GW01406659

Biotechnology of Micrc

Production, Biocatalysis, and Industrial

Biotechnology of Microbial Enzymes

Production, Biocatalysis, and Industrial Applications

Second Edition

Edited by

Goutam Brahmachari
Laboratory of Natural Products & Organic Synthesis,
Department of Chemistry, Visva-Bharati (A Central University),
Santiniketan, West Bengal, India

ELSEVIER

ACADEMIC PRESS
An imprint of Elsevier

Academic Press is an imprint of Elsevier
125 London Wall, London EC2Y 5AS, United Kingdom
525 B Street, Suite 1650, San Diego, CA 92101, United States
50 Hampshire Street, 5th Floor, Cambridge, MA 02139, United States
The Boulevard, Langford Lane, Kidlington, Oxford OX5 1GB, United Kingdom

Copyright © 2023 Elsevier Inc. All rights reserved.

No part of this publication may be reproduced or transmitted in any form or by any means, electronic or mechanical, including photocopying, recording, or any information storage and retrieval system, without permission in writing from the publisher. Details on how to seek permission, further information about the Publisher's permissions policies and our arrangements with organizations such as the Copyright Clearance Center and the Copyright Licensing Agency, can be found at our website: www.elsevier.com/permissions.

This book and the individual contributions contained in it are protected under copyright by the Publisher (other than as may be noted herein).

Notices
Knowledge and best practice in this field are constantly changing. As new research and experience broaden our understanding, changes in research methods, professional practices, or medical treatment may become necessary.

Practitioners and researchers must always rely on their own experience and knowledge in evaluating and using any information, methods, compounds, or experiments described herein. In using such information or methods they should be mindful of their own safety and the safety of others, including parties for whom they have a professional responsibility.

To the fullest extent of the law, neither the Publisher nor the authors, contributors, or editors, assume any liability for any injury and/or damage to persons or property as a matter of products liability, negligence or otherwise, or from any use or operation of any methods, products, instructions, or ideas contained in the material herein.

ISBN: 978-0-443-19059-9

For Information on all Academic Press publications
visit our website at https://www.elsevier.com/books-and-journals

Publisher: Mica H. Haley
Acquisitions Editor: Peter B. Linsley
Editorial Project Manager: Michaela Realiza
Production Project Manager: Fahmida Sultana
Cover Designer: Greg Harris

Typeset by MPS Limited, Chennai, India

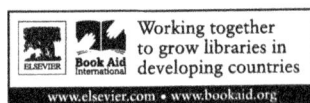

Working together to grow libraries in developing countries
www.elsevier.com • www.bookaid.org

Dedication

In memory of Dr. Arnold L. Demain (United States).

Goutam Brahmachari

Contents

List of contributors

Komal Agrawal Bioprocess and Bioenergy Laboratory (BPEL), Department of Microbiology, Central University of Rajasthan, Bandarsindri, Kishangarh, Ajmer, Rajasthan, India; Department of Microbiology, School of Bioengineering and Biosciences, Lovely Professional University, Phagwara, Punjab, India

Erika Cristina G. Aguieiras Federal University of Rio de Janeiro, Department of Biochemistry, Rio de Janeiro, Brazil; Federal University of Rio de Janeiro, Campus UFRJ - Duque de Caxias Prof. Geraldo Cidade, Duque de Caxias, Rio de Janeiro, Brazil

Hiroshi Amesaka Department of Biomolecular Chemistry, Kyoto Prefectural University, Kyoto, Japan

K.S. Anantharaju Department of Chemistry, Dayananda Sagar College of Engineering, Bangalore, Karnataka, India

Miguel Arroyo Department of Biochemistry and Molecular Biology, Faculty of Biology, University Complutense of Madrid, Madrid, Spain

Prashant S. Arya Department of Microbiology and Biotechnology, School of Sciences, Gujarat University, Ahmedabad, Gujarat, India

Emanueli Backes State University of Maringá, Maringá, Brazil

Dhritiksha M. Baria Department of Microbiology and Biotechnology, University School of Sciences, Gujarat University, Ahmedabad, Gujarat, India

José Luis Barredo Curia, Parque Tecnológico de León, León, Spain

Carlos Barreiro Department of Molecular Biology, Area of Biochemistry and Molecular Biology, Faculty of Veterinary, University of León, León, Spain

Sudhanshu S. Behera Department of Biotechnology, National Institute of Technology Raipur, Raipur, Chhattisgarh, India; Centre for Food Biology & Environment Studies, Bhubaneswar, Odisha, India

Reeta Bhati Amity Institute of Microbial Technology, Amity University, Noida, Uttar Pradesh, India

Kanishk Bhatt Department of Biotechnology and Bioinformatics, Jaypee University of Information Technology, Waknaghat, Solan, Himachal Pradesh, India

Elba P.S. Bon Federal University of Rio de Janeiro, Department of Biochemistry, Rio de Janeiro, Brazil

Adelar Bracht State University of Maringá, Maringá, Brazil

Goutam Brahmachari Laboratory of Natural Products & Organic Synthesis, Department of Chemistry, Visva-Bharati (a Central University), Santiniketan, West Bengal, India

Filipe Carvalho Faculty of Engineering, Universidade Lusófona de Humanidades e Tecnologias, Lisboa, Portugal; iBB—Institute for Bioengineering and Biosciences, Instituto Superior Técnico, Universidade de Lisboa, Lisboa, Portugal; Associate Laboratory i4HB—Institute for Health and Bioeconomy at Instituto Superior Técnico, Universidade de Lisboa, Lisboa, Portugal

Servio Tulio Alves Cassini Federal University of Espírito Santo, Vitoria, Espírito Santo, Brazil; Center for Research, Innovation and Development of Espírito Santo, CPID, Cariacica, Espírito Santo, Brazil

Chiu-Wen Chen Department of Marine Environmental Engineering, National Kaohsiung University of Science and Technology, Kaohsiung City, Taiwan

Rubia Carvalho Gomes Corrêa UniCesumar, Cesumar University Centre, Maringá, Brazil

Cristina Coscolín Systems Biotechnology Group, ICP, CSIC, Madrid, Spain

Ayla Sant'Ana da Silva National Institute of Technology, Ministry of Science, Technology and Innovation, Avenida Venezuela, Rio de Janeiro, Brazil; Federal University of Rio de Janeiro, Department of Biochemistry, Rio de Janeiro, Brazil

Bruna Polacchine da Silva UniSalesiano, Catholic Salesian Auxilium University Centre, Araçatuba, Brazil

Thais de Andrade Silva Federal University of Espírito Santo, Vitoria, Espírito Santo, Brazil

Isabel de la Mata Department of Biochemistry and Molecular Biology, Faculty of Biology, University Complutense of Madrid, Madrid, Spain

Jairo Pinto de Oliveira Federal University of Espírito Santo, Vitoria, Espírito Santo, Brazil; Center for Research, Innovation and Development of Espírito Santo, CPID, Cariacica, Espírito Santo, Brazil

Ronaldo Rodrigues de Sousa National Institute of Technology, Ministry of Science, Technology and Innovation, Avenida Venezuela, Rio de Janeiro, Brazil; Federal University of Rio de Janeiro, Department of Biochemistry, Rio de Janeiro, Brazil

Marcella Fernandes de Souza Ghent University, Faculty of Bioscience Engineering, Gent, Belgium

Cheng-Di Dong Department of Marine Environmental Engineering, National Kaohsiung University of Science and Technology, Kaohsiung City, Taiwan

Jaqueline Greco Duarte Federal University of Rio de Janeiro, Department of Biochemistry, Rio de Janeiro, Brazil; SENAI Innovation Institute for Biosynthetics and Fibers, SENAI CETIQT, Rio de Janeiro, Brazil

Roberta Pereira Espinheira National Institute of Technology, Ministry of Science, Technology and Innovation, Avenida Venezuela, Rio de Janeiro, Brazil; Federal University of Rio de Janeiro, Department of Biochemistry, Rio de Janeiro, Brazil

Mariana de Oliveira Faber National Institute of Technology, Ministry of Science, Technology and Innovation, Avenida Venezuela, Rio de Janeiro, Brazil; Federal University of Rio de Janeiro, Department of Biochemistry, Rio de Janeiro, Brazil

Daniel Oluwagbotemi Fasheun National Institute of Technology, Ministry of Science, Technology and Innovation, Avenida Venezuela, Rio de Janeiro, Brazil; Federal University of Rio de Janeiro, Department of Biochemistry, Rio de Janeiro, Brazil

Pedro Fernandes Faculty of Engineering, Universidade Lusófona de Humanidades e Tecnologias, Lisboa, Portugal; iBB—Institute for Bioengineering and Biosciences, Instituto Superior Técnico, Universidade de Lisboa, Lisboa, Portugal; Associate Laboratory i4HB—Institute for Health and Bioeconomy at Instituto Superior Técnico, Universidade de Lisboa, Lisboa, Portugal

Viridiana S. Ferreira-Leitão National Institute of Technology, Ministry of Science, Technology and Innovation, Avenida Venezuela, Rio de Janeiro, Brazil; Federal University of Rio de Janeiro, Department of Biochemistry, Rio de Janeiro, Brazil

Manuel Ferrer Systems Biotechnology Group, ICP, CSIC, Madrid, Spain

Niyonzima Francois Department of Biotechnologies, Faculty of Applied Fundamental Sciences, INEs Ruhengeri, Rwanda

Denise M.G. Freire Federal University of Rio de Janeiro, Department of Biochemistry, Rio de Janeiro, Brazil

José Luis García Centro de Investigaciones Biológicas Margarita Salas CSIC, Madrid, Spain

Carlos García-Estrada Department of Biomedical Sciences, University of León, León, Spain

Vishal A. Ghadge Natural Products & Green Chemistry Division, CSIR-Central Salt and Marine Chemicals Research Institute (CSIR-CSMCRI), Council of Scientific and Industrial Research (CSIR), Bhavnagar, Gujarat, India; Academy of Scientific and Innovative Research (AcSIR), Ghaziabad, Uttar Pradesh, India

Peter N. Golyshin Centre for Environmental Biotechnology, Bangor University, Bangor, United Kingdom

Carolina Reis Guimarães National Institute of Technology, Ministry of Science, Technology and Innovation, Avenida Venezuela, Rio de Janeiro, Brazil

Venkatesh S. Joshi School of Basic and Applied Sciences, Dayananda Sagar University, Bangalore, Karnataka, India

Shigenori Kanaya Department of Material and Life Science, Graduate School of Engineering, Osaka University, Osaka, Japan

Camila Gabriel Kato Federal University of Mato Grosso do Sul, Campo Grande, Brazil

Ankush Kerketta Department of Biotechnology, National Institute of Technology Raipur, Raipur, Chhattisgarh, India

Yuichi Koga Department of Biotechnology, Graduate School of Engineering, Osaka University, Osaka, Japan

Chandrakant Kokare Department of Pharmaceutics, Sinhgad Technical Education Society, Sinhgad Institute of Pharmacy, Narhe, Pune, Maharashtra, India

Bikash Kumar Bioprocess and Bioenergy Laboratory (BPEL), Department of Microbiology, Central University of Rajasthan, Bandarsindri, Kishangarh, Ajmer, Rajasthan, India; Department of Biosciences and Bioengineering, Indian Institute of Technology Guwahati, Guwahati, Assam, India

Pankaj Kumar Natural Products & Green Chemistry Division, CSIR-Central Salt and Marine Chemicals Research Institute (CSIR-CSMCRI), Council of Scientific and Industrial Research (CSIR), Bhavnagar, Gujarat, India; Academy of Scientific and Innovative Research (AcSIR), Ghaziabad, Uttar Pradesh, India

Paloma Liras Department of Molecular Biology, University of León, León, Spain

Xiangyang Liu UNT System College of Pharmacy, University of North Texas Health Science Center, Fort Worth, TX, United States

Juan F. Martín Department of Molecular Biology, University of León, León, Spain

Patricia Molina-Espeja Systems Biotechnology Group, ICP, CSIC, Madrid, Spain

Sunil S. More School of Basic and Applied Sciences, Dayananda Sagar University, Bangalore, Karnataka, India

Veena S. More Department of Biotechnology, Sapthagiri College of Engineering, Bangalore, Karnataka, India

Shivangi Mudaliar Bioprocess and Bioenergy Laboratory (BPEL), Department of Microbiology, Central University of Rajasthan, Bandarsindri, Kishangarh, Ajmer, Rajasthan, India

Ashok Kumar Nadda Department of Biotechnology and Bioinformatics, Jaypee University of Information Technology, Waknaghat, Solan, Himachal Pradesh, India

Ajay Nair School of Basic and Applied Sciences, Dayananda Sagar University, Bangalore, Karnataka, India

Lakshana Nair (G) Bioprocess and Bioenergy Laboratory (BPEL), Department of Microbiology, Central University of Rajasthan, Bandarsindri, Kishangarh, Ajmer, Rajasthan, India

Nandini Amrutha Nandyal School of Basic and Applied Sciences, Dayananda Sagar University, Bangalore, Karnataka, India

Francois N. Niyonzima Department of Math, Science and PE, CE, University of Rwanda, Rwamagana, Rwanda

Florien Nsanganwimana Department of Math, Science and PE, CE, University of Rwanda, Rwamagana, Rwanda

Rakeshkumar R. Panchal Department of Microbiology and Biotechnology, University School of Sciences, Gujarat University, Ahmedabad, Gujarat, India

Ashok Pandey CSIR-Indian Institute for Toxicological Research, Lucknow, Uttar Pradesh, India

Dimple S. Pardhi Department of Microbiology and Biotechnology, University School of Sciences, Gujarat University, Ahmedabad, Gujarat, India

Anil Kumar Patel Department of Marine Environmental Engineering, National Kaohsiung University of Science and Technology, Kaohsiung City, Taiwan

Nidhi Y. Patel Department of Microbiology and Biotechnology, University School of Sciences, Gujarat University, Ahmedabad, Gujarat, India

Rosane Marina Peralta State University of Maringá, Maringá, Brazil

Laura Marina Pinotti Federal University of Espírito Santo, São Mateus, Espírito Santo, Brazil

K.R. Pooja School of Basic and Applied Sciences, Dayananda Sagar University, Bangalore, Karnataka, India

Kiransinh N. Rajput Department of Microbiology and Biotechnology, University School of Sciences, Gujarat University, Ahmedabad, Gujarat, India

Archana S. Rao School of Basic and Applied Sciences, Dayananda Sagar University, Bangalore, Karnataka, India

Meena R. Rathod Natural Products & Green Chemistry Division, CSIR-Central Salt and Marine Chemicals Research Institute (CSIR-CSMCRI), Council of Scientific and Industrial Research (CSIR), Bhavnagar, Gujarat, India

Vikram H. Raval Department of Microbiology and Biotechnology, School of Sciences, Gujarat University, Ahmedabad, Gujarat, India; Department of Microbiology and Biotechnology, University School of Sciences, Gujarat University, Ahmedabad, Gujarat, India

Ramesh C. Ray Centre for Food Biology & Environment Studies, Bhubaneswar, Odisha, India

Diana Rocha Instituto de Investigaciones Biomédicas, Universidad Nacional Autónoma de México (UNAM), México Distrito Federal, México

Romina Rodríguez-Sanoja Instituto de Investigaciones Biomédicas, Universidad Nacional Autónoma de México (UNAM), México Distrito Federal, México

Alba Romero Instituto de Investigaciones Biomédicas, Universidad Nacional Autónoma de México (UNAM), México Distrito Federal, México

Beatriz Ruiz-Villafán Instituto de Investigaciones Biomédicas, Universidad Nacional Autónoma de México (UNAM), México Distrito Federal, México

Harshal Sahastrabudhe Natural Products & Green Chemistry Division, CSIR-Central Salt and Marine Chemicals Research Institute (CSIR-CSMCRI), Council of Scientific and Industrial Research (CSIR), Bhavnagar, Gujarat, India; Academy of Scientific and Innovative Research (AcSIR), Ghaziabad, Uttar Pradesh, India

Hima A. Salu School of Basic and Applied Sciences, Dayananda Sagar University, Bangalore, Karnataka, India

Igor Carvalho Fontes Sampaio Center for Research, Innovation and Development of Espírito Santo, CPID, Cariacica, Espírito Santo, Brazil

Sergio Sánchez Instituto de Investigaciones Biomédicas, Universidad Nacional Autónoma de México (UNAM), México Distrito Federal, México

Vinícius Mateus Salvatore Saute State University of Maringá, Maringá, Brazil

Flávio Augusto Vicente Seixas State University of Maringá, Maringá, Brazil

Ayushi Sharma Department of Biotechnology and Bioinformatics, Jaypee University of Information Technology, Waknaghat, Solan, Himachal Pradesh, India

Shagun Sharma Department of Biotechnology and Bioinformatics, Jaypee University of Information Technology, Waknaghat, Solan, Himachal Pradesh, India

Pramod B. Shinde Natural Products & Green Chemistry Division, CSIR-Central Salt and Marine Chemicals Research Institute (CSIR-CSMCRI), Council of Scientific and Industrial Research (CSIR), Bhavnagar, Gujarat, India; Academy of Scientific and Innovative Research (AcSIR), Ghaziabad, Uttar Pradesh, India

Rahul Shrivastava Department of Biotechnology and Bioinformatics, Jaypee University of Information Technology, Waknaghat, Solan, Himachal Pradesh, India

Rajni Singh Amity Institute of Microbial Technology, Amity University, Noida, Uttar Pradesh, India

Sanju Singh Natural Products & Green Chemistry Division, CSIR-Central Salt and Marine Chemicals Research Institute (CSIR-CSMCRI), Council of Scientific and Industrial Research (CSIR), Bhavnagar, Gujarat, India; Academy of Scientific and Innovative Research (AcSIR), Ghaziabad, Uttar Pradesh, India

Reeta Rani Singhania Department of Marine Environmental Engineering, National Kaohsiung University of Science and Technology, Kaohsiung City, Taiwan

Swati Srivastava Amity Institute of Microbial Technology, Amity University, Noida, Uttar Pradesh, India

Kazufumi Takano Department of Biomolecular Chemistry, Kyoto Prefectural University, Kyoto, Japan

Shun-ichi Tanaka Ritsumeikan Global Innovation Research Organization, Ritsumeikan University, Shiga, Japan; Department of Biomolecular Chemistry, Kyoto Prefectural University, Kyoto, Japan; Department of Biotechnology, College of Life Science, Ritsumeikan University, Shiga, Japan

Ricardo Sposina Sobral Teixeira Federal University of Rio de Janeiro, Department of Biochemistry, Rio de Janeiro, Brazil

Marina Cristina Tomasini National Institute of Technology, Ministry of Science, Technology and Innovation, Avenida Venezuela, Rio de Janeiro, Brazil; Federal University of Rio de Janeiro, Department of Biochemistry, Rio de Janeiro, Brazil

Thaís Marques Uber State University of Maringá, Maringá, Brazil

Ryo Uehara Division of Cancer Cell Regulation, Aichi Cancer Center Research Institute, Nagoya, Japan; Ritsumeikan Global Innovation Research Organization, Ritsumeikan University, Shiga, Japan

Pradeep Verma Bioprocess and Bioenergy Laboratory (BPEL), Department of Microbiology, Central University of Rajasthan, Bandarsindri, Kishangarh, Ajmer, Rajasthan, India

Shivani M. Yagnik Department of Microbiology, Christ College, Rajkot, Gujarat, India

Julio Pansiere Zavarise Federal University of Espírito Santo, São Mateus, Espírito Santo, Brazil

About the editor

Born on April 14, 1969 in Barala, a village in the district of Murshidabad (West Bengal, India), Goutam Brahmachari had his early education in his native place. He received his high school degree in scientific studies in 1986 at Barala R. D. Sen High School under the West Bengal Council of Higher Secondary Education (WBCHSE). Then he moved to Visva-Bharati (a Central University founded by Rabindranath Tagore at Santiniketan, West Bengal, India) to study chemistry at the undergraduate level. After graduating from this university in 1990, he completed his master's in 1992 with a specialization in organic chemistry. After that, receiving his Ph.D. in 1997 in chemistry from the same university, he joined his alma mater the very next year and currently holds the position of a full professor of chemistry since 2011. The research interests of Prof. Brahmachari's group include natural products chemistry, synthetic organic chemistry, green chemistry, and the medicinal chemistry of natural and natural product-inspired synthetic molecules. With more than 24 years of experience in teaching and research, he has produced about 250 scientific publications, including original research papers, review articles, books, and invited book chapters in the field of natural products and green chemistry. He has already authored/edited 26 books published by internationally reputed major publishing houses, namely, Elsevier Science (The Netherlands), Academic Press (Oxford), Wiley-VCH (Germany), Alpha Science International (Oxford), De Gruyter (Germany), World Scientific (Singapore), CRC Press (Taylor & Francis Group, USA), Royal Society of Chemistry (Cambridge), etc. Prof. Brahmachari serves as a life member for the Indian Association for the Cultivation of Science (IACS), Indian Science Congress Association (ISCA), Kolkata, Indian Chemical Society (ICS), Kolkata, and Chemical Research Society of India (CRSI), Bangalore. He has also been serving as a co-editor for *Current Green Chemistry*.

Prof. Brahmachari serves as the founder series editor of Elsevier Book Series' *Natural Product Drug Discovery*. He is an elected fellow of the Royal Society of Chemistry, and a recipient of CRSI (Chemical Research Society of India) Bronze Medal, 2021 (contributions to research in chemistry), Dr. Basudev Banerjee Memorial Award, 2021 (scholastic contribution to the field of chemical sciences), INSA (Indian National Science Academy) Teachers Award, 2019, Dr. Kalam Best Teaching Faculty Award, 2017, and Academic Brilliance Award, 2015 (Excellence in Research). Prof. Brahmachari was featured in the World's Top 2% Scientists (organic chemistry category) in 2020 and 2021 and the AD Scientific Index 2022 World Ranking of Scientists, 2022.

Preface

The success of the first edition of the book titled *Biotechnology of Microbial Enzymes: Production, Biocatalysis, and Industrial Applications*, coedited with Dr. Arnold L. Demain and Dr. Jose L. Adrio, prompted me to plan this second edited version. The other two major reasons behind this planning were to keep pace with the rapid progress of this remarkable field and to offer a tribute to the legendary microbiologist, Dr. Demain, who left us on April 3, 2020. The book's first edition was developed due to his keen interest, organization, and encouragement.

Since its first publication in 2017, the book has been well acclaimed among readers of all sections. Hence, upgrading the book has become quite rational, and this newly revised second edition satisfies the demand. This enlarged volume, comprising a total of 26 chapters, is an endeavor to focus on the recent cutting-edge research advances in the biotechnology of microbial enzymes, mainly focusing on their productions, modifications, and industrial applications. These selectively screened chapters are contributed by active researchers and leading experts in microbial enzymes from several countries in response to my invitation. In addition to the new groups of authors, a good number of contributors from the book's first edition also participated in this revised version. While a few earlier chapters are dropped, the others are thoroughly revised, and a considerably good number of new chapters are included. I am most grateful to all the contributors for their generous and timely response despite their busy and tight schedules with academics, research, and other responsibilities.

Enzymes are potential biocatalysts produced by living cells to bring about specific biochemical reactions linked with the metabolic processes of the cells. Due to the unique biochemical properties of enzymes, such as high specificity, fast action, and biodegradability, the demand for industrial enzymes is on a continuous rise, driven by a growing need for sustainable solutions. Microorganisms are the primary source of enzymes because they are cultured in large quantities in a short time. These enzymes have found practical and industrial applications from ancient times dating back many centuries. The use of barley malt for starch conversion in brewing and dung for bating of hides in the leather making is just a couple of examples of early use of microbial enzymes. Microorganisms have thus provided and continued to offer an impressive amount of such biocatalysts with a wide range of applications across several industries such as food, animal feed, household care, technical industries, biofuels, fine chemicals, and pharmaceuticals. The beneficial characteristics of microbial enzymes, such as thermotolerance, thermophilic nature, and tolerance to a varied range of pH and other harsh reaction conditions, are exploited for their commercial interest and industrial applications. However, natural enzymes do not often fulfill all process requirements despite these advantages and need further tailoring or redesign to fine-tune fundamental catalytic properties.

Recent advances in "*omics*" technologies (e.g. genomics, metagenomics, and proteomics), efficient expression systems, and emerging recombinant DNA techniques facilitate the discovery of new microbial enzymes either from nature or by creating (or evolving) enzymes with improved catalytic properties. The implementation of genetic manipulations on bacterial cells can also enhance enzyme production. Besides, recently several lines of study have been initiated to isolate new bacterial and fungal strains, which may render new types of enzymes with remarkable properties and efficacies. Combinations of newly isolated, engineered and de novo designed enzymes coupled with chemistry have been successful in generating more (and even new) chemicals and materials from cheaper and renewable resources, thereby opening a new window to establishing a bio-based economy and achieving low-carbon green growth. The ongoing progress and interest in enzymes provide further success in many areas of industrial biocatalysis. Besides, applying one-pot multistep reactions using multifunctional catalysts or new and improved enzyme immobilization techniques is also receiving growing interest in biocatalysis.

Many of the technologies and strategies mentioned above are gathered together in this book. A variety of 26 chapters brings together an overview of current discoveries and trends in this remarkable area. Chapter 1 presents an overview of the book and summarizes the contents of all technical chapters to offer glimpses of the subject matter covered to the readers before they go in for a detailed study. Chapters 2–26 are devoted to exploring the ongoing innovative ideas and tools directed toward the fruitful production, modifications/tailoring, and applications of microbial enzymes in both academic and industrial sectors.

This timely revised volume encourages interdisciplinary work among synthetic and natural product chemists, medicinal chemists, green chemistry practitioners, pharmacologists, biologists, and agronomists interested in microbial enzymes. Representation of facts and their discussions in each chapter are exhaustive, authoritative, and deeply informative. The broad interdisciplinary approach in this book would surely make the work much more attractive to the scientists deeply engaged in the research and/or use of microbial enzymes. I would like to thank all the contributors again for their excellent reviews on this remarkable area. Their participation made my effort to organize such a book possible. Their masterly accounts will surely provide the readers with a strong awareness of current cutting-edge research approaches in this remarkable field. In continuation to its first edition, this thoroughly revised second edition would also serve as a key reference for recent developments in the frontier research on the biotechnological developments of microbial enzymes and their prospective industrial applications. It would also motivate young scientists in the dynamic field of biotechnology of microbial enzymes.

I would also like to express my deep sense of appreciation to all of the editorial and publishing staff members associated with Elsevier Inc., for their keen interest in taking initiation for the second edition and publishing the works, and also for their all-round help to ensure that the highest standards of publication have been maintained in bringing out this book.

June 22, 2022

Goutam Brahmachari

Chapter 1

Biotechnology of microbial enzymes: production, biocatalysis, and industrial applications—an overview

Goutam Brahmachari
Laboratory of Natural Products & Organic Synthesis, Department of Chemistry, Visva-Bharati (a Central University), Santiniketan, West Bengal, India

1.1 Introduction

The second edition of this book, titled *Biotechnology of Microbial Enzymes: Production, Biocatalysis, and Industrial Applications* is an endeavor to have vivid information on the ongoing developments and recent cutting-edge research advances in the field of microbial enzymes regarding the identification of their source microbes, isolation, purification, biocatalysis, and multifaceted applications in various industrial sectors, including agricultural, chemical, pharmaceuticals, textile, paper, bioremediation, biorefineries, biofuels, and bioenergy. The enlarged second edition of the book encourages more interdisciplinary works among chemists, pharmacologists, clinicians, technologists, biologists, botanists, and agronomists interested in microbes and microbial enzymes. This edition comprising 25 technical chapters, offers recent updates on microbial enzymatic research with an intention to unravel their production, biocatalysis and industrial applicability to a greater extent to maintain the environmental sustainability.

Enzymes are potential biocatalysts produced by living cells to bring about specific biochemical reactions linked with the metabolic processes of the cells. Due to unique biochemical properties of enzymes, the demand for industrial enzymes is on a continuous rise driven by a growing need for sustainable solutions. Microorganisms have provided and continue to provide an impressive amount of such biocatalysts with a wide range of applications across several industries such as food, animal feed, household care, technical

Biotechnology of Microbial Enzymes. DOI: https://doi.org/10.1016/B978-0-443-19059-9.00026-8
© 2023 Elsevier Inc. All rights reserved.

industries, biofuels, fine chemicals, and pharmaceuticals. As mentioned, the unique properties of enzymes such as high specificity, fast action and biodegradability allow enzyme-assisted processes in industry to run under milder reaction conditions, with improved yields and reduced waste generation. However, despite these advantages, natural enzymes do not often fulfill all process requirements and need further tailoring or redesign in order to fine-tune key catalytic properties.

This introductory chapter (Chapter 1) presents an overview of the book and summarizes the contents and subject matter of each chapter to offer certain glimpses of the coverage of discussion to the readers before they go for a detailed study.

1.2 An overview of the book

The second edition of this book contains 25 technical chapters—Chapters 2–26. This section summarizes the contents and subject matter of each of them.

1.2.1 Chapter 2

Chapter 2 by Sánchez and coauthors underlines the usefulness of microbial enzymes focusing on their industrial applications, improvements, and discovery of newer versions. This chapter aims to prepare the readers to go in-depth into the book.

1.2.2 Chapter 3

Chapter 3 by Singhania and coauthors offers an overview of the production, purification, and application of microbial enzymes in industrial sectors. Enzyme technology is an ever-evolving branch of science and technology, and with the intervention and influence of biotechnology and bioinformatics, continuously novel or improved applications of enzymes are emerging. Screening for new and improved enzymes, selection of microorganisms and strain improvement for qualitative and quantitative enhancement, fermentation for enzyme production, large-scale enzyme purifications, and formulation of enzymes for sale are the key aspects of enzyme technology that enable industries and consumers to replace processes using aggressive chemicals with mild and environment-friendly enzymatic processes. Enzymes, being highly specific in nature, can revolutionize the whole industrial sector. The authors presented a detailed discussion on these aspects in their chapter that is anticipated to be much helpful to the biologists working in enzyme production, purification, and improvement.

1.2.3 Chapter 4

Chapter 4 by Ray and his group offers a thorough account of solid-state fermentation (SSF) for the production of microbial cellulases. Cellulolytic enzymes convert lignocellulosic biomass into products with high added value. The SSF of cellulosic biomass involving cellulases and cellulolytic microorganisms is currently a hot area of biotechnological research. This chapter focuses on the importance of SSF and its comparative aspects with submerged fermentation processes in cellulase production. The authors addressed the method of extraction of microbial cellulases and the measurement of their enzymatic activity in SSF. The overview covers a detailed discussion on the lignocellulosic residues/solid substrates in SSF, pretreatment of agricultural residues, environmental factors affecting cellulase production, and strategies to improve the production of microbial cellulases. This illustrative review would create enthusiasm among the readers.

1.2.4 Chapter 5

Chapter 5 by Tanaka and coauthors presents an in-depth overview of the unique maturation, stabilization mechanisms, and applications of the hyperthermophilic subtilisin-like proteases from *Thermococcus kodakarensis*. Hyperthermophiles, known for their exceptional tolerance against chemical and thermal denaturation, are attractive sources of enzymes. The genome of a hyperthermophilic archaeon, *T. kodakarensis* KOD1, contains three genes encoding subtilisin-like serine proteases. Two proteases, Tk-subtilisin and Tk-SP, have biochemically and structurally been characterized. Tk-subtilisin and Tk-SP exhibit extraordinarily high stability compared with their mesophilic counterparts. Thus these two proteases find potential biotechnological applications. The authors underlined all these issues in a clear view in their chapter, and this exhaustive review would offer huge relevant information to the readers.

1.2.5 Chapter 6

Chapter 6 by Peralta and coauthors offers an excellent overview of the enzymes from Basidiomycetes, a fascinating group of fungi that act as natural lignocellulose destroyers and can accommodate themselves to detrimental conditions of the environment. Basidiomycetes are considered one of the most peculiar and efficient tools for biotechnology. The ability of basidiomycetes to degrade the complex structure of lignocellulose makes them potentially useful in exploring the lignocellulosic biomass for producing fuel ethanol and other value-added commodity chemicals. In their chapter, the authors documented a general panorama of the enzymes involved in the capability of these fungi to degrade vegetal biomass and their industrial and biotechnological applications. This thorough and illuminating review of a specific class of fungi would be much helpful to the readers.

1.2.6 Chapter 7

Chapter 7 by Ferrer and coauthors overviews the impact of metagenomics and new enzymes on the bioeconomy. Metagenomics refers to the application of genomics to study microbial communities and enzymes directly extracted from the environment without the need for culturing. Proper actions in searching for new enzymes and designing technologies are required to achieve environmental and circular economy goals, following circular economy criteria, new products and processes that are more environment-friendly and sustainable. For this purpose, metagenomics tools, resources, approaches, results, and practical applications can be much helpful. The present chapter gives an insight into this remarkable area of interest.

1.2.7 Chapter 8

Chapter 8 by Singh and her group presents an overview of the enzymatic biosynthesis of β-lactam antibiotics, a group of bactericidal drugs that inhibit bacterial growth by obstructing penicillin-binding proteins responsible for the transpeptidation/cross-linking process during cell-wall biosynthesis. β-Lactam antibiotics consist of four major classes: penicillin derivatives, cephalosporins, monobactams, and carbapenems. Most β-lactams are produced *via* fermentation or modification of fermented intermediates except for carbapenems and aztreonam. The β-lactam biosynthesis generally follows nonoxidative reactions. However, enzymatic synthesis is also important by using enzymes like 2-oxoglutarate (2OG)-dependent oxygenases, isopenicillin N-synthase (IPNS), clavaminic acid synthase, β-lactam synthetases (BLS), nonribosomal peptide synthetases, etc. The authors summarized such available understanding and knowledge of the enzymology leading to the biosynthesis of β-lactam antibiotics and their effective derivatives. The readers will be benefitted largely with these insightful discussions.

1.2.8 Chapter 9

Chapter 9 by Martín and coauthors extends the discussion on the β-lactam antibiotics by offering insights into molecular mechanisms of β-lactam antibiotic synthesizing and modifying enzymes in fungi. The authors reviewed the molecular mechanism of the core enzymes such as ACV synthetase, IPNS, and isopenicillin N-acyltransferase, including details on their structures. Detailed analyses of recent findings on the transport of intermediates through organelles and the controversial mechanisms of penicillin secretion through the cell membrane in *Penicillium chrysogenu* are discussed.

In addition, the authors also afforded available information on studying different penicillin acylases used for the industrial production of semisynthetic β-lactam antibiotics. This thorough and explicit discussion on β-lactam antibiotics would interest the readers.

1.2.9 Chapter 10

Chapter 10 by Shinde and his group is devoted to the role of glycosyltransferases in the biosynthesis of antibiotics required for everyday life functions. Most of them are naturally decorated with various sugars. The importance of glycosylated antibiotics in treating infections and chronic diseases has motivated the researchers to explore and have a better understanding of glycosylation. Glycosyltransferase enzymes catalyze glycosidic bond formation between a donor sugar molecule and a hydroxyl group of an acceptor molecule. The structural basis of enzymes provides an excellent opportunity for the genetic engineering of these enzymes. Diversifying natural products through glycosyltransferases catalyzed by glycosylations is exceptional, which is practically impossible to achieve using chemical synthesis. The authors furnished an insightful discussion on the effects of classifications of glycosyltransferase enzymes, their role in glycosylated antibiotics, and different strategies employed to carry out glycosylation.

1.2.10 Chapter 11

Chapter 11 by Sanchez and his group deals with the relevance of microbial glucokinases, widely distributed in all domains of life. This group of microbial enzymes is responsible for glucose phosphorylation utilizing diverse phosphoryl donors such as ATP, ADP, and/or polyphosphate. Apart from glucose phosphorylation, some glucokinases also present a regulatory role. Glucokinases, especially those that are thermostable, find industrial applications, taking advantage of their phosphorylating activity. In their present chapter, the authors outlined the physicochemical and biochemical characteristics of glucokinases and their potential applications that will invoke much interest in the readers.

1.2.11 Chapter 12

Chapter 12 by Shrivastava and coauthors is devoted to the microbial enzyme, *Mycobacterium tuberculosis* DapA as a target for antitubercular drug design. The enzymes involved in mycobacterial cell-wall biosynthesis are usually targeted to design better and more efficacious antitubercular drugs to meet the ever-increasing challenges of tuberculosis. Despite the growing research, tuberculosis eradication is still a worldwide challenge. Research findings indicate that the diaminopimelate (DAP) pathway enzymes are indispensable for the growth and survival of *M. tuberculosis*; hence, inhibiting this pathway in mammals can provide an effective target for the bacteria to discover anti-tubercular drugs. The present chapter describes the DAP pathway that leads to lysine production and provides an overview of the studies about inhibiting DAP pathway enzymes. The chapter also provides a systematic review of the effects of inhibitors reported against *M. tuberculosis* DapA.

1.2.12 Chapter 13

Chapter 13 by Brahmachari portrays an updated review of the lipase-catalyzed organic transformations. In the recent past, lipase has emerged as one of the most promising enzymes for broad practical applications in organic synthesis, with a remarkable ability to carry out a wide variety of chemo-, regio-, and enantioselective transformations and very broad substrate specificity. This chapter is an updated version of the earlier edition highlighting the lipase-catalyzed organic reactions reported from 2013 to 2021. This overview reflects the biocatalytic efficacy of the enzyme in carrying out various types of organic reactions, including esterification, transesterification, additions, ring-closing, oxidation, reduction, and amidation. The overview is anticipated to boost ongoing research in chemoenzymatic organic transformations, particularly the biocatalytic applications of lipases.

1.2.13 Chapter 14

Chapter 14 by Nadda and coauthors describes the fundamentals and applications of two important microbial enzymes, tyrosinase and oxygenase. The authors offered detailed mechanistic aspects of catalytic activity of both tyrosinase and oxygenase, and also multifaceted uses of tyrosinase in various sectors such as the food, textile, cosmetic, bioremediation and medical sectors, and the significance of oxygenase in cleaving aromatic wastes.

1.2.14 Chapter 15

Chapter 15 by Barredo and coauthors deals with applying microbial enzymes as drugs in human therapy and healthcare. The application of microbial enzymes is an emerging alternative for treating a wide range of human diseases. In their review, the authors provided a good deal of examples of microbial enzymes for the effective treatment of different disease conditions. The authors also addressed the recent research outcomes in this area, including microbial enzymes useful as "clot busters" or digestive aids, for the treatment of congenital and infectious diseases, burn debridement and fibroproliferative diseases, and the treatment of cancer and other health disorders.

1.2.15 Chapter 16

Chapter 16 by Raval and coauthors highlights the application of microbial enzymes in the pharmaceutical industry. Due to their efficacy coupled with specificity, microbial enzymes find immense applications as biocatalysts for synthesizing active pharmaceutical products. The authors offered a vivid description of therapeutic enzymes having productive applications in the pharmaceutical industry for drug development. This illustrative review is worthy enough to attract the attention of the readers.

1.2.16 Chapter 17

Chapter 17 by Liu and Kokare underlines the production and impacts of microbial enzymes for use in industrial sectors. The enormous diversity of microbial enzymes makes them an exciting group of chemical entities for application in many industrial sectors such as chemical, pharmaceutical, food processing, textile, wood processing, and cosmetics. The authors herein presented various classifications, resources, production, and applications of a group of industrially used microbial enzymes.

1.2.17 Chapter 18

Chapter 18 by Fernandes and Carvalho categorically overviews microbial enzymes used in the food industry. Enzymes used in food production and processing have a long history and tradition. This trend is directly related to the biocompatible nature of these biocatalysts and their selective nature and ability to operate under mild conditions. The authors offered a comprehensive overview of the different applications of enzymes in food production and processing, highlighting the role of enzymes, their sources, and particular features and formulations required for targeted applications.

1.2.18 Chapter 19

Chapter 19 by More and coauthors deals with the biocatalytic applications of carbohydrase enzymes. Carbohydrases hydrolyze complex carbohydrates into simple sugars. Carbohydrase enzymes such as maltases, amylases, xylanases, mannanases, glucanases, etc., are used in several industrial steps offering multifaceted benefits. Industries such as food and detergents utilize these enzymes as potent catalysts and product ingredients. In this chapter, the authors extensively reviewed the origin and potential applications of carbohydrases.

1.2.19 Chapter 20

Chapter 20 by Raval and coauthors delineated the role of microbial enzymes in the agricultural industry. The importance and application of enzymes in agriculture, particularly those microbial enzymes found in soil, have been increasing steadily. Soil enzymes are necessary for organic matter transformations, nutrient cycle, and uptake. Soil microbes and their enzymes are frequently utilized in agronomy as accurate markers of soil health, soil fertility, and crop health and yield. The authors provided detailed information on the current state of the eco-friendly use of microbial enzymes in the agricultural industry.

1.2.20 Chapter 21

Chapter 21 by Ferreira-Leitão and coauthors overviews the opportunities and challenges for producing fuels and chemicals and the impacts of microbial enzymes in addressing these challenges for biorefineries. Sustainability linked to using renewable materials for industrial production is considered an unavoidable path. Integrating biofuels and biomass chemicals stimulates the transition to the inevitable bioeconomy era. The authors depicted the current situation of the two main biofuels in Brazil: ethanol and biodiesel. They also explored the opportunities and bottlenecks in exploiting lignocellulosic and oleaginous materials, focusing on the vital role of enzymatic and microbial processes in supporting a sustainable industry.

1.2.21 Chapter 22

Chapter 22 by de Oliveira and coauthors is devoted to using lipases to produce biofuels, highlighting the major advances in lipases for the catalysis of biodiesel, the production methods, immobilization strategies, and raw materials used. The authors also underlined the current limitations and the main challenges to be met for attaining further progress in this demanding field.

1.2.22 Chapter 23

Chapter 23 by More and coauthors enlightens on using microbial enzymes in the textile industry. The most commonly utilized microbial enzymes in textile industries are amylases, peroxidases, catalases, cellulases, and laccases. They can remove the starchy soils; degrade excess hydrogen peroxide and lignin; and take part in de-sizing, scouring, bleaching, garment washing, denim washing, dyeing, and biofinishing in a more effective and nontoxic manner. Enzymes are utilized in the textile industries to make the environment safe and the textile manufacturing processes cost-effective. The present chapter deals with how to produce microbial enzymes used in textile industries in cost-effective and considerable amounts to replace optimally the chemicals used in them. The isolation and identification of microorganisms that produce significant quantities of textile enzymes are also addressed, emphasizing genetic material manipulation.

1.2.23 Chapter 24

Chapter 24 by Yagnik and coauthors presents an excellent overview of microbial enzymes used in bioremediation. The deposition of environmental pollutants like xenobiotic chemicals such as plastics, insecticides, hydrocarbon-containing substances, heavy metals, synthetic dyes, pesticides, and chemical fertilizers has reached an alarming level in recent years due to urbanization growth and industrial expansion. Enzyme-based bioremediation is considered a viable, cost-effective, and eco-friendly solution among modern remediation technologies. The authors summarized the bacterial strains and their enzymes involved in the bioremediation of toxic, carcinogenic, and hazardous environmental contaminants, including industrial bioremediation.

1.2.24 Chapter 25

Chapter 25 by Verma and coauthors overviews the role of microbes and their enzymes for bioelectricity generation. Microbial enzymes and related products form the foundation of bio-based technologies. Bio-based methods should be best exploited to produce various value-added products/chemicals and biofuels. Hydrolytic enzymes play a vital role in bio-based refineries. Recently, the production of bioelectricity by using microorganisms in microbial fuel cells has been receiving attention globally as it can be an efficient source for a steady supply of sustainable energy. The present chapter offers an insight into this area of tremendous interest and future applications.

1.2.25 Chapter 26

Chapter 26 by Verma and his team offers an excellent review of the discovery of untapped nonculturable microbes based on an advanced next-generation metagenomics approach for exploring novel industrial enzymes. Man has been harnessing enzymes from microbes to meet industrialization growth, and the yield from conventional methods could be consumable as microbes continually modify their characteristics. Consequently, the search for new advanced techniques is warranted. Next-generation sequencing and metagenomics have already been found effective in identifying and exploiting several novel enzymes from unculturable microbes. Many untouched aspects will be dealt with in the coming future to explore more about unculturable microbes with the advancements in metagenomics. This approach could better understand the uncultured microbes in the environment and their possible applications in the near future. The authors of the present chapter enlightened the readers on this spectacular aspect of the microbial world.

1.3 Concluding remarks

This introductory chapter summarizes each technical chapter of the book for which the representation of facts and their discussions are exhaustive, authoritative, and deeply informative. The readers would find interest in each of the chapters, which practically cover a broad spectrum of microbial enzymes in terms of sources, production, purification, and applications in various industrial sectors, including agricultural, chemical, pharmaceuticals, textile, paper, bioremediation, biorefineries, biofuels, and bioenergy. The enlarged second edition of this book encourages more interdisciplinary works among chemists, pharmacologists, clinicians, technologists, biologists, botanists, and agronomists interested in microbes and microbial enzymes. Hence, the present book would surely serve as a key reference in this domain.

Chapter 2

Useful microbial enzymes—an introduction

Beatriz Ruiz-Villafán, Romina Rodríguez-Sanoja and Sergio Sánchez
Instituto de Investigaciones Biomédicas, Universidad Nacional Autónoma de México (UNAM), México Distrito Federal, México

2.1 The enzymes: a class of useful biomolecules

According to the International Union of Biochemistry, and based on the nature of their reaction, enzymes are divided into six classes: oxidoreductases, transferases, hydrolases, lyases, isomerases, and ligases. The use of enzymes in industrial processes has been crucial since they can eliminate the use of organic solvents, high temperatures, or extreme pH values. At the same time, they offer high substrate specificity, low toxicity, product purity, reduced environmental impact, and ease of termination of activity (Jemli et al., 2016). Microorganisms constitute the major source of enzymes as they produce high concentrations of extracellular enzymes. Screening for the best enzymes is simple, allowing the examination of thousands of cultures in a short time. Microorganisms used for enzyme production include around 50 Generally Recognized as Safe (GRAS) bacteria and fungi. Bacteria are mainly represented by *Bacillus subtilis, Bacillus licheniformis*, and various *Streptomyces* species, while fungi are represented by *Aspergillus*, *Mucor*, and *Rhizopus*.

Microorganisms can be cultured in large quantities relatively fast by well-established fermentation methods. Microbial enzyme production is economical on a large scale due to inexpensive culture media and short fermentation cycles.

More than 3000 different enzymes are known, but only 5% are commercially used (Parameswaran et al., 2013), and more than 500 commercial products are manufactured using enzymes (Kumar et al., 2014). Its global figures depend on the consulted source regarding the total enzyme market. In one case, the market reached $9.9 billion in 2019 and is predicted to rise 7.1% per annum to grasp $14.9 billion in 2027 (Grand View Research, 2020). A second report estimated $8.47 billion in 2019 and is predicted to

Biotechnology of Microbial Enzymes. DOI: https://doi.org/10.1016/B978-0-443-19059-9.00024-4
© 2023 Elsevier Inc. All rights reserved.

reach $11.63 billion by 2026 (Suncoast News Network, 2021). The major technical enzymes are used in bulk to manufacture detergents, textiles, leather, pulp, paper, and biofuels. The market for these enzymes reached $1.2 billion in revenues in 2011 and is still rising. Other enzymatic applications include household care, foods, animal feed, fine chemicals, and pharmaceuticals. Enzymes have unique properties such as rapid action, high specificity, biodegradability, high yields, ability to act under mild conditions, and reduction in the generation of waste materials. These properties offer flexibility for operating conditions in the reactor.

Enzymes are used to increase nutrient digestibility and degrade unacceptable feed components. Proteases, phytases, glucanases, alpha-galactosidases, alpha-amylases, and polygalacturonases are utilized in the poultry and swine industry. Recent emphasis has been on developing heat-stable enzymes, economic and more reliable assays, improvement of activity, and discovery of new nonstarch polysaccharide-degrading enzymes.

Enzymes for food and beverage manufacture constitute a significant part of the industrial enzyme market, reaching almost $2.2 billion in 2021 (Markets and Markets 2021). Lipases include a substantial portion of the usage, targeting fats and oils. To maximize flavor and fragrance, control of lipase concentration, pH, temperature, and emulsion content are necessary. Lipases are potentially helpful as emulsifiers for foods, pharmaceuticals, and cosmetics. *Aspergillus oryzae* is used as a cloning host to produce fungal lipases, as also obtained from *Rhizomucor miehei*, *Thermomyces lanuginosus*, and *Fusarium oxysporum*.

Fundamental detergent additives include proteases, lipases, oxidases, amylases, peroxidases, and cellulases, the catalytic activity of which begins upon the addition of water. The useful ones are active at thermophilic temperatures (c. 60°C) and alkalophilic pH (9–11) and in the presence of components of washing powders.

Over 60% of the worldwide enzyme market is devoted to proteases. These enzymes are involved in manufacturing foods, pharmaceuticals, leather, detergents, silk, and agrochemicals. Their use in laundry detergents constitutes 25% of global enzyme sales. They include (1) the *B. licheniformis* alcalase Biotex, (2) the first recombinant detergent lipase called Lipolase, made by cloning the lipase from *T. lanuginosa* into *A. oryzae*, (3) the *Pseudomonas mendocina* lipase (Lumafast), and (4) the *Pseudomonas alcaligenes* lipase (Lipomax).

Natural enzymes are often unsuitable as industrial biocatalysts and need modifications. Genetic manipulation usually modifies the production strains to improve their properties, including high production levels. With recombinant DNA technology, it has been possible to clone genes encoding enzymes from microbes and express them at levels tens to hundreds of times higher than those produced by unmodified microorganisms. Because of this, the enzyme industry rapidly accepted the technology and moved enzyme production from strains not suited for the sector into industrial strains (Galante and Formantici, 2003). Genomics, metagenomics, proteomics, and recombinant DNA technology are employed to discover new enzymes from microbes in nature and create or evolve improved enzymes. Several unique and valuable enzymes have been obtained by metagenomics (Adrio and Demain, 2014; Thies et al., 2016, Robinson et al., 2021).

Directed protein evolution includes several methodologies such as DNA shuffling, whole-genome shuffling, heteroduplexing, and transient template shuffling. Additionally, there are the techniques of engineered oligonucleotide assembly, mutagenic and unidirectional reassembly, exon shuffling, Y-ligation-based block shuffling, nonhomologous recombination, and the combination of rational design with directed evolution (Arnold, 2018; Bershstein and Tewfic, 2008). Currently, machine learning is used to improve the quality and diversity of solutions for protein engineering problems (Wu et al., 2019). Directed evolution has increased enzyme activity, stability, solubility, and specificity. For example, it increased the activity of glyphosate-*N*-acetyltransferase 7000-fold and, at the same time, its thermostability by 2- to 5-fold (Siehl et al., 2005).

2.2 Microbial enzymes for industry

According to their applications, microbial enzymes have been applied to numerous biotechnology products and in processes commonly encountered in the production of laundry, food and beverages, paper and textile industries, clothing, biorefinery, etc. *Bacilli* are very useful for enzyme production, especially *B. subtilis*, *B. amyloliquefaciens*, and *B. licheniformis*. This is due to their excellent fermentation properties, high product yields (23–25 g/L), and lack of toxic by-products (Schallmey et al., 2004).

The use of enzymes as detergent additives represents a major application of industrial enzymes. The detergent market for enzymes has grown enormously in the last 25 years (around $1.3 billion in 2017). The first detergent containing a bacterial protease was introduced in 1956, and in 1960, Novo Industry A/S introduced an alkaline protease produced by *B. licheniformis* (Biotex). Proteases are the major enzymes used for detergent preparation, with a market value of around $0.71 billion in 2020. The global protease market is expected to reach $3.35 billion by 2028. The protease market is estimated to represent 72% of the global market for detergent enzymes (Maurer, 2015). Only in Europe in 2013, proteases for the manufacture of detergents had a production level of 900 tons per year (van Dijl and Hecker, 2013).

Cellulase from *Bacillus* sp. KSM-635 has been used in detergents because of its alkaline pH optimum and insensitivity to components in laundry detergents (Ozaki et al., 1990). Later, Novozyme launched a detergent using a cellulase complex isolated from *Humicola insolens* (Celluzyme).

Certain microorganisms called extremophiles grow under extreme conditions such as 100°C, 4°C, 250 atm, pH 10, or 5% NaCl. Their enzymes that act under such extreme conditions are known as extremozymes. Cellulase 103 is an extremozyme isolated from an alkaliphile and commercialized because of its ability to break down microscopic lint from cellulose fibers that trap dirt in cotton fabric. It has been used for over 10 years in detergents to return the "newness" of cotton clothes, even after many washes. As early as the mid-1990s, virtually, all laundry detergents contained genetically engineered enzymes (Adrio and Demain, 2014). Over 60% of the enzymes used in detergents are of recombinant origin (Adrio and Demain, 2010).

Enzymes for food manufacture constitute a significant part of the industrial enzyme market. Their global market was valued at $2.75 billion in 2019.

Fungal alpha-amylase, glucoamylase, and bacterial glucose isomerase are used to produce "high-fructose corn syrup" from starch in a $1 billion-a-year business. Fructose syrups are also made from glucose using a "glucose isomerase" (actually xylose isomerase) at an annual level of 15 million tons per year. The food industry also uses invertase from *Kluyveromyces fragilis*, *Saccharomyces cerevisiae* and *S. carlsbergensis* to manufacture candy and jam. Beta-galactosidase (lactase), produced by *Kluyveromyces lactis*, *K. fragilis* or *Candida pseudotropicalis*, is used to hydrolyze lactose from milk or whey. Alpha-galactosidase from *S. carlsbergensis* is employed to crystallize beet sugar.

Microbial lipases catalyze the hydrolysis of triacylglycerol to glycerol and fatty acids. They are commonly used in producing various products ranging from fruit juices, baked foods, pharmaceuticals, and vegetable fermentations to dairy enrichment. The microbial lipase market was estimated at $425.0 million in 2018, and it is projected to reach $590.2 million by 2023 (Chandra et al., 2020). Fats, oils, and related compounds are the main targets of lipases in food technology. Accurate control of lipase concentration, pH, temperature, and emulsion content is required to maximize flavor and fragrance production. The lipase mediation of carbohydrate esters of fatty acids offers a potential market as emulsifiers in foods, pharmaceuticals, and cosmetics.

Another application of increasing importance involves lipases to remove pitch (hydrophobic components of wood, mainly triglycerides and waxes).

Nippon Paper Industries use lipase from *Candida rugosa* to remove up to 90% of these compounds (Jaeger and Reetz, 1998). The use of enzymes as an alternative to chemicals in leather processing has proved successful in improving its quality and reducing the environmental pollution. Alkaline lipases from *Bacillus* strains, which grow under high alkaline conditions in combination with other alkaline or neutral proteases, are currently being used in this industry. Lipases are also used in detergent formulations to remove lipid stains, greasy food stains, and sebum from fabrics (Hasan et al., 2010). Alkaline yeast lipases are preferred because they can work at lower temperatures than bacterial and fungal lipases. Cold-active lipase detergent formulation is used for cold washing, reducing energy consumption, and wear on textile fibers. It is estimated that about 1000 tons of lipases are added to approximately $13 billion tons of detergents (Zaitsev et al., 2019).

The major application of proteases in the dairy industry is for cheese manufacturing. Food and Drug Administration has approved four recombinant proteases for cheese production. Calf rennin had been preferred in cheese-making due to its high specificity. Still, microbial proteases produced by GRAS microorganisms such as *Rhizomucor miehei*, *R. pusilis*, *B. subtilis*, and *Endothia parasitica* are gradually replacing them. The primary function of these enzymes in cheese-making is to hydrolyze the specific peptide bond (Phe105-Met106) that generates para-k-casein and macropeptides. *A. oryzae* produces nearly 40,000 U/g of milk-clotting activity at 120 h by solid-state fermentation (Vishwanatha et al., 2009). For many years, proteases have also been used to produce low allergenic milk proteins used as ingredients in baby milk formulas (Gupta et al., 2002).

Proteases can also be used for the synthesis of peptides in organic solvents. Thermolysin is used to make aspartame (Alsoufi and Aziz, 2019). Aspartame sold for $1.5 billion in 2003 (Baez-Viveros et al., 2004). In 2004 its production amounted to 14,000 metric tons (Adrio and Demain, 2014). The global sugar substitute market is the fastest growing sector of the sweetener market.

In other enzyme applications, laccases oxidize phenolic and nonphenolic lignin-related compounds and environmental pollutants (Kunamneni et al., 2008; Rodríguez-Couto and Toca-Herrera, 2006). They are used to detoxify effluents from the paper and pulp, textile, and petrochemical industries, bioremediation of herbicides, pesticides, explosives in soil, cleaning agents for water purification systems, as catalysts in drug manufacturing, and as ingredients in cosmetics.

Enzymes are also used in a wide range of agro-biotechnological processes, and the major application is the production of food supplements to improve feed efficiency. A recent advance in food enzymes involves the application of phytases in agriculture as an ingredient in animal feed and in food to enhance the absorption of phosphorus from plants by monogastric animals (Vohra and Satyanarayana, 2003). Phytate phosphorus is often unavailable to farm animals and chelates valuable minerals. Phytase allows phosphorus liberation from plant feedstuffs, which contain about two-thirds of their phosphorus as phytate. Hydrolysis of phytate prevents its passage via manure to the soil, which would be hydrolyzed by microbes from soil and water, causing eutrophication. Therefore phytase in the food industry involves removing phytic acid, an antinutritional factor. The annual market for phytase is about $500 million. The enzyme is produced by many bacteria, yeasts, and filamentous fungi. Production is controlled by phosphate.

Cloning the phytase-encoding gene *phyA* from *Aspergillus niger* var. *awamori* and reintroduction at a higher dosage increased phytase production by 7-fold (Piddington et al., 1993). Recombinant *Hansenula polymorpha* produced 13 g/L of phytase (Mayer et al., 1999). New fungal phytases with higher specific activities or improved thermostability have been recently identified (Haefner et al., 2005).

In the paper and textile industries, enzymes are increasingly used to develop cleaner processes and reduce raw materials and waste production. An alternative enzymatic process developed in the manufacture of cotton is based on a pectate lyase.

Removing pectin and other hydrophobic materials from cotton fabrics is performed at much lower temperatures and with less water than the classical method. Single-site mutants with improved thermotolerance were isolated by Gene Site Saturation Mutagenesis technology applied on DNA encoding pectinolytic enzymes. In addition, variants with improved thermotolerance were produced by Gene Reassembly technology (Solvak et al., 2005). The best performing variant (CO14) contained eight mutations and showed a melting temperature 16°C higher than the wild-type enzyme while retaining the same specific activity at 50°C. The optimal temperature of the evolved enzyme was 70°C, which is 20°C higher than the wild type. Scouring results obtained with the evolved enzyme was significantly better than those obtained with chemical scouring, making it possible to replace the conventional and environmentally harmful chemical scouring process (Solvak et al., 2005). Furthermore, alkaline pectinases are used to treat pectic wastewaters, degumming of plant bast fibers, paper making, and coffee and tea fermentations (Hoondal et al., 2002).

In the chemical industry, enzymes are used to replace chemical processes if they compete successfully for cost. Enzymes require less energy, yield a higher titer with enhanced catalytic efficiency, produce less waste and catalyst by-products, and lower volumes of wastewater streams. They often involve hydrolases and ketoreductases that are stable in organic solvents. Enzymes also can be used to produce valuable compounds such as L-amino acids. For example, L-tyrosine has been made from phenol, pyruvate, pyridoxal phosphate, and ammonium chloride with chemo- and thermostable tyrosine phenol lyase from *Symbiobacterium toebii*. With continuous substrate feeding, the amino acid was produced at 130 g/L in 30 h. About 150 biocatalytic processes are used in the chemical industry, and this number will increase with the application of genomics and protein engineering.

Enzymes are also important in the pharmaceutical industry (Anbu et al., 2015). They are used in the preparation of beta-lactam antibiotics such as semisynthetic penicillins and cephalosporins. This antibiotic group is extremely important, making up 60%–65% of the total antibiotic market. Enzymes are also involved in the preparation of chiral medicines, that is, complex chiral pharmaceutical intermediates. For example, esterases, proteases, lipases, and ketoreductases are used to prepare chiral alcohols, carboxylic acids, amines, and epoxides (Craik et al., 2011).

Recently, biochemical and biological conversion platforms that convert lignocellulosic residues into biofuels for more sustainable production and the energy industry have drastically increased since the effects of climate change are becoming more prevalent (De Buck et al., 2020).

Microbial enzymes as tools for biotechnological processes were reviewed by Adrio and Demain (2014) in a book titled "Biotechnology of Microbial Enzymes: Production, Biocatalysis and Industrial Applications" (Brahmachari et al., 2017).

2.3 Improvement of enzymes

Certain enzymes face problems, such as poor stability, substrate/product inhibition, narrow substrate specificity, or enantioselectivity. Genetic modification is often carried out using recombinant DNA techniques to face these problems. In some cases, this has improved activity by 100-fold. The enzyme is modified by (1) rational redesign of the biocatalyst and/or by (2) combinatorial methods in which the desired functionality is searched in randomly generated libraries. The rational design approach is carried out by site-directed mutagenesis to target amino acid substitutions (Yang et al., 2014). It requires knowledge about the three-dimensional structure of the enzyme and the chemical mechanism of the reaction. This approach often fails, although successes have been achieved (Wu et al., 2017; Chen et al., 2018). Combinatorial methods include directed evolution which does not require extensive knowledge about the enzyme. Here, many variants are created for screening for catalytic efficiency, enantioselectivity, solubility, catalytic rate, specificity, and enzyme stability. The directed evolution method is rapid and inexpensive (Rubin-Pitel and Zhao, 2006). It includes a range of molecular biological methods which allow the achievement of genetic diversity, mimicking mechanisms of evolution in nature. Random mutagenesis of the protein-encoding gene is carried out by various techniques such as (1) error-prone polymerase chain reaction (PCR), (2) repeated oligonucleotide-directed mutagenesis, or (3) action of chemical agents. Error-prone PCR introduces random point mutations in a population of enzymes. Molecular breeding techniques, such as DNA shuffling, allow in vitro random homologous recombination, usually between parental genes with similarity above 70% (Ness et al., 2000). After cloning and inducing protein expression, an extensive collection of enzyme variants, that is, 10^4-10^6, is generated and subjected to screening or selection (Liu et al., 2013).

Protein engineering is used to change protein sequence rationally or combinatorially. Rational methods include site-directed mutagenesis to target amino acid substitutions. A small number of variants are produced, which are then screened.

Improvement of enzymes and whole-cell catalysis has been reviewed by De Carvalho (2011). As a result of such improvement, certain enzymes have achieved large markets. For example, the Taq DNA polymerase isolated from *Thermus aquaticus* had sales of $500 million in 2009 (De Carvalho, 2011).

2.4 Discovery of new enzymes

Screening natural microbes for enzymes suffers from the fact that less than 1% of the microbes inhabiting the biosphere can be cultivated in the laboratory by standard techniques. Genomics, metagenomics, proteomics, and recombinant DNA technology are now employed to facilitate the discovery of new enzymes from microbes in nature and create or evolve improved enzymes (Oates et al., 2021).

Several new and valuable enzymes have been obtained by metagenomics (Thies et al., 2016). Metagenomic screening (Madhavan et al., 2017) involves the preparation of a genomic library from environmental DNA and the systematic screening of the library for open reading frames (ORFs) potentially encoding novel enzymes (Uchiyama and Miyazaki, 2009; Gilbert and Dupont, 2011). Metagenomic screening of particular habitats such as Arctic tundra, cow rumen, volcanic vents, marine environments, and termite guts has yielded valuable enzymes. Examples include lipases, oxidoreductases, amidases, amylases, nitrilases, decarboxylases, epoxide hydrolases, and beta-glucosidases. Although *Escherichia coli* has been the usual host for screening foreign genes, the system has been improved by using alternative hosts and expression systems such as *Streptomyces lividans*, *Pseudomonas putida*, and *Rhizobium leguminosarum*.

Genome mining involves exploring genome sequence databases for genes encoding new enzymes. An example of a useful database is the National Center for Biotechnology Information database (NCBI Microbial Genomes, https://www.ncbi.nlm.nih.gov/genome) which includes more than 100,000 genome sequences and draft assemblies (Klemetsen et al., 2018). Two methods are used for the discovery of new enzymes. One of them, genome hunting, involves the search for ORFs in the genome of a particular microbe. Those annotated sequences as putative enzymes are subjected to subsequent cloning, overexpression, and activity screening. A second approach called data mining is based on homology alignment among all sequences deposited in databases. Using bioinformatics tools such as basic local alignment search tool, the search for conserved regions between sequences yields orthologous protein sequences that are then considered candidates for further study.

In cooperation with the Pfizer pharmaceutical corporation, enzyme evolution was used by the Codexis Corporation to produce (*R*)-2-methylpentanol, an essential intermediate for the manufacture of pharmaceuticals and liquid crystals (Gooding et al., 2010). Codexis has also developed enzymatic processes to replace and improve chemical transformations for the production of sitagliptin (Saviole et al., 2010), montelukast (Singulair) (Liang et al., 2009a), and sulopenem (Liang et al., 2009b).

Extremophiles can survive under extreme conditions. These include temperature (−2°C to 12°C, 60°C to 110°C) pressure, radiation, salinity (2−5 M NaCl), and pH (<2, >9). These microorganisms contain extremely stable enzymes. Genera, such as *Clostridium*, *Thermotoga*, *Thermus*, and *Bacillus*, contain thermophiles growing at 60°C−80°C, whereas hyperthermophiles are the members of *Archaea*, for example, *Pyrococcus*, *Methanopyrus*, and *Thermococcus*. An example of a useful enzyme is the maltogenic amylase of *Bacillus stearothermophilus*. This enzyme, sold as Novamyl (Novozymes), is used in the bakery industry for improved freshness and other bread qualities (Sarmiento et al., 2015). The industry already uses thermophilic cellulases, amylases, and proteases.

Psychrophiles already supply cold-active enzymes such as proteases, amylases, and lipases for the future development of detergents to reduce the wear of textile fibers. In producing second-generation biofuels via saccharification of pretreated lingocellulosic biomass, cold-active cellulases and xylanases are of interest in the pulp and paper industry. They are also potentially useful for extracting and clarifying fruit juices, improving bakery products, bioremediation of waters contaminated with hydrocarbons or oils, and polishing and stonewashing textiles. Halophilic xylanases, proteases, amylases, and lipases have been isolated from several halophiles such as *Halobacillus*, *Halobacterium*, and *Halothermothrix* (Van den Burg, 2003).

Also of interest are microbes surviving under extreme pH conditions, which could be useful for isolating thermoalkaliphilic proteases and lipases as additives in laundry and dishwashing detergents (Shukla et al., 2009).

2.5 Concluding remarks

Industrial product makers have long used microbial enzymes as major catalysts to transform raw materials into specific products. Over 500 commercial products are made using enzymes. They are economically produced by different microorganisms and are quickly broken down upon completing their job. New technical tools to use enzymes as crystalline catalysts to recycle cofactors and engineering enzymes to function in various solvents with multiple activities are important technological developments which will steadily create new applications.

The industrial enzyme market will grow steadily mainly due to improved production efficiency, resulting in cheaper enzymes, new application fields, new enzymes from screening programs, and engineering properties of traditional enzymes. Tailoring enzymes for specific applications will be a future trend with continuously improving tools, further understanding of structure–function relationships, and increased searching for enzymes from exotic environments. New applications are to be expected in textiles and new animal diets such as ruminant and fish feed. It can be expected that breakthroughs in pulp and paper applications will materialize. The use of cellulases to convert waste cellulose into sugars and further to ethanol or butanol by fermentative organisms has been a major topic of study for years. Increasing environmental pressures and energy prices will make this application a real possibility in the future.

Enzymes should never be considered alone but rather part of a biocatalyst technology. Recent developments in genetic engineering and protein chemistry are bringing ever more powerful means of analysis to study enzyme structure and function. These developments will undoubtedly lead to the rational modification of enzymes to match specific requirements and the design of new enzymes with novel properties. The techniques such as protein engineering, gene shuffling, and directed evolution will enable enzymes to be better suited to industrial environments (Kumar and Singh, 2013). These tools will also allow the synthesis of new biocatalysts for completely novel applications, resulting in the production and commercialization of new enzymes, thus seeding a second explosive expansion to the current multibillion-dollar enzyme industry.

Acknowledgments

We are indebted to Marco A. Ortíz-Jiménez for his assistance during this manuscript edition. Part of this work was supported by the CONACYT grant AS1–9143. This work was supported by the NUATEI program from Instituto de Investigaciones Biomédicas, UNAM.

Abbreviations

GRAS Generally Recognized as Safe
PCR polymerase chain reaction
ORF open reading frame

References

Adrio, J.L., Demain, A.L., 2010. Recombinant organisms for production of industrial products. Bioengineered Bugs 1 (2), 116–131. Available from: https://doi.org/10.4161/bbug.1.2.10484.

Adrio, J.L., Demain, A.L., 2014. Microbial enzymes: tools for biotechnological processes. Biomolecules 4, 117–139. Available from: https://doi.org/10.3390/biom4010117.

Alsoufi, M.A., Aziz, R.A., 2019. Production of aspartame by immobilized thermolysin. Iraqi J. Sci. 60, 1232–1239. Available from: https://doi.org/10.24996/ijs.2019.60.6.6.

Anbu, P., Gopinath, S.C.B., Chaulagain, B.P., Tang, T.-H., Citartan, M., 2015. Microbial enzymes and their applications in industries and medicine. 2014. BioMed. Res. Int. 2015, 816419. Available from: https://doi.org/10.1155/2015/816419.

Arnold, F.H., 2018. Directed evolution: bringing new chemistry to life. Angew. Chem. 57 (16), 4143–4148. Available from: https://doi.org/10.1002/anie.201708408.

Baez-Viveros, J.L., Osuna, J., Hernández-Chávez, G., Soberón, X., Bolívar, F., Gosset, G., 2004. Metabolic engineering and protein directed evolution increase the yield of L-phenylalanine synthesized from glucose in *Escherichia coli*. Biotechnol. Bioeng. 87, 516–524.

Bershstein, S., Tewfic, D.S., 2008. Advances in laboratory evolution of enzymes. Curr. Opin. Chem. Biol. 12, 151–158.

Brahmachari, G., Demain, A.L., Adrio, J.L. (Eds.), 2017. Biotechnology of Microbial Enzymes: Production, Biocatalysis and Industrial Applications. Amsterdam, Elsevier/Academic Press.

Chandra, P., Enespa, Singh, R., Arora, P.K., 2020. Microbial lipases and their industrial applications: a comprehensive review. Microb. Cell Factories 19, 169. Available from: https://doi.org/10.1186/s12934-020-01428-8.

Chen, H., Li, M., Liu, C., Zhang, H., Xian, Mo, Liu, H., 2018. Enhancement of the catalytic activity of isopentenyl diphosphate isomerase (IDI) from Saccharomyces cerevisiae through random and site-directed mutagenesis. Microb. Cell Factories 17, 65. Available from: https://doi.org/10.1186/s12934-018-0913-z.

Craik, C.S., Page, M.J., Madison, E.L., 2011. Proteases as therapeutics. Biochemical J. 435, 1–16.

De Buck, V., Polanska, M., Van Impe, J., 2020. Modeling biowaste biorefineries: a review. Front. Sustain. Food Syst. 4, 11. Available from: https://doi.org/10.3389/fsufs.2020.00011.

De Carvalho, C.C., 2011. Enzymatic and whole cell catalysis: Finding new strategies for old processes. Biotechnol. Adv. 29, 75–83.

Grand View Research, 2020. Enzymes market size, share & trends analysis report. Market Analysis Report. https://www.grandviewresearch.com/industry-analysis/enzymes-industry

Galante, Y.M., Formantici, C., 2003. Enzyme applications in detergency and in manufacturing industries. Curr. Org. Chem. 13, 1399–1422.

Gilbert, J.A., Dupont, C.L., 2011. Microbial metagenomics: beyond the genome. Annu. Rev. Mar. Sci. 3, 347–371.

Suncoast News Network, 2021. Global "Enzymes Market" (2016–2027). https://www.snntv.com/story/44244479/global-enzymes-market-value-and-size-expected-to-reach-usd-11630-million-growing-at-cagr-of-46-forecast-period-2021-2027.

Gooding, O.W., Voladr, R., Bautista, A., Hopkins, T., 2010. Development of a practical biocatalytic process for (R)-2-methylpentanol. Org. Process. Res. & Dev. 74, 119–126.

Gupta, R., Beg, Q.K., Lorenz, P., 2002. Bacterial alkaline proteases: molecular approaches and industrial applications. Appl. Microbiology Biotechnol. 59, 15–32.

Haefner, S., Knietsch, A., Scholton, E., Braun, J., Lohscheidt, M., Zelder, O., 2005. Biotechnological production and applications of phytases. Appl. Microbiology & Biotechnol. 68, 588–597.

Hasan, F., Shah, A.A., Javed, S., Hameed, A., 2010. Enzymes used in detergents: lipases. Afr. J. Biotechnol. 9, 4836–4844.

Hoondal, G.S., Tiwari, R.P., Tewari, R., Dahiya, N., Beg, Q.K., 2002. Microbial alkaline pectinases and their industrial applications: a review. Appl. Microbiology Biotechnol. 59 (4–5), 409–418. Available from: https://doi.org/10.1007/s00253-002-1061-1.

Jaeger, K.-E., Reetz, M.T., 1998. Microbial lipases form versatile tools for biotechnology. TibTechnology 16, 396–403.

Jemli, S., Ayadi-Zouari, D., Hlima, H.B., Bejar, S., 2016. Biocatalysts: application and engineering for industrial purposes. Crit. Rev. Biotechnol. 36, 246–255.

Klemetsen, T., Raknes, I.A., Fu, J., Agafonov, A., Balasundaram, S.V., Tartari, G., et al., 2018. The MAR databases: development and implementation of databases specific for marine metagenomics. Nucleic acids Res. 46, D692–D699. Available from: https://doi.org/10.1093/nar/gkx1036.

Kumar, D., Parshad, D., Gupta, V.K., 2014. Application of a statistically enhanced, novel, organic solvent stable lipase from *Bacillus safensis* DVL-43. Int. J. Biol. Macromolecules 66, 97107.

Kumar, A., Singh, S., 2013. Directed evolution: tailoring biocatalysts for industrial applications. Crit. Rev. Biotechnol. 36, 365–378.

Kunamneni, A., Plou, F.J., Ballesteros, A., Alcalde, M., 2008. Laccases and their applications: a patent review. Recent. Pat. Biotechnol. 2, 10–24.

Liang, J., Lalonde, J., Borup, B., Mitchell, V., Mundorff, E., Trinh, N., et al., 2009a. Development of a biocatalytic process as an alternative to the (−)-DIP-Cl-mediated asymmetric reduction of a key intermediate of montelukast. Org. Process. Res. Dev. 14, 193–198.

Liang, J., Mundorff, E., Voladri, R., Jenne, S., Gilson, L., Conway, A., et al., 2009b. Highly enantioselective reduction of a small heterocyclic ketone: biocatalytic reduction of tetrahydrothiophene-3-one to the corresponding (R)-alcohol. Org. Process. Res. Dev. 14, 188–192.

Liu, L., Yang, H., Shin, H.D., Chen, R.R., Li, J., Du, G., et al., 2013. How to achieve high-level expression of microbial enzymes: strategies and perspectives. Bioengineered. 4, 212–223.

Madhavan, A., Sindhu, R., Parameswaran, B., Parameswaran, B., Sukumaran, R.K., Pandey, A., 2017. Metagenome analysis: a powerful tool for enzyme bioprospecting. Appl. Biochem. Biotechnol. 183 (636–651), 2017. Available from: https://doi.org/10.1007/s12010-017-2568-3.

Markets and Markets, 2021. Available from: https://www.marketsandmarkets.com/Market-Reports/food-enzymes-market-800.html?gclid = Cj0KCQjwrs2XBhDjARIsAHVymmTvRHFPJv4ot28n_c80S5sYCZC32ht8-LfMxo_ygeTld2iNjC8quVkaAl9fEALw_wcB (2021) (accessed 10.08.22).

Maurer, K.-H., 2015. Detergent proteases. In: Grunwald, P. (Ed.), Industrial Biocatalysis. CRC Press, Taylor & Francis Group, Boca Raton, Fl. USA, pp. 949–984.

Mayer, A.F., Hellmuth, K., Schlieker, H., López-Ubarri, R., Oertel, S., Dahlems, U., et al., 1999. An expression system matures: a highly efficient and cost-effective process for phytase production by recombinant strains of *Hansenula polymorpha*. Biotechnol. Bioeng. 63, 373–381.

Ness, J.E., del Cardayre, S.B., Minshull, J., Stemmer, W.P., 2000. Molecular breeding: The natural approach to protein design. Adv. ProteChem. 55, 261–292.

Oates, N.C., Abood, A., Schirmacher, A.M., Alessi, A.M., Bird, S.M., Bennett, J.- P., et al., 2021. A multi-omics approach to lignocellulolytic enzyme discovery reveals a new ligninase activity from *Parascedosporium putredinis* NO1. Proc. Natl Acad. Sci. 118 (18). Available from: https://doi.org/10.1073/pnas.2008888118e2008888118.

Ozaki, K., Shikata, S., Kaway, S., Ito, S., Okamoto, K., 1990. Molecular cloning and nucleotide sequence of a gene for alkaline cellulase from *Bacillus* sp. KSM-635. J. Gen. Microbiology 136, 1327–1334.

Parameswaran, B., Palkhiwala, P., Gaikaiwari, R., Nampoothiri, K.M., Duggal, A., Dey, K., et al., 2013. Industrial enzymes - present status & future perspectives for India. J. Sci. Ind. Res. 72, 271–286.

Piddington, C.S., Houston, C.S., Paloheimo, M., Cantrell, M., Miettinen-Oinonen, A., Nevalainen, H., et al., 1993. The cloning and sequencing of the genes encoding phytase (*phy*) and pH 2.5-optimum acid phosphatase (*aph*) from *Aspergillus niger* var. *awamori*. Gene. 133, 55–62.

Robinson, S.L., Piel, J., Sunagawa, S., 2021. A roadmap for metagenomic enzyme discovery. Nat. Products Reports. Advance Artic. Available from: https://doi.org/10.1039/d1np00006c.

Rodríguez-Couto, S., Toca-Herrera, J.L., 2006. Industrial and biotechnological applications of laccases: A review. Biotechnol. Adv. 24, 500–513.

Rubin-Pitel, S.B., Zhao, H., 2006. Recent advances in biocatalysis by directed enzyme evolution. Combinatorial Chem. High. Throughput Screen. 9, 247–257.

Sarmiento, F., Peralta, R., Blamey, J.M., 2015. Cold and hot extremozymes: industrial relevance and current trends. Front. Bioeng. Biotechnol. 3, 148. Available from: https://doi.org/10.3389/fbioe.2015.00148.

Saviole, C.K., Janey, J.M., Mundorff, E.C., Moore, J.C., Tam, S., Jarvis, W.R., et al., 2010. Biocatalytic asymmetric synthesis of chiral amines from ketones applied to sitagliptin manufacture. Science 329, 305–309.

Schallmey, M., Singh, A., Ward, O.P., 2004. Developments in the use of *Bacillus* species for industrial production. Canadian J. Microbiology 50 (1), 1–17. Available from: https://doi.org/10.1139/w03-076. PMID: 15052317.

Shukla, A., Rana, A., Kumar, L., Singh, B., Ghosh, D., 2009. Assessment of detergent activity of *Streptococcus* sp. AS02 protease isolated from soil of Sahastradhara, Doon Valley, Uttarakhand. Asian J. Microb. Biotechnol. Environ. Sci. 11, 587–591.

Siehl, D.L., Castle, L.A., Gorton, R., Chen, Y.H., Bertain, S., Cho, H.J., et al., 2005. Evolution of a microbial acetyl transferase for modification of glyphosate: a novel tolerance strategy. Pest. Manag. Sci. 2005 (61), 235–240.

Solvak, A.I., Richardson, T.H., McCann, R.T., Kline, K.A., Bartnek, F., Tomlinson, G., et al., 2005. Discovery of pectin-degrading enzymes and directed evolution of a novel pectate lyase for processing cotton fabric. J. Biol. Chem. 280, 9431–9438.

Thies, S., Rausch, S.C., Kovacic, F., Schmidt-Thaler, A., Wilhelm, S., Rosenau, F., et al., 2016. Metagenomic discovery of novel enzymes and biosurfactants in a slaughterhouse biofilm microbial community. Sci. Rep. 6, 27035. Available from: https://doi.org/10.1038/srep27035.

Uchiyama, T., Miyazaki, K., 2009. Functional metagenomics for enzyme discovery: Challenges to efficient screening. Curr. Opin. Biotechnol. 20, 616–622.

Van den Burg, B., 2003. Extremophiles as a source for novel enzymes. Curr. Opin. Microbiology 6, 213–218.

van Dijl, J.M., Hecker, M., 2013. *Bacillus subtilis*: from soil bacterium to super-secreting cell factory. Microb. Cell Factories 12, 3–9.

Vishwanatha, K.S., Appu Rao, A.G., Singh, S.A., 2009. Production and characterization of a milk-clotting enzyme from *Aspergillus oryzae* MTCC 5341. Appl. Microbiology Biotechnol. 85, 1849–1859.

Vohra, A., Satyanarayana, T., 2003. Phytases: microbial sources, production, purification, and potential biotechnological applications. Crit. Rev. Biotechnol. 23, 29–60.

Wu, Z., Kan, S.B.J., Lewis, R.D., Wittman, B.J., Arnold, F.H., 2019. Machine learning-assisted directed protein evolution with combinatorial libraries. Proc. Natl Acad. Sci. 116 (18), 8852–8858. Available from: https://doi.org/10.1073/pnas.1901979116.

Wu, X., Tian, Z., Jiang, X., Zhang, Q., Wang, L., 2017. Enhancement in catalytic activity of *Aspergillus niger* XynB by selective site-directed mutagenesis of active site amino acids. Appl. Microbiology Biotechnol. 102, 249–260.

Yang, H., Li, J., Shin, H.D., Du, G., Liu, L., Chen, J., 2014. Molecular engineering of industrial enzymes: recent advances and future prospects. Appl. Microbiology Biotechnol. 98, 23–29.

Zaitsev, S.Y., Savina, A.A., Zaitsev, I.S., 2019. Biochemical aspects of lipase immobilization at polysaccharides for biotechnology. Adv. Colloid Interface Sci. 272, 102016.

Chapter 3

Production, purification, and application of microbial enzymes

Anil Kumar Patel[1], Cheng-Di Dong[1], Chiu-Wen Chen[1], Ashok Pandey[2] and Reeta Rani Singhania[1]

[1]*Department of Marine Environmental Engineering, National Kaohsiung University of Science and Technology, Kaohsiung City, Taiwan,* [2]*CSIR-Indian Institute for Toxicological Research, Lucknow, Uttar Pradesh, India*

3.1 Introduction

Nature always uses natural catalysts for carrying out chemical conversions of its substances to speed up the reaction and/or control the process. These natural catalysts are the enzymes that are active proteins (except RNAse) capable of catalyzing biochemical reactions. These are biomolecules required for both syntheses and breakdown reactions by living organisms. All living organisms are built up and maintained by these enzymes, which are genuinely termed biological catalysts that can convert a specific compound (as substrate) into products at a higher reaction rate. These are biological and stable, hence making the most eco-friendly catalysts. Like other catalysts, an enzyme increases the reaction rate by lowering its activation energy (E_a). Thus products are formed faster, and reactions rapidly reach equilibrium. The rates of most of the enzymatic reactions are millions of times faster than that of the uncatalyzed reactions. They can perform a specific conversion in minutes or even in seconds which otherwise may take hundreds of years (Dalby, 2003; Otten and Quax, 2005). Enzymes are known to catalyze about 4000 biochemical reactions in living beings (Bairoch, 2000). For example, lactase is a glycoside hydrolase, which can hydrolyze lactose (milk sugar) into constituent galactose and glucose monomers. It is produced by various microorganisms in the small intestine of humans and other mammals, helping to digest milk. Enzymes are also enantioselective catalysts, which can either separate enantiomers from racemic mixture or synthesize chiral compounds.

Biotechnology of Microbial Enzymes. DOI: https://doi.org/10.1016/B978-0-443-19059-9.00019-0
© 2023 Elsevier Inc. All rights reserved.

Humans recognized the importance of enzymes thousands of years ago; clarification and filtration of wines and beer are the earliest examples of the application of industrial enzymes. Since prehistoric times, enzymes have been used in brewing, baking, and alcohol production. However, they did not call them enzymes! One of the earliest written references to enzymes is found in Homer's Greek epic poems dating from about 800 BCE, where it has been mentioned that enzymes were used in producing cheese. For more than a thousand years, the Japanese have also used naturally occurring enzymes in making fermented products such as Sake, the Japanese schnapps brewed from rice. Nature has designed some enzymes to form complex molecules from simpler ones, while others have been designed for breaking up the complex molecules into simpler ones. These reactions involve the making and breaking of the chemical bonds within these components. Owing to their "specificity," a property of enzymes that allows them to recognize a particular substrate, they may be designed to target specifics. Hence, they are useful for industrial processes capable of catalyzing reactions between certain chemicals even though they are present in mixtures with many chemicals. Natural enzymes are environmentally safe and applied very safely in food and even pharmaceutical industries. Still, enzymes, being proteins, can cause allergic reactions; hence, protective measures are necessary during their production and applications.

Enzyme technology is an ever-evolving branch of "science and technology." With the intervention and influence of biotechnology and bioinformatics, continuously novel or improved applications of enzymes are emerging. With somewhat novel applications, the need for enzymes with improved properties is also cooccurring. The development of commercial enzymes is a specialized business that is usually undertaken by companies possessing high skills in:

- screening for new and improved enzymes,
- selection of microorganisms and strain improvement for qualitative and quantitative improvement,
- fermentation for enzyme production,
- large-scale enzyme purifications, and
- formulation of enzymes for sale.

Enzyme technology allows industries and consumers to replace processes using aggressive chemicals with mild and environment-friendly enzyme processes. About 3000 enzymes exist, out of which only 150–170 are being exploited industrially. Currently, only 5% of the chemical products are produced through the biological routes in this green era. With time, enzymatic processes are emerging as economically feasible and eco-friendly alternatives to physicochemical and mechanical processes. Based on different application sectors, industrial enzymes can be classified as: (1) enzymes in the food industry, (2) enzymes for processing aids, (3) enzymes as industrial biocatalysts, (4) enzymes in genetic engineering, and (5) enzymes in cosmetics.

Today, the enzymes are envisaged as the bread and butter of biotechnology because they are the main tools for several biotechnological techniques (gene restriction, ligation and cloning, etc.) and bioprocesses (fermentation and cell culture) and analytics in human and animal therapy as medicines or as drug targets. Furthermore, they find applications in several other industries such as food and feed, textiles, effluent and waste treatment, paper, tannery, baking, brewing, dairy, pharmaceuticals, confectionery, etc. (Pandey et al., 2006, Patel et al., 2016).

The enzymes utilized today are also found in animals (pepsin, trypsin, pancreatin, and chymosin) and plants (papain, bromelain, and ficin), but most are microbial origin such as glucoamylase, alpha-amylase, pectinases, etc. The advantage of using microbes for enzyme production is based on their higher growing abilities, higher productivity, easier genetic manipulation for enhanced enzyme production, etc. Enzymes produced from the microbial origin are termed microbial enzymes. Microbes are mainly exploited in industries for enzyme production. Moreover, microbial enzymes are ample supplied, well standardized, and marketed by several competing companies worldwide. Depending on the type of process, enzymes can be used in soluble (animal proteases and lipases in tannery) and immobilized forms (isomerization of glucose to fructose by glucose isomerase).

3.2 Production of microbial enzymes

Naturally occurring microorganisms produce most industrial enzymes. The industry has exploited this knowledge for more than 50 years. Bacteria and filamentous fungi are the microorganisms best suited to the industrial production of enzymes. They are easy to handle, can be grown in huge tanks without light, and have a very high growth rate.

Bacterium *Bacillus subtilis* and the fungus *Aspergillus oryzae* are the most employed microorganisms for enzyme production by Novozymes. Both have an immense capacity for producing enzymes and are considered completely harmless for humans (http://www.novozymes.com/en/about-us/our-business/what-are-enzymes/Pages/creating-the-perfect-enzyme.aspx). The ideal characteristics of a microorganism include fast growth and the ability to produce high titers of the desired enzyme at mild temperature while consuming inexpensive nutrients. However, it is usually challenging to get through the ideal microorganism naturally. Most microorganisms found naturally might not suit well to culture in large fermentation tanks. Some only produce tiny quantities of enzyme or take a long time to grow. Others can produce undesired by-products that would disturb industrial processes, especially in downstream processing. So, a perfect microorganism is a foremost requirement for industrial production. Table 3.1 shows the microorganisms involved in producing enzymes and their industrial applications.

TABLE 3.1 Industrial application of enzymes.

Enzyme	Source organism	Method of production	Industrial application
Amylase (α and gluco)	Bacteria (*Bacillus amyloliquefaciens, Bacillus licheniformis, Bacillus coagulans*)	SmF	1. Mashing for beer making 2. Sugar recovery from scrap candy in the candy industry 3. Starch modification for paper coating in the paper industry 4. Cold swelling laundry starch in starch and syrup industry 5. Wallpaper removal 6. Desizing of fabrics in textiles 7. Degradation of protein, causing stains in the detergent industry
	Fungi (*Aspergillus oryzae, Aspergillus niger, Rhizopus* sp.)	SSF and SmF	1. Precooked baby foods and breakfast foods in the cereals industry 2. Sugar recovery from scrap candy in the candy industry 3. Removal of starch, clarification, oxygen removal for flavor enhancement 4. Starch removal from pectin in fruits and fruit juices 5. Corn syrup in starch and syrup industry 6. Production of glucose in starch and syrup industry 7. Bread baking in the baking and milling industry 8. Digestive aids in clinics and pharmaceuticals

(*Continued*)

TABLE 3.1 (Continued)

Enzyme	Source organism	Method of production	Industrial application
			9. Liquefying purees and soups
Protease	Bacteria (*B. amyloliquefaciens*)	SmF	1. Chill proofing in the beer industry 2. For condiment in the food industry 3. Milk protein hydrolysate making in the dairy industry 4. Unhairing and bating in the leather industry 5. Recovery of silver from films in photography 6. Degradation of fat, causing stains in the detergent industry
	Fungi (*A. oryzae, A niger, Pseudomonas* sp., *Penicillium chrysosporium, Rhizopus oligosporus, Actinomycetes strain)*	SSF and SmF	1. Bread baking in the baking and milling industry 2. Chill proofing in the beer industry 3. For condiment in the food industry 4. Milk protein hydrolysate making in the dairy industry 5. Evaporated milk stabilization in the dairy industry 6. Spot removal in dry cleaning, the laundry industry 7. Digestive aids in clinics and pharmaceuticals 8. Unhairing and bating in the leather industry 9. Meat tendering, tenderizing casings, condensed fish soluble

(*Continued*)

TABLE 3.1 (Continued)

Enzyme	Source organism	Method of production	Industrial application
			10. Resolution racemic mixture of amino acids
Glucose oxidase	Fungi (*A. niger* and *Penicillum* sp.)	SSF and SmF	1. Oxygen removal in beer and beverages 2. Dried milk, oxygen removal in the dairy industry 3. Oxygen and oxygen removal, mayonnaise in dried and egg industry 4. Paper test strips for diabetic glucose in pharmaceuticals 5. Oxygen removal for flavors enhancement in fruits and juices 6. In toothpaste to convert glucose into gluconic acid and hydrogen peroxide as both act as a disinfectant
Pectinases	Fungi (*A. niger, Penicillium* sp.)	SSF and SmF	1. Coffee bean fermentation, coffee concentrates in the coffee industry 2. Clarification, filtration, concentration in fruit and fruit juices 3. Pressing, clarification, filtration in the wine industry
Lactase	Yeast (*Kluyveromyces*)	SmF	1. Whole milk concentrates 2. Ice cream and frozen desserts 3. Whey concentrates 4. Lactose hydrolysis in the dairy industry

(*Continued*)

TABLE 3.1 (Continued)

Enzyme	Source organism	Method of production	Industrial application
Cellulase	Fungi (*Trichoderma reesei, Trichodermaviride, Penicillium* sp., *Humicola grisea, Aspergillus* sp., *Chrysosporium lucknowense, Acremonium* sp.)	SSF and SmF	1. Deinking of papers for recycling in the paper and pulp industry 2. Bio-stonewashing denim in textile industry 3. Hydrolyzing cellulosic biomass to generate glucose for ethanol production in biofuel industry 4. Loosening of cellulose fibers to easily remove dirt and color in the detergent industry
Xylanase	Fungi (*Myceliophthora thermophila, Bacillus* sp., *A. oryzae, Trichoderma* sp.)	SSF and SmF	1. Biobleaching in paper and pulp industry 2. Fiber solubility in animal feed industry
Lipase and proteinase	*A. oryzae, Aspergillus terreus, Pseudomonas* sp., *Alcaligenes* sp., *Staphylococcus* sp., *Candida albicans, Rhizopus* sp., *Mucor*	SSF	1. Contact lens cleaning 2. Brightening in the detergent industry 3. Ripening of cheese in the dairy industry
Phytase	*Aspergillus* sp., *Aspergillus ficuum, Penicillium funiculosum, Bacillus* sp., *Pseudomonas, Xanthomonas oryzae*	SSF	1. Release of phosphate in the animal feed industry
Dextrinase	Fungi	SSF	2. Corn syrup preparation in starch and syrup making
Invertase	yeast (*Saccharomyces*)	SmF	1. Soft center candies and fondants 2. High test molasses

(*Continued*)

TABLE 3.1 (Continued)

Enzyme	Source organism	Method of production	Industrial application
Laccases and peroxidase	*Aspergillus nidulans*, *Aspergillus* sp., *Basidiomycetes*	SSF	1. Polymerize materials with wood based fibers in paper and pulp industry
Hemicellulase	Fungi (*A. niger, T. reesei, Penicillium* sp.)	SSF and SmF	1. Coffee concentrates 2. Hydrolyzing hemicellulosic biomass to generate glucose for ethanol production in the biofuel industry
Catalase	*Aspergillus* sp.		1. Biopolishing and bleach cleanup in textile industry

SmF, Submerged fermentation; *SSF*, Solid-state fermentation.

3.2.1 Enzyme production in industries

Different microorganisms have been employed for industrial enzyme production, varying from eukaryotic systems such as yeast and fungi to prokaryotic systems involving Gram-positive and Gram-negative bacteria.

The first enzyme industry was developed for producing subtilisin, an alkaline protease naturally secreted by *Bacillus licheniformis* to break down proteinaceous substrate used in detergent. An industry for alpha-amylase production was also based on *B. licheniformis*, which naturally secretes a highly thermostable alpha-amylase capable of breaking down starch to easily digestible oligosaccharides. Hence, strains of *Bacillus* were regarded as workhorses of enzyme production for decades because of their ability to overproduce subtilisin and alpha-amylase. Amylase from *Bacillus* is being used for a long period to liquefy starch. Another enzyme called glucoamylases is required for a complete breakdown of starch into its monosaccharide. The most widely used glucoamylases for the hydrolysis of starch into glucose are produced by fungal strains of the genus *Aspergillus* (Sonenshein et al., 1993). Overproducing strains have been isolated over the years for glucoamylase. Likewise, an acidic cellulase complex is secreted by a fungal strain of *Trichoderma*. This enzyme complex was assumed to convert cellulosic substrate to glucose, similar to the starch-degrading enzyme, but this application was not initially commercialized due to slow action. Instead, it has found its application in the treatment of textiles and as an additive in detergents.

The potential of cellulases for biomass hydrolysis for bioethanol production was lost in oblivion. But in recent times, its potential for cellulose hydrolysis regained attention from researchers worldwide, mainly for applying bioethanol production using cellulosic biomass, which is the most abundant material available for use to mankind. Several efforts were made to improve the efficiency of the enzyme complex and its expression. Presently, several enzyme companies, such as "Genencor" and "Novozymes" are preparing cellulase cocktails for bioethanol application. These are produced from filamentous fungi such as *Trichoderma*, *Aspergillus*, *Penicillium*, etc.

Glucose isomerase catalyzes the conversion of glucose into fructose, resulting in a product sweeter in taste. This particular enzyme was produced by species of *Streptomyces* which led to the development of the food industry. Fructose has the highest relative sweetness and, at the same time, has the lowest glycemic index among all the naturally occurring sugars (Kumar et al., 2020).

All of the above strains employed for industrial applications are capable of differentiation; for example, *Bacillus* shows the tendency to survive adverse environmental conditions by forming dormant yet viable spores. The spore remains dormant until it reaches a favorable environment where it can germinate and multiply. This differentiation is highly associated with the regulation of enzyme production by microorganisms. This highly complex behavior makes modeling studies difficult for process development. Protease production by *Bacillus* is highly regulated with differentiation, similar to *Aspergillus* and *Trichoderma*. *Streptomyces* does not truly sporulate though it differentiates by forming filaments, unlike isolated single cells. This property exerts an effect on the production and physical properties of fermentation broth. An organism can be considered as a metabolic system capable of utilizing substrates to produce cell mass and by-products. Enzymes catalyze different reactions vital for organism's growth and metabolic activities. Each cell is equipped with a mechanism that economically regulates the synthesis of enzymes, enabling the cell to respond adequately to environmental changes.

The basic mechanism of enzyme synthesis includes transcription, translation, and posttranslational processing, which is highly conserved (Rehm and Reed, 1985). However, several differences exist between various organisms and some fundamental differences between prokaryotic and eukaryotic organisms. The enzymes themselves differ in their molecular structure, the number of a polypeptide chain, degree of glycosylation, and isoelectric point. Though all the differences influence the synthetic pattern, the basic enzyme synthesis mechanisms are similar enough to allow a general treatment of the microbiological production process. Differences exist between the production kinetics of different enzymes by different microorganisms because of their varied physical characteristics and growth pattern, which necessitates optimizing each production process separately (Patel et al., 2016).

3.2.2 Industrial enzyme production technology

Fermentation technologies have been employed exclusively to produce industrial enzymes, preferably by microorganisms such as bacteria or fungi, under carefully controlled conditions due to their ease of multiplication and handling. Microorganisms employed are GRAS (Generally Recognized as Safe) strains due to their application in food and feed industries (Singhania et al., 2010; Pandey et al., 2008). Researchers are currently isolating extremophile organisms from different parts of the world, ranging from rainforests to arid regions and ocean bottom. These isolates produce enzymes with a promising industrial nature. In practice, most microbial enzymes come from a minimal number of genera, of which *Aspergillus*, *Trichoderma*, *Penicillium*, *Bacillus*, and *Streptomyces* and *Kluyveromyces* species predominate. Most of the strains used have been employed by the food industry for many years or derived from such strains by mutation and selection (Sarrouh et al., 2012).

Selection of the strain for industrial enzyme production is a significant factor in leading a successful industrial process. Ideally, the strain producing extracellular enzyme should be selected as it makes the purification and recovery far more accessible than when had intracellularly. Different organisms may also differ in their suitability for fermentation. The process characteristics, such as viscosity, recoverability, and legal clearance of the organism, should also be considered before selection. Industrial strains typically produce up to 50 g/L of extracellular protein. Filamentous fungi are known to secrete up to 100 g/L protein in an industrial fermentation process, making them highly suitable for the commercial production of enzymes (Cherry and Fidantsef, 2003). Fermentation process design is interdisciplinary and requires knowledge of chemical engineering and microbial physiology to scale up. Submerged fermentation (SmF) and solid-state fermentation (SSF) are the two crucial fermentation technologies available. Both of these technologies offer several benefits and have their limitations. Most industries employ SmF for enzyme production; however, there is a resurgence in the popularity of SSF for a few applications and specific industries (Pandey, 2003; Singhania et al., 2009; Thomas et al., 2013a,b).

3.2.2.1 *Submerged fermentation*

Fermentation done in excess free water is termed SmF. Using submerged aerobic culture in a stirred-tank reactor is a typical industrial process for enzyme production involving microorganisms that produce extracellular enzymes. It is the preferred technology for industrial enzyme production due to its easy handling at a large scale compared to SSF. Large-scale fermenters, varying in volume from thousands to hundred thousand liters for SmF, are well developed and offer online control over several parameters such as pH, temperature, DO (dissolved oxygen), and foam formation. Moreover, there is no problem with mass transfer and heat removal. Thus these are some of the benefits which make this production technology superior to SSF and widely accepted for industrial metabolite production. The medium in the SmF is liquid which remains in contact with the microorganisms. A supply of oxygen is essential in the SmF, which is done through a sparger. Stirrer and impellers play a vital role in mixing gas, biomass, and suspended particles in these fermenters.

There are four main ways of growing the microorganisms in SmF: batch culture, fed-batch culture, perfusion batch culture, and continuous culture. In the batch culture, the microorganisms are inoculated in a fixed volume of the medium. In the case of fed-batch culture, the concentrated components of the nutrient are gradually added to the batch culture. In the perfusion batch culture, adding the culture and withdrawing an equal volume of used cell-free medium is performed. In the continuous culture, a fresh medium is added to the batch system at the exponential phase of the microbial growth with a corresponding withdrawal of the product's medium. Continuous cultivation gives a near-balanced growth, with slight fluctuation of the nutrients, metabolites, cell numbers, or biomass. Some enzymes are produced more as a secondary metabolite, and specific productivity may then be an inverse function of growth rate, that is, nongrowth-associated production. Here a recycling reactor may be most suitable. A recycling reactor is similar to a continuous culture, but a device is added to return a significant fraction of the cells to the reactor. Low growth rates with high cell concentration can often be achieved in such systems.

In practice, scale-up effects are more pronounced for the aerobic process than the anaerobic process. So aeration and agitation are maintained during the fermenter's scale-up to have a constant oxygen supply. Scale-up complications arise from cell response to distributed DO values, temperature, pH, and nutrients. For enzyme production, the economy of scale leads to the use of fermenters with a volume of 20–200 m^3. The simultaneous mass and heat transfer problems are usually neglected in small fermenters and low cell densities. However, industrial microbiology must consider transport processes with the abovementioned fermenter volumes and the economic necessity of using the highest possible cell densities. These can limit the metabolic rates, such as oxygen limitation, which leads the microorganisms to respond with physiological patterns. In these conditions, the desired control of microbial metabolism is lost. In the controlled operation of an industrial process, metabolic rates must be limited to a level just below the transport capacity of the fermenter. Therefore the highest possible productivity in a fermenter is obtained at maximal transport capacity.

3.2.2.2 *Solid-state fermentation*

Current developments in biotechnology are yielding new applications for enzymes. SSF holds tremendous potential for the production of enzymes. It can be of special interest in those processes where the crude fermented products may be used directly as enzyme sources. This system offers numerous advantages over SmF system, including high titer, relatively higher concentration of the products, less effluent generation, the requirement for simple fermentation equipment, less trained labor, etc.(Pandey et al., 2007).

Many microorganisms, including bacteria, yeast, and fungi, produce different groups of enzymes. However, selecting a particular strain remains a tedious task, especially when commercially competent enzyme yields are to be achieved. For example, it has been reported that while a strain of *Aspergillus niger* produced 19 types of enzymes, alpha-amylase was being produced by as many as 28 microbial cultures (Pandey et al., 1999). Thus selecting a suitable strain for the required purpose depends upon several factors, particularly the nature of the substrate and environmental conditions. Generally, hydrolytic enzymes, for example, cellulases, xylanases, pectinases, etc., are produced by fungal cultures since such enzymes are used in nature by fungi for their growth. *Trichoderma* spp. and *Aspergillus* spp. have most widely been used for these enzymes. Amylolytic enzymes too are commonly produced by filamentous fungi, and the preferred strains belong to the species of *Aspergillus* and *Rhizopus*. Although commercial production of amylases is carried out using both fungal and bacterial cultures, bacterial alpha-amylase is generally preferred for starch liquefaction due to its high-temperature stability. Genetically modified strains would hold the key to enzyme production to achieve high productivity with less production cost.

Agro-industrial residues are generally considered the best substrates for the SSF processes, and the use of SSF for the production of metabolites, either enzymes or even secondary metabolites (Kumar et al., 2021). Several such substrates have been employed to cultivate microorganisms to produce enzymes. Some of the substrates that have been used included sugarcane bagasse, wheat bran, rice bran, maize bran, gram bran, wheat straw, rice straw, rice husk, soy hull, sago hampers, grapevine trimming dust, sawdust, corncobs, coconut coir pith, banana waste, tea waste, cassava waste, palm oil mill waste, aspen pulp, sugar beet pulp, sweet sorghum pulp, apple pomace, peanut meal, rapeseed cake, coconut oil cake, mustard oil cake, cassava flour, wheat flour, cornflour, steamed rice, steam pretreated willow, starch, etc. (Pandey et al., 1999; Kumar et al., 2021). Wheat bran, however, holds the key and has most commonly been used in various processes. The selection of a substrate for enzyme production in an SSF process depends upon several factors, mainly related to the cost and availability of the substrate, and thus may involve screening of several agro-industrial residues. In an SSF process, the solid substrate supplies the nutrients to the microbial culture and serves as an anchorage for the cells. The substrate that provides all the needed nutrients for growing microorganisms is considered an ideal substrate (Costa et al., 2018). However, some nutrients may be available in suboptimal concentrations or even absent in the substrates. In such cases, it would become necessary to supplement them externally. It has also been a practice to pretreat (chemically or mechanically) some of the substrates before using them in SSF processes (e.g., lignocellulose), thereby making them more easily accessible for microbial growth.

Among the several factors that are important for microbial growth and enzyme production using a particular substrate; particle size, initial moisture level and water activity are the most critical. Generally, smaller substrate particles provide a larger surface area for microbial attack and, thus, are desirable. However, too small a substrate particle may result in substrate agglomeration, interfering with microbial respiration/aeration and resulting in poor growth. In contrast, larger particles provide better respiration/aeration efficiency (due to increased interparticle space) but offer a limited surface for microbial attack. This necessitates a compromised particle size for a particular process. Over the years, different types of fermenters (bioreactors) have been employed in SSF systems. The aspects of the design of fermenters in SSF processes have been discussed elaborately (Pandey, 1995; Krishania et al., 2018). Laboratory studies are generally carried out in Erlenmeyer flasks, beakers, Petri dishes, Roux bottles, jars, and glass tubes (as column fermenters). Large-scale fermentation has been carried out in-tray-, drum-, or deep-trough-type fermenters. Developing a practical and straightforward fermenter with automation is yet to be achieved for the SSF processes.

SSF processes are distinct from SmF culturing since microbial growth and product formation occur at or near the surface of the solid substrate particle having low moisture contents. Thus it is crucial to provide optimized water content and control the fermenting substrate's water activity (a_w) as water availability in lower or higher concentrations affects microbial activity adversely. Moreover, water has a profound impact on the solids' physicochemical properties, which, in turn, affects the overall process productivity. The major factors that affect the microbial synthesis of enzymes in an SSF system include the selection of a suitable substrate and microorganism, pretreatment of the substrate, particle size (interparticle space and surface area) of the substrate, water content and a_w of the substrate, relative humidity, type and size of the inoculum, control of the temperature of fermenting matter/removal of metabolic heat, period of cultivation, maintenance of uniformity in the environment of SSF system, and the gaseous atmosphere, that is, oxygen consumption rate, and carbon dioxide evolution rate. Ideally, almost all the known microbial enzymes can be produced under SSF systems. Literature survey reveals that much work has been carried out on the production of enzymes of industrial importance, such as proteases, cellulases, ligninases, xylanases, pectinases, amylases, glucoamylases, etc. Attempts are also being made to study SSF processes for the production of inulinases, phytases, tannases, phenolic acid esterases, microbial rennets, aryl-alcohol oxidases, oligosaccharide oxidases, tannin acyl hydrolase, α-L-arabinofuranosidase, etc. using SSF systems.

3.3 Strain improvements

It is well recognized that a large majority of the naturally occurring microorganisms do not produce enzymes at industrially appreciable quantities or often do not possess desirable properties for applications. Hence, tremendous efforts have been made to improve the strains using classical or molecular tools to obtain hyper-producing strains or develop required characteristics (Pandey et al., 2010).

3.3.1 Mutation

Most of the strains used for industrial enzyme production have been improved by classical selection. There are four classes of mutations: (1) spontaneous mutations (molecular decay), (2) mutations due to error-prone replication by-pass of naturally occurring DNA damage (also called error-prone translation synthesis), (3) errors introduced during DNA repair, and (4) induced mutations caused by mutagens. Scientists may also deliberately introduce mutant sequences through DNA manipulation for scientific experimentation. Mutagenesis by UV radiation or chemical mutagens has been applied to find useful variants quickly. Many cells are subjected to mutation, and the resulting mutants are selected for the desired combination of traits. Usually, the mutation causes changes in protein structure via changing the DNA sequence, which results in deterioration of function. Changes in structural components by mutation rarely result in improvements unless the specific loss of function is required for production purposes, for example, when a loss of regulatory function results in enhanced enzyme production. Mutation and selection are directed primarily toward higher overall productivity rather than mutation of a specific function, but a loss of regulatory function is highly probable. The report says that 8% of overall mutations resulted in being useful. There are several examples of mutant strains known as hyperproducers, such as *Trichoderma reesei* RUT C30, one of the best cellulase producers for decades. A mutation is preferred over other strain improvement methods as it is more natural. Each living organism undergoes mutation over time; however, we expedite it by using mutagens. Most of the commercial producers of enzymes are mutants.

3.3.2 Recombinant DNA technology

Some microorganisms have the capability of producing the perfect enzyme. Others could win the Olympic gold medal in growth and enzyme production! Combining the best from each organism could obtain a microorganism that proliferates on inexpensive nutrients while producing large quantities of the suitable enzyme. This is done by identifying the gene that codes for the desired enzyme and transferring it to a production organism known to be a good enzyme producer. Industrial enzymes need to be perfectly suited to their tasks, but sometimes they are impossible to find the perfect enzyme for a specific job. However, this does not mean that we cannot make an enzyme for the job. Usually, our scientists can discover suitable naturally occurring enzymes that can also be upgraded to the desired efficiency using modern biotechnology. This is done by altering small parts of the microorganism genes that code for the enzyme's production. These tiny alterations only change the enzyme's structure very slightly, but this usually is enough to make a good enzyme into a perfect enzyme.

Microorganisms isolated from diverse environments represent a source of enzymes that can be used for industrial process chemistry. Though high-throughput screening methods have enabled us to find novel and potent enzymes from microorganisms, many of those microorganisms are not easily cultivated in laboratory conditions, or their enzyme yield is too low for economical use. Using recombinant DNA technology, cloning the genes encoding these enzymes and heterologous expression for commonly used industrial strains has become a common practice. The novel enzymes suitable for specific conditions may be obtained by genetically modifying the microorganism. The industrial production of insulin is produced by genetically modified *Escherichia coli*. Recombinant DNA technology enables the production of enzymes at levels 100-fold greater than the native expression, making them available at low cost and in large quantities (Shu-Jen, 2004). As a result, several important food processing enzymes such as amylases and lipases with properties tailored to particular food applications have become available. Several microbial strains have been engineered to increase the enzyme yield by deleting native genes encoding the extracellular proteases. Moreover, certain fungal production strains have been modified to reduce or eliminate their potential for the production of toxic secondary metabolites (Olempska-Beer et al., 2006). Although recombinant DNA technology significantly lowers the cost of enzyme production, the applications of enzymes are still limited. Most chemicals with industrial interest are not natural substrates for these enzymes. If the desired enzyme activity is found, the yield is often low. Moreover, enzymes are not usually stable in harsh reaction conditions, such as pH higher or lower than physiological pH 7, high temperature, or the presence of organic solvents required to solubilize many substrates. This approach precludes the transfer of any extraneous or unidentified DNA from the donor organisms to the production strain.

3.3.3 Clustered regularly interspaced short palindromic repeats-Cas9 technology

Genomes of prokaryotic organisms, such as bacteria and archaea, harbor a family of DNA sequences called CRISPR (clustered regularly interspaced short palindromic repeats) (Barrangou, 2015). Cas9 is an enzyme that uses

CRISPR sequences as a guide to recognize and cleave specific strands of DNA that are complementary to the CRISPR sequence. CRISPR-Cas9 technology can be used to edit genes within organisms (Zhang et al., 2014). Even in filamentous fungi, this technique has been successfully employed for strain improvement to enhance cellulase production. Liu et al. (2015) demonstrated the establishment of a CRISPR-Cas9 system in *T. reesei* by in vitro RNA transcription and specific codon optimization. Site-specific mutations were generated through efficient homologous recombination in target genes, even using short homology arms. This system provides a promising and applicable approach to target multiple genes simultaneously. *T. reesei* RUT-C30 was engineered by using CRISPR-Cas9 technology, and six genetic modifications were introduced, which resulted in a significant enhancement in protein secretion. The engineered *T. reesei* RUT-C30 overcame the deficiency of β-glucosidase (BGL) and was able to utilize sucrose for its growth, eliminating the need for inducers for enzyme production. The development of the CRISPR-Cas9 genome editing technique was awarded with the Nobel Prize in chemistry in 2020, which was awarded to Emmanuelle Charpentier and Jennifer Doudna.

3.3.4 Protein engineering

Recent advances in polymerase chain reaction technology, site-specific and random mutagenesis are readily available to improve enzyme stability in the broader range of pH and temperature and tolerance to various organic solvents. Since a large quantity of enzyme can be obtained by recombinant expression, X-ray crystallography can facilitate understanding the tertiary structure of an enzyme and its substrate-binding/recognition sites. This information may assist a rational design of the enzyme, predicting amino acid changes for altering substrate specificity, catalytic rate, and enantioselectivity (in the case of chiral compound synthesis). Two approaches are presently available to engineer a commercially available enzyme to be a better industrial catalyst: a random method called directed evolution and a protein engineering method called rational design.

Protein engineering is a method of changing a protein sequence to achieve the desired result, such as a change in substrate specificity, increased temperature stability, organic solvents, and/or extremes of pH. Many specific methods for protein engineering exist, but they can be grouped into two major categories: those involving the rational design of the protein changes and the combinatorial methods which make changes more randomly. Protein engineering or rational methods, such as site-directed mutagenesis, require targeted amino acid substitutions. Therefore a large body of knowledge about the biocatalyst is being improved, including the three-dimensional structure and the chemical mechanism of the reaction. The main advantage of the rational design is that a minimal number of protein variants are created, meaning that minimal effort is necessary to screen for the improved properties.

On the other hand, the combinatorial methods create many variants that must be assayed; however, they have the advantage of not requiring such extensive knowledge about the protein. In addition, nonobvious changes in the protein sequence often lead to significant improvements in their properties, which are extremely hard to predict rationally. Thus they can only be identified by combinatorial methods.

Several enzymes have already been engineered to function better in industrial processes. These include the proteinases, lipases, cellulases, alpha-amylases, and glucoamylases. Xylanase is a good example of an industrial enzyme, which needs to be stable at high temperatures and active at the physiological temperature and pH when used as the feed additive and in the alkaline conditions when used in the bleaching in the pulp and paper industry. One of the industrial production organisms of the xylanases is *Trichoderma* sp. Its xylanase has been purified and crystallized. By the designed mutagenesis, its thermal stability has been increased by about 15°C. The mutational changes increased the half-life in the thermal inactivation of this enzyme from approximately 40 s to approximately 20 min at 65°C, and from less than 10 s to approximately 6 min at 70°C (Fenel et al., 2004).

By designed mutagenesis, its thermal stability has been increased about 2000 times at 70°C, and its pH optimum shifted toward the alkaline region by one pH unit. The most successful strategies to improve the stability of the *Trichoderma* xylanase include the stabilization of the alpha-helix region and the N-terminus. The abovementioned strategies for strain improvement and production process optimization may reduce the cost of enzymes to an extent. However, downstream processing and product formulation for reasonably purified and highly stable enzymes are the key steps toward the success of any enzyme industry.

3.4 Downstream processing/enzyme purification

The goal of the fermentation process is to produce a final formulated enzyme product, and it also includes many postfermentation unit operations. Still, the maximum production rate could be the most important factor. However, the lowest unit production cost could also be an important driving force. Optimization of each unit operation does not always lead to optimal overall process performance, especially when there are strong interactions between unit operations (Groep et al., 2000). Understanding these interactions is crucial to overall process optimization. For instance, product concentration or purity in the fermentation broth can significantly impact downstream purification unit operation. If the fermentation is optimized for productivity without considering its effect on the purification step, the overall process productivity can be negatively affected. Antifoaming agents in the fermentation process are another example of such a trade-off. By reducing foaming in the fermentation, a higher working volume can be used to optimize the fermentation unit operations. Care has to be taken as antifoams could have an adverse effect on filtration unit membranes. Thus it is important to know how the fermentation process will affect the other unit operations of downstream processing.

Usually, the purification process aims to achieve the maximum possible yield, maximum catalytic activity, and the maximum possible purity. Most of the industrial enzymes produced are extracellular, and the first step in their purification is the separation of cells from the fermentation broth. For intracellular enzymes, disruption of cells by mechanical or nonmechanical methods is required. Filtration, centrifugation, flocculation, floatation, and concentration methods lead to the development of a concentrated product. Salting-out solvent precipitation methods could be employed for protein concentration in industries; acetone precipitation is a popular method of protein concentration in industries as acetone can be recycled. Ultrafiltration, electrophoresis, and chromatography lead to a highly purified product. Fig. 3.1 shows the basic steps followed during the microbial enzyme production process downstream and verifies that extracellular enzymes are more desirable for industrial applications, being more economical than downstream processes. However, the number of steps and economic viability is highly associated with the degree of purity required, related to the enzyme's end application.

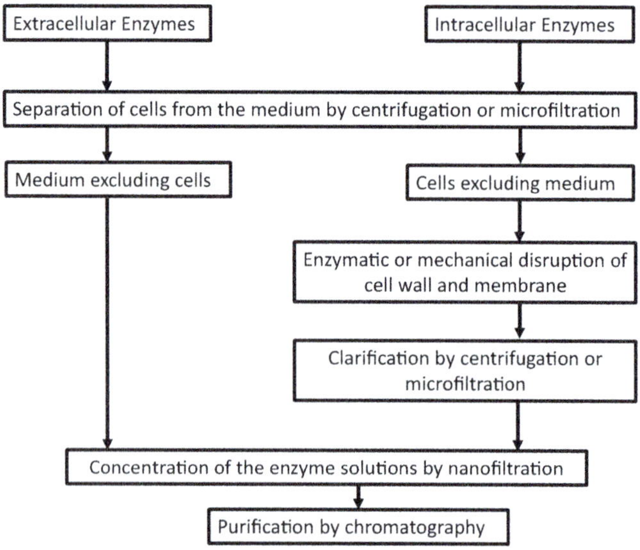

FIGURE 3.1 Downstream processing of industrial enzymes.

Thus it is evident that the extent of purification required is based on its end application. For pharmaceutical and food industries, purification is critical, whereas in the textile, detergent, biofuel industries, generally cocktails are preferred. It would be challenging to discuss purification steps, their principle, and relevance to a particular enzyme product, etc. There could be a set of steps for the purification of a particular enzyme, and also, a particular set of steps can be employed for several enzymes. High-volume low-value enzymes should minimize the extent of purification steps to be economically viable. Cellulase is among many industrial enzymes and has several applications, including in the textile industry, detergent industry, and cellulosic biomass degradation for bioethanol production. Cellulases are available commercially in the liquid stage as well as powder stage, and both were found to be quite stable. These are normally produced by SmF and SSF as well and are generally concentrated by acetone precipitation. After concentration, principal separation methods are employed based on the properties of the enzyme to be separated, which are enlisted in Table 3.2.

3.5 Product formulations

The primary objective of a commercial formulation is to minimize losses in enzymatic activity during transport, storage, and usage. So enzymes are sold as stabilized liquid concentrates or as particulate solids. Enzymes are often exposed to humid, hot, or oxidative environments in industrial applications such as detergents, textile formulations, and food and beverage processing. Chemical stabilizers are available to protect the thermolabile enzymes thermally and chemically to an extent. So, it is preferred to select structurally more stable or resistant to oxidation enzymes during screening itself.

Formulations enhance stability by counteracting the primary forces of deactivation: denaturation, catalytic-site deactivation, and proteolysis (Becker et al., 1997). Denaturation occurs by physically unfolding an enzyme's tertiary protein structure under thermal or chemical stress. Once an enzyme begins to unfold, it becomes dramatically more vulnerable to deactivation and proteolysis. To minimize unfolding, the formulator can alter the protein's environment so as to induce a compact protein structure. This is done most effectively by adding water associating compounds such as sugars, polyhydric alcohols, and lyotropic salts that detach water molecules from the protein surface via "preferential exclusion." The best ways to combat active site inactivation are to ensure sufficient levels of required cofactors, add reversible inhibitors, and exclude reactive or oxidizing species from the formulation. There are several key secondary requirements besides enzymatic stability a formulation should meet, including preservation against microbial contamination, avoidance of physical precipitation or haze formation, minimization of the formation of sensitizing dusts or aerosols, and the optimization of esthetic criteria such as color and odor.

TABLE 3.2 Principal separation methods used in the purification of enzymes.

Property	Method	Scale
Size or mass	Centrifugation Gelfiltration Dialysis, ultrafiltration	Large or small Generally small Generally small
Polarity • Charge • Hydrophobic character	Ion-exchange chromatography Chromatofocusing Electrophoresis Isoelectric focusing Hydrophobic chromatography	Large or small Generally small Generally small Generally small Generally small Generally small
Solubility	Change in pH Change in ionic strength Decrease in dielectric constant	Generally large Large or small Generally large
Specific binding sites or structural features	Affinity chromatography Immobilized metal ion chromatography Affinity elution Dye-ligand chromatography Immunoadsorption Covalent chromatography	Generally small Generally small Large or small Large or small Generally small Generally small

Many of these problems are best addressed by focusing as far "upstream" as possible, including the choice of raw materials in the fermentation or enzyme recovery process. Downstream operations such as diafiltration, adsorption, chromatography, crystallization, and extraction can be used to remove impurities responsible for color, odor, and precipitation (Becker, 1995). The risk of physical precipitation could be minimized by formulating the enzyme near its isoelectric point with hydrophilic solvents such as glycerol or propylene glycol. A moderate amount of solvating salts can also be added to avoid either salting-out or "reverse salting-in." A combination of filtration, acidification, and the minimization of free water, biocides could be effective in preventing microbial contamination. The range of acceptable chemicals for controlling or killing microbes should only be used, circumscribed by health and safety regulations. Dry granular enzyme formulations for powdered laundry detergents and textile formulations result in workers' safety because enzyme granules have become increasingly resistant to physical breakage and the formation of airborne dust upon handling. Two processes producing the most attrition-resistant granules to date are high-shear granulation and fluidized-bed spray coating. These processes use various binders, coatings, and particle morphologies to make nonfriable particles that protect enzymes during storage and, at the same time, allow for their ready release in solution during application.

3.6 Global enzyme market scenarios

According to the report on industrial enzymes (Global Industry Analysts, Inc., 2011), the global market for the industrial enzyme was pretty immune to the turmoil in the global economy and grew moderately during 2008–09. In matured economies such as the United States, Western Europe, Japan, and Canada was relatively stable, while in developing economies of Asia-Pacific, Eastern Europe, and Africa emerged as the fastest growing markets for industrial enzymes. The United States and Europe collectively command a major share of the global market of industrial enzymes as they are the major producers of pharmaceutical, medicinal, and cosmetics products, whereas Asia-Pacific growth was stagnant with just an 8% compounded annual growth rate in 2008–12 (Sarrouh et al., 2012).

In 2019 the global industrial enzymes market was valued at 5.6 billion USD and expected to grow at a compound annual growth rate (CAGR) of 6.4% from 2020 to 2027, as shown in Fig. 3.2. Increased demand from the

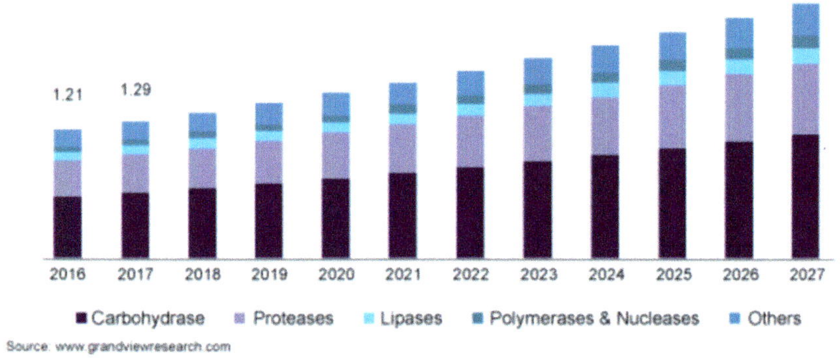

FIGURE 3.2 Expected growth of industrial enzyme market from 2020 to 2027.

end-use industries, such as home cleaning, animal feed, food, and beverage, and biofuel, has led the industry growth over the forecast period. Increasing demand for carbohydrase and proteases in the food and beverage applications in the emerging economies of Asia-Pacific, such as China, India, and Japan, has fueled the growth of the industry. Industrialization and advancements in the nutraceutical sector can be attributed to increasing growth in developed economies. Moreover, people are getting more health-conscious, which has led to increased demand for functional foods causing increased demand for enzymes. Such factors have bolstered the product demand on a large scale (https://www.grandviewresearch.com/industry-analysis/industrial-enzymes-market). Fig. 3.3 shows the industrial enzyme market size in the United States, based on products from 2016 to 2017 (USD Billions). The market is highly competitive, dominated by large companies, and highly sensitive to production costs. http://www.idiverse.com/html/target_industrial_enzymes.htm. Fig. 3.4 shows the proportion of industrial enzymes needed by sector based on application.

Major enzyme producers are located in Europe, the United States, and Japan. Major players in enzyme market are Novozymes (45%) and Danisco (17%) in Denmark, Genencor in United States, DSM in the Netherlands, and BASF in Germany (Binod et al., 2008; Binod et al., 2013; BCC-Business Communications Company, Inc., 2009). The pace of development in emerging markets suggests that companies from India and China can join this restricted party in the very near future [Research and Markets, 2011a (India) and Research and Markets, 2011b (China), Chandel et al., 2007, Carrez and Soetaert, 2005]. Another published research report on enzymes market (Global Industry Analysts, Inc., 2011) highlighted the fact that proteases constitute the largest product segment in the global industrial enzymes market, and the carbohydrases market was projected to be the fastest growing product segment, with a CAGR of more than 7.0% over the analysis period.

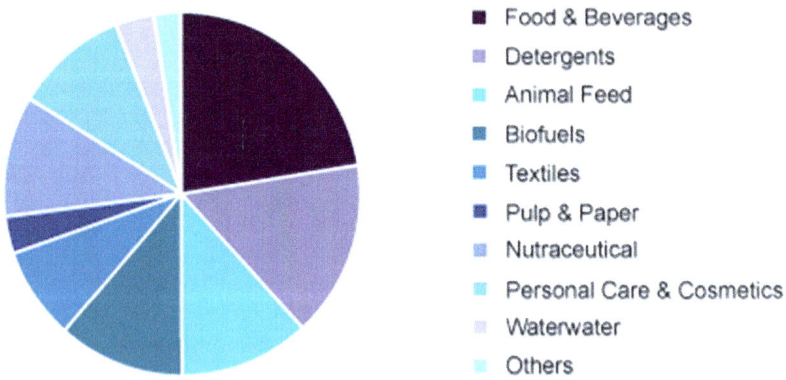

FIGURE 3.3 Industrial enzyme sales by sector in 2016–2017.

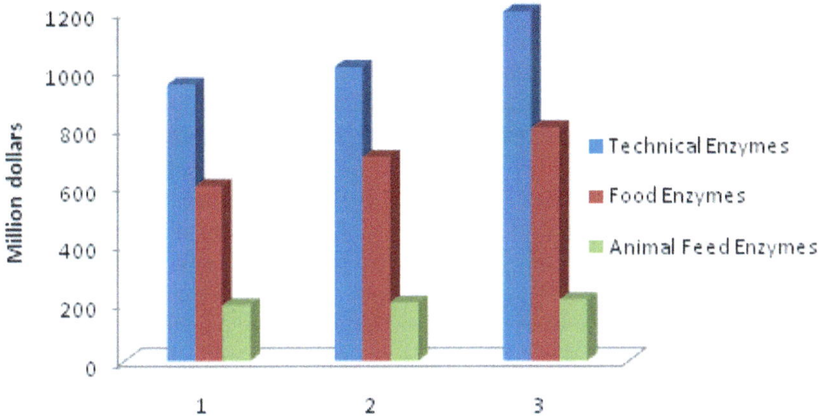

FIGURE 3.4 Global enzyme market by different application sectors.

Lipases represent the other major product segment in the global industrial enzymes market with high growth potential. Sectors such as pharmaceuticals and bioethanol have succeeded in drawing significant attention of the investors and are self-sufficient in undertaking new product development activities and in launching novel and unique products in the market, thus offering new opportunities to the industrial enzyme manufacturers.

Market researchers highlight the fact that industrial demands for enzymes is being driven by new enzyme technologies and the increased use of organic compounds in place of petrochemical-based ingredients in cosmetics and other products (Pitman, 2011). The global industrial enzymes market consists of various enzymes, which can be broadly categorized as carbohydrases, proteases, lipases, and others. Carbohydrases industrial enzymes are further classified as amylases, cellulases, and other carbohydrases, which are projected to grab the market at a fast pace (Industrial enzymes market, 2014). This projection came out so true that carbohydrases alone accounted for more than 47.0% share of the global revenue in 2019 (https://www.grandviewresearch.com/industry-analysis/industrial-enzymes-market).

3.7 Industrial applications of enzymes

It was very well realized that by applying the enzymes to biological processes, the reaction rate could be enhanced, and the production process could be performed within a fraction of time at lower temperatures and pressure with cheaper raw materials. The enzyme industry is searching for sustainable processes that enable higher yields with improved efficiency and dynamic nature. From creating lactose-free dairy products to fast-acting laundry detergents, innovation is the key in engineering; improved, cost-effective end-products for textiles, foods, detergents, animals, biofuels, and more. Table 3.1 shows large-scale enzyme applications; however, there are other applications too which are not enlisted in the table, such as therapeutic and speciality enzymes, which are not required in bulk. Still, purity is the primary concern as these need to be free from other enzyme activities.

3.7.1 Food industry

3.7.1.1 Starch industry

Starch-degrading enzymes were the first large-scale application of microbial enzymes in the food industry. Mainly two enzymes carry out the conversion of starch to glucose: alpha-amylase and glucoamylase (Pandey et al., 2000, Pandey, 1995). Sometimes additional debranching enzymes such as pullulanase are added to improve the glucose yield. Beta-amylase is commercially produced from barley grains and used for the production of the disaccharide maltose (Selvakumar et al., 1996). Studies have been carried out on the application of transglutaminase as a texturing agent in the processing of sausages, noodles and yoghurt, where cross-linking of proteins provides improved viscoelastic properties of the products (Kuraishi et al., 2001). In the United States, large volumes of glucose syrups are converted by glucose isomerase after Ca^{2+} removal (alpha-amylase needs Ca^{2+} for activity, but it inhibits glucose isomerase) to fructose-containing syrup. This is done by bacterial enzymes, which need Mg^{2+} ions for activity. Fructose is separated from glucose by large-scale chromatographic separation and crystallized. Alternatively, fructose is concentrated to 55% and used as a high-fructose corn syrup in the soft drink industry.

3.7.1.2 Baking industry

Alpha-amylases have been most widely studied in connection with improved bread quality and increased shelf life. Both fungal and bacterial amylases are used. The added amount needs to be controlled as overdosage may lead to a sticky dough. One of the motivations to study the effect of enzymes on dough and bread qualities comes from the pressure to reduce other additives. In addition to starch, flour typically contains minor amounts of cellulose, glucans, and hemicelluloses such as arabinoxylan and arabinogalactan. There is evidence that the use of xylanases decreases water absorption and thus reduces the quantity of water added needed in baking. This leads to a more stable dough. Primarily xylanases are used in wholemeal rye baking and dry crisps common in Scandinavia. Proteinases can be added to improve dough-handling properties; glucose oxidase has been used to replace chemical oxidants and lipases to strengthen gluten, which leads to a more stable dough and better bread quality.

3.7.1.3 Brewing industry

Enzymes have many applications in the drink industry. Chymosin is used in cheese making to coagulate milk protein. Another enzyme used in the milk industry is β-galactosidase or lactase, which splits milk-sugar lactose into glucose and galactose. This process is used for milk products that lactose-intolerant consumers consume.

3.7.1.4 Fruit juice industry

Enzymes are also used in fruit juice manufacturing. The addition of pectinase, xylanase, and cellulase improves the liberation of the juice from the pulp. Pectinases and amylases are used in juice clarification. Similarly, enzymes are widely used in wine production to better extract the necessary components and thus improving the yield. Enzymes hydrolyze high-molecular-weight substances such as pectin. Enzymes can be used to help the starch hydrolysis (typically alpha-amylases), solve filtration problems caused by beta-glucans present in malt (beta-glucanases), hydrolyze proteins (neutral proteinase), and control haze during maturation, filtration, and storage (papain, alpha-amylase, and beta-glucanase).

3.7.2 Textile industry

The use of enzymes in the textile industry is one of the most rapidly growing fields in industrial enzymology. Amylases are used for desizing of textile fibers. Another important enzyme used in the textile industry is cellulases. Because of their ability to modify the cellulosic fibers in a controlled and desired manner to improve fabrics, these (neutral or acidic) cellulases offer a great alternative to stonewashing of blue denim garments, eliminating the disadvantages associated with stones, such as damage to the washers, safety issues, and handling issues. The enzymatic stonewashing allows up to 50% higher jean load and yields the desired look and softer finish. The neutral cellulase is the enzyme of choice for stonewashing because of the reduction in back staining and its broader pH profile. This latter property reduces the need for the rigid pH control of the wash, resulting in a more reproducible finish from wash to wash. Fuzz formation and pilling are the common problems associated with the fabric using cotton or other natural fibers; the cellulases are utilized to digest the small thread ends protruding from the fabric, resulting in a better finish.

Catalase is used to degrade excess peroxide, as hydrogen peroxides are used as bleaching agents to replace chlorine-based chemicals. Another recent approach is to use oxidative enzymes directly to bleach textiles. Laccase—a polyphenol oxidase from fungi—is a new candidate in this field. It is a copper-containing enzyme oxidized by oxygen, and in its oxidized state, it can oxidatively degrade many different types of molecules such as dye pigments.

3.7.3 Detergent industry

The detergent industry is the largest single industry for enzymes, using about 25%–30% of the total industrial enzyme. Nearly half of the detergents available in the market contain enzymes in their formulations; however, rarely the information is published about the formulations.

Dirt in clothes could be proteins, starches, or lipids in nature. It is possible to remove most types of dirt using detergents in water at high temperatures with vigorous mixing, but the cost of heating the water is high and lengthy mixing or beating would be required, which shortens the life of cloths. The use of enzymes allows lower temperatures to be employed, and shorter periods of agitation are needed, often after a preliminary soaking period. In general, enzyme detergents remove protein from clothes soiled with blood, milk, sweat, grass, etc., far more effectively than nonenzyme detergents. Cellulases are employed to loosen the fibers to remove the dirt easily and give a finishing touch by digesting fine fibers during washing. At present, only proteases, amylases, cellulases, and lipases are commonly used in the detergent industry.

Enzymes are used in surprisingly small amounts in most detergent preparations, only 0.4%–0.8% crude enzyme by weight (about 1% by cost). The ability of enzymes to withstand the conditions of use is a more important criterion than its cost. Now second-generation detergent enzymes are available, having enhanced activity at low temperature and alkaline pH.

3.7.4 Pulp and paper industry

Intensive studies have been carried out during the last 20 years to apply many different enzymes in the pulp and paper industry. Xylanases are applied in pulp bleaching, which liberates lignin fragments by hydrolyzing residual xylan (Thomas et al., 2013a). This reduces the need for chlorine-based bleaching chemicals considerably. Cellulases are used for deinking the cellulose fibers during recycling. In papermaking, amylases are used especially in the modification of starch, which improves paper's strength, stiffness, and erasability. The starch suspension must have a certain viscosity, which is achieved by adding amylase enzymes in a controlled process. Pitch is a sticky substance composed of lipids present mainly in softwoods. It causes a problem for the paper machine when mechanical pulps of red pine are used as a raw material that lipases can remove.

3.7.5 Animal feed industry

Enzyme addition in the animal feed has been intensively started in the 1980s. Such application reduces viscosity, increases absorption of nutrients, liberates nutrients either by hydrolysis of nondegradable fibers or by releasing nutrients blocked by these fibers, and reduces the number of feces. They are added as enzyme premixes (enzyme-flour mixture) during the feed manufacturing process, which involves the extrusion of wet feed mass at high temperatures (80°C–90°C). Therefore the feed enzymes need to be thermotolerant during the feed manufacturing and operative in the animal body temperature. The first commercial success was the addition of β-glucanase into barley-based feed diets. Barley contains β-glucan, which causes high viscosity in the chicken gut. The net effect of enzyme usage in feed has been increased animal weight gain with the same amount of barley, resulting in an increased feed conversion ratio. The addition of xylanase to wheat-based broiler feed is capable of increasing the available metabolizable energy to 7%–10%.

Another important feed enzyme is the phytase, which is a phosphoesterase and liberates the phosphate from the phytic acid. Phytic acid is commonly present in plant-based feed materials. The supplementation of the phytase results in a reduced amount of phosphorous in the feces, resulting in reduced environmental pollution. It also minimizes the need to add phosphorus to the feed. Currently, phytase from fungal sources is a potent feed enzyme (Pandey et al., 2001). Usually, a feed-enzyme preparation is a multi-enzyme cocktail containing glucanases, xylanases, proteinases, and amylases.

3.7.6 Leather industry

The leather industry uses proteolytic and lipolytic enzymes in leather processing. The use of these enzymes is associated with the structure of animal skin as a raw material. Enzymes are used to remove unwanted parts. Alkaline proteases are added in the soaking phase. This improves water uptake by the dry skins, removal and degradation of protein, dirt, and fats and reduces the processing time. In some cases, pancreatic trypsin is also used in this phase. Proteases are used in de-hairing and de-wooling leather and improving its quality (cleaner and stronger surface, softer leather, fewer spots). Lipases are used in this phase or in bating phase to remove grease precisely. The use of lipases is a reasonably new development in the leather industry.

3.7.7 Biofuel from biomass

Perhaps the most important emerging application of the enzymes currently being investigated actively is in the utilization of the lignocellulosic biomass for the production of biofuel (Patel et al., 2014). Biomass represents the most abundant renewable resource available to mankind for effective utilization. However, the lack of cost-effective enzyme conversion technologies made it difficult to realize, which is due to the high cost of the cellulases and the lack of specificities for various lignocellulosic substrates (Patel et al., 2019). The strategy employed currently in the bioethanol production from the biomass is a multistep process in which its enzymatic hydrolysis is a crucial step (Singhania et al., 2015). In the effort to develop efficient technologies for biofuel production, significant research has been directed toward identifying efficient cellulase systems and process conditions, besides studies directed at the biochemical and genetic improvement of the existing organisms utilized in the process (Singhania et al., 2021b). Dupont and Novozymes have been actively involved in cellulase research and significantly reduced the cost of the enzyme and improved the efficiency of the cellulolytic enzyme by exploring novel enzyme components such as BGL and LPMOs (lytic polysaccharide monooxygenases), leading to the development of economically feasible and efficient enzyme production. Commercially available cellulase cocktail for biomass hydrolysis is enriched with LPMOs, and it has found a place among key components of biomass-degrading enzymatic cocktails (Singhania et al., 2021a).

In recent times, bioethanol from biomass via the enzymatic route has taken the shape of reality, with several biofuel industries existing and coming up in the current time (Patel et al., 2019). Advanced Biofuels, LLC, in Scotland (SD, USA, since 2008), BetaRenewables in Rivalta (Italy, since 2009), Inbicon in Kalundborg (Denmark, since 2009), and Clariant in Munich (Germany, since 2009) are some of the proofs of a dream turned to reality. Beta Renewables constructed the world's first commercial-scale cellulosic ethanol plant in Crescentino, Italy, with a capacity of 20 MGY. This plant began operation at the end of 2012, and it uses the PROESA process to convert agricultural nonfood wastes to ethanol (Gusakov, 2013). Few of the above stopped functioning due to logistics and economic reasons as the cost of petroleum fuel dropped massively; however, technology was appreciated.

3.7.8 Enzyme applications in the chemistry and pharma sectors

An important issue in the pharma sector is the large number of compounds that must be tested for biological activity to find a single promising lead. The combinatorial biocatalysis has received much attention here. It could add a level of complexity to the diversity of existing chemical libraries or could be used to produce the libraries de novo (Rich et al., 2002). An example is the use of glycosyltransferases to change the glycosylation pattern of the bioactive compounds. Only a few commodity chemicals, such as acrylamide, are now produced by enzyme technology (annual production scale 40,000 tons). Nonetheless, this success has demonstrated that bioconversion technology can be scaled up. Many other chemicals, including chiral compounds (Jaeger et al., 2001), are also produced by biocatalysis on a multiton scale.

3.7.8.1 Speciality enzymes

In addition to significant volume enzyme applications, there are a large number of speciality applications for enzymes. These include enzymes in clinical analytical applications, flavor production, protein modification, personal care products, DNA technology, and fine chemical production. Contrary to bulk industrial enzymes, these enzymes need to be free from side activities, emphasizing an elaborate purification process. Alkaline phosphatase and peroxidases are used for immunoassays. A significant development in analytical chemistry is biosensors. The most widely used application is a glucose biosensor involving glucose oxidase catalyzed reaction:

$$\text{Glucose} + O_2 + H_2O \rightarrow \text{gluconic acid} + H_2O_2$$

Several commercial instruments apply this principle to measure molecules such as glucose, lactate, lactose, sucrose, ethanol, methanol, cholesterol, and some amino acids.

3.7.8.2 Enzymes in personal care products

Personal care products are a relatively new area for enzymes, and the amounts used are small but worth mentioning as a future growth area. One application is contact lens cleaning. Proteinase- and lipase-containing enzyme solutions are used for this purpose. Hydrogen peroxide is used in the disinfections of contact lenses. After disinfection, the residual hydrogen peroxide can be removed by a heme-containing catalase enzyme, which degrades hydrogen peroxide. Glucoamylase and glucose oxidase are used in some toothpaste, as glucoamylase liberates glucose from starch-based oligomers produced by alpha-amylase and glucose oxidase converts glucose to gluconic acid and hydrogen peroxide, both of which function as disinfectants. Dentures can be cleaned with protein degrading enzyme solutions. Enzymes such as chitinase are also studied for applications in skin and hair care products.

3.7.8.3 Enzymes in DNA technology

DNA-modifying enzymes play a crucial role in DNA technology, which has revolutionized traditional and modern biotechnology. It can be divided into two classes:

1. Restriction enzymes: It recognizes specific DNA sequences and cut the chain at these recognition sites.
2. DNA-modifying enzymes: These synthesize nucleic acids, degrade them, join pieces together, and remove parts of the DNA.

Restriction enzymes produce cleavage after recognizing a specific code sequence in the DNA. These enzymes are essential in gene technology. DNA-polymerases synthesize new DNA-chains using a model template which they copy. Nucleases hydrolyze the phosphodiester bonds between DNA sugars. Kinases add phosphate groups and phosphatases remove them from the end of DNA chain. Ligases join adjacent nucleotides together by forming phosphodiester bonds between them. These enzymes are involved in

DNA replication, foreign DNA degradation, repairing mutated DNA, and recombining different DNA molecules in the cell. The enzymes used in gene technology are produced like any other enzyme, but their purification needs extra attention. The use of enzymes in industrial applications has been limited by several factors, such as the high cost of the enzymes, availability in small amounts, and instability. Also, the enzymes are soluble in aqueous media, and it is difficult and expensive to recover them from reactor effluents at the end of the catalytic process. This limits the use of soluble enzymes to batch operations, followed by the spent enzyme-containing solvent disposal.

3.8 Concluding remarks

Though enzyme technology is a well-established branch of science, still it is passing through a continuous phase of evolution. Our society is moving toward an eco-friendly and sustainable "biorefinery" replacing "petrorefinery" to protect our environment for the future generation. Searches for novel enzymes based on a potential application for a known enzyme are moving ahead simultaneously. Enzymes have already shown a tremendous capacity to guide us toward biological processes as biocatalysts. Biological processes have replaced several chemical processes with benefits such as mild operating conditions, specificity, and environmental feasibility. Enzymes have interventions in almost all the major commercial sectors. Enzymes will continue with their potential and beneficial roles in a more intensified manner in the days coming ahead.

Abbreviations

SSF	solid-state fermentation
SmF	submerged fermentation
GRAS	Generally Recognized as Safe
DO	dissolved oxygen
CAGR	compound annual growth rate
MGY	mega gallons per year
CRISPR	clustered regularly interspaced short palindromic repeats
BGL	β-Glucosidase
LPMOs	lytic polysaccharide monooxygenases

References

Bairoch, A., 2000. The ENZYME database in 2000 (PDF). Nucleic Acids Res. 28 (1), 304–305.

Barrangou, R., 2015. The roles of CRISPR-Cas systems in adaptive immunity and beyond. Curr. Opin. Immunology 32, 36–41. Available from: https://doi.org/10.1016/j.coi.2014.12.008.

BCC-Business Communications Company, Inc., 2009, In: Report FOD020C-World markets for fermentation ingredients. Wellesley, MA 02481.

Becker, T., 1995. Separation and purification processes for recovery of industrial enzymes. In: Singh, R.K., Rizvi, S.S.H. (Eds.), Bioseparation Processes in Foods. Marcel Dekker, New York, pp. 427–445.

Becker, T., Park, G., Gaertner, A.L., 1997. Formulation of detergent enzymes. In: Van, E.E.J.H., Misset, O., Baas, E. (Eds.), Enzymes in Detergency. Marcel Dekker, New York, pp. 299–325.

Binod, P., Singhania, R.R., Soccol, C.R., Pandey, A., 2008. Industrial enzymes. Advances in Fermentation Technology. Asiatech Publishers, New Delhi, India, pp. 291–320.

Binod, P., Palkhiwala, P., Gaikaiwari, R., Nampoothiri, K.M., Duggal, A., Dey, K., et al., 2013. Industrial enzymes: present status and future perspectives for India. J. Sci. Ind. Res. 72 (5), 271–286.

Carrez, D., Soetaert, W., 2005. Looking ahead in Europe: white biotech by 2025. Ind. Biotechnol. 1, 95–101.

Chandel, A.K., Rudravaram, R., Rao, L.V., Ravindra, P., Narasu, M.L., 2007. Industrial enzymes in bioindustrial sector development: an Indian perspective. J. Commer. Biotechnol. 13, 283–291.

Cherry, J.R., Fidantsef, A.L., 2003. Directed evolution of industrial enzymes: an update. Curr. Opin. Biotechnol. 14 (4), 438–443. ISSN 0958-1669. Available from: https://doi.org/10.1016/S0958-1669(03)00099-5.

Costa, J.A.V., Treichel, H., Kumar, V., Pandey, A., 2018. Chapter 1 - Advances in solid-state fermentation. In: Pandey, A., Larroche, C., Ricardo Soccol, C. (Eds.), Current Developments in Biotechnology and Bioengineering. Elsevier, pp. 1–17. Available from: https://doi.org/10.1016/B978-0-444-63990-5.00001-3.

Dalby, P.A., 2003. Optimizing enzyme functions by directed evolution. Curr. Opin. Struct. Biol. 13, 500–505.

Fenel, F., Leisola, M., Janis, J., Turunen, O., 2004. A *de novo* designed *N*-terminal disulphide bridge stabilizes the *Trichoderma reesei endo*-1,4-beta-xylanase II. J. Biotechnol. 108 (2), 137–143.

Global Industry Analysts, Inc., 2011. In: Report- Global strategic business.

Groep, M.E., Gregory, M.E., Kershenbaum, L.S., Bogle, I.D.L., 2000. Performance modeling and simulation of biochemical process sequences with interacting unit operations. Biotechnol. Bioeng. 67, 300–311.

Gusakov, A.V., 2013. Cellulases and hemicellulases in the 21st century race for cellulosic ethanol. Biofuels 4, 567–569.

http://www.idiverse.com/html/target_industrial_enzymes.htm (date 19/04/2015).

http://www.novozymes.com/en/about-us/our-business/what-are-enzymes/Pages/creating-the-perfect-enzyme.aspx (date 19/04/2015).

https://www.grandviewresearch.com/industry-analysis/industrial-enzymes-market (date 13/04/2021).

marketsandmarkets.com, March 2014. Industrial enzymes market by types (carbohydrase, protease, lipase), applications (food &beverages, cleaning agents, bio-fuel, animal feed), & geography - Global trends &forecasts to 2018, report code: FB 2277.

Jaeger, K.E., Eggert, T., Eipper, A., Reetz, M.T., 2001. Directed evolution and the creation of enantioselective biocatalysts. Appl. Microbiol. Biotechnol. 55, 519–530.

Krishania, M., Sindhu, R., Binod, P., Ahluwalia, V., Kumar, V., Pandey, A., et al., 2018. Design of bioreactors in solid-state fermentation. In: Pandey, A., Larroche, C., Soccol, C.R. (Eds.), Current Developments in Biotechnology and Bioengineering. Elsevier, pp. 85–96. ISBN:9780444639905. Available from: https://doi.org/10.1016/B978-0-444-63990-5.00001-3.

Kumar, S., Sharma, S., Kansal, S.K., Elumalai, S., 2020. Efficient conversion of glucose into fructose via extraction-assisted isomerization catalyzed by endogenous polyamine spermine in the aqueous phase. ACS Omega 2020 5 (5), 2406–2418. Available from: https://doi.org/10.1021/acsomega.9b03918.

Kumar, V., Ahluwalia, V., Saran, S., Kumar, J., Patel, A.K., Singhania, R.R., 2021. Recent developments on solid-state fermentation for production of microbial secondary metabolites: challenges and solutions. Bioresour. Technol. 323, 124566.

Kuraishi, C., Yamazaki, K., Susa, Y., 2001. Transglutaminase: its utilization in the food industry. Food Rev. Int. 17, 221–246.

Liu, R., Chen, L., Jiang, Y., Zhou, Z., Zou, G., 2015. Efficient genome editing in filamentous fungus *Trichoderma reesei* using the CRISPR/Cas9 system. Cell Discov. 1, 15007.

Olempska-Beer, Z.S., Merker, R.I., Ditto, M.D., DiNovi, M.J., 2006. Food-processing enzymes from recombinant microorganisms-a review. Regul. Toxicol. Pharm. 45, 144–158.

Otten, L.G., Quax, W.J., 2005. Directed evolution: selecting today's biocatalysts. Biomol. Eng. 22, 1–9.

Pandey, A., 1995. Glucoamylase research: an overview. Starch/Starke 47 (11), 439–445.

Pandey, A., 2003. Solid-state fermentation. Biochem. Eng. J. 13 (2–3), 81–84.

Pandey, A., Selvakumar, P., Soccol, C.R., Nigam, P., 1999. Solid-state fermentation for the production of industrial enzymes. Curr. Sci. 77 (1), 149–162.

Pandey, A., Singhania, R.R., 2008. Production and application of industrial enzymes. Chem. Ind. Dig. 21, 82–91.

Pandey, A., Nigam, P., Soccol, C.R., Singh, D., Soccol, V.T., Mohan, R., 2000. Advances in microbial amylases. Biotechnol. Appl. Biochem. 31, 135–152.

Pandey, A., Szakacs, G., Soccol, C.R., Rodriguez-Leon, J.A., Soccol, V.T., 2001. Production, purification and properties of microbial phytases. Bioresour. Technol. 77 (3), 203–214.

Pandey, A., Webb, C., Soccol, C.R., Larroche, C. (Eds.), 2006. Enzyme Technology. Springer Science, USA, p. 740.

Pandey, A., Soccol, C.R., Larroche, C. (Eds.), 2007. Current Developments in Solid-state Fermentation. Springer, USA & Asiatech Publishers, Inc., New Delhi, p. 517.

Pandey, A., Larroche, C., Soccol, C.R., Dussap, C.G. (Eds.), 2008. Advances in Fermentation Technology. Asiatech Publishers, Inc., New Delhi, p. 672.

Pandey, A., Binod, P., Ushasree, M.V., Vidya, J., 2010. Advanced strategies for improving industrial enzymes. Chem. Ind. Dig. 23, 74–84.

Patel, A.K., Singhania, R.R., Pandey, A., 2014. Biofuels from biomass. In: Agarwal, A., Pandey, A., Gupta, A., Aggarwal, S., Kushari, A. (Eds.), Novel Combustion Concepts for Sustainable Energy Development. Springer, New Delhi. Available from: https://doi.org/10.1007/978-81-322-2211-8_3.

Patel, A.K., Singhania, R.R., Pandey, A., 2016. Novel enzymatic processes applied to the food industry. Curr. Opin. Food Sci. 7, 64–72.

Patel, A.K., Dixit, P., Pandey, A., Singhania, R.R., 2019. Promising enzymes for biomass processing. In: Singh, S.P., Singhania, R.R., Kondo, A., Larroche, C. (Eds.), Advances in Enzyme Catalysis and Technologies. Elsevier, pp. 245–264. ISBN:978-0-12-819820-9.

Pitman, S., 2011. Growth in enzyme market driven by cosmetic demand. Online news, Cosmetics design.

Rehm, H.J., Reed, G. (Eds.), 1985. Biotechnology, a Comprehensive Treatise in Eight Volumes. VCH Verlagsgesellschaft, Weinheim, Germany.

Research and Markets, 2011a. In: Report - Indian Industrial Enzymes Market.

Research and Markets, 2011b. In: Report - Chinese Markets for Enzymes.

Rich, J.O., Michels, P.C., Khmelnitsky, Y.L., 2002. Combinatorial biocatalysis. Curr. Opin. Chem. Biol. 6, 161–167.

Sarrouh, B., Santos, T.M., Miyoshi, A., Dias, R., Azevedo, V., 2012. Up-to-date insight on industrial enzymes applications and global market. J. Bioprocess. Biotech. S4, 002. Available from: https://doi.org/10.4172/2155-9821.S4-002.

Selvakumar, P., Ashakumary, L., Pandey, A., 1996. Microbial synthesis of starch saccharifying enzyme in solid-state fermentation. J. Sci. Ind. Res. 55 (5–6), 443–449.

Shu-Jen, C., 2004. Strain improvement for fermentation and biocatalysis processes by genetic engineering technology. J. Ind. Microbiol. Biotechnol. 31, 99–108.

Singhania, R.R., Patel, A.K., Soccol, C.R., Pandey, A., 2009. Recent advances in solid-state fermentation. Biochem. Eng. J. 44 (1), 13–18.

Singhania, R.R., Patel, A.K., Pandey, A., 2010. The industrial production of enzymes. In: Soetaert, W., Vandamme, E.J. (Eds.), Industrial Biotechnology. Wiley-VCH Verlag, Weinheim, Germany, pp. 207–226.

Singhania, R.R., Saini, R., Adsul, M., Saini, J.K., Mathur, A., Tuli, D., 2015. An integrative process for bio-ethanol production employing SSF produced cellulase without extraction. Biochem. Eng. J. 102, 45–48. Available from: https://doi.org/10.1016/j.bej.2015.01.002.

Singhania, R.R., Dixit, P., Patel, A.K., Kuo, C.H., Chen, C.W., Dong, C.D., 2021a. LPMOs: a boost to catalyse lignocellulose deconstruction. Bioresour. Technol. 335, 125261.

Singhania, R.R., Ruiz, H.A., Awasthi, M.K., Dong, C.D., Patel, A.K., 2021b. Challenges in cellulase bioprocess for biofuel applications. Renew. Sust. Energ. Rev. 151, 111622.

Sonenshein, A.L., Hoch, J.A., Losick, R. (Eds.), 1993. *Bacillus subtilis* and Other Gram-Positive Bacteria: Biochemistry, Physiology and Molecular Genetics. ASM Press, Washington DC.

Thomas, L., Arumugam, M., Pandey, A., 2013a. Production, purification characterization and over-expression of xylanases from actinomycetes. Ind. J. Exp. Biol. 51 (6), 875–884.

Thomas, L., Larroche, C., Pandey, A., 2013b. Current developments in solid-state fermentation. Biochem. Eng. J. 81, 146–161.

Zhang, F., Wen, Y., Guo, X., 2014. CRISPR/Cas9 for genome editing: progress, implications and challenges. Hum. Mol. Genet. 23 (R1), R40–R46. Available from: https://doi.org/10.1093/hmg/ddu125.

Chapter 4

Solid-state fermentation for the production of microbial cellulases

Sudhanshu S. Behera[1,2], Ankush Kerketta[1] and Ramesh C. Ray[2]
[1]*Department of Biotechnology, National Institute of Technology Raipur, Raipur, Chhattisgarh, India,* [2]*Centre for Food Biology & Environment Studies, Bhubaneswar, Odisha, India*

4.1 Introduction

Lignocellulosic biomass, which is created from biological photosynthesis, is a readily available sustainable and clean resource for the development of fine chemicals and fuels with no additional carbon emissions (Velvizhi et al., 2022). Furthermore, the syntheses of lignocellulose-derived chemicals and fuels offer significant potential to lower our reliance on fossil fuels, enhance the country's economic growth, and prevent the deterioration of the environment (Xia et al., 2016; Velvizhi et al., 2022). Typical sources of lignocellulosic biomass include grasses (e.g., maize stalks and wheat straw), softwoods (white pine and larch), and hardwoods (camphor and birch) (Xia et al., 2016; Velvizhi et al., 2022; Yuan et al., 2022). In terms of biochemistry, lignocellulosic biomass is a collection of materials high in hemicellulose (15%–30%), cellulose (35%–50%), and lignin (15%–30%) (Yuan et al., 2022). Each of the lignocellulose's core ingredients can be transformed into high-value compounds and foundation molecules. However, before lignocellulosic biomass or cellulose can be used as a carbon feedstock for the production of high-value products, they must be pretreated and degraded by enzymes (Han et al., 2019; Yuan et al., 2022).

Enzymes, particularly cellulase, serve a significant part in the industrial sector as it converts lignocellulosic biomass (or cellulose) to simple glucose feedstock. Fungi, actinomycetes, and bacteria synthesize cellulose when they grow in lignocellulosic biomass, but fungi have been the most prevalent producers (Ejaz et al., 2021). Cellulases from bacteria and fungi typically include two or more structural and functional regions interconnected by a peptide spacer (Sakka et al., 2000). Cellulase seems to be involved in cellulose breakdown by hydrolyzing the β-1,4-glycosidic linkages. β-Glucosidase

Biotechnology of Microbial Enzymes. DOI: https://doi.org/10.1016/B978-0-443-19059-9.00012-8
© 2023 Elsevier Inc. All rights reserved.

(BGL), endo-1,4-D-glucanase (endoglucanase), and exo-1,4-D-glucanase (exoglucanase) are the three enzymes that makeup cellulose. These three enzymes work together to hydrolyze cellulose in a coordinated manner for effective and successful cellulose breakdown. Endoglucanase attacks oligosaccharide interior regions in amorphous cellulose, carboxymethyl cellulose, and cellooligosaccharides. Exoglucanase catalyzes the breakdown of crystalline cellulose at the nonreducing ends to produce glucose or cellobiose (Fig. 4.1). BGL targets nonreducing ends of cellodextrin and cellobiose (Patel et al., 2019).

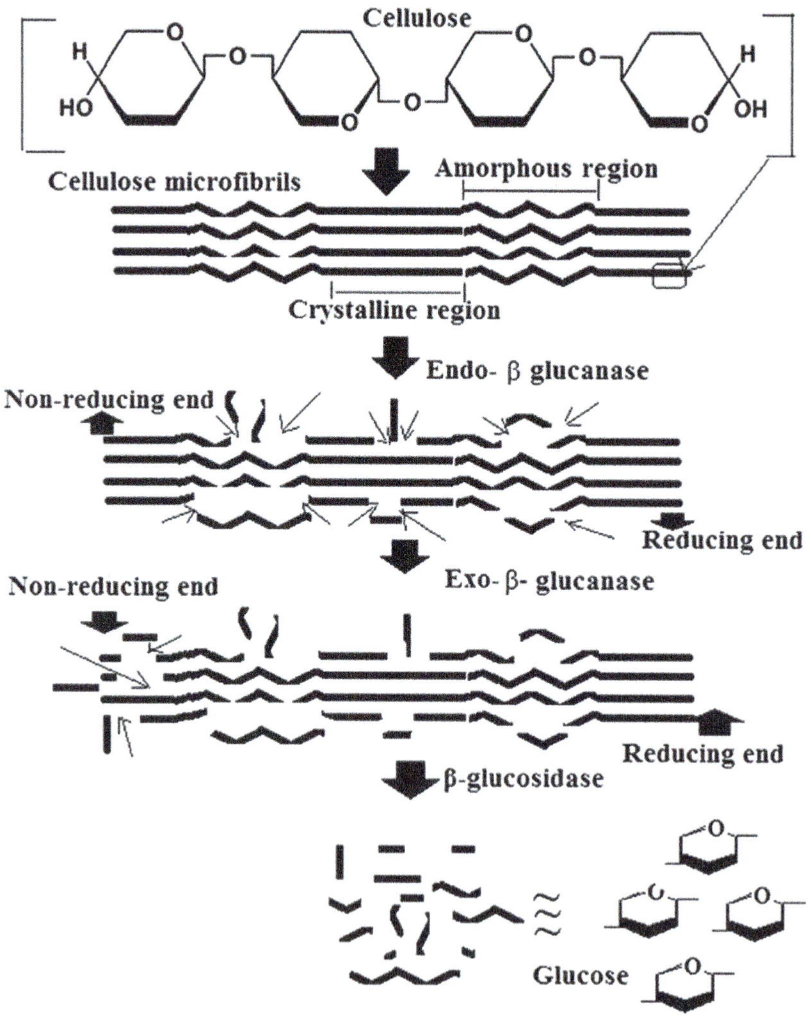

FIGURE 4.1 Schematic representation of sequential stages in cellulolysis.

4.2 Solid-state fermentation

Solid-state fermentation (SSF) is found to be a promising strategy for synthesizing cellulose by microorganisms (Gad et al., 2022). It is a fermentation method where microorganisms grow on solid substrates without the availability of free liquid. Employing solid substrates is perhaps the old way to make microbes serve humans (Bhargav et al., 2008). SSF replicates the local habitat of the majority of microorganisms, primarily fungus and molds. It is less prone to bacterial contamination and allows for higher enzyme performance for several enzymes. Nevertheless, the substrate, microorganisms, and operating parameters all show the impact of the targeted enzyme synthesis (Chilakamarry et al., 2022). SSF has demonstrated great potential in developing cellulose production in the past few years.

4.2.1 Comparative aspects of solid-state and submerged fermentations

In recent years, fermentation technology and biochemical engineering elements of enzyme production have advanced dramatically, including advancements in production tank design for solid substrate (SSF) and liquid-phase/submerged fermentation (SmF) configurations. A comparative analysis of the two fermentation methods is provided below:

- Industrially, SmF is the most widely used technology for synthesizing enzymes since it offers benefits, including a well-developed bioreactor with control and monitoring tools; nevertheless, the SmF method has some disadvantages, such as a moderate mass yield, massive cost, and the generation of large amounts of wastewater (Chakraborty et al., 2019).
- SSF uses an inexpensive substrate, including lignocellulosic biomass, to offer a framework for enhanced microorganism–substrate interaction, resulting in increased enzyme concentration. Its fermentation waste can be recycled and used in other operations (Chakraborty et al., 2019).
- SSF is problematic for microorganisms with a greater water function demand, including bacteria. The majority of SSFs utilized in the enzyme industry employ fungi as native producers but are not used to synthesize recombinant enzymes (Ouedraogo and Tsang, 2021).
- SSF resembles the filamentous fungi's native habitat significantly better than submerged culture, which may justify their outstanding output. Compared to bacteria and other unicellular microorganisms, fungal filaments in solid fermentation facilitate significant penetration into the material and have excellent osmotic resistance (Juturu and Wu, 2014).

SSF is cost-effective in the mentioned industrial sectors (low water requirement, cheaper carbon source, and simple and environment-friendly method).

Many different researchers investigated the development of microbial cellulases in SmF and SSF. Ramamoorthy et al. (2019) examined a mixture of packaging cardboard and surgical cotton waste as a carbon source for cellulase production by *Trichoderma harzanium* ATCC 20846. The results showed that cellulase synthesized through an SSF (FPase activities = 3.230 IU/mL) exhibits 21% higher enzyme activity compared to cellulase synthesized with an SmF (FPase activities = 1.94 IU/mL). Singh et al. (2021) employed rice straw as a substrate for fungal growth during SSF and SmF environments. *Aspergillus heteromorphus* significantly increased cellulase activity of 6.4 IU/g FPase and 125 IU/g CMCase under SSF environments and 3.8 IU/g FPase and 94 IU/g CMCase under SmF environments. Likewise, *Aspergillus niger* showed FPase and CMCase activity (CA) of 5.8 and 113 IU/g, correspondingly, during SSF, whereas FPase activity and CA were 3.5 and 88 IU/g during SmF.

4.2.2 Cellulase-producing microorganisms in solid-state fermentation

Presently, a wide range of microorganisms has been utilized in industrial cellulase production via fermentation. Cellulases are synthesized in high or low titers by a variety of fungi and bacteria (Singh et al., 2021). Cellulase is produced by bacteria in the form of single or multisubunit enzyme cellulosomes, which include a combination of cellulolytic enzymes attached to a scaffold protein; however, fungi generate cellulases as liberated molecules (Jayasekara and Ratnayake, 2019). Table 4.1 lists many cellulase makers from various microorganisms.

4.2.3 Extraction of microbial cellulase in solid-state fermentation

Different investigators have described the extraction of cellulase from microbiological cells (bacteria, fungi, and actinomycetes) after SSF (Amadi et al., 2022; Fadel et al., 2021; Moran-Aguilar et al., 2021; Srivastava et al., 2022). In short, a crude enzyme present in fermented substrate had been extracted with a sodium citrate buffer (50 mM, 4.8 pH). However, simplified distilled water may be used for extraction (Fadel et al., 2021). A mixture of 10 mL of buffer/g fermented material was stirred at 150 rpm for 30 min. The enzyme extract subsequently passed through muslin cloth, and the filtrate was ultracentrifuged. The supernatant was recovered by ultracentrifugation and had used effectively in enzyme experiments.

TABLE 4.1 Major microorganisms[a] employed in cellulase production via solid-state fermentation.

Major group	Microorganisms	
	Genus	**Species**
Fungi	*Aspergillus*	*A. niger, A. nidulans, A, oryzae, A. fumigatus,A. phoenicis, A. ibericus, A. penicillioides, A. ibericus, A. terreus, A. unguis*
	Bjerkandera	*B. adusta*
	Ceriporiopsis	*C. subvermispora*
	Cerrena	*C. maxima*
	Cladosporium	*C. cladosporioides, C. sphaerospermum*
	Coriolus	*C. versicolor*
	Coriolopsis	*C. polyzona*
	Eupenicillium	*E. crustaceum*
	Emericella	*E. variecolor*
	Fomes	*F. fomentarius*
	Funalia	*F. trogi*
	Fusarium	*F. oxysporum*
	Geotrichum	*G. candidum*
	Gloeophyllum	*G. trabeum*
	Humicola	*H. insolens, H. grisea*
	Irpex	*I. lacteus*
	Laetiporus	*L. sulphureus*
	Lentinus	*L. strigosus, L. edodes, L. tigrinus*
	Melanocarpus	*M. albomyces*
	Myceliophthora	*M. thermophila*
	Microporus	*M. xanthopus*
	Mucor	*M. indicus, M. hiemalis*
	Neurospora	*N. crassa*
	Paecilomyces	*P. themophila*
	Psedotremella	*P. gibbosa*

(Continued)

TABLE 4.1 (Continued)

Major group	Microorganisms	
	Genus	**Species**
	Piptoporus	*P. betulinus*
	Penicillium	*P. oxalicum, P. citrinum, P. brasilianum, P. decumbans, P. occitanis*
	Phanerochaete	*P. chrysosporium*
	Phomopsis	*P. stipata*
	Phaseolus	*P. coccineus*
	Pleurotus	*P. ostreatus, P. dryinus, P. tuberregium, P. sajor-caju, P. pulmonarius*
	Pycnoporus	*P. sanguineus, P. coccineus*
	Rhizopus	*R. oryzae*
	Saccharomyces	*S. cerevisiae*
	Schizophyllum	*S. commune*
	Trichaptum	*T. biforme*
	Trichoderma	*T. longibrachiatum, T. afroharzianum, T. asperellum, T. reesei, T. stromaticum, T. koningii, T. viridae, T koningii, T. harzianum*
	Trametes	*T. versicolor, T. trogii, T. pubescens, T. hirsute,T. ochrace*
	Wolfiporia	*W. cocos*
Bacteria	*Acidothermus*	*A. cellulolyticus*
	Bacillus	*B. subtilis, B. pumilus*
	Clostridium	*C. acetobutylicum, C. thermocellum*
	Cellulomonas	*C. fimi, C. bioazotea, C. uda*
Actinomycetes	*Streptomyces*	*S. enissocaesilis, S. viridochromogenes, S. philanthi*
	Thermomonospora	*T. fusca, T. curvata*

[a]*Updated from Ray and Behera (2017).*

4.2.4 Measurement of cellulase activity in solid-state fermentation

4.2.4.1 Filter paper activity (FPase)

Fundamentally, the filter paper activity (FPase) has been measured utilizing filter paper as a substrate (Singh et al., 2021) and is calculated as the total amount of reducing sugar which are produced after the 1 mL enzyme assay (Eveleigh et al., 2009; Singh et al., 2021). Filter paper offers the benefit of employing an easily accessible and repeatable substrate that is neither too sensitive nor too resistant, which could be determined by unit area, eliminating the boredom of weighting a solid or attempting to evenly dispense a solid suspension (Eveleigh et al., 2009). In short, the filter paper was incubated with a dilute solution of cellulase in citrate buffer at 50°C for 30 min. The released reducing sugars in the reaction were measured using Miller's dinitro-salicylic acid test (1959) (Singh et al., 2021). The amount of enzyme which generates 1 mmoL of glucose in 1 min has been expressed in international units (IU).

4.2.4.2 Carboxymethyl cellulase activity (CMCase)

Carboxymethyl cellulase (CMCase) activity is frequently used to describe single endoglucanase activity. CA, which mainly demonstrates endoglucanase activity, may be utilized to represent total free or bound cellulase activity via cellulose enzymatic hydrolysis for particular cellulose–cellulase reaction system (Zhou et al., 2004). In brief, a fixed quantity of the crude enzyme was introduced to a CMC solution made in sodium citrate buffer (1% w/v) (50 mM, pH 5.0). This combination was incubated for 30 min at 50°C. A conventional 3′,5′-dinitrosalicylic acid (DNS or DNSA) technique (1959) was used to detect the reducing sugars liberated from the substrates after cellulose–cellulase reaction. The quantity of enzyme which generates 1 mmol of reducing sugars per minute is calculated as one unit of enzyme activity (U) (Boontanom and Chantarasiri, 2021; Taherzadeh-Ghahfarokhi et al., 2021).

4.2.4.3 Xylanase activity

Usually, xylanase activity is reported by researchers based on the liberation of reducing sugars from semi-soluble xylan substrates (Bailey et al., 1992). The activity of xylanase was determined by employing a substrate of 1% (wt/v) xylan in sodium phosphate buffer (100 mM, pH 7.0) (Moran-Aguilar et al., 2021; Yadav et al., 2018). The amount of xylose generated from xylan was determined using the 3′ 5′ dinitrosalicylic acid (DNSA) technique with xylose as a reference. Endo-1, 4–xylanase units were established as the quantity of enzyme needed to generate 1 mol of xylose per minute under normal assay conditions.

4.2.4.4 β-Glucosidase activity

BGLs are a crucial element of the cellulase system (cellulose metabolizing enzymes) that facilitate the final and most crucial step in cellulose hydrolysis. Cellulase enzymes break down cellulose into cellobiose and various short oligosaccharides that are then broken down further by BGL into glucose (Singh et al., 2016). The rate of *p*-nitrophenyl β-D glucopyranoside (PNG) hydrolysis was used to measure BGL activity (Alarcón et al., 2021). This was accomplished by combining 20 μL of PNG with 980 μL of crude enzymatic extract and incubating it at 50°C for 60 min. The solution was suspended in 1 mL of 0.5 M NaOH following incubation. The reaction solution was spectrophotometrically determined at a wavelength of 412 nm.

4.3 Lignocellulosic residues/wastes as solid substrates in solid-state fermentation

The concept of adopting alternate substrates stems from the desire to establish more economical methods, lowering the end by-products' price. The utilization of lignocellulosic feedstock, mostly agro-industrial leftovers, to synthesize cellulase and other lignocellulolytic enzymes is a practical method of lowering process input costs (Siqueira et al., 2020).

Agricultural waste remains among the most acceptable substrates for enzyme synthesis for SSF. Agro-waste substrates potentially offer the nutrients necessary for microbial growth, resulting in high enzyme output. Wheat straw, wheat bran, rice straw, maize, grains, brans, coconut coir and pith, tea, coffee, and cassava waste, oil mill waste, sugarcane bagasse and sorghum pulp waste, coconut oil cakes, mustard, soybean, peanut, and cassava oil cakes are some of the most commonly used enzyme substrates (Chakraborty et al., 2019; Leite et al., 2021).

SSF is recognized to have several upsides over SmF (or liquid fermentation), including increased yield and production output, increased product stability, reduced production expense, relatively low protein breakdown (that is particularly important if an enzyme is the intended outcome), fairly low contamination risk, lesser energy need, lesser energy expenses for sterilization, lesser fermenter volume, and lower (or absence of) contamination risk. SSF also expands the range of agro-industrial wastes and by-products that can be used as raw materials in their natural state (nonpretreated) which is a salient feature in aspects of economic viability, as feedstock is frequently noted as among the main operational charges in enzyme production processes (Ramamoorthy et al., 2019).

4.4 Pretreatment of agricultural residues

Cellulose and hemicellulose hydrolysis by cellulase in lignocellulosic biomass is limited by a variety of possible factors. For instance, minerals, sugars, lignin, oils, particle size, and the partly crystalline structure of cellulose might indeed influence the hydrolysis being restricted (Beig et al., 2021). Because of such a significant hurdle, lignocellulose requires pretreatment. The pretreatment makes the substrate readily available to microorganisms and improves enzyme activity, resulting in the production of more enzymes. The fine feedstock was preferable as they give more outstanding production; nevertheless, there are several techniques for altering substrate before actual employment in fermentation, each of which has a distinct effect on the lignocellulose contents (Fatima, 2021). As a result, distinct procedures are developed and used depending on the fermentation phases needed. Because it saves time and effort required for fermentation, it simplifies the system and is more cost-effective due to lower energy use.

4.4.1 Physical pretreatments

Mechanical comminution is described as physical pretreatment via combining chipping, grinding, and milling. Physical pretreatment is, in fact, a precondition for subsequent chemical or biological treatment. This decrement in size corresponds to the greater surface area, reduced crystalline structure, and lignin structural alteration, providing it ideal for future processing. Physical pretreatments of lignocellulosic material need a certain amount of energy that corresponds to the desired fine particle size and crystallinity. The size of the substrate was typically 10–30 mm and 0.2–2 mm after chipping and milling or grinding, respectively. An additional advantage would be that the disintegration of the structure might reduce by-product generation throughout fermentation, which optimizes the downstream process encouraging enzymatic transformation. One thing to consider has been that excessive depolymerization and extremely small, fine materials will result in clumping. An SSF for cellulase synthesis employing pretreated biomass (waste surgical cotton and packaging cardboard) has 23% higher enzyme activity compared to cellulase generated utilizing an untouched substrate composition (Behera and Ray, 2016; Ramamoorthy et al., 2019).

4.4.2 Physiochemical pretreatment

One of the best effective pretreatments includes physicochemical pretreatment, which mixes physical or mechanical operations and chemical reactions to break down the cell walls of biomass. Lignocellulose does have the capacity to withstand deconstruction, which must be overcome for SSF to operate efficiently. Physiochemical approaches such as a steam explosion, liquid hot water, and ammonia fiber explosion can be used to break down this barrier.

Under steam explosion pretreatment, biomass is processed with high saturated steam before the pressure is rapidly decreased, causing the materials to decompress explosively. The steam explosion has been outlined as a huge interruption of lignocellulosic structure is facilitated by heat from steam (Thermo), shear factors due to moisture increase (mechano), and hydrolysis of glycosidic bonds (chemical), which results in cleavages of some available glycosidic links, β-ether linkages of lignin and lignin–carbohydrate complex bonds, and minor chemical modifications of lignin and carbohydrates. Temperatures vary from 190°C to 270°C, with exposure time ranging from 1 to 10 min. The starting material determines the amount of time and temperature required. It allows for a high cellulose yield.

Ammonia fiber expansion is another potential and leading physicochemical procedure for substantially pretreating lignocellulosic biomass. For pretreatment, lignocellulosic materials are impregnated with liquid ammonia before being heated to approximately 90°C. The generated gas ammonia reacts with biomass at pressure before being quickly released, resulting in cellulose de-crystallization, hemicellulose prehydrolysis, and lignin structural changes.

4.4.3 Chemical pretreatments

Chemical pretreatment has been extensively used by corporations since it is the more cost-effective approach (Trivedi et al., 2015b). It assures improved output generation and fewer breakdowns, releases significantly higher rates of sugar, and reduces the requirement for enzymes and solvent loading, usually combined with other procedures. Acids, alkalis, ammonia, ethylamine, ozone, and other compounds are commonly used in chemical procedures (Ramamoorthy et al., 2019; Trivedi et al., 2015b). According to the nature of the substrate required for SSF, these compounds are employed in several ways.

The acid approach results in hemicelluloses and lignin reduction, whereas the alkaline approach transforms lignocellulose into alcohol and soap while also changing the composition of lignin.

4.4.4 Biological pretreatment

Biological pretreatment uses microorganisms in the pretreated feedstock to boost the enzymatic accessibility of residual solids. The microorganisms used can generally degrade lignin and carbohydrate polymers, with white-, brown-, and soft-rot fungi attracting the most consideration. These fungi could make lignocellulolytic enzymes that act together to break down plant cell walls. Furthermore, it has been demonstrated that several white- and brown-rot fungi generate hydrogen peroxide, which undergoes the Fenton reaction and results in the production of free hydroxyl radicals (•OH). These free radicals break polysaccharides and lignin in lignocellulosic materials in a nonspecific fashion, causing cleavages that allow lignocellulolytic enzymes to invade more easily. A portion of the lignin and hemicelluloses were destroyed after white-rot fungus pretreatment, leading to an improvement in SSA and pore size of the feedstock; nevertheless, the crystallinity index (CrI) of the feedstock does not quite alter much.

4.5 Environmental factors affecting microbial cellulase production in solid-state fermentation

Apart from the feedstock and microorganisms utilized, the efficiency of SSF is influenced by procedural and environmental parameters. Moisture content, pH, temperature, aeration, fermentation length, substrate nature (including particle size), inducer, and nutrients are all critical parameters that have a significant impact on enzyme synthesis. These elements are essential for creating a link between microbial growth and creating appropriate models. Nonetheless, the key issues that prevent SSF from being used broadly are monitoring and controlling the process parameters (Singhania et al., 2017).

4.5.1 Water activity/moisture content

Moisture represents a factor that is closely linked to the concept of SSF as well as the biological material's features. The significance of water in the system arises from the premise that the vast proportion of viable cells has a moisture content of 70%–80%. Based on this basic observation, it is straightforward to conclude that a specific amount of water is required for new cell production. The importance of moisture in SSF operations was already widely recognized. In principle, it has been proven that the moisture content of the solid matrix had to be greater than 70% in the scenario of bacteria. The moisture range for yeast might be as large as 72%–60%, whereas the range for fungus can be as broad as 70%–20%. This assurance might be a significant benefit in creating a specific SSF procedure. This might act as a natural barrier to pollutants if the optimal moisture level for the growth of the utilized microorganism is comparatively lower. In this instance, sterilization treatment may be less stringent or perhaps unnecessary.

4.5.2 Temperature

Biological processes are distinguished by the fact that they occur within a very small temperature range. In several cases, high-temperature thresholds for growth were identified at temperatures between 60°C and 80°C, with some strains reaching 120°C. The importance of temperature in the evolution of a biological process has been that it may influence things such as protein denaturation, enzyme inhibition, encouragement or inhibition of metabolite synthesis, cell death, and so on. The importance of temperature in microbial development leads to a categorization based on temperature, or the temperature range wherein the microbe develops. Microorganisms that are thermophiles may grow at temperatures ranging from 80°C to 95°C., including an optimal growth temperature of 45°C–70°C. Mesophiles are the microorganisms that can only survive at temperatures below 50°C, and temperatures of 30°C–45°C are ideal for their growth. Microorganisms that are psychrophiles typically withstand temperatures below 0°C, with the optimal temperature range being 10°C–20°C.

4.5.3 Mass transfer processes: aeration and nutrient diffusion

4.5.3.1 Gas diffusion

When the gas transfer occurs in a bioreactor, the primary aspects are (1) O_2 transmission and (2) CO_2 evolution. The activity of the fungal culture in SSF may be immediately shown by measuring microbial activity, such as O_2 utilization, CO_2, and heat generation during SSF. Smaller particle size offers advantages for heat transfer and exchange of O_2 and CO_2 between the air and the solid surface.

4.5.3.2 Nutrient diffusion

The transport of nutrients and enzymes within solid substrate particles is referred to as nutrient diffusion. When the water activity is low, the dispersion of nutrients through the solid matrix is reduced, but the substrate particles are compacted when the water activity is high. As a result, adequate water activity and moisture levels in the solid substrate are the critical components of SSF processes.

4.5.4 Substrate particle size

The substrate grain size is very important for substrate characterization and system ability to exchange with microbial growth, heat, and mass transfer throughout the SSF process. It establishes numerous characteristics that are used to characterize a heterogeneous system.

4.5.5 Other factors

Microbial growth in SSF has also been affected by pH, inoculums size, and sterilization of the system. The inoculum size regulates microbial multiplication in the fermentation cycle; nevertheless, a small proportion of inoculum might not have been sufficient to initiate microbe growth, while a high inoculum causes challenges such as mass transfer limitations. The concentration of inoculum is an essential parameter in the SSF (Pandey, 2003, Singhania et al., 2017; Thomas et al., 2013). In the SSF system, pH is a critical and highly influencing growth parameter for microbial metabolism and development. The substrate employed in SSF has a buffering effect owing to its complex chemical composition; nevertheless, due to the heterogeneous characteristics of the substrate and the lack of free water, it is hard to observe the changes. Bacterial contamination in fungus and yeasts may be avoided by cultivating them at a pH that is not ideal for bacteria since fungi and yeast have a wide pH range for growth that can be utilized (Lopez-Calleja et al., 2012). In SSF, ammonium salts were combined with urea or nitrate salts to balance the acidity and alkalization effects (Singhania et al., 2017). Microorganisms can cause illness and are often prevalent in the surroundings. These are the main sources of pollution and the ones to blame. Sterilization has the goal of removing or destroying them from a substance or surface. Sterilization is a crucial operational necessity for any biotech firm to preserve pure culture and make the industrial fermentation process effective (Deindoerfer, 1957). Sterilization can be accomplished in a variety of ways, including using a combination of heat, chemicals, irradiation, high pressure, and filtration, or using dry heat, ultraviolet (UV) radiation, gas vapor sterilants, chlorine dioxide gas, and other methods (Liu et al., 2013).

4.6 Strategies to improve production of microbial cellulase

Several attempts have recently been concentrated on improving cellulase production via improved cellulolytic activities and the creation of predicted enzyme characteristics. Numerous techniques, such as heterologous or recombinant cellulase expression, coculture, or integrated metabolic engineering procedures, are gaining importance for better cellulase production (Table 4.2).

4.6.1 Metabolic engineering and strain improvement

For optimum cellulase production, growth conditions are optimized, and strain enhancement procedures are applied (Vu et al., 2011). Many studies believe that mutagenesis substances can be used to boost strains for microbial cellulase production (El-Ghonemy et al., 2014; Li et al., 2010). Several

TABLE 4.2 Spectrum of microbial cultures employed for the production of *cellulases* in solid-state fermentation (SSF) systems.

Types of wastes/substrate	Inoculant	Nutrient and culture conditions	Activity (cellulase production)	References
Substrate as carbon source				
Oil palm trunk	*Aspergillusfumigatus* SK1	PDA (30°C, 5 d) + 0.75 g/L peptone and 2 mL/L Tween 80	CMCase: 54.27 U/gds; FPase: 3.36 U/gds; β-glucosidase: 4.54 U/gds	Ang et al. (2013)
Wheat bran	*Chrysoporthe cubensis*	125 mL + 5 gds + 12 mL of culture media	Endoglucanase: 33.84 U/gds; FPase: 2.52 U/gds; β-glucosidase: 21.55 U/gds and Xylanase: 362.38 U/gds	Falkoski et al. (2013)
Rice grass (*Spartina* spp.)	*Aspergillus* sp. SEMCC-3.248	Rice grass 2.5 g, Wheat bran 1.5 g, 4 mL of nutrient; Spore suspension (1×10^7 spores/mL) Moisture content 70%, pH 5.0, 32°C and 5 d	Cellulase: 1.14 FPU/gds	Liang et al. (2012)
Green seaweed (*Ulva fasciata*)	*Cladosporium sphaerospermum*	PDA (4°C); 250 mL + 10 g DPS + mineral salt medium	CMCase: 0.20 ± 0.40 U/gds;FPase: 9.60 ± 0.64 U/gds	Trivedi et al. (2015a)
Sugarcane bagasse and corn steep liquor	*Bacillus* sp. SMIA-2	pH 6.5–8.0 and 70°C	Avicelase: 0.83 U/mL; CMCase: 0.29 U/mL	Ladeira et al. (2015)

Mixed or coculture system				
Agricultural residues (cauliflower waste, kinnow pulp, rice straw, pea-pod waste, wheat bran)	*Aspergillus niger* and *Trichoderma reesei*	MWM; tray fermentation (120 h)	FPase: 24.17 IU/gds; β-glucosidase: 24.54 IU/gds	Dhillon et al. (2011)
Wheat straw	*T. reesei* RUT-C30 + *Aspergillus saccharolyticus* AP, *Aspergillus carbonarius* ITEM 5010 or *A. niger* CBS 554.65	10% glycerol, −80°C; 25.6% (w/v) wheat bran; 1 mL of 5×10^6/mL spore suspension (25°C, 10 d)	80% more avicelase activity	Kolasa et al. (2014)
Reed	*T. reesei* RUT-C30 ATCC 56765 + *Clostridium acetobutylicum* ATCC824	Glucose, 40.01 g/L + xylose, 3.55 g /L; pH 6.5; 2 M NaOH; carbon source and yeast extract (3 g/L) and sterilized (121°C, 15 min)	–	Zhu et al. (2015)
Comparative studies of SmF and SSF				
Wheat bran, corn bran, and kinnow peel	*A. niger* NCIM 548	30°C, 170 rpm, media components g/L [$(NH_4)_2SO_4$, $MgSO_4$, $FeSO_4$, $7H_2O$, and KH_2PO_4]	Pectinase, 7.13 and cellulase, 1.95 times higher in SSF than in SmF	Kumar et al. (2011)
Sugarcane bagasse	*T. reesei* RUT-C30	Conidia suspension of 10^7 spores/mL; 100 mL; glucose, 30 g/L	4.2-folds more in SSF than in SmF	Florencio et al. (2015)
Metabolic engineering and strain improvements				
Sugarcane bagasse	*Chaetomium cellulolyticum* NRRL 18756	Peptone 1% (w/w), 2.5 mM MgSO4, 0.05% (v/w) Tween 80; moisture content 40%(v/w) and pH 5.0–6.5 (40°C 4d)	CMCase yield 4-folds more than wild-type strain	Fawzi and Hamdy (2011)
Castor bean meal (*Ricinus communis* L.)	*Aspergillus japonicus* URM5620	Spores (10^7) + malt extract agar plates + 0.05 M citrate buffer; pH 3–10 and 30°C–50°C	β-glucosidase: 88.3U/g; FPase: 953.4 U/g; CMCase: 191.6 U/g	Herculano et al. (2011)

(*Continued*)

TABLE 4.2 (Continued)

Types of wastes/substrate	Inoculant	Nutrient and culture conditions	Activity (cellulase production)	References
Sequential cultivation methodology				
Sugarcane bagasse	*A. niger* A12	Shake flasks + bubble column bioreactor; PDA (32°C, 7d)	Endoglucanase: 57 ± 13 IU/L/h	Cunha et al. (2014)
Sugarcane bagasse	*T. reesei* RUT-C30	Conidia suspension (10^7 spores/mL); Glucose, 30 g/L; 72 h, 30°C, and 200 rpm	4.2-fold improvement compared with SmF	Florencio et al. (2015)
Bioreactor systems				
Soybean hulls	*T. reesei* + *Aspergillus oryzae*	Static tray fermenters; 250-mL; PDA, 100 mL, Tween-80, 0.1 mL; 30°C, 5–6 d	FPase: 10.7 FPU/gds; β-glucosidase: 10.7 IU/gds	Brijwani et al. (2010)
Sugarcane bagasse	*T. reesei* NRRL-6156	Packed-bed bioreactor; medium for preinoculum (g/L): $(NH_4)_2SO_4$, $MgSO_4$, $FeSO_4 \cdot 7H_2O$, $MnSO_4 \cdot 7H_2O$, Tween 80	Higher for hydrolysis (229 g/kg); 43.4°C and enzymatic extract of 18.6% (w/v)	Gasparotto et al. (2015)

DPS, Dried and powdered seaweed; *FPU*, filter paper units; gds, gram dry substrate; *MWM*, Mendel–Weber medium; *PDA*, potato dextrose agar; *SmF*, submerged fermentation; *U*, units, *BGL*, β-glucosidase.

mutagenic agents, such as UV irradiation, gamma irradiation (Co60-rays), ethyl-methane sulfonate (EMS), and *N*-methylN0-nitro-*N*-nitrosoguanidine, have been used on several fungal strains to increase cellulase production (El-Ghonemy et al., 2014; Fawzi and Hamdy, 2011; Mostafa, 2014). Dhillon et al. (2012) used a multistep change approach to develop a changed strain of *Aspergillus oryzae* NRRL 3484 with increased extracellular cellulase production.

Vu et al. (2011) found that cellulose production was 8.5-fold greater utilizing mutagenized strain (*Aspergillus* sp. SU-M15) fermented onto SSF medium than wild-type strain fermented onto wheat bran medium under better fermentation conditions and SSF medium. When *Chaetomium cellulolyticum* NRRL 18756 was exposed to numerous doses of gamma light (0.5 KGy), it produced up to 1.6 times more CMCase than wild-type strains (Fawzi and Hamdy, 2011). *Trichoderma viride* FCBP-142, a wild-type strain, was used to establish hyperproducer strains for cellulases production and was subjected to UV and EMS mutations, according to Shafique et al. (2011). *Trichoderma asperellum* RCK2011 was mutagenized for high cellulase output by UV irradiation; however, it was less susceptible to catabolite suppression (Raghuwanshi et al., 2014).

4.6.2 Recombinant strategy (heterologous cellulase expression)

"Consolidated bioprocessing (CBP)," a technology with tremendous potential to generate cellulase, has been the subject of various studies or investigations so far. CBP for cellulose production is described by Geddes et al. (2011) as "cellulase production integrating cellulose hydrolysis and fermentation in a single process without the need for enzymes provided externally."

Current CBP research focuses on the use of microbial cellulase, which is supported by designed (recombinant) noncellulolytic bacteria that produce large profitability and quantity to deliver a heterologous cellulase system (Khattak et al., 2013; Ward, 2012). Furthermore, recombinant DNA technology allows for cellulase new gene discovery, large-scale cellulase manufacturing, and the creation of customized enzymes through genetic engineering for a variety of applications (Van Zyl et al., 2014).

4.6.3 Mixed-culture (coculture) systems

Several studies have been conducted to use a particular pure culture strain for biomass biotransformation. Nonetheless, substrate consumption by a single strain in pure culture-based methods was already confined to a narrow range of biologically generated materials such as starch (Lin et al., 2011). Still, a wide range of microorganisms and natural inocula are employed to structure mixed cultures that allow for the bioconversion of heterogeneous biomass under nonsterile conditions without compromising strain stability.

The use of integrated microbial communities to degrade lignocellulosic agricultural waste could provide a well-ordered biologically generated matter breakdown and subsequent adaptation to high-value products. Wongwilaiwalin et al. (2010) developed active heat-tolerant lignocellulose degrading microbial consortium comprising eight coexisting primary microorganisms. For the degradation of cellulose-rich feedstock from high-temperature sugarcane bagasse compost, the diverse set of microorganisms includes aerobic/facultative anaerobic (including *Rhodocyclaceae bacterium*), anaerobic bacterial genera (e.g., *Thermoanaero* and *Clostridium bacterium*), and nonculture bacteria. The findings demonstrated that in the biotechnology-based industrial sector, the lignocellulose-based enzyme system is suited for the breakdown and transformation of biologically created materials.

Furthermore, under unfavorable environmental conditions, cellulase bacteria activity and development increased in the mixed system found in bacterial communities. The effect of a noncellulolytic bacteria W2−10 (*Geobacillus* sp.) on the cellulase activity of a cellulolytic CTI-6 (*Clostridium thermocellum*) was demonstrated to rise from 0.23 to 0.47 U/mL in coculture using a straw (cellulose materials) and paper as a substrate in a peptone cellulose solution medium (Lü et al., 2013).

4.7 Fermenter (bioreactor) design for cellulase production in solid-state fermentation

Various types of fermenters (bioreactors) have been used for a variety of applications over the years, including cellulase production in SSF systems (Arora et al., 2018). The cellulase synthesis through SSF has been employed in batch, fed-batch, or continuous modes, with batch being the most prevalent. The most often utilized laboratory size bioreactors for SSF include tray bioreactors (Arora et al., 2018), packed bed bioreactors (Castro et al., 2015), rotary drum bioreactors (RDBs) (Soccol et al., 2017), and fluidized bed bioreactors (FBBs) (Moshi et al., 2014; Arora et al., 2018).

4.7.1 Tray bioreactor

A bioreactor is a method that enables the ideal conditions for microorganisms to proliferate, resulting in maximum productivity. Tray bioreactors have been used to make fermented foods, mostly in Asian nations (Arora et al., 2018). Surprisingly, many commercial operations used tray bioreactors in various sectors for distinct uses (Thomas et al., 2013). A tray bioreactor consists of flat trays made of wood, metal, or plastic that can be perforated and packed with substrate support that creates a 1.5 or 2 cm thick layer. These reactors were stacked on top of one another and housed in a temperature and humidity-controlled chambers (Arora et al., 2018; Behera and Ray, 2016). In the scenario of tray bioreactors, typically, the surface area or the number of trays is extended to achieve high amounts of the output (Arora et al., 2018). Brijwani et al. (2010) used a tray bioreactor to produce cellulases using a mixed culture fermentation approach with *Trichoderma reesei* and *A. oryzae* cultured on soybean meal and wheat bran (4:1), with a 1 cm bed height and optimal operational parameters. A proper C:N ratio in the feedstock had been kept, which facilitated enhanced cellulase titers and resulted in the balanced production of all three enzymes, which is essential for full biomass conversion in biofuel production. Dhillon et al. (2011) found higher amounts of BGL when fermenting apple pomace in a tray bioreactor with *A. niger* and *T. reesei*. Although the tray bioreactor is widely used in the fermentation industry for SSF operations, it has several drawbacks, including poor heat transmission owing to reduced thermal conductivity and low bed height (Arora et al., 2018).

4.7.2 Packed bed reactor

A packed bed reactor (PBR) has the property of forced aeration through a static bed, which helps to replenish O_2 and moisture while reducing heat and CO_2 buildup. When mixing is undesired or harmful to microbial development, a PBR is used to provide greater control. The substrate is usually housed in a cylindrical glass or metal tube/drum, and the walls of the cylinder or drum may be jacketed, which can be furnished with cooling plates within the bed to allow effective heat transmission (Arora et al., 2018). In a study, *A. niger* FTCC 5003 was grown for 7 days on palm kernel cake as a substrate in the manufacture of cellulase, with a cellulase yield of 244.53 U/gds, utilizing 100 g of palm kernel cake (Abdeshahian et al., 2011). PBR was utilized to produce hydrolases utilizing babassu cake as a substrate, high in carbohydrates and known for causing *Aspergillus awamori* to induce and secrete an enzyme pool. Except for xylanases, higher enzyme yields have been recorded for cellulases, proteases, endoamylases, and exoamylases (Castro et al., 2015). The major source of heat creation in a PBR is the inefficient elimination of biologically produced heat, caused by the substrate's poor effective thermal conductivity and the moderate airflow rates used.

4.7.3 Rotary drum bioreactor

RDBs are made up of a horizontal cylinder. The solid medium tumbles owing to stirring, which is aided by baffles on the inside wall of the revolving drum, which may or may not be perforated (Soccol et al., 2017). Since there is no agitator, a better baffle design offers gentle homogeneous mixing

(Behera and Ray, 2016). *Trichoderma harzianum* was utilized to test cellulase production in terms of FPase activity. At the same time, *Penicillium verruculosum* COKE4E alkali-treated empty palm fruit bunch fiber was employed as the only carbon source in an RDB (Alam et al., 2009; Kim and Kim, 2012). The RDB reactor's technical principles, particularly its air circulation and constant mixing, make it suitable for biofuel production on a pilot or lab scale utilizing cellulosic feedstock.

4.7.4 Fluidized bed reactor

The FBB is a form of SSF bioreactor where the substrate and microorganisms are kept in a fluidized condition by supplying upward air movement (Arora et al., 2018). The substrate availability rises because of its fluidized condition, resulting in extremely efficient microbial growth. Air velocities cause more turbulence on the surface, which aids in effective heat transmission all across the reactor. While cassava starch had been hydrolyzed employing coimmobilized glucoamylase in an SSF process alongside concurrent fermentation utilizing *Saccharomyces cerevisiae* in pectin gel in an FBR, ethanol production increased to 11.7 g/L/h (Trovati et al., 2009). FBR was utilized to design a technique for producing high bioethanol titers from wild, nonedible cassava *Manihot glaziovii* through fed-batch as well as simultaneously saccharification and fermentation. This enabled the fermentation of up to 390 g/L of starch-derived glucose, exhibiting yields of roughly 94% of the theoretical value and a high bioethanol percentage of up to 190 g/L (24% v/v) (Moshi et al., 2014).

4.8 Biomass conversions and application of microbial cellulase

Several research organizations have studied enzymes in recent years due to their growing market in a variety of sectors (Behera and Ray 2016; Ray and Behera, 2017). Many scientists have concentrated on learning about their structure, manufacture, features, and qualities in the recent past and the elements that influence their efficiency and applications (Kuhad et al., 1994). Because the yearly sale of cellulase has been expected to surpass the protease customer base in the long term, taking account 10% of the global industrial enzyme supply (Singh et al., 2016), and has already gone up to 8% of the overall enzyme industry, better and more efficient methods of producing such enzymes has become the subject of numerous studies (Horn et al., 2012). Several organizations and scientists have investigated and financed renewable energy sources, resulting in biomass being used to generate electricity. Biomass usage has resurfaced due to environmental difficulties (Ramos and Wilhelm, 2005; Zaldivar et al., 2001). The major goal of this project was to extract monosaccharides from biomass using cellulase for liquid fuel generation and other chemicals (Bozell and Petersen, 2010).

Following the oil crisis of the 1970s, cellulase is now employed to help in the manufacturing of biofuels.

4.8.1 Textile industry

According to recent reports, cellulase consumption in the textile sector is increased by 13.77% in 2016. (Jayasekara and Ratnayake, 2019). It could be utilized either finishing, manufacturing cellulose-containing fabric, or developing novel fabric by upgrading fundamental processes in the textile industry, which does not require additional apparatus or machinery. It dates from the 1980s when biostoning (Arja, 2007; Mondal and Khan, 2014) was first introduced. Previously, amylase and pumice stonewash (Menon and Rao, 2012) was employed, but it had various drawbacks, including the fact that it had to be utilized on bulk cloth, that it might produce tear effects, and that the stone had to be physically eliminated. Cellulase was later employed to generate a stonewashed look and biofinishing of cellulase-based fabric, making the fabric more wearable and of higher quality. Small fibers protrude from the yarn's surface, which functions as a catalyst for cellulase to release the dye and turn it into water-soluble sugars, making it easier to eliminate during washing (Galante et al., 1998; Singh et al., 2007; Sukumaran et al., 2005). *T. reesei* is a suitable alternative (Miettinen-Oinonen, 2004). Cotton and linen, for example, are cellulose-based textiles that frequently develop fuzz on their surfaces. This fuzz is made up of loosely connected fibers that form pilling, which is little balls. This provides the cloth with an unappealing appearance and feel. Fabric is coated with cellulose throughout the wet manufacturing steps to avoid this. This acidic cellulase eliminates microfibers and improves fabric suppleness while also providing the surface with a lustrous look (Sreenath et al., 1996). This improves the fabric's hand feel as well as its water absorption qualities (Bhat, 2000). Cellulase, particularly endoglucanases-rich cellulase, restores and brightens color, softens the fabric, enhances cellulosic fabric durability, and eliminates excess dye (Ibrahim et al., 2011). Cellulase, for example, is employed in the textile sector since it is simple to use and does not require additional equipment; instead, it may be used with existing equipment. They are also environment-friendly since they are biodegradable and cost-effective because they save energy and chemicals, lowering final consumer prices.

4.8.2 Laundry and detergent

The utilization of enzymes in biological detergents has been used in the laundry industry since the 1960s, but the use of cellulase, including other enzymes such as proteases and lipases, is a relatively new technique (Singh et al., 2007). Based on a survey, the detergent business ranked first in the industry for the enzyme in 2014, with overall sales of 25%–30% (Jayasekara and Ratnayake, 2019). Alkaline cellulase is the latest cellulase development in the laundry business. This is currently being looked into (Singh et al., 2007). Alkaline cellulase binds to the cellulose in the fabric and dissolves the soil inside the fibrils. However, all of this occurs in the presence of additional detergent components (Sukumaran et al., 2005). Employing cellulase has several advantages, the most notable of which are the cleaning and textile care advantages. It helps improve and brighten the color of faded garments, giving them a better appearance by removing fuzz and stray strands from the cloth (Maurer, 2004). In the detergent business, *Trichoderma* species, *A. niger*, and a few *Humicola* species have been examined.

4.8.3 Paper and pulp industry

The paper industry is the world's biggest manufacturing industry and that has proceeded to expand (Nagar et al., 2011). Cellulase consumption has now risen from 320 to 395 million tons (Mai et al., 2004; Przybysz Buzała et al., 2018). Enzymes can be used in this business for a variety of operations over the years, including deinking for recycling, pulping, and strength qualities. Xylanase, hemicellulase, laccase, lipase (Demuner et al., 2011), and cellulase have been employed (Garg et al., 2011; Subramaniyan and Prema, 2002). Cellulase is primarily utilized in the pulping operation to recover cellulose from raw materials and remove contaminants before paper production. This procedure can be performed either mechanically or biochemically (Bajpai, 1999). Nevertheless, there are certain disadvantages in doing it mechanically. It entails grinding woody raw materials despite the end product's high bulk, fineness, and stiffness. It also costs a significant amount of energy to do so. Using enzymes, including such cellulase for biomechanical pulping, on the other hand, not just conserves energy by 20%–40% (Bhat, 2000; Demuner et al., 2011) but also enhances paper quality. This tends to make it much more viable for businesses since it reduces energy usage and overall costs while also improving pulp quality (Demuner et al., 2011).

Deinking is yet another procedure that uses cellulase. Many chemicals had been employed in the conventional approach to eliminate the ink from the paper; this produced yellowing, dullness in the paper, increased process costs, and caused pollution. A small quantity of enzyme is employed to break down the surface cellulose to release the ink. Cellulase and xylanase mixtures may be employed in biobleaching (Kumar and Satyanarayana, 2012). It improves the appearance of paper by making it brighter, whiter, and cleaner while also reducing pollutants.

4.8.4 Bioethanol and biofuel production

As the globe develops, so does the energy demand. Researchers and scientists have already investigated alternative means of providing adequate energy to meet this criterion. Much new research on lignocellulosic waste for biofuel generation has subsequently been published (Binod et al., 2010; Liang et al., 2014; Zhu et al., 2015).

Cellulase breaks down the pretreated lignocellulosic waste into sugars, which are then turned into bioethanol (Ahmed and Bibi, 2018). It has the energy-producing potential to fulfil the demands of today's fast-paced, energy-driven world (Anwar et al., 2014). Cellulase has been shown to transform lignocellulose feedstock into ethanol in various investigations effectively. It increases and improves the grade of ethanol generated when it is made from SSF (Shrestha et al., 2010).

Because cellulase aids in the production of ethanol, it also aids in the reduction of environmental pollution generated by the burning of fossil fuels (Horn et al., 2012). In the 1970s a global energy crisis prompted not just researchers but also governments to hunt for alternate energy sources. Biofuels received special consideration as a result of the joining of forces. Because biofuels are renewable and sustainable, cellulase aids the environment and the energy crisis (Srivastava et al., 2022).

4.8.5 Food industry

Cellulase has an important position in the food business. The presence of orange juice on breakfast tables and a bottle of olive oil in the kitchen is attributable to the refining of food by cellulase. Many fruits and plants employ cellulase to improve flavor, texture, and scent (Baker and Wicker, 1996). It clarifies and stabilizes juices and increases the volume of specific juices produced (De Carvalho et al., 2008). Utilizing cellulase, several nectars and purees of fruits such as peaches, mango, plum, and papaya have been stabilized, and a better texture is achieved with the correct thickness (Bhat, 2000; De Carvalho et al., 2008). It improves the flavor of baked products as well as the nutritional content of food for cattle and other animals.

4.8.6 Agriculture

Cellulase is used in agriculture for a variety of reasons. Cellulase is used to treat plant illnesses and infections that might lead to further difficulties. Combined with other enzymes, cellulases also contribute to the production of healthier crops (Bhat, 2000). It is recognized for increasing soil quality by breaking down straw that is placed into the soil as a substitute for synthetic fertilizers. Exogenous cellulase may hydrolyze the cellulose in straw, causing it to decompose and release sugars, increasing soil fertility.

4.9 Concluding remarks

The application of enzymes on an industrial scale still presents a challenge. Cellulolytic enzymes are used to convert lignocellulosic biomass into products with high added value. SSF has been researched for a long time in the industrial processes and is of economic importance. Apart from the purification of cellulase and separation of end products, limited knowledge of protein engineering and its implementation also adds an extra obstacle to the overall bioconversion process. In light of the aforementioned challenges, more and more research is needed to develop and optimize cost-effective cellulases employing SSF and biomass as a low-cost method. Additionally, research on (1) cheaper technologies for pretreatment of cellulosic biomass providing for a better microbial attack, (2) designing of bioreactors, (3) process optimization leading to higher cellulase yields, (4) treatment of biomass for production of hydrolytic products, and (5) protein engineering to improve cellulase qualities may also support to achieve novel and more economical processes for cellulase production compared to the existing process technologies.

Abbreviations

OH	hydroxyl radicals
CBP	consolidated bioprocessing
CMCase	carboxymethyl cellulase
CrI	crystallinity index
DNS	3′,5′-dinitrosalicylic acid
EMS	ethyl-methane sulfonate
FBB	fluidized bed bioreactor
FPase	filter paper hydrolase
IU/mL	international units per milliliter
PBR	packed bed reactor
PNG	*p*-nitrophenyl β-D glucopyranoside
RDBs	rotary drum bioreactors
SmF	submerged fermentation
SSF	solid-state fermentation
UV	ultraviolet

References

Abdeshahian, P., Samat, N., Hamid, A.A., Yusoff, W.M.W., 2011. Solid substrate fermentation for cellulase production using palm kernel cake as a renewable lignocellulosic source in packed-bed bioreactor. Biotechnol. Bioprocess. Eng. 16 (2), 238–244.

Ahmed, A., Bibi, A., 2018. Fungal cellulase; production and applications: mini review. Int. J. Health Life Sci. 4 (1), 19–36.

Alam, M.Z., Mamun, A.A., Qudsieh, I.Y., Muyibi, S.A., Salleh, H.M., Omar, N.M., 2009. Solid state bioconversion of oil palm empty fruit bunches for cellulase enzyme production using a rotary drum bioreactor. Biochemical Eng. J. 46 (1), 61–64.

Alarcón, E., Hernández, C., García, G., Ziarelli, F., Gutiérrez-Rivera, B., Musule, R., et al., 2021. Changes in chemical and structural composition of sugarcane bagasse caused by alkaline pretreatments [Ca (OH) 2 and NaOH] modify the amount of endoglucanase and β-glucosidase produced by *Aspergillus niger* in solid-state fermentation. Chem. Eng. Commun. 1–13.

Amadi, O.C., Awodiran, I.P., Moneke, A.N., Nwagu, T.N., Egong, J.E., Chukwu, G.C., 2022. Concurrent production of cellulase, xylanase, pectinase and immobilization by combined Cross-linked enzyme aggregate strategy-advancing tri-enzyme biocatalysis. Bioresour. Technol. Rep. 101019.

Ang, S.K., Shaza, E.M., Adibah, Y., Suraini, A.A., Madihah, M.S., 2013. Production of cellulases and xylanase by *Aspergillus fumigatus* SK1 using untreated oil palm trunk through solid state fermentation. Process. Biochem. 48, 1293–1302.

Anwar, Z., Gulfraz, M., Irshad, M., 2014. Agro-industrial lignocellulosic biomass a key to unlock the future bio-energy: a brief review. J. Radiat. Res. Appl. Sci. 7 (2), 163–173.

Arja, M.O., 2007. Cellulases in the textile industry. In: Polaina, J., MacCabe, A.P. (Eds.), Industrial Enzymes: Structure, Function and Applications. Springer.

Arora, S., Rani, R., Ghosh, S., 2018. Bioreactors in solid state fermentation technology: design, applications and engineering aspects. J. Biotechnol. 269, 16–34.

Bailey, M.J., Biely, P., Poutanen, K., 1992. Interlaboratory testing of methods for assay of xylanase activity. J. Biotechnol. 23 (3), 257–270.

Bajpai, P., 1999. Application of enzymes in the pulp and paper industry. Biotechnol. Prog. 15 (2), 147–157.

Baker, R.A., Wicker, L., 1996. Current and potential applications of enzyme infusion in the food industry. Trends Food Sci. & Technol. 7 (9), 279–284.

Behera, S.S., Ray, R.C., 2016. Solid state fermentation for production of microbial cellulases: recent advances and improvement strategies. Int. J. Biol. macromolecules 86, 656–669.

Beig, B., Riaz, M., Naqvi, S.R., Hassan, M., Zheng, Z., Karimi, K., et al., 2021. Current challenges and innovative developments in pretreatment of lignocellulosic residues for biofuel production: a review. Fuel 287, 119670.

Bhargav, S., Panda, B.P., Ali, M., Javed, S., 2008. Solid-state fermentation: an overview. Chem. Biochemical Eng. Q. 22 (1), 49–70.

Bhat, M.K., 2000. Cellulases and related enzymes in biotechnology. Biotechnol. Adv. 18 (5), 355–383.

Binod, P., Sindhu, R., Singhania, R.R., Vikram, S., Devi, L., Nagalakshmi, S., et al., 2010. Bioethanol production from rice straw: an overview. Bioresour. Technol. 101 (13), 4767–4774.

Boontanom, P., Chantarasiri, A., 2021. Diversity and cellulolytic activity of culturable bacteria isolated from the gut of higher termites (*Odontotermes* sp.) in eastern Thailand. Biodiversitas J. Biol. Diversity 22 (8).

Bozell, J.J., Petersen, G.R., 2010. Technology development for the production of biobased products from biorefinery carbohydrates—the US Department of Energy's "Top 10" revisited. Green. Chem. 12 (4), 539–554.

Brijwani, K., Oberoi, H.S., Vadlani, P.V., 2010. Production of a cellulolytic enzyme system in mixed-culture solid-state fermentation of soybean hulls supplemented with wheat bran. Process. Biochem. 45 (1), 120–128.

Castro, A.M., Castilho, L.R., Freire, D.M., 2015. Performance of a fixed-bed solid-state fermentation bioreactor with forced aeration for the production of hydrolases by *Aspergillus awamori*. Biochemical Eng. J. 93, 303–308.

Chakraborty, S., Yadav, G., Saini, J.K., Kuhad, R.C., 2019. Comparative study of cellulase production using submerged and solid-state fermentation. New and Future Developments in Microbial Biotechnology and Bioengineering. Elsevier, pp. 99–113.

Chilakamarry, C.R., Sakinah, A.M., Zularisam, A.W., Sirohi, R., Khilji, I.A., Ahmad, N., et al., 2022. Advances in solid-state fermentation for bioconversion of agricultural wastes to value-added products: opportunities and challenges. Bioresour. Technol. 343, 126065.

Cunha, F.M., Kreke, T., Badino, A.C., Farinas, C.S., Ximenes, E., Ladisch, M.R., 2014. Liquefaction of sugarcane bagasse for enzyme production. Bioresour. Technol. 172, 249–252.

De Carvalho, L.M.J., De Castro, I.M., Da Silva, C.A.B., 2008. A study of retention of sugars in the process of clarification of pineapple juice (*Ananas comosus*, L. Merril) by micro-and ultra-filtration. J. Food Eng. 87 (4), 447–454.

Deindoerfer, F.H., 1957. Calculation of heat sterilization times for fermentation media. Appl. microbiology 5 (4), 221–228.

Demuner, B.J., Pereira Junior, N., Antunes, A., 2011. Technology prospecting on enzymes for the pulp and paper industry. J. Technol. Manag. & Innov. 6 (3), 148–158.

Dhillon, G.S., Kaur, S., Brar, S.K., Verma, M., 2012. Potential of apple pomace as a solid substrate for fungal cellulase and hemicellulase bioproduction through solid-state fermentation. Ind. Crop. Products 38, 6–13.

Dhillon, G.S., Oberoi, H.S., Kaur, S., Bansal, S., Brar, S.K., 2011. Value-addition of agricultural wastes for augmented cellulase and xylanase production through solid-state tray fermentation employing mixed-culture of fungi. Ind. Crop. Products 34 (1), 1160–1167.

Ejaz, U., Sohail, M., Ghanemi, A., 2021. Cellulases: from bioactivity to a variety of industrial applications. Biomimetics 6 (3), 44.

El-Ghonemy, D.H., Ali, T.H., Moharam, M.E., 2014. Optimization of culture conditions for the production of extracellular cellulases via solid state fermentation. Br. Microbiology Res. J. 4 (6), 698.

Eveleigh, D.E., Mandels, M., Andreotti, R., Roche, C., 2009. Measurement of saccharifying cellulase. Biotechnol. Biofuels 2 (1), 1–8.

Fadel, M., Hamed, A.A., Abd-Elaziz, A.M., Ghanem, M.M., Roshdy, A.M., 2021. Cellulases and animal feed production by solid-state fermentation by *Aspergillus fumigatus* NRCF-122 mutant. Egypt. J. Chem. 64 (7), 3511–3520.

Falkoski, D.L., Guimarães, V.M., de Almeida, M.N., Alfenas, A.C., Colodette, J.L., de Rezende, S.T., 2013. *Chrysoporthe cubensis*: a new source of cellulases and hemicellulases to application in biomass saccharification processes. Bioresour. Technol. 130, 296–305.

Fatima, M., 2021. Recent Insight into production of cellulase by fungi and its industrial applications. Ann. Romanian Soc. Cell Biol. 25 (7), 1444–1472.

Fawzi, E.M., Hamdy, H.S., 2011. Improvement of carboxymethyl cellulase production from *Chaetomium cellulolyticum* NRRL 18756 by mutation and optimization of solid state fermentation. Bangladesh J. Bot. 40, 139–147.

Florencio, C., Cunha, F.M., Badino, A.C., Farinas, C.S., 2015. Validation of a novel sequential cultivation method for the production of enzymatic cocktails from *Trichoderma* strains. Appl. Biochem. Biotechnol. 175, 1389–1402.

Gad, A.M., Suleiman, W.B., El-Sheikh, H.H., Elmezayen, H.A., Beltagy, E.A., 2022. Characterization of cellulase from *Geotrichum candidum* strain Gad1 approaching bioethanol production. Arab. J. Sci. Eng. 1–14.

Galante, Y.M., DE Conti, A.L.B.E.R.T.O., Monteverdi, R., 1998. Application of *Trichoderma* enzymes. *Trichoderma* and Gliocladium. Enzymes, Biol. Control. Commercial Appl. (2), 327.

Garg, G., Mahajan, R., Kaur, A., Sharma, J., 2011. Xylanase production using agro-residue in solid-state fermentation from *Bacillus pumilus* ASH for biodelignification of wheat straw pulp. Biodegradation 22 (6), 1143–1154.

Gasparotto, J.M., Werle, L.B., Foletto, E.L., Kuhn, R.C., Jahn, S.L., Mazutti, M.A., 2015. Production of cellulolytic enzymes and application of crude enzymatic extract for saccharification of lignocellulosic biomass. Appl. Biochem. Biotechnol. 175, 560–572.

Geddes, C.C., Nieves, I.U., Ingram, L.O., 2011. Advances in ethanol production. Curr. Opin. Biotechnol. 22 (3), 312–319.

Han, X., Guo, Y., Liu, X., Xia, Q., Wang, Y., 2019. Catalytic conversion of lignocellulosic biomass into hydrocarbons: a mini review. Catal. Today 319, 2–13.

Herculano, P.N., Porto, T.S., Moreira, K.A., Pinto, G.A., Souza-Motta, C.M., Porto, A.L.F., 2011. Cellulase production by *Aspergillus japonicus* URM5620 using waste from castor bean (*Ricinus communis* L.) under solid-state fermentation. Appl. Biochem. Biotechnol. 165, 1057–1067.

Horn, S.J., Vaaje-Kolstad, G., Westereng, B., Eijsink, V., 2012. Novel enzymes for the degradation of cellulose. Biotechnol. biofuels 5 (1), 1–13.

Ibrahim, N.A., El-Badry, K., Eid, B.M., Hassan, T.M., 2011. A new approach for biofinishing of cellulose-containing fabrics using acid cellulases. Carbohydr. Polym. 83 (1), 116–121.

Jayasekara, S., Ratnayake, R., 2019. Microbial cellulases: an overview and applications. Cellulose 22.

Juturu, V., Wu, J.C., 2014. Microbial cellulases: engineering, production and applications. Renew. Sustain. Energy Rev. 33, 188–203.

Khattak, W.A., Khan, T., Ha, J.H., Ul-Islam, M., Kang, M.K., Park, J.K., 2013. Enhanced production of bioethanol from waste of beer fermentation broth at high temperature through consecutive batch strategy by simultaneous saccharification and fermentation. Enzyme Microb. Technol. 53 (5), 322–330.

Kim, S., Kim, C.H., 2012. Production of cellulase enzymes during the solid-state fermentation of empty palm fruit bunch fiber. Bioprocess. Biosyst. Eng. 35 (1), 61–67.

Kolasa, M., Ahring, B.K., Lubeck, P.S., Lubeck, M., 2014. Co-cultivation of *Trichoderma reesei* RutC30 with three black *Aspergillus* strains facilitates efficient hydrolysis of pretreated wheat straw and shows promises for on-site enzyme production. Bioresour. Technol. 169, 143–148.

Kuhad, R.C., Kumar, M., Singh, A., 1994. A hypercellulolytic mutant of *Fusarium oxysporum*. Lett. Appl. microbiology 19 (5), 397–400.

Kumar, S., Sharma, H.K., Sarkar, B.C., 2011. Effect of substrate and fermentation conditions on pectinase and cellulase production by *Aspergillus niger* NCIM 548 in submerged (SmF) and solid state fermentation (SSF). Food Sci. Biotechnol. 20, 1289–1298.

Kumar, V., Satyanarayana, T., 2012. Thermo-alkali-stable xylanase of a novel polyextremophilic *Bacillus halodurans* TSEV1 and its application in biobleaching. Int. Biodeterior. & Biodegrad. 75, 138–145.

Ladeira, S.A., Cruz, E., Delatorre, A.B., Barbosa, J.B., Leal Martins, M.L., 2015. Cellulase production by thermophilic *Bacillus* sp.: SMIA-2 and its detergent compatibility. Electron. J. Biotechnol. 18, 110–115.

Leite, P., Sousa, D., Fernandes, H., Ferreira, M., Costa, A.R., Filipe, D., et al., 2021. Recent advances in production of lignocellulolytic enzymes by solid-state fermentation of agro-industrial wastes. Curr. Opin. Green. Sustain. Chem. 27, 100407.

Li, X.H., Yang, H.J., Roy, B., Park, E.Y., Jiang, L.J., Wang, D., et al., 2010. Enhanced cellulase production of the *Trichoderma viride* mutated by microwave and ultraviolet. Microbiological Res. 165 (3), 190–198.

Liang, S., Liang, K., Luo, L., Zhang, Q., Wang, C., 2014. Study on low-velocity impact of embedded and co-cured composite damping panels with numerical simulation method. Composite Struct. 107, 1–10.

Liang, X., Huang, Y., Hua, D., Zhang, J., Xu, H., Li, Y., et al., 2012. Cellulase production by *Aspergillus* sp. on rice grass (*Spartina* spp.) under solid-state fermentation. Afr. J. Microbiol. Res. 6, 6785–6792.

Lin, C.W., Wu, C.H., Tran, D.T., Shih, M.C., Li, W.H., Wu, C.F., 2011. Mixed culture fermentation from lignocellulosic materials using thermophilic lignocellulose-degrading anaerobes. Process. Biochem. 46 (2), 489–493.

Liu, X., Hong, F., Guo, Y., Zhang, J., Shi, J., 2013. Sterilization of *Staphylococcus aureus* by an atmospheric non-thermal plasma jet. Plasma Sci. Technol. 15 (5), 439.

Lopez-Calleja, A.C., Cuadra, T., Barrios-González, J., Fierro, F., Fernandez, F.J., 2012. Solid-state and submerged fermentations show different gene expression profiles in cephalosporin C production by *Acremonium chrysogenum*. Microb. Physiol. 22 (2), 126–134.

Lü, Y., Li, N., Yuan, X., Hua, B., Wang, J., Ishii, M., et al., 2013. Enhancing the cellulose-degrading activity of cellulolytic bacteria CTL-6 (*Clostridium thermocellum*) by co-culture with non-cellulolytic bacteria W2–10 (*Geobacillus* sp.). Appl. Biochem. Biotechnol. 171 (7), 1578–1588.

Mai, C., Kües, U., Militz, H., 2004. Biotechnology in the wood industry. Appl. microbiology Biotechnol. 63 (5), 477–494.

Maurer, K.H., 2004. Detergent proteases. Curr. Opin. Biotechnol. 15 (4), 330–334.

Menon, V., Rao, M., 2012. Trends in bioconversion of lignocellulose: biofuels, platform chemicals & biorefinery concept. Prog. energy Combust. Sci. 38 (4), 522–550.

Miettinen-Oinonen, A., 2004. Trichoderma reesei strains for production of cellulases for the textile industry. VTT Publications.

Mondal, M.I.H., Khan, M.M.R., 2014. Characterization and process optimization of indigo dyed cotton denim garments by enzymatic wash. Fash. Text. 1 (1), 1–12.

Moran-Aguilar, M.G., Costa-Trigo, I., Calderón-Santoyo, M., Domínguez, J.M., Aguilar-Uscanga, M.G., 2021. Production of cellulases and xylanases in solid-state fermentation by different strains of *Aspergillus niger* using sugarcane bagasse and brewery spent grain. Biochemical Eng. J. 172, 108060.

Moshi, A.P., Crespo, C.F., Badshah, M., Hosea, K.M., Mshandete, A.M., Mattiasson, B., 2014. High bioethanol titre from *Manihot glaziovii* through fed-batch simultaneous saccharification and fermentation in Automatic Gas Potential Test System. Bioresour. Technol. 156, 348–356.

Mostafa, A.A., 2014. Effect of gamma irradiation on *Aspergillus niger* DNA and production of cellulases enzymes. J. Am. Sci. 10 (5), 152–160.

Nagar, S., Mittal, A., Kumar, D., Kumar, L., Kuhad, R.C., Gupta, V.K., 2011. Hyper production of alkali stable xylanase in lesser duration by *Bacillus pumilus* SV-85S using wheat bran under solid state fermentation. N. Biotechnol. 28 (6), 581–587.

Ouedraogo, J.P., Tsang, A., 2021. Production of native and recombinant enzymes by fungi for industrial applications.

Pandey, A., 2003. Solid-state fermentation. Biochemical Eng. J. 13 (2–3), 81–84.

Patel, A.K., Singhania, R.R., Sim, S.J., Pandey, A., 2019. Thermostable cellulases: current status and perspectives. Bioresour. Technol. 279, 385–392.

Przybysz Buzała, K., Kalinowska, H., Borkowski, J., Przybysz, P., 2018. Effect of xylanases on refining process and kraft pulp properties. Cellulose 25 (2), 1319–1328.

Raghuwanshi, S., Deswal, D., Karp, M., Kuhad, R.C., 2014. Bioprocessing of enhanced cellulase production from a mutant of *Trichoderma asperellum* RCK2011 and its application in hydrolysis of cellulose. Fuel 124, 183–189.

Ramamoorthy, N.K., Sambavi, T.R., Renganathan, S., 2019. A study on cellulase production from a mixture of lignocellulosic wastes. Process. Biochem. 83, 148–158.

Ramos, L.P., Wilhelm, H.M., 2005. Current status of biodiesel development in Brazil. Appl. Biochem. Biotechnol. 123 (1), 807–819.

Ray, R.C., Behera, S.S., 2017. Solid state fermentation for production of microbial cellulases. Biotechnology of microbial enzymes. Academic Press, pp. 43–79.

Sakka, K., Kimura, T., Karita, S., Ohmiya, K., 2000. Molecular breeding of cellulolytic microbes, plants, and animals for biomass utilization. J. Biosci. Bioeng. 90, 227–233.

Shafique, S., Bajwa, R., Shafique, S., 2011. Strain improvement in *Trichoderma viride* through mutation for overexpression of cellulase and characterization of mutants using random amplified polymorphic DNA (RAPD). Afr. J. Biotechnol. 10 (84), 19590–19597.

Shrestha, P., Khanal, S.K., Pometto III, A.L., van Leeuwen, J.H., 2010. Ethanol production via in situ fungal saccharification and fermentation of mild alkali and steam pretreated corn fiber. Bioresour. Technol. 101 (22), 8698–8705.

Singh, A., Bajar, S., Devi, A., Bishnoi, N.R., 2021. Evaluation of cellulase production from *Aspergillus niger* and *Aspergillus heteromorphus* under submerged and solid-state fermentation. Environ. Sustainability 4 (2), 437–442.

Singh, A., Bajar, S., Devi, A., Pant, D., 2021. An overview on the recent developments in fungal cellulase production and their industrial applications. Bioresour. Technol. Rep. 14, 100652.

Singh, A., Kuhad, R.C., Ward, O.P., 2007. Industrial application of microbial cellulases. In: Kuhad, R.C., Singh, A. (Eds.), Lignocellulose Biotechnologgy: Future Prospects. I.K. International Publishing House, pp. 345–358.

Singh, G., Verma, A.K., Kumar, V., 2016. Catalytic properties, functional attributes and industrial applications of β-glucosidases. 3 Biotech. 6 (1), 1–14.

Singhania, R.R., Patel, A.K., Thomas, L., Pandey, A., 2017. Solid-state fermentation. Ind. Biotechnology: Products Processes 4, 187–204.

Siqueira, J.G.W., Rodrigues, C., de Souza Vandenberghe, L.P., Woiciechowski, A.L., Soccol, C.R., 2020. Current advances in on-site cellulase production and application on lignocellulosic biomass conversion to biofuels: a review. Biomass Bioenergy 132, 105419.

Soccol, C.R., Costa, E.D., Letti, L.A.J., Karp, S.G., Woiciechowski, A.L., Vandenberghe, L.D.S., 2017. Recent developments and innovations in solid state fermentation. Biotechnol. Res. Innov. 1, 52–71.

Sreenath, H.K., Shah, A.B., Yang, V.W., Gharia, M.M., Jeffries, T.W., 1996. Enzymatic polishing of jute/cotton blended fabrics. J. Fermentation Bioeng. 81 (1), 18–20.

Srivastava, N., Singh, R., Mohammad, A., Pal, D.B., Syed, A., Elgorban, A.M., et al., 2022. Graphene oxide mediated enhanced cellulase production using pomegranate waste following co-cultured condition with improved pH and thermal stability. Fuel 312, 122807.

Subramaniyan, S., Prema, P., 2002. Biotechnology of microbial xylanases: enzymology, molecular biology, and application. Crit. Rev. Biotechnol. 22 (1), 33–64.

Sukumaran, R.K., Singhania, R.R., Pandey, A., 2005. Microbial cellulases-production, applications and challenges. J. Sci. Res.

Taherzadeh-Ghahfarokhi, M., Panahi, R., Mokhtarani, B., 2021. Medium supplementation and thorough optimization to induce carboxymethyl cellulase production by Trichoderma reesei under solid state fermentation of nettle biomass. Prep. Biochem. & Biotechnol. 1–8.

Thomas, L., Larroche, C., Pandey, A., 2013. Current developments in solid-state fermentation. Biochemical Eng. J. 81, 146–161.

Trivedi, N., Gupta, V., Reddy, C.R.K., Jha, B., 2015a. Detection of ionic liquid stable cellulase produced by the marine bacterium *Pseudoalteromonas* sp. isolated from brown alga *Sargassum polycystum* C. Agardh. Bioresour. Technol. 132, 313–319.

Trivedi, N., Reddy, C.R.K., Radulovich, R., Jha, B., 2015b. Solid state fermentation (SSF)-derived cellulase for saccharification of the green seaweed Ulva for bioethanol production. Algal Res. 9, 48–54.

Trovati, J., Giordano, R.C., Giordano, R.L., 2009. Improving the performance of a continuous process for the production of ethanol from starch. Appl. Biochem. Biotechnol. 156 (1), 76–90.

Van Zyl, J.H.D., Den Haan, R., Van Zyl, W.H., 2014. Over-expression of native *Saccharomyces cerevisiae* exocytic SNARE genes increased heterologous cellulase secretion. Appl. microbiology Biotechnol. 98 (12), 5567–5578.

Velvizhi, G., Goswami, C., Shetti, N.P., Ahmad, E., Pant, K.K., Aminabhavi, T.M., 2022. Valorisation of lignocellulosic biomass to value-added products: Paving the pathway towards low-carbon footprint. Fuel 313, 122678.

Vu, V.H., Pham, T.A., Kim, K., 2011. Improvement of fungal cellulase production by mutation and optimization of solid state fermentation. Mycobiology 39 (1), 20–25.

Ward, O.P., 2012. Production of recombinant proteins by filamentous fungi. Biotechnol. Adv. 30 (5), 1119–1139.

Wongwilaiwalin, S., Rattanachomsri, U., Laothanachareon, T., Eurwilaichitr, L., Igarashi, Y., Champreda, V., 2010. Analysis of a thermophilic lignocellulose degrading microbial consortium and multi-species lignocellulolytic enzyme system. Enzyme Microb. Technol. 47 (6), 283–290.

Xia, Q., Chen, Z., Shao, Y., Gong, X., Wang, H., Liu, X., et al., 2016. Direct hydrodeoxygenation of raw woody biomass into liquid alkanes. Nat. Commun. 7, 11162.

Yadav, P., Maharjan, J., Korpole, S., Prasad, G.S., Sahni, G., Bhattarai, T., et al., 2018. Production, purification, and characterization of thermostable alkaline xylanase from *Anoxybacillus kamchatkensis* NASTPD13. Front. Bioeng. Biotechnol. 6, 65.

Yuan, Z., Dai, W., Zhang, S., Wang, F., Jian, J., Zeng, J., et al., 2022. Heterogeneous strategies for selective conversion of lignocellulosic polysaccharides. Cellulose 1–19.

Zaldivar, J., Nielsen, J., Olsson, L., 2001. Fuel ethanol production from lignocellulose: a challenge for metabolic engineering and process integration. Appl. microbiology Biotechnol. 56 (1), 17–34.

Zhou, X., Chen, H., Li, Z., 2004. CMCase activity assay as a method for cellulase adsorption analysis. Enzyme Microb. Technol. 35 (5), 455–459.

Zhu, Y., Luo, B., Sun, C., Li, Y., Han, Y., 2015. Influence of bromine modification on collecting property of lauric acid. Miner. Eng. 79, 24–30.

Chapter 5

Hyperthermophilic subtilisin-like proteases from *Thermococcus kodakarensis*

Ryo Uehara[1,2], Hiroshi Amesaka[3], Yuichi Koga[4], Kazufumi Takano[3], Shigenori Kanaya[5] and Shun-ichi Tanaka[2,3,6]

[1]*Division of Cancer Cell Regulation, Aichi Cancer Center Research Institute, Nagoya, Japan,* [2]*Ritsumeikan Global Innovation Research Organization, Ritsumeikan University, Shiga, Japan,* [3]*Department of Biomolecular Chemistry, Kyoto Prefectural University, Kyoto, Japan,* [4]*Department of Biotechnology, Graduate School of Engineering, Osaka University, Osaka, Japan,* [5]*Department of Material and Life Science, Graduate School of Engineering, Osaka University, Osaka, Japan,* [6]*Department of Biotechnology, College of Life Science, Ritsumeikan University, Shiga, Japan*

5.1 Introduction

Hyperthermophiles are microorganisms that optimally grow at temperatures above 80°C (Stetter, 2013) or above 90°C (Adams and Kelly, 1998). These microorganisms, predominantly distributed in the domain Archaea, have been isolated from a large variety of natural thermal environments, including hot springs and volcanoes and artificial environments. These organisms' proteins, DNAs, RNAs, and cytoplasmic membranes have adapted to the high temperatures necessary to thrive in such harsh environments (Imanaka, 2011; Stetter, 2013). Enzymes from hyperthermophiles generally display superior stability for their mesophilic homologs. They, therefore, have been regarded not only as excellent models for studying protein stabilization mechanisms but also as potential candidates for industrial applications. Thermostable enzymes are useful for industrial enzymatic processes performed at high temperatures (de Miguel Bouzas et al., 2006). Performing enzymatic reactions at high temperatures increases reaction rates and solute solubility while simultaneously reducing the risk of contamination, resulting in increased product yields. Furthermore, enhanced thermal stability often confers resistance to chemicals such as surfactants, denaturants, and oxidative reagents. Thus hyperthermophilic enzymes allow us to apply enzymatic catalytic techniques to a wide variety of environments.

Biotechnology of Microbial Enzymes. DOI: https://doi.org/10.1016/B978-0-443-19059-9.00003-7
© 2023 Elsevier Inc. All rights reserved.

Proteases that catalyze the hydrolysis of peptide bonds in proteins and peptides are ubiquitous. Proteolytic activity is essential for many biological processes, including development, differentiation, and immune response. Furthermore, proteases are the most important industrial enzymes, accounting for at least 65% of sales in the world market (de Miguel Bouzas et al., 2006; Rao et al., 1998). Proteases are used in many market sectors, including detergents, food, pharmaceuticals, leather, diagnostics, waste management, and silver recovery from X-ray film. Notably, the largest share of the enzyme market is held by bacterial alkaline proteases used as detergent additives. These are highly active at alkaline pHs and enhance the washing ability of detergents by degrading protein stains. However, the stability of these proteases deteriorates in the presence of surfactants, which are usually present in detergents, and therefore they are destabilized and degraded by their activities. Searches for more stable, thermophilic proteases have helped solve this issue and develop protein engineering methods, which can improve the stability of existing enzymes. Hyperthermophiles are additional, attractive sources of novel proteases, which are sufficiently stable even in the presence of surfactants.

Serine proteases are classified based on their amino acid sequence. The SB clan, known as the subtilisin-like protease (subtilase) superfamily, is diverse and found in archaea, bacteria, fungi, and higher eukaryotes (Siezen and Leunissen, 1997, Siezen et al., 2007). The SB clan is further divided into six families, subtilisin, thermitase, proteinase K, pyrolysin, kexin, and lantibiotic peptidase. These proteases all possess a catalytic triad consisting of serine, histidine, and aspartate amino acids. The serine triggers catalytic hydrolysis through nucleophilic attack of the peptide bond carbonyl group. Histidine is the general base that extracts a proton from the hydroxyl group of serine. At the same time, aspartate stabilizes the protonated imidazole group of histidine and orients it into the correct position. Most subtilases display maximum activity at alkaline pHs, because deprotonation of the imidazole group is required for catalysis. Prokaryotic subtilases are generally secreted into the extracellular environment and provide nutrients by degrading extracellular proteins, whereas eukaryotic subtilases play essential roles in many varied cellular events. Examples in human cells include PCSK9 (Awan et al., 2014) and furin (Seidah et al., 2008), responsible for cholesterol metabolism and the maturation of hormone precursors, respectively. Prokaryotic subtilases are present not only in mesophiles, but also in thermophiles (Catara et al., 2003; Choi et al., 1999; Gödde et al., 2005; Jang et al., 2002; Kannan et al., 2001; Kluskens et al., 2002; Kwon et al., 1988; Li and Li, 2009, Li et al., 2011; Matsuzawa et al., 1988; Sung et al., 2010; Wu et al., 2004) and psychrophiles (Arnórsdottir et al., 2002; Davail et al., 1994; Dong et al., 2005; Kulakova et al., 1999; Kurata et al., 2007; Morita et al., 1998; Narinx et al., 1997). The crystal structures of several have been reported (Almog et al., 2003, 2009; Arnórsdóttir et al., 2005;

Kim et al., 2004; Smith et al., 1999; Teplyakov et al., 1990). Extracellular subtilases are often used for various biotechnological applications. The detergent industry has widely used the subtilisins from mesophilic *Bacillus* species due to broad substrate specificity and ease of large-scale active enzyme preparation (Saeki et al., 2007).

Subtilisin E from *Bacillus subtilis* (Stahl and Ferrari, 1984), subtilisin BPN′P from *Bacillus amyloliquefaciens* (Wells et al., 1983), and subtilisin Carlsberg from *Bacillus licheniformis* (Jacobs et al., 1985) are representative bacterial subtilisins. The structure and maturation of these three subtilisins have been extensively studied (Bryan, 2002; Chen and Inouye, 2008; Eder et al., 1993; Eder and Fersht, 1995; Fisher et al., 2007; Hu et al., 1996; Li et al., 1995; Shinde and Inouye, 2000; Takagi et al., 1988; Yabuta et al., 2001). Bacterial subtilisins are synthesized as an inactive precursor, termed prepro-subtilisin, containing a signal peptide (pre) and a propeptide (pro) at the *N*-terminus of the subtilisin domain. The presequence assists in the translocation of subtilisin to the extracellular environment and is cleaved off through the secretion process. The subtilisin precursor is then secreted in pro-form (pro-subtilisin) and subsequently matures following three steps: (1) folding of pro-subtilisin, (2) autoprocessing of propeptide, and (3) degradation of propeptide (Fig. 5.1). The propeptide is essential for maturation owing to its dual function as an intramolecular chaperone that assists folding the subtilisin domain and as an inhibitor that forms an inactive complex with subtilisin upon autoprocessing. The subtilisin domain alone cannot fold into its native structure; subtilisin folding stops at a molten globule-like intermediate, because its folding rate is extremely low in the absence of propeptide (Eder et al., 1993). In the propeptide:subtilisin complex structure, the propeptide interacts with two surface helixes and the active-site cleft of subtilisin. The *C*-terminus region of the propeptide binds to the active site in a substrate-like manner and competitively inhibits activity upon autoprocessing (Li et al., 1995). The propeptide of mesophilic subtilisins is an intrinsically unstructured protein (Hu et al., 1996; Huang et al., 1997; Wang et al., 1998) and therefore cleaved by subtilisin immediately upon its dissociation. The dissociation of the first propeptide molecule initiates a cascade, in which the entire propeptide in the complex is destroyed *in trans*, and subsequently, active proteases are released. Hence, the release of the propeptide is the rate-determining step in the maturation reaction. This propeptide-mediated bacterial subtilisin maturation is a general maturation model for not only subtilases (Baier et al., 1996; Basak and Lazure, 2003; Jia et al., 2010; Marie-Claire et al., 2001; Shinde and Thomas, 2011) but also for other extracellular proteases (Grande et al., 2007; Marie-Claire et al., 1999; Nemoto et al., 2008; Schilling et al., 2009; Salimi et al., 2010; Tang et al., 2002; Winther and Sørensen, 1991).

Subtilisin stability is highly dependent on Ca^{2+} ions. Bacterial subtilisins bind Ca^{2+} ions at two binding sites, a high-affinity site (the A or Ca1 site) and a low-affinity site (the B or Ca2 site), according to the crystal structures (Bott et al., 1988; Jain et al., 1998; McPhalen and James, 1988). Both sites

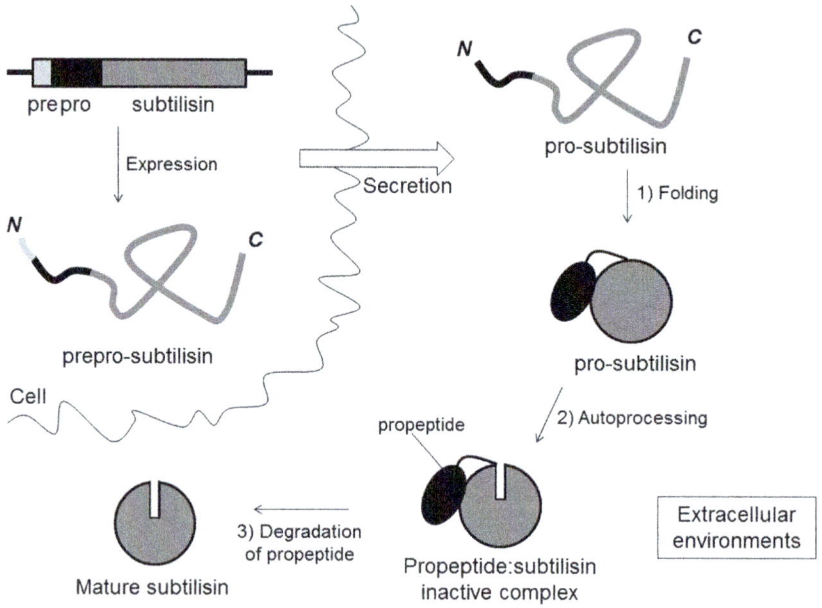

FIGURE 5.1 Maturation of bacterial subtilisin in the extracellular environment.

are located far from the active site, and the Ca^{2+} ions are not involved in catalysis. The two Ca^{2+} ions in bacterial subtilisins contribute to the enzyme's stability, and their removal significantly destabilizes the structures (Gallagher et al., 1993). In contrast, pro-subtilisin can mature without Ca^{2+} ions, indicating that Ca^{2+} ions are not required for the maturation of bacterial subtilisins (Yabuta et al., 2002).

As mentioned above, the structure and maturation of subtilisins from mesophilic bacteria have been studied in great detail. However, our understanding of hyperthermophilic subtilisins is still quite limited, even though they should prove to be promising candidates for many industrial applications. Our recent studies have revealed the mechanism by which two hyperthermophilic subtilisins mature effectively and acquire their stable structures in an extremely hot environment. This chapter summarizes the unique maturation and stabilization mechanisms of two subtilisin-like proteases from a hyperthermophilic archaeon. We also discuss the prospect of their application in medical and industrial fields.

5.2 Two Subtilisin-like proteases from *Thermococcus Kodakarensis* KOD1

Thermococcus kodakarensis KOD1 is a hyperthermophilic archaeon isolated from a solfatara at a wharf on Kodakara Island, Kagoshima, Japan (Morikawa et al.,

1994; Atomi et al., 2004). The strain grows at temperatures ranging from 60°C to 100°C, with an optimum growth temperature of 85°C. This strain offers a promising source of a wide range of commercially valuable enzymes, as represented by KOD polymerase (Toyobo Co., Ltd., Japan), which has exceptionally high fidelity and efficiency in polymerase chain reaction applications, and its hyperstable thiol protease (Morikawa et al., 1994). The *T. kodakarensis* genome contains three genes encoding subtilisin-like proteases, Tk-1675 (NCBI YP_184088), Tk-1689 (NCBI YP_184102), and Tk-0076 (NCBI YP_182489) (Fukui et al., 2005). Tk-1675 and Tk-1689 are designated prepro-Tk-subtilisin (Kannan et al., 2001) and prepro-Tk-SP (Foophow et al., 2010b), respectively. These two proteases are enzymatically active when overproduced in *Escherichia coli*. Furthermore, *T. kodakarensis* cells produce these enzymes as active extracellular proteases when the genes are expressed under a strong constitutive promoter (Takemasa et al., 2011). In contrast, an active form of Tk-0076 has not yet been obtained.

The amino acid sequences of the three subtilisins from *T. kodakarensis* are compared with that of prepro-subtilisin E (Protein ID AAA22742) from *B. subtilis* in Fig. 5.2. Prepro-subtilisin E consists of a signal peptide [Met(−106)−Ala(−78)], a propeptide [Ala(−77)−Tyr(−1)], and a mature domain, subtilisin E (Ala1−Gln275). Similarly, prepro-Tk-subtilisin consists of a signal peptide [Met(−24)−Ala(−1)], a propeptide, Tkpro (Gly1−Leu69), and a mature domain, Tk-subtilisin (Gly70−Gly398). Prepro-Tk-SP consists of a signal peptide [Met(−23)−Ala(−1)], a propeptide, proN (Ala1−Ala113), and a mature domain, Tk-SP (Val114−Gly640).

The pro-forms of Tk-subtilisin and Tk-SP are designated Pro-Tk-subtilisin and Pro-Tk-SP, respectively. The three amino acid residues of the subtilisin catalytic triad, and the asparagine residue that is required for the formation of an oxyanion hole, are fully conserved in the sequences. Comparison of the Tk-subtilisin and Tk-SP sequences with that of bacterial subtilisin prepro-subtilisin E shows that Tk-subtilisin contains three insertion sequences, termed IS1 (Gly70−Pro82), IS2 (Pro207−Asp226), and IS3 (Gly346−Ser358) and that Tk-SP has a long *C*-terminal extension (Ala422−Gly640). Except for these inserted sequences, Tk-subtilisin and Tk-SP show 31% identity to each other, and both of them show approximately 40% identity to bacterial prepro-subtilisin E.

5.3 TK-subtilisin

5.3.1 Ca^{2+}-dependent maturation of Tk-subtilisin

The primary structures of Pro-Tk-subtilisin (Gly1−Gly398), Tk-subtilisin (Gly70−Gly398), Tkpro (Gly1−Leu69), and their derivatives are schematically shown in Fig. 5.3. When Pro-Tk-subtilisin is overproduced in *E. coli* under the T7 promoter, it accumulates in inclusion bodies. The protein is solubilized in the presence of 8M urea, purified in a denatured form, and then

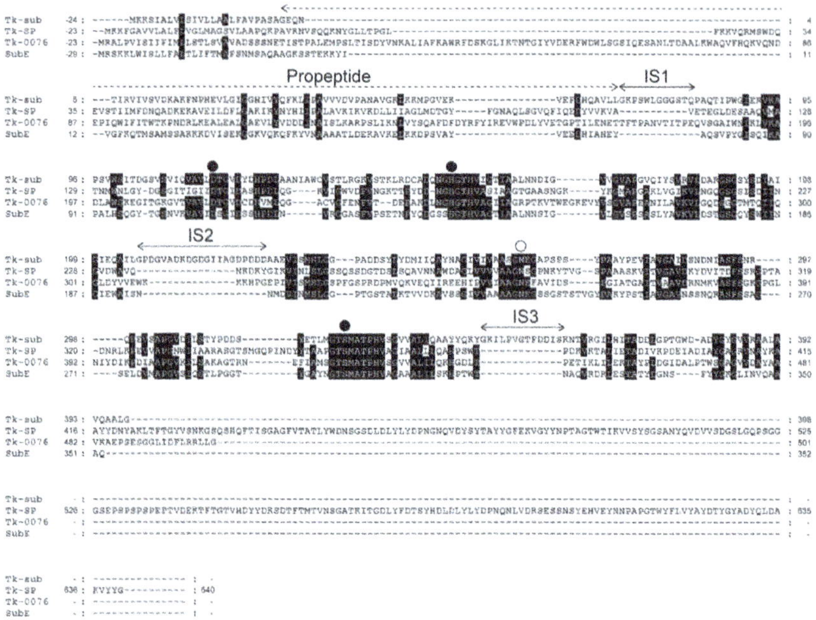

FIGURE 5.2 Sequence alignment of subtilisin-like serine proteases. The amino acid sequences of Prepro-Tk-subtilisin (Tk-sub), Prepro-Tk-SP (Tk-SP), Tk-0076, and prepro-subtilisin E (SubE) are aligned; NCBI accession numbers: YP_184088 (Tk-sub), YP_184102 (Tk-SP), YP_182489 (Tk-0076), and AAA22742 (subE). Highly conserved residues are indicated with white letters. Amino acid residues that form the catalytic triad, and the asparagine residue that forms the oxyanion hole, are indicated by solid and open circles, respectively. The propeptide domain of the proteases and the insertion sequences (IS1–IS3) of Pro-Tk-subtilisin are shown above the sequences.

refolded by removing the urea. The refolded protein yields an approximately 45 kDa protein band on sodium dodecyl sulfate polyacrylamide gel electrophoresis (SDS-PAGE) (Fig. 5.4, 0 min) and remains inactive unless incubated in the presence of Ca^{2+} ions. When Pro-Tk-subtilisin (Gly1–Gly398, 45 kDa) is incubated at 80°C and pH 9.5 in the presence of 5 mM $CaCl_2$, it starts autoprocessing into Tk-subtilisin (Gly70–Gly398, 37 kDa) and Tkpro (Gly1–Leu69, 8 kDa) within 2 min. Then Tkpro is gradually degraded by Tk-subtilisin (Fig. 5.4). Tk-subtilisin activity increases as the amount of Tkpro decreases, indicating that Tkpro acts as an inhibitor, which forms an inactive complex with Tk-subtilisin upon autoprocessing. Therefore Tk-subtilisin is fully activated by the eventual degradation of Tkpro. The autoprocessing of Pro-Tk-subtilisin does not occur in the absence of Ca^{2+} ions. This result suggests that Pro-Tk-subtilisin requires Ca^{2+} ions for correct folding. Surprisingly, Tk-subtilisin is correctly refolded into its native structure even in the absence of Tkpro, as long as sufficient Ca^{2+} ions are present, although the refolding yield is quite lower than that in the presence of

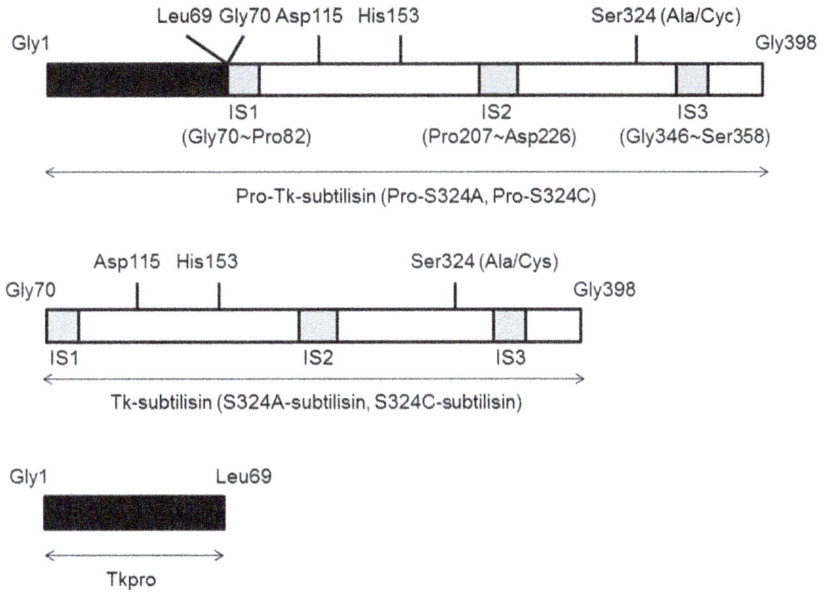

FIGURE 5.3 Schematic representation of the Pro-Tk-subtilisin, Tk-subtilisin, and Tkpro primary structures. The dark box represents the propeptide domain, the open box represents the subtilisin domain, and gray boxes represent the insertion sequences (IS1–IS3). The locations of the catalytic triad, Asp115, His153, and Ser324 (alanine in Pro-S324A and S324A-subtilisin, and cysteine in Pro-S324C and S324C-subtilisin), and the *N*- and *C*-terminal residues of each domain are shown. These proteins were separately produced in *Escherichia coli*. Pro-Tk-subtilisin (Pro-S324A and Pro-S324C) and Tk-subtilisin (S324A-subtilisin) accumulate in inclusion bodies upon overproduction and are purified in a denatured form. S324C-subtilisin is obtained when Pro-S324C is autoprocessed into S324C-subtilisin and Tkpro in the presence Ca^{2+} ions. Tkpro is produced in a soluble form and purified.

Tkpro (Pulido et al., 2006; Tanaka et al., 2008). Bacterial subtilisins, such as pro-subtilisin E, are refolded, autoprocessed, and then activated in the absence of Ca^{2+} ions, because they do not require Ca^{2+} ions for folding, only needing the Ca^{2+} ions for stability (Yabuta et al., 2002). When pro-subtilisin E is overproduced in *E. coli*, it matures in the cells and exhibits serious toxicity (Li and Inouye, 1994). However, Pro-Tk-subtilisin does not correctly fold in *E. coli* cells, probably due to insufficient Ca^{2+} concentration, and accumulates in inclusion bodies. These results suggest that the maturation of Tk-subtilisin is initiated when its precursor is secreted into the Ca^{2+}-rich extracellular environment so as not to degrade cellular proteins.

The maturation rate of Pro-Tk-subtilisin greatly depends on temperature. Pro-Tk-subtilisin fully matures within 30 min at 80°C (Fig. 5.4) and 3 h at 60°C, while the maturation is not complete even after 4 h when incubated at 40°C and below (Pulido et al., 2006). Pro-Tk-subtilisin is immediately autoprocessed, even at 20°C, in the presence of Ca^{2+} ions. Nevertheless, the

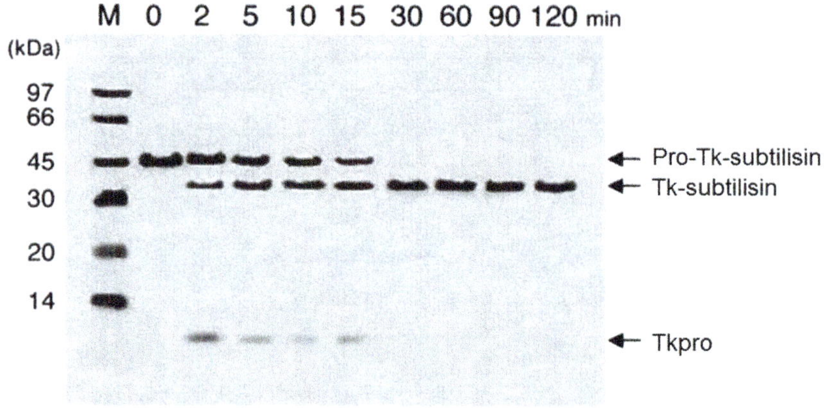

FIGURE 5.4 Autoprocessing and degradation of Tkpro at 80°C. Pro-Tk-subtilisin (300 nM) was incubated in 50 mM CAPS-NaOH (pH 9.5) containing 5 mM $CaCl_2$ at 80°C for the time indicated at the top of the gel. Upon incubation, all proteins were precipitated by addition of trichloroacetic acid (TCA), and subjected to 15% Tricine-SDS-PAGE. Arrows indicate Pro-Tk-subtilisin (45 kDa), Tk-subtilisin (38 kDa), and Tkpro (7 kDa), from top to bottom.

subsequent degradation of Tkpro does not occur at this temperature. Circular dichroism (CD) spectroscopy shows that Tkpro is highly structured in an isolated form, unlike the propeptide of bacterial subtilisins, which are unstructured upon dissociation from the mature domain. Therefore Tkpro is tolerant to proteolysis and requires a high temperature for its degradation. In conclusion, Tk-subtilisin matures from its pro-form by the same three steps, folding, autoprocessing, and degradation of Tkpro, as do bacterial subtilisins. Still, the maturation mechanism is unique in that it requires Ca^{2+} ions for folding, instead of its cognate propeptide (Tkpro), as well as a high temperature for the degradation of Tkpro.

5.3.2 Crystal structures of Tk-subtilisin

The crystal structures of Tk-subtilisin in the three maturation steps (unautoprocessed, autoprocessed, and mature) were determined using mutant enzymes, in which the catalytic serine (Ser324) was replaced with either alanine or cysteine (Tanaka et al., 2006; Tanaka et al., 2007a,b). Pro-Tk-subtilisin with the Ser324→Ala mutation, termed Pro-S324A, represents the unautoprocessed form of Tk-subtilisin. The Ser324→Ala mutation completely abolishes enzymatic activity, and therefore Pro-S324A is not autoprocessed upon folding. In contrast, the Ser324→Cys mutation greatly reduces activity such that Pro-S324C is autoprocessed into Tkpro and S324C-subtilisin upon folding, but further degradation of Tkpro does not occur. Pro-S324C forms a stable Tkpro: S324C-subtilisin complex after autoprocessing and therefore represents the autoprocessed form. The maturation of Pro-Tk-subtilisin prepares the matured form in the presence of Ca^{2+} ions at 80°C. Matured Tk-subtilisin is inactivated by modifying the catalytic serine with a specific inhibitor, diisopropyl fluorophosphate, to prevent the self-degradation of Tk-subtilisin during crystallization processes. It is likely that this modification does not seriously affect the overall structure of Tk-subtilisin. Thus the resultant monoisopropylphospho-Tk-subtilisin (MIP-Tk-subtilisin) represents the mature form of Tk-subtilisin. S324A-subtilisin, that is Tk-subtilisin with the Ser324→Ala mutation, was refolded in the presence of Ca^{2+} ions, and the structure was determined as another mature form (mature form 2). The preparation of these proteins is schematically described in Fig. 5.5A. These structures are shown in Fig. 5.5B–E, compared with pro-S221C-subtilisin E, which represents the autoprocessed form of pro-subtilisin E (Fig. 5.5F).

The overall structure of autoprocessed Pro-S324C (Fig. 5.5C) highly resembles that of autoprocessed pro-S221C-subtilisin E (Fig. 5.5F), except that the mature domain contains seven, instead of two, Ca^{2+} ions, and three unique surface loops, which are formed by the insertion sequences (IS1, IS2, and IS3). The number of Ca^{2+} ions in Tk-subtilisin is the highest among those so far reported for various subtilases (Almog et al., 2003, 2009; Arnórsdóttir et al., 2005; Betzel et al., 1988; Bode et al., 1987; Bott et al., 1988; Gros et al., 1988; Helland et al., 2006; Henrich et al., 2003; Jain et al., 1998; Kim et al., 2004; McPhalen and James, 1988; Smith et al., 1999; Teplyakov et al., 1990; Vévodová et al., 2010). The Ca1 site is conserved in the members of the subtilisin, thermitase, and kexin families, while the other six, with a few exceptions for Ca6 (Almog et al., 2003, 2009), are unique in the structure of Tk-subtilisin. All Ca^{2+} binding sites are located on surface loops far from the active site. The Ca2–Ca5 sites are localized in the unique Ca^{2+} binding loop, which is mostly formed by IS2 (Fig. 5.5G), whereas the Ca1, Ca6, and Ca7 sites are located on different surface loops.

The entire structure of the unautoprocessed form (Fig. 5.5B) and autoprocessed form (Fig. 5.5C) of Pro-Tk-subtilisin is similar, indicating that the autoprocessing reaction does not cause significant conformational changes. In contrast, the crystal structure of the unautoprocessed form of pro-subtilisin E has not yet been determined because it is highly unstable owing to the absence of the Ca1 site (Yabuta et al., 2002). The Ca1 site of subtilisin E is formed only when the peptide bond between the propeptide and the mature domain is cleaved, and the *N*-terminus of the mature domain leaves the active site, such that Gln2 at the *N*-terminus of the mature domain can directly coordinate the Ca^{2+} ion at the Ca1 site. The corresponding residue of Tk-subtilisin, Gln84, coordinates the Ca1 ion in the unautoprocessed form owing to the extension loop between the propeptide and the mature domain (Fig. 5.5B). Tk-subtilisin has a 13 amino acid residue insertion sequence (IS1, Gly70–Pro82) at the *N*-terminus. IS1 forms a long surface loop and allows Gln84 to directly coordinate the Ca1 ion in the unautoprocessed form.

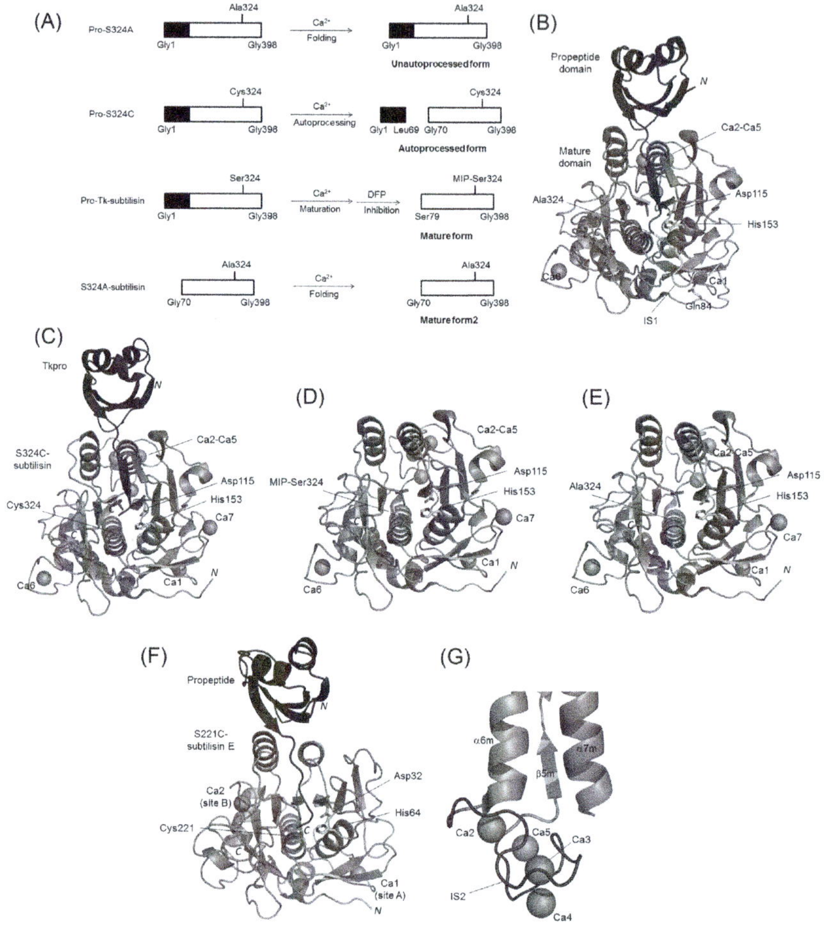

FIGURE 5.5 (A) Schematic representation of protein preparation. Filled and open boxes represent the propeptide and mature domains, respectively. MIP-Ser represents monoisopropylphospho-Ser. (B–G) Crystal structures of Tk-subtilisin and subtilisin E. The propeptide and subtilisin domains are colored black and gray, respectively. The catalytic triad, Asp115, His153, and Ser324 (or Ala324, Cys324, and MIP-Ser324) are indicated with stick models. The Ca^{2+} ions are displayed as gray spheres. N and C represent the *N*- and *C*-termini, respectively. (B) Structure of the unautoprocessed form (Pro-S324A). (C) Structure of the autoprocessed form (Pro-S324C). (D) Structure of the mature form (MIP-Tk-subtilisin). Tk-subtilisin is matured from Pro-Tk-subtilisin by incubation at 80°C in the presence of 10 mM $CaCl_2$ and inactivated with diisopropyl fluorophosphate. (E) Structure of S324A-subtilisin. S324A-subtilisin is refolded in the presence of Ca^{2+} ions and absence of Tkpro. Misfolded and unfolded proteins are removed by proteolysis. (F) Structure of autoprocessed pro-subtilisin E (pro-S221C-subtilisin E). (G) The Ca^{2+} binding loop of Tk-subtilisin IS2 is highlighted with dark gray.

Hence, Tk-subtilisin nearly completes folding before autoprocessing. The structure of MIP-Tk-subtilisin is almost identical to that of S324A-subtilisin (Fig. 5.5D and E). S324A-subtilisin is refolded in the presence of Ca^{2+} ions and the absence of Tkpro. This result also suggests that Tk-subtilisin can fold into its native structure without the assistance of Tkpro.

The Ca7 ion is missing in the unautoprocessed form, whereas the other six Ca^{2+} ions are observed in all three forms (Fig. 5.5B). The missing Ca^{2+} ion is removed from the Ca7 site when the unautoprocessed form is dialyzed against a Ca^{2+}-free buffer. The Ca7 site is stable in the autoprocessed form and mature form such that the Ca^{2+} ion is bound to these structures even upon dialysis (Fig. 5.5C and D). The Ca7 ion is observed in the unautoprocessed form when the crystal is soaked with 10 mM $CaCl_2$ solution (Tanaka et al., 2007a). These results suggest that the Ca7 ion weakly binds to the unautoprocessed form, but this binding site is stabilized upon autoprocessing.

5.3.3 Requirement of Ca^{2+}-binding loop for folding

The structure of Pro-Tk-subtilisin significantly changes upon binding to Ca^{2+} ions. Pro-Tk-subtilisin is folded into an inactive form with a molten globule-like structure, highly susceptible to proteolysis without Ca^{2+} ions. In contrast, it is folded into its native structure in the presence of Ca^{2+} ions (Tanaka et al., 2007b). Tk-subtilisin binds seven Ca^{2+} ions. Of those, four Ca^{2+} ions (Ca2–Ca5) continuously bind to a single surface loop, termed the Ca^{2+} binding loop, which primarily consists of the second Tk-subtilisin's unique insertion sequence (IS2, Pro207-Asp226). This Ca^{2+} binding loop contains a Dx[DN]xDG motif, which is recognized as Ca^{2+} binding motif in a variety of Ca^{2+} binding proteins (Rigden et al., 2011). Removal of the Ca^{2+} binding loop from Tk-subtilisin completely abolishes its ability to fold into its native structure (Takeuchi et al., 2009). The loop-deleted mutant (Δloop-Pro-S324A) is trapped into a molten globule-like structure even in the presence of Ca^{2+} ions. Similarly, the mutation that removes either the Ca2 or Ca3 site substantially reduces the refolding rate of Pro-S324A (Takeuchi et al., 2009). We kinetically analyzed the refolding of Pro-S324A and its mutants, ΔCa2-Pro-S324A and ΔCa3-Pro-S324A, which lack the Ca2 and Ca3 ions, respectively, by initiating and terminating the refolding reaction by the addition of $CaCl_2$ and ethylenediaminetetraacetic acid (EDTA), respectively, with appropriate intervals, and determining the amount of the correctly folded protein by SDS-PAGE (Fig. 5.6). The rate constants for refolding the proteins in the presence of 10 mM $CaCl_2$ at 30°C are 1.3 min^{-1} for Pro-S324A, 0.020 min^{-1} for ΔCa2-Pro-S324A, and 0.019 min^{-1} for ΔCa3-Pro-S324A. Our crystallographic analysis revealed that the overall structures of ΔCa2-Pro-S324A and ΔCa3-Pro-S324A are nearly identical to that of Pro-S324A, except that these structures lack the Ca2 and Ca3 ions, respectively (Takeuchi et al., 2009). Also, the thermal

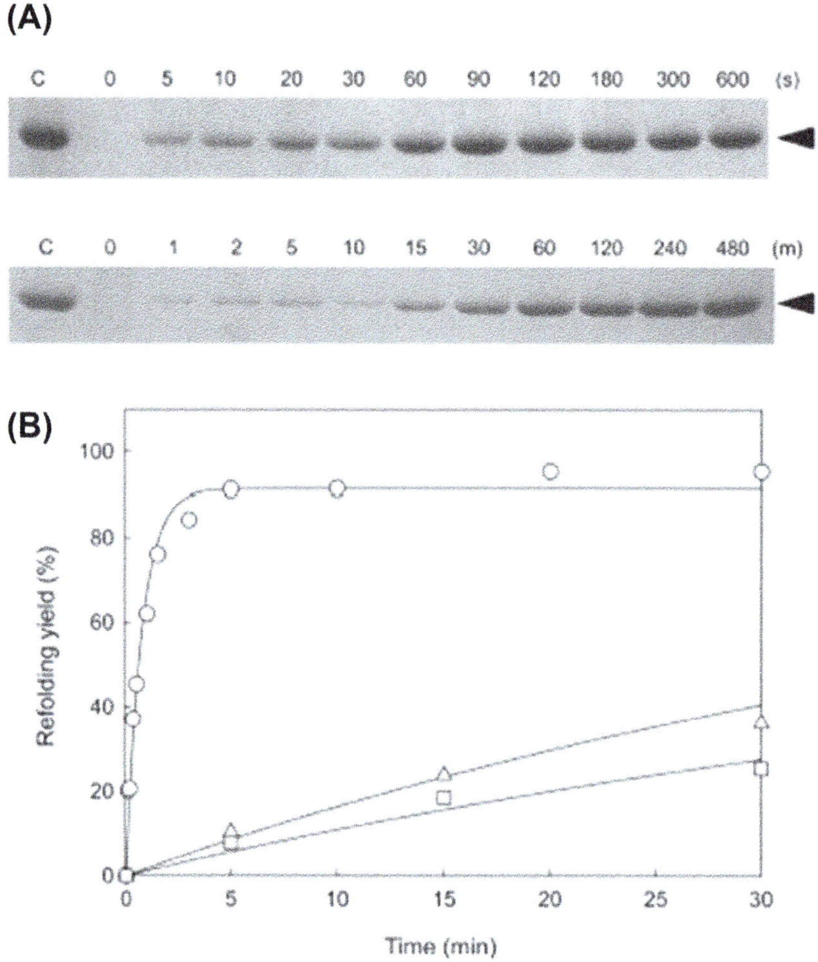

FIGURE 5.6 Refolding rates of Pro-S324A and its derivatives. (A) SDS-PAGE of refolded proteins. Pro-S324A (top) and ΔCa2-Pro-S324A (bottom) in a Ca^{2+}-free form were denatured using 6M GdnHCl, diluted 100-fold with 50 mM Tris–HCl (pH 8.0) containing 10 mM $CaCl_2$ and 1 mM DTT, and then incubated at 30°C for refolding. At appropriate intervals (shown at the top of the gel), incorrectly folded proteins were digested with chymotrypsin, and refolded proteins were analyzed with 12% SDS-PAGE. (B) Refolding curves. The refolding yields of Pro-S324A (open circle), ΔCa2-Pro-S324A (open triangle), and ΔCa3-Pro-S324A (open square) are shown as a function of incubation time. The refolding yields were calculated by estimating the amount of correctly refolded protein from the intensity of the band visualized with CBB staining following SDS-PAGE. Lines represent the optimal fit to the data. *CBB*, Coomassie brilliant blue.

stabilities of these proteins are not reduced compared with that of Pro-S324A. These results suggest that the Ca2 and Ca3 ions are responsible for the Ca^{2+}-dependent folding of Tk-subtilisin, but not for the stability gained upon folding. Likewise, the mutation that removes the Ca4 or Ca5 site also greatly reduces the refolding rate of Pro-S324A without seriously affecting its thermal stability, indicating that the Ca4 and Ca5 ions also contribute to the folding of Tk-subtilisin, although the crystal structures of these mutants remain to be determined (Takeuchi, unpublished data).

The Ca^{2+} binding loop is inserted in between the sixth α-helix of the mature domain (α6m-helix), and the fifth β-strand of the mature domain (β5m-strand), which together compose the central αβα substructure along with the seventh α-helix of the mature domain (α7m-helix), as shown in Fig. 5.5G. This loop is rich in aspartate and probably unstructured in the absence of Ca^{2+} ions because of extensive negative charge repulsions. The Ca^{2+} binding loop is only folded into its correct structure when Ca^{2+} ions bind to the loop, permitting a proper substructure arrangement. Bryan et al. (1992,1995) proposed that the formation of the αβα substructure is crucial for folding bacterial subtilisins. Subsequently, the Ca^{2+} binding loop is probably required to initiate the folding of Tk-subtilisin by stabilizing the αβα substructure and most likely acts as an intramolecular chaperone for Tk-subtilisin. These results suggest that Tk-subtilisin has two intramolecular chaperones, its unique Ca^{2+} binding loop and its cognate propeptide (Tkpro). The propeptides of extracellular proteases, such as bacterial subtilisins, assist folding by reducing the high kinetic barrier between unfolded and folded states, ensuring proteolytic stability and allowing these proteases to function in harsh environments (Jaswal et al., 2002). The kinetic barrier of folding may be much higher for Tk-subtilisin than for mesophilic proteases because *T. kodakarensis* grows in an extremely hot environment. Hence, Tk-subtilisin might require the chaperone function of Tkpro and the folding of its Ca^{2+} binding loop to overcome this high kinetic barrier.

5.3.4 Ca^{2+} ion requirements for hyperstability

The other three Ca^{2+} ions (Ca1, Ca6, and Ca7) bind to the different surface loops located far from one another (Fig. 5.5C–E). The Ca1 site is conserved among bacterial subtilisins and other subtilases (Fig. 5.5F). This site in bacterial subtilisin is essential for structural stability (Bryan et al., 1992; Pantoliano et al., 1989; Voordouw et al., 1976). The Ca1 ion is relatively buried inside the protein molecule according to the crystal structure. Therefore the dissociation of the Ca^{2+} ion from the Ca1 site is very slow (Bryan et al., 1992), resulting in a high-affinity Ca^{2+} binding site. We constructed Pro-Tk-subtilisin and Pro-S324A derivatives lacking the Ca1, Ca6, and Ca7 ions by removing the binding loop or mutating the aspartate that directly coordinates the Ca^{2+} ions to examine the role of these three Ca^{2+} ions in Tk-subtilisin. These Pro-Tk-subtilisin derivatives (ΔCa1-Pro-Tk-subtilisin, ΔCa6-Pro-Tk-subtilisin, and ΔCa7-Pro-Tk-subtilisin) were used for maturation analysis, and the Pro-S324A derivatives (ΔCa1-Pro-S324A,

ΔCa6-Pro-S324A, and ΔCa7-Pro-S324A) were used for analyses of stability and structure.

Differential scanning calorimetry analysis showed that removing the Ca1, Ca6, or Ca7 ions reduces Pro-S324A's thermal stability T_m value by 26.6°C, 11.7°C, or 4.0°C, respectively. The crystal structure of ΔCa6-Pro-S324A revealed that the overall structure is nearly identical to that of Pro-S324A, except that the surface loop containing the Ca6 site is disordered owing to the absence of the Ca^{2+} ion. Although the crystal structures of ΔCa1-Pro-S324A and ΔCa7-Pro-S324A have not been determined, the secondary structure features of these mutants are indistinguishable from that of Pro-S324A in CD spectroscopy. Additionally, the refolding rates of these mutants are not reduced compared with Pro-S324A. These results suggest that the Ca1, Ca6, and Ca7 ions stabilize the structure of Tk-subtilisin, and especially that the Ca1 ion is critical for stability. The Pro-Tk-subtilisin derivatives lacking these Ca^{2+} ions are autoprocessed in the presence of Ca^{2+} ions. Still, the subsequent degradation of Tkpro does not occur for ΔCa1-Pro-Tk-subtilisin, while the other two mutants are effectively matured with rates comparable to that of Pro-Tk-subtilisin (Fig. 5.7). The active form of ΔCa1-Tk-subtilisin, which is obtained by refolding the mature domain alone in the presence of Ca^{2+} ions exhibits approximately 50% of the activity of Tk-subtilisin at 60°C, and less than 5% at 80°C (Uehara et al., 2012a). The reduced proteolytic activity of ΔCa1-Tk-subtilisin probably causes insufficient degradation

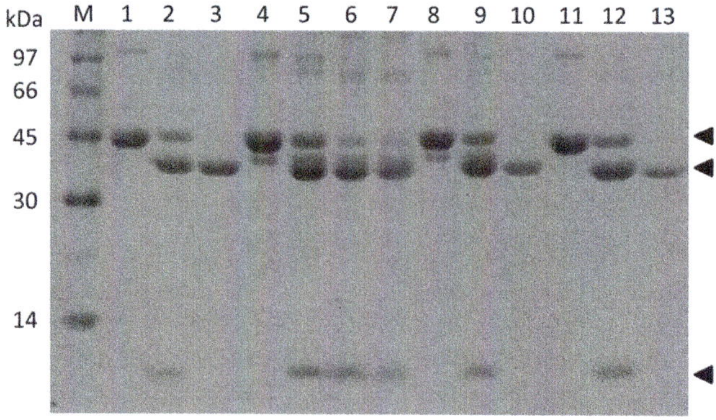

FIGURE 5.7 Maturation of Pro-Tk-subtilisin Ca^{2+}-deleted mutants. Pro-Tk-subtilisin (lanes 1–3), ΔCa1-Pro-Tk-subtilisin (lanes 4–7), ΔCa6-Pro-Tk-subtilisin (lanes 8–10), and ΔCa7-Pro-Tk-subtilisin (lanes 11–13) (0.3 μM each) were incubated in 1 mL of 50 mM CAPS-NaOH (pH 9.5) containing 10 mM $CaCl_2$ and 1 mM DTT at 80°C for 2 min. Lanes 2, 5, 9, and 12 were incubated for 30 min; lanes 3, 6, 10, and 13 for 60 min; lane 7 was precipitated by the addition of 120 μL of TCA, and subjected to 12% Tricine-SDS-PAGE. Lanes 1, 4, 8, and 11 contained the sample without exposure to the buffer containing Ca^{2+} ions. Lane M contained the low-molecular-weight markers (GE Healthcare). Proteins were stained with CBB. The arrowheads indicate Pro-Tk-subtilisin or its derivative, Tk-subtilisin or its derivative, and Tkpro from top to bottom, respectively. *CBB*, Coomassie brilliant blue.

of Tkpro. Hence, the Ca1 ion is most likely crucial for Tk-subtilisin activity at high temperatures and thereby required for maturation. ΔCa6-Tk-subtilisin and ΔCa7-Tk-subtilisin exhibit activity comparable with that of Tk-subtilisin, but the optimum temperature for proteolytic activity is decreased by 10°C. The Ca6 site is not conserved among bacterial subtilisins, but is found in several subtilases, such as mesophilic subtilisin, sphericase (Sph) from *Bacillus sphaericus* (Almog et al., 2003), and psychrophilic subtilisin S41 from Antarctic *Bacillus* TA41 (Almog et al., 2009). Because removal of the Ca6 ion increases the flexibility of the binding loop as evidenced by the ΔCa6-Pro-S324A structure, the Ca6 site may be important in protecting the surface loop from self-degradation. The Ca7 ion is observed in the autoprocessed form and the mature form, but missing in the unautoprocessed form. This autoprocessing, as described above, likely increases the affinity of Ca^{2+} ions at this site. Removal of the Ca7 ion destabilizes Pro-S324A by only 4°C, probably because it weakly binds to the unautoprocessed form. Destabilization by removal of the Ca7 ion appears to be more serious in the mature form than in the unautoprocessed form, such that the half-life of ΔCa7-Tk-subtilisin at 95°C is decreased fourfold compared with that of Tk-subtilisin. Interestingly, the *N*-terminal region of the mature domain (Leu75–Thr80) is disordered when the Ca7 ion is binding to the unautoprocessed form. Gln81 changes the conformation so much that it can stabilize the Ca7 site upon autoprocessing. Therefore it is tempting to speculate whether the Ca7 ion promotes the autoprocessing reaction by shifting the equilibrium between the unautoprocessed and autoprocessed forms. However, ΔCa7-Pro-Tk-subtilisin is autoprocessed at a similar rate as Pro-Tk-subtilisin (Uehara et al., 2012a). Thus the Ca7 ion must contribute to the stabilization of the autoprocessed form and the mature form but is likely not responsible for the autoprocessing efficiency.

5.3.5 Role of Tkpro

Unlike bacterial subtilisins, Tkpro is not required for the folding of Tk-subtilisin. The crystal structure of the active-site mutant of Tk-subtilisin (S324A-subtilisin), which is refolded in the presence of Ca^{2+} ions and the absence of Tkpro, is essentially the same as that of the mature form (Fig. 5.5E). In contrast, the refolding rate of S324A-subtilisin is significantly increased in the presence of Tkpro. Tkpro increases the refolding rate of S324A-subtilisin by more than 100-fold in the presence of 1 mM $CaCl_2$, while it is further increased 10-fold in the presence of 10 mM $CaCl_2$, suggesting that the chaperone function of Tkpro is dependent on Ca^{2+} ion concentration and more predominant at lower concentrations (Tanaka et al., 2009). Hence, Tkpro probably acts as an intramolecular chaperone in an auxiliary manner.

Tkpro interacts with Tk-subtilisin predominantly at three regions, according to the crystal structure of the Tkpro and S324A-subtilisin complex (Tanaka et al., 2007a). These regions are the *C*-terminal extended region (His64–Leu69), which binds to the active-site cleft in a substrate-like manner; Glu61 and Asp63, which interact with the *N*-termini of two surface helices (α6m and α7m) of the αβα substructure by hydrogen bonds; and the hydrophobic loop containing three hydrophobic residues (Phe33, Leu35, and Ile36) between the second and third β-strand of Tkpro (β2p and β3p strands), which interact with Glu201 by hydrogen bonds. Glu201 is located in the α6m-helix. Furthermore, three homologous acidic amino acids are highly conserved among bacterial subtilisins and form similar hydrogen bonds. Either the truncation of the Tkpro *C*-terminal extended region or the combination of this truncation plus the double mutations substituting Glu61 and Asp63 to alanine reduces the chaperone function and binding ability of folded S324A-subtilisin but does not abolish them. This suggests that the interactions in these two regions are not essential for the chaperone function of Tkpro. However, these interactions seem to accelerate propeptide-catalyzed folding by promoting binding to a folding intermediate with a native-like structure.

In contrast, the mutation at Glu201 almost fully abolishes the chaperone function of Tkpro, but nevertheless, Tkpro retains the significant binding ability to E201/S324A-subtilisin with a native structure. These results suggest that the hydrophobic loop of Tkpro, which interacts with Glu201 through hydrogen bonds, is required to initiate propeptide-catalyzed folding by promoting binding to a folding intermediate with a nonnative structure. Because the hydrophobic loop is tightly packed with two surface helices (α6m and α7m), it seems likely that Glu201-mediated interactions promote an association between the two helices and the subsequent formation of the central αβα substructure. Hydrophobic effects with Tkpro may also further stabilize these structures. Therefore Glu201-mediated interactions might be the first folding step promoted by Tkpro (Tanaka et al., 2009). The glutamate residue corresponding to Glu201 of Tk-subtilisin is well conserved in members of the subtilisin and proteinase K families, although it is occasionally replaced by glutamine, aspartate, or asparagine. Likewise, the hydrophobic loop and two acidic residues (Glu61 and Asp63) are located in the conserved motif of the propeptide (N2 motif). These results suggest that their cognate propeptides catalyze the folding of these subtilases through similar mechanisms. The propeptide of aqualysin, a member of the proteinase K family, can function as a chaperone for subtilisin (Marie-Claire et al., 2001). Hence, the interactions required for the propeptide-catalyzed folding of Tk-subtilisin might be shared with all members of the subtilisin and proteinase K families.

Tkpro remains bound to the mature domain after autoprocessing and forms an inactive complex. According to the crystal structure, the *C*-terminal four residues of Tkpro (Ala66–Leu69) are located in the Tk-subtilisin substrate-binding pockets (S1–S4 subsites). Therefore Tkpro competitively inhibits Tk-subtilisin activity when added *in trans* (Pulido et al., 2006, Pulido et al., 2007a,b; Uehara et al., 2013b). Inhibition progress curves of mature Tk-subtilisin by Tkpro revealed a hyperbolic pattern, indicating that Tkpro acts as a slow binding inhibitor, similar to the propeptides of bacterial subtilisins. The inhibition potency of Tkpro is higher than those of bacterial propeptides, such that the concentration of propeptide required for complete inhibition is ~50 nM for Tkpro and 0.5–5 μM for bacterial propeptides at low or middle temperatures. Tkpro is well structured even in an isolated form, while bacterial propeptides are intrinsically unstructured proteins (Hu et al., 1996, Huang et al., 1997; Wang et al., 1995, 1998). Because of its defined structure, Tkpro is effectively degraded by Tk-subtilisin only at high temperatures (≥60°C), where the stability and binding affinity of Tkpro decrease, and the activity of Tk-subtilisin simultaneously increases. Tk-subtilisin can be refolded in the absence of Tkpro. However, the yield of mature Tk-subtilisin when refolding Tk-subtilisin in the presence of Ca^{2+} ions and the absence of Tkpro is considerably lower than with Tkpro (≤5%), because the mature Tk-subtilisin molecules that refold earlier attack the other molecules destined to refold later. Hence, Tkpro is required not only for promoting refolding, but also for increasing the yield of mature Tk-subtilisin by protecting the precursor from auto-degradation.

5.3.6 Role of the insertion sequences

Tk-subtilisin has three insertion sequences (IS1, IS2, and IS3). All three insertions form long surface loops on the structure. IS2 contains four Ca^{2+} binding sites and comprises the unique Ca^{2+} binding loop required for folding. In contrast, the other two loops do not bind Ca^{2+} ions. Because we have already reviewed the role of IS2 in the aforementioned section, this section concentrates on the role of the other two insertions. Our crystal structure of the unautoprocessed form revealed that IS1 forms an *N*-terminal extension loop and allows Gln84 to directly coordinate the Ca1 ion before autoprocessing. This loop is nearly disordered in the autoprocessed form, and most of it (Gly70–Gly78) is cleaved away in the mature form (Tanaka et al., 2007a). This suggests that IS1 specifically stabilizes the structure of the unautoprocessed form by promoting the formation of the Ca1 site. We constructed IS1-deleted mutants (ΔIS1-Pro-Tk-subtilisin, ΔIS1-Pro-S324A, and ΔIS1-Pro-S324C) and characterized the resultant mutant proteins. As described previously for Pro-S324A and Pro-S324C, ΔIS1-Pro-S324A and ΔIS1-Pro-S324C represent the unautoprocessed and autoprocessed forms ΔIS1-Pro-Tk-subtilisin, respectively. Thermal denaturation analyses showed that ΔIS1-Pro-S324A is less stable than ΔIS1-Pro-S324C by a 26.3°C lower T_m value, whereas the thermal stability of Pro-S324A is comparable with that of Pro-S324C (Uehara et al., 2012b). 8-Anilinonaphthalene-1-sulfonic acid fluorescence spectra demonstrated that the ΔIS1-Pro-S324A structure is not fully folded, and that the interior hydrophobic region is partially exposed. These results suggest that the covalent bond between Leu69 and Ala83 in ΔIS1-Pro-S324A causes a strain, which disrupts the structure of the *N*-terminal region of the mature domain. Gln84 cannot coordinate Ca^{2+} ions in this structure because of the large distance between its active and Ca1 sites. The *N*-terminus is released from steric strain by autoprocessing and moves to its original position where Gln84 can coordinate the Ca1 ion. Hence, IS1 is required to form the Ca1 site before autoprocessing.

ΔIS1-Pro-Tk-subtilisin matures without decreasing the yield of the mature protein, but its autoprocessing rate is significantly less than that of Pro-Tk-subtilisin. The structure around the scissile peptide bond between Tkpro and the mature domain may change owing to a strain caused by the connection between Ala83 and Leu69 in the ΔIS1-Pro-Tk-subtilisin unautoprocessed form. This structural change may be responsible for the slow autoprocessing and maturation of ΔIS1-Pro-Tk-subtilisin. We note that the source organism, *T. kodakarensis*, optimally grows at 85°C (Atomi et al., 2004), yet Pro-Tk-subtilisin might not fold without IS1 at this temperature because of the instability of the unautoprocessed form. In fact, ΔIS1-Pro-Tk-subtilisin cannot mature at 80°C. Hence, IS1 is required for the maturation of Pro-Tk-subtilisin in its native environment because it stabilizes the structure of the unautoprocessed form and promotes maturation.

Bacterial subtilisins also have the Ca1 site. The crystal structure of the unautoprocessed form has not yet been determined because of the structural instability. We constructed a stable unautoprocessed form of pro-subtilisin E by inserting the 13 amino acid residues of IS1 between the propeptide and the mature domain (IS1-pro-S221A-subtilisin E) and determined its crystal structure (Uehara et al., 2013a). In this structure, IS1 forms a surface loop and the Ca1 ion is bound to the protein. This result suggests that the *N*-terminal extension sequence stabilizes the unautoprocessed form of subtilisins containing a Ca1 site by supporting the formation of the Ca1 site. As described above, IS1 increases the autoprocessing rate of Pro-Tk-subtilisin. Likewise, IS1-pro-S221C-subtilisin E is autoprocessed more rapidly than pro-S221C-subtilisin E. When we overproduce pelB-pro-subtilisin E (a pelB leader attached to the *N*-terminus of pro-subtilisin E) in *E. coli*, it accumulates in inclusion bodies, whereas pelB-IS1-pro-subtilisin E does not. IS1-pro-subtilisin E may rapidly fold and mature in the cells. Li and Inouye (1994) has reported that subtilisin E activity is cytotoxic to host cells. The rapid activation of IS1-pro-subtilisin E may cause serious damage to the cells. This may be the reason why pro-subtilisin E and its homologs do not have insertion sequences corresponding to IS1. In contrast, Pro-Tk-subtilisin requires IS1 to mature at the high temperatures where the source organism grows. The maturation of Pro-Tk-subtilisin is regulated by Ca^{2+} ions and is initiated only when Pro-Tk-subtilisin is secreted into the external medium where Ca^{2+} ions are abundant. Therefore the stabilization of the unautoprocessed form of Pro-Tk-subtilisin by IS1 may not cause serious damage to the *T. kodakarensis* cells.

Henrich et al. (2005) has proposed, based on structural models, that the unautoprocessed form of pro-furin, which has a short six residue insertion sequence between its propeptide and mature domain, assumes an incompletely folded structure, in which the Ca1 site is not formed. Meanwhile, the unautoprocessed forms of other pro-protein convertases (pro-PCs) with a long 12 residues insertion sequence assume a fully folded structure, in which the Ca1 site is formed. These results suggest that the maturation rates of the pro-PCs are controlled by the presence or absence of a long insertion sequence between the propeptide and mature domain.

We have recently investigated the role of IS3 (Gly346–Ser358) in the folding and maturation of Tk-subtilisin (Uehara et al., 2021). IS3 is not responsible for binding any Ca^{2+} ions, but this insertion seems likely to be important for the molecular architecture of the mature domain. The IS3-deleted mutant (ΔIS3-Pro-Tk-subtilisin) does not undergo autoprocessing even in the presence of Ca^{2+} ions. However, the structure is significantly changed upon binding to the Ca^{2+} ions as analyzed by CD spectroscopy. The far-UV CD spectrum of the Ca^{2+}-bound form of ΔIS3-Pro-Tk-subtilisin showed an intermediate feature between those of the Ca^{2+}-free molten globule state and the Ca^{2+}-bound native state of Pro-S324A. Interestingly, the intermediate state contains a well-folded proteolysis-resistant core structure, including the central αβα substructure and not well-folded *N*- and *C*-terminal subdomains. Several studies on bacterial subtilisins have proposed that the folding of subtilisin domain initially occurs in the αβα substructure with the assistance of propeptide as an intramolecular chaperone, and the folding subsequently propagates into the terminal subdomains (Gallagher et al., 1995; Kim et al., 2020). ΔIS3-Pro-Tk-subtilisin has an ability to fold the central αβα substructure that is initiated by Ca^{2+} ion binding to IS2 but fails to proceed to the propagation step, arresting Tk-subtilisin folding at the intermediate state. These results suggest that IS3 is a key element for the efficient folding of Tk-subtilisin. Furthermore, mutational analysis on amino acid residues within IS3, which are highly conserved among hyperthermophilic subtilases, revealed that Asp356 is a critical residue responsible for the IS3-mediated Tk-subtilisin folding. The crystal structure of Tk-subtilisin reveals that the side chain of Asp356 serves as a center of intraloop hydrogen-bonding interactions (Fig. 5.8). Deletion of the hydrogen bonds by alanine-substitution of Asp356 compromises the folding ability of Pro-S324A in a temperature-dependent manner as the refolding rate was decreased with temperature increased. Consequently, D356A mutant of Pro-Tk-subtilisin does not become effectively mature at 80°C. In contrast to its importance for Tk-subtilisin folding, D356A mutation does not seriously affect the thermal stability of the folded native structure. Taking together, the Asp356-mediated

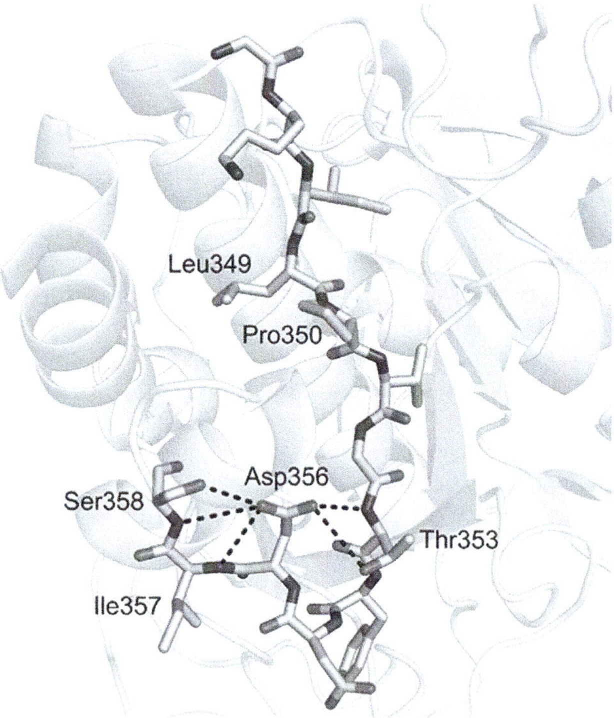

FIGURE 5.8 Structure of IS3. The residues of IS3 are shown as stick models. Hydrogen-bonding network mediated by Asp356 are shown as black dashed lines.

intraloop interactions within IS3 are critical for Tk-subtilisin folding under the high-temperature environments but are not maintained upon folding.

Database searches indicate that, in addition to Tk-subtilisin, only subtilases from hyperthermophilic archaea, such as *Thermococcus onnurineus* NA1 (NCBI YP_002308296), *Pyrococcus* sp. NA2 (NCBI YP_004424756), *Ferroglobus placidus* DSM 10642 (NCBI YP_003436500), *Pyrolobus fumarii* 1A (NCBI YP_004781243), and *Aeropyrum pernix* K1 (NCBI NP_147093), contain these insertion sequences. Among these hyperthermostable proteases, the IS2 and IS3 sequences are highly conserved, while IS1 sequence similarities are poor. This may be because IS1 functions as an extension loop without specific interactions with other residues. Although the insertion sequences of the other subtilases have not been characterized, Catara et al. (2003) reported that the enzymatic activity of *A. pernix* K1 subtilase (pernisine) is greatly reduced by EDTA or ethylene glycol tetraacetic acid treatment. Thus Ca^{2+}-dependent folding may be a common feature among these hyperthermophilic subtilisins. Our studies on the Tk-subtilisin insertion sequences suggest that the acquisition of these insertion sequences is a strategy that hyperthermophilic subtilases have evolved to adapt to the extremely high-temperature environment in which they function.

5.3.7 Cold-adapted maturation through Tkpro engineering

Tk-subtilisin effectively matures only at temperatures ≥ 80°C, because Tkpro forms a stable, inactive complex with Tk-subtilisin and is barely degraded at mild temperatures. *E. coli* HB101 transformed by a plasmid containing the entire Prepro-Tk-subtilisin gene do not form halos, that is, clear zones around colonies, at ≤ 70°C, but do form halos at 80°C on casein-LB-plates. We screened for cold-adapted Pro-Tk-subtilisin mutants, which exhibited the halo-forming activity at ≤ 70°C, upon random mutagenesis in the propeptide region (Ala1–Leu69). A single Gly56 → Ser mutation was identified that greatly accelerated the maturation of Pro-Tk-subtilisin, using this method. The mutant protein (G56S-Pro-Tk-subtilisin) matured within 1 h at 60°C without reducing the mature form yield, whereas Pro-Tk-subtilisin does not complete maturation within 3 h at this temperature. Tkpro containing the G56S mutation (G56S-Tkpro) was overproduced, purified, and characterized to examine why this mutation accelerates the maturation of Pro-Tk-subtilisin. G56S-Tkpro is unstructured in an isolated form, and its stability and inhibitory potency are greatly decreased. In contrast, G56S-Tkpro retains the ability to bind Tk-subtilisin and is folded into a compact structure as revealed by the crystal structure of G56S-Tkpro:S324A-subtilisin complex, even though the hydrophobic core of G56S-Tkpro is partially exposed (Pulido et al., 2007b). Gly56 is in the hydrophobic core of Tkpro and assumes a right-handed conformation. When serine is substituted into this position, it assumes a left-handed conformation and leads to a conformational change of the other hydrophobic residues in the core region. Several researchers have reported the relationship between the stability and the inhibitory potency of the propeptide for bacterial subtilisins (Huang et al., 1997; Kojima et al., 2001; Wang et al., 1995, 1998). These propeptides are almost entirely disordered in their isolated form but adopt a significant secondary structure in the presence of 10% glycerol. The association constant of the interaction between subtilisin E and its cognate propeptide is increased 10-fold in the presence of 10% glycerol, indicating that the propeptide folded in an isolated form is a more potent inhibitor of subtilisins. Pro-Tk-subtilisin presumably matures at temperatures ≥ 80°C in its native environment, and the independent folding of Tkpro is probably required to strongly bind Tk-subtilisin at high temperatures. We also introduced a Phe17 → His mutation in Tkpro to examine whether a nonpolar-to-polar amino acid substitution at a different position in the hydrophobic core would accelerate the maturation of Pro-Tk-subtilisin in a similar manner. The maturation rate of the Pro-Tk-subtilisin Phe17 → His mutant (F17H-Pro-Tk-subtilisin) greatly increased compared with that of Pro-Tk-subtilisin, such that F17H-Pro-Tk-subtilisin effectively matures even at temperatures as low as 40°C. This mutation nearly abolishes Tkpro secondary structure in its isolated form and increases its sensitivity to chymotryptic digestion (Yuzaki et al., 2013). These results suggest that destabilization of the Tkpro hydrophobic core disrupts the structure in its isolated form, thereby increasing Pro-Tk-subtilisin's maturation rate due to the rapid degradation of Tkpro. In contrast, these mutations do not seriously affect the refolding rate of Pro-Tk-subtilisin nor the yield of mature Tk-subtilisin. Tkpro is covalently linked to the mature domain until it is autoprocessed. Hence, the mutant propeptides might be folded by interactions with Tk-subtilisin and become fully functional when they act as an intramolecular chaperone. In fact, F17H-Tkpro and G56S-Tkpro are mostly folded in the crystal structures of Pro-F17H/S324A and G56S-Tkpro:S324A-subtilisin complex, respectively. We propose that destabilization of the Tkpro hydrophobic core by a single mutation is an effective way to promote the degradation of Tkpro, the rate-determining step of maturation, and thereby accelerate the maturation of Pro-Tk-subtilisin without seriously affecting the yield of the mature protein.

The third mutation for cold-adapted maturation was found in the Tkpro *C*-terminus. This Leu69→Pro mutation increases the Pro-Tk-subtilisin maturation rate as much as the G56S mutation does (Fig. 5.9). Unlike the aforementioned mutants, L69P-Tkpro is fully folded in its isolated form and retains stability to proteolysis by other proteases even at high temperatures. Nevertheless, L69P-Tkpro is rapidly degraded by Tk-subtilisin. A kinetic binding analysis showed that Tkpro and L69P-Tkpro bind to Tk-subtilisin at similar rates but that L69P-Tkpro dissociates faster than Tkpro. In addition, the inhibitory potency of L69P-Tkpro is reduced, especially at high

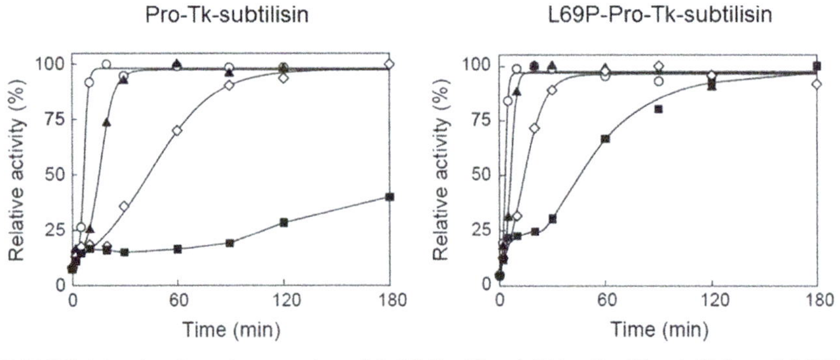

FIGURE 5.9 Accelerated maturation of L69P-Pro-Tk-subtilisin. Pro-Tk-subtilisin and L69P-Pro-Tk-subtilisin were incubated at 50°C (closed square), 60°C (open diamond), 70°C (closed triangle), and 80°C (open circle) for maturation. After incubation at these temperatures, an aliquot was withdrawn and enzymatic activity was determined at 20°C using Suc-AAPF-*p*NA as a substrate.

temperatures, although L69P-Tkpro retains its secondary structure at these temperatures. These results suggest that the L69P mutation accelerates maturation by reducing the binding ability to Tk-subtilisin. The Tkpro *C*-terminal extended region binds to the active-site cleft of Tk-subtilisin, and Ala66–Leu69 are located in the Tk-subtilisin substrate-binding pockets (S1–S4 subsites). The crystal structure of the L69P-Tkpro:S324A-subtilisin complex revealed that the conformation of the L69P-Tkpro *C*-terminal region is shifted away from the substrate-binding pockets compared with that of Tkpro (Uehara et al., 2013b). This conformational change is probably caused by the restricted conformation of the *C*-terminal proline residue (Pro69). The proline cyclic side chain limits the backbone dihedral φ angle to a small range, and therefore proline acts as a structural disruptor in secondary structure elements. The extended Tkpro *C*-terminal region assumes a β-strand conformation, which forms an antiparallel β-sheet with another β-strand of the mature domain. The hydrogen bonds formed between the *C*-terminal region and the mature domain are mostly disrupted in the structure of L69P-Tkpro:S324A-subtilisin, indicating that the *C*-terminal proline reduces interactions between Tkpro and Tk-subtilisin. This mutation does not seriously affect the chaperone function of Tkpro, because the Glu201-mediated interactions, which are essential for the chaperone function of Tkpro, are completely preserved in this structure. We emphasize, however, that the accelerated maturation mechanism of the L69P mutation is different from those of the F17H and G56S mutations. This mutation does not affect structure and stability but does reduce binding ability. Therefore introducing mutations in the Tkpro *C*-terminal region may effectively alter inhibitory potency without affecting stability and chaperone function, thereby modulating Pro-Tk-subtilisin's maturation rate.

5.3.8 Degradation of PrP^{Sc} by Tk-subtilisin

Prion diseases are fatal neurodegenerative disorders that include Creutzfeldt–Jakob disease (CJD), Gerstmann–Sträussler–Scheinker syndrome, fatal familial insomnia, and kuru in humans, bovine spongiform encephalopathy in cattle, and scrapie in sheep (Prusiner, 1998; Wissmann, 2004). These diseases are associated with an abnormal isoform of the prion protein-rich with β-sheets (PrP^{Sc}), which is converted from the normal form containing an α-helix rich conformation (PrP^{C}). The pathogenic PrP^{Sc} propagates by promoting the conversion of cellular PrP^{C} to the abnormal form by an unknown mechanism. Therefore decontamination of PrP^{Sc} attached to medical instruments is essential to prevent secondary infection. However, PrP^{Sc} is highly resistant to general medical instrument sterilization and decontamination methods. The World Health Organization infection control guidelines recommend autoclaving or strong chemical treatment using a high concentration of sodium hydroxide or sodium hypochlorite to inactivate

PrPSc-contaminated reusable instruments completely. Although effective for removing infectivity, these procedures are not applicable to delicate medical devices. Hence, PrP^{Sc} decontamination methods with sufficient potency and safety have been eagerly anticipated.

Tk-subtilisin is highly active and stable at a wide range of temperatures and pHs. Tk-subtilisin exhibits the highest activity toward protein substrates at 90°C and a mildly alkaline pH. In contrast, subtilisin E shows its most increased activity at 60°C. The maximum specific activity of Tk-subtilisin is approximately sevenfold higher than that of subtilisin E (Pulido et al., 2006). Furthermore, Tk-subtilisin retains almost 100% activity even after 1 h of incubation in the presence of 5% SDS, 8M urea, and 6M guanidine hydrochloride (GdnHCl), whereas most commercial proteases are immediately denatured under these conditions. Because of its hyperstability and high activity, Tk-subtilisin is expected to degrade proteolysis-resistant proteins such as PrP^{Sc} under severe physical and chemical conditions where most other proteases would be denatured. We tested whether Tk-subtilisin can be applied to PrP^{Sc} decontamination. Brain homogenates of terminally diseased mice infected with the Candler scrapie strain (SBH) were treated with Tk-subtilisin at various concentrations. Western blot analysis of the proteolysis products showed that Tk-subtilisin is capable of degrading PrP^{Sc} to an undetectable level (Koga et al., 2014). Tk-subtilisin's PrP^{Sc} decontamination activity is much higher than that of proteinase K, a versatile commercial protease; 90 mU of Tk-subtilisin degrades PrP^{Sc} in brain homogenates from patients having sporadic CJD more effectively than 16,700 mU of proteinase K (Fig. 5.10). Tk-subtilisin activity is further increased by the presence of SDS and other industrial surfactants. These results suggest that Tk-subtilisin would be a powerful tool for reducing prion infectivity and applied in many different conditions. Bacterial subtilisins have been used as detergent additives, which improve washing ability. Tk-subtilisin may be useful for universal detergents and special medical detergents used for PrP^{Sc} decontamination.

5.3.9 Tk-subtilisin pulse proteolysis experiments

Determining protein stability and folding kinetics are essential procedures for studying protein structure and energetics. Optical spectroscopy, such as CD and fluorescence, has been conventionally used to characterize the chemical- and thermal-unfolding/refolding properties of proteins, to evaluate ΔG values of stability, and to determine k_f and k_u values for folding reactions. The pulse proteolysis method has recently been developed for taking thermodynamic measurements of protein stability/folding, and for the determination of folding/unfolding pathways without biophysical instrumentation (Park and Marqusee, 2005; Okada et al., 2012). Pulse proteolysis is designed to digest only the unfolded regions of a mixture of folded and unfolded proteins using

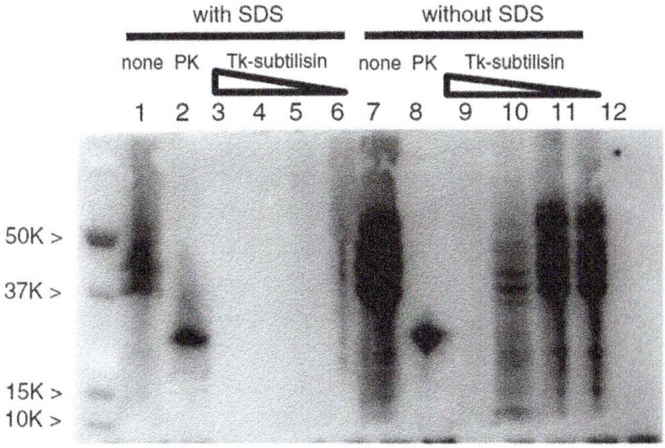

FIGURE 5.10 Proteolysis of PrP^{Sc} in the brain homogenate of a patient having sporadic CJD. A sporadic CJD brain homogenate was digested with proteinase K (PK) and Tk-subtilisin in the presence and absence of SDS. Lane 1 and 7 contained no enzyme. Lanes 2 and 8 contained 14 μg (16.7 units) of proteinase K. Lanes 3–6 and 9–12 contained 9100, 910, 91, and 9.1 μU Tk-subtilisin, respectively. *CJD*, Creutzfeldt–Jakob disease.

a protease in a very short incubation time. After proteolysis, the fraction of folded protein versus digested product is determined by SDS-PAGE. The method is simple and can be applied to high-throughput systems or for the measurement of an unpurified protein's stability/folding. However, chemically and thermally stable proteases are required, because the proteolysis reactions are often performed in conditions that would unfold most proteins.

Tk-subtilisin retains high stability and activity in high temperatures as well as in the presence of chemical denaturants. We examined the unfolding pathway of ribonuclease H2 from *T. kodakarensis* (Tk-RNase H2) using pulse proteolysis with Tk-subtilisin (Okada et al., 2012). Tk-subtilisin retains activity in highly concentrated GdnHCl; Tk-RNase H2 unfolds in this situation. The Tk-subtilisin pulse proteolysis degradation products of Tk-RNase H2 during its unfolding were detected with tricine-SDS-PAGE (Fig. 5.11). The intact Tk-RNase H2 band and several cleavage products were observed. The intact band was reduced by more than half, and the presence of bands representing 20 and 22 kDa fragments developed at early time points. However, the amount of intact protein, and of the 20 and 22 kDa fragments, was less at this early incubation time of 0.5 min, than it was at intermediate incubation times of 2–16 min. These bands gradually disappeared, and then a 9 kDa fragment appeared (fragment 9) over time. At 120 min, only the faint bands of the intact protein and of fragment 9 were observed. Pulse proteolysis using Tk-subtilisin enabled us to detect the unfolding intermediates of a hyperthermophilic protein. This is exceptional. The intermediates were further examined and identified by *N*-terminal sequencing, mass

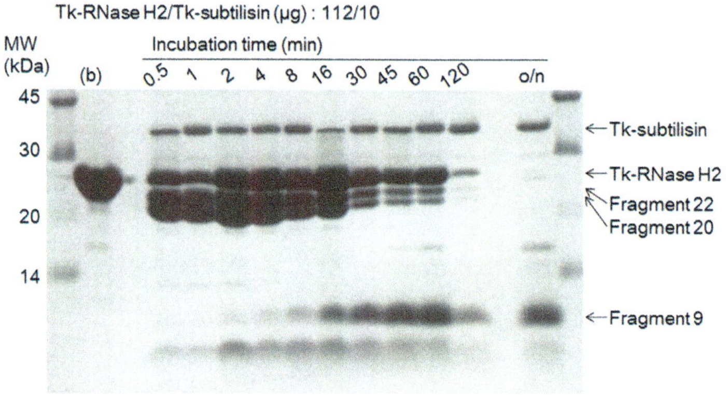

FIGURE 5.11 Tk-subtilisin pulse proteolysis in the kinetic unfolding of Tk-RNase H2. Lane b represents Tk-RNase H2 (112 μg). Tk-RNase H2 was unfolded by adding 4 M GdnHCl. At each time point (0.5–120 min and overnight), the sample was dispensed into tubes, and proteolysis was performed by adding Tk-subtilisin (10 μg) and incubating for 45 s. Proteolysis was quenched by adding 10% TCA, and the products were quantified using tricine-SDS-PAGE. Bands corresponding to Tk-subtilisin, Tk-RNase H2, and the cleavage products are indicated.

spectrometry, and protein engineering. We found that the Tk-RNase H2 native state (N-state) changes to an I_A-state, which is digested by Tk-subtilisin in the early stage of unfolding. The I_A-state shifts to two intermediate forms, the I_B-state and I_C-state, and the I_B-state is digested by Tk-subtilisin in the *C*-terminal region, but the I_C-state is a Tk-subtilisin-resistant form. These states gradually unfold through the I_D-state. These results show that pulse proteolysis by the superstable protease, Tk-subtilisin, is a suitable strategy and an effective tool for analyzing the intermediate structure of proteins.

5.4 Tk-SP

5.4.1 Maturation of Pro-Tk-SP

Pro-Tk-SP is a subtilisin-like serine protease encoded in the gene Tk-1689 (NCBI YP_184102), of *T. kodakarensis*. The primary structures of Pro-Tk-SP (Ala1–Gly640) and its derivatives are schematically shown in Fig. 5.12. Asp147, His180, and Ser359 form the catalytic triad. When Pro-Tk-SP (68 kDa) is overproduced in *E. coli*, 55 and 44 kDa proteins accumulate and show proteolytic activity in zymographic analysis, although the production levels of these proteins are too low to be detected by Coomassie brilliant blue (CBB) staining of the gel following SDS-PAGE (Fig. 5.13). In contrast, the active-site mutant Pro-S359A, in which the catalytic serine (Ser359) is replaced with alanine, accumulates in a soluble form as a 65 kDa protein.

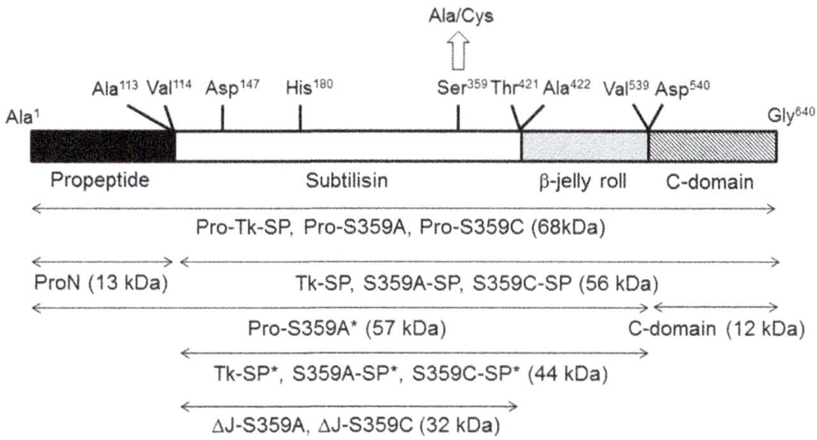

FIGURE 5.12 Schematic representation of the primary structure of Pro-Tk-SP and its derivatives. The dark box represents the propeptide domain (proN); the open box represents the subtilisin domain; the gray box represents the β-jelly roll domain; and the hatched box represents the C-domain. The locations of the three active-site residues, Asp147, His180, and Ser359 (alanine and cysteine for the active-site mutants), and the *N*- and *C*-terminal residues of each domain are shown. Each region and the terminology associated with the recombinant proteins are shown with their molecular masses.

These results suggest that Pro-Tk-SP, like bacterial subtilisins, start maturation in *E. coli* cells, and that its proteolytic activity prevents further protein production owing to serious cytotoxicity. The 55 and 44 kDa proteins cannot be separated using normal purification procedures, such as column chromatography. However, the 44 kDa protein fraction gradually increases during purification and eventually becomes the only protein fraction, as the 55 kDa fraction decreases upon heat treatment at 80°C. Therefore Pro-Tk-SP (68 kDa) must be autoprocessed into the 44 kDa protein through the 55 kDa protein, and this autoprocessing reaction is promoted at high temperatures.

Matrix-assisted laser desorption/ionization time-of-flight mass spectrometry revealed that the 44 kDa protein is a Pro-Tk-SP derivative (Val114–Val539) lacking the *N*-terminal (Ala1–Ala113) and *C*-terminal (Asp540–Gly640) regions (Foophow et al., 2010b). The 55 kDa protein was identified to be the intermediate (Val114–Gly640) lacking the *N*-terminal region (Ala1–Ala113). These results suggest that Pro-Tk-SP is matured through the autoprocessing of both the *N*-terminal and *C*-terminal domains, and that the first autoprocessing event occurs at the *N*-termini. The 55 kDa protein, 44 kDa protein, *N*-terminal domain, and *C*-terminal domain are designated the Tk-SP, Tk-SP*, ProN, and C-domains, respectively, as shown in Fig. 5.12. Because Tk-SP is correctly refolded into its native structure in the absence of Ca^{2+} ions and exhibits significant activity in gel assays, Tk-SP requires neither propeptide nor Ca^{2+} ions for folding.

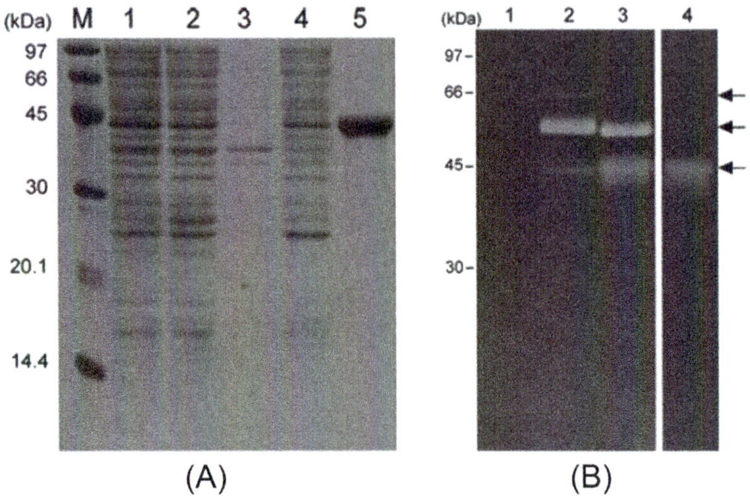

FIGURE 5.13 Overproduction and purification of Pro-Tk-SP. (A) *Escherichia coli* BL21-Codonplus (DE3)-RIL cells transformed by a pET25b derivative designed for the overproduction of Pro-Tk-SP were subjected to 15% SDS-PAGE. The gel was stained with CBB following electrophoresis. Lane M: low-molecular-weight marker; lane 1: whole cell extract without IPTG induction; lane 2: whole cell extract with IPTG induction (1 mM); lane 3: insoluble fraction after sonication lysis of the cells; lane 4: soluble fractions after sonication; lane 5: purified Tk-SP protein (4 μg). (B) Activity staining of gel. The gel contained 0.1% gelatin and SDS. After electrophoresis, the gel was washed with 2.5% (v/v) Triton X-100, incubated in 50 mM Tris–HCl (pH 9.0) at 80°C for 2 h and stained with CBB. Lanes 1, 2, and 3 are the same samples as those loaded onto lanes 1, 2, and 4 in panel (A), respectively; lane 4 is the purified Tk-SP protein (1 μg). The arrows indicate the position of the 65, 55, and 44 kDa proteins from top to bottom. *CBB*, Coomassie brilliant blue.

5.4.2 Crystal structure of Pro-S359A*

We successfully determined the crystal structure of Pro-S359A* at 2.0 Å resolution. The structure consists of the propeptide domain (proN, Lys4–Ala113), the subtilisin domain (Val114–Thr421), and the β-jelly roll domain (Ala422–Pro522), as shown in Fig. 5.14. The structure of Pro-S359A* is similar to that of Pro-S324A (unautoprocessed form of Pro-Tk-subtilisin), but Pro-S359A* has an additional helix (α1p) and a long extension loop, which extends to the β-jelly roll domain across the subtilisin domain, at the *N*-terminus. The subtilisin domain lacks four α-helices (α1m–α4m of Pro-S324A) and Ca^{2+} ions. The root mean square deviation values between Pro-S359A* and Pro-S324A is 1.5 Å for both the propeptide and subtilisin domains.

The β-jelly roll domain is not observed in the Pro-S324A structure. It is composed of nine β-strands and contains two Ca^{2+} ions. The Ca1 site of Pro-S359A* is located in the β-jelly roll domain, while the Ca2 site is located at the interface between the subtilisin and β-jelly roll domains.

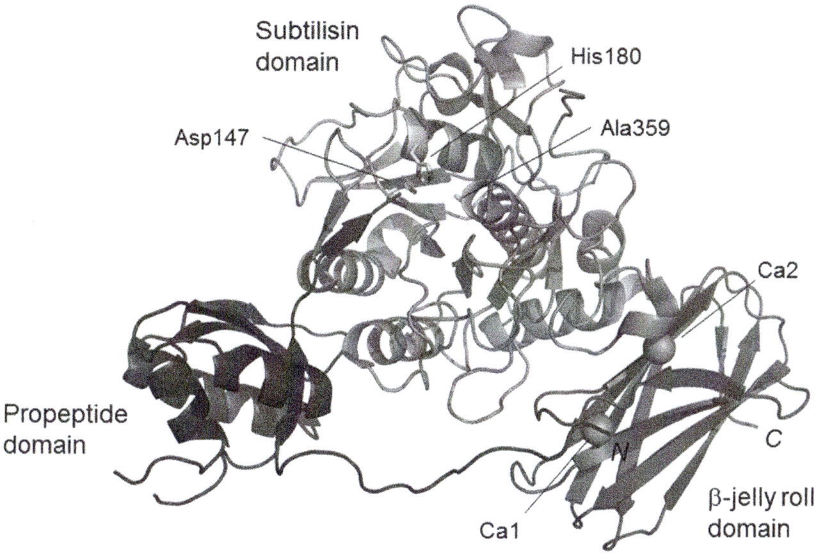

FIGURE 5.14 Crystal structure of Pro-S359A*. The propeptide, subtilisin and β-jelly roll domain are colored black, gray, and dark gray, respectively. The two active-site residues and Ala359, which was substituted for the catalytic serine residue, are indicated with stick models. Two Ca^{2+} ions (Ca1 and Ca2) are shown as gray spheres. N and C represent the *N*- and *C*-terminus, respectively.

These Ca^{2+} binding sites are relatively conserved in the β-jelly roll-like domains of subtilisin-like alkaline serine protease Kp-43 from *Bacillus* sp. (Nonaka et al., 2004), kexin-like proteases (Henrich et al., 2003; Holyoak et al., 2003; Kobayashi et al., 2009), and tomato subtilase 3 (Ottmann et al., 2009). However, the amino acid sequence similarities of the β-jelly roll domains between Pro-S359A* and these proteases are poor, and the site corresponding to the Ca2 site of Pro-S359A* is not located at the interface between the subtilisin and β-jelly roll-like domains of these proteases owing to the different arrangement of the β-jelly roll-like domain relative to the subtilisin domain.

We constructed an S359A-SP* derivative lacking the β-jelly roll domain (ΔJ-S359A-SP*) to examine whether the β-jelly roll domain is important for the stability of Tk-SP. The far- and near-UV CD spectra of ΔJ-S359A-SP* are similar to those of S359A-SP*. Thermal denaturation curves of these proteins were measured by monitoring the change in CD values at 222 nm as the temperature increases, in the presence of either 10 mM $CaCl_2$ or EDTA. The midpoints of the transition of these thermal denaturation curves, their T_m values, were 88.3°C ± 0.86°C, 58.8°C ± 0.93°C, 58.9°C ± 1.3°C, and 58.4°C ± 0.93°C for S359A-SP^{*Ca}, S359A-SP^{*EDTA}, ΔJ-S359A-SP^{*Ca}, and ΔJ-S359A-SP^{*EDTA}, respectively (Foophow et al., 2010a). The T_m value of

ΔJ-S359A-SP^{*Ca}, which was measured in the presence of 10 mM $CaCl_2$, is lower than that of S359A-SP^{*Ca} by 29.4°C, whereas those of S359A-SP^{*EDTA} and ΔJ-S359A-SP^{*EDTA} are similar, suggesting that the β-jelly roll domain contributes to the stabilization of Tk-SP by binding with Ca^{2+} ions.

5.4.3 Role of proN

The *N*-terminal propeptide domain (proN), similar to the propeptides of other subtilases, binds to the subtilisin domain in a substrate/product-like manner (Fig. 5.14). This suggests that proN inhibits the activity of Tk-SP as a competitive inhibitor. In fact, proN inhibits the activity of Tk-SP* when added *in trans* although the progress curve for the inhibition does not show a clear hyperbolic pattern (Yamanouchi, unpublished data). ProN is not required for the folding of these proteins, because Tk-SP and Tk-SP* exhibit activity in gel assays. However, the possibility that proN accelerates the folding rate of Tk-SP and Tkpro cannot be excluded. The critical interactions of the Glu201 residue for propeptide-catalyzed folding observed in Tk-subtilisin are conserved in the crystal structure of Pro-S359A*, for the most part, indicating that proN may function as an intramolecular chaperone. Further study will be required to understand the chaperone function of proN thoroughly.

5.4.4 Role of the C-domain

Attempts to obtain S359A-SP and Pro-S359A crystals containing the C-domain have remained elusive, and therefore the structure of the C-domain is unknown. When Pro-Tk-SP is overproduced in *E. coli* and purified, the C-domain is cleaved by Tk-SP during the purification procedures. We constructed the active-site mutant of Tk-SP, S359C-SP, to examine the autoprocessing of the C-domain in more detail. The activity of S359C-SP was, similar to S324C-subtilisin, substantially reduced. When S359C-SP was incubated at 80°C in the absence of Ca^{2+}, the C-domain was autoprocessed, whereas the autoprocessing reaction was not observed in the presence of 10 mM $CaCl_2$ (Sinsereekul et al., 2011). Similarly, the isolated C-domain, which was produced in His-tagged form using an *E. coli* expression system, was susceptible to proteolytic degradation by Tk-SP in the absence of Ca^{2+}, but was resistant in the presence of Ca^{2+}. Therefore the Ca^{2+}-bound form is most likely stable and resistant to autoprocessing owing to a conformational change of the C-domain induced by Ca^{2+} binding. In fact, the far-UV CD spectra of the His-tagged C-domain indicated increased secondary structure in the presence of Ca^{2+}. Tk-SP's source organism, *T. kodakarensis*, was isolated from sediments and seawater samples from a solfatara at a wharf of Kodakara Island (Kagoshima, Japan) (Morikawa et al., 1994). Therefore Pro-Tk-SP must mature in its native growth environment where Ca^{2+} ions are enriched (approximately 10 mM in seawater). We propose that Tk-SP is the mature form, rather than Tk-SP*, because the C-domain is unlikely autoprocessed under natural conditions.

Thermal denaturation curves of S359C-SP and S359C-SP* show that S359C-SP is more stable than S359C-SP* by 7.5°C in the presence of Ca^{2+}, and 25.9°C in the absence of Ca^{2+}. These results suggest that the C-domain contributes to the stabilization of Tk-SP by Ca^{2+} binding, although why the C-domain contributes more to the stabilization of Tk-SP in the absence of Ca^{2+} than in the presence of Ca^{2+} remains to be elucidated.

5.4.5 PrP^{Sc} degradation by Tk-SP

Tk-SP is a highly thermostable enzyme with a half-life of 100 min at 100°C and exhibits its highest activity at 100°C. It is also resistant to treatment with chemical denaturants, detergents, and chelating agents. Therefore Tk-SP is, like Tk-subtilisin, a promising candidate as a novel detergent enzyme. We tested whether Tk-SP can degrade PrP^{Sc} in scrapie-infected mouse brain homogenates using a combination of chemical treatments (Hirata et al., 2013). Western blot analysis revealed that PrP^{Sc} is completely degraded by Tk-SP in both the absence and presence of 1% SDS (Fig. 5.15). These results suggest that Tk-SP has potential application as a detergent additive for decreasing the infectivity of PrP^{Sc}. Further quantitative assessment of both Tk-SP and Tk-subtilisin for decontaminating PrP^{Sc} will be required because the minimum amount of protease for complete PrP^{Sc} degradation in various individual conditions has not been determined.

5.5 Concluding remarks

Our studies have revealed the unique maturation and stabilization mechanisms of two hyperthermophilic proteases. Tk-subtilisin has a Ca^{2+}-dependent maturation mechanism characterized by its unique Ca^{2+} binding loop, which acts as an intramolecular chaperone, and also by its Ca^{2+}-dependent stabilization mechanism. Tk-SP requires neither a propeptide nor Ca^{2+} ions for its maturation and stabilization mechanisms and is characterized by a β-jelly roll domain bound to Ca^{2+} ions. These mechanisms have evolved as successful strategies in hyperthermophilic proteases to adapt to high-temperature environments.

Tk-subtilisin and Tk-SP both exhibit superior stability against heat, detergents, and denaturants. Thus they are potentially applicable to industrial and medical technologies for degrading persistent proteins under harsh conditions where most other proteins would be denatured. In fact, both proteases effectively degrade infectious prion proteins (PrP^{Sc}) from human and mouse brain homogenates in combination with SDS. Our results indicate the great potential of these proteases as versatile detergent enzymes, not only for household

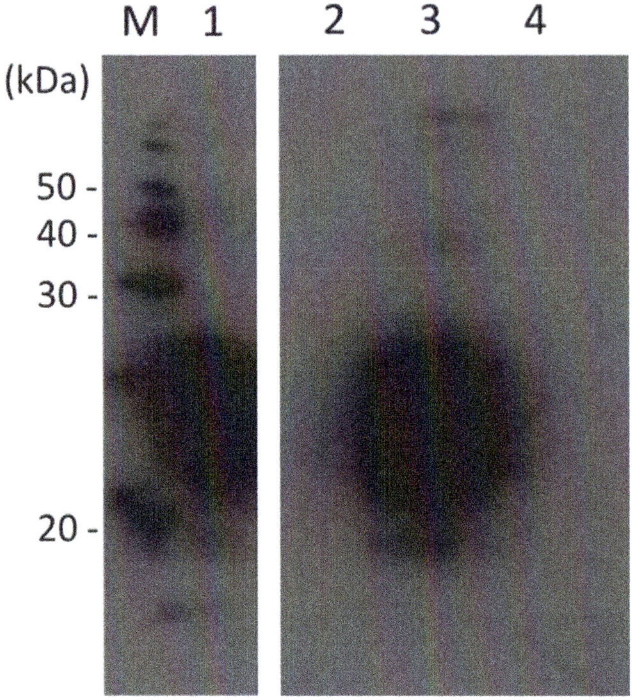

FIGURE 5.15 Western blot analysis of PrP^{Sc} Tk-SP digests. 1% mouse brain homogenate was subjected to digestion at 100°C for 1 h with buffer only (lane 1), 0.02 mg/mL (0.4 mM) Tk-SP (lane 2), 1% SDS (lane 3), or 0.02 mg/mL (0.4 mM) Tk-SP plus 1% SDS (lane 4).

use but also for the decontamination of infectious materials on medical instruments.

Acknowledgments

We thank Dr. Tita Foophow, Dr. Marian Pulido, Mr. Yuki Takeuchi, Mr. Nitat Sinsereekful, Ms. Mai Yamanouchi, Mr. Kenji Saito, Mr. Kota Yuzaki, Dr. Clement Angkawidjaja, and Dr. You Dong-Ju for technical assistance and helpful discussion. We also thank Dr. Takuya Yoshizawa, Dr. Hiroyoshi Matsumura, Dr. Mitsuru Haruki, Dr. Masaaki Morikawa, and Dr. Tadayuki Imanaka for helpful discussions and encouragement.

Abbreviations

TCA	trichloroacetic acid
CD	circular dichroism
SDS	sodium dodecyl sulfate
PAGE	polyacrylamide gel electrophoresis
GdnHCl	guanidine hydrochloride
EDTA	ethylenediaminetetraacetic acid

Suc-AAPF-*p*NA	*N*-succinyl-Ala-Ala-Pro-Phe-*p*-nitroanilide
CBB	Coomassie brilliant blue

References

Adams, M.W., Kelly, R.M., 1998. Finding and using hyperthermophilic enzymes. Trends Biotechnol. 16, 329–332.

Almog, O., González, A., Godin, N., de Leeuw, M., Mekel, M.J., Klein, D., et al., 2009. The crystal structures of the psychrophilic subtilisin S41 and the mesophilic subtilisin Sph reveal the same calcium-loaded state. Proteins 74, 489–496.

Almog, O., González, A., Klein, D., Greenblatt, H.M., Braun, S., Shoham, G., 2003. The 0.93 Å crystal structure of sphericase: a calcium-loaded serine protease from *Bacillus sphaericus*. J. Mol. Biol. 332, 1071–1082.

Arnórsdóttir, J., Kristjánsson, M.M., Ficner, R., 2005. Crystal structure of a subtilisin-like serine proteinase from a psychrotrophic *Vibrio* species reveals structural aspects of cold adaptation. FEBS J. 272, 832–845.

Arnórsdottir, J., Smáradóttir, R.B., Magnússon, O.T., Thorbjarnardóttir, S.H., Eggertsson, G., Kristjánsson, M.M., 2002. Characterization of a cloned subtilisin-like serine proteinase from a psychrotrophic *Vibrio* species. Eur. J. Biochem. 269, 5536–5546.

Atomi, H., Fukui, T., Kanai, T., Morikawa, M., Imanaka, T., 2004. Description of *Thermococcus kodakaraensis* sp. nov., a well studied hyperthermophilic archaeon previously reported as *Pyrococcus* sp. KOD1. Archaea 1, 263–267.

Awan, Z., Baass, A., Genest, J., 2014. Proprotein convertase subtilisin/kexin type 9 (PCSK9): lessons learned from patients with hypercholesterolemia. Clin. Chem. 60, 1380–1389.

Baier, K., Nicklisch, S., Maldener, I., Lockau, W., 1996. Evidence for propeptide-assisted folding of the calcium-dependent protease of the cyanobacterium *Anabaena*. Eur. J. Biochem. 241, 750–755.

Basak, A., Lazure, C., 2003. Synthetic peptides derived from the prosegments of proprotein convertase 1/3 and furin are potent inhibitors of both enzymes. Biochem. J. 373, 231–239.

Betzel, C., Pal, G.P., Saenger, W., 1988. Three-dimensional structure of proteinase K at 0.15-nm resolution. Eur. J. Biochem. 178, 155–171.

Bode, W., Papamokos, E., Musil, D., 1987. The high-resolution X-ray crystal structure of the complex formed between subtilisin Carlsberg and eglin c, an elastase inhibitor from the leech *Hirudo medicinalis*. Structural analysis, subtilisin structure and interface geometry. Eur. J. Biochem. 166, 673–692.

Bott, R., Ultsch, M., Kossiakoff, A., Graycar, T., Katz, B., Power, S., 1988. The three-dimensional structure of *Bacillus amyloliquefaciens* subtilisin at 1.8 Å and an analysis of the structural consequences of peroxide inactivation. J. Biol. Chem. 263, 7895–7906.

Bryan, P.N., 2002. Prodomain and protein folding catalysis. Chem. Rev. 102, 4805–4816.

Bryan, P.N., Alexander, P., Strausberg, S., Schwarz, F., Lan, W., Gilliland, G., et al., 1992. Energetics of folding subtilisin BPN'. Biochemistry 31, 4937–4945.

Bryan, P.N., Wang, L., Hoskins, J., Ruvinov, S., Strausberg, S., Alexander, P., et al., 1995. Catalysis of a protein folding reaction: mechanistic implications of the 2.0 Å structure of the subtilisin-prodomain complex. Biochemistry 34, 10310–10318.

Catara, G., Ruggiero, G., La Cara, F., Digilio, F.A., Capasso, A., Rossi, M., 2003. A novel extracellular subtilisin-like protease from the hyperthermophile *Aeropyrum pernix* K1: biochemical properties, cloning, and expression. Extremophiles 7, 391–399.

Chen, Y.J., Inouye, M., 2008. The intramolecular chaperone-mediated protein folding. Curr. Opin. Struct. Biol. 18, 765–770.

Choi, I.G., Bang, W.G., Kim, S.H., Yu, Y.G., 1999. Extremely thermostable serine-type protease from *Aquifex pyrophilus*. Molecular cloning, expression, and characterization. J. Biol. Chem. 274, 881–888.

Davail, S., Feller, G., Narinx, E., Gerday, C., 1994. Cold adaptation of proteins. Purification, characterization, and sequence of the heat-labile subtilisin from the antarctic psychrophile *Bacillus* TA41. J. Biol. Chem. 269, 17448–17453.

de Miguel Bouzas, T., Barros-Velázquez, J., Villa, T.G., 2006. Industrial applications of hyperthermophilic enzymes: a review. Protein Pept. Lett. 13, 645–651.

Dong, D., Ihara, T., Motoshima, H., Watanabe, K., 2005. Crystallization and preliminary X-ray crystallographic studies of a psychrophilic subtilisin-like protease Apa1 from Antarctic *Pseudoalteromonas* sp. strain AS-11. Acta Crystallogr. Sect. F. Struct. Biol. Cryst. Commun. 61, 308–311.

Eder, J., Fersht, A.R., 1995. Pro-sequences assisted protein folding. Mol. Microbiol. 16, 609–614.

Eder, J., Rheinnecker, M., Fersht, A.R., 1993. Folding of subtilisin BPN′: characterization of a folding intermediate. Biochemistry 32, 18–26.

Fisher, K.E., Ruan, B., Alexander, P.A., Wang, L., Bryan, P.N., 2007. Mechanism of the kinetically-controlled folding reaction of subtilisin. Biochemistry 46, 640–651.

Foophow, T., Tanaka, S., Angkawidjaja, C., Koga, Y., Takano, K., Kanaya, S., 2010a. Crystal structure of a subtilisin homologue, Tk-SP, from *Thermococcus kodakaraensis*: requirement of a C-terminal beta-jelly roll domain for hyperstability. J. Mol. Biol. 400, 865–877.

Foophow, T., Tanaka, S., Koga, Y., Takano, K., Kanaya, S., 2010b. Subtilisin-like serine protease from hyperthermophilic archaeon *Thermococcus kodakaraensis* with N- and C-terminal propeptides. Protein Eng. Des. Sel. 23, 347–355.

Fukui, T., Atomi, H., Kanai, T., Matsumi, R., Fujiwara, S., Imanaka, T., 2005. Complete genome sequence of the hyperthermophilic archaeon *Thermococcus kodakaraensis* KOD1 and comparison with *Pyrococcus* genomes. Genome Res. 15, 352–363.

Gallagher, T., Bryan, P., Gilliland, G.L., 1993. Calcium-independent subtilisin by design. Proteins 16, 205–213.

Gallagher, T., Gilliland, G., Wang, L., Bryan, P., 1995. The prosegment–subtilisin BPN′ complex: crystal structure of a specific 'foldase'. Structure 3, 907–914.

Gödde, C., Sahm, K., Brouns, S.J., Kluskens, L.D., van der Oost, J., de Vos, W.M., et al., 2005. Cloning and expression of islandisin, a new thermostable subtilisin from *Fervidobacterium islandicum*, in *Escherichia coli*. Appl. Environ. Microbiol. 71, 3951–3958.

Grande, K.K., Gustin, J.K., Kessler, E., Ohman, D.E., 2007. Identification of critical residues in the propeptide of LasA protease of *Pseudomonas aeruginosa* involved in the formation of a stable mature protease. J. Bacteriol. 189, 3960–3968.

Gros, P., Betzel, C., Dauter, Z., Wilson, K.S., Hol, W.G., 1988. Molecular dynamics refinement of a thermitase-eglin-c complex at 1.98 Å resolution and comparison of two crystal forms that differ in calcium content. J. Mol. Biol. 210, 347–367.

Helland, R., Larsen, A.N., Smalås, A.O., Willassen, N.P., 2006. The 1.8 Å crystal structure of a proteinase K-like enzyme from a psychrotroph *Serratia* species. FEBS J. 273, 61–71.

Henrich, S., Cameron, A., Bourenkov, G.P., Kiefersauer, R., Huber, R., Lindberg, I., et al., 2003. The crystal structure of the proprotein processing proteinase furin explains its stringent specificity. Nat. Struct. Biol. 10, 520–526.

Henrich, S., Lindberg, I., Bode, W., Than, M.E., 2005. Proprotein convertase models based on the crystal structures of furin and kexin: explanation of their specificity. J. Mol. Biol. 345, 211–227.

Hirata, A., Hori, Y., Koga, Y., Okada, J., Sakudo, A., Ikuta, K., et al., 2013. Enzymatic activity of a subtilisin homolog, Tk-SP, from *Thermococcus kodakarensis* in detergents and its ability to degrade the abnormal prion protein. BMC Biotechnol. 13, 19.

Holyoak, T., Wilson, M.A., Fenn, T.D., Kettner, C.A., Petsko, G.A., Fuller, R.S., et al., 2003. 2.4 Å resolution crystal structure of the prototypical hormone-processing protease Kex2 in complex with an Ala-Lys-Arg boronic acid inhibitor. Biochemistry 42, 6709–6718.

Hu, Z., Haghjoo, K., Jordan, F., 1996. Further evidence for the structure of the subtilisin propeptide and for its interactions with mature subtilisin. J. Biol. Chem. 271, 3375–3384.

Huang, H.W., Chen, W.C., Wu, C.Y., Yu, H.C., Lin, W.Y., Chen, S.T., et al., 1997. Kinetic studies of the inhibitory effects of propeptides subtilisin BPN' and Carlsberg to bacterial serine proteases. Protein Eng. 10, 1227–1233.

Imanaka, T., 2011. Molecular bases of thermophily in hyperthermophiles. Proc. Jpn. Acad. Ser. B Phys. Biol. Sci. 87, 587–602.

Jacobs, M., Eliasson, M., Uhlen, M., Flock, J.I., 1985. Cloning, sequencing and expression of subtilisin Carlsberg from *Bacillus licheniformis*. Nucleic Acids Res. 13, 8913–8926.

Jain, S.C., Shinde, U., Li, Y., Inouye, M., Berman, H.M., 1998. The crystal structure of an autoprocessed Ser221Cys-subtilisin E-propeptide complex at 2.0 Å resolution. J. Mol. Biol. 284, 137–144.

Jang, H.J., Lee, C.H., Lee, W., Kim, Y.S., 2002. Two flexible loops in subtilisin-like thermophilic protease, thermicin, from *Thermoanaerobacter yonseiensis*. J. Biochem. Mol. Biol. 35, 498–507.

Jaswal, S.S., Sohl, J.L., Davis, J.H., Agard, D.A., 2002. Energetic landscape of α-lytic protease optimizes longevity through kinetic stability. Nature 415, 343–346.

Jia, Y., Liu, H., Bao, W., Weng, M., Chen, W., Cai, Y., et al., 2010. Functional analysis of propeptide as an intramolecular chaperone for in vivo folding of subtilisin nattokinase. FEBS Lett. 584, 4789–4796.

Kannan, Y., Koga, Y., Inoue, Y., Haruki, M., Takagi, M., Imanaka, T., et al., 2001. Active subtilisin-like protease from a hyperthermophilic archaeon in a form with a putative prosequence. Appl. Environ. Microbiol. 67, 2445–2452.

Kim, S.G., Chen, Y.J., Falzon, L., Baum, J., Inouye, M., 2020. Mimicking cotranslational folding of prosubtilisin E in vitro. J. Biochem. 167, 473–482.

Kim, J.S., Kluskens, L.D., de Vos, W.M., Huber, R., van der Oost, J., 2004. Crystal structure of fervidolysin from *Fervidobacterium pennivorans*, a keratinolytic enzyme related to subtilisin. J. Mol. Biol. 335, 787–797.

Kluskens, L.D., Voorhorst, W.G., Siezen, R.J., Schwerdtfeger, R.M., Antranikian, G., van der Oost, J., et al., 2002. Molecular characterization of fervidolysin, a subtilisin-like serine protease from the thermophilic bacterium *Fervidobacterium pennivorans*. Extremophiles 6, 185–194.

Kobayashi, H., Utsunomiya, H., Yamanaka, H., Sei, Y., Katunuma, N., Okamoto, K., et al., 2009. Structural basis for the kexin-like serine protease from *Aeromonas sobria* as sepsis-causing factor. J. Biol. Chem. 284, 27655–27663.

Koga, Y., Tanaka, S., Sakudo, A., Tobiume, M., Aranishi, M., Hirata, A., et al., 2014. Proteolysis of abnormal prion protein with a thermostable protease from *Thermococcus kodakarensis* KOD1. Appl. Microbiol. Biotechnol. 98, 2113–2120.

Kojima, S., Yanai, H., Miura, K., 2001. Accelerated refolding of subtilisin BPN′ by tertiary-structure-forming mutants of its propeptide. J. Biochem. (Tokyo) 130, 471–474.

Kulakova, L., Galkin, A., Kurihara, T., Yoshimura, T., Esaki, N., 1999. Cold-active serine alkaline protease from the psychrotrophic bacterium *Shewanella* strain Ac10: gene cloning and enzyme purification and characterization. Appl. Environ. Microbiol. 65, 611–617.

Kurata, A., Uchimura, K., Shimamura, S., Kobayashi, T., Horikoshi, K., 2007. Nucleotide and deduced amino acid sequences of a subtilisin-like serine protease from a deep-sea bacterium, *Alkalimonas collagenimarina* AC40(T). Appl. Microbiol. Biotechnol. 77, 311–319.

Kwon, S.T., Terada, I., Matsuzawa, H., Ohta, T., 1988. Nucleotide sequence of the gene for aqualysin I (a thermophilic alkaline serine protease) of *Thermus aquaticus* YT-1 and characteristics of the deduced primary structure of the enzyme. Eur. J. Biochem. 173, 491. 197.

Li, Y., Hu, Z., Jordan, F., Inouye, M., 1995. Functional analysis of the propeptide of subtilisin E as an intramolecular chaperone for protein folding. Refolding and inhibitory abilities of propeptide mutants. J. Biol. Chem. 270, 25127–25132.

Li, Y., Inouye, M., 1994. Autoprocessing of prothiolsubtilisin E in which active-site serine 221 is altered to cysteine. J. Biol. Chem. 269, 4169–4174.

Li, A.N., Li, D.C., 2009. Cloning, expression and characterization of the serine protease gene from *Chaetomium thermophilum*. J. Appl. Microbiol. 106, 369–380.

Li, A.N., Xie, C., Zhang, J., Zhang, J., Li, D.C., 2011. Cloning, expression, and characterization of serine protease from thermophilic fungus *Thermoascus aurantiacus* var. *levisporus*. J. Microbiol. 49, 121–129.

Marie-Claire, C., Ruffet, E., Beaumont, A., Roques, B.P., 1999. The prosequence of thermolysin acts as an intramolecular chaperone when expressed in trans with the mature sequence in *Escherichia coli*. J. Mol. Biol. 285, 1911–1915.

Marie-Claire, C., Yabuta, Y., Suefuji, K., Matsuzawa, H., Shinde, U.P., 2001. Folding pathway mediated by an intramolecular chaperone: the structural and functional characterization of the aqualysin I propeptide. J. Mol. Biol. 305, 151–165.

Matsuzawa, H., Tokugawa, K., Hamaoki, M., Mizoguchi, M., Taguchi, H., Terada, I., et al., 1988. Purification and characterization of aqualysin I (a thermophilic alkaline serine protease) produced by *Thermus aquaticus* YT-1. Eur. J. Biochem. 171, 441–447.

McPhalen, C.A., James, M.N., 1988. Structural comparison of two serine proteinase-protein inhibitor complexes: Eglin-c-subtilisin Carlsberg and CI-2-subtilisin Novo. Biochemistry 27, 6582–6598.

Morikawa, M., Izawa, Y., Rashid, N., Hoaki, T., Imanaka, T., 1994. Purification and characterization of a thermostable thiol protease from a newly isolated hyperthermophilic *Pyrococcus* sp. Appl. Environ. Microbiol. 60, 4559–4566.

Morita, Y., Hasan, Q., Sakaguchi, T., Murakami, Y., Yokoyama, K., Tamiya, E., 1998. Properties of a cold-active protease from psychrotrophic *Flavobacterium balustinum* P104. Appl. Microbiol. Biotechnol. 50, 669–675.

Narinx, E., Baise, E., Gerday, C., 1997. Subtilisin from psychrophilic antarctic bacteria: characterization and site-directed mutagenesis of residues possibly involved in the adaptation to cold. Protein Eng. 10, 1271–1279.

Nemoto, T.K., Ohara-Nemoto, Y., Ono, T., Kobayakawa, T., Shimoyama, Y., Kimura, S., et al., 2008. Characterization of the glutamyl endopeptidase from *Staphylococcus aureus* expressed in *Escherichia coli*. FEBS J. 275, 573–587.

Nonaka, T., Fujihashi, M., Kita, A., Saeki, K., Ito, S., Horikoshi, K., et al., 2004. The crystal structure of an oxidatively stable subtilisin-like alkaline serine protease, KP-43, with a C-terminal β-barrel domain. J. Biol. Chem. 279, 47344–47351.

Okada, J., Koga, Y., Takano, K., Kanaya, S., 2012. Slow unfolding pathway of the hyperthermophilic Tk-Rnase H2 examined by pulse proteolysis using the stable protease Tk-subtilisin. Biochemistry 51, 9178–9191.

Ottmann, C., Rose, R., Huttenlocher, F., Cedzich, A., Hauske, P., Kaiser, M., et al., 2009. Structural basis for Ca^{2+}-independence and activation by homodimerization of tomato subtilase 3. Proc. Natl. Acad. Sci. USA 106, 17223–17228.

Pantoliano, M.W., Whitlow, M., Wood, J.F., Dodd, S.W., Hardman, K.D., Rollence, M.L., et al., 1989. Large increases in general stability for subtilisin BPN' through incremental changes in the free energy of unfolding. Biochemistry 28, 7205–7213.

Park, C., Marqusee, S., 2005. Pulse proteolysis: A simple method for quantitative determination of protein stability and ligand binding. Nat. Methods 2, 207–212.

Prusiner, S.B., 1998. Prions. Proc. Natl. Acad. Sci. USA 95, 13363–13383.

Pulido, M.A., Koga, Y., Takano, K., Kanaya, S., 2007a. Directed evolution of Tk-subtilisin from a hyperthermophilic archaeon: identification of a single amino acid substitution responsible for low-temperature adaptation. Protein Eng. Des. Sel. 20, 143–153.

Pulido, M.A., Saito, K., Tanaka, S., Koga, Y., Morikawa, M., Takano, K., et al., 2006. Ca^{2+}-dependent maturation of subtilisin from a hyperthermophilic archaeon, *Thermococcus kodakaraensis*: the propeptide is a potent inhibitor of the mature domain but is not required for its folding. Appl. Environ. Microbiol. 72, 4154–4162.

Pulido, M.A., Tanaka, S., Sringiew, C., You, D.-J., Matsumura, H., Koga, Y., et al., 2007b. Requirement of left-handed glycine residue for high stability of the Tk-subtilisin propeptide as revealed by mutational and crystallographic analyses. J. Mol. Biol. 374, 1359–1373.

Rao, M.B., Tanksale, A.M., Ghatge, M.S., Deshpande, V.V., 1998. Molecular and biotechnological aspects of microbial proteases. Microbiol. Mol. Biol. Rev. 62, 597–635.

Rigden, D.J., Woodhead, D.D., Wong, P.W., Galperin, M.Y., 2011. New structural and functional contexts of the Dx[DN]xDG linear motif: insights into evolution of calcium-binding proteins. PLoS One 6, e21507.

Saeki, K., Ozaki, K., Kobayashi, T., Ito, S., 2007. Detergent alkaline proteases: enzymatic properties, genes, and crystal structures. J. Biosci. Bioeng. 103, 501–508.

Salimi, N.L., Ho, B., Agard, D.A., 2010. Unfolding simulations reveal the mechanism of extreme unfolding cooperativity in the kinetically stable α-lytic protease. PLoS Comput. Biol. 6, e1000689.

Schilling, K., Körner, A., Sehmisch, S., Kreusch, A., Kleint, R., Benedix, Y., et al., 2009. Selectivity of propeptide-enzyme interaction in cathepsin L-like cysteine proteases. Biol. Chem. 390, 167–174.

Seidah, N.G., Mayer, G., Zaid, A., Rousselet, E., Nassoury, N., Poirier, S., et al., 2008. The activation and physiological functions of the proprotein convertases. Int. J. Biochem. Cell Biol. 40, 1111–1125.

Shinde, U.P., Inouye, M., 2000. Intramolecular chaperones: Polypeptide extensions that modulate protein folding. Semin. Cell Dev. Biol. 11, 35–44.

Shinde, U.P., Thomas, G., 2011. Insights from bacterial subtilases into the mechanisms of intramolecular chaperone-mediated activation of furin. Methods Mol. Biol. 768, 59–106.

Siezen, R.J., Leunissen, J.A., 1997. Subtilases: the superfamily of subtilisin-like serine proteases. Protein Sci. 6, 501–523.

Siezen, R.J., Renckens, B., Boekhorst, J., 2007. Evolution of prokaryotic subtilases: genome-wide analysis reveals novel subfamilies with different catalytic residues. Proteins 67, 681–694.

Sinsereekul, N., Foophow, T., Yamanouchi, M., Koga, Y., Takano, K., Kanaya, S., 2011. An alternative mature form of subtilisin homologue, Tk-SP, from *Thermococcus kodakaraensis* identified in the presence of Ca^{2+}. FEBS J. 278, 1901–1911.

Smith, C.A., Toogood, H.S., Baker, H.M., Daniel, R.M., Baker, E.N., 1999. Calcium-mediated thermostability in the subtilisin superfamily: the crystal structure of *Bacillus* Ak.1 protease at 1.8 Å resolution. J. Mol. Biol. 294, 1027–1040.

Stahl, M.L., Ferrari, E., 1984. Replacement of the *Bacillus subtilis* subtilisin structural gene with an *in vitro*-derived deletion mutation. J. Bacteriol. 158, 411–418.

Stetter, K.O., 2013. A brief history of the discovery of hyperthermophilic life. Biochem. Soc. Trans. 41, 416–420.

Sung, J.H., Ahn, S.J., Kim, N.Y., Jeong, S.K., Kim, J.K., Chung, J.K., et al., 2010. Purification, molecular cloning, and biochemical characterization of subtilisin JB1 from a newly isolated *Bacillus subtilis* JB1. Appl. Biochem. Biotechnol. 162, 900–911.

Takagi, H., Morinaga, Y., Ikemura, H., Inouye, M., 1988. Mutant subtilisin E with enhanced protease activity obtained by site-directed mutagenesis. J. Biol. Chem. 263, 19592–19596.

Takemasa, R., Yokooji, Y., Yamatsu, A., Atomi, H., Imanaka, T., 2011. *Thermococcus kodakarensis* as a host for gene expression and protein secretion. Appl. Environ. Microbiol. 77, 2392–2398.

Takeuchi, Y., Tanaka, S., Matsumura, H., Koga, Y., Takano, K., Kanaya, S., 2009. Requirement of a unique Ca^{2+}-binding loop for folding of Tk-subtilisin from a hyperthermophilic archaeon. Biochemistry 48, 10637–10643.

Tanaka, S., Matsumura, H., Koga, Y., Takano, K., Kanaya, S., 2007a. Four new crystal structures of Tk-subtilisin in unautoprocessed, autoprocessed and mature forms: insight into structural changes during maturation. J. Mol. Biol. 372, 1055–1069.

Tanaka, S., Matsumura, H., Koga, Y., Takano, K., Kanaya, S., 2009. Identification of the interactions critical for propeptide-catalyzed folding of Tk-subtilisin. J. Mol. Biol. 394, 306–319.

Tanaka, S., Saito, K., Chon, H., Matsumura, H., Koga, Y., Takano, K., et al., 2006. Crystallization and preliminary X-ray diffraction study of an active-site mutant of pro-Tk-subtilisin from a hyperthermophilic archaeon. Acta Crystallogr. Sect. F. Struct. Biol. Cryst. Commun. 62, 902–905.

Tanaka, S., Saito, K., Chon, H., Matsumura, H., Koga, Y., Takano, K., et al., 2007b. Crystal structure of unautoprocessed precursor of subtilisin from a hyperthermophilic archaeon: evidence for Ca^{2+}-induced folding. J. Biol. Chem. 282, 8246–8255.

Tanaka, S., Takeuchi, Y., Matsumura, H., Koga, Y., Takano, K., Kanaya, S., 2008. Crystal structure of Tk-subtilisin folded without propeptide: requirement of propeptide for acceleration of folding. FEBS Lett. 582, 3875–3878.

Tang, B., Nirasawa, S., Kitaoka, M., Hayashi, K., 2002. The role of the N-terminal propeptide of the pro-aminopeptidase processing protease: refolding, processing, and enzyme inhibition. Biochem. Biophys. Res. Commun. 296, 78–84.

Teplyakov, A.V., Kuranova, I.P., Harutyunyan, E.H., Vainshtein, B.K., Frömmel, C., Höhne, W. E., et al., 1990. Crystal structure of thermitase at 1.4 Å resolution. J. Mol. Biol. 214, 261–279.

Uehara, R., Angkawidjaja, C., Koga, Y., Kanaya, S., 2013a. Formation of the high-affinity calcium binding site in pro-subtilisin E with the insertion sequence IS1 of pro-Tk-subtilisin. Biochemistry 52, 9080–9088.

Uehara, R., Dan, N., Amesaka, H., Yoshizawa, T., Koga, Y., Kanaya, S., et al., 2021. Insertion loop-mediated folding propagation governs efficient maturation of hyperthermophilic Tk-subtilisin at high temperatures. FEBS Lett. 595, 452–461.

Uehara, R., Takeuchi, Y., Tanaka, S., Takano, K., Koga, Y., Kanaya, S., 2012a. Requirement of Ca^{2+} ions for the hyperthermostability of Tk-subtilisin from *Thermococcus kodakarensis*. Biochemistry 51, 5369–5378.

Uehara, R., Tanaka, S., Takano, K., Koga, Y., Kanaya, S., 2012b. Requirement of insertion sequence IS1 for thermal adaptation of Pro-Tk-subtilisin from hyperthermophilic archaeon. Extremophiles 16, 841–851.

Uehara, R., Ueda, Y., You, D.-J., Koga, Y., Kanaya, S., 2013b. Accelerated maturation of Tk-subtilisin by a Leu→Pro mutation at the C-terminus of the propeptide, which reduces the binding of the propeptide to Tk-subtilisin. FEBS J. 280, 994–1006.

Vévodová, J., Gamble, M., Künze, G., Ariza, A., Dodson, E., Jones, D.D., et al., 2010. Crystal structure of an intracellular subtilisin reveals novel structural features unique to this subtilisin family. Structure 18, 744–755.

Voordouw, G., Milo, C., Roche, R.S., 1976. Role of bound calcium ions in thermostable, proteolytic enzymes. Separation of intrinsic and calcium ion contributions to the kinetic thermal stability. Biochemistry 15, 3716–3724.

Wang, L., Ruan, B., Ruvinov, S., Bryan, P.N., 1998. Engineering the independent folding of the subtilisin BPN' pro-domain: correlation of pro-domain stability with the rate of subtilisin folding. Biochemistry 37, 3165–3171.

Wang, L., Ruvinov, S., Strausberg, S., Gallagher, D.T., Gilliland, G., Bryan, P.N., 1995. Prodomain mutations at the subtilisin interface: correlation of binding energy and the rate of catalyzed folding. Biochemistry 34, 15415–15420.

Wells, J.A., Ferrari, E., Henner, D.J., Estell, D.A., Chen, E.Y., 1983. Cloning, sequencing, and secretion of *Bacillus amyloliquefaciens* subtilisin in *Bacillus subtilis*. Nucleic Acids Res. 11, 7911–7925.

Winther, J.R., Sørensen, P., 1991. Propeptide of carboxypeptidase Y provides a chaperone-like function as well as inhibition of the enzymatic activity. Proc. Natl. Acad. Sci. USA 88, 9330–9334.

Wissmann, C., 2004. The state of the prion. Nat. Rev. Microbiol. 2, 861–871.

Wu, J., Bian, Y., Tang, B., Chen, X., Shen, P., Peng, Z., 2004. Cloning and analysis of WF146 protease, a novel thermophilic subtilisin-like protease with four inserted surface loops. FEMS Microbiol. Lett. 230, 251–258.

Yabuta, Y., Subbian, E., Takagi, H., Shinde, U.P., Inouye, M., 2002. Folding pathway mediated by an intramolecular chaperone: dissecting conformational changes coincident with autoprocessing and the role of Ca^{2+} in subtilisin maturation. J. Biochem. 131, 31–37.

Yabuta, Y., Takagi, H., Inouye, M., Shinde, U.P., 2001. Folding pathway mediated by an intramolecular chaperone: Propeptide release modulates activation precision of pro-subtilisin. J. Biol. Chem. 276, 44427–44434.

Yuzaki, K., Sanda, Y., You, D.-J., Uehara, R., Koga, Y., Kanaya, S., 2013. Increase in activation rate of Pro-Tk-subtilisin by a single nonpolar-to-polar amino acid substitution at the hydrophobic core of the propeptide domain. Protein Sci. 22, 1711–1721.

Chapter 6

Enzymes from basidiomycetes—peculiar and efficient tools for biotechnology

Thaís Marques Uber[1], Emanueli Backes[1], Vinícius Mateus Salvatore Saute[1], Bruna Polacchine da Silva[2], Rubia Carvalho Gomes Corrêa[3], Camila Gabriel Kato[4], Flávio Augusto Vicente Seixas[1], Adelar Bracht[1] and Rosane Marina Peralta[1]

[1]*State University of Maringá, Maringá, Brazil,* [2]*UniSalesiano, Catholic Salesian Auxilium University Centre, Araçatuba, Brazil,* [3]*UniCesumar, Cesumar University Centre, Maringá, Brazil,* [4]*Federal University of Mato Grosso do Sul, Campo Grande, Brazil*

6.1 Introduction

Fungi are the only group of organisms that occupy a kingdom all to themselves, the kingdom fungi. At least 100,000 different species of fungi have been identified, and recent estimates based on high-throughput sequencing methods suggest that as many as 5.1 million fungal species exist (Blackwell, 2011). Recent studies, however, have considered this number an overestimate of the fungal richness by 1.5- to 2.5-fold (Hawksworth and Lücking, 2017; Tedersoo et al., 2014). The Ascomycotina, called "sac fungi" (ascomycetes), with over 60,000 described species, and the Basidiomycotina, called "club fungi" (basidiomycetes), with over 30,000 described species, are the largest groups of known fungi. Basidiomycetes are called "club fungi" because their spores are attached to a club-shaped structure named *basidium* (pl. *basidia*). Basidiomycetous fungi include edible and medicinal mushrooms, pathogens for plants and animals, symbionts and endophytes in lichens, plant root mycorrhizas, leaves and needles, and saprotrophytes.

6.2 Brown- and white-rot fungi

Basidiomycetes and ascomycetes play a crucial role in the balance of ecosystems. They are the major decomposers of lignocellulosic material in several ecosystems and play an essential role in cycling carbon and other nutrients.

Biotechnology of Microbial Enzymes. DOI: https://doi.org/10.1016/B978-0-443-19059-9.00023-2
© 2023 Elsevier Inc. All rights reserved.

A wide variety of wood types and trees in different stages of decomposition can be colonized by these fungi. The wood-decaying fungi have specific preferences for certain host species and stages of wood decay. Besides, those wood-decaying fungi that are plant pathogens are adapted to survive against the plant defenses with particular abilities to defeat antifungal substances such as phenolics, tannins, and alkaloids (Maciel et al., 2012).

Wood is a carbon-rich substrate with low nitrogen and other essential nutrients. Lignocellulose, the primary wood component, is a complex mixture of polymers that includes mainly cellulose, hemicellulose, and lignin (Fig. 6.1).

Lignocellulose is the major renewable organic matter in nature. It has been estimated that there is an annual worldwide production of 50×10^9 tons of lignocellulose, accounting for about half of the global biomass yield (Lynch, 1987; Meng et al., 2021). However, lignocellulose is recalcitrant to degradation. The recalcitrance of lignocellulose, which confers protection against microbial attack and enzymatic action, derives mainly from lignin, an irregular and nonrepeating polymer. Its biosynthesis results from oxidative polymerization of several phenyl-propanoid precursors, such as coniferyl alcohol, sinapyl alcohol, and *p*-coumaryl alcohol. Polymerization occurs randomly by various carbon–carbon and ether bonds resulting in an irregular structure that is impossible to hydrolyze under natural conditions. The proportions and location of these polymers (cellulose, hemicelluloses, and lignin) vary among plant groups or species. To use woody materials as substrates and get access to its limited nutrients, the species of wood-decaying fungi have developed distinct mechanisms of growth as well as metabolic and enzymatic abilities dependent on environmental conditions such as temperature, humidity, and availability of food resources. Based on the ability to degrade lignin along or not with cellulose and hemicellulose, wood decay has traditionally been divided into white rot and brown rot, mainly performed by basidiomycetes, and soft rot, mainly brought about by ascomycetes.

Many of the brown-rot fungi produce bracket-shaped fruiting bodies on the trunks of dead trees. Still, the characteristic feature of these fungi is that the decaying wood is brown and shows brick-like cracking—a result of the uneven pattern of decay, causing the wood to split along the lines of weakness. The term "brown rot" refers to the characteristic color of the decayed wood, because most of the cellulose and hemicelluloses are degraded, leaving the lignin more or less intact as a brown, chemically modified framework (Fig. 6.2A).

The term white rot is related to the bleached (white) appearance frequently observed on wood attacked by these fungi (Fig. 6.2B). A common characteristic among white-rot fungi is the ability to degrade all main components of the plant cell wall: cellulose, hemicelluloses, and lignin. The decomposition of lignocelluloses is achieved by a series of enzymatic

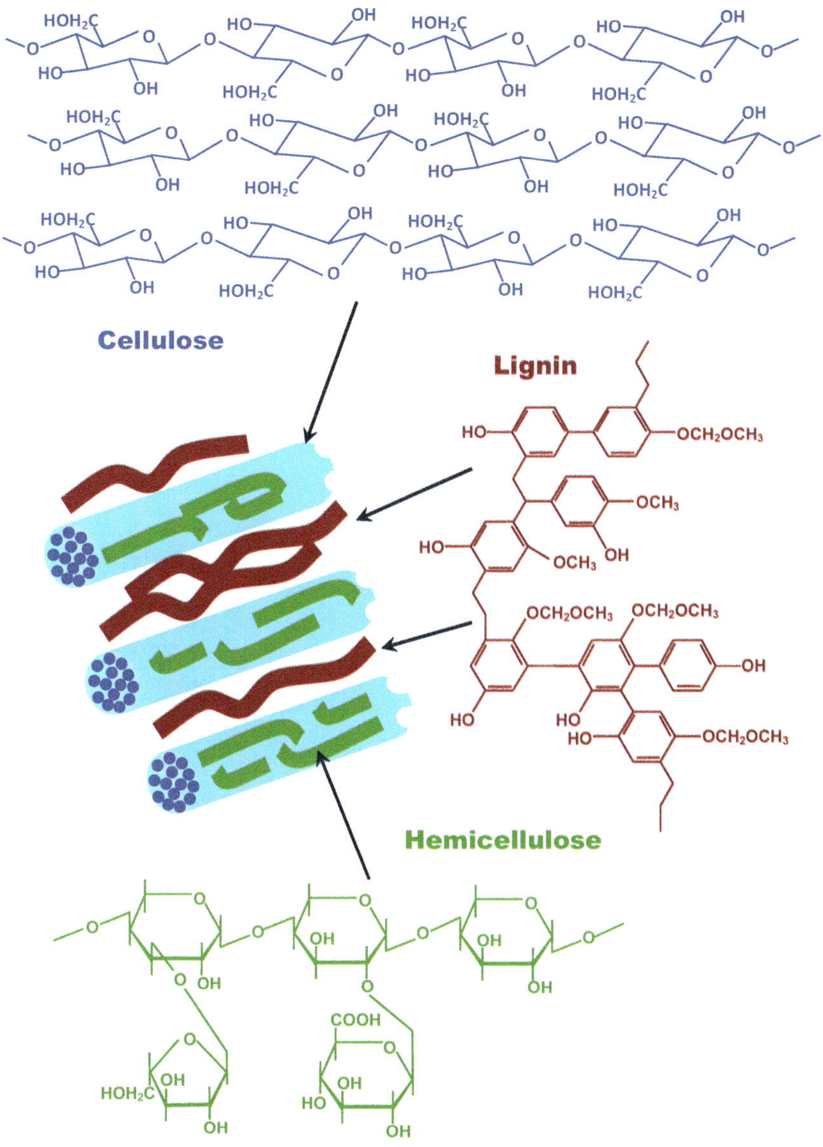

FIGURE 6.1 Main components of lignocellulosic material.

(hydrolases and oxidoreductases) and nonenzymatic mechanisms, and the production of ligninolytic enzymes is typical of this group of fungi.

Less than 10% of the wood-rotting basidiomycete species are brown rots, but they are prevalent in nature. Typical species of brown-rot fungi are *Gloeophyllum trabeum*, *Serpula lacrymans*, *Coniophora puteana*, known as

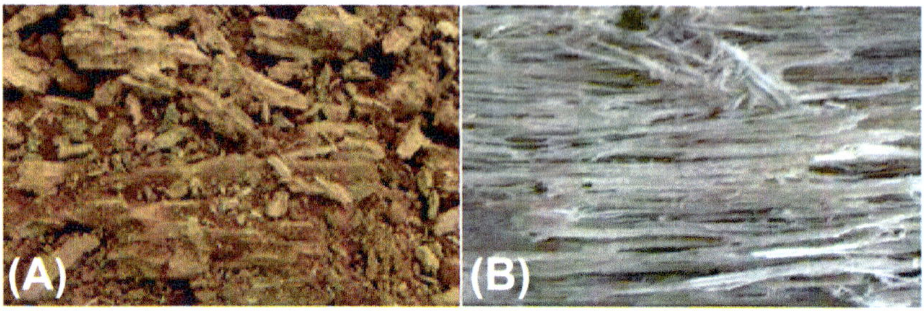

FIGURE 6.2 Wood decay caused by brown-rot fungi (A) and white-rot fungi (B).

FIGURE 6.3 Brown-rot basidiomycetes. (A) *Gloeophyllum trabeum*; (B) *Serpula lacrymans*; (C) *Coniophora puteana*; (D) *Schizophyllum commune;* (E) *Postia placenta;* (F) *Fomes fomentarius.*

the "cellar fungus," *Schizophyllum commune*, *Postia placenta*, and *Fomes fomentarius* (Fig. 6.3).

The brown-rot fungi can degrade cellulose and hemicelluloses, but they can only modify lignin, which remains a polymeric residue in the decaying wood (Arantes and Goodell, 2014). The main lignin modification carried out by brown-rot fungi is a demethylation reaction. The majority of the brown-rot basidiomycetes have long been thought to lack the processive cellulases, especially exocellulases. This makes the generation of hydroxyl radicals through Fenton-based reactions ($Fe^{2+} + H_2O_2 \rightarrow Fe^{3+} + {}^{\bullet}OH + OH^{-}$), which depolymerize polysaccharides, highly important. However, brown-rot fungi such as *Fomitopsis palustris* (Yoon et al., 2007) and *G. trabeum* (Cohen et al., 2005) produce endoglucanase and can degrade microcrystalline cellulose. A study of the genome sequence of *P. placenta* revealed the absence of exocellulases and abundance of genes involved in reactive oxygen species formation (Martinez et al., 2009), reinforcing the idea that brown-rot fungi can depolymerize cellulose via a combination of oxidative reactions and endocellulases (Cohen et al., 2005).

White-rot fungi are a diverse and abundant group classified into the Agaricomycetes class. These fungi can be found colonizing either living trees (e.g., heart rot) or dead wood (e.g., logs, stumps) from temperate to tropical climates, presenting a variety of morphologies such as caps, brackets or resupinaceous (corticioid) basidiomes. There are about 10,000 species of white-rot fungi, with varying capacities to degrade lignin, cellulose, and hemicelluloses. However, only a few dozen have been adequately studied. The most commonly studied species of white-rot fungi are subdivided into six families: Phanerochaetaceae (e.g., *Phanerochaete chrysosporium*), Polyporaceae (e.g., *Trametes versicolor* and *Pycnoporus sanguineus*), Marasmiaceae (e.g., *Lentinula edodes*), Pleurotaceae (oyster mushrooms such as *Pleurotus ostreatus* and *Pleurotus pulmonarius*), Hymenochaetaceae (e.g., *Inonotus hispidus* and *Phellinus igniarius*), Ganodermataceae (e.g., *Ganoderma lucidum* and *Ganoderma applanatum*), Meruliaceae (e.g., *Bjerkandera adusta*, *Irpex lacteus*, and *Phlebia radiata*) (Fig. 6.4).

In nature, several microorganisms can produce hydrolases (cellulases and hemicellulases) capable of hydrolyzing to a full extent all polysaccharide components of wood (cellulose and hemicelluloses) into monosaccharides. However, when these polysaccharides are complexed with lignin, they are resistant to enzymic hydrolytic breakdown. Therefore lignin appears to inhibit hydrolytic activity. This is one reason, among others, why research on biotransformation of lignocellulose takes a long time to develop. The degradation of lignin is a critical step for efficient carbon cycling. Thus, as organisms capable of completely metabolizing lignin, white-rot fungi are an essential part of forest ecosystems.

The patterns and rates of wood and woody material degradation vary among white-rot fungi. Usually, the degradation systems can be divided into two subtypes: (1) those producing oxidative cleavage of lignin and structural polysaccharides at similar rates, leading to progressive erosion and thinning of wood cell walls, often referred to as simultaneous degraders, and (2) those capable of removing lignin in advance of cellulose and hemicelluloses which are called selective delignifiers. The first subtype is considered the normal or the most common process of wood degradation and is used by the majority of white-rot fungi (Skyba et al., 2013). In some cases, a slight preference for the removal of lignin in advance of carbohydrates may occur, but an extensive loss of carbohydrates usually appears simultaneously or immediately after lignin removal. In selective delignification, the white-rot fungi preferentially remove lignin from wood, causing moderate losses of hemicelluloses and leaving cellulose practically intact or only slightly degraded. Selective decayed wood presents white pockets consisting entirely of cellulose.

FIGURE 6.4 White-rot basidiomycetes. (A) *Phanerochaete chrysosporium*; (B) *Trametes versicolor*; (C) *Lentinula edodes*; (D) *Pleurotus ostreatus*; (E) *Pleurotus pulmonarius*; (F) *Inonotus hispidus*; (G) *Phellinus igniarius*; (H) *Ganoderma lucidum*; (I) *Ganoderma applanatum*; (J) *Bjerkandera adusta;* (K) *Irpex lacteus*; (L) *Phlebia radiata*.

Although rare, selective delignification can occur under natural conditions. *G. applanatum* has been associated with the formation of "palo podrido" wood in the evergreen rainforests of Southern Chile, a highly delignified type of wood where cellulose remains intact while hemicellulose and lignin are degraded (Dill and Kraepelin, 1986). *Ceriporiopsis subvermispora, Phlebia* spp., and *Physisporinus rivulosus* are some examples of white-rot fungi that selectively attack lignin, while *T. versicolor, P. chrysosporium,* and *I. lacteus* simultaneously degrade all cell wall components (Maciel et al., 2012). The white-rot fungi capable of preferentially removing lignin from lignocellulosic materials have received increasing attention for their applicability in industrial processes such as biopulping and bioethanol production as well as in the formulation of cellulose-enriched products in animal feed (Koutrotsios et al., 2014). However, the application of white-rot fungi in selective delignification may depend on strain, cultivation methods, and lignocellulosic material composition (Salmones et al., 2005). In contrast, much variation exists in the ability of certain fungi to cause selective delignification and/or simultaneous rot.

The first determined genomes of the white-rot fungus *P. chrysosporium* (Martinez et al., 2004) and the brown-rot fungus *P. placenta* (Martinez et al., 2009) revealed a gene complement consistent with their respective modes of wood decay. Recently, the analysis of more than 30 genomes, however, has led to the view that the ability of basidiomycetes to degrade wood should be reclassified based on a continuum, rather than on only two groups (Riley et al., 2014). The argument in favor of a new classification of wood decay is based on the fact that some white-rot species, such as *Botryobasidium botryosum* and *Jaapia argillacea,* lack the peroxidases involved in lignin degradation, thus resembling brown-rot fungi. Still, they do possess the cellulose-degrading apparatus typical of white-rot fungi.

6.3 Isolation and laboratory maintenance of wood-rot basidiomycetes

Over several decades the basidiomycetes have been explored as biofactories of novel bioactive substances with great potential for biotechnological applications. In fact, basidiomycetes represent a reservoir for discovering new compounds, such as antibiotics, enzymes, antioxidants, immunomodulators, and anticancer and antiparasitic compounds, for use in the pharmaceutical and agrochemical industries (Erjavec et al., 2012). For this reason, there has been a growing interest in screening for new species and strains. A classical strategy is to collect basidioma or mycelia of white-rot fungi in forests, dead trees, and lignocellulosic crop residues showing signs of attack by fungi. Samples of basidioma or mycelia are aseptically transferred onto potato dextrose agar or malt extract agar and subcultured until pure mycelia are obtained. Identification is based on morphological, physiological, biochemical, and genetic characteristics of the basidioma, hyphae, and spores. Molecular biology techniques are also being used in the identification of new isolates. Advancement in molecular methods has permitted a more rational study of the phylogenetic relationships within the various microorganisms. Noncoding internal transcribed spacer regions (ITS1 and ITS2) of the ribosomal DNA seem to be one of the most frequently used analytical tools (Prewitt et al., 2008).

There are several options for maintaining brown and white-rot basidiomycetes in the laboratory. Continuous growth methods of preservation, in which the fungi are grown on agar (e.g., malt extract agar, potato dextrose agar, and yeast extract agar), are typically used for short-term storage. For long-term storage, preservation in distilled water (called the Castellani method) and anhydrous silica gel are some of the most indicated methods. All are considered low-cost methods, but none of them is considered permanent. Lyophilization and liquid nitrogen refrigeration (cryo-preservation) are expensive methods but are considered permanent.

6.4 Basidiomycetes as producers of enzymes involved in the degradation of lignocellulose biomass

As stated above, the lignocellulosic biomass consists mainly of three types of polymers, that is, lignin, cellulose, and hemicellulose interlinked in a hetero-matrix. Consequently, its complete degradation requires the synergistic action of many oxidative, hydrolytic, and nonhydrolytic enzymes. A panoramic view of enzymes involved in the complex processes of lignocellulose degradation will be presented here. Those enzymes typically found in basidiomycetes with potential biotechnological applications will be described in more detail.

6.4.1 Enzymes involved in the degradation of cellulose and hemicelluloses

The classical model for degradation of cellulose into glucose involves the cooperative action of endocellulases (EC 3.2.1.4), exocellulases (cellobiohydrolases, CBH, EC 3.2.1.91) glucanohydrolases (EC 3.2.1.74), and beta-glucosidases (EC 3.2.1.21). Endocellulases hydrolyze internal glycosidic linkages in a random fashion, which results in a rapid decrease in polymer length and a gradual increase in the reducing sugar concentration. Exocellulases hydrolyze cellulose chains by removing mainly cellobiose either from the reducing or the nonreducing ends, which leads to a rapid release of reducing sugars but little change in polymer length. Endocellulases and exocellulases act synergistically on cellulose to produce cello-oligosaccharides and cellobiose, cleaved by beta-glucosidases to glucose. It has been shown that some fungi, including *Hypocrea jecorina* (syn. *Trichoderma reesei*), also produce a class of oxidative enzymes, known as polysaccharide mono-oxygenases, which directly cleave cellulose chains through an oxidative mechanism, and appear to act synergistically with the traditional hydrolytic enzymes (Harris et al., 2010; Hansson et al., 2017).

Cellulose is particularly resistant to degradation and requires several different enzymatic attack modes to degrade effectively. In cellulose fibers, crystalline and amorphous regions alternate. The amorphous regions are formed by cellulose chains with weaker organization, more accessible to enzymatic attack. On the other hand, the crystalline regions are very cohesive, with a rigid structure formed by the parallel juxtaposition of linear chains, resulting in intermolecular hydrogen bonds. This characteristic contributes to the insolubility and low reactivity of cellulose.

Hydrolysis of hemicelluloses involves enzymes such as glycoside hydrolases, carbohydrate esterases, polysaccharide lyases, *endo*-hemicellulases, and others. The concerted action hydrolyzes glycosidic bonds and ester bonds and removes the substituent chains or side chains. These include *endo*-1,4-β-xylanase (EC 3.2.1.8), β-xylosidase (EC 3.2.1.37), β-mannanase (EC 3.2.1.78), β-mannosidase, (EC 3.2.1.25), α-glucuronidase (EC 3.2.1.1), α-L-arabinofuranosidase (EC 3.2.1.55), acetyl xylan esterase (EC 3.1.1.72), *p*-coumaric and ferulic acid esterases (EC 3.1.1.1), and feruloyl esterases (EC 3.1.1.73). The last one can specifically cleave the ester bond between ferulic acid and arabinose, acting synergistically with cellulases, xylanases, and other hemicellulases in the saccharification of lignocellulose.

Considering the catalytic mechanisms and enzymatic specificities, the carbohydrate active enzymes are divided into families and subfamilies and compiled in the knowledge-based CAZy database (http://www.cazy.org) together with information about enzymes with auxiliary activities (Kues, 2015; Levasseur et al., 2013).

Filamentous fungi are considered good sources of hydrolytic enzymes, which find application in various fields, from process industries to diagnostic laboratories. Traditionally, enzymes from basidiomycetes involved in the degradation of plant cell wall polysaccharides have been receiving less attention than those of soft-rot ascomycetes and deuteromycetes. *T. reesei* (anamorph of *H. jecorina*) is the best known mesophilic soft-rot fungus producer of cellulases and hemicellulases (He et al., 2014; Martinez et al., 2008; Schuster and Schmoll, 2010). It is widely employed to produce enzymes for applications in the pulp and paper, food, feed, and textile industries, and, currently, with increasing importance, in biorefining (Kuhad et al., 2011; Seiboth et al., 2011). The genera *Aspergillus*, *Fusarium*, *Humicola*, *Rhizopus*, *Alternaria*, *Monilia*, *Mucor*, and *Penicillium* are also considered great producers of cellulases and hemicellulases, including xylanases and pectinases (Juturu and Wu, 2013; Polizeli et al., 2005). The genera are also explored as the producers of amylases and proteases (Payne et al., 2015). These enzymes have been commercially available for more than 30 years and represent a target for both academic and industrial research. Basic and applied studies on cellulolytic and xylanolytic enzymes have demonstrated their biotechnological potential in various fields, including food, animal feed, brewing, wine making, biomass refining, textile, pulp and paper industries, as well as in agriculture and laundry.

The brown-rot fungi, *C. puteana*, *Lanzites trabeum*, *Poria placenta*, *Tyromyces palustris*, *F. palustris*, and *Piptoporus betulinus*, and the white fungi *P. chrysosporium*, *Sporotrichum thermophile*, *T. versicolor*, *Agaricus arvensis*, *P. ostreatus*, and *Phlebia gigantea* are among those most studied concerning their cellulase and xylanase enzymatic complexes (Cohen et al., 2005; Daniel, 2014; Kuhad et al., 2011; Valásková and Baldrian, 2006; Yoon et al., 2007; Zilly et al., 2012).

Considering that the brown- and white-rot basidiomycetes are able to completely degrade the lignocellulosic materials, it is reasonable to suppose that they are able to produce at least some of the hydrolytic enzymes involved in the degradation of the polymers cellulose and hemicellulose (cellulase and xylanase complexes). Again, the most studied species in this respect is *P. chrysosporium*. The number of extracellular hydrolytic enzymes described in *P. chrysosporium* is elevated and includes several proteases, amylases, xylanases, and other carbohydrases (Dey et al., 1991; Ishida et al., 2007). *T. versicolor*, one of the most studied species, can produce hydrolytic enzymes (cellulases and xylanases especially) involved in the degradation of lignocellulosic biomass in both submerged and solid-state cultures (Tisma et al., 2021). *G. applanatum* also produces both cellulose and xylanases in liquid cultures (Salmon et al., 2014). Bentil et al. (2018) reviewed the production of cellulose and xylanase by several white-rot fungi. They concluded that submerged cultures are a more suitable technology than solid-state cultivation (SSC) in production yields and enzyme recovery. In a general way, however, the hydrolases of white-rot basidiomycetes are not yet commercially explored. This low interest in the basidiomycete hydrolytic enzymes is easy to understand. First, the soft-rot ascomycetes and deuteromycetes can be easily cultured to produce hydrolytic enzymes with high productivity and low cost. Second, when compared with the ascomycete and deuteromycete hydrolytic enzymes, the white-rot basidiomycete hydrolytic enzymes do not possess the desired special characteristics such as thermal stability, tolerance to high temperatures, and stability over a large range of pH values. Also, they do not retain their activity under severe reaction conditions, such as the presence of metals and organic solvents (Nigam, 2013).

6.4.2 Enzymes involved in lignin degradation

Lignin degradation by white-rot basidiomycetes involves a set of enzymes called lignin-modifying enzymes (LMEs). Most LMEs are secreted as multiple isoforms by many different species of white-rot fungi under various conditions. The set of LMEs comprises a phenoloxidase, laccase (Lcc, EC

1.10.3.2), and three peroxidases (high oxidation potential class II peroxidases), lignin peroxidase (LiP, EC 1.11.1.14), manganese peroxidase (MnP, EC 1.11.1.13), and versatile peroxidase (VP, 1.11.1.16). Laccase, which was first described over 128 years ago, is one of the oldest known enzymes. LiP and MnP were initially discovered in *P. chrysosporium*. VP was also added to the group of LMEs and was originally discovered in a strain of *Pleurotus eryngii*.

White-rot fungi usually secrete one or more LMEs in different combinations. The distribution of white-rot fungi into groups according to their enzymatic systems has been undertaken. This general classification is based on the capacity of different fungi to produce one or a combination of peroxidases and laccase. Generically, white-rot fungi can be distributed into four groups, according to their ability to produce laccases and peroxidases (LiP, MnP, and VP): (1) laccase and MnP and LiP (*T. versicolor, B. adusta*); (2) laccase and at least one of the peroxidases (*L. edodes*, *P. eryngii*, and *C. subvermispora*); (3) only laccase (*S. commune*); (4) only peroxidases (*P. chrysosporium*). The most frequently observed LMEs among the white-rot basidiomycetes are laccases and MnP, and the least ones are LiP and VP (Maciel et al., 2012).

The absence or nondetection of these enzymes in some white-rot fungi, the sequencing of white-rot fungal genomes, and the discovery of new enzymes can lead to the establishment of different groups. For example, the white-rot fungus *C. subvermispora* delignifies lignocellulose with high selectivity. Still, up to a few years ago, it appeared to lack the specialized peroxidases, termed LiPs and VPs, which are generally thought important for ligninolysis (Lobos et al., 1994). The recently sequenced *C. subvermispora* genome was screened for genes that encode peroxidases with a potential ligninolytic role (Fernandez-Fueyo et al., 2012). Among 26 peroxidase genes, two newly discovered *C. subvermispora* peroxidases are functionally competent LiPs, phylogenetically and catalytically intermediate between classical LiPs and VPs. These observations offer new insights into selective lignin degradation by *C. subvermispora*.

In addition to the peroxidases and laccases, fungi produce other accessory process enzymes, unable to degrade lignin on their own, but necessary to complete the process of lignin and/or xenobiotic degradation: aryl-alcohol dehydrogenase (AAD, EC 1.1.1.90), glyoxal oxidase (GLOX, EC 1.2.3.5), quinone reductase (QR, EC 1.1.5.1), cellobiose dehydrogenase (CDH, EC 1.1.99.18), superoxide dismutase (SOD, EC 1.15.1.1), glucose 1-oxidase (GOX, EC 1.1.3.4), pyranose 2-oxidase (P_2Ox, EC 1.1.3.4), and methanol oxidase (EC 1.1.3.13). These are mostly oxidases generating H_2O_2, dehydrogenases of lignin, and many xenobiotics (Maciel et al., 2012).

Cytochrome P450 mono-oxygenases (CYPs, EC 1.14.14.1) are also significant components involved in the degradation of lignin and chemically associated xenobiotics (Coelho-Moreira et al., 2013a; Ning and Wang,

2012). These unspecific mono-oxygenases are intracellular heme-thiolate-containing oxidoreductases acting on a wide range of substrates in stereo-selective and regio-selective manners under consumption of O_2. Activated by a reduced heme iron, these enzymes add one atom of molecular oxygen to a substrate, usually by a hydroxylation reaction. A series of other reactions can occur, including epoxidation, sulfoxidation, and dealkylation (Kues, 2015).

Recent additions to the enzymatic systems of white-rot fungi include dye-decolorizing peroxidases (DyP), involved in the oxidation of synthetic high redox-potential dyes and nonphenolic lignin model compounds (Liers et al., 2010) and aromatic peroxygenases that catalyze diverse oxygen transfer reactions which can result in the cleavage of ethers (Hofrichter et al., 2010).

6.5 Production of ligninolytic enzymes by basidiomycetes: screening and production in laboratory scale

The potential application of ligninolytic enzymes in biotechnology has stimulated investigations on selecting promising enzyme producers and finding convenient substrates to obtain large amounts of low-cost enzymes. First, screening of fungal species and their variants is important for selecting suitable LME-producing organisms. For this reason, one usually relies on the use of inexpensive, rapid, and sensitive testing methods. The screening strategy must aim at identifying fungal strains and enzymes that will work under industrial conditions. The discovery of novel ligninolytic enzymes with different substrate specificities and improved stabilities is important for industrial applications. Fungi that produce ligninolytic enzymes have been screened either by visual detection of ligninolytic enzymes on solid media containing colored indicator compounds or, alternatively, by enzyme activity measurements in samples obtained from liquid cultivations. The use of colored indicators is generally simpler as no sample handling and measurement are required. As ligninolytic enzymes oxidize various substrates, several different compounds have been used as indicators of laccase production. The traditional screening reagents, tannic acid and gallic acid (Bavendam reaction), have mainly been replaced by synthetic phenolic reagents, such as guaiacol, or by the anthraquinonic dye Remazol Brilliant Blue R (RBBR) and the polymeric dye Poly R-478. These dyes are decolorized by ligninolytic fungi, and the production of enzymes is observed as a colorless halo around the microbial growth. With guaiacol, a positive reaction is indicated by forming a reddish-brown halo. In contrast, the positive reaction with tannic acid and gallic acid is seen as a dark-brown-colored zone (Bazanella et al., 2013; Ryu et al., 2003).

In general, industrial enzymes on a large scale have been carried out using well-established submerged systems where the fungi are grown in a fully liquid system, allowing control over process parameters, such as pH, temperature, and aeration. Submerged cultures of several white-rot basidiomycetes have been conducted in Erlenmeyer flasks and bioreactors with high volumetric capacity. The production of ligninolytic enzymes by the white-rot basidiomycetes depends on the bioavailable nitrogen and carbon concentrations. For several fungal species, such as *P. chrysosporium* producing LiP and MnP under nutrient-limited conditions, ligninolytic enzyme activities are suppressed by high nitrogen concentrations in the medium. For other species such as *P. ostreatus* and *Trametes trogii*, high amounts of laccase and MnP are produced in the presence of high concentrations of nutrient nitrogen, while in *Dichomitus squalens* cultures, MnP is secreted under both high- and low-nitrogen conditions (Janusz et al., 2013). The addition of inducers can obtain overproduction of ligninolytic enzymes. The list of inducers is large and includes oxalic acid, veratryl alcohol, 2,6-dimethoxyphenol, 2,5-xylidine, ferulic acid, vanillin, ethanol, copper, manganese, among others (Piscitelli et al., 2013; Souza et al., 2004). The highest production of LMEs by different white wood-rot fungi may require a different set of conditions that include composition of the growth medium, carbon/nitrogen ratio, pH, temperature, presence of inducers, growing time, and agitation. All aspects must be considered with care since inappropriate screening procedures and/or assays of enzyme activity may discard strains of potential interest.

Cultivation of white-rot fungi in solid-state systems appears as a promising technology for producing ligninolytic enzymes on a large scale (Martani et al., 2017). SSC is defined as the cultivation process in which microorganisms grow on solid materials without free liquid. SSC using different agro-industrial residues appears to be a good technique for culturing basidiomycetes. The solid-state cultures reproduce the natural living conditions of these microorganisms, leading to the production of enzymes with high productivity and low cost. The selection of adequate support for performing SSC is significant since the success of the process depends on it (Soccol et al., 2017).

The selection of agro-industrial residues for utilization in SSC depends on physical parameters, such as particle size, moisture level, intraparticle spacing, and nutrient composition within the substrate. Wheat bran is the most commonly used substrate for cultivating white-rot fungi. However, the list of possibilities is extensive. It includes several lignocellulolytic wastes such as sugarcane bagasse, corn cob, wheat straw, oat straw, rice straw, and food processing wastes such as banana, kiwi fruit, and orange wastes, cassava bagasse, sugar beet pulp/husk, oil cakes, apple pomace, grape juice, grape seed, coffee husk, coir pith, and others (Couto, 2008; Holker et al., 2004; Zilly et al., 2012). Even so, it is worth searching for new substrates, especially if they are available in large amounts, allowing the growth of white-rot fungi without further supplementation and facilitating the obtainment of valuable products. Additionally, cost and availability are important factors to consider when choosing a residue as substrate or support in SSC (Soccol et al., 2017).

After successful production, enzymes can be separated or purified depending on the field of application, depending on whether the desired enzymes can be effectively separated and purified. This final step is commonly known as downstream processing or bio-separation, accounting for up to 60% of the total production costs, excluding the purchased raw materials. The downstream processing includes extraction, concentration, purification, and stabilization methods.

6.6 General characteristics of the main ligninolytic enzymes with potential biotechnological applications

A general description of the mechanisms and functions of the main ligninolytic enzymes is presented here.

6.6.1 Laccases

Laccases are copper oxidases that catalyze the one-electron oxidation of phenolics, aromatic amines, and other electron-rich substrates with a concomitant reduction of O_2 to H_2O (Fig. 6.5).

The enzyme is widely distributed in higher fungi and has also been found in insects and bacteria. The majority of laccases characterized so far have been derived from white-rot fungi, the efficient lignin degraders. Many fungi contain laccase-encoding genes, but their biological roles are not well understood. *Agaricus bisporus*, *Botrytis cinerea*, *B. adusta*, *Coprinus cinereus*, *G. lucidum*, *P. radiata*, *P. ostreatus*, *P. pulmonarius*, *Rigidoporus lignosus*, and *T. versicolor* are examples of basidiomycetes that produce laccases (Viswanath et al., 2014). Fungal laccases are mainly extracellular glycoproteins with molecular weights between 50 and 130 kDa, and acid pI values of 3–6 (Agrawal et al., 2018; Lundell et al., 2010). Fungal laccases often occur as isozymes with monomeric or dimeric protein structures, all showing a similar architecture consisting of three sequentially arranged domains of a β-barrel-type structure (Fig. 6.6).

4 benzenediol + O_2 → 4 benzo-semiquinone + 2 H_2O

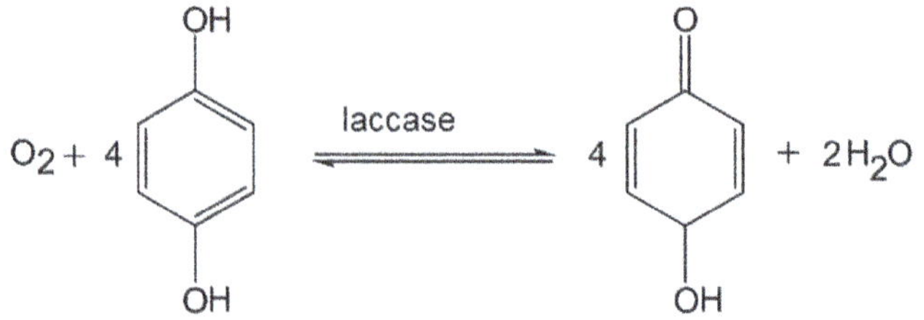

FIGURE 6.5 General reaction catalyzed by laccase.

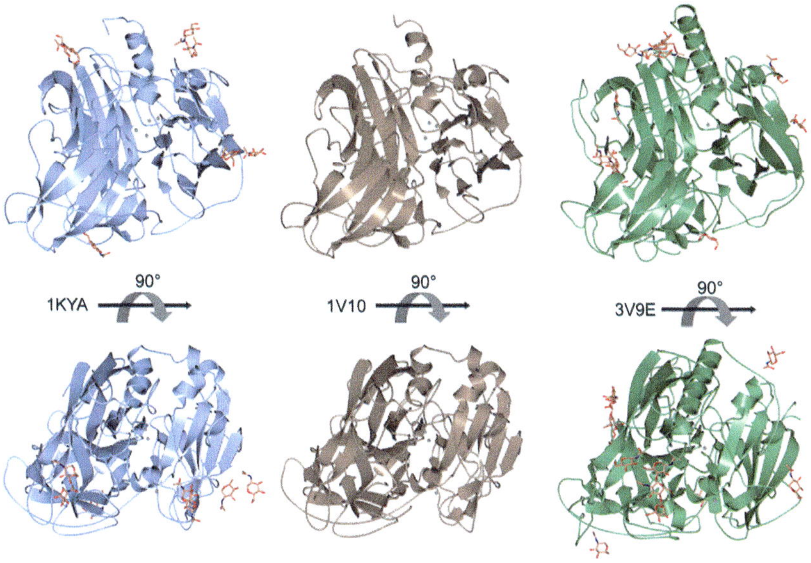

FIGURE 6.6 Ribbon models of laccases from *Trametes vesicolor* (pdbid: 1KYA); *Rigidoporus lignosus* (pdbid: 1V10); and *Botrytis aclada* (pdbid: 3V9E) showing a similar fold motif. Crystallographic data evidence carbohydrates (coral color) at glycosylation binding sites and copper ions (gray spheres) at redox centers.

Laccase attacks the phenolic subunits of lignin, leading to Cα oxidation, Cα–Cβ cleavage, and aryl–alkyl cleavage. This oxidation results in an oxygen-centered free radical, which can then be converted into quinone by a second enzyme-catalyzed reaction. The quinone and the free radicals can then undergo polymerization (Christopher et al., 2014). Most monomeric laccase molecules contain four copper atoms in their structure that can be classified into three groups using UV/visible and electron paramagnetic resonance (EPR) spectroscopy. The type I copper (T1) is responsible for the intense blue color of the enzymes (with an absorption peak at 600 nm) and is EPR-detectable. The type II copper (T2) is colorless but is EPR-detectable, and the type III copper (T3) consists of a pair of copper atoms that give a weak absorbance near the UV spectrum but no EPR signal. The T2 and T3 copper sites are located close to each other and form a trinuclear center that is involved in the catalytic mechanism of the enzyme. Based on the type T1–T3 copper properties, laccases can be categorized into enzymes with high (0.6–0.8 V) or low (0.4–0.6 V) redox potential (Christopher et al., 2014). The catalytic efficiency of laccases depends on their redox potential, which explains the interest in laccases with high redox potential, such as the laccases from *T. versicolor*, *Pycnoporus coccineus*, and *P. sanguineus* (Morozova et al., 2007; Uzan et al., 2010).

Due to their specificity for phenolic subunits in lignin, restricted access to lignin in the fiber wall, and low redox potential, laccases have a limited potential to oxidize lignin and for being used in biotechnological applications. Laccases possess relatively low redox potentials ($\leq$0.8 V) compared to ligninolytic peroxidases (> 1 V). Thus their action would be restricted to the oxidation of the phenolic lignin moiety (less than 20% of lignin polymers). Nonphenolic substrates, having redox potentials above 1.3 V, cannot be directly oxidized by laccases. Nevertheless, this limitation has been overcome through nature mimicking, that is, by using redox mediators in the so-called laccase-mediator system (LMS). These small natural and synthetic low-molecular-weight compounds with higher redox potential than laccase itself (> 0.9 V), called mediators, may be used to oxidize the nonphenolic part of lignin (Bibi et al., 2011; Camarero et al., 2004; Christopher et al., 2014; Johannes and Majcherczyk, 2000; Poppins-Levlin et al., 1999). In the last few years, discovering new and efficient synthetic mediators has extended laccase catalysis toward several xenobiotic substrates (Camarero et al., 2005; Canã and Camarero, 2010). A mediator is continuously oxidized by laccase and subsequently reduced by the substrate. The mediator acts as a carrier of electrons between the enzyme and the substrate (Fig. 6.7).

Phenolic products generated during lignin degradation by white-rot fungi, fungal metabolites such as 2,2′-azino-bis(3-ethylbenzothiazoline-6-sulfonic acid) (ABTS), violuric acid (VLA), 1-hydroxybenzotriazole (HBT), acetosyringone, syringaldehyde, *p*-coumaric acid, vanillin, and 4-hydroxybenzoic acid are considered as potential mediators. Fig. 6.8 shows the chemical structures of some largely used natural and synthetic mediators.

The applicability and effectiveness of the LMS depend on the choice of proper mediators (Jeon et al., 2008). It should be noted that mediators are also substrate-selective in the same way as enzymes. The mediator HBT, for example, efficiently improves the laccase-induced transformation of some dyes such as Reactive Black 5, Bismarck Brown R, and Lanaset Grey G, but it is not effective in the decoloration of RBBR (Daâssi et al., 2014). The ideal mediator should be nontoxic, of low-cost and efficient, with stable oxidized and reduced forms (Morozova et al., 2007) and should be able to maintain a continuous cyclic redox conversion (Christopher et al., 2014). The utilization of synthetic mediators, such as ABTS and HBT, in

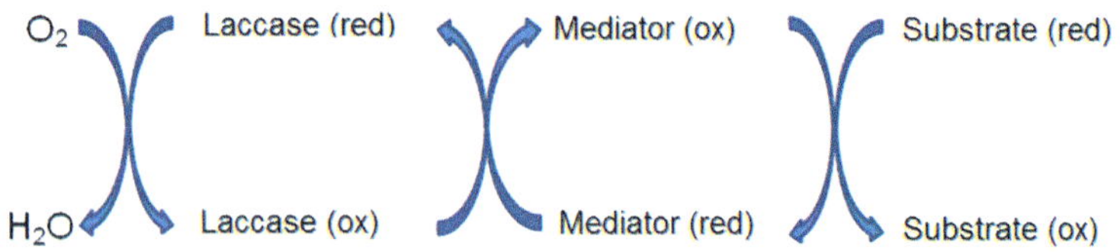

FIGURE 6.7 Schematic representation of laccase-catalyzed cycles for substrate oxidation in the presence of a chemical mediator.

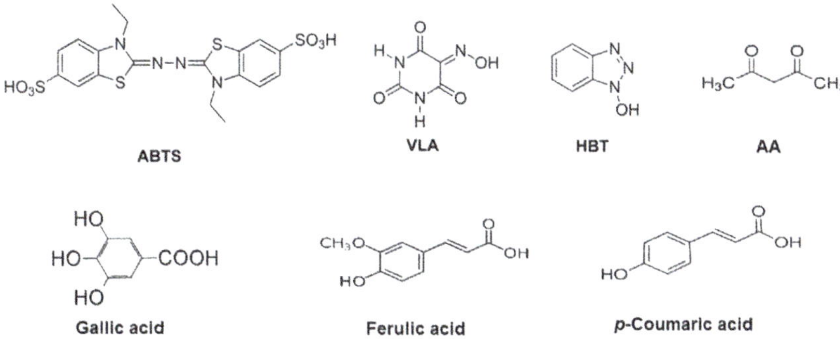

FIGURE 6.8 Chemical structures of natural and synthetic laccase mediators. *ABTS*, 2,2′-azino-bis(3-ethylbenzothiazoline-6-sulfonicacid); *VLA*, violuric acid; *HBT*, 1-hydroxybenzotriazole; *AA*, acetylacetone.

industrial processes is additionally hindered by their high cost, toxicity, and recycling-associated problems (Canã and Camarero, 2010). Low-cost and environmentally benign mediators are still highly needed to facilitate the application of laccases in various biotechnological processes in wastewater treatment. Acetylacetone (2,4-pentanedione), denoted as AA in Fig. 6.7, is an inexpensive small-molecular diketone that presents low toxicity (Ballantyne and Cawley, 2001). It has been demonstrated that AA could act as a mediator for laccase from *P. coccineus* and *Myceliophthora thermophila* to initiate free radical polymerization of acrylamide (Hollmann et al., 2008). AA was also a highly effective mediator in the *T. versicolor* laccase-induced grafting copolymerization of acrylamide and chitosan, as well as in the decolorization of the dye malachite green (Yang et al., 2015).

6.6.2 Peroxidases

Three peroxidases involved in lignin degradation are produced by white-rot fungi. LiP is characterized by its ability to oxidize high redox-potential aromatic compounds (including veratryl alcohol), whereas MnP requires Mn^{2+} to complete the catalytic cycle and forms Mn^{3+} chelates acting as diffusing oxidizers. LiP and MnP were first described as true ligninases because of their high redox potential. The third peroxidase, VP, can oxidize Mn^{2+} as well as nonphenolic aromatic compounds, phenols, and dyes (Martínez, 2002).

The lignin-modifying peroxidases (LiP, MnP, and VP), belonging to class II of fungal heme peroxidases, the so-called LiPs, catalyze the oxidation of various nonphenolic aromatic compounds as well as phenolic aromatic compounds such as veratryl alcohol, which is a metabolite produced by *P. chrysosporium*. Under ligninolytic conditions, veratryl alcohol can also act as a mediator (Fig. 6.9).

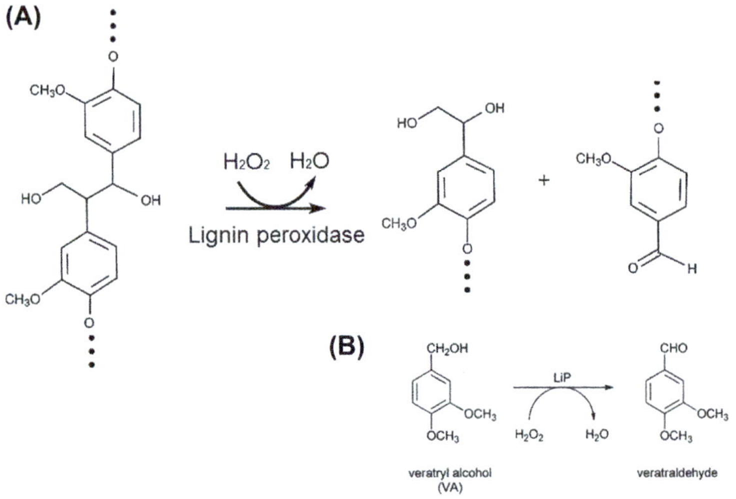

FIGURE 6.9 General reaction catalyzed by lignin peroxidase. (A) Cleavage of C–C of lignin; (B) oxidation of veratryl alcohol is generally used to estimate the lignin peroxidase activity.

MnP is the most common lignin-modifying peroxidase produced by almost all wood-colonizing basidiomycetes causing white rot. Multiple forms of this glycosylated heme protein with molecular weights normally from 40 to 50 kDa are secreted by ligninolytic fungi into their microenvironment. MnP preferentially oxidizes Mn^{2+}, always present in woods and soils, into the highly reactive Mn^{3+}, which is stabilized by fungal chelators such as oxalic acid. Since Mn^{3+} is unstable in aqueous media, MnP releases it as a Mn^{3+}-carboxylic acid chelate. There is a variety of carboxylic acid chelators, including oxalate, malonate, tartrate, and lactate; however, oxalate is the most common. Chelated Mn^{3+}, in turn, acts as a low-molecular-weight, diffusible redox mediator that attacks phenolic lignin structures, resulting in the formation of unstable free radicals that tend to disintegrate spontaneously. The enzyme has extraordinary potential to oxidize a range of different phenolic and nonphenolic complex compounds (Asgher and Iqbal, 2011). MnP is capable of oxidizing and depolymerizing natural and synthetic lignins and entire lignocelluloses (milled straw or wood, pulp) in cell-free systems (in vitro). In vitro depolymerization is enhanced in the presence of cooxidants such as thiols (e.g., glutathione) or unsaturated fatty acids and their derivatives (e.g., Tween 80).

Classical LiP producers are *B. adusta*, *P. chrysosporium*, *Trametes cervina*, and *T. versicolor*. The occurrence of MnP is higher than LiP. Typical MnP producers are *C. subvermispora*, *D. squalens*, *I. lacteus*, *L. edodes*, *P. chrysosporium*, *P. ostreatus*, *P. pulmonarius*, *and T. versicolor*.

The existence of a VP, a peroxidase with ability to oxidize both Mn^{2+} and aromatic compounds, was first reported in *P. eryngii*, and subsequently, other VPs were isolated from *P. pulmonarius*, *P. ostreatus*, *B. adusta*, and *Bjerkandera* sp. (Hernández-Bueno et al., 2021). These peroxidases, similarly to MnP, are able to oxidize Mn^{2+} to Mn^{3+} and to oxidize veratryl alcohol and *p*-dimethoxybenzene to veratraldehyde and *p*-benzoquinone, respectively, as reported for LiP.

The catalytic cycle of three LiPs, consisting of the resting peroxidase and compounds I (two-electron oxidized form) and II (one-electron oxidized form), is common to other peroxidases. The unique catalytic properties of ligninolytic peroxidases are provided by the heme environment, conferring high redox potential to the oxo-ferryl complex ($\geq$1.0 V) and by the existence of specific binding sites (and mechanisms) for oxidation of their characteristic substrates (Anastasi et al., 2013). These include nonphenolic aromatics in the case of LiP, manganous ion in the case of MnP, and both types of compounds in the case of the VP (Martínez et al., 2005). Similar heme environments in the above three peroxidases are located at the central region of the protein (Fig. 6.10).

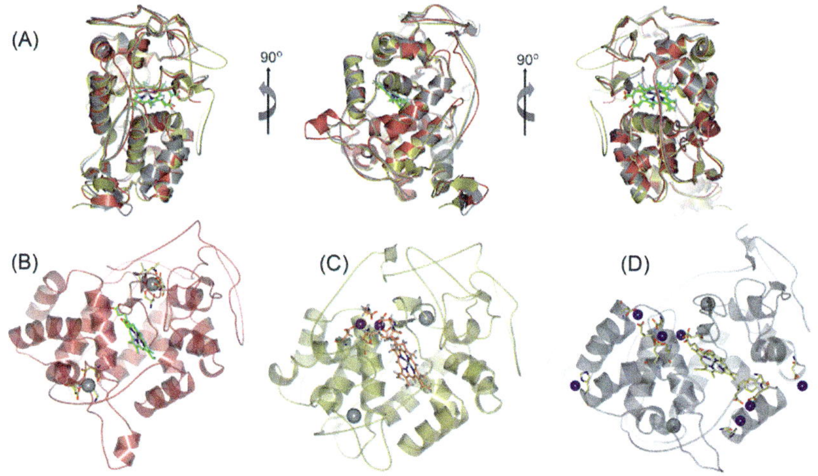

FIGURE 6.10 Ribbon models of peroxidases from *Trametes cervina* (vine color, pdbid: 3Q3U); *Ceriporiopsis subvermispora* (gold color, pdbid: 4CZO); and *Pleurotus eryngii* (gray color, pdbid: 3FKG). (A) The overlay fit of these three enzymes evidences a similar fold motif detaching a heme group (green) common in this protein family (heme peroxidase, pfam id: PF00141). (B) A lignin peroxidase from *T. cervina* detaching two calcium ions (gray spheres) bonded at their sites (yellow sticks). (C) Manganese peroxidase from *C. subvermispora* detaching two manganese ions (violet spheres) bonded at their sites (orange sticks). (D) Versatile peroxidase from *P. eryngii* detaching zinc ions (blue spheres) bonded at their sites (gold sticks). In the case of lignin peroxidase and versatile peroxidase, the heme group is in close contact (within 4.0 Å) with manganese or zinc ions.

6.7 Industrial and biotechnological applications of ligninolytic enzymes from basidiomycetes

A considerable number of reviews detailing the numerous applications of ligninolytic enzymes have been published during the last few years. Ligninolytic enzymes are applied or present potential application in biofuel production, bioremediation of several xenobiotics, detoxification of wastewater, organic synthesis, food industry, and pharmaceutical and cosmetic industries, among others (Table 6.1).

Among the ligninolytic enzymes, laccases seem to be the most suitable to be used on a large scale due to the existence of a considerable number of microbial producers, ease of production in both submerged cultivation and SSC, broad substrate specificity, and their ability to use atmospheric oxygen as electron donor compared to the H_2O_2 requirement of the peroxidases. These facts explain the higher number of research and review articles on laccase when compared to those on peroxidases. Below, a brief description of some important industrial and biotechnological applications of ligninolytic enzymes is presented.

6.7.1 Application of ligninolytic enzymes in delignification of vegetal biomass and biological detoxification for biofuel production

The increasing demand for energy and the depletion of fossil fuel reserves urge us to find large quantities of alternative precursors for the petrol-based chemical industry and transportation sectors. Cellulosic biomass, derived from nonfood sources, such as trees and grasses, is being explored as a feedstock for cellulosic ethanol production (Bilal et al., 2018). As mentioned at the beginning of this chapter, lignocellulosic residues to produce bioethanol are hindered by the presence of lignin. The highly recalcitrant lignin structure makes the enzymatic and chemical degradation highly problematic. For this reason, the previous degradation of lignin is a prerequisite for saccharification of polysaccharides in biomasses. Pretreatments are required to remove or modify lignin into a lignocellulosic fiber structure to facilitate the hydrolytic enzymes access to the polysaccharides. Ideally, these pretreatments should predominantly modify lignin without causing major breaks in the structural carbohydrates, making the latter available for fermentation processes. Mild pretreatments, avoiding the generation of waste and pollutants, are desirable. Different pretreatment techniques have been developed in the last few decades and can be divided into four groups—physical, chemical, physicochemical, and biological processes. Several chemical and physicochemical pretreatment processes, such as acid pretreatment, alkaline pretreatment, steam explosion, and ammonia fiber explosion, have been used to enhance the enzymatic hydrolysis of lignocellulose (Baruah et al., 2018). These processes usually require high temperatures and pressures, resulting in

TABLE 6.1 **Last 10-year reviews of industrial and biotechnological applications of ligninolytic enzymes.**

Enzyme (s) and application fields	References
Addressing ligninolytic enzymes	
The role of WRF and their ligninolytic enzymes in bioremediation/biodegradation of synthetic dyes.	Bazanella et al. (2013)
The capability of laccase, lignin peroxidase, manganese-dependent peroxidase, and versatile peroxidase in the degradation of different xenobiotics (heavy metals, polychlorinated biphenyls, petroleum hydrocarbons, pesticides, and phenolic derivatives) was discussed in this review.	Dhagat and Jujjavarapu (2021)
General aspects as well as the potential uses of ligninolytic enzymes in different industrial processes (food, pharmaceutical, textile, pulp and paper, environmental and bioenergy) are discussed in this review.	Vilar et al. (2021)
Discussion about mixed enzyme systems (ligninolytic and hydrolytic) for delignification of lignocellulosic biomass.	Woolridge (2014)
The application of ligninolytic enzymes in the production of second-generation ethanol was reviewed. The analysis includes an evaluation of the biochemical process, feedstocks, and ethanol production.	Placido and Capareda (2015)
The authors discuss the WRF and their enzymes in the bio-delignification of lignocellulosic biomass, enzymatic hydrolysis, and fermentation of hydrolyzed feedstock. Metabolic engineering, enzymatic engineering, and synthetic biology aspects for ethanol production and platform chemicals production are also reviewed.	Bilal et al. (2018)
Physical, chemical, and enzymatic pretreatments for exploitation of grass lignocellulosic biomass are discussed with special emphasis on fungal ligninolytic enzymes and the most recent findings and developments in their current application, issues, and perspectives.	Bilal and Iqbal (2020)
Different techniques, including heterologous gene expression, mutagenesis, and coculturing for improving production and catalytic and stability properties of ligninolytic enzymes are discussed in this review.	Paramjeet et al. (2018)
Pretreatment of recalcitrant lignocellulosic biomass for biofuel production and other industrial and environmental applications such as paper industry, textile industry, wastewater treatment, and the degradation of herbicides.	Abdel-Hamid et al. (2013)

(*Continued*)

TABLE 6.1 (Continued)

Enzyme (s) and application fields	References
PAHs degradation by fungi belonging to different ecophysiological groups (white-rot and litter-decomposing fungi) under submerged cultivation and during mycoremediation of PAH-contaminated soils. The possible functions of ligninolytic enzymes of these fungi in PAH degradation are discussed.	Pozdnyakova (2012)
Degradation of persistent organic pollutants in waste waters, emphasizing the utilization of immobilized enzymes, especially laccases.	Kues (2015)
Degradation of herbicides by white-rot fungi, emphasizing ligninolytic enzymes and cytochrome P450.	Coelho-Moreira et al. (2013a,b)
The role of mushrooms in mycoremediation, emphasizing their capability of biodegradation, bioaccumulation, and bioconversion.	Kulshreshtha et al. (2014)
Descriptive information on the several enzymes from various microorganisms, including ligninolytic enzymes, involved in the biodegradation of a wide range of pollutants, applications, and suggestions of how to overcome the limitations of their efficient use.	Karigar and Rao (2011)
WRF and their ligninolytic enzymes have been widely applied in the removal of PAHs, PhACs, EDCs, pesticides, synthetic dyes, and other environmental pollutants, wherein promising results have been achieved.	Zhuo and Fan (2021)
Application of ligninolytic enzymes in the food industry, including stabilization of wine, beer and fruit juices as well as in baking and sugar beet pectin gelation.	Chowdhary et al. (2019a,b)
The application of ligninolytic enzymes in the degradation and detoxification of xenobiotics released from various types of industries, wastewater treatment, decolorization of dark color of the effluent, and in soil treatment is discussed in this review.	Zainith et al. (2020)
Addressing laccases	
Use of laccase in several fields including pulp and paper industry, textile industry, food industry, pharmaceutical and cosmetic industries, organic synthesis, biofuel cells, and biosensing.	Piscitelli et al. (2013)
Application of laccases in the decolorization of dyes, detoxification of industrial effluents, wastewater treatment, paper and pulp industries, xenobiotic degradation, bioremediation and in biosensors. The review also compares several techniques such as micropatterning, self-assembled monolayer, and layer-by-layer techniques, which immobilize laccases and preserve their enzymatic activity.	Viswanath et al. (2014)

(Continued)

TABLE 6.1 (Continued)

Enzyme (s) and application fields	References
Application of laccases in pulp and paper industry.	Virk et al. (2012)
Laccase-mediator systems and their application in areas such as bioremediation and lignocellulose biorefineries.	Canã and Camarero (2010)
Recent progress in lignin degradation with laccase-mediator systems.	Christopher et al. (2014)
Application of enzymes in bioremediation: degradation of xenobiotics, decolorization of dyes, effluent treatment, among others.	Viswanath et al. (2014)
Use of laccase in the environmental area, emphasizing detoxification and bioremediation of polluted wastewaters and soils.	Strong and Clauss (2011)
The focus of this work was to review different types of laccase immobilization followed by a discussion of the results obtained in the application of immobilized laccase for water purification in recent years.	Zhou et al. (2021)
The current uses of laccases in food processing were the focus of this review. Application of laccases in the synthesis of new compounds with functional properties such as antioxidant and antimicrobial activities as well as recent developments in the field of cross-linking of polymers (proteins and polysaccharides) using laccases were also discussed.	Backes et al. (2021)
The authors discuss the bioenergy generation using laccase in the pretreatment of biomass. The work provides an overview of the biological delignification and detoxification through whole-cell and enzymatic methods, use of laccase-mediator systems, and immobilized laccases.	Malhotra and Suman (2021)
Addressing lignin peroxidase and manganese peroxidase	
This review article is focused on the sources, catalytic reaction mechanisms, and different biotechnological applications of manganese peroxidase, such as in alcohol, pulp and paper, biofuel, agriculture, cosmetic, textile, and food industries.	Chowdhary et al. (2019a,b)
General aspects of production, structure, and action mechanism of lignin peroxidase are described in this review. Contemporary (delignification of feedstock for ethanol production, textile effluent treatment and dye decolorization, coal depolymerization and degradation of other xenobiotics) and prospective (prospects in drug discovery) functionalities of lignin peroxidase are also discussed.	Falade et al. (2017)

(*Continued*)

TABLE 6.1 (Continued)

Enzyme (s) and application fields	References
The authors discuss the best lignin peroxidase (LiP) producers and the molecular properties of LiP. The main focus of the review is the wide variety of current and future lignin peroxidase applications, such as effluent treatment, dye decolorization, catalytic elimination of pharmaceutical and endocrine-disrupting compounds. These issues were discussed with suitable examples.	Singh et al. (2021)

EDCs, Endocrine disruptor compounds; *PAHs*, polycyclic aromatic hydrocarbon; *PhACs*, pharmaceutically active compounds; *WRF*, white rot fungi.

high costs and undesirable products (Agbor et al., 2011). However, pretreatments with dilute acids are more suitable at an industrial scale as they bring about the conversions in an economical and environment-friendly manner (Kumari and Singh, 2018; Solarte-Toro et al., 2019). Nonconventional pretreatment methods such as extrusion, microwave, supercritical fluid, and deep eutectic solvents were recently revised and considered by the authors as novel, cost-effective, energy-efficient, and eco-friendly pretreatment approaches to be used in association with the existing pretreatment processes (Mankar et al., 2021).

Biological pretreatment of lignocellulosic residues can be done using ligninolytic fungi (Castoldi et al., 2014; Singh et al., 2008) or their ligninolytic enzymes. LMS, peroxidases, or mixtures of two or three ligninolytic enzymes have been used in biological delignification of lignocellulose (Placido and Capareda, 2015; Rico et al., 2014; Toco et al., 2021). Ligninolytic enzymes can also be useful in the detoxification of vegetal biomass after conventional pretreatments such as steam explosion for removing toxic phenolics (Jonsson et al., 2013; Placido and Capareda, 2015). Several soluble phenolics derived from lignin, such as vanillin, syringaldehyde, trans-cinnamic acid, and hydroxybenzoic acid, inhibit cellulases, thereby reducing the efficiency of saccharification of biomass (Ximenes et al., 2010). Biological pretreatments present interesting perspectives, especially due to low energy requirements and no generation of toxic compounds. However, their disadvantages, such as low efficiency, a considerable loss of carbohydrates, and long residence periods, are obstacles for using them on a large industrial scale (Zheng et al., 2014). Recently, the use of combined biological and chemical/physicochemical pretreatments has been considered for improving the fiber degradation, sugar yield, and final biofuel production (bioethanol or biogas) (Meenakshisundaram et al., 2021).

6.7.2 Application of ligninolytic enzymes in the degradation of xenobiotic compounds

White-rot fungi and their ligninolytic enzymes have been demonstrated to be capable of transforming and/or degrading a wide range of xenobiotic compounds, including aromatic amines, a wide number of phenolic compounds, including chlorophenols, secondary aliphatic polyalcohols, polycyclic aromatic hydrocarbons, herbicides, and pesticides, among others. Two mechanisms or systems have been proposed. The first consists of transformation in the extracellular space and it involves lignin-degrading enzymes. This powerful capability of white-rot fungi resides in the fact that many pollutants have structural similarities to lignin and because ligninolytic enzymes are nonspecific, that is, they can also act on the pollutant molecules. Furthermore, the transformation of some compounds can be enhanced by using mediators, which can extend the reactivity of enzymes toward the substrates. In recent years, the capability of white-rot fungi and their enzymes to biodegrade several xenobiotics and recalcitrant pollutants has generated considerable research interest in the area of industrial/environmental microbiology. As a consequence, a considerable number of reviews detailing the numerous characteristics and applications of ligninolytic enzymes have been published (Table 6.1). The participation of extracellular enzymes in the transformation of several xenobiotics by white-rot fungi has been conclusively demonstrated by studies performed with purified enzymes.

The second system of white-rot fungi involved in xenobiotic transformation is an intracellular enzymatic mechanism, represented mainly by cytochrome P450. Purification of fungal cytochrome P450, in order to obtain conclusive data, has been accomplished in only a few studies due to the difficulties in maintaining the activation of the enzymes during microsome preparation. Hence, most conclusions were drawn from the results of indirect experiments consisting in the addition of specific cytochrome P450 inhibitors to the culture medium, such as piperonyl butoxide and 1-aminobenzotriazole (Ning and Wang, 2012; Coelho-Moreira et al., 2013a,b). Direct evidence is also available. Some experiments were carried out with the microsomal fraction isolated from *P. ostreatus* (Jauregui et al., 2003). In this work, the investigators found that the microsomes transformed the pesticides in vitro in a NADPH-dependent reaction.

6.7.3 Application of ligninolytic enzymes in the degradation of textile dyes

More than 0.7 million tons of dyes and pigments are produced annually worldwide, presenting more than 10,000 different chemical structures (Young and Yu, 1997). Several of these dyes are conducive to light, temperature, and microbial attack, making them recalcitrant compounds. Dyes can obstruct the passage of sunlight through water resources, reducing photosynthesis by aquatic plants coupled to consequent decreases of the concentration of dissolved oxygen and diminution of the biodegradation of organic matter. Presently, the removal of color from the colored effluents is conducted using physical and chemical methods such as adsorption, precipitation, coagulation–flocculation, oxidation, filtration, and photodegradation, which present advantages and disadvantages (Ferreira-Leitão et al., 2007; Robinson et al., 2001). Many works have reported dye degradation by the use of ligninolytic enzymes. Different dye classes, such as heterocyclic, polymeric, triphenymethane, azo, indigo, anthraquinones, and phtalocyanin, were degraded by ligninolytic enzymes (Neifar et al., 2011; Zilly et al., 2002, 2011). Although most studies on dye decolorization are carried out using spectrophotometric analysis, this technique is limited as an analytical tool since it allows only the evaluation of the chemical modifications that occur in the chromophore groups. However, more recently, new methodologies such as liquid chromatography-mass spectrometry, ^{13}C-nuclear magnetic resonance, high-performance liquid chromatography systems equipped with a diode array detector, LC-Ms-electrospray ionization, and EPR spectroscopy have been introduced to analyze the decomposition of dyes by ligninolytic enzymes (Baratto et al., 2015; Hsu et al., 2012; Michniewicz et al., 2008; Murugesan et al., 2009; Zhao and Hardin 2007; Zille et al., 2005). These methodologies allow the identification of the metabolite products and the proposition of catalytic transformation mechanisms. Peroxidases are applied in the degradation of dyes (Xu et al., 2017). Still, free and immobilized laccases associated or not to natural and synthetic mediators are the most common ligninolytic enzymes used in these studies (Iark et al., 2019; Uber et al., 2020). The immobilization of laccase results in several improvements for its application on a large scale, including an increase in storage and operational stabilities, better control of the enzymatic reaction in aqueous solution, and the possibility of repeated use (Brugnari et al., 2018; Dai et al., 2016; Uber et al., 2020; Zhou et al., 2021).

6.7.4 Application of ligninolytic enzymes in pulp and paper industry

The manufacture of pulp, paper, and paper products ranks among the world's largest industries. The traditional pulp and paper production process is based on chemicals and mechanical processing, which consumes large amounts of raw materials, water, and energy and creates considerable pressure on the environment (Rosenfeld and Feng, 2011). A range of enzyme applications in the pulp and paper industry, including ligninolytic enzymes can be useful for reducing the environmental impacts caused by this important economic activity (Jegannathana and Nielsen, 2013). LiPs are useful in bleaching pulp (Bajpai et al., 2006; Sigoillot et al., 2005), and both LiP and MnP have been shown to be efficient in the decolorization of kraft pulp mill effluents (Ferrer et al., 1991; Moreira et al., 2003).

Laccase can be used in the pulp and paper industry in a number of ways, including lignin degradation, deinking, pitch control, grafting on fibers to improve properties, and pulp and paper mill effluent detoxification (Virk et al., 2012). Lignozym-process, which refers to the LMS employing mediators, such as ABTS and HBT, can remove lignin from pulp. This improves the brightness of pulp, thereby making the paper "white" (Call and Mucke, 1997). The combination of laccase mediators with cellulases and hemicellulases has also been considered for deinking to produce pulps with improved physical and optical properties. The combination of hydrolytic and oxidative enzymes has been described as highly efficient with lower environmental impacts (Woolridge, 2014; Singh et al., 2016; Zainith et al., 2020).

6.8 Concluding remarks

Basidiomycetes represent a reservoir of important bioactive compounds. In this chapter, efforts were made to present a general panorama of the enzymes involved in the capability of these fungi to degrade vegetal biomass and their industrial and biotechnological applications. Basidiomycete enzymes involved in the degradation of plant cell wall polysaccharides have been receiving less attention than those of soft-rot ascomycetes and deuteromycetes. However, their ligninolytic systems (peroxidases and laccases) have great biotechnological importance. Ligninolytic enzymes have been used especially in the so-called white biotechnology, where vegetal biomass can be useful as an alternative to fossil resources for producing chemicals such as biofuels and biopolymers. Ligninolytic enzymes also have several environmental applications; thanks to their nonspecificity, these enzymes are able to degrade several xenobiotics and recalcitrant pollutants, including pesticides, herbicides, and textile dyes.

Additionally, ligninolytic enzymes present applications in the food, medical, pharmaceutical, cosmetic, and nanotechnological areas. After describing so many enzymes, the present trend is to characterize synergism as a real possibility of enhancing efficiency. The final goal is to use highly efficient enzymatic cocktails for industrial purposes.

Acknowledgments

R. M. Peralta and A. Bracht are fellowship holders of the Conselho Nacional de Desenvolvi-mento Científico e Tecnológico (CNPq). This work was supported by a grant from CNPq.

Abbreviations

AA acetylacetone
AAD aryl-alcohol dehydrogenase
ABTS 2,2′-Azino-bis(3-ethylbenzothiazoline-6-sulfonic acid)
APD aromatic peroxygenases
CyPs cytochrome P450 mono-oxygenases
DyP dye-decolorizing peroxidases
EDCs endocrine disruptor compounds
EPR electron paramagnetic resonance
GLOX glyoxal oxidase
GOX glucose 1-oxidase
HBT 1-hydroxybenzotriazole
Lcc laccase
LiP lignin peroxidase
LMEs lignin-modifying enzymes
LMS laccase-mediator system
LS-Ms liquid chromatography-mass spectrometry
MnP manganese peroxidase
PhACs pharmaceutically active compounds
P2Ox pyranose 2-oxidase
PAHs polycyclic aromatic hydrocarbons
QR quinone reductase
RBBR Remazol Brilliant Blue R
SSC solid-state cultivation
VLA violuric acid
VP versatile peroxidase

References

Abdel-Hamid, A.M., Solbiati, J.O., Cann, I.K.O., 2013. Insights into lignin degradation and its potential industrial applications. Adv. Appl. Microbiol. 82, 1–28.

Agbor, V.B., Cicek, N., Sparling, R., Berlin, A., Levin, D.B., 2011. Biomass pretreatment: fundamentals toward application. Biotechnol. Adv. 29, 675–685.

Agrawal, K., Chaturvedi, V., Verma, P., 2018. Fungal laccase discovered but yet undiscovered. Bioresour. Bioprocess. 5, article number 4.

Anastasi, A., Tigini, V., Varese, G.C., 2013. The bioremediation potential of different ecophysiological groups of fungi. In: Goltapeh, E., Mohammadi, D., Younes, R., Varma, A. (Eds.), Fungi as Bioremediators. Springer-Verlag, Berlin Heidelberg, pp. 29–49.

Arantes, V., Goodell, B., 2014. Current understanding of brown rot fungal biodegradation mechanisms: a review. In: Schultz, T.P., Goodell, B., Nicholas, D.D. (Eds.), Deterioration and Protection of Sustainable Biomaterials. ACS Publication, USA, pp. 3–21.

Asgher, M., Iqbal, H.M.N., 2011. Characterization of a novel manganese peroxidase purified from solid state culture of *Trametes versicolor* IBL-04. BioResources 6, 4317–4330.

Backes, E., Kato, C.G., Corrêa, R.C.G., Peralta Muniz Moreira, R.F., Peralta, R.A., Barros, L. Ferreira, et al., 2021. Laccases in food processing: current status, bottlenecks and perspectives. Trends Food Sci. & Technol. 115, 445–460.

Bajpai, P., Ananad, A., Bajpai, P.K., 2006. Bleaching with lignin oxidizing enzymes. Biotechnol. Annu. Rev. 12, 349–378.

Ballantyne, B., Cawley, T.J., 2001. 2,4-Pentanedione. J. Appl. Toxicol. 21, 165–171.

Baratto, M.C., Juarez-Moreno, K., Pogni, R., Basosi, R., Vazquez-Duhalt, R., 2015. EPR and LC-MS studies on the mechanism of industrial dye decolorization by versatile peroxidase from *Bjerkandera adusta*. Environ. Sci. Pollut. Res. Int. 22, 8683–8692.

Baruah, J., Nath, B.K., Sharma, R., Kumar, S., Deka, R.C., Baruah, D.C., et al., 2018. Recent trends in the pretreatment of lignocellulosic biomass for value-added products. Front. Energy Res. 6, article 141.

Bazanella, G.C.S., Araújo, C.A.V., Castoldi, R., Maciel, G.M., Indácio, F.D., Souza, C.G.M., et al., 2013. Ligninolytic enzymes from white-rot fungi and application in the removal of synthetic dyes. In: Polizeli, M.L.T.M., Rai, M. (Eds.), Fungal Enzymes. CRC Press, Boca Raton, pp. 258–279.

Bentil, J.A., Thygesen, A., Mensah, M., Lange, L., Meyer, A.S., 2018. Cellulase production by white-rot basidiomycetous fungi: solid-state vs submerged cultivation. Appl. Microbiol. Biotechnol. 102, 5827–5839.

Bibi, I., Bhati, H.N., Asgher, M., 2011. Comparative study of natural and synthetic phenolic compounds as efficient laccase mediators for the transformation of cationic dye. Biochem. Eng. J. 56, 225–231.

Bilal, M., Iqbal, H.M.N., 2020. Ligninolytic enzymes mediated ligninolysis: an untapped biocatalytic potential to deconstruct lignocellulosic molecules in a sustainable manner. Catal. Lett. 150, 524–543.

Bilal, M., Nawaz, M.Z., Iqbal, H.M.N., Hou, J.L., Mahboob, S., Al-Ghanim, K.A., et al., 2018. Engineering ligninolytic consortium for bioconversion of lignocelluloses to ethanol and chemicals. Protein Pept. Lett. 25, 108–119.

Blackwell, M., 2011. The fungi: 1,2,3...0.5.1 million species? Am. J. Bot. 98, 426–438.

Brugnari, T., Pereira, M.G., Bubna, G.A., Freitas, E.N., Contato, A.G., Corrêa, R.C., et al., 2018. A highly reusable MANAE-agarose-immobilized *Pleurotus ostreatus* laccase for degradation of bisphenol A. Sci. Total. Envirn 634, 1346–1351.

Call, H.P., Mucke, I., 1997. History, overview and applications of mediated. Biotechnol. 53, 163–202.

Camarero, S., Garcia, O., Vidal, T., Colom, J., del Rio, J.C., Gutierrez, A., et al., 2004. Efficient bleaching of non-wood high-quality paper pulp using laccase-mediator system. Enzyme Microbiol. Tech. 35, 113–120.

Camarero, S., Ibarra, D., Martinez, M.J., Martinez, A.T., 2005. Lignin-derived compounds as efficient laccase mediators for decolorization of different types of recalcitrant dyes. Appl. Environ. Microbiol. 71, 1775–1784.

Canã, A.I., Camarero, S., 2010. Laccases and their natural mediators: biotechnological tools for sustainable eco-friendly processes. Biotechnol. Adv. 28, 694–705.

Castoldi, R., Bracht, A., Morais, G.R., Baesso, M.L., Correa, R.C.G., Peralta, R.A., et al., 2014. Biological pretreatment of *Eucalyptus grandis* sawdust with white-rot fungi: study of degradation patterns and saccharification kinetics. Chem. Eng. J. 258, 240–246.

Chowdhary, P., More, N., Yadav, A., Bharagava, R.N., 2019a. Ligninolytic enzymes: an introduction and applications in the food industry. Enzymes Food Biotechnology, Production, Applications, Future Prospect. Academic Press, pp. 181–195. Available from: doi:10.1016/b978-0-12-813280-7.00012-8.

Chowdhary, P., Shukla, G., Raj, G., Ferreira, L.F.R., Bharagava, R.N., 2019b. Microbial manganese peroxidase: a ligninolytic enzyme and its ample opportunities in research. SN Appl. Sci. 1, article number 45.

Christopher, L.P., Yao, B., Ji, Y., 2014. Lignin biodegradation with laccase-mediator systems. Front. Energy Res. 2. Available from: https://doi.org/10.3389/fenrg.2014.00012article 12, 13 pages.

Coelho-Moreira, J.S., Bracht, A., Souza, A.C.S., Oliveira, R.F., Sá-Nakanishi, A.B., Souza, C.G. M., et al., 2013a. Degradation of diuron by *Phanerochaete chrysosporium*: role of ligninolytic enzymes and cytochrome P450. BioMed. Res. Int. 9, Article ID 251354.

Coelho-Moreira, J.S., Maciel, G.M., Castoldi, R., Mariano, S.S., Inácio, F.D., Bracht, A., et al., 2013b. Involvement of lignin-modifying enzymes in the degradation of herbicides. In: Price, A. (Ed.), Herbicides - Advances in Research. InTech, pp. 165–187. ISBN: 978-953-51-1122-1.

Cohen, R., Suzuki, M.R., Hammel, K.E., 2005. Processive endoglucanase active in crystalline cellulose hydrolysis by the brown rot basidiomycete *Gloeophyllum trabeum*. Appl. Environ. Microbiol. 71, 2412–2417.

Couto, S.R., 2008. Exploitation of biological wastes for the production of value-added products under solid state fermentation conditions. Biotechnol. J. 3, 859–870.

Daâssi, D., Rodriguez-Couto, S., Nasri, M., Mechichi, T., 2014. Biodegradation of textile dyes by immobilized laccase from *Coriolopsis gallica* into Ca-alginate beads. Int. Biodeter. Biodegrad. 90, 71–78.

Dai, J., Wang, H., Chi, H., Wang, Y., Zhao, J., 2016. Immobilization of laccase from *Pleurotus ostreatus* on magnetic separable SiO_2 support and excellent activity towards azo dye decolorization. J. Environ. Chem. Eng. 4, 2585–2591.

Daniel, G., 2014. Fungal and bacterial biodegradation: white-rots, brown-rots, soft rots, and bacteria. In: Schultz, T.P., Goodell, B., Nicholas, D.D. Deterioration and Protection of Sustainable Biomaterials. ACS Symposium Series, 1158, pp. 23–58.

Dey, S., Maiti, T.K., Saha, N., Banerjee, R., Bhattacharyya, B.C., 1991. Extracellular protease and amylase activities in ligninase-producing liquid culture of *Phanerochaete chrysosporium*. Proc. Biochem. 26, 325–329.

Dhagat, S., Jujjavarapu, S.E., 2021. Utility of lignin-modifying enzymes: a green technology for organic compound mycodegradation. J. Chem. Technol. Biotechnol. Available from: https://doi.org/10.1002/jctb.6807 *In press*.

Dill, I., Kraepelin, G., 1986. Palo podrido: model for extensive delignification of wood by *Ganoderma applanatum*. Appl. Environ. Microbiol. 52, 1305–1312.

Erjavec, J., Kos, J., Ravnikar, M., Dreo, T., Sabotic, J., 2012. Proteins of higher fungi-from forest to application. Trends Biotechnol. 30, 259–273.

Falade, C., Green, E., Mabinva, L.V., Okoh, A.I., 2017. Lignin peroxidase functionalities and prospective applications. MicrobiologyOpen. 6, e00394.

Fernandez-Fueyo, E., Ruiz-Dueñas, F.J., Miki, Y., Martinez, M.J., Hammel, K.E., Martinez, A. T., 2012. Lignin-degrading peroxidases from genome of selective ligninolytic fungus *Ceriporiopsis subvermispora*. J. Biol. Chem. 287, 16903–16916.

Ferreira-Leitão, V.S., Carvalho, M.E., Bon, E.P.S., 2007. Lignin peroxidase efficiency for methylene blue decoloration: comparison to reported methods. Dye. Pigment. 74, 230–236.

Ferrer, I., Dezotti, M., Durán, N., 1991. Decolorization of Kraft effluent by free and immobilized lignin peroxidases and horseradish peroxidase. Biotechnol. Lett. 13, 577–582.

Hansson, H., Karkehabadi, S., Mikkelsen, N., Douglas, N.R., Kim, S., Lam, A., et al., 2017. High-resolution structure of a lytic polysaccharide monooxygenase from *Hypocrea jecorina* reveals a predicted linker as an integral part of the catalytic domain. J. Biol. Chem. 292, 19099–19109.

Harris, P.V., Welner, D., McFarland, K.C., Re, E., Poulsen, J.N., Brown, K., et al., 2010. Stimulation of lignocellulosic biomass hydrolysis by proteins of glycoside hydrolase family 61: structure and function of a large, enigmatic family. Biochem. (NY.) 49, 3305–3316.

Hawksworth, D.L., Lücking, R., 2017. Fungal diversity revisited: 2.2 to 3.8 million species. Microbiol. Spectr. 5.

He, J., Wu, A., Chen, D., Yu, B., Mao, X., Zheng, P., et al., 2014. Cost-effective lignocellulolytic enzyme production by *Trichoderma reesei* on a cane molasses medium. Biotechnol. Biofuels 7, 43–51.

Hernández-Bueno, N.S., Suárez-Rodríguez, R., Balcázar-López, E., Folch-Mallol, J.L., Ramírez-Trujillo, J.A., Iturriaga, G., 2021. A versatile peroxidase from the fungus *Bjerkandera adusta* confers abiotic stress tolerance in transgenic tobacco. Plants 10, 859–876.

Hofrichter, M., Ullrich, R., Pecyna, M.J., Liers, C., Lundell, T., 2010. New and classic families of secreted fungal heme peroxidases. Appl. Microbiol. Biotechnol. 87, 871–897.

Holker, U., Hofer, M., Lenz, J., 2004. Biotechnological advantages of laboratory-scale solid-state fermentation with fungi. Appl. Microbiol. Biotechnol. 64, 175–186.

Hollmann, F., Gumulya, Y., Tölle, C., Liese, A., Thum, O., 2008. Evaluation of the laccase from *Myceliophthora thermophila* as industrial biocatalyst for polymerization reactions. Macromolecules 41, 8520–8524.

Hsu, C.A., Wen, T.N., Su, Y.C., Jiang, Z.B., Chen, C.W., Shyur, L.F., 2012. Biological degradation of anthraquinone and azo dyes by a novel laccase from *Lentinus* sp. Environ. Sci. Technol. 46, 5109–5117.

Iark, D., Buzzo, A.J.R., Garcia, J.A.A., Correa, V.G., Helm, C.V., Corrêa, R.C.G., et al., 2019. Enzymatic degradation and detoxification of azo dye Congo red by a new laccase from *Oudemansiella canarii*. Bioresour. Technol. 289, 121655.

Ishida, T., Yaoi, K., Hiyoshi, A., Igarashi, K., Samejima, M., 2007. Substrate recognition by glycoside hydrolase family 74 xyloglucanase from the basidiomycete *Phanerochaete chrysosporium*. FEBS J. 274, 5727–5736.

Janusz, G., Kucharzyk, K.H., Pawlik, A., Staszczak, M., Paszczynski, A.J., 2013. Fungal laccase, manganese peroxidase and lignin peroxidase: Gene expression and regulation. Enzyme Microb. Tech. 52, 1–12.

Jauregui, J., Valderrama, B., Albores, A., Vazquez-Duhalt, R., 2003. Microsomal transformation of organophosphorus pesticides by white rot fungi. Biodegradation 14, 397–406.

Jegannathana, K.R., Nielsen, P.H., 2013. Environmental assessment of enzyme use in industrial production – a literature review. J. Clean. Prod. 42, 228–240.

Jeon, J.R., Murugesan, K., Kim, Y.M., Kim, E.J., Chang, Y.S., 2008. Synergistic effect of laccase mediators on pentachlorophenol removal by *Ganoderma lucidum* laccase. Appl. Microbiol. Biotechnol. 81, 783–790.

Johannes, C., Majcherczyk, A., 2000. Natural mediators in the oxidation of polycyclic aromatic hydrocarbons by laccase mediator systems. Appl. Environ. Microbiol. 66, 524–528.

Jonsson, L., Abriksson, B., Nilvebrant, N.-O., 2013. Bioconversion of lignocellulose: inhibitors and detoxification. Biotechnol. Biofuels 6, 16–26.

Juturu, V., Wu, J.C., 2013. Insight into microbial hemicellulases other than xylanases: a review. J. Chem. Technol. Biotechnol. 88, 353–363.

Karigar, C.S., Rao, S.S., 2011. Role of microbial enzymes in the bioremediation of pollutants: a review. Enzyme Res. 11. Article ID 805187.

Koutrotsios, G., Mountzouris, K.C., Chatzipavlidis, I., Zervakis, G.I., 2014. Bioconversion of lignocellulosic residues by *Agrocybe cylindracea* and *Pleurotus ostreatus* mushroom fungi: assessment of their effect on the final product and spent substrate properties. Food Chem. 161, 127–135.

Kues, U., 2015. Fungal enzymes for environmental management. Cur. Opin. Biotechnol. 33, 268–278.

Kuhad, R.C., Gupta, R., Singh, A., 2011. Microbial cellulases and their industrial applications. Enzyme Res. 10. article ID 280696.

Kulshreshtha, S., Mathur, N., Bhatnagar, P., 2014. Mushroom as a product and their role in mycoremediation. AMB. Express 4, 29–36.

Kumari, D., Singh, R., 2018. Pretreatment of lignocellulosic wastes for biofuel production: a critical review. Renew. Sustain. Energy Rev. 90, 877–891.

Levasseur, A., Drula, E., Lombard, V., Coutinho, P.M., Henrissat, B., 2013. Expansion of the enzymatic repertoire of the CAZy database to integrate auxiliary redox enzymes. Biotechnol. Biofuels 6, 41–54.

Liers, C., Bobeth, C., Pecyna, M., Ullrich, R., Hofrichter, M., 2010. DyP-like peroxidases of the jelly fungus *Auricularia auricula-judae* oxidize bnonphenolic lignin model compounds and high-redox potential dyes. Appl. Microbiol. Biotechnol. 85, 1869–1879.

Lobos, S., Larraín, J., Salas, L., Cullen, D., Vicuña, R., 1994. Isoenzymes of manganese-dependent peroxidase and laccase produced by the lignin-degrading basidiomycete *Ceriporiopsis subvermispora*. Microbiology 140, 2691–2698.

Lundell, T.K., Makela, M.R., Hildén, K., 2010. Lignin-modifying enzymes in filamentous basidiomycetes- ecological, functional and phylogenetic review. J. Bas. Microbiol. 50, 5–20.

Lynch, J.M., 1987. Utilisation of lignocellulosic wastes. J. Appl. Bact. Symp. Suppl. 71S-83S.

Maciel, G.M., Bracht, A., Souza, C.G.M., Costa, A.M., Peralta, R.M., 2012. Fundamentals, diversity and application of white rot fungi. In: Silva, A.P., Sol, M. (Eds.), Fungi: Types, Environmental Impact and Role in Disease. Nova Science Publishers Inc, New York, pp. 409–457.

Malhotra, M., Suman, S.K., 2021. Laccase-mediated delignification and detoxification of lignocellulosic biomass: removing obstacles in energy generation. Environ. Sci. Pollut. Res. Available from: https://doi.org/10.1007/s11356-021-13283-0.

Mankar, A.R., Pandey, A., Modak, A., Pant, K.K., 2021. Pretreatment of lignocellulosic biomass: a review on recent advances. Bioresour. Technol. 334, article 125235.

Martani, F., Beltrametti, F., Porro, D., Branduardi, P., Lotti, M., 2017. The importance of fermentative conditions for the biotechnological production of lignin modifying enzymes from white-rot fungi. FEMS Microbiol. Lett. 364, fnx134.

Martínez, A.T., 2002. Molecular biology and structure function of lignin-degrading heme peroxidases. Enzyme Microb. Technol. 30, 425–444.

Martinez, D., et al., 2009. Genome, transcriptome, and secretome analysis of wood decay fungus *Postia placenta* supports unique mechanisms of lignocellulose conversion. Proc. Natl. Acad. Sci. USA 106, 1954–1959.

Martinez, D., Larrondo, L.F., Putnam, N., Gelpke, M.D., Huang, K., Chapman, J., et al., 2004. Genome sequence of the lignocellulose degrading fungus *Phanerochaete chrysosporium* strain RP78. Nat. Biotechnol. 22, 695–700.

Martínez, A.T., Speranza, M., Ruiz-Dueñas, F.J., Ferreira, P., Camarero, S., Guillén, F., et al., 2005. Biodegradation of lignocellulosics: microbial, chemical, and enzymatic aspects of the fungal attack of lignin. Int. Microbiol. 8, 195–2004.

Martinez, D., Berka, R.M., Henrissat, B., Saloheimo, M., Arvas, M., Baker, S.E., et al., 2008. Genome sequencing and analysis of the biomass-degrading fungus *Trichoderma reesei* (syn. *Hypocrea jecorina*). Nat. Biotechnol. 26, 553–560.

Meenakshisundaram, S., Fayeulle, A., Leonard, E., Ceballos, C., Pauss, A., 2021. Fiber degradation and carbohydrate production by combined biological and chemical/physicochemical pretreatment methods of lignocellulosic biomass – a review. Bioresour. Technol. 331, article number 125053.

Meng, Q., Yan, J., Wu, R., Liu, H., Sun, Y., Wu, N., et al., 2021. Sustainable production of benzene from lignin. Nat. Comm. 12, 4534.

Michniewicz, A., Ladakowicz, S., Ullrich, R., Hofrichter, M., 2008. Kinetics of the enzymatic decolorisation of textile dyes by laccase from *Cerrena unicolor*. Dye. Pigment. 77, 295–302.

Moreira, M.T., Feijoo, G., Canaval, J., Lema, J.M., 2003. Semipilot-scale bleaching of Kraft pulp with manganese peroxide. Wood Sci. Technol. 37, 117–123.

Morozova, O.V., Shumakovich, G.P., Gornbacheva, M.A., Shleev, S.V., Yaropolov, A.I., 2007. Blue laccases. Biochemistry 72, 1136–1150.

Murugesan, K., Yang, I.-H., Kim, Y.-M., Jeon, J.-R., Chang, Y.-S., 2009. Enhanced transformation of malachite green by laccase of *Ganoderma lucidum* in the presence of natural phenolic compounds. Appl. Microbiol. Biotechnol. 82, 341–350.

Neifar, M., Jaouani, A., Kamoun, A., Ellouze-Ghorbel, R., Ellouze-Chaabouni, S., 2011. Decolorization of solophenyl red 3BL polyazo dye by laccase-mediator system: optimization through response surface methodology. Enz. Res. Article ID 179050.

Nigam, P.S., 2013. Microbial enzymes with special characteristics for biotechnology applications. Biomolecules 3, 597–611.

Ning, D., Wang, H., 2012. Involvement of cytochrome P450 in pentachlorophenol transformation in a white rot fungus *Phanerochaete chrysosporium*. PLoS One 7, e45887.

Paramjeet, S., Manasa, P., Korrapati, N., 2018. Biofuels: Production of fungal-mediated ligninolytic enzymes and the modes of bioprocesses utilizing agro-based residues. Biocatal. Agric. Biotechnol. 14, 57–71.

Payne, C.M., Knott, B.C., Mayes, H.B., Hansson, H., Himmel, M.E., Sandgren, M., et al., 2015. Fungal cellulases. Chem. Rev. 115, 1308–1448.

Piscitelli, A., Pezzella, C., Lettera, V., Giardina, P., Faraco, V., Sannia, G., 2013. Fungal laccases: structure, function and application. In: Polizeli, M.L.T.M., Rai, M. (Eds.), Fungal Enzymes. CRC Press, pp. 113–151.

Placido, J., Capareda, S., 2015. Ligninolytic enzymes: a biotechnological alternative for bioethanol production. Bioresour. Bioproc. 2, 23–34.

Polizeli, M.L.T.M., Rizzatti, A.C., Monti, R., Terenzi, H.F., Jorge, J.A., Amorim, D.S., 2005. Xylanases from fungi: properties and industrial application. Appl. Microbiol. Biotechnol. 67, 577–591.

Poppins-Levlin, K., Wang, W., Tamminen, T., Hortling, B., Viikari, L., Nikiu-Paavola, M.L., 1999. Effects of laccase/HBT treatment on pulps and lignin structures. J. Pulp Pap. Sci. 25, 90–94.

Pozdnyakova, N.N., 2012. Involvement of the ligninolytic system of white-rot and litter-decomposing fungi in the degradation of polycyclic aromatic hydrocarbons. Biotechnol. Res. Int. 20. Article ID 243217.

Prewitt, M.L., Diehl, S., McElroy, T.C., Diehl, W.J., 2008. Comparison of general fungal and basidiomycetes-specific ITS primers for identification of wood decay fungi. For. Prod. J. 58, 68–71.

Rico, A., Rencoret, J., del Rio, J.C., Martinez, A.T., Gutierrez, A., 2014. Pretreatment with laccase and a phenolic mediator degrades lignin and enhances saccharification of *Eucalyptus* feedstock. Biotechnol. Biofuels 7, 6–20.

Riley, R., et al., 2014. Extensive sampling of basidiomycete genomes demonstrates inadequacy of the white-rot/brown-rot paradigm for wood decay fungi. Proc. Natl. Acad. Sci. U.S.A. 111, 9923–9928.

Robinson, T., McMullan, G., Marchant, R., Nigam, P., 2001. Remediation of dyes in textile effluent: a critical review on current treatment technologies with a proposed alternative. Bioresour. Technol. 77, 247–255.

Rosenfeld, E.P., Feng, L.G.H., 2011. The paper and pulp industry. In: Rosenfeld, P.E., Feng, L. (Eds.), Risks of Hazardous Wastes. Elsevier, Oxford.

Ryu, W.Y., Jang, M.Y., Cho, M.H., 2003. The selective visualization of lignin peroxidase, manganese peroxidase and laccase produced by white rot fungi on solid media. Biotechnol. Bioproc. Eng. 8, 130–134.

Salmon, D.N.X., Spier, M.R., Soccol, C.R., Vanderberghe, L.P.S., Montibeller, V.W., Bier, M.C. J., et al., 2014. Analysis of inducers of xylanase and cellulase activities production by *Ganoderma applanatum* LPB MR-56. Fungal Biol. 118, 655–662.

Salmones, D., Mata, G., Waliszewski, K.N., 2005. Comparative culturing of *Pleurotus* spp on coffee pulp and wheat straw: biomass production and substrate biodegradation. Bioresour. Technol. 96, 537–544.

Schuster, A., Schmoll, M., 2010. Biology and biotechnology of *Trichoderma*. Appl. Microbiol. Biotechnol. 87, 787–799.

Seiboth, B., Ivanova, C., Seidl-Seiboth, V., 2011. *Trichoderma reesei:* a fungal enzyme producer for cellulosic biofuels. In: Bernardes, M.A.S. (Ed.), Biofuel Production: Recent Developments and Prospects. InTech, pp. 309–340. , 105772/16848.

Sigoillot, C., Camarero, S., Vidal, T., Record, E., Asther, M., Pérez-Boada, M., et al., 2005. Comparison of different fungal enzymes for bleaching high-quality paper pulps. J. Biotechnol. 115, 333–343.

Singh, P., Suman, A., Tiwari, P., 2008. Biological pretreatment of sugarcane trash for its conversion to fermentable sugars. World J. Microbiol. Biotechnol. 24, 667–673.

Skyba, O., Douglas, C.J., Mansfiled, S.D., 2013. Syringyl-rich lignin renders poplars more resistant to degradation by wood decay fungi. Appl. Environ. Microbiol. 79, 2560–2571.

Singh, G., Capalash, N., Kaur, K., Puri, S., Sharma, P., 2016. Enzymes: applications in pulp and paper industry. enzymes. Agro-Industrial Wastes Feedstock for Enzyme Production. Academic Press, pp. 157–172. Available from: doi:10.1016/b978-0-12-802392-1.00007-1.

Singh, A.K., Bilal, M., Iqbal, M.N., Raj, A., 2021. Lignin peroxidase in focus for catalytic elimination of contaminants — a critical review on recent progress and perspectives. Int. J. Biol. Macromol. 177, 58–82.

Soccol, C.R., da Costa, E.S.F., Letti, L.A.J., Karp, S.G., Woiciechowski, A.L., Vandenberghe, L. P.S., 2017. Recent developments and innovations in solid state fermentation. Biotechnol. Res. Innov. 1, 52–71.

Solarte-Toro, J.C., Romero-Garcia, J.M., Martínez-Patino, J.C., Ruiz-Ramos, E., Castro-Galiano, E., Cardona-Alzate, A., 2019. Acid pretreatment of lignocellulosic biomass for energy vectors production: a review focused on operational conditions and techno-economic assessment for bioethanol production. Renew. Sustain. Energy Rev. 107, 587–601.

Souza, C.G.M., Tychanowicz, G.K., Souza, D.F., Peralta, R.M., 2004. Production of laccase isoforms by *Pleurotus pulmonarius* in response to presence of phenolic and aromatic compounds. J. Bas. Microbiol. 44, 129–136.

Strong, P.J., Claus, H., 2011. Laccase: a review of its past and its future in bioremediation. Crit. Rev. Environ. Sci.Technol. 41, 373–434.

Tedersoo, L., et al., 2014. Fungal biogeography. Global diversity and geography of soil fungi. Science 346 (6213), 1256688.

Tisma, M., Žnidaršič-Plazl, Selo, G., Tolj, I., Speranda, M., Bucic-Kojic, A., et al., 2021. *Trametes versicolor* in lignocellulose-based bioeconomy: state of the art, challenges and opportunities. Bioresour. Technol. 330, article number 124997.

Toco, D., Carucci, C., Monduzzi, M., Salis, A., Sanjust, E., 2021. Recent developments in the delignification and exploitation of grass lignocellulosic biomass. ACS Sustain. Chem. Eng. 9, 2412–2432.

Uber, T.M., Buzzo, A.J.R., Scaratti, G., Amorim, S.M., Helm, C.V., Maciel, G.M., et al., 2020. Comparative detoxification of Remazol brilliant blue R by free and immobilized laccase of *Oudemansiella canarii*. Biocatal.Biotransformation. Available from: https://doi.org/10.1080/10242422.2020.1835873.

Uzan, E., Nousiarnen, P., Balland, V., Sipila, J., Piumi, F., Navarro, D., et al., 2010. High redox potential laccases from the ligninolytic fungi *Pycnoporus coccineus* and *Pycnoporus sanguineus* suitable for white biotechnology: from gene cloning to enzyme characterization and application. J. Appl. Microbiol. 108, 2199–2213.

Valášková, V., Baldrian, P., 2006. Degradation of cellulose and hemicelluloses by the brown rot fungus *Piptoporus betulinus*: production of extracellular enzymes and characterization of the major cellulases. Microbiology 152, 3613–3622.

Vilar, D.S., Bilal, M., Bharagava, R.N., Kumar, A., Nadda, A.K., Salazar-Banda, G.R., et al., 2021. Lignin-modifying enzymes: a green and environmental responsive technology for organic compound degradation. J. Chem. Technol. Biotechnol. Available from: https://doi.org/10.1002/jctb.6751.

Virk, A.P., Sharma, P., Capalash, N., 2012. Use of laccase in pulp and paper industry. Biotechnol. Progr. 28, 21–32.

Viswanath, B., Rajesh, B., Janardhan, A., Kumar, A.P., Narasimha, G., 2014. Fungal laccases and their applications in bioremediation. Enzyme Res. 21. Article ID 163242.

Woolridge, E.M., 2014. Mixed enzyme systems for delignification of lignocellulosic biomass. Catalysts 4, 1–34.

Ximenes, E., Kim, Y., Mosier, N., Dien, B., Ladish, M., 2010. Inhibition of cellulases by phenols. Enzyme Microb. Technol. 46, 170–176.

Xu, H., Guo, M.-Y., Gao, Y.-H., Bai, X.-H., Zhou, X.-W., 2017. Expression and characteristics of manganese peroxidase from *Ganoderma lucidum* in *Pichia pastoris* and its application in the degradation of four dyes and phenol. BMC Biotechnol. 17, article number 19.

Yang, H., Sun, H., Zhang, S., Wu, B., Pan, B., 2015. Potential of acetylacetone as a mediator for *Trametes versicolor* laccase in enzymatic transformation of organic pollutants. Environ. Sci. Pollut. Res. Available from: https://doi.org/10.1007/s11356-015-4312-2.

Yoon, J.J., Cha, C.J., Kim, Y.S., Son, D.W., Kim, Y.K., 2007. The brown rot basidiomycete *Fomitopsis palustris* has the *endo*-glucanases capable of degrading microcrystalline cellulose. J. Microbiol. Biotechnol. 17, 800–805.

Young, L., Yu, J., 1997. Ligninase-catalyzed decolorization of synthetic dyes. Water Res. 31, 1187–1193.

Zainith, S., Chowdhary, P., Mani, S., Mishra, S., 2020. Microbial ligninolytic enzymes and their role in bioremediation. Microorg. Sustain. Environ. Health 179–203. Available from: https://doi.org/10.1016/B978-0-12-819001-2.00009-7.

Zhao, X., Hardin, I.R., 2007. HPLC and spectrophotometric analysis of biodegradation of azo dyes by *Pleurotus ostreatus*. Dye. Pigment. 73, 322–325.

Zheng, Y., Zhao, J., Xu, F., Li, Y., 2014. Pretreatment of lignocellulosic biomass for enhanced biogas production. Prog. Energy Combust. Sci. 42, 35–53.

Zhuo, R., Fan, F., 2021. A comprehensive insight into the application of white rot fungi and their lignocellulolytic enzymes in the removal of organic pollutants. Sci. Total. Environ. 778, article number 146132.

Zhou, W.T., Zhang, W.X., Cai, Y.P., 2021. Laccase immobilization for water purification: a comprehensive review. Chem. Eng. J. 403, article number 126272.

Zille, A.B., Górnacka, B., Rehorek, A., Cavaco-Paulo, A., 2005. Degradation of azo dyes by *Trametes villosa* laccase over long periods of oxidative conditions. Appl. Environ. Microbiol. 71, 6711–6718.

Zilly, A., Souza, C.G., Barbosa-Tessmann, I.P., Peralta, R.M., 2002. Decolorization of industrial dyes by a Brazilian strain of *Pleurotus pulmonarius* producing laccase as the sole phenol-oxidizing enzyme. Folia Microbiol. 47, 273–277.

Zilly, A., Coelho-Moreira, J.S.C., Bracht, A., Souza, C.G.M., Carvajal, A.E., Koehnlein, E.A., et al., 2011. Influence of NaCl and Na_2SO_4 on the kinetics and dye decolorization ability of crude laccase from *Ganoderma lucidum*. Int. Biodeter. Biodegr 65, 340–344.

Zilly, A., Bazanella, G.C.S., Helm, C.V., Araújo, C.A.V., Souza, C.G.M., Bracht, A., et al., 2012. Solid-state bioconversion of passion fruit waste by white-rot fungi for production of oxidative and hydrolytic enzymes. Food Bioproc. Technol. 5, 1573–1580.

Chapter 7

Metagenomics and new enzymes for the bioeconomy to 2030

Patricia Molina-Espeja[1], Cristina Coscolín[1], Peter N. Golyshin[2] and Manuel Ferrer[1]
[1]Systems Biotechnology Group, ICP, CSIC, Madrid, Spain, [2]Centre for Environmental Biotechnology, Bangor University, Bangor, United Kingdom

7.1 Introduction

Enzymes are proteins, that is, polymers made up of amino acids, which catalyze chemical reactions in all living organisms. For example, a single medium-size genome of a bacterium such as *Escherichia coli* contains approximately 607 enzymes that support more than 700 reactions (Ouzounis and Karp, 2000). The pivotal assets provided by the use of enzymes in industrial processes and products are (1) lower energy footprint, (2) reduced waste production and chemical consumption, (3) safer process conditions, and (4) the use of renewable feedstocks (Bommarius, 2015; Pellis et al., 2018; Sheldon and Woodley, 2018; Hodgson, 2019; Sheldon and Brady, 2019).

Hence, replacing chemicals (including chemical catalysts) with enzymes in industrial processes or products is expected to positively impact greenhouse gas emissions (reported savings from 0.3 to 990 kg CO_2 equivalent/kg product; Jegannathan and Nielsen, 2013) and global warming issues by reducing water and energy consumption (estimates: 6000 million (mln) m^3 and 167 TWh, equivalent to 100 mln barrels of oil) (OECD, 2011; Timmis et al., 2014). This is why enzymes can strengthen the market positioning and competitiveness of multiple industries (Global Index, 2018). However, there is strong evidence that the different enzymes on the market are not in line with current and future stricter environmental regulations (FMCG Gurus, 2019). For example, the Paris Climate Conference (COP21) has prescribed a reduction of the emissions of greenhouse gases to 55 Gt of CO_2 equivalent ($GtCO_2eq$) in 2030 (equivalent to driving 2,036,650 km in an average car or 55 times around the world), and most known enzymes cannot meet these

Biotechnology of Microbial Enzymes. DOI: https://doi.org/10.1016/B978-0-443-19059-9.00013-X
© 2023 Elsevier Inc. All rights reserved.

demands. This is important as we have agreed on a number of steps to promote climate neutrality by 2050 while enhancing economic competitiveness in a world facing major environmental threats. And more recently, these objectives are marked by the recovery after the COVID-19 crisis, for which many countries have agreed on Recovery, Transformation and Resilience Mechanisms. In addition to all of this, there is also strong evidence, through socioeconomic assessments evaluating the consumer's perception of the environmental impacts of daily life habits, that (1) c. 90% of consumers have a more positive image of a company that supports biotechnology and will buy a product with an environmental benefit, (2) 50% of European consumers are willing to recognize a green premium for a more sustainable greener alternative, and (3) changes in the consumer behavior can significantly decrease environmental impacts (FMCG Gurus, 2019) (Box 7.1).

7.2 Metagenomics

The search for new enzymes has undergone several waves related to advances in protein structure determination (the 1980s), rational design and directed evolution (the 1990s and 2000s), and the ab initio design of artificial enzymes (late 2010–20s) with desired properties (Bornscheuer et al., 2012; Hodgson, 2019; Sheldon and Brady, 2019). All these advances, and the strategies developed, have been transferred to the synthesis of new molecules (Sheldon and Brady, 2019). We can envision the importance of some of these advances by the recognition given to directed evolution, recipient of the 2018 Nobel Prize in chemistry (Arnold, 2019).

In genetic engineering, we start a priori from an industrial problem that responds to increasing some enzymatic property that leads to scientific advances (Arnold, 2019). These increases can be achieved, for example, through directed evolution and iterative cycles of mutagenesis, the search for better mutants and testing of the property sought, and through rational design in which a computational analysis suggests mutations that are tested

BOX 7.1 Technological innovations for implementing new enzymes

The establishment or implementation of new enzyme-based processes and products may not depend on the enzyme's price but on whether the industry is and will be capable of searching and implementing new enzymes in timeframes not exceeding the climate and environmental policy commitments and serving the present and future recovery, transformation, resilience, and socioeconomic demands in multiple sectors. Clearly, any action addressing a technology that can help in this direction, which can be of interest, and can be transferred to industry, is of great importance.

individually or jointly. Recent examples include substrate-driven screening to reach high activity levels of an enzyme through a rational and directed evolution refinement and an initial smart selection of amino acid hotspots derived from extensive enzyme-substrate molecular models (Mateljak et al., 2019; Roda et al., 2021). For increasing enzyme stability, theoretical–experimental approaches combining thermodynamic predictors (FoldX, Rosetta), the clustering and validation of promising positions (top 50–100), and machine learning into additive terms can also be applied (Roda et al., 2021). Due to its high speed and screening capacity, this tool will be the preferred one to promote the performance of enzymes for which crystal structures are available or high-quality homology models can be generated. Indeed, the possibility of accessing high-quality, three-dimensional models in hours/minutes using the AlphaFold/RoseTTAFold (Senior et al., 2020; Jumper et al., 2021; Tunyasuvunakool et al., 2021) allows protein engineering without the need of crystal structures.

In both cases, enzymes with known sequences and structures are used. This does not occur through the application of the so-called genomic or metagenomic techniques in which completely novel enzymes can be searched by simply isolating DNA from a single microbial culture or a microbial community inhabiting an environmental sample that can then be further cloned and traced by activity assays or sequenced directly to search for enzymes with the help of computers (Ferrer et al., 2019; Ye et al., 2019) (Box 7.2).

Metagenomics is a culture-independent technique that makes it theoretically possible to study any type of sample and offers us the possibility of studying DNA from an entire organismal community and screening new

BOX 7.2 Metagenomics for exploring microbial enzymes

The term "metagenomics" was introduced as the application of genomics to the study of microbial communities and enzymes directly extracted from the environment without the need for culturing (Handelsman et al., 1998). Metagenomics is analogous to genomics, but the genome is not derived from a single organism but, rather, from a whole community. This is why metagenomics has proved in the last two decades to be an effective method for exploring the microbial diversity directly from environmental sources, without the need for culturing (Ferrer et al., 2019). By applying bioinformatics and experimental methods, this methodology contributed to disclosing part of the microorganisms' biodiversity and their related activities (Handelsman, 2004). Indeed, using this technique, the search for new enzymes can be performed via either massive searches through direct sequencing of environmental genomic material or a selection of interesting activities in clone libraries created from environmental DNA (Mende et al., 2012; Ferrer et al., 2019; Ye et al., 2019).

enzymes that can be accessed with no limit (Berini et al., 2017; Verma et al., 2021). Both function- and DNA sequence-based metagenomic methods are complementary, with each having advantages and disadvantages. Bioinformatics methods allow a rapid process of enzyme searching; however, in prokaryotic genomes, >30% of genes remain annotated as "hypothetical, conserved hypothetical, or with general prediction", and large numbers of genes may have nonspecific annotations (such as putative hydrolases). Tools such as MetaEuk for high-throughput, reference-based discovery, and annotation of protein-coding genes in eukaryotic metagenomic contigs have been designed to solve this problem (Levy Karin et al., 2020). The analysis of biochemical functions, by the meaning of naïve screens, is likely to provide a superior approach to avoid this limitation, especially when screening novel enzymes. However, only a few hundred specific enzymatic assays exist, with a limited number of them applied in a high-throughput manner for the naïve screening of metagenomic libraries.

The application of these three techniques (directed evolution, rational design, and metagenomics) (Fig. 7.1) is the key today as ~90% of industrial processes use a major catalytic step, and between 20% and 30% of these catalysts will be enzymatic because they can make some industrial processes more sustainable and cleaner and thus solve global problems (Bommarius, 2015; Pellis et al., 2018; Sheldon and Woodley, 2018; Hodgson, 2019;

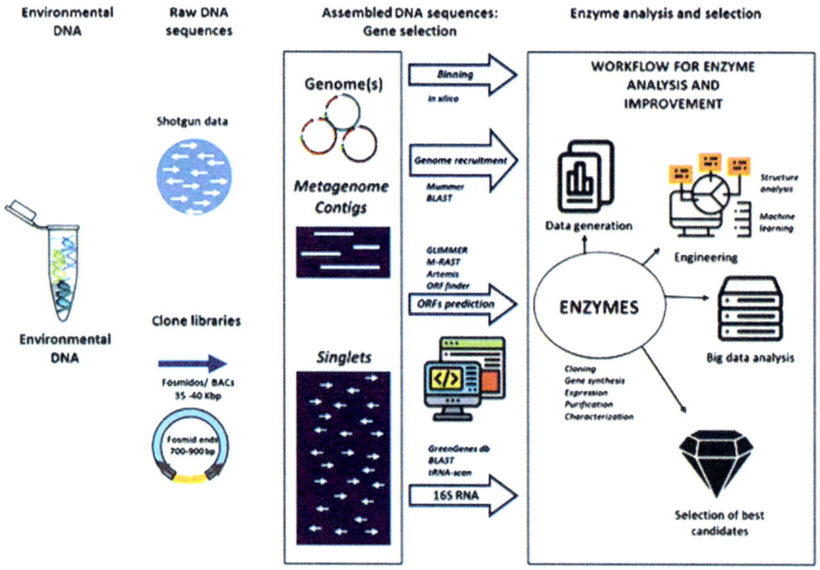

FIGURE 7.1 Workflow for enzyme screening and implementation. The figure summarizes the different steps covering the steps and tools applied to screen enzymes starting from the DNA from microbial communities inhabiting an environmental sample (the metagenome) to their further engineering and selection of best candidates after big data analysis.

Sheldon and Brady, 2019). Increasing advances are being achieved (1) in the generation of better biocatalytic systems with new properties (such as Synthetic enzymes, NanoZymes, and PluriZymes) and with new reactivities (such as the design of enzymes for reactions reserved for organic chemistry: C-Si, C-B, etc.); (2) in computational, structural, and bioinformatics techniques [including computational and structural techniques such as molecular dynamics, XFEL (X-ray electron free laser), and biodynamic nuclear magnetic resonance (NMR) for "molecular movies" during catalysis, electric field rearrangement, massive analyses, ancestral reconstructions, OMICS, etc.]; and (3) automation (high-throughput screening systems, in vitro translation, gene synthesis, etc.) (Devine et al., 2018; Gumulya et al., 2018; Welborn et al., 2018; Arnold, 2019; Alonso et al., 2020; Bell et al., 2021).

However, the growing use of enzymes as a greener alternative to chemical catalysts demands constant innovation, and the access to new enzymes as a starting point for future developments continues to be essential.

7.3 Activity-based methods for enzyme search in metagenomes

Since a gene function is often manifested by the direct activity of its translated protein, the analysis of protein biochemical function is likely to provide a superior approach for elucidating gene function compared to the analysis of a sequence (Editorial, 2018) and to screen enzymatic activity in clone metagenome libraries (Ferrer et al., 2019). Multiple protocols for constructing metagenomic small-insert and large-insert libraries in several vectors have been described (Simon and Daniel, 2017). These techniques can be applied to screen enzymes such as esterases, lipases, proteases, and glycosidases, to cite some, that can be easily screened because of the availability of multiple screen methods at hand (Reyes-Duarte et al., 2012; Popovic et al., 2017; Martínez-Martínez et al., 2018). However, the design of new screen methods allowed the possibility to also extend the screen possibilities to other enzymes such as hydrogenases (Adam and Perner, 2017), rhodopsins (Pushkarev and Béjà, 2016), enzyme-producing biosurfactants (Williams et al., 2019), pollutant-degrading enzymes (Ufarté et al., 2015), amine transferases (Ferrandi et al., 2017; Coscolín et al., 2019), and, more recently, plastic-degrading enzymes (Hajighasemi et al., 2018; Pérez-García et al., 2021; Karunatillaka et al., 2022). These techniques can be applied to DNA from microbial communities inhabiting multiple environments, including extreme and common environments (Popovic et al., 2017; Martínez-Martínez et al., 2018; Wohlgemuth et al., 2018; Ferrer et al., 2019).

Screening of metagenomic libraries offers access to novel enzymes with new activity and stability features. Still, the probability of identifying a certain gene depends on multiple factors that must be taken into account during the planning of a metagenomic study. In this direction, several issues limit the use of metagenomics to identify new enzyme activities. Recent developments are underway to reduce the time required for enzyme identification to its application in technological developments (Jemli et al., 2014). Although a successful analysis requires the combination of an appropriate DNA extraction method, a suitable host–vector system, and an effective screening pipeline, the most significant technological challenges can be summarized as follows:

(1) Understanding the complexity of the environmental material, as due to the high diversity within microbial communities, target genes encoding for novel enzymes represent a small fraction of the total nucleic acid sample (Guazzaroni et al., 2015); (2) the isolation of pure and high-molecular-weight DNA and the construction of a library proportionally representing the composition of microbial community; library size is also a critical factor for success: to obtain a representative coverage of the diversity in soil microbes, 10^6-10^7 bacterial artificial chromosome clones (100 kb insert size) are required, if all species are assumed to be equally abundant (Handelsman et al., 1998); (3) the low proportion of enzymes selected under the conditions required in industrial processes (Martínez-Martínez et al., 2013); (4) the lack of relevant industrial substrates for functional screens (Fernández-Arrojo et al., 2010); (5) the low efficiency of screening methods for rare activities (Ferrer et al., 2016); (6) the low yield of enzymes isolated by these techniques under unnatural conditions (Fernández-Arrojo et al., 2010); (7) the high number of enzymes identified that are inactive after expression in *E. coli* (Loeschcke et al., 2013); (8) the lack of reliable bioinformatics sources for the analysis of bulk sequencing data (Nyyssönen et al., 2013); and (9) the lack of reliable systems for predicting enzymatic activities in hypothetical protein-coding sequences (Bastard et al., 2014).

To solve the abovementioned issues, advances have been made in several lines, including, the design and use of novel vectors to clone the DNA fragments and improve the expression possibilities of genes included in the DNA fragments (Weiland-Bräuer et al., 2017; Cheng et al., 2014), and the design or use of multiple surrogate hosts (Terrón-González et al., 2013; Katzke et al., 2017; Williams et al., 2019; Cecchini et al., 2022). However, how many enzyme activities can be measured using actual screening methods? There are several thousands of different enzyme-catalyzed reactions encoded in each genome or metagenome. However, there are a few hundreds of specific enzymatic assays, including spectrophotometric, fluorimetric, and calorimetric chemiluminescent, radiometric, chromatographic, mass spectrometry, NMR, etc. (Kuznetsova et al., 2005). Some of these methods can be extended for their use in screening new enzymes in clone libraries. The design of new fluorogenic substrates (Nasseri et al., 2018) compatible with agar plate screening and high-throughput robotic screening systems (Smart et al., 2017; Chuzel et al., 2021), including microfluidics (Neun et al., 2020; Cecchini et al., 2022) have also been subjected to investigation to promote the success rate of enzyme discovery (Box 7.3).

BOX 7.3 Supercomputers for enzyme search

Despite the advances in screening assays, the design of large supercomputers and the development of bioinformatics and computational techniques, with the help of massive analysis and highly accurate protein structure prediction tools, are shifting the search for new enzymes towards in silico enzyme searches.

BOX 7.4 Computing resources minimize the time to search enzymes

Imagine, for example, that one wants to screen by homoloy search a metagenome of 15 Mbp and 53,000 open reading frames of more than 50 amino acids, using the Lipase Engineering Database. This database contains more than 280,638 sequences encoding lipases and related proteins sharing the same α/β hydrolase fold (Bauer et al., 2020). If one uses a personal computer with single-core at 3.6 GHz, the search may take about 142 min using Diamond (as fastest search standard). The same analysis using a computing cluster takes about 18 min using the minimum configuration with a single node composed of 40 cores at 2.5 GHz, from a hypothetical maximum of up to 134 nodes. Finally, if one has access to cloud resources (with up-to-date hardware), the search for a single genome usually takes 1–5 min.

7.4 Computers applied to metagenomic enzyme search

A wealth of genetic information is available in public databases. About 180 million protein sequences are in databases, and only around 170,000 protein structures are in the Protein Data Bank (Velankar et al., 2021). In addition, from 1982 to the present, the number of bases in GenBank has doubled approximately every 18 months, with 420,000 formally described species (Sayers et al., 2020). These sequences and others in private repositories, many of which can potentially encode prominent enzymes, can be screened by applying homology/bioinformatics-based approaches with the help of curated databases with representative target sequences (Box 7.3). The speed of the homology-based search process depends on the computing facilities, the size of the reference database (containing sequences of target enzymes) and, even more, the size of the sequence repository (or genome or metagenome) to screen. It also depends on whether one uses Diamond blastp (a faster blast implementation) (Buchfink et al., 2015) or hmmer3 (for Hidden Markov Model profiling) (Borchert et al., 2021; Fernandez-Lopez et al., 2021) (Box 7.4).

The sequences obtained using massive sequencing methods can be filtered according to their similarities with databases of general sequences, such as UniProt (UniProt Consortium, 2021), and databases with conserved domain sequences such as Pfam (Mistry et al., 2021). They can also be filtered using specific databases containing sequences and biochemical information for certain groups of enzymes, such as the Carbohydrate-Active enZYme database (Cantarel et al., 2009; Tasse et al., 2010), which is used to identify glycosidases, and a database for identifying laccases (Sirim et al., 2011), peroxidases (Fawal et al., 2013), oxidoreductases (Duarte et al., 2014), lipases and esterases (Bauer et al., 2020), epoxide hydrolases and haloalkane dehalogenases (Barth et al., 2004), and halohydrin dehalogenases (Schallmey et al., 2014). More recently, a curated database with diverse protein sequences featuring enzyme families relevant to bone degradation, including glycosidases and peptidases, has been established to identify by homology BLAST-search similar sequences in genomes, metagenome-assembled genomes, and metagenome reads sequences encoding targeted bone-degrading enzymes (Fernandez-Lopez et al., 2021). Prediction tools such as antiSMASH have been recently developed for the search of genes coding for enzymes for the synthesis of secondary or bioactive metabolites, such as lactones, bacteriocins, siderophores, ectoines, and polyketides (Blin et al., 2013, 2014).

Whatever the speed of the in silico search process, the output will be a list of sequences encoding enzymes, the activity, specificity, and stability of which could not be inferred by any known method, and each enzyme encoded by selected sequences needs to be expressed and characterized to find whether its properties fit those requested; this is a time-consuming process with a low rate of success (Ferrer et al., 2019). Having said that, it is now accepted that a current main bottleneck in the implementation of enzymes in greener industry processes and products, hindering economic competitiveness and greater sustainability, is that, although bioprospecting and engineering technologies are enjoying a high level of sophistication (Pellis et al., 2018; Arnold, 2018; Ferrer et al., 2019), we are unable to accurately predict enzyme parameters from a protein sequence (Editorial, 2018). This limits the exploitation of the existing large sequence databases for searching enzymes with industrial and manufacturing requirements. Succeeding in this undertaking would revolutionize the possibilities of the industry to better recover the needed enzymes from the ever-increasing amount of sequencing data, which can be generated at an ever-lower price compared to any other method (Editorial, 2018).

The second bottleneck is that if someone intends to use known enzymes, most current engineering efforts come up short when addressing the desired industrial needs in terms of activity and stability; this also occurs when applying current methods to new enzymes (Editorial, 2018). The third crucial problem is poor enzyme productivity and high development, production, and formulation costs once an enzyme is selected and implemented (Ferrer et al., 2016). Another major drawback preventing the acquisition of enzymes by

BOX 7.5 Challenges for replacing chemicals by enzymes

The challenges for replacing chemical counterparts with enzymes are manifold: first, finding and customizing novel enzymes occur in a linear, stepwise, and iterative manner. Second, the process is costly and time-consuming, with enzyme screening alone mounting up to 30 k€ and taking up to 15 months through iterative optimization cycles and scale-up productions. Third, the process often provides gradual improvements. However, most of the current engineering efforts applied to known enzymes fall short of the need for their use in future products and processes, and concerns remain about using current machine learning algorithms to improve new enzymes. Fourth is the limited incorporation of artificial intelligence for enzyme discovery. For example, millions of sequences (each representing an enzyme) are available in public databases, and thus new enzymes encoded by those can be cheaply obtained. However, little use has been made of them due to the lack of appropriate computational resources (hardware, software) to screen them and find those with potential use in the industry while improving the success rate (Boulund et al., 2017; Kusnezowa and Leichert, 2017). Fifth, the lack of automation, artificial intelligence, and modeling of technologies prevent industrial scalability: the low productivity of enzymes, and the high cost of development, production, and formulation once an enzyme has been selected and applied. Sixth, various and complex regulatory and safety aspects are often road blockers for commercial developments.

the market is of limited knowledge about enzymes and their benefits among manufacturers, policymakers, and consumers. For decades, the limited repertoire of available enzymes and lack of reference enzymes have remained a common challenge (Box 7.5).

7.5 Concluding remarks

Microbes are involved in many processes, including the carbon and nitrogen cycles, and are responsible for using and producing greenhouse gases such as carbon dioxide and methane. Microbes can have positive and negative responses to temperature, making them an important component of climate change models (Cavicchioli et al., 2019). At the same time, they are hosts for enzymes that may have the potential to be implemented in industrial products and processes. Such enzymes will respond to the industry's and public's demand for more eco-friendly, efficient, and durable products and processes. How we are and will be able to find and develop such enzymes will be key to the future of the bioeconomy. Besides the broad range of working conditions microbes operate on, it is estimated that 1 trillion (10^{12}) microbial species coexist (Locey and Lennon, 2016). This number can make us think about the astronomic possibilities we have to solve climate change issues using their enzymes as present in nature after their adaptation by engineering. With the help of supercomputers, bioinformatics, computational, and accurate protein structure prediction tools, metagenomic techniques will help accessing and further expand such enzymatic diversity.

Acknowledgments

This study was conducted under the auspices of the FuturEnzyme Project funded by the European Union's Horizon 2020 Research and Innovation Programme under Grant Agreement No. 101000327. We also acknowledge financial support under Grants PCIN-2017–078 (within the Marine Biotechnology ERA-NET, GA No. 604814) (M.F.), BIO2017–85522-R (M.F.), PID2020–112758RB-I00 (M.F.), and PDC2021–121534-I00 (M.F.) from the Ministerio de Economía, Industria y Competitividad, Ministerio de Ciencia e Innovación, Agencia Estatal de Investigación (AEI) (Digital Object Identifier 10.13039/501100011033), Fondo Europeo de Desarrollo Regional (FEDER) and the European Union ("NextGenerationEU/PRTR"), and Grant 2020AEP061 (M.F.) from the Agencia Estatal CSIC.

P.N.G. acknowledges Sêr Cymru programme part-funded by ERDF through Welsh Government for the support of the project BioPOL4Life, the project "Plastic Vectors" funded by the Natural Environment Research Council UK (NERC), Grant No. NE/S004548/N, the Centre for Environmental Biotechnology Project cofunded by the European Regional Development Fund (ERDF) through the Welsh Government.

References

Adam, N., Perner, M., 2017. Activity-based screening of metagenomic libraries for hydrogenase enzymes. Methods Mol. Biol. 1539, 261–270.

Alonso, S., Santiago, G., Cea-Rama, I., Fernandez-Lopez, L., Coscolín, C., Modregger, J., et al., 2020. Genetically engineered proteins with two active sites for enhanced biocatalysis and synergistic chemo- and biocatalysis. Nat. Catal. 3, 319–328.

Arnold, F.H., 2018. Directed evolution: bringing new chemistry to life. Angew. Chem. Int. (Ed.) Engl. 57, 4143–4148.

Arnold, F.H., 2019. Innovation by evolution: bringing new chemistry to life (Nobel lecture). Angew. Chem. Int. (Ed.) Engl. 58, 14420–14426.

Barth, S., Fischer, M., Schmid, R.D., Pleiss, J., 2004. The database of epoxide hydrolases and haloalkane dehalogenases: one structure, many functions. Bioinformatics 20, 2845–2847.

Bastard, K., Smith, A.A., Vergne-Vaxelaire, C., Perret, A., Zaparucha, A., De Melo-Minardi, R., et al., 2014. Revealing the hidden functional diversity of an enzyme family. Nat. Chem. Biol. 10, 42–49.

Bauer, T.L., Buchholz, P.C.F., Pleiss, J., 2020. The modular structure of α/β-hydrolases. FEBS J. 287, 1035–1053.

Bell, E.L., Finnigan, W., France, S.P., Green, A.P., Hayes, M.A., Hepworth, L.J., et al., 2021. Biocatalysis. Nat. Rev. Methods Prim. 1, 46.

Berini, F., Casciello, C., Marcone, G.L., Marinelli, F., 2017. Metagenomics: novel enzymes from non-culturable microbes. FEMS Microbiol. Lett. 364, 21.

Blin, K., Kazempour, D., Wohlleben, W., Weber, T., 2014. Improved lanthipeptide detection and prediction for antiSMASH. PLoS One 9, e89420.

Blin, K., Medema, M.H., Kazempour, D., Fischbach, M.A., Breitling, R., Takano, E., et al., 2013. antiSMASH 2.0—a versatile platform for genome mining of secondary metabolite producers. Nucleic Acids Res. 41, W204—W212.

Bommarius, A.S., 2015. Biocatalysis: a status report. Annu. Rev. Chem. Biomol. Eng. 6, 319—345.

Borchert, E., García-Moyano, A., Sanchez-Carrillo, S., Dahlgren, T.G., Slaby, B.M., Bjerga, G. E.K., et al., 2021. Deciphering a marine bone-degrading microbiome reveals a complex community effort. mSystems 6, e01218—e01220.

Bornscheuer, U.T., Huisman, G.W., Kazlauskas, R.J., Lutz, S., Moore, J.C., Robins, K., 2012. Engineering the third wave of biocatalysis. Nature 485, 185—194.

Boulund, F., Berglund, F., Flach, C.F., Bengtsson-Palme, J., Marathe, N.P., Larsson, D.G.J., et al., 2017. Computational discovery and functional validation of novel fluoroquinolone resistance genes in public metagenomic data sets. BMC Genomics 18, 682.

Buchfink, B., Xie, C., Huson, D.H., 2015. Fast and sensitive protein alignment using DIAMOND. Nat. Methods 12, 59—60.

Cantarel, B.L., Coutinho, P.M., Rancurel, C., Bernard, T., Lombard, V., Henrissat, B., 2009. The Carbohydrate-Active EnZymes database (CAZy): an expert resource for glycogenomics. Nucleic Acids Res. 37, D233—D238.

Cavicchioli, R., Ripple, W.J., Timmis, K.N., Azam, F., Bakken, L.R., Baylis, M., et al., 2019. Scientists' warning to humanity: microorganisms and climate change. Nat. Rev. Microbiol. 17, 569—586.

Cecchini, D.A., Sánchez-Costa, M., Orrego, A.H., Fernández-Lucas, J., Hidalgo, A., 2022. Ultrahigh-throughput screening of metagenomic libraries using droplet microfluidics. Methods Mol. Biol. 2397, 19—32.

Cheng, J., Pinnell, L., Engel, K., Neufeld, J.D., Charles, T.C., 2014. Versatile broad-host-range cosmids for construction of high quality metagenomic libraries. J. Microbiol. Methods 99, 27—34.

Chuzel, L., Fossa, S.L., Boisvert, M.L., Cajic, S., Hennig, R., Ganatra, M.B., et al., 2021. Combining functional metagenomics and glycoanalytics to identify enzymes that facilitate structural characterization of sulfated N-glycans. Microb. Cell Fact. 20, 162.

Coscolín, C., Katzke, N., García-Moyano, A., Navarro-Fernández, J., Almendral, D., Martínez-Martínez, M., et al., 2019. Bioprospecting reveals class III ω-transaminases converting bulky ketones and environmentally relevant polyamines. Appl. Env. Microbiol. 85, e02404—e02418.

Devine, P.N., Howard, R.M., Kumar, R., Thompson, M.P., Truppo, M.D., Turner, N.J., 2018. Extending the application of biocatalysis to meet the challenges of drug development. Nat. Rev. Chem. 2, 409—421.

Duarte, M., Jauregui, R., Vilchez-Vargas, R., Junca, H., Pieper, D.H., 2014. AromaDeg, a novel database for phylogenomics of aerobic bacterial degradation of aromatics. Database bau. 118.

Editorial, 2018. On advances and challenges in biocatalysis. Nat. Catal. 1, 635—636.

Fawal, N., Li, Q., Savelli, B., Brette, M., Passaia, G., Fabre, M., et al., 2013. PeroxiBase: a database for large-scale evolutionary analysis of peroxidases. Nucleic Acids Res. 37, D441—D444.

Fernandez-Lopez, L., Sanchez-Carrillo, S., García-Moyano, A., Borchert, E., Almendral, D., Alonso, S., et al., 2021. The bone-degrading enzyme machinery: from multi-component understanding to the treatment of residues from the meat industry. Comput. Struct. Biotechnol. J. 19, 6328—6342.

Fernández-Arrojo, L., Guazzaroni, M.E., López-Cortés, N., Beloqui, A., Ferrer, M., 2010. Metagenomic era for biocatalyst identification. Curr. Opin. Biotechnol. 21, 725—733.

Ferrandi, E.E., Previdi, A., Bassanini, I., Riva, S., Peng, X., Monti, D., 2017. Novel thermostable amine transferases from hot spring metagenomes. Appl. Microbiol. Biotechnol. 101, 4963–4979.

Ferrer, M., Martínez-Martínez, M., Bargiela, R., Streit, W.R., Golyshina, O.V., Golyshin, P.N., 2016. Estimating the success of enzyme bioprospecting through metagenomics: current status and future trends. Microb. Biotechnol. 9, 22–34.

Ferrer, M., Méndez-García, C., Bargiela, R., Chow, J., Alonso, S., García-Moyano, A., et al.,The INMARE Consortium 2019. Decoding the ocean's microbiological secrets for marine enzyme biodiscovery. FEMS Microbiol. Lett. 366, fny285.

FMCG Gurus, 2019. Global and Regional Sustainability Survey, Q3–2019.

Global Index, 2018, Industrial enzymes - a global market overview; Ipsos MORI, Fashion Revolution. Fashion Revolution Consumer Survey Report, 2018.

Guazzaroni, M.E., Silva-Rocha, R., Ward, R.J., 2015. Synthetic biology approaches to improve biocatalyst identification in metagenomic library screening. Microb. Biotechnol. 8, 52–64.

Gumulya, Y., Baek, J.M., Wun, S.J., Thomson, R.E.S., Harris, K.L., Hunter, D.J.B., et al., 2018. Engineering highly functional thermostable proteins using ancestral sequence reconstruction. Nat. Catal. 1, 878–888.

Hajighasemi, M., Tchigvintsev, A., Nocek, B., Flick, R., Popovic, A., Hai, T., et al., 2018. Screening and characterization of novel polyesterases from environmental metagenomes with high hydrolytic activity against synthetic polyesters. Env. Sci. Technol. 52, 12388–12401.

Handelsman, J., 2004. Metagenomics: application of genomics to uncultured microorganisms. Microbiol. Mol. Biol. Rev. 68, 669–685.

Handelsman, J., Rondon, M.R., Brady, S.F., Clardy, J., Goodman, R.M., 1998. Molecular biological access to the chemistry of unknown soil microbes: a new frontier for natural products. Chem. Biol. 5, R245–R249.

Hodgson, J., 2019. Biotech's baby boom. Nat. Biotechnol. 37, 502–512.

OECD, 2011. Industrial Biotechnology and Climate Change.

Jegannathan, K.R., Nielsen, P.H., 2013. Environmental assessment of enzyme use in industrial production: a literature review. J. Clean. Prod. 42, 228–240.

Jemli, S., Ayadi-Zouari, D., Hlima, H.B., Bejar, S., 2014. Biocatalysts: application and engineering for industrial purposes. Crit. Rev. Biotechnol. 6, 1–13.

Jumper, J., Evans, R., Pritzel, A., Green, T., Figurnov, M., Ronneberger, O., et al., 2021. Highly accurate protein structure prediction with AlphaFold. Nature 596, 583–589.

Karunatillaka, I., Jaroszewski, L., Godzik, A., 2022. Novel putative polyethylene terephthalate (PET) plastic degrading enzymes from the environmental metagenome. Proteins 90, 504–511.

Katzke, N., Knapp, A., Loeschcke, A., Drepper, T., Jaeger, K.E., 2017. Novel tools for the functional expression of metagenomic DNA. Methods Mol. Biol. 1539, 159–196.

Kusnezowa, A., Leichert, L.I., 2017. In silico approach to designing rational metagenomic libraries for functional studies. BMC Bioinforma. 18, 267.

Kuznetsova, E., Proudfoot, M., Sanders, S.A., Reinking, J., Savchenko, A., Arrowsmith, C.H., et al., 2005. Enzyme genomics: application of general enzymatic screens to discover new enzymes. FEMS Microbiol. Rev. 29, 263–279.

Levy Karin, E., Mirdita, M., Söding, J., 2020. MetaEuk-sensitive, high-throughput gene discovery, and annotation for large-scale eukaryotic metagenomics. Microbiome 8, 48.

Locey, K.J., Lennon, J.T., 2016. Scaling laws predict global microbial diversity. Proc. Natl Acad. Sci. U.S.A 113, 5970–5975.

Loeschcke, A., Markert, A., Wilhelm, S., Wirtz, A., Rosenau, F., Jaeger, K.E., et al., 2013. TREX: a universal tool for the transfer and expression of biosynthetic pathways in bacteria. ACS Synth. Biol. 2, 22–33.

Martínez-Martínez, M., Alcaide, M., Tchigvintsev, A., Reva, O., Polaina, J., Bargiela, R., et al., 2013. Biochemical diversity of carboxyl esterases and lipases from Lake Arreo (Spain): a metagenomic approach. Appl. Env. Microbiol. 79, 3553–3562.

Martínez-Martínez, M., Coscolín, C., Santiago, G., Chow, J., Stogios, P.J., Bargiela, R., et al., The Inmare Consortium 2018. Determinants and prediction of esterase substrate promiscuity patterns. ACS Chem. Biol. 13, 225–234.

Mateljak, I., Monza, E., Lucas, M.F., Guallar, V., Aleksejeva, O., Ludwig, R., et al., 2019. Increasing redox potential, redox mediator activity, and stability in a fungal laccase by computer-guided mutagenesis and directed evolution. ACS Catal. 9 (5), 4561–4572.

Mende, D.R., Waller, A.S., Sunagawa, S., Järvelin, A.I., Chan, M.M., Arumugam, M., et al., 2012. Assessment of metagenomic assembly using simulated next generation sequencing data. PLoS one 7, e31386.

Mistry, J., Chuguransky, S., Williams, L., Qureshi, M., Salazar, G.A., Sonnhammer, E.L.L., et al., 2021. Pfam: the protein families database in 2021. Nucleic Acids Res. 49 (D1), D412–D419.

Nasseri, S.A., Betschart, L., Opaleva, D., Rahfeld, P., Withers, S.G., 2018. A mechanism-based approach to screening metagenomic libraries for discovery of unconventional glycosidases. Angew. Chem. Int. (Ed.) Engl. 57, 11359–11364.

Neun, S., Zurek, P.J., Kaminski, T.S., Hollfelder, F., 2020. Ultrahigh throughput screening for enzyme function in droplets. Methods Enzymo 643, 317–343.

Nyyssönen, M., Tran, H.M., Karaoz, U., Weihe, C., Hadi, M.Z., Martiny, J.B.H., et al., 2013. Coupled high-throughput functional screening and next generation sequencing for identification of plant polymer decomposing enzymes in metagenomic libraries. Front. Microbiol. 4, 282.

Ouzounis, C.A., Karp, P.D., 2000. Global properties of the metabolic map of *Escherichia coli*. Genome Res. 10, 568–576.

Pellis, A., Cantone, S., Ebert, C., Gardossi, L., 2018. Evolving biocatalysis to meet bioeconomy challenges and opportunities. N. Biotechnol. 40 (Pt A), 154–169.

Popovic, A., Hai, T., Tchigvintsev, A., Hajighasemi, M., Nocek, B., Khusnutdinova, A.N., et al., 2017. Activity screening of environmental metagenomic libraries reveals novel carboxylesterase families. Sci. Rep. 7, 44103.

Pushkarev, A., Béjà, O., 2016. Functional metagenomic screen reveals new and diverse microbial rhodopsins. ISME J. 10, 2331–2335.

Pérez-García, P., Danso, D., Zhang, H., Chow, J., Streit, W.R., 2021. Exploring the global metagenome for plastic-degrading enzymes. Methods Enzymol. 648, 137–157.

Reyes-Duarte, D., Ferrer, M., García-Arellano, H., 2012. Functional-based screening methods for lipases, esterases, and phospholipases in metagenomic libraries. Methods Mol. Biol. 861, 101–113.

Roda, S., Fernandez-Lopez, L., Cañadas, R., Santiago, G., Ferrer, M., Guallar, V., 2021. Computationally driven rational design of substrate promiscuity on serine ester hydrolases. ACS Catal. 11 (6), 3590–3601.

Sayers, E.W., Cavanaugh, M., Clark, K., Ostell, J., Pruitt, K.D., Karsch-Mizrachi, I., 2020. GenBank. Nucleic Acids Res. 48 (D1), D84–D86.

Schallmey, M., Koopmeiners, J., Wells, E., Wardenga, R., Schallmey, A., 2014. Expanding the halohydrin dehalogenase enzyme family: identification of novel enzymes by database mining. Appl. Env. Microbiol. 80, 7303–7315.

Senior, A.W., Evans, R., Jumper, J., Kirkpatrick, J., Sifre, L., Green, T., et al., 2020. Improved protein structure prediction using potentials from deep learning. Nature 577, 706–710.

Sheldon, R.A., Brady, D., 2019. Broadening the scope of biocatalysis in sustainable organic synthesis. ChemSusChem 12, 2859–2881.

Sheldon, R.A., Woodley, J.M., 2018. Role of biocatalysis in sustainable chemistry. Chem. Rev. 118, 801–838.

Simon, C., Daniel, R., 2017. Construction of small-insert and large-insert metagenomic libraries. Methods Mol. Biol. 1539, 1–12.

Sirim, D., Wagner, F., Wang, L., Schmid, R.D., Pleiss, J., 2011. The Laccase Engineering Database: a classification and analysis system for laccases and related multicopper oxidases. Database 2011, bar006.

Smart, M., Huddy, R.J., Cowan, D.A., Trindade, M., 2017. Liquid phase multiplex high-throughput screening of metagenomic libraries using p-nitrophenyl-linked substrates for accessory lignocellulosic enzymes. Methods Mol. Biol. 1539, 219–228.

Tasse, L., Bercovici, J., Pizzut-Serin, S., Robe, P., Tap, J., Klopp, C., et al., 2010. Functional metagenomics to mine the human gut microbiome for dietary fiber catabolic enzymes. Genome Res. 20, 1605–1612.

Terrón-González, L., Medina, C., Limón-Mortés, M.C., Santero, E., 2013. Heterologous viral expression systems in fosmid vectors increase the functional analysis potential of metagenomic libraries. Sci. Rep. 3, 1107.

Timmis, K., de Lorenzo, V., Verstraete, W., Garcia, J.L., Ramos, J.L., Santos, H., et al., 2014. Pipelines for new chemicals: a strategy to create new value chains and stimulate innovation-based economic revival in Southern European countries. Env. Microbiol. 16, 9–18.

Tunyasuvunakool, K., Adler, J., Wu, Z., Green, T., Zielinski, M., Žídek, A., et al., 2021. Highly accurate protein structure prediction for the human proteome. Nature 596, 590–596.

Ufarté, L., Laville, E., Duquesne, S., Potocki-Veronese, G., 2015. Metagenomics for the discovery of pollutant degrading enzymes. Biotechnol. Adv. 33, 1845–1854.

UniProt Consortium, 2021. UniProt: the universal protein knowledgebase in 2021. Nucleic Acids Res. 49 (D1), D480–D489.

Velankar, S., Burley, S.K., Kurisu, G., Hoch, J.C., Markley, J.L., 2021. The Protein Data Bank Archive. Methods Mol. Biol. 2305, 3–21.

Verma, S., Meghwanshi, G.K., Kumar, R., 2021. Current perspectives for microbial lipases from extremophiles and metagenomics. Biochimie 182, 23–36.

Weiland-Bräuer, N., Langfeldt, D., Schmitz, R.A., 2017. Construction and screening of marine metagenomic large insert libraries. Methods Mol. Biol. 1539, 23–42.

Welborn, V.V., Ruiz Pestana, L., Head-Gordon, T., 2018. Computational optimization of electric fields for better catalysis design. Nat. Catal. 1, 649–655.

Williams, W., Kunorozva, L., Klaiber, I., Henkel, M., Pfannstiel, J., Van Zyl, L.J., et al., 2019. Novel metagenome-derived ornithine lipids identified by functional screening for biosurfactants. Appl. Microbiol. Biotechnol. 103, 4429–4441.

Wohlgemuth, R., Littlechild, J., Monti, D., Schnorr, K., van Rossum, T., Siebers, B., et al., 2018. Discovering novel hydrolases from hot environments. Biotechnol. Adv. 36, 2077–2100.

Ye, S.H., Siddle, K.J., Park, D.J., Sabeti, P.C., 2019. Benchmarking metagenomics tools for taxonomic classification. Cell 178, 779–794.

Chapter 8

Enzymatic biosynthesis of β-lactam antibiotics

Swati Srivastava, Reeta Bhati and Rajni Singh
Amity Institute of Microbial Technology, Amity University, Noida, Uttar Pradesh, India

8.1 Introduction

The β-lactam antibiotics (BLAs) family is the most important class of clinically used antibiotics, with more than half of the global antibiotic market share. Transpeptidase enzymes involved in bacterial cell wall biosynthesis are the targets of BLAs. Peptidoglycan cross-linking is catalyzed by the transpeptidases or penicillin-binding protein (PBP) enzymes. It is acceptable to modern medicinal approaches complementing the immense structure–activity relationship datasets obtained in the period leading up to the 1990s. β-Lactam shows its antibiotic response by satirizing the natural D-Ala-D-Ala substrate of the PBPs, responsible for cross-linking peptidoglycan components of the bacterial cell wall. BLAs have specifically earmarked their structure and absolute stereochemistry against PBPs, the members of the single clan of serine hydrolases. The PBPs, particularly the D,D-transpeptidases induce the typical 4–3 cross-linking of muramyl peptide stems throughout bacterial cell wall (peptidoglycan) biosynthesis (Townsend, 2016). Lee et al. (2001) reported the crystal structure of a cephalosporin derivative connected to a bifunctional carboxypeptidase/transpeptidase from *Streptomyces* sp. strain *R61* responsible for inhibiting the transpeptidation activity and disturbing the cell wall integrity, which eventually results in cell lysis (Worthington and Melander, 2013). There are various classes of BLAs comprising penicillins, cephalosporins, carbapenems, and monobactams (Fig. 8.1) and the evolution of new BLAs through side chain modification is continuous development.

In the early 20th century, the discovery of penicillin entered the modern era with the treatment of infectious diseases, leading to antibiotic control and abolishing infections that would otherwise be uncontrollable. At present, also when our dependence upon antibiotics is alarming by the ever-increasing menace of antibiotic resistance, the β-lactam class of compounds holds for beyond 50% of all antibiotic prescriptions. Penicillins initially manifested

Biotechnology of Microbial Enzymes. DOI: https://doi.org/10.1016/B978-0-443-19059-9.00007-4
© 2023 Elsevier Inc. All rights reserved.

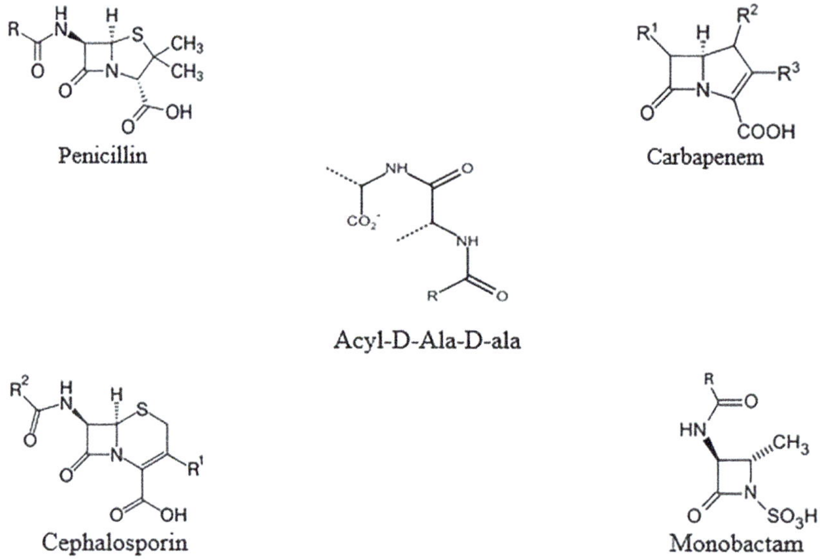

FIGURE 8.1 Classes of β-lactam antibiotics.

little activity against Gram-negative pathogens, later conquered by aminopenicillins, active against *Escherichia coli*, *Shigella*, and *Salmonella* species but not against *Pseudomonas aeruginosa* or *Klebsiella* species. The substitution of the amino group of aminopenicillin with a carboxyl group, inducing carboxypenicillins, provided the β-lactams effective against *P. aeruginosa* as an outcome of their low affinity for the AmpC β-lactamase. It is repeatedly noticed that resistant strains are developed shortly after the establishment of any new antibiotics (Singh et al., 2005).

The cephalosporin class of BLAs was discovered in the 1940s, which was stable for staphylococcal β-lactamase, a severe clinical problem initially, but later various generations of semisynthetic cephalosporins were established. Initially, for treating infections caused by Gram-negative bacteria, cephalosporins were proved beneficial except for *P. aeruginosa*, which was successfully served with third-generation cephalosporins, including cefoperazone and ceftazidime for many years. With the discovery of carbapenem and monobactam classes of BLAs, there was a rise in the increased therapeutic alternatives for bacterial infections that had become intractable to treatment with other β-lactams, consequently with the introduction of imipenem in 1985 and aztreonam in 1986. In the past, carbapenems have been moderately used for extremely severe infections caused by Gram-negative bacteria. However, resistance is now extensive in Enterobacteriaceae, *P. aeruginosa* and *Acinetobacter* spp., predominantly due to the increasing prevalence of carbapenemases (Sreedharan and Singh, 2019). β-Lactamases are enzymes responsible for the degradation of the lactam rings and are divided into two types on a structural basis: (1) serine β-lactamases and 2) metallo-β-lactamases (MBL). Serine-β-lactamases comprise extended-spectrum β-lactamases (ESBL) that mainly hydrolyze later-generation cephalosporins and carbapenemases such as *Klebsiella pneumoniae* carbapenemases that hydrolyze carbapenem antibiotics, besides later-generation cephalosporins. MBLs are Zn(II)-dependent enzymes that can adjust most β-lactams in their active site and hydrolyze nearly all BLAs involving carbapenems (Sharma et al., 2021a, b).

Primarily resistance to BLAs arises from one of the two mechanisms: (1) β-lactamase production, that is, the most frequent resistance mechanism in Gram-negative bacteria, and (2) an altered PBP production with a lower affinity for most BLAs (Sreedharan et al., 2019). Many new β-lactams were available in response to the spread of β-lactam resistance. Although both TEM (refers to an extended mutated enzyme that confers BLA resistance to Gram-negative bacteria) and AmpC (mediate resistance to most penicillins) hydrolyze the first-generation cephalosporins, such as cephaloridine and cefazolin, initially the reported literature suggested that both cephamycins (such as cefoxitin) and the third-generation cephalosporins containing an oxyimino side chain (such as cefotaxime) are resistant to both types of enzymes (Saxena and Singh, 2010). Although TEM was inactive against a few Gram-negative bacteria such as *Enterobacter* and *Serratia*, AmpC was efficient in killing these organisms. Their inception in the clinics resulted in resistant strains that showed the overproduction of the chromosomal AmpC enzyme (Bajpai et al., 2017). These compounds have shallow *KM* values and relatively high Vmax/*KM* values with respect to the AmpC enzymes (Gupta et al., 2015a, b). The AmpC enzyme that needs to be persuaded as third-generation cephalosporins are chosen for constitutive mutants of *ampC* gene, and the third-generation cephalosporins were effective against these bacteria simply because they were ineffective inducers of this enzyme (Priya et al., 2021). Moreover, lately in species that do not express the chromosomally coded *ampC*, strong expression of plasmid-coded AmpC has been perceived (Nikaido, 2009).

Eventually, fourth-generation cephalosporins (cefepime, cefpirome) have been developed that are more resistant to hydrolysis by the AmpC enzyme. The selection of plasmids created mutants of common enzymes, such as TEM or its relative sulfydryl variable, which can now hydrolyze third-generation and sometimes even fourth-generation cephalosporins due to the continued selective pressure. These enzymes are known as ESBL (Robin et al., 2007). Among the ESBL enzymes, the mainly troublesome ones are known as CTX-M (preferential hydrolytic activity against cefotaxime) (Cantón et al., 2012). The chromosome of a rarely encountered Gram-negative bacterium, *Kluyvera*, apparently ought to be the origin of the gene coding for these enzymes and have transferred to *R*-plasmids (Sharma et al., 2021a, b). This transfer or mobilization unusually appears several times, and subsequently, the enzyme rapidly became widespread among *R*-plasmid-containing pathogenic bacteria. In terms of effectiveness, β-lactams with a newly discovered nucleus, such as carbapenems (e.g., imipenem), continue to be effectual. Still, their use may result in the expanded presence of enzymes competent for hydrolyzing these compounds (Bharathala et al., 2020). Multidrug-resistant (MDR) bacteria constitute a major threat to all fields of medical science as its repercussions can lead to treatment failure, which can have severe outcomes, especially in the case of scathing patients. Considerably, extremely drug-resistant *Mycobacterium tuberculosis, P. aeruginosa, E. coli, Acinetobacter baumannii*, methicillin-resistant *Staphylococcus aureus* (MRSA), *K. pneumoniae* bearing NDM-1 (New Delhi metallo-beta-lactamase-1), vancomycin-resistant MRSA, and vancomycin-resistant enterococci (Singh et al., 2014) are some of the severe MDR organisms nowadays.

8.2 Enzymes involved in the biosynthesis of β-lactam antibiotics

β-Lactams are chemically very distinctive as antibacterials, and the reactions involved in β-lactam biosynthesis are striking. The β-lactam biosynthesis generally follows nonoxidative reactions, but other enzymatic-dependent reactions such as the ferrous iron moiety, 2-oxoglutarate (2OG)-dependent oxygenases, and allied oxidase isopenicillin N synthase (IPNS)-catalyzed reactions are also of specific interest (Fig. 8.2). The 2OG oxygenases were first discovered during collagen biosynthesis, and at that time, their participation in β-lactam biosynthesis was unexpected (Gupta et al., 2015a, b). The nonpenicillin β-lactam biosynthetic pathways of 2OG oxygenases play roles in diversifying the chemistry and, consequently, the activities of the

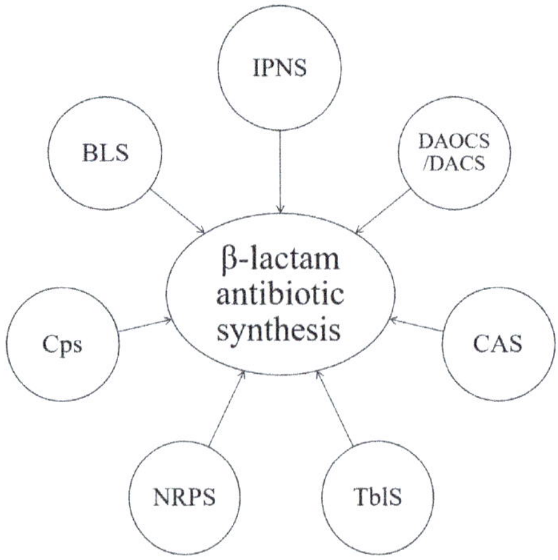

FIGURE 8.2 Enzymes involved in the synthesis of β-lactam antibiotic.

β-lactams formed by non-oxygen-dependent pathways. In the course of biosynthesis of the clavam and the carbapenem bicyclic ring systems, β-lactam development is induced by asparagine synthetase-related enzymes. The genesis of the β-lactam ring from relevant β-amino acid precursors is induced by β-lactam synthetases (BLSs), persuading the reverse of β-lactamase catalysis using ATP to activate the carboxylic acid (Table 8.1). In clavam biosynthesis, a monocyclic β-lactam ring is formed, which acts as a precursor for the bicyclic clavam ring in reactions induced by the 2OG oxygenase clavaminic acid synthase (CAS). During the synthesis of carbapenem, (3*S*,5*S*)-carbapenam ring system induced by BLSs, which is epimerized and desaturated by a 2OG oxygenase to carbapen-2-em-3-carboxylate, that is, CarC, to give a (5*R*)-carbapenem. Thus the involvement of 2OG oxygenases in these pathways indicates the roles of these enzymes, both in forming a bicyclic β-lactam ring system (clavams) and in altering a bicyclic β-lactam ring system (carbapenems) to construct antibacterials. The action of nonribosomal peptide synthetases (NRPSs) catalyze the formation of the N1−C4 β-lactam bond from synthetase-bound precursor peptide, which is involved in the formation of some monocyclic β-lactams, for example, the nocardicins and monobactams. In a few cases, 2OG oxygenases are entailed in modifications resulting in β-lactam formation, for instance, monobactams and tabtoxin.

8.2.1 Isopenicillin N synthase

IPNS is a nonheme iron protein belonging to the 2-oxoglutarate (2OG)-dependent dioxygenase oxidoreductase family. The formation of isopenicillin N from δ-(L-α-aminoadipoyl)-L-cysteinyl-D-valine is catalyzed by this enzyme. Most IPNSs contain an iron ion in their active sites and have a densely functionalized heterocyclic penicillin ring system for oxidative closure. IPNS catalyzes the formation of fused β-lactam and thiazolidine core of penicillin scaffold by four-electron oxidation of peptide precursor with reduction of dioxygen to two water molecules (Rabe et al., 2018). Based on extensive spectroscopic, computational, crystallographic, and substrate analog studies, the current, thorough understanding of IPNS mechanism is analyzed. The first crystal structure of IPNS from *Aspergillus nidulans* was procured in complex with manganese replacing iron. This structure disclosed that single active site metal is synchronized by two histidine (His214 and His270) and one aspartate residue (Asp216) as well as by the side chain of Gln330, in the resting enzyme state (Zhang et al., 2019). The distorted double-stranded β-helix (DSBH or "jelly roll") core is the basis of an overall fold of IPNS supporting a triad of Fe(II) binding residues and was eventually conserved in 2OG oxygenases, including deacetoxycephalosporin C synthase (DAOCS). The binding of NO (acting as a dioxygen analog) adjacent to the

TABLE 8.1 β-Lactam antibiotic synthesizing enzymes and their functions and microbial sources.

β-Lactam antibiotic	Enzyme involved in synthesis	Enzyme functionality	Microorganism involved	References
Penicillin	IPNS	Enables oxidative ring closure to the densely functionalized heterocyclic penicillin ring system.	*Aspergillus nidulans, Penicillium chrysogenum, Aspergillus oryzae, Trichophyton verrucosum, Trichophyton tonsurans, Trichophyton rubrum, Arthroderma benhamiae, Epidermophyton, Malbranchea, Pleurophomopsis, Aspergillus fumigatus (Sartorya), Polypaecilum*	Tahlan and Jensen (2013); Hamed et al. (2013); Raber et al. (2009); Chapman and Rutledge (2021)
Cephalosporins	IPNS, DAOCS/ DACS	DAOCS catalyzes oxidative ring expansion reaction involving conversion of penicillin β-sulfoxide to cephalosporin. DACS catalyzed hydroxylation reaction is 2OG-dependent oxygenase-type reaction.	*Xanthothecium (Anixiopsis), Acremonium chrysogenum, Arachnomyces, Scopulariopsis, Spiroidium, Diheterospora (Verticillium), Kallichroma Tethys*	
Cephamycins	IPNS	Same as above.	*Streptomyces microflavus (lipmanii), Streptomyces clavuligerus, Streptomyces cattleya, Streptomyces jumonjinensis, Streptomyces griseus, Lactamdurans, Amycolatopsis (Nocardia), Streptomyces panayensis, Streptomyces fimbriatus, Streptomyces viridochromogenes, Streptomyces wadayamensis, Streptomyces hygroscopicus, Streptomyces heteromorphus, Streptomyces sulfonofaciens*	

Cephabacins	IPNS	Same as above.	*Flavobacterium sp., Lysobacter lactamgenus, Xanthomonas lactamgena*
Clavams (Clavulanic acid)	CAS, BLS, BLS 2	BLS catalyzes the two-stage reaction in which N^2-(2-carboxyethyl)-L-arginine is first adenylated, and then undergoes intramolecular ring closure.	*Saccharomonospora viridis, S. clavuligerus, Streptomyces jumonjinensis, Streptomyces flavogriseus, Anoxybacillus flavithermus, Streptomyces katsurahamanus*
Clavams (5S Clavams)	CAS, BLS, BLS 1, BLS 3	CAS catalyzes three distinct oxidations, that is, hydroxylation, cyclisation, and desaturation reactions for clavaminic acid synthesis, which acts as branch point for diverging pathways yielding clavulanic acid and the (5*S*)-clavams.	*Streptomyces hygroscopicus, Streptomyces antibioticus, Streptomyces microflavus (lipmanii), S. clavuligerus*
Carbapenems (carbapenem-3-carboxylic acid)	BLS, carbapenam synthetase (Cps)	Cps catalyzes both epimerization and desaturation of a carbapenam in an unusual process catalyzed by an iron- and 2-oxoglutarate-dependent oxygenase.	*Dickeya zeae, Pectobacterium carotovorum, Pantoea sp., Photorhabdus luminescens, Serratia sp., Pelosinus fermentans*
Carbapenems (Thienamycin/olivanic acid-type)	Cps	Same as above.	*Streptomyces flavogriseus, Streptomyces olivaceus, Streptomyces cattleya*
Monocyclic β-lactams (sulfazecin-type monobactams)	NRPS	A modular enzyme catalyzes the synthesis of important peptide products from various standard and nonproteinogenic amino acid substrates.	*Pseudomonas acidophila, Gluconobacter* spp.*, Chromobacterium violaceum, Pseudomonas mesoacidophila, Acetobacter* spp.*, Rhizobium radiobacter*

(*Continued*)

TABLE 8.1 (Continued)

β-Lactam antibiotic	Enzyme involved in synthesis	Enzyme functionality	Microorganism involved	References
Monocyclic β-lactams (tabtoxin-type)	Tabtoxinine β-lactam synthetase (Tbl S)	Hinders glutamine synthetase.	*Streptomyces sp., Pseudomonas syringae*	
Monocyclic β-lactams (nocardicin A)	NRPS	Same as above.	*Actinosynnema mirum, Microtetraspora caesia, Nocardia uniformis, Nocardiopsis atra*	
Monamphilectine	ND		*Hymeniacidon* sp.	
Pachystermine	ND		*Pachysandra terminalis*	

BLS, β-Lactam synthetase; *CAS*, clavaminic acid synthase; *DACS*, deacetylcephalosporin C synthase; *DAOCS*, deacetoxycephalosporin C synthase; *IPNS*, isopenicillin N synthase; *ND*, not determined; *NRPS*, nonribosomal peptide synthetases.

cysteinyl thiol with the oxygen atom of the complexed NO directed toward the cysteinyl β-carbon was manifested by the crystallographic analysis of IPNS.ACV.Fe(II).NO complex.

Crystallographic study of IPNS with the substrate analog δ-(L-α-aminoadipoyl)-L-(cysteinyl)-S-methyl-D-cysteine supported for initial β-lactam ring formation under anaerobic conditions then exposed to high pressure of oxygen. These joint investigations extend the information of earlier spectroscopic analyses and IPNS turnover studies of this major protein and NHIO enzyme family (Chapman and Rutledge 2021).

8.2.2 β-Lactam synthetase

BLS is an expanding class of enzymes that construct the overcritical β-lactam ring in clavam and carbapenem antibiotics. BLS is an ATP/Mg^{2+}-dependent enzyme that induces the reaction in two stages in which N^2-(2-carboxyethyl)-L-arginine is first adenylated, and then undergoes intramolecular ring closure (Raber et al., 2009).This intramolecular amide bond origination was previously unknown and has led to the appointment of a BLS. Adenylation of the CEA β-carboxylate is a simple substitution reaction. No net charge is generated in this step at physiological pH, occurring in the controlled displacement of phosphoric anhydride of ATP by a carboxylate oxygen atom. The drift of negative charge is created from the β carboxylate to PPi. β-Lactam formation is more complex as intramolecular nucleophilic addition of the α-amine of CEA to the activated β-carboxylethyladenylate would consequently lead to the formation of an oxyanion intermediate or transition state, which undergoes α-elimination, expelling AMP/Mg^{2+} to form the β-lactam ring (Miller et al., 2002).

In clavulanic acid biosynthesis, a synthetase enzyme induces the formation of the monocyclic β-lactam ring (McNaughton et al., 1998). The ATP-utilizing enzyme BLS synthesizes 2-azetidinone ring of the Class A and D β-lactamase inhibitor clavulanic acid. The effectiveness of this compound is strengthened by the addition of hydroxyethyl group to C-6 of clavulanic acid in (S) configuration against Class C β-lactamases (Labonte et al., 2012). *Orf3*, a clavulanic acid gene cluster component having deduced amino acid similarity to a subset of amidotransferases, encodes BLS. The location of *Orf3* is approximately 1 kb upstream of *orf5* that encodes clavaminate synthase, a later enzyme in the pathway that is accountable for the genesis of the bicyclic nucleus of clavulanic acid (Gupta et al., 2017).

A single enzyme catalyzes the conversion of CEA to deoxyguanidinoproclavaminic acid (DGPC) in the presence of ATP/Mg^{2+}. Complete blocking of clavulanic acid production and CEA accumulation was observed due to targeted disruption of *orf3* in the wild-type *Streptomyces clavuligerus*. Clavulanic acid production was restored via chemical complementation of *orf3* mutants with genuine DGPC. After IPNS, BLS is the second enzyme that has shown the ability to induce the formation of β-lactam ring using an unnatural substrate (Sleeman et al., 2002).

8.2.3 Carbapenam synthetase (Cps)

McGowan et al. (1996) described the gene cluster accountable for the biosynthesis of simplest carbapenem (5*R*)-carbapen-2-em-3-carboxylic acid from *Erwinia carotovora* (now *Pectobacterium carotovorum*). Carbapenem synthase (CarC) is an uncommon member of α-ketoglutarate (α-KG)-dependent dioxygenases.

The nonheme iron oxygenases use reliable His-X-Asp/Glu-X_n-His motif in conjunction with bi-dentate coordination of α-KG, to bind Fe^{II}, withdrawing the sixth and sole remaining open site on iron available to coordinate dioxygen. Oxidative decarboxylation of α-KG generates CO_2, succinate and highly reactive iron^{IV}-oxo ($Fe^{IV}=O$) intermediate by dioxygen binding (Phelan and Townsend, 2013).

CarC is suggested to induce both epimerization and desaturation of (3*S*,5*S*)-carbapenam-3-carboxylic acid, forming (5*R*)-carbapen-2-em-3-carboxylic acid [as well as (3*S*,5*R*)-carbapenam-3-carboxylic acid]. In accordant with solution studies, crystal structures of CarC complexed with Fe(II) and 2-oxoglutarate disclose it to be hexameric (space group *C*222). The monomers of CarC contain a double-stranded β-helix core that supports ligands binding a single Fe(II) to which 2-oxoglutarate complexes in a bi-dentate manner (Clifton et al., 2003).

8.2.4 Tabtoxinine β-lactam synthetase (Tbl S)

Tabtoxinine-β-lactam (TβL), phytotoxin produced by plant pathogenic strains of *Pseudomonas syringae*, hinders glutamine synthetase (GS), unlike other BLAs, inhibiting PBPs. Identification of biosynthetic precursors of tabtoxinine as done by incorporating ^{13}C-labeled compounds. The side chain comprises L-threonine and L-aspartate, whereas pyruvic acid and the methyl group of L-methionine make up β-lactam moiety.

It has the biosynthetic cluster of 31-kb, which consists of:

- TabA (P31851), an enzyme that is related to lysA (diaminopimelate decarboxylase);
- TabB (P31852), an enzyme that is related to dapD (THDPA succinyl-CoA succinyltransferase, THDPA-ST); and
- TblA (P31850), an enzyme that has no close paralogs (identified as a member of SAMe-dependent methyltransferase superfamily by InterPro).

Tabtoxinine hydrolyzes to release TβL for aminopeptidase activity with the help of zinc. Traditional BLAs are mechanically distinct from TβL, for example, penicillin, which covalently inhibits transpeptidases in the serine hydrolase superfamily by a β-lactam ring-opening acylation mechanism. During GS inhibition, the **TβL** β-lactam ring remains flawless as a tetrahedral template that matches the conformation, polarity and chirality of GS transition state (Patrick et al., 2018).

8.2.5 Deacetoxycephalosporin C synthase and deacetylcephalosporin C synthase

The committed step in cephalosporin biosynthesis is induced by DAOCS. The eukaryotic microorganisms have a single bifunctional 2OG oxygenase DAOCS/deacetylcephalosporin C synthase (DACS), which shows the activity of both DAOCS and DACS. In comparison, prokaryotes have two distinct enzymes that are highly, but incompletely, selective for the penam expansion (DAOCS) and the DAOC hydroxylation steps (DACS).

One of the 2OG oxygenases is involved in 7α-functionalization of cephalosporin. Two types of known 7α-functionalized cephalosporins with potent antibiotic activity are cephamycins with a 7α-methoxy group, and cephabacins with a 7α-formylamino group. Though the biosynthesis of the 7α-formylamino group has not been elucidated, the methyl group and the oxygen of the 7α-methoxy group are derived from methionine and dioxygen, respectively. Studies using *O*-carbamoyl DAC as a substrate with cell-free extracts of *S. clavuligerus* manifested the process to be dependent on Fe(II), 2OG, and *S*-adenosylmethionine. Isotopic labeling experiments, spectroscopy and crystallography techniques provided insight into DAOCS/DACS mechanisms. The DACS-induced hydroxylation reaction is a 2OG-dependent oxygenase-type reactions occurring at an allylic position. In comparison, to date, in enzymology, the DAOCS-induced oxidative ring expansion reaction is very typical in which a penicillin β-sulfoxide is transformed into a cephalosporin where the (*pro-S*) β-methyl group of penicillin is included into dihydrothiazine/cephem ring. Both in vitro and in vivo studies have exhibited that during cephalosporin biosynthesis, the β-methyl group of penicillin N creates endocyclic C-3 of DAOC (Hamed et al., 2013).

8.2.6 Clavaminic acid synthase

The conversion of DGPC to clavaminic acid involves four steps. Three steps are induced by a single 2OG oxygenase, CAS. The initial steps in clavam biosynthesis occur via a conserved pathway leading to (3*S*,5*S*)-clavaminic acid, which acts as a bifurcating point for diverging pathways generating (3*R*,5*R*)-clavulanic acid and the (5*S*)-clavams. The 2OG oxygenase CAS induces three distinct oxidations, that is, hydroxylation, cyclization, and desaturation reactions. In the biosynthesis of clavam, the reaction of L-arginine with glyceraldehyde-3-phosphate forms β-amino acid N^2-(2-carboxyethyl) arginine (CEA), which is induced by carboxyethylarginine synthase (CEAS, thiamine pyrophosphate enzyme). An ATP-dependent reaction induced by BLS creates (3*S*,5*S*)-β-lactam core of clavams. In a usual 2OG oxygenases reaction, CAS induces hydroxylation of deoxyguanidinoproclavaminic acid generating guanidino-proclavaminic acid. Previous oxidation by CAS leads to the formation of proclavaminic acid as proclavaminate amidino hydrolase inducing hydrolysis of guanidine group of guanidinoproclavaminic acid. Further two more reactions are induced by CAS: first unusual bicyclization of proclavaminic acid to form dihydroclavaminic acid, then a desaturation reaction to form clavaminic acid. The substitution of an amino group of (3*S*,5*S*)-clavaminic acid with a hydroxyl group and epimerization of two stereocenters leads to clavulanic acid. There is the involvement of NADPH/NADH-dependent reduction of (3*R*,5*R*)-clavaldehyde to give clavulanic acid in the concluding step of the pathway. Elson et al. (1987) developed the production of clavaminic acid and proclavaminic acid by a clavulanate-producing strain of *S. clavuligerus* while exploring the cell-free production of clavulanic acid. The mechanism of CAS involves combined crystallographic and solution studies. A crystal structure of the complex CAS1 (one of two CAS isozymes in *Streptomyces* spp.) with Fe(II) and 2OG revealed characteristic 2OG oxygenase core fold constructed on a biased double-stranded β-helix fold, comprising eight β-strands enveloped by two α-helical regions. This enzyme contains nonheme iron, and the iron-binding is mediated by the side chains of His144, Glu146, and His279.

8.2.7 Nonribosomal peptide synthetases

The biosynthesis of monocyclic β-lactams is taking attention to interesting chemical mechanisms of cyclisation. Although β-lactam rings of both nocardicins and sulfazecin are formed in reactions induced by NRPS, still the exact mechanisms of formation differ for these two BLAs as described below.

Biosynthesis of nocardicin is initiated by NRPS proteins NocA and NocB. A pentapeptide product/intermediate was predicted with a sequence of L-pHPG-L-Arg-D-pHPG-L-Ser-L-pHPG (pHPG, *p*-hydroxyphenylglycine), on the basis of five adenylation domains of these two proteins. The formation of β-lactam occurs as the nascent peptide chain covalently attaches to NocB. Its origin involves dehydration of serine residue (as part of a NocB tetrapeptide peptide chain) bound to peptidyl carrier protein (PCP) domain PCP_4, before inclusion of C-terminal pHPG residue. The α-amino group of pHPG residue can attach with protein-linked dehydroalanine residue when bound to PCP domain 5 as a thioester. The consequent secondary amino group of PCP_5-bound pHPG then attack thioester link to PCP_4, establishing β-lactam and liberating the nascent peptide chain from PCP_4. Ultimately, after thioesterase domain-mediated hydrolysis, pro-nocardicin G is procured and further a hydrolytic step and tailoring reactions lead toward nocardicin A.

TABLE 8.2 Types of semisynthetic β-lactam derivatives.

Antibiotic group	Subgroups	Functions	Semisynthetic derivatives	References
Penicillin	Aminopenicillins	Inhibition of cell wall synthesis by binding to specific PBPs leading to cell lysis by autolysins.	Amoxycillin	Miller et al. (2002); Chapman and Rutledge (2021); Tahlan and Jensen (2013)
			Ampicillin	
			Hetacillin	
			Pivampicillin	
	Antistaphylococcal penicillins	Same as above.	Cloxacillin	
			Dicloxacillin	
			Nafcillin	
			Oxacillin	
			Methicillin	
			Flucloxacillin	
	Antipseudomonal antibiotics (carboxypenicillins)	Same as above.	Carbenicillin	
			Ticarcillin	
	(Ureidopenicillins)		Mezlocillin	
			Piperacillin	
			Azlocillin	
	(β-lactamase inhibitor)	Binding and inhibiting β-lactamase produced by bacterial cells to prevent it from reducing antibiotic activity.	Clavulanic acid	
			Sulbactam	
			Tazobactam	

(*Continued*)

TABLE 8.2 (Continued)

Antibiotic group	Subgroups	Functions	Semisynthetic derivatives	References
Carbapenem	–	Inhibition of cell wall by binding to specific PBPs of cell wall.	Imipenem	
			Meropenem	
			Ertapenem	
			Doripenem	
			Panipenem	
			Tebipenem	
Cephalosporin	First generation	Inhibition of cell wall synthesis by binding to specific PBPs at third and last stage of bacterial cell wall synthesis leading to cell lysis by autolysins.	Cefalexin	
			Cefadroxil	
			Cefazolin	
			Cefapirin	
			Cefacetrile	
			Cefaloglycin	
			Cefaloridine	
			Cefalotin	
			Cefazaflur	
			Cefradine	
			Cefroxadine	

Second generation	Inhibits PBPs interfering with the formation and remodeling of the cell wall structure ultimately causing cell lysis mediated by autolysins.	Cefuroxime
		Cefprozil
		Cefaclor
		Cefonicid
		Cefuzonam
		Cefoxitin
		Cefotetan
		Cefmetazole
		Cefotiam
		Loracarbef
Third generation	Inhibition of cell wall synthesis via binding to PBPs. Penetrates the bacterial cell wall by combating inactivation by beta-lactamase enzymes and inactivating PBPs, leading to interfere in transpeptidation, eventually causing cell lysis.	Cefdinir
		Ceftriaxone
		Ceftazidime
		Cefixime
		Cefpodoxime
		Cefotaxime
		Ceftizoxime
		Cefditoren
		Ceftibuten
		Cefdaloxime
		Cefmenoxime

(*Continued*)

TABLE 8.2 (Continued)

Antibiotic group	Subgroups	Functions	Semisynthetic derivatives	References
			Cefpimizole	
			Cefteram	
			Ceftiolene	
	Fourth generation	Inhibiting cell wall synthesis by binding to PBPs ultimately causing cell death. In this class mechanism of action is not yet known properly.	Cefepime	
			Cefiderocol	
			Cefquinome	
			Cefclidine	
			Cefluprenam	
			Cefoselis	
	Fifth generation	Antibacterial activity mediated through binding to PBPs leading to inhibition of cell wall synthesis.	Ceftaroline	
			Ceftolozane	
			Ceftobiprole	
Monobactams	–	Inhibition of cell wall synthesis by binding to specific PBPs by inhibiting the third and last stage of bacterial cell wall leading to cell lysis by autolysins.	Aztreonam	
			Tigemonam	
			Nocardicin A	
			Tabtoxin	

PBPs, Penicillin-binding proteins.

The occurrence of β-lactam ring of sulfazecin is directed through a stimulant of an NRPS thioesterase domain. An enzyme-bound tripeptide intermediate consists D-Glu residue (linked via its γ-carboxyl group), a D-Ala residue and an L-2,3 diaminopropionate (Dap or β-aminoalanine) residue via NRPSs SulI and SulM. SulN induces a 3-amino group of Dap which undergoes an unusual N-sulfonation reaction using 3′-phosphoadenosine 5′-phosphosulfate as cosubstrate. The tripeptide is then transferred to thioesterase domain, causing a nucleophilic attack of sulfonated amino group on thioester carbonyl, forming β-lactam.

8.3 Semisynthetic β-lactam derivatives

BLAs are among the oldest antibiotics group, which are being used extensively nowadays because of their renewability. Different semisynthetic derivatives of the four major classes of BLAs have been developed (Table 8.2). These constantly generating new semisynthetic derivatives are one of the primary reasons for their extensive use in clinical prescriptions. Apart from those annotated in the table, there are a few other semisynthetic derivatives (drugs) that are not discussed here as they are not yet fully annotated and are considered a stub. Enzymes play a vital role in forming these derivatives and make them very specific, potent, and safe coupled.

8.4 Concluding remarks

This chapter offers the enzymology leading to the biosynthesis of BLAs and their effective derivatives. In the family of antimicrobial agents, BLAs are the largest and the most extensively used antibiotics. The BLAs were conventionally manufactured using synthetic chemical routes, which are now slowly getting replaced with enzyme biocatalysis. The expansion of this enzymatic synthesis is growing fast because of several benefits such as cost-effectiveness, minimization of wastes, and generation of less harmful effluent impurities, thereby satisfying certain green chemistry aspects (Saxena and Singh, 2011). Generally, kinetically controlled methods with activated acyl-substrates are used to synthesize semisynthetic antibiotics (amoxicillin, cephalexin, ampicillin, etc.) in good yields (Ulijn et al., 2002). Also, enzymatic synthesis has allowed the synthesis of potentially effective BLAs and more effective β-lactamase inhibitors.

Abbreviations

BLA β-Lactam antibiotic
PBP penicillin-binding protein
MBL metallo-β-lactamases
ESBL extended-spectrum β-lactamases

MDR multidrug-resistant
MRSA methicillin-resistant *Staphylococcus aureus*
NDM-1 New Delhi metallo-beta-lactamase-1
2OG 2-Oxoglutarate-dependent oxygenases
IPNS isopenicillin N synthase
DAOCS deacetoxycephalosporin C synthase
DACS deacetylcephalosporin C synthase
CAS clavaminic acid synthase
BLS β-Lactam synthetases
NRPS nonribosomal peptide synthetases
Cps carbapenam synthetase
CarC Carbapenem synthase
CEA N^2-(2-carboxyethyl)arginine
DGPC deoxyguanidinoproclavaminic acid
αKG α-Ketoglutarate
TβL tabtoxinine-β-lactam
Tbl S tabtoxinine-β-lactam synthetase
GS glutamine synthetase
CEAS carboxyethylarginine synthase
pHPG *p*-Hydroxyphenylglycine
PCP peptidyl carrier protein
CTX-M cefotaxime
KM Michaelis constant
TEM extended mutated β-lactam enzyme
AmpC class C beta-lactamase

References

Bajpai, T., Pandey, M., Varma, M., Bhatambare, G.S., 2017. Prevalence of TEM, SHV, and CTX-M Beta-Lactamase genes in the urinary isolates of a tertiary care hospital. Avicenna J Clin Med. 7 (01), 12–16.

Bharathala, S., Singh, R., Sharma, P., 2020. Controlled release and enhanced biological activity of chitosan-fabricated carbenoxolone nanoparticles. Int. J. Biol. Macromol. 164, 45–52.

Cantón, R., González-Alba, J.M., Galán, J.C., 2012. CTX-M enzymes: origin and diffusion. Front. Microbial. 3, 110.

Chapman, N.C., Rutledge, P.J., 2021. Isopenicillin N synthase: crystallographic studies. Chembiochem 22 (10), 1687–1705.

Clifton, I.J., Doan, L.X., Sleeman, M.C., Topf, M., Suzuki, H., Wilmouth, R.C., et al., 2003. Crystal structure of carbapenem synthase (CarC). J. Biol. Chem. 278 (23), 20843–20850.

Elson, S.W., Baggaley, K.H., Gillett, J., Holland, S., Nicholson, N.H., Sime, J.T., et al., 1987. Isolation of two novel intracellular β-lactams and a novel dioxygenase cyclising enzyme from *Streptomyces clavuligerus*. ChemComm. 22, 1736–1738.

Gupta, S., Nigam, A., Singh, R., 2015a. Purification and characterization of a *Bacillus subtilis* keratinase and its prospective application in feed industry. Acta Biol. Szeged. 59 (2), 197–204.

Gupta, S., Singh, S.P., Singh, R., 2015b. Synergistic effect of reductase and keratinase for facile synthesis of protein-coated gold nanoparticles. J. Microbiol. Biotechnol. 25 (5), 612–619.

Gupta, S., Tewatia, P., Misri, J., Singh, R., 2017. Molecular modeling of cloned *Bacillus subtilis* keratinase and its insinuation in psoriasis treatment using docking studies. Indian J. Microbiol. 57 (4), 485–491.

Hamed, R.B., Gomez-Castellanos, J.R., Henry, L., Ducho, C., McDonough, M.A., Schofield, C. J., 2013. The enzymes of β-lactam biosynthesis. Nat. Prod. Rep. 30 (1), 21–107.

Labonte, J.W., Kudo, F., Freeman, M.F., Raber, M.L., Townsend, C.A., 2012. Engineering the synthetic potential of β-lactam synthetase and the importance of catalytic loop dynamics. MedChemComm. 3 (8), 960–966.

Lee, W., McDonough, M.A., Kotra, L.P., Li, Z.H., Silvaggi, N.R., Takeda, Y., et al., 2001. A 1.2-Å snapshot of the final step of bacterial cell wall biosynthesis. Proc. Natl. Acad. Sci. USA 98 (4), 1427–1431.

McGowan, S.J., Sebaihia, M., Porter, L.E., Stewart, G.S.A.B., Williams, P., Bycroft, B.W., et al., 1996. Analysis of bacterial carbapenem antibiotic production genes reveals a novel β-lactam biosynthesis pathway. Mol. Microbiol. 22 (3), 415–426.

McNaughton, H., Thirkettle, J., Schofield, C., Jensen, S., 1998. β-Lactam synthetase: implications for β-lactamase evolution. ChemComm. 21, 2325–2326.

Miller, M.T., Bachmann, B.O., Townsend, C.A., Rosenzweig, A.C., 2002. The catalytic cycle of β-lactam synthetase observed by x-ray crystallographic snapshots. Proc. Natl. Acad. Sci. 99 (23), 14752–14757.

Nikaido, H., 2009. Multidrug resistance in bacteria. Annu. Rev. Biochem. 78, 119–146.

Patrick, G.J., Fang, L., Schaefer, J., Singh, S., Bowman, G.R., Wencewicz, T.A., 2018. Mechanistic basis for ATP-dependent inhibition of glutamine synthetase by tabtoxinine-β-lactam. Biochemistry 57 (1), 117–135.

Phelan, R.M., Townsend, C.A., 2013. Mechanistic insights into the bifunctional non-heme iron oxygenase carbapenem synthase by active site saturation mutagenesis. J. Am. Chem. Soc. 135 (20), 7496–7502.

Priya, T., Sharma, C., Singh, R., 2021. Comparative in-silico analysis of structural and functional attributes of microbial lipases: structural and functional attributes of microbial lipases. SPAST Abstracts. 1 (01).

Rabe, P., Kamps, J.J., Schofield, C.J., Lohans, C.T., 2018. Roles of 2-oxoglutarate oxygenases and isopenicillin N synthase in β-lactam biosynthesis. Nat. Prod. Rep. 35 (8), 735–756.

Raber, M.L., Castillo, A., Greer, A., Townsend, C.A., 2009. A conserved lysine in β-lactam synthetase assists ring cyclization: implications for clavam and carbapenem biosynthesis. Chembiochem 10 (18), 2904–2912.

Robin, F., Delmas, J., Schweitzer, C., Tournilhac, O., Lesens, O., Chanal, C., et al., 2007. Evolution of TEM-type enzymes: biochemical and genetic characterization of two new complex mutant TEM enzymes, TEM-151 and TEM-152, from a single patient. Antimicrob. Agents Chemother. 51 (4), 1304–1309.

Saxena, R., Singh, R., 2010. Statistical optimization of conditions for protease production from Bacillus sp. Acta Biol. Szeged. 54 (2), 135–141.

Saxena, R., Singh, R., 2011. Amylase production by solid-state fermentation of agro-industrial wastes using Bacillus sp. Braz. J. Microbiol. 42 (4), 1334–1342.

Sharma, C., Nigam, A., Singh, R., 2021a. Computational-approach understanding the structure-function prophecy of Fibrinolytic Protease RFEA1 from *Bacillus cereus* RSA1. PeerJ. 9, e11570.

Sharma, C., Osmolovskiy, A., Singh, R., 2021b. Microbial fibrinolytic enzymes as anti-thrombotics: production, characterisation and prodigious biopharmaceutical applications. Pharmaceutics. 13 (11), 1880.

Singh, R., Jain, A., Panwar, S., Gupta, D., Khare, S.K., 2005. Antimicrobial activity of some natural dyes. Dyes Pigm. 66 (2), 99–102.

Singh, R., Smitha, M.S., Singh, S.P., 2014. The role of nanotechnology in combating multi-drug resistant bacteria. J. Nanosci. Nanotechnol. 14 (7), 4745–4756.

Sleeman, M.C., MacKinnon, C.H., Hewitson, K.S., Schofield, C.J., 2002. Enzymatic synthesis of monocyclic β-lactams. Bioorg. Med. Chem. Lett. 12 (4), 597–599.

Sreedharan, S.M., Singh, R., 2019. Ciprofloxacin functionalized biogenic gold nanoflowers as nanoantibiotics against pathogenic bacterial strains. Int. J. Nanomedicine 14, 9905.

Sreedharan, S.M., Singh, S.P., Singh, R., 2019. Flower shaped gold nanoparticles: biogenic synthesis strategies and characterization. Indian J. Microbiol. 59 (3), 321–327.

Tahlan, K., Jensen, S.E., 2013. Origins of the β-lactam rings in natural products. J. Antibiot. Res. 66 (7), 401–410.

Townsend, C.A., 2016. Convergent biosynthetic pathways to β-lactam antibiotics. Curr. Opin. Chem. Biol. 35, 97–108.

Ulijn, R.V., De Martin, L., Halling, P.J., Moore, B.D., Janssen, A.E.M., 2002. Enzymatic synthesis of β-lactam antibiotics via direct condensation. J. Biotechnol. 99 (3), 215–222.

Worthington, R.J., Melander, C., 2013. Overcoming resistance to β-lactam antibiotics. J. Org. Chem. 78 (9), 4207–4213.

Zhang, H., Che, S., Wang, R., Liu, R., Zhang, Q., Bartlam, M., 2019. Structural characterization of an isopenicillin N synthase family oxygenase from *Pseudomonas aeruginosa* PAO1. Biochem. Biophys. Res. Commun. 514 (4), 1031–1036.

Chapter 9

Insights into the molecular mechanisms of β-lactam antibiotic synthesizing and modifying enzymes in fungi

Juan F. Martín[1], Carlos García-Estrada[2] and Paloma Liras[1]
[1]*Department of Molecular Biology, University of León, León, Spain,* [2]*Department of Biomedical Sciences, University of León, León, Spain*

9.1 Introduction

The discovery of penicillins and later cephalosporin C in fungi, and cephamycin C in some actinobacteria, motivated a great interest in finding novel molecules containing the β-lactam ring. These include, in addition to the classical β-lactam, the cephabacins, monobactams, olivanic acids, thienamycin, clavulanic acid, the antifungal clavams, and the toxic tabtoxin (Liras and Martín, 2009), among others. Out of these β-lactams, penicillins, cephalosporins, and clavulanic acid have contributed significantly to combat bacterial infections and still are among the most important selling drugs (Demain, 2014).

In nature, the ability to produce β-lactam antibiotics is distributed among fungi and Gram-positive (and some Gram-negative) bacteria. However, the genetic information required is frequently restricted to a few species within a genus or even to certain strains within a particular species (Laich et al., 2002; Kim et al., 2003; Houbraken et al., 2011). The distribution in nature of β-lactam-producing microorganisms has been reviewed earlier (Aharonowitz et al., 1992), and its possible transmission by horizontal gene transfer has been proposed (Landan et al., 1990; Peñalva et al., 1990; Liras et al., 1998). Recent evidence on the genomes of many filamentous fungi supports the horizontal transfer theory (Martín and Liras., 2019). The biosynthesis of metabolites containing the β-lactam ring is an important source of information on novel enzymes with interesting molecular mechanisms of catalysis. Particularly well-known are the enzymes involved in penicillin biosynthesis. These are excellent models for understanding the biochemistry of related

Biotechnology of Microbial Enzymes. DOI: https://doi.org/10.1016/B978-0-443-19059-9.00015-3
© 2023 Elsevier Inc. All rights reserved.

secondary metabolites. However, some of the enzymes involved in cephalosporin biosynthesis are not so well known [e.g., the two-component isopenicillin N (IPN) epimerization system].

This article is focused on the molecular characterization of the enzymes involved in penicillin biosynthesis and penicillin acylases for the industrial production of semisynthetic β-lactams.

9.1.1 Penicillin and cephalosporin biosynthesis: a brief overview

Penicillins and cephalosporins are nonribosomal peptide antibiotics containing the β-lactam ring formed by cyclization of a linear tripeptide. The biosynthesis of these antibiotics starts with the condensation of the amino acids L-α-aminoadipic acid, L-cysteine, and L-valine to form the tripeptide δ-(L-α-aminoadipyl)-L-cysteinyl-D-valine (ACV). Activation of the three amino acids and the condensation reactions are catalyzed by a nonribosomal peptide synthetase (NRPS) named ACV synthetase (ACVS). In a second step, the ACV tripeptide is cyclized to IPN by the enzyme IPN synthase (IPNS), an iron-dependent oxidase. In *Penicillium chrysogenum* NRRL1951 (syn. *Penicillium rubens*), the IPN is converted to hydrophobic penicillins, such as benzylpenicillin or phenoxymethylpenicillin, by IPN acyltransferase (IAT), an enzyme with amidase, acyltransferase, and transacylase activities. In *Acremonium chrysogenum* and actinomycetes, IPN is later isomerized to penicillin N, a step required for the last cephalosporin reactions to occur. The epimerization reaction in filamentous fungi is still poorly understood (Ullan et al., 2002; Martín et al., 2004), whereas, in the case of *Streptomyces clavuligerus* and *Amycolatopsis lactamdurans*, it is mediated by the pyridoxal phosphate-dependent IPN epimerase (Usui and Yu 1989; Laiz et al., 1990). The penicillin N is then converted by a ring expanding oxygenase that uses 2-oxoglutarate as cosubstrate to deacetoxycephalosporin C (DAOC) and deacetylcephalosporin C (DAC) and finally transformed to cephalosporin C by acetylation via the DAC acetyltransferase (Brakhage, 1998; Martín et al., 2012a).

Penicillin biosynthesis is compartmentalized between cytosol and peroxisomes in *P. chrysogenum*. Whereas ACVS and IPNS are cytosolic enzymes, IAT is localized in peroxisomes together with the phenylacetyl-CoA-ligase involved in the activation of the side chain. Similar compartmentalization of intermediates occurs in *A. chrysogenum*, with ACVS and IPNS being cytosolic enzymes. Two enzymes IPN-CoA ligase and IPN-CoA epimerase that catalyze the conversion of IPN in penicillin N are located in peroxisomes. Two genes encoding major facilitator superfamily (MFS) transporters, *cefP* and *cefM*, are present in the early cephalosporin gene cluster. These transporters have been localized in peroxisomes by confocal fluorescence microscopy. Regarding cephalosporin secretion from the producer cells, the third gene of *A. chrysogenum*, *cefT*, encodes an MFS protein, targeted to the cell membrane. This protein appears to be involved in cephalosporin secretion, although *cefT*-disrupted mutants are still able to export cephalosporin by redundant transporters (Martín, 2020).

9.1.2 Genes involved in penicillin and cephalosporin biosynthesis

The central enzymes of the penicillin biosynthetic pathway in *P. chrysogenum* are encoded by a set of three genes linked together in a cluster (the *pen* cluster) in chromosome I of *P. chrysogenum* or in chromosome VI of *Aspergillus nidulans* (Montenegro et al., 1992). The core of the penicillin biosynthetic pathway is encoded by three genes, *pcbAB* (11.5 kbp, encoding the ACVS), *pcbC* (1.0 kbp, encoding the IPNS), and *penDE* (1.2 kbp, encoding the IAT).

The *P. chrysogenum pcbAB* gene is a large open reading frame (ORF) (11376 bp). It lacks introns and encodes a protein of 3791 amino acids (Díez et al., 1990). The orthologous gene of *A. chrysogenum* is very similar (54.9% identity) to that of *Penicillium* (Gutiérrez et al., 1992). The *pcbAB* gene is linked to the *pcbC* and *penDE* genes, clustered together with other nonessential ORFs in a 56.8 kb DNA region amplified in tandem repeats in high penicillin-producing strains (Fierro et al., 1995; 2006; van den Berg et al., 2007). In wild-type *P. chrysogenum*, the penicillin (*pen*) gene cluster is present in a single copy. The number of copies of this amplifiable unit can be changed by recombination (Fierro et al., 1995) and even removed completely, resulting in a nonproducer strain (Fierro et al., 1995, 1996; Harris et al., 2009).

In *A. chrysogenum*, the first two genes of the pathway (*pcbAB* and *pcbC*) are linked to the genes *cefD1* and *cefD2*, encoding the two components of the epimerization system, and to two MFS transporter-encoding genes for cell membrane and organelle (peroxisome) transport systems. The two late genes of the pathway, *cefEF* and *cefG*, encoding, respectively, the ring expandase (DAOC synthase/hydroxylase), and the DAC acetyltransferase, form a separate gene cluster located on a different chromosome (Gutiérrez et al., 1992; Martín et al., 2012a).

The basic molecular genetics of penicillin and cephalosporin biosynthesis is well known (Aharonowitz et al., 1992; Brakhage et al., 2004; Martín, 2000a; Liras and Martín, 2009; Liras and Demain, 2009) and is not detailed here.

9.2 ACV synthetase

ACVS is one of the first known fungal NRPSs and serves as a model of more complex peptide synthetases (Aharonowitz et al., 1993; Zhang and Demain, 1992; Byford et al., 1997). NRPSs are multimodular proteins involved in the biosynthesis of thousands of microbial and plant secondary metabolites (Marahiel et al., 1997). The large size of the NRPSs and their multiple enzyme activities have made difficult the characterization of their catalytic sites. The in vitro activities of these enzymes are usually very poor

(Theilgaard et al., 1997; van Liempt et al., 1989) because they may lose their native structure or suffer alterations, for example, dissociation of NRPS-ancillary proteins that improve substrate affinity in vivo (Baltz, 2011; Walsh et al., 2001). However, molecular genetic approaches allow their characterization in vivo (Wu et al., 2012).

In the cells, ACV can be found either in a reduced monomeric form or as an oxidized dimer (bis-ACV) (Theilgaard and Nielsen, 1999). In the presence of oxygen, bis-ACV is the predominant form released to the culture broth (López-Nieto et al., 1985; García-Estrada et al., 2007). Intracellularly, the ACV is kept in the reduced form by the thioredoxin and thioredoxin reductase system (Cohen et al., 1994) as only the reduced monomer is used as a substrate for the IPNS (Aharonowitz et al., 1992; Theilgaard and Nielsen 1999). The bis-ACV secreted to the culture broth appears to be a waste product of the β-lactam pathway because it is not reintroduced into the producer cells (García-Estrada et al., 2007) and does not serve as an inducer in a quorum-sensing mechanism (Martín et al., 2011). Its secretion reflects the presence of limiting steps in the middle and late parts of the penicillin or cephalosporin pathways (e.g., the *A. chrysogenum* N2 strain, Shirafuji et al., 1979; Ramos et al., 1986), but the mechanism by which the ACV tripeptide is secreted is still unknown.

9.2.1 The ACV assembly line

As indicated above, ACV is synthesized by condensation of three amino acids: L-α-aminoadipic acid, which is an intermediate in the lysine biosynthetic pathway, L-cysteine, and L-valine (Byford et al., 1997; Zhang and Demain, 1992; Martín, 2000b). Amino acid condensation is catalyzed by the nonribosomal multidomain ACVS (Zhang and Demain, 1992; Kallow et al., 2000). The ACVS is a large cytosolic protein with a molecular mass of about 415 kDa that is not located in peroxisomes, unlike the IPN acyltransferase (see below). However, the recent finding of a vacuolar membrane transporter that affects ACV biosynthesis may imply that ACVS is a cytosolic protein associated with vacuoles (Fernández-Aguado et al., 2013; Martín and Liras, 2021), as also proposed for the more complex cyclosporin synthetase (Hopper et al., 2001).

The ACVS contains three different modules, each of approximately 1000 amino acids, and can catalyze multiple reactions, including substrate amino acid activation by adenylation, translocation to the phosphopantetheine arm, peptide bond formation, epimerization, and tripeptide release by an integrated thioesterase (Martín, 2000b; Aharonowitz et al., 1993; Byford et al., 1997). The domains of each NRPSs module are partially conserved, giving a reiterated structure (Zucher and Keller, 1997). Each ACVS module contains adenylate-forming (designated A for activation), aminoacyl (or peptidyl) carrier PCP (also abbreviated T, for thiolation), and condensation (C) domains (Marahiel et al., 1997). In the third module, there is an epimerization domain

(E). These domains are arranged in the characteristic order ATC ATC ATE. The A domain is involved in ATP binding and amino acyl-adenylate formation and determines substrate specificity (Rausch et al., 2005; Röttig et al., 2011; Prieto et al., 2012). The aminoacyl/peptidyl carrier domain contains a conserved serine residue that binds a thiol-containing phosphopantetheine cofactor through its phosphate group (Baldwin et al., 1991). Phosphopantetheine is added to the apo-ACVS by the enzyme 4′-phosphopantetheine transferase (PPTase), encoded by the *pptA* gene (Keszenman-Pereyra et al., 2003; García-Estrada et al., 2008a; Márquez-Fernández et al., 2007). The C domain of ACVSs is involved in the condensation of two amino acids activated on adjacent modules and catalyzes the peptide elongation reaction (Crécy-Lagard et al., 1995).

At the end of the third module, there is an epimerase (E) domain (365 amino acids), which was proposed to catalyze the conversion of the precursor amino acid L-valine into D-valine to form the tripeptide LLD-ACV (Díez et al., 1990; Gutiérrez et al., 1991; Stachelhaus and Marahiel, 1995). Several conserved motifs (E1 to E7) have been found within this domain in the ACVSs from different fungi, including *P. chrysogenum, A. chrysogenum, A. nidulans* (Kallow et al., 2000), *Aspergillus oryzae, Kallychroma tethys* (Kim et al., 2003), and in the actinobacteria *A. lactamdurans* and *S. clavuligerus* (Coque et al., 1991; Martín, 2000b). In all these microorganisms, the consensus sequence for the E4 motif (EGHGRE) is fully conserved as well as in the epimerization domains of other NRPS (Martín, 2000b).

9.2.2 The cleavage function of the integrated thioesterase domain

Next to the epimerase domain and integrated into the C-terminal region of the ACVS, there is a thioesterase domain (designated TE in Fig. 9.1) of about 230 amino acids. Within the thiosterase region, the fungal and bacterial ACVSs contain a motif (GXSXG) homologous to the amino acid sequence present in the active center of the oleyl-ACP thioesterase I of vertebrate fatty acid synthetases (Díez et al., 1990; Theilgaard et al., 1997). This domain was proposed to catalyze the hydrolysis of the thioester bond between the nascent ACV tripeptide and the enzyme-bound phosphopantetheine.

The cleavage function of the integrated thioesterase domain has been clearly established by directed in vitro mutation of the serine residue at the thioesterase active center (Kallow et al., 2000; Wu et al., 2012) and by deletion of the entire TE domain (Wu et al., 2012). In both studies, mutants in the thioesterase domain could not produce significant amounts of penicillin. When the thioesterase domain of *A. nidulans* ACVS was modified, the resulting mutant showed more than a 95% decrease in penicillin production. Still, some ACV was released unexpectedly, part of which had the LLL configuration (Kallow et al., 2000). These results suggest an interaction between the epimerase and thioesterase domains adjacent in the C-terminal regions of

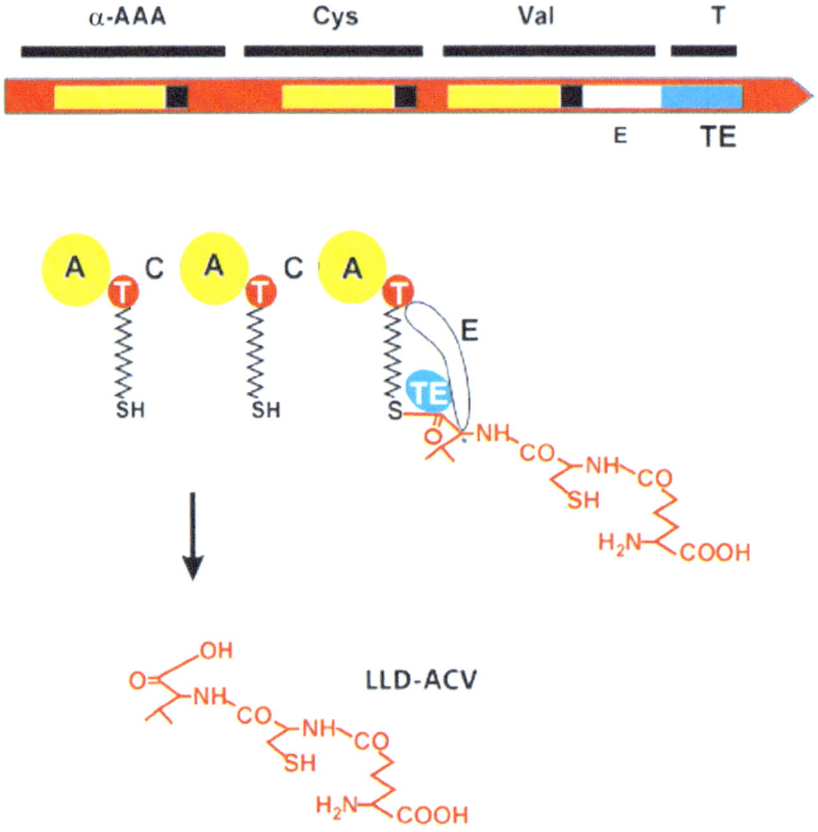

FIGURE 9.1 Scheme of the modules and domains of the ACVS showing the interaction of the epimerization domain (E) with the nascent peptide while it is attached to the pantetheine of the third module, and with the thiosterase (TE) domain (see text for details). A, activation; T, thiolation (peptidyl carrier); C, condensation domains. The C-2 carbon atom of valine that is epimerized to the D configuration is indicated by an asterisk. *ACVS*, ACV synthetase.

the ACVSs (Fig. 9.1). Increasing evidence indicated that the thioesterase domain at the carboxyl-terminal end of the ACVS has a proofreading role in removing the release of peptides that are not properly converted to the LLD-form (Martín and Liras, 2017). Recent evidence is that some NRPSs are modified by acetylation/acylation of key amino acids. Still, there is no clear information on whether these posttranslational modifications affect the ACVS activity (Martín et al., 2021).

9.3 Isopenicillin N synthase

The second enzyme of the penicillin and cephalosporin pathway was purified, characterized in the 1980s from extracts of *P. chrysogenum* (Ramos et al.,

1985) and *Cephalosporium acremonium* (syn. *Acremonium chysogenum)* (Pang et al., 1984). These early biosynthetic studies have been reviewed (Aharonowitz et al., 1992; Martín, 1998) and are not discussed here. Studies in the last decade have provided considerable insight into the mechanisms of IPN biosynthesis.

The IPNS is a mononuclear iron-containing nonheme oxidase, which may also work as an oxygenase, related evolutionarily to other nonheme oxygenases. When acting on its natural substrate ACV, this enzyme functions as an oxidase consuming a whole molecule of O_2 and removing four H atoms from the substrate to form two molecules of H_2O (Feig and Lippard, 1994). This oxidative reaction results in the closure of the β-lactam and thiazolidine rings, characteristic of penicillins. The IPNS belongs to a group of iron enzymes that bind O_2 to a ferrous iron-containing center and activate O_2 to react with the organic substrate (Feig and Lippard, 1994; Solomon et al., 2000). These enzymes contain a catalytic center consisting of two histidine residues and an acidic amino acid carboxylate group (the so-called facial triad). Most of these enzymes work as oxygenases, but two of them, IPNS and the closely related 1-aminocyclopropane 1-carboxylic acid oxidase, exhibit a reaction mechanism different from that of oxygenases and act as oxidases.

The O_2 cleavage mechanism by iron-dependent oxygenases was believed to be a homolytic cleavage, in which the Fe^{II}-hydroperoxide complex formed after binding of O_2, follows a Fenton reaction resulting in the cleavage of the O–O bond and formation of a hydroxyl radical and a Fe^{III}-oxo species. However, in the IPNS, cleavage of the O_2 molecule was found to be heterolytic with an entirely different mechanism.

9.3.1 Binding and lack of cyclization of the LLL-ACV

During early studies on the substrate specificity of IPNS, a large number of possible substrates were tested (Abraham, 1986; Baldwin and Bradley, 1990; Huffman et al., 1992). In those studies on the substrate specificity of IPNS, it was observed that this enzyme is unable to convert the LLL-ACV tripeptide to a penicillin molecule. It was concluded that only the LLD-ACV estereoisomer, the fully modified product ACV, served as a substrate of IPNS. However, those studies revealed that the LLL-ACV stereoisomer interferes with the IPNS cyclization activity. Later, Howard-Jones et al. (2005), using crystallographic studies with the LLL-tripeptide analogue δ-(L-α-aminoadipyl)-L-cysteinyl-L-3,3,3,3′,3′,3′-hexafluorylvaline, proved that the LLL-tripeptide binds strongly to the active center of IPNS as does the LLD-ACV stereoisomer, but the configuration of the valine seems to prevent the correct cyclization.

9.3.2 The iron-containing active center

Initial studies by Randall et al. (1993) identified the ferrous active site of IPNS. Using spectroscopic techniques, the coordination of Fe^{II} at the active site with O_2, NO (a dioxygen structural analogue), and the ACV substrate was identified, leading to the proposal of a Fe^{II}-ACV-[NO] complex at the active center (Brown et al., 2007). The crystal structure of IPNS complexed with Fe^{II}-ACVS and NO revealed a six-coordinate iron center.

9.3.3 The crystal structure of isopenicillin N synthase

The first crystal structure of IPNS was provided by Roach et al. (1995) using recombinant *A. nidulans* enzyme complexed with manganese (instead of Fe^{II}) at a resolution of 2.4 Å. These authors reported that (1) the protein has a "jelly-roll"-like structure and (2) the active center is deeply buried inside the protein structure, lined by a set of hydrophobic amino acids. The secondary structure of IPNS consists of 10 α-helices and 16 β-strands. Eight of the β-strands fold into the so-called jelly-roll structure. Several β-strands are combined to give a large sheet on each side of the roll, and the active center is buried inside the β-barrel. The C-terminal region of the protein (amino acids 324–331), extending from the last α-helix (α10), enters between the two faces of the jelly-roll barrel allowing the glutamine residue conserved in all sequenced IPNSs (Gln^{330} in the *P. chrysogenum* amino acid sequence) to bind to the metal in the active center.

The manganese (or iron in the native enzyme) is bound as a six-coordinate nucleus to four protein ligands (His^{214}, Asp^{216}, His^{270}, and Gln^{330}) and to two molecules of water (Roach et al., 1995). These authors propose that the substrate ACV and O_2 bind to the coordination sites initially occupied by the two water molecules and to the Gln^{330}.

In further work, the *A. nidulans* IPNS was crystallized complexed with ferrous iron and ACV at a resolution of 1.3 Å, and a mechanism of ACV cyclization was proposed (Roach et al., 1997). These studies suggest that the reaction is initiated by binding the thiolate of ACV to the Fe^{II} at the active center. The binding of the substrate thiolate to Fe^{II} creates a vacant coordination site in the Fe^{II} center into which the O_2 molecule is bound. According to the proposed model, the cyclization then proceeds by removing the four H atoms from ACV to form the IPN double ring structure, converting the O_2 into two water molecules. This model was supported by studies using the O_2 structural analogue molecule NO and the substrate ACV (Roach et al., 1997). This cyclization model proposes that the IPNS activity participates in the early stages of the reaction to create a Fe^{II}-oxygen species that then cyclizes the substrate without the further direct involvement of the protein ligands in the transfer of the H atoms.

In summary, the accumulated evidence indicates that the cyclization step is initiated by the formation of a bond between the hydroperoxide in the nucleus complex and the N atom of the amide bond of Cys-Val at the ACV.

9.3.4 Recent advances in the cyclization mechanism

Functional Density Theory studies have contributed to refining the previous ACV cyclization mechanism, proposing that the initial iron (Fe^{III}) atom at the active center abstracts an H atom from the cysteine β-carbon of ACV. This H atom abstraction includes an electron transfer from ACV to the oxygen yielding the known Fe^{II}-hydroperoxide complex and a double bond between the S atom of the L-cysteine and the adjacent carbon atom of cysteine in the ACV. The Fe^{II}-hydroperoxide complex deprotonates either the amide NH (Ge et al., 2008; MacFaul et al., 1998; Long et al., 2005) or the iron-bound H_2O molecule (Lundberg and Morokuma, 2007; Lundberg et al., 2009), resulting in the heterolytic cleavage of the O–O bond (Brown-Marshall et al., 2010).

9.4 Acyl-CoA ligases: a wealth of acyl-CoA ligases activate penicillin side-chain precursors

In addition to hydrophobic penicillins containing aromatic side chains, *P. chrysogenum* produces a variety of natural penicillins containing saturated (6–8 atom carbon) or unsaturated fatty acids. Indeed, the IAT in vitro is able to accept a variety of acyl-CoA molecules as substrates, including hexanoyl-CoA, hex-3-enoyl-CoA, and octanoyl-CoA, among others (Luengo et al., 1986; Aplin et al., 1993). The use of side-chain precursors for penicillin biosynthesis requires their activation as thioesters, usually as CoA derivatives. However, other thioesters, for example, acyl-glutathione, are also used in vitro by the IAT (Alvarez et al., 1993).

The phenylacetyl-CoA-ligase-encoding gene (*phlA*) was first cloned by Lamas-Maceiras et al. (2006). The encoded protein, PhlA (also known as PCL) belongs to a well-known family of acyl-CoA synthetases that use ATP to activate fatty acids and organic acids as acyl-AMP intermediates. The PhlA protein belongs to the subfamily of *p*-coumaroyl-CoA ligases that are widely used in the biosynthesis of plant and microbial secondary metabolites. Detailed analysis of the *P. chrysogenum* genome revealed the presence of at least five genes encoding enzymes of this subfamily in *P. chrysogenum* (Table 9.1) (Martín et al., 2012b). The PhlA protein contains a canonical PTS1 peroxisomal targeting sequence that supports the biochemical description of a peroxisomal acyl-CoA ligase (Gledhill et al., 2001). Disruption of the *phlA* gene reduced the production of benzylpenicillin by about 50%, and, interestingly, the cell extract of the disrupted mutant still contained 60% of phenylacetyl-CoA-ligase activity (Lamas-Maceiras et al., 2006), leading to the conclusion that *P. chrysogenum* contains other acyl-CoA ligases involved in penicillin biosynthesis.

Later, Wang et al. (2007) described one of these redundant acyl-CoA ligases, named PhlB, as an isozyme involved in activating phenylacetic acid.

TABLE 9.1 **Phenylacetyl-CoA, adipyl-CoA and related aryl-CoA ligases in the genome of *Penicillium chrysogenum*.**

Gene	*P. chrysogenum* gene number	PST1 targeting sequence (C-terminal)	Enzyme	References
phlA	Pc22g14900	SKI	Phenylacetate-CoA ligase	Lamas-Maceiras et al. (2006)
phlB	Pc22g202700	AKL	Adipyl and fatty acyl-CoA ligase	Wang et al. (2007); Koetsier et al. (2009, 2010)
phlC	Pc13g12270	AKL	Aryl (coumaroyl, phenylacetyl)-CoA ligase	Yu et al. (2011)
ary1	Pc22g24780	AKL	Similar to *Arabidopsis thaliana* 4-coumaroyl-CoA ligase	This work
ary2	Pc21g22010	SKL	Similar to *A. thaliana* 4-coumaroyl-CoA ligase	This work
ary3	Pc06g01160	TKI	Similar to *A. thaliana* 4-coumaroyl-CoA ligase	This work
ary4	Pc21g20650	ARL	Similar to *A. thaliana* 4-coumaroyl-CoA ligase	This work
ary5	Pc21g07810	Lacks PTS1	Similar to *A. thaliana* 4-coumaroyl-CoA ligase	This work

Note: The *ary1* to *ary5* genes found in *P. chrysogenum* genome (van den Berg et al., 2008) have not been studied biochemically.

However, Koetsier and coworkers (2009, 2010) purified this second acyl-CoA ligase and concluded that it activates medium- and long-chain fatty acids but has essentially no activity on phenylacetic acid. This second enzyme activates adipic acid and might be involved in producing some natural nonaromatic penicillins and in the synthesis of adipyl-7-aminodeacetoxycephalosporin C in engineered *P. chrysogenum* strains. The PhlB protein has also a peroxisomal targeting sequence (Wang et al., 2007; Koetsier et al., 2009) and is proposed to be located in the peroxisomes.

Expression of *phlA* gene is strongly induced by phenylacetic acid (Lamas-Maceiras et al., 2006; Harris et al., 2009), whereas expression of *phlB* is induced by adipic acid (Koetsier et al., 2010).

A third acyl-CoA ligase-encoding gene, *phlC*, has been cloned from *P. chrysogenum*, and the encoded enzyme has been characterized (Yu et al., 2011). This third acyl-CoA ligase is an orthologue of a putative *p*-coumaroyl-CoA ligase of *Aspergillus fumigatus* (73% identical amino acids). In contrast, it has only 37% and 38% identity with PhlA and PhlB. The PhlC protein also contains the peroxisomal targeting sequence in its C-terminus (Table 9.1). Although inefficiently, the recombinant enzyme expressed in *Escherichia coli* showed broad substrate specificity and activated phenylacetic acid. It was more active on caproic acid (C6) and the aromatic cinnamic and coumaric acids. Although PhlC can activate phenylacetic acid, the reaction turnover was very low, and therefore, the contribution of PhlC to penicillin biosynthesis appears to be very limited (Yu et al., 2011). Unfortunately, no mutants lacking the *phlC* gene have been obtained, and, therefore, it is impossible to estimate the actual contribution of this enzyme to penicillin biosynthesis. Other putative acyl-CoA ligases identified in the *P. chrysogenum* genome by bioinformatic analysis remain to be characterized (Table 9.1). In summary, several acyl-CoA ligases with characteristic substrate specificity are involved in the biosynthesis of different natural penicillins. Some of them, for example, PhlB, may have to be overexpressed to enhance the production of adipyl-containing recombinant β-lactams.

9.5 Isopenicillin N acyltransferase (IAT)

The initial IAT purification studies (Alvarez et al., 1987) revealed that IAT may use either IPN or 6-APA as substrate in the acylation/transacylation reaction. The activity on 6-APA, that is, phenylacetyl-CoA:6-APA acyltransferase, is very high as compared to the activity on IPN, and this led to the initial proposal that there might be two enzymes involved in the conversion of IPN to benzylpenicillin which might require two encoding genes, *penD* and *penE*. However, we found that a single gene, *penDE*, (Barredo et al., 1989; Montenegro et al., 1992), encodes five related enzyme activities in the IAT protein, namely, IPN aminohydrolase, IPN acyltransferase, 6-APA acyltransferase, benzylpenicillin (PenG) acylase, and PenG/PenV transacylase (Alvarez et al., 1993).

9.5.1 Posttranslational maturation of the IAT

The mature IAT, as occurs with other penicillin and cephalosporin acylases (CA), is a heterodimeric protein consisting of two subunits: α (11 kDa) and β (29 kDa). Both subunits derive from a pro-IAT precursor protein by cleavage of the $gly^{102}-cys^{103}$ bond of the precursor protein (Barredo et al., 1989; Aplin et al., 1993; Tobin et al., 1995), although in some analyses, the Thr^{104} was identified in the N-terminus of the β subunit (Barredo et al., 1989).

Several amino acid residues of the IAT were shown to be important for IAT activity (Tobin et al., 1994, 1995). One of the critical amino acid residues in the IAT is Cys^{103}, which after cleavage of the IAT precursor (pro-IAT) becomes the N-terminal amino acid of the β-subunit. This finding and the similarity with the maturation mechanisms of self-processing penicillin G acylases (PGA), penicillin V acylases (PVA), and CA of *E. coli* and other Gram-negative bacteria indicate that the IAT belongs to the Ntn (N-terminal nucleophile) hydrolase family (Brannigan et al. 1995). The *E. coli* PGA has a serine (Ser^{264}) as the nucleophile and, indeed, it is an autoprocessing enzyme that cleaves itself at the 263–264 peptide bond (Duggleby et al., 1995), the so-called excisite (excision site). However, IAT has a low overall identity to PGA (10% identity) or PVA (11% identity), indicating that IAT belongs to a different class of Ntn proteins (see below). Although different members of the Ntn protein superfamily do not show extensive amino acid similarity, they all share the same folding structure composed of two layers of β-sheets sandwiched between two layers of α-helices ($\alpha\beta\beta\alpha$ structure).

An interesting finding on both the fungal IAT and the bacterial PGA is that the nucleophile residue at the active center is involved in both the self-processing during maturation of the pro-enzyme and the cleavage of the substrate (IPN for IAT, PenG for PGA, or PenV for PVA). Hewitt et al. (2000) reported that a slow self-processing PGA variant is also less active in penicillin G hydrolysis. Similarly, we observed that the *A. nidulans* IAT (less active than the *P. chrysogenum* IAT) is a very slow self-processing enzyme (Fernández et al., 2003), presumably due to differences in some amino acid residues of these two fungal enzymes. The amino acids present at the cleavage site of both the *P. chrysogenum* and *A. nidulans* IATs are conserved (Fig. 9.2), but other important residues are different (Montenegro et al., 1992).

The fungal IATs have a cysteine nucleophile, whereas other NTN enzymes have a serine or threonine at that position (Murzin et al., 1995; Andreeva et al., 2004). During autoprocessing, the side chain of the cysteine, serine, or threonine nucleophile attacks the carbonyl (CO) group of the preceding peptide bond and cleaves it. Directed mutation of the Cys^{103} to ser or to ala makes the pro-IAT unprocessable (Tobin et al., 1994; García-Estrada

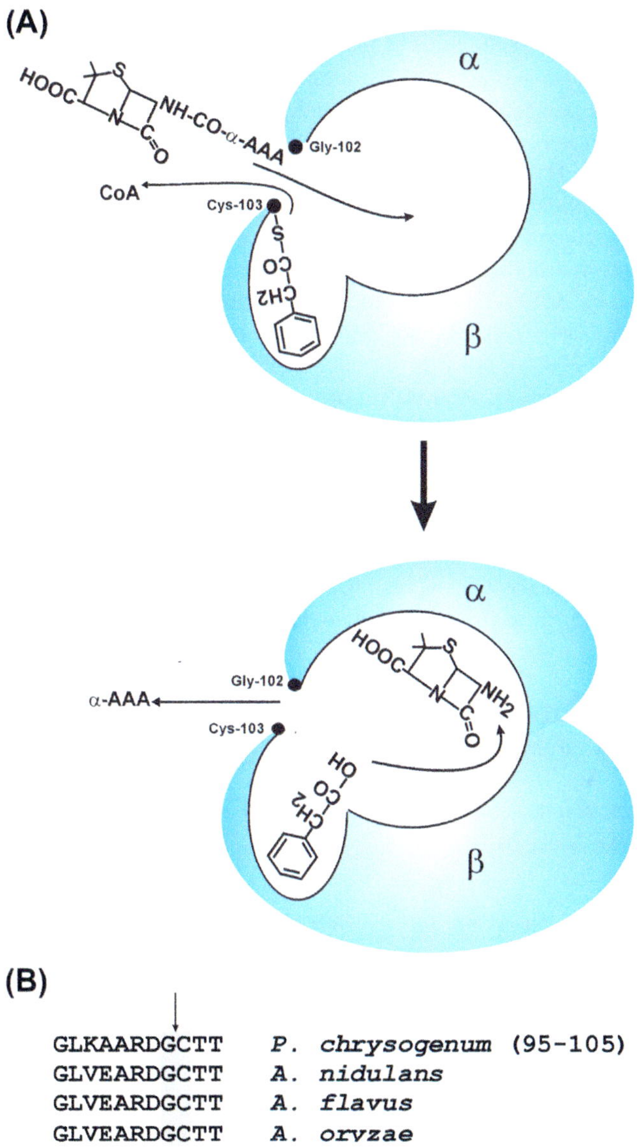

FIGURE 9.2 (A) Model of the reactions occurring in the cavity between the α and β subunits of IAT. Upper drawing: Entry of IPN (with α-AAA as arrowhead) into the cavity and binding of phenylacetic acid to the SH of Cys^{103} as thioester. Lower drawing: Cleavage of the CO-NH bond between α-AAA and 6-APA by the nucleophile attack and exit of α-aminoadipic acid; subsequently, the phenylacetic acid is attached to the 6-APA in the 6-APA binding pocket of the cavity. (B) Comparison of the amino acids forming the lid (residues 95–102 in *Penicillium chrysogenum* enzyme) that closes the entrance to the cavity in *P. chrysogenum* and *Aspergillus nidulans*, *Aspergillus flavus*, and *Aspergillus oryzae* IATs. The cleavage site G^{102}-C^{103} is indicated by a vertical arrow. Note the small sequence differences between the lids of *Penicillium* and *Aspergillus*.

et al., 2008b; Bokhove et al., 2010). The $Cys^{103}Ser$ mutant has not sufficient nucleophilicity, as compared to Cys^{103} in the wild type, to attack the peptide bond. The inducer molecule, or cellular condition that activates the enzyme to trigger the nucleophilic attack, is unknown, but it may be related to its localization in peroxisomes. The higher pH environment of the peroxisomal matrix (above pH 7.0), as compared to the cytosol, and the accumulation of hydrogen peroxide and catalase in these organelles may be relevant to triggering the nucleophile attack on the $Gly^{102}-Cys^{103}$ peptide bond and, therefore, for the hydrolytic activity on IPN. Using the unprocessable $Cys^{103}Ser$ ($C^{103}S$) variant enzyme, we proved that the unprocessed IAT is perfectly targeted to the peroxisomes and transported into these organelles (García-Estrada et al., 2008b), but it is yet unclear if the wild-type enzyme is cleaved or not in the preperoxisomal traffic of the protein from the ER to the peroxisomes (Roze et al., 2011; Martín et al., 2012b; 2013).

Hensgens and coworkers (2002) and Bokhove et al. (2010) crystallized the IAT and studied its autoprocessing mechanism. They proposed that the peptide extending from positions 91–02 of the pro-IAT acts as a lid that closes the access of IPN to the substrate-binding pocket in the precursor IAT. When the Gly^{102}-Cys^{103} bond is cleaved, the C-terminal peptide of the α-subunit (amino acids 95–102) departs from the $\alpha-\beta$ heterodimeric structure, opening the substrate-binding pocket in the precursor protein. The 95–102 peptide's proposed flexibility is based on models deduced from electron density maps of the mature IAT and its $C^{103}A$ variant after crystallization. The distance between residue 101 in the C-terminus of the α-subunit and the Cys^{103} (β subunit) in the mature enzyme was over 25 Ao. The Gly^{102} C-terminal residue of the α subunit does not show up in the electron density maps, indicating that it is not in a rigid permanent position (Bokhove et al., 2010).

9.5.2 The IPN/6-APA/PenG substrate-binding pocket

The IAT crystallographic studies have shown a large cavity between the α- and β-subunits with a narrow entrance in which the nucleophilic Cys^{103} interacts with the IPN substrate (Bokhove et al., 2010). Cleavage of the $Gly^{102}-Cys^{103}$ bond and opening of the lid (amino acids 95–102) do not drastically change the cavity structure and, therefore, the uncapped narrow entrance to the cavity still poses a limitation for bulky substrates such as benzylpenicillin, or phenoxymethylpenicillin. Based on the data of amino acids lining the cavity, Bokhove et al. (2010) proposed a tetrahedral reaction intermediate. In this model, IPN enters the cavity as an arrow with the α-aminoadipic acid chain as arrowhead establishing bonds through the α-amino and carboxylic groups. Following the nucleophilic attack, cleavage of the amide bond that releases the α-aminoadipic moiety occurs. α-Aminoadipic acid has to leave the cavity, providing room for the entry of the aromatic acyl side-chain precursor. These authors propose that the 6-APA formed in the hydrolytic reaction is temporarily stored in a substrate subpocket of the cavity, allowing the released α-aminoadipic acid to leave the reaction cavity. The exit of α-aminoadipic acid might be a limiting step. This model explains our early observation that the IPN amidohydrolase reaction is more inefficient (probably rate-limiting in the IAT) than the 6-APA acyltransferase reaction (Alvarez et al., 1993).

9.5.3 A transient acyl-IAT intermediate

Alvarez et al. (1987, 1993) observed that the IAT has a thioesterase site (GXS^{309}XG) close to the C-terminus of the β-subunit that appears to be involved in recognition of the phenylacetyl-CoA (or related acyl-CoA thioesters) and hydrolysis of the thioester bond transferring the phenylacetic acid moiety to a conserved cysteine residue in the IAT.

The serine residue in the GXS^{309}XG sequence is essential for IAT activity (Tobin et al., 1994). Although no enzymatic analysis has been made of the putative thioesterase activity, Aplin et al. (1993), using electrospray protein mass spectrometry, concluded that an acyl-IAT intermediate with a mass increase of 116 Da is formed in phenylacetic acid-supplemented culture medium that appears to correspond to a phenylacetyl-IAT transient intermediate. The mass increase was different when the culture was supplemented with phenoxyacetic acid. Similar transient mass increases were observed in in vitro reactions when the enzymatic reaction was carried out in the presence of other acyl-CoAs that correlate with the acyl group mass (Aplin et al., 1993). Bokhove et al. (2010) proposed that the transient acyl-enzyme intermediate is formed at the SH group of Cys^{103} at the entrance of the substrate pocket. This allows attack of the thioester bond of the acyl-enzyme intermediate by the amino group of 6-APA, stored temporarily at the "parking subpocket," resulting in the formation of the final product of the pathway, benzylpenicillin, which leaves the cavity. When 6-APA, instead of IPN, is used as substrate in vitro, the 6-APA acyltransferase reaction is facilitated by the absence of α-aminoadipic acid. The IAT could be crystallized, complexed with 6-APA that interacts efficiently with amino acids lining the reaction pocket (Bokhove et al., 2010).

9.5.4 The origin of IAT: an homologous AT in many fungal genomes

The *penDE* gene, encoding IAT, contains three introns in *P. chrysogenum*, *A. nidulans*, and *A. oryzae*, and its origin is different from that of *pcbAB* and *pcbC* lacks introns and appears to be of bacterial origin (Landan et al., 1990; Peñalva et al., 1990; Liras et al., 1998). The *penDE* gene is only present in penicillin-producing fungi but not in the cephalosporin producer *A. chrysogenum* (Martín and Liras, 2019). It is closely related to the orthologous *penDE* gene (also named *aat*) in other penicillin-producing fungi, and its origin is intriguing.

The IAT belongs to a group of related fungal Ntn amidase/acylases enoded by genes designated as *ial* (for IAT-like) that occur in many ascomycetes (García-Estrada et al., 2009). The *ial* gene of *P. chrysogenum* was initially described during analysis of the *P. chrysogenum* genome (van den Berg et al., 2008), and the orthologous gene of *A. nidulans* (named *aatB*) was then identified (Spröte et al., 2008). The IAL protein shares 54% similarity (34% identity) with *P. chrysogenum* IAT and 52% similarity (35% identity) with *A. nidulans* IAT. This similarity is higher than that of bacterial PGA or PVAs (10% identity). IAL has an orthologous gene in most ascomycetes that contains 2 or 3 introns. They encode proteins with similarities ranging from 76% to 81% with proteins of unknown function (members of the IAL subfamily).

The *P. chrysogenum* and *A. nidulans* IAL proteins contain the characteristic Ntn motifs of self-processing enzymes but lack the peroxisomal targeting sequence ARL or ANI that occurs in the IAT. Using enzyme purified after expression of *ial* in *E. coli*, it was confirmed that the 40 kDa pro-IAL is processed into two subunits α (about 12 kDa) and β (28 kDa) by self-cleavage, presumably at the G^{105}–C^{106} bond, as occurs with the IAT. The *ial* gene is located away from the penicillin gene cluster, and mutants lacking the *penDE* gene (e.g., *npe*10 AB-C) but containing a functional *ial* gene were unable to produce penicillin, indicating that the *ial* gene does not encode a second IAT enzyme. Conversely, mutants deleted in the *ial* gene, but conserving the penicillin gene cluster, produce normal amounts of benzylpenicillin (García-Estrada et al., 2009). Since the IAL protein lacks the peroxisomal targeting signal, an engineered protein carrying the canonical C-terminal ARL amino acids was used in complementation studies. Even when targeted to peroxisomes, the IAL failed to show in vivo IPN-amidolyase acid or IPN acyltransferase activities. Detailed analysis of the protein concluded that the lack of IPN acyltransferase activity in IAL is probably due to the lack of the thioesterase center (at Ser^{309} in the IAT) that participates in the cleavage of phenylacetyl-CoA and the transfer of phenylacetic acid to form the phenylacetyl-enzyme intermediate in the acyltransferase reaction. Indeed, the *A. nidulans* IAL enzyme (named AatB), which conserves the Ser^{309} amino acid, shows some penicillin biosynthesis activity (Spröte et al., 2008), even though the protein lacks the canonical peroxisomal targeting signal.

In summary, differences in the protein sequence explain the different activities of IAT and IAL proteins, but the actual substrate(s) of IAL in all ascomycetes remains unknown. Although the initial observation of *ial* genes in β-lactam-producing fungi led to speculation that the *penDE* and *ial* genes might be related, their low overall identity and the lack of contribution of *ial* to penicillin biosynthesis suggest that the role of IAL in fungal cells is different from that of IAT. However, both genes may derive from a common acyltransferase ancestor gene. Further biochemical characterization of the IAL substrates and reaction products will clarify the relationship between these two enzymes.

9.6 Transport of intermediates and penicillin secretion

9.6.1 Transport of isopenicillin N into peroxisomes

An MFS transporter was found in *P. chrysogenum*. This *P. chrysogenum* protein, named PenM (for microbodies), is encoded by Pc21g09220 gene. PenM has 12 transmembrane spanning domains and a size of 508 amino acids (Fernández-Aguado et al., 2014).

A *penM* overexpressing strain produces increased levels of penicillin ranging from 169% to 236% concerning the parental strain, *P. chrysogenum* Wis54−1255. These results indicate that transport of IPN into peroxisomes is rate-limiting for penicillin biosynthesis in a complex production medium. Several mutants silenced in the expression of the *penM* gene of *P. chrysogenum* Wis54−1255 showed reduced benzylpenicillin production. The IAT activity of the *penM*-silenced transformants is still normal, as shown by immunoblotting assays of IAT and in vitro determination of its enzyme activity. These results demonstrate that the *penM*-silenced strain is deficient in the transport of intermediates, but the mutation does not affect the normal incorporation of IAT into peroxisomes.

Confocal microscopy studies using labeled Pen-MDsRed protein and the peroxisomal control marker protein EFG-SKL showed that both proteins colocalize in peroxisomes. This evidence supports the conclusion that PenM is a peroxisomal membrane protein. Experiments with the PenM-silenced transformant using increasing concentrations of phenylacetic acid and/or 6-APA show that in this strain, its low penicillin biosynthesis is independent of the concentration of phenylacetic acid provided, suggesting that the transport of phenylacetic acid is not mediated by the PenM transporter but is performed by a different phenylacetic acid carrier.

9.6.2 IAT is easily accessible to external 6-APA

As indicated above, a mutant deficient in the PenM transporter could not synthesize benzylpenicillin even in a medium supplied with phenylacetic acid (Fernández-Aguado et al., 2014). Surprisingly, this mutant produced normal amounts of benzylpenicillin when 6-APA was supplied extracellularly to the cells (Fernández-Aguado et al., 2014). The formation in vivo of benzylpenicillin in cultures supplemented with 6-APA was linearly dependent on the amount of 6-APA supplied to the cells. This result suggests that 6-APA has easy access to the IAT in the peroxisomes. Alternatively, the IAT may contact the 6-APA at the cell membrane or in endosomes or traffic vesicles.

9.6.3 Intracellular traffic of intermediates and secretion of penicillins

The optimal function of the β-lactam biosynthetic enzymes relies on a sophisticated temporal and spatial organization of the enzymes, the intermediates, and the final products. As described above, the first and second enzymes of the penicillin pathway, ACVS and IPNS, in *P. chrysogenum* and *A. nidulans* are cytosolic. In contrast, the last two enzymes of the penicillin pathway, phenylacetyl-CoA-ligase and IAT, are located in peroxisomes working in tandem at their optimal pH, which coincides with the peroxisomes pH. Two MFS transporters, PenM and PaaT, have been found to be involved in the import of the intermediate IPN and phenylacetic acid, respectively, into peroxisomes (Fernández-Aguado et al., 2013, 2014).

The secretion of penicillin from peroxisomes to the extracellular medium is still unclear. Attempts have been made to identify a gene encoding the penicillin transporter proteins among the 48 ABC-transporters of *P. chrysogenum*. The highly efficient secretion system that exports penicillin against a concentration gradient may involve active penicillin extrusion systems mediated by vesicles that fuse to the cell membrane (Martín, 2020). In summary, the penicillin biosynthesis finding shows that it is essential to target enzymes to organelles adequately to increase the biosynthesis of novel secondary metabolites.

9.7 Production of semisynthetic penicillins by penicillin acylases

Production of specific penicillins, such as benzylpenicillin (penicillin G) or phenoxymethylpenicillin (penicillin V), for industrial purposes, requires the addition of specific side-chain precursors to the fermentation tank; that is, phenylacetic acid for benzylpenicillin and phenoxyacetic acid for phenoxymethylpenicillin (García-Estrada and Martín, 2019). These two penicillins are dubbed natural (biosynthetic) penicillins and represent the precursor molecules for the production of semisynthetic penicillins (antistaphylococcal, aminopenicillins, carboxypenicillins, ureidopenicillins, and β-lactamase-resistant penicillins), which show better properties in terms of potency, the spectrum of activity, stability, pharmacokinetics, and safety than biosynthetic penicillins. They have contributed to mitigating the important problem of adaptive microbial resistance to antibiotics (Oshiro, 1999).

9.7.1 Molecular mechanisms of penicillin acylases

Natural (biosynthetic) penicillins undergo enzymatic hydrolysis for the production of semisynthetic penicillins, thus giving rise to the antibiotic nucleus (6-APA), followed by further amidation of this nucleus with different carboxylic acyl donor chains (Volpato et al., 2010). The enzymes catalyzing the cleavage of the acyl side chains (deacylation) that are present in penicillins, thus yielding 6-APA and the corresponding organic acid, are called penicillin acylases (E.C. 3.5.1.11) or penicillin amidohydrolases (Shewale and SivaRaman, 1989). These enzymes have contributed to the transformation of the industrial production of 6-APA, which was traditionally based on chemical synthesis. This chemical process, which began around 1970 and required the use of hazardous chemicals and solvents for the one-pot deacylation of benzylpenicillin produced in fermentation, remained for 15–20 years until it was widely replaced by the use of these enzymes (Arroyo et al., 2003).

Based on substrate specificity, penicillin acylases are classified into two main groups: PGAs, preferentially hydrolyzing benzylpenicillin, and PVAs, preferentially hydrolyzing phenoxymethyl penicillin, although other minority groups could also be considered depending on the activity over other substrates (Deshpande et al., 1994; Arroyo et al., 2003). These enzymes are naturally produced intra- and extracellularly by fungi and bacteria, showing differences in their structural nature and substrate spectrum, thus offering a prominent diversity in their source and localization. For example, PGA from *E. coli* is targeted at the periplasmic space, whereas this enzyme is extracellular in *Bacillus megaterium* and intracellular in *Bacillus badius*. In the same way, periplasmic signal sequences have been found in the PVAs from Gram-negative bacteria. In contrast, this enzyme is cytoplasmic in *Bacillus subtilis* and can be found both intra- and extracellularly in fungi (e.g., *P. chrysogenum*) (reviewed by Avinash et al., 2016). Although they have an in vivo role not yet fully characterized, it has been suggested that it is related to the metabolization of aromatic compounds (Valle et al., 1991) or to the bacterial signaling phenomenon quorum sensing (Avinash et al., 2016).

From the biochemical point of view, penicillin acylases are Ntn hydrolases with a catalytic N-terminal nucleophile residue and an αββα-fold that undergo posttranslational processing leading to an autocatalytically active enzyme (Brannigan et al., 1995; Oinonen and Rouvinen, 2000). PGAs and PVAs differ in structure, PGAs being usually heterodimeric, unlike PVAs, which usually form homotetramers (Suresh et al., 1999; Avinash et al., 2016). After processing, the α-subunit of PGAs varies between 20 and 24 kDa, with a β-subunit of about 60–65 kDa (Rajendhran and Gunasekaran, 2004). Regarding PVAs, the size of the monomer ranges between 30 and 35 kDa (Shewale and Sudhakaran, 1997). The catalytic reaction mechanism is also different between PGAs and PVAs (Avinash et al., 2016). While Ser is the N-terminal nucleophile in PGAs, Cys plays this role in PVAs (Oinonen and Rouvinen, 2000). The N-terminal serine (βSer1) of PGAs acts as a nucleophile in the cleavage of the Thr-Ser bond of the precursor, thus removing the spacer peptide and generating the active mature enzyme, and undergoing a nucleophilic attack on the carbonyl carbon of the amide bond of the substrate benzylpenicillin, with the subsequent formation of a tetrahedral intermediate stabilized by oxyanion hole residues, which helps in the release of the product (Kasche et al., 1999;

McVey et al., 2001). It has been proposed that βGln^{23} and βAsn^{241} residues also play a role in the complex interaction (Zhiryakova et al., 2009), whereas βArg^{145} and αArg^{263} are involved in the substrate binding and in some way influence the stereoselectivity of the enzyme (Guncheva et al., 2004). On the other hand, Cys^{1} acts as the nucleophile in PVAs, where residue Asp^{274} can interact with the extra oxygen in penicillin V, enabling substrate recognition over penicillin G (Chandra et al., 2005). In addition, Asn^{82} and Arg^{228}, present in the PVAs catalytic site, have been suggested to stabilize the transition state region (Lodola et al., 2012).

9.7.2 Novel developments in industrial applications of penicillin acylases

There are many industrial processes catalyzed by penicillin acylases, such as the industrial production of semisynthetic penicillins and cephalosporins, protection of amino and hydroxyl groups in peptide synthesis, as well as the resolution of racemic mixtures of chiral compounds (Arroyo et al., 2003). Enzymatic production of 6-APA by penicillin acylases is only economically viable when immobilized biocatalysts are used. However, immobilization of enzyme on a carrier results in the loss of enzyme activity (Rajendhran and Gunasekaran, 2007), which has promoted the development of very robust biocatalysts (carrier-free immobilized enzymes) such as cross-linked enzymes, cross-linked enzyme crystals, and cross-linked enzyme aggregates (Arroyo et al., 2003).

Large-scale biotechnological production of penicillin acylases is interesting from the industrial point of view and is achieved by strain improvement of the producer microorganisms, fermentation optimization, and the use of heterologous hosts for expression (Arroyo et al., 2003; Srirangan et al., 2013). The catalytic activity and stability of PGAs from *E. coli* and *Alcaligenes faecalis* have been the subject of study in order to improve them by protein engineering (Tishkov et al., 2010). In addition, immobilization of free PGA enzyme or whole cells has been performed to reduce manufacturing costs and mass transfer limitations to increase effectiveness and competitiveness as industrial biocatalysts (Tischer and Kasche 1999; Li et al., 2020). On the contrary, the application of PVAs has been limited by slightly higher costs of penicillin V and low PVA activity, despite it has been reported that the combination between PVA and penicillin V could be more advantageous due to the resistance of this enzyme to acid conditions, its availability with high substrate concentration, its high stability, and higher yield of 6-APA (Shewale and Sudhakaran, 1997). Overall, penicillin V is responsible for only 15% of all manufactured 6-APA (Demain, 2000), which has reached an estimated annual worldwide production of more than 20,000 tons (Maresova et al., 2014).

Much of the effort made to achieve a competitive biocatalytic process has been focused on the development of new penicillin acylases with better properties. As a result, many different enzymes from different microorganisms have been tested as biocatalysts with exciting results, such as the PVA from *Streptomyces lavendulae* (Torres-Bacete et al., 2000) or the PGA from *Achromobacter* sp. CCM 4824 (Bečka et al., 2014), among others. The use of mutagenesis has also been employed to improve the properties of already known enzymes. For example, saturation site-directed mutagenesis of residues αR^{145} and αF^{146} of PGA from *E. coli* increased the synthetic rate over the hydrolytic rate 5- to 15-fold, thus improving the synthetic yield by a 50% (Jager et al., 2008). In another report, a triple mutant PVA ($T^{63}S/N^{198}Y/S^{110}C$) from *Bacillus sphaericus* (BspPVA) was developed by directed evolution and showed 12.4-fold specific activity and 11.3-fold catalytic efficiency higher than the wild-type enzyme, with a 98% conversion yields of 6-APA using 20% (w/v) penicillin V as substrate (Xu et al., 2018).

9.8 Concluding remarks

Considerable progress has been made on understanding the structure and function of the main enzymes involved in penicillin formation and the early steps of cephalosporin biosynthesis. Still, our knowledge of the structure–activity relationship of the ACVS is very limited and needs to be studied in more detail. The protein–protein interactions between ACVS and IPNS, suggested many years ago, have not been confirmed experimentally. These two proteins are encoded by linked genes in all penicillin-, cephalosporin-, and cephamycin-producing microorganisms. In fungi, both genes *pcbAB* and *pcbC* are expressed from a promoter region as divergent transcripts, which facilitates their coordinate regulation (Brakhage, 1998; Martín 2000a). The availability of advanced proteomic and mass-spectrometry tools should be explored to gain insight into their protein–protein interactions.

In bacterial β-lactam gene clusters, the *pcbAB* and *pcbC* genes are in a tail to head organization, and both genes are expressed as a polycistronic transcript from a promoter located upstream of *pcbAB*. The knowledge of the β-lactam biosynthetic enzymes is useful for further improvement of β-lactam antibiotic production. For instance, the adipyl-CoA ligase activity encoded by *phlB* is helpful for synthesizing adipyl-containing cephalosporins in engineered *P. chrysogenum* strains. Similarly, some members of the Ntn acylases, for example, PGA and CA (also named glutaryl acylase) (Velasco et al., 2000) are beneficial for semisynthetic penicillin and cephalosporin production and may be modified to improve their enzyme kinetics and thermal stability. Advances in developing novel penicillin acylases with better properties, together with the design of new routes and better substrates, and the improvement in immobilization protocols and product recovery, can lead to successful use of enzymes in new areas in the antibiotics industry.

Knowledge of the protein structure, substrate specificity, cofactor requirements, and possible modifications of the kinetic parameters by targeted mutation of the encoding genes will facilitate the application of these enzymes for specific substrate conversions in vitro (Barends et al., 2004).

Abbreviations

ACV L-α-aminoadipyl-L-cysteinyl-D-valine
ACVS ACV synthetase
IPN isopenicillin N
IPNS IPN synthase
IAT IPN acyltransferase
IAL IAT-like
DAC deacetylcephalosporin C
DAOC deacetoxycephalosporin C
NRPS nonribosomal peptide synthetase
PCP peptidyl carrier protein
6-APA 6-Aminopenicillanic acid
PGA penicillin G acylase
PVA penicillin V acylase

References

Abraham, E.P., 1986. enzymes involved in penicillin and cephalosporin formation. In: Kleinkauf, H., von Döhren, H., Dornauer, H., Neseman, G. (Eds.), Regulation of Secondary Metabolites Formation. VCH, Weinheim, pp. 115–132.

Aharonowitz, Y., Cohen, G., Martín, J.F., 1992. Penicillin and cephalosporin biosynthetic genes: structure, organization, regulation, and evolution. Annu. Rev. Microbiol. 46, 461–495.

Aharonowitz, Y., Bergmeyer, J., Cantoral, J.M., Cohen, G., Demain, A.L., Fink, U., et al., 1993. Delta-(L-alpha-aminoadipyl)-L-cysteinyl-D-valine synthetase, the multienzyme integrating the four primary reactions in beta-lactam biosynthesis, as a model peptide synthetase. Biotechnology (N.Y.) 11, 807–810.

Alvarez, E., Cantoral, J.M., Barredo, J.L., Díez, B., Martín, J.F., 1987. Purification to homogeneity and characterization of the acyl-CoA:6-APA acyltransferase of *Penicillium chrysogenum*. Antimicrob. Agents Chemother. 31, 1675–1682.

Alvarez, E., Meesschaert, B., Montenegro, E., Gutiérrez, S., Díez, B., Barredo, J.L., et al., 1993. The isopenicillin N acyltransferase of *Penicillium chrysogenum* has isopenicillin N amidohydrolase, 6-aminopenicillanic acid acyltransferase and penicillin amidase acitivies, all of which are encoded by the single *pen*DE gene. Eur. J. Biochem. 215, 323–332.

Andreeva, A., Howorth, D., Brenner, S.E., Hubbard, T.J., Chothia, C., Murzin, A.G., 2004. SCOP database in 2004: refinements integrate structure and sequence family data. Nucleic Acid Res. 32, D226–D229.

Aplin, R.T., Baldwin, J.E., Roach, P.L., Robinson, C.V., Schofield, C.J., 1993. Investigations into the post-translational modification and mechanism of isopenicillin N:acyl-CoA acyltransferase using electrospray mass spectrometry. Biochem. J. 1294, 357–363.

Arroyo, M., de la Mata, I., Acebal, C., Castillón, M.P., 2003. Biotechnological applications of penicillin acylases: state-of-the-art. Appl. Microbiol. Biotechnol. 60, 507–514.

Avinash, V.S., Pundle, A.V., Ramasamy, S., Suresh, C.G., 2016. Penicillin acylases revisited: importance beyond their industrial utility. Crit. Rev. Biotechnol. 36 (2), 303–316.

Baldwin, J.E., Bradley, M., 1990. Isopenicillin N synthase: mechanistic studies. Chem. Rev. 90, 1079–1088.

Baldwin, J.E., Bird, J.W., Field, R.A., O'Callaghan, N.M., Schofield, C.J., Willis, A.C., 1991. Isolation and partial purification of ACV synthetase from *Cephalosporium acremonium* and *Streptomyces clavuligerus*: evidence for the presence of phosphopantothenate in ACV synthetase. J. Antibiot. 44, 241–248.

Baltz, R.H., 2011. Function of MbtH homologs in nonribosomal peptide biosynthesis and applications in secondary metabolite discovery. J. Ind. Microbiol. Biotechnol. 38, 1747–1760.

Barends, T.R.M., Yoshida, H., Dijkstra, B.W., 2004. Three-dimensional structures of enzymes useful for β-lactam antibiotic production. Curr. Opin. Biotechnol. 15, 356–363.

Barredo, J.L., van Solingen, P., Díez, B., Alvarez, E., Cantoral, J.M., Kattevilder, A., et al., 1989. Cloning and characterization of the acyl-coenzyme A:6-aminopenicillanic-acid-acyltransferase gene of *Penicillium chrysogenum*. Gene 83, 291–300.

Bečka, S., Štěpánek, V., Vyasarayani, R.W., Grulich, M., Maršálek, J., Plháčková, K., et al., 2014. Penicillin G acylase from *Achromobacter* sp. CCM 4824: an efficient biocatalyst for syntheses of beta-lactam antibiotics under conditions employed in large-scale processes. Appl. Microbiol. Biotechnol. 98, 1195–2103.

Bokhove, M., Yoshida, H., Hensgens, C.M., van der Laan, J.M., Sutherland, J.D., Dijkstra, B. W., 2010. Structures of an isopenicillin N converting Ntn-hydrolase reveal different catalytic roles for the active site residues of precursor and mature enzyme. Structure 18, 301–308.

Brakhage, A.A., 1998. Molecular regulation of beta-lactam biosynthesis in filamentous fungi. Microbiol. Mol. Biol. Rev. 62, 547–585.

Brakhage, A.A., Spröte, P., Qusai Al-Abdallah, Q., Gehrke, A., Plattner, H., Tüncher, A., 2004. Regulation of penicillin biosynthesis in filamentous fungi. Adv. Biochem. Eng./Biotechnol. 88, 45–90.

Brannigan, J.A., Dodson, G., Duggleby, H.J., Moody, P.C., Smith, J.L., Tomchick, D.R., et al., 1995. A protein catalytic framework with an N-terminal nucleophile is capable of self-activation. Nature 378, 416–419.

Brown, C.D., Neidig, M.L., Neibergall, M.B., Lipscomb, J.D., Solomon, E.I., 2007. VTVH-MCD and DFT studies of thiolate bonding to [FeNO]7/[FeO2]8 complexes of isopenicillin N synthase: substrate determination of oxidase vs oxygenase activity in nonheme Fe enzymes. J. Am. Chem. Soc. 129, 7427–7438.

Brown-Marshall, C.D., Diebold, A.R., Solomon, E.I., 2010. Reaction coordinate of isopenicillin N synthase: oxidase vs oxygenase activity. Biochemistry 49, 1176–1182.

Byford, M.F., Baldwin, J.E., Shiau, C.Y., Schofield, C.J., 1997. The mechanism of ACV synthetase. Chem. Rev. 97, 2631–2650.

Chandra, P.M., Brannigan, J.A., Prabhune, A., Pundle, A., Turkenburg, J.P., Dodson, G.G., et al., 2005. Cloning, preparation and preliminary crystallographic studies of penicillin V acylase autoproteolytic processing mutants. Acta Crystallogr. F 61, 124–127.

Cohen, G., Argaman, A., Schreiber, R., Mislovati, M., Aharonowitz, Y., 1994. The thioredoxin system of *Penicillium chrysogenum* and its possible role in penicillin biosynthesis. J. Bacteriol. 176, 973–984.

Coque, J.J., Martín, J.F., Calzada, J.G., Liras, P., 1991. The cephamycin biosynthetic genes *pcbAB*, encoding a large multidomain peptide synthetase, and *pcbC* of *Nocardia lactamdurans* are clustered together in an organization different from the same genes in *Acremonium chrysogenum* and *Penicillium chrysogenum*. Mol. Microbiol. 5, 1125–1133.

Crécy-Lagard, V., Marliere, P., Saurin, W., 1995. Multienzymatic non-ribosomal peptide biosynthesis: identification of the functional domains catalyzing peptide elongation and epimerization. Comptes R. Acad. Sci. III 318, 927–936.

Demain, A.L., 2000. Small bugs, big business: the economic power of the microbe. Biotechnol. Adv. 18, 499–514.

Demain, A.L., 2014. Valuable secondary metabolites from fungi. In: Martín, J.F., García-Estrada, C., Zeilinger, S. (Eds.), Biosynthesis and Molecular Genetics of Fungal Secondary Metabolites. Springer Verlag, New York, pp. 1–15.

Deshpande, B.S., Ambedkar, S.S., Sudhakaran, V.K., Shewale, J.G., 1994. Molecular biology of beta-lactam acylases. World J Microbiol Biotechnol 10, 129–138.

Díez, B., Gutiérrez, S., Barredo, J.L., van Solingen, P., van der Voort, L.H., Martín, J.F., 1990. The cluster of penicillin biosynthetic genes. Identification and characterization of the *pcbAB* gene encoding the alpha-aminoadipyl-cysteinyl-valine synthetase and linkage to the *pcbC* and *penDE*. genes. J. Biol. Chem. 265, 16358–16365.

Duggleby, H.J., Tolley, S.P., Hil,l, C.P., Dodson, E.J., Dodson, G., Moody, P.C., 1995. Penicillin acylase has a single-amino-acid catalytic centre. Nature 373, 264–268.

Feig, A.L., Lippard, S.J., 1994. Reactions of non-heme iron (II) centres with dioxygen in biology and chemistry. Chem. Rev. 94, 759–805.

Fernández, F.J., Cardoza, R.E., Montenegro, E., Velasco, J., Gutiérrez, S., Martín, J.F., 2003. The isopenicillin N acyltransferase of *Aspergillus nidulans* and *Penicillium chrysogenum* differ in their ability to maintain the 40 kDa $\alpha\beta$ heterodimer in an undissociated form. Eur. J. Biochem. 270, 1958–1968.

Fernández-Aguado, M., Ullán, R.V., Teijeira, F., Rodríguez-Castro, R., Martín, J.F., 2013. The transport of phenylacetic acid across the peroxisomal membrane is mediated by the PaaT protein in *Penicillium chrysogenum*. Appl. Microbiol. Biotechnol. 97, 3073–3084.

Fernández-Aguado, M., Martín, J.F., Rodríguez-Castro, R., García-Estrada, C., Albillos, S.M., Teijeira, F., et al., 2014. New insights into the isopenicillin N transport in *Penicillium chrysogenum*. Metab. Eng. 22, 89–103.

Fierro, F., Barredo, J.L., Díez, B., Gutiérrez, S., Fernández, F.J., Martín, J.F., 1995. The penicillin gene cluster is amplified in tandem repeats linked by conserved hexanucleotide sequences. Proc. Natl. Acad. Sci. USA 92, 6200–6204.

Fierro, F., Montenegro, E., Gutiérrez, S., Martín, J.F., 1996. Mutants blocked in penicillin biosynthesis show a deletion of the entire penicillin gene cluster at a specific site within a conserved hexanucleotide sequence. Appl. Microbiol. Biotechnol. 44, 597–604.

Fierro, F., García-Estrada, C., Castillo, N.I., Rodríguez, R., Velasco-Conde, T., Martín, J.F., 2006. Transcriptional and bioinformatic analysis of the 56.8 kb DNA region amplified in tandem repeats containing the penicillin gene cluster in *Penicillium chrysogenum*. Fungal Genet. Biol. 43, 618–629.

García-Estrada, C., Vaca, I., Lamas-Maceiras, M., Martín, J.F., 2007. *In vivo* transport of the intermediates of the penicillin biosynthetic pathway in tailored strains of *Penicillium chrysogenum*. Appl. Microbiol. Biotechnol. 76, 169–182.

García-Estrada, C., Ullán, R.V., Velasco-Conde, T., Godio, R.P., Teijeira, F., Vaca, I., et al., 2008a. Post-translational enzyme modification by the phosphopantetheinyl transferase is required for lysine and penicillin biosynthesis but not for roquefortine or fatty acid formation in *Penicillium chrysogenum*. Biochem. J. 415, 317–324.

García-Estrada, C., Vaca, I., Fierro, F., Sjollema, K., Veenhuis, M., Martín, J.F., 2008b. The unprocessable isopenicillin N acyltransferase (IATC103S) of *Penicillium chrysogenum* is located in peroxisomes and regulates the processing of the wild-type preprotein. Fungal Gen. Biol. 45, 1043–1052.

García-Estrada, C., Vaca, I., Ullán, R.V., van den Berg, M.A., Bovenberg, R.A.L., Martín, J.F., 2009. Molecular characterization of a fungal gene paralogue of the penicillin *penDE* gene of *Penicillium chrysogenum*. BMC Microbiol. 9, 104.

García-Estrada, C., Martín, J.F., 2019. Penicillins and cephalosporins. In: Moo-Young, M. (Ed.), Comprehensive Biotechnology, 3. Elsevier, Pergamon, pp. 283–296.

Ge, W., Clifton, I.J., Stok, J.E., Adlington, R.M., Baldwin, J.E., Rutledge, P.J., 2008. Isopenicillin N synthase mediates thiolate oxidation to sulfenate in a depsipeptide substrate analogue: implications for oxygen binding and a link to nitrile hydratase? J. Am. Chem. Soc., 130. pp. 10096–10102.

Gledhill, L., Greaves, P.A., Griffin, J.P., 2001. Phenylacetyl-CoA ligase from *Penicillium chrysogenum*. US Patent 6245524.

Guncheva, M., Ivanov, I., Galunsky, B., Stambolieva, N., Kaneti, J., 2004. Kinetic studies and molecular modelling attribute a crucial role in the specificity and stereoselectivity of penicillin acylase to the pair ArgA145-ArgB263. Eur. J. Biochem. 271, 2272–2279.

Gutiérrez, S., Díez, B., Montenegro, E., Martín, J.F., 1991. Characterization of the *Cephalosporium acremonium pcb*AB gene encoding α-aminoadiyl-cysteinyl-valine synthetase, a large multidomain peptide synthetase: Linkage to the *pcb*C gene as a cluster of early cephalosporin-biosynthetic genes and evidence of multiple functional domains. J. Bacteriol. 173, 2354–2365.

Gutiérrez, S., Velasco, J., Fernández, F.J., Martín, J.F., 1992. The *cef*G gene of *Cephalosporium acremonium* is linked to the *cef*EF gene and encodes a deacetylcephalosporin C acetyltransferase closely related to homoserine O-acetyltransferase. J. Bacteriol. 174, 3056–3064.

Harris, D.M., van der Krogt, Z.A., Klaassen, P., Raamsdonk, L.M., Hage, S., van den Berg, M. A., et al., 2009. Exploring and dissecting genome-wide gene expression responses of *Penicillium chrysogenum* to phenylacetic acid consumption and penicillin G production. BMC Genomics 10, 75.

Hensgens, C.M., Kroezinga, E.A., van Montfort, B.A., van der Laan, J.M., Sutherland, J.D., Dijkstra, B.W., 2002. Purification, crystallization and preliminary X-ray diffraction of Cys103Ala acyl coenzyme A: isopenicillin N acyltransferase from *Penicillium chrysogenum*. Acta Crystallogr. D Biol. Crystallogr. 58, 716–718.

Hewitt, L., Kasche, V., Lummer, K., Lewis, R.J., Murshudov, G.N., Verma, C.S., et al., 2000. Structure of a slow processing precursor penicillin acylase from *Escherichia coli* reveals the linker peptide blocking the active-site cleft. J. Mol. Biol. 301, 887–898.

Hopper, M., Gentzsch, C., Schorgendorfer, K., 2001. Structure and localization of cyclosporine synthetase, the key enzyme of cyclosporine biosynthesis in *Tolypocladium inflatum*. Arch. Microbiol. 176, 285–293.

Houbraken, J., Frisvad, J.C., Samson, R.A., 2011. Fleming's penicillin-producing strain is not *Penicillium chrysogenum* but *P. rubens*. IMA Fungus. 2, 87–95.

Howard-Jones, A.R., Rutledge, P.J., Clifton, I.J., Adlington, R.M., Baldwin, J.E., 2005. Unique binding of a non-natural L,L,L-substrate by isopenicillin N synthase. Biochem. Biophys. Res. Commun. 336, 702–708.

Huffman, G.W., Gesellchen, P.D., Turner, J.R., Rothenberger, R.B., Osborne, H.E., Miller, F.D., et al., 1992. Substrate specificity of isopenicillin N synthase. J. Med. Chem. 35, 1897–1914.

Jager, S.A., Shapovalova, I.V., Jekel, P.A., Alkema, W.B., Svedas, V.K., Janssen, D.B., 2008. Saturation mutagenesis reveals the importance of residues alphaR145 and alphaF146 of penicillin acylase in the synthesis of beta-lactam antibiotics. J. Biotechnol. 133, 18–26.

Kasche, V., Lummer, K., Nurk, A., Piotraschke, E., Rieks, A., Stoeva, S., et al., 1999. Intramolecular autoproteolysis initiates the maturation of penicillin amidase from *Escherichia coli*. Biochim. Biophys. Acta 1433, 76–86.

Kallow, W., Kennedy, J., Arezi, B., Turner, G., von Dohren, H., 2000. Thioesterase domain of delta-(L-alpha-aminoadipyl)-L-cysteinyl-D-valine synthetase: alteration of stereospecificity by site-directed mutagenesis. J. Mol. Biol. 297, 395–408.

Keszenman-Pereyra, D., Lawrence, S., Twfieg, M.E., Price, J., Turner, G., 2003. The *npgA/cfwA* gene encodes a putative 4'-phosphopantetheinyl transferase which is essential for penicillin biosynthesis in *Aspergillus nidulans*. Curr. Genet. 43, 186–190.

Kim, C.F., Lee, S.K., Price, J., Jack, R.W., Turner, G., Kong, R.Y., 2003. Cloning and expression analysis of the *pcbAB-pcbC* beta-lactam genes in the marine fungus *Kallichroma tethys*. Appl. Environ. Microbiol. 69, 1308–1314.

Koetsier, M.J., Jekel, P.A., van den Berg, M.A., Bovenberg, R.A., Janssen, D.B., 2009. Characterization of a phenylacetate-CoA ligase from *Penicillium chrysogenum*. Biochem. J. 417, 467–476.

Koetsier, M.J., Gombert, A.K., Fekken, S., Bovenberg, R.A., van den Berg, M.A., Kiel, J.A., et al., 2010. The *Penicillium chrysogenum aclA* gene encodes a broad-substrate-specificity acyl-coenzyme A ligase involved in activation of adipic acid, a side-chain precursor for cephem antibiotics. Fungal Genet. Biol. 47, 33–42.

Laich, F., Fierro, F., Martín, J.F., 2002. Production of penicillin by fungi growing on food products: Identification of a complete penicillin gene cluster in *Penicillium griseofulvum* and a truncated cluster in *Penicillium verrucosum*. Appl. Environ. Microbiol. 68, 1211–1219.

Laiz, L., Liras, P., Castro, J.M.y, Martín, J.F., 1990. Purification and characterization of the isopenicillin N epimerase from *Nocardia lactamdurans*. J. Gen. Microbiol. 136, 663–671.

Lamas-Maceiras, M., Vaca, I., Rodríguez, E., Casqueiro, J., Martín, J.F., 2006. Amplification and disruption of the phenylacetyl-CoA ligase gene of *Penicillium chrysogenum* encoding an aryl-capping enzyme that supplies phenylacetic acid to the isopenicillin N aceyltransferase. Biochem. J. 395, 147–155.

Landan, G., Cohen, G., Aharonowitz, Y., Shuali, Y., Graur, D., Shiffman, D., 1990. Evolution of isopenicillin N synthase genes may have involved horizontal gene transfer. Mol. Biol. Evol. 7, 399–406.

Li, K., Mohammed, M.A.A., Zhou, Y., Tu, H., Zhang, J., Liu, C., et al., 2020. Recent progress in the development of immobilized penicillin acylase for chemical and industrial applications: A review. Polym. Adv. Technol. 31, 368–388.

Liras, P., Rodríguez-García, A., Martín, J.F., 1998. Evolution of the clusters of genes for ß-lactam antibiotics: a model for evolutive combinatorial assembling of new ß-lactams. Int. Microbiol. 1, 271–278.

Liras, P., Demain, A.L., 2009. Enzymology of β-lactam compounds with cephem structure produced by actinomycete. In: Hopwood, D.A. (Ed.), Methods in Enzymology, 458. Academic Press, Burlington, pp. 401–429.

Liras, P., Martín, J.F., 2009. β-Lactam Antibiotics. In: Schaechter, M. (Ed.), Encyclopedia of Microbiology, Third Edition Elsevier Ltd., Oxford (U.K.), pp. 274–289.

Lodola, A., Branduardi, D., De Vivo, M., Capoferri, L., Mor, M., Piomelli, D., et al., 2012. A catalytic mechanism for cysteine N-terminal nucleophile hydrolases, as revealed by free energy simulations. PLoS One 7, e32397.

Long, A.J., Clifton, I.J., Roach, P.L., Baldwin, J.E., Rutledge, P.J., Schofield, C.J., 2005. Structural studies on the reaction of isopenicillin N synthase with the truncated substrate analogues delta-(L-alpha-aminoadipoyl)-L-cysteinyl-glycine and delta-(L-alpha-aminoadipoyl)-L-cysteinyl-D-alanine. Biochemistry 44, 6619–6628.

López-Nieto, M.J., Ramos, F.R., Luengo, J.M., Martín, J.F., 1985. Characterization of the biosynthesis *in vivo* of α-aminoadipyl-cysteinyl-valine in *Penicillium chrysogenum*. Appl. Microbiol. Biotechnol. 22, 343–351.

Luengo, J.M., Iriso, J.L., López-Nieto, M.J., 1986. Direct enzymatic synthesis of natural penicillins using phenylacetyl-CoA: 6-APA phenylacetyl transferase of *Penicillium chrysogenum*: minimal and maximal side chain length requirements. J Antibiot. 39, 1754–1759.

Lundberg, M., Morokuma, K., 2007. Protein environment facilitates O-2 binding in non-heme iron enzyme. An insight from ONIOM calculations on isopenicillin N synthase (IPNS). J. Chem. Phys. B 111, 9380–9389.

Lundberg, M., Kawatsu, T., Morokuma, K., 2009. Transition states in a protein environment-ONIOM QM:MM modeling of isopenicillin N synthesis. J. Chem. Theory Comput. 5, 222–234.

MacFaul, P.A., Wayner, D.D.M., Ingold, K.U., 1998. A radical account of "oxygenated Fenton chemistry.". Acc. Chem. Res. 31, 159–162.

Marahiel, M.A., Stachelhaus, T., Mootz, H.D., 1997. Modular peptide synthetases involved in nonribosomal peptide synthesis. Chem. Rev. 97, 2651–2674.

Maresova, H., Plackova, M., Grulich, M., Kyslik, P., 2014. Current state and perspectives of penicillin G acylase-based biocatalyses. Appl. Microbiol. Biotechnol. 98, 2867–2879.

Márquez-Fernández, O., Trigos, A., Ramos-Balderas, J.L., Viniegra-González, G., Deising, H.B., Aguirre, J., 2007. Phosphopantetheinyl transferase CfwA/NpgA is required for *Aspergillus nidulans* secondary metabolism and asexual development. Eukaryot. Cell 6, 710–720.

Martín, J.F., 1998. New aspects of genes and enzymes for ß-lactam antibiotic biosynthesis. Appl. Microbiol. Biotechnol. 50, 1–15.

Martín, J.F., 2000a. Molecular control of expression of penicillin biosynthesis genes in fungi: Regulatory proteins interact with a bidirectional promoter region. J. Bacteriol. 182, 2355–2362.

Martín, J.F., 2000b. α-Aminoadipyl-cysteinyl-valine synthetases in β-lactam producing organisms. From Abraham's discoveries to novel concepts of non-ribosomal peptide synthesis. J. Antibiot. 53, 1008–1021.

Martín, J.F., 2020. Transport systems, intracellular traffic of intermediates and secretion of β-lactam antibiotics in fungi. Fungal Biol Biotechnol. 7, 6.

Martín, J.F., Liras, P., 2017. Insights into the structure and molecular mechanisms of b-lactam synthesizing enzymes in fungi. In: Brahmachati, G., Demain, A.L., Adrio, J.L. (Eds.), Biotechnology of Microbial Enzymes. Elsevier, New York, pp. 215–241.

Martín, J.F., Liras, P., 2019. Transfer of secondary metabolite gene clusters: assembly and reorganization of the β-lactam gene cluster from bacteria to fungi and arthropods. In: Villa, T. G., Viñas, M. (Eds.), Horizontal Gene Transfer. Springer Nature Switzerland, pp. 337–359.

Martín, J.F., Liras, P., 2021. The PenV vacuolar membrane protein that controls penicillin biosynthesis is a putative member of a subfamily of stress-gated transient receptor calcium channels. Current Res. Biotechnol. 3, 317–322.

Martín, J.F., Ullán, R.V., Casqueiro, F.J., 2004. Novel Genes Involved In System. In: Brakhage, A. (Ed.), Advances in Biochemical Engineering-Biotechnology. Springer-Verlag, Berlin, pp. 91–109.

Martín, J., García-Estrada, C., Rumbero, A., Recio, E., Albillos, S.M., Ullán, R.V., et al., 2011. Characterization of an autoinducer of penicillin biosynthesis in *Penicillium chrysogenum*. Appl. Env. Microbiol. 77, 5688–5696.

Martín, J.F., García-Estrada, C., Ullán, R., 2012a. Genes encoding penicillin and cephalosporin biosynthesis in *Acremonium chrysogenum*: two separate clusters are required for cephalosporin production. In: Gupta, V.J., Ayyachamy, M. (Eds.), Biotechnology of Fungal Genes. CRC Press, pp. 113–138.

Martín, J.F., Ullán, R.V., García-Estrada, C., 2012b. Role of peroxisomes in the biosynthesis and secretion of β-lactams and other secondary metabolites. J. Ind. Microbiol. Biotechnol. 39, 367–382.

Martín, J.F., García-Estrada, C., Ullán, R.V., 2013. Transport of substrates into peroxisomes: the paradigm of β-lactam biosynthetic intermediates. Bio. Mol. Concepts 4, 197–211.

Martín, J.F., Liras, P., Sánchez, S., 2021. Modulation of gene expression in actinobacteria by translational modification of transcriptional factors and secondary metabolite biosynthetic enzymes. Front Microbiol. 12, 630694.

McVey, C.E., Walsh, M.A., Dodson, G.G., Wilson, K.S., Brannigan, J.A., 2001. Crystal structures of penicillin acylase enzyme-substrate complexes, structural insights into the catalytic mechanism. J. Mol. Biol. 313, 139–150.

Montenegro, E., Fierro, F., Fernández, F.J., Gutiérrez, S., Martín, J.F., 1992. Resolution of chromosomes III and VI of *Aspergillus nidulans* by pulsed-field gel electrophoresis shows that the penicillin biosynthetic pathway genes *pcb*AB, *pcb*C, and *pen*DE are clustered on chromosome VI (3.0 megabases). J. Bacteriol. 174, 7063–7067.

Murzin, A.G., Brenner, S.E., Hubbard, T.J., Chothia, C., 1995. SCOP: a structural classification of protein database for the investigation of sequences and structures. J. Mol. Biol. 247, 536–540.

Oinonen, C., Rouvinen, J., 2000. Structural comparison of Ntn-hydrolases. Prot Sci 9, 2329–2337.

Oshiro, B.T., 1999. The semisynthetic penicillins. Prim. Care Update Ob Gyns 6, 56–60.

Pang, C.P., Chakravarti, B., Adlington, R.M., Ting, H.H., White, R.L., Jayatilake, G.S., et al., 1984. Purification of isopenicillin N synthetase. Biochem. J. 222, 789–795.

Peñalva, M.A., Moya, A., Dopazo, J., Ramón, D., 1990. Sequences of isopenicillin N synthetase genes suggest horizontal gene transfer from prokaryotes to eukaryotes. Proc. Biol. Sci. 241164–241169.

Prieto, C., García-Estrada, C., Lorenzana, D., Martín, J.F., 2012. NRPSsp: non-ribosomal peptide synthase substrate predictor. Bioinformatics 28, 426–427.

Rajendhran, J., Gunasekaran, P., 2004. Recent biotechnological interventions for developing improved penicillin G acylases. J. Biosci. Bioeng. 97, 1–13.

Rajendhran, J., Gunasekaran, P., 2007. Application of cross-linked enzyme aggregates of *Bacillus badius* penicillin G acylase for the production of 6-aminopenicillanic acid. Lett. Appl. Microbiol. 44 (1), 43–49.

Ramos, F.R., López-Nieto, M.J., Martín, J.F., 1985. Isopenicillin N synthetase of *Penicillium chrysogenum*, an enzyme that converts δ-(L-α-aminoadipyl)-L- cysteinyl-D-valine to isopenicillin N. Antimicrob. Agents Chemother. 27, 380–387.

Ramos, F.R., López-Nieto, M.J., Martín, J.F., 1986. Coordinate increase of isopenicillin N synthetase, isopenicillin N epimerase and deacetoxycephalosporin C synthetase in a high cephalosporin-producing mutant of *A. chrysogenum* and simultaneous loss of the three enzymes in a non-producing mutant. FEMS Microbiol. Lett. 35, 123–127.

Randall, C.R., Zang, Y., True, A.E., Que Jr., L., Charnock, J.M., Garner, C.D., et al., 1993. X-ray absorption studies of the ferrous active site of isopenicillin N synthase and related model complexes. Biochemistry 32, 6664–6673.

Rausch, C., Weber, T., Kohlbacher, O., Wohlleben, W., Huson, D.H., 2005. Specificity prediction of adenylation domains in nonribosomal peptide synthetases (NRPS) using transductive support vector machines (TSVMs). Nucleic Acids Res. 33, 5799–5808.

Roach, P.L., Clifton, I.J., Fülöp, V., Harlos, K., Barton, G.J., Hajdu, J., et al., 1995. Crystal structure of isopenicillin N synthase is the first from a new structural family of enzymes. Nature 375, 700–704.

Roach, P.L., Clifton, I.J., Hensgens, C.M., Shibata, N., Schofield, C.J., Hajdu, J., et al., 1997. Structure of isopenicillin N synthase complexed with substrate and the mechanism of penicillin formation. Nature 387, 827–830.

Röttig, M., Medema, M.H., Blin, K., Weber, T., Rausch, C., Kohlbacher, O., 2011. NRPSpredictor2–a web server for predicting NRPS adenylation domain specificity. Nucleic Acids Res. 39 (Web Server issue), W362–W367.

Roze, L.V., Chanda, A., Linz, J.E., 2011. Compartmentalization and molecular traffic in secondary metabolism: a new understanding of established cellular processes. Fungal Genet. Biol. 48, 35–48.

Shewale, J.G., SivaRaman, H., 1989. Penicillin acylase: enzyme production and its application in the manufacture of 6-APA. Process Biochem. 24, 146–154.

Shewale, J.G., Sudhakaran, V.K., 1997. Penicillin V acylase: its potential in the production of 6-amonipenicillanic acid. Enzyme Microb. Tech. 20, 402–410.

Shirafuji, H., Fujisawa, Y., Makoto Kida, M., Toshihiko Kanzaki, T., Yoneda, M. (Eds.), 1979. Accumulation of tripeptide derivatives by mutants of *Cephalosporium acremonium*. Agric. Biol. Chem. 43, 155–160.

Solomon, E.I., Brunold, T.C., Davis, M.I., Kemsley, J.N., Lee, J.N., Lehnert, S.K., et al., 2000. Geometric and electronic structure/function correlations in non-heme iron enzymes. Chem. Revs. 100, 235–349.

Spröte, P., Hynes, M.J., Hortschansky, P., Shelest, E., Scharf, D.H., Wolke, S.M., et al., 2008. Identification of the novel penicillin biosynthesis gene *aatB* of *Aspergillus nidulans* and its putative evolutionary relationship to this fungal secondary metabolism gene cluster. Mol. Microbiol. 70, 445–461.

Srirangan, K., Orr, V., Akawi, L., Westbrook, A., Moo-Young, M., Chou, C.P., 2013. Biotechnological advances on penicillin G acylase: pharmaceutical implications, unique expression mechanism and production strategies. Biotechnol. Adv. 31, 1319–1332.

Stachelhaus, T., Marahiel, M.A., 1995. Modular structure of genes encoding multifunctional peptide synthetases required for non-ribosomal peptide synthesis. FEMS Microbiol. Lett. 125, 3–14.

Suresh, C.G., Pundle, A.V., SivaRaman, H., Rao, K.N., Brannigan, J.A., McVey, C.E., et al., 1999. Penicillin V acylase crystal structure reveals new Ntnhydrolase family members. Nat. Struct. Biol. 6, 414–416.

Theilgaard, H.A., Nielsen, J., 1999. Metabolic control analysis of the penicillin biosynthetic pathway: the influence of the LLD-ACV:bisACV ratio on the flux control. Ant. v. Leeuwenhoek 75, 145–154.

Theilgaard, H.B., Kristiansen, K.N., Henriksen, C.M., Nielsen, J., 1997. Purification and characterization of delta-(L-alpha-aminoadipyl)-L-cysteinyl-D-valine synthetase from *Penicillium chrysogenum*. Biochem. J. 327, 185–191.

Tischer, W., Kasche, V., 1999. Immobilized enzymes: crystals or carriers? Trends Biotechnol. 17, 326–334.

Tishkov, V.I., Savin, S.S., Yasnaya, A.S., 2010. Protein engineering of penicillin acylase. Acta Naturae 2, 47–61.

Tobin, M.B., Cole, S.C., Kovacevic, S., Miller, J.R., Baldwin, J.E., Sutherland, J.D., 1994. Acyl-coenzyme A: isopenicillin N acyltransferase from *Penicillium chrysogenum*: effect of amino acid substitutions at Ser^{227}, Ser^{230} and Ser^{309} on proenzyme cleavage and activity. FEMS Microbiol. Lett. 121, 39–46.

Tobin, M.B., Cole, S.C., Miller, J.R., Baldwin, J.E., Sutherland, J.D., 1995. Amino-acid substitutions in the cleavage site of acyl-coenzyme A: isopenicillin N acyltransferase from *Penicillium chrysogenum*: effect on proenzyme cleavage and activity. Gene 162, 29–35.

Torres-Bacete, J., Arroyo, M., Torres-Guzman, R., de la Mata, I., Castillon, M.P., Acebal, C., 2000. Optimization of 6-APA production by using a new immobilized penicillin acylase. Biotechnol. Appl. Biochem. 32, 173–177.

Ullan, R.V., Casqueiro, J., Banuelos, O., Fernandez, F.J., Gutierrez, S., Martín, J.F., 2002. A novel epimerization system in fungal secondary metabolism involved in the conversion of isopenicillin N into penicillin N in *Acremonium chrysogenum*. J. Biol. Chem. 277, 46216–46225.

Usui, S., Yu, C.A., 1989. Purification and properties of isopenicillin N epimerase from *Streptomyces clavuligerus*. Biochim. Biophys. Acta 999, 78–85.

Valle, F., Balbas, P., Merino, E., Bolivar, F., 1991. The role of penicillin amidases in nature and in industry. Trends Biochem. Sci. 16, 36–40.

van den Berg, M.A., Westerlaken, I., Leeflang, C., Kerkman, R., Bovenberg, R.A., 2007. Functional characterization of the penicillin biosynthetic gene cluster of *Penicillium chrysogenum* Wisconsin 54–1255. Fungal Genet. Biol. 44, 830–844.

van den Berg, M.A., Albang, R., Albermann, K., Badger, J.H., Daran, J.M., Driessen, A.J., et al., 2008. Genome sequencing and analysis of the filamentous fungus *Penicillium chrysogenum*. Nat. Biotechnol. 26, 1161–1168.

van Liempt, H., von Döhren, H., Kleinkauf, H., 1989. Delta-(L-alpha-aminoadipyl)-L-cysteinyl-D-valine synthetase from *Aspergillus nidulans*. The first enzyme in penicillin biosynthesis is a multifunctional peptide synthetase. J. Biol. Chem. 264, 3680–3684.

Velasco, J., Adrio, J.L., Moreno, M.A., Díez, B., Soler, G., Barredo, J.L., 2000. Environmentally safe production of 7-aminodeacetoxycephalosporanic acid (7-ADCA) using recombinant strains of Acremonium chrysogenum. Nat. Biotechnol. 18, 857–861.

Volpato, G., Rodrigues, R.C., Fernandez-Lafuente, R., 2010. Use of enzymes in the production of semi-synthetic penicillins and cephalosporins: drawbacks and perspectives. Curr. Med. Chem. 17, 3855–3873.

Walsh, C.T., Chen, H., Keating, T.A., Hubbard, B.K., Losey, H.C., Luo, L., et al., 2001. Tailoring enzymes that modify nonribosomal peptides during and after chain elongation on NRPS assembly lines. Curr. Opin. Chem. Biol. 5, 525–534.

Wang, F.Q., Liu, J., Dai, M., Ren, Z.H., Su, C.Y., He, J.G., 2007. Molecular cloning and functional identification of a novel phenylacetyl-CoA ligase gene from *Penicillium chrysogenum*. Biochem. Biophys. Res. Commun. 360, 453–458.

Wu, X., García-Estrada, C., Vaca, I., Martín, J.F., 2012. Motifs in the C-terminal region of the *Penicillium chrysogenum* ACV synthetase are essential for valine epimerization and processivity of tripeptide formation. Biochimie 94, 354–364.

Xu, G., Zhao, Q., Huang, B., Zhou, J., Cao, F., 2018. Directed evolution of a penicillin V acylase from *Bacillus sphaericus* to improve its catalytic efficiency for 6-APA production. Enzyme Microb. Technol. 119, 65–70.

Yu, Z.L., Liu, J., Wang, F.Q., Dai, M., Zhao, B.H., He, J.G., et al., 2011. Cloning and characterization of a novel CoA-ligase gene from *Penicillium chrysogenum*. Folia Microbiol. 56, 246–252.

Zhang, J., Demain, A.L., 1992. ACV synthetase. Crit. Rev. Biotechnol. 12, 245–260.

Zhiryakova, D., Ivanov, I., Ilieva, S., Guncheva, M., Galunsky, B., Stambolieva, N., 2009. Do N-terminal nucleophile hydrolases indeed have a single amino acid catalytic centre? FEBS J. 276, 2589–2598.

Zucher, Z., Keller, U., 1997. Thiol template peptide synthesis systems in bacteria and fungi. Adv. Microb. Physiol. 38, 85–131.

Chapter 10

Role of glycosyltransferases in the biosynthesis of antibiotics

Pankaj Kumar[1,2], Sanju Singh[1,2], Vishal A. Ghadge[1,2], Harshal Sahastrabudhe[1,2], Meena R. Rathod[1] and Pramod B. Shinde[1,2]

[1]*Natural Products & Green Chemistry Division, CSIR-Central Salt and Marine Chemicals Research Institute (CSIR-CSMCRI), Council of Scientific and Industrial Research (CSIR), Bhavnagar, Gujarat, India,* [2]*Academy of Scientific and Innovative Research (AcSIR), Ghaziabad, Uttar Pradesh, India*

10.1 Introduction

Natural antimicrobials that can circumvent drug resistance and provide new mechanisms of action are in urgent demand given the prevalence of new arising infective agents. The current discovery of antibiotics is not being able to cope with their high demand while maintaining their efficiency. Hence, the generation of new chemical entities is required more preferably from the natural resources. On average, it takes around 10 to 15 years to discover and develop new drugs and get them into the market. Most antibiotics such as spiramycin, vancomycin, avermectin B, amphotericin B, nystatin A, and fidaxomicin, which currently rule the market, are naturally glycosylated with various sugars. Sugars are a basic structure used in structural biology and immunology along with other chemical moieties for everyday life functions. Glycosylation is a vital modification reaction in the biosynthesis of natural products, resulting in enhanced solubility, bioavailability, stability, and bioactivity of diverse natural products (Liang et al., 2015). Glycosyltransferases (GTs) are the enzymes that catalyze the glycosidic bond formation between a donor sugar molecule and a hydroxyl group of acceptor molecules. GTs form an integral part of life in all domains and are involved in synthesizing glycans, glycolipids, and glycoproteins, playing an essential role in determining the structure of cell walls in eukaryotes and prokaryotes. Besides basic housekeeping, GTs are also biotechnologically important enzymes pertaining to their flexibility for substrates. The structural basis of the enzyme provides an excellent opportunity for the genetic engineering of these enzymes. The typical process for synthesizing sugar-containing antibiotics is the assembly

Biotechnology of Microbial Enzymes. DOI: https://doi.org/10.1016/B978-0-443-19059-9.00022-0
© 2023 Elsevier Inc. All rights reserved.

of an aglycone, a sugar moiety, and transferring the sugar moiety on the aglycone catalyzed by GTs (Pandey and Sohng, 2016). Representative examples of glycosylated antibiotics are shown in Fig. 10.1. Erythromycin was the first macrolide glycoside used for medical purposes, and it consists of two sugar molecules L-cladinose and D-desoamine. Azithromycin and clarithromycin are semisynthetic antibiotics that contain the same sugar moieties and show similar antibacterial properties. Glycosylation is not only limited to the macrolide class, but GTs also glycosylate antibiotics of polyene, terpenoids, nonribosomal peptides, and glycolipid class. Vancomycin is the first glycopeptide (containing sugars, vancosamine and glucose) that was available in the market to treat multidrug resistance infection. Similarly, fidaxomicin is another frontline antibiotic for multidrug-resistant *Clostridium difficile* and is glycosylated with two modified D-rhamnose moieties.

Aminoglycosides such as kanamycin, gentamicin, apramycin, and neomycin are another class of antibiotics, the biosynthesis of which is mainly

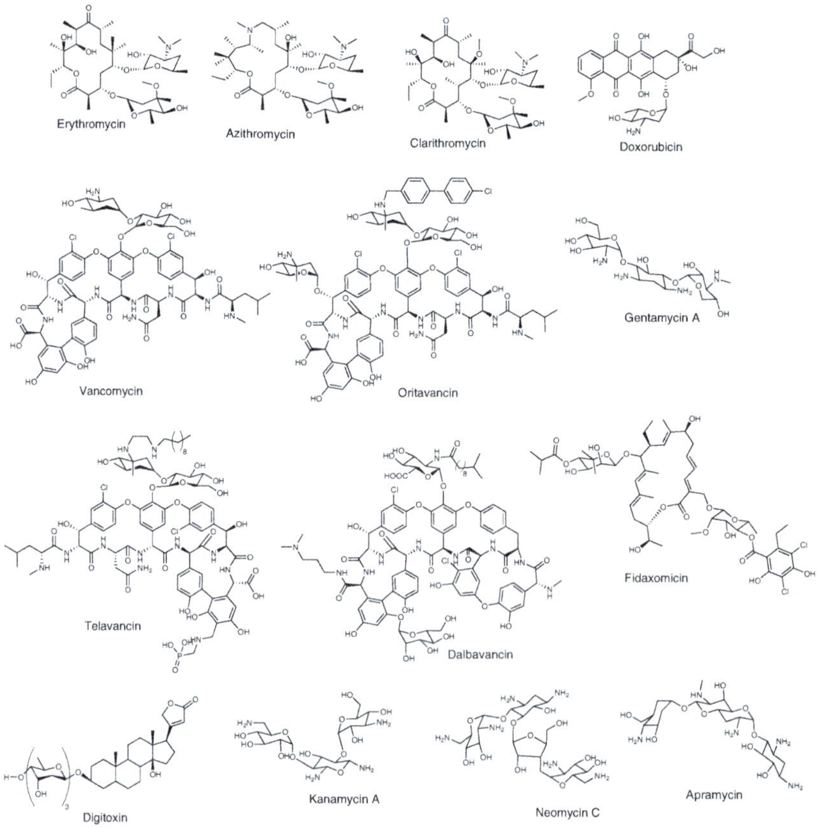

FIGURE 10.1 Examples of glycosylated antibiotics.

governed by GTs. The generation of a library of glycosylated natural products is done with the help of multiple strategies such as chemical synthesis, glycorandomization, or combinatorial biosynthesis (Blanchard and Thorson, 2006). Combinatorial biosynthesis of natural products heavily relies upon the coexpression of sugar cassette genes and their GTs in the heterologous host containing the substrate aglycone. Enzymatic approaches have excelled their way in both in vivo and in vitro synthesis for the diversification of chemical scaffolds. The importance of glycosylated antibiotics in treating infections and chronic diseases has made it highly urgent to gain knowledge of glycosylation so that it can be further used to improve the efficacy of antibiotics. This chapter conveys the classification of GTs, their role in glycosylated antibiotics, and different strategies employed to carry out glycosylation.

10.2 Classification and structural insights of glycosyltransferases

GTs utilize active donor substrates for the reaction that mainly have phosphate leaving groups such as nucleoside diphosphate (NDP) sugars and may be extended to nucleoside monophosphate sugars, lipid phosphates, or unsubstituted phosphates. Glycosylation generally occurs at the O-linked position, but in the case of natural products, N-, S-, and C-linked glycosylation can also be formed. GTs facilitate glycosidic bond formation by either retention or inversion of the anomeric configuration with respect to the donor sugar molecule (Gloster, 2014) (Fig. 10.2).

To better understand the role of GTs in the structural diversity of antibiotics, understanding different types of GTs and their structural specialties is primary. Structural characterization of GTs is challenging pertaining to low isolation yields, which are not enough for crystallographic studies and size precluding analysis by nuclear magnetic resonance (NMR) spectroscopy. Moreover, the high number of flexible loops in the structure results in low electron density, posing difficulties in revealing the catalytic domains of the enzyme (Schmid et al., 2016). However, the first X-Ray crystal structure for GTs was reported in 1994, describing a T4-glucosyltransferase from a bacteriophage. The structural basis for enzyme classification is based on the folding patterns and is divided into three forms: GT-A, GT-B, and GT-C (Taujale et al., 2021). Classification based on amino acid sequence similarities is compiled by Carbohydrate Active Enzymes (CAZy) database into 114 GT families (Drula et al., 2022). Out of these, 34 families account for GT-A type, 32 families for GT-B type, and approximately 10 families for GT-C category (Fig. 10.3). Some enzymes are still not characterized and cannot fit into these families.

The GT-A enzymes are generally dependent on divalent metal ions. They comprise two $\alpha/\beta/\alpha$ sandwich Rossmann-like domains, that is a classical structural motif (a seven-stranded β-sheet with 3214657 topology in which

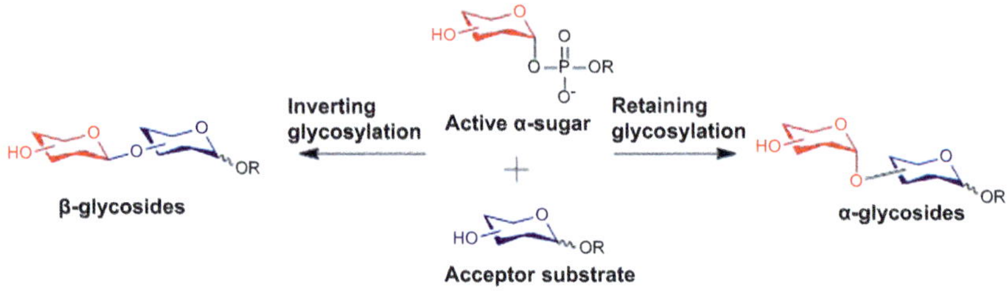

FIGURE 10.2 Two types of glycosylation reactions based on the stereochemistry of sugar.

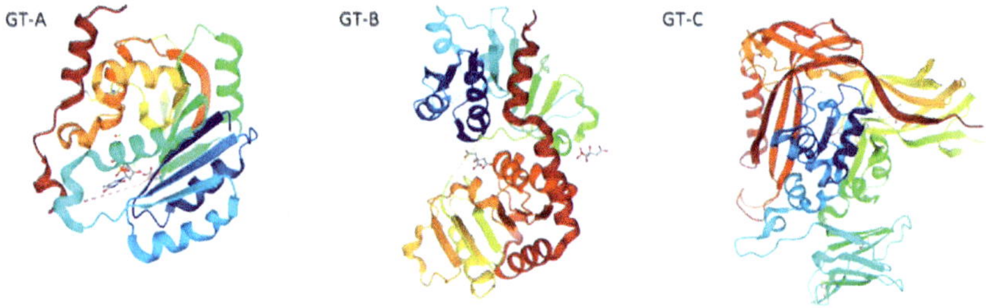

FIGURE 10.3 Representative 3D structure of three main categories of GTs. GT-A fold is represented by inverting type SpsA from *Bacillus subtilis* subsp. subtilis str. 168, involved in spore coat formation of microbe; CAZy database: GT-2, PDB accession no. 1H7L. GT-B fold is represented by the structure of a YjiC GTs from *B. subtilis* subsp. subtilis str. 168, involved in glycosylation of minor secondary metabolite and have diverse substrate spectrum, placed in GT-1 family of GTs in the CAZy database, PDB accession no. 6KQW. GT-C fold is represented by is oligosaccharyltransferase Stt3 from *Pyrococcus furiosus* DSM 3638, CAZy database: GT-66, PDB accession no. 2ZAG. *3D*, Three-dimensional; *GT*, glycosyltransferase.

strand 6 is antiparallel to the rest) consisting of 100–120 amino acids ending with a highly conserved DXD motif (Aspartate-X-Aspartate) within the active site. The DXD motif primarily interacts with the phosphate groups of nucleotide donors (mostly ribose) through the coordination of a divalent cation, typically Mn^{2+}. Some crystal structure indicates the presence of specific amino acids in the β_1 and β_4 strands of the C-terminal that interact with uridine diphosphate (UDP). Reaction strategies between aspartate residues and metal ions differ in the case of retaining and inverting enzymes (Persson et al., 2001; Tarbouriech et al., 2001). Although the C-terminal region is highly flexible for recognizing acceptors, a common structural motif in the conserved region β6–α4–α5 forms a part of the active site that reacts with both donor and sugar acceptor (Breton et al., 2006).

GTs involved in the biosynthesis of therapeutically important secondary metabolites predominantly belong to the GT-B superfamily (Park et al., 2009b). The structure of GT-B enzymes has a fold of two β/α/β Rossmann-like domains linked together that face each other. GT-B are generally independent of metal ions, and the active site is spatially located between the two folds. The C-terminal domain that corresponds to nucleotide-binding sites is structurally conserved in the GT-B superfamily. A higher number of variations are observed to occur in N-terminal domains consisting of flexible loops and folds forming the active site ultimately evolved to accommodate a large variety of acceptors. The characteristic peptide motifs of the GT-B fold are notably a glutamate residue and glycine-rich loops that interact with the ribose and phosphate moieties of nucleotide donor (Wrabl and Grishin, 2001). Even though the catalytic activity of GT-B is independent of metal ions, some GT-B members do show enhanced activity in the presence of divalent cations (Williams and Thorson, 2009).

The recently discovered GT-C enzymes isolated from the *Campylobacter jejuni* as sialyltransferases (Cst II) are hydrophobic integral membrane proteins having a different type of α/β/α sandwich (a seven-stranded parallel β sheet with 8712456 topology) and a modified DXD signature in the first extracellular loop. It mainly uses lipid phosphate-linked sugar donors (Schmid et al., 2016; Gloster, 2014). These are mainly found in the plasma membrane or on the endoplasmic reticulum, nearly having 8–13 transmembrane helices with probable active site location in the soluble C-terminal region (Lairson et al., 2008).

Bacterial GTs show very few sequence similarities but exhibit comparatively higher structural similarities. GTs from the higher group of organisms have stringent regiospecificity, and hence, their use is limited to the biosynthesis of a specific product. In contrast, the bacterial GTs can recognize a wide variety of acceptor substrates because of their relatively larger acceptor binding pockets. For example, the OleD GTs from *Streptomyces antibiotics* not only execute glycosylation on flavones, stilbenes, indole alkaloids, and steroids but can also process O-, N-, and S-glycosylations. A database provides an extensive list of the known three-dimensional (3D) structures of GTs (http://www.cermav.cnrs.fr/glyco3d). The GTs are classified according to not only CAZy systematics (Coutinho et al., 2003) but also to the organism of origin, the linkage formed by the enzymatic reaction, or the protein fold.

10.3 Role of glycosylation in enhancing bioactivity

Enhancing the bioactivity of existing antimicrobials is a great challenge and also a requirement because of the increase in resistance by pathogens. The attachment of a suitable sugar to a drug candidate can significantly alter its pharmacological and pharmacokinetic properties. Glycosylation increases the polarity of natural products, which improves their solubility in different mediums (Park et al., 2009a; Weymouth-Wilson, 1997). For example, nadifloxacin is a fluoroquinolone antibiotic having broad-spectrum activity against Gram-positive, Gram-negative aerobes and

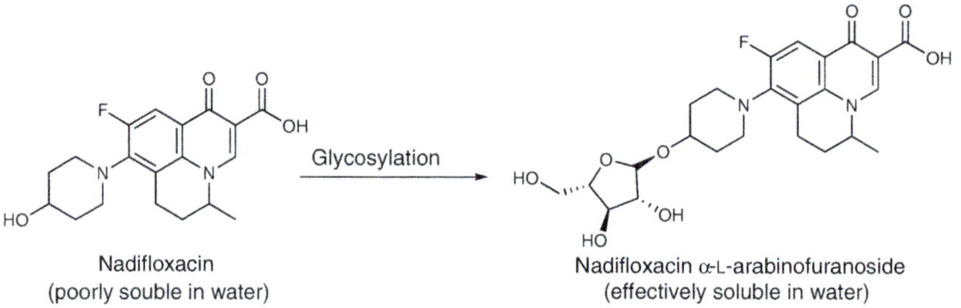

FIGURE 10.4 Improving the solubility of nadifloxacin through glycosylation.

anaerobes. But the poor solubility of nadifloxacin limits its use only as a topical ointment for skin infections. Hutchins et al. (2022) performed the glycosylation of nadifloxacin to overcome the solubility issues and synthesized nadifloxacin-α-L-arabinofuranoside with increased solubility and efficiency against MRSA infection (Fig. 10.4). As a result of the synthesis of the analogs, the orally administrable formulation of nadifloxacin for antibacterial infections, including methicillin-resistant *Staphylococcus aureus* (MRSA), became possible.

Glycosylation of natural products can also enhance hydrogen bond formation capability, which helps in recognizing specific drug targets and increases the efficiency of antibiotics. Most 14- and 16-membered macrolide antibiotics such as erythromycin and tylosin have a sugar moiety appended at the C-5 position. Dimethylamino group and 2′-OH of sugar moiety attached at C-5 are essential for effectiveness and binding to the biological target (Janas and Przybylski, 2019). A crystal structure study revealed the mechanism of binding macrolides with ribosomes that indicated the importance of the saccharide chain at C-5 position in macrolides. Macrolides inhibit protein synthesis by blocking peptide chain elongation. The saccharide chain provides steric hindrance to the peptidyl transferase center, due to which the peptide chain stops growing further. Binding of 16-membered macrolides to the ribosome is initiated by the interaction of C-5 saccharide chain and a complementary binding site in the ribosome's peptide elongation tunnel. Length of the saccharide chain directly affects steric hindrance at the peptide exit tunnel, and steric hindrance can also result in the termination of the peptide chain prematurely (Hansen et al., 2002). Such examples of the mechanism of action of glycosylated antibiotics signify the role of sugar moieties in bioactivity. The role of glycosylation or GTs in bioactivity can be understood by examining the effect of modifications such as detachment and replacement of the sugar moieties naturally found in a bioactive molecule (Liang et al., 2015). Further, to better understand the glycosylation-driven effect on bioactivity, some examples of glycosylated derivatives are discussed in the following sections in detail.

10.3.1 Vancomycin

Vancomycin is an antibiotic of glycopeptide class known for its activity against multidrug-resistant Gram-positive bacteria such as MRSA. Vancomycin binds with the precursor *N*-acyl-D-Ala-D-Ala and inhibits the synthesis of the cell wall by blocking trans-glycosylation/transpeptidation activity. The antibiotic contains two sugar moieties attached as L-vancosaminyl-1,2-D-glucosyl disaccharide to the heptapeptide chain. The glycosylation of the vancomycin aglycone to synthesize vancomycin was catalyzed by enzymes GtfE and GtfD, facilitating the addition of UDP-glucose and thymidine diphosphate (TDP)-vancosamine, respectively. The enzymes β-1,4-galactosyltransferase and α-2,3-sialyltransferases catalyzed galactose and sialic acid transfer on the pseudo vancomycin. Although only the aglycon part binds with the precursor of the cell wall synthesis, the sugar moiety contributes significantly to the bioactivity. The aglycone of vancomycin showed reduced activity in comparison to the glycosylated vancomycin. The alkylated vancosamine also exhibited activity against the resistant strains, and many glycosylated analogs were produced, such as oritavancin, dalbavancin, and telavancin (Williams and Thorson, 2009).

10.3.2 Tiacumicin B

Tiacumicin B, also known as lipiarmycin, fidaxomicin or difimicin, is a macrolide glycoside antibiotic isolated from *Dactylosporangium aurantiacum* actinobacteria and shows bioactivity against Gram-positive pathogens, especially against hospital-acquired *C. difficile*-mediated diarrhea. It is a member of 18-membered macrolide glycosylated with two modified D-rhamnose moieties. In the process of identification of the gene cluster responsible for the biosynthesis of fidaxomicin, different analogs of fidaxomicin were generated. TiaG1 and TiaG2 enzymes characterized as the 5-C-methyl-D-rhamnosyl-transferase and the 2-*O*-methyl-D-rhamnosyltransferase, respectively, are the two GTs responsible for the glycosylation of tiacumicin. The bioactivity of glycosylated tiacumicin and its aglycone formed by making its deletion mutant was compared. The activity of aglycone reduced significantly against *S. aureus* and *Enterococcus faecalis* compared to glycosylated analogs. TiaG1 and TiaG2 enzymes are regioselective and also extend their substrate flexibility to mannose. The mutation studies can also result in the knowledge of the catalytic order of two enzymes present simultaneously (Xiao et al., 2011).

10.3.3 Amycolatopsins

Amycolatopsis sp. MST-108494 produces amycolatopsins A − C which display antimycobacterial activity and cytotoxicity against colon and lung cancer cell lines. These glycosylated polyketides show cytotoxicity dependent on the length of the sugar chain. Amycolatopsin C, which has only one sugar molecule attached at the C-19 position of carbon, showed the least activity against tested cancer cell lines in comparison to amycolatopsin A and B, which has a trisaccharide attached at the C-19 position and showed better activity against the same cancer cell lines. The addition of hydroxyl group at C-6 and more glycosylation at C-19 position of amycolatopsis can improve their bioactive potential (Khalil et al., 2017).

10.3.4 Digitoxin

A cardiac glycoside from the plant used for treating congestive heart failure is another example exhibiting the importance of glycosylation for its anticancer activity. Experimental chemical synthesis of digitoxin monosaccharide analogs was done using a palladium catalyst to add sugars β-D-digitoxose, α-L-rhamnose and α-L-amicetose. These analogs were able to exhibit better apoptotic activity than the original digitoxin and its aglycone digitoxigenin. Thus it can be inferred from the example that glycosylation plays a crucial role in bioactivity, whether it is chemically or biologically synthesized. Further, using compatible host bacterial strains, GTs, and genetic engineering, the work can be tried to be replicated in microbial systems (Wang et al., 2011).

10.3.5 Aminoglycosides

They have a broad range of biological activities such as antibacterial, antiinfective (gentamicin, kanamycin), antidiabetic (acarbose), and crop protection (trehalase inhibition by validamycin) and cytotoxicity (pactamycin). The structural diversity of the aminoglycosides is due to the action of different sugar modifying enzymes such phosphotransferases, acetyltransferases, and nucleotidyltransferases. They functionalize the sugar molecules and improve the bioactive range of the antibiotics. The role of promiscuous GTs is not studied so prominently in this class of antibiotics. Such a study will readily enhance the diversity and library of aminoglycoside class of the molecules.

10.4 Engineering biosynthetic pathway of antibiotics by altering glycosyltransferases

The production of secondary metabolites in microbes is often a factor in their environmental conditions. Stressful conditions accelerate the synthesis of various metabolites that support stress tolerance and successful survival. Genetic engineering is such a tool through which one can manage the synthesis of metabolites by altering the expression of genes. Gene cluster-specific secondary metabolite biosynthesis contains various genes required for its modification.

Genetic engineering helps improve the yield of metabolites and facilitates the biosynthesis of diverse/modified analogs. Biosynthetic pathway engineering for the production of glycosylated derivatives can be done by mutating codons of amino acids in GTs that are responsible for specificity among different substrates. Glycosylated derivatives can also be obtained by inactivating genes accountable for sugar synthesis. The absence of parent sugar paves the way for the formation of glycosides using sugar intermediates. It can also lead to the generation of novel glycosides formed by using sugar substrates that can be used by enzymes available in the host. Inherent promiscuity of GTs can be exploited to generate glycosylated analogs by either deleting genes of native sugar molecules and feeding other sugars or transforming the host–microbe with plasmids for expression of active sugar donors not already present in the microbe.

Spectinomycin is an aminoglycoside class of antibiotics made up of actinamine (an aminocyclitol monomer) and a sugar moiety actinospectose (3-keto-4,6-dideoxy-glucose). SpcG, a putative GT is responsible for the reaction of these two units during the biosynthesis of spectinomycin. *Streptomyces venezuelae* is a well-developed model organism for genetic engineering experiments and especially for the heterologous expression of secondary metabolites. Lamichhane et al., 2014 performed the heterologous expression of spectinomycin genes in *S. venezuelae*. A vector that contains genes for the synthesis of actinamine and SpcG GT was transformed into *S. venezuelae* and was named pSM5. pSM5 has genes for SpcA (myo-inositol monophosphatase), SpcB (dehydrogenase), SpcS2 (aminotransferase), and SpcM (methyltransferase), which are the enzymes involved in the biosynthesis of actinamine. Sugar moiety acinospectose is the intermediate of desosamine biosynthetic pathway. SpcG is a unique member of GT-A superfamily, which can form two bonds between two sugar moieties. This type of heterologous expression of aminoglycosides can be extended to mix two different biosynthetic pathways of aminoglycosides and make new analogs. This chapter will further discuss some more interesting approaches for making glycosidic derivatives.

10.4.1 Combinatorial biosynthesis

Earlier glycosylation engineering solely meant the insertion or deletion of enzymatic genes to produce glycosyl-modified metabolites. However, in 1997 an attempt to apply heterologous GT genes for the production of hybrid glycopeptide antibiotics was successful, and the process was termed combinatorial biosynthesis. To simplify, combinatorial biosynthesis modifies biosynthetic pathways by combining natural and engineered biosynthetic enzymes from disparate sources. It is an efficient tool for increasing the diversity, novelty, and yield of natural products. GTs exhibit substrate promiscuity toward acceptor/donor sugar substrates that facilitate the combinatorial synthesis of glycosylated metabolites. Moreover, the wide availability of GT genes among microbes has been exploited for the synthesis of glycosylated derivatives of erythromycin, pikromycin, rebeccamycin, daunorubicin, etc. The most common and well-studied example is erythromycin and generating its library of 6-deoxyerythronolide B (6-DEB) analogs using combinatorial synthesis (Sun et al., 2015).

Urdamycin is an angucycline class antibiotic that shows antibacterial and antitumor activity. UrdGT1b and UrdGT1c are GTs involved in the biosynthesis of urdamycin in *Streptomyces fradiae* Tü2717. These enzymes have strict sugar and acceptor substrate specificity provided by a specific set of amino acids. These amino acids in both the enzymes were identified, and their random combination was performed to generate a library of GTs with different sets of amino acids capable of performing glycosylations. An engineered glycosyltransferase GT1707 was formed, which has the capability to produce a novel glycosylated derivative urdamycin P containing a branched saccharide chain (Fig. 10.5). The branched saccharide chain in urdamycin P has D-olivose linked by β-(1→4) glycosidic bond and L-rhodinose linked by α-(1→3)-glycosidic bond with D-olivose of 12-β-derhodinosylurdamycin G linked by C-glycosidic bond to the aglycon (Hoffmeister et al., 2002)

Sipanmycins are 26-membered macrolactams glycosylated with two disaccharides and are known for antibacterial and cytotoxic activities. Disaccharide moiety is composed of β-D-xylosamine and β-D-sipanose, which are crucial for antibacterial and antitumor activity. Both of these sugar molecules are glycosylated by two different pairs of GTs. Glycosylation of UDP-D-xylosamine to aglycon required SipS5/SipS14 and TDP-D-sipanose's attachment to β-D-xylosamine occurs in the presence of a pair of SipS9/SipS14 GTs. To generate glycosylation derivatives of sipanmycin, plasmids containing genes for the biosynthesis of L-digitoxose, L-amicetose, and D-olivose were introduced in wild-type strain *Streptomyces* sp. strain CS149.

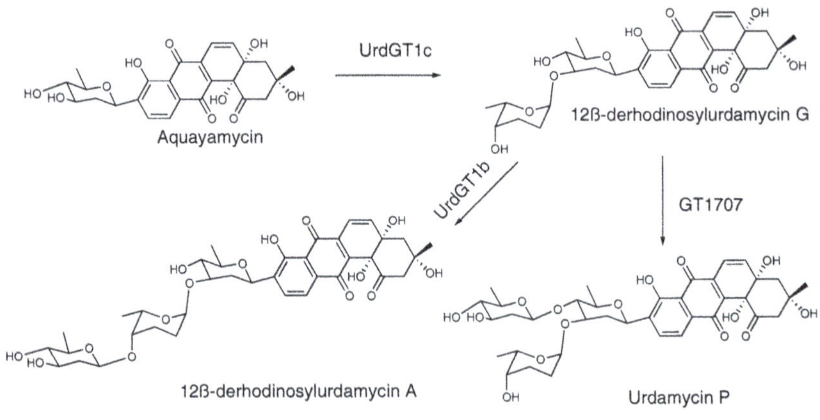

FIGURE 10.5 Schematic representation of the production of branched saccharide derivative of urdamycin.

Engineered sipanmycins showed replacement of β-D-sipanose by α-L-digitoxose, β-D-olivose, β-D-olivomycose, and β-D-forosamine (Fig. 10.6). The GTs pair SipS9/SipS14 has a wide range of substrate flexibility and can be utilized to incorporate different sugar molecules into the sipanmycin (Malmierca et al., 2020).

Another example of biosynthetic pathway engineering is shown by Trefzer et al., 2001 in which LanGT1 and LanGT2 GTs involved in the biosynthesis of landomycin A were cloned and expressed in *S. fradiae* Tu2717, which is originally a producer strain of urdamycin A. GTs of urdamycin were deleted in the host strain to understand the functioning of LanGT1 and LanGT2. This led to the synthesis of novel glycosylated urdamycin derivatives.

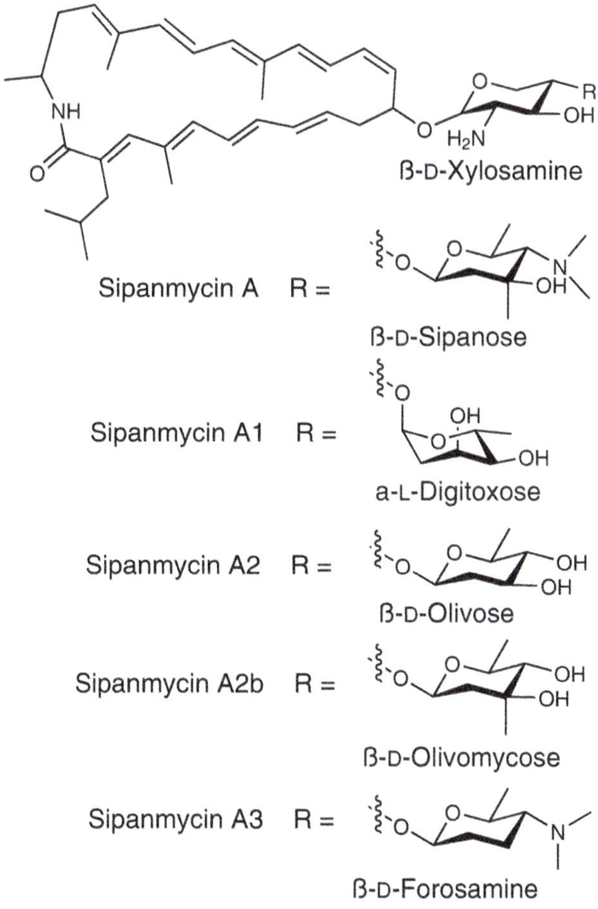

FIGURE 10.6 Glycosylated derivatives of sipanmycin generated by combinatorial biosynthesis.

The use of promiscuous GTs and techniques of genetic engineering such as combinatorial biosynthesis can fill the gap in new and effective antibiotics against emerging multidrug resistance microbes. KanF is one of the GTs of kanamycin biosynthetic pathway, which catalyze ligation of 2-*N*-acetylglucosamine and aglycon 2-deoxystreptamine. Nepal et al., 2010 demonstrated that KanF could attach 2-*N*-acetylglucosamine to streptamine. This combinatorial biosynthesis approach resulted in the biosynthesis of a new kanamycin derivative, oxykanamycin C. Oxykanamycin C is produced by combining biosynthetic pathways of kanamycin and spectinomycin in *S. venezuelae* YJ003. Yet another example of GTs importance in antibiotics is in improving the yield of products. In a study by Wu et al., 2017 production of gentamicin is enhanced by overexpressing the GTs from kanamycin and gentamicin biosynthetic pathways (KanM1 and GenM2) in *Micromonospora echinospora* JK4. The strategy successfully increased the yield of gentamicin B by 54% and gentamicin C1a by 45%.

10.4.2 Glycorandomization

Creating a library of randomly glycosylated natural products is important to harness more benefits of glycosylation. The generation of randomized glycosylated libraries can be accomplished in two different ways: the first is a chemical method called neoglycorandomization (NGR) and the second is chemoenzymatic glycorandomization (CGR). NGR uses a nonactive reducing sugar and aglycon with alkoxyamine linker to form a glycosidic bond. Monosaccharides used in NGR can either be a natural sugar or a chemical one, synthesized by a catalyzed aldol reaction. The reaction can be catalyzed by dihydroxyacetone phosphate, deoxyribose-5-phosphate aldolase (DERA), proline organocatalyst, etc. Detailed information about this approach is covered in detail by Langenhan et al. in a review published in 2005.

CGR is a process of glycosylation using substrate flexible GTs and random nucleotide sugars modified by anomeric kinase with nucleotidylyltransferases (NTs). This process is divided into two different approaches; in vitro and in vivo. Synthesis of active sugar donors using enzymes and their glycosylation by GTs falls in the category of in vitro approach. The process is advantageous, being flexible, regioselective, stereoselective, and a one-pot synthesis. In vitro CGR is further divided into three steps. The first step is the phosphorylation of sugar, a product created by anomeric kinases. The utilization of enzymes in the process increases the probability of improving enzymatic activity. Galactokinase (GalK) is a model anomeric kinase isolated from *E. coli*. Site-directed mutation led to an exchange of single amino acid in the active site, resulting in promiscuity for different sugar molecules for the GTs. NTs carry out the second step for the synthesis of NDP sugar using nucleotide triphosphate and sugar-phosphate substrates.

α-D-glucopyranosyl phosphate thymidylyltransferase (TTs) produced by *Salmonella enterica* LT2 is an example of NTs that facilitates the generation of a library of active sugars for GTs (Barton et al., 2002). Further, a point mutation at the substrate-binding pocket of NTs increases its substrate flexibility to recognize sugar phosphates (Thorson et al., 2004). Three NTs identified by Timmons et al. (2007). Cps2L (*Streptococcus pneumoniae* R6), RmlA (*Streptococcus mutans* UA159), and RmlA3 (*Aneurinibacillus thermoaerophilus*) (DSM 10155) have greater flexibility for recognition of sugar 1-phosphates (both α-D- and β-L-) and nucleotide phosphates (thymidine triphosphate, uridine triphosphate, adenosine triphosphate, guanosine triphosphate, and cytidine triphosphate). The third step of in vitro CGR involves glycosylation by GTs aglycon scaffold using activated sugars. GTs used in this step should be promiscuous for different structures of aglycon and NDP sugar (Fig. 10.7). GTs Bs-YjiC from *Bacillus subtilis* 168 and VinC from *Streptomyces halstedii* HC-34 are the examples of enzymes producing stereo- and region-specific O-, N-, S-glycosides using a diverse class of aglycone and NDP sugars.

In vivo glycorandomization is a process that involves microbes to generate glycosylated libraries either by pathway engineering or bioconversion. Microbes harboring inherent promiscuous NTs and GTs are best suited for this approach. There are three different alternatives to perform in vivo glycorandomization: (1) Addition of monosaccharides in the culture media for generating active sugar libraries using NTs, aglycon, and GTs already present in the microbe. (2) Addition of different aglycons as substrate and multiple monosaccharides in the culture of microbes expressing anomeric kinase, NTs and GTs (natural or engineered), and (3) Microbes that express GTs and produce active sugar donor are fed with aglycon to generate glycoside derivatives.

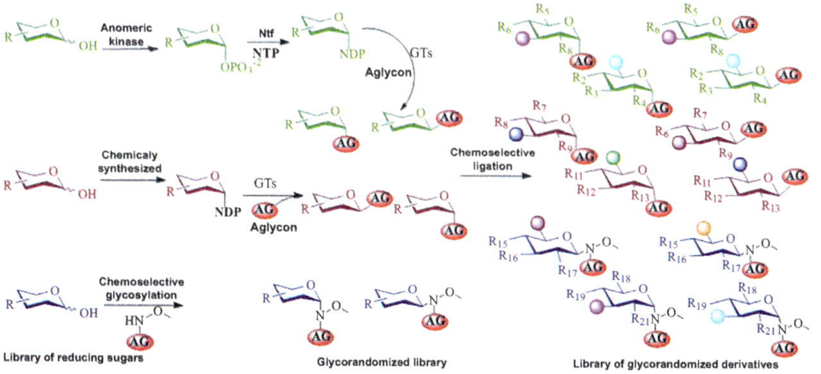

FIGURE 10.7 General representation of glycorandomization scheme (Langenhan et al., 2005).

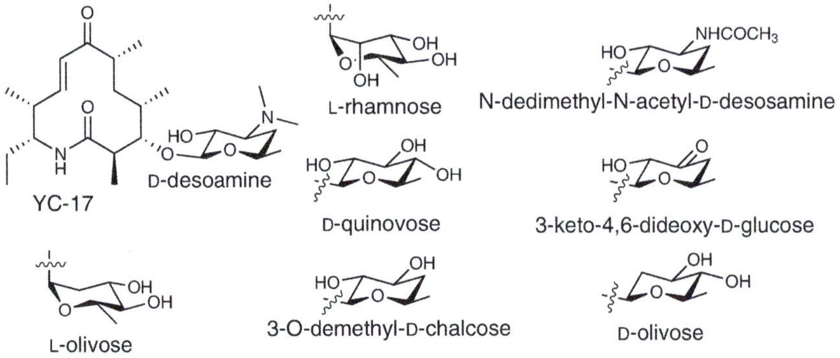

FIGURE 10.8 Glycosylated derivatives of YC-17 antibiotics.

Inherent or engineered enzymes that are promiscuous in nature form the first choice to perform glycorandomization. *E. coli* and *Streptomyces* strains are the favorable hosts pertaining to their feasibility for genetic engineering. *E. coli* BL21 (DE3) strain is a prototype strain for type II in vivo glycorandomization, wherein it was engineered to harbor TDP16 (an improved version of OleD GTs), GalK M173L/Y371H (promiscuous anomeric kinase), and RmlA L89T (promiscuous nucleotidyltransferase) (Williams et al., 2011). *S. venezuelae* YJ028 strain having genes deletion for polyketide synthase (PKS) and active sugar donor (gene cluster of pikromycin and desosamine) was used as a host to perform Type III in vivo glycorandomization to generate glycosylated analogs of YC-17 (Fig. 10.8). After deletion of wild-type PKS it was transformed with a plasmid encoded with DesII/DesIII GTs famous for sugar promiscuity. Thus the engineered strain was further modified to express sugar cassette for the production of active sugar donor when fed with chemically synthesized aglycon of YC-17. Analysis of antimicrobial activity of thus generated glycosylated derivatives reveals that the two analogs with L-rhamnose and D-quinovose attached to YC-17 are active against erythromycin-resistant clinical isolates *E. faecium* P00558 and *S. aureus* P00740 (Shinde et al., 2015). Such a mix and match of sugar cassettes, engineered hosts and enzymes with enhanced promiscuity can help to generate highly diverse glycosylated libraries of different classes of molecules.

10.5 Identification of glycosyltransferases and glycosylated molecules using bioinformatics

The bioinformatic analysis facilitates the identification of GTs and glycosylated secondary metabolites using whole-genome sequences. Comprehensive databases of amino acid sequence and protein structures of GTs help in the identification on the basis of homology between 3D structure and deoxyribose nucleic acid sequences of enzymes. Accessibility to whole-genome sequencing technology is essential for bioinformatic identification of GTs and glycosylated secondary metabolites. A major tool used for the identification of secondary metabolites from the entire genome sequence of a microbe is antiSMASH. This web-based tool identifies the glycosylated compounds and the GTs involved in the biosynthesis by possible gene clusters of secondary metabolites that are being predicted (Blin et al., 2019). PRediction Informatics for Secondary Metabolome (PRISM) is another software for predicting secondary metabolite gene clusters from the whole-genome sequence of microbes. PRISM can identify GTs and biosynthetic pathways of sugars attached to secondary metabolites (Skinnider et al., 2015; Skinnider et al., 2020). antiSMASH and PRISM are comprehensive tools for the identification of secondary metabolite gene clusters. A dedicated tool for GTs of secondary metabolites was developed in 2005 called as SEARCHGTr. It could analyze the sequence of GTs to predict the specificity of enzymes for acceptor secondary metabolite and donor sugar molecules (Kamra et al., 2005). Generally, GTs are found in the same gene cluster for transcription at the same time, along with the core molecule. In addition to GTs, genes responsible for the synthesis of active sugar molecules are also found in the same gene cluster, which further eases the possibility of predicting the sugar and promiscuity of GTs.

Stambomycins are 51-membered macrolide glycosides having anticancer properties. Genome-based study of *Streptomyces ambofaciens* ATCC23877 leads to detecting a silent gene cluster of huge Type I PKS. Analysis of SEARCHGTr identified *samR0481* as a putative GTs capable of attaching mycaminosyl to the product of PKS. Overexpression of a regulatory gene found within the gene cluster led to identifying the product of this giant modular PKS (Laureti et al., 2011).

Catenulisporolide is a glycosylated triene macrolide isolated from the actinomycete *Catenulispora* reported for the first time in 2018. Analysis of the whole-genome sequence of the producing strains using antiSMASH can readily identify the gene cluster involved in the biosynthesis. The sequence of gene clusters retrieved from antiSMASH and the function of individual genes was deduced from the BlastP search. Cat8 and Cat23 are the putative GTs for glycosylation of D-olivose and D-oleandrose on the metabolite's aglycon. The importance of whole-genome sequence and bioinformatic analysis in the identification of GTs was shown in the biosynthetic pathway of pulvomycin (Moon et al., 2020). PulQ is the GT identified by antiSMASH and BlastP, involved in the glycosylation of di-*O*-methyl-D-fucose to pulvomycin aglycons. The gene responsible for the formation of associated sugar can be identified along with the gene cluster of secondary metabolites. The information about GTs and sugar modifying genes can also help to identify sugar and to plan the biosynthesis of glycosylated macrolides by deleting genes involved in sugar biosynthesis.

10.6 Concluding remarks

Sugar molecules are an important part of life. They are involved in the cell wall formation of microbes, cell membrane glycolipids of eukaryotes, the structural backbone of genetic material, and antibodies. These sugar molecules are attached to their respective sites on different types of molecules by GTs. GT enzymes play an integral part in the formation of natural products and their derivatives. The attachment of sugar molecules to antibiotics after administration to the patient by pathogenic GTs can alter an effective antibiotic into an ineffective drug. At the same time, a similar family of enzymes can attach a sugar to an antibiotic and inhibit the pathogen more effectively. Because there are separate domains for recognizing the target aglycon and donor sugar molecule, GTs can be engineered to attach desirable sugar to the target. However, excellent knowledge and skills are required for successful modification of the enzymes and antibiotics. Whole-genome sequence-based tools can identify glycosylated secondary metabolites, and GTs involved in the process. Readily identification of GTs enhances the chances of discovering novel GTs and their glycosylated natural products. GT-driven library of glycosylated natural products and antibiotics derivatives will provide an edge for tackling the existing and forthcoming antimicrobial resistance and pandemics.

Abbreviations

GTs	glycosyltransferases
UDP	uridine diphosphate
CAZy	carbohydrate Active Enzymes
TDP	thymidine diphosphate
NGR	neoglycorandomization
CGR	chemoenzymatic glycorandomization
NT	nucleotidylyltransferase
TT	thymidylyltransferase
3D	three dimensional
antiSMASH	antibiotics and Secondary Metabolite Analysis Shell
PRISM	PRediction Informatics for Secondary Metabolome
PKS	polyketide synthase
BlastP	Basic Local Alignment Search Tool for Protein

References

Barton, W.A., Biggins, J.B., Jiang, J., Thorson, J.S., Nikolov, D.B., 2002. Expanding pyrimidine diphosphosugar libraries via structure-based nucleotidylyltransferase engineering. Proc. Nat. Acad. Sci. USA 99 (21), 13397–13402.

Blanchard, S., Thorson, J.S., 2006. Enzymatic tools for engineering natural product glycosylation. Curr. Opin. Chem. Biol. 10 (3), 263–271.

Blin, K., Shaw, S., Steinke, K., Villebro, R., Ziemert, N., Lee, S.Y., et al., 2019. antiSMASH 5.0: updates to the secondary metabolite genome mining pipeline. Nuc. Acids Res. 47 (W1), W81–W87.

Breton, C., Šnajdrová, L., Jeanneau, C., Koča, J., Imberty, A., 2006. Structures and mechanisms of glycosyltransferases. Glycobiol 16 (2), 29R–37R.

Coutinho, P.M., Deleury, E., Davies, G.J., Henrissat, B., 2003. An evolving hierarchical family classification for glycosyltransferases. J. Mol. Biol. 328 (2), 307–317.

Drula, E., Garron, M.L., Dogan, S., Lombard, V., Henrissat, B., Terrapon, N., 2022. The carbohydrate-active enzyme database: functions and literature. *Nucleic acids research, 50* (D1), pp.D571-D577. Available from: http://www.cazy.org/.

Gloster, T.M., 2014. Advances in understanding glycosyltransferases from a structural perspective. Curr. Opin. Struc. Biol. 28, 131–141.

Hansen, J.L., Ippolito, J.A., Ban, N., Nissen, P., Moore, P.B., Steitz, T.A., 2002. The structures of four macrolide antibiotics bound to the large ribosomal subunit. Mol. Cell 10 (1), 117–128.

Hoffmeister, D., Wilkinson, B., Foster, G., Sidebottom, P.J., Ichinose, K., Bechthold, A., 2002. Engineered urdamycin glycosyltransferases are broadened and altered in substrate specificity. Chem. Biol. 9 (3), 287–295.

Hutchins, M., Bovill, R.A., Stephens, P.J., Brazier, J.A., Osborn, H., 2022. Glycosides of nadifloxacin-synthesis and antibacterial activities against methicillin-resistant *Staphylococcus aureus*. Molecules 27 (5), 1504.

Janas, A., Przybylski, P., 2019. 14- and 15-membered lactone macrolides and their analogues and hybrids: structure, molecular mechanism of action and biological activity. Eur. J. Med. Chem. 182, 111662.

Kamra, P., Gokhale, R.S., Mohanty, D., 2005. SEARCHGTr: a program for analysis of glycosyltransferases involved in glycosylation of secondary metabolites. Nuc. Acids Res 33 (suppl_2), W220–W225.

Khalil, Z.G., Salim, A.A., Vuong, D., Crombie, A., Lacey, E., Blumenthal, A., et al., 2017. Amycolatopsins A–C: antimycobacterial glycosylated polyketide macrolides from the Australian soil *Amycolatopsis sp*. MST-108494. J. Antibiot. 70 (12), 1097–1103.

Lairson, L.L., Henrissat, B., Davies, G.J., Withers, S.G., 2008. Glycosyltransferases: structures, functions, and mechanisms. Annu. Rev. Biochem. 77, 521–555.

Lamichhane, J., Jha, A.K., Singh, B., Pandey, R.P., Sohng, J.K., 2014. Heterologous production of spectinomycin in *Streptomyces venezuelae* by exploiting the dTDP-D-desosamine pathway. J. Biotechnol. 174, 57–63.

Langenhan, J.M., Griffith, B.R., Thorson, J.S., 2005. Neoglycorandomization and chemoenzymatic glycorandomization: two complementary tools for natural product diversification. J. Nat. Prod. 68 (11), 1696–1711.

Laureti, L., Song, L., Huang, S., Corre, C., Leblond, P., Challis, G.L., et al., 2011. Identification of a bioactive 51-membered macrolide complex by activation of a silent polyketide synthase in *Streptomyces ambofaciens*. Proc. Nat. Acad. Sci. USA 108 (15), 6258–6263.

Liang, D.M., Liu, J.H., Wu, H., Wang, B.B., Zhu, H.J., Qiao, J.J., 2015. Glycosyltransferases: mechanisms and applications in natural product development. Chem. Soc. Rev. 44 (22), 8350–8374.

Malmierca, M.G., Pérez-Victoria, I., Martín, J., Reyes, F., Méndez, C., Salas, J.A., et al., 2020. New sipanmycin analogues generated by combinatorial biosynthesis and mutasynthesis approaches relying on the substrate flexibility of key enzymes in the biosynthetic pathway. Appl. Environ. Microbiol. 86 (3), e02453 – 19.

Moon, K., Cui, J., Kim, E., Riandi, E.S., Park, S.H., Byun, W.S., et al., 2020. Structures and biosynthetic pathway of Pulvomycins b–d: 22-membered macrolides from an estuarine *Streptomyces* sp. Org. Lett. 22 (14), 5358–5362.

Nepal, K.K., Yoo, J.C., Sohng, J.K., 2010. Biosynthetic approach for the production of new aminoglycoside derivative. J. Biosci. Bioeng. 110 (1), 109–112.

Pandey, R.P., Sohng, J.K., 2016. Glycosyltransferase-mediated exchange of rare microbial sugars with natural products. Front. Microbiol. 7, 1849.

Park, S.H., Park, H.Y., Cho, B.K., Yang, Y.H., Sohng, J.K., Lee, H.C., et al., 2009a. Reconstitution of antibiotics glycosylation by domain exchanged chimeric glycosyltransferase. J. Mol. Catal. B Enzym. 60, 29–35.

Park, S.H., Park, H.Y., Sohng, J.K., Lee, H.C., Liou, K., Yoon, Y.J., et al., 2009b. Expanding substrate specificity of GT-B fold glycosyltransferase via domain swapping and high-throughput screening. Biotechnol. Bioeng. 102 (4), 988–994.

Persson, K., Ly, H.D., Dieckelmann, M., Wakarchuk, W.W., Withers, S.G., Strynadka, N.C., 2001. Crystal structure of the retaining galactosyltransferase LgtC from *Neisseria meningitidis* in complex with donor and acceptor sugar analogs. Nat. Struct. Biol. 8 (2), 166–175.

Schmid, J., Heider, D., Wendel, N.J., Sperl, N., Sieber, V., 2016. Bacterial glycosyltransferases: challenges and opportunities of a highly diverse enzyme class toward tailoring natural products. Front. Microbiol. 7, 182.

Shinde, P.B., Oh, H.S., Choi, H., Rathwell, K., Ban, Y.H., Kim, E.J., et al., 2015. Chemoenzymatic synthesis of glycosylated macrolactam analogues of the macrolide antibiotic YC-17. Adv. Synth. Catal. 357 (12), 2697–2711.

Skinnider, M.A., Dejong, C.A., Rees, P.N., Johnston, C.W., Li, H., Webster, A.L., et al., 2015. Genomes to natural products prediction informatics for secondary metabolomes (PRISM). Nuc. Acids Res. 43 (20), 9645–9662.

Skinnider, M.A., Johnston, C.W., Gunabalasingam, M., Merwin, N.J., Kieliszek, A.M., MacLellan, R.J., et al., 2020. Comprehensive prediction of secondary metabolite structure and biological activity from microbial genome sequences. Nat. Commun. 11 (1), 1–9.

Sun, H., Liu, Z., Zhao, H., Ang, E.L., 2015. Recent advances in combinatorial biosynthesis for drug discovery. Drug. Des. Devel. Ther. 9, 823.

Tarbouriech, N., Charnock, S.J., Davies, G.J., 2001. Three-dimensional structures of the Mn and Mg dTDP complexes of the family GT-2 glycosyltransferase SpsA: a comparison with related NDP-sugar glycosyltransferases. J. Mol. Biol. 314 (4), 655–661.

Taujale, R., Zhou, Z., Yeung, W., Moremen, K.W., Li, S., Kannan, N., 2021. Mapping the glycosyltransferase fold landscape using interpretable deep learning. Nat. Commun. 12 (1), 1–12.

Thorson, J.S., Barton, W.A., Hoffmeister, D., Albermann, C., Nikolov, D.B., 2004. Structure-based enzyme engineering and its impact on in vitro glycorandomization. Chem. Biochem. 5 (1), 16–25.

Timmons, S.C., Mosher, R.H., Knowles, S.A., Jakeman, D.L., 2007. Exploiting nucleotidylyltransferases to prepare sugar nucleotides. Org. Lett. 9 (5), 857–860.

Trefzer, A., Fischer, C., Stockert, S., Westrich, L., Künzel, E., Girreser, U., et al., 2001. Elucidation of the function of two glycosyltransferase genes (lanGT1 and lanGT4) involved in landomycin biosynthesis and generation of new oligosaccharide antibiotics. Chem. Biol. 8 (12), 1239–1252.

Wang, H.Y.L., Xin, W., Zhou, M., Stueckle, T.A., Rojanasakul, Y., O'Doherty, G.A., 2011. Stereochemical survey of digitoxin monosaccharides. ACS Med. Chem. Lett. 2 (1), 73–78.

Weymouth-Wilson, A.C., 1997. The role of carbohydrates in biologically active natural products. Nat. Prod. Rep. 14 (2), 99–110.

Williams, G.J., Yang, J., Zhang, C., Thorson, J.S., 2011. Recombinant *E. coli* prototype strains for in vivo glycorandomization. ACS Chem. Biol. 6 (1), 95–100.

Williams, G.J., Thorson, J.S., 2009. Natural product glycosyltransferases: properties and applications. Adv. Enzymol. Relat. Areas Mol. Biol. 76, 55.

Wrabl, J.O., Grishin, N.V., 2001. Homology between O-linked GlcNAc transferases and proteins of the glycogen phosphorylase superfamily. J. Mol. Biol. 314 (3), 365–374.

Wu, Z., Gao, W., Zhou, S., Wen, Z., Ni, X., Xia, H., 2017. Improving gentamicin B and gentamicin C1a production by engineering the glycosyltransferases that transfer primary metabolites into secondary metabolites biosynthesis. Microbiol. Res. 203, 40–46.

Xiao, Y., Li, S., Niu, S., Ma, L., Zhang, G., Zhang, H., et al., 2011. Characterization of tiacumicin B biosynthetic gene cluster affording diversified tiacumicin analogues and revealing a tailoring dihalogenase. J. Am. Chem. Soc. 133 (4), 1092–1105.

Chapter 11

Relevance of microbial glucokinases

Beatriz Ruiz-Villafán, Diana Rocha, Alba Romero and Sergio Sánchez
Instituto de Investigaciones Biomédicas, Universidad Nacional Autónoma de México (UNAM), México Distrito Federal, México

11.1 Introduction

The transferase family (EC 2) includes a group of enzymes capable of transferring phosphorus-containing groups to an alcohol group as an acceptor (EC 2.7.1). Inside this broad group are the glucokinases (Glks). These enzymes are responsible for glucose phosphorylation, and the product of this reaction, glucose-6-phosphate (G6P), may follow different fates. At least, in yeast, the role of hexokinases in glucose uptake/oxidation and controlling the pentose phosphate pathway and energy metabolism has been demonstrated (Gao and Leary, 2003). Glks (D-glucose-6-phosphotransferases) transfer the group γ-phosphate to the OH group from the C_6 of D-glucose, using various phosphate donors such as ATP (EC: 2.7.1.1 and EC: 2.7.1.2), ADP (EC: 2.7.1.147), and polyphosphate (PolyP) (EC: 2.7.1.63) (Kyoto Encyclopedia of Genes and Genomes). Glks are present both in prokaryotes and eukaryotes. However, the phosphorylation of glucose in eukaryotes is mostly achieved by ATP-dependent kinases and these enzymes show broad substrate specificity for hexoses and are called hexokinases (HKs).

In contrast to HKs, bacterial Glks usually show high specificity for glucose. Based on their primary structure, Glks/HKs from eukaryotes or prokaryotes are broadly classified into two distinct nonhomologous families: HK and ribokinase (RK) (Kawai et al., 2005). The RK family comprises Glks from euryarchaeota and eukaryotes (mammals). Glks in the RK family show homology in primary and tertiary structures, whereas the HK members only have a few conserved motifs, initially identified in conserved tertiary structures. Glks in the HK family are subgrouped in HK, A and B. The HK group consists entirely of HKs from eukaryotes. Group A is composed of Glks from Gram-negative bacteria, cyanobacteria, and amitochondriate protists. Group B includes HKs from Crenarchaeota, Glks from

Biotechnology of Microbial Enzymes. DOI: https://doi.org/10.1016/B978-0-443-19059-9.00011-6
© 2023 Elsevier Inc. All rights reserved.

Gram-positive bacteria, some PolyP-Glks, and also proteins belonging to the ROK family (repressor- Open Reading Frame-kinase). Based on their primary amino acid structure, another previous classification divided microbial Glks into three distinct families (Lunin et al., 2004): (1) Glks from archaea, (2) ATP-Glks without the ROK motif, and (3) ATP-Glks belonging to the ROK family.

1. The first family involves ADP-dependent Glks (ADP-Glks) from archaea and higher eukaryotes (Fig. 11.1). Nowadays, approximately 1005 sequences belonging to this family are compiled in Pfam: PF04587 (Protein family database, Mistry et al., 2021). These enzymes are involved in a modified Embden-Meyerhof pathway in archaea requiring ADP as the phosphoryl group donor, instead of ATP (Siebers and Schönheit, 2005; Hansen et al., 2002). In euryarchaeota, two types of glucose-phosphorylating enzymes have been reported: (i) the ADP-Glks from the hyperthermophilic euryarchaea *Pyrococcus furiosus* (Kengen et al., 1995; Koga et al., 2000; Tuininga et al., 1999), *Thermococcus litoralis* (Koga et al., 2000), and *Archaeoglobus fulgidus* strain 7324 (Labes and Schonheit, 2003) and (ii) the ATP-dependent glucose-phosphorylating enzymes.

   Also, a bifunctional ADP-Glk/phosphofructokinase has been described in *Methanococcus jannaschii* (Sakuraba et al., 2002). As can be seen in Fig. 11.1, in the Pfam database, a greater number of ADP-dependent Glks from eukaryotes were reported, followed by bacteria and archaea. Likewise, 20 of such Glks were reported in firmicutes, 45 in actinobacteria, 11 in proteobacteria, and 1 in chloroflexi.
2. The second family (Pfam: PF02685) groups together ATP-Glks that do not contain the ROK motif. The Pfam database currently has approximately 3082 full or partial protein sequences belonging to this family (Fig. 11.2). Most family members are of bacterial origin, 230 from eukaryotes and 3 from archaea. The main members of bacteria belong to proteobacteria and cyanobacteria (COG0837).
3. The third family (ATP-Glks belonging to the ROK family) Pfam: PF00480, essentially comprises ATP-Glks from both archaea and bacteria (primarily Gram-positive) with an ROK motif. Even though a vast number of ROK members are of prokaryotic origin, there are proteins with ROK domains in all branches of life (Conejo et al., 2010) (Fig. 11.3). Sugar kinases that are classified as members of the ROK family have been found in many bacterial species constituting the largest family of bacterial Glks. Approximately 33024 ROK-family proteins have been identified so far, mostly in prokaryotes, but family members are found in all kingdoms of life (Świątek et al., 2013). Sugar kinases that are classified as members of the ROK family have been found in many bacterial species and constitute the most prominent family of bacterial Glks, with approximately 3600 members in Pfam. Most are

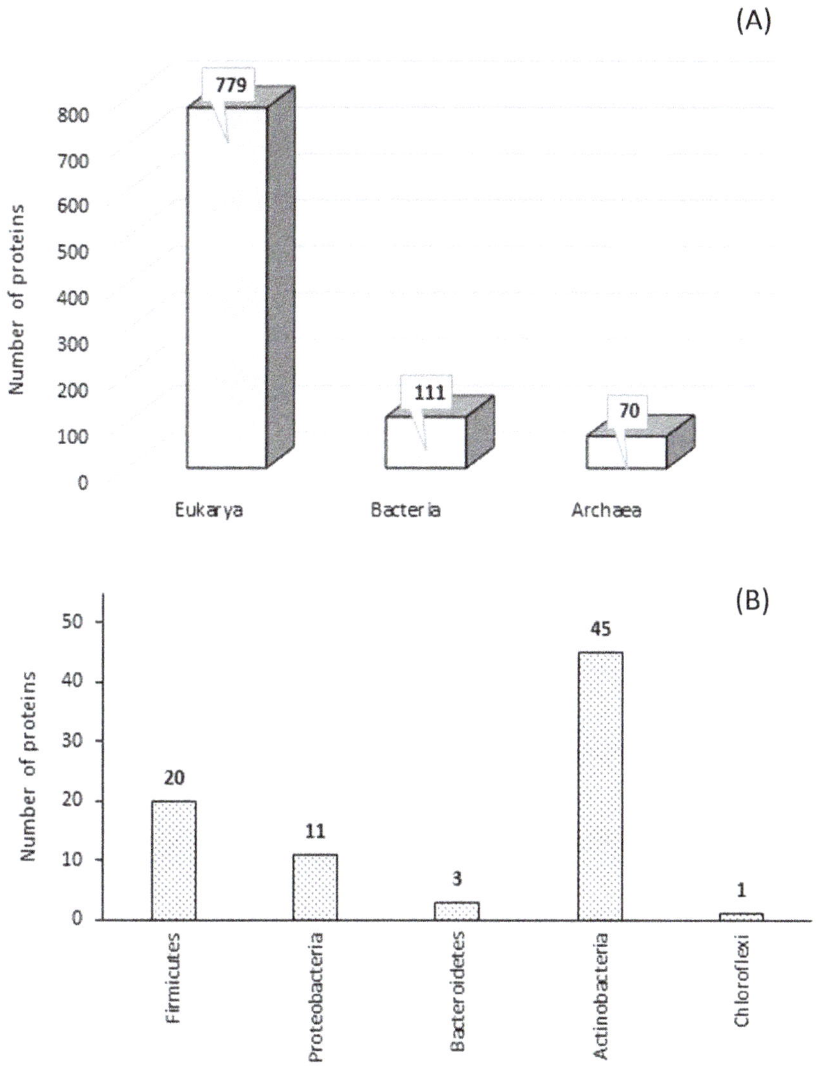

FIGURE 11.1 Distribution of ADP-dependent Glks (ADP-Glks) (Pfam: PF04587) in the three domains of life (Panel A) and bacterial phyla (Panel B). The presence of ADP-Glks was searched in the Pfam database, thus the numbers indicate how many proteins belonging to PF04587 are reported.

present in firmicutes from these members, followed by proteobacteria and actinobacteria (Fig. 11.3).

This family also includes many Glks that use inorganic PolyP as a phosphate donor (PolyP-Glks). PolyP can be found in organisms representing

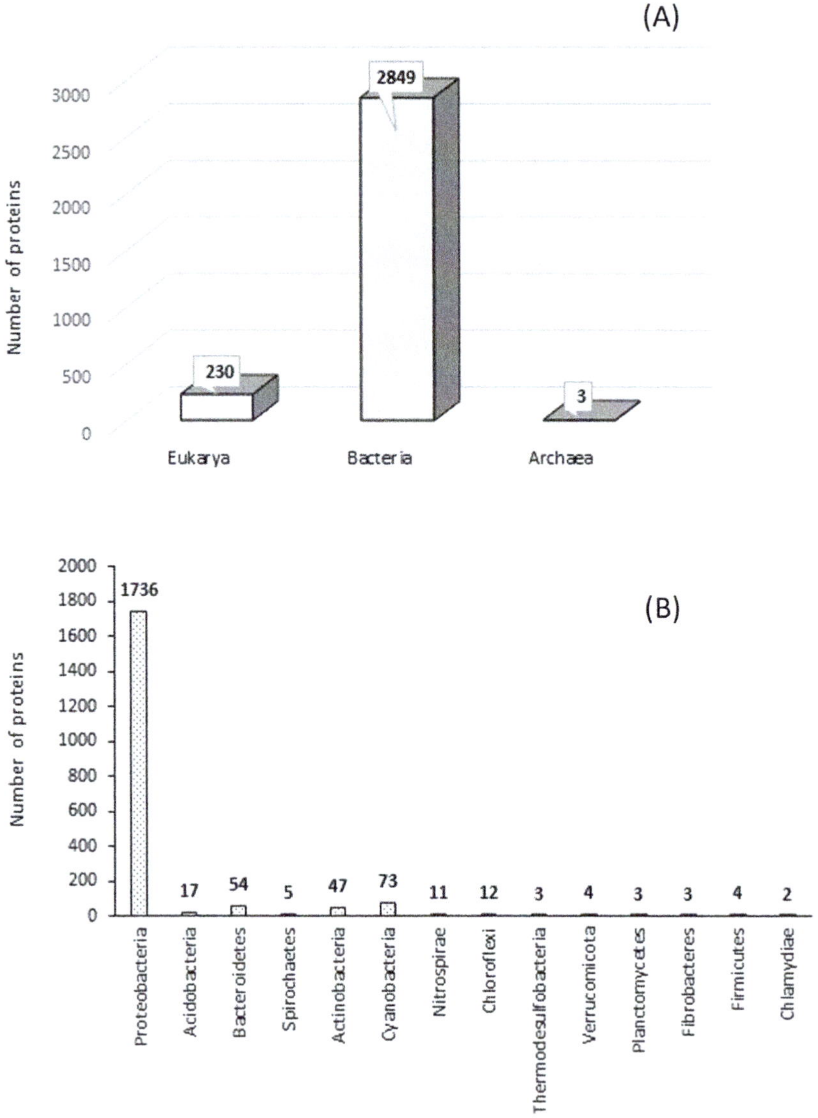

FIGURE 11.2 Distribution of ATP-dependent Glks (ATP-Glks) (Pfam: PF02685) in the three domains of life (Panel A) and in bacterial Phyla (Panel B). The presence of ATP-Glks was searched in the Pfam database, thus the numbers indicate how many proteins belonging to PF02685 are reported.

species from each domain in nature: Eukarya, Archaea, and Bacteria. Among other functions in prokaryotes, PolyP and its associated enzymes play a crucial role in basic metabolism and stress responses (Rao et al., 2009; Whitehead et al., 2013). PolyP-Glks were first reported in the

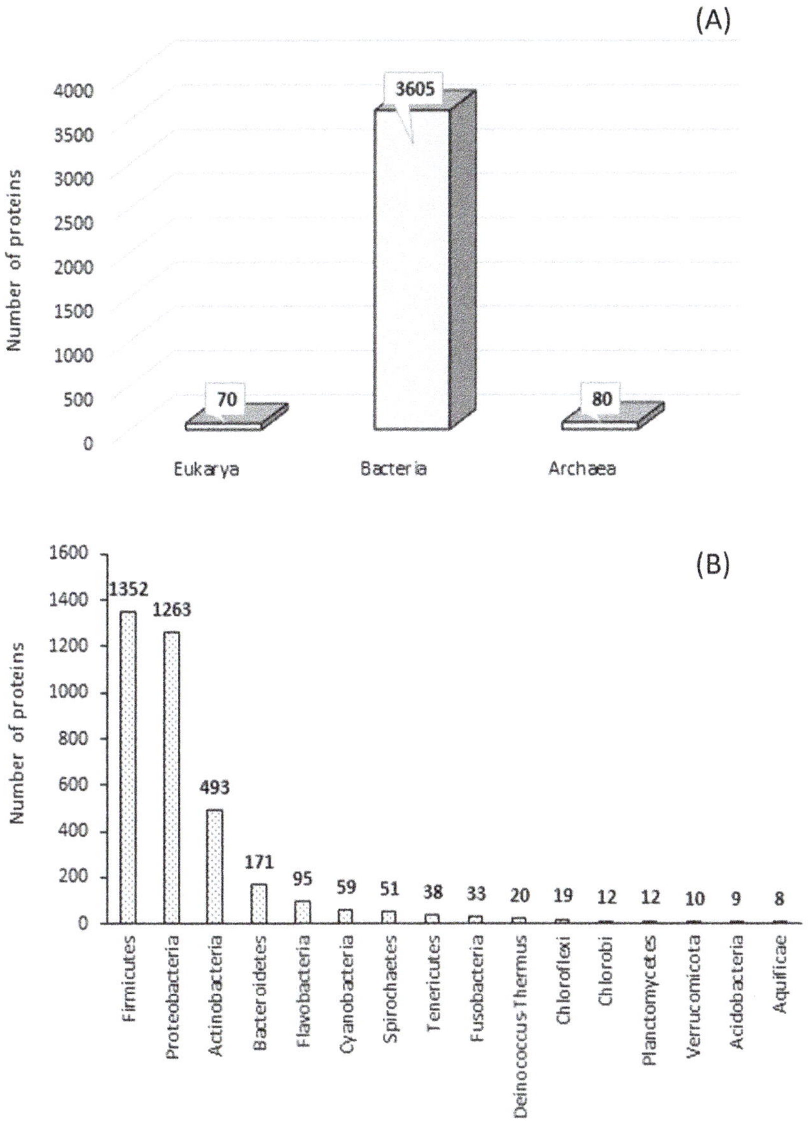

FIGURE 11.3 Distribution of ATP-dependent Glks (ATP-Glks) with the ROK motif (Pfam: PF00480) in the three domains of life (Panel A) and in bacterial Phyla (Panel B). The presence of ATP-Glks was searched in the Pfam database, thus the numbers indicate how many proteins belonging to PF00480 with the ROK motif are reported.

actinobacterium *Mycobacterium phlei* (Szymona and Ostrowski, 1964), although there are many other reports in other species of actinobacteria. This Glk uses PolyP as the phosphoryl donor, as well as ATP. Inorganic PolyP is an energy- and phosphorus-rich biopolymer that is present in a variety of organisms. The energy contained in the phosphodiester bonds of PolyP is thermodynamically equivalent to the energy obtained from ATP and can be utilized directly or indirectly for the phosphorylation of cellular molecules (Rao et al., 2009; Mukai et al., 2004). Furthermore, two strictly PolyP-Glks that utilize PolyP as the sole phosphoryl donor was reported in *Microlunatus phosphovorus* and *Thermobifida fusca* (Tanaka et al., 2003; Liao et al., 2012). Evolutionarily, PolyP/ATP-Glk and PolyP/ATP-NAD kinases are presumed to be intermediate enzymes between an ancient PolyP-specific kinase and a present-day ATP-specific kinase family (Kawai et al., 2005).

As outlined before, Glks and HKs are widely distributed among practically all living beings. Regarding eukaryotes, mammals have four HK isoenzymes, while the yeast *Saccharomyces cerevisiae* contains three enzymes that catalyze the phosphorylation of glucose, that is, HKs1 and 2 (HxK1 and Hxk2) and Glk (GlK1). Nevertheless, HxK2 is the isoenzyme that predominates when this microorganism is grown on glucose (Rodriguez et al., 2001).

11.2 Synthesis, biochemical properties, and regulation

In general, Glks from Gram-positive and Gram-negative bacteria have been cloned and expressed in *Escherichia coli* to determine their biochemical characteristics. Some of these recombinant Glks and their properties are listed in Table 11.1. Additionally, kinetic properties for different yeast systems have also been summarized (Gao and Leary, 2003). The biochemical properties and substrate specificities of HK isoforms can be used to distinguish between pathogenic and nonpathogenic species. For instance, *Entamoeba histolytica*, responsible for amebic colitis, is morphologically indistinguishable from the nonpathogenic *Entamoeba dispar*, but due to the HK isoenzyme patterns, it is possible to distinguish between both (Pineda et al., 2015).

For many years, Glks and other glycolytic enzymes were viewed just as "housekeeping" proteins with the functional purpose of fueling more important and complex biochemical processes (Kim and Dang, 2005). Nevertheless, evidence is emerging to support the unexpected multifunctional roles of glycolytic proteins.

Particularly, the Glk from *Streptomyces coelicolor* (Angell et al., 1994; Mahr et al., 2000; van Wezel et al., 2007), *Streptomyces peucetius* var. *caesius* (Rocha-Mendoza et al., 2021), and the Hxk2 from *S. cerevisiae* (Rodriguez et al., 2001) have been implicated in glucose regulation and carbon catabolite repression (CCR) (Ruiz-Villafán et al., 2021).

To understand the Glk/HK functions and signaling properties, it is critical to have insight into its regulation to manipulate them or their metabolic pathways eventually.

TABLE 11.1 Some biochemical characteristics of Glks.

Microorganism	Phosphoryl donor	km (mM)	V_{max} for glucose	Source	References
Sporolactobacillus inulinus Y2–8	ATP	ATP 1.03 Glc 4.26	62.3 U/mg	Overexpressed in *Escherichia coli*	Zheng et al. (2012)
Leptospira interrogans	ATP	ATP 1.011 Glc 0.43	NR	Overexpressed in *E. coli*	Zhang et al. (2011)
Bacillus subtilis	ATP	ATP 0.77 Glc 0.24	93 U/mg	Overexpressed in *E. coli*	Skarlatos and Dahl (1998)
Methylocrobium alcaliphilum 20Z	ATP	ATP 0.37 Glc 0.32	216.7 U/mg	Overexpressed in *E. coli*	Mustakhimov et al. (2017)
Thermus caldophilus	ATP	ATP 0.77 Glc 0.13	196 U/mg	Overexpressed in *E. coli*	Bae et al. (2005)
Sulfolobus tokodaii	ATP	ATP 0.12 Glc 0.05	67 U/mg	Overexpressed in *E. coli*	Nishimasu et al. (2006) Nishimasu et al. (2007)
Pyrobaculum calidifontis	ATP	ATP 0.9 Glc 0.66	550 U/mg	Overexpressed in *E. coli*	Bibi et al. (2018)
Thermotoga maritima	ATP	ATP 0.36 Glc 1	365 U/mg	Overexpressed in *E. coli*	Hansen and Schönheit (2003)
Methanococcus jannaschii	ADP	ADP 0.032 Glc 1.6	21.5 U/mg	Overexpressed in *E. coli*	Sakuraba et al. (2002)

(*Continued*)

TABLE 11.1 (Continued)

Microorganism	Phosphoryl donor	km (mM)	V_{max} for glucose	Source	References
Pyrococcus furiosus	ADP	ADP 0.45 Glc 2.61	1740 U/mg	Overexpressed in *E. coli*	Verhees et al. (2002)
Thermococcus kodakarensis	ADP	ADP 0.1 Glc 0.48	343 U/mg	Overexpressed in *E. coli*	Shakir et al. (2021)
Corynebacterium glutamicum	ATP/PP	ATP 6 PP 1 Glc 21.1	21.1 U/mg	Overexpressed in *E. coli*	Lindner et al. (2010)
Streptomyces coelicolor	ATP/PP	PoliP 3.87X 10^{-3} Glc 1.24X10^{-2}	NR	Overexpressed in *E. coli*	Koide et al. (2013)
Yeast	ATP	ATP 0.235 Glc 2.857	NR	Yeast Sigma	Socorro et al. (2000)
Saccharomyces cerevisiae	ATP	ATP 0.21 Glc 0.071	NR	Self-expression	Golbik et al. (2001)
Schizosaccharomyces pombe	ATP	ATP 0.0886 Glc 0.1097	NR	Self-expression	Tsai and Chen (1998)

In bacteria from the genus *Streptomyces*, like *S. coelicolor*, Glk seems to be constitutively expressed. At the same time, its kinase activity largely depends on the carbon source present due to a possible posttranscriptional mechanism (van Wezel et al., 2007). In *S. peucetius* var. *caesius*, the ATP and PolyP-Glk activities are both induced by glucose (Ruiz-Villafán et al., 2014). In the yeast *S. cerevisiae*, the principal Glk, HxK2, has long been implicated in the CCR process. Initially, this property was attributed to its catalytic activity but now seems to result from an interaction with other regulatory proteins (Pérez et al., 2014). For instance, HxK2 from *S. cerevisiae* could interact with the transcriptional repressor Mig1 and the Snf1 kinase (Rodríguez-Saavedra et al., 2021). Hxk2 interacts with Mig1 when the yeast grows in a high glucose concentration, preventing its phosphorylation at serine 311 by Snf1, avoiding Mig1 nuclear export and de-repression of genes subjected to CCR (Ahuatzi et al., 2004, 2007).

11.3 Structure

The tertiary structure of Glk consists of two domains (a large and a small). Between them, there is a deep cleft, where the substrate-binding site is formed (Fig. 11.4).

The binding of glucose causes movement of the two domains to get close to the cleft. It leads to a change in the conformation of the enzyme, a phenomenon known as induced fit. The domains acquire a closed and open conformation in the presence and absence of glucose, respectively (Fig. 11.5).

The bound glucose molecule is in a chair conformation and adopts the β-anomeric configuration. The glucose molecule participates in an extensive hydrogen bonding network within the active site pocket.

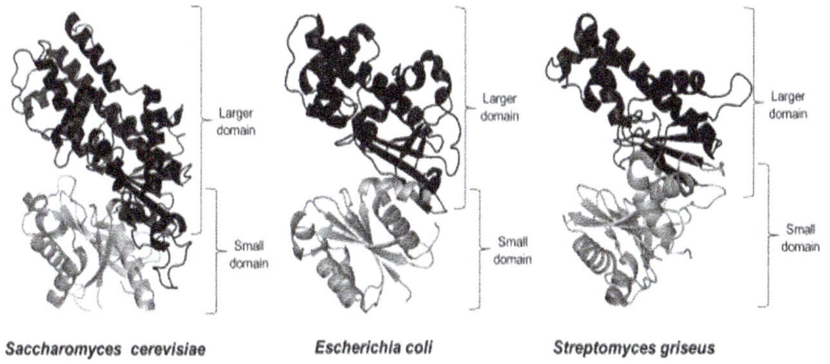

FIGURE 11.4 Ribbon model of the glucokinase monomer. The larger (black) and small domain (gray) for the hexokinase I from *Saccharomyces cerevisiae* (PDB, 3B8A), Glk from *Escherichia coli* (PDB, 1Q18) and ATP-dependent Glk from *Streptomyces griseus* (PDB, 3VGK).

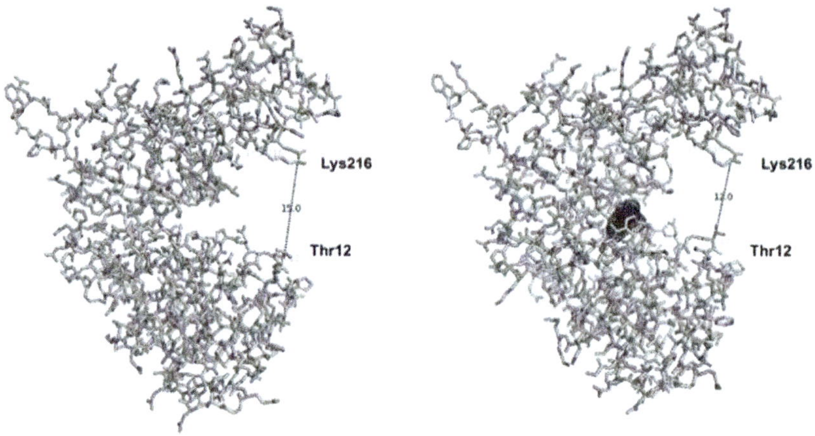

FIGURE 11.5 Comparison of open (left) and closed (right) structure of glucokinase "induced fit" of ATP-dependent Glk from *Streptomyces griseus* (PDB 3VGK apo form and 3VGM glucose complex).

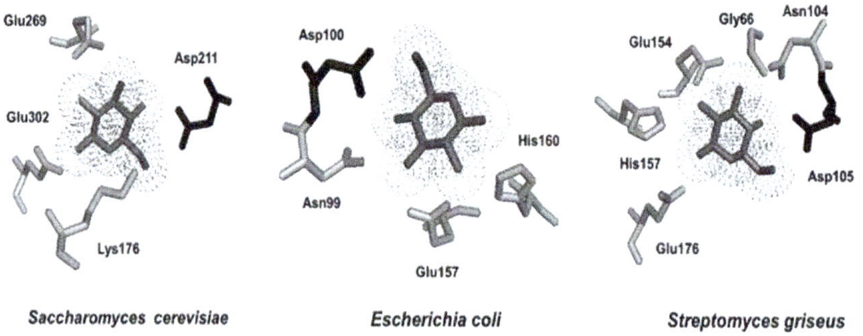

FIGURE 11.6 Structure of the active site region of hexokinase I from *Saccharomyces cerevisiae* (PDB, 3B8A), Glk from *Escherichia coli* (PDB, 1SZ2) and ATP-dependent Glk from *Streptomyces griseus* (PDB, 3VGM). Showing electron density for glucose (dark gray), amino acid residues that link glucose with hydrogen bonds (gray) and the catalytic residue Asp (black).

Glks contain an N-terminal ATP-binding motif denoted by the sequence LXXDXGGTNXRXXL. The PP-Glk has two putative sequences for PP-binding: (1) TXGTGIGXA (correlated with the signature for ROK proteins) and (2) SXXX-W/Y-A (Ruiz-Villafán et al., 2014).

In the ATP-Glk, the main conserved amino acids binding the glucose-forming hydrogen bonds are Gly, Asn, Asp, Glu, and His (Fig. 11.6). The large number of hydrogen bonds formed between the enzyme and glucose contribute to the stability of the closed structure. The putative catalytic amino acid that acts as a base in the reaction mechanism of Glk is well conserved: Asp100 *E. coli*, Asp189 *S. cerevisiae*, Asp451 *T. litoralis*, Asp440 *P. furiosus, and* Asp443 *Pyrococcus horikoshii*.

11.4 Catalytic mechanism

Although the presence of glucose can induce a conformational change in microbial HKs and Glks, their mechanism can be different. In the case of ADP-dependent Glk enzymes of Archaea such as *T. litoralis, P. horikoshii*, and *Aeropyrum pernix*, a sequential mechanism has been observed, where a ternary complex must be formed before the products are released. In this sequential mechanism, Mg•ADP⁻ is the first substrate to bind the catalytic site, and Mg•AMP is the final released product.

The most striking structural aspect of the nucleotide and D-glucose binding to the active site is that both binding events indicate a sequential conformational change. ThusMg•ADP⁻ triggers the first structural change (semiclosed conformation), which favors the oncoming of the small domain toward the large domain. This is followed by the entry of D-glucose, which in turn leads to a ternary complex formation and closure of the total domain. This sequential conformational change strongly suggested an induced fit mechanism (Rivas-Pardo et al., 2013).

The crystal structure of ADP-Glk from *P. furiosus* revealed that 17 water molecules were confined entirely in the active site cleft and formed a hydrophilic pocket on the closing of the domain. This is the only report of ADP-Glk from Archaea, the native structure of which corresponds to a homodimer (Ito et al., 2003).

Tokarz et al. (2018) cocrystallized the ADP-Glk from *Methanocaldococcus jannaschii* with the inhibitor 5-iodotubercidin. Surprisingly, they found an intermediate where there is a phosphate ion trapped between the D-glucose and the inhibitor. The phosphate ion is stabilized by its interaction with magnesium and a guanidine group of Arg197. This study revealed the position of the magnesium ion that appears to be essential for stabilization of the transition state during phosphate transfer.

Analogous to other bacterial kinases, the ATP chemical mechanism of Glk catalysis in *E. coli* is a S_N2 nucleophilic attack over the glucose O6 atom on the electropositive P atom of the γ-phosphoryl group of ATP (Fig. 11.7). Initial abstraction of the proton from the CH_2OH group of O6 is presumably performed by an Asp, acting as a general base (Lunin et al., 2004).

Kinases and almost all phosphate-transferring enzymes have been shown to have Mg^{2+} in their active sites. They interact with both the β- and γ-phosphate to assist the reaction by orienting and stabilizing the terminal phosphate during its transfer to an acceptor (Ito et al., 2001).

The kinetic data support the postulation of a sequential mechanism for the Glk reaction. These data are consistent with an ordered type of

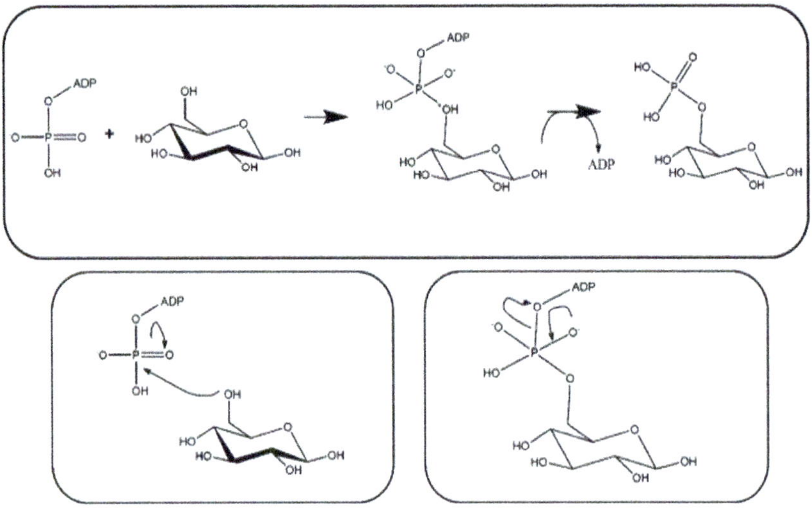

FIGURE 11.7 Mechanism of Glk activities. Simultaneous binding of ATP and glucose to the enzyme provides the proximity for the nucleophilic attack of the 6-OH of glucose on the terminal phosphoryl of ATP. Electron rearrangement leads to the production of glucose-6-phosphate and ADP.

mechanism in which the glucose binding is initially observed, ending with G6P dissociation.

Studies of the kinetic mechanisms of both PolyP and ATP of Glk of *Mycobacterium tuberculosis* indicate that the activity follows a steady-state ordered Bi-Bi mechanism in both PolyP- and ATP-dependent reactions. Product and dead-end inhibition studies suggest that PolyP binds to the free enzyme as the first substrate and is released as the first product after the terminal phosphate is transferred to glucose, in the ordered Bi-Bi mechanism. Comparison of efficiencies suggests that PolyP is favored over ATP as the phosphoryl donor for Glk in *M. tuberculosis*. The mechanism of PolyP utilization is nonprocessive, since it requires its dissociation from the enzyme prior to complete utilization (Hsieh et al., 1996).

It is known that *S. cerevisiae* has three distinct enzymes for glucose phosphorylation, the HKs (1 and 2), and Glk. All three enzymes display a broad specificity towards the sugar substrate, except fructose, which is not phosphorylated by the yeast Glk. Both HKs are dimers of subunits with approximately 52 kDa in molecular weight and share between themselves considerable structural similarity. The molecular weight of the Glk subunit was found to be 51 kDa (Maitra and Lobo, 1977) (Table 11.2).

HK binds glucose, mannose, or xylose. Xylose induces the conformation change in the active site, necessary for the interaction of ATP-Mg and the formation of catalytic ternary complexes. The mechanism of the reaction

TABLE 11.2 **The molecular dimensions of Glks.**

Taxon	Organism	Enzyme	Molecular mass (kDa) Subunit	Native	References
Archaea	*Thermococcus litoralis*	ADP-Glk (Ribokinase family)	54	Monomeric	Ito et al. (2001)
	Pyrococcus horikoshii	ADP-Glk (Ribokinase family)	52	Monomeric	Tsuge et al. (2002)
	Pyrococcus furiosus	ADP-Glk (Ribokinase family)	51	Dimeric	Ito et al. (2003)
	Aeropyrum pernix	ATP-Glk (Ribokinase family)	36	Monomeric	Hansen et al. (2002)
	Thermococcus kodakarensis	ADP-Glk (Ribokinase family)	50	Monomeric	Shakir et al. (2021)
Bacteria	*Escherichia coli*	ATP-Glk (Hexokinase family, A subgroup)	35	Dimeric	Lunin et al. (2004)
	Streptomyces griseus	ATP-Glk (Hexokinase family, B subgroup)	33	Tetrameric	Miyazono et al. (2012)
	Thermus thermophilus HB8	ATP-Glk (Hexokinase family, B subgroup)	31	Tetrameric	Nakamura et al. (2012)
	Methylomicrobium alcaliphilum 20Z	ATP-Glk (Hexokinase family, B subgroup)	35.4	Dimeric	Mustakhimov et al. (2017)
	Pyrobaculum calidifontis	ATP-Glk (Hexokinase family, B subgroup)	31	Monomeric	Bibi et al. (2018)
	Arthrobacter sp. KM	ATP/PP-GMK (Hexokinase family, B subgroup)	30	Monomeric	Mukai et al. (2004)

(Continued)

TABLE 11.2 (Continued)

			Molecular mass (kDa)		
Taxon	**Organism**	**Enzyme**	**Subunit**	**Native**	**References**
	Corynebacterium glutamicum	ATP/PP-Glk (Hexokinase family, B subgroup)	27	Dimeric	Lindner et al. (2010)
	Mycobacterium tuberculosis	ATP/PP-Glk (Hexokinase family, B subgroup)	33	Dimeric	Hsieh et al. (1993)
	Microlunatus phosphovorus	PP-Glk (Hexokinase family, B subgroup)	32	Dimeric	Tanaka et al. (2003)
	Streptomyces coelicolor	ATP/PP-Glk (Hexokinase family, B subgroup)	27	Aggregates	Koide et al. (2013)
Fungi	*Saccharomyces cerevisiae*	PI, PII (Hexokinase family, Hk subgroup)	51	Monomeric or dimeric	Schmidt and Colowick (1973)
	S. cerevisiae	ATP-Glk (Hexokinase family, Hk subgroup)	51	Aggregates	Maitra and Lobo (1977)
	Hansenula polymorpha	Hexokinase Hexokinase family, Hk subgroup	54.2	Monomeric	Karp et al. (2004)
Protozoa	*Trypanosoma brucei*	Hexokinase (Hexokinase family, A subgroup)	51	Aggregates	Misset et al. (1986)
	Trypanosoma cruzi	Glk (Hexokinase family, A subgroup)	43	Monomeric or dimeric	Cáceres et al. (2007)
	Leishmania spp.	Glk (Hexokinase family, A subgroup)	46	Monomeric or dimeric	Cáceres et al. (2007)

catalyzed by yeast HK seems essentially ordered (Steitz et al., 1981). The sugar substrate must first bind at its specific site, and then the conformational change induced allows ATP-Mg to interact at the nucleotide site (Roustan et al., 1974, Willson and Perie, 1999).

Trypasosomatids are the only organisms that possess both HKs and group A Glk enzymes. The *Trypanosoma cruzi* Glk (42 kDa) and HK (50 kDa) differ considerably in their substrate affinity; their km values for glucose are 0.7 mM and 0.06 mM, respectively. The crystallographic structure of Glk shows a homodimer in the asymmetric unit. Each monomer forms a complex to β-D-glucose and ADP (Cordeiro et al., 2007) (Table 11.2).

11.5 Production

Despite their industrial relevance, as well as their importance for clinical analyses and research, to our knowledge, very few methodologies for large-scale production of Glks have been described and patented. One of the best known is the production of a thermostable Glk by *Bacillus stearothermophilus* UK 788, which is about 10 microns longer than *B. stearothermophilus* IAM 11001. The new strain, having cells that settle smoothly and an easily breakable cell wall, was isolated due to screening naturally occurring micro-organisms by Koch's plate culture. The new strain was isolated from manure in Ogura, Uji, Kyoto, Japan. In addition to discovering a new strain, the patent provided a process for the large-scale industrial production of useful enzymes such as a heat-resistant polynucleotide phosphorylase, heat-resistant maleate dehydrogenase, heat-resistant Glk, heat-resistant G6P dehydrogenase (G6PDH), and heat-resistant pyruvate kinase. In brief, cells in the last stage of the logarithmic growth phase were collected in batch culture in a 30 jar fermentor. Chemostatic fermentation was performed, supplying fresh medium (Nakajima et al., 1982) to the fermentor and withdrawing its content with a metering pump at a rate of 24 L per h. The physical parameters were: temperature, 60°C; pH, 6.8–7.0, air supply rate, 20 L per min; and stirring speed, 600 rpm. Throughout the continuous fermentation that lasted for about 4 h, the cell concentration was maintained at the level present at the start of the fermentation (5.8 g of wet cells per liter or 0.75 g of dry cells per liter), and 550 g of wet cells were centrifuged from 96 L of the fermentation liquor. To collect intracellular protein, cells were subjected to ultrasonic treatment and protein was determined by the biuret method. In this way, the yield was 12.8 U of heat-resistant Glk per gram of wet cells (0.16 U per mg of protein) (Nakajima et al., 1982). Later, considering the high energy needed to industrially produce the thermostable Glk due to its high growth temperature (50°C–60°C) and also to produce larger amounts, another invention was patented. The goal was to provide genetic material for genetic engineering to produce the thermostable Glk from *B. stearothermophilus* at a larger scale in a mesophilic bacterium such as *E. coli*, lowering the production cost. The authors isolated and cloned on a commercial vector, like pUC29, a gene encoding a thermostable Glk from *B. stearothermophilus* and the construct was introduced in *E. coli*. About cloning, commercial vectors such as pUC19, pKK223−3′, and pPL-Qt may be used, although the authors suggest using a vector constructed by combining the ori and tac promoters originating in a multicopy vector such as pUC19. The host strains are preferably *E. coli* TG1 or BL21. In detail, the gene isolation was conducted using a chromosomal DNA library from *B. stearothermophilus*. The full-length gene was successfully amplified by PCR using a combination of two primers capable of amplifying the N-terminal and C-terminal parts of the Glk gene. The Glk gene-containing DNA fragments were mixed with vector plasmid fragments and subjected to ligation by using a T4 phage-derived DNA ligase at 16°C for 30 min.

The plasmid with the Glk gene was transformed in *E. coli* TG1 competent cells prepared by the calcium method. The transformant *E. coli* was inoculated into 300 mL of medium containing 50 μg/mL of ampicillin and cultured at 37°C overnight. The overnight culture was inoculated into 20 L of medium supplemented with 50 μg/mL of ampicillin and cultured at 37°C for 10 h. Then,1 mM isopropyl β-thiogalactopyranoside was added, and the culturing was continued for an additional 15 h. The cells were collected and showed 1000,000 U of Glk activity. The intracellular proteins were obtained by ultrasonic disruption in 25 mM phosphate buffer (pH 8.0). The Glk was purified by affinity chromatography with Blue Sepharose and ion exchange chromatography using DEAE-Sepharose. In this way, 360,000 U of Glk were recovered, which was 50 times as much as the amount of Glk obtained by culturing 20 L *B. stearothermophilus* UK-563 (Kawase and Kurosaka, 2003). This production method reflects a clear cost reduction.

Most probably, heterologous expression may be an important alternative to increase Glk yield, avoiding regulation and energetic cost associated with thermophilic bacteria. For instance, in Table 11.3, examples of cloned genes for enzyme expression of microbial Glk by PCR can be seen.

11.6 Potential applications in industrial processes

Glk functions as the "glucose sensor" in pancreatic β-cells regulating the glucose-stimulated insulin secretion. Therefore its structural integrity is determinant for maintaining normal glucose homeostasis. In addition, there are several variants of congenital pancreatic diseases affecting Glk, causing either hipo (Njølstad et al., 2001) or hiperinsulinism (Gilis-Januszewska et al., 2021). According to WHO, the global prevalence of diabetes in 2014 was estimated to be 8.5% among adults, and in 2019, diabetes caused 1.5 million deaths worldwide. Therefore Glk has attracted attention as a diagnostic and therapeutic target for diabetes and is of great significance for industrial applications and medical purposes (Guzmán and Gurrola-Díaz 2021).

TABLE 11.3 Examples of cloned bacterial Glks.

Gram-positive bacteria

				Cloning		Subcloning		
Name	**Enzyme**	**Size (bp)**	**Accession number**	**Vector**	***Escherichia coli* Host strain**	**Vector**	***E. coli* Host strain**	**References**
Bacillus subtilis	ATP-Glk	966	AL009126 region 2570606–2571571	pMD492[b]	UE26	pQE9[c]	RB791	Skarlatos and Dahl (1998)
Corynebacterium glutamicum	Pp/ATP-Glk	753	NC_006958 region 1980681–1981433	pEKEx3	LJ142[a]	pET16b[c]	BL21 (DE3)	Lindner, et al. (2010)
Bacillus sphaericus C3–41	ATP Glk	876	EF065663	pUC18	ZSC13[a]	pET28a[c]	BL21	Han, et al. (2007)
Sporolactobacillus inulinus Y2–8	ATP-Glk	975	JN860435	pMD18	DH5α	pET28a[c]	BL21 (DE3)	Zheng, et al. (2012)
Streptomyces coelicolor A3(2)	Pp-Glk	741	NC_003888 region 5499055–5499795	pMD19	DH5α	pColdI[c]	BL21 (DE3)	Koide, et al. (2013)
S. coelicolor A3(2)	ATP-Glk	954	NC_003888 region 2285983–2286936	pET15b	DH5α	pFT61[b,c]	FT1 (pLysS)	Mahr, et al. (2000)
Thermobifida fusca YX	Pp-Glk	789	CP000088 region 2111660–2112448	pCG	DH5α	pC-ppgk[b,c]	BL21 Star (DE3)	Liao, et al. (2012)
Thermotoga maritime MSB8	ATP-Glk	954	NC_000853 region 1481349–1482302	pET19b	JM109 BL21 -CodonPlus (DE3)-RILc	–	–	Hansen and Schönheit (2003)

(*Continued*)

TABLE 11.3 (Continued)

Gram-positive bacteria								
				Cloning		Subcloning		
Name	Enzyme	Size (bp)	Accession number	Vector	*Escherichia coli* Host strain	Vector	*E. coli* Host strain	References
Thermococcus kodakarensis	ADP-Glk	1362	BAD85299	pTZ57R/T	XL1-Blue	pET21a (+)	BL21 -CodonPlus (DE3)-RIL	Shakir et al. (2021)
Gram-negative bacteria								
E. coli O157:H7 str. EDL933	ATP-Glk	966	NC_002655.2 region 3306555–3307520	pET15[c]	*E. coli* BL21 (DE3)	–	–	Lunin, et al. (2004)
Leptospira interrogans	ATP-Glk	897	NC_004342 region 1438799–1439695	pET28b[c]	*E. coli* BL21 (DE3)	–	–	Zhang, et al. (2011)

[1]For Genomic DNA library construction.
[a]*Strain without pts and glk genes for complementation analysis.*
[b]*Plasmid with cloned glk gene.*
[c]*For protein expression.*

Glucose monitoring in diabetes. One of the challenges for diabetic patients is the regular monitoring of glucose levels without constant finger needle pricks. Thus, there is an urgent need to develop technology for the in vivo measurement of glucose (D'Auria et al., 2002; Hussain et al., 2005; Pickup et al., 2005a). One of the first assays was the glucose determination by a spectrophotometric method based on NADPH production by the coupled action of the enzymes Glk and G6P-DH. The formation of NADPH is therefore proportional to the amount of glucose present in the assay, which is measured as a change in absorbance at 340 nm (Fig. 11.8). For this purpose, the Glk of *B. stearothermophilus* has attracted attention because of its thermal stability, which allows its use for about one1 month at room temperature (Tomita et al., 1995). A novel, thermostable adaptation of the coupled-enzyme assay (Glk-G6PDH) for monitoring glucose concentrations was recently developed. This thermostable enzyme complex was isolated from the marine hyperthermophile *Thermotoga maritima*, and it works at 85°C (McCarthy et al., 2012).

Otherwise, the quantification of glucose has also been performed using biosensors. Biosensors have been defined as "analytical devices or units, which incorporate a biological or biologically-derived sensitive recognition element integrated to or associated with a physicochemical transducer" (Yoo and Lee, 2010). Biosensors are composed basically of (1) a recognition element of biological origin (receptors, enzymes, antibodies, nucleic acids, microorganisms or lectins) able to differentiate the target molecules in the presence of other chemical agents; (2) a transducer (electrochemical, optical, thermometric, piezoelectric, or magnetic) that converts the recognition of the target molecule into a measurable signal; and (3) a signal processing system

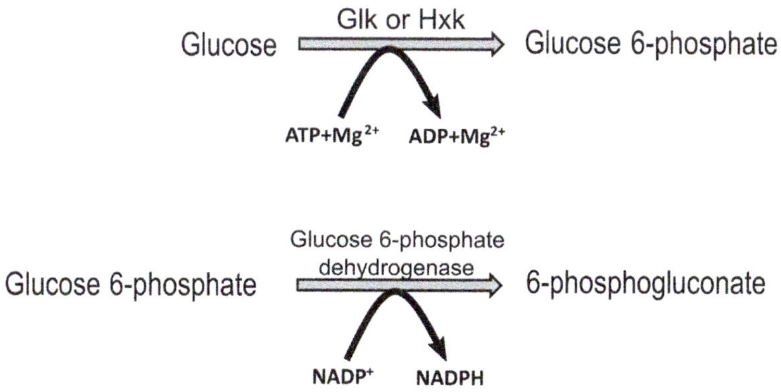

FIGURE 11.8 Diagram of the coupled reaction of Glk/Hxk and glucose-6-phosphate dehydrogenase for the quantification of glucose. *Modified from Tomita, K., Nomura, K., Kondo, H., Nagata, K., Tsubota, H., 1995. Stabilized enzymatic reagents for measuring glucose, creatine kinase and gamma-glutamyltransferase with thermostable enzymes from a thermophile, Bacillus stearothermophilus. J. Pharmaceut. Biomed. 13, 477–481.*

that converts the signal into a reading (Yoo and Lee, 2010). The majority of the current glucose biosensors are of the electrochemical type due to their better sensitivity, reproducibility, easy maintenance, and low cost (Yoo and Lee, 2010). Electrochemical sensors may be subdivided into potentiometric, amperometric or conductometric types. Amperometric sensors monitor currents generated when electrons are exchanged directly or indirectly between a biological system and an electrode. Enzymatic amperometric glucose biosensors are the most commercially available devices mainly based on glucose oxidase (GOD) enzyme activity (Hussain et al., 2005; Yoo and Lee, 2010).

One disadvantage of continuous glucose sensors based on needle-type amperometric enzyme electrodes with immobilized GOD is the need for frequent calibration to compensate for impaired responses and signal drift in vivo; for that reason, new glucose-sensing approaches are being explored (Hussain et al., 2005).

An example of this is the sol−gel coimmobilization of GOD and the Hxk from yeast to develop an amperometric biosensor for the simultaneous detection of glucose and ATP (Liu and Sun, 2007).

There is one patent for Glk immobilization from *B. stearothermophilus* since this enzyme is quite specific for glucose with very low interference derived from other monosaccharides. Besides, the use of this Glk avoids the problems associated with the use of the GOD enzyme, such as low sensitivity due to low oxygen solubility (Iida and Kawabe 1990).

Preliminary studies on the use of Glk for glucose determination through changes in fluorescence were performed using the Glk from *B. stearothermophilus* (BSGlk) (D'Auria et al., 2002). In this report, they compared the stability in the liquid between the BSGlk and the HXK from yeast. The enzyme from yeast proved to have poor stability over time at room temperature, unlike that of the BSGlk, which showed 100% of activity after 20 days of incubation. However, a disadvantage of the system was the poor fluorescence obtained by the authors with the fluorophore 2-(4 (iodoacetoamido) aniline) naphthalene-6-sulfonic acid (IA-ANS). Despite the low specificity for glucose of the HXK from yeast, it is possible to monitor its conformational changes when bound to the substrate using fluorescence (Pickup et al., 2005b; Hussain et al., 2005). Each subunit of yeast HXK has four tryptophan residues, two surface residues, one quenchable residue in the cleft, and one buried (Fig. 11.6). Approximately at an excitable wavelength of 300 nm, both monomers and dimers have a steady-state fluorescence emission maximum at about 300 nm, attributable to tryptophan fluorescence (Pickup et al., 2005b). To avoid problems associated with low fluorescence due to the binding of glucose to the enzyme (Pickup et al., 2005b) and those associated with interferences due to the nature of the serum sample, Hussain et al. (2005) conducted a HXK immobilization in a sol−gel matrix. This development is promising for the in vivo measurement of glucose in diabetic patients.

The Proassay Glucokinase kit from ProteinOne is a convenient, high-throughput method for the enzyme-luminescence detection of glucokinase activity, which can be used for the screening of glucokinase modulators (http://www.proteinone.com/glucokinase-kit-proassay-glucokinase-kit-2-10-137.php).

Glk act as a glucose sensor regulating insulin and glucagon secretion in the pancreas. Mutations in the gene encoding Glk cause a decrease in its activity and, therefore, an increase in blood glucose levels (Gao et al., 2021). Recent efforts are performed to develop safety Glk activators (GKA) to treat type-2 diabetes (Matschinsky et al., 2011).

Some clinical trials with the GKAs piragliatin and MK-0941 presented the incidence of hypoglycemia. However, a complete clinical trial in phase III of the GKA dorzagliatin showed a significant decrease in blood glucose levels with a low incidence of hypoglycemia (Gao et al., 2021). TransTech Pharma has developed several compounds covered by the composition of matter patents and applications, which is the case of its lead compound TTP399. Currently, this compound is in Phase-II clinical trials against type-2 diabetes mellitus (adjunctive treatment) in the United States (PO) (NCT02405260).

On the other hand, a human ADP-dependent Glk is located in the endoplasmic reticulum. It was suggested that this ADP-Glk links hyperglycemia caused by oxidative stress and an upregulated immune response ($TCD8^+$ and $TCD4^+$) in patients with type-2 diabetes (Imle et al., 2019). Therefore, the search for ADP-Glk inhibitors is of current interest.

Other clinical analyses. The determination of magnesium in serum and urine could be utilized for the diagnosis of renal diseases and gastrointestinal disorders. For this assay, it is possible to perform a reaction with Glk as stated in Fig. 11.8, as the magnesium ion is required in complex with ATP to carry out the phosphorylation of glucose. Thus the magnesium concentration can be determined spectrophotometrically by the increase of NADPH at 340 nm. This method has proven to be linear up to 100 mg/mL, better than that obtained in the colorimetric Xylidyl Blue method (Tabata et al., 1986; Tomita et al., 1990).

Serum creatine kinase (SCK) activity has been used for a long time to diagnose myocardial infarction or progressive muscular dystrophy (Shiraishi et al., 1991, Tomita et al., 1995). Glk could be used to determine SCK activity in a coupled reaction (Fig. 11.9), measuring at 340 nm the appearance of NADPH.

The determination of alpha-amylase activity in serum and urine is used to diagnose pancreatic and parotid diseases (Kondo et al., 1988, Tomita et al., 1990). Alternative methods have presented several problems, such as time-consuming assays or some interference from the production of maltose during the action of alpha-amylase. The alpha-amylase assay occurs in two groups of reactions (Fig. 11.10). In the main group of reactions, the action of the alpha-amylase on its substrate, the maltopentaose (G_5), produces maltose

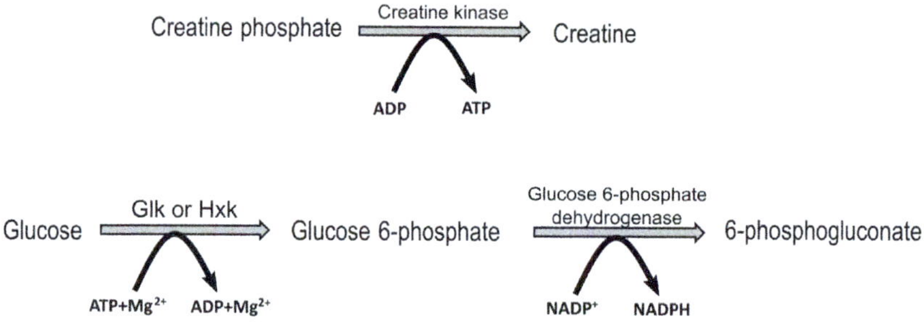

FIGURE 11.9 Diagram of the coupled reaction of CK and Glk/Hxk for measuring serum creatine kinase activity. *CK*, Creatine kinase. *Modified from Tomita, K., Nomura, K., Kondo, H., Nagata, K., Tsubota, H., 1995. Stabilized enzymatic reagents for measuring glucose, creatine kinase and gamma-glutamyltransferase with thermostable enzymes from a thermophile, Bacillus stearothermophilus. J. Pharmaceut. Biomed. 13, 477–481.*

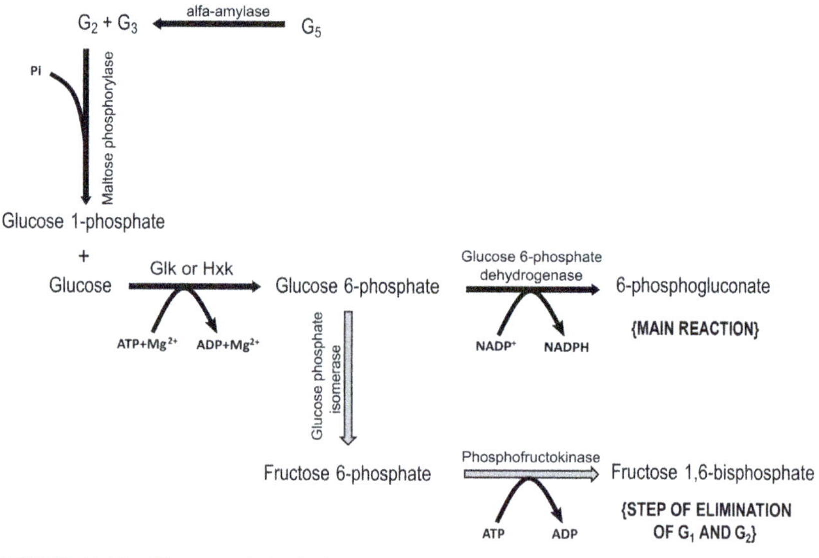

FIGURE 11.10 Diagram of the basis of the assay for alpha-amylase activity determination using Glk. Darker arrows show the main reaction while lighter arrows show the step of elimination of glucose (G_1) and maltose (G_2). G_5 is for maltopentaose, and G_3 for maltotriose. *Modified from Kondo, H., Shiraishi, T., Nagata, K., Tomita, K., 1988. An enzymatic method for the alpha-amylase assay which comprises a new procedure for eliminating glucose and maltose in biological fluids. Clin. Chim. Acta 172, 131–140.*

(G_2), which is converted to glucose by maltose phosphorylase (MP). Then, glucose is converted to 6-phosphogluconate (6-P gluconate) by the action of G6PDH producing NADPH, which can be measured by the change in absorbance at 340 nm (Kondo et al., 1988; Tomita et al., 1990). The elimination reaction converts glucose and maltose to fructose-1,6-biphosphate by the action of the enzymes MP, glucose phosphate isomerase (Glc-P isomerase), and phosphofructokinase (PFK). In the first step, glucose and maltose present in the serum samples are removed. Then, an inhibitor or MP is added, and subsequently, the alpha-amylase assay is initiated (Kondo et al., 1988).

Hxks of yeast origin, or Glks from *B. stearothermophilus*, were used in the aforementioned clinical applications. Currently, there is an increased interest in Glks capable of using PolyP instead of more labile and unstable ATP. One option is the Pp-Glk from *T. fusca*, which has been cloned, expressed, and purified (Liao et al., 2012). It has been mentioned that the low-cost generation of G6P by using PolyP would be advantageous to produce metabolites of interest such as hydrogen (Liao et al., 2012).

Other applications requiring immobilized Glks have used immobilization matrices resistant to continuous utilization. In this sense, Liu and Sun (2007) created an electrode based on a silicate hybrid sol–gel membrane to measure glucose and ATP. As the new matrix was biocompatible, the HXK and Glc6P-DH from Baker's yeast and the glucose oxidase from *Aspergillus niger* were immobilized. The resultant electrode showed high sensitivity, fast response, and good stability (Liu and Sun, 2007).

Nonclinical applications. The utilization of less costly PolyP rather than ATP as the phosphate donor for Glk activity is potentially attractive to produce high-yield hydrogen at a low cost without ATP (Liao et al., 2012). Another potential application may be cell-free protein synthesis, which requires much ATP input. By integration of PolyP-Glk that can produce G6P from low-cost PolyP with enzymes in the glycolysis pathway in the *E. coli* cell lysate (Calhoun and Swartz 2005; Wang and Zhang 2009), it could be possible to synthesize proteins from low-cost substrates rather than from costly substrates such as creatine phosphate, PEP, and acetate phosphate.

A process was recently described for G6P production coupled with an adenosine triphosphate (ATP) regeneration system. In this process, glucose is phosphorylated using Glk and acetyl phosphate to produce G6P with a conversion yield greater than 97% at 37°C for 1 h (Yan et al., 2014).

It is possible to lower costs in biofuel production by using cellulose as a substrate. However, the cellulose must be degraded by an enzymatic process prior to fermentation. To monitor glucose release during the latter approach, McCarthy et al. (2003) developed a continuous assay using the thermostable Glk and Glc6P-DH from *T. maritima*, which can withstand the conditions used to hydrolyze cellulose. The coupled reaction included 1,4-β-D-glucan glucohydrolase (for cellobiose hydrolysis), Glk, and Glc6P-DH and was performed at 85°C and a pH range of 7–8.5. This assay proved to be fast and straightforward and could also be coupled to glucose oxidase.

The emerging markets for polylactic acid are likely to stimulate a significantly increased demand for high optical purity D-lactic acid. *Sporolactobacillus inulinus*, a homofermentative lactic acid bacterium, is widely used for the industrial production of D-lactic acid of high optical purity; it efficiently ferments glucose exclusively to D-lactic acid (Zheng et al., 2010). Phosphorylation of glucose is the first step of glycolysis for D-lactic acid production. It was found that the Glk pathway was the major route for glucose uptake and phosphorylation in *S. inulinus* D-lactate production. Glk was prominently upregulated, followed by a large transmembrane proton gradient, while the phosphotransferase system pathway was completely repressed (Zheng et al., 2012).

11.7 Concluding remarks

Glucokinases are responsible for glucose phosphorylation, and aside from their catalytic activity, some Glks also present a regulatory role. They are produced by an incredible array of microbial systems, including bacteria, fungi, and other eukaryotes. Some Glks, especially those thermostable enzymes, have been envisaged for industrial purposes, taking advantage of their phosphorylating activity. In clinical analyses, they have been used to quantify glucose in diabetic patients accurately. In a coupled reaction with Glc6P-DH, Glk has been used to determine SCK activity to diagnose myocardial infarction. In another coupled reaction with Glc6P-DH, Glk has been utilized to monitor glucose release in biofuel production from cellulose as a substrate. Glk has also been used for D-lactic acid production of high optical purity. Finally, polyphosphate Glks have great potential for generating G6P and high-yield hydrogen based on low-cost polyphosphate.

Acknowledgments

We are indebted to Marco A. Ortiz Jimenez and Betsabé Linares for helping in the manuscript preparation. This work was partially supported by grants DGAPA, PAPIIT, IN202216 and IN-205922, UNAM and CONACYT, and A-S1-9143.

References

Angell, S., Lewis, C.G., Buttner, M.J., Bibb, M.J., 1994. Glucose repression in *Streptomyces coelicolor* A3 (2): a likely regulatory role for glucose kinase. Mol. Gen. Genet. 244, 135–143.

Ahuatzi, D., Herrero, P., de la Cera, T., Moreno, F., 2004. The glucose-regulated nuclear localization of hexokinase 2 in Saccharomyces cerevisiae is Mig1-dependent. J. Biol. Chem. 279 (14), 14440–14446. Available from: https://doi.org/10.1074/jbc.M313431200.

Ahuatzi, D., Riera, A., Pelaez, R., Herrero, P., Moreno, F., 2007. Hxk2 regulates the phosphorylation state of Mig1 and therefore its nucleocytoplasmic distribution. J. Biol. Chem. 282, 4485–4493.

Bae, J., Kim, D., Coi, Y., Koh, S., Park, J.E., Kim, J.S., et al., 2005. A hexokinase with broad sugar specificity from a thermophilic bacterium. Biochem. Biophys. Res. Comm. 334, 754–763.

Bibi, T., Ali, M., Rashid, N., Muhammad, M.A., Akhtar, M., 2018. Enhancement of gene expression in *Escherichia coli* and characterization of highly stable ATP-dependent glucokinase from *Pyrobaculum calidifontis*. Extremophiles 22, 247–257.

Cáceres, A.J., Quiñones, W., Gualdrón, M., Cordeiro, A., Avilán, L., Michels, P.A.M., et al., 2007. Molecular and biochemical characterization of novel glucokinases from *Trypanosoma cruzi* and *Leishmania* spp. Mol. Biochem. Parasitol. 156, 235–245.

Calhoun, K.A., Swartz, J.R., 2005. Energizing cell-free protein synthesis with glucose metabolism. Biotechnol. Bioeng. 90, 606–613.

Conejo, M.S., Thompson, S.M., Miller, B.G., 2010. Evolutionary bases of carbohydrate recognition and substrate discrimination in the ROK protein family. J. Mol. Evol. 70, 545–556.

Cordeiro, A.T., Cáceres, A.J., Vertommen, D., Concepción, J.L., Michels, P., et al., 2007. The crystal structure of *Trypanosoma cruzi* glucokinase reveals features determining oligomerization and anomer specificity of hexose-phosphorylating enzymes. J. Mol. Biol. 372, 1215–1226.

D'Auria, S., DiCesare, N., Staiano, M., Gryczynski, Z., Rossi, M., Lakowicz, J.R., 2002. A novel fluorescence competitive assay for glucose determinations by using a thermostable glucokinase from the thermophilic microorganism *Bacillus stearothermophilus*. Anal. Biochem. 303, 138–144.

Gao, H., Leary, J.A., 2003. Multiplex inhibitor screening and kinetic constant determinations for Gao using mass spectrometry based assays. J. Am. Soc. Mass Spectrom. 14, 173–181.

Gao, Q., Zhang, W., Li, T., Yang, G., Zhu, W., Chen, N., et al., 2021. The efficacy and safety of glucokinase activators for the treatment of type-2 diabetes mellitus: a *meta*-analysis. Medicine 100 (40), e27476.

Gilis-Januszewska, A., Boguslawska, A., Kowalik, A., Rzepka, E., Soczowka, K., Przybylik-Mazurek, E., et al., 2021. Hyperinsulinemic hypoglycemia in three generations of a family with glucokinase activating mutation, c-295T > C (p.Trp99Arg), Genes (Basel), 12. p. 1566.

Golbik, R., Naumann, M., Otto, A., Muller, E.C., Behlke, J., Reuter, R., et al., 2001. Regulation of phosphotransferase activity of hexokinase 2 from *Saccharomyces cerevisiae* by modification at serine-14. Biochemistry 40, 1083–1090.

Guzmán, T.J., Gurrola-Díaz, C.M., 2021. Glucokinase activation as antidiabetic therapy: effect of nutraceuticals and phytochemicals on glucokinase gene expression and enzymatic activity. Arch. Physiol. Biochem. 127, 182–193.

Han, B., Liu, H., Hu, X., Cai, Y., Zheng, D., Yuan, Z., 2007. Molecular characterization of a glucokinase with broad hexose specificity from *Bacillus sphaericus* strain C3–41. Appl. Environ. Microbiol. 73, 3581–3586.

Hansen, T., Reichstein, B., Schmid, R., Schönheit, P., 2002. The first archaeal ATP-dependent glucokinase, from the hyperthermophilic crenarchaeon *Aeropyrum pernix*, represents a monomeric, extremely thermophilic ROK glucokinase with broad hexose specificity. J. Bacteriol. 184, 5955–5965.

Hansen, T., Schönheit, P., 2003. ATP-dependent glucokinase from the hyperthermophilic bacterium *Thermotoga maritima* represents an extremely thermophilic ROK glucokinase with high substrate specificity. FEMS Microbiol. Lett. 226, 405–411.

Hsieh, P.C., Shenoy, B.C., Jentoft, J.E., Phillips, N.F.B., 1993. Purification of polyphosphate and ATP glucose phosphotransferase from *Mycobacterium tuberculosis* H37Ra: evidence that poly(P) and ATP glucokinase activities are catalyzed by the same enzyme. Protein Express. Purif. 4, 76–84.

Hsieh, P.C., Kowalczyk, T.H., Phillips, N.F.B., 1996. Kinetic mechanisms of polyphosphate glucokinase from *Mycobacterium tuberculosis*. Biochemistry 35, 9772–9781.

Hussain, F., Birch, D.J.S., Pickup, J.C., 2005. Glucose sensing based on the intrinsic fluorescence of sol-gel immobilized yeast hexokinase. Anal. Biochem. 339, 137–143.

Iida, T., Kawabe, T., 1990. Enzyme sensor using immobilized glucokinase. U.S. patent 4, 900, 423. Feb. 13.

Ito, S., Fushinobu, S., Yoshioka, I., Koga, S., Matsuzawa, H., Wakagi, T., 2001. Structural basis for the ADP-specificity of a novel glucokinase from a hyperthermophilic Archaeon. Structure 9, 205–214.

Imle, R., Wang, B.T., Stützenberger, N., Birkenhagen, J., Tandon, A., Carl, M., et al., 2019. ADP-dependent glucokinase regulates energy metabolism via ER-localized glucose sensing. Scientific Reports 9 (1), 14248. Available from: https://doi.org/10.1038/s41598-019-50566-6.

Ito, S., Fushinobu, S., Jeong, J.J., Yoshioka, I., Koga, S., Shoun, H., et al., 2003. Crystal structure of an ADP-dependent glucokinase from *Pyrococcus furiosus*: implications for a sugar-induced conformational change in ADP-dependent kinase. J. Mol. Biol. 331, 871–883.

Kawai, S., Mukai, T., Mori, S., Mikami, B., Murata, K., 2005. Hypothesis: structures, evolution, and ancestor of glucose kinases in the hexokinase family. J. Biosc. Bioeng. 99, 320–330.

Karp, H., Järviste, A., Kriegel, T.M., Alamäe, T., 2004. Cloning and biochemical characterization of hexokinase from the methylotrophic yeast *Hansenula polymorpha*. Curr. Genet. 44, 268–276.

Kawase, S., Kurosaka, K., 2003. Gene for thermostable glucokinase, recombinant vector containing the same, transformant containing the recombinant vector and process for producing thermostable glucokinase using the transformant. U.S. patent 6,566,109 b2.

Kengen, S.W., Tuininga, J.E., De Bok, F.A., Stams, A.J., De Vos, W.M., 1995. Purification and characterization of a novel ADP-dependent glucokinase from the hyperthermophilic archaeon *Pyrococcus furiosus*. J. Biol. Chem. 270, 30453–30457.

Kim, J.W., Dang, C.V., 2005. Multifaceted roles of glycolytic enzymes. Trends Biochem. Sci. 30, 142–150.

Koga, S., Yoshioka, I., Sakuraba, H., Takahashi, M., Sakasegawa, S., Shimizu, S., et al., 2000. Biochemical characterization, cloning, and sequencing of ADP-dependent (AMP-forming) glucokinase from two hyperthermophilic archaea, *Pyrococcus furiosus* and *Thermococcus litoralis*. J. Biochem. 128, 1079–1085.

Koide, M., Miyanaga, A., Kudo, F., Eguchi, T., 2013. Characterization of polyphosphate glucokinase SCO5059 from *Streptomyces coelicolor* A3(2). Biosci. Biotechnol. Biochem. 77, 2322–2324.

Kondo, H., Shiraishi, T., Nagata, K., Tomita, K., 1988. An enzymatic method for the alpha-amylase assay which comprises a new procedure for eliminating glucose and maltose in biological fluids. Clin. Chim. Acta 172, 131–140.

Labes, A., Schonheit, P., 2003. ADP-dependent glucokinase from the hyperthermophilic sulfate-reducing archaeon *Archaeoglobus fulgidus* strain 7324. Arch. Microbiol. 180, 69–75.

Liao, H., Myung, S., Zhang, Y.H.P., 2012. One-step purification and immobilization of thermophilic polyphosphate glucokinase from *Thermobifida fusca* YX: glucose-6-phosphate generation without ATP. Appl. Microbiol. Biotechnol. 93, 1109–1117.

Lindner, S.N., Knebel, S., Pallerla, S.R., Schoberth, S.M., Wendisch, V.F., 2010. Cg2091 encodes a polyphosphate/ATP-dependent glucokinase of *Corynebacterium glutamicum*. Appl. Microbiol. Biotechnol. 87, 703–713.

Liu, S., Sun, Y., 2007. Co-immobilization of glucose oxidase and hexokinase on silicate hybrid sol-gel membrane for glucose and ATP detections. Biosens. Biolectron. 22, 905–911.

Lunin, V.V., Li, Y., Schrag, J.D., Lannuzzi, P., Cygler, M., Matte, A., 2004. Crystal structures of *Escherichia coli* ATP-dependent glucokinase and its complex with glucose. J. Bacteriol. 186, 6915–6927.

Mahr, K., Wezel, G., Svensson, C., Krengel, U., Bibb, M., Titgemeyer, F., 2000. Glucose kinase of *Streptomyces coelicolor* A3(2): large-scale purification and biochemical analysis. Anton. Leeuw. 78, 253–261.

Maitra, P.K., Lobo, Z., 1977. Molecular properties of yeast glucokinase. Mol. Cell .

Matschinsky, F.M., Zelent, B., Grimsby, J., 2011. Glucokinase activators for diabetes therapy. Diabetes Care 34 (Suppl 2), S236–S243.

McCarthy, J.K., O'Brien, C.E., Eveleigh, D.E., 2003. Thermostable continuous coupled assay for measuring glucose using glucokinase and glucose-6-phosphate dehydrogenase from the marine hyperthermophile *Thermotoga maritima*. Anal. Biochem. 318, 196–203.

McCarthy, J.K., O'Brien, C.E., Eveleigh, D.E., 2012. Thermostable continuous coupled assay for measuring glucose using glucokinase and glucose-6-phosphate dehydrogenase from the marine hyperthermophile *Thermotoga maritima*. Anal. Biochem. 318 (2), 196–203.

Misset, O., Bos, O.J.M., Opperdoes, F.R., 1986. Glycolytic enzymes of *Trypanosoma brucei*. Simultaneous purification, intraglycosomal concentrations and physical properties. Eur. J. Biochem. 157, 441–453.

Mistry, J., Chuguransky, S., Williams, L., Qureshi, M., Salazar, G.A., Sonnhammer, E.L.L., et al., 2021. Pfam: the protein families database in 2021. Nucleic Acids Res. 49, D412–D419.

Miyazono, K.-I., Tabei, N., Morita, S., Ohnishi, Y., Horinouchi, S., Tanokura, M., 2012. Substrate recognition mechanism and substrate-dependent conformational changes of an ROK family glucokinase from *Streptomyces griseus*. J. Bacteriol. 194, 607–616.

Mukai, T., Kawai, S., Mori, S., Mikami, B., Murata, K., 2004. Crystal structure of bacterial inorganic polyphosphate/ATP-glucomannokinase. Insights into kinase evolution. J. Biol. Chem. 279, 50591–50600.

Mustakhimov, I.I., Rozova, O.N., Solntseva, N.P., Khmelenina, V.N., Reshetnikov, A.S., Trotsenko, Y.A., 2017. The properties and potential metabolic role of glucokinase in halotolerant obligate methanotroph *Methylomicrobium alcaliphilum* 20Z. Antonie Van Leeuwenhoek 110, 375–386.

Nakajima, H., Nagata, K., Kageyama, M., Suga, T., Suzuki, T., Motosugi, K., 1982. *Bacillus stearothermophilus* strain UK 788 and process for producing a useful enzyme. US Patent 4,331,762.

Nakamura, T., Kashima, Y., Mine, S., Oku, T., Uegaki, K., 2012. Characterization and crystal structure of the thermophilic ROK hexokinase from *Thermus thermophilus*. J. Biosc. Bioeng. 114, 150–154.

Nishimasu, H., Fushinobu, S., Shoun, H., Wakagi, T., 2006. Identification and characterization of an ATP-dependent hexokinase with broad substrate specificity from the hyperthermophilic archaeon Sulfolobus tokodaii. J. Bacteriol. 188 (5), 2014–2019. Available from: https://doi.org/10.1128/JB.188.5.2014-2019.2006.

Nishimasu, H., Fushinobu, S., Shoun, H., Wakagi, T., 2007. Crystal structures of an ATP-dependent hexokinase with broad substrate specificity from the hyperthermophilic Archaeon *Sulfolobus tokodaii*. J. Biol. Chem. 282, 9923–9931.

Njølstad, P.R., Sovic, O., Cuesta-Munoz, A., Bjørkhaug, L., Massa, O., Barbetti, F., et al., 2001. Neonatal diabetes mellitus due to complete glucokinase deficiency. N. Engl. J. Med. 344, 1588–1592.

Pickup, J.C., Hussain, F., Evans, N.D., Sachedina, N., 2005a. *In vivo* glucose monitoring: the clinical reality and the promise. Biosens. Bioelectron. 20, 1897–1902.

Pickup, J.C., Hussain, F., Evans, N.D., Rolinski, O.J., Birch, D.J., 2005b. Fluorescence-based glucose sensors. Biosens. Bioelectron. 20, 2555–2565.

Pineda, E., Encalada, R., Vázquez, C., González, Z., Moreno-Sánchez, R., Saavedra, E., 2015. Glucose metabolism and its controlling mechanisms in *Entamoeba histolytica*. In: Nozaki, T., Bhattacharya, A. (Eds.), Amebiasis. Springer, Tokyo. Available from: https://doi.org/10.1007/978-4-431-55200-0_20.

Pérez, E., Fernández, F., Fierro, F., Mejia, A., Marcos, A., Martin, J., et al., 2014. Yeast HXK2 gene reverts glucose regulation mutation of penicillin biosynthesis in *P. chrysogenum*. Brazilian J. Microbiol. 45, 873–883.

Rao, N.N., Gómez-García, M.R., Kornberg, A., 2009. Inorganic polyphosphate: essential for growth and survival. Annu. Rev. Biochem. 78, 605–647.

Rivas-Pardo, J., Herrera-Morande, A., Castro-Fernández, V., Fernández, F.J., Vega, C.M., Guixé, V., 2013. Crystal structure, SAXS and kinetic mechanism of hyperthermophilic ADP-dependent glucokinase from *Thermococcus litoralis* reveal a conserved mechanism for catalysis. PLoS One 8 (6), e66687.

Rocha-Mendoza, D., Manzo-Ruiz, M., Romero-Rodríguez, A., Ruiz-Villafán, B., Rodríguez-Sanoja, R., Sánchez, S., 2021. Dissecting the role of the two *Streptomyces peucetius* var. *caesius* glucokinases in the sensitivity to carbon catabolite repression. J. Ind. Microbiol. Biotechnol. 48. Available from: https://doi.org/10.1093/jimb/kuab047 kuab047.

Rodriguez, A., De La Cera, T., Herrero, P., Moreno, F., 2001. The hexokinase 2 protein regulates the expression of the GLK1, HXK1 and HXK2 genes of *Saccharomyces cerevisiae*. Biochem. J. 355, 625–631.

Rodríguez-Saavedra, C., Morgado-Martínez, L.E., Burgos-Palacios, A., King-Díaz, B., López-Coria, M., Sánchez-Nieto, S., 2021. Moonlighting proteins: the case of hexokinases. Front. Mol. Biosci. 8, 701975.

Roustan, C., Brevet, A., Pradel, L., van Thoai, N., A., 1974. Yeast hexokinase: interaction with substrates and analogs studied by difference spectrophotometry. Eur. J. Biochem. 44, 353–358.

Ruiz-Villafán, B., Rodríguez-Sanoja, R., Aguilar-Osorio, G., Gosset, G., Sánchez, S., 2014. Glucose kinases from *Streptomyces peucetius* var. *caesius*. Appl. Microbiol. Biotechnol. 98, 6061–6071.

Ruiz-Villafán, B., Cruz-Bautista, R., Manzo-Ruiz, M., Kumar Passari, A., Villarreal-Gómez, K., Rodríguez-Sanoja, R., et al., 2021. Carbon catabolite regulation of secondary metabolite formation, an old but not well-established regulatory system. Microb. Biotechnol. 14, 2021. ISSN: 1751-7915.

Sakuraba, H., Yoshioka, I., Koga, S., Takahashi, M., Kitahama, Y., Satomura, T., et al., 2002. ADP-dependent glucokinase/ phosphofructokinase, a novel bifunctional enzyme from the hyperthermophilic archaeon *Methanococcus jannaschii*. J. Biol. Chem. 277, 12495–12498.

Schmidt, J.J., Colowick, S.P., 1973. Chemistry and subunit structure of yeast hexokinase isoenzymes. Arch. Biochem. Biophys. 158, 458–470.

Shakir, N.A., Aslam, M., Bibi, T., Rashid, N., 2021. ADP-dependent glucose/glucosamine kinase from Thermococcus kodakarensis: Cloning and characterization. Int. J. Biol. Macro. 173, 168–179. Available from: https://doi.org/10.1016/j.ijbiomac.2021.01.019.

Shiraishi, T., Kondo, H., Tsubota, H., 1991. A new method for measuring serum pyruvate kinase and creatine kinase activities using a thermostable glucokinase. Jpn. J. Clin. Chem. 20, 235–241.

Siebers, B., Schönheit, P., 2005. Unusual pathways and enzymes of central carbohydrate metabolism in Archaea. Curr. Opin. Microbiol. 8, 695–705.

Skarlatos, P., Dahl, M.K., 1998. The glucose kinase of *Bacillus subtilis*. J. Bacteriol. 180, 3222–3226.

Socorro, J.M., Olmo, R.C., Blanco, M.D., Teijon, J.M., 2000. Analysis of aluminum–yeast hexokinase interaction: modifications on protein structure and functionality. J. Protein Chem. 19, 199–208.

Steitz, T.A., Shoham, M., Bennett Jr., W.S., 1981. Structural dynamics of yeast hexokinase during catalysis. Phil. Trans. R. Soc. Lond. 293, 43–52.

Świątek, M.A., Gubbens, J., Bucca, G., Song, E., Yang, Y.-H., Laing, E., et al., 2013. The ROK family regulator Rok7B7 pleiotropically affects xylose utilization, carbon catabolite repression, and antibiotic production in *Streptomyces coelicolor*. J. Bacteriol. 195, 1236–1248.

Szymona, M., Ostrowski, W., 1964. Inorganic polyphosphate glucokinase of *Mycobacterium phlei*. Biochim. Biophys. Acta 85, 283–295.

Tabata, M., Kido, T., Totani, M., Murachi, T., 1986. Usefulness of glucokinase from *Bacillus stearothermophilus* for the enzymatic measurement of magnesium in human serum. Agric. Biol. Chem. 50, 1909–1910.

Tanaka, S., Lee, S.O., Hamaoka, K., Kato, J., Takiguchi, N., Nakamura, K., et al., 2003. Strictly polyphosphate-dependent glucokinase in a polyphosphate-accumulating bacterium, *Microlunatus phosphovorus*. J. Bacteriol. 185, 5654–5656.

Tokarz, P., Wisniewska, M., Kaminski, M.M., Dubin, G., Grudnik, P., 2018. Crystal structure of ADP-dependent glucokinase from *Methanocaldococcus jannaschii* in complex with 5-iodotubercidin reveals phosphoryl transfer mechanism. Protein Sci. 27, 790–797.

Tomita, K., Nagata, K., Kondo, H., Shiraishi, T., Tsubota, H., Suzuki, H., et al., 1990. Thermostable glucokinase from *Bacillus stearothermophilus* and its analytical application. Ann. NY Acad. Sci. 613, 421–425.

Tomita, K., Nomura, K., Kondo, H., Nagata, K., Tsubota, H., 1995. Stabilized enzymatic reagents for measuring glucose, creatine kinase and gamma-glutamyltransferase with thermostable enzymes from a thermophile, *Bacillus stearothermophilus*. J. Pharmaceut. Biomed. 13, 477–481.

Tsai, S.C., Chen, Q., 1998. Purification and kinetic characterization of hexokinase and glucose-6-phosphate dehydrogenase from *Schizosaccharomyces pombe*. Biochem. Cell Biol. 76, 107–113.

Tsuge, H., Sakuraba, H., Kobe, T., Kujime, A., Katunuma, N., Ohshima, T., 2002. Crystal structure of the ADP-dependent glucokinase from *Pyrococcus horikoshii* at 2.0-Å resolution: A large conformational change in ADP-dependent glucokinase. Prot. Sci. 11, 2456–2463.

Tuininga, J.E., Verhees, C.H., van der Oost, J., Kengen, S.W., Stams, A.J., de Vos, W.M., 1999. Molecular and biochemical characterization of the ADP-dependent phosphofructokinase from the hyperthermophilic archaeon *Pyrococcus furiosus*. J. Biol. Chem. 274, 21023–21028.

van Wezel, G., Konig, M., Mahr, K., Nothaft, H., Thomae, A., Bibb, M., et al., 2007. A new piece of an old jigsaw: glucose kinase is activated postrtranslationally in a glucose transport-dependent manner in *Streptomyces coelicolor*A3(2). J. Mol. Microbiol. Biotechnol. 12, 67–74.

Verhees, C.H., Koot, D.G., Ettema, T.J., Dijkema, C., de Vos, W.M., van der Oost, J., 2002. Biochemical adaptations of two sugar kinases from the hyperthermophilic archaeon *Pyrococcus furiosus*. Biochem. J. 366, 121–127.

Wang, Y., Zhang, Y.-H.P., 2009. Overexpression and simple purification of the *Thermotoga maritime* 6-phosphogluconate dehydrogenase in *Escherichia coli* and its application for NADPH regeneration. Microb. Cell Fact. 8, 30.

Whitehead, M., Hooley, P., Brown, M., 2013. Horizontal transfer of bacterial polyphosphate kinases to eukaryotes: implications for the ice age and land colonisation. BMC Res. Notes 6, 221.

Willson, M., Perie, J., 1999. Inhibition of yeast hexokinase: a kinetic and phosphorous nuclear magnetic resonance study. Spectrochimica Acta Part A 55, 911–917.

Yan, B., Ding, Q., Ou, L., Zou, Z., 2014. Production of glucose-6-phosphate by glucokinase coupled with an ATP regeneration system. World J. Microbiol. Biotechnol. 30, 1123–1128.

Yoo, E.-H., Lee, S.-Y., 2010. Glucose biosensors: an overview of use in clinical practice. Sensors-basel 10, 4558–4576.

Zhang, Q., Zhang, Y., Zhong, Y., Peng, M.J., Cao, N., Yang, X., et al., 2011. *Leptospira interrogans* encodes an ROK family glucokinase involved in a cryptic glucose utilization pathway. Acta Biochim. Biophys. Sin. 43, 618–629.

Zheng, L., Bai, Z., Xu, T., He, B., 2012. Glucokinase contributes to glucose phosphorylation in D-lactic acid production by *Sporolactobacillus inulinus* Y2–8. J. Ind. Microbiol. Biotechnol. 39, 1685–1692.

Zheng, H.J., Gong, J.X., Chen, T., Chen, X., Zhao, X.M., 2010. Strain improvement of *Sporolactobacillus inulinus* ATCC 15538 for acid tolerance and production of D-lactic acid by genome shuffling. Appl. Microbiol. Biotechnol. 85, 1541–1549.

Chapter 12

Myctobacterium tuberculosis DapA as a target for antitubercular drug design

Ayushi Sharma, Ashok Kumar Nadda and Rahul Shrivastava
Department of Biotechnology and Bioinformatics, Jaypee University of Information Technology, Waknaghat, Solan, Himachal Pradesh, India

12.1 Introduction

Tuberculosis (TB), a communicable disease caused by the notorious agent *Mycobacterium tuberculosis* (MTB), requires a long-term (at least 6 months) administration of an antibiotic cocktail (ethambutol, isoniazid, rifampicin, and pyrazinamide) for its treatment (Burley et al., 2020). The United Nation declarations acknowledge it among the top 10 lethal diseases worldwide (Shetye et al., 2020). TB is an airborne disease spread by individuals through coughing and expelling bacteria into the surrounding air. It typically attacks human lungs, causing pulmonary TB, besides exhibiting the potential of infecting alternative sites and causing extrapulmonary TB. Poverty and lack of sanitation contribute extensively to the dissemination of the disease. Other determinants for TB comprise human immunodeficiency virus (HIV) infection, undernutrition, diabetes, smoking, alcohol abuse, renal disease, organ transplantation, malignancies, and air pollution (Singh et al., 2020).

Decreased permeability of the mycobacterial cell wall to foreign hydrophilic molecules is an important aspect favoring MTB survival inside hosts (Bhat et al., 2017). Exploring this feature, numerous mycobacterial cell wall biosynthesis enzymes have been popularly targeted by clinical drug discovery groups to design novel TB therapeutics. The chapter provides an analysis of DapA as a drug target involved in the cell wall biosynthesis of MTB. It focuses on the associated mechanisms and an array of cell wall target(s) with potential chemical inhibitors. Previous reports targeting MTB DAP pathway enzymes and the significance of the identified inhibitors has also been systematically reviewed.

Biotechnology of Microbial Enzymes. DOI: https://doi.org/10.1016/B978-0-443-19059-9.00008-6
© 2023 Elsevier Inc. All rights reserved.

12.1.1 Tuberculosis: global epidemiology

As per World Health Organization (WHO) report (WHO, 2020), around 10 million cases (3.2 million women, 5.6 million men, and 1.2 million children) and 1.4 million deaths (encompassing 208,000 people coinfected with HIV) were recorded worldwide, in the year 2019 alone. 87% cases were reported from the 30 high-TB burden countries, where 8 out of these 30 countries accounted for two-thirds of the total patients. The year 2019 counts for the highest 44% new cases in the South-East Asian region, followed by 25% new cases in the African areas and 18% new cases in the Western Pacific region. Worldwide, India ranked "number one" with the highest cases (26%), followed by different South-Asian countries (China, Philippines, Pakistan, Indonesia, Bangladesh), and South Africa.

The WHO report (2020) indicated a 10% increase in rifampicin-resistant, or multidrug resistant (MDR) TB cases in 2019, as compared to the count from the year 2018. A total of 186,883 cases of MDR TB were reported in 2018, with the count increasing to 206,030 by the year 2019. Fig. 12.1 represents countries with high-TB cases in 2019. Fifty-four countries from the WHO European and WHO American region reported a relatively lower fraction of TB cases (<10 cases per 100,000 population/year). Around 150–400 cases were recorded for most of the high-burden countries per 100,000 population, while greater than 500 cases were from South Africa, Central African Republic, Democratic Republic of Korea, Philippines, and Lesotho. A fraction of 8.2% of TB patients was found to be coinfected with HIV, with the highest ratio (around 50%) from the WHO African regions.

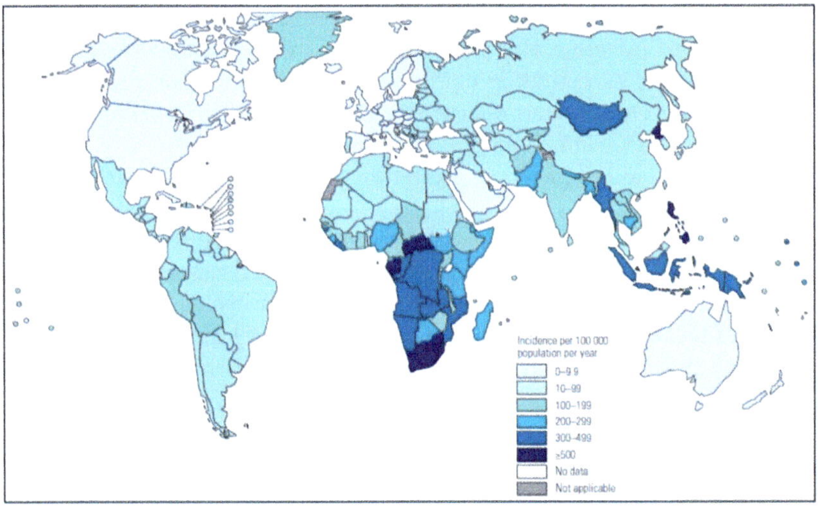

FIGURE 12.1 Global tuberculosis incidence (per 100,000 population), 2019 (WHO, 2020).

12.2 Challenges encountered by the scientific communities

The WHO regards MTB as a global health emergency. Since the advent of MTB, the mycobacterial cell wall has been widely exploited as a target fueling research for new and effective TB therapeutics (Abrahams and Besra, 2018). However, the cell wall inhibitors currently used as treatment options for TB [ethionamide and isoniazid (mycolic acid inhibitor), ethambutol (arabinogalactan inhibitor), and cycloserine (peptidoglycan inhibitor)] are not effective in shortening the TB-treatment duration (Mdluli and Spigelman, 2006). The mycolyl-arabinogalactan-peptidoglycan (mAGP) structure of the MTB cell wall is not being targeted by current TB therapeutics, owing to the presence of β-lactam nullifying enzyme β-lactamase (Catalão et al., 2019). On the counter-side, Bacillus Calmette–Guerin inefficacy, increasing MDR and extensively drug-resistant MTB isolates, and its catastrophic synergistic budding with HIV fuel the global TB pandemic and pose a significant challenge in the treatment of the disease (Shafiani et al., 2005). Slow growth rate, potential to exist in multiple environments (granulomas, macrophages, and aerobic and anaerobic conditions), drug resistance, and persistence add to a load of challenges (Abrahams and Besra, 2020). Rifampicin, an antimycobacterial drug, is progressively showing patient incompliance owing to its adverse effects on the treated individuals (Amir et al., 2014). The use of second-line anti-TB drugs such as kanamycin, cycloserine, capreomycin, para-aminosalicylate, fluoroquinolones, and ethionamide present several serious complications. Further, treatment difficulties are observed due to dormant, metabolically silent, and persistent bacilli within-host lesions (Bushra and Adem, 2016). TB drug pipeline hence requires identification and validation of novel targets, potentially effective chemical inhibitors, new drug regimen, cost and time-efficient chemotherapy, better diagnostics (Shetye et al., 2020), and a significantly reduced target mutation frequency for combating upsurge resistance (Abrahams and Besra, 2020).

12.3 MTB cell wall: a source of drug targets

MTB cell envelope is about 10,000 times more resistant to nutrient penetration when compared to the outer coating of *Escherichia coli* (Nieto et al., 2017). MTB cell wall (Fig. 12.2) is a three-dimensional complex formed by covalent interlinking of outer mycolic acids (long-chain), middle arabinogalactan polysaccharides (branched-chain), and inner polymeric peptidoglycans (interconnected) (Shetye et al., 2020). The mycolic acid layer is surrounded by an outer layer (capsule) that is composed of lipids (diacyl trehaloses, phosphatidyl-myo-inositol mannosides, phosphatidylethanolamine, and phthiocerol dimycocerosates), polysaccharides, and proteins (Shaku et al., 2020).

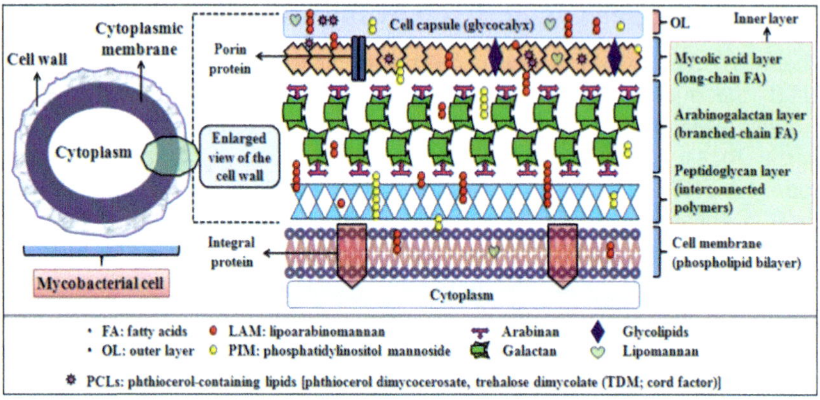

FIGURE 12.2 Structure of the mycobacterial cell envelope.

The lipid-rich coating protects MTB bacilli from macrophage defense actions such as the secretion of hydrolytic enzymes and the generation of toxic radicals (Bushra and Adem, 2016). MTB cell wall enzymes involved in the biosynthesis of the MTB cell wall can be exploited as targets for the design and development of new antitubercular therapeutics since the wall is an essential structure for MTB survival and pathogenesis (Moraes et al., 2015; Shetye et al., 2020). Moreover, it helps maintain turgor, resists antibiotic pressure, and modulates the host's innate immune responses (Catalão et al., 2019). The peptidoglycan layer itself maintains the integrity of the cell envelope and is involved in MTB virulence (Alderwick et al., 2015; Maitra et al., 2019). The absence of peptidoglycan biosynthesis in humans efficiently helps in targeting the nonhomologous sequences for drug generation. The uniqueness of the biosynthetic pathway to the pathogen opens possibilities of exploiting the concerned enzymes for novel drug discovery (Bushra and Adem, 2016).

12.3.1 Targeting MTB cell wall enzymes

Previous studies have suggested numerous MTB cell wall components as potential drug targets. A list of the identified targets and their inhibitors is provided in Table 12.1.

12.4 The diaminopimelate (DAP) pathway (lysine synthesis pathway)

Diaminopimelic acid/diaminopimelate (DAP) synthesis occurs through three pathways: succinylase, acetylase, and the dehydrogenase pathways, all of which can synthesize *L*-2,3,4,5-tetrahydrodipicolinic acid (*L*-THDP)

TABLE 12.1 *Mycobacterium tuberculosis* **(MTB) cell wall target(s) and their inhibitors (Mdluli and Spigelman, 2006; Abrahams and Besra, 2018; Catalão et al., 2019; Shetye et al., 2020; Shaku et al., 2020).**

MTB cell wall layer	Target(s)	Chemical inhibitor
Mycolic acid layer	InhA (2-trans-enoyl-ACP reductase)	Isoniazid, GSK693 (thiadiazoles), triclosan, ethionamide, pyridomycin, 4-hydroxy-2-pyridines (NITD-113, NITD-916), diazaborines (AN12855, AN12541), 2-(o-tolyloxy)-5-hexylpnenols (PT70)
	KasA (β-ketoacyl-ACP synthase)	GSK724 (indazole sulfonamide), GSK3011724A
	β-ketoacyl synthases (KasA, KasB)	Thiolactomycin (TLM), cerulenin, platensimycin
	Methoxy and keto mycolic acid inhibitor (actual target unknown)	Delamanid, pretomanid
	MmpL3 (transmembrane transport protein-3)	THPP and Spiro, AU1235 (adamantyl urea derivative), NITD-304 and NITD-349 (indolcarboxamides), SQ109 (1,2-ethylene diamine), BM212 and BM635 (1,5-diarylpyrrole derivative), C215, PIPD1 (a piperidinol-containing molecule)
	FabH (β-ketoacyl-ACP synthase III)	TLM analogs
	MabA (β-ketoacyl-ACP reductase)	Pteleoellagic acid, anthranilic acid analogs
	HadAB/BC [(3)-hydroxyacyl-ACP dehydratase subunit A/B/C]	Thiacetazone, isoxyl
	FabD32 (fatty-acid-AMP ligase)	Diarylcoumarin
	Pks13 (polyketide synthase-13)	Benzofuran (TAM 16), thiophenes, β-lactones (EZ120), coumestans

(Continued)

TABLE 12.1 (Continued)

MTB cell wall layer	Target(s)	Chemical inhibitor
	Antigen 85	I3-AG85 (2-amino-6-propyl-4,5,6,7-tetrahydro-1-benzothiphene-3-carbonitrile), cyclipostins and its analogs
	MmaA1 (mycolic acid methyltransferase)	3-(2-Morpholinoacetamido)-N-(1,4-dihydro-4-oxoquinazolin-6-yl) benzamide
	EchA6 (probable enoyl-CoA hydratase)	THPPs [tetrahydropyrazo(1,5-a) pyrimidine-3-carboxamides]
Arabinogalactan layer	DprE1 (decaprenylphosphoryl-β-d-ribose 2′-epimerase) (covalent inhibitors)	BTZ-043 (nitrobenzothiazinone), PBTZ169 (macozinone, piperazinobenzothiazinone), benzothiazole
	DprE1 (non-covalent inhibitors)	Pyrrole-benzothiazinone (PyrBTZ01), TCA1 (thiophene), OPC-167832, TBA7371
	WecA (decaprenyl-phosphate–GlcNAc-1-phosphate transferase)	Caprazamycin, CPZEN-45 and X-J99620886 (caprazene nucleoside), Tunicamycin
	Galactofuranosyltransferases (GlfT1 and GlfT2)	Uridine diphosphate (UDP)-Gal*f* derivatives
	UbiA (decaprenol-1-phosphate-5-phosphoribosyltransferase)	KRT2029
	Arabinofuranosyltransferases (AftA, AftB, AftC, AftD)	Decaprenylphosphoryl-D-arabinose (DPA) analogs
	Emb arabinosyltransferases (EmbA and EmbB)	Ethambutol and analogs (SQ109, SQ775)

Peptidoglycan layer	L,D-transpeptidases	Carbapenem β-lactam (meropenem)
	L,D-transpeptidases and β lactamase	Meropenem and clavulanate
	MurX/MraY (muramic acid residue X/Y)	Capuramycin, capuramycin analogs (U T-01320,SQ641, X-J99620886), liposidomycin, muramycin, caprazamycin, tunicamycin
	Lipid II	Teixobactin, ramoplanin, enduracidin
	Ddl (*D*-Ala-*D*-Ala ligase)	*D*-cycloserine
	MurG (muramic acid residue G)	Uridine-linked transition state analogs
	GlmU (bifunctional acetyltransferase/uridyltransferase)	Glucosamine-1-phosphate (GlcN-1-P) substrate analogs, 4-aminoquinazolines (compounds HMP-15 and HMP-05)
	MurB (muramic acid residue B)	Dioxopyrazolidines
	Alr (alanine racemase)	*D*-cycloserine, thiadiazolidinones
	BlaC (β-lactamase)	Diazabicyclooctanes (zidebactam and nacubactam), sulbactam, tazobactam, avibactam, clavulanate
	PknB (protein kinase B)	5-Substituted pyrimidine analogs
	Carboxypeptidase	Meropenem

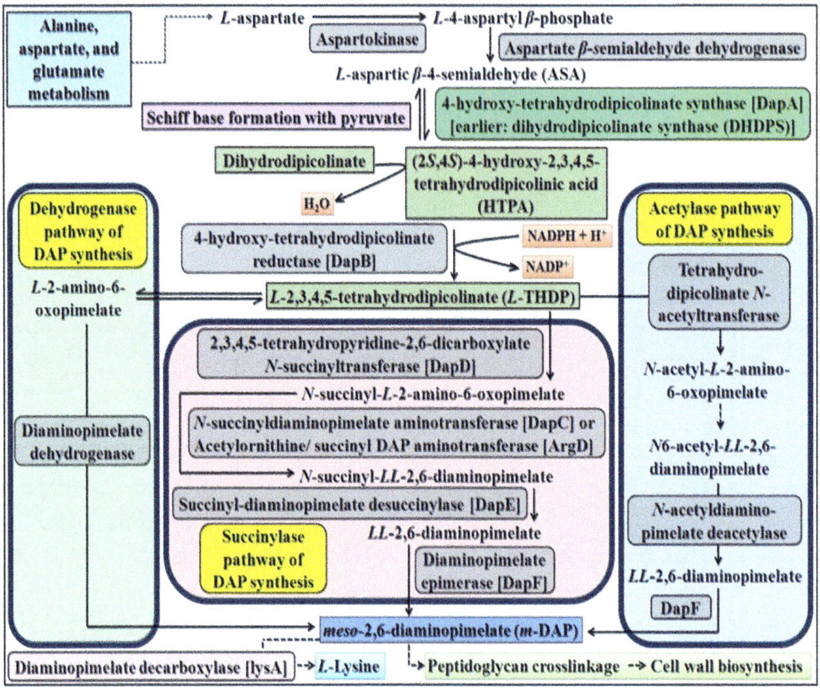

FIGURE 12.3 *meso*-DAP biosynthesis pathway. *Mycobacteria* (especially MTB) specifically utilizes succinylase pathway of DAP biosynthesis.

as an intermediate (Usha et al., 2016). Fig. 12.3 pictorially describes the chemical reactions undergoing in *meso*-DAP biosynthesis. The most common pathway synthesizing lysine in bacterial species is the succinylase pathway. It is inherent to *E. coli*. The acetylase pathway proceeds via four steps, incorporating *N*-acetyl moieties instead of *N*-succinyl groups. It is observed in bacteria belonging to *Bacillus* species (*Bacillus subtilis*, *Bacillus anthracis*). It begins with tetrahydrodipicolinate *N*-acetyltransferase mediated catalysis of *L*-THDP to *N*-acetyl-*L*-2-amino-6-oxopimelate, followed by generation of *N*6-acetyl-*LL*-2,6-diaminopimelate. The substrate is further acted upon by *N*-acetyldiamino-pimelate deacetylase to generate *LL*-2,6-diaminopimelate. DapF then catalyzes the conversion of *LL*-2,6-diaminopimelate to *meso*-DAP, in a manner as in the succinylase pathway. Another subpathway, the dehydrogenase route to *meso*-DAP generation, is common in particular *Bacillus* and *Corynebacterium* species. The pathway proceeds via a single step. *L*-THDP is reversibly converted to *L*-2-amino-6-oxopimelate. This substrate is further converted to *meso*-DAP by the NADPH-dependent enzyme diaminopimelate dehydrogenase (Dogovski et al., 2012).

Mycobacterial *meso*-DAP synthesis occurs in eight steps via the succinylase pathway of DAP biosynthesis. The first step proceeds through aspartokinase-mediated *L*-aspartate phosphorylation (by adenosine triphosphate) into *L*-4-aspartyl *β*-phosphate. Its conversion to *L*-aspartic *β*-4-semialdehyde (ASA) is then catalyzed by aspartate *β*-semialdehyde dehydrogenase (Usha et al., 2012). The reaction is common to threonine, methionine, isoleucine, and lysine biosynthesis (Dogovski et al., 2009). DapA then converts ASA to 4-hydroxy-tetrahydrodipicolinate synthase (HTPA) (earlier: 2,3-dihydrodipicolinic acid) through aldol condensation with pyruvate. In the next step, 4-hydroxy-tetrahydrodipicolinate reductase (DapB) (earlier: dihydrodipicolinate reductase) carries out a pyridine nucleotide-dependent reaction for converting dihydrodipicolinate to tetrahydrodipicolinate (Pavelka et al., 1997). DapD further converts cyclic compound *L*-THDP into *N*-succinyl-*L*-2-amino-6-oxopimelate (acyclic compound) using succinyl-CoA. DapC or ArgD catalyzes the sixth step, transferring amino group from *L*-glutamate, subsequently converting the substrate to *N*-succinyl-*LL*-2,6-diaminopimelate. A pyridoxal phosphate cofactor is used for mediating the reaction. Zinc containing metallohydrolase DapE hydrolyzes *N*-succinyl-*LL*-2,6-diaminopimelate to succinate and *LL*-2,6-diaminopimelate (Usha et al., 2016). The eighth step of the DAP biosynthesis pathway utilizes DapF for generating *meso*-2,6-diaminopimelate (*m*-DAP) through a racemization reaction (Munshi et al., 2013).

All the three DAP biosynthetic pathways converge to generate lysine and carbon dioxide from *meso*-DAP (Dogovski et al., 2012). MTB *lysA* encodes pyridoxal-5′-phosphate-dependent diaminopimelate decarboxylase enzyme that is essential for lysine biosynthesis in bacteria. LysA facilitates the last step in lysine biosynthesis, converting *meso*-DAP to *L*-lysine. Lysine, the end product, is essential for bacterial viability and growth (Gokulan et al., 2003). *lysA* mutation has previously been observed to be bacteriocidal in mycobacteria *Mycobacterium smegmatis* (Pavelka et al., 1997). Lysine precursor *meso*-DAP contributes significantly to mycobacterial cell wall development by establishing peptidoglycan cross-linkage. These linkages provide cell wall stability and aid in resisting osmotic pressure prevalent within cells (Gokulan et al., 2003).

12.5 Dihydrodipicolinate synthase (DapA)

DapA catalyzes the reaction between ASA and pyruvate, generating HTPA through aldol condensation (Shrivastava et al., 2016). Kefala and Weiss (2006) cloned MTB DapA in *E. coli*, solving its structure to 2.28 Å. Analytical ultracentrifugation studies showed the enzyme exists as a homotetramer of 120 kDa (approximately). According to Kefala et al. (2008), it consists of two independent tetramers that constitute the asymmetric unit. Both the tetramers were believed to form the functional enzyme unit. Each subunit of the enzyme is composed of 300 amino acids (Garg et al., 2010), 1–233 of which belong to the N-terminal $(\beta/\alpha)_8$ barrel, and 234–300 belong to C-terminal α-helical domain. A variant of MTB DapA, A204R, was proposed by Evans et al. (2011) with a novel view of designing inhibitors against the enzyme. The approach focused on disruption of the native tetramer structure by targeting dimer–dimer interface of DapA, suggesting the nonessentiality of tetrameric MTB DapA.

12.5.1 Structure of MTB DapA

MTB DapA is an *N*-acetylneuraminate lyase $(\beta/\alpha)_8$ (Shrivastava et al., 2016). Its structure is a homotetramer (a dimer of tightly clustered dimers) at 2.28 Å resolution as studied using X-ray diffraction. The four identical monomers show a D_2 symmetrical arrangement. Each monomer consists of $(\beta/\alpha)_8$ TIM barrel at the N-terminal that acts as the catalytic core. Apart from this, the enzyme is composed of three α-helices making the C-terminal domain. Two tetramers, however, come together to form the asymmetrical crystallographic unit (Singh et al., 2012).

Phyre2 (Kelley et al., 2015) was used for generating the three-dimensional ribbon model of MTB DapA monomer (Fig. 12.4). The input amino acid sequence was taken from the UniProt database [4-hydroxy-tetrahydrodipicolinate synthase, *dapA* gene, *M. tuberculosis* (strain ATCC 25618/ H37Rv)].

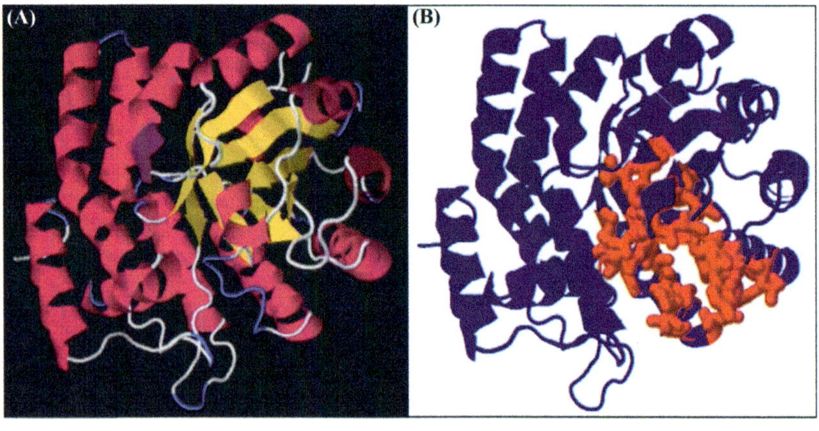

FIGURE 12.4 Phyre2 predicted three-dimensional ribbon model of MTB DapA monomer (PDB ID: 1xxx); chain A, domain (1) based on homology modeling. (A) The model was generated with 100% confidence using 99% of the input sequence (296 residues). Model dimensions (Å): **X**:51.540; **Y**:44.954; **Z**:50.722. Alpha-helices are shown colored in pink, beta-strands are colored yellow, and the coils are colored both white and blue. (B) Largest binding pocket (colored in red) shows the enzyme active site. Illustrations are depicted using JSmol interface of Phyre2.

12.5.2 Action mechanism of MTB DapA

DapA-catalyzed conversion has been widely studied and characterized in *E. coli*. It proceeds via the mechanistic ping-pong kinetics. The first step begins with pyruvate (DapA substrate) condensation with ε-amino group of Lys 161, *E. coli* DapA active site residue. ASA then undergoes hydrogen bonding to Arg 138 residue at the entrance of the active site. This reaction leads to Schiff-base (imine) formation. Further, aldol-type reaction and tautomerization of pyruvate-bound enzyme with ASA lead to the generation of an enzyme-tethered intermediate of acyclic nature that undergoes transamination to generate an unstable heterocyclic compound HTPA. Aldol condensation is carried out by a proton relay motif consisting of three residues, Tyr 107, Thr 44, and Tyr 133. HTPA is then released from the enzyme active site, succeeding protonated water molecule release and providing the product dihydrodipicolinate (Dogovski et al., 2009). MTB DapA follows a similar mechanism of action, except that the amino acid residues in function are Tyr 143, Thr 54, Arg 148, Thr 55, and Lys 171.

12.5.3 Active site of MTB DapA

The active site of MTB DapA resides at C-terminal $(\beta/\alpha)_8$ barrel domain and consists of residues from two subunits (at their interface) (Kefala et al., 2008). It comprises the amino acid residues Tyr 143, Thr 54, Arg 148, Thr 55, and Lys 171 (Singh et al., 2012). These residues are mainly conserved among DapA homologs, albeit the proton relay motif (Tyr 117, Thr 54, Tyr 143) appears disrupted (Kefala et al., 2008). Lys 171 is particularly associated with the binding of substrate and catalytic reactions. Lys 171 forms Schiff-base with pyruvate (Singh et al., 2012). The residue resides central to each monomer, within the $(\beta/\alpha)_8$ domain, facing the central tetramer cavity (Garg et al., 2010). Thr 54, Tyr 143, and Tyr 117 form a conserved catalytic triad. Two adjacent monomers share the Tyr 117 residue at their interface. Residues lining active site of the MTB enzyme, located opposite to Lys 171, contain a cystein (Cys 248) residue, and a methionine (Met 251) residue (Kefala et al., 2008).

12.5.4 Kinetic parameters of MTB DapA

Kefala et al. performed a kinetic study of MTB DapA. MTB DapA shows the following kinetic parameters: K_M (pyruvate) = 0.17 ± 0.01 mM, K_M (ASA) = 0.43 ± 0.02 mM, and V_{max} = 4.42 ± 0.08 μmol/s/mg. Values of MTB DapA are found to be six times higher than that of *E. coli* DapA (Kefala et al., 2008).

12.5.5 Regulation of MTB DapA activity

MTB DapA has been reported as insensitive to (*S*)-lysine feedback inhibition, owing to nonconservation of (S)-lysine allosteric binding site, except when used at a high concentration of 50 mM and/or above. MTB DapA is an exception to all the structurally resolved DapA/dihydrodipicolinate synthase (DHDPS) enzymes that have the nonconserved asparagine residue of (*S*)-lysine-binding site substituted by a tyrosine residue, hampering the binding of (*S*)-lysine and consequent shutdown of the feedback inhibition loop (Kefala et al., 2008).

12.5.6 Inhibitors against MTB DapA

Garg et al. (2010) identified potential antitubercular molecules targeting MTB DapA through virtual screening. Three virtual screening protocols were used. The combinatorial library consisted of pyruvate (DapA substrate) analogs. Another procedure comprised a flexible three-dimensional similarity search against PubChem and National Cancer Institute databases to obtain molecules structurally similar to pyruvate. The third approach collected 3847 antiinfective molecules from PubChem, and their filtering using Lipinski's five filters. Inhibitors exhibiting favorable interactions were selected after docking the shortlisted molecules.

A molecular docking study was performed exploiting MTB DapA as the target, and checking its interaction with the antibacterial 2-methylheptyl isonicotinate, from *Streptomyces* sp. 201. The interaction was analyzed at three DapA binding sites. Favorable interactions were observed at all the three sites under study when compared to the interactions observed with experimentally validated DapA inhibitors: piperidine-2,6-dicarboxylic acid, dimethyl-1,4-dihydro-4-oxopyridine-2,6-dicarboxylate, dimethylpiperidine-2,6-dicarboxylate, 1,4-dihydro-4-oxopyridine-2,6-dicarboxylic acid, and pyridine-2,6-dicarboxylic acid (Singh et al., 2012).

An 88% inhibition of MTB DapA activity was observed with α-ketopimelic acid. Besides, an IC_{50} value of 21 μM was recorded when α-ketopimelic acid was used along with 400 μM ASA and 500 μM pyruvate. Its structural analog α-ketoadipic acid also showed potential as MTB DapA inhibitor, owing to an inhibition percentage of 40% (Shrivastava et al., 2016).

12.6 Previous experiments targeting MTB Dap pathway enzymes

Transposon mutagenesis in MTB established essentiality of all the DAP pathway enzymes [Ask (aspartokinase), Asd (aspartate-semialdehyde dehydrogenase), DapA, DapD, DapC, DapE, and DapF], for mycobacterial growth (Shrivastava et al., 2016), excluding only DapB. MTB *dapB* was cloned for the purpose of determining cofactor selectivity information of the enzyme DapB (Pavelka et al., 1997). Cloning, expression, and characterization of MTB Asd was done in *E. coli*. The enzyme exhibited two domains, dimerization and N-terminal NADP binding domain (Shafiani et al., 2005).

MTB Asd contains Cys 130 active site within its catalytic residue. The residue shows conservation in all bacteria. It usually forms an acyl intermediate while generating *L*-β-aspartate semialdehyde from β-aspartyl phosphate. Other catalytic residues comprise His 256 and Gln 157. Homology modeling studies propose His 256 and Cys 130 as the key residues for developing selective inhibitors against MTB Asd (Singh et al., 2008). Ajijur et al. (2021) recently predicted AP0600 and AP0639 as novel lead molecules against MTB aspartate β-semialdehyde dehydrogenase through a combinatorial approach. The DAP pathway substrate β-aspartyl phosphate was used as the template for designing novel lead molecules, facilitating understanding of MTB aspartate β-semialdehyde dehydrogenase inhibition mechanisms. The enzyme is a validated drug target (Shafiani et al., 2005).

M. smegmatis is a fast growing mycobacterial species commonly used as a model organism. The aspartate amino acid pathway was shown to be essential for *M. smegmatis* survival. Mutation in aspartokinase, the enzyme catalyzing the first step for conversion of *L*-aspartate to *L*-4-aspartyl β-phosphate, yielded nonviable *M. smegmatis* (Pavelka and Jacobs, 1996). The first auxotrophic mutant of DAP pathway, generated in *M. smegmatis* by disrupting the *ask* gene, resulted in immediate lysis upon deprivation of *meso*-DAP. Complementation, in this case, was done with a functional *ask* gene (wild-type) (Pavelka and Jacobs, 1996; Pavelka et al., 1997).

MTB DapB was observed as a homotetramer of 100 kDa (Kefala et al., 2005). The enzyme shows the presence of ternary structures complexed with NADH/NADPH and the inhibitor pyridine-2,6-dicarboxylic acid (Cirilli et al., 2003). MTB DapB can use either NADPH or NADH as a cofactor with equal efficiency (Janowski et al., 2010; Sagong and Kim, 2016). N-terminal of DapB consists of a catalytic Rossmann-like fold and a tetramerization domain at the C-terminal, facilitating interdomain flexibility (Kefala et al., 2005). MTB DapB exhibited a much larger binding pocket for substrates or inhibitors than its *E. coli* counterpart, owing to the differences in pocket shape at β8 N-terminal end (β9 in *E. coli*) (Cirilli et al., 2003).

MTB DapD structure showed its existence as a homotrimer, each monomer of which consisted of three domains—α/β globular domain (at the N-terminal), a parallel β-helix (left-handed), and a much smaller domain at the C-terminal (Dogovski et al., 2012). Gly 222 and Glu 199 residues of the enzyme were shown to be important for the catalytic function. Mn^{2+}, Ca^{2+}, and Mg^{2+} were suggested as the key activators of the enzyme, whereas Zn^{2+} and Co^{2+} as its inhibitors (Schuldt et al., 2009).

MTB DapC was resolved to a three-dimensional structure of 2.0 Å by X-ray diffraction analysis. It displayed a characteristic S-shape, known to be exhibited by pyridoxal-5′-phosphate (PLP)-binding proteins (Class I) (Weyand et al., 2007). The lethality of *M. smegmatis* DapE deletion mutants confirmed the essentiality of DapE for the growth and proliferation of the bacteria (Pavelka and Jacobs, 1996). Usha et al., 2016 cloned and purified

MTB DapE as a fusion protein having N-terminal hexahistidine containing a single zinc ion per monomer of DapE. DAP pathway was accordingly redesigned, and MTB DapE characterization was done with *N*-succinyl-*LL*-2,6-diaminopimelate and its derivatives (as substrates). It showed insensitivity towards *L*-captopril, the available DapE inhibitor.

X-ray diffraction analysis resolved the MTB DapF unliganded crystal structure to 2.6 Å by X-ray diffraction analysis. The enzyme shows the presence of two pseudo-symmetrical α/β domains. MTB DapF backbone was found to be stabilized by a mycobacterial DAP epimerase-specific tyrosine residue. Subtle changes in interactions facilitated by the residue were hypothesized to destabilize DapF, thereby suggesting a clear rationale for designing MTB specific DapF inhibitors (Usha et al., 2009).

12.7 Significance of inhibitors against MTB Dap pathway enzymes

Metabolic pathways are generally targeted for rational drug design since each step is a validated progression assisting bacterial survival (Bushra and Adem, 2016). DAP biosynthesis occurs specifically in plants and bacteria, altogether being absent in humans. Humans neither produce nor require DAP (Shafiani et al., 2005), suggesting that inhibitors of DAP pathway might provide novel antimicrobial compounds exhibiting minimum mammalian toxicity (Dogovski et al., 2009). The pathway leads to the synthesis of *L*-lysine precursor *meso*-DAP (*D,L*-DAP) (Usha et al., 2016). Lysine and *meso*-DAP are both essential for bacterial viability. Lysine facilitates protein synthesis, whereas *meso*-DAP (a well-recognized virulence factor) establishes peptidoglycan crosslinks (Shrivastava et al., 2016). Both components provide rigidity and strength to the bacterial cell wall (Singh et al., 2012). All enzymes of the DAP pathway have been exploited as potential drug targets (Maitra et al., 2019). De novo lysine biosynthesis facilitates MTB survival during infection (Ajijur et al., 2021). Mycobacteria has DAP incorporated into peptides forming peptidoglycan layer of the cell wall (Pavelka et al., 1997). DAP-negative bacteria are prone to immediate lysis, owing to cell wall fragility (Shafiani et al., 2005). Enzymes involved in the biosynthesis of *meso*-DAP (covering both protein and peptidoglycan synthesis) can hence be attractively exploited as potent antimicrobial targets (Usha et al., 2016). Moreover, since DapA and DapB catalyze the steps prior to the junction initiating other subpathways (dehydrogenase and acetylase) generating *meso*-DAP, they exhibit significant possibilities of being exploited to develop broad-spectrum antimicrobials (Impey and Soares da Costa, 2018).

12.8 Concluding remarks

TB drug discovery requires increased efforts to develop novel and better therapeutics. The present TB drug pipeline offers numerous targets, continuously being exploited for drug-inhibitor studies to achieve higher cure rates. Despite the growing research, TB eradication is still a worldwide challenge (Chauhan et al., 2014). Studies leading to the identification of novel drug targets are hence warranted. Genes essential for mycobacterial replication, pathogenesis, and survival during infection need to be exploited to a greater extent. Since the mycobacterial cell wall is an essential component for MTB survival, pathogenesis, and virulence, it offers numerous targets for drug discovery. The present chapter suggests an immense possibility of designing new inhibitors with potential as antimycobacterial compounds targeting DAP pathway enzymes. In vivo studies, including generation and study of knockout for DAP pathway enzymes, are required for further validation of the targets, which would undoubtedly open new prospects for the development of antitubercular therapeutics, aiding in holistically managing the MTB consequences.

Acknowledgment

We would like to acknowledge Jaypee University of Information Technology, Waknaghat, Solan (Himachal Pradesh, India) administration for providing infrastructure and research facilities to the authors.

Abbreviations

ASA	*L*-aspartate-*β*-semialdehyde
DAP	diaminopimelate pathway
DHDPS	dihydrodipicolinate synthase
HIV	human immunodeficiency virus
HTPA	4-Hydroxy-tetrahydrodipicolinate synthase
***L*-THDP**	*L*-2,3,4,5-Tetrahydrodipicolinic acid
***m*-DAP**	*meso*-diaminopimelate
MDR	multidrug resistant
MTB	*Mycobacterium tuberculosis*
NADPH	nicotinamide adenine dinucleotide phosphate
TB	tuberculosis
TLM	thiolactomycin
WHO	World Health Organization

References

Abrahams, K.A., Besra, G.S., 2018. Mycobacterial cell wall biosynthesis: a multifaceted antibiotic target. Parasitology. 145 (2), 116–133.

Abrahams, K.A., Besra, G.S., 2020. Mycobacterial drug discovery. RSC Med. Chem. 11 (12), 1354–1365.

Ajijur, R., Salman, A., Ahmad, K.M., 2021. Combinatorial design to decipher novel lead molecule against *Mycobacterium tuberculosis*. Biointerface Res. Appl. Chem. 11 (5), 12993–13004.

Alderwick, L.J., Harrison, J., Lloyd, G.S., Birch, H.L., 2015. The mycobacterial cell wall-peptidoglycan and arabinogalactan. Cold Spring Harb. Perspect. Med. 5 (8), a021113.

Amir, A., Rana, K., Arya, A., Kapoor, N., Kumar, H., Siddiqui, M.A., 2014. *Mycobacterium tuberculosis* H37Rv: *in silico* drug targets identification by metabolic pathways analysis. Int. J. Evolut. Biol. 2014, 284170.

Bhat, Z.S., Rather, M.A., Maqbool, M., Lah, H.U., Yousuf, S.K., Ahmad, Z., 2017. Cell wall: a versatile fountain of drug targets in *Mycobacterium tuberculosis*. Biomed. Pharmacother. 95, 1520–1534.

Burley, K.H., Cuthbert, B.J., Basu, P., Newcombe, J., Irimpan, E.M., Quechol, R., et al., 2020. Structural and molecular dynamics of *Mycobacterium tuberculosis* malic enzyme, a potential anti-TB drug target. ACS Infect. Dis. 7 (1), 174–188.

Bushra, E., Adem, J., 2016. Mycobacterial metabolic pathways as drug targets: a review. Int. J. Microbiology Res. 7 (3), 74–87.

Catalão, M.J., Filipe, S.R., Pimentel, M., 2019. Revisiting anti-tuberculosis therapeutic strategies that target the peptidoglycan structure and synthesis. Front. Microbiol. 10, 190.

Chauhan, R.S., Chanumolu, S.K., Rout, C., Shrivastava, R., 2014. Can mycobacterial genomics generate novel targets as speed-breakers against the race for drug resistance. Curr. Pharm. Des. 20 (27), 4319–4345.

Cirilli, M., Zheng, R., Scapin, G., Blanchard, J.S., 2003. The three-dimensional structures of the *Mycobacterium tuberculosis* dihydrodipicolinate reductase – NADH – 2,6-PDC and – NADPH – 2,6-PDC complexes. Structural and mutagenic analysis of relaxed nucleotide specificity. Biochemistry. 42 (36), 10644–10650.

Dogovski, C., Atkinson, S.C., Dommaraju, S.R., Downton, M., Hor, L., Moore, S., et al., 2012. Enzymology of bacterial lysine biosynthesis. Biochemistry IntechOpen..

Dogovski, C., Atkinson, S.C., Dommaraju, S.R., Hor, L., Dobson, R.C., Hutton, C.A., et al., 2009. Lysine biosynthesis in bacteria: an unchartered pathway for novel antibiotic design. Encycl. Life Support. Syst. 11, 116–136.

Evans, G., Schuldt, L., Griffin, M.D., Devenish, S.R., Pearce, F.G., Perugini, M.A., et al., 2011. A tetrameric structure is not essential for activity in dihydrodipicolinate synthase (DHDPS) from *Mycobacterium tuberculosis*. Arch. Biochem. Biophysics 512 (2), 154–159.

Garg, A., Tewari, R., Raghava, G.P., 2010. Virtual screening of potential drug-like inhibitors against Lysine/DAP pathway of *Mycobacterium tuberculosis*. BMC Bioinforma. 11 (1), 1–9.

Gokulan, K., Rupp, B., Pavelka Jr, M.S., Jacobs Jr, W.R., Sacchettini, J.C., 2003. Crystal structure of *Mycobacterium tuberculosis* diaminopimelate decarboxylase, an essential enzyme in bacterial lysine biosynthesis. J. Biol. Chem. 278 (20), 18588–18596.

Impey, R.E., Soares da Costa, T.P., 2018. Review: targeting the biosynthesis and incorporation of amino acids into peptidoglycan as an antibiotic approach against gram negative bacteria. EC Microbiology. 14, 200–209.

Janowski, R., Kefala, G., Weiss, M.S., 2010. The structure of dihydrodipicolinate reductase (DapB) from *Mycobacterium tuberculosis* in three crystal forms. Acta Crystallogr. Sect. D: Biol. Crystallography 66 (1), 61–72.

Kefala, G., Evans, G.L., Griffin, M.D., Devenish, S.R., Pearce, F.G., Perugini, M.A., et al., 2008. Crystal structure and kinetic study of dihydrodipicolinate synthase from *Mycobacterium tuberculosis*. Biochemical J. 411 (2), 351–360.

Kefala, G., Janowski, R., Panjikar, S., Mueller-Dieckmann, C., Weiss, M.S., 2005. Cloning, expression, purification, crystallization and preliminary X-ray diffraction analysis of DapB (Rv2773c) from *Mycobacterium tuberculosis*. Acta Crystallogr. Sect. F: Struct. Biol. Crystallization Commun. 61 (7), 718–721.

Kefala, G., Weiss, M.S., 2006. Cloning, expression, purification, crystallization and preliminary X-ray diffraction analysis of DapA (Rv2753c) from *Mycobacterium tuberculosis*. Acta Crystallogr. Sect. F: Struct. Biol. Crystallization Commun. 62 (11), 1116–1119.

Kelley, L.A., Mezulis, S., Yates, C.M., Wass, M.N., Sternberg, M.J., 2015. The Phyre2 web portal for protein modeling, prediction and analysis. Nat. Protoc. 10 (6), 845–858.

Maitra, A., Munshi, T., Healy, J., Martin, L.T., Vollmer, W., Keep, N.H., et al., 2019. Cell wall peptidoglycan in *Mycobacterium tuberculosis*: an Achilles' heel for the TB-causing pathogen. FEMS Microbiology Rev. 43 (5), 548–575.

Mdluli, K., Spigelman, M., 2006. Novel targets for tuberculosis drug discovery. Curr. Opin. Pharmacology 6 (5), 459–467.

Moraes, G.L., Gomes, G.C., De Sousa, P.R., Alves, C.N., Govender, T., Kruger, H.G., et al., 2015. Structural and functional features of enzymes of *Mycobacterium tuberculosis* peptidoglycan biosynthesis as targets for drug development. Tuberculosis 95 (2), 95–111.

Munshi, T., Gupta, A., Evangelopoulos, D., Guzman, J.D., Gibbons, S., Keep, N.H., et al., 2013. Characterisation of ATP-dependent Mur ligases involved in the biogenesis of cell wall peptidoglycan in *Mycobacterium tuberculosis*. PLoS One 8 (3), e60143.

Nieto, L.M., Mehaffy, C., Dobos, K.M., 2017. The physiology of *Mycobacterium tuberculosis* in the context of drug resistance: a system biology perspective. Mycobacterium-Research Development. IntechOpen.

Pavelka Jr, M.S., Jacobs Jr, W.R., 1996. Biosynthesis of diaminopimelate, the precursor of lysine and a component of peptidoglycan, is an essential function of *Mycobacterium smegmatis*. J. Bacteriol. 178 (22), 6496.

Pavelka Jr, M.S., Weisbrod, T.R., Jacobs Jr, W.R., 1997. Cloning of the dapB gene, encoding dihydrodipicolinate reductase, from *Mycobacterium tuberculosis*. J. Bacteriol. 179 (8), 2777.

Sagong, H.Y., Kim, K.J., 2016. Structural insight into dihydrodipicolinate reductase from *Corynebacterium glutamicum* for lysine biosynthesis. J. Microbiol. Biotechnol. 26 (2), 226–232.

Schuldt, L., Weyand, S., Kefala, G., Weiss, M.S., 2009. The three-dimensional structure of a mycobacterial DapD provides insights into DapD diversity and reveals unexpected particulars about the enzymatic mechanism. J. Mol. Biol. 389 (5), 863–879.

Shafiani, S., Sharma, P., Vohra, R.M., Tewari, R., 2005. Cloning and characterization of aspartate-β-semialdehyde dehydrogenase from *Mycobacterium tuberculosis* H37Rv. J. Appl. Microbiol. 98 (4), 832–838.

Shaku, M., Ealand, C., Kana, B.D., 2020. Cell surface biosynthesis and remodeling pathways in mycobacteria reveal new drug targets. Front. Cell. Infect. Microbiol. 10, 700.

Shetye, G.S., Franzblau, S.G., Cho, S., 2020. New tuberculosis drug targets, their inhibitors, and potential therapeutic impact. Transl. Res. 220, 68–97.

Shrivastava, P., Navratna, V., Silla, Y., Dewangan, R.P., Pramanik, A., Chaudhary, S., et al., 2016. Inhibition of *Mycobacterium tuberculosis* dihydrodipicolinate synthase by alpha-ketopimelic acid and its other structural analogues. Sci. Rep. 6 (1), 1–7.

Singh, S.P., Bora, T.C., Bezbaruah, R.L., 2012. Molecular interaction of novel compound 2-methylheptyl isonicotinate produced by *Streptomyces* sp. 201 with dihydrodipicolinate synthase (DHDPS) enzyme of *Mycobacterium tuberculosis* for its antibacterial activity. Indian. J. Microbiol. 52 (3), 427–432.

Singh, R., Dwivedi, S.P., Gaharwar, U.S., Meena, R., Rajamani, P., Prasad, T., 2020. Recent updates on drug resistance in *Mycobacterium tuberculosis*. J. Appl. Microbiol. 128 (6), 1547–1567.

Singh, A., Kushwaha, H.R., Sharma, P., 2008. Molecular modelling and comparative structural account of aspartyl β-semialdehyde dehydrogenase of *Mycobacterium tuberculosis* (H37Rv). J. Mol. Modeling 14 (4), 249–263.

Usha, V., Dover, L.G., Roper, D.I., Fuetterer, K., Besra, G.S., 2009. Structure of the diaminopimelate epimerase DapF from *Mycobacterium tuberculosis*. Acta Crystallogr. Sect. D: Biol. Crystallography 65 (4), 383–387.

Usha, V., Lloyd, A.J., Lovering, A.L., Besra, G.S., 2012. Structure and function of *Mycobacterium tuberculosis* meso-diaminopimelic acid (DAP) biosynthetic enzymes. FEMS Microbiol. Lett. 330 (1), 10–16.

Usha, V., Lloyd, A.J., Roper, D.I., Dowson, C.G., Kozlov, G., Gehring, K., et al., 2016. Reconstruction of diaminopimelic acid biosynthesis allows characterisation of *Mycobacterium tuberculosis* N-succinyl-L, L-diaminopimelic acid desuccinylase. Sci. Rep. 6 (1), 1–10.

Weyand, S., Kefala, G., Weiss, M.S., 2007. The three-dimensional structure of N-succinyldiaminopimelate aminotransferase from *Mycobacterium tuberculosis*. J. Mol. Biol. 367 (3), 825–838.

World Health Organization, 2020. Global tuberculosis report 2020. Geneva, Switzerland: World Health Organization. https://www.who.int/news-room/fact-sheets/detail/tuberculosis.

Chapter 13

Lipase-catalyzed organic transformations: a recent update ☆

Goutam Brahmachari

Laboratory of Natural Products & Organic Synthesis, Department of Chemistry, Visva-Bharati (a Central University), Santiniketan, West Bengal, India

13.1 Introduction

The development of new synthetic and catalytic methods to access new classes of chemical entities and their various analogues, emerging as the key target of molecular research in the drug discovery process, is of prime interest and importance to organic and bioorganic chemists. With the advent of green and sustainable chemistry, chemists are now also to take care of the rising constraints imposed by environmental concerns while incorporating new catalysts in designing new synthetic protocols (Brahmachari, 2015, 2021). This is the fact that many of the new reagents and catalysts that have benefited organic synthesis in the last several years contain transitional metals or heavy elements. In cases where they can be environmentally acceptable, their uses sometimes impose several disadvantages concerning handling and disposal issues. Replacement of such metals and their slats with catalysts with better environmental acceptability would almost always be an advantage. The design of synthetic processes may also suffer from several constraints arising from volatile and toxic organic solvents. Thus water as a solvent for reactions has already become an attractive and encouraging option. In addition, chemical syntheses with classical catalysts frequently result in undesirable mixtures of racemic components. However, optically pure isomers are particularly important in synthesizing biologically active compounds and other useful purposes.

Chemists are dedicated to searching for novel alternatives to overcome all these constraints. It has already been established that enzymes possess

☆ This chapter is dedicated to Professor Maya Shankar Singh on the occasion of his 63rd birthday.

Biotechnology of Microbial Enzymes. DOI: https://doi.org/10.1016/B978-0-443-19059-9.00005-0
© 2023 Elsevier Inc. All rights reserved.

enormous possibility and potentiality in emerging as the most suited catalysts (*called biocatalysts*) to solve many aspects concerned (Halgas, 1992; Tawaki and Klibanov, 1992; Koskinen and Klibanov, 1996; Roberts, 1998, 1999, 2000; Gotor, 1999; Davis and Boyer, 2001; Klibanov, 2001; Liese et al., 2006; Vasic-Racki, 2006; Gröger, 2010; Paravidino et al., 2010; Kamerlin and Warshel, 2010; Faber, 2011; Drauz et al., 2012; Ranoux and Hanefeld, 2013, Kumar et al., 2016). Enzymes are, in general, natural proteins that catalyze chemical reactions. These are intrinsically eco-friendly materials that operate best in water and organic solvents or under solvent-free conditions (Martinek et al., 1986; Gupta, 1992; Vulfson et al., 2001; Gupta and Roy, 2004, Sheldon and Woodley, 2018; Tan and Dou, 2020). The fact that many enzymes are reported to possess activity against nonnatural substrates in organic media has invoked interest in their potential use to carry out synthetic transformations. The use of enzymes in fine organic synthesis is most feasible when the molecules subjected to chemical rearrangement are rather complex and contain chemically similar bonds, only one (or several) of which is to be modified without the need for protection of any functional group. That is why enzymes are indispensable for the synthesis of the derivatives of natural products with complex structures (Kobayashi et al., 2003; Träff et al., 2008; Choi et al., 2009; Leijondahl et al., 2009; Krumlinde et al., 2009, 2010; Han et al., 2010a, b; Hoye et al., 2010; Andrushko and Andrushko, 2013; Wever et al., 2015). In a true sense, enzymes nowadays offer a vital part of the spectrum of catalysts available to synthetic chemistry.

Enzyme catalysis is characterized by two main factors: specificity and rate acceleration. Chemoenzymatic syntheses provide enantioselective products in most cases, which are much more demanding. The active sites of enzymes are chiral and contain moieties, namely, amino acid residues and, in the case of some enzymes, *cofactors*, that are responsible for these properties of an enzyme (Rahman and Shah, 1993; O'Brien and Herschlag, 1999; Tsai and Huang, 1999; Arora et al., 2014a). Enzyme catalysis offers several benefits at the same time, such as wide applicability, mild reaction conditions required for complex and chemically unstable molecules, low catalyst loading, good and effective reusability of biocatalyst, desired biodegradability of enzyme (catalyst) to promote green chemistry, its safe and eco-friendly nature, the possibility to reduce or to eliminate reaction by-products, and carrying out a conventional multistage reaction via a single-stage process without going for protection/deprotection steps.

In the recent past, lipase (triacylglycerol acyl hydrolase EC 3.1.1.3) has emerged as one of the most promising enzymes for broad practical application in organic synthesis (Theil, 1995; Itoh et al., 1997; Andersch et al., 1997; Schmid and Verger, 1998; Davis and Boyer, 2001; Fan et al., 2012; Adlercreutz, 2013; Stergiou et al., 2013). Unlike many other enzymes, it has extremely broad substrate specificity. It can use a wide range of structurally diverse nonnatural compounds as substrates in research laboratories and industry. So, lipases have been extensively utilized to synthesize many biologically active compounds and natural products. In general, lipases are the most frequently used biocatalysts in organic synthesis. There are various natural sources of lipases, such as plants, animals, and microbes (mainly bacteria and fungi). The lipases available from various sources have considerable variation in their reaction specificities (referred to as enzyme specificity). There are several good reviews and book chapters on their sources, enzyme specificity, biotechnological developments, immobilization of lipases, and reusability (Copping et al., 1990; Tramper et al., 1992; Cedrone et al., 2000; Sharma et al., 2001; Patel, 2000, 2007; Raillard et al., 2001; Schmid et al., 2001; Bolon et al., 2002; Krishna, 2002; Saxena et al., 2003; Hasan et al., 2006; Ran et al., 2008; Zhao, 2010; Faber, 2011; Sanchez and Demain, 2011; Bornscheuer et al., 2012; Sharma and Kanwar, 2012; Adrio and Demain, 2014); hence, these are not the subject matters of this present article. As mentioned, during the last decade, lipases have been finding enormous applications in organic synthesis, and the experimental outcomes have also been summarized on a regular basis in literature (Saxena et al., 1999; Davis and Boyer, 2001; Hassan et al., 2013; Adlercreutz, 2013, Kumar et al., 2016; Sheldon and Woodley, 2018; Tan and Dou, 2020). This chapter is an updated version of the earlier edition highlighting the lipase-catalyzed organic transformations of interest reported from 2013 to 2021.

13.2 Chemoenzymatic applications of lipases in organic transformations: a recent update

As discussed in the earlier section that lipases find enormous biocatalytic applications in organic synthesis. In the current scenario of chemoenzymatic organic reactions, lipases stand among the most important biocatalysts carrying out such novel organic transformations with efficiency under eco-friendly conditions. This section updates the useful catalytic applications of lipases in synthetic organic chemistry reported from 2013 to 2021.

Among various types of lipases, lipase B from *Candida antarctica* (CAL-B) is regarded as a promising biocatalyst by organic chemists. Branneby et al. (2003) demonstrated the first example of carbon−carbon bond formation by a CAL-B-catalyzed aldol addition of hexanal. They observed that the mutant Ser l05Ala catalyzed this reaction faster than the wild-type enzyme. Since then, various groups of researchers have reported on the direct application of this hydrolytic enzyme in the formation of Michael-type adducts (Torre et al., 2004; Carlqvist et al., 2005; Svedendahl et al., 2005; Lou et al., 2008; Strohmeier et al., 2009). CAL-B belongs to the folding family of α/β hydrolases, and the three amino acid residues, the so-called catalytic triad 187Asp−224 His−105 Ser, are supposed to play a key role in the catalytic process, including normal and promiscuous reactions (Bornscheuer and Kazlauskas, 2004; Busto et al., 2010; Wu et al., 2010; Humble and

Berglund, 2011). From these reports, it can be suggested that the carbonyl function of a substrate is activated through an oxyanion hole, and proton transfer gets facilitated through the basic nitrogen of the Asp-His pair during the catalytic process. However, such successful examples of CAL-B-catalyzed C–C bond formations via Michael or Michael-type additions were mainly restricted to the reactions between α,β-unsaturated carbonyl compounds and activated carbon nucleophiles, such as acetylacetone, acetoacetates, and methyl nitroacetate. Less-activated carbon nucleophiles such as cyclohexanone were found to be challenging to add to the Michael acceptor. Chen et al. (2013) were successful in carrying out such a transformation between a series of aromatic and heteroaromatic nitroolefins (**1**) as the Michael acceptors and less-activated cyclic and acyclic ketones (**2**) as the Michael donors in the presence of (CAL-B)/acetamide as a cocatalytic system to obtain the corresponding Michael adducts **3** (Scheme 13.1). Interestingly, it was also revealed that neither CAL-B nor acetamide could independently catalyze the reaction to any appreciable extent. The experimental outcomes confirmed the involvement of hydrogen bonds between acetamide and oxo functionalities for the observed activation, and a new input was offered on the mechanistic insights into CAL-B/acetamide cocatalysis (Chen et al., 2013).

In addition to the esterification and transesterification reactions, lipases are reported to catalyze amidation reactions via ammonolysis and aminolysis (Zaks and Klibanov, 1985; Adamczyk and Grote, 1996, 1997, 1999). Gotor-Fernández et al. (2006) published an informative review on CAL-B-catalyzed ammonolysis and aminolysis. In recent times, aminolyses of carboxylic esters with propargyl amine containing an alkynyl moiety that can be functionalized judiciously were explored for the first time by Hassan et al. (2013). Propargyl amides are particularly interesting as biologically active functionalities and as synthetic building blocks (Garg et al., 2005; Merkul and Müller, 2006; Bonger et al., 2008; Sanda et al., 2008; Merkul et al., 2009; Grotkopp et al., 2011; Gruit et al., 2011; Arkona and Rademann, 2013, Borowiecki and Dranka, 2019). The investigators demonstrated that *C. antarctica* lipase B (CAL-B) immobilized on an acrylic resin (Novozyme 435) smoothly catalyzes the aminolysis of methyl esters (**4**) with

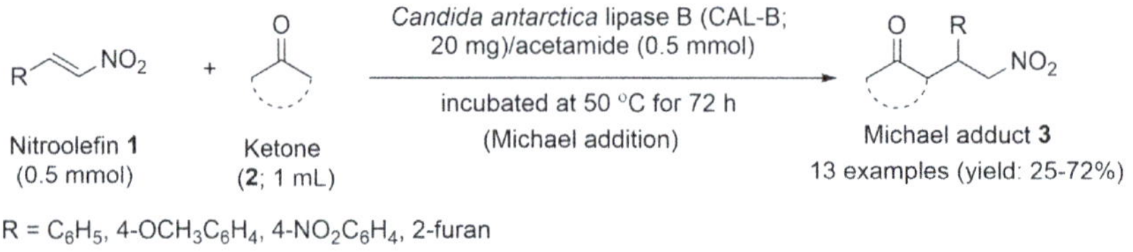

SCHEME 13.1 Michael addition between nitroolefins and less-activated ketones under cocatalytic system.

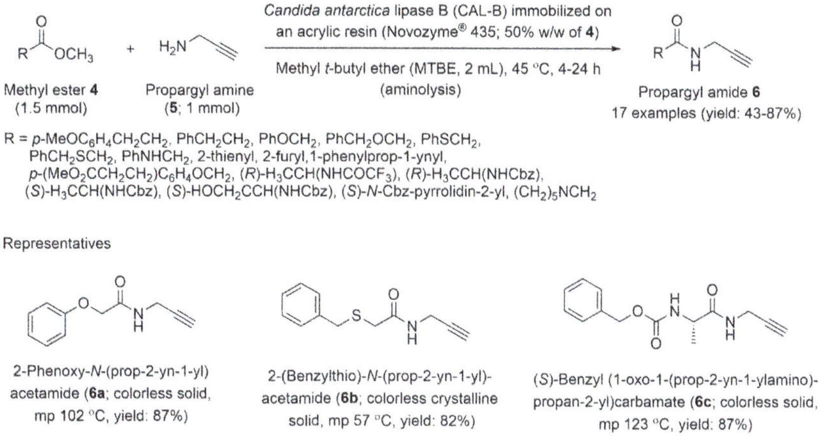

SCHEME 13.2 Aminolysis of methyl esters with propargyl amine.

propargyl amine (**5**), affording a series of diverse propargyl amides (**6**) with moderate to good yields (Scheme 13.2). Additionally, the investigators utilized these propargyl derivatives **6** in the synthesis of amide ligated 1,2,3-triazoles in a Cu(I)-catalyzed azide-alkyne cycloaddition (CuAAC) click reaction in good to excellent yields in the same reaction vessel (Hassan et al., 2013).

The same group of investigators (Hassan et al., 2015) further extended their works in this direction. They disclosed for the first time consecutive three-component syntheses of (hetero)arylated propargyl amides **10** by CAL-B (Novozyme 435)-assisted aminolysis–Sonogashira coupling sequence from the reaction of substituted esters (**7**), propargyl amine (**4**) and alkyl/aryl/heteroaryl iodides (**9**) (Scheme 13.3). The propargyl amides (**8**) formed in the first step upon CAL-B-catalyzed aminolysis of ester **7**, subsequently underwent Sonogashira coupling to finally afford the desired product of (hetero)arylated propargyl amides (**10**) in a one-pot. This combination of enzyme-metal catalyzed methodology may find useful application to more sophisticated peptides and aryl halides as a bioorganic tool for the efficient generation of peptidomimetics in a one-pot fashion.

Amidation of organic acids is an important organic transformation, and substituted amides of phenolic and cinnamic acids are known to exhibit a wide range of biological activities, including antioxidant (Moon and Terao, 1998; Pérez-Alvarez et al., 2001; Spasova et al., 2007), antiinflammatory (Sudina et al., 1993), antimutagenic (Namiki, 1990), and antihyperlipidimic (Lee et al., 2007; Stankova et al., 2009) activities. These types of scaffolds are also reported to possess antibacterial, antifungal, antiviral, insecticidal, nematicidal, and herbicidal properties (Torres et al., 2004; Christodoulopoulou et al., 2005; Lamberth et al., 2006; Cohen et al., 2008; Hu et al., 2008;

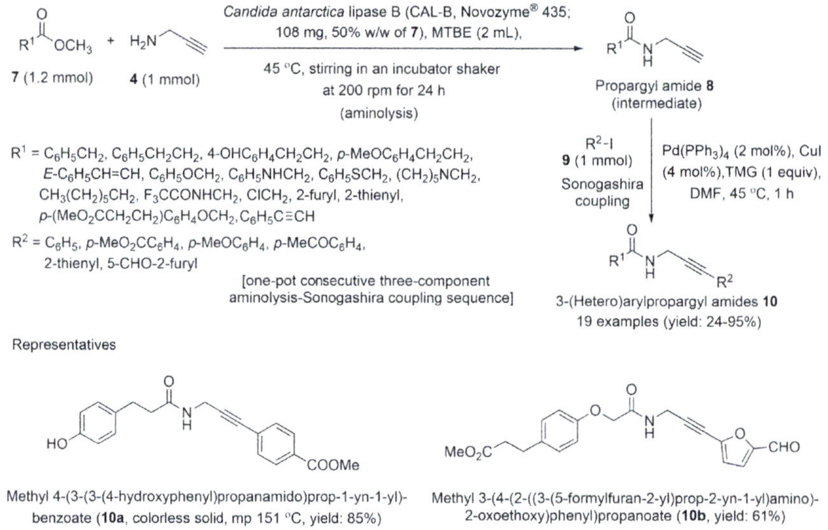

SCHEME 13.3 Chemoenzymatic regioselective acetylation of one of the two diastereotopic hydroxymethyl functions in 3-*O*-benzyl-4-*C*-hydroxymethyl-1,2-*O*-isopropylidene-α-D-ribofuranose.

Debonsi et al., 2009; Vishnoi et al., 2009; Bose et al., 2010). That is why there are so many methods available in the literature for their synthesis. With the advent of enzyme-catalyzed organic reactions that have provided a great impetus to organic synthesis during the past two decades, Kaushik et al. (2015) have recently envisioned the use of *Candida anatrctica* lipase B (CAL-B) to carry out biocatalytic one-pot amidation of a variety of phenolic/cinnamic acids (**11/12**) with alkyl amines (**13**) in bulk at 60°C−90°C under solvent-free conditions in vacuum (Scheme 13.4). A series of *N*-alkyl-substituted amides (**14/15**) were synthesized with good yields (75.6%−83.5%). The present enzymatic procedure offers some useful advantages over conventional catalysts, which demand large-scale applications in industrial processes.

An elegant example of the chemoenzymatic application of lipase for regioselective acetylation of diol sugars on a multiple-gram scale was demonstrated by Sharma et al. (2014a,b) (Scheme 13.5). Novozyme 435 (*C. antarctica* lipase B) was found to catalyze regioselective acetylation of the two diastereotopic hydroxymethyl functions in 3-*O*-benzyl-4-*C*-hydroxymethyl-1,2-*O*-isopropylidene-α-D-ribofuranose (**16**) with vinyl acetate in diisopropyl ether (DIPE) at 45°C (incubator temperature) in an efficient manner to produce the monoacetate derivative, 5-*O*-acetyl-3-*O*-benzyl-4-*C*-hydroxymethyl-1,2-*O*-isopropylidene-α-D-ribofuranose (**17**) in quantitative yield. The investigators also exploited this enzymatic methodology successfully for the first time in the convergent synthesis of bicyclic nucleosides (LNA monomers); T, U, A, and C with a relatively shorter route and with significant

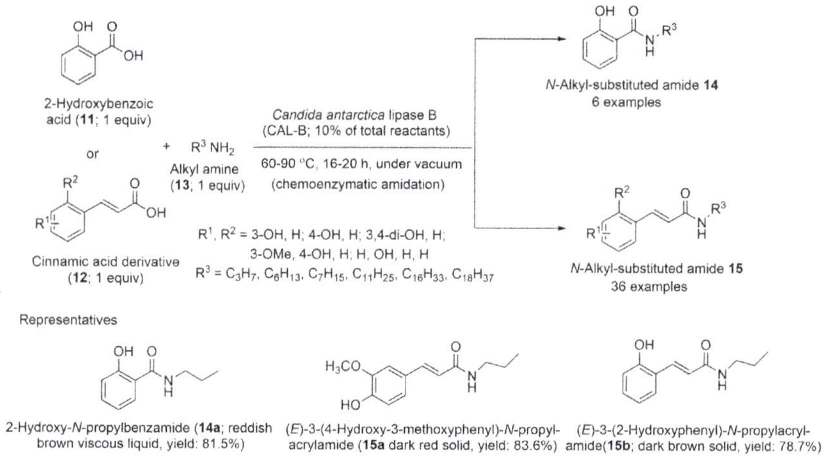

SCHEME 13.4 Consecutive three-component synthesis of (hetero)arylated propargyl amides by chemoenzymatic aminolysis–Sonogashira coupling sequence.

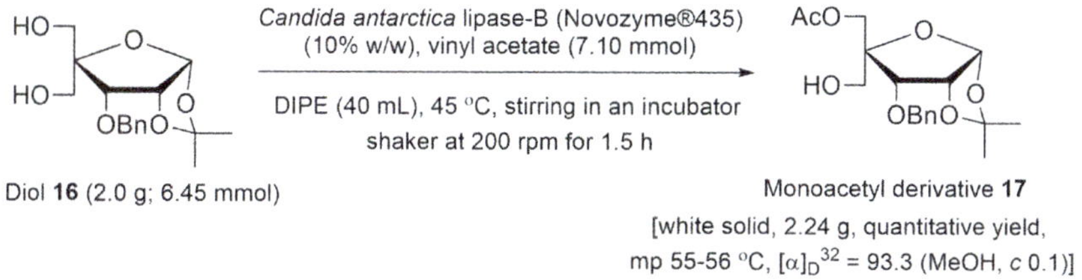

SCHEME 13.5 Solvent-free amidation of phenolic acids.

improvement in overall yields. They screened lipases from different sources, such as *C. antarctica* lipase B (Novozyme 435; CAL-B), *Thermomyces lanuginosus* lipase immobilized on silica (Lipozyme TL IM), *Candida rugosa* lipase (CRL), and porcine pancreatic lipase (PPL) in five sets of organic solvents (tetrahydrofuran, acetonitrile, toluene, diisopropyl ether, acetone) using both acetic anhydride and vinyl acetate as acetyl donor at 45°C and 200 rpm in an incubator shaker for the study of selective acetylation of one over the other primary hydroxyl function in diol **16**. Among them, Novozyme 435 (CAL-B) came out as the best one to effect the transformation in diisopropyl ether (DIPE) with vinyl acetate as the acetyl donor (Scheme 13.3). They have also demonstrated that the biocatalyst can be used for ten cycles of the acylation reaction without losing selectivity and efficiency. The newly developed methodology demands useful applications for the commercial synthesis of LNA monomers of current pharmaceutical promise (Koshkin et al., 1998; Wengel, 1999; Hildebrandt-Eriksen et al., 2012; Watts, 2013; Gebert et al., 2014; Sharma et al., 2014a,b).

In another report, Hoang and Matsuda (2015) successfully utilized immobilized *C. antarctica* lipase B (Novozyme 435) as a useful biocatalyst for transesterification of *rac*-alcohols in liquid carbon dioxide medium in a batch reactor (Scheme 13.6). The investigators also successfully performed a large-scale kinetic-resolution of secondary alcohols by immobilizing lipase with a continuous packed-column reactor that afforded corresponding enantiopure products, thereby minimizing waste.

In the previous year, Janssen et al. (2014) also demonstrated the preparation of octyl formate (**23**) *via* immobilized lipase-catalyzed transesterification of ethyl formate (**21**) with 1-octanol (**22**) (Scheme 13.7). Although some formyl esters are valuable compounds, *e.g.*, ingredients in flavors and fragrances (Herrmann, 2007), the formation of formic acid esters, despite several promising early contributions (Bevinakatti and Newadkar, 1989), has so far failed to be the focus of research. The present method offers simple access to a hydrophobic formic acid ester, which can be used as a reactive organic phase in biocatalytic redox reactions (Churakova et al., 2014). In addition, ethyl formate added in surplus to shift the reaction equilibrium could be partially recovered for subsequent reactions, and the biocatalyst could also be reused at least 27 times (Janssen et al., 2014).

Lipase-catalyzed synthesis of kojic acid ricinoleate was also reported in the recent past. Kojic acid (**24**; 5-hydroxy-2-(hydroxymethyl)-1,4-pyrone) is widely used as a food additive to prevent the browning reaction or in cosmetics as a skin whitening agent (Bentley, 2006; Chang, 2009; Kang et al., 2009). The main shortcomings of using kojic acid for industrial purposes are its water-solubility and instability at high temperatures. Its long-term storage and direct use in incorporating in oil base cosmetic products suffer from a

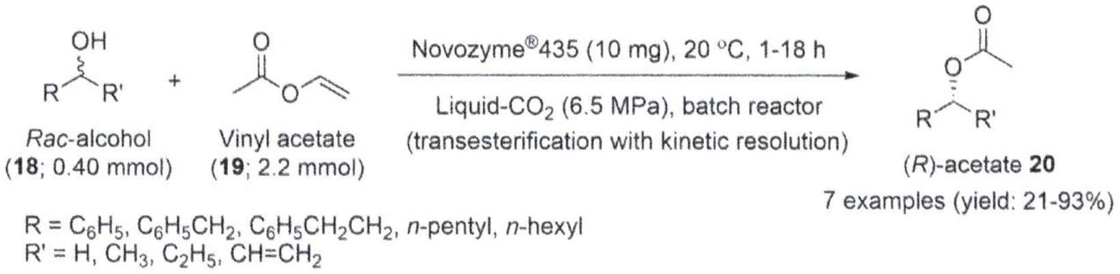

SCHEME 13.6 Transesterification of *rac*-alcohols with vinyl acetate with a kinetic resolution.

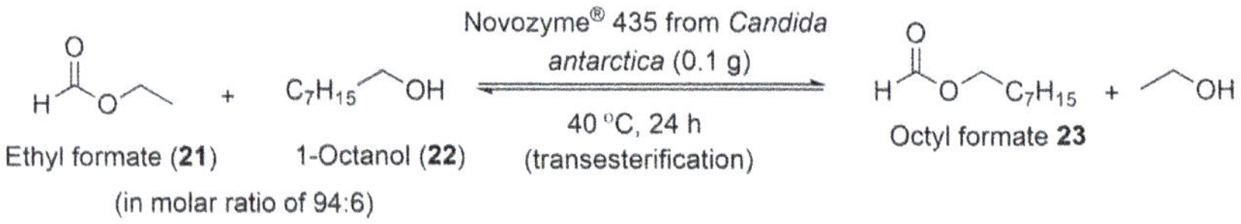

SCHEME 13.7 Transesterification of ethyl formate to octyl formate.

major problem. To improve the kojic acid properties, such as storage stability, compatibility and oil-solubility, many kojic acid derivatives have been synthesized by different groups, usually by modifying the C-5 hydroxyl function to form hydroxyphenyl ethers or esters or by using this function to form glycosides or peptide derivatives (Nishimura et al., 1995; Kadokawa et al., 2003; Kim et al., 2004; Hsieh et al., 2007). Kojic acid possesses two different hydroxyl groups: the secondary hydroxyl group at C-5 position and the primary hydroxyl group at C-7. The hydroxyl group at the C-5 position of kojic acid is essential to the radical scavenging activity and tyrosinase interference activity (Raku and Tokiwa, 2003). But in practice, the esterification protocol of kojic acid with long-chain fatty acids in the presence of acid or alkaline catalysts usually results in a complex mixture. It makes accessible the formation of esters at C-5, the secondary hydroxyl group of kojic acid. However, immobilized lipases offer a solution to these inherent problems associated with chemical catalysts. Liu and Shaw (1998) improved the lipophilic property of kojic acid by lipase-catalyzed acylation with lauric and oleic acids in the presence of acetonitrile as solvent. In this case, the acylation was also carried out at the C-5 hydroxyl group. Subsequently, Khamaruddin et al. (2009) tried to improve Liu yields by esterifying kojic acid and oleic acid using lipase from *Candida rugosa* and *Aspergillus niger* organic media. The maximum yield was not exceeded 45%. Optimized enzymatic synthesis of kojic acid monooleate was reported in the same year by Ashari et al. (2009) but with an unsatisfactory yield (40% after 48 h reaction time). In both cases, kojic acid was also esterified in the C-5 hydroxyl group. Under this background, El-Boulifi et al. (2014) have demonstrated a modified protocol for the enzymatic esterification of kojic acid (**24**) for the first time with hydroxyl-fatty acid, ricinoleic acid (*cis*-12-hydroxy-9-octadecenoic acid; **25**) using Novozyme 435, in a solvent-free system. This lipase-catalyzed esterification took place regioselectively with the C-7 hydroxy group affording kojic acid ricinoleate (**26**) with a good yield of 68% (Scheme 13.8).

The unsaturated chiral γ-lactone, (*S*)-γ-hydroxymethyl-α,β-butenolide (HBO) is an important intermediate for the synthesis of many drugs (such as Burseran or Isostegane) (Tomioka et al., 1979; Enders et al., 2002), flavors (Takashi et al., 1990), and antiviral agents against human immunodeficiency

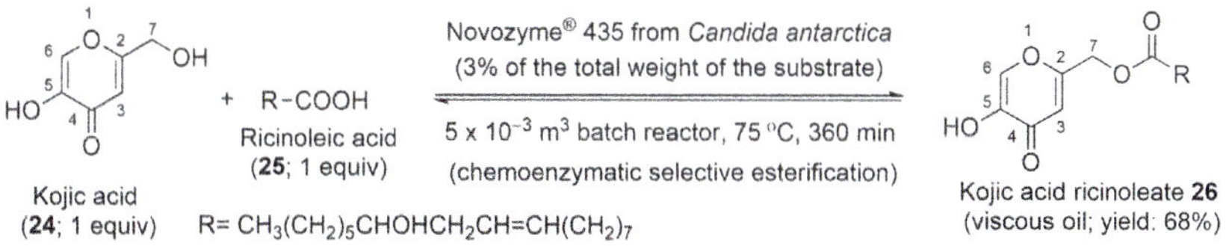

SCHEME 13.8 Solvent-free synthesis of kojic acid ricinoleate, a novel hydroxyl-fatty acid derivative of kojic acid.

virus (HIV) or hepatitis B virus (Hawakami et al., 1990; Diaz-Rodriguez et al., 2009; Flores et al., 2011). In 2015, Flourat et al. (2015) synthesized this key intermediate from (−)-levoglucosenone (LGO) using a two-step sequence involving a lipase-mediated Baeyer–Villiger oxidation and acid hydrolysis (Scheme 13.9). This chemoenzymatic synthetic protocol offers a cost-effective, less toxic, and greener alternative.

Arora et al. (2014b) reported that lipase could also catalyze the Cannizzaro-type reaction of substituted benzaldehydes in an aqueous medium at 30°C without adding any external redox reagent (Scheme 13.10).

Zhou et al. (2014) reported the first-time enzyme-catalyzed direct vinylogous Michael addition reaction of electron-deficient vinyl malononitriles to nitroalkenes. A series of nitroalkenes (**34**) underwent smooth Michael addition with varying vinyl malononitriles (**33**) to generate the corresponding products **35** with moderate to high yields (57%–94%) in the presence of Lipozyme (immobilized lipase from *Mucor miehei*) with excellent diastereoselectivities (Scheme 13.11).

Wang and his group (2014) also synthesized a series of bis-lawsone derivatives, 3,3′-(arylmethylene)bis(2-hydroxynaphthalene-1,4-diones) (**38**), from the reaction of 2-hydroxy-1,4-naphthoquinone (**36**) with aromatic aldehydes (**37**) using lipase as green and inexpensive biocatalyst in excellent yields (Scheme 13.12). This new protocol has several advantages over the earlier reported methods, particularly in terms of green chemistry aspects.

Kumar et al. (2011) first demonstrated the application of tandem catalysis in a cross-aldol reaction of aromatic aldehydes with enol acetate using the commercial lipase (Novozym 435) and triethylamine (TEA) as

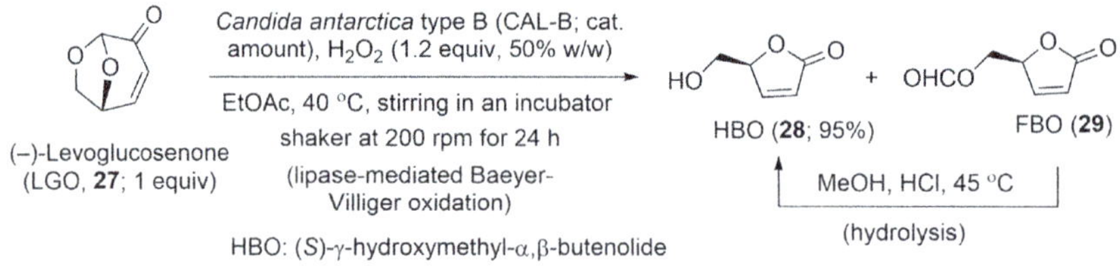

SCHEME 13.9 Chemoenzymatic synthesis of (*S*)-γ-hydroxymethyl-α,β-butenolide via lipase-mediated Baeyer–Villiger oxidation of (−)-levoglucosenone.

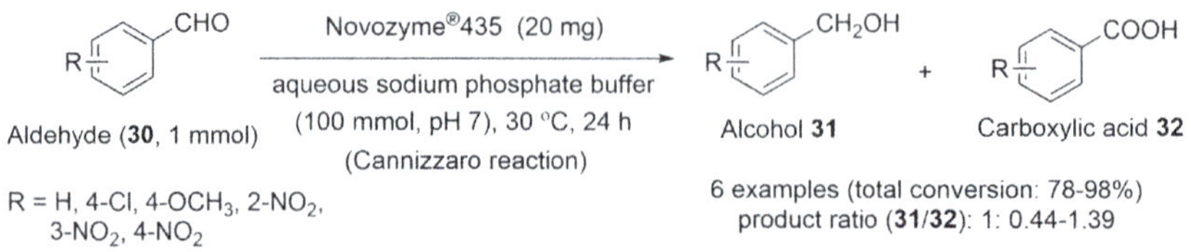

SCHEME 13.10 Cannizzaro-type reaction of substituted benzaldehydes in an aqueous medium.

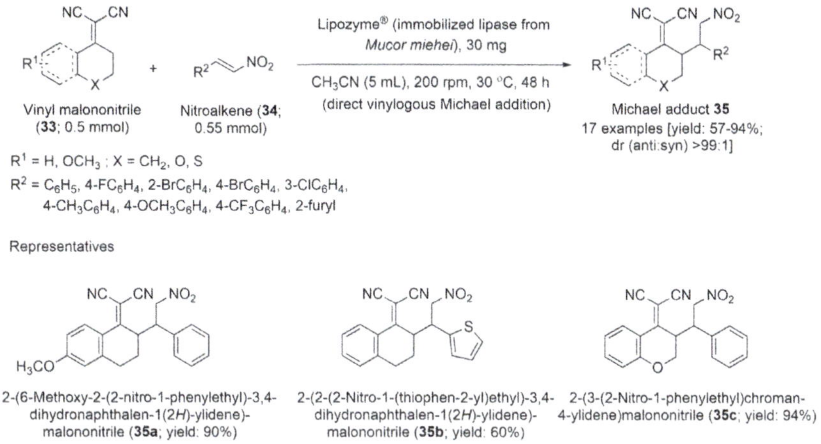

SCHEME 13.11 Lipase-catalyzed direct vinylogous Michael addition reaction.

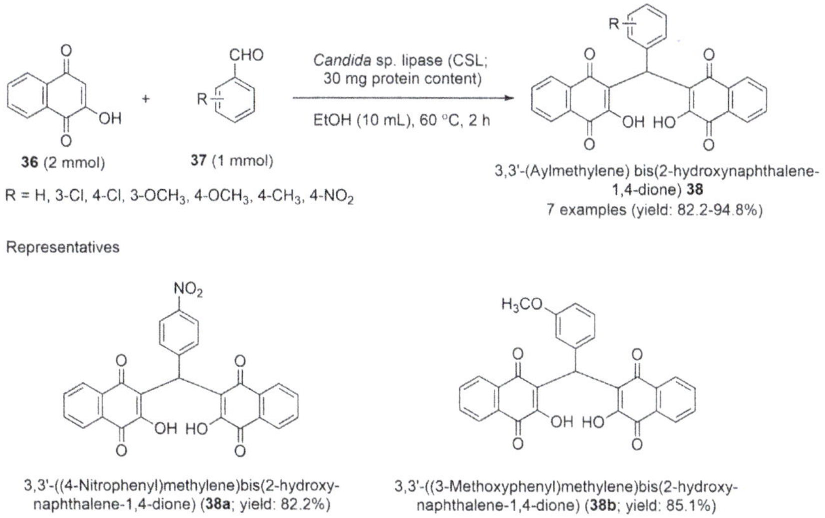

SCHEME 13.12 Synthesis of functionalized 3,3′-(arylmethylene) bis(2-hydroxynaphthalene-1,4-diones).

an organocatalyst (Scheme 13.13). The reaction was facilitated through the lipase-catalyzed in situ generations of acetaldehyde/acetone from commercial vinyl/isopropenyl acetate (**39**). The key advantages of the present methodology include the mild and facile reaction conditions, renewability of the lipase, comparatively high yields, and minimal side product formation. Still, the method suffers from a long reaction time.

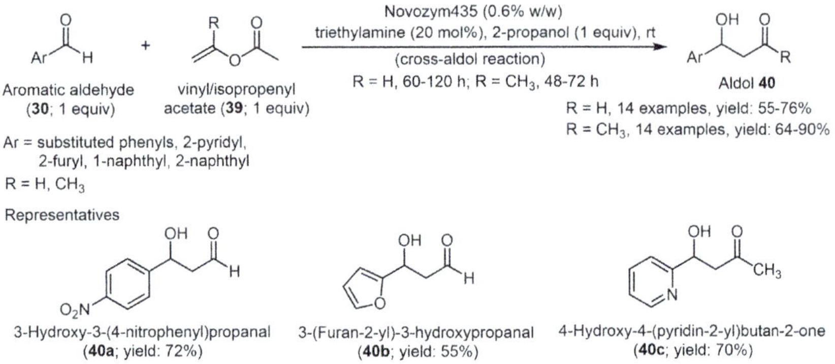

SCHEME 13.13 Lipase-catalyzed cross-aldol reaction of aromatic aldehydes with vinyl acetate and isopropenyl acetate.

Among the heterocyclic compounds, thiazole derivatives are also regarded as useful chemical entities in organic and medicinal chemistry for their varied biological activities, such as antitumor, antifungal, antibiotic, and antiviral properties (Alcaide et al., 2007; Diness et al., 2011; Mouri et al., 2012; Bevk et al., 2013). Very particularly, 2,4-disubstituted thiazoles demand special mention for exhibiting multifarious bioactivities, and hence synthetic methods have been reported continually (Bharti et al., 2010; Chimenti et al., 2011; Naveena et al., 2013). Zhang (2014) developed a novel enzyme-catalyzed multicomponent synthetic method for a series of new 2,4-disubstituted thiazoles (**45**) from the reaction of aldehydes (**41**), amines (**42**), thioacetic acid (**43**), and methyl 3-(dimethyl amino)-2-isocyanoacrylate (**44**) under the catalysis of lipase from porcine pancreas (PPL) in methanol under mild conditions with good yields (Scheme 13.13). This one-pot enzymatic multicomponent conversion method provides a novel strategy and useful tool for synthesizing 2,4-disubstituted thiazoles and satisfies certain green chemistry perspectives (Scheme 13.14).

1,3-Oxathiolan-5-ones and their related derivatives are also attractive heterocyclic targets for their existence in numerous natural products and broad biological activities and their importance as intermediates for highly successful and valuable pharmaceuticals. These chemical scaffolds are reported to possess inhibitory activities toward human type-II (nonpancreatic) secretory phospholipase A2 (PLA2) (Higashiya et al., 1998). The oxathionyl nucleosides, emtricitabine (Coviracil) and lamivudine (3TC), are two of the most potent antiviral drugs as nucleoside reverse transcriptase inhibitors for the treatment of diseases, such as HIV or hepatitis B (Beach et al., 1992; Jeong et al., 1992). Since the initial discovery of the antiviral activity of this motif, the synthesis of enantiomerically pure 1,3-oxathiolan-5-one derivatives has received significant attention (Chu et al., 1991; Humber et al., 1992; Roy et al., 2009). Zhang et al. (2014) successfully demonstrated a lipase-catalyzed dynamic covalent kinetic resolution protocol based on reversible hemithioacetal (**48**) formation

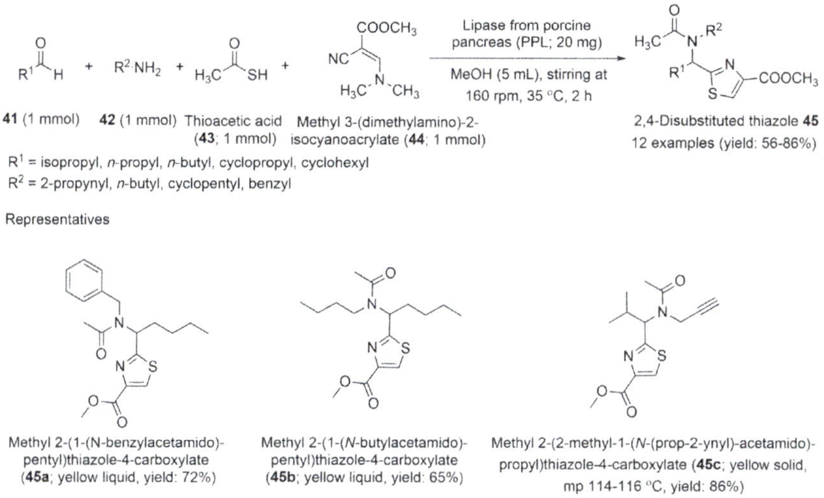

SCHEME 13.14 One-pot multicomponent synthesis of 2,4-disubstituted thiazoles.

for the asymmetric synthesis of 1,3-oxathiolan-5-one derivatives (**49**). Among various lipases, lipase B from *C. antarctica* (CAL-B) was found to show the most potent efficiency in carrying out the transformation in good yields with moderate to good enantiomeric excess (ee) for the final products. The investigators exploited such lipase-catalyzed resolution to obtain selective γ-lactonization of the hemithioacetal (**48**) arising out of the reaction between the aldehydic substrates (**46**) and methyl 2-sulfanylacetate (**47**) in a one-pot process (Scheme 13.15). Furthermore, some of the synthesized 1,3-oxathiolan-5-one derivatives showed potential for simple access to the core structure of active pharmaceutical nucleoside analogues.

Yang et al. (2015) reported on the development of a facile and simple method for the synthesis of biologically relevant benzo[*g*]chromenes (**51**) using lipase as an efficient biocatalyst from the one-pot multicomponent reaction (MCR) of aromatic aldehydes (**30**), malononitrile (**50**), and 2-hydroxy-1,4-naphthoquinone (**36**) in ethanol medium at 55°C (Scheme 13.16). Benzo[*g*]chromenes are known to exhibit a wide range of bioactivities, including antimicrobial, antiproliferative, and antitumor properties (Mohr et al., 1975; Coudert et al., 1988; Tandon et al., 1991; Zamocka et al., 1992; Brunavs et al., 1994; El-Agrody et al., 2000). In their experiment, the present investigators also examined several kinds of lipases such as *C. antarctica* lipase B (CAL-B), Porcine pancreas lipase (PPL), *Candida* sp. lipase (CSL), *Candida rugosa* lipase (CRL), *Pseudomonas fluorescens* lipase (PFL), *Pseudomonas* sp. lipase (PSL), and *Bacillus subtilis* lipase (BSL2) to catalyze this MCR, and *Candida* sp. lipase (CSL) came out with the highest catalytic activity, thereby suggesting that the catalytic activities depend

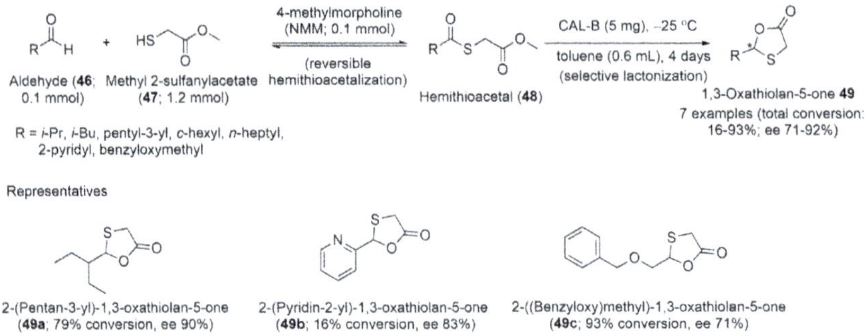

SCHEME 13.15 Asymmetric synthesis of 1,3-oxathiolan-5-one derivatives.

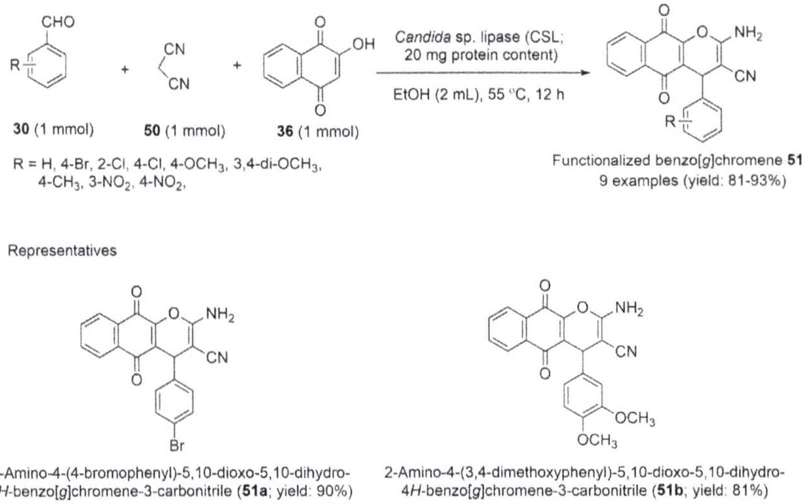

SCHEME 13.16 Synthesis of functionalized benzo[*g*]chromenes.

mainly on the lipase origin. When the denatured lipase or bovine serum albumin was used as the catalyst, almost no product could be detected, which suggested a unique active conformation of the enzyme playing a crucial role in effecting this MCR. Synthesis of such compounds was reported earlier by using nonenzymatic catalysts (Wang et al., 2009; Yao et al., 2009; Khurana et al., 2010, 2012; Dekamin et al., 2013; Azizi and Heydari, 2014), but this chemoenzmatic route is more advantageous over the existing protocols in terms of atom economy, environmental friendliness, and operational simplicity.

Recently, Zhang et al. (2017) accomplished a straightforward Biginelli approach via a one-pot tandem MCR, affording a series of 3,4-dihydropyrimidin-2(1 *H*)-ones (**54**) in good yields, using CAL-B as an economical and eco-friendly initiator (Scheme 13.17). This tandem strategy involves two

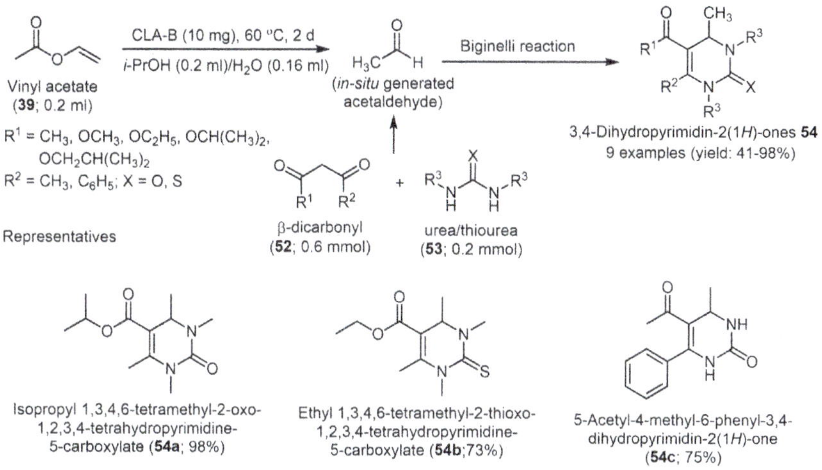

SCHEME 13.17 Lipase-catalyzed tandem Biginelli reactions.

steps, in situ generation of acetaldehyde from vinyl acetate by lipase catalysis at first, followed by a Biginelli reaction. The investigators showed the structurally diverse molecular libraries as luminogens in the solid state.

Biphenyl esters are a class of chemical agents with diverse pharmaceutical and technical relevance (Baheti et al., 2009; Sharma et al., 2017; Kwong et al., 2017; Li et al., 2018). Ehlert et al. (2019) recently accomplished selective ester monohydrolysis of biphenyl esters using lipases under mild conditions. This strategy is advantageous over the conventional basic-hydrolysis protocol limited by competing elimination reactions under the reaction conditions and chemoselectivity (Niwayama, 2000; Nicolaou et al., 2005; Niwayama et al., 2008). The investigators showed that *ortho*-substituents within the biphenyl moiety profoundly affect the chemoselectivity of the ester hydrolysis (Scheme 13.18).

13.3 Concluding remarks

Lipases are among the most important biocatalysts that are in use to carry out a broad spectrum of organic transformations in both aqueous and nonaqueous media to generate biologically relevant organic molecules of potential practical interest, both in research laboratories and in industry. Lipases have the remarkable ability to carry out a wide variety of chemo-, regio-, and enantioselective transformations and very broad substrate specificity. The present chapter offers a recent update on the lipase-catalyzed organic transformations reported during 2013–21. This overview reflects the biocatalytic efficacy of the enzyme in carrying out various types of organic reactions, including esterifications, transesterifications, additions, ring-closing,

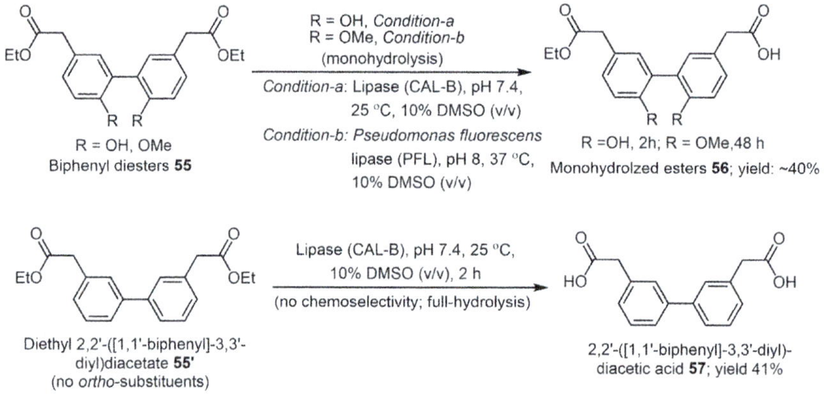

SCHEME 13.18 Lipase-catalyzed ester hydrolysis of biphenyl esters.

oxidation, reduction, amidation, and many others. Ease of handling, broad substrate tolerance, high stability toward temperatures and solvents, high enantioselectivity, convenient commercial availability and reusability are the key advantages of choosing lipase as a biocatalyst in a huge number of organic transformations. The author hopes that this overview should boost ongoing research in chemoenzymatic organic transformations, particularly the biocatalyic applications of lipases. It is noteworthy that each lipase has its own unique properties, and that fine-tuning of any methodology employing lipases to suit the individual enzyme is to be screened carefully. We need to explore detailed mechanisms behind the lipase-catalyzed reactions in more depth for broader usage of this beneficial and effective enzyme. Protein engineering of lipases and the further improvement of lipase preparations and reaction methodology will offer an excellent potential to generate even better bioconversions in the future.

References

Adamczyk, M., Grote, J., 1996. *Pseudomonas cepacia* lipase mediated amidation of benzyl esters. Tetrahedron Lett. 37, 7913–7916.

Adamczyk, M., Grote, J., 1997. Stereoselective *Pseudomonas cepacia* lipase mediated synthesis of α-hydroxyamides. Tetrahedron: Asymmetry 8, 2509–2512.

Adamczyk, M., Grote, J., 1999. *N*-Phosphonoalkyl-5-aminomethylquinoxaline-2,3-diones: *In vivo* active AMPA and NMDA(glycine) antagonists. Bioorg. Med. Chem. Lett. 9, 245–248.

Adlercreutz, P., 2013. Immobilisation and application of lipases in organic media. Chem. Soc. Rev. 42, 6406–6436.

Adrio, J.L., Demain, A.L., 2014. Microbial enzymes: tools for biotechnological processes. Biomolecules 4, 117–139.

Alcaide, B., Almendros, P., Redondo, M.C., 2007. N1–C4 β-lactam bond cleavage in the 2-(trimethylsilyl)thiazole addition to β-lactam aldehydes: asymmetric synthesis of spiranic and tertiary α-alkoxy-γ-keto acid derivatives. Eur. J. Org. Chem. 22, 3707–3710.

Andersch, P., Berger, M., Hermann, J., Laumen, K., Lobell, M., Seemayer, R., et al., 1997. Ester synthesis *via* acyl transfer (transesterification). Methods Enzymol. 286, 406–443.

Andrushko, V., Andrushko, N. (Eds.), 2013. Stereoselective Synthesis of Drugs and Natural Products, 2 vol. set. John Wiley & Sons, Inc, Hoboken.

Arkona, C., Rademann, J., 2013. Propargyl amides as irreversible inhibitors of cysteine proteases – a lesson on the biological reactivity of alkynes. Angew. Chem. Int. (Ed.) 52, 8210–8212.

Arora, B., Mukherjee, J., Gupta, M.N., 2014a. Enzyme promiscuity: using the dark side of enzyme specificity in white biotechnology. Sus. Chem. Proc. 2, 25. Available from: https://doi.org/10.1186/s40508-014-0025-y.

Arora, B., Pandey, P.S., Gupta, M.N., 2014b. Lipase catalyzed Cannizzaro-type reaction with substituted benzaldehydes in water. Tetrahedron Lett. 55, 3920–3922.

Ashari, S.E., Mohamad, R., Ariff, A., Basri, M., Salleh, A.B., 2009. Optimization of enzymatic synthesis of palm-based kojic acid ester using response surface methodology. J. Oleo. Sci. 58, 501–510.

Azizi, K., Heydari, A., 2014. A simple, green, one-pot synthesis of magnetic-nanoparticle-supported proline without any source of supplemental linkers and application as a highly efficient base catalyst. RSC Adv. 4, 6508–6512.

Baheti, A., Tyagi, P., Thomas, K.R.J., Hsu, Y.C., Lin, J.T.S., 2009. Simple triarylamine-based dyes containing fluorene and biphenyl linkers for efficient dye-sensitized solar cells. J. Phys. Chem. C. 113, 8541–8547.

Beach, J.W., Jeong, L.S., Alves, A.J., Pohl, D., Kim, H.O., Chang, C.N., et al., 1992. Synthesis of enantiomerically pure (2′*R*,5′*S*)-(−)-1-(2-hydroxymethyloxathiolan-5-yl)cytosine as a potent antiviral agent against hepatitis B virus (HBV) and human immunodeficiency virus (HIV.). J. Org. Chem. 57, 2217–2219.

Bentley, R., 2006. From miso, saké and shoyu to cosmetics: a century of science for kojic acid. Nat. Prod. Rep. 23, 1046–1062.

Bevinakatti, H.S., Newadkar, R.V., 1989. Lipase catalysis in organic solvents transesterification of *O*-formyl esters of secondary alcohols. Biotechnol. Lett. 11, 785–788.

Bevk, D., Marin, L., Lutsen, L.L., Vanderzande, D., Maes, W., 2013. Thiazolo[5,4-*d*]thiazoles – promising building blocks in the synthesis of semiconductors for plastic electronics. RSC Adv. 3, 11418–11431.

Bharti, S.K., Tilak, G.R., Singh, S.K., 2010. Synthesis, anti-bacterial and anti-fungal activities of some novel Schiff bases containing 2,4-disubstituted thiazole ring. Eur. J. Med. Chem. 45, 651–660.

Bolon, D.N., Voigt, C.A., Mayo, S.L., 2002. *De novo* design of biocatalysts. Curr. Opin. Struct. Biol. 6, 125–129.

Bonger, K.M., van den Berg, R.J.B.H.N., Knijnenburg, A.D., Heitman, L.H., Ijzerman, A.P., Oosterom, J., et al., 2008. Synthesis and evaluation of homodimeric GnRHR antagonists having a rigid bis-propargylated benzene core. Bioorg. Med. Chem. 16, 3744–3758.

Bornscheuer, U.T., Huisman, G.W., Kazlauskas, R.J., Lutz, S., Moore, J.C., Robins, K., 2012. Engineering the third wave of biocatalysis. Nature 485, 185–194.

Bornscheuer, U.T., Kazlauskas, R.J., 2004. Catalytic promiscuity in biocatalysis: using old enzymes to form new bonds and follow new pathways. Angew. Chem. Int. (Ed.) 116, 6032–6040.

Borowiecki, P., Dranka, M., 2019. A facile lipase-catalyzed KR approach toward enantiomerically enriched homopropargyl alcohols. Bioorg. Chem. 93, 102754.

Bose, A., Shakil, N.A., Pankaj Kumar, J., Singh, M.K., 2010. Biocatalytic amidation of carboxylic acids and their antinemic activity. J. Environ. Sci. Health B 45, 254–261.

2nd edn. Brahmachari, G. (Ed.), 2021. Green Synthetic Approaches for Biologically Relevant Heterocycles, Vol. 1 and 2. Elsevier, Amsterdam.

Brahmachari, G., 2015. Room Temperature Organic Synthesis, 1st edn. Elsevier, Amsterdam.

Branneby, C., Carlqvist, P., Magnusson, A., Hult, K., Brinck, T., Berglund, P., 2003. Carbon-carbon bonds by hydrolytic enzymes. J. Am. Chem. Soc. 125, 874–875.

Brunavs, M., Dell, C.P., Owton, W.M., 1994. Direct fluorination of the anthraquinone nucleus: scope and application to the synthesis of novel rhein analogues. J. Fluor. Chem. 68, 201–203.

Busto, E., Gotor-Fernández, V., Gotor, V., 2010. Hydrolases: catalytically promiscuous enzymes for non-conventional reactions in organic synthesis. Chem. Soc. Rev. 39, 4504–4523.

Carlqvist, P., Svedendahl, M., Branneby, C., Hult, K., Brinck, T., Berglund, P., 2005. Exploring the active-site of a rationally redesigned lipase for catalysis of Michael-type additions. ChemBioChem 6, 331–336.

Cedrone, F., Menez, A., Quemeneur, E., 2000. Tailoring new enzyme functions by rational redesign. Curr. Opin. Struct. Biol. 10, 405–410.

Chang, T.S., 2009. An updated review of tyrosinase inhibitors. Int. J. Mol. Sci. 10, 2440–2475.

Chen, X.-Y., Chen, G.-J., Wang, J.-L., Wu, Q., Lin, X.-F., 2013. Lipase/acetamide-catalyzed carbon-carbon bond formations: a mechanistic view. Adv. Synth. Catal. 355, 864–868.

Chimenti, F., Bizzarri, B., Bolasco, A., Secci, D., Chimenti, P., Granese, A., et al., 2011. Synthesis and biological evaluation of novel 2,4-disubstituted-1,3-thiazoles as anti-*Candida* spp. Agents. Eur. J. Med. Chem. 46, 378–382.

Choi, Y.K., Kim, Y., Han, K., Park, J., Kim, M.-J., 2009. Synthesis of optically active amino acid derivatives *via* dynamic kinetic resolution. J. Org. Chem. 74, 9543–9545.

Christodoulopoulou, L., Tsoukatou, M., Tziveleka, L.A., Vagias, C., Petrakis, P.V., Roussis, V., 2005. Piperidinyl amides with insecticidal activity from the maritime *Plantotanthus maritimus*. J. Agric. Food Chem. 53, 1435–1439.

Churakova, E., Tomaszewski, B., Buehler, K., Schmid, A., Arends, I.W.C.E., Holl-mann, F., 2014. The taming of oxygen: biocatalytic oxyfunctionalisations. Top. Catal. 57, 385–391.

Chu, C.K., Beach, J.W., Jeong, L.S., Choi, B.G., Comer, F.I., Alves, A.J., et al., 1991. Enantiomeric synthesis of (+)-BCH-189 [(+)-(2S,5R)-1-[2-(hydroxymethyl)-1,3-oxathiolan-5-yl]cytosine] from D-mannose and its anti-HIV activity. J. Org. Chem. 56, 6503–6505.

Cohen, Y., Rubin, A., Gotlieb, D., 2008. Activity of carboxylic acid amide (CAA) fungicides against *Bremia lactucae*. Eur. J. Plant. Pathol. 122, 169–183.

Copping, L.G., Martin, R., Pickett, J.A., Bucke, C., Bunch, A.W. (Eds.), 1990. Opportunities in Biotransformations. Elsevier, London.

Coudert, P., Coyquelet, J.M., Bastide, J., Marion, Y., Fialip, J., 1988. Synthesis and anti-allergic properties of *N*-arylnitrones with furo-pyran structure. Ann. Pharm. Fr. 46, 91–96.

Davis, B.G., Boyer, V., 2001. Biocatalysis and enzymes in organic synthesis. Nat. Prod. Rep. 18, 618–640.

Debonsi, H.M., Miranda, J.E., Murata, A.T., Bortoli, S.A., Kato, M.J., Bolzani, V.S., et al., 2009. Isobutyl amides—potent compounds for controlling *Diatraea saccharalis*. Pest. Manag. Sci. 65, 47–51.

Dekamin, M.G., Eslami, M., Maleki, A., 2013. Potassium phthalimide-*N*-oxyl: a novel, efficient, and simple organocatalyst for the one-pot three-component synthesis of various 2-amino-4*H*-chromene derivatives in water. Tetrahedron 69, 1074–1085.

Diaz-Rodriguez, A., Sanghvi, Y.S., Fernandez, S., Schinazi, R.F., Theodorakis, E.A., Ferreroa, M., et al., 2009. Synthesis and anti-HIV activity of conformationally restricted bicyclic hexahydroisobenzofuran nucleoside analogs. Org. Biomol. Chem. 7, 1415–1423.

Diness, F., Nielsen, D.S., Fairlie, D.P., 2011. Synthesis of the thiazole–thiazoline fragment of largazole analogues. J. Org. Chem. 76, 9845–9851.

Drauz, K., Gröger, H., May, O. (Eds.), 2012. Enzyme Catalysis in Organic Synthesis. 3rd edn. Wiley-VCH Verlag & Co. KGaA, Weinheim.

Ehlert, J., Kronemann, J., Zumbrägel, N., Preller, M., 2019. Lipase-catalyzed chemoselective ester hydrolysis of biomimetically coupled aryls for the synthesis of unsymmetric biphenyl esters. Molecules 24, 4272.

El-Agrody, A.M., El-Hakim, M.H., Abd El-Latif, M.S.A., Fakery, A.H., El-Sayed, M., El-Ghareab, K.A., 2000. Synthesis pyrano[2,3-d]pyrimidine and pyrano[3,2-e][1,2,4]triazolo [2,3-c]pyrimidine derivatives with promising antibacterial activities. Acta Pharm. 50, 111–120.

El-Boulifi, N., Ashari, S.E., Serrano, M., Aracil, J., Martínez, M., 2014. Solvent-free lipase-catalyzed synthesis of a novel hydroxyl-fatty acid derivative of kojic acid. Enzyme Microb. Tech. 55, 128–132.

Enders, D., Lausberg, V., Del Signore, G., Berner, O.M., 2002. A general approach to the asymmetric synthesis of lignans: (−)methyl piperitol, (−)sesamin, (−)-aschantin, (+)-yatein, (+)-dihydroclusin, (+)burseran, and (−)-isostegane. Synthesis 515–522.

Faber, K., 2011. Biotransformations in Organic Chemistry, 6th edn. Springer, Heidelberg.

Fan, X., Niehus, X., Sandoval, G., 2012. Lipases as biocatalyst for biodiesel production. Methods Mol. Biol. 861, 471–483.

Flores, R., Rustullet, A., Alibes, R., Alvarez-Larena, A., de March, P., Figueredo, M., et al., 2011. Synthesis of purine nucleosides built on a 3-oxabicyclo[3.2.0]heptane scaffold. J. Org. Chem. 76, 5369–5383.

Flourat, A.L., Peru, A.A.M., Teixeira, A.R.S., Brunissena, F., Allais, F., 2015. Chemo-enzymatic synthesis of key intermediates (*S*)-γ-hydroxymethyl-α,β-butenolide and (*S*)-γ-hydroxymethyl-γ-butyrolactone *via* lipase-mediated Baeyer–Villiger oxidation of levoglucosenone. Green. Chem. 17, 404–412.

Garg, N.K., Woodroofe, C.C., Lacenere, C.J., Quake, S.R., Stoltz, B.M., 2005. A ligand-free solid-supported system for Sonogashira couplings: applications in nucleoside chemistry. Chem. Commun. 4551–4553.

Gebert, L.F., Rebhan, M.A., Crivelli, S.E., Denzler, R., Stoffel, M., Hall, J., 2014. Miravirsen (SPC3649) can inhibit the biogenesis of miR-122. Nucleic Acids Res. 42, 609–621.

Gotor, V., 1999. Non-conventional hydrolase chemistry: amide and carbamate bond formation catalyzed by lipases. Bioorg. Med. Chem. 7, 2189–2197.

Gotor-Fernández, V., Busto, E., Gotor, V., 2006. *Candida antarctica* lipase B: an ideal biocatalyst for the preparation of nitrogenated organic compounds. Adv. Synth. Catal. 348, 797–812.

Gröger, H., 2010. In: Ojima, I. (Ed.), Catalytic Asymmetric Synthesis, 3rd edn. John Wiley & Sons, Inc, Hoboken, pp. 269–341. Ch. 6.

Grotkopp, O., Ahmad, A., Frank, W., Müller, T.J.J., 2011. Blue-luminescent 5-(3-indolyl)oxazoles via microwave-assisted three-component coupling–cycloisomerization–Fischer indole synthesis. Org. Biomol. Chem. 9, 8130–8140.

Gruit, M., Davtyan, A.P., Beller, M., 2011. Platinum-catalyzed cyclization reaction of alkynes: synthesis of azepino[3,4-*b*]indol-1-ones. Org. Biomol. Chem. 9, 1148–1159.

Gupta, M.N., Roy, I., 2004. Enzymes in organic media: forms, function and applications. Eur. J. Biochem. 271, 2575–2583.

Gupta, M.N., 1992. Enzyme function in organic solvents. Eur. J. Biochem. 203, 25–31.

Halgas, J., 1992. Biocatalysts in organic synthesis, Studies in Organic Chemistry, Vol. 46. Elsevier, Amsterdam.

Han, K., Kim, C., Park, J., Kim, M.-J., 2010a. Chemoenzymatic synthesis of rivastigmine *via* dynamic kinetic resolution as a key step. J. Org. Chem. 75, 3105–3108.

Han, K., Kim, Y., Park, J., Kim, M.-J., 2010b. Chenoenzymatic synthesis of the calcimimetics (+)-NPS R-568 *via* asymmetric reductive acylation of ketoxime intermediate. Tetrahedron Lett. 51, 3536–3537.

Hasan, F., Shah, A.A., Hameed, A., 2006. Industrial applications of microbial lipases. Enzyme Microb. Tech. 39, 235–251.

Hassan, S., Tschersich, R., Müller, T.J.J., 2013. Three-component chemoenzymatic synthesis of amide ligated 1,2,3-triazoles. Tetrahedron Lett. 54, 4641–4644.

Hassan, S., Ullrich, A., Müller, T.J.J., 2015. Consecutive three-component synthesis of (hetero) arylated propargyl amides by chemoenzymatic aminolysis–Sonogashira coupling sequence. Org. Biomol. Chem. 13, 1571–1576.

Hawakami, H., Ebata, T., Koseki, K., Statsumoto, K., Matsushita, H., Naoi, Y., et al., 1990. Stereoselectivities in the coupling reaction between silytated pyrimidine bases and 1-halo-2,3-dideoxyribose. Heterocycles 31, 2041–2054.

Herrmann, A., 2007. Controlled release of volatiles under mild reaction conditions: From Nature to everyday products. Angew. Chem. Int. (Ed.) 46, 5836–5863.

Higashiya, S., Narizuka, S., Konno, A., Maeda, T., Momota, K., Fuchigami, T., 1998. Electrolytic partial fluorination of organic compounds. 30. Drastic improvement of anodic monofluorination of 2-substituted 1,3-oxathiolan-5-ones using the novel fluorine source $Et_4NF \cdot 4HF$. J. Org. Chem. 64, 133–137.

Hildebrandt-Eriksen, E.S., Aarup, V., Persson, R., Hansen, H.F., Munk, M.E., Orum, H., 2012. A locked nucleic acid oligonucleotide targeting microRNA 122 is well-tolerated in cynomolgus monkeys. Nucleic Acid. Ther. 22, 152–161.

Hoang, H.N., Matsuda, T., 2015. Liquid carbon dioxide as an effective solvent for immobilized *Candida antarctica* lipase B catalyzed transesterification. Tetrahedron Lett. 56, 639–641.

Hoye, T.R., Jeffrey, C.S., Nelson, D.P., 2010. Dynamic kinetic resolution during a vinylogous Pane rearrangement: a concise synthesis of the polar pharmacophore subunit of (+)-scyphostatin. Org. Lett. 12, 52–55.

Hsieh, H., Giridhar, R., Wu, W., 2007. Regioselective formation of kojic acid-7-*O*-alpha-dglucopyranoside by whole cells of mutated *Xanthomonas campestris*. Enzyme Microb. Tech. 40, 324–328.

Humber, D.C., Jones, M.F., Payne, J.J., Ramsay, M.V.J., Zacharie, B., Jin, H., et al., 1992. Expeditious preparation of (−)-2′-deoxy-3′-thiacytidine (3TC). Tetrahedron Lett. 33, 4625–4628.

Humble, M.S., Berglund, P., 2011. Biocatalytic promiscuity. Eur. J. Org. Chem. 19, 3391–3401.

Hu, D., Wan, Q., Yang, S., Song, B., Bhadury, P.S., Jin, L., et al., 2008. Synthesis and antiviral activities of amide derivatives containing the α-aminophosphonate moiety. J. Agric. Food Chem. 56, 998–1001.

Itoh, T., Takagi, Y., Tsukube, H., 1997. Synthesis of chiral building blocks for organic synthesis via lipase-catalyzed reaction: New method of enhancing enzymatic reaction enantioselectivity. J. Mol. Catal. B- Enzym. 3, 259–270.

Janssen, L.M.G., van Oosten, R., Paul, C.E., Arends, I.W.C.E., Hollmann, F., 2014. Lipase-catalyzed transesterification of ethyl formate to octyl formate. J. Mol. Catal. B- Enzym. 105, 7–10.

Jeong, L.S., Alves, A.J., Carrigan, S.W., Kim, H.O., Beach, J.W., Chu, C.K., 1992. An efficient synthesis of enantiomerically pure (+)-(2*S*,5*R*)-1-[2-(hydroxymethyl)-1,3-oxathiolan-5-yl] cytosine [(+)-BCH-189] from d-galactose. Tetrahedron Lett. 33, 595–598.

Kadokawa, J., Nishikura, T., Muraoka, R., Tagaya, H., Fukuoka, N., 2003. Synthesis of kojic acid derivatives containing phenolic hydroxy groups. Synth. Commun. 33, 1081–1086.

Kamerlin, S.C., Warshel, A., 2010. At the dawn of the 21st century: Is dynamics the missing link for understanding enzyme catalysis? Prots: Struct., Funct., Bioinformat 78, 1339–1375.

Kang, S.S., Kim, H.J., Jin, C., Lee, Y.S., 2009. Synthesis of tyrosinase inhibitory (4-oxo-4*H*-pyran-2-yl)acrylic acid ester derivatives. Bioorg. Med. Chem. Lett. 19, 188–191.

Kaushik, P., Shakil, N.A., Kumar, J., Singh, B.B., 2015. Lipase-catalyzed solvent-free amidation of phenolic acids. Synth. Commun. 45, 569–577.

Khamaruddin, N.H., Basri, M., Lian, E.G.C., Salleh, A., Rahman, A.S.R., Ariff, A., et al., 2009. Enzymatic synthesis and characterization of palm-based kojic acid ester. J. Oil Palm. Res. 20, 461–469.

Khurana, J.M., Magoo, D., Chaudhary, A., 2012. Efficient and green approaches for the synthesis of 4*H*-benzo[g]chromenes in water, under neat conditions, and using task-specific ionic liquid. Synth. Commun. 42, 3211–3219.

Khurana, J.M., Nand, B., Saluja, P., 2010. DBU: a highly efficient catalyst for one-pot synthesis of substituted 3,4-dihydropyrano[3,2-c]chromenes, dihydropyrano[4,3-b]pyranes, 2-amino-4H-benzo[h]chromenes and 2-amino-4H benzo[g]chromenes in aqueous medium. Tetrahedron 66, 5637–5641.

Kim, H., Choi, J., Cho, J.K., Kim, S.Y., Lee, Y.S., 2004. Solid-phase synthesis of kojic acidtri-peptides and their tyrosinase inhibitory activity, storage stability, and toxicity. Bioorg. Med. Chem. Lett. 14, 2843–2846.

Klibanov, A.M., 2001. Improving enzymes by using them in organic solvents. Nature 409, 241–246.

Kobayashi, S., Fujikawa, S., Ohmae, M., 2003. Enzymatic synthesis of chondroitin and its derivatives catalyzed by hyaluronidase. J. Am. Chem. Soc. 125, 14357–14369.

Koshkin, A.A., Singh, S.K., Nielson, P., Rajwanshi, V.K., Kumar, R., Meldgaard, M., et al., 1998. LNA (Locked Nucleic Acids): synthesis of the adenine, cytosine, guanine, 5-methylcytosine, thymine and uracil bicyclonucleoside monomers, oligomerization, and unprecedented nucleic acid recognition. Tetrahedron 1998 (54), 3607–3630.

Koskinen, A.M.P., Klibanov, A.M. (Eds.), 1996. Enzymatic Reactions in Organic Media. Blackie Academic & Professional, London.

Krishna, S.H., 2002. Developments and trends in enzyme catalysis in nonconventional media. Biotechnol. Adv. 20, 239–267.

Krumlinde, P., Bogár, K., Bäckvall, J.-E., 2009. Synthesis of a neonicotinoide pesticide derivative *via* chemoenzymatic dynamic kinetic resolution. J. Org. Chem. 74, 7407–7410.

Krumlinde, P., Bogár, K., Bäckvall, J.-E., 2010. Asymmetric synthesis of bicyclic diol derivative through metal and enzyme catalysis: application to the formal synthesis of sertraline. Chem. Eur. J. 16, 4031–4036.

Kumar, A., Khan, A., Malhotra, S., Mosurkal, R., Dhawan, A., Pandey, M.K., et al., 2016. Synthesis of macromolecular systems via lipase catalyzed biocatalytic reactions: applications and future perspectives. Chem. Soc. Rev. 45, 6855–6887.

Kumar, M., Shah, B.A., Taneja, S.C., 2011. Tandem catalysis by lipase in a vinyl acetate-mediated cross-aldol reaction. Adv. Synth. Catal. 353, 1207–1212.

Kwong, H.C., Chidan Kumar, C.S., Mah, S.H., Chia, T.S., Quah, C.K., Loh, Z.H., et al., 2017. Novel biphenyl ester derivatives as tyrosinase inhibitors: synthesis, crystallographic, spectral analysis and molecular docking studies. PLoS ONE 12, e0170117.

Lamberth, C., Cederbaum, F., Jeanguenat, A., Kempf, H., Zeller, M., Zeun, R., 2006. Synthesis and fungicidal activity of *N*-2-(3-methoxy-4-propargyloxy)phenethyl amides. Part II: Anti-oomycetic mandelamides. Pest. Manag. Sci. 62, 446–451.

Lee, S., Lee, C.H., Kim, E., Jung, S.H., Lee, H.K., 2007. Hydroxylated hydrocinnamides as hypocholesterolemic agents. Bull. Korean Chem. Soc. 28, 1787–1791.

Leijondahl, K., Borén, L., Braun, R., Bäckvall, J.-E., 2009. Enzyme- and ruthenium-catalyzed dynamic kinetic asymmetric transformation of 1,5-diols. Applications to the synthesis of (+)-solenopsin A. J. Org. Chem. 74, 1988–1993.

Liese, A., Seelbach, K., Wandrey, C. (Eds.), 2006. Industrial Biotransformations. 2nd edn. Wiley-VCH Verlag GmbH, Weinheim.

Liu, K.J., Shaw, J.F., 1998. Lipase-catalyzed synthesis of kojic acid esters in organic solvents. J. Am. Oil Chem. Soc. 75, 1507–1511.

Li, X., Deng, Y., Zheng, Z., Huang, W., Chen, L., Tong, Q., et al., 2018. Corilagin, a promising medicinal herbal agent. Biomed. Pharmacother. 99, 43–50.

Lou, F.W., Liu, B.K., Wu, Q., Lv, D.S., Lin, X.F., 2008. *Candida antarctica* lipase B (CAL-B)-catalyzed carbon-sulfur bond addition and controllable selectivity in organic media. Adv. Synth. Catal. 350, 1959–1962.

Martinek, K., Levashov, A.V., Klyachko, N., Khmelnitsky, Y.L., Berezin, I.V., 1986. Micellar enzymology. Eur. J. Biochem. 155, 453–468.

Merkul, E., Boersch, C., Frank, W., Müller, T.J.J., 2009. Three-component synthesis of *N*-boc-4-iodopyrroles and sequential one-pot alkynylation. Org. Lett. 11, 2269–2272.

Merkul, E., Müller, T.J.J., 2006. A new consecutive three-component oxazole synthesis by an amidation–coupling–cycloisomerization (ACCI) sequence. Chem. Commun. 4817–4819.

Mohr, S.J., Chirigos, M.A., Fuhrman, F.S., Pryor, J.W., 1975. Pyran copolymer as an effective adjuvant to chemotherapy against a murine leukemia and solid tumor. Cancer Res. 35, 3750–3754.

Moon, J.H., Terao, J.J., 1998. Antioxidant activity of caffeic acid and dihydrocaffeic acid in lard and human low-density lipoprotein. J. Agric. Food Chem. 46, 5062–5065.

Mouri, K., Saito, S., Yamaguchi, S., 2012. Highly flexible π-expanded cyclooctatetraenes: cyclic thiazole tetramers with head-to-tail connection. Angew. Chem. Int. (Ed.) 51, 5971–5975.

Namiki, M., 1990. Antioxidants/antimutagens in food. Crit. Rev. Food Sci. 29, 273–300.

Naveena, C.S., Poojary, B., Arulmoli, T., Manjunatha, K., Prabhu, A., Kumari, N.S., 2013. Synthesis and evaluation of biological and nonlinear optical properties of some novel 2,4-disubstituted [1,3]-thiazoles carrying 2-(aryloxymethyl)-phenyl moiety. Med. Chem. Res. 22, 1925–1937.

Nicolaou, K.C., Estrada, A.A., Zak, M., Lee, S.H., Safina, B.S., 2005. A mild and selective method for the hydrolysis of esters with trimethyltin hydroxide. Angew. Chem. Int. (Ed.) 44, 1378–1382.

Nishimura, T., Kometani, T., Takii, H., Terada, Y., Okada, S., 1995. Comparison of some properties of kojic acid glucoside with kojic acid. Nippon. Shokuhin Kagaku Kogaku Kaishi 42, 602–606.

Niwayama, S., 2000. Highly efficient selective monohydrolysis of symmetric diesters. J. Org. Chem. 65, 5834–5836.

Niwayama, S., Cho, H., Lin, C., 2008. Highly efficient selective monohydrolysis of dialkyl malonates and their derivatives. Tetrahedron Lett. 49, 4434–4436.

O'Brien, P.J., Herschlag, D., 1999. Catalytic promiscuity and the evolution of new enzymatic activities. Chem. Biol. 6, R91–R105.

Paravidino, M., Sorgedrager, M.J., Orru, R.V.A., Hanefeld, U., 2010. Activity and enantioselectivity of the hydroxynitrile lyase MeHNL in dry organic solvents. Chem. Eur. J. 16, 7596–7604.

Patel, R.N. (Ed.), 2000. Stereoselective Biocatalysis. Marcel Dekker, New York.

Patel, R.N. (Ed.), 2007. Biocatalysis in the Pharmaceutical and Biotechnology Industries. CRC Press, Boca Raton.

Pérez-Alvarez, V., Bobadilla, R.A., Muriel, P., 2001. Structure–hepatoprotective activity relationship of 3,4-dihydroxycinnamic acid (caffeic acid) derivatives. J. Appl. Toxicol. 21, 527–531.

Rahman, A.-u, Shah, Z., 1993. Stereoselective Synthesis in Organic Chemistry. Springer-Verlag, New Jersey.

Raillard, S., Krebber, A., Chen, Y., Ness, J.E., Bermudez, E., Trinidad, R., et al., 2001. Novel enzyme activities and functional plasticity revealed by recombining highly homologous enzymes. Chem. Biol. 8, 891–898.

Raku, T., Tokiwa, Y., 2003. Regioselective synthesis of kojic acid esters by *Bacillus subtilis* protease. Biotechnol. Lett. 25, 969–974.

Ranoux, A., Hanefeld, U., 2013. Improving transketolase. Top. Catal. 56, 750–764.

Ran, N., Zhao, L., Chen, Z., Tao, J., 2008. Recent applications of biocatalysis in developing green chemistry for chemical synthesis at the industrial scale. Green. Chem. 10, 361–372. 112.

Roberts, S.M., 1998. Preparative biotransformations: the employment of enzymes and whole-cells in synthetic organic chemistry. J. Chem. Soc., Perkin Trans. 1, 157–170.

Roberts, S.M., 1999. Preparative biotransformations. J. Chem. Soc., Perkin Trans. 1, 1–22.

Roberts, S.M., 2000. Preparative biotransformations. J. Chem. Soc., Perkin Trans. 611–633.

Roy, B.N., Singh, G.P., Srivastava, D., Jadhav, H.S., Saini, M.B., Aher, U.P., 2009. A novel method for large-scale synthesis of lamivudine through cocrystal formation of racemic lamivudine with (*S*)-(−)-1,1′-bi(2-naphthol) [(*S*)-(BINOL)]. Org. Process. Res. Dev. 13, 450–455.

Sanchez, S., Demain, A.L., 2011. Enzymes and bioconversions of industrial, pharmaceutical, and biotechnological significance. Org. Process. Res. Dev. 15, 224–230.

Sanda, F., Fujii, T., Tabei, J., Shiotsuki, M., Masuda, T., 2008. Synthesis of hydroxy group-containing poly(*N*-propargylamides): examination of the secondary structure and chiral-recognition ability of the polymers. Macromol. Chem. Phys. 209, 112–118.

Saxena, R.K., Ghosh, P.K., Gupta, R., Davidson, W.S., Bradoo, S., Gulati, R., 1999. Microbial lipases: potential biocatalysts for the future industry. Curr. Sci. 77, 101–115.

Saxena, R.K., Sheoran, A., Giri, B., Davidson, W.S., 2003. Purification strategies for microbial lipases. J. Microbiol. Methods 52, 1–18.

Schmid, A., Dordick, J.S., Hauer, B., Kiener, A., Wubbolts, M., Witholt, B., 2001. Industrial biocatalysis today and tomorrow. Nature 409, 258–268.

Schmid, R.D., Verger, R., 1998. Lipases: interfacial enzymes with attractive applications. Angew. Chem. Int. (Ed.) 37, 1609–1633.

Sharma, R., Chisti, Y., Banerjee, U.C., 2001. Production, purification, characterization, and applications of lipases. Biotech. Adv. 19, 627–662.

Sharma, C.K., Kanwar, S.S., 2012. Purification of a novel thermophilic lipase from *B. licheniformis* MTCC-10498. Int. Res. J. Biol. Sci. 1, 43–48.

Sharma, V.K., Kumar, M., Olsen, C.E., Prasad, A.K., 2014a. Chemoenzymatic convergent synthesis of 2′-O,4′-C-methyleneribonucleosides. J. Org. Chem. 79, 6336–6341.

Sharma, P., McClees, S.F., Afaq, F., 2017. Pomegranate for prevention and treatment of cancer: An update. Molecules 2017 (22), 177.

Sharma, V.K., Rungta, P., Prasad, A.K., 2014b. Nucleic acid therapeutics: basic concepts and recent developments. RSC Adv. 4, 16618–16631.

Sheldon, R.A., Woodley, J.M., 2018. Role of biocatalysis in sustainable chemistry. Chem. Rev. 118, 801–838.

Spasova, M., Ivanova, G., Weber, H., Ranz, A., Lankmayr, E., Milkova, T., 2007. Amides of substituted cinnamic acids with aliphatic monoamines and their antioxidative potential. Oxid. Commun. 30, 803–813.

Stankova, I., Chuchkov, K., Shishkov, S., Kostova, K., Mukova, L., Galabov, A.S., 2009. Synthesis, antioxidative and antiviral activity of hydroxycinnamic acid amides of thiazole containing amino acid. Amino Acids 37, 383–388.

Stergiou, P.Y., Foukis, A., Filippou, M., Koukouritaki, M., Parapouli, M., Theodorou, L.G., et al., 2013. Advances in lipase-catalyzed esterification reactions. Biotechnol. Adv. 3, 1846–1859.

Strohmeier, G.A., Sović, T., Steinkellner, G., Hartner, F.S., Andryushkova, A., Purkarthofer, T., et al., 2009. Investigation of lipase-catalyzed Michael-type carbon–carbon bond formations. Tetrahedron 65, 5663–5668.

Sudina, F., Mirzoeva, K., Pushkareva, A., Korshunova, A., Sumbutya, V., Vartolomeev, D., 1993. Caffeic acid phenethyl ester as a lipoxygenase inhibitor with antioxidant properties. FEBS Lett. 329, 21–24.

Svedendahl, M., Hult, K., Berglund, P., 2005. Fast carbon-carbon bond formation by a promiscuous lipase. J. Am. Chem. Soc. 127, 17988–17989.

Takashi, E., Katsuya, M., Hajime, Y., Koseki, K., Kawakami, H., Matsushita, H., 1990. Synthesis of (+)-*trans*-whisky lactone, (+)-*trans*-cognac lactone and (+)-eldanolide. Heterocycles 31, 1585–1588.

Tandon, V.K., Vaish, M., Jain, S., Bhakuni, D.S., Srimal, R.C., 1991. Synthesis, carbon-13 NMR and hypotensive action of 2,3-dihydro-2,2-dimethyl-4*H*-naphtho[1,2-*b*]pyran-4-one. Indian. J. Pharm. Sci. 53, 22–23.

Tan, J.-N., Dou, Y., 2020. Deep eutectic solvents for biocatalytic transformations: focused lipase-catalyzed organic reactions. Appl. Microbiol. Biotechnol. 104, 1481–1496.

Tawaki, S., Klibanov, A.M., 1992. Inversion of enzyme enantioselectivity mediated by the solvent. J. Am. Chem. Soc. 114, 1882–1884.

Theil, F., 1995. Lipase-supported synthesis of biologically active compounds. Chem. Rev. 95, 2203–2227.

Tomioka, K., Ishiguro, T., Koga, K., 1979. Asymmetric total synthesis of the antileukaemic lignans (+)-trans-burseran and (−)-isostegane. J. Chem. Soc. Chem. Commun. 652–653.

Torres, J.M., Cabrera, C.J.S., Salinas, C.A., Chvez, E.R., 2004. Fungistatic and bacteriostatic activities of alkamides from heliopsis longipes roots: affinin and reduced amides. J. Agric. Food Chem. 52, 4700–4704.

Torre, O., Alfonso, I., Gotor, V., 2004. Lipase catalyzed Michael addition of secondary amines to acrylonitrile. Chem. Commun. 1724–1725.

Träff, A., Bogár, K., Warner, M., Bäckvall, J.-E., 2008. Highly efficient route for enatioselective preparation of chlorohydrins via dynamic kinetic resolution. Org. Lett. 10, 4807–4810.

Biocatalysis in Non-conventional Media. In: Tramper, J., Vermue, M.H., Beeftink, H.H., von Stockar, U. (Eds.), Progress in Biotechnology, vol 8. Elsevier, Amsterdam.

Tsai, S., Huang, C.M., 1999. Enantioselective synthesis of (*S*)-ibuprofen ester prodrugs by lipases in cyclohexane. Enzyme Microb. Technol. 25, 682–688.

Vasic-Racki, D., 2006. History of industrial biotransformations – dreams and realities. In: Liese, A., Seelbach, K., Wandrey, C. (Eds.), Industrial Biotransformations, 2nd edn. Wiley-VCH Verlag GmbH, Weinheim.

Vishnoi, S., Agrawal, V., Kasana, V.K., 2009. Synthesis and structure – activity relationships of substituted cinnamic acids and amide analogues: a new class of herbicides. J. Agric. Food Chem. 57, 3261–3265.

Vulfson, E.N., Halling, P.J., Holland, H.L., 2001. Enzymes in Nonaqueous Solvents: Methods and Protocol. Humana Press, New Jersey.

Wang, H., Wang, Z., Wang, C., Yang, F., Zhang, H., Yue, H., et al., 2014. Lipase catalyzed synthesis of 3,3′-(arylmethylene)bis(2-hydroxynaphthalene-1,4-dione). RSC Adv. 4, 35686–35689.

Wang, X.H., Zhang, X.H., Tu, S.J., Shi, F., Zou, X., Yan, S., et al., 2009. A facile route to the synthesis of 1,4-pyranonaphthoquinone derivatives under microwave irradiation without catalyst. J. Heterocycl. Chem. 46, 832–836.

Watts, J.K., 2013. Locked nucleic acid: tighter is different. Chem. Commun. 49, 5618–5620.

Wengel, J., 1999. Synthesis of 3′-*C*- and 4′-*C*-branched oligodeoxynucleotides and the development of locked nucleic acid (LNA). Acc. Chem. Res. 32, 301–310.

Wever, W.J., Bogart, J.W., Baccile, J.A., Chan, A.N., Schroeder, F.C., Bowers, A.A., 2015. Chemoenzymatic synthesis of thiazolyl peptide natural products featuring an enzyme catalyzed formal [4 + 2] cycloaddition. J. Am. Chem. Soc. 137, 3494–3497.

Wu, Q., Liu, B.K., Lin, X.F., 2010. Enzymatic promiscuity for organic synthesis and cascade process. Curr. Org. Chem. 14, 1966–1988.

Yang, F., Wang, H., Jiang, L., Yue, H., Zhang, H., Wang, Z., et al., 2015. A green and one-pot synthesis of benzo[*g*]chromene derivatives through a multicomponent reaction catalyzed by lipase. RSC Adv. 5, 5213–5216.

Yao, C.S., Yu, C.X., Li, T.J., Tu, S.J., 2009. An efficient synthesis of 4*H*-benzo[*g*]chromene-5,10-dione derivatives through triethylbenzylammonium chloride catalyzed multicomponent reaction under solvent-free conditions. Chin. J. Chem. 27, 1989–1994.

Zaks, A., Klibanov, A.M., 1985. Enzyme-catalyzed processes in organic solvents. Proc. Natl. Acad. Sci. USA 82, 3192–3196.

Zamocka, J., Misikova, E., Durinda, J., 1992. Study of the preparation, properties and effect of (5-hydroxy)- or (5-methoxy-4-oxo-4*H*-pyran-2-yl)methyl-2-alkoxycarbanilates. Cesk-Farm (Ceska a Slovenska Farmacie) 41, 170–172.

Zhang, H., 2014. A novel one-pot multicomponent enzymatic synthesis of 2,4-disubstituted thiazoles. Catal. Lett. 144, 928–934.

Zhang, Y., Schaufelberger, F., Sakulsombat, M., Liu, C., Ramström, O., 2014. Asymmetric synthesis of 1,3-oxathiolan-5-one derivatives through dynamic covalent kinetic resolution. Tetrahedron 70, 3826–3831.

Zhang, W., Wang, N., Yang, Z.-J., Li, Y.-R., Yu, Y., Pu, X.-M., et al., 2017. Lipase-initiated tandem Biginelli reactions via in situ-formed acetaldehydes in one pot: discovery of single-ring deep blue luminogens. Adv. Synth. Catal. 359, 3397–3406.

Zhao, H., 2010. Methods for stabilizing and activating enzymes in ionic liquids – a review. J. Chem. Technol. Biotechnol. 85, 891–907.

Zhou, L.-H., Wang, N., Chen, G.-N., Yang, Q., Yang, S.-Y., Zhang, W., et al., 2014. Lipase-catalyzed highly diastereoselective direct vinylogous Michael addition reaction of α, α-dicyanoolefins to nitroalkenes. J. Mol. Catal. B- Enzym. 109, 170–177.

Chapter 14

Tyrosinase and Oxygenases: Fundamentals and Applications

Shagun Sharma, Kanishk Bhatt, Rahul Shrivastava and Ashok Kumar Nadda

Department of Biotechnology and Bioinformatics, Jaypee University of Information Technology, Waknaghat, Solan, Himachal Pradesh, India

14.1 Introduction

Tyrosinase and oxygenase are important enzymes with immense applications in various fields. Tyrosinase is a copper-containing metalloenzyme of type 3, with two copper ions in its active center (Decker et al., 2006). Both enzymes belong to polyphenol oxidases family. Other vital metalloenzymes of this family are catechol oxidase, phenoloxidase, and hemocyanin. This copper-containing tyrosinase enzyme is present as a 75 kDa glycoprotein known as the t4 molecule. It is ubiquitous in nature and found in microbes and animals, including mammals. Tyrosinase plays an essential for pigmentation and melanogenesis. Tyrp1 and Tyrp2 (two tyrosinase-related proteins) and tyrosinase are essential for melanin biosynthesis in the melanocytes' melanosomes through an enzymatically driven process. Epithelial, mucosal, retinal, and ciliary body melanocytes cells synthesize tyrosinase and are stored in cytoplasmic organelles to aid melanin synthesis further. The tyrosinase and tyrosinase-like protein synthesis occur in the rough endoplasmic reticulum transported through the Golgi apparatus to small vesicles. This further passes through the endosomal–lysosomal compartment and fuses with the melanosomes. Tyrosinase plays an essential function in transforming l-tyrosine to dopaquinone through l-dopa (Afshin et al., 2017; Alberti et al., 2006). Tyrosinase activity monitoring and pharmacological regulation can aid in detecting melanoma and Parkinson's disease and treat diseases like albinism, vitiligo, etc. Increased melanin deposition in the human body causes spots of brown colour and freckles, and it is also linked to fruit and vegetable browning and insect moulting (Qu et al., 2020). Oxygenase is a group of oxidative enzymes that catalyze the oxidative reaction by incorporating an oxygen atom into various substrates (Hayaishi, 2004). These enzymes are also as ubiquitous in nature as that tyrosinase. They play a role in the transformation reactions of

Biotechnology of Microbial Enzymes. DOI: https://doi.org/10.1016/B978-0-443-19059-9.00014-1
© 2023 Elsevier Inc. All rights reserved.

various substrates. Oxidase and oxygenase are frequently confused as the same protein, and both of these catalyze the oxidation-reduction reactions. Still, oxygenase specifically incorporates molecular oxygen into different substrates, whereas oxidases are responsible for catalyzing the dehydrogenation of substrates and electron transfer reactions. Oxygenases are mostly engaged in the anabolism and catabolism of biological and synthetic substances, whereas oxidases are primarily concerned with energy metabolism. The oxygenase is divided into two subtypes, i.e., 'mono' and 'di' oxygenases. Incorporating one oxygen atom/mole of a substrate can be catalyzed by monooxygenase. In contrast, the inclusion of two oxygen atoms/mole of a substrate can be catalyzed by dioxygenase (Hayaishi, 2004).

Oxygenases carry out the breakdown and purification of a variety of substances. Mineralization of xenobiotic compounds is carried out with the help of oxygenases. Even though these enzymes serve a critical function, little is known. Cytochrome p450 oxidases are among the most significant monooxygenases in the body, and they are accountable for the breakdown of a variety of substances. Dioxygenases are also known as oxygen transferases. They are responsible for integrating the two atoms of molecular oxygen into the reaction substrate.

Along with molecular oxygen, oxygenase requires the presence of a co-substrate that can donate a pair of electrons, allowing the second atom of oxygen to be reduced to water. Different compounds act as a co-substrate for the monooxygenase, for example, pyridine nucleotides, flavins, ferredoxins, hydroquinone, ascorbate, etc. On the other hand, dioxygenases generally do not often include heme, and their non-heme iron-sulfur group makes up the active site. Here, in this chapter, detailed information on the structure, mechanism of action, and applications of tyrosinase and oxygenase have been given to better understand the role of these enzymes in the living system.

14.2 Origin and Sources

14.2.1 Tyrosinase

Tyrosinase enzyme is a widely utilized enzyme amongst microbes (mostly bacteria) and in the kingdom of Fungi, Plantae and Animalia. As a result, tyrosinases may be easily isolated, refined, and examined for their functional determination from various sources (Table 14.1).

14.2.1.1 Bacterial source

One of the bacterial tyrosinase from genus *Streptomyces* is well-reported. Bacterial tyrosinase is an extracellular enzyme that participates in melanin formation. Various genera, including *Rhizobium, Marinomonas, Pseudomonas, Serratia*, and *Bacillus* can produce melanin. Some of the examples of bacterial

TABLE 14.1 Bacterial tyrosinase sources and their induction.

Bacterial species	Inducing condition	References
Bacillus megaterium	Tyrosine and copper	(Shuster, 2009)
Bacillus thuringiensis	Heat (42°C)	(Liu et al., 2004)
Pseudomonas sp. Dsm13540	L-tyrosine	(Lantto et al., 2002)
Streptomyces antibioticus	L-methionine	(Katz, 1998)
Streptomyces castaneoglobisporus	Methionine, copper	(Ikeda et al., 1996)
Streptomyces glaucescens	L-tyrosine, l- methionine	(Lerch and Ettinger, 1972)
Streptomyces michiganensis	Copper	(Held and Kutzner, 1990)
Vibrio cholera	Osmotic stress, heat (>30°C)	(Coyne et al., 1992)

genera are *Bacillus thuringiensis* (molecular weight of 14kda) (Liu et al., 2004), *Pseudomonas putida*, *Streptomyces castaneoglobisporus*, *Streptomyces antibioticus* (molecular weight of 30 kDa (Liu et al., 2004) *Ralstonia solanacearum*, and *Verrucomicrobium spinosum*, etc.

14.2.1.2 Fungal source

Tyrosinase has been extracted and isolated from various fungi, including *Agaricusbisporus*, *Amanita muscaria*, *Neurospora crassa*, *Aspergillus oryzae*, *Pycnoporus-sanguineus*, *Lentinulaedodes*, *Trichoderma reesei*, *Portabella* mushrooms, and *Lentinula boryana*.

14.2.1.3 Plant and animal source

Tyrosinase has been examined in various plant sources, including the Monastrellgrapes, sunflower seed, apple, and *Solanum melongena*. Tyrosinase is present in plant tissues' chloroplasts, and the vacuole contains the substrate. Tyrosinase is also abundant in *Portulaca grandiflora* (Portulacaceae). It induces unfavorable enzymatic browning of plants, resulting in a major reduction in farm goods' nutritional and commercial value. As an animal source, tyrosinase is found in human melanocytes, responsible for melanin biosynthesis. The absence of tyrosinase can inhibit melanin synthesis leading to albinism and other diseases. (TABLE 14.2)

TABLE 14.2 Various tyrosinase sources and parameters that control the activity.

Source	Molecular weight (kDa)	Optimum temperature	Optimum pH
Bacterial			
Aeromonas media	58	50°C	8.0
Bacillus megaterium	31	50°C	7.0
Bacillus thuringiensis	16.8	75°C	9.0
Streptomyces glaucescens	3.09	-	-
Pseudomonas putida	36–39	30°C	7.0
Fungi			
Lentinula boryana	20–40	50°C	6.0
Lentinula edodes	70–105	-	6.5
Neurospora crassa	46	-	5.0
Agaricus bisporus	112800	25°C	7.0
Aspergillus oryzae	67	-	-
Pycnoporus sanguineus	45	25°C	6.5
Trichoderma reesei	43.2	30°C	9.0
Plant			
Beta vulgaris	41	25°C	6.0

14.2.2 Oxygenase

Oxygenase is classified into two categories depending on whether either one or two oxygen atoms are present per mole of a substrate. (TABLE 14.3)

14.2.2.1 Monooxygenase

On the active sites of metal-containing monooxygenase enzymes, there is either heme iron, non-heme iron, or copper. Considering reversible di-oxygen transporting proteins, there exist three monooxygenase enzymes. It shows that the initial step in the enzymatic pathways is di-oxygen binding to the metalloenzyme in its reduced form, followed by further events that result in substrate oxygenation. The monooxygenase enzymes are: (i) cytochrome p-450, a heme-containing protein; (ii) cytochrome p-450, a heme-containing protein; (iii) cytochrome p-450, a heme-containing protein, present in all animals,

TABLE 14.3 The basic difference between monooxygenase and dioxygenase.

Monooxygenases	Dioxygenases
It incorporates one atom from oxygen into a substrate and reduces the other atom to water.	2 atoms of oxygen are incorporated into a substrate.
For example- most of the hydrolases except for lysine and proline.	Example- tryptophan pyrolase, homogentisic acid oxidase, cysteine dioxygenases, catechol dioxygenases.

plants, fungi, and microorganisms. Cyp enzymes are membrane-bound and are present in mammals and may be found in all organs, with the largest concentrations in the liver and small intestine. Cyps are also found in the inner membranes of mitochondria in steroidogenic tissue such as the testis, adrenal cortex, breast, ovary, and placenta, where they are involved in the production and breakdown of endogenous steroid hormones. (ii) methane monooxygenase: two non-heme iron ions are near together in this enzyme. In terms of spectroscopic characteristics, it is similar to hemerythrin.

14.2.2.1.1 Sub-types of monooxygenases

1. Cyclo-oxygenase (cox) or prostaglandin synthase cox-1 (constitutive) and cox-2 (non-constitutive) are the two isoforms that are expressed in cells (inducible). The rate-limiting step catalyzed by these is the creation of prostaglandins.
2. Hemeoxygenase: heme oxygenase, often known as ho, is a catalytic enzyme that catalyzes the breakdown of heme. It produces biliverdin by cleaving the heme ring at the -methene bridge. Biliverdin is then converted to bilirubin, iron, and carbon monoxide via bilirubin reductase.

14.2.2.2 Dioxygenases

Dioxygenase enzymes are oxidoreductases that include copper or manganese, heme and non-heme iron. These enzymes catalyze many oxygenations of the available substrates and metal-binding sites. As a result, numerous unconnected pathways might be at work behind the hood of these enzymes.

14.2.2.3 Sub-types of dioxygenases

14.2.2.3.1 Intradiol catechol dioxygenases

Degradation of catechol derivatives to muconic acids is done with the help of these non-heme iron-containing enzymes. When the bacteria's only carbon sources are aromatic compounds, they are triggered. Protocatechuate-3,4-dioxygenase (pcd) and catechol 1, 2-dioxygenase (ctd) are the two most well-studied members of this class.

14.2.2.3.2 Indoleamine-2,3-dioxygenase

It is a 45 kDa monomeric, contains heme oxidoreductase. It catalyzes the oxidative cleavage of l-tryptophan into N-formyl-kynurenine, which is the rate-limiting initial step in the kynurenine pathway. Lipoxygenases (lox): belong to the non-heme iron dioxygenases family. They have a role in the synthesis and metabolism of fatty acid hydroperoxidases. Lipoxygenase isozymes are divided into six categories: 5-lox, 12s-lox, 12r-lox, e-lox 15-lox-1, 15-lox-2.

14.3 Molecular Structure of Tyrosinase and Oxygenase

14.3.1 Molecular structure of Tyrosinase

The tyrosinase active site comprises two copper ions binding to the histidine residues. The copper ion in the center is linked by a coordination bridge which is endogenous in nature (Qu et al., 2020). The enzyme's active site forms a compound with the hydroxyl group of tyrosine or other chemicals.

The catalytic site is also classified into three states. The main difference between the three types is the copper ion and the structure of the active center. Eoxy is square in shape made of two copper(II) atoms, where NH ligand is attached to each copper atom. The Cu-Cu has a weaker axial bond, whereas the NH ligands are attached with strong equatorial bonds to each Cu atom. By producing peroxides, the oxygen binds and connects the copper sites. Cu-Cu bonds have a length of roughly 0.35 nm. The active center of the eoxy may be represented as [Cu (II)-O_2-Cu (II), which represents a combination of oxygen molecules that results in structure creation. The electrical structure of peroxides is important for eoxy's biological effects. Emet is similar to eoxy in that it has two tetragonal copper(II) atoms connected by a bridge endogenous in nature. The distinguishable feature is that instead of peroxide, hydroxide is the bridging ligand between copper ions. Eoxy and emet are also different in terms of oxidative characteristics. Emet cannot oxidize monophenolic compounds, whereas eoxy can oxidize both monophenols and diphenols at the same time. In the absence of substrates, emet is the predominant form found in organisms, and a gradual oxidation mechanism breaks it down. Same as the deoxyhemocyanin, deoxytyrosinase (edeoxy) possess symmetric structure [(Cu(I)-Cu(I)]. Because since the ligands that cause bridging between binuclear copper and peroxide or hydroxide, OH in water is a crucial ligand bridging (Qu et al., 2020) (Fig. 14.1).

There is a four-helix bundle in the domain that binds copper, which provides residues of histidine that are six in number to the copper ions at the active site. Tyrosinase and tyrosinase-like proteins have very similar crystal structures. The signal peptide having *N*-terminal, a sole *trans*-membrane helix plus a short, an intra-melanosomal domain, flexible *C*-terminal cytoplasmic domains are all conserved.

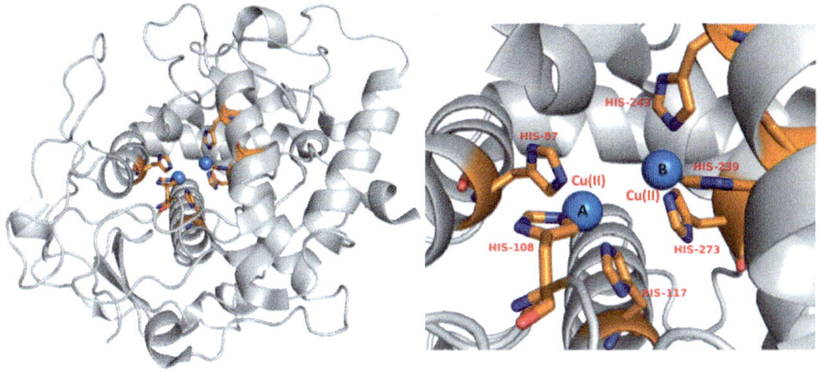

FIGURE 14.1 Crystal structure of tyrosinase the figure has been reproduced with permission from (Pillaiyar et al., 2017).

14.3.2 Oxygenase

14.3.2.1 Heme oxygenase (HO)

The entire human HO-1 is 288 amino acids long; however, the crystallized core does not have the carboxy-terminal 55 amino acids, making it 233 amino acids long. This carboxy-terminal deletion had a slight impact on catalytic activity and did not affect regiospecificity (Schuller et al., 1998). The initial nine and last ten amino acids of the shortened protein are not visible in the configuration due to the irregularity of these residues in the electron density maps (Montellano, 2000). The configuration of human HO-1 is made largely of helices and represents new protein folds. The heme is maintained among 2 helices by the carboxylate groups, which engage with arg183, lys22, lys18, lys179, and tyr134 at the molecular surface (Schuller et al., 1998). The residue his25 represents the proximal heme iron ligand. Three glycines provide the distal helix kinks right above the heme and flaccidity in the extremely preserved pattern gly139–asp–leu–ser–gly–gly144. The heme group is in direct interaction with gly139 and gly143. Such observations, combined with the distal helix's substantial crystallographic thermal factors, clearly suggest that the distal helix's elasticity permits the heme crevice to open and close in order to bind to the heme substrate and dissociate the biliverdin product. At a distance of 1.8 from the iron, a distal water ligand is identified in the crystal structure.

Furthermore, there are no polar side chains in near proximity to directly bind hydrogen to this water molecule, specifically no histidines. There are nonpolar residues that might obstruct the catalytic process. The -*meso*-heme side of the protein is present on the surface, whereas the other three sides are buried within. A hydrophobic barrier composed of phe214, met34, and phe37 supports the actual oxidized -*meso* site. Above the -*meso* border, nevertheless, a massive water-filled hollow channel extends into the protein core. The space supplied by this cavity for evacuation of the carbon monoxide created in the reaction may be significant in avoiding self-inhibition of the enzyme, in combination with the higher selectivity for the attachment of O_2 over carbon monoxide in HO-1 vs myoglobin (Montellano, 2000).

Although HO-1 lacks cysteines, HO-2 has two preserved cysteine residues. In HO-2, cysteine residues are found as cys–pro pairs separated by phenylalanine, a configuration Zhang and Guarente observed in different proteins as a heme-binding regulatory motif (HRM). It has been proposed that heme bound to these sites works as a nitric oxide sink, but there is no evidence for this, or indeed any unique role for the HRMin HO-2. (Montellano, 2000).

14.3.2.2 Cytochrome p450

External monooxygenases are p450s that require an external electron transfer partner protein for activation. Because of the existence of the heme moiety as a prosthetic group, the term cytochrome was coined. The letter p in p450 holds for pigment. The number 450 refers to these enzymes' ability to produce a 'Soret absorption peak' after reduction and bonding with carbon monoxide (CO) at 450 nm (Omura and Sato, 1964). However, as the fifth ligand of the heme-iron, they all utilize a cysteine thiolate group, so they have been dubbed heme-thiolate proteins. (Hannemann et al., 2007). Cyps are 400–500 amino acid proteins (Montellano et al., 2005). Low spin and high spin are two spin states of ferric iron (Fe^{3+}) (Shannon and Prewitt, 1970). A sixth axial ligand is created by water molecules of the Fe^{3+} in the substrate free form, maintaining the ls state of the ion, according to spectral, NMR, and crystallographic evidence (Poulos et al., 1986; Groves and Watanabe, 1988). When substrates bind to the enzyme, the iron-water molecule is displaced, shifting the Fe^{3+} coordination state from six to five, causing the Fe^{3+} to move out of the plane of the heme ring (Raag & Poulos, 1989). The above-mentioned residues bind the redox partner cytochrome b5. The eukaryotic accessory flavoprotein NADPH-p450 reductase is thought to bind to the same basic region (Estrada et al., 2013). In humans, the active center volume ranges from 190 å3 (Porubsky et al., 2008) to an estimated 1438 å3 (Schoch et al., 2008), however, in bacteria, the active site volume is reported to be 2446 å3 (Takahashi et al., 2014). The sizes of p450's active center stay consistent with familiar substrate selectivity over availability, although the active site shape varies. For example, the huge cavity of p450 2c8 is relatively "l-shaped," whereas the cavity of p450 3a4 is more "open," indicating broader catalytic specificity.

14.4 Mechanism of Catalytic Action

14.4.1 Tyrosinase: mechanism of the reaction

Tyrosinase is a crucial biocatalyst in the manufacture of melanin. It is the essential pigment, produced by a physiological technique known as melanogenesis known as melanosomes, which are still available in black; melanocytes are skin dendritic cells. Their activity is the primary determinant of hair and skin color. By absorbing UV radiation and eliminating reactive oxygen species, they perform a critical function in shielding the skin from harmful UV light. The number of melanosomes grows due to exposure to sunshine, increasing their melanin concentration and allowing them to transfer to keratinocytes.

The development of an efficient coordination link between the OH group on the substrate and the active center of the tyrosinase is the main mechanism by which tyrosinase and related substrates react. Monophenolase and diphenolase activity in mammals is triggered by eoxy and appropriate substrates. Monophenols (L-tyrosine) are oxidized by monophenolase to generate *o*-quinones (*o*-dopaquinone), a key progenitor of melanin, eoxy, and emetandedeoxy. Oxidised-diphenols (L-dopa) can create *o*-dopaquinone amid diphenolase action. Researchers typically believe this process because it most correctly reflects the kinetic features of tyrosinase, in which the monophenol cycle is the rate-limiting step in melanin formation (Qu et al., 2020). (Fig. 14.2)

14.4.2 Oxygenase

14.4.2.1 Regulation mechanisms of heme oxygenase-1

Heme oxygenase (HO) is a microsomal enzyme that is divided into two forms: HO-1 and HO-2. These have indistinguishable substrate specificities and cofactors in humans, with about 40% amino acid sequence homology.

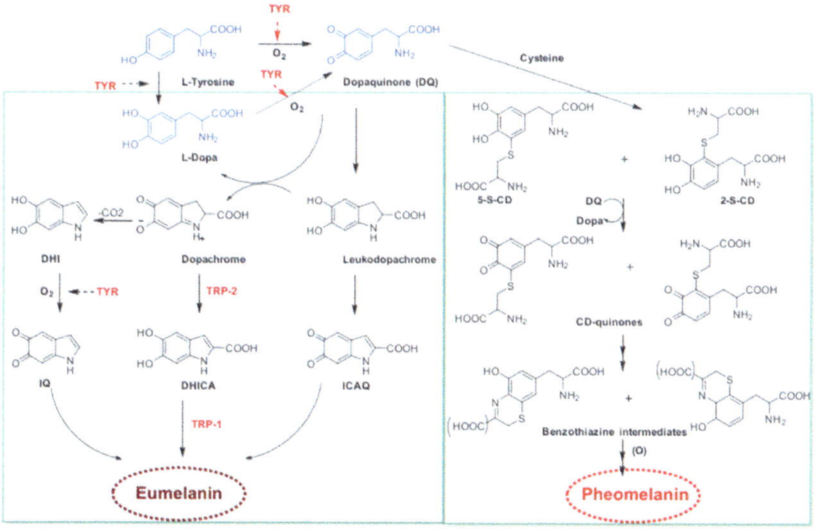

FIGURE 14.2 Melanin biosynthesis using tyrosinase figure has been reproduced from (Pillaiyar et al., 2017) with permission.

The instability of the proline- glutamic acid- a serine-threonine region in the carboxy terminus makes HO-1 vulnerable to breakdown. The decomposition of cellular heme is catalyzed by ho and NADPH cytochrome p450 reductase, which produces ferrous iron, biliverdin, and CO. The biliverdin reductase then converts biliverdin to bilirubin, one of the bile pigments. It is not a heme protein in and of itself, and it does utilize heme as an active center and a substrate.

The number of O_2 molecules consumed is three, and seven electrons are also consumed in the preceding phase. NADPH–cytochrome-p450-reductase provides these electrons in humans, but different electron-donating groups are used in microbes and plants. The heme is initially oxidized to -meso-hydroxyheme, which possesses free radical properties and interacts with oxygen to form verdoheme and CO in its deprotonated state. This reaction does not require any external reduction equivalents to take place. NADPH–cytochrome-p450-reductase and O_2, converts the verdoheme to biliverdin with free iron. The release of biliverdin is a relatively slow process. (Fig. 14.3)

14.4.2.2 Catalytic reaction by cytochrome p450

A wide range of redox reactions is catalyzed by the cyps, including dealkylation, heteroatom oxygenation, hydroxylation, desaturation, epoxidation, heme degradation, and many others (Guengerich, 2008; Isin & Guengerich, 2007; Hrycay & Bandiera, 2015). Monooxygenation processes driven by cyp entail the inclusion of one atom of oxygen into the substrate and the water reduction (Isin & Guengerich, 2007). The energy necessary to break the connection between the oxygen molecules is provided by NADPH cytochrome p450 reductase or cytochrome b5. Cytochrome b5 functions as a redox partner and hence as an allosteric modulator for several cyps implicated in mammalian steroidogenesis control (Storbeck et al., 2015).

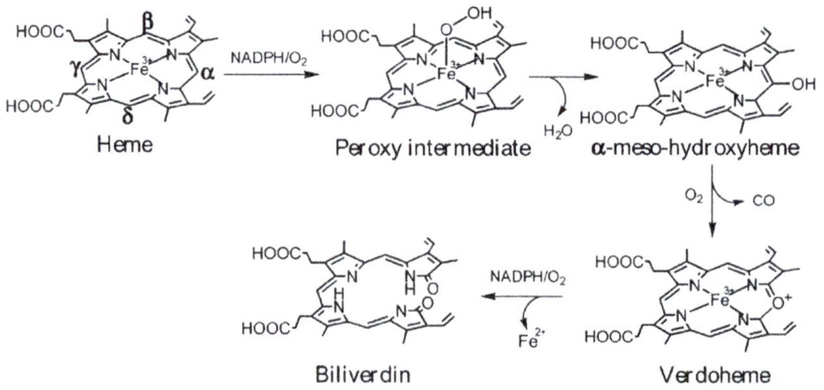

FIGURE 14.3 Mechanism of producing biliverdin through heme oxygenase-1 catalysis. (David J. Schuller, 1998).

Cytochrome b5 is accountable for providing the second electron to specific cyps faster than cyp reductase in the smooth endoplasmic reticulum (Im &Waskell, 2011). While electrons are transported from NADPH to redoxin via redoxin reductase and subsequently to cyp in the mitochondria (Guengerich et al., 2016; Gonzalez, 1990). In the absence of NADPH, cyp enzymes use a range of donors (oxygen atom), including peracids, chlorite, iodosobenzene, perborate, hydroperoxides, percarbonate, periodate, and n-oxides (Guengerich, 2008).

The removal of a hydrogen atom from the substrate during cyp-catalyzed monooxygenation results in the synthesis of a porphyrin radical ferryl intermediate or rather its ferryl radical resonance form, known as compound 1. Because compound 1 is very sensitive and unstable, it produces compound 2, a protonated intermediate with a carbon-centered alkyl radical that rebounds onto the ferryl hydroxyl molecule to produce the hydroxylated substrate (Hrycay & Bandiera, 2015).

14.5 Applications of Tyrosinase and Oxygenase

Tyrosinases are regarded as essential enzymes involved in a wide range of biological functions and defense systems. Tyrosine-related melanogenesis is liable for mammalian coloration of hair, skin, and eyes, as coloration is an important part of skin protection against UV radiation. Faulty melanin can come from abnormalities at any place in the biosynthetic route, such as defective tyrosinase or inadequacies produced by the migration of melanosomes to keratinocytes. Tyrosinases are type-3 copper proteins that are important in the first phase of melanin synthesis.

The economically accessible tyrosinase is derived from *agaricus bisporus* and *Streptomyces* spp., which are edible mushrooms (Agarwal et al., 2019). Due to enzymatic browning issues throughout postharvest storage, the best characterized tyrosinase comes from the mushroom *Bisporus*. Tyrosinases are enzymes found in fungi that are involved in the production and stabilization of spores. They are also important for defense, virulence processes, browning, and coloring. Tyrosinases present in a variety of different fungi have lately been discovered to have significant and beneficial features for various applications of biotechnology.

Because indigenous fungal tyrosinases are mainly intracellular and generated in small quantities, their uses are also limited (Halaouli et al., 2006). Enzyme demonstrates effective bioremediation of toxic phenolic compounds (Montiel et al., 2004; Ikehata and Nicell, 2000; Lee et al., 1996; Nagatsua and Sawadab, 2009) as biosensing for phenol tracking; in the therapy of parkinson's disease *via* synthesis of l-dopafrom l-tyrosine; and in the food and cosmetic sectors, biosynthesizing melanin (Obata et al., 2004; Konishi et al., 2007; Rao et al., 2011).

Oxygenases are widely involved in organic synthesis. They also show chemo, regio and enantio-selectivity. The insertion of a single (monooxygenases) or both (dioxygenases) oxygen atoms into an organic substrate, via molecular oxygen (O_2) as oxygen donor, is usually catalyzed by oxygenase. Mostly oxygenase utilizes inorganic co-factors, for example, flavins, metal-ions and hemes. This is done for transferring electrons to molecular oxygen for their activation. (Fig. 14.4)

14.5.1 Biological applications

In mammals, the tyrosinase enzyme is important in producing melanin in melanosomes, which causes the coloration of the skin, hair, and eyes. As a result, UV protection is provided, as well as the risk of skin cancer is reduced. The enzyme also plays an important role in an immune reaction and wound repair in plants, different invertebrates, and sponges due to its fungistatic, bacteriostatic, and antiviral capabilities (Ashida and Brey, 1995; Soderhall, 1998). While in the form of a cream or ointment, melanin precursor and tyrosinase work as natural antibacterial agents. Vital metabolites, including amino acids, lipids, carbohydrates, porphyrins, vitamins, and

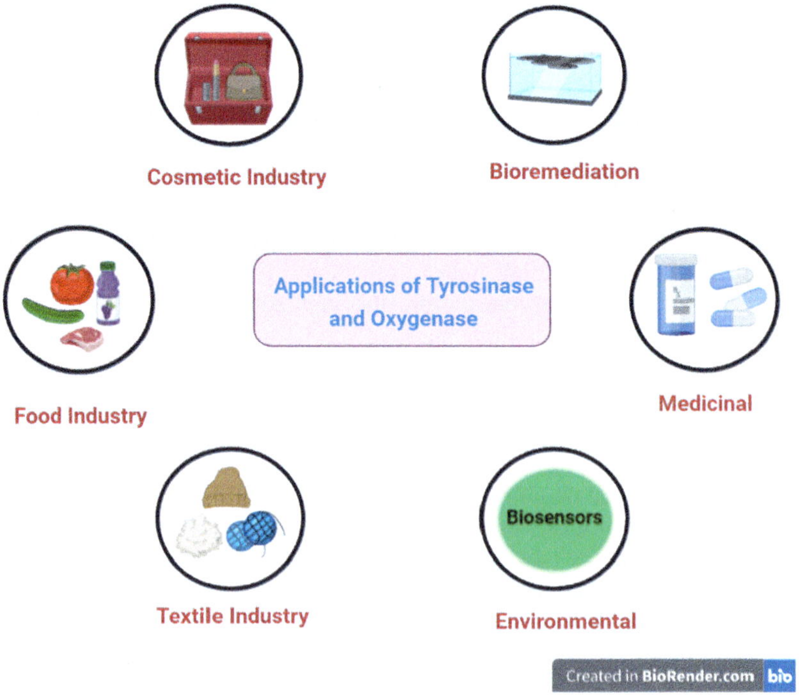

FIGURE 14.4 Various applications of tyrosinase and oxygenase.

hormones are biosynthesized, transformed, and degraded by oxygenase. They are also important in the metabolic elimination of foreign substances, including medications, pesticides, and toxins.

Furthermore, they have a significant role in the natural breakdown of different native and manufactured substances by soil and airborne microbes, making them important in environmental studies. Ribulose 1,5-bisphosphate carboxylase/oxygenase (rubisco), for example, catalyzes both CO_2 fixation and dioxygenase reactions. L-formyl kynurenine is produced by tryptophan 2,3-dioxygenase (tryptophan pyrrolase), which catalyzes the pyrrole ring cleavage by inserting two oxygen atoms. This enzyme is found in the liver and is highly selective for the amino acid l-tryptophan.

14.5.2 Applications in food industry

Enzyme tyrosinase can produce food additives like hydroxyl tyrosol. This can be utilized in the food industry, for example, to create aflavins, which are a group of chemicals found in black tea that have anticancer characteristics (Valipour & Burhan, 2016). Several food sectors employ various forms of biopolymers. The biopolymers being used must possess various capabilities such as forming a gel, good texture, and binding properties, which are important for some meat products. Different enzymes like transglutaminase, lactase, and tyrosinase are productively utilized to test pork and chicken proteins (Selinheimo et al., 2008). The enzyme mechanism is also used as emulsifiers that can be used for manufacturing less fat and low-calorie foods.

14.5.3 Applications in bioremediation

Serious environmental and health problems emerge due to the accumulation of phenols and dye in the ecosystem. Paper, chemical, textile, mining, coal, and petroleum industries, among others, generate wastewater, including phenols and their derivatives, such as chlorophenols (Marino et al., 2011). The tyrosinase enzyme can oxidize phenols into insoluble substances (quinones), making further elimination by precipitating or filtration comparatively easy (Faria et al., 2007). Tyrosine is a xenobiotic detoxification agent. Different sectors, such as leather, textile, pharmaceutical, and paper, are responsible for producing numerous dyes, the majority of which are azo dyes, which are detrimental to the environment. As a result, bacteria decolorize them first and use the tyrosinase enzyme to break them down. (Fig. 14.5)

Oxygenase has an essential part in the organic compound's metabolism through expanding reactivity or solubility of water or causing cleavage of an aromatic ring. Oxygenases are effective towards various chemicals, like chlorinated aliphatic, further having a vast substrate range. The cleavage of aromatic rings is brought about through the influx of oxygen atoms into the organic molecule. This provides us with a linear, non/less toxic chain. (TABLE 14.4)

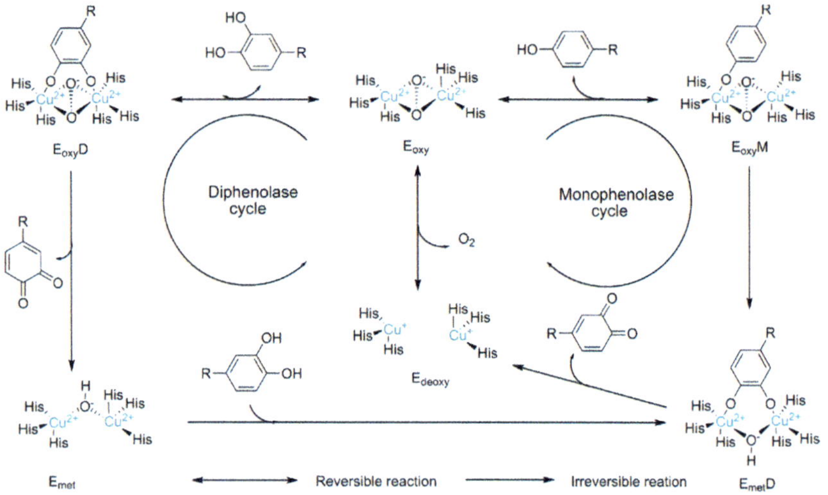

FIGURE 14.5 Monophenol cycle and bisphenol cycle catalytic cycle employed for bioremediation (Qu, et al., 2020).

TABLE 14.4 Different applications of tyrosinase.

Fields	Applications	References
Food industry	In cereal processing to improve baking. In dairy processing to link dairy proteins. In meat processing to improve gelation.	Facio, 2011 Selinheimo, 2008 Selinheimo, 2008
Medical fields	As a pro-drug in immunoassays and antibody microarrays	Selinheimo, 2008; Valipor & Buran, 2016; Zaidi et al., 2014
Textile industry	To modify the wool fibres and production of different dyes	Selinheimo, 2008; valipor & burhan, 2016
Cosmetic industry	As a self-tanning agent	Selinheimo, 2008; Valipor & Burhan, 2016
Environmental significance	As a biosensor	Selinheimo, 2008; Singh & Singh, 2002

14.5.4 Medicinal applications

Medical application of tyrosinases includes the synthesis of melanin for therapeutic purposes and the production of L-dopa, the choice of drug to treat

Parkinson's disease. Mushroom tyrosinase is employed to treat vitiligo and synthesis natural compounds with estrogenic activity. Other applications include the manufacture of antibiotic lincomycin and treating different neurological diseases (Valipour & Burhan, 2016).

14.5.5 Industrial applications

Tyrosinase has been considered as a possible prodrug in the treatment of melanoma patients who have responded well to tyrosinase activity. Sericin is peptide in silk textile companies used to make conjugates by microbial tyrosinase. Tyrosinase protein called spinosum is being utilized for making custom melanin and other polyphenolic substances from diverse phenols and catechol as raw materials. The materials get a vast range of applications, including organic semiconductors and photovoltaics. The tyrosinase enzyme is utilized to make proteins that are cross-linked, making it possible to reuse enzyme biocatalysts like lipase.

14.6 Concluding Remarks

Despite their widespread acceptance in the global market, there is an increasing need for diverse enzymes in various sectors. Tyrosinase is widely used in industrial operations such as medicines, cosmetics, and food production. It also has a lot of potential in enzyme medicine, agriculture, and various environmental applications. Synthetic melanin is also used, which acts as an emission shield and can be employed to play the role of cation exchangers, immunogens, medication transporters, antioxidants, and antiviral agents. Enzymatic procedures that utilize immobilized oxidoreductases (oxygenase and tyrosinases) have various advantages, including low cost, long-term sustainability, and gentle process conditions. Because of their poor substrate specificity, immobilized oxidoreductases may effectively convert a wide range of phenol and phenolic derivatives, including medicines, estrogens, bisphenols, and dyes, with clearance rates typically surpassing 90%. Thus more applications of tyrosinase and oxygenase need to be explored and used for solving the problems of mankind.

Acknowledgement

The authors thankfully acknowledge the basic facilities and infrastructure provided by Jaypee University of Information Technology (JUIT) Waknaghat, Solan, H.P. India.

Abbreviations

CYP Cytochrome p-450
COX Cylco-oxygenase
HO Hemeoxygenase

PCD Protocatechuate-3,4-dioxygenase
CTD catechol 1, 2-dioxygenase
LOX Lipooxygenases
CO Carbon monoxide
HRM Heme-binding regulatory motif

References

Afshin, A., et al., 2017. Health Effects of Overweight and Obesity in 195 Countries over 25 Years. N. Engl. J. Med. 377 (1), 13–27.

Agarwal, P., Singh, M., Singh, J., Singh, R., 2019. Microbial Tyrosinases: A Novel Enzyme, Structural Features, and Applications. Appl. Microbiology Bioeng. 3–19.

Alberti, K.G., Zimmet, P., Shaw, J., 2006. Metabolic syndrome--a new world-wide definition. A Consensus Statement from the International Diabetes Federation. Diabet. Med. 23 (5), 469–480. Available from: https://doi.org/10.1111/j.1464-5491.2006.01858.x. PMID: 16681555.

Ashida, M., Brey, P.T., 1995. Role of the integument in insect defense: pro-phenol oxidase cascade in the cuticular matrix. Proc. Natl Acad. Sci. U S A. 92 (23), 10698–10702. Available from: https://doi.org/10.1073/pnas.92.23.10698. PMID: 11607587; PMCID: PMC40679.

Coyne, V.E., a.-H. L., 1992. Induction of melanin biosynthesis in Vibrio cholerae. Appl. Env. Microbiol. 2861–2865.

Decker, H., Schweikardt, T., Tuczek, F., 2006. The first crystal structure of tyrosinase: all questions answered? Angew. Chem. Int. Ed. Engl. 45 (28), 4546–4550.

Estrada, D., F. Laurence, J.S., Scott, E.E., 2013. Substrate-modulated Cytochrome P450 17A1 and Cytochrome b5 Interactions Revealed by NMR. J. Biol. Chem. 288 (23), 17008–17018.

Gonzalez, F., 1990. Molecular genetics of the P-450 superfamily. Pharmacol. Ther. 45, 1–38.

Groves, J., Watanabe, Y., 1988. Reactive iron porphyrin derivatives related to the catalytic cycles of cytochrome P-450 and peroxidase. Studies of the mechanism of oxygen activation. J. Am. Chem. Soc. 110, 8433–8452.

Guengerich, F., 2008. Cytochrome P450 and chemical toxicology. Chem. Res. Toxicol. 21 (1), 70–83.

Guengerich, F., Waterman, M., Egli, M., 2016. Recent structural insights into cytochrome P450 function. Trends Pharmacol. Sci. 37 (8), 625–640.

Halaouli, S., Asther, M., Sigoillot, J.C., Hamdi, M., Lomascolo, A., 2006. Fungal tyrosinases: new prospects in molecular characteristics, bioengineering and biotechnological applications. J. Appl. Microbiol. 100 (2), 219–232. Available from: https://doi.org/10.1111/j.1365-2672.2006.02866.x. PMID: 16430498.

Hannemann, F., Ewen, K., et al., 2007. Cytochrome P450 systems-biological variations of electron transport chains. Biochim. Biophys. Acta 1770, 330–344.

Hayaishi, O., 2004. Oxygenases, Volume 3. *Elsevier*, pp. 178–182.

Held, T., Kutzner, J.H., 1990. Transcription of the tyrosinase gene in Streptomyces michiganensis DSM 40015 is induced by copper and repressed by ammonium. Microbiology 2413–2419.

Hrycay, E., Bandiera, S., 2015. Monooxygenase, peroxidase and peroxygenase properties and reaction mechanisms of cytochrome P450 enzymes. Adv. Exp. Med. Biol. 1–61.

Ikeda, K., Masujima, T., Sugiyama, M., 1996. Effects of methionine and Cu2+ on the expression of tyrosinase activity in Streptomyces castaneoglobisporus. J. Biochem. 1141–1145.

Ikehata, K., Nicell, J.A., 2000. Color and toxicity removal following tyrosinase-catalyzed oxidation of phenols. Biotechnol. Prog. 16 (4), 533–540. Available from: https://doi.org/10.1021/bp0000510. PMID: 10933824.

Im, S., Waskell, L., 2011. The interaction of microsomal cytochrome P450 2B4 with its redox partners, cytochrome P450 reductase and cytochrome b5. Arch. Biochem. Biophys. 507 (1), 144–153.

Isin, E., Guengerich, F., 2007. Complex reactions catalyzed by cytochrome P450 enzymes. Biochim. Biophys. Acta 1770 (3), 314–329.

Lantto, R., Niku-Paavola, M., Schonberg, C., Buchert, J., 2002. Atyrosinase enzyme. Intern. Publ. Num. WO 02/14484 A1.

Katz, E., B.A., 1988. Induction of tyrosinase by l-methionine in Streptomyces antibioticus. Can. J. Microbiol. 1297–1303.

Lerch, K., Ettinger, L., 1972. Purification and characterization of a tyrosinase from Streptomyces glaucescens. Eur. J. Biochem. 31 (3), 427–437. Available from: https://doi.org/10.1111/j.1432-1033.1972.tb02549.x. PMID: 4631007.

Liu, N., Zhang, T., Wang, Y.J., Huang, Y.P., Ou, J.H., Shen, P., 2004. A heat inducible tyrosinase with distinct properties from Bacillus thuringiensis. Lett. Appl. Microbiol. 39 (5), 407–412. Available from: https://doi.org/10.1111/j.1472-765X.2004.01599.x. PMID: 15482430.

Montellano, P.R., 2000. The mechanism of heme oxygenase. 2000. Curr Opin Chem Biol. 4, 221–227.

Montellano, O.D., James, J., De Voss, Paul, R., 2005. Substrate Oxidation by Cytochrome P450 Enzymes. Cytochrome P450: Structure, Mechanism, and Biochemistry 183–245. New York: Kluwer Academic/Plenum Publishers.

Montiel, A.M., Fernández, F.J., Marcial, J., Soriano, J., Barrios-González, J., Tomasini, A., 2004. A fungal phenoloxidase (tyrosinase) involved in pentachlorophenol degradation. Biotechnol. Lett. 26 (17), 1353–1357. Available from: https://doi.org/10.1023/B:BILE.0000045632.36401.86. PMID: 15604763.

Nagatsua, T., Sawadab, M., 2009. L-dopa therapy for Parkinson's disease: past, present, and future. Parkinsonism Relat. Disord. 15 (Suppl 1), S3–S8. Available from: https://doi.org/10.1016/S1353-8020(09)70004-5. PMID: 19131039.

Obata, H., Ishida, H., Hata, Y., Kawato, A., Abe, Y., Akao, T., et al., 2004. Cloning of a novel tyrosinase-encoding gene (melB) from Aspergillus oryzae and its overexpression in solid-state culture (Rice Koji). J. Biosci. Bioeng. 97 (6), 400–405. Available from: https://doi.org/10.1016/S1389-1723(04)70226-1. PMID: 16233650.

Omura, T., Sato, R., 1964. The carbon monoxide-binding pigment of liver microsomes: I. Evidence for its hemoprotein nature. J. Biol. Chem. 239, 2370–2378.

Pillaiyar, T., Manickam, M., Namasivayam, V., 2017. Skin whitening agents: medicinal chemistry perspective of tyrosinase inhibitors. J. enzyme inhibition Med. Chem. 32 (1), 403–425.

Porubsky, P.R., Meneely, K.M., Scott, E.E., 2008. Structures of human cytochrome P-450 2E1. Insights into the binding of inhibitors and both small molecular weight and fatty acid substrates. J. Biol. Chem. 283 (48), 33698–33707.

Poulos, T.L., Finzel, B.C., Howard, A.J., 1986. Crystal structure of substrate-free Pseudomonas putida cytochrome P-450. Biochemistry 25 (18), 5314–5322.

Qu, Y., Zhan, Q., e, a, 2020. Catalysis-based specific detection and inhibition of tyrosinase and their application. J. Pharm. Anal. 414–425.

Raag, R., Poulos, T., 1989. Crystal structure of the carbon monoxidesubstrate-cytochrome P-450CAM ternary complex. Biochemistry 28 (19), 7586–7592.

Rao, A., Pimprikar, P., Bendigiri, C., Kumar, A.R., Zinjarde, S., 2011. Cloning and expression of a tyrosinase from Aspergillus oryzae in Yarrowia lipolytica: application in L-DOPA biotransformation. Appl. Microbiol. Biotechnol. 92 (5), 951–959. Available from: https://doi.org/10.1007/s00253-011-3400-6. Epub 2011 Jun 18. PMID: 21687965.

Schoch, G.A., et al., 2008. Determinants of cytochrome P450 2C8 substrate binding: structures of complexes with montelukast, troglitazone, felodipine, and 9-cis-retinoic acid. J. Biol. Chem. 283 (25), 17227–17237.

Schuller, D.J., Wilks, A., Ortiz de Montellano, P., Poulos, T.L., 1998. Crystallization of recombinant human heme oxygenase-1. Protein Sci. 7 (8), 1836–1838. Available from: https://doi.org/10.1002/pro.5560070820. PMID: 10082382; PMCID: PMC2144071.

Shannon, R.D., Prewitt, C.T., 1970. Revised values of effective ionic radii. Acta Cryst. 26 (7), 1046–1048.

Shuster, V., F.A., 2009. Isolation, cloning and characterization of a tyrosinase with improved activity in organic solvents from Bacillus megaterium. J. Mol. Microbiol. Biotechnol. 188–200.

Soderhall, K., C.L., 1998. Role of the prophenoloxidase activating system in invertebrate immunity. Curr. Opin. Immunol. 10, 23–28.

Storbeck, K., Swart, A., et al., 2015. Cytochrome b5 modulates multiple reactions in steroidogenesis by diverse mechanisms. J. Steroid Biochem. Mol. Biol. 151, 66–73.

Takahashi, S., Nagano, S., et al., 2014. Structure-function analyses of cytochrome P450revI involved in reveromycin A biosynthesis and evaluation of the biological activity of its substrate, reveromycin T. J. Biol. Chem. 289 (47), 32446–32458.

Zaidi, K.U., Ayesha, S.Ali, Sharique, A.Ali, Ishrat, Naaz, 2014. Microbial Tyrosinases: Promising Enzymes for Pharmaceutical, Food Bioprocessing, and Environmental Industry. Biochem. Res. Int. 1–16.

Chapter 15

Application of microbial enzymes as drugs in human therapy and healthcare

Miguel Arroyo[1], Isabel de la Mata[1], Carlos Barreiro[2], José Luis García[3] and José Luis Barredo[4]

[1]*Department of Biochemistry and Molecular Biology, Faculty of Biology, University Complutense of Madrid, Madrid, Spain,* [2]*Department of Molecular Biology, Area of Biochemistry and Molecular Biology, Faculty of Veterinary, University of León, León, Spain,* [3]*Centro de Investigaciones Biológicas Margarita Salas CSIC, Madrid, Spain,* [4]*Curia, Parque Tecnológico de León, León, Spain*

15.1 Introduction

Many enzymes can be used to treat different human diseases, ranging from inherited congenital disorders to cancer (Cioni et al., 2021; Labrou, 2019; Tandon et al., 2021; Vachher et al., 2021; Vellard, 2003). For example, enzyme replacement therapy (ERT) is a crucial therapeutic option for controlling severe disorders triggered by the absence or deficiency of specific enzymes (de la Fuente et al., 2021; Li, 2018). Besides, many enzymatic drugs are used as food supplements and cosmetic formulations. That would explain how the demand for enzymes in the health-care market has increased steadily, with an anticipated value of $10,519 million by 2024 (Tandon et al., 2021). This chapter overviews the available microbial therapeutic enzymes (Fig. 15.1) and evaluates current discoveries and innovations.

15.2 Manufacture of therapeutic enzymes

15.2.1 Production and purification

Most therapeutic enzymes were traditionally isolated from human or animal tissues or even from undesirable sources such as urine (e.g., urokinase) or snake venom (e.g., ancrod, a thrombin-like serine protease). The advances in genetic engineering and recombinant DNA technology have allowed heterologous protein expression using different host organisms, thus enabling the

Biotechnology of Microbial Enzymes. DOI: https://doi.org/10.1016/B978-0-443-19059-9.00002-5
© 2023 Elsevier Inc. All rights reserved.

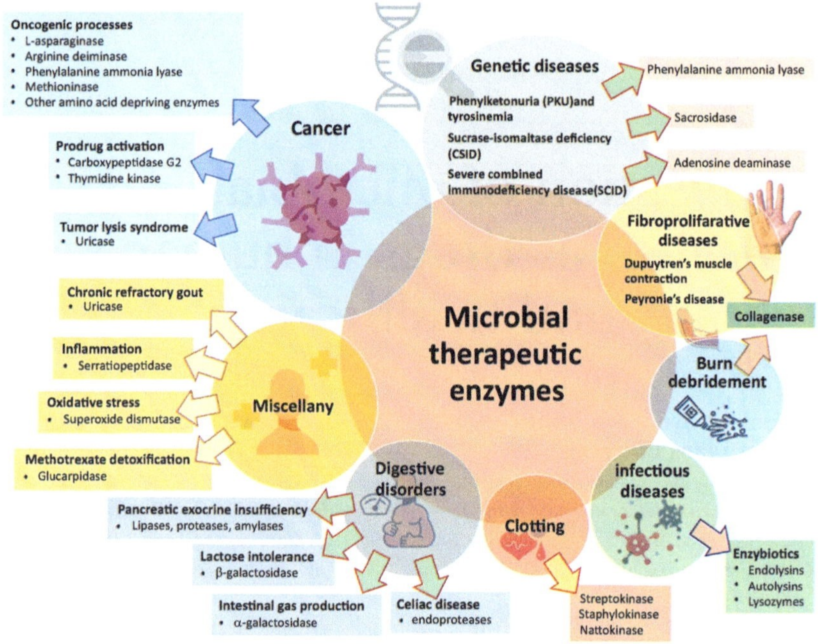

FIGURE 15.1 Disorders and diseases that can be treated with microbial enzymes.

industrial production of many therapeutic enzymes from animals, plants, or microbial origins. Furthermore, recombinant production of relevant human enzymes within Chinese hamster ovary (CHO) cells has enabled production without the risk of disease transmission through the infected source material, such as bovine spongiform encephalopathy. Transgenic animals, plants, and insects can produce recombinant enzymes with complex structures and post-translational modifications.

Nevertheless, recombinant microbial enzymes are particularly attractive since they are cheaper to produce, display higher catalytic efficiency, and have a more comprehensive pH range of activity and stability than their animal counterparts (Singh et al., 2016). Industrial production of therapeutic enzymes via microbial origin has been achieved using different hosts (Gupta and Shukla, 2017), such as bacteria (e.g., *Escherichia coli*, *Bacillus subtilis*, etc.) and yeasts (e.g., *Pichia pastoris* (*Komagataella pastoris*), *Saccharomyces cerevisiae*). *E. coli* is the most common and least expensive expression system, but the conventional use of this bacterium is limited to nonglycosylated proteins. In addition, high-level expression of recombinant enzymes in *E. coli* often results in protein aggregation as inclusion bodies, which imposes a tremendous hurdle in production and purification. In contrast, yeasts show several advantages as cloning hosts since they proliferate rapidly in simple cultivation media. After posttranslational modifications, they may produce glycosylated proteins and may even secrete heterologous proteins extracellularly. In this sense, extracellular enzyme production is advantageous since cell disruption to release the enzyme of interest is obviated. Industrial production of microbial enzymes can be performed employing either submerged fermentation (SmF) or solid-state fermentation (SSF) (Ashok and Kumar, 2017; Meghwanshi et al., 2020). Although most industries have adopted SmF for enzyme production, there is a renewed interest in SSF technology (Arora et al., 2018).

In all of the current production schemes, protein purification is considered the bottleneck in the manufacturing processes of therapeutic enzymes. As a rule of thumb, enzymes administered intravenously must be purified to homogeneity before their final formulation. This process includes multiple high-resolution chromatographic steps employing packed-bed columns. Moreover, the whole downstream process must comply with the principles and guidelines of the present-day "good manufacturing practices". Therapeutic enzymes for intravenous administration are often marketed as lyophilized preparations, including the pure protein, biocompatible buffering salts, and diluents to reconstitute the aqueous enzymatic solution (Gurung et al., 2013).

15.2.2 Preparation of "single-enzyme nanoparticles": SENization

The main limitations of therapeutic enzymes are the risk of unleashing an immune response, low stability, and lack of suitable delivery. In this context, the term "SENization" can be defined as the process of transforming a single enzyme into a "single-enzyme nanoparticle" (SEN), maintaining its original enzymatic function with improved half-life and utility (Kim et al., 2020). One of the most popular methods of SENization is PEGylation, which consists of the chemical modification of enzymes by site-specific or random covalent conjugation with polyethylene glycol (PEG). This particular approach renders PEGylated enzymes with lower immunogenicity, higher stability, and enhanced solubility (Dozier and Distefano, 2015; Turecek et al., 2016). In this sense, PEGylation increases the molecular weight and volume of small therapeutic proteins. Therefore the size of PEGylated enzymes is increased, resulting in a prolonged circulatory time by reducing renal clearance (Veronese and Mero, 2008). In addition, PEG interacts with those residues located at epitopic sites and makes them resistant to the action of proteases as well.

However, random PEGylation may interfere negatively with protein structural dynamics, decreasing enzyme catalytic activity (Rodríguez-Martínez et al., 2009). As reported, this shortcoming can be overcome by introducing cysteine residues at specific locations through site-directed mutagenesis, thereby reducing the degree of PEGylation, retaining the catalytic activity of the modified enzyme, and simultaneously modulating immunogenicity and proteolysis (Ramírez-Paz et al., 2018). However, recent studies have suggested that PEGylated proteins could induce an immune response, and therefore their bioactivity could be significantly decreased (Wan et al., 2017). Combined conjugation of PEGylated enzymes with arabinogalactan has been proposed as an alternative to solve the abovementioned problem (Qi et al., 2019). On the other hand, enzyme modification can also be performed by introducing other biocompatible moieties such as monoclonal antibodies, therefore the resulting SENs could be delivered to specific lesions or cells (Sharma and Bagshawe, 2017a). Other SENization strategies have been implemented to enhance the delivery of therapeutic enzymes to specific intended targets, such as glycosylation (e.g., with mannose-6-phosphate) (Kang et al., 2021) and the production of recombinant fusion proteins that contain the enzymatic domain and an affinity peptide that enhances the attachment to the cell surface (Ghosh et al., 2018; Van Nguyen et al., 2018).

15.2.3 Oral enzyme therapy

Oral administration of therapeutic enzymes is a noninvasive, patient-friendly, and highly desirable option, but enzymes are prone to deactivation in the gastrointestinal tract. In this sense, oral ERT has achieved a certain degree of success due to different stabilization strategies such as protein modifications or formulation approaches (Fuhrmann and Leroux, 2014). The primary sequence of enzymes can be manipulated by DNA recombinant technology to reduce their propensity to unfold in acidic environments and decrease their susceptibility to pepsin-mediated digestion. In addition, to altogether avoid digestion in the stomach at a low acidic pH (1.0–2.5), enzymes can be encapsulated in gastro-resistant polymer-coated tablets or microparticles. These coatings, prepared with cellulose acetate phthalate or methacrylate copolymers, are dissolved in a higher pH environment of the small intestine, thereby releasing the therapeutic enzymes. Currently, oral administration of exogenous enzymes is undertaken for exocrine pancreatic insufficiency, lactose and sucrose intolerance, phenylketonuria, and celiac disease.

15.3 Examples of microbial enzymes aimed at human therapy and healthcare

15.3.1 "Clot buster" microbial enzymes

When homeostasis fails, a blood clot or thrombus developed in the circulatory system might cause vascular blockage leading to ischemic stroke, pulmonary embolism, deep vein thrombosis, and acute myocardial infarction. Such pathologies require a rapid clinical intervention consisting of intravenous administration of thrombolytic agents that include tissue plasminogen activator (tPA), urokinase (UK), as well as microbial streptokinase and staphylokinase. These agents lead to the conversion of the zymogen plasminogen (PG) to plasmin (PN), an active protease that dissolves the fibrin clots. On the other hand, nattokinase dissolves blood clots by directly hydrolyzing fibrin and plasmin. Although the most relevant agents to dissolve clots are recombinant human-like tPAs (e.g., alteplase, reteplase, and tenecteplase) (Baldo, 2015), "clot buster" microbial enzymes show less undesirable side effects and lower expenses (Sharma et al., 2021).

15.3.1.1 Streptokinase

Streptokinase (SK, EC 3.4.99.22) is an extracellular protein with a molecular weight of 47.0 kDa, produced by various strains of β-hemolytic *Streptococci*. Unlike UK and tPA, SK has no proteolytic activity on its own but activates PG to PN indirectly. It first forms a high-affinity equimolar complex with PG (Fig. 15.2A), which results in conformational expression of an active catalytic site on this zymogen without the usual strict requirement for peptide bond cleavage (Boxrud et al., 2004). Although streptokinase is an efficient thrombolytic agent, conventional formulations (Streptase, Kabikinase) (Hermentin et al., 2005) have limited activity and several shortcomings such as protein immunogenicity, short half-life, lack of tissue targeting, and peripheral bleeding. Modified SKs, obtained by chemical modification or protein engineering, do not overcome these problems since SK domains responsible for antigenicity, stability, and plasminogen activation appear to overlap to some degree (Banerjee et al., 2004). Compared to recombinant human-like tPAs, microbial SK is the least expensive and thus remains an affordable therapy for poor health-care systems (Kunamneni et al., 2007). *Streptococcus equisimilis* H46A was initially considered the best SK

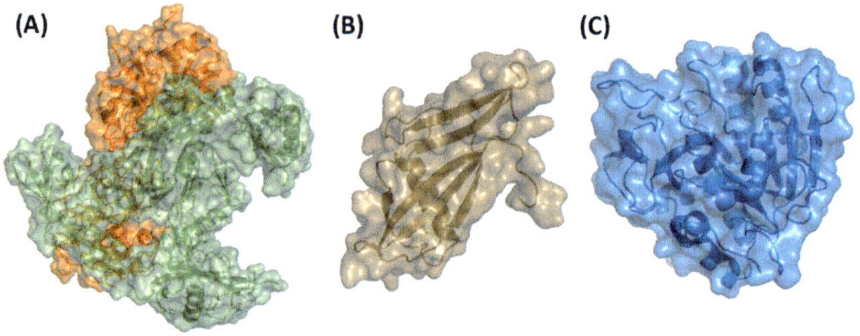

FIGURE 15.2 3D structures of clot buster microbial enzymes: (A) Complex of the catalytic domain of human plasmin (in orange) and streptokinase (in green) (PDB ID 1BML); (B) Staphylokinase sakSTAR variant (PDB ID 2SAK); (C) Nattokinase from *B. subtilis* natto (PDB ID 4DWW).

producer. Further cloning and expression of the enzyme encoding gene in *E. coli* has allowed the development of a fed-batch process for SK production (Aghaeepoor et al., 2019).

15.3.1.2 Staphylokinase

Despite the potent clot lysis obtained with recombinant tPAs and SK, these agents show several drawbacks: high production cost, systemic bleeding, intracranial hemorrhage, vessel reocclusion by platelet-rich, and retracted secondary clots and nonfibrin specificity. Staphylokinase (SAK, EC 3.4.24.29) from *Staphylococcus aureus* (Bahareh et al., 2017) is an indirect PG activator (Okada et al., 2001) of 17.0 kDa that hinders the systemic degradation of fibrinogen and reduces the risk of severe hemorrhage. SAK achieves its function primarily by forming plasminogen activating complex together with plasmin itself (Fig. 15.2B). Cost-effective recombinant wild-type SAK (SakSTAR) production (Faraji et al., 2017), and the design of mutant variants with fewer side effects have positioned this enzyme as a promising thrombolytic agent (Nedaeinia et al., 2020). Published reports have suggested that encapsulated SAK variant K35R into poly(lactic-*co*-glycolic acid) microspheres (He et al., 2006), as well as PEGylated SAK variant C104R (Xu et al., 2017), could be suitable for clinical applications. Likewise, alternative modification of SAK with an arabinogalactan-PEG conjugate has improved its therapeutic efficacy (Qi et al., 2019).

15.3.1.3 Nattokinase

Nattokinase (NK, EC 3.4.21.62) is a 27.7 kDa serine protease of the subtilisin family that is produced during the fermentation of soybeans by *B. subtilis* var. *natto* in the preparation of the Japanese food natto (Ali and Bavisetty, 2020). Compared to other fibrinolytic enzymes (UK, t-PA, and SK), NK exhibits the advantages of having no side effects, low cost, and long half-life. Western medicine has recently recognized its potential to be used as a therapeutic enzyme for treating cardiovascular disease, and it might be served as a functional food additive. The enzyme, the 3D structure (Fig. 15.2C) of which is almost identical to that of subtilisin E, can also be produced by recombinant techniques (Cai et al., 2017). In practice, NK production can be performed by recombinant *B. subtilis* using batch cultures rather than from bacteria using traditional extraction (Cho et al., 2010). Studies also indicate that NK can ameliorate other diseases, such as hypertension, stroke, Alzheimer's disease, and atherosclerosis. NK is currently undergoing a clinical trial study for atherothrombotic prevention (Weng et al., 2017), and there are some available food supplements in the market such as Nattiase, NattoZyme, and NattoMax, among others.

15.3.2 Microbial enzymes as digestive aids

The application of enzymes as drugs begins with their use as digestive aids to treat gastrointestinal diseases. For such purposes, the amylase preparation Taka-Diastase was launched by Parke-Davis & Company in the late 19th century (Patil, 2012), and many other enzymatic preparations have been marketed since then.

15.3.2.1 Lactose intolerance: β-galactosidase

Reduction in β-galactosidase (lactase, EC 3.2.1.108) activity during adulthood affects 70% of the world adult population. It can cause severe digestive disorders that are signs of lactose intolerance (Corgneau et al., 2017). Enzyme supplements containing microbial lactase can be used prior to or added to milk/dairy-containing meals to aid lactose digestion. However, commercial oral capsules or tablets under different brand names (Lactaid, Dairy-Care, Lacteeze, Silact, and Lifeplan, among others) contain fungal enzymes with characteristics not ideally suited to lactose hydrolysis. Therapeutic lactases should be resilient to digestive proteases and gastrointestinal acidic environments. In this sense, some lactases with these properties have been described to alleviate lactose intolerance, such as β-galactosidases from *Aspergillus niger* van Tiegh (Hu et al., 2010; O'Connell and Walsh, 2010) and *Aspergillus carbonarius* ATCC 6276 (O'Connell and Walsh, 2008).

15.3.2.2 Pancreatic exocrine insufficiency: cocktail of pancreatic enzymes

Pancreatic exocrine insufficiency (PEI) occurs in patients suffering cystic fibrosis, advanced chronic pancreatitis, or surgical pancreas resection. Nowadays, an oral cocktail of digestive enzymes (amylase, lipase, and protease) from the porcine pancreas (pancreatin/pancrelipase) is the only option for PEI treatment. Since 2009 there are currently six US-FDA-approved products under the brand names, Creon, Zenpep, Pancreaze, Ultresa, Viokace, and Pertzye which have been considered effective, safe, and of sufficient quality. However, there is a considerable interest in finding microbial enzymes that cope with those nutritional, allergenic, and even religious issues derived from their porcine origin. In this sense, extracellular lipase (EC 3.1.1.3) from *Yarrowia lipolytica* (YLLIP2) possesses several biochemical properties like human pancreatic lipase, such as a high activity on long-chain triglycerides at pH 6.0 in the absence of inhibition by bile salts, the generation of 2-monoglycerides and free fatty acids which are absorbed at the intestinal level, and high stability at low pH values at 37°C (Aloulou et al., 2007).

In a minipig model of PEI, oral administration of milligrams of YLLIP2 significantly increased the coefficient of fat absorption, similar to that obtained with 1.2 g pancreatin (Aloulou et al., 2015). Nevertheless, further studies with humans are needed to consider YLLIP2 as a therapeutic lipase either alone or in combination with proteases and/or amylases. Finally, it is worth mentioning the use of immobilized lipases to improve fat absorption in patients who receive enteral nutrition (EN). In this sense, lipases from *Chromobacterium viscosum*, *Pseudomonas fluorescens*, *Burkholderia cepacia*, and *Rhizopus oryzae* have been attached to a single-use cartridge (RELiZORB, Alcresta Therapeutics) that can be connected in-line with EN feeding sets. FDA has approved devices for pediatric and adult patients to hydrolyze up to 90% of the fats in enteral preparations (Stevens et al., 2018).

15.3.2.3 Celiac disease: endoproteases

Celiac disease (CeD) is triggered and maintained by ingesting wheat, barley, and rye gluten proteins. CeD affects approximately 1% of most world populations, causing a wide range of symptoms, including diarrhea, abdominal pain, and bloating. To date, a strict gluten-free diet is the only clinical solution for CeD since no effective or approved treatment is currently available. Oral administration of gluten-degrading enzymes is a promising therapeutic approach, but such enzymes must be active in gastroduodenal conditions and quickly neutralize the T-cell activating gluten peptides. Low stability and autodegradation hamper their therapeutic application, but different strategies have been devised to overcome these drawbacks (Wei et al., 2020). As a result, some microbial enzymes have been reported to cleave human digestion-resistant gluten peptides, including prolyl endopeptidase (PEP, EC 3.4.21.26) from *Sphingomonas capsulata* (SC-PEP) (Kabashima et al., 1998), recombinant glutamine-specific cysteine endoprotease from barley (EP-B2) produced in *E. coli* (Bethune et al., 2006; Vora et al., 2007), and engineered serine endoprotease from *Alicyclobacillus sendaiensis* (Gordon et al., 2012) (KumaMax). The combination of EP-B2 and SC-PEP in a 1:1 ratio (Latiglutenase) (Gass et al., 2007) has shown promising results in clinical studies.

15.3.2.4 Intestinal gas production: α-galactosidase

α-Galactosidase (α-Gal, EC 3.2.1.22) catalyzes the hydrolysis of α-1,6-linked terminal galactose residues of RFOs (raffinose family oligosaccharides such as raffinose, melibiose, stachyose) (Fig. 15.3), which cannot be digested in the small intestine due to the absence of endogenous α-galactosidase. Consequently, large amounts of carbon dioxide, hydrogen, and small quantities of methane and short-chain fatty acids are produced. Oral administration of exogenous α-Gal (Nutritek, Beano, Vitacost Gas Enzyme, and Veganzyme, among others) improves digestion of RFOs from legumes, minimizing bloating, and preventing flatulence (Di Stefano et al.,

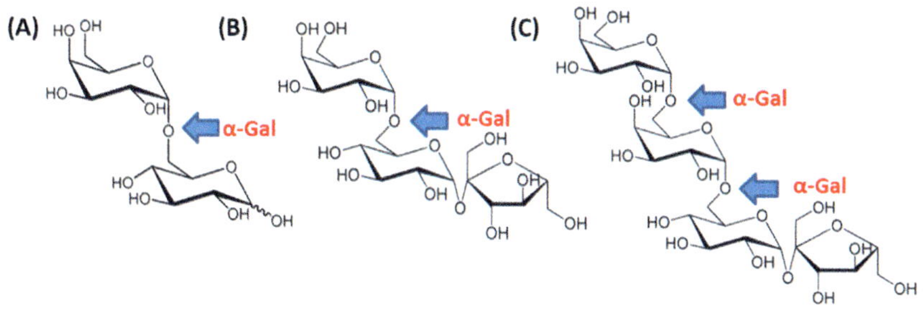

FIGURE 15.3 Hydrolysis of α-1,6 glycosidic linkages of RFOs catalyzed by α-galactosidase (α-Gal): (A) melibiose; (B) raffinose; (C) stachyose. *RFOs*, raffinose family oligosaccharides.

2007). *A. niger* has been used as the primary microbial source to produce α-Gal on a large scale (Bhatia et al., 2020; Katrolia et al., 2014).

15.3.3 Microbial enzymes for the treatment of congenital diseases

15.3.3.1 Phenylketonuria (PKU) and tyrosinemia: phenylalanine ammonia lyase

Phenylalanine ammonia lyase (EC 4.3.1.24), hereafter abbreviated as PAL, has emerged as an important therapeutic enzyme for the treatment of phenylketonuria (PKU), a rare autosomal recessive disorder resulting from the deficit of hepatocellular phenylalanine hydroxylase (PAH) (Kawatra et al., 2020). PAL catalyzes the nonoxidative deamination of L-phenylalanine (L-Phe) to *trans*-cinnamic acid and ammonia (Fig. 15.4C), reducing the enhanced L-Phe level that alters cognitive functions. Since PAL is absent in mammals, the interest of researchers has been focused on searching microbial PALs for biomedical applications. This enzyme can be found in some microorganisms such as fungi, yeasts, cyanobacteria, and bacteria (including actinomycetes). In this sense, large quantities of recombinant PAL from *Rhodotorula toruloides* (*Rt*PAL, formerly *Rhodosporidium toruloides*) were first obtained to meet the amounts required for PKU treatment (Sarkissian et al., 1999). However, the use of such enzyme (Fig. 15.5C) (Calabrese et al., 2004) showed certain limitations (reduced specific activity, short half-life, and proteolysis at neutral pH) when administered intraperitoneally and enterally. This led to studies that were focused on finding novel PALs such as those from the cyanobacteria *Anabaena variabilis* (*Av*PAL) (Fig. 15.5D) (Moffitt et al., 2007) and the yeast *Rhodotorula glutinis* (*Rg*PAL) (Zhu et al., 2013) and improving their properties by different approaches such as site-directed mutagenesis, directed enzyme evolution, and PEGylation. As a result, a PEGylated *Av*PAL double mutant (C503S, C565S) (pegvaliase, Palynziq) has been recently approved in 2018 (US-FDA) and 2019 (EMA) for combating PAH deficiency and treating PKU (Kawatra et al., 2020).

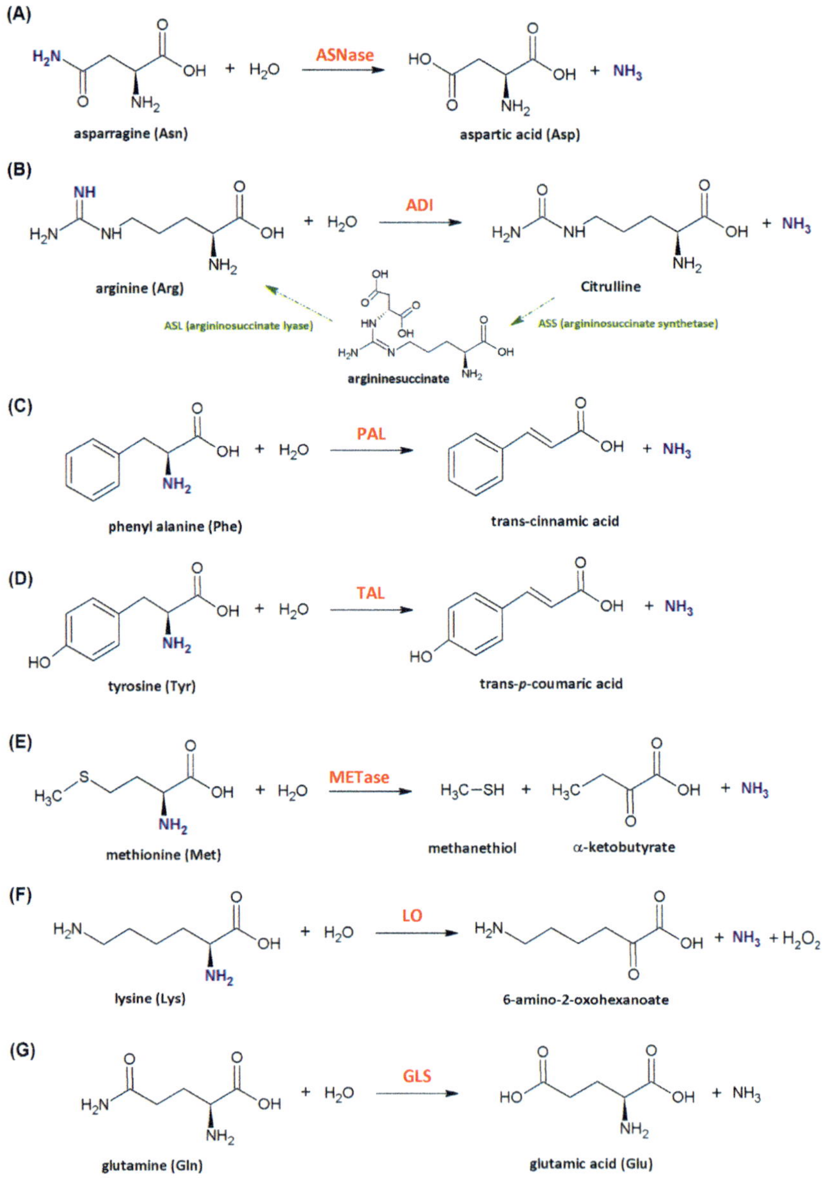

FIGURE 15.4 Reactions catalyzed by amino acid depriving enzymes: (A) L-ASNase: L-asparaginase; (B) ADI: arginine deiminase; (C) PAL: phenylalanine ammonia lyase; (D) TAL: tyrosine ammonia lyase; (E) METase: methionine gamma lyase (methioninase); (F) LO: L-lysine-α-oxidase; (G) GLS: glutaminase.

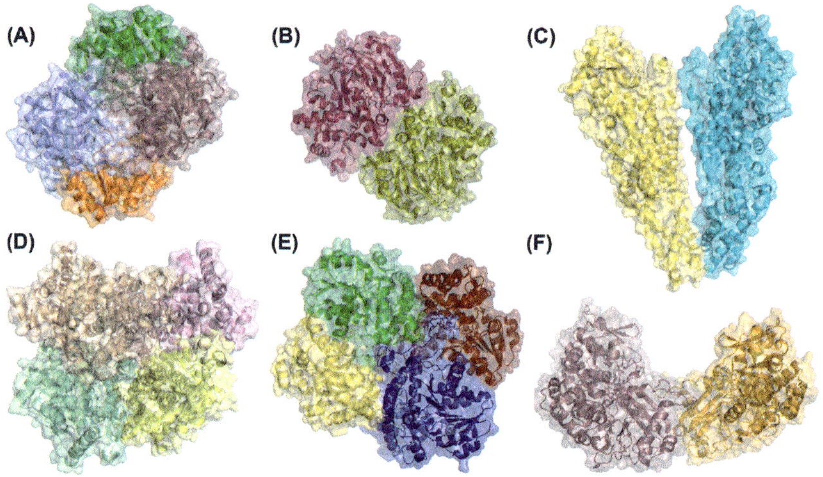

FIGURE 15.5 Available structures of amino acid depriving enzymes: (A) L-asparaginase type II from *Escherichia coli* (PDB ID 3ECA); (B) ADI: arginine deiminase from *Mycoplasma arginini* (PDB ID 1LXY); (C) Phenylalanine ammonia lyase from *R. toruloides* (PDB ID 1T6J); (D) Phenylalanine ammonia lyase from *A. variabilis* (PDB ID 2NYN); (E) Methioninase from *Pseudomonas putida* (PDB ID 1GC2); (F) L-lysine-α-oxidase from *Trichoderma viride* (PDB ID 3X0V).

Conversely, the treatment of hereditary tyrosinemia type 1 (HT1) with enzymes has been seriously considered since some PALs can also recognize L-tyrosine (L-Tyr) as a substrate and transform it into trans-*p*-coumaric acid (MacDonald and D'Cunha, 2007), therefore showing tyrosine ammonia lyase (TAL) activity (Fig. 15.4D). These enzymes with dual activity are often referred to as PAL/TAL enzymes (EC 4.3.1.25) (Vannelli et al., 2007). People suffering from HT1 disease are highly vulnerable to hepatocellular carcinoma, and therapy currently involves the administration of nitisinone combined with a lifetime restricted from L-Phe and L-Tyr. However, this drug has been shown to disturb the homeostasis of cognitive hormones (such as serotonin, dopamine, noradrenaline, etc.). Therefore complementary enzyme therapy is worth considering for relieving the burden of such a strict diet for HT1 patients. In this sense, an ancestral variant of *Rg*PAL and PAL/TAL from *Trichosporon cutaneum* was found to exhibit enhanced therapeutic efficacy and high potency to cure tyrosinemia (Hendrikse et al., 2020).

15.3.3.2 Sucrase-isomaltase deficiency: sacrosidase

Patients with inherited congenital sucrase-isomaltase deficiency cannot digest disaccharide sucrose, leading to abdominal distension, bloating, and watery diarrhea. Since a strict diet without sucrose-containing food is difficult to follow, especially in children, the clinical option to treat the disease consists of the oral administration of yeast extract (preferably on a filled stomach) supplemented with a yeast-derived sacrosidase solution (Sucraid, approved by US-FDA in 1998) (Treem et al., 1999). Sacrosidase (invertase, EC 3.2.1.26) is exceptionally resistant to acidic environments due to enzyme glycosylation and its use at high concentrations (Treem et al., 1993).

15.3.3.3 Severe combined immunodeficiency disease: adenosine deaminase

Adenosine deaminase (ADA, EC 3.5.4.4) is an important enzyme in the purine degradation and salvage pathway, and its complete or near-complete absence to abnormal DNA synthesis and repair enhanced apoptosis of cells and impaired intracellular signaling. Inherited defects in the ADA gene cause severe combined immune deficiency, which has been treated with PEGylated bovine-extracted ADA (pegademase, Adagen) since its approval by US-FDA in 1990 (Hershfield et al., 1987). To overcome the reliance on bovine products and the associated risk of infections, a recombinant ADA mutant (C74S) was expressed in *E. coli* and conjugated to PEG using succinimidyl carbonate as a linker (Murguia-Favela et al., 2020). This new PEGylated enzyme (elapegademase, Revcovi) was approved by US-FDA in 2018 for ADA-deficient patients, and Adagen has been discontinued in North America since then.

15.3.4 Microbial enzymes for the treatment of infectious diseases: enzybiotics

Nowadays, there is an urgent and increasing necessity for new innovative antibiotics due to the quick emergence of antibiotic resistance worldwide and the scarce development of new antimicrobial molecules (Barreiro and Barredo, 2021). Recently, "enzybiotics" derived from endolysins have arisen as an alternative to conventional antibiotics (Ho et al., 2021). Endolysins (EC 3.2.1.17) are phage-derived enzymes that degrade cell wall peptidoglycan, thereby killing bacteria by osmotic imbalance (Domingo-Calap and Delgado-Martínez, 2018). These enzymes (such as Cpl-1, Pal, Cpl-7, etc.) usually contain an *N*-terminal enzymatically active domain and a *C*-terminal binding domain that attaches specifically to teichoic acids unique to the pneumococcal surface (Hermoso et al., 2003). Their high specificity reduces the probability of developing bacterial resistance, and their low stability and lack of solubility have been overcome by protein engineering (São-José, 2018). In addition, protein-engineered endolysins have also been designed to improve bacterial lysis (Love et al., 2020). For instance, Cpl-711 combines endolysins Cpl-1 and Cpl-7S from Cp-1 and Cp-7 pneumococcal bacteriophages against

Streptococcus pneumoniae and their multiresistant strains (Díez-Martínez et al., 2015), although other examples of chimeric lysins can be found in the literature (Yang et al., 2019a).

Another key example is the recent development of enzybiotics against Gram-negative pathogens in which the exogenous application of endolysins is hindered by the outer membrane (OM). These novel antimicrobials (termed Artilysins) are based on a fusion of a selected endolysin and a specific OM-peptide (Briers et al., 2014; Gerstmans et al., 2016; Rodríguez-Rubio et al., 2016). As a result of this research, some commercial enzybiotics have been developed for human healthcare, such as StaphEfekt (Micreos) (Totté et al., 2017) that specifically kills *S. aureus* (including methicillin-resistant *Staphylococcus aureus*; MRSA) on intact skin, as well as Artilysins (Lysando AG) for wound care sprays to combat several pathogenic Gram-positive and Gram-negative bacteria. Currently, several lysins have progressed to human clinical trials for the treatment of *S. aureus* infections, such as natural lysin CF-301 (NCT03163446), chimeric lysin P128 (NCT01746654), and recombinant endolysin SAL-1 (N-Rephasin SAL200 (Kim et al., 2018), NCT03089697).

Other enzybiotics include autolysins (*N*-acetylmuramoyl L-alanine amidases, EC 3.4.24.38) such as LytA amidase from *S. pneumoniae* (Rodríguez-Cerrato et al., 2007), and also lysozymes (β-1,4-*N*-acetylmuramidases, EC 3.2.1.17) (Masschalck and Michiels, 2003). Due to its mild antibacterial effect, lysozyme is included in topical formulations to cure and prevent acne and bedsores, wound dressings, and mouthwash for patients suffering from xerostomia. Currently, most of these health-care products (e.g., Reflap, Biotene) are based on hen egg-white lysozyme (HWL) due to its availability and low-cost production. However, human lysozyme (hLYZ) is safer and has two and a half times higher antimicrobial activity than HWL. In this sense, optimization of recombinant hLYZ production using *Kluyveromyces lactis* K7 would allow to reduce its price and be more affordable (Ercan and Demirci, 2015; Ercan and Demirci, 2014). On the other hand, antivirulence approaches can reduce resistance development, targeting pathogenicity without exerting a bacteriostatic or bactericidal effect. In this sense, *quorum sensing* (QS) systems control the expression of virulence factors and biofilm formation in many pathogens, and interference with QS (termed *quorum quenching*, QQ) has been proposed as a strategy for the reduction of pathogenicity and prevention of bacterial infections. *N*-Acyl homoserine lactones (AHLs) are important QS signal molecules that modulate the virulence of Gram-negative pathogenic bacteria and QQ enzymes involved in AHL degradation (Chen et al., 2013) [mainly microbial AHL acylases (Serrano-Aguirre et al., 2021; Velasco-Bucheli et al., 2020) and AHL lactonases (Kyeong et al., 2015)] have been proposed as novel agents to treat infectious diseases (Basavaraju et al., 2016). However, further research is needed to prove their therapeutic potential.

15.3.5 Microbial enzymes for burn debridement and fibroproliferative diseases: collagenase

Enzymatic debridement entails the controlled digestion and removal of necrotic tissues from skin ulcers, burns, and wounds, employing proteases from different biological sources. Consequently, scars rapidly heal, and patients do not suffer surgical debridement of surrounding healthy tissues. Although proteolytic enzymes extracted from fruits such as papain from papaya (Debridace) and bromelain from pineapple (Nexobrid) have been positively used in burn injuries (Heitzmann et al., 2020), the efficacy of bacterial collagenase (EC 3.4.24.3) has also been proven to be very useful. Ointments containing collagenase from *Clostridium histolyticum* (such as Santyl) have been successfully used in clinical practice for many years as debriding agents in chronic wounds (Sheets et al., 2016). Recently, a novel collagenase from *Vibrio alginolyticus* has been developed as an improved debriding agent, which is much gentler on perilesional, healthy skin than *C. histolyticum* collagenase (Di Pasquale et al., 2019). Apart from its topical applications, intralesional injection of *C. histolyticum* collagenase (Xiaflex, Xiapex) is currently used for the treatment of Dupuytren's muscle contraction (Kaplan, 2011) and Peyronie's disease (Dhillon, 2015), fibroproliferative disorders of the tunica albuginea of the palmar fascia of the hand and penis, respectively.

15.3.6 Enzymes for the treatment of cancer

15.3.6.1 Enzymes against oncogenic processes

Targeted cancer therapies are based on metabolic variations, which differentiate normal from neoplastic cells. In this case, some auxotrophic tumors require a high metabolic input of some amino acids to support the proliferative ability of their cells. Consequently, amino acid deprivation therapy (AADT) is emerging as a promising strategy for developing novel therapeutics against cancer (Dhankhar et al., 2020). The enzymes used in AADT are mostly of microbial origin for easy production and availability. Many have been crystallized, and their 3D structures have been elucidated (Fig. 15.5). This section will overview the most important microbial amino acid depriving enzymes.

15.3.6.1.1 L-Asparaginase (L-ASNase)

L-Asparagine (L-Asn) is a nonessential amino acid in normal cells but essential to certain malignant cells due to a deficiency in the expression of asparagine synthetase. Since L-asparaginase (L-ASNase, EC 3.5.1.1) catalyzes the conversion of L-asparagine to aspartic acid and ammonia (Fig. 15.4A), the clinical use of this enzyme was based on depriving such malignant cells of their source of L-Asn to survive. Therefore L-ASNase has been used as the main therapeutic agent to treat some types of blood cancers, for example, acute lymphoblastic leukemia (ALL), acute myeloid leukemia, and non-Hodgkin lymphoma (Chand et al., 2020). Current commercial L-ASNases for human therapy are derived from two microbial sources: ASNase type II from *E. coli* (hereafter EcAII) (de Moura et al., 2020) and ASNase from *Dickeya dadantii* (hereafter EcA since the microbial source was formerly described as *Erwinia chrysanthemi*) (Aghaiypour et al., 2001). In addition, ASNase from *Erwinia carotovora* (hereafter EwA) (Krasotkina et al., 2004) has emerged as a potential therapeutic enzyme due to its increased glutaminase activity since glutamine can recover asparagine-deprived cells through asparagine regeneration via a transamidation chemical reaction.

L-ASNases consist of four identical subunits and 140–150 kDa of molecular mass (Fig. 15.5A). The *N*- and *C*-terminal domains of two adjacent monomers are involved in forming each active site, which contains the catalytic nucleophile Thr15, common to all ASNases. Clinical use of microbial ASNases has been challenging because proteases break down these enzymes in the blood, and consequently, their half-lives ($t_{1/2}$) are short. The sensitivity of EwA to protease inactivation and its low half-life has hindered its clinical application. Currently, few commercial ASNases (native, recombinant, and PEGylated) have been produced industrially since their US-FDA or EMA approval (Table 15.1). All the available ASNases products should be used with care because of the possibility of severe reactions, including anaphylaxis and sudden death. Among all of them, PEGylated EcAII (pegaspargase, Oncaspar) is currently the most effective oncology product for any ALL treatment since 2006. Additionally, EcA (crisantaspase, Erwinase) is the second-line drug in case of allergic reaction to native or PEGylated EcAII. However, EcA shows very poor stability in the blood stream compared to PEGylated EcAII ($t_{1/2} = 6–16$ h vs 6 days, respectively). It is worth mentioning that a new modified ASNase was obtained replacing the succinimidyl succinate (SS) linker in PEGylated EcAII with a succinimidyl carbamate linker (Angiolillo et al., 2014), creating a more stable ASNase (calaspargase pegol, Asparlas) that the US-FDA approved as part of a multiagent chemotherapeutic regimen for ALL in patients aged from 1 month to 21 years.

15.3.6.1.2 Arginine deiminase (ADI)

L-ASNase is probably the most relevant growth-inhibitory enzyme for treating acute leukemia and certain lymphomas. However, alternative treatments have been sought due to their serious side effects, including anaphylactic shock, coagulopathies, and liver and pancreatic toxicity. In this sense, arginine

TABLE 15.1 Available commercial ASNases for human therapy.

Enzyme	Microbial source	Year of approval[a]	Commercial trade name	Manufacturer
Native ASNase	*Escherichia coli*	1978 (US-FDA)	Elspar	Ovation Pharmaceuticals
			Leukanase	Sanofi-Aventis
			Kidrolase	Sanofi-Aventis, EUSA Pharma
SS-PEGylated ASNase	*E. coli*	1994 (US-FDA)	Oncaspar	Enzon Pharmaceuticals
SC-PEGylated ASNase	*E. coli*	2018 (US-FDA)	Asparlas	Servier Pharmaceuticals
Native recombinant ASNase	*E. coli*	2016 (EMA)	Spectrila	Medac Gelleschaft
Native recombinant ASNase	*D. dadantii* (formerly *E. chrysanthemi*)	2011 (US-FDA)	Erwinase, Erwinaze	Porton Biopharma, EUSA Pharma

[a]*Food and Drug Administration from the United States of America (US-FDA); European Medicine Agency (EMA).*

deiminase (ADI, EC 3.5.3.6) has been suggested to be a potentially better therapeutic agent than L-ASNase (Gong et al., 2000). Specifically, the enzyme ADI has been shown to degrade dietary arginine and result in enhanced killing in select tumor cells that lack argininosuccinate synthetase (ASS, EC 6.3.4.5), the rate-limiting step in the synthesis of arginine from citrulline (Fig. 15.4B). Therefore arginine deprivation has been shown to decrease cancer cell survival and induce autophagy and later cell death via caspase-independent apoptosis. And although ADI is synthesized by several microorganisms (*Pseudomonas* spp., *Streptococcus* spp., etc.) (Han et al., 2016), the highest amount of enzyme is produced by *Mycoplasma arginini* (*Ma*ADI). *Ma*ADI is composed of two identical monomeric subunits of 45.0 kDa according to its 3D structure (Das et al., 2004) (Fig. 15.5B). Modification of *Mycoplasma* ADI with 20,000 molecular weight PEG (ADI-PEG20) has resulted in PEGylated enzyme with longer half-life and reduced immunogenicity in animal models, but ongoing clinical trials with humans (Table 15.2) must confirm its potential for cancer treatment of ASS-deficient tumors (Abou-Alfa et al., 2018; Harding et al., 2018; Tomlinson et al., 2015).

TABLE 15.2 Clinical trials with microbial enzymes for amino acid deprivation cancer therapy.

Enzyme	Microbial source	Status	ClinicalTrials.gov identifier
PAL (phenylalanine ammonia lyase)	PEG-recombinant *Anabaena variabilis* PAL	Under phase II clinical trial	NCT00925054
L-ASNase (L-asparaginase)	PEG-recombinant *Erwinia* asparaginase	Under phase III clinical trial	NCT03150693
	Recombinant *Escherichia coli* asparaginase	Completed phase III clinical trial	NCT00784017
ADI (Arginine deiminase)	PEG-*Mycoplasma* arginine deiminase (ADI-PEG 20)	Completed phase III clinical trial	NCT01287585
METase (Methioninase)	Recombinant *Salmonella enterica,* serotype *typhimurium* (SGN1) expressing METase	Under phase I clinical trial	NCT05038150

15.3.6.1.3 Phenylalanine ammonia lyase

Apart from its clinical application in PKU treatment (Section 15.3.3.1), PAL has demonstrated effectiveness in regressing L-Phe auxotrophic tumors by rapidly depleting this amino acid (Fig. 15.4C) (Babich et al., 2013; Yang et al., 2019b). Normal cells avert the L-Phe scarcity by using phenylpyruvate as a substitute for growth, whereas malignant cells cannot do so (MacDonald and D'Cunha, 2007). The chemotherapeutic potential of PALs from *A. variabilis* (*Av*PAL), *R. toruloides* (*Rt*PAL), and *R. glutinis* (*Rg*PAL) has been confirmed in vitro and in vivo using a variety of cancer cell lines and animal models (Abell et al., 1973; Babich et al., 2013; Yang et al., 2019b). However, further research and clinical trials on PALs are needed to warrant its development as an alternative for anticancer therapy (Table 15.2).

15.3.6.1.4 Methioninase (METase)

Methioninase or methionine gamma lyase (METase, MGL, EC 4.4.1.11) is an PLP-dependent enzyme that catalyzes the α,γ elimination of L-Met to α-ketobutyrate, methanethiol and ammonia (Fig. 15.4E). Application of this enzyme as a therapeutic agent has spawned interest in methioninase, since deprivation of L-Met can cause cell cycle arrest in late S/G2 phase in several cancers, such as primary ductal carcinoma, melanoma, glioma, osteosarcoma,

ALL, non-small lung cancer, and glioblastoma (Cavuoto and Fenech, 2012). In this case, recombinant METase from *P. putida* (*Pp*MGL) (Fig. 15.5E) (Sharma et al., 2014) has shown no clinical toxicity in patients with several advanced cancers (Hoffman et al., 2019). Likewise, a recombinant strain of *Salmonella enterica* (serotype *typhimurium* (SGN1) that expresses METase) is under phase I clinical trial to assess its safety and tolerability as a therapeutic bacterium that can preferentially replicate and accumulate in tumors and starve them of essential L-Met (Table 15.2).

15.3.6.1.5 Other amino acid depriving enzymes

Other amino acid depriving enzymes of different microbial sources have also sparked interest as potential anticancer agents, such as L-lysine-α-oxidases (LO, EC 1.4.3.14) (Fig. 15.4F) from *T. viride* (*Tv*LO) (Fig. 15.5F) (Amano et al., 2015) and *Trichoderma harzianum* (*Th*LO) (Treshalina et al., 2000) and L-glutaminase (L-glutamine amidohydrolase, GLS, EC 3.5.1.2) (Fig. 15.4G) from *Aeromonas veroni* (Jesuraj et al., 2017), *Aspergillus flavus* (Abu-Tahon and Isaac, 2019), *Bacillus cereus* (Singh and Banik, 2013), *Halomonas meridiana* (Mostafa et al., 2021), and *Streptomyces canarius* (Reda, 2015). However, the role of these enzymes in tumor suppression is controversial and requires further analysis.

15.3.6.2 Enzymes for prodrug activation

Directed enzyme prodrug therapy (DEPT) is a chemotherapy strategy that involves delivering prodrug-activating enzymes directly to a tumor site before administering a nontoxic prodrug, thereby enabling its conversion into the desired anticancer therapeutics and allowing its administration in smaller quantities (Yari et al., 2017). Several delivery vectors have been explored for DEPT application, including antibodies (ADEPT) (Sharma and Bagshawe, 2017a), genes (GDEPT) (Dachs et al., 2009), and gold-coated magnetic nanoparticles (MNDEPT) (Gwenin et al., 2011).

15.3.6.2.1 Antibody-directed enzyme prodrug therapy

SENization by introducing monoclonal antibodies has allowed a route to obtain modified enzymes which can be targeted to specific tumors. ADEPT allows the administered prodrugs to be metabolized into their active forms only at the extracellular areas of the tumor (Fig. 15.6). This approach has been widely reported by many groups using a wide variety of microbial enzymes and prodrugs (Schellmann et al., 2010; Sharma and Bagshawe, 2017b). However, it should be noted that application of ADEPT is still scarce, only being employed in clinical studies performed with carboxypeptidase G2 from *Pseudomonas* sp. (*Ps*CPG2, EC 3.4.17.11) (Springer et al., 1990) conjugated to antibodies directed at carcinoembryonic antigen (CEA) in combination with several mustard prodrugs such as CMDA (Martin et al.,

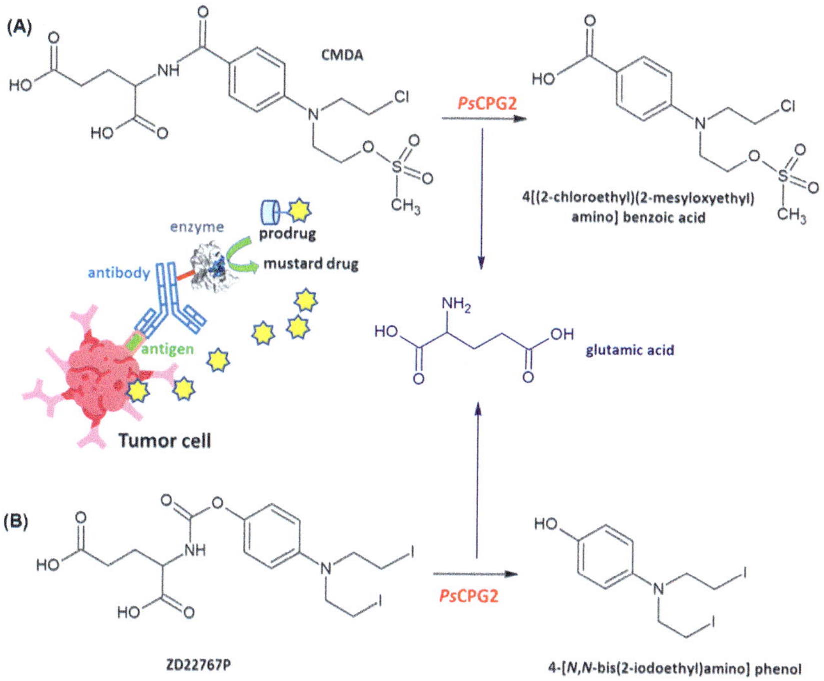

FIGURE 15.6 Reactions catalyzed by carboxypeptidase G2 from *Pseudomonas* sp. (*Ps*CPG2) employing mustard prodrugs as substrates for ADEPT: (A) CMDA: 4-[(2-chloroethyl)(2-mesyloxyethyl) amino] benzoyl-L-glutamic acid; (B) ZD2767P: *N*-{4-[*N*,*N*-bis(2-iodoethyl)amino]-phenoxycarbonyl}-L-glutamic acid.

1997) and ZD2767P (Francis et al., 2002; Mayer et al., 2006) (Fig. 15.6). Although *Ps*CPG2 can be conjugated to humanized or fully human antibodies, immunogenicity has been established as the major limitation of *Ps*CPG2. Therefore any new ADEPT system would need to address this drawback.

15.3.6.2.2 Gene-directed enzyme prodrug therapy

GDEPT is a two-step gene therapy approach where the gene for a nonendogenous enzyme is delivered to tumors. In contrast to ADEPT extracellular approach, enzymes are expressed intracellularly in targeting malignant cells where they can activate subsequently administered nontoxic prodrugs to cytotoxic drugs (Denny, 2003; Niculescu-Duvaz and Springer, 2005). The gene encoding the prodrug-activating enzyme needs to be expressed selectively and efficiently in tumor cells to spare normal tissue from damage. The application of viral vectors has been established as the most advanced system for GDEPT (sometimes categorized as VDEPT or viral DEPT) (Schepelmann and Springer, 2006) compared to gene expression within bacterial vectors (BDEPT). Best examples of GDEPT are herpes simplex virus thymidine kinase (HSV-tk, EC 2.7.1.21) in combination with ganciclovir (Sawdon et al., 2021), cytosine deaminase (CD, EC 3.5.4.5) from bacteria or yeast with 5-fluorocytosine (5-FC) (Horo et al., 2020; Warren et al., 2020), and nitroreductase from *E. coli* (NTR, E.C.1.6.99.7) with 5-(azaridin-1-yl)-2,4-dinitrobenzamide (NfnB-CB1954 combination) (Williams et al., 2015) for the treatment of different tumors. However, GDEPT has not yet reached the clinic despite enormous efforts, promising preliminary results, and more than 25 years since this approach was first mentioned (Alekseenko et al., 2021).

15.3.6.2.3 Magnetic nanoparticle-directed enzyme prodrug therapy

MNDEPT is a novel approach that might overcome the inherent drawbacks of the traditional biological methods used to deliver the prodrug-activating enzymes to the tumor site. In this scenario, gold-coated magnetic nanoparticles have been used as the delivery system of several enzymes, including novel nitroreductases from *B. cereus* (YdgI_Bc and YfkO_Bc) (Gwenin et al., 2015) and *Bacillus licheniformis* (YfkO) (Ball et al., 2021), as well as xenobiotic reductases from *P. putida* (XenA and XenB) (Ball et al., 2020), which can be directed to a solid tumor using an external magnetic field.

15.3.7 Other enzymes for the treatment of other health disorders

15.3.7.1 Uricase

Urate oxidase (uricase, EC 1.7.3.3) converts uric acid to 5-hydroxyisourate and H_2O_2, leading to the formation of water-soluble allantoin, which is then readily excreted by the kidneys. Therefore uricase can be used to treat diseases such as gout, hyperuricemia, and osteoporosis. There are several clinically approved uricases, namely Krystexxa (pegloticase, a PEGylated recombinant porcine uricase produced in *E. coli* launched by Savient Pharmaceuticals, USA), for the treatment of chronic refractory gout (Hershfield et al., 2010; Keenan et al., 2021) and Fasturtec/Elitek (rasburicase, a recombinant uricase from *A. flavus* expressed in *S. cerevisiae*) for tumor lysis syndrome (Bayol et al., 2002).

15.3.7.2 Serratiopeptidase

Serratiopeptidase (serrapeptase, EC 3.4.24.40) from *Serratia marcescens* (formerly *Serratia* E15, isolated in the silkworm *Bombyx mori*) has emerged as a potent antiinflammatory agent (Jadhav et al., 2020). As a result of its exceptional ability to dissolve dead and damaged tissues without harming the healthy ones, this protease is often prescribed in orthopedics, surgery, gynecology, otorhinolaryngology, and dentistry. It might also be used in atherosclerosis management since it shows caseinolytic and fibrinolytic properties

(Bhagat et al., 2013). Due to its hazardous origin, the recombinant production of serratiopeptidase by *E. coli* C43(DE3) has allowed the elimination of proteins associated with toxicity (Srivastava et al., 2019). The FDA has approved its use in dietary supplements for the US market. In contrast, the same organism has been approved for using serratiopeptidase in India as a pharmaceutical agent to treat acute pain in combination with nonsteroidal antiinflammatory drugs (NSAID) such as diclofenac (Volvib, Edase).

15.3.7.3 Superoxide dismutase

Low and diminished superoxide dismutase (SOD, EC 1.15.1.1) activity has been associated with a significant risk of oxidative stress that might result in various illnesses, such as hypertension, hypercholesterolemia, atherosclerosis, diabetes, heart failure, stroke, and other cardiovascular diseases. Thus SOD supplementation has been suggested to treat different pathophysiological conditions, from protecting the immune system to preventing ageing (Rosa et al., 2021; Younus, 2018). Some nonmicrobial SODs have gradually been introduced in some drug preparations, such as the veterinary NSAID product Palosein (orgotein from bovine source) and the antioxidant food supplements for sensitive skin GliSODin and SOD B (extracted from *Cucumis melo*), but with limited success. Recently, a highly stable manganese SOD (Mn-SOD) from a mutant thermoresistant strain (Ms-AOE) has shown promising results as a potential therapeutic agent to treat intestinal mucositis (Yan et al., 2020), a common side effect of anticancer chemotherapy.

The antioxidative effect of orally administered enteric-coated Mn-SOD from *Bacillus amyloliquefaciens* has also been investigated using γ-radiation- and dextransulfate sodium (DSS)-induced oxidative models in mice (Kang et al., 2018). Other SODs were found in different microbial sources such as *Corynebacterium glutamicum* (El Shafey et al., 2008) and the cyanobacteria *Nostoc* PCC 7120 (El Shafey et al., 2008). However, their potential application as therapeutic agents has not been assessed yet.

15.3.7.4 Glucarpidase

Glucarpidase (carboxypeptidase G2 from *Pseudomonas* sp. RS-16, *Ps*CPG2) is an FDA-approved recombinant enzyme (Voraxaze, BTG Specialty Pharmaceuticals) that rapidly hydrolyzes methotrexate (MTX) in patients with renal dysfunction during high-dose MTX treatment (Howard et al., 2016). As mentioned in Section 15.3.6.2.1, conjugation of *Ps*CPG2 to specific monoclonal antibodies has allowed its application in ADEPT. Despite the effectiveness of *Ps*CPG2 in cancer therapy and MTX detoxification, this enzyme has shown some drawbacks that include immunogenicity, protease susceptibility, as well as thermal instability, limiting its therapeutic application. Recently, a novel carboxypeptidase G2 from *Acinetobacter* sp. (*Ac*CPG2) has been kinetically characterized and compared with *Ps*CPG2

(Sadeghian and Hemmati, 2021), showing similar features but better pH versatility and higher thermostability that is beneficial during purification, formulation, transport, and storage of this enzyme variant.

15.3.7.5 Angiotensin-converting enzyme 2

Located on the surface of human vascular endothelium, respiratory epithelium, and other cell types, angiotensin-converting enzyme 2 (ACE2, EC 3.4.17.23) is a metalloprotease of 805 amino acids that cleaves angiotensin II (Ang II) into heptapeptide angiotensin-(1−7). As a key enzyme in the renin-angiotensin system, ACE2 negatively regulates vasoconstriction, proliferation, fibrosis, and proinflammation. It is closely connected to cardiovascular, kidney, and lung diseases (Li et al., 2017; Tan et al., 2018). Likewise, it has been reported that SARS-CoV2 bind to the peptidase domain of ACE2 through its receptor-binding domain of spike protein and participate in the viral infection process (Medina-Enríquez et al., 2020). Therefore the therapeutic application of soluble ACE2 as an antiviral agent was proposed for the treatment of patients with COVID-19, since the enzyme might act as a decoy to block the interaction between the spike protein and the surface-bound cellular ACE2 (Batlle et al., 2020; Marquez et al., 2021). Consequently, a soluble recombinant form of human ACE2 (rhACE2) was developed by Apeiron Biologics (APN01, alunacedase alfa), and it has successfully completed phase 2 clinical trials in severe diseases such as pulmonary arterial hypertension (NCT03177603) and COVID-19 (NCT04335136). Production of APN01 is carried out by CHO cells, which can express a glycosylated dimer of peptide chains that form the extracellular domain of ACE2 and several intra-subunit disulfide bridges. Nevertheless, a truncated rhACE2 (30−356aa) has been effectively expressed in *E. coli*, providing a novel, useful tool for the study of SARS-CoV2 infection mechanism, as well as for antiviral drug screening and development of new effective antibodies or fusion proteins against COVID-19 infection (Gao et al., 2021).

15.4 Concluding remarks

The application of microbial enzymes is an emerging alternative for treating a wide range of human diseases. Thanks to recent advances in recombinant DNA technology, the cost of microbial therapeutic enzymes and their improved mutant variants has been reduced, and their unwanted effects (immunogenicity, poor bioavailability, etc.) have been partially overcome by different approaches (such as PEGylation or encapsulation) in order to preserve both therapeutic efficacy and safety. Consequently, many commercial enzymes are now in the market and are currently used as drugs. Ongoing research on novel microbial enzymes will broaden their applicability to a broader spectrum of pathologies.

Abbreviations

AADT	amino acid deprivation therapy
ACE2	angiotensin-converting enzyme 2
ADA	adenosine deaminase
ADEPT	antibody-directed enzyme prodrug therapy
ADI	arginine deiminase
AHL	*N*-Acyl homoserine lactone
ALL	acute lymphoblastic leukemia
CeD	celiac disease
CHO	Chinese hamster ovary
COVID-19	Coronavirus disease 2019
DEPT	directed enzyme prodrug therapy
EC	Enzyme Commission
EcA	L-Asparaginase from *Dickeya dadantii*
EcAII	L-Asparaginase type II from *Escherichia coli*
EMA	European Medicines Agency
EN	enteral nutrition
ERT	enzyme replacement therapy
EwA	L-Asparaginase from *Erwinia carotovora*
GDEPT	gene-directed enzyme prodrug therapy
GLS	glutaminase
hLYZ	human lysozyme
HT1	hereditary tyrosinemia type 1
HWL	hen egg-white lysozyme
L-ASNase	L-Asparaginase
LO	L-Lysine-α-oxidase
METase	methioninase
MNDEPT	magnetic nanoparticle-directed enzyme prodrug therapy
MTX	methotrexate
NCT	National Clinical Trial
NK	nattokinase
NSAID	nonsteroidal antiinflammatory drug
OM	outer membrane
PAL	phenylalanine ammonia lyase
PEG	polyethylene glycol
PEI	pancreatic exocrine insufficiency
PEP	prolyl endopeptidase
PG	plasminogen
PKU	phenylketonuria
PN	plasmin
QQ	quorum quenching
QS	quorum sensing
RFOs	raffinose family oligosaccharides
rhACE2	recombinant human angiotensin-converting enzyme 2
SAK	staphylokinase
SEN	single-enzyme nanoparticle
SK	streptokinase

SmF	submerged fermentation
SOD	superoxide dismutase
SSF	solid-state fermentation
TAL	tyrosine ammonia lyase
tPA	tissue plasminogen activator
UK	urokinase
US-FDA	United States Food and Drug Administration

References

Abell, C.W., Hodgins, D.S., Stith, W.J., 1973. *In vivo* evaluation of chemotherapeutic potency of phenylalanine ammonia-lyase. Cancer Res. 33, 2529–2532.

Abou-Alfa, G.K., Qin, S., Ryoo, B.Y., Lu, S.N., Yen, C.J., Feng, Y.H., et al., 2018. Phase III randomized study of second line ADI-PEG 20 plus best supportive care vs placebo plus best supportive care in patients with advanced hepatocellular carcinoma. Ann. Oncol. 29, 1402–1408.

Abu-Tahon, M.A., Isaac, G.S., 2019. Purification, characterization and anticancer efficiency of L-glutaminase from *Aspergillus flavus*. J. Gen. Appl. Microbiol. 65, 284–292.

Aghaeepoor, M., Akbarzadeh, A., Kobarfard, F., Shabani, A.A., Dehnavi, E., Aval, S.J., et al., 2019. Optimization and high level production of recombinant synthetic streptokinase in *E. coli* using response surface methodology. Iran. J. Pharm. Res. 18, 961–973.

Aghaiypour, K., Wlodawer, A., Lubkowski, J., 2001. Structural basis for the activity and substrate specificity of *Erwinia chrysanthemi* L-asparaginase. Biochemistry 40, 5655–5664.

Alekseenko, I., Kuzmich, A., Kondratyeva, L., Kondratieva, S., Pleshkan, V., Sverdlov, E., 2021. Step-by-step immune activation for suicide gene therapy reinforcement. Int. J. Mol. Sci. 22, 9376.

Ali, A.M.M., Bavisetty, S.C.B., 2020. Purification, physicochemical properties, and statistical optimization of fibrinolytic enzymes especially from fermented foods: a comprehensive review. Int. J. Biol. Macromol. 163, 1498–1517.

Aloulou, A., Rodriguez, J.A., Puccinelli, D., Mouz, N., Leclaire, J., Leblond, Y., et al., 2007. Purification and biochemical characterization of the LIP2 lipase from *Yarrowia lipolytica*. Biochim. Biophys. Acta: Mol. Cell Biol. Lipids 1771, 228–237.

Aloulou, A., Schue, M., Puccinelli, D., Milano, S., Delchambre, C., Leblond, Y., et al., 2015. *Yarrowia lipolytica* lipase 2 is stable and highly active in test meals and increases fat absorption in an animal model of pancreatic exocrine insufficiency. Gastroenterology 149, 1910–1919.

Amano, M., Mizuguchi, H., Sano, T., Kondo, H., Shinyashiki, K., Inagaki, J., et al., 2015. Recombinant expression, molecular characterization and crystal structure of antitumor enzyme, L-lysine alpha-oxidase from *Trichoderma viride*. J. Biochem. 157, 549–559.

Angiolillo, A.L., Schore, R.J., Devidas, M., Borowitz, M.J., Carroll, A.J., Gastier-Foster, J.M., et al., 2014. Pharmacokinetic and pharmacodynamic properties of calaspargase pegol *Escherichia coli* L-asparaginase in the treatment of patients with acute lymphoblastic leukemia: results from children's oncology group study AALL07P4. J. Clin. Oncol. 32, 3874–3882.

Arora, S., Rani, R., Ghosh, S., 2018. Bioreactors in solid state fermentation technology: design, applications and engineering aspects. J. Biotechnol. 269, 16–34.

Ashok, A., Kumar, D.S., 2017. Different methodologies for sustainability of optimization techniques used in submerged and solid state fermentation. 3 Biotech. 7, 301.

Babich, O.O., Pokrovsky, V.S., Anisimova, N.Y., Sokolov, N.N., Prosekov, A.Y., 2013. Recombinant L-phenylalanine ammonia lyase from *Rhodosporidium toruloides* as a potential anticancer agent. Biotechnol. Appl. Biochem. 60, 316–322.

Bahareh, V., Navid, N., Manica, N., Maryam, Y., Bijan, Z., Younes, G., 2017. Staphylokinase enzyme: an overview of structure, function and engineered forms. Curr. Pharm. Biotechnol. 18, 1026–1037.

Baldo, B.A., 2015. Enzymes approved for human therapy: indications, mechanisms and adverse effects. Biodrugs 29, 31–55.

Ball, P., Halliwell, J., Anderson, S., Gwenin, V., Gwenin, C., 2020. Evaluation of two xenobiotic reductases from *Pseudomonas putida* for their suitability for magnetic nanoparticle-directed enzyme prodrug therapy as a novel approach to cancer treatment. MicrobiologyOpen 9, e1110.

Ball, P., Hobbs, R., Anderson, S., Thompson, E., Gwenin, V., Von Ruhland, C., et al., 2021. The YfkO nitroreductase from *Bacillus licheniformis* on gold-coated superparamagnetic nanoparticles: towards a novel directed enzyme prodrug therapy approach. Pharmaceutics 13, 517.

Banerjee, A., Chisti, Y., Banerjee, U.C., 2004. Streptokinase: a clinically useful thrombolytic agent. Biotechnol. Adv. 22, 287–307.

Barreiro, C., Barredo, J.L., 2021. Worldwide clinical demand for antibiotics: is it a real countdown? In: Barreiro, C., Barredo, J.L. (Eds.), Antimicrobial Therapies - Methods and Protocols, 2296. Humana Press, New York, NY, pp. 3–15.

Basavaraju, M., Sisnity, V.S., Palaparthy, R., Addanki, P.K., 2016. Quorum quenching: signal jamming in dental plaque biofilms. J. Dental Sci. 11, 349–352.

Batlle, D., Wysocki, J., Satchell, K., 2020. Soluble angiotensin-converting enzyme 2: a potential approach for coronavirus infection therapy? Clin. Sci. 134, 543–545.

Bayol, A., Capdevielle, J., Malazzi, P., Buzy, A., Bonnet, M.C., Colloc'h, N., et al., 2002. Modification of a reactive cysteine explains differences between rasburicase and Uricozyme®, a natural *Aspergillus flavus* uricase. Biotechnol. Appl. Biochem. 36, 21–31.

Bethune, M.T., Strop, P., Tang, Y., Sollid, L.M., Khosla, C., 2006. Heterologous expression, purification, refolding, and structural-functional characterization of EP-B2, a self-activating barley cysteine endoprotease. Chem. Biol. 13, 637–647.

Bhagat, S., Agarwal, M., Roy, V., 2013. Serratiopeptidase: a systematic review of the existing evidence. Int. J. Surg. 11, 209–217.

Bhatia, S., Singh, A., Batra, N., Singh, J., 2020. Microbial production and biotechnological applications of alpha-galactosidase. Int. J. Biol. Macromol. 150, 1294–1313.

Boxrud, P.D., Verhamme, I.M., Bock, P.E., 2004. Resolution of conformational activation in the kinetic mechanism of plasminogen activation by streptokinase. J. Biol. Chem. 279, 36633–36641.

Briers, Y., Walmagh, M., Grymonprez, B., Biebl, M., Pirnay, J.-P., Defraine, V., et al., 2014. Art-175 is a highly efficient antibacterial against multidrug-resistant strains and persisters of *Pseudomonas aeruginosa*. Antimicrob. Agents Chemother. 58, 3774–3784.

Cai, D., Zhu, C., Chen, S., 2017. Microbial production of nattokinase: current progress, challenge and prospect. World J. Microbiol. Biotechnol. 33, 84.

Calabrese, J.C., Jordan, D.B., Boodhoo, A., Sariaslani, S., Vannelli, T., 2004. Crystal structure of phenylalanine ammonia lyase: multiple helix dipoles implicated in catalysis. Biochemistry 43, 11403–11416.

Cavuoto, P., Fenech, M.F., 2012. A review of methionine dependency and the role of methionine restriction in cancer growth control and life-span extension. Cancer Treat. Rev. 38, 726–736.

Chand, S., Mahajan, R.V., Prasad, J.P., Sahoo, D.K., Mihooliya, K.N., Dhar, M.S., et al., 2020. A comprehensive review on microbial l-asparaginase: bioprocessing, characterization, and industrial applications. Biotechnol. Appl. Biochem. 67, 619–647.

Chen, F., Gao, Y.X., Chen, X.Y., Yu, Z.M., Li, X.Z., 2013. Quorum quenching enzymes and their application in degrading signal molecules to block quorum sensing-dependent infection. Int. J. Mol. Sci. 14, 17477–17500.

Cho, Y.-H., Song, J.Y., Kim, K.M., Kim, M.K., Lee, I.Y., Kim, S.B., et al., 2010. Production of nattokinase by batch and fed-batch culture of *Bacillus subtilis*. N. Biotechnol. 27, 341–346.

Cioni, P., Gabellieri, E., Campanini, B., Bettati, S., Raboni, S., 2021. Use of exogenous enzymes in human therapy: approved drugs and potential applications. Curr. Med. Chem. 29, 411–452.

Corgneau, M., Scher, J., Ritie-Pertusa, L., Le, D.T.L., Petit, J., Nikolova, Y., et al., 2017. Recent advances on lactose intolerance: tolerance thresholds and currently available answers. Crit. Rev. Food Sci. Nutr. 57, 3344–3356.

Dachs, G.U., Hunt, M.A., Syddall, S., Singleton, D.C., Patterson, A.V., 2009. Bystander or no bystander for gene directed enzyme prodrug therapy. Molecules 14, 4517–4545.

Das, K., Butler, G.H., Kwiatkowski, V., Clark, A.D., Yadav, P., Arnold, E., 2004. Crystal structures of arginine deiminase with covalent reaction intermediates: implications for catalytic mechanism. Structure 12, 657–667.

de la Fuente, M., Lombardero, L., Gomez-Gonzalez, A., Solari, C., Angulo-Barturen, I., Acera, A., et al., 2021. Enzyme therapy: current challenges and future perspectives. Int. J. Mol. Sci. 22, 9181.

de Moura, W.A.F., Schultz, L., Breyer, C.A., de Oliveira, A.L.P., Tairum, C.A., Fernandes, G. C., et al., 2020. Functional and structural evaluation of the antileukaemic enzymel-asparaginase II expressed at low temperature by different *Escherichia coli* strains. Biotechnol. Lett. 42, 2333–2344.

Denny, W.A., 2003. Prodrugs for gene-directed enzyme-prodrug therapy (suicide gene therapy). J. Biomed. Biotechnol. 1, 48–70.

Dhankhar, R., Gupta, V., Kumar, S., Kapoor, R.K., Gulati, P., 2020. Microbial enzymes for deprivation of amino acid metabolism in malignant cells: biological strategy for cancer treatment. Appl. Microbiol. Biotechnol. 104, 2857–2869.

Dhillon, S., 2015. Collagenase *Clostridium histolyticum*: a review in Peyronie's disease. Drugs 75, 1405–1412.

Di Pasquale, R., Vaccaro, S., Caputo, M., Cuppari, C., Caruso, S., Catania, A., et al., 2019. Collagenase-assisted wound bed preparation: an *in vitro* comparison between *Vibrio alginolyticus* and *Clostridium histolyticum* collagenases on substrate specificity. Int.Wound J. 16, 1013–1023.

Di Stefano, M., Miceli, E., Gotti, S., Missanelli, A., Mazzocchi, S., Corazza, G.R., 2007. The effect of oral alpha-galactosidase on intestinal gas production and gas-related symptoms. Dig. Dis. Sci. 52, 78–83.

Díez-Martínez, R., De Paz, H.D., García-Fernández, E., Bustamante, N., Euler, C.W., Fischetti, V.A., et al., 2015. A novel chimeric phage lysin with high *in vitro* and *in vivo* bactericidal activity against *Streptococcus pneumoniae*. J. Antimicrob. Chemother. 70, 1763–1773.

Domingo-Calap, P., Delgado-Martínez, J., 2018. Bacteriophages: protagonists of a post-antibiotic era. Antibiotics 7, 66.

Dozier, J.K., Distefano, M.D., 2015. Site-specific PEGylation of therapeutic proteins. Int. J. Mol. Sci. 16, 25831–25864.

El Shafey, H.M., Ghanem, S., Merkamm, M., Guyonvarch, A., 2008. *Corynebacterium glutamicum* superoxide dismutase is a manganese-strict non-cambialistic enzyme *in vitro*. Microbiol. Res. 163, 80–86.

Ercan, D., Demirci, A., 2014. Enhanced human lysozyme production in biofilm reactor by *Kluyveromyces lactis* K7. Biochem. Eng. J. 92, 2–8.

Ercan, D., Demirci, A., 2015. Enhanced human lysozyme production by *Kluyveromyces lactis* K7 in biofilm reactor coupled with online recovery system. Biochem. Eng. J. 98, 68–74.

Faraji, H., Ramezani, M., Sadeghnia, H.R., Abnous, K., Soltani, F., Mashkani, B., 2017. High-level expression of a biologically active staphylokinase in *Pichia pastoris*. Prep. Biochem. Biotechnol. 47, 379–387.

Francis, R.J., Sharma, S.K., Springer, C., Green, A.J., Hope-Stone, L.D., Sena, L., et al., 2002. A phase I trial of antibody directed enzyme prodrug therapy (ADEPT) in patients with advanced colorectal carcinoma or other CEA producing tumours. Br. J. Cancer 87, 600–607.

Fuhrmann, G., Leroux, J.C., 2014. Improving the stability and activity of oral therapeutic enzymes-recent advances and perspectives. Pharm. Res. 31, 1099–1105.

Gao, X., Liang, K., Mei, S., Peng, S., Vong, E.G., Zhan, J., 2021. An efficient system to generate truncated human angiotensin converting enzyme 2 (hACE2) capable of binding RBD and spike protein of SARS-CoV2. Protein Expr. Purif. 184, 105889.

Gass, J., Bethune, M.T., Siegel, M., Spencer, A., Khosla, C., 2007. Combination enzyme therapy for gastric digestion of dietary gluten in patients with celiac sprue. Gastroenterology 133, 472–480.

Gerstmans, H., Rodriguez-Rubio, L., Lavigne, R., Briers, Y., 2016. From endolysins to Artilysin®s: novel enzyme-based approaches to kill drug-resistant bacteria. Biochem. Soc. Trans. 44, 123–128.

Ghosh, D., Peng, X., Leal, J., Mohanty, R.P., 2018. Peptides as drug delivery vehicles across biological barriers. Int. J. Pharm. Investig. 48, 89–111.

Gong, H., Zölzer, F., von Recklinghausen, G., Havers, W., Schweigerer, L., 2000. Arginine deiminase inhibits proliferation of human leukemia cells more potently than asparaginase by inducing cell cycle arrest and apoptosis. Leukemia 14, 826–829.

Gordon, S.R., Stanley, E.J., Wolf, S., Toland, A., Wu, S.J., Hadidi, D., et al., 2012. Computational design of an α-gliadin peptidase. J. Am. Chem. Soc. 134, 20513–20520.

Gupta, S.K., Shukla, P., 2017. Sophisticated cloning, fermentation, and purification technologies for an enhanced therapeutic protein production: a review. Front. Pharmacol. 8, 419.

Gurung, N., Ray, S., Bose, S., Rai, V., 2013. A broader view: microbial enzymes and their relevance in industries, medicine, and beyond. Biomed. Res. Int. 2013, 329121.

Gwenin, V.V., Gwenin, C.D., Kalaji, M., 2011. Colloidal gold modified with a genetically engineered nitroreductase: toward a novel enzyme delivery system for cancer prodrug therapy. Langmuir 27, 14300–14307.

Gwenin, V.V., Poornima, P., Halliwell, J., Ball, P., Robinson, G., Gwenin, C.D., 2015. Identification of novel nitroreductases from *Bacillus cereus* and their interaction with the CB1954 prodrug. Biochem. Pharmacol. 98, 392–402.

Han, R.Z., Xu, G.C., Dong, J.J., Ni, Y., 2016. Arginine deiminase: recent advances in discovery, crystal structure, and protein engineering for improved properties as an anti-tumor drug. Appl. Microbiol. Biotechnol. 100, 4747–4760.

Harding, J.J., Do, R.K., Dika, I.E., Hollywood, E., Uhlitskykh, K., Valentino, E., et al., 2018. A phase 1 study of ADI-PEG 20 and modified FOLFOX6 in patients with advanced hepatocellular carcinoma and other gastrointestinal malignancies. Cancer Chemother. Pharmacol. 82, 429–440.

He, J.T., Su, H.B., Li, G.P., Tao, X.M., Mo, W., Song, H.Y., 2006. Stabilization and encapsulation of a staphylokinase variant (K35R) into poly(lactic-co-glycolic acid) microspheres. Int. J. Pharm. 309, 101–108.

Heitzmann, W., Fuchs, P.C., Schiefer, J.L., 2020. Historical perspectives on the development of current standards of care for enzymatic debridement. Medicina 56, 706.

Hendrikse, N.M., Holmberg Larsson, A., Svensson Gelius, S., Kuprin, S., Nordling, E., Syrén, P. O., 2020. Exploring the therapeutic potential of modern and ancestral phenylalanine/tyrosine ammonia-lyases as supplementary treatment of hereditary tyrosinemia. Sci. Rep. 10, 1315.

Hermentin, P., Cuesta-Linker, T., Weisse, J., Schmidt, K.-H., Knorst, M., Scheld, M., et al., 2005. Comparative analysis of the activity and content of different streptokinase preparations. Eur. Heart J. 26, 933–940.

Hermoso, J.A., Monterroso, B., Albert, A., Galán, B., Ahrazem, O., García, P., et al., 2003. Structural basis for selective recognition of pneumococcal cell wall by modular endolysin from phage Cp-1. Structure 11, 1239–1249.

Hershfield, M.S., Buckley, R.H., Greenberg, M.L., Melton, A.L., Schiff, R., Hatem, C., et al., 1987. Treatment of adenosine deaminase deficiency with polyethylene glycol-modified adenosine deaminase. N. Engl. J. Med. 316, 589–596.

Hershfield, M.S., Roberts, L.J., Ganson, N.J., Kelly, S.J., Santisteban, I., Scarlett, E., et al., 2010. Treating gout with pegloticase, a PEGylated urate oxidase, provides insight into the importance of uric acid as an antioxidant *in vivo*. Proc. Natl. Acad. Sci. USA 107, 14351–14356.

Ho, M.K.Y., Zhang, P.F., Chen, X., Xia, J., Leung, S.S.Y., 2021. Bacteriophage endolysins against gram-positive bacteria, an overview on the clinical development and recent advances on the delivery and formulation strategies. Crit. Rev. Microbiol. 1–24.

Hoffman, R.M., Tan, Y., Li, S., Han, Q., Zavala, J., Zavala, J., 2019. Pilot phase I clinical trial of methioninase on high-stage cancer patients: rapid depletion of circulating methionine. In: Hoffman, R.M. (Ed.), Methionine dependence of cancer and aging: Methods and Protocols. Springer, NY, pp. 231–242.

Horo, H., Porathoor, S., Anand, R., Kundu, L.M., 2020. A combinatorial approach involving *E. coli* cytosine deaminase and 5-fluorocytosine-nanoparticles as an enzyme-prodrug therapeutic method for highly substrate selective *in situ* generation of 5-fluorouracil. J. Drug. Deliv. Sci. Technol. 58, 101799.

Howard, S.C., McCormick, J., Pui, C.H., Buddington, R.K., Harvey, R.D., 2016. Preventing and managing toxicities of high-dose methotrexate. Oncologist 21, 1471–1482.

Hu, X., Robin, S., O'Connell, S., Walsh, G., Wall, J.G., 2010. Engineering of a fungal β-galactosidase to remove product inhibition by galactose. Appl. Microbiol. Biotechnol. 87, 1773–1782.

Jadhav, S.B., Shah, N., Rathi, A., Rathi, V., Rathi, A., 2020. Serratiopeptidase: insights into the therapeutic applications. Biotechnol. Rep. 28, e00544.

Jesuraj, S.A.V., Sarker, M.M.R., Ming, L.C., Praya, S.M.J., Ravikumar, M., Wui, W.T., 2017. Enhancement of the production of L-glutaminase, an anticancer enzyme, from *Aeromonas veronii* by adaptive and induced mutation techniques. PLoS One 12, 17.

Kabashima, T., Fujii, M., Meng, Y., Ito, K., Yoshimoto, T., 1998. Prolyl endopeptidase from *Sphingomonas capsulata*: isolation and characterization of the enzyme and nucleotide sequence of the gene. Arch. Biochem. Biophys. 358, 141–148.

Kang, J.-E., Kim, H.-D., Park, S.-Y., Pan, J.-G., Kim, J.H., Yum, D.-Y., 2018. Dietary supplementation with a *Bacillus* superoxide dismutase protects against γ-radiation-induced oxidative stress and ameliorates dextran sulphate sodium-induced ulcerative colitis in mice. J. Crohns Colitis 12, 860–869.

Kang, J.Y., Choi, H.Y., Kim, D.I., Kwon, O., Oh, D.B., 2021. *In vitro* N-glycan mannosyl-phosphorylation of a therapeutic enzyme by using recombinant Mnn14 produced from *Pichia pastoris*. J. Microbiol. Biotechnol. 31, 163–170.

Kaplan, F.T.D., 2011. Collagenase *Clostridium histolyticum* injection for the treatment of Dupuytren's contracture. Drugs Today 47, 653–667.

Katrolia, P., Rajashekhara, E., Yan, Q., Jiang, Z., 2014. Biotechnological potential of microbial α-galactosidases. Crit. Rev. Biotechnol. 34, 307–317.

Kawatra, A., Dhankhar, R., Mohanty, A., Gulati, P., 2020. Biomedical applications of microbial phenylalanine ammonia lyase: current status and future prospects. Biochimie 177, 142–152.

Keenan, R.T., Botson, J.K., Masri, K.R., Padnick-Silver, L., LaMoreaux, B., Albert, J.A., et al., 2021. The effect of immunomodulators on the efficacy and tolerability of pegloticase: a systematic review. Semin. Arthritis Rheum. 51, 347–352.

Kim, D.H., Lee, H.S., Kwon, T.W., Han, Y.M., Kang, N.W., Lee, M.Y., et al., 2020. Single enzyme nanoparticle, an effective tool for enzyme replacement therapy. Arch. Pharm. Res. 43, 1–21.

Kim, N.H., Park, W.B., Cho, J.E., Choi, Y.J., Choi, S.J., Jun, S.Y., et al., 2018. Effects of phage endolysin SAL200 combined with antibiotics on *Staphylococcus aureus* infection. Antimicrob. Agents Chemother. 62, e00731-18.

Krasotkina, J., Borisova, A.A., Gervaziev, Y.V., Sokolov, N.N., 2004. One-step purification and kinetic properties of the recombinant L-asparaginase from *Erwinia carotovora*. Biotechnol. Appl. Biochem. 39, 215–221.

Kunamneni, A., Abdelghani, T.T.A., Ellaiah, P., 2007. Streptokinase: the drug of choice for thrombolytic therapy. J. Thromb. Thrombolysis 23, 9–23.

Kyeong, H.H., Kim, J.H., Kim, H.S., 2015. Design of N-acyl homoserine lactonase with high substrate specificity by a rational approach. Appl. Microbiol. Biotechnol. 99, 4735–4742.

Labrou, N. (Ed.), 2019. Therapeutic Enzymes: Function and Clinical Implications. Springer Nature Singapore Pte Ltd.

Li, M.D., 2018. Enzyme replacement therapy: a review and its role in treating lysosomal storage diseases. Pediatr. Ann. 47, e191–e197.

Li, X.C., Zhang, J.F., Zhuo, J.L., 2017. The vasoprotective axes of the renin-angiotensin system: physiological relevance and therapeutic implications in cardiovascular, hypertensive and kidney diseases. Pharmacol. Res. 125, 21–38.

Love, M.J., Abeysekera, G.S., Muscroft-Taylor, A.C., Billington, C., Dobson, R.C.J., 2020. On the catalytic mechanism of bacteriophage endolysins: opportunities for engineering. Biochim. Biophys. Acta Proteins Proteom. 1868, 140302.

MacDonald, M.J., D'Cunha, G.B., 2007. A modern view of phenylalanine ammonia lyase. Biochem. Cell Biol. 85, 273–282.

Marquez, A., Wysocki, J., Pandit, J., Batlle, D., 2021. An update on ACE2 amplification and its therapeutic potential. Acta Physiol. 231, e13513.

Martin, J., Stribbling, S.M., Poon, G.K., Begent, R.H.J., Napier, M., Sharma, S.K., et al., 1997. Antibody-directed enzyme prodrug therapy: pharmacokinetics and plasma levels of prodrug and drug in a phase I clinical trial. Cancer Chemother. Pharmacol. 40, 189–201.

Masschalck, B., Michiels, C.W., 2003. Antimicrobial properties of lysozyme in relation to foodborne vegetative bacteria. Crit. Rev. Microbiol. 29, 191–214.

Mayer, A., Francis, R.J., Sharma, S.K., Tolner, B., Springer, C.J., Martin, J., et al., 2006. A phase I study of single administration of antibody-directed enzyme prodrug therapy with the recombinant anti-carcinoembryonic antigen antibody-enzyme fusion protein MFECP1 and a bis-iodo phenol mustard prodrug. Clin. Cancer Res. 12, 6509–6516.

Medina-Enríquez, M.M., López-León, S., Carlos-Escalante, J.A., Aponte-Torres, Z., Cuapio, A., Wegman-Ostrosky, T., 2020. ACE2: the molecular doorway to SARS-CoV-2. Cell Biosci. 10, 148.

Meghwanshi, G.K., Kaur, N., Verma, S., Dabi, N.K., Vashishtha, A., Charan, P.D., et al., 2020. Enzymes for pharmaceutical and therapeutic applications. Biotechnol. Appl. Biochem. 67, 586–601.

Moffitt, M.C., Louie, G.V., Bowman, M.E., Pence, J., Noel, J.P., Moore, B.S., 2007. Discovery of two cyanobacterial phenylalanine ammonia lyases: kinetic and structural characterization. Biochemistry 46, 1004–1012.

Mostafa, Y.S., Alamri, S.A., Alfaifi, M.Y., Alrumman, S.A., Elbehairi, S.E.I., Taha, T.H., et al., 2021. L-Glutaminase synthesis by marine *Halomonas meridiana* isolated from the red sea and its efficiency against colorectal cancer cell lines. Molecules 26, 1963.

Murguia-Favela, L., Min, W., Loves, R., Leon-Ponte, M., Grunebaum, E., 2020. Comparison of elapegademase and pegademase in ADA-deficient patients and mice. Clin. Exp. Immunol. 200, 176–184.

Nedaeinia, R., Faraji, H., Javanmard, S.H., Ferns, G.A., Ghayour-Mobarhan, M., Goli, M., et al., 2020. Bacterial staphylokinase as a promising third-generation drug in the treatment for vascular occlusion. Mol. Biol. Rep. 47, 819–841.

Niculescu-Duvaz, I., Springer, C.J., 2005. Introduction to the background, principles, and state of the art in suicide gene therapy. Mol. Biotechnol. 30, 71–88.

O'Connell, S., Walsh, G., 2008. Application relevant studies of fungal β-galactosidases with potential application in the alleviation of lactose intolerance. Appl. Biochem. Biotechnol. 149, 129–138.

O'Connell, S., Walsh, G., 2010. A novel acid-stable, acid-active β-galactosidase potentially suited to the alleviation of lactose intolerance. Appl. Microbiol. Biotechnol. 86, 517–524.

Okada, K., Ueshima, S., Fukao, H., Matsuo, O., 2001. Analysis of complex formation between plasmin(ogen) and staphylokinase or streptokinase. Arch. Biochem. Biophys. 393, 339–341.

Patil, P.N. (Ed.), 2012. Discoveries in Pharmacological Sciences. World Scientific Publishing Co Pte Ltd.

Qi, F., Qi, J., Hu, C., Shen, L., Yu, W., Hu, T., 2019. Conjugation of staphylokinase with the arabinogalactan-PEG conjugate: Study on the immunogenicity, *in vitro* bioactivity and pharmacokinetics. Int. J. Biol. Macromol. 131, 896–904.

Ramírez-Paz, J., Saxena, M., Delinois, L.J., Joaquín-Ovalle, F.M., Lin, S., Chen, Z., et al., 2018. Thiol-maleimide poly(ethylene glycol) crosslinking of L-asparaginase subunits at recombinant cysteine residues introduced by mutagenesis. PLoS One 13, e0197643.

Reda, F.M., 2015. Kinetic properties of *Streptomyces canarius* L-glutaminase and its anticancer efficiency. Braz. J. Microbiol. 46, 957–968.

Rodríguez-Cerrato, V., Garcia, P., Huelves, L., García, E., del Prado, G., Gracia, M., et al., 2007. Pneumococcal LytA autolysin, a potent therapeutic agent in experimental peritonitis-sepsis caused by highly beta-lactam-resistant *Streptococcus pneumoniae*. Antimicrob. Agents Chemother. 51, 3371–3373.

Rodríguez-Martínez, J.A., Rivera-Rivera, I., Solá, R.J., Griebenow, K., 2009. Enzymatic activity and thermal stability of PEG-α-chymotrypsin conjugates. Biotechnol. Lett. 31, 883–887.

Rodríguez-Rubio, L., Chang, W.L., Gutiérrez, D., Lavigne, R., Martínez, B., Rodríguez, A., et al., 2016. 'Artilysation' of endolysin λSa2lys strongly improves its enzymatic and antibacterial activity against streptococci. Sci. Rep. 6, 35382.

Rosa, A.C., Corsi, D., Cavi, N., Bruni, N., Dosio, F., 2021. Superoxide dismutase administration: a review of proposed human uses. Molecules 26, 1844.

Sadeghian, I., Hemmati, S., 2021. Characterization of a stable form of carboxypeptidase G2 (Glucarpidase), a potential biobetter variant, from *Acinetobacter* sp. 263903-1. Mol. Biotechnol. 63, 1155–1168.

São-José, C., 2018. Engineering of phage-derived lytic enzymes: improving their potential as antimicrobials. Antibiotics (Basel) 7, 29.

Sarkissian, C.N., Shao, Z., Blain, F., Peevers, R., Su, H., Heft, R., et al., 1999. A different approach to treatment of phenylketonuria: phenylalanine degradation with recombinant phenylalanine ammonia lyase. Proc. Natl. Acad. Sci. USA 96, 2339–2344.

Sawdon, A.J., Zhang, J., Peng, S., Alyami, E.M., Peng, C.A., 2021. Polymeric nanovectors incorporated with ganciclovir and HSV-tk encoding plasmid for gene-directed enzyme prodrug therapy. Molecules 26, 1759.

Schellmann, N., Deckert, P.M., Bachran, D., Fuchs, H., Bachran, C., 2010. Targeted enzyme prodrug therapies. Mini Rev. Med. Chem. 10, 887–904.

Schepelmann, S., Springer, C.J., 2006. Viral vectors for gene-directed enzyme prodrug therapy. Curr. Gene Ther. 6, 647–670.

Serrano-Aguirre, L., Velasco-Bucheli, R., García-Álvarez, B., Saborido, A., Arroyo, M., de la Mata, I., 2021. Novel bifunctional acylase from *Actinoplanes utahensis*: a versatile enzyme to synthesize antimicrobial compounds and use in quorum quenching processes. Antibiotics (Basel) 10, 922.

Sharma, B., Singh, S., Kanwar, S.S., 2014. L-Methionase: a therapeutic enzyme to treat malignancies. Biomed. Res. Int. 2014, 506287.

Sharma, C., Osmolovskiy, A., Singh, R., 2021. Microbial fibrinolytic enzymes as anti-thrombotics: production, characterisation and prodigious biopharmaceutical applications. Pharmaceutics 13, 1880.

Sharma, S.K., Bagshawe, K.D., 2017a. Antibody directed enzyme prodrug therapy (ADEPT): trials and tribulations. Adv. Drug. Deliv. Rev. 118, 2–7.

Sharma, S.K., Bagshawe, K.D., 2017b. Translating antibody directed enzyme prodrug therapy (ADEPT) and prospects for combination. Expert. Opin. Biol. Ther. 17, 1–13.

Sheets, A.R., Demidova-Rice, T.N., Shi, L., Ronfard, V., Grover, K.V., Herman, I.M., 2016. Identification and characterization of novel matrix-derived bioactive peptides: a role for collagenase from Santyl® ointment in post-debridement wound healing? PLoS One 11, e0159598.

Singh, P., Banik, R.M., 2013. Biochemical characterization and antitumor study of L-glutaminase from *Bacillus cereus* MTCC 1305. Appl. Biochem. Biotechnol. 171, 522–531.

Singh, R., Kumar, M., Mittal, A., Mehta, P.K., 2016. Microbial enzymes: industrial progress in 21st century. 3 Biotech. 6, 174.

Springer, C.J., Antoniw, P., Bagshawe, Searle, F., Bisset, G.M.F., Jarman, M., 1990. Novel prodrugs which are activated to cytotoxic alkylating agents by carboxypeptidase G2. J. Med. Chem. 33, 677–681.

Srivastava, V., Mishra, S., Chaudhuri, T.K., 2019. Enhanced production of recombinant serratiopeptidase in *Escherichia coli* and its characterization as a potential biosimilar to native biotherapeutic counterpart. Microb. Cell Fact. 18, 215.

Stevens, J., Wyatt, C., Brown, P., Patel, D., Grujic, D., Freedman, S.D., 2018. Absorption and safety with sustained use of RELiZORB evaluation (ASSURE) study in patients with cystic fibrosis receiving enteral feeding. J. Pediatr. Gastroenterol. Nutr. 67, 527–532.

Tan, W.S.D., Liao, W.P., Zhou, S., Mei, D., Wong, W.S.F., 2018. Targeting the renin-angiotensin system as novel therapeutic strategy for pulmonary diseases. Curr. Opin. Pharmacol. 40, 9–17.

Tandon, S., Sharma, A., Singh, S., Sharma, S., Sarma, S.J., 2021. Therapeutic enzymes: discoveries, production and applications. J. Drug. Deliv. Sci. 63, 102455.

Tomlinson, B.K., Thomson, J.A., Bomalaski, J.S., Diaz, M., Akande, T., Mahaffey, N., et al., 2015. Phase I trial of arginine deprivation therapy with ADI-PEG 20 plus docetaxel in patients with advanced malignant solid tumors. Clin. Cancer Res. 21, 2480–2486.

Totté, J.E.E., van Doorn, M.B., Pasmans, S.G.M.A., 2017. Successful treatment of chronic *Staphylococcus aureus*-related dermatoses with the topical endolysin Staphefekt SA.100: a report of 3 cases. Case Rep. Dermatol. 9, 19–25.

Treem, W.R., Ahsan, N., Sullivan, B., Rossi, T., Holmes, R., Fitzgerald, J., et al., 1993. Evaluation of liquid yeast-derived sucrase enzyme replacement in patients with sucrase-isomaltase deficiency. Gastroenterology 105, 1061–1068.

Treem, W.R., McAdams, L., Stanford, L., Kastoff, G., Justinich, C., Hyams, J., 1999. Sacrosidase therapy for congenital sucrase-isomaltase deficiency. J. Pediatr. Gastroenterol. Nutr. 28, 137–142.

Treshalina, H.M., Lukasheva, E.V., Sedakova, L.A., Firsova, G.A., Guerassimova, G.K., Gogichaeva, N.V., et al., 2000. Anticancer enzyme L-lysine alpha-oxidase: properties and application perspectives. Appl. Biochem. Biotechnol. 88, 267–273.

Turecek, P.L., Bossard, M.J., Schoetens, F., Ivens, I.A., 2016. PEGylation of biopharmaceuticals: a review of chemistry and nonclinical safety information of approved drugs. J. Pharm. Sci. 105, 460–475.

Vachher, M., Sen, A., Kapila, R., Nigam, A., 2021. Microbial therapeutic enzymes: A promising area of biopharmaceuticals. Curr. Res. Biotechnol. 3, 195–208.

Van Nguyen, T., Shin, M.C., Min, K.A., Huang, Y., Oh, E., Moon, C., 2018. Cell-penetrating peptide-based non-invasive topical delivery systems. J. Pharm. Investig. 48, 77–87.

Vannelli, T., Xue, Z.X., Breinig, S., Qi, W.W., Sariaslani, F.S., 2007. Functional expression in *Escherichia coli* of the tyrosine-inducible tyrosine ammonia-lyase enzyme from yeast *Trichosporon cutaneum* for production of p-hydroxycinnamic acid. Enzyme Microb. Technol. 41, 413–422.

Velasco-Bucheli, R., Hormigo, D., Fernández-Lucas, J., Torres-Ayuso, P., Alfaro-Ureña, Y., Saborido, A.I., et al., 2020. Penicillin acylase from *Streptomyces lavendulae* and aculeacin A acylase from *Actinoplanes utahensis*: two versatile enzymes as useful tools for quorum quenching processes. Catalysts 10, 730.

Vellard, M., 2003. The enzyme as drug: application of enzymes as pharmaceuticals. Curr. Opin. Biotechnol. 14, 444–450.

Veronese, F.M., Mero, A., 2008. The impact of PEGylation on biological therapies. Biodrugs 22, 315–329.

Vora, H., McIntire, J., Kumar, P., Deshpande, M., Khosla, C., 2007. A scalable manufacturing process for pro-EP-B2, a cysteine protease from barley indicated for celiac sprue. Biotechnol. Bioeng. 98, 177–185.

Wan, X., Zhang, J., Yu, W., Shen, L., Ji, S., Hu, T., 2017. Effect of protein immunogenicity and PEG size and branching on the anti-PEG immune response to PEGylated proteins. Process. Biochem. 52, 183–191.

Warren, T.D., Patel, K., Rivera, J.L., Eshleman, J.R., Ostermeier, M., 2020. Comprehensive mutagenesis on yeast cytosine deaminase yields improvements in 5-fluorocytosine toxicity in HT1080 cells. Aiche J. 66, e16688.

Wei, G.X., Helmerhorst, E.J., Darwish, G., Blumenkranz, G., Schuppan, D., 2020. Gluten degrading enzymes for treatment of celiac disease. Nutrients 12, 2095.

Weng, Y., Yao, J., Sparks, S., Wang, K.Y., 2017. Nattokinase: an oral antithrombotic agent for the prevention of cardiovascular disease. Int. J. Mol. Sci. 18, 523.

Williams, E.M., Little, R.F., Mowday, A.M., Rich, M.H., Chan-Hyams, J.V.E., Copp, J.N., et al., 2015. Nitroreductase gene-directed enzyme prodrug therapy: insights and advances toward clinical utility. Biochem. J. 471, 131–153.

Xu, Y., Shi, Y., Zhou, J., Yang, W., Bai, L., Wang, S., et al., 2017. Structure-based antigenic epitope and PEGylation improve the efficacy of staphylokinase. Microb. Cell Fact. 16, 197.

Yan, X.X., Li, H.L., Zhang, Y.T., Wu, S.Y., Lu, H.L., Yu, X.L., et al., 2020. A new recombinant MS-superoxide dismutase alleviates 5-fluorouracil-induced intestinal mucositis in mice. Acta Pharmacol. Sin. 41, 348–357.

Yang, H., Gong, Y.J., Zhang, H.D., Etobayeva, I., Miernikiewicz, P., Luo, D.H., et al., 2019a. ClyJ is a novel pneumococcal chimeric lysin with a cysteine- and histidine-dependent amidohydrolase/peptidase catalytic domain. Antimicrob. Agents Chemother. 63, e02043-18.

Yang, J., Tao, R., Wang, L., Song, L.J., Wang, Y., Gong, C.Y., et al., 2019b. Thermosensitive micelles encapsulating phenylalanine ammonia lyase act as a sustained and efficacious therapy against colorectal cancer. J. Biomed. Nanotech 15, 717–727.

Yari, M., Ghoshoon, M.B., Vakili, B., Ghasemi, Y., 2017. Therapeutic enzymes: applications and approaches to pharmacological improvement. Curr. Pharm. Biotechnol. 18, 531–540.

Younus, H., 2018. Therapeutic potentials of superoxide dismutase. Int. J. Health Sci. 12, 88–93.

Zhu, L., Cui, W., Fang, Y., Liu, Y., Gao, X., Zhou, Z., 2013. Cloning, expression and characterization of phenylalanine ammonia-lyase from *Rhodotorula glutinis*. Biotechnol. Lett. 35, 751–756.

Chapter 16

Microbial enzymes in pharmaceutical industry

Nidhi Y. Patel[1], Dhritiksha M. Baria[1], Dimple S. Pardhi[1], Shivani M. Yagnik[2], Rakeshkumar R. Panchal[1], Kiransinh N. Rajput[1] and Vikram H. Raval[1]
[1]*Department of Microbiology and Biotechnology, University School of Sciences, Gujarat University, Ahmedabad, Gujarat, India,* [2]*Department of Microbiology, Christ College, Rajkot, Gujarat, India*

16.1 Introduction

The microbial system possesses diverse and vast pools of unique enzymes which play a vital role in many industrial processes. A microbial system includes unique metabolic pathways which can be regulated with the help of enzymes. Microbial enzymes are a valuable resource for medicinal treatments due to their robust nature, metabolic capability, and ease of cultivation. The major benefit of microbial enzyme production is that it produces higher yields on low-cost media in a shorter duration. Recent studies (Patel et al., 2022) show the considerable potential of diverse microbial enzymes in the medicine and biopharmaceutical sectors, resulting in a significant increase in the use of these enzymes for the treatment of many diseases (Taipa et al., 2019). Enzymes play a crucial role in human welfare due to their effectiveness and catalytic abilities. Several indications of medicinal properties are mentioned in available literature (Patel, 2020). Several medical problems or conditions linked to or associated with enzyme deficiency are a concern in the current scenario.

Enzymes are of many types based on catalytic and functional attributes (Raval et al., 2018). The enzymes used for treatment purposes are referred to belong to a class of therapeutic enzymes. Therapeutic enzymes are biocatalysts that are used to treat a variety of disorders. Enzymes have been utilized for therapeutics since the last few decades when they were first employed as a digestive aid. Several other therapeutic enzymes, such as L-glutaminase, L-asparaginase, hydrolases, amidases, phenylalanine ammonia-lyase (PAL), and uricase, have been used in the pharmaceutical industry over the last

Biotechnology of Microbial Enzymes. DOI: https://doi.org/10.1016/B978-0-443-19059-9.00025-6
© 2023 Elsevier Inc. All rights reserved.

100 years. Therapeutic enzymes bind selectively to their targets, distinguishing them from nonenzymatic medicines. These benefits, however, have been accompanied by several drawbacks, including high production costs, targeted enzyme delivery, elicitation of an immunological response against it, and a shorter in vivo half-life, compared to others (Reshma, 2019).

Apart from their therapeutic role, certain enzymes (viz., serratiopeptidase, collagenase, superoxide dismutase, etc.) have also been explored for their antiinflammatory ability. Additionally, enzybiotics are also cited for playing a crucial role in the pharmaceutical sector (viz., lysins, autolysin, bacteriocin, lysozyme, etc.). However, exploring a novel microbe that can be utilized to synthesize therapeutic enzymes cost-effectively and with lesser side effects is difficult in commercializing therapeutic enzymes (Sugathan et al., 2017). The most promising challenges in today's scenario are the discovery and manufacturing of novel enzymes and innovative alterations to existing enzymes. It plays a vital role in controlling various diseases and improving health care. It is necessary to evaluate novel enzymes and upgrade and enhance current ones. These microbial enzymes should be explored in multiple ways to treat human diseases (Vachher et al., 2021). The present chapter deals with microbial-based therapeutic enzymes in allied pharmaceutical sectors.

16.2 Cataloging of hydrolases used in pharmaceutical industry

Hydrolases are the ubiquitous enzymes that result in the hydrolysis of a chemical bond, thereby dividing a large molecule into smaller ones. The examples of hydrolases group involve amidase group which has application in the field of therapeutics, additionally cataloging of enzyme provides detailed insights (Table 16.1) into each enzymatic catalytic ability and its role and various applications as far as a pharmaceutical is concerned.

16.3 Microbial enzymes in pharmaceutical processes

16.3.1 Therapeutics

16.3.1.1 L-Asparaginase

L-Asparaginase amidohydrolase (L-ASNase) is an enzyme belonging to class EC 3.5.1.1, which causes the breakdown of amino acid, asparagine into aspartic acid and ammonia (Fig. 16.1). The L-ASNase sourced from microbial sources is successfully administered with a low level of nonessential amino acid requirements. Also, it rapidly stops the proliferation of cancerous cells, which needs a high amount of asparagine. Presently, it is the cornerstone drug applied in treating acute lymphoblastic leukemia (ALL), pediatric cancer, etc. However, though these lymphoblastic cells lack the expression

TABLE 16.1 Cataloging of hydrolases group of enzymes.

Enzymes	EC no.	Properties	References
L-Asparaginase	EC 3.5.1.1	• Systematic name: L-asparagine amidohydrolase. • Catalytic reaction: L-asparagine + H_2O = L-aspartate + NH_3. • It is involved in aspartate and asparagine metabolism. • It plays major role in pediatric and adult leukemia.	Chakravarty et al. (2022)
L-Glutaminase	EC 3.5.1.2	• Systematic name: L-glutamine amidohydrolase. • Catalytic reaction: L-glutamine + H_2O = L-glutamate + NH_3. • It is majorly involved in glutamate metabolism. • Glutaminase C is the first enzyme in glutaminolysis • It has role in pulmonary hypertension disease as well as in neuroinflammatory associated depression and other type of cancer cell proliferation.	Patel et al. (2021); Wang et al. (2022)
Omega amidase	EC 3.5.1.3	• Systematic name: omega-amidodicarboxylate amidohydrolase. • Catalytic reaction: a monoamide of a dicarboxylate + H_2O = a dicarboxylate + NH_3. • It involves asparagine coupled with glutamine transamination pathway. • It helps in alleviate risk for hyperammonemia.	Mao et al. (2022)
Acylamidase	EC 3.5.1.4	• Systematic name: acylase amidohydrolase • Catalytic reaction: a monocarboxylic acid amide + H_2O = a monocarboxylate + NH_3 • Involved in arginine, proline, and phenylalanine metabolism.	Reina et al. (2022)
Urease	EC 3.5.1.5	• Systematic name: urea amidohydrolase. • Catalytic reaction: urea + H_2O = CO_2 + $2NH_3$ • It plays major role in removing urea from blood. • Its deficiency leads to kidney failure.	Basso and Serban (2019)

(Continued)

TABLE 16.1 (Continued)

Enzymes	EC no.	Properties	References
β-Ureidopropionase	EC 3.5.1.6	• Systematic name: *N*-carbamoyl-beta-alanine amidohydrolase. • Catalytic reaction: *N*-carbamoyl-beta-alanine + H_2O = beta-alanine + CO_2 + NH_3 • They work on carbon–nitrogen bonds except the peptide bonds, precisely in linear amides. • Contributes in beta-alanine and pyrimidine metabolism as well as coenzyme A and pantothenate biosynthesis. • Involved in degradation of pyrimidines and pyrimidine-based anticancer drugs.	Maurer et al. (2018)
Formyl aspartate deformylase	EC 3.5.1.8	• Systematic name: *N*-formyl-L-aspartate amidohydrolase. • Catalyzed reaction: *N*-formyl-L-aspartate + H_2O = formate + L-aspartate. • It is mainly involved in histidine, glyoxylate, dicarboxylate, alanine, aspartic acid, and asparagine metabolism. • Its deficiency causes histidinemia in newborn babies.	Dayan (2019)
Penicillin amidase	EC 3.5.1.11	• Systematic name: penicillin amidohydrolase. • Catalytic reaction: penicillin + H_2O = a carboxylate + 6-aminopenicillanate. • Penicillin and cephalosporin biosynthesis pathway is involved in the enzyme. • Used in stimuli-sensitive drug delivery systems for site-specific antibiotic release and for other active pharmaceutical chemical synthesis.	Nazir et al. (2022)
Biotinidase	EC 3.5.1.12	• Systematic name: biotin amide amidohydrolase. • Catalytic reaction: biotin amide + H_2O = biotin + NH_3. • Involves in biotin metabolism, biotinylating proteins. • Its deficiency causes neurological problems.	Canda et al. (2020)
N-Acetyl-amino-hydrolase	EC 3.5.1.14	• Common name: aminoacylase. • Catalytic reaction: *N*-acetyl-amino acid + H_2O = carboxylate + L-amino acid. • This enzyme generally participates in amino acid metabolism and urea cycle.	Hernick and Fierke (2010)

Fatty acylamidase	EC 3.5.1.99	• Systematic name: fatty acylamide amidohydrolase. • Catalytic reaction: anandamide + H_2O = arachidonic acid + ethanolamine. • Oleamide + H_2O = oleic acid + NH_3. • Anandamide degradation pathway is involved. • It is a promising target for neuropathic pain and other CNS-related disorders.	Jaiswal et al. (2022)
Deaminase	–	• Causes deamination, that is, removal of the amine group from a compound. • Such enzymes are named based on the substrate they utilize, such as adenosine deaminase (EC 3.5.4.4), cytidine deaminase (EC 3.5.4.1), guanine deaminase (EC 3.5.4.3), etc. • They are generally involved with purine metabolism.	Ireton et al. (2002)

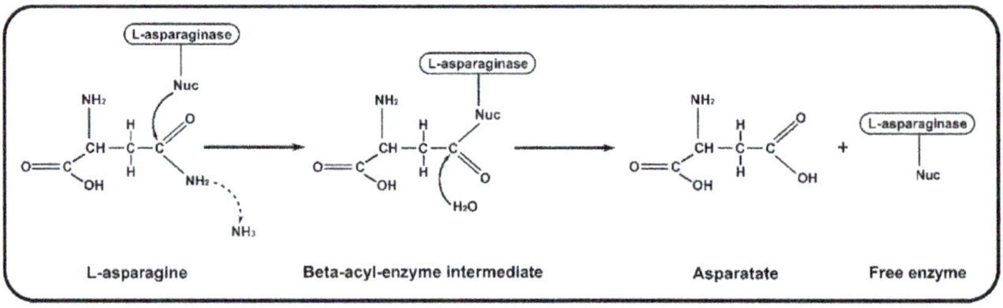

FIGURE 16.1 Schematic representation of the catalytic mechanism of L-asparaginase.

level of asparagine synthetase, these cells utilize extracellular asparagine for proliferation and survival. Current reports demonstrate the potential application of L-ASNase in treating other types of cancer, namely, hematological or solid types of cancers. There is a limitation of the said enzyme due to various immunogenic and adverse side effects during clinical trials, which lead to discontinuation of the use of the enzyme for treatment. To overcome these limitations, the newer formulation of L-ASNase is desirable. Additionally, the multiple other mechanisms, that is, ASNS promoter reactivation and desensitization may play an important role in research area that fuels the newer combination of formulations to overcome chemoresistance (Van Trimpont et al., 2022).

Long back, in 1904, L-asparaginase was initially described as an enzyme of nonhuman origin that hydrolyzes L-asparagine to L-aspartic acid with the release of ammonia in various bovine tissues. Kidd (1953) reported that serum content from guinea pigs could inhibit the growth of modified lymphomas in mice. Still, after about 10 years, Broome reported the inhibitory effect of the L-ASNase enzyme (Broome, 1963). Later on, in 1966, the initial use of purified serum for the treatment of ALL in the patient was reported by researchers (Dolowy et al., 1966). The higher costs and lesser availability of sources of guinea pig serum, there is a constant need to move to an alternative to get a large-scale production level in pharmaceuticals for cancer treatments due to several side effects of the drug in clinical trials, such as higher toxicity profiling, including immunological ill-effects and other nonimmunological toxicities, namely, pancreatitis, liver toxicities, coagulopathy, neurotoxicity, etc (Cachumba et al., 2016).

16.3.1.2 L-Glutaminase

The investigative research on the amidohydrolase group of enzymes was initiated in 1956, and the prominence of L-glutaminase in primary gout was mentioned by Alexander B. Gutman (Gutman, 1963). L-Glutaminase amidohydrolase (L-GLNase), EC 3.5.1.2, is a crucial and peculiar enzyme belonging to serine-dependent β-lactamases and penicillin-binding proteins (Patel et al., 2022) due to its higher affinity to cause alteration in peptidoglycan

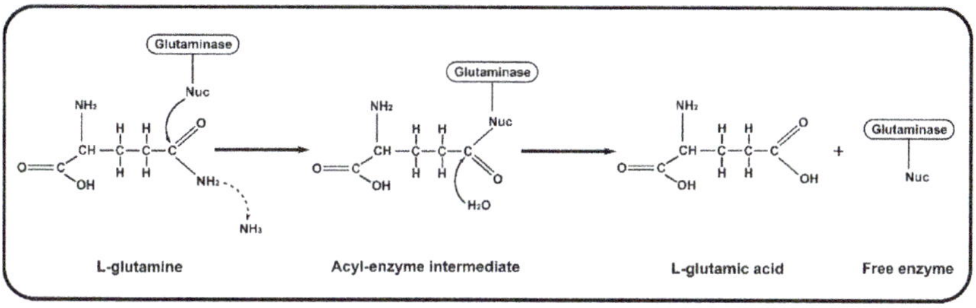

FIGURE 16.2 Schematic representation of the catalytic mechanism of L-glutaminase.

biosynthesis which requires for making bacterial cell wall. Additionally, based on catalytic efficacy, L-glutaminase is further cataloged as the type of proteolytic endopeptidase (Patel et al., 2020), responsible for cleaving peptide linkages with the release of ammonia and glutamate as a by-product. (Fig. 16.2). Though other sources of L-glutaminase are available, the microbial ones are preferable compared to plant- or animal-based due to ease of cultivation and less toxicity.

L-Glutaminase is majorly associated with nitrogen-related metabolism. It imparts particular umami flavor content to fermented food such as soy sauce, miso, sufu, etc. Due to this, it finds suitable application in the food industry as a flavor-enhancing agent (Rastogi and Bhatia, 2019). It can act as a potent in vivo antioxidant for glutathione synthesis. For the biosynthesis of L-theanine, a nutraceutical is used in biotechnological processes and for de novo biosynthesis of vitamin B_6, L-glutaminase is preferable (Amobonye et al., 2022). L-Glutaminase has been reported to inhibit cancer cells in vitro, due to which it finds suitable applications in the pharmaceutical sector in anticancer drug development. The inhibitory effect of L-glutaminase on various cancerous cell lines, such as Vero, HeLa, MCF7, JURKAT, HCT-116 RAW, etc., is cited by many researchers (Patel et al., 2021; Gomaa, 2022). The L-glutaminase also possesses antibacterial, antiviral properties in addition to antioxidants. L-Glutaminase, in combination with other enzymes, namely, glucose oxidase, lactate oxidase, and glutamate oxidase, used as a biosensor for measuring glutamine level in vitro cell line study and hybridoma, the measurement for various biological fluids.

16.3.1.3 L-Methioninase

L-Methioninase EC 4.4.1.11 (L-methionine-a-deamino-c-mercaptomethane-lyase) is a type of multifunctional enzyme pyridoxal-L-phosphate-dependent enzyme that causes the breakdown of L-methionine to methanethiol, a ketobutyrate along with the release of ammonia by oxidative deamination and demethylation mechanism (Fig. 16.3). Except for a few mammals, this enzyme is ubiquitous in occurrence. The methionine dependency was cited

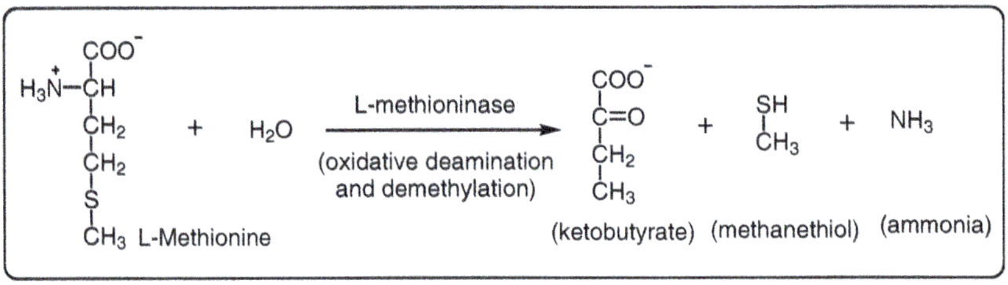

FIGURE 16.3 Schematic representation of the catalytic mechanism of L-methioninase.

(Goseki et al., 1995) as an efficient biochemical phenomenon among various cancerous cells likely seen in normal cells. For triggering the majority of cancerous cells, L-methioninase is applied. L-Methioninase is used as a chemotherapeutic agent in combination with other aspects, clearly showing its therapeutic potential. L-Methionine is considered an essential amino acid that plays a crucial role in the biosynthesis of protein, glutathione, and polyamines methylation of DNA, thus regulating gene expression. In combination with chemotherapeutic drugs, the methionine-free diet is effective against gastric cancers. Additionally, L-methioninase has been proven effective against breast, kidney, colon, lung, and prostate cancer cell lines (Yamamoto et al., 2022).

16.3.1.4 CGTase

Cyclodextrin glucanotransferase (CGTase), EC 2.4.1.19, is an important extracellular enzyme that belongs to the α-amylase glycoside hydrolase family. It converts starch and other 1,4-linked α-glucans to nonreducing maltooligosaccharides, called cyclodextrins (CDs), by an intramolecular transglycosylation (cyclization) reaction. They are also capable of cleaving cyclodextrin ring and transferring it to a linear acceptor substrate as well as converting two linear oligosaccharides into linear oligosaccharides of different sizes by an intermolecular transglycosylation reaction, coupling and disproportionation, respectively. Moreover, they have a weak hydrolyzing activity in which water is the glycosyl acceptor (Rajput et al., 2016).

CDs consist of 6–12 glucose units linked by α-1,4- glycosidic bonds, among which the most common types are α (cyclohexamilose $_cG_6$), β (cycloheptamilose $_cG_7$), and γ (cyclooctamilose $_cG_8$) (Fig. 16.4). The key difference between them is the apolar cavity size and the solubility in water (de Freitas et al., 2004). Generally, CGTases produce a mixture of α-, β-, and γ-cyclodextrins in different ratios depending on the reaction conditions and the type of CGTase producing microorganisms. In contrast, CD production mainly depends on the methods and incubation conditions used. Among the three CDs, usually, β-CD was reported with a maximum concentration of approximately 80% of the mixture (Sian et al., 2005).

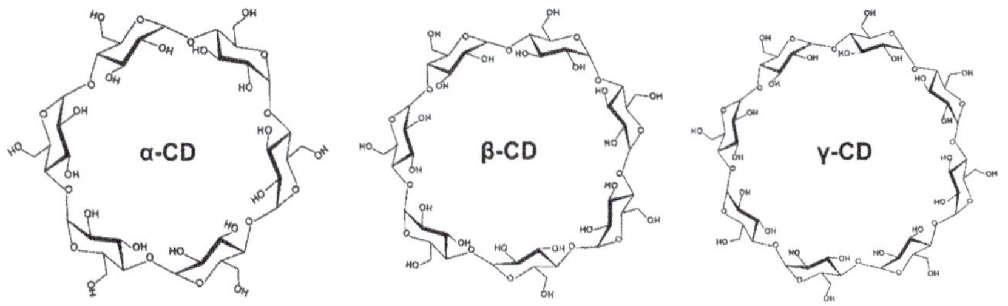

FIGURE 16.4 Structures of different cyclodextrins.

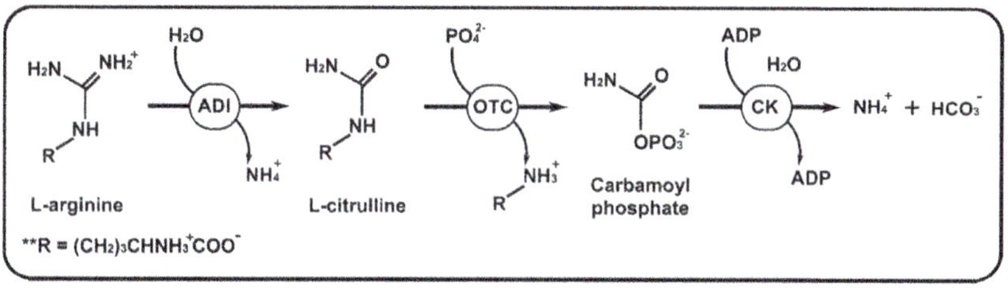

FIGURE 16.5 Schematic representation of arginine deaminase pathway.

Majorly cyclodextrins and their derivatives are related to the drug delivery systems such as making inclusion complexes. Currently, microbial approaches replace synthetic cyclodextrin production because of the highly active CGTases. A cyclodextrin derived from *Bacillus* sp. CD 18 was found to be compatible and increase the solubility of paracetamol and aceclofenac (Kaur et al., 2014). Studies about the immediate release of oral dosage forms revealed that β-CDs, when complexed with Irbesartan, work best for drug dissolution and release (Rajesh et al., 2015).

16.3.1.5 Arginine-deiminase

Arginine-deiminase is one of the important parts of the protein superfamily which is involved in arginine catabolism. Among the various pathways, it extensively uses the arginine dihydrolase pathway for microbial degradation of arginine. This pathway comprises three different enzymes (Fig. 16.5):

- Arginine dihydrolase (ADI): hydrolyzes arginine to citrulline and ammonia, EC 3.5.3.6
- Ornithine transcarbamoylase (OTC): forms carbamoyl phosphate and ornithine, EC 2.1.3.3
- Carbamate kinase (CK): phosphorylate ADP to ATP, bicarbonate/CO_2, ammonia from carbamoyl phosphate, EC 2.7.2.2

The ADI pathway generates energy in the form of ATP (one mole) by consuming arginine (one mole). The phylogenetic analysis of these enzymes proposes that this pathway generally accumulated based on enzyme recruitment. Bacterial cells contain clusters of genes for such enzymes and an amino acid transporter gene. The ADI enzyme is the first and an attractive target for designing antimicrobial drugs among these three. It generally occurs in archaea, bacteria, and anaerobic eukaryotes but does not found in higher eukaryotes or humans. Besides energy production, the ADI pathway also produces ammonia that aids acid tolerance in several human pathogens. The structural and pharmacological characteristics of ADI were studied from several bacterial cells, but the most frequent ADI enzyme was found from *Mycoplasma arginine* with a recognizable therapeutic perspective.

The attention turned around therapeutic applications of ADI after it was successfully used in inhibition of human tumor cell lines and some murein in both in vitro and in vivo through the exhaustion of ARG supply in the late 1990s (Takaku et al., 1992; Takaku et al., 1995). After that, many researchers purified ADI from *Mycoplasma* sp. and reported its potential as an antitumor agent for different types of tumors. In addition, ADI was also found to be a possible candidate for angiogenesis inhibition that regulates neovascularization. The recombinant ADI was demonstrated for inhibiting the microvessel tube-like structure of human umbilical vein endothelial (HUVE) cells on Matrigel-coated surfaces as well as migration inhibition in the scratch wounded area in HUVE cell monolayers in in vitro because of an antiangiogenic property (Beloussow et al., 2002). ADI effectively cure several tumors and other diseases because of its supportive antiangiogenic and antiproliferative properties.

16.3.1.6 Lysine oxidase

Lysine oxidase (LO) or more precisely known as L-lysine α-oxidase (LO, EC 1.4.3.14) is one of the best studied enzymes belonging to the L-amino acid oxidases, which showed an exclusive substrate specificity and deaminated L-lysine to produce ammonia, α-keto-ε-aminocaproate, and H_2O_2 (Fig. 16.6). Microbial strains such as *Trichoderma viride*, *Streptomyces lividans* TK24, and *Escherichia coli* were extensively studied to analyze the

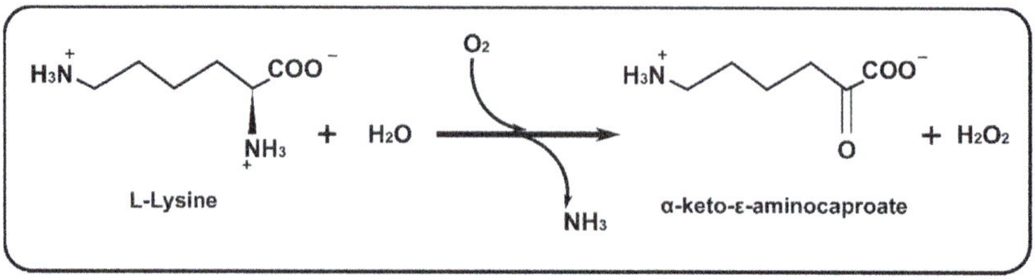

FIGURE 16.6 Schematic representation of the catalytic reaction of L-lysine α-oxidase.

structural configurations of LOs. This enzyme is generally obtained in crystal forms with a molecular weight ranging from 100 to 110 kDa (Lukasheva et al., 2021).

Most cancer cells are susceptible to essential growth factors and cannot tolerate their deficiency, including amino acids; hence, L-lysine reduction becomes the primary stage that contributes to the anticancer effect of LOs. The LOs have cytotoxic, antitumor, and antimetastatic properties are proven by multiple in vitro and in vivo practices. As shown in Fig. 16.6, LOs work on a dual mechanism of action that targets the tumor cells, that is, (1) depletion of L-lysine and (2) formation of H_2O_2. A *Trichoderma* cf. *aureoviride* Rifai VKM F-4268D derived LO showed a dose-dependent cytotoxic activity against different cancer cell lines in an in vitro experiment. The cells of breast cancer MCF7, erythromyeloblastic leukemia K562, and colon cancer LS174T with IC_{50} of 8.4×10^{-7}, 3.2×10^{-8}, and 5.6×10^{-7} mg/mL are found to be the highly sensitive one toward LOs isolated from this source (Pokrovsky et al., 2013). The LOs isolated from *Amanita phalloides* (ApLAO) and *Clitocybe geotropa* (CgLAO) showed cytotoxicity to Jurkat and MCF7 cells and were found to be involved in the multiple signaling pathways that trigger apoptosis (Pišlar et al., 2016). Sabotič et al. (2020) identified a potential protein L-amino acid oxidases (LAOs) from fruiting bodies of *A. phalloides* (*Ap*LAO) and *Infundibulicybe geotropa* (*Cg*LAO) that exhibits antibacterial activity against *Ralstonia solanacearum*, a plant pathogen.

16.3.1.7 Phenylalanine ammonia-lyase

The enzyme PAL has lately gained attention as a vital therapeutic enzyme with a plethora of medicinal applications. This enzyme converts L-phenylalanine to *trans*-cinnamate and ammonia (Fig. 16.7). PAL is found in higher plants, algae, ferns, and microbes, but it is not found in vertebrates. Although microbial PAL has been widely used to produce industrially significant metabolites in the past, its high substrate specificity and catalytic activity have recently sparked interest in its medicinal uses (Kawatra et al., 2020).

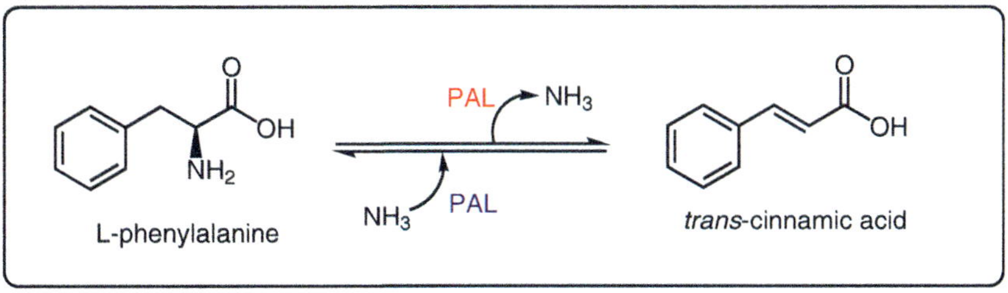

FIGURE 16.7 Schematic representation showed deamination of L-phenylalanine into *trans*-cinnamic acid and ammonia.

16.3.1.7.1 Phenylalanine ammonia-lyase in cancer treatment

Targeted cancer treatments are based on metabolic differences that separate normal cells from cancerous cells. The latter demands a large metabolic intake of amino acids to sustain their proliferative capabilities. Auxotrophic cancer cells require only a few amino acids, including L-phenylalanine. By targeting the metabolism of its particular amino acids and quickly reducing the quantity of exogenous phenylalanine in the cells, PAL has been proven to be beneficial in regressing cancers (Pokrovsky et al., 2019; Dhankhar et al., 2020). Normal cells overcome phenylalanine shortage by using phenylpyruvate as a growth replacement, but malignant cells cannot overcome it. The first report to show the considerable chemotherapeutic activity of PAL from *Rhodotorula glutinis* in mice tumors in vitro and in vivo was published in 1973 (Abell et al., 1973). RgPAL significantly reduced the plasma concentrations of L-Phe, suggesting that PAL might be used to treat leukemia. PAL was shown to be effective against breast cancer and prostate cancer, as evidenced by a substantial decrease in cell viability during the cytotoxicity experiment. Furthermore, it was deduced that infusing PAL at a lower concentration into these two malignant cell lines in vitro, resulted in higher cytotoxicity than increasing the PAL dosage. Again, when PAL was compared to other anticancer enzymes such as *E. coli* L-asparaginase, it demonstrated similar cytotoxicity (Costa and Silva 2021). As a result, further study on PAL is needed before it can be developed as a cancer-curing alternative.

16.3.1.7.2 Phenylketonuria treatment

Phenylketonuria (PKU) is a rare autosomal recessive condition caused by a lack of the hepatocellular enzyme phenylalanine hydroxylase (PAH). This enzyme is responsible for controlling L-phenylalanine levels in the blood (Levy et al., 2018). Increased quantities of neurotoxic phenylalanine occur from the inherent lag in its action. Seizures, tremors, autism, and persistent psychiatric abnormalities are the symptoms of increased amino acid levels in the brain (Xu et al., 2021). Sapropterin dihydrochloride, an FDA-approved oral cofactor for the enzyme PAH, was previously used to treat the recessive condition (Van Spronsen et al., 2021). However, there was a drawback because it could only be given to patients who had some residual PAH activity. Hence, it was not adequate for patients entirely devoid of PAH activity. Large quantities of PAL necessary for PKU treatment were generated using recombinant DNA technology. In a mouse model of PKU, parenteral PAL formulations also demonstrated promising benefits, drastically lowering blood phenylalanine levels in vivo. Repeated injections triggered an immunological response that rendered the enzyme inactive. PEGylation and genetic manipulation approaches have also been used to modify PAL's enzymatic properties for effective treatment. Recent research suggests that PAL might be used to remove phenylalanine from commercial protein supplements (Kawatra et al., 2020).

16.3.1.7.3 Phenylalanine ammonia-lyase in tyrosinemia therapy

PAL benefits in treating tyrosine-related metabolic disorders, such as tyrosinemia (Hendrikse et al., 2020). Due to the functional impairment of tyrosine metabolizing enzymes, these disorders are characterized by elevated levels of tyrosine and its metabolites in blood plasma. Increased tyrosine levels have been linked to liver cirrhosis, cognitive impairment, and renal failure. Type 1 tyrosinemia, also known as hereditary tyrosinemia, is the most lethal to the liver (Demirbas et al., 2018). Hepatocellular carcinoma is a severe threat to the person suffering from these diseases. In combination with a lifelong limited phenylalanine and tyrosine diet, a drug, nitisinone, is the current therapy approach for this disease. However, by preventing the degradation of L-tyrosine, this substance has been demonstrated to disrupt the regulation of cognitive hormones such as serotonin, dopamine, and noradrenaline (Demirbas et al., 2018). Furthermore, the risk of developing hepatocellular carcinoma is not diminished completely; hence, this warrants searching for alternative treatments for tyrosinemia. PAL produced by some fungi such as *Trichosporon cutaneum*, *Sporobolomyces roseus*, *R. glutinis*, and photosynthetic bacteria *Rhodobacter capsulatus* utilizes tyrosine as a substrate. As a result, either directly catabolizing tyrosine or addressing the depletion of phenylalanine might help to reduce the excessive expression of tyrosine. Recently, ancestral sequence reconstruction, an enzyme engineering technique, was used to improve the therapeutic capabilities of PAL for treating tyrosinemia, precisely its stability and reduced substrate specificity (Demirbas et al., 2018; Kawatra et al., 2020).

16.3.1.7.4 Health supplement production

These novel extremophilic PALs are the ideal candidates for synthesizing medically relevant, optically pure natural and artificial compounds because of their structural stability and catalytic effectiveness. Amino acids and their derivatives are the building blocks of life. PAL activity may be used to biosynthesize various D and L-configured amino acids in vitro. Pure amino acid synthesis is essential for manufacturing a variety of therapeutic peptides and analgesics. Previously, hydrolytic enzymes such as lipases were utilized to make chiral amino acids (Sarmah et al., 2018). The enzyme PAL's biocompatibility with the aqueous environment and absence of any external cofactor need to make it a suitable catalyst for synthesizing high yield amino acids. Moreover, PAL transforms L-phenylalanine to *trans*-cinnamic acid and ammonia, pure L-α-phenylalanine can be produced by reversing the process by adding ammonia to an unsaturated acid in a stereospecific manner. PAL also efficiently catalyzes derivatives of cinnamic acid, resulting in a wide spectrum of phenylalanine analogues (Boros et al., 2021). Several biological metabolites such as L-DOPA (L-3,4-dihydroxyphenylalanine), fluoroepinephrine, and anticancer peptides are synthesized using these derivatives. Many artificial D-α-amino acids are produced using the stereo-destructive character of PALs, in addition to the L-α-amino acid and analogues. These D-arylalanines are used to make a variety of medications, including analgesics, stress relievers, and anticoagulants (Zhou et al., 2020a). PAL might be used to produce aspartame, an ester derivative of phenylalanine that is an important artificial sweetener (Kawatra et al., 2020).

16.3.1.7.5 Antimicrobial production

The advent of resistant "superbugs" regularly represents a severe danger to global health care. As a result, there is a constant demand for particular, nontoxic "designed therapies" that are possibly active against multiple-drug-resistant (MDR) strains. Biocatalysis of PAL has been exploited in producing effective antimicrobial therapeutics. Antimicrobial peptides (AMPs) (such as enterocin) are one such strategy for dealing with multidrug-resistant bacteria. Enterocin is a cationic bacteriostatic peptide that is very selective. According to previous research, the synthesis of enterocin by *Sporobolus maritimus* is regulated by a new PAL encoded by the bacterial gene encP (Zhou et al., 2020a). Antibiotic production was decreased when the bacterial gene encP was disrupted, but cinnamic acid and benzoic acid supplementation restored it. This shows that PAL's bioactivity is essential for enterocin synthesis. Ammonia lyases have also been utilized to make *p*-hydroxycinnamate, a molecule with antiviral, antibacterial, and antioxidant properties (Khatkar et al., 2017).

16.3.1.8 Uricase

Uricase (UC), also known as urate oxidase, is an enzyme that catalyzes the conversion of uric acid to 5-hydroxyisourate and hydrogen peroxide (H_2O_2) (Fig. 16.8). Allantoin is formed after hydrolysis and decarboxylation, and it is readily eliminated by the kidney in most bacterial and eukaryotic organisms. During hominoid evolution, however, all five taxa of hominoids (human, chimp, gorilla, orangutan, and gibbon) lost uricase activity. Hyperuricemia is the condition in which excess uric acid in the blood can cause urate crystal deposition, which is linked to gout, chronic renal disease, tumor lysis syndrome (TLS), hypertension, and a variety of cardiovascular disorders (Piani and Johnson, 2021). Gout is a painful inflammatory disorder caused by the accumulation of uric acid crystals (monosodium urate crystals) within joints. Hyperuricemia and osteoporosis are two other medical conditions resulting from excess uricase release in the bloodstream. Increased uricase levels in the

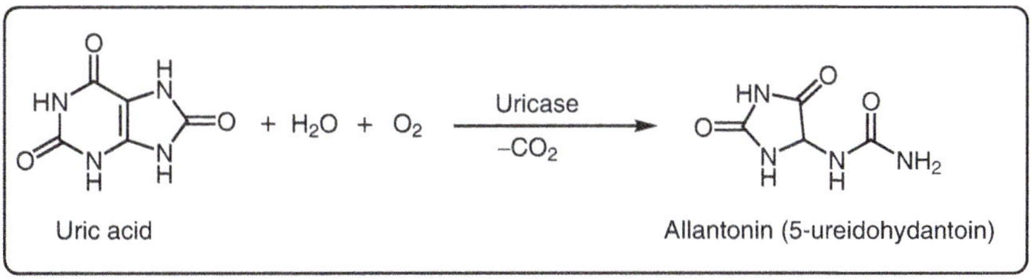

FIGURE 16.8 Schematic representation of the conversion of uric acid into allantoin by uricase.

blood have long been linked to cardiometabolic illnesses, kidney stones, Lesch–Nyhan syndrome, hypertension, and type 2 diabetes. As a result, the usage of the uricase enzyme is seen to be the most effective therapy choice for such conditions. Hypouricemic agents such as allopurinol, urate oxidase, etc. and diuretics are also used to treat TLS, a severe oncological disorder. *Arthrobacter globiformis* Uricase (AgUricase) is a homotetrameric enzyme comprising four identical 34 kDa subunits with high expression levels in *E. coli* and great activity at neutral pH, suggesting that it might be used as a therapeutic. To provide appropriate therapeutic effectiveness, this enzyme must have great thermostability and a long half-life under physiological conditions (Zhu et al., 2022).

16.3.2 Antiinflammatory

16.3.2.1 Serratiopeptidase

The enzyme serratiopeptidase belongs to class EC 3.4.24.40 and is included in the group Serralysin belonging to the trypsin family with about 45–60 kDa molecular weight. Initially, it originated from *Serratia marcescens* isolated from the intestinal portion of a silkworm named *Bombyx mori L.* It is a type of leading therapeutic enzyme which has immense applications as an antiinflammatory, antibiofilm, analgesic, antiedemic, and fibrinolytic effects. There are many in vitro, in vivo, and clinical-based reports on the therapeutic applicability of the enzyme (Kordi et al., 2022). In 1957 Japan was the first to explore Serratiopeptidase for its antiinflammatory effects. Serratiopeptidase possesses tremendous scope to overcome the symptoms of inflammation (Jadhav et al., 2020). However, the Serine proteases bear a higher affinity toward cyclooxygenase (COX-I and COX-II), an important enzyme associated with the secretion of various inflammatory mediators, namely, interleukins (IL), prostaglandins (PGs), and thromboxane (TXs), etc. In the current scenario, the most common inflammatory disease that affects entire human being is arthritis, sinusitis, bronchitis, fibrocystic breast disease, carpal tunnel syndrome, etc. (Tiwari, 2017). The most basic drugs, namely, nonsteroidal antiinflammatory drugs (NSAIDs) either or in combination with other drugs can be used to treat symptoms of inflammation. In the case of chronic inflammation, NSAIDs are used in combination with steroidal drugs, that is, autoimmune disorders. This major adverse effect of using this drug is seen as adverse drug reaction, and to overcome this drawback and other complexities, the new upcoming trend is based on the enzyme-based drug for inflammation. The inflammation serves as a protective mechanism against injury and various infections. The immune system responds quickly to any foreign substances and tissue injury by conscripting immunogenic mediators and various immune cells toward the target site. Thereby inflammation is regarded as a cleaning process of the body that leads to balance homeostasis. Depending upon the pathological conditions of tissue and intensity, the inflammation can be acute or chronic. Apart from its antiinflammatory feature, Serratiopeptidase is explored for other therapeutic abilities. Though the antiinflammatory property of enzymes is explored still, there is a constant need to optimize a particular dose depending on its applicability.

16.3.2.2 Collagenase

Collagenase is a therapeutic enzyme involved in the cleavage of peptide bonds found in collagen. Collagenase is a Zn^{2+}-dependent matrix metalloproteinase that is a fundamental component of the extracellular matrix (ECM). It features a saddle-shaped tertiary structure, and the active site contains a zinc moiety. The side chains containing two histidines, one glutamate, and a water residue coordinate the zinc moiety in a tetrahedral fashion, with the water molecule hydrogen linked to another glutamate residue. Bacterial collagenases are metalloproteases that may break down the ECM, making them essential virulence agents (Vachher et al., 2021). *Clostridium histolyticum*-produced collagenase is used to disintegrate burn scars instead of harsh surgical debridement. Debridement is the process of removing damaged or dead tissue to provide enhanced treatments and active healing. Surgical and mechanical debridement procedures are less precise and painful. This enzyme destroys necrotic dead tissues painlessly and selectively. It promotes wound healing by releasing collagen-derived peptides, which boost macrophage chemotaxis and cytokine release, promoting wound healing. Collagenase A widely established enzymatic debridement regimen is Santyl ointment, which contains clostridial collagenase and a few other generic proteases (Sheets et al., 2016). It is administered topically to burns or ulcers on the skin to speed up healing and remove dead tissue. In the United States, clostridial collagenase ointment is the only FDA-approved treatment for enzymatic wound debridement in severe burns, shortening wound healing time, and discomfort while lowering infection risk (Pham et al., 2019).

There are mainly four basic types of collagenases, that is, collagenase I, II, III, and IV. Each of them acts on a specific substrate, and their absence or nonfunctionality may play a key role in several disorders. Collagenases have been utilized as a therapeutic enzyme to treat multiple diseases and medical problems, including cartilage repair, Dupuytren's disease (DD), cellulite therapy, glaucoma diseases, keloid disease, etc. DD is a fibroproliferative disorder defined by the development and progressive contracture of isolated portions of the palmar aponeurosis, resulting in considerable limitations in hands functioning in advanced stages of the disease (Hartig-Andreasen et al., 2019). Deposition of collagen-rich cords occurs, as well as gradual finger flexion. It is a devastating lifelong condition with various symptoms and clinical manifestations. Overproduction and low production of collagenase in the body are the most common causes of these disorders. *C. histolyticum* collagenase injections have been recommended as a potential first-line therapy for DD (Hartig-Andreasen et al., 2019).

Peyronie's disease (PD) in adult males is also treated with intralesional injections of a combination of class I and class II collagenases from *C. histolyticum* (Xiaflex and Xiapex) in various countries. It has been demonstrated to be effective in disrupting type I and type III collagen, the most common collagen types seen in PD plaques. It spared type IV collagen in nearby connective tissue arteries and veins, lowering PD plaques without damaging axon myelin sheaths, elastic tissue, and vasculature. According to new research, collagenase is a well-tolerated, painless, and effective therapy for various diseases (Hoy, 2020).

16.3.2.3 Superoxide dismutase

Superoxide dismutase (SOD), EC 1.15.1.1, catalyzes the dismutation of superoxide (O_2^-) to oxygen (O_2) and hydrogen peroxide (H_2O_2), which avoids the production of more reactive hydroxyl radicals (•OH) via the Haber–Weiss reaction. They are mostly found in aerobic microbes and play a vital role in the oxidative stress defense system. SODs are metalloenzymes that significantly contain Cu^{2+}, Fe^{2+}, Mn^{2+}, and Zn^{2+} in the active sites of their chemical moieties, which influences and facilitates the dismutation process (Mittra et al., 2017). Based on these metal ion cofactors, SODs are classified into three isozymes: Cu/Zn-SOD, Fe-SOD, and Mn-SOD. All these isozymes have varying degrees of sensitivity toward the H_2O_2 and cyanide (CN), which eventually helps to identify them. The Cu/Zn-SOD is highly sensitive to both H_2O_2 and CN, whereas Mn-SOD is recognized as insensitive to both. In contrast to this, Fe-SOD has sensitivity against H_2O_2 but not for CN. Moreover, chloroform-ethanol showed no effect on Cu/Zn-SOD but was found inhibitory for Fe-SOD and Mn-SOD. The increased SOD production is resembling with the increased tolerance of organisms against oxidative stress caused by biotic aggressors (Roychowdhury et al., 2019).

SOD can prevent precancerous cells, cystic fibrosis, aging, ischemia, rheumatoid arthritis, diabetes, and neurodegenerative diseases but plays a major role in antiinflammation. Porfire et al. (2014) investigated the role of CU/Zn-SOD in antiinflammatory and oxidative systems involved with peritonitis in rats. A recombinant chimeric SOD2/3 showed improved protection in various animals against inflammation and/or ischemia (Hernandez-Saavedra et al., 2005). Exogenous SOD with H_2O_2 can induce apoptotic neutrophils and regulate neutrophil-mediated tissue injury (Yasui et al., 2005). Ghio et al. (2002) explored the O_2^- inhibition by overexpression of extracellular SOD to prevent lung inflammation in transgenic mice.

16.3.3 Enzybiotics

16.3.3.1 Lysins

Lysins or endolysins are special hydrolytic enzymes produced by bacteriophages having double-stranded DNA. These enzymes are sometimes also known as murein hydrolases. Their main target is to cleave the covalent bonds of peptidoglycan or murein of the host bacterial cell during the last phase of the lytic cycle and release progeny virions (Santos et al., 2019). Some specialized pseudomurein-cleaving lysins are also studied to lyse the cell wall-containing archaea (Visweswaran et al., 2010). Lysins are generally used to replace antibiotics that are vulnerable to bacterial resistance, as they are highly effective and specific as compared to the other antibacterial agents. Five functionally different catalytic domains help the lysins to cleave the peptidoglycan bonds (Fischetti, 2008):

- Endopeptidases: hydrolyze any peptide bonds present within the amino acids
- *Endo*-β-*N*-acetylglucosaminidases (endoglycosidase H, EC 3.2.1.96): cleaves *N*-acetylglucosamine (NAG)
- *N*-Acetylmuramidases (Lysozymes, EC 3.2.1.17): cleaves *N*-acetylmuramic acid (NAM)
- *N*-acetylmuramoyl-L-alanine amidases (T7-like, EC 3.5.1.28): cleaves the amide bond between amino acid and sugar moieties
- γ-D-glutaminyl-L-lysine endopeptidase (EC 3.4.14.13): cleaves the gamma bond occurred between the L-lysine and D-glutamine residues

Endolysins can easily access the peptidoglycans and subsequent destruction of Gram-positive bacteria because they lack the outer cell membrane. In future, Lysins may procure a major impact on controlling the bacterial pathogens related to humans' health without disturbing the normal flora. An endolysin LysIME-EF1 was found with highly efficient bactericidal activities against 29 multidrug-resistant clinical strains of *Enterococcus faecalis* (Zhou et al., 2020b). TSPphg, a new lysin isolated from an extremophilic *Thermus* phage TSP4, showed a bactericidal effect on *Klebsiella pneumonia* and *Staphylococcus aureus*, responsible for skin infections (Wang et al., 2020). Oliveira et al. (2015) studied the endolysin PlyPI23 with great potential to regulate the American foulbrood disease-causing *Paenibacillus larvae*. The methicillin-resistant pathogens *S. aureus* and *Streptococcus pyogenes* were controlled by PlySs2, an lysin derived from the *Streptococcus suis* phage (Gilmer et al., 2013). Similarly, a recombinant cytolysin P128 showed lethal actions against *S. aureus* to control bacteraemia in the mouse model (Channabasappa et al., 2018).

16.3.3.2 Autolysins

Another class of lytic enzymes that could be used as enzybiotics is autolysins, also known as *N*-acetylmuramoyl L-alanine amidases. They are a fascinating group of amidases that act on the producer bacterial cell and cleave the amide link between *N*-acetyl muramic acid and L-alanine in the peptidoglycan layer of the bacterial cell wall. They may be bacterial in origin and are involved in various important bacterial processes such as cell growth and division, protein secretion, cell wall turnover, and maturation (Xu et al., 2021). There are conserved amidase domains of bacterial amidases, eukaryotic glutathionylspermidine amidase, NLP/P60 family proteins, and glutamyl L-diamino endopeptidases have been found in phage-encoded autolysins that act as peptidoglycan hydrolases. LytA amidase from *Streptococcus pneumoniae* strains was the first autolysin to be employed as an antibacterial agent (Vachher et al., 2021).

16.3.3.3 Bacteriocins

Bacteriocins are potent small AMPs about 30–60 amino acids synthesized by Gram-positive and Gram-negative bacteria in addition to antibiotics. Bacteriocins form a network in the target cell membrane, leading to lower weight leakage due to proton motive force, resulting in disruption. However, as far as AMPs are concerned, they are secreted heterogeneous in size, structure, mechanisms, spectrum, target cell receptor, etc. A broad range of antimicrobial strategies is adopted by the immune system of many life forms ranging from insects to humans, represented as the synthesis of AMPs. In the case of microbial systems, it is found that the majority of the Gram-positive and archaeal systems secrete at least one AMP for self-preservation which bears an advantage in the ecological niche. In bacteria, bacteriocins are found in lactic acid bacteria (Fernandes and Jobby, 2022). There is extensive research done on bacteriocins which shows the potency as an antibacterial substance that is found to be associated with strain-producing species. However, due to these properties, bacteriocins are referred to as Generally Recognized as Safe status as they can degrade gastrointestinal enzymes. There are differences in the antibacterial spectrum which may be due to the occurrence of amphiphilic helices. As far as the food industry is concerned, the bacteriocins may be used directly or incorporated into food while cultivating it with the help of bacteriocin-producing strains. However, the majority of the bacteriocins gain applicability in the health care allied sector as antibacterial and anticancer agents (Gradisteanu Pircalabioru et al., 2021). One such example of the bacteriocins such as nisin shows very specificity in its action, namely, antibacterial activity toward multidrug-resistant strains of Gram-positive bacteria, that is, methicillin-resistant *S. aureus*.

A few studies demonstrated the benefits of incorporating antibacterial nanoparticles with bacteriocins from Gram-positive bacteria in exhibiting the growth-inhibitory efficacy of Gram-negative bacteria, such as *Pseudomonas aeruginosa*. As adjuvant for bioactive materials nanoparticle (Naskar & Kim, 2021) employed, a study is quite promiscuous to overcome troubleshoot. Major reasons behind using nanomaterials are broad-spectrum activity, physiological stability, higher surface area-to-volume ratio compared to others, very easy to synthesize with economic feasibility, and nontoxicity. With this respect, presently most promiscuous use of bacteriocins is in biomedical properties for anticancer studies. Despite various advantages, there are a few shortcomings that restrict the biomedical applications of bacteriocins such as less yield, vulnerability to intracellular proteases, higher production costs, and higher dosage forms for MDR bacteria. However, based on limitations, only three FDA- approved bacteriocins, namely, nisin (precursor peptide secreted by *Lactococcus lactis* with 34 mature forms of amino acids which are further processed by posttranslational modifications), pediocin (mature 44 amino acids' form is not further processed by posttranslational modifications), and micocin, are available in processes related to the food sector. Another bacteriocin sourced from Gram-negative bacteria *E. coli* is colicin, namely, higher molecular-weight protein and used to prevent or inhibit the growth of Gram-negative bacteria. On the contrary, there is another bacteriocin, that is, microcin sourced from bacteria *E. coli* has its property similar to Gram-positive bacteriocins in terms of pH and thermal stability and protease resistance. Other Gram-negative bacteria *P. aeruginosa* secretes pyocins or aeruginocins, and *K. pneumonia* secretes klebicin, a type of bacteriocins. Such bacteriocins may be advantageous as it depends upon a particular purpose. Due to a thick multilayered peptidoglycan wall instead of an outer membrane, Gram-positive bacteria have a broad antibacterial spectrum.

The bacteriocins possess antibacterial activity owing to their biomedical applicability in vitro; however, they possess low in vivo stability and are subject to degradation by proteolytic enzymes, limiting their clinical application (Tang et al., 2022). The bacteriocins are labile in tissue, serum, and organs, namely, liver and kidneys; thereby, it needs modification to ensure their stability. Various functionalization approaches have been employed to show the stability feature of bacteriocins, that is, cyclization by incorporating D forms of amino acids. However, the bacteriocins such as nicins are generally used in combination with gellan gum which enhances efficacy. In a few studies, bacteriocins from *Enterococcus faecium* are immobilized into cellulose nanocrystals which leads to enhance stability. Other substances such as liposomes, double hydroxide nanoparticles, chitosan nanoparticles, and nanovesicles were used to encapsulate bacteriocins which enhanced their antibacterial features and led to wide applicability. Also, by employing bioengineering approaches, physicochemical and biological attributes of bacteriocins can be enhanced. Many other bacteriocins have been cited to show cytotoxic/ anticancer activity against various cancer cell lines which is due to apoptosis or depolarization. For example, few studies highlight the cytotoxic and apoptotic effects of azurin named bacteriocin from *P. aeruginosa* without harming normal cells. Few bacteriocins such as nisin and fermenticin have shown spermicidal ability by reducing spermatozoa motility, thus they can also be used as contraceptives. The majority of the anticancer studies of bacteriocins have been executed in vitro, and there is a constant urge to validate in vivo studies.

16.3.3.4 Lysozymes

Lysozymes, also known as 1, 4-*N*-acetylmuramidases, are hydrolytic enzymes that hydrolyzed 1, 4-glycosidic linkages in the peptidoglycan layer between the *N*-acetylmuramic acid and *N*-acetylglucosamine moieties. Lysozymes are ubiquitously expressed in the plant and animal worlds and play an essential role in defense. Various microorganisms, including bacteria and viruses, produce lysozymes. Bacteria, such as *Arthrobacter crystalopoites, S. aureus*, *Bacillus subtilis*, *Streptomyces griseus, B. thuringiensis*, and *Enterococcus hirae*, have lysozymes (Vachher et al., 2021). Lysozymes are one of a kind since they have antiviral, anticancer, antiinflammatory, and immunomodulatory properties and antibacterial properties. They are widely recognized for their capacity to catalyze the cleavage of peptidoglycan, as previously stated. Lysozymes, on the other hand, may lyse bacteria by a variety of nonenzymatic methods. They can also cause the plasma membrane to become unstable by eliminating divalent ions from the membrane's surface (Pizzo et al., 2018; Kundu, 2020).

Gel formulations for topical administration on wounds, control of infections originating from skin piercing, acne therapy, and aerosols for treating tracheitis, amygdalitis, pneumonia, and faucitis are among the most current protocols for using lysozymes in medicine (Portilla et al., 2020). Lysozyme is also found in oral health products such as mouthwashes (Sajedinejad et al., 2018), as it can destroy a variety of germs. It has also been demonstrated that lysozyme may be used as a new antimicrobial treatment for accurately delivering phenolic antimicrobial drugs (triclosan) into bacterial cells (Vachher et al., 2021). Recent studies showed that the lysozyme-loaded polyurethane dressing could supply sufficient wound hydration while preventing bacterial infection, indicating that this is a future development direction for wound dressings (Xiao et al., 2021).

16.4 Concluding remarks

Microbial systems synthesize various enzymes that have the potential to be used in the pharmaceutical industry. Enzymes are the essential part of all metabolic processes that occur directly or indirectly in a living system. The applicability of biocatalysts for synthesizing active pharmaceutical products has increased rapidly in the current scenario. The physiological functions of the enzyme in the living system are digestion, metabolism, immune function, and reproduction. Many enzymes are available; among them, the microbial source is preferably used for biotechnological processes due to ease of cultivation and economic feasibility. Enzyme-based therapeutics is more preferable over chemicals due to specificity, selectivity, and safety. The enormous, diverse microbial enzymes make an interesting group of products for application in the particularly pharmaceutical sector. The current chapter deals with major emphasis on the therapeutic enzyme that has potential application in the pharmaceutical industry. Particularly by employing molecular approaches, the enhancement in features of the microbial system may be possible through recombinant technology. The therapeutic enzyme seems to possess catalytic proficiencies that have broad applicability in the pharmaceutical industry for drug development. Furthermore, employing the metagenome approach paves the way for in-depth knowledge of novel enzymes that have plenty of scope in future.

Abbreviations

ADI	arginine dihydrolase
AgUricase	*Arthrobacter globiformis* Uricase
ALL	acute lymphoblastic leukemia
AMPs	antimicrobial peptides
CDs	cyclodextrins
CGTase	cyclodextrin glucanotransferase
CK	carbamate kinase
COX	cyclooxygenase
DD	Dupuytren's disease
ECM	extracellular matrix
H_2O_2	hydrogen peroxide
HeLa	Henrietta lacks
HUVE	human umbilical vein endothelial
IC 50	maximal Inhibitory Concentration
IT	interleukins
LAOs	L-Amino acid oxidases
L-ASNase	L-Asparaginase amidohydrolase

L-DOPA	L-3, 4-Dihydroxyphenylalanine
L-GLNase	L-Glutaminase amidohydrolase
LO	lysine oxidase
MCF7	Michigan Cancer Foundation-7
MDR	multiple-drug resistant
MRSA	methicillin-resistant *Staphylococcus aureus*
NAM	*N*-Acetylmuramic acid
NSAIDs	nonsteroidal antiinflammatory drugs
OTC	ornithine transcarbamoylase
PAH	phenylalanine hydroxylase
PAL	phenylalanine ammonia-lyase
PD	Peyronie's disease
PGs	prostaglandins
PKU	phenylketonuria
SOD	superoxide dismutase
TLS	tumor lysis syndrome
TXs	thromboxane
Vero	Verda Reno

References

Abell, C.W., Hodgins, D.S., Stith, W.J., 1973. An *in-vivo* evaluation of the chemotherapeutic potency of phenylalanine ammonia-lyase. Cancer Res. 33 (10), 2529–2532.

Amobonye, A., Singh, S., Mukherjee, K., Jobichen, C., Qureshi, I.A., Pillai, S., 2022. Structural and functional insights into fungal glutaminase using a computational approach. Process. Biochem. 117, 76–89. Available from: https://doi.org/10.1016/j.procbio.2022.03.019.

Basso, A., Serban, S., 2019. Industrial applications of immobilized enzymes—a review. Mol. Catal. 479, 110607. Available from: https://doi.org/10.1016/j.mcat.2019.110607.

Beloussow, K., Wang, L., Wu, J., Ann, D., Shen, W.C., 2002. Recombinant arginine deiminase as a potential antiangiogenic agent. Cancer Lett. 183, 155–162. Available from: https://doi.org/10.1016/S0304-3835(01)00793-5.

Boros, K., Moisă, M.E., Nagy, C.L., Paizs, C., Toşa, M.I., Bencze, L.C., 2021. Robust, site-specifically immobilized phenylalanine ammonia-lyases for the enantioselective ammonia addition of cinnamic acids. Catal. Sci. Technol. 11 (16), 5553–5563. Available from: https://doi.org/10.1039/D1CY00195G.

Broome, J.D., 1963. Evidence that the L-asparaginase of guinea pig serum is responsible for its antilymphoma effects: I. properties of the L-asparaginase of guinea pig serum in relation to those of the antilymphoma substance. J. Exp. Med. 118 (1), 99–120. Available from: https://doi.org/10.1084/jem.118.1.99.

Cachumba, J.J., Antunes, F.A., Peres, G.F., Brumano, L.P., Santos, J.C., Da Silva, S.S., 2016. Current applications and different approaches for microbial L-asparaginase production. Braz. J. Microbiol. 47 (Suppl. S1), 77–85. Available from: https://doi.org/10.1016/j.bjm.2016.10.004.

Canda, E., Uçar, S.K., Çoker, M., 2020. Biotinidase deficiency: prevalence, impact and management strategies. Pediatric Health Med. Ther. 11, 127–133. Available from: https://doi.org/10.2147/PHMT.S198656.

Chakravarty, N., Mathur, A., Singh, R.P., 2022. L-asparaginase: insights into the marine sources and nanotechnological advancements in improving its therapeutics. In: Sarma, H., Gupta, S.,

Narayan, M., Prasad, R., Krishnan, A. (Eds.), Engineered Nanomaterials for Innovative Therapies and Biomedicine. Nanotechnology in the Life Sciences. Springer, Cham. Available from: https://doi.org/10.1007/978-3-030-82918-6_4.

Channabasappa, S., Chikkamadaiah, R., Durgaiah, M., Kumar, S., Ramesh, K., Sreekanthan, A., et al., 2018. Efficacy of chimeric ectolysin P128 in drug-resistant *Staphylococcus aureus* bacteraemia in mice. J. Antimicrob. Chemothe 73 (12), 3398–3404. Available from: https://doi.org/10.1093/jac/dky365.

Costa, M.N., Silva, R.N., 2021. Cytotoxic activity of L-lysine alpha-oxidase against leukaemia cells. Seminars in Cancer Biology. Academic Press. Available from: https://doi.org/10.1016/j.semcancer.2021.09.015.

Dayan, F.E., 2019. Current status and future prospects in herbicide discovery. Plants 8 (9), 341. Available from: https://doi.org/10.3390/plants8090341.

de Freitas, T.L., Monti, R., Contiero, J., 2004. Production of CGTase by a *Bacillus alkalophilic* CGII strain isolated from wastewater of a manioc flour industry. Braz. J. Microbiol. 35, 255–260. Available from: https://doi.org/10.1590/S1517-83822004000200015.

Demirbas, D., Brucker, W.J., Berry, G.T., 2018. Inborn errors of metabolism with hepatopathy: metabolism defects of galactose, fructose, and tyrosine. Pediatr. Clin. 65 (2), 337–352. Available from: https://doi.org/10.1016/j.pcl.2017.11.008.

Dhankhar, R., Gupta, V., Kumar, S., Kapoor, R.K., Gulati, P., 2020. Microbial enzymes for deprivation of amino acid metabolism in malignant cells: a biological strategy for cancer treatment. App. Microbiol. Biotechnol. 104 (7), 2857–2869. Available from: https://doi.org/10.1007/s00253-020-10432-2.

Dolowy, W.C., Henson, D., Cornet, J., Sellin, H., 1966. Toxic and antineoplastic effects of L-asparaginase. Study of mice with lymphoma and normal monkeys and report on a child with leukaemia. Cancer 19, 1813–1819. Available from: https://doi.org/10.1002/1097-0142(196612)19:12%3C1813::aid-cncr2820191208%3E3.0.co;2-e.

Fernandes, A., Jobby, R., 2022. Bacteriocins from lactic acid bacteria and their potential clinical applications. Appl. Biochem. Biotechnol. 1–23. Available from: https://doi.org/10.1007/s12010-022-03870-3.

Fischetti, V.A., 2008. Bacteriophage lysins as effective antibacterials. Curr. Opin. Microbiol. 11 (5), 393–400. Available from: https://doi.org/10.1016/j.mib.2008.09.012.

Ghio, A.J., Suliman, H.B., Carter, J.D., Abushamaa, A.M., Folz, R.J., 2002. Overexpression of extracellular superoxide dismutase decreases lung injury after exposure to oil fly ash. Am. J. Physiol. Lung Cell Mol. Physiol. 283, 211–218. Available from: https://doi.org/10.1152/ajplung.00409.2001.

Gilmer, D.B., Schmitz, J.E., Euler, C.W., Fischetti, V.A., 2013. Novel bacteriophage lysin with broad lytic activity protects against mixed infection by *Streptococcus pyogenes* and methicillin-resistant *Staphylococcus aureus*. Antimicrob. Agents Chemother. 57, 2743–2750. Available from: https://doi.org/10.1128/AAC.02526-12.

Gomaa, E.Z., 2022. Production, characterization, and antitumor efficiency of L-glutaminase from halophilic bacteria. Bull. Natl Res. Cent. 46 (1), 1–11. Available from: https://doi.org/10.1186/s42269-021-00693-w.

Goseki, N., Yamazaki, S., Shimojyo, K., Kando, F., Maruyama, M., Endo, M., et al., 1995. Synergistic effect of methionine-depleting total parenteral nutrition with 5-fluorouracil on human gastric cancer. A randomized, prospective clinical trial. Jpn. J. Cancer Res. 86, 484–489. Available from: https://doi.org/10.1111/j.1349-7006.1995.tb03082.x.

Gradisteanu Pircalabioru, G., Popa, L.I., Marutescu, L., Gheorghe, I., Popa, M., Czobor Barbu, I., et al., 2021. Bacteriocins in the era of antibiotic resistance: rising to the challenge. Pharmaceutics 13 (2), 196. Available from: https://doi.org/10.3390/pharmaceutics13020196.

Gutman, A.B., 1963. An abnormality of glutamine metabolism in primary gout. Am. J. Med. 35 (6), 820–831. Available from: https://doi.org/10.1016/0002-9343(63)90244-4.

Hartig-Andreasen, C., Schroll, L., Lange, J., 2019. *Clostridium histolyticum* as first-line treatment of Dupuytren's disease. Dan. Med. J. 66 (2).

Hendrikse, N.M., Holmberg Larsson, A., Svensson Gelius, S., Kuprin, S., Nordling, E., Syrén, P. O., 2020. Exploring the therapeutic potential of modern and ancestral phenylalanine/tyrosine ammonia-lyase as supplementary treatment of hereditary tyrosinemia. Sci. Rep. 10 (1), 1–13. Available from: https://doi.org/10.1038/s41598-020-57913-y.

Hernandez-Saavedra, D., Zhou, H., McCord, J.M., 2005. Anti-inflammatory properties of a chimeric recombinant superoxide dismutase: SOD2/3. Biomed. Pharmacother. 59 (4), 204–208. Available from: https://doi.org/10.1016/j.biopha.2005.03.001.

Hernick, M., Fierke, C., 2010. Mechanisms of metal-dependent hydrolases in metabolism, Comprehensive Natural Products II, 8. Elsevier, pp. 547–581. Available from: https://doi.org/10.1016/B978-008045382-8.00178-7.

Hoy, S.M., 2020. Collagenase *Clostridium histolyticum*: A review in Peyronie's disease. Clin. Drug. Invest. 40, 83–92. Available from: https://doi.org/10.1007/s40261-019-00867-5.

Ireton, G.C., McDermott, G., Black, M.E., Stoddard, B.L., 2002. The structure of *Escherichia coli* cytosine deaminase. J. Mol. Biol. 315 (4), 687–697. Available from: https://doi.org/10.1006/jmbi.2001.5277.

Jadhav, S.B., Shah, N., Rathi, A., Rathi, V., Rathi, A., 2020. Serratiopeptidase: insights into the therapeutic applications. Biotechnol. Rep. 28, e00544. Available from: https://doi.org/10.1016/j.btre.2020.e00544.

Jaiswal, S., Uniyal, A., Tiwari, V., Ayyannan, S.R., 2022. Synthesis and evaluation of dual fatty acid amide hydrolase-monoacylglycerol lipase inhibition and antinociceptive activities of 4-methylsulfonylaniline-derived semicarbazones. Bioorg. Med. Chem. 116698. Available from: https://doi.org/10.1016/j.bmc.2022.116698.

Kaur, S., Kaur, S., Kaur, K., 2014. Studies on β-cyclodextrin production from CGTase producing bacteria and its effect on drug solubility. Int. J. Pharma. Pharma. Sci. 6 (10), 383–387.

Kawatra, A., Dhankhar, R., Mohanty, A., Gulati, P., 2020. Biomedical applications of microbial phenylalanine ammonia-lyase: current status and future prospects. Biochimie 177, 142–152. Available from: https://doi.org/10.1016/j.biochi.2020.08.009.

Khatkar, A., Nanda, A., Kumar, P., Narasimhan, B., 2017. Synthesis, antimicrobial evaluation and QSAR studies of p-coumaric acid derivatives. Arab. J. Chem. 10, S3804–S3815. Available from: https://doi.org/10.1016/j.arabjc.2014.05.018.

Kidd, J.G., 1953. Regression of transplanted lymphomas induced *in-vivo* by means of normal guinea pig serum: II. Studies on the nature of the active serum constituent: histological mechanism of the regression: tests for effects of guinea pig serum on lymphoma cells *in-vitro*: Discussion. J. Exp. Med. 98 (6), 583–606. Available from: https://doi.org/10.1084/jem.98.6.583.

Kordi, M., Salami, R., Bolouri, P., Delangiz, N., Asgari Lajayer, B., van Hullebusch, E.D., 2022. White biotechnology and the production of bio-products. Syst. Microbiology Biomanufacturing 1–17. Available from: https://doi.org/10.1007/s43393-022-00078-8.

Kundu, R., 2020. Cationic amphiphilic peptides: synthetic antimicrobial agents inspired by nature. ChemMedChem 15 (20), 1887–1896. Available from: https://doi.org/10.1002/cmdc.202000301.

Levy, H.L., Sarkissian, C.N., Scriver, C.R., 2018. Phenylalanine ammonia-lyase (PAL): From discovery to enzyme substitution therapy for phenylketonuria. Mol. Genet. Metab. 124 (4), 223–229. Available from: https://doi.org/10.1016/j.ymgme.2018.06.002.

Lukasheva, E.V., Babayeva, G., Karshieva, S.S., Zhdanov, D.D., Pokrovsky, V.S., 2021. L-lysine α-oxidase: enzyme with anticancer properties. Pharm. (Basel) 14 (11), 1070. Available from: https://doi.org/10.3390/ph14111070.

Mao, X., Chen, H., Lin, A., Kim, S., Burczynski, M.E., Na, E., et al., 2022. Glutaminase 2 knockdown reduces hyperammonemia and associated lethality of urea cycle disorder mouse model. J. Inherit. Metab. D. Available from: https://doi.org/10.1002/jimd.12474.

Maurer, D., Lohkamp, B., Krumpel, M., Widersten, M., Dobritzsch, D., 2018. Crystal structure and pH-dependent allosteric regulation of human β-ureidopropionase, an enzyme involved in anticancer drug metabolism. Biochem. Eng. J. 475 (14), 2395–2416. Available from: https://doi.org/10.1042/BCJ20180222. s.

Mittra, B., Laranjeira-Silva, M.F., Miguel, D.C., de Menezes, J.P.B., Andrews, N.W., 2017. The iron-dependent mitochondrial superoxide dismutase SODA promotes *Leishmania* virulence. J. Biol. Chem. 292 (29), 12324–12338. Available from: https://doi.org/10.1074/jbc.M116.772624.

Naskar, A., Kim, K.S., 2021. Potential novel food-related and biomedical applications of nanomaterials combined with bacteriocins. Pharmaceutics 13 (1), 86. Available from: https://doi.org/10.3390/pharmaceutics13010086.

Nazir, F., Tabish, T.A., Tariq, F., Iftikhar, S., Wasim, R., Shahnaz, G., 2022. Stimuli-sensitive drug delivery systems for site-specific antibiotic release. Drug. Discovery Today . Available from: https://doi.org/10.1016/j.drudis.2022.02.014.

Oliveira, A., Leite, M., Kluskens, L.D., Santos, S.B., Melo, L.D., Azeredo, J., 2015. Correction: the first *Paenibacillus* larvae bacteriophage endolysin (PlyPl23) with high potential to control American foulbrood. PLoS one 10 (8), e0136331. Available from: https://doi.org/10.1371/journal.pone.0136331.

Patel, N.Y., 2020. Exploring the Potential of Marine Bacteria for L-Glutaminase Biosynthesis and its Biochemical Profiling (M.Phil. thesis). Gujarat University.

Patel, N.Y., Baria, D.M., Arya, P.S., Rajput, K.N., Panchal, R.R., Raval, V.H., 2020. L-glutaminase biosynthesis from marine bacteria. Biosci. Biotechnol. Res. Commun. 13 (1), 67–72. Available from: http://bbrc.in/bbrc/wp-content/uploads/2020/05/FINAL-PDF-Special-Issue-Vol-13-No-1-2020.pdf.

Patel, N.Y., Baria, D.M., Yagnik, S.M., Rajput, K.N., Panchal, R.R., Raval, V.H., 2021. Bioprospecting the future in perspective of amidohydrolase L-glutaminase from marine habitats. Appl. Microbiol. Biotechnol. 105 (13), 5325–5340. Available from: https://doi.org/10.1007/s00253-021-11416-6.

Patel, N.Y., Yagnik, S.M., Baria, D.M., Raval, V.H., 2022. Application of extremophiles in medicine and pharmaceutical industries. Physiology, Genomics, Biotechnological Applications of Extremophiles. IGI Global, pp. 260–285. Available from: https://doi.org/10.4018/978-1-7998-9144-4.ch013.

Pham, C.H., Collier, Z.J., Fang, M., Howell, A., Gillenwater, T.J., 2019. The role of collagenase ointment in acute burns: a systematic review and *meta*-analysis. J. Wound Care 28 (Sup2), S9–S15. Available from: https://doi.org/10.12968/jowc.2019.28.Sup2.S9.

Piani, F., Johnson, R.J., 2021. Does gouty nephropathy exist, and is it more common than we think? Kidney Int. 99 (1), 31–33. Available from: https://doi.org/10.1016/j.kint.2020.10.015.

Pišlar, A., Sabotič, J., Šlenc, J., Brzin, J., Kos, J., 2016. Cytotoxic L-amino-acid oxidases from *Amanita phalloides* and *Clitocybe geotropa* induce caspase-dependent apoptosis. Cell Death Discovery 2, 16021. Available from: https://doi.org/10.1038/cddiscovery.2016.21.

Pizzo, E., Cafaro, V., Di Donato, A., Notomista, E., 2018. Cryptic antimicrobial peptides: identification methods and current knowledge of their immunomodulatory properties. Curr. Pharm. Des. 24 (10), 1054–1066. Available from: https://doi.org/10.2174/1381612824666180327165012.

Pokrovsky, V.S., Chepikova, O.E., Davydov, D.Z., Zamyatnin Jr, A.A., Lukashev, A.N., Lukasheva, E.V., 2019. Amino acid degrading enzymes and their application in cancer therapy. Curr. Med. Chem. 26 (3), 446–464. Available from: https://doi.org/10.2174/0929867324666171006132729.

Pokrovsky, V.S., Treshalina, H.M., Lukasheva, E.V., Sedakova, L.A., Medentzev, A.G., Arinbasarova, A.Y., et al., 2013. Enzymatic properties and anticancer activity of L-lysine α-oxidase from *Trichoderma* cf. *aureoviride* Rifai BKMF-4268D. Anticancer. Drugs 846–851. Available from: https://doi.org/10.1097/CAD.0b013e328362fbe2.

Porfire, A.S., Leucuţa, S.E., Kiss, B., Loghin, F., Pârvu, A.E., 2014. Investigation into the role of Cu/Zn-SOD delivery system on its antioxidant and anti-inflammatory activity in rat model of peritonitis. Pharmacol. Rep. 66, 670–676. Available from: https://doi.org/10.1016/j.pharep.2014.03.011.

Portilla, S., Fernández, L., Gutiérrez, D., Rodríguez, A., García, P., 2020. Encapsulation of the antistaphylococcal endolysin LysRODI in pH-sensitive liposomes. Antibiotics 9 (5), 242. Available from: https://doi.org/10.3390/antibiotics9050242.

Rajesh, Y., Narayanan, K., Reddy, M.S., Bhaskar, V.K., Shenoy, G.G., Subrahmanyam, V.M., et al., 2015. Production of β -cyclodextrin from pH and thermostable cyclodextrin glycosyltransferase, obtained from *Arthrobacter mysorens* and its evaluation as a drug carrier for Irbesartan. Curr. Drug. Deliv. 12 (4), 444–453. Available from: https://doi.org/10.2174/1567201812666150422163531.

Rajput, K.N., Patel, K.C., Trivedi, U.B., 2016. Screening and selection of medium components for cyclodextrin glucanotransferase production by new alkaliphile *Microbacterium terrae* KNR 9 using Plackett-Burman design. Biotec. Res. Intern. Available from: https://doi.org/10.1155/2016/3584807.

Rastogi, H., Bhatia, S., 2019. Future prospectives for enzyme technologies in the food industry. Enzymes in food biotechnology. Academic Press, pp. 845–860. Available from: https://doi.org/10.1016/B978-0-12-813280-7.00049-9.

Raval, V.H., Bhatt, H.B., Singh, S.P., 2018. Adaptation strategies in halophilic bacteria. Extremophiles. CRC Press, pp. 137–164. Available from: https://doi.org/10.1201/9781315154695-7.

Reina, J.C., Pérez, P., Llamas, I., 2022. Quorum quenching strains isolated from the microbiota of sea anemones and holothurians attenuate *Vibrio corallilyticus* virulence factors and reduce mortality in *Artemia salina*. Microorganisms 10 (3), 631. Available from: https://doi.org/10.3390/microorganisms10030631.

Reshma, C.V., 2019. Microbial enzymes: therapeutic applications. Microbiol. Res. J. Int. 1, 1–8. Available from: https://doi.org/10.9734/MRJI/2019/v27i230093.

Roychowdhury, R., Khan, M.H., Choudhury, S., 2019. Physiological and molecular responses for metalloid stress in rice-a comprehensive overview. Chapter 16. Advances in Rice Research for Abiotic Stress Tolerance. Woodhead Publishing, pp. 341–369. Available from: https://doi.org/10.1016/B978-0-12-814332-2.00016-2.

Sabotič, J., Brzin, J., Erjavec, J., Dreo, T., Tušek Žnidarič, M., Ravnikar, M., et al., 2020. L-amino acid oxidases from mushrooms show antibacterial activity against the phytopathogen *Ralstonia solanacearum*. Front. Microbiol. 11, 977. Available from: https://doi.org/10.3389/fmicb.2020.00977.

Sajedinejad, N., Paknejad, M., Houshmand, B., Sharafi, H., Jelodar, R., Shahbani Zahiri, H., et al., 2018. *Lactobacillus salivarius* NK02: a potent probiotic for clinical application in mouthwash. Probiotics Antimicrob. Proteins. 10 (3), 485–495. Available from: https://doi.org/10.1007/s12602-017-9296-4.

Santos, S.B., Oliveira, A., Melo, L.D.R., Azeredo, J., 2019. Identification of the first endolysin cell-binding domain (CBD) targeting *Paenibacillus larvae*. Sci. Rep. 9, 2568. Available from: https://doi.org/10.1038/s41598-019-39097-2.

Sarmah, N., Revathi, D., Sheelu, G., Yamuna Rani, K., Sridhar, S., Mehtab, V., et al., 2018. Recent advances in sources and industrial applications of lipases. Biotechnol. Prog. 34 (1), 5–28. Available from: https://doi.org/10.1002/btpr.2581.

Sheets, A.R., Demidova-Rice, T.N., Shi, L., Ronfard, V., Grover, K.V., Herman, I.M., 2016. Identification and characterization of novel matrix-derived bioactive peptides: a role for collagenase from Santyl® ointment in post-debridement wound healing. PLoS one 11 (7), e0159598. Available from: https://doi.org/10.1371/journal.pone.0159598.

Sian, H.K., Said, M., Hassan, O., Kamaruddin, K., Ismail, A.F., Rahman, R.A., et al., 2005. Purification and characterization of cyclodextrin glucanotransferase from alkalophilic *Bacillus* sp. G1. Process. Biochem. 4, 1101–1111. Available from: https://doi.org/10.1016/j.procbio.2004.03.018.

Sugathan, S., Pradeep, N.S., Abdulhameed, S. (Eds.), 2017. Bioresources and Bioprocess in Biotechnology: Volume 2: Exploring Potential Biomolecules. Springer. Available from: https://doi.org/10.1007/978-981-10-4284-3.

Taipa, M.A., Fernandes, P., de Carvalho, C., 2019. Production and purification of therapeutic enzymes. Adv. Exp. Med. Biol. 1148, 1–24. Available from: https://doi.org/10.1007/978-981-13-7709-9_1.

Takaku, H., Matsumoto, M., Misawa, S., Miyazaki, K., 1995. Antitumor-activity of arginine deiminase from *Mycoplasma arginini* and its growth-inhibitory mechanism. Jap. J. Cancer Res. 86, 840–846. Available from: https://doi.org/10.1111/j.1349-7006.1995.tb03094.x.

Takaku, H., Takase, M., Abe, S., Hayashi, H., Miyazaki, K., 1992. *In-vivo* antitumor-activity of arginine deiminase purified from *Mycoplasma arginini*. Int. J. Cancer 51, 244–249. Available from: https://doi.org/10.1002/ijc.2910510213.

Tang, H.W., Phapugrangkul, P., Fauzi, H.M., Tan, J.S., 2022. Lactic acid bacteria bacteriocin, an antimicrobial peptide effective against multidrug resistance: a comprehensive review. Int. J. Pept. Res. Ther. 28 (1), 1–14. Available from: https://doi.org/10.1007/s10989-021-10317-6.

Tiwari, M., 2017. The role of serratiopeptidase in the resolution of inflammation. Asian J. Pharm. Sci. 12 (3), 209–215. Available from: https://doi.org/10.1016/j.ajps.2017.01.003.

Vachher, M., Sen, A., Kapila, R., Nigam, A., 2021. Microbial therapeutic enzymes: a promising area of biopharmaceuticals. CRBIOTECH 3, 195–208. Available from: https://doi.org/10.1016/j.crbiot.2021.05.006.

Van Spronsen, F.J., Blau, N., Harding, C., Burlina, A., Longo, N., Bosch, A.M., 2021. Phenylketonuria. Nat. Rev. Dis. Prim. 7 (1), 1–19. Available from: https://doi.org/10.1038/s41572-021-00267-0.

Van Trimpont, M., Peeters, E., De Visser, Y., Schalk, A.M., Mondelaers, V., De Moerloose, B., et al., 2022. Novel insights on the use of L-asparaginase as an efficient and safe anti-cancer therapy. Cancers 14 (4), 902. Available from: https://doi.org/10.3390/cancers14040902.

Visweswaran, G.R., Dijkstra, B.W., Kok, J., 2010. Two major archaeal pseudomurein endoisopeptidases: PeiW and PeiP. Archaea 480492. Available from: https://doi.org/10.1155/2010/480492.

Wang, F., Ji, X., Li, Q., Zhang, G., Peng, J., Hai, J., et al., 2020. TSPphg lysin from the extremophilic thermus bacteriophage TSP4 as a potential antimicrobial agent against both Gram-negative and Gram-positive pathogenic bacteria. Viruses 12, 192. Available from: https://doi.org/10.3390/v12020192.

Wang, S., Yan, Y., Xu, W.J., Gong, S.G., Zhong, X.J., An, Q.Y., et al., 2022. The role of glutamine and glutaminase in pulmonary hypertension. Front. Cardiovasc. Med. 461. Available from: https://doi.org/10.3389/fcvm.2022.838657.

Xiao, L., Ni, W., Zhao, X., Guo, Y., Li, X., Wang, F., et al., 2021. A moisture balanced antibacterial dressing loaded with lysozyme possesses antibacterial activity and promotes wound healing. Soft Matter 17 (11), 3162–3173. Available from: https://doi.org/10.1039/D0SM02245D.

Xu, H.M., Huang, H.L., Zhou, Y.L., Zhao, H.L., Xu, J., Shou, D.W., et al., 2021. Faecal microbiota transplantation: a new therapeutic attempt from the gut to the brain. Gastroenterol. Res. Pract. Available from: https://doi.org/10.1155/2021/6699268.

Yamamoto, J., Inubushi, S., Han, Q., Tashiro, Y., Sugisawa, N., Hamada, K., et al., 2022. Linkage of methionine addiction, histone lysine hypermethylation and malignancy. iScience 104162. Available from: https://doi.org/10.21873/cgp.20299.

Yasui, K., Kobayashi, N., Yamazaki, T., Agematsu, K., Matsuzaki, S., Ito, S., et al., 2005. Superoxide dismutase (SOD) is a potential inhibitory mediator of inflammation via neutrophil apoptosis. Free. Radic. Res. 39 (7), 755–762. Available from: https://doi.org/10.1080/10715760500104066.

Zhou, B., Zhen, X., Zhou, H., Zhao, F., Fan, C., Perčulija, V., et al., 2020b. Structural and functional insights into a novel two-component endolysin encoded by a single gene in *Enterococcus faecalis* phage. PLOS Pathog. 16 (3), e1008394. Available from: https://doi.org/10.1371/journal.ppat.1008394.

Zhou, Y., Sekar, B.S., Wu, S., Li, Z., 2020a. Benzoic acid production via cascade biotransformation and coupled fermentation-biotransformation. Biotechnol. Bioeng. 117 (8), 2340–2350. Available from: https://doi.org/10.1002/bit.27366.

Zhu, T.T., Chen, H.N., Yang, L., Liu, Y.B., Li, W., Sun, W.X., 2022. Characterization and Cys-directed mutagenesis of urate oxidase from *Bacillus subtilis* BS04. Biologia 77 (1), 291–301. Available from: https://doi.org/10.1007/s11756-021-00941-4.

Chapter 17

Microbial enzymes of use in industry

Xiangyang Liu[1] and Chandrakant Kokare[2]
[1]*UNT System College of Pharmacy, University of North Texas Health Science Center, Fort Worth, TX, United States,* [2]*Department of Pharmaceutics, Sinhgad Technical Education Society, Sinhgad Institute of Pharmacy, Narhe, Pune, Maharashtra, India*

17.1 Introduction

Many microbes such as bacteria, actinomycetes, fungi, and yeast extracellularly or intracellularly produce a group of versatile and attractive enzymes with a wide variety of structures and commercial applications. Many microbial enzymes, such as amylases, proteases, pectinases, lipases, xylanases, cellulases, and laccases, are extracellularly produced. Some enzymes, such as catalase from *Saccharomyces cerevisiae* and *Aspergillus niger*, are intracellular (Fiedurek and Gromada, 2000; Venkateshwaran et al., 1999). As biocatalytic molecules, microbial enzymes are ecologically effective and highly specific, resulting in stereo- and region-chemically defined reaction products with a rate acceleration of 10^5-10^8 (Gurung et al., 2013; Koeller and Wong, 2001). Among the industrial enzymes, 50% are made by fungi and yeast, 35% are from bacteria, and 15% are from plants (Saranraj and Naidu, 2014). When compared to animal and plant enzymes, microbial enzymes have several advantages: first, microbial enzymes are more active and stable than plant and animal enzymes. By developing fermentation processes, mainly selected strains can produce purified, well-characterized enzymes on a large scale. Second, enzymes produced by microorganisms have high yields and are easy for product modification and optimization owing to the biochemical diversity and susceptibility to gene manipulation. Engineering techniques have been applied to microorganisms to improve the production of enzymes and alter the properties of enzymes by protein engineering (Gurung et al., 2013). Third, microbes represent a rich source for discovering microbial enzymes through modern techniques such as metagenome screening, genome mining, and exploring the diversity of extremophiles (Adrio and Demain, 2014; Zhang and Kim, 2010).

Biotechnology of Microbial Enzymes. DOI: https://doi.org/10.1016/B978-0-443-19059-9.00021-9
© 2023 Elsevier Inc. All rights reserved.

Currently, approximately 200 types of microbial enzymes from 4000 known enzymes are used commercially. However, only about 20 enzymes are produced on a truly industrial scale. About 12 major producers and 400 minor suppliers meet the worldwide enzyme demand. Three top enzyme companies make nearly 75% of the total enzymes, that is, Denmark-based Novozymes, US-based DuPont (through the May 2011 acquisition of Denmark-based Danisco), and Switzerland-based Roche. The highly competitive market has small profit margins and is technologically intensive (Li et al., 2012). An increasing number of industrial enzymes can be supplied with an improved understanding of microbial recombination, metagenome mining, fermentation processes, and recovery methods. For example, recombinant DNA technology can be applied to microorganisms to produce enzymes commercially that could not be produced previously. Approximately 90% of industrial enzymes are recombinant versions (Adrio and Demain, 2014).

The industrial applications for microbial enzymes have grown immensely in recent years. For example, the estimated value of worldwide sales of industrial enzymes for 2012, 2013, and 2015 is $1 million, $3 billion, and $3.74 billion, respectively (Deng et al., 2010; Godfrey and West, 1996). Protease sales represent more than 60% of all industrial enzyme sales globally (Rao et al., 1998) and still constitute the most extensive product segment in the 2015 global industrial enzyme market. Amylases comprise about 30% of the world's enzyme production. Lipases represent the other significant product segment in such a market. Geographically, demand for industrial enzymes in matured economies such as the United States, Western Europe, Japan, and Canada was relatively stable, while developing economies of Asia-Pacific, Eastern Europe, Africa, and Middle East regions emerged fastest growing markets for industrial enzymes. Based on the application, commercial applications of enzymes can be divided into nine broad categories, including food and feed, detergents, etc. (Sharma et al., 2010). About 150 industrial processes use enzymes or whole microbial cell catalysts (Adrio and Demain, 2014). Food and feed represent the largest segment for industrial enzymes, followed by detergents, which constitute another significant segment for industrial enzymes (Arora et al., 2020). This chapter covers biotechnologically and industrially valuable microbial enzymes' classification, resource, production, and applications.

17.2 Classification and chemical nature of microbial enzymes

Based on the catalyzed reaction, microbial enzymes can be classified into six types: oxidoreductases (EC 1, catalyze oxidation/reduction reactions), transferases (EC 2, transfer a functional group), hydrolases (EC 3, catalyze the hydrolysis of various bonds), lyases (EC 4, cleave multiple bonds by means other than hydrolysis and oxidation), isomerases (EC 5, catalyze isomerization changes within a single molecule), and ligases (EC 6, join two molecules with covalent bonds) (http://www.chem.qmul.ac.uk/iubmb/enzyme/). Currently, there are 510 commercially valuable microbial enzymes in the metagenomics database (Sharma et al., 2010). Of the industrial enzymes, 75% are hydrolytic (Li et al., 2012).

17.2.1 Amylases

Amylases catalyze the hydrolysis of starch into sugars such as glucose and maltose (Sundarram and Murthy, 2014). Amylases are divided into three subclasses (α, β, and γ) according to the type of bond/link they can cleave (Fig. 17.1). α-Amylases (EC 3.2.1.1) catalyze the hydrolysis of internal α-1,4-*O*-glycosidic bonds in polysaccharides with the retention of α-anomeric configuration in the products. Most of the α-amylases are metalloenzymes, which require calcium ions (Ca^{2+}) for their activity, structural integrity, and stability. They belong to family 13 (GH-13) of the glycoside hydrolase (GH) group of enzymes. β-Amylases (EC 3.2.1.2) are *exo*-hydrolase enzymes that act from the nonreducing end of a polysaccharide chain by hydrolyzing α-1,4-glucan linkages to yield successive maltose units. Since it cannot cleave branched linkages present in branched polysaccharides such as glycogen or amylopectin, the hydrolysis is incomplete, and dextrin units remain. γ-Amylases (EC 3.2.1.3) cleave α(1−6)glycosidic linkages, in addition to cleaving the last α(1−4)glycosidic linkages at the nonreducing end of amylose and amylopectin, unlike the other forms of amylase, yielding glucose. α-Amylase is produced by several bacteria, fungi, and genetically modified species of microbes. The most widely used source among the bacterial species is *Bacillus* spp., *B. amyloliquefaciens* and *B. licheniformis*. Fungal sources of α-amylase are mostly *Aspergillus* species and only a few *Penicillium* species, *P. brunneum* being one of them.

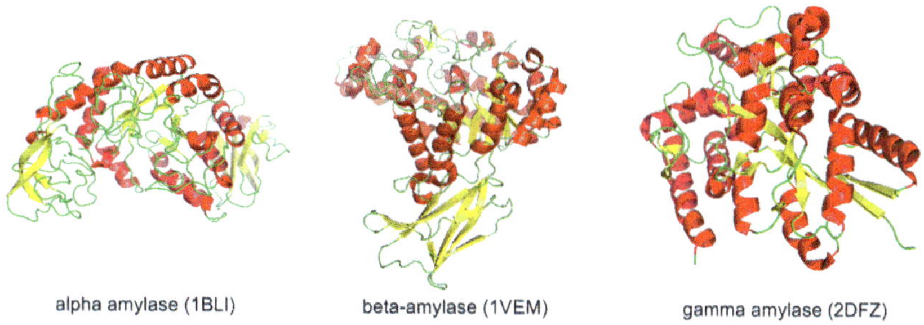

FIGURE 17.1 Structures of selected microbial amylases.

17.2.2 Catalases

Catalases (EC 1.11.1.6) are antioxidant enzymes that catalyze hydrogen peroxide to water and molecular oxygen. According to the structure and sequence, catalases can be divided into three classes (Fig. 17.2): monofunctional catalase or typical catalase, catalase-peroxidase, and pseudo-catalase or Mn-catalase (Zhang et al., 2010). Currently, at least eight strains can produce catalases (Zhang et al., 2010): *Penicillium variabile*, *A. niger*, *S. cerevisiae*, *Staphylococcus*, *Micrococcus lysodeikticus*, *Thermoascus aurantiacus*, *Bacillus subtilis*, and *Rhizobium radiobacter*. Catalases are used in several industrial applications such as food or textile processing to remove hydrogen peroxide that is used for sterilization or bleaching.

17.2.3 Cellulases

Cellulases are enzymes that hydrolyze β-1,4 linkages in cellulose chains. The catalytic modules of cellulases have been classified into numerous families based on their amino acid sequences and crystal structures (Henrissat, 1991). Cellulases contain noncatalytic carbohydrate-binding modules and/or other functionally known or unknown modules located at the N- or C-terminus of a catalytic module. In nature, complete cellulose hydrolysis is mediated by a combination of three main types of cellulases (Juturu and Wu, 2014; Kuhad et al., 2011; Sukumaran et al., 2005; Yang et al., 2013) (Fig. 17.3). These are

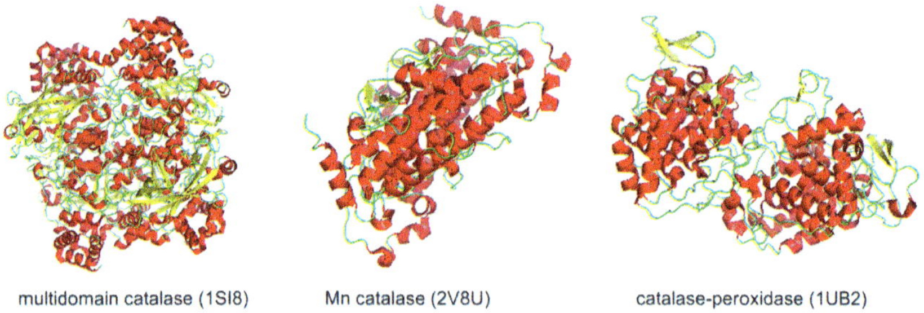

FIGURE 17.2 Structures of selected microbial catalases.

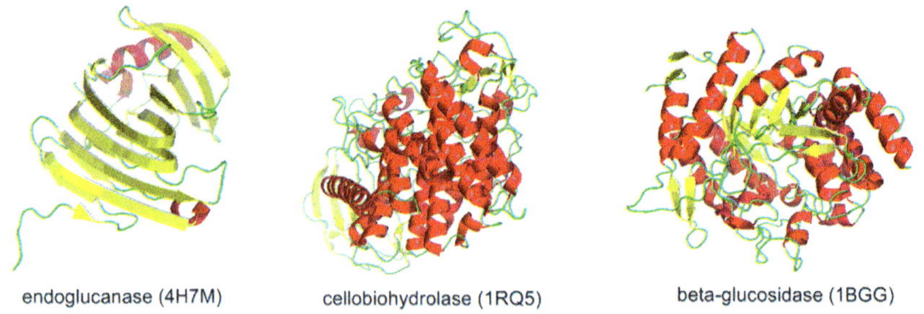

FIGURE 17.3 Structures of selected microbial cellulases.

endoglucanase (EC 3.2.1.4) exoglucanase (EC 3.2.1.91), and glucosidases (EC 3.2.1.21). Endoglucanases hydrolyze glycosidic bonds at the amorphous regions of the cellulose generating long-chain oligomers (nonreducing ends) for the action of exoglucanases or cellobiohydrolases, which cleave the long-chain oligosaccharides generated by the activity of endoglucanases into short-chain oligosaccharides. There are two types of exoglucanases, acting unidirectionally on the long-chain oligomers either from the reducing (EC 3.2.1.176) or nonreducing ends (EC 3.2.1.91) liberating cellobiose, which is further hydrolyzed to glucose by β-glucosidases (EC 3.2.1.21)(Juturu and Wu, 2014). Cellulases are inducible enzymes synthesized by many microorganisms, including fungi and bacteria, during their growth on cellulosic materials (Ma et al., 2013; Quintanilla et al., 2015). These microorganisms can be aerobic, anaerobic, mesophilic, or thermophilic. The genera of *Clostridium, Cellulomonas, Thermomonospora, Trichoderma*, and *Aspergillus* are the most extensively studied cellulose producer (Kuhad et al., 2011).

17.2.4 Lipases

Lipases (triacylglycerol acyl hydrolases, EC 3.1.1.3) catalyze the hydrolysis of triacylglycerols to glycerol, diacylglycerols, mono glycerol, and free fatty acids (Treichel et al., 2010). Bacterial lipases are classified into eight families (families I–VIII) based on differences in their amino acid sequences and biological properties (Arpigny and Jaeger, 1999). The family I of true lipases is the most represented one and can be further divided into *Pseudomonas* lipase subfamily, *Bacillus* lipase subfamily, *Staphylococcal* lipase subfamily, etc. (Fig. 17.4). Lipases belong to the class of serine hydrolases. Therefore lipases do not need any cofactor. Lipases catalyze the hydrolysis of ester bonds at the interface between an insoluble substrate phase and the aqueous phase where the enzymes remain dissolved. Lipases do not hydrolyze dissolved substrates in the bulk fluid. In nature, lipases have considerable variations in their reaction specificities. From the fatty acid side, some lipases have an affinity for short-chain fatty acids (C2, C4,

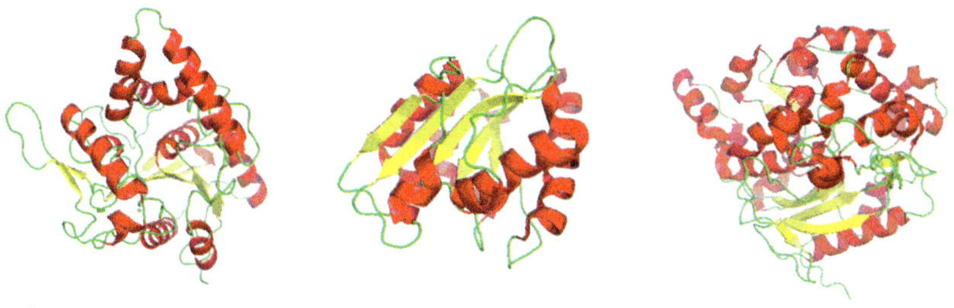

Pseudomonas lipase (2LIP) *Bacillus* lipase (1ISP) *Staphylococcal* lipase (1KU0)

FIGURE 17.4 Structures of selected microbial lipases.

C6, C8, and C10), some have a preference for unsaturated fatty acids (oleic, linoleic, linolenic, etc.) while many others are nonspecific and randomly split the fatty acids from the triglycerides. Some of the most commercially important lipase-producing fungi belong to the genera *Rhizopus* sp., *Aspergillus*, *Penicillium*, *Geotrichum*, *Mucor*, and *Rhizomucor* (Gupta et al., 2004; Treichel et al., 2010). The main terrestrial species of yeasts found to produce lipases are (Treichel et al., 2010) *Candida spp.* such as *Candida rugosa*, *Candida tropicalis*, *Candida antarctica*, *Candida cylindracea*, *Candida parapsilosis*, *Candida deformans*, *Candida curvata*, and *Candida valida*, furthermore, *Yarrowia lipolytica*, *Rhodotorula glutinis*, *Rhodotorula pilimornae*, *Pichia* spp. (*Pichia bispora*, *Pichia mexicana*, *Pichia silvicola*, *Pichia xylose*, and *Pichia burtonii*), *Saccharomycopsis crataegenesis*, *Torulaspora globosa*, and *Trichosporon asteroids*. Among bacterial lipases being exploited, *Bacillus* exhibit exciting properties that make them potential candidates for biotechnological applications (Gupta et al., 2004; Treichel et al., 2010). *Bacillus* spp., *B. subtilis*, *Bacillus pumilus*, *B. licheniformis*, *Bacillus coagulans*, *Bacillus stearothermophilus*, and *Bacillus alcalophilus* are the most common lipase-producing strains. In addition, *Pseudomonas* sp., *Pseudomonas aeruginosa*, *Burkholderia multivorans*, *Burkholderia cepacia*, and *Staphylococcus caseolyticus* are also reported as bacterial lipase producers (Gupta et al., 2004).

17.2.5 Pectinases

Pectinases are a group of enzymes that catalyze pectic substance degradation through depolymerization (hydrolases and lyases) and de-esterification (esterases) reaction (Pedrolli et al., 2009) (Fig. 17.5). According to the cleavage mode and specificity, pectic enzymes are divided into three major types (Fig. 17.5): pectinesterases (PE), depolymerizing enzymes, and cleaving (Tapre and Jain, 2014). These types can be further divided into 13 groups: protopectinases, pectin methyl esterases (PME), pectin acetyl esterases (PAE), polymethyl galacturonases (PMG),

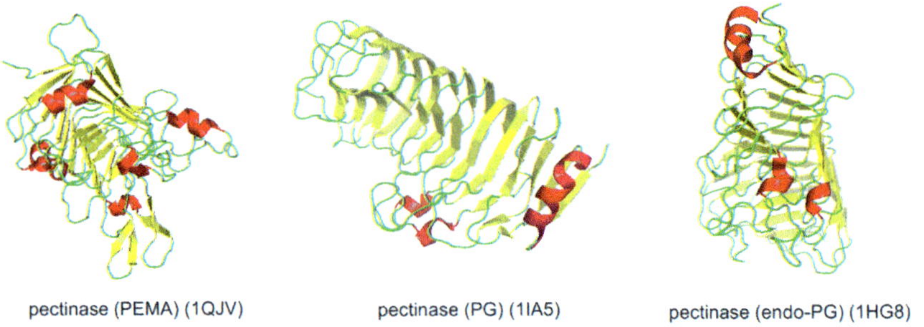

FIGURE 17.5 Structures of selected microbial pectinases.

polygalacturonases (PG), polygalacturonate lyases (PGL), pectin lyases (PL), rhamnogalacturonan rhamnohydrolases, rhamnogalacturonan galacturonohydrolases, rhamnogalacturonan hydrolases, rhamnogalacturonan lyases, rhamnogalacturonan acetyl esterases, and xylogalacturonase (Pedrolli et al., 2009). For example, PME or pectinesterase (EC 3.1.1.11) catalyzes de-esterification of the methoxyl group of pectin, forming pectic acid and methanol. The enzyme acts preferentially on a methyl ester group of galacturonate units next to an on-esterified galacturonate unit (Kashyap et al., 2001). PAE (EC 3.1.1.6) hydrolyzes the acetyl ester of pectin, forming pectic acid and acetate (Shevchik and Hugouvieux-Cotte-Pattat, 1997). PG catalyzes the hydrolysis of α-1,4-glycosidic linkages in polygalacturonic acid-producing D-galacturonate. Both groups of hydrolase enzymes (PMG and PG) can act in an *endo-* or *exo*-mode. *Endo*-PG (EC 3.2.1.15) and *endo*-PMG catalyze random cleavage of the substrate, whereas *exo*-PG (EC 3.2.1.67) and *exo*-PMG catalyze hydrolytic cleavage at substrate nonreducing end producing monogalacturonate ordigalacturonate in some cases (Kashyap et al., 2001). Homogalacturonan-degrading enzymes are well known among them (Pedrolli et al., 2009). It has been reported that microbial pectinases account for 25% of the global food enzyme sales and 10% of global industrial enzymes produced (Ceci and Lozano, 1998; Jayani et al., 2005; Saranraj and Naidu, 2014). Pectinase production has been reported from bacteria, including actinomycetes, yeast, and fungi (Murad and Azzaz, 2011; Saranraj and Naidu, 2014). However, almost all the commercial preparations of pectinases are produced from fungal sources (Singh et al., 1999). *A. niger* is the most commonly used fungal species for the industrial production of pectinolytic enzymes (Gummadi and Panda, 2003).

17.2.6 Proteases

Proteases (EC 3:4, 11–19, 20–24, 99) (peptidase or proteinase) constitute a very large and complex group of enzymes that catalyze the hydrolysis of covalent peptide bonds. Proteases can be classified based on pH, substrate specificity, similarity to well-characterized enzymes, and the active site amino acid (Ellaiah et al., 2002). Based on the pH optima, they are referred to as acidic, neutral, or alkaline proteases (Rao et al., 1998). Based on their site of action on protein substrates, proteases are broadly classified as *endo-* or *exo*-enzymes (Rao et al., 1998). They are further categorized as serine proteases, aspartic proteases, cysteine proteases, or metallo proteases-depending on their catalytic mechanism (Jisha et al., 2013) (Fig. 17.6). Microorganisms account for a two-thirds share of commercial protease production in the enzyme market across the world. Alkaline serine proteases are the most dominant proteases produced by bacteria, fungi, yeast, and actinomycetes. Currently, at least 29 *Bacillus* species and 17

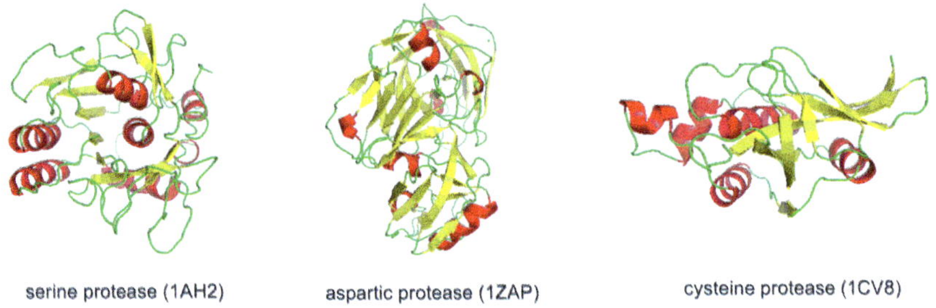

FIGURE 17.6 Structures of selected microbial proteases.

fungal producers have been reported to produce alkaline proteases (Jisha et al., 2013). Commercial producers of alkaline proteases include protein-engineered *B. licheniformis*, alkalophilic *Bacillus* sp., and *Aspergillus* sp. (Ellaiah et al., 2002).

17.2.7 Xylanases

Xylanases are among the xylanolytic enzyme system that includes endoxylanase, β-xylosidase, α-glucuronidase, α-arabinofuranosidase, and acetyl xylan esterase (Juturu and Wu, 2012). Xylanases are a group of GH enzymes that degrade the linear polysaccharide xylan into xylose by catalyzing the hydrolysis of the glycosidic linkage (β-1,4) of xylosides. Xylanases have been classified in at least three ways: based on the molecular weight and isoelectric point (Wong et al., 1988), the crystal structure (Jeffries, 1996) and kinetic properties, or the substrate specificity and product profile (Motta et al., 2013). The good system for the classification of xylanases is based on the primary structure and comparison of the catalytic domains (Collins et al., 2005; Henrissat and Coutinho, 2001). According to the CAZy database (http://www.cazy.org), xylanases (EC 3.2.1.8) are related to GH families 5, 7, 8, 9, 10, 11, 12, 16, 26, 30, 43, 44, 51, and 62. Among those, xylanases in GH 10 and 11 are the two families that have been thoroughly studied (Fig. 17.7). GH family 10 is composed of *endo*-1,4-β-xylanases and *endo*-1,3-β-xylanases (EC 3.2.1.32) (Motta et al., 2013). Members of this family can hydrolyze the aryl β-glycosides of xylobiose and xylotriose at the aglyconic bond. Furthermore, these enzymes are highly active on short xylooligosaccharides, indicating small substrate-binding sites. Family 11 is composed only of xylanases (EC 3.2.1.8), leading to their consideration as "true xylanases," as they are exclusively active on D-xylose-containing substrates. Among all xylanases, endoxylanases are the most important due to their direct involvement in cleaving the glycosidic bonds and liberating short xylooligosaccharides (Collins et al., 2005). However, several *Bacillus* species secrete high levels of extracellular xylanase (Beg et al., 2001) and

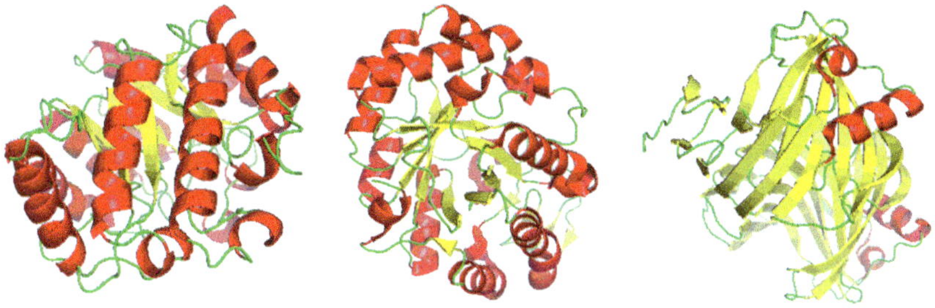

FIGURE 17.7 Structures of selected microbial xylanases.

filamentous fungi secret high amounts of extracellular xylanase, which are often accompanying cellulolytic enzymes—for example, as in species of *Trichoderma*, *Penicillium*, and *Aspergillus* (Kohli et al., 2001; Polizeli et al., 2005; Wong and Saddler, 1992).

17.2.8 Other enzymes

Carbohydrates serve as the multifunctional and primary energy source in all food products. Carbohydrates chemically can vary into a few subtypes, such as monosaccharides, disaccharides, oligosaccharides, and polysaccharides. Fructans are multifunctional fructose-based oligo- and polysaccharides. These serve as precursors/substrates to various metabolic reactions in vivo. They depict excellent physicochemical properties such as chemical inertness, water solubility, and low calorific value. The latter property can be used to our advantage to develop natural alternatives to synthetic sweeteners for diabetic patients. Additionally, fructans have significant prebiotic properties, immunomodulatory properties, wound-healing properties, antitumor properties, cryoprotective properties, and drug delivery applications. Fructans derived from microbial sources are obtained via extracellular enzymes called fructosyltransferases, namely, levansucrase (EC 2.4.1.10) and inulosucrase (EC 2.4.1.9). Biosynthetic pathways depict a series of reactions for levansucrase and inulosucrase, yielding levanbiose and inulobiose, respectively. Moreover, fructosyltransferases also function in the synthesis of fructan oligosaccharide intermediates. Fructosyltransferase and levansucrase are obtained from numerous microbes, namely, *Bacillus subtilis, B. megaterium, B. methylotrophicus, Gluconacetobacter diazotrophicus, Erwinia amylovora*, and *Leuconostoc mesenteroides*. Inulosucrase production has been reported using *Streptomyces turgidiscabies, Bacillus agaradhaerens, Lactobacillus johnsonii, Streptococcus mutans*, and *Leuconostoc citreum*. Certain fungi species have also been reported used in the production of fructans, such as

Aspergillus spp. *(Aspergillus japonicus, A. ficuum, A. oryzae*, and *A. kawachii), Schwanniomyces occidentalis, S. cerevisiae, Xanthophyllomyces dendrorhous, Aureobasidium pullulans, Scopulariopsis brevicaulis*, and *Penicillium expansum*. Interestingly, *Aspergillus japonicus* has also been reported in the production of fructan oligosaccharides via β-fructofuranosidase. Fructans can be produced enzymatically via several methods such as whole-cell synthesis, production via the isolated enzyme, and enzyme immobilization technology. Microbial fructosyltransferase, that is, levansucrase, inulosucrase, and β-fructofuranosidase, mainly belongs to the GH68 and GH32 enzyme family, which are responsible for the majority of transfructosylation reactions.

Fructosyltransferases act via a mechanism called double displacement, which enables hydrolysis and the formation of a fructosyl-enzyme intermediate. The majority of levansucrase depict/follow Michaelis–Menten kinetics, thereby retaining substrate saturation. On the contrary, inulosucrase does not show/follow Michaelis–Menten kinetics. During the synthesis of fructan polymer, the polymer remains attached to levansucrase, undergoing chain elongation. This type of reaction is called a proportionate reaction.

On the other hand, in a nonproportionate reaction, short fructan chains are released after completing the transfructosylation reaction. Moreover, β-fructofuranosidase (invertase) also obeys Michaelis–Menten kinetics and is responsible for the reversible synthesis of fructan oligosaccharides. However, the yield is relatively low and mainly depends on the equilibrium state between free enzyme and enzyme–fructan complex. Production, efficiency, and efficacy of fructosyltransferases are affected by several factors such as their physicochemical properties, molecular weight, pH, temperature, substrate affinity, presence of cofactor, ionic strength, presence of solvents, inhibitors to the stop reaction and reaction time (Tezgel et al., 2020).

Chitinases have been divided into two main groups: endochitinases (EC 3.2.1.14) and *exo*-chitinases (Fig. 17.8). The endochitinases randomly split

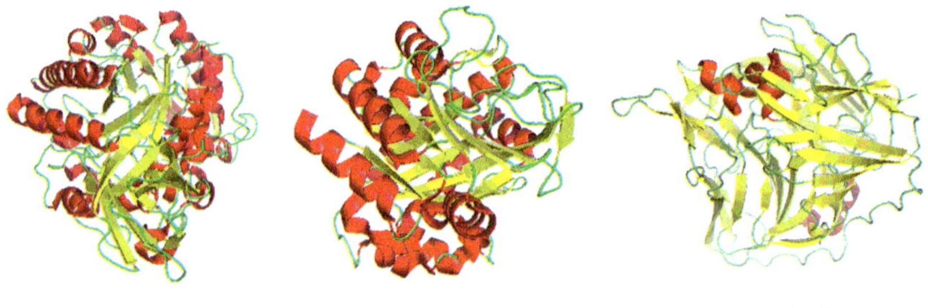

FIGURE 17.8 Structures of selected microbial chitinases and laccase.

chitin at internal sites, forming the dimer di-acetylchitobiose and soluble low-molecular mass multimers of GlcNAc such as chitotriose and chitotetraose. The exochitinases have been further divided into two subcategories: chitobiosidases (EC 3.2.1.29), which are involved in catalyzing the progressive release of di-acetylchitobiose starting at the nonreducing end of the chitin microfibril, and 1–4-β-glucosaminidase (EC 3.2.1.30), cleaving the oligomeric products of endochitinases and chitobiosidases, thereby generating monomers of *N*-acetylglucosamine (GlcNAc). Chitinases (EC 3.2.1.14) can catalyze the hydrolysis of chitin to its monomer *N*-acetyl-D-glucosamine. Chitinases are widely distributed in bacteria such as *Serratia, Chromobacterium*, *Klebsiella*, *Pseudomonas*, *Clostridium*, *Vibrio, Arthrobacter*, *Beneckea*, *Aeromonas*, and *Streptomyces*. They are also found in fungi such as *Trichoderma*, *Penicillium*, *Lecanicillium*, *Neurospora*, *Mucor*, *Beauveria*, *Lycoperdon*, *Aspergillus*, *Myrothecium*, *Conidiobolus*, *Metarhizium*, *Stachybotrys*, and *Agaricus* (Felse and Panda, 2000; Islam and Datta, 2015; Matsumoto, 2006).

Laccases (benzenediol: oxygen oxidoreductase, EC 1.10.3.2) are multicopper oxidases that participate in cross-linking of monomers, degradation of polymers, and ring cleavage of aromatic compounds. These enzymes contain 15%–30% carbohydrate and have a molecule mass of 60–90 kDa(Shraddha et al., 2011) (Fig. 17.8). Laccases contain four copper atoms termed Cu T1 (where the reducing substrate binds) and trinuclear copper cluster T2/T3 (electron transfer from type I Cu to the type II Cu and type III Cu trinuclear cluster/reduction of oxygen to water at the trinuclear cluster) (Gianfreda et al., 1999). These four copper ions are classified into three categories: Type 1 (T1), Type 2 (T2), and Type 3 (T3). Laccases carry out one-electron oxidation of phenolic and its related compound and reduce oxygen to water. When a laccase oxidizes the substrate, it loses a single electron and usually forms a free radical which may undergo further oxidation or nonenzymatic reactions, including hydration, disproportionation, and polymerization. Most laccases are extracellularly produced by fungi (Agematu et al., 1993; Brijwani et al., 2010; Chandra and Chowdhary, 2015; Mougin et al., 2003). The production of laccase can also be seen by soil and some freshwater *Ascomycetes* species (Banerjee and Vohra, 1991; Junghanns et al., 2005; Rodríguez et al., 1996; Scherer and Fischer, 1998). In addition, laccases are also produced by *Gaeumannomyces graminis*, *Magnaporthe grisea*, *Ophiostoma novoulmi*, *Marginella*, *Melanocarpus albomyces*, *Monocillium indicum*, *Neurospora crassa*, and *P. anserine* (Binz and Canevascini, 1997; Edens et al., 1999; Froehner and Eriksson, 1974; Iyer and Chattoo, 2003; Kiiskinen et al., 2002; Molitoris and Esser, 1970; Palonen et al., 2003; Thakker et al., 1992).

The production of native laccases is hindered due to low productivity, high cost, and a commercially undesirable mixture of isoenzymes. Laccases have excellent catalytic activity due to the presence of four Cu^{2+} coordinated atoms capable of an oxidizing variety of aromatic pollutants (anthracene, bisphenol-A, and triclosan). Since its inception by Stanley Cohen and Herbert Boyer in 1973, recombinant DNA technology has been proven an efficient and effective tool in various fields of biotechnology and medicine. Laccase produced by rDNA technology overcomes its industrial limitations and improves its productivity. Recombinant laccases have been used as a decolorizing agent to remove azo dyes under alkaline conditions and as a polymerizing agent. They also exuberate multifarious medicinal properties and are used in the treatment of viral infections and cancer. Recombinant laccases are produced simply by combining the desired gene from a source and then expressing it in a targeted vector by using plasmid DNA. Various sources have been reported to obtain laccase gene such as *Trametes trogii, Trametes versicolor, Pleurotus ostreatus, Pycnoporus* spp. *(Pycnoporus cinnabarinus* and *Pycnoporus eryngii), Hypsizygus ulmarius, Volvariella volvacea, Lentinula edodes, Rigidoporus lignosus, Auricularia auricula-judae, Streptomyces* spp. *(Streptomyces ipomoea, Streptomyces coelicolor,* and *Streptomyces cyaneus), Bacillus* spp.*(B. licheniformis, B. subtilis, Bacillus* sp. HR03, and *Bacillus halodurans* C125*), Ochrobactrum* sp., and *Lactobacillus plantarum* J16. The isolated genes are expressed in suitable vectors such as *Escherichia coli* (BL21, DH5α, and XL1), *Kluyveromyces lactis, Pichia* spp. *(Pichia pastoris* and *Pichia methanolica), Aspergillus* spp. (*A. oryzae* and *A. sojae*), and transgenic maize. Recombinant laccase shows several advantages, such as, it provides saturation mutagenesis, which leads to increased tolerance to a variety of solvents. Furthermore, it offers random and site-directed mutagenesis, thereby aiding in varying optimum conditions (pH and temperature), improved kinetic stability, decreasing barrier for electron transfer, and increased enzyme production. Moreover, it provides a shift in stop codon site synergistically affecting enzyme activity, tagging laccase improves catalytic activity, and enhances thermal stability by nitrogen implantation. Significant differences between engineered recombinant laccase and native laccase make computer-aided laccase engineering an efficient and excellent tool for industrial and commercial applications (Preethi et al., 2020).

Cytochromes P450 (CYPs) catalyze various reactions such as hydroxylation, epoxidation, alcohol and aldehyde oxidation, *O*-dealkylation, *N*-dealkylation, oxidative dehalogenation, and oxidative C−C bond cleavage (Sakaki, 2012). Among these, regio- and enantioselective hydroxylation by P450 is quite attractive as a bioconversion process. There are 10 classes of CYPs(Kelly and Kelly, 2013). Most bacterial CYPs are class I and driven by ferredoxin and ferredoxin reductases. CYPs have potential applications in bioconversion processes, biosensors, and bioremediation due to their regio- and enantioselective hydroxylation, which is difficult for chemical synthesis. *Streptomyces carbophilus* CYP105A3 is a CYP that is successfully applied in bioconversion to produce pravastatin (Watanabe et al., 1995).

17.3 Production of microbial enzymes

Fermentation is the technique of the biological conversion of complex substrates into simple compounds by various microorganisms. It has been widely used for the production of many microbial enzymes (Aehle, 2007). Much work has been focused on the screening of enzyme-producing microorganisms, physiological optimizations for substrates, carbon source and nitrogen source, pH of the media, and the cultivation temperature during the fermentation process (Ellaiah et al., 2002; Juturu and Wu, 2014; Sundarram and Murthy, 2014).

17.3.1 Fermentation methods

There are two cultivation methods for all microbial enzymes: submerged fermentation (SMF) and solid-state fermentation (SSF). SMF involves the nurturing of microorganisms in high oxygen concentrated liquid nutrient medium. The viscosity of broth is the major problem associated with the fungal SMFs. When fungal cells grow and a mycelium is produced, this hinders impeller action due to oxygen and mass transfer limitations. SSF is suitable for producing enzymes using natural substrates such as agricultural residues because they mimic the conditions the fungi grow naturally.

Advancement in biotechnology over the last decade has led to novel production methods of numerous enzymes via SSF. These advancements led to the reutilization of several process by-products and agricultural wastes as precursors/substrates for microbial growth. This reutilization had a considerable impact on waste management, especially by reducing vast amounts of wastewater disposal. Several agriculture residues, such as straw from corn, soy, cotton, rice, and wheat; baggase from sugarcane and orange, with several other bakery-derived wastes, serve as a potential food source for rich microbes in carbohydrates. These carbohydrates can then be digested by several enzymes such as glucanase, polygalacturonase, xylanase, and amylase. Some agricultural wastes contain lignin, which is a complex, irregular, and insoluble polymer. Laccases, oxidases, and peroxidases observed effective degradation of lignin with several other intracellular metabolites and reactive oxygen species. Bacteria need an environment and, more specifically, water for growth and proliferation, making it unlikely to be used in SSF. On the other hand, yeast and fungi are the most suitable microbes for SSF.

Nowadays, there's an increase in demand for high-quality food products with enhanced flavor, nutrition, taste, and natural-based products. This surge led to the development of enzymes such as glucanase, amylase, and xylanase in food processing, especially in juice products. Amylase also finds its applications in the textile industry and others, such as catalase, peroxidase, pectinase, lipase, laccase, and glucanase, all obtained via fungi-based SSF. These enzymes have enhanced the production of threads, fabrics, and yarn from various natural and synthetic fibers. Xylanase produced via SSF has applications in the paper and pulp industry. Xylanase utilization has led to a reduction in chemical pollutants from pulp bleaching, prebleaching stage, increased density of paper, and reduction in contaminants and waste generated in the recycling process. Phytase is a nutritional enzyme that improves the digestion of celluloses in animal feed, thereby improving its nutritional value. Another diverse yet important area is sustainable renewable resource development, which utilizes lipases, xylanases, amylases, and glucanases to produce biofuels such as biodiesel and ethanol (Londoño-Hernandez et al., 2020).

Since SSF involves relatively little liquid when compared with SMF, downstream processing from SSF is theoretically simpler and less expensive. During the past 10 years, a renewed interest in SSF has developed due, in part, to the recognition that many microorganisms, including genetically modified organisms, may produce their products more effectively by SSF (Singh et al., 2008). SSF has three significant advantages: (1) high volumetric productivity, (2) relatively higher concentration of the products, (3) less effluent generation, the requirement for simple fermentation equipment, etc. Moreover, the biosynthesis of microbial enzymes in the SMF process is of economic importance because it is strongly affected by catabolic and end-product repressions. The amenability of the SSF technique to use up to 20%–30% substrate, in contrast, to the maximum of 5% in the SMF process, has been documented (Pamment et al., 1978).

17.3.2 Purification methods

Enzymes are manufactured in bioreactors for commercial use. These enzymes are in crude form and have to be purified for further use. The purification processes follow the extraction methods. There are mainly three primary purification methods for microbial enzymes: (1) based on ionic properties of enzymes, (2) based on the ability to get adsorbed, and (3) based on the difference in the size of molecules. Special procedures employed for enzyme purification are crystallization, electrophoresis, and chromatography. The main applications of industrial-scale chromatography were the desalination of enzyme solutions using highly cross-linked gels such as Sephadex G-25 and batch separations using ion exchangers such as DEAE-Sephadex A-50. The stability and hydraulic properties of chromatographic media have been improved so that these techniques are now used on a production scale. Important parameters for the scale-up of chromatographic systems are the height of the column, the linear flow rate, and the ratio of sample volume to bed volume. Zone spreading interferes with the performance of the column. Factors contributing to zone spreading are longitudinal diffusion in the column, incomplete equilibration, and inadequate column packing. Longitudinal diffusion can be minimized by using a high flow rate. On the other hand, equilibration between the stationary and the mobile phases is optimal at low flow rates. A compromise must be made because a good process economy depends largely on the flow rate. In addition, the flow rate is also dependent on particle size; the decisive factor is usually the pressure drop along with large columns. Although the optimal resolution is obtained only with the smallest particles, the gel must have a particle size that favors a good throughput and reduces processing times. The use of segmented columns prevents a significant pressure drop in the column. Above all, the column must be uniformly packed so that the particle-size distribution, porosity, and resistance to flow are the same throughout the column. If this is not done, viscous protein solutions can give an uneven elution profile, which would lead to zone bleeding. The design of the column head is important for the uniform distribution of the applied sample. This is generally achieved by the symmetrical arrangement of several inlets and perforated inserts for good liquid distribution. The outlet of the column must have the minimum volume to prevent back-mixing of the separated components (Aehle, 2007).

17.4 Applications of microbial enzymes

Microbial enzymes are of great importance in the development of industrial bioprocesses. The end-use market for industrial enzymes is prevalent with numerous industrial applications (Adrio and Demain, 2014). Over 500 industrial products are made using different microbial enzymes (Kumar and Singh, 2013). The demand for industrial enzymes is continuously growing, driven by a growing need for sustainable solutions.

Microbes are one of the most significant and valuable sources of many enzymes (Demain and Adrio, 2008). A large number of new enzymes have been designed with the input of protein engineering and metagenomics. Various molecular techniques have also been applied to improve the quality and performance of microbial enzymes for their broader applications in many industries (Chirumamilla et al., 2001; Nigam, 2013). Many microorganisms, including bacteria, actinomycetes, and fungi, have been globally studied to synthesize economically viable preparations of various enzymes for commercial applications (Pandey et al., 1999). The unique characteristics of enzymes are exploited for their commercial interest and industrial applications (Table 17.1), including thermotolerance, tolerance to a wide range of pH, and stability of enzyme activity over a harsh reaction conditions.

The majority of currently used industrial enzymes are hydrolytic, and they are used to degrade various natural substances. Proteases are one of the most important enzymes for the detergent and dairy industries. Carbohydrases, primarily amylases, and cellulases, used in starch, textile, detergent, and baking industries, represent the second largest group (Underkofler et al., 1958). The fastest growth over the past decade has been seen in the baking and animal feed industries, but growth is also being

TABLE 17.1 Microbial enzymes and their applications.

Industry	Enzyme	Applications	References
Food, dairy and beverage	Protease, lipase, lactase, pectinase, amylase, laccase, amyloglucosidase, phospholipase, elagitanase	Degradation of starch and proteins into sugars, production of low caloric beer, fruit juice processing, cheese production, glucose production from lactose, dough stability and conditioning.	Gurung et al. (2013); Nigam and Singh (1995); Londoño-Hernandez et al. (2020)
Biopesticide biosensor	Chitinase, peroxidase	Obtained from *Penicillium ochrochloron MTCC 517* proved an effective and efficient alternative against chemical pesticides. Obtained from *Phanerochaete chrysosporium* employed in the development of stimuli-responsive drug delivery and several biosensor applications.	Londoño-Hernandez et al. (2020)
Detergents	Amylase, cellulase, lipase, protease, mannanase	Remove protein after staining, cleaning agents, removing insoluble starch, fats and oils, to increase the effectiveness of detergents.	Pandey et al. (2000a); Wintrode et al. (2000)
Textiles	Amylase, cellulase, pectinase, catalase, protease, peroxidase, keratinase, polygalacturonase	Fabric finishing in denims, wool treatment, degumming of raw silk (biopolishing), cotton softening,	Liu et al. (2013); Saha et al. (2009); Londoño-Hernandez et al. (2020)
Animal feed	Phytase, Xylanase	Increase total phosphorus content for growth, Digestibility	Mitidieri et al. (2006); Tomschy et al. (2000)
Ethanol production	Cellulase, ligninase, mannanase	Formation of ethanol	Jolly (2001)

Biofuel and biocatalyst	Lipase	Used in catalysis and biotransformation reactions to produce eco-friendly and energy-efficient high-energy products	Kumar et al. (2020)
Paper and pulp	Amylase, lipase, protease, cellulase, hemicellulase, esterase, ligninase, xylanase, laccase, and mannanase	Degrade starch to lower viscosity, aiding sizing, deinking, and coating paper. Cellulase and hemicellulase smooth fibers enhance water drainage and promote ink removal. Lipases reduce pitch, and ligninase removes lignin to soften the paper.	Kirk et al. (2002); Kohli et al. (2001); Polizeli et al. (2005); Angural, et al. (2020)
Plastic Industry (PET)-Polyethylene terephthalate	PETase, MHETase, cutinase, esterase, hydrolase, arylesterase, carboxylesterase	Studies have reported excellent catalytic activity (over 90% of PET degradation), enhanced degradation rate, enhanced enzyme affinity, and enhanced melting temperatures.	Samak, et al. (2020)
Leather	Protease, lipase	Unhearing, bating, depicking	Parameswaran et al. (2013); Saha et al. (2009)
Pharmaceuticals	Penicillin acylase, peroxidase, aminoacylase Asparaginase, argininedeiminase, methionase, lysine oxidase, glutaminase, phenylalanine ammonia lyase	Synthesis of semisynthetic antibiotics, Antimicrobials Offers edge over conventional chemotherapy. Used as amino acid deprivation therapy against cancers such as auxotrophic tumors, leukemia, hepatocellular carcinoma, Walker-256 carcinoma, colorectal cancer, etc.	(Gurung et al., 2013); (Roberts et al., 2010) Londoño-Hernandez et al. (2020); Dhankhar et al. (2020)
Molecular biology	DNA ligase, restriction enzymes, polymerase	Manipulate DNA in genetic engineering. DNA restriction and the polymerase chain reaction, Important in forensic science.	Nigam (2013); Roberts et al. (2010)

generated from applications established in a wealth of other industries spanning from organic synthesis to paper and pulp and personal care. The use of enzymes in animal nutrition is essential and growing, especially for pig and poultry nutrition.

Enzymes play critical roles in numerous biotechnology products and processes that are commonly encountered in the production of food and beverages, detergents, clothing, paper products, transportation fuels, pharmaceuticals, and monitoring devices (Gurung et al., 2013). As the industrial enzyme market has expanded at about 10% annually, microbial enzymes have largely replaced the traditional plant and animal enzymes. DNA technology has been used to modify substrate specificity and improve the stability properties of enzymes for increasing yields of enzyme-catalyzed reactions. Enzymes can display regional stereospecificity, properties that have been exploited for asymmetric synthesis and racemic resolution. Chiral selectivities of enzymes have been employed to prepare enantiomerically pure pharmaceuticals, agrochemicals, chemical feedstock, and food additives.

17.4.1 Plastic/polymer biodegradation

PET is aromatic polyester synthesized from ethylene glycol, and terephthalic acid is highly resistant to degradation. Hence, it becomes a prime concern to innovate and implement more green methods for its degradation. Despite its high resistance, recent advances in polymer biodegradation have shown promising areas for PET and other plastics. Owing to the complexity of the polymer, its biodegradation is affected by several factors, such as, size, polarity/nature, charge, shape, crystallinity, porosity, and surface characteristics. For example, polymer crystallites can retard moisture penetration, slowing its degradation. Depending upon these factors, one can design a strategy to break down the polymer into its constituent oligomers and monomers cleaved via either enzymatic catalysts or hydrolysis or a combination of both (Khot, et al., 2021).

In the case of PET biodegradation, a strategy involving enzymes derived from microbes or microbes as a whole can be employed to ensure its complete biodegradation. The process usually shortens the length of the polymer and breaks it into its constituent oligomers, dimers, and monomers, which serve as carbon/energy sources for the microbes. Recent studies reported using some saprophytic scavengers such as certain species of fungi and bacteria are helpful in PET biodegradation resulting in monomers, namely, mono and bis(2-hydroxyethyl)terephthalate. The use of whole organisms instead of specific enzymes may prove effective and efficient by reducing the enzyme purification process and overall degradation time. Examples of such whole organisms are, *Comamonas testosteroni, Ideonella sakaiensis, Thermomyces lanuginosus, C. antarctica, Triticum aestivum,* and

Burkholderia spp. Studies have reported two particular enzymes: PETase and MHETase [mono (2-hydroxyethyl) terephthalate] for PET biodegradation.

Moreover, PETase showed significantly higher hydrolyzing activity toward PET at low-temperature conditions than cutinase (*Fusarium solani*) and hydrolase (*Thermobifida fusca*). Esterase has been found to cleave ester bonds in PET monomers and is responsible for PET polyester surface alterations. Lipases also show similar characteristics with significant hydrolyzing activity toward PET. Leaf branch cutinases are lipolytic enzymes with the ability to ester bonds. Cutinases show slightly different catalytic activity, that is, affinity toward soluble and emulsified substrates of PET. Apart from the abovementioned microorganisms, there are quite a few alternatives for PET biocatalysts, such as *E. coli, P. pastoris*, microalgae *(Phaeodactylum tricornutum and Chlamydomonas. reinhardtii), Clostridium thermocellum, Rhodococcus* sp., *Acetobacterium woodii, Pseudomonas* spp. *(P. aeruginosa* and *P. putida)* (Samak et al., 2020).

17.4.2 Food and beverage

In the 20th century, enzymes began to be isolated from living cells, which led to their large-scale commercial production and broader application in the food industry. Food and beverage enzymes constitute the largest segment of industrial enzymes, with revenues of nearly $1.2 billion in 2011, which is expected to grow to $2.0 billion by 2020. Enzymes used in food can be divided into food additives and processing aids. Most food enzymes are considered processing aids used during the manufacturing process of foodstuffs (Saha et al., 2009), with only a few used as additives, such as lysozyme and invertase. The applications of different enzymes in the food industry are shown in Table 17.2.

Amylases are the essential enzymes in the industrial starch conversion process. Amylolytic enzymes act on starch and related oligo- and polysaccharides (Pandey et al., 2000a). The application of these enzymes has been established in starch liquefaction, paper, food, sugar, and pharmaceutical industries. In the food industry, amylolytic enzymes have a large scale of applications, such as the production of glucose syrups and maltose syrup, reduction of viscosity of sugar syrups, to produce clarified fruit juice for longer shelf life, and solubilization of starch in the brewing industry (Pandey et al., 2000b). The baking industry uses amylases to delay the staling of bread and other baked products.

The major application of proteases in the dairy industry is for manufacturing cheese. Calf rennin (chymosin) had been preferred in cheese making due to its high specificity, but microbial proteases are also used. Chymosin is an aspartic acid protease that causes the coagulation of milk. The primary function of these enzymes in cheese making is to hydrolyze the specific peptide bond that generates para-k-casein and macropeptides (Rao et al., 1998).

TABLE 17.2 Applications of enzymes in the food industry.

Process	Enzyme	Applications	References
Baking	Amylase, protease	Conversion of sugar into ethanol and CO_2, To prepare bread	Collar et al. (2000)
Brewing	Amylase, protease	Conversion of sugar into ethanol and CO_2, To prepare the alcoholic drink	Pandey et al. (2000a)
Corn syrup	Amylase, glucoamylase	Preparation of low dextrose equivalent syrups Production of starch	Kirk et al. (2002); Londoño-Hernandez et al. (2020)
Cheese making	Rennin, lipases	Milk clotting, favor production	Okanishi et al. (1996)
Baby foods	Trypsin	Digestion	Parameswaran et al. (2013)
Coffee	Pectinase	Coffee bean fermentation	Kirk et al. (2002)
Dairy industry	Protease, lactase lactoperoxidase, β-galactosidase, β-glucosidase, protease	Preparation of protein hydrolysates; preparation of milk and ice cream; cold sterilization of milk improved production of milk and milk products by implementing paradigm-shifting techniques	Tucker and Woods (1995); Akdeniz and Akalın (2020)
Fruit juices	Glucose oxidase, pectinase	Oxygen removal, clarification of fruit juices	Godfrey and West (1996)
Soft drinks	Glucose oxidase	Stabilization	Kirk et al. (2002)
Meat and fish industries	Proteinase	To tenderize the meat and solubilize fish products	Saha and Demirjian (2000)

The calf rennin (chymosin) in recombinant *A. niger* var *awamori* amounted to about 1 g/L after nitrosoguanidine mutagenesis and selection for 2-deoxyglucose resistance. FDA has approved four recombinant proteases for cheese production (Bodie et al., 1994; Pariza and Johnson, 2001).

The application of enzymes (proteases, lipases, esterases, lactase, and catalase) in dairy technology is well established. Rennets (rennin) are used for the coagulation of milk in the first stage of cheese production. Proteases of various kinds are used to accelerate cheese ripening and modify functional properties and milk proteins to reduce the allergenic properties of cow milk products for infants. Lipases are primarily used in cheese ripening for the development of lipolytic flavors. Lactase is used to hydrolyze lactose to glucose and galactose as a digestive aid and improve the solubility and sweetness of various dairy products. Lactose hydrolysis helps these lactose-intolerant people to drink milk and eat multiple dairy products (Tucker and Woods, 1995). Lactases have also been used to process dairy wastes and as a digestive aid taken by humans in tablet form when consuming dairy products. Recently three novel thermophilic xylanases (XynA, B, C) have been characterized (Du et al., 2013). These were produced by *Humicola* sp. for their potential applications in the brewing industry. This XynA also possessed higher catalytic efficiency and specificity for a range of substrates.

Proteinases, either indigenous (cathepsin) or those obtained from plants and microorganisms, are used in the meat and fish industries to tenderize the meat and solubilize fish products. Tenderization of meat can be achieved by keeping the rapidly chilled meat at 1°C–2°C to allow proteolysis by indigenous enzymes. Enzymes are also used to separate hemoglobin from blood proteins and remove meat from bones. Proteases hydrolyze minced meat or meat by-products to produce a liquid meat digest or slurry with a much lower viscosity (Saha and Demirjian, 2000). Fish protein concentrates are generally prepared by treating groundfish parts with a protease.

In the baking industry, there is an increasing focus on lipolytic enzymes. Recent findings suggest that phospholipases can be used to substitute or supplement traditional emulsifiers, as the enzymes degrade polar wheat lipids to produce emulsifying lipids. Also, research is currently devoted to further understanding bread staling and the mechanisms behind the enzymatic prevention of staling in the presence of α-amylases and xylanases (Andreu et al., 1999). Lipases are commonly used to produce various products, ranging from fruit juices, baked foods, and vegetable fermentations. Fats, oils, and related compounds are the main targets of lipases in food technology. Accurate control of lipase concentration, pH, temperature, and emulsion content is required to maximize the production of flavor and fragrance. The lipase mediation of carbohydrate esters of fatty acids offers a potential market for emulsifiers in foods and pharmaceuticals. There are three recombinant fungal lipases currently used in the food industry: *Rhizomucor miehei*, *T. lanuginosus*, and *F. oxysporum* (Mendez and Salas, 2001).

Enzymes can play essential roles in preparing and processing various fruit and vegetable juices such as apple, orange, grapefruit, pineapple, carrot, lemon, etc. Fruits and vegetables are particularly rich in pectic substances. Pectin, a hydrocolloid, has a great affinity for water and can form gels under certain conditions. The addition of pectinases, pectin lyase, pectin esterase, and polygalacturonase, reduces viscosity and improves possibility as the pectin gel collapses (Tucker and Woods, 1995). Hemicellulases and amylases can be used with pectinases for the complete liquefaction of fruits and vegetable juices. Flavoprotein glucose oxidase is used to scavenge oxygen in fruit juice and beverages to prevent color and taste changes upon storage. Glucose oxidase is produced by various fungi such as *A. niger* and *P. purpurogenum* (Godfrey and West, 1996).

A therapeutically important class of mono- and sesquiterpenes is found in many plant-derived essential oils with various applications. Studies have shown that certain microbial strains such as *E. coli* and *Y. lipolytica* can also produce compounds, for example, limonene. Monoterpenes are also helpful as flavoring agents in the food industry. One such example is α-terpineol which is produced with the help of *Sphingobium* sp. as a biocatalyst. Aromatic sesquiterpenes are valued for their aroma and fragrance across the cosmetic and food industry. An example of such is valencene produced by valencene synthase using *S. cerevisiae*. Nootkatone is another example of valuable sesquiterpene. Both valencene and nootkatone are biosynthesized using genetically engineered *Y. lipolytica strain* coexpressing (d)-valencene synthase, (d)-valencene oxidase, and NADPH-cytochrome P450 reductase. This *Y. lipolytica strain*, along with *E. coli* has also been reported to produce some important sesquiterpenes farnesene and santalol. De novo synthetic approach has also been employed in the synthesis of isomers of santalene and santalol from engineered *S. cerevisiae*, along with *Santalum album* and *Clausena lansium*. Certain terpenic esters of citronellol and geraniol were reported for commercial production using lipases from *C. antarctica* and *T. lanuginosus*.

Another class of aromatic compounds, namely, lactones exhibit a peculiar fruity essence. These lactones have been biosynthesized using *Y. lipolytica*. Aromatic compounds used as flavoring agents are vanillin, 2-phenyl ethanol, and raspberry ketone. Isoeugenol monooxygenase produced by *E. coli* BL21 *strain* was utilized in the biosynthesis of vanillin from isoeugenol. Additionally, there are various enzyme substrates for producing vanillins such as feruloyl-CoA synthetase, enoyl-CoA hydratase, ferulic acid decarboxylase, carotenoid cleavage oxygenase, lipoxygenase, and eugenol oxidase. Another interesting aromatic compound found in coffee and chocolate is 2,3,5,6-tetramethylpyrazine, reported for its several medicinal properties such as antiinflammatory, antidiabetic, etc. Enzymes produced by *B. licheniformis strains* such as α-acetolactate decarboxylase and α-acetolactate synthase have been reported in the biosynthesis of 2,3,5,6-tetramethylpyrazine (Paulino et al., 2020).

17.4.3 Detergents

The detergent industry occupies about 30% of the entire industrial enzyme market. The application of enzymes in detergents enhances the detergents' ability to remove tough stains and makes detergent eco-friendly. Constantly, new and improved engineered versions of the "traditional" detergent enzymes, proteases and amylases, are developed. These new second- and third-generation enzymes are optimized to meet the requirements for performance in detergents. Over half of the laundry detergents contain protease, amylase, lipase, and cellulase enzymes. These enzymes must be very efficient in laundry detergent environments, work at alkaline pH conditions and high temperatures, be stable in the presence of chelating agents and surfactants, and possess long storage stability. Proteases are the most widely used enzymes in the detergent industry. DNA technology has been used extensively to modify the protein catalysts, primarily for increasing stability properties (Bisgaard-Frantzen et al., 1999; Wintrode et al., 2000). These detergent enzymes (serine proteases) are produced by fermentation of *B. licheniformis*, *B. amyloliquefaciens*, or *Bacillus* sp.

Novo Industries A/S produces and supplies three proteases, Alcalase, from *B. licheniformis*, Esperase, from an alkalophilic strain of a *B. licheniformis*, and Savinase, from an alkalophilic strain of *B. amyloliquefaciens*. Gist-Brocades produce and supply Maxatase from *B. licheniformis*. Alcalase and Maxatase are recommended for use at 10°C–65°C and pH 7–10.5. Savinase and Esperase may be used at up to pH 11 and 12, respectively.

Amylases are the second type of enzyme used in the detergent formulation, and 90% of all liquid detergents contain these enzymes (Mitidieri et al., 2006). These enzymes are used for laundry and automatic dishwashing to clean up residues of starchy foods such as mashed potato, custard, oatmeal, and other small oligosaccharides. The α-amylase supplied for detergent use is Termamyl, the enzyme from *B. licheniformis*, which is also used to produce glucose syrups. α-Amylase is particularly useful in dishwashing and de-starching detergents.

Lipases facilitate removing fatty stains such as lipsticks, frying fats, butter, salad oil, sauces, and tough stains on collars and cuffs. Recently, an alkali-stable fungal cellulase preparation has been introduced to wash cotton fabrics. Treatment with these cellulase enzymes removes the tiny fibers extending from the fabric, without apparently damaging the major fibers, and restores the fabric by improving color brightness and enhancing softness feel. Cellulases are used in textile manufacturing to partially remove dye (indigo) from denim, producing a stone-washed appearance. Bleach-stable enzymes (amylase, protease) are now available for use in automatic dishwashing detergents. The most commercially important field of application for hydrolytic lipases is their addition to detergents used mainly in household and industrial laundry and in household dishwashers. To improve detergency, modern types of heavy-duty powder detergents and automatic dishwasher detergents usually contain one or more enzymes, such as protease, amylase, cellulase, and lipase (Ito et al., 1998).

17.4.4 Removal of pollutants

The important pollutants that end up in soil and water are hydrocarbons from crude oil and laundry waste. This causes harmful physical and chemical changes in the soil affecting its porosity, texture, and waterlogging. Lipids are another waste that can decrease soil fertility by reducing water holding capacity. Therefore removing such oils and lipids is essential, and it is achieved by using lipase-containing microbes such as *Pseudomonas sp.*, with lipolytic activity for bioremediation of soil contaminants. Other microbes, namely, *Klebsiella sp., Rhodopseudomonas palustris, Alicyclobacillus tengchogenesis, B. licheniformis, B. cereus, Brevibacillus* sp., *A. niger*, and *Nephotettix cincticeps*, with similar activity have also been reported to remove several soil contaminants.

Lipases have found diverse applications via energy-efficient production of high-value-added products. They are indispensable in bioremediation and the removal of pollutants from wastewater. Lipases serve as alternative chassis for synthesis and degradation of plastics and biopolymers, biosensors in detecting environmental pollutants. Moreover, ester bonds containing plastics, insecticides, pesticides, and parabens can be removed via hydrolysis using lipases, significantly tackling global waste management. Furthermore, lipases will serve as energy-saving and bioenergy production (Kumar et al., 2020).

17.4.5 Textiles

Enzymes are being used increasingly in textile processing, mainly in the finishing of fabrics and garments. Some of the more critical applications are desizing and jeans finishing. The use of enzymes in the textile industry allows the development of environment-friendly technologies in fiber processing and strategies to improve the final product quality. The consumption of energy as well as increased awareness of environmental concerns related to the use and disposal of chemicals into landfills, water, or release into the air during chemical processing of textiles are the principal reasons for the application of enzymes in finishing textile materials (O'Neill et al., 2007).

Enzymes have been used increasingly in the textile industry since the late 1980s. Many of the enzymes developed in the last 20 years can replace chemicals used by mills. The first breakthrough was when enzymes were introduced for stonewashing jeans in 1987. Within a few years, most denim finishing laundries had switched from pumice stones to enzymes.

The main enzymes used in the textile industry are hydrolases and oxidoreductases. The group of hydrolases includes amylases, cellulases, proteases, pectinases, lipases, and esterases. Amylases were the only enzymes applied in textile processing until the 1980s. These enzymes are still used to remove starch-based sizes from fabrics after weaving. Nowadays, amylases are commercialized and preferred for desizing due to their high efficiency and specificity, completely removing the size without harming the fabric (Etters and Annis, 1998). Cellulases have been employed to enzymatically remove fibrils and fuzz fibers and have also successfully been introduced to the cotton textile industry. Further applications have been found for these enzymes to produce the aged look of denim and other garments. The potential of proteolytic enzymes was assessed for the removal of wool fiber scales, resulting in improved antifelting behavior. Esterases have been successfully studied for the partial hydrolysis of synthetic fiber surfaces, improving their hydrophilicity and aiding further finishing steps. Besides hydrolytic enzymes, oxidoreductases have also been used as powerful tools in various textile processing steps. Catalases have been used to remove H_2O_2 after bleaching and to reduce water consumption. In the textile industry, lipases are used to remove size lubricants, which increases the absorbance ability of fabrics for improved levelness in dyeing (Raja et al., 2012). In the denim abrasion systems, it is used to lessen the frequency of cracks and streaks.

17.4.6 Animal feed

Animal feed is the largest livestock and poultry production cost, accounting for 60%–70% of total expenses. To save on costs, many producers supplement feed with enzyme additives, enabling them to produce more meat cheaper and faster. Found in all living cells, enzymes catalyze chemical processes that convert nutrients into energy and new tissue. They do this by binding to substrates in the feed and breaking them down into smaller compounds. For example, proteases break down proteins into amino acids, carbohydrases split carbohydrates into simple sugars, and lipases take apart lipids into fatty acids and glycerol.

Animal feed is composed mainly of plant materials such as cereals, agricultural and grain milling by-products, and agricultural waste residues. These contain nonstarch polysaccharides, protein, and phytic acid. Monogastric animals generally cannot fully digest and utilize the fiber-rich feedstuffs. Due to the complex nature of the feed materials, starch sequestered by β-glucans and pentosans is also not digestible. Feed enzymes can increase the digestibility of nutrients, leading to greater efficiency in feed utilization (Choct, 2006). Currently, feed enzymes commercially available are phytases, proteases, α-galactosidases, glucanases, xylanases, α-amylases, and polygalacturonases (Selle and Ravindran, 2007). The use of enzymes as feed additives is restricted in many countries by local regulatory authorities (Pariza and Cook, 2010), and applications may vary from country to country.

During recent years focus has been on utilizing natural phosphorus bound in phytic acid in cereal-based feed for monogastrics. Phytic acid forms complexes with metal ions such as calcium, magnesium, iron, and zinc, thus preventing their assimilation by the animal. Better utilization of total plant phosphorus, of which 85%–90% is bound in phytic acid, is only obtained by adding the enzyme phytase to the feed. Microbial phytase liberates part of the bound phosphorus and makes it possible to reduce the number of supplements (phosphorus, calcium, and other nutrients) added to the animal diet. Phytase in animal feed can alleviate environmental pollution from bound phosphorous in animal waste and develop dietary deficiencies in animals (Lei and Stahl, 2000). The most common source of microbial phytase is *A. ficuum*.

Protein utilization from vegetables can be enhanced by using microbial proteases. Thus feed utilization and digestion by animals can be enhanced by adding enzymes to the feed (Lehmann et al., 2000). Various microbial enzymes are now used as feed enhancers and hold the prospect of serving larger animal and poultry production roles. Commercially available enzymes can be derived from plants and animals (e.g., actinidin from kiwi and rennet from calf stomachs) as well as microorganisms (e.g., amylase from *Bacillus* and lactase from *Aspergillus*).

17.4.7 Ethanol production

In the alcohol industry, enzymes for fermentable sugars from starch are also well established. Making ethanol from starch involves basic steps such as preparing the glucose feedstock, fermentation of glucose to ethanol, and recovery of ethanol. Enzymes have an important role in preparing the feedstock to convert starch into fermentable sugar, and glucose. Corn kernels contain 60%–70% starch, and it is the dominant source (97%) of starch feedstock used for ethanol production.

Over the past decade, there has been an increasing interest in fuel alcohol due to increased environmental concerns and higher crude oil prices. Therefore intense efforts are currently being undertaken to develop improved enzymes that can enable the utilization of cheaper and partially utilized substrates such as lignocellulose to make bio-ethanol more competitive with fossil fuels (Wheals et al., 1999; Zaldivar et al., 2001).

Two methods are used industrially to process corn to make starch accessible to enzymes in subsequent treatment. Corn is steeped in acidic water solutions in the wet-milling process, and the oil, protein, and fiber fractions are successively removed as products leaving the starch fraction. Enzymatic liquefaction and saccharification of the starch fraction are then carried out for the production of glucose. Microbial enzymes have replaced the traditional hydrolytic enzymes formerly supplied by adding malt. The traditional yeast *S. cerevisiae* ferments glucose to ethanol, which can be recovered by distillation. In beverage ethanol processes, the beer may be treated with acetolactate decarboxylase from *Bacillus brevis* or *Lactobacillus* sp. to convert acetolactate into acetoin via nonoxidative decarboxylation. Saccharification and fermentation steps can also be carried out concurrently in a process known as simultaneous saccharification and fermentation. In the United States, most ethanol (over 80%) from corn is produced from corn processed through dry grind facilities because of the lower capital investment required compared to wet mills. In the typical dry grind process, corn is mechanically milled to coarse flour. Following liquefaction, enzymatic saccharification using glucoamylase and fermentation using the conventional yeast are carried out simultaneously (Taylor et al., 2000; Taylor et al., 2001). The addition of protein-splitting enzymes (proteases) releases soluble nitrogen compounds from the fermentation mash and promotes yeast growth, decreasing fermentation time. The residue left after fermenting the sugars is known as distiller grains used as animal feed. Typically, large-scale industrial fermentation processes provide 12%–15% (v/v) ethanol with an ethanol yield as high as 95% of theoretical, based on starch feedstock.

17.4.8 Other applications

In recent years, tremendous research efforts have been made to reduce chlorine used for bleaching kraft pulp after pulping. Environmental regulatory pressures have prompted the pulp and paper industry to adapt new technology to eliminate various contaminants in the bleaching plant effluents. The main constituents of wood are cellulose, hemicellulose, and lignin. Research in the use of enzymes in pulp manufacture involves the degradation or modification of hemicellulose and lignin without disturbing the cellulose fibers.

Xylanase preparations used for wood processing in the paper industry should be free of cellulose activity. Cellulase-free xylanase preparations have applications to provide brightness to the paper due to their preferential solubilization of xylans in plant materials and selective removal of hemicelluloses from the kraft pulp. The production of cellulase-free extracellular *endo*-1,4-β-xylanase has been studied at a higher temperature of 50°C and pH 8.5, employing a strain of *Thermoactinomyces thalophilus* (Kohli et al., 2001). The paper and pulp industry requires separating and degrading lignin from plant material. The pretreatment of wood pulp using ligninolytic enzymes is essential for a milder and cleaner strategy of lignin removal compared to chemical bleaching. Bleach enhancement of mixed wood pulp has been achieved using coculture strategies through the combined activity of xylanase and laccase (Dwivedi et al., 2010). The ligninolytic enzyme system is used in the bio-bleaching of kraft pulp and other industries. Fungi are the most potent producers of lignin-degrading enzymes. The use of laccase to promote lignin degradation and bleaching of pulp has attracted considerable interest as a cost-effective replacement for chlorine bleaches. Thermophilic laccase enzyme is of particular use in the pulping industry. Recently, the biophysical characterization of thermophilic laccase isoforms has been reported (Kumar and Srikumar, 2013).

The removal of a pitch by chemical pulping or bleaching is not efficient. Pitch is the sticky, resinous material in wood. Treatment with lipases helps reduce pitch deposits since lipases hydrolyze the triglycerides in the wood resin to fatty acids and glycerol, making the material less viscous. The enzyme does not affect the cellulose quality. Removal of ink is an essential part of waste paper processing. Conventional deinking involves pulping the paper in a highly alkaline solution. It has been reported that cellulase enzymes can increase the efficiency of the deinking process. The coating treatment makes the paper's surface smooth and firm to improve the writing quality of the paper. For paper sizing, the viscosity of natural enzyme is too high, and this can be changed by partially degrading the polymer with α-amylases in a batch or continuous processes (van der Maarel et al., 2002). Starch is considered an excellent sizing agent for finishing paper, improving the quality and reusability, besides being a good coating for the paper.

In the leather industry, skins are soaked initially to clean them and to allow rehydration. Proteolytic enzymes effectively facilitate the soaking process. Lipases have also been used to dissolve and remove fat. Dehairing is then carried out using alkaline protease, such as subtilisin. Alkaline conditions swell the hair roots, easing the hair removal by allowing proteases to selectively attack the protein in the hair follicle. Conventional dehairing processes require harsh chemicals such as slaked lime and sodium sulfide, which essentially swell the hide, loosen, and damage the hair (Godfrey and West, 1996). Enzyme-based dehairing has led to much lower pollution emissions from tanneries.

Enzymes are used in various analytical methods for medical and nonmedical purposes. Immobilized enzymes are used as biosensors to analyze organic and inorganic compounds in biological fluids. A glucose biosensor consists of a glucose oxidase membrane and an oxygen electrode, while a biosensor for lactate consists of immobilized lactate oxidase and an oxygen electrode. The lactate sensor functions by monitoring the decrease in dissolved oxygen, resulting from the oxidation of lactate in the presence of lactate oxidase (Saha et al., 2009). The amperometric determination of pyruvate can be carried out with the pyruvate oxidase sensor. A bioelectrochemical system for total cholesterol estimation was developed based on a double-enzymatic method. An immobilized enzyme reactor containing cholesterol esterase and cholesterol oxidase is coupled with an amperometric detector system in this system. An amino acid electrode to determine total amino acids has also been developed using the enzymes L-glutamate oxidase, L-lysine oxidase, and tyrosinase. Enzyme electrodes are used for continuous control of fermentation processes.

Successful application of enzymatic processes in the chemical industry depends mainly on cost competitiveness with the existing and well-established chemical methods (Tufvesson et al., 2011). However, new scientific developments in genomics and protein engineering facilitate the tailoring of enzyme properties to increase that number significantly (Jackel and Hilvert, 2010; Lutz, 2010). An enzymatic conversion was devised to produce the amino acid L-tyrosine. Phenol, pyruvate, pyridoxal phosphate, and ammonium chloride are converted to L-tyrosine using a thermostable and chemostable tyrosine phenol lyase obtained from *Symbiobacterium toebii* (Kim do et al., 2007; Sanchez and Demain, 2011).

The numerous biocatalytic routes scaled up for pharmaceutical manufacturing have been recently reviewed (Bornscheuer et al., 2012), showing the competitiveness of enzymes versus traditional chemical processes. Enzymes help prepare β-lactam antibiotics such as semisynthetic penicillins and cephalosporins (Volpato et al., 2010). The semisynthetic penicillins have largely replaced natural penicillins, and about 85% of penicillins marketed for medicinal use are semisynthetic. 6-Aminopenicillanic acid is obtained by the hydrolysis of the amide bond of the naturally occurring penicillin with the enzyme penicillin amidase, which, unlike chemical hydrolysis, does not open the β-lactam ring. The most critical biocatalysis applications are synthesizing complex chiral pharmaceutical intermediates efficiently and economically. Esterases, lipases, proteases, and ketoreductases are widely applied in the preparation of chiral alcohols, carboxylic acids, amines, or epoxides (Zheng and Xu, 2011). Kinetic resolution of racemic amines is a common method used in the synthesis of chiral amines. Acylation of a primary amine moiety by a lipase is used by BASF for the resolution of chiral primary amines in a multi-thousand-ton scale (Sheldon, 2008). Atorvastatin, the active ingredient of Lipitor, a cholesterol-lowering drug, can be produced enzymatically. The process is based on three enzymatic activities: ketone reductase, glucose dehydrogenase, and halohydrin dehalogenase. Several iterative rounds of DNA shuffling for these three enzymes led to a 14-fold reduction in reaction time, a 7-fold increase in substrate loading, a 25-fold reduction in enzyme use, and a 50% improvement in isolated yield (Ma et al., 2010).

Therapeutic enzymes have various specific uses, such as on colytics, thrombolytics, or anticoagulants and as replacements for metabolic deficiencies. Enzymes are being used to treat many diseases such as cancer, cardiac problems, cystic fibrosis, dermal ulcers, inflammation, digestive disorders, etc. Proteolytic enzymes serve as good antiinflammatory agents. The action of collagenases is highly specific on native collagen and does not hydrolyse other structural proteins. This specific action of collagenase has been used for treatment of dermal ulcers and burns. Papain has been shown to produce a marked reduction of obstetrical inflammation and edema in dental surgery. Deoxyribonuclease is used as a mucolytic agent for use in patients with chronic bronchitis. Trypsin and chymotrypsin have been successfully used in the treatment of athletic injuries and postoperative hand trauma. Hyaluronidase has hydrolytic activity on chondroitin sulfate and may help regenerate damaged nerve tissue (Moon et al., 2003). Lysozyme hydrolyzes the chitins and mucopeptides of bacterial cell walls. Hence, it is used as an antibacterial agent, usually in combination with standard antibiotics. Lysozyme has also been found to have activity against HIV, as RNase A and urinary RNase U present selectively degrade viral RNA (Lee-Huang et al., 1999), showing possibilities for the treatment of HIV infection.

Cancer research has some good instances of the use of enzyme therapeutics. Recent studies have proved that arginine-degrading enzyme (PEGylated arginine deaminase) can inhibit human melanoma and hepatocellular carcinoma (Ensor et al., 2002). Another PEGylated enzyme, Oncaspar1 (pegaspargase), has shown promising results for treating children newly diagnosed with acute lymphoblastic leukemia. The further application of enzymes as therapeutic agents in cancer is described by antibody-directed enzyme prodrug therapy. A monoclonal antibody carries an enzyme specific to cancer cells where the enzyme activates a prodrug and destroys cancer cells but not normal cells. This approach is being utilized to discover and develop cancer therapeutics based on tumor-targeted enzymes that activate prodrugs. Certain enzymes such as L-asparaginase help treat cancer. L-asparaginase by lowering the concentration of asparagine retards the growth of cancer cells. It has proven particularly useful in treating lymphoblastic leukemia and certain forms of lymphomas and acute lymphoblastic leukemia.

Auxotrophic tumors are the type of tumors that cannot produce essential amino acids for growth and proliferation. Researchers have been investigating amino acid deprivation therapy as a promising strategy in the treatment of such tumors. This strategy utilizes enzymes from microbial sources, namely, L-asparaginase, arginase, lysine oxidase, phenylalanine ammonia lyase, arginine deiminase, glutaminase, and methionase for several anticancer treatments. There are two commercial enzymes available Kidrolase and Erwinase obtained from *E. coli* and *Erwinia chrysanthemi*, respectively, which are used to treat acute myeloid leukemia. Asparaginase obtained from *E. coli* exists in two distinct forms asparaginase-I and II. Both are produced in different locations inside the bacteria and also differ in their affinity toward cancer cells.

L-Arginine is a semiessential amino acid synthesized in the urea cycle by two enzymes, argininosuccinate synthetase and argininosuccinate lyase. Numerous cancers such as malignant melanoma and hepatocellular carcinoma exhibit arginine auxotrophy due to the downregulation of argininosuccinate synthetase. Arginine deiminase was first found in *Bacillus pyocyaneus*, and it was evaluated against various cancer lines in 1990. Later, this enzyme was produced in several other species such as *Pseudomonas putida, Enterococcus spp., Halobacterium salinarum, Mycoplasma arginini, Mycoplasma hominis, P. aeruginosa, Lactococcus lactis* ssp., *P. plecoglossicida* and *P. furukawaii*. The primary anticancer mechanisms exhibited by arginine deiminase are protein synthesis inhibition, reduced angiogenesis, and induction of apoptosis. PEGylated arginine deiminase has been investigated and proven successful in phase II (NCT02006030) and III (NCT01287585) clinical trials for hepatocellular carcinoma and phase II (NCT00450372) trial for melanoma.

Also known as methionine-γ-lyase, methioninase belongs to a family of pyridoxal-L-phosphate-dependent enzymes. It is one of the essential amino acids in humans involved in protein synthesis. It serves as a precursor for several biologically important molecules, polyamines, glutathione, and S-adenosyl methionine. Despite being essential, studies have shown that long-term deprivation of methionine is not fatal, which can be due to alternative options to synthesize it from enzymes such as L-methionine synthase and betaine-homocysteine methyltransferase. Due to the inherent inability to produce methionine cancer types such as primary ductal carcinoma, triple-negative breast cancer, osteosarcoma, non-small-cell lung carcinoma, melanoma, and certain gliomas undergo enzymatic depletion of methionine, resulting in cell cycle arrest at late "S" or "G2" phase. Interestingly, except mammalian species, certain eubacteria, archaea, plants, fungi, and protozoans can produce methionase. Species that produce methionase are *Aeromonas sp., Clostridium sporogenes, P. putida, E. coli*, and *Brevibacterium linens.*

L-Glutamine is a nonessential amino acid responsible for maintaining acid-base equilibrium, protein, and DNA synthesis. L-Glutamine amidohydrolase a.k.a. glutaminase causes hydrolysis of glutamine thereby depriving cells of performing cellular functions. Interestingly, there is an enormous requirement of glutamine by cancer cells as compared to normal cells. The mechanism here is the same as all the other amino acid deprivation therapies. The inherent inability of cancer cells to produce L-glutamine synthetase relies on an external source. L-Glutaminase can be obtained from *Achromobacter* sp. and *Alcaligenes faecalis* and have been reported for antitumor and cytotoxic activity against hepatocellular carcinoma (Hep-G2) and HeLa cell lines (Dhankhar et al., 2020).

Genetic engineering involves taking the relevant gene from the microorganism that naturally produces a particular enzyme (donor) and inserting it into another microorganism to make the enzyme more efficient (host). The first step is to cleave the DNA of the donor cell into fragments using restriction enzymes. The DNA fragments with the code for the desired enzyme are then placed, with the help of ligases, in a natural vector called a plasmid that can be transferred to the host bacterium or fungus. In recombinant DNA technology, restriction enzymes recognize specific base sequences in double-helical DNA and bring out the cleavage of both strands of the duplex in regions of the defined line. Restriction enzymes cleave foreign DNA molecules. The term *restriction endonuclease* comes from observing that certain bacteria can block virus infections by specifically destroying the incoming viral DNA (Adrio and Demain, 2014). Such bacteria are known as restricting hosts since they restrict the expression of foreign DNA. Specific nicks in duplex DNA can be sealed by an enzyme-DNA ligase which generates a phosphodiester bond between a 5′-phosphoryl group and a directly adjacent 3′-hydroxyl, using either ATP or NAD^+ as an external energy source.

17.5 Future of microbial enzymes

Enzymes are critical biomolecules with a wide range of applications in the industrial and biomedical fields. Today, it is one of the essential molecules widely used in every sector: dairy, industrial, agriculture, or pharmaceutical. The global market for industrial enzymes was estimated at $3.3 billion in 2010 and is expected to reach $5.0 billion by 2020. The market segmentation for various application areas shows that 34% of the market is for food and animal feed, followed by detergent and cleaners (29%). Paper and pulps share 11% market, while 17% of the market is captured by the textile and leather industries (Parameswaran et al., 2013).

The ongoing progress and interest in enzymes provide further success in industrial biocatalysis. There is a need for exciting developments in biotransformation and molecular biology. Many factors influence the growing interest in biocatalysts, including enzyme promiscuity, robust computational methods combined with directed evolution, and screening technologies to improve enzyme properties to meet process prospects (Adrio and Demain, 2014).

Recent advances in genomics, proteomics, efficient expression systems, and emerging recombinant DNA techniques have facilitated the discovery of new microbial enzymes from nature or by creating enzymes with improved catalytic properties. The future trend is to develop more effective systems that use much smaller quantities of chemicals and less energy to attain maximum product yield. Modern biotechnology will lead to the development of enzyme products with improved effects on diverse physiological conditions. Biotechnology offers an increasing potential for producing goods to meet various human needs. Enzyme technology is a subfield of biotechnology. New processes have been developed and are still evolving to manufacture bulk and value-added products utilizing enzymes as biocatalysts to meet needs such as food, fine chemicals, agriculture, and pharmaceuticals.

Enzymes contributed to more environmentally adapted clean and green technology due to their biodegradable nature. It can be used to develop environment-friendly alternatives to chemical processes in almost all steps of textile fiber processing (Araujo et al., 2008). Further research is required to implement commercial enzyme-based strategies for the biomodification of synthetic and natural fibers. An active field of study is searching for new enzyme-producing microorganisms and enzymes extracted from extremophilic microorganisms (Schumacher et al., 2001).

During the last two decades, enzyme applications continuously increased with high research and development-oriented activity covering various scientific and technological issues. Many enzymes need rigorous research and development to explore commercially through fundamental research in enzymology and process engineering. The functional understanding of different enzyme classes will likely provide new applications in the future. Multidisciplinary research involving industries is required to develop application-oriented research on enzymes. Over the past 10 years, significant advances in DNA technologies and bioinformatics have provided critical support to the field of biocatalysis. These tools have promoted the discovery of novel enzymes in natural resources and have substantially accelerated the redesign of existing biocatalysts. Next-generation DNA sequencing technology has allowed parallel sequence analysis on a massive scale and at dramatically reduced cost (Bornscheuer et al., 2012).

New and exciting enzyme applications are likely to benefit other areas such as less harm to the environment, greater efficiency, lower cost, lower energy consumption, and the enhancement of product properties. New enzyme molecules capable of achieving this will be developed through protein engineering and recombinant DNA techniques. Industrial biotechnology has a vital role in the way modern foods are processed. New ingredients and alternative solutions to current chemical processes will challenge the enzyme industry. Compared with chemical reactions, the more specific and cleaner technologies made possible by enzyme-catalyzed methods will promote the continued trend toward natural processes in food production.

17.6 Concluding remarks

The enzyme industry is one of the world's major industries, and there is a great market for enzymes. Enzymes are used in several different industrial products and processes, and new applications are constantly being added because of advances in modern biotechnology. Microorganisms provide an impressive amount of catalysts with a wide range of applications across several industries such as food, animal feed, technical industries, paper, fine chemicals, and pharmaceuticals. The unique properties of enzymes, such as high specificity, fast action, and biodegradability, allow enzyme-assisted processes in the industry to run under milder reaction conditions, with improved yields and reduced waste generation. Naturally occurring enzymes are often modified by molecular biology techniques to redesign the enzyme itself to fine-tune substrate specificity activity and thermostability. Enzyme technology offers great potential for many industries to meet the challenges in the future with the help of recombinant technology.

References

Adrio, J.L., Demain, A.L., 2014. Microbial enzymes: tools for biotechnological processes. Biomolecules 4, 117–139.

Aehle, W., 2007. Enzymes in industry - production and applications. Wiley-VCH Verlag GmbH & Co., Weinheim, Germany.

Agematu, H., Tsuchida, T., Kominato, K., Shibamoto, N., Yoshioka, T., Nishida, H., et al., 1993. Enzymatic dimerization of penicillin X. J. Antibiot. (Tokyo) 46, 141–148.

Akdeniz, V., Akalın, A.S., 2020. Recent advances in dual effect of power ultrasound to microorganisms in dairy industry: activation or inactivation. Crit. Rev. Food Sci. Nutr. 1–16.

Andreu, P., Collar, C., Martínez-Anaya, M.A., 1999. Thermal properties of doughs formulated with enzymes and starters. Eur. Food Res. Technol. 209, 286–293.

Angural, S., Rana, M., Sharma, A., Warmoota, R., Puri, N., Gupta, N., 2020. Combinatorial biobleaching of mixedwood pulp with lignolytic and hemicellulolytic enzymes for paper making. Indian. J. Microbiol. 60 (3), 383–387.

Araujo, R, Casal, M, Cavaco-Paulo, A, 2008. Application of enzymes for textile fibres processing. Biocatalysis and Biotransformation 26 (5), 332–349.

Arora, N.K., Mishra, J., Mishra, V., 2020. Microbial Enzymes: Roles and Applications in Industries. Springer, Singapore.

Arpigny, J.L., Jaeger, K.E., 1999. Bacterial lipolytic enzymes: classification and properties. Biochem. J. 343 (Pt 1), 177–183.

Banerjee, U.C., Vohra, R.M., 1991. Production of laccase by *Curvularia* sp. Folia Microbiol. (Praha) 36, 343–346.

Beg, Q., Kapoor, M., Mahajan, L., Hoondal, G., 2001. Microbial xylanases and their industrial applications: a review. Appl. Microbiol. Biotechnol. 56, 326–338.

Binz, T., Canevascini, G., 1997. Purification and partial characterization of the extracellular laccase from *Ophiostoma novo-ulmi*. Curr. Microbiol. 35, 278–281.

Bisgaard-Frantzen, H., Svendsen, A., Norman, B., Pedersen, S., Kjaerulff, S., Outtrup, H., et al., 1999. Development of industrially important α-amylases. J. Appl. Glycosci. 46, 199–206.

Bodie, E.A., Armstrong, G.L., Dunn-Coleman, N.S., 1994. Strain improvement of chymosin-producing strains of *Aspergillus niger* var. *awamori* using parasexual recombination. Enzyme Microb. Technol. 16, 376–382.

Bornscheuer, U.T., Huisman, G.W., Kazlauskas, R.J., Lutz, S., Moore, J.C., Robins, K., 2012. Engineering the third wave of biocatalysis. Nature. 485, 185–194.

Brijwani, K., Rigdon, A., Vadlani, P.V., 2010. Fungal laccases: production, function, and applications in food processing. Enzyme Res. 2010, 149748.

Ceci, L., Lozano, J., 1998. Determination of enzymatic activities of commercial pectinases for the clarification of apple juice. Food Chem. 61, 237–241.

Chandra, R., Chowdhary, P., 2015. Properties of bacterial laccases and their application in bioremediation of industrial wastes. Env. Sci. Process. Impacts 17, 326–342.

Chirumamilla, R.R., Muralidhar, R., Marchant, R., Nigam, P., 2001. Improving the quality of industrially important enzymes by directed evolution. Mol. Cell Biochem. 224, 159–168.

Choct, M., 2006. Enzymes for the feed industry. World Poult. Sci. 62, 5–15.

Collar, C., Martinez, J.C., Andreu, P., Armero, E., 2000. Effects of enzyme associations on bread dough performance. a response surface analysis. Food Sci. Technol. Int. 6, 217–226.

Collins, T., Gerday, C., Feller, G., 2005. Xylanases, xylanase families and extremophilic xylanases. FEMS Microbiol. Rev. 29, 3–23.

Demain, A.L., Adrio, J.L., 2008. Contributions of microorganisms to industrial biology. Mol. Biotechnol. 38, 41–55.

Deng, A., Wu, J., Zhang, Y., Zhang, G., Wen, T., 2010. Purification and characterization of a surfactant-stable high-alkaline protease from Bacillus sp. B001. Bioresour. Technol. 101 (7111–7117).

Dhankhar, R., Gupta, V., Kumar, S., Kapoor, R.K., Gulati, P., 2020. Microbial enzymes for deprivation of amino acid metabolism in malignant cells: biological strategy for cancer treatment. Appl. Microbiol. Biotechnol. 104 (7), 2857–2869.

Du, Y., Shi, P., Huang, H., Zhang, X., Luo, H., Wang, Y., et al., 2013. Characterization of three novel thermophilic xylanases from Humicolainsolens Y1 with application potentials in the brewing industry. Bioresour. Technol. 130, 161–167.

Dwivedi, P., Vivekanand, V., Pareek, N., Sharma, A., Singh, R.P., 2010. Bleach enhancement of mixed wood pulp by xylanase-laccase concoction derived through co-culture strategy. Appl. BiochemBiotechnol 160, 255–268.

Edens, W.A., Goins, T.Q., Dooley, D., Henson, J.M., 1999. Purification and characterization of a secreted laccase of *Gaeumannomycesgraminis var. tritici.* Appl. Env. Microbiol. 65, 3071–3074.

Ellaiah, P., Srinivasulu, B., Adinarayana, K., 2002. A review on microbial alkaline proteases. J. Sci. Ind. Res. 61, 690–704.

Ensor, C.M., Holtsberg, F.W., Bomalaski, J.S., Clark, M.A., 2002. PEGylated arginine deiminase (ADI-SS PEG20,000 mw) inhibits human melanomas and hepatocellular carcinomas in vitro and in vivo. Cancer Res. 62, 5443–5450.

Etters, J.N., Annis, P.A., 1998. Textile enzyme use: a developing technology. Am. Dyest. Report. 5, 18–23.

Felse, P.A., Panda, T., 2000. Production of microbial chitinases – a revisit. Bioprocess. Eng. 23, 127–134.

Fiedurek, J., Gromada, A., 2000. Production of catalase and glucose oxidase by Aspergillus niger using unconventional oxygenation of culture. J. Appl. Microbiol. 89, 85–89.

Froehner, S.C., Eriksson, K.E., 1974. Purification and properties of *Neurospora crassa* laccase. J. Bacteriol. 120, 458–465.

Gianfreda, L., Xu, F., Bollag, J.M., 1999. Laccases: a useful group of oxidoreductive enzymes. Biorem J. 3, 1–26.

Godfrey, T., West, S.I., 1996. Introduction to industrial enzymology. In: Godfrey, T., West, S. (Eds.), Industrial Enzymology. Macmillan Press, London, pp. 1–8.

Godfrey, R., West, M., 1996. Industrial Enzymology. Mac Millan Publishers Inc, New York.

Gummadi, S.N., Panda, T., 2003. Purification and biochemical properties of microbial pectinases—a review. Process. Biochem. 38, 987–996.

Gupta, R., Gupta, N., Rathi, P., 2004. Bacterial lipases: an overview of production, purification and biochemical properties. Appl. Microbiol. Biotechnol. 64, 763–781.

Gurung, N., Ray, S., Bose, S., Rai, V., 2013. A broader view: microbial enzymes and their relevance in industries, medicine, and beyond. Biomed. Res. Int. 2013, 1–18.

Henrissat, B., 1991. A classification of glycosyl hydrolases based on amino acid sequence similarities. Biochem. J. 280, 309–316.

Henrissat, B., Coutinho, P.M., 2001. Classification of glycoside hydrolases and glycosyltransferases from hyperthermophiles. In: Michael, W.W., Adams, R.M.K. (Eds.), Methods in Enzymology. Academic Press, pp. 183–201.

Islam, R., Datta, B., 2015. Diversity of chitinases and their industrial potential. Int. J. Appl. Res. 1, 55–60.

Ito, S., Kobayashi, T., Ara, K., Ozaki, K., Kawai, S., Hatada, Y., 1998. Alkaline detergent enzymes from alkaliphiles: enzymatic properties, genetics, and structures. Extremophiles. 2, 185–190.

Iyer, G., Chattoo, B.B., 2003. Purification and characterization of laccase from the rice blast fungus, *Magnaporthe grisea*. FEMS Microbiol. Lett. 227, 121–126.

Jackel, C., Hilvert, D., 2010. Biocatalysts by evolution. CurrOpinBiotechnol 21, 753–759.

Jayani, R.S., Saxena, S., Gupta, R., 2005. Microbial pectinolytic enzymes: a review. Process. Biochem. 40, 2931–2944.

Jeffries, T.W., 1996. Biochemistry and genetics of microbial xylanases. CurrOpinBiotechnol 7, 337–342.

Jisha, V.N., Smitha, R.B., Pradeep, S., Sreedevi, S., Unni, K.N., Sajith, S., et al., 2013. Versatility of microbial proteases. Adv. Enzyme Res. 1, 39–51.

Jolly, L., 2001. The commercial viability of fuel ethanol from sugar cane. Int. Sugar J. 103, 17–143.

Junghanns, C., Moeder, M., Krauss, G., Martin, C., Schlosser, D., 2005. Degradation of the xenoestrogen nonylphenol by aquatic fungi and their laccases. Microbiology. 151, 45–57.

Juturu, V., Wu, J.C., 2012. Microbial xylanases: engineering, production and industrial applications. Biotechnol. Adv. 30, 1219–1227.

Juturu, V., Wu, J.C., 2014. Microbial cellulases: engineering, production and applications. Renew. Sust. Energ. Rev. 33, 188–203.

Kashyap, D.R., Vohra, P.K., Chopra, S., Tewari, R., 2001. Applications of pectinases in the commercial sector: a review. Bioresour. Technol. 77, 215–227.

Kelly, S.L., Kelly, D.E., 2013. Microbial cytochromes P450: biodiversity and biotechnology. Where do cytochromes P450 come from, what do they do and what can they do for us? Philos. Trans. R. Soc. Lond. B Biol. Sci. 368, 20120476.

Khot, S., Rawal, S.U., Patel, M.M., 2021. Dissolvable-soluble or biodegradable polymers. In: Chappel, E. (Ed.), Drug Delivery Devices and Therapeutic Systems. Academic Press, pp. 367–394.

Kiiskinen, L.L., Viikari, L., Kruus, K., 2002. Purification and characterisation of a novel laccase from the ascomycete *Melanocarpus albomyces*. Appl. Microbiol. Biotechnol. 59, 198–204.

Kim do, Y., Rha, E., Choi, S.L., Song, J.J., Hong, S.P., Sung, M.H., et al., 2007. Development of bioreactor system for L-tyrosine synthesis using thermostable tyrosine phenol-lyase. J. Microbiol. Biotechnol. 17, 116–122.

Kirk, O., Borchert, T.V., Fuglsang, C.C., 2002. Industrial enzyme applications. CurrOpinBiotechnol 13, 345–351.

Koeller, K.M., Wong, C.H., 2001. Enzymes for chemical synthesis. Nature. 409, 232–240.

Kohli, U., Nigam, P., Singh, D., Chaudhary, K., 2001. Thermostable, alkalophilic and cellulase free xylanase production by *Thermoactinomyces thalophilus* subgroup C. Enzyme Microb. Technol. 28, 606–610.

Kuhad, R.C., Gupta, R., Singh, A., 2011. Microbial cellulases and their industrial applications. Enzyme Res. 2011, 280696.

Kumar, A., Gudiukaite, R., Gricajeva, A., Sadauskas, M., Malunavicius, V., Kamyab, H., et al., 2020. Microbial lipolytic enzymes–promising energy-efficient biocatalysts in bioremediation. Energy. 192, 116674.

Kumar, A., Singh, S., 2013. Directed evolution: tailoring biocatalysts for industrial applications. Crit. Rev. Biotechnol. 33, 365–378.

Kumar, G., Srikumar, K., 2013. Biophysical characterization of thermophilic laccase from the xerophytes: *Cereus pterogonus* and *Opuntia vulgaris*. Cellulose. 20, 115–125.

Lee-Huang, S., Huang, P.L., Sun, Y., Kung, H.F., Blithe, D.L., Chen, H.C., 1999. Lysozyme and RNases as anti-HIV components in beta-core preparations of human chorionic gonadotropin. Proc. Natl Acad. Sci. U S A 96, 2678–2681.

Lehmann, M., Kostrewa, D., Wyss, M., Brugger, R., D'Arcy, A., Pasamontes, L., et al., 2000. From DNA sequence to improved functionality: using protein sequence comparisons to rapidly design a thermostable consensus phytase. Protein Eng. 13, 49–57.

Lei, X.G., Stahl, C.H., 2000. Nutritional benefits of phytase and dietary determinants of its efficacy. J. Appl. Anim. Res. 17, 97–112.

Liu, B., Zhang, J., Li, B., Liao, X., Du, G., Chen, J., 2013. Expression and characterization of extreme alkaline, oxidation-resistant keratinase from *Bacillus licheniformis* in recombinant *Bacillus subtilis* WB600 expression system and its application in wool fiber processing. World J. Microbiol. Biotechnol. 29, 825–832.

Li, S., Yang, X., Yang, S., Zhu, M., Wang, X., 2012. Technology prospecting on enzymes: application, marketing and engineering. Comput. Struct. Biotechnol. J. 2, e201209017.

Londoño-Hernandez, L., Ruiz, H.A., Toro, C.R., Ascacio-Valdes, A., Rodriguez-Herrera, R., Aguilera-Carbo, A., et al., 2020. Advantages and progress innovations of solid-state fermentation to produce industrial enzymes. In: Arora, N., Mishra, J., Mishra, V. (Eds.), Microbial Enzymes: Roles and Applications in Industries. Springer, Singapore, pp. 87–113.

Lutz, S., 2010. Biochemistry. Reengineering enzymes. Science. 329, 285–287.

van der Maarel, M.J., van der Veen, B., Uitdehaag, J.C., Leemhuis, H., Dijkhuizen, L., 2002. Properties and applications of starch-converting enzymes of the alpha-amylase family. J. Biotechnol. 94, 137–155.

Matsumoto, K.S., 2006. Fungal chitinases. In: Guevara-González, R.G., Torres-Pacheco, I. (Eds.), Advances in Agricultural and food biotechnology. Research Signpost, Kerala, India, pp. 289–304.

Ma, S.K., Gruber, J., Davis, C., Newman, L., Gray, D., Wang, A., et al., 2010. A green-by-design biocatalytic process for atorvastatin intermediate. Green. Chem. 12, 81–86.

Ma, L., Li, C., Yang, Z., Jia, W., Zhang, D., Chen, S., 2013. Kinetic studies on batch cultivation of *Trichoderma reesei* and application to enhance cellulase production by fed-batch fermentation. J. Biotechnol. 166, 192–197.

Mendez, C., Salas, J.A., 2001. Altering the glycosylation pattern of bioactive compounds. Trends Biotechnol. 19, 449–456.

Mitidieri, S., Martinelli, Souza, Schrank, A.H., Vainstein, M.H., A., 2006. Enzymatic detergent formulation containing amylase from *Aspergillus niger*: a comparative study with commercial detergent formulations. Bioresour. Technol. 97, 1217–1224.

Molitoris, H.P., Esser, K., 1970. The phenoloxidases of the ascomycete *Podospora anserina*. V. properties of laccase I after further purification. Arch. Mikrobiol. 72, 267–296.

Moon, L.D., Asher, R.A., Fawcett, J.W., 2003. Limited growth of severed CNS axons after treatment of adult rat brain with hyaluronidase. J. Neurosci. Res. 71, 23–37.

Motta, F.L., Andrade, C.C.P., Santana, M.H.A., 2013. A Review of xylanase production by the fermentation of xylan: classification, characterization and applications. In: Chandel, A.K., Silvério da Silva, S. (Eds.), Sustainable degradation of lignocellulosic biomass - techniques, applications and commercialization. InTech, p. 284.

Mougin, C., Jolivalt, C., Briozzo, P., Madzak, C., 2003. Fungal laccases: from structure-activity studies to environmental applications. Env. Chem. Lett. 1, 145–148.

Murad, H.A., Azzaz, H.H., 2011. Microbial pectinases and ruminant nutrition. Res. J. Microbiol. 6, 246–269.

Nigam, P., Singh, D., 1995. Enzyme and microbial systems involved in starch processing. Enzyme Microb. Technol. 17, 770–778.

Nigam, P.S., 2013. Microbial enzymes with special characteristics for biotechnological applications. Biomolecules. 3, 597–611.

Okanishi, M., Suzuki, N., Furuta, T., 1996. Variety of hybrid characters among recombinants obtained by interspecific protoplast fusion in *Streptomycetes*. BiosciBiotechnolBiochem 60, 1233–1238.

O'Neill, A., Araújo, R., Casal, M., Guebitz, G., Cavaco-Paulo, A., 2007. Effect of the agitation on the adsorption and hydrolytic efficiency of cutinases on polyethylene terephthalate fibres. Enzyme Microb. Technol. 40, 1801–1805.

Palonen, H., Saloheimo, M., Viikari, L., Kruus, K., 2003. Purification, characterization and sequence analysis of a laccase from the ascomycete *Mauginiella* sp. Enzyme Microb. Technol. 33, 854–862.

Pamment, N., Robinson, C.W., Hilton, J., Moo-Young, M., 1978. Solid-state cultivation of *Chaetomium cellulolyticum* on alkali-pretreated sawdust. BiotechnolBioeng 20, 1735–1744.

Pandey, A., Nigam, P., Soccol, C.R., Soccol, V.T., Singh, D., Mohan, R., 2000a. Advances in microbial amylases. Biotechnol. Appl. Biochem. 31 (Pt 2), 135–152.

Pandey, A., Selvakumar, P., Soccol, C.R., Nigam, P., 1999. Solid-state fermentation for the production of industrial enzymes. Curr. Sci. 77, 149–162.

Pandey, A., Soccol, C.R., Nigam, P., Soccol, V.T., Vandenberghe, L.P.S., Mohan, R., 2000b. Biotechnological potential of agro-industrial residues. II: cassava bagasse. Bioresour. Technol. 74, 81–87.

Parameswaran, B., Piyush, P., Raghavendra, G., Nampoothiri, K.M., Duggal, A., Dey, K., et al., 2013. Industrial enzymes - present status and future perspectives for India. J. Sci. Ind. Res. 72, 271–286.

Pariza, M.W., Cook, M., 2010. Determining the safety of enzymes used in animal feed. RegulToxicolPharmacol 56, 332–342.

Pariza, M.W., Johnson, E.A., 2001. Evaluating the safety of microbial enzyme preparations used in food processing: update for a new century. RegulToxicolPharmacol 33, 173–186.

Paulino, B.N., Sales, A., de Oliveira Felipe, L., Pastore, G.M., Molina, G., Bicas, J.L., 2020. Recent advances in the microbial and enzymatic production of aroma compounds. CurrOpin Food Sci. 37, 98–106.

Pedrolli, D.B., Monteiro, A.C., Gomes, E., Carmona, E.C., 2009. Pectin and pectinases: production, characterization and industrial application of microbial pectinolytic enzymes. Open. Biotechnol. J. 3, 9–18.

Polizeli, M.L., Rizzatti, A.C., Monti, R., Terenzi, H.F., Jorge, J.A., Amorim, D.S., 2005. Xylanases from fungi: properties and industrial applications. Appl. Microbiol. Biotechnol. 67, 577–591.

Preethi, P.S., Gomathi, A., Srinivasan, R., Kumar, J.P., Murugesan, K., Muthukailannan, G.K., 2020. Laccase: recombinant expression, engineering and its promising applications. In: Arora, N., Mishra, J., Mishra, V. (Eds.), Microbial Enzymes: Roles and Applications in Industries. Springer, Singapore, pp. 63–85.

Quintanilla, D., Hagemann, T., Hansen, K., Gernaey, K.V., 2015. Fungal morphology in industrial enzyme production-modelling and monitoring. In: Krull, R., Bley, T. (Eds.), Filaments In Bioprocesses. Springer International Publishing, Switzerland, pp. 29–54.

Raja, K.S., Vasanthi, N.S., Saravanan, D., Ramachandran, T., 2012. Use of bacterial lipase for scouring of cotton fabrics. Indian. J. Fibre Text. Res. 37, 299–302.

Rao, M.B., Tanksale, A.M., Ghatge, M.S., Deshpande, V.V., 1998. Molecular and biotechnological aspects of microbial proteases. Microbiol. Mol. Biol. Rev. 62, 597–635.

Roberts, R.J., Vincze, T., Posfai, J., Macelis, D., 2010. REBASE–a database for DNA restriction and modification: enzymes, genes and genomes. Nucleic Acids Res. 38, D234–D236.

Rodríguez, A., Falcón, M.A., Carnicero, A., Perestelo, F., la Fuente, G.D., Trojanowski, J., 1996. Laccase activities of *Penicillium chrysogenum* in relation to lignin degradation. Appl. Microbiol. Biotechnol. 45, 399–403.

Saha, B.C., Demirjian, D.C., 2000. Applied Biocatalysis in Specialty Chemicals and Pharmaceuticals. American Chemical Society, Washington, DC.

Saha, B.C., Jordan, D.B., Bothast, R.J., 2009. Enzymes, industrial (overview). Encyclopedia of Microbiology. Academic Press, Oxford.

Sakaki, T., 2012. Practical application of cytochrome. Biol. Pharm. Bull. 35, 844–P849.

Samak, N.A., Jia, Y., Sharshar, M.M., Mu, T., Yang, M., Peh, S., et al., 2020. Recent advances in biocatalysts engineering for polyethylene terephthalate plastic waste green recycling. Env. Int. 145, 106144.

Sanchez, S., Demain, A.L., 2011. Enzymes and bioconversions of industrial, pharmaceutical, and biotechnological significance. Org. Process. Res. Dev. 15, 224–230.

Saranraj, P., Naidu, M.A., 2014. Microbial pectinases: a review. Glob. J. Trad. Med. Sys 3, 1–9.

Scherer, M., Fischer, R., 1998. Purification and characterization of laccase II of *Aspergillus nidulans*. Arch. Microbiol. 170, 78–84.

Schumacher, K., Heine, E., Hocker, H., 2001. Extremozymes for improving wool properties. J. Biotechnol. 89, 281–288.

Selle, P.H., Ravindran, V., 2007. Microbial phytase in poultry nutrition. Anim. Feed. Sci. Technol. 135, 1–41.

Sharma, V.K., Kumar, N., Prakash, T., Taylor, T.D., 2010. MetaBioME: a database to explore commercially useful enzymes in metagenomic datasets. Nucleic Acids Res. 38, D468–D472.

Sheldon, R.A., 2008. E factors, green chemistry and catalysis: an odyssey. Chem. Commun. (Camb.) 3352–3365.

Shevchik, V.E., Hugouvieux-Cotte-Pattat, N., 1997. Identification of a bacterial pectin acetyl esterase in *Erwinia chrysanthemi* 3937. Mol. Microbiol. 24, 1285–1301.

Shraddha, Shekher, R., Sehgal, S., Kamthania, M., Kumar, A., 2011. Laccase: microbial sources, production, purification, and potential biotechnological applications. Enzyme Res. 2011, 217861.

Singh, S.A., Ramakrishna, M., Appu Rao, A.G., 1999. Optimisation of downstream processing parameters for the recovery of pectinase from the fermented bran of *Aspergillus carbonarius*. Process. Biochem. 35, 411–417.

Singh, S., Sczakas, G., Soccol, C., Pandey, A., 2008. Production of enzymes by solid-state fermentation. In: Pandey, A., Soccol, C., Larroche, C. (Eds.), Current Developments in Solid-State Fermentation. Springer, New York, pp. 183–204.

Sukumaran, R.K., Singhania, R.R., Pandey, A., 2005. Microbial cellulases-production, applications and challenges. J. Sci. Ind. Res. 64, 832–844.

Sundarram, A., Murthy, T.P.K., 2014. α-amylase production and applications: a review. J. Appl. Env. Microbiol. 2, 166–175.

Tapre, A.R., Jain, R.K., 2014. Pectinases: enzymes for fruit processing industry. Int. Food Res. J. 21, 447–453.

Taylor, F., Kurantz, M.J., Goldberg, N., McAloon, A.J., Craig Jr., J.C., 2000. Dry-grind process for fuel ethanol by continuous fermentation and stripping. Biotechnol. Prog. 16, 541–547.

Taylor, F., McAloon, A.J., Craig Jr., J.C., Yang, P., Wahjudi, J., Eckhoff, S.R., 2001. Fermentation and costs of fuel ethanol from corn with Quick-Germ process. Appl. BiochemBiotechnol 94, 41–49.

Tezgel, N., Kırtel, O., Van den Ende, W., Oner, E.T., 2020. Fructosyltransferase enzymes for microbial fructan production. In: Arora, N., Mishra, J., Mishra, V. (Eds.), Microbial Enzymes: Roles and Applications in Industries. Springer, Singapore, pp. 1–39.

Thakker, G., Evans, C., Rao, K.K., 1992. Purification and characterization of laccase from *Monocillium indicum* Saxena. Appl. Microbiol. Biotechnol. 37, 321–323.

Tomschy, A., Tessier, M., Wyss, M., Brugger, R., Broger, C., Schnoebelen, L., et al., 2000. Optimization of the catalytic properties of *Aspergillus fumigatus* phytase based on the three-dimensional structure. Protein Sci. 9, 1304–1311.

Treichel, H., de Oliveira, D., Mazutti, M., Di Luccio, M., Oliveira, J.V., 2010. A review on microbial lipases production. Food Bioprocess. Technol. 3, 182–196.

Tucker, G.A., Woods, L.F.J., 1995. Enzymes in Food Processing. Blackie Academic, Bishopbriggs, Glasgow, UK.

Tufvesson, P., Lima-Ramos, J., Nordblad, M., Woodley, J.M., 2011. Guidelines and cost analysis for catalyst production in biocatalytic processes. Org. Process. Res. Dev. 15, 266–274.

Underkofler, L.A., Barton, R.R., Rennert, S.S., 1958. Production of microbial enzymes and their applications. Appl. Microbiol. 6, 212–221.

Venkateshwaran, G., Somashekar, D., Prakash, M.H., Agrawal, R., Basappa, S.C., Joseph, R., 1999. Production and utilization of catalase using *Saccharomyces cerevisiae*. Process. Biochem. 34, 187–191.

Volpato, G., Rodrigues, R.C., Fernandez-Lafuente, R., 2010. Use of enzymes in the production of semi-synthetic penicillins and cephalosporins: drawbacks and perspectives. Curr. Med. Chem. 17, 3855–3873.

Watanabe, I., Nara, F., Serizawa, N., 1995. Cloning, characterization and expression of the gene encoding cytochrome P-450sca-2 from *Streptomyces carbophilus* involved in production of pravastatin, a specific HMG-CoA reductase inhibitor. Gene. 163, 81–85.

Wheals, A.E., Basso, L.C., Alves, D.M., Amorim, H.V., 1999. Fuel ethanol after 25 years. Trends Biotechnol. 17, 482–487.

Wintrode, P.L., Miyazaki, K., Arnold, F.H., 2000. Cold adaptation of a mesophilic subtilisin-like protease by laboratory evolution. J. Biol. Chem. 275, 31635–31640.

Wong, K.K.Y., Saddler, J.N., 1992. Trichoderma xylanases, their properties and application. Crit. Rev. Biotechnol. 12, 413–435.

Wong, K.K., Tan, L.U., Saddler, J.N., 1988. Multiplicity of beta-1,4-xylanase in microorganisms: functions and applications. Microbiol. Rev. 52, 305–317.

Yang, S.T., El-Enshasy, H.A., Thongchul, N., 2013. Cellulases: characteristics, sources, production and applications. In: Yang, S.T., El-Enshasy, H.A., Thongchul, N. (Eds.), Bioprocessing Technologies in Biorefinery for Sustainable Production of Fuels, Chemicals, and Polymers. John Wiley & Sons, Inc., Hoboken, New Jersey.

Zaldivar, J., Nielsen, J., Olsson, L., 2001. Fuel ethanol production from lignocellulose: a challenge for metabolic engineering and process integration. Appl. Microbiol. Biotechnol. 56, 17–34.

Zhang, D.X., Du, G.C., Chen, J., 2010. Fermentation production of microbial catalase and its application in textile industry. Chin. J. Biotech. 26, 1473–1481.

Zhang, C., Kim, S.K., 2010. Research and application of marine microbial enzymes: status and prospects. Mar. Drugs 8, 1920–1934.

Zheng, G.W., Xu, J.H., 2011. New opportunities for biocatalysis: driving the synthesis of chiral chemicals. CurrOpinBiotechnol 22, 784–792.

Chapter 18

Microbial enzymes used in food industry

Pedro Fernandes[1,2,3] and Filipe Carvalho[1,2,3]
[1]*Faculty of Engineering, Universidade Lusófona de Humanidades e Tecnologias, Lisboa, Portugal,* [2]*iBB—Institute for Bioengineering and Biosciences, Instituto Superior Técnico, Universidade de Lisboa, Lisboa, Portugal,* [3]*Associate Laboratory i4HB—Institute for Health and Bioeconomy at Instituto Superior Técnico, Universidade de Lisboa, Lisboa, Portugal*

18.1 Introduction

18.1.1 A global perspective on the use of enzymes in the food industry

For thousands of years, man has relied on microorganisms and enzymes for food production. Typical examples include the making of beer, bread, cheese, or wine, where enzymes were unknowingly used for thousands of years. Currently, enzymes either obtained by microbial fermentation or extracted from plants or animals are used for the production and/or processing of foods, leading to improved or new processes or goods. Enzymes used in food and beverages have a significant market share of the global enzyme market (Bilal and Iqbal, 2020), roughly 20% by 2019, according to Fortune Business Insight, corresponding to USD 1.69 billion in 2019. Moreover, the food enzyme market is expected to reach USD 2.39 billion by 2027 while displaying a CAGR (compound annual growth rate) of 4.70%. (https://www.fortunebusinessinsights.com/food-enzymes-market-102835; https://www.fortunebusinessinsights.com/industry-reports/enzymes-market-100595). A recent report by MarketsandMarkets is even more optimistic, as it values the food enzymes market at USD 2.2 billion in 2021 and foresees a value of USD 3.1 billion by 2026, while exhibiting a CAGR of 6.4% (https://www.marketsandmarkets.com/Market-Reports/food-enzymes-market-800.html). Different enzyme classes find application in the production and processing of food and beverages because diverse sets of reactions are required, although hydrolases among these carbohydrases visibly stand out (Madhavan et al., 2021) (Table 18.1). Besides specificity, enzymes may be required to present

Biotechnology of Microbial Enzymes. DOI: https://doi.org/10.1016/B978-0-443-19059-9.00009-8
© 2023 Elsevier Inc. All rights reserved.

TABLE 18.1 A brief overview of some major roles played by enzymes within the scope of food processing.

Class	Enzyme	Role
Hydrolases	Amylases	Starch liquefaction and saccharification. Production of starch-derived goods, processing of bread and juices, brewing
	Asparaginase	Decrease of acrylamide, which may occur in baked, roasted, and fried starch-rich products
	Chitinases, chitosanases	Production of chitooligosaccharides (nutraceuticals)
	Glucanases	Breakdown of cell walls of cereals, brewing
	Glucoamylases	Saccharification of starch, production of sugar syrups
	Glutaminase	Hydrolysis of glutamine into ammonia and glutamate, the later used as a flavor enhancer
	Glycosidases	Hydrolysis of β-glycosidic bond. Release of aroma compounds in wine production
	Inulinase	Production of fructooligosaccharides (prebiotics), high-fructose syrups
	Invertase	Sucrose hydrolysis, production of invert sugar syrup
	Lactase (β-galactosidase)	Lactose hydrolysis (Lactose-free milk), whey hydrolysis, synthesis of galactooligosaccharides
	Lipase	Cheese flavor, in situ emulsification for dough conditioning, support for lipid digestion in young animals, synthesis of aromatic molecules

	Naringinase	Removal of bitterness from citrus juices
	Proteases (viz., alcalase, bromelain, chymosin, chymotrypsin, cyprosin, papain, pepsin, trypsin)	Protein hydrolysis, milk clotting, low allergenic infant food formulation, enhanced digestibility and utilization, flavor improvement in milk and cheese, meat tenderizer, prevention of haze formation in brewing, dough conditioning, production of bioactive peptides
	Pectinase	Viscosity reduction, clarification of fruit juices, fruit peeling
	Peptidase	Cheese ripening
	Phospholipase	In situ emulsification for dough conditioning
	Phytases	Feed processing through the release of phosphate from phytate, enhanced digestibility
	Polygalacturonase	Production of pectic oligosaccharides (food additives)
	Pullulanase	Saccharification, complementary to glucoamylases
	Xylanases	Viscosity reduction, enhanced digestibility, dough conditioning, production of xylooligosaccharides (food additives and nutraceuticals)
Isomerases	Arabinose isomerase	Isomerization of D-galactose to D-tagatose (rare, low-calorie sweetener)
	Cellobiose-2-epimerase	Isomerization of lactose to lactulose
	Rhamnose isomerase	Isomerization of L-mannose and L-rhamnose to L-fructose (rare, low-calorie sweetener) and L-rhamnulose (rare sugar, provider of caramel-liked aroma), respectively
	Tagatose-3-epimerase	Isomerization of D-fructose to D-psicose (D-allulose) (rare, low-calorie sweetener)
	Xylose (glucose) isomerase	Isomerization of D-glucose to D-fructose (production of high-fructose corn syrup), isomerization of D-psicose to D-allose (rare, low-calorie sweetener)

(*Continued*)

TABLE 18.1 (Continued)

Class	Enzyme	Role
Lyases	Acetolactate decarboxylase	Fastening beer maturation, through the conversion of acetolactate to acetoin
	Alginate lyase	Additive to modify the texture, increased viscosity of ice cream
	Pectinolytic lyase	Production of pectin oligosaccharides (food additives)
Oxidoreductases	Glucose oxidase	Improved dough strength and handling properties, removal of molecular oxygen (from air-tight packages), removal of glucose (suppress Maillard reaction)
	Laccases	Stabilization of color in wine making, dough strengthener, increased storage life of beer, improved flavor of vegetable oils, preparation of cork stoppers
	Lipoxygenase	Dough strengthening, bread whiting
	Peroxidases	Fruit juice clarification
Transferases	Cyclodextrin glycosyltransferases	Production of cyclodextrins, production of high intensity sweeteners (steviol glycoside based)
	Fructosyltransferases (levansucrases)	Synthesis of fructooligosaccharides (production of prebiotics)
	Glycosyltransferases	Synthesis of glucooligosaccharides and galactooligosaccharides (production of prebiotics)
	Transglutaminase	Cross-linking role, modification of viscoelastic and water binding properties. Dough processing, meat processing
Ligases	Butelase, peptiligase, sortase	Peptides as food additives

diverse, if not opposite, operational requirements because of the role intended. Thus high thermal and operational stabilities are required for glucose isomerase for the isomerization of glucose to fructose in the production of high-fructose corn syrup, to allow for high volumetric productivity and adequate shift of the thermodynamic equilibrium toward the product (DiCosimo et al., 2013). Similar features are required for the production of D-tagatose through the isomerization of D-galactose catalyzed by L-arabinose isomerase (Oh, 2007; Xu et al., 2018). In both cases, operation in a slightly acidic environment is desired to minimize by-product formation (Xu et al., 2014, 2018).

On the other hand, cold-active pepsins are looked after for the riddling process in caviar production (Guérard et al., 2005), as operation at low temperatures reduces the risk of microbial contamination and thermal degradation of food. Also, cold-active proteases are looked after for meat tenderization, as these enzymes display high activity at low temperatures, and are denatured during cooking, therefore ruling out the risk of overtenderization (Zhao et al., 2012; Mageswari et al., 2017). Additionally, cold-active proteases have also been shown to reduce the increased pH of stored meat samples compared to untreated meat samples. As the authors tentatively assigned such pH increase to food spoilage, they concluded that enzyme treatment could minimize such deleterious effects while performing the intended role as a meat tenderizer (Mageswari et al., 2017). A cold-active recombinant *Alteromonas* sp. ML117 β-galactosidase was also shown to hydrolyze 86% of milk lactose at 10°C, over 24 h. This behavior is appealing since operation at such a temperature contributes to retaining the original taste and nutritional value of milk (Yao et al., 2019).

18.1.2 Identification/improvement of the right biocatalyst

Enzymes that display the required features for the application intended may be identified through proper screening. Improvement or modification of enzymes, aiming for optimization toward the specific goal aimed, may be achieved through protein engineering using directed evolution, semirational design, computer-assisted design (CAD), machine learning, and synthesis and/or adequate formulation, namely immobilization, alongside with other forms of solid-state and liquid enzyme formulations (Madhavan et al., 2017; Bernal et al., 2018; Manning et al., 2018; Chowdhury and Maranas, 2020; Mikl et al., 2020; Rodríguez-Núñez et al., 2021; Saito et al., 2021; Xiong et al., 2021; Wu et al., 2022). Some of the key features involved in protein engineering are summarized in Table 18.2. Despite their acknowledged biocatalytic functionalities, most native enzymes lack the catalytic efficiency and stability to cope with the harsh operational conditions typical of industrial environments and process demands (Bilal and Iqbal, 2020; Wiltschi et al., 2020). To tackle these issues, several enzymatic features have to be

TABLE 18.2 Key issues of the different methodologies for protein engineering (Chen et al., 2018; Steiner and Schwab, 2012; Singh et al., 2013; Saito et al., 2021; Xiong et al., 2021; Wu et al., 2022).

	Rational design	Directed evolution	Semirational design
Protein structure	Detailed knowledge needed	No information needed	Partial information required
Catalytic mechanisms	Detailed knowledge needed	No information needed	Partial information required
Assay system needed	Sensitive method required, no need for HTP method	Cheap, fast, and reliable HTP method	Sensitive method required, HTP method helpful Increasingly complex computational need
Synergistic effects of neighboring mutations	Common	Low probability	Intermediate, possible identification of effects missed in rational design

HTP, High throughput.

tuned to increase the successful application in industrial processes, such as catalytic power, enantioselectivity, specificity, stability, and tolerance toward substrate, product, and solvents used as well as to inhibitors that may be present, substrate scope, and high space-time yield (Singh et al., 2013; Bilal and Iqbal, 2020; Wiltsch et al., 2020; Madhavan et al., 2021). Protein engineering can be implemented through rational design, directed evolution (random design), or semirational design. In rational design, site-specific mutations are introduced to replace given residues in an enzyme molecule with carefully selected residues to achieve the intended goal with a minimal number of variants. This approach thus requires detailed knowledge of the structure of the active site and structural–functionality interactions. A recent example involved the construction of endoglucanase mutants from *Chaetomium thermophilum* with improved thermal stability and activity. Two single-point mutants obtained complied with both goals. In contrast, a double-point mutant exhibited higher stability at high temperatures but did not display the increased catalytic efficiency of the single mutant counterparts (Chen et al., 2018). Conversely, directed evolution is carried out through random modifications in protein structure due to either haphazard changes in single protein sequences (viz. error-prone polymerase chain reaction) or accidental recombination of a set of related sequences (viz. gene shuffling). In each round, the mutants are screened based on the intended trait and selected for further runs. The process typically requires several rounds since each step often results in small, not all positive, changes and results in a large number of variants. Directed evolution allows the combination of useful mutations, often required to produce improved biocatalysts as single-point mutations may fail to deliver the proper outcome. Considering the large library of mutants obtained, high-throughput technologies are advised to handle the screening requirements in a timely manner. A recent example of this strategy involves the catalytic improvement of α-1,3-fucosyltransferase for the synthesis of fucosylated glycoconjugates toward the production of prebiotic supplements, which was achieved through a seven-point mutant (Tan et al., 2019). Semirational methods rely on protein structure, sequence, and function information, along with computational tools, for example, molecular dynamics using software such as Amber, CAVER, GROMACS, IPRO, MAESTRO, Osprey, Transcent, and Rosetta for protein engineering. A recent example involves the construction of mutants α-rhamnosidase mutants with improved catalytic activity and thermostability. Energy variety analysis enabled the prediction of the mutant with better thermostability, and molecular dynamics simulations highlighted the impact of the mutations on the interactions between the catalytic domain and the substrate (Li et al., 2021). Significant improvements in computational power and developments in protein design algorithms have enabled predicting the changes in free energy of the substrate and coenzyme binding in the engineered protein, typically determined through Monte-Carlo approaches, enabling the CAD of proteins. As a natural follow-up of such computing developments, *de novo* design of enzymes has emerged. Rather than modifying the binding pocket of the substrate, the active site is designed from the score to include the aminoacid residues that stabilize the transition state and lower the activation energy of the reaction to be catalyzed. Hence, this approach requires detailed insight into the molecular mechanisms of the reaction and typically involves in silico modeling to establish the proper positioning of residues. The new active site can then be inserted in a suitable scaffold, selected from the Protein Data Bank (Reetz et al., 2006; Lutz, 2010; Steiner and Schwab, 2012; Bornscheuer, 2013; Madhavan et al., 2017, 2021; Chowdhury and Maranas, 2020; Saito et al., 2021; Xiong et al., 2021; Wu et al., 2022). Even in this semirational method, not all the consensus mutations contribute positively to the envisaged goal, namely, stability, and may even impact other properties of the enzyme, namely, activity (Steiner and Schwab, 2012). The roots of the implementation of protein engineering methods in enzymes for industrial applications can be somehow traced back to food enzymes, namely, glucose isomerase and the modification of a specific residue to improve the thermostability of this particular enzyme (Sicard et al., 1990). Given the relevance of the role of glucose isomerase in the industrial production of sweeteners, it is only natural that many engineered mutants have been produced, aiming: to improve its activity, thermal stability, and weak acidity (Xu et al., 2014; Jin et al., 2021a); to increase the catalytic activity of a thermostable enzyme at the operational temperature for industrial glucose isomerization and in a mildly acidic environment (Sriprapundh et al., 2003); and to decrease the requirements/inhibition regarding metal ions (Hlima et al., 2012). Such variants were obtained using either site-directed mutagenesis (Hlima et al., 2012; Xu et al., 2014; Jin et al., 2021a, b) or directed evolution (Sriprapundh et al., 2003). Other recent examples of protein engineering to improve the performance of isomerases include the semirational design based on molecular dynamics simulations upon point mutations of a cellobiose-2-epimerase to improve its thermostability and isomerization activity (Chen et al., 2020); the semirational design of a triple point mutant sucrose isomerase with increased specificity for the synthesis of isomaltulose from sucrose (Pilak et al., 2020); the rational design of a mutant L-arabinose isomerase through single-point mutation to increase galactose isomerase activity in mesophilic environments (Jayaraman et al., 2021) The different approaches for protein engineering have been seminal in gaining insight on the design of more efficient glucansucrases, which can be advantageously used for the tailor-made synthesis of oligosaccharides with a prebiotic role. The protein engineering techniques allowed for identifying the relevant residues for catalysis and substrate specificity and provided relevant information for understanding the sequence–structure–function relationships of the enzymes. Moreover, it also paved the way for a more rational design of improved glucansucrases (Daudé et al., 2014). It broadened the range of glucan and oligosaccharide amenable to catalysis by engineering linkage specificity of glucansucrases (Molina et al., 2021). Notwithstanding, site-directed mutagenesis (double-point mutation) has been used to improve the thermal stability of a dextransucrase from *Leuconostoc mesenteroides* with no impact in specificity (Li et al., 2018a). Amylases have also been focused on obtaining engineered variants displaying modification in given properties, namely, substrate specificity and cleavage pattern, thermal and pH stability, pH/activity profile, pH/stability profile, and metal ion dependency (Andersen et al., 2013). Using a semirational method, where functionally correlated variation sites of proteins are used as hotspot sites to construct focused mutant libraries, allowed the production of α-amylase mutants with improved thermal stability by 8°C as compared to the native enzymes. Such a goal proved unfeasible with rational single and double-point mutations, requiring a relatively small library (Wang et al., 2012). More recently, site-directed mutation of an α-amylase enhanced the optimum temperature from 75°C to 80°C and the half-life at 90°C roughly doubled as compared to the wild type (Zhang et al., 2021a). Protein engineering has also been used to render more cost-effective the production of food enzymes. Thus a hybrid β-galactosidase was produced that combined the intracellular β-galactosidase of *Kluyveromyces lactis*, of particular interest for lactose hydrolysis and galactooligosaccharide production (Panesar et al., 2010b), with its extracellular homologue from *Aspergillus niger*, including a heterologous signal peptide for secretion at the *N*-terminus of the recombinant protein. The resulting variant displayed improved secretion to the fermentation medium, easing downstream processing and making industrial production more competitive. Moreover, the hybrid β-galactosidase displayed an increase in the optimal temperature and enhanced thermal stability, an affinity for natural (lactose) and synthetic (ONPG, *o*-nitrophenyl-β-D-galactopyranoside) substrates, as well as a shift in the optimum pH from 7.0 to 6.5, when compared with the native enzyme from *K. lactis*, making the variant a more effective biocatalyst for processing lactose and whey (Rodríguez et al., 2006). To overcome the sensitiveness of β-galactosidase to galactose, one of the end products that hampers lactose hydrolysis, Liu and coworkers (2021a) constructed a mutant by semirational design displaying a semirational design decreased affinity toward galactose. Molecular docking identified galactose binding residues, and the replacement of asparagine at position 148 with aspartic acid led to a mutant with increased hydrolytic efficiency compared to the wild type.

18.1.3 Enzyme sources and safety issues

Given the particular nature of the applications of enzymes in food industry, with clear implications for public health, these industrially produced enzymes are assessed for safety by regulatory agencies, although the defined safety criteria are far from universal (Olempska-Beer et al., 2006; Magnuson et al., 2013; Farias et al., 2014; Singh et al., 2019; Srivastava, 2019; Deckers et al., 2020). The need for such monitoring has become more noticeable since the late 1980s, with the introduction of protein and genetic engineering techniques and recombinant DNA technology into enzyme production, additionally coupled with the intensive screening of enzymes, particularly from extremophiles (Akanbi et al., 2020; Chowdhury and Maranas, 2020; Deckers et al., 2020; Fasim et al., 2021; Saito et al., 2021; Xiong et al., 2021; Wu et al., 2022). Particular care has thus been given to the selection of enzyme sources. Hence, within the vast array of microorganisms used to produce enzymes for industrial applications, those labeled as GRAS (Generally Regarded as Safe) are particularly favored within the scope of food industry (Adrio and Demain, 2010; Liu et al., 2013; Srivastava, 2019; Deckers et al., 2020). GRAS labeling is either related to a long track record of safe use or compliance with the outcome of a set of scientific procedures based on FDA (Food and Drug Administration) regulations. Still, with the relatively recent onset of genetic manipulation of these microbial strains, long-term effects on the health and environment of the engineered strains remain yet to be established. Regulatory agencies such as EFSA (European Food Safety Agency) are currently developing efforts to thoroughly evaluate the safety of enzymes used in food production and processing. Strains that secrete enzymes from cells are again favored as downstream processing complexity, and costs are lower than intracellular produced enzymes. The capacity to achieve a reasonable extracellular protein concentration, namely, over 50 g/L, is examined. Microbial enzyme producers on an industrial scale that comply with these demands are typically a few strains, such as *A. oryzae*, *A. niger*, *Bacillus subtilis*, or *B. licheniformis*, that have furthermore proven adequate hosts for the expression of homologous and heterologous enzymes (Sarrouh et al., 2012; Liu et al., 2013; Adrio and Demain, 2014; Fasim et al., 2021).

The present chapter offers an overview of the microbial enzymes currently finding potential use in food industry.

18.2 Microbial enzymes in food industry

18.2.1 Production of enzymes for food processing

Recombinant DNA technology and genetic engineering have enabled: (1) the production by industrial microorganisms of enzymes obtained originally from pathogenic microorganisms or producers of toxins, challenging to grow or even considered unculturable. An increasing number of these enzymes are obtained from extremophiles. Since they are obtained from microorganisms that thrive in harsh environments, extremozymes are expected to display high thermostability and endure pH extremes denaturing agents, for example, chaotropic agents or organic solvents. Hence, extremozymes can be of particular interest in processes where high temperatures are required, for example, starch hydrolysis or real-time prevention of acrylamide formation during cooking of starchy foods; removal of lactose from milk at low temperatures is envisaged not to tamper with organoleptic features could be addressed by psychrophilic enzymes; preparation of salty foods is performed, for example, soy sauce or sauerkraut, which could be improved using halophilic enzymes. However, the use of extremozymes in food production/processing is still far from achieving its full potential (Akanbi et al., 2020); (2) enhanced enzyme productivity using adequate promoters, signal sequences, and multiple gene copies (Adrio and Demain, 2010; Dalmaso et al., 2015; Liu et al., 2013; Neifar et al., 2015). Systems biology has now been gaining relevance as a most valuable tool for the consistent integration of multiome data. As an outcome, detailed insights into the metabolism of microbial industrial workhorses, namely, *A. niger* and *B. subtilis*, enable a more rational and cost-efficient approach for massive enzyme production (Zhu et al., 2012; Vongsangnak and Nielsen, 2013; Brandl and Andersen, 2015). Protein engineering has largely contributed to the design and production of enzymes with improved activity, specificity, or stability (Singh et al., 2013; Damborsky and Brezovsky, 2014;

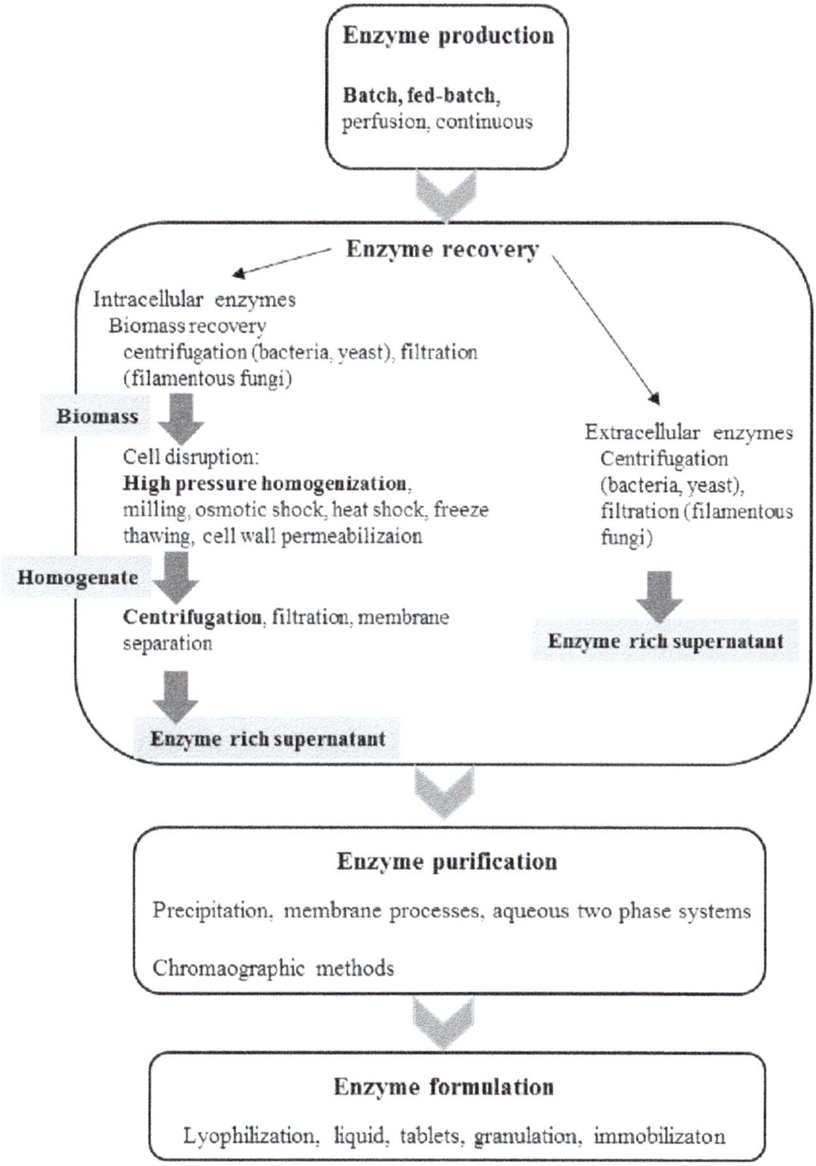

FIGURE 18.1 A simplified flow-sheet of enzyme production.

Zhang et al., 2018; Li et al., 2022). The overall production process for enzyme production is depicted in Fig. 18.1, involving enzyme synthesis through fermentation, recovery of the enzyme from the fermentation medium, purification of the enzyme, to remove unwanted contaminants, and formulation according to the intended use (Panesar et al., 2010a; Ramos et al., 2013). Production of enzymes has been mostly performed through submerged fermentation, although solid-state fermentation (SSF) has been gaining relevance in recent years (Singhania et al., 2010; Thomas et al., 2013; Niyonzima, 2019; Singh and Kumar, 2019; Tarafdar et al., 2021). Submerged fermentation takes place in vessels of up to 200 m^3, with a wide range of substrates, from defined ingredients, such as dextrose, ammonia, and urea to undefined ingredients, often by-products from the food industry such as molasses, whey, soybean, fish meal and yeast extract, and minerals, such as carbonates and phosphates (van den Berg et al., 2010; Ramos et al., 2013). Fermentation is implemented under monitoring and controlling variables such as temperature, pH, and dissolved oxygen tension. The batch mode of operation is widely used since it is a well-established methodology. Yet, productivity is limited, often because of substrate inhibition, a drawback overcome by the fed-batch approach. In the later method, after a given period of batch cultivation, nutrients are fed to the bioreactor according to a given pattern and up to a final volume. This mode of cultivation is considered particularly suitable for producing enzymes, as the operation is relatively straightforward and metabolic responses of the producing cells can be controlled and has thus been gaining relevance (Illanes, 2008; Chisti, 2010; Ramos et al., 2013). Perfusion culture, where a cell-free product stream is continuously harvested from the bioreactor, while an equal volume of medium is added to the bioreactor, is also an alternative since it reduces or prevents the accumulation of inhibitory metabolites within the bioreactor. However, it is somehow cumbersome and expensive for large-scale production of low-value products as industrial enzymes (Singhania et al., 2010; Salehmin et al., 2013; Spohner et al., 2015). Continuous production is hardly considered a realistic alternative. Despite allowing for steady-state operation, with concomitantly near-balanced growth and slight fluctuation in operational parameters, it is most susceptible to microbial contamination (Chisti, 2010).

SSF is carried out in the (near)-absence of free water. Yet, care has to be taken to ensure that the substrate possesses the moisture required to support the growth and metabolic activity of the microorganism used for enzyme production. In SSF, microorganisms grow in a solid matrix that is either the source of nutrients or an inert support material impregnated with a growth solution. Some relevant features of SSF fermentation are summarized in Table 18.3. Given the particular features of SSF processes, fungal cultures are deemed the most suitable for growth in such an environment. Accordingly, several relevant enzymes in food processing have been produced by SSF (Table 18.4) from bench to large-scale (Pandey et al., 2008; Thomas et al., 2013; Balakrishnan et al., 2021). After fermentation, the enzyme must be recovered. In the case of SSF, enzymes are typically excreted. The extraction is carried out with either a suitable buffer or water,

TABLE 18.3 Some significant aspects of solid-state fermentation (Barrios-González, 2012; Ramos et al., 2013; Niyonzima, 2019; Singh and Kumar, 2019).

Concentrated fermentation media, resulting in small reactor volume and low capital investment
Possible use of agro-industrial wastes and low-cost material as substrates, namely, defatted soybean cake, rice bran, or wheat bran
Low risk of contamination due to low moisture content and substrate complexity
Simple technology and low production of effluents
High product yield and eased downstream processing
Limitations in heat and mass transfer

TABLE 18.4 Some relevant enzymes in the food industry produced by solid-state fermentation. At least, α-amylase, glucoamylase, lipase, pectinase, phytase, and protease have been produced in large scale (Thomas et al., 2013; Niyonzima, 2019).

Enzyme	Microbial source	References
α-Amylase	*Aspergillus oryzae, Bacillus* sp., *Bacillus subtilis,*	Balakrishnan et al. (2021); Almanaa et al. (2020); Dey and Banerjee (2012); Derakhti et al. (2012); Saxena and Singh (2011); Rajagopalan and Krishnan (2010)
Glucoamylase	*Aspergillus awamori, Aspergillus sp., A. oryzae, Aspergillus niger; Penicillium javanicum, Rhizopus K1*	El-Gendy and Alzahrani (2020); Meng et al. (2019); Tang et al. (2018); Kiran et al. (2014); El-Gendy (2012); Parbat and Singhal (2011); Slivinski et al. (2011)
Inulinase	*Aspergillus brasiliensis, Aspergillus flavus var. Flavus, Aspergillus tamarii, B. subtilis, Kluyveromyces marxianus, Penicillium oxalicum, Saccharomyces* sp., *Penicillium amphipolaria*	Das et al. (2020); Garuba and Onilude (2020); Das et al. (2019); Singh et al. (2018); Dilipkumar et al. (2013); Onilude et al. (2012); Mazutti et al. (2010)

(*Continued*)

TABLE 18.4 (Continued)

Enzyme	Microbial source	References
Invertase	*A. brasiliensis, A. niger, A. tamarii, B. subtilis, Saccharomyces cerevisiae*	Guerrero-Urrutia et al. (2021); de Oliveira et al. (2020); Lincoln and More (2018); Al-Hagar et al. (2015); AL-Sa'ady (2014); Kumar and Kesavapillai (2012)
Lipase	*Aspergillus ibericus, A. niger, Rhizopus homothallicus, Yarrowia lipolytica*	de Souza et al. (2019); Nema et al. (2019); Oliveira et al. (2017); Farias et al. (2014); Velasco-Lozano et al. (2012)
Naringinase	*A. niger,* marine fungi	de Oliveira et al. (2022); Shehata and El Aty (2014); Shanmugaprakash et al. (2011)
Pectinase	*A. niger, B. subtilis*	Ahmed et al. (2021); Kaur and Gupta (2017); Alcântara and da Silva (2014); Heerd et al. (2012); Ruiz et al. (2012)
Pectin lyase	*B. subtilis*	Kaur and Gupta (2017)
Phytase	*A. niger, Acremonium zeae B, B. subtilis, Pichia membranifaciens* S3, *Thermomyces lanuginosus*	Singh et al. (2021); Soman et al. (2020); Pires et al. (2019); Berikten and Kivanc (2014); Rodríguez-Fernández et al. (2013)
Protease	*Aspergillus* sp., *Penicillium citrinum, Rhizopus stolonifer*	Usman et al. (2021); Xiao et al. (2015); Kranthi et al. (2012)

followed by solid–liquid separation (viz. centrifugation) to remove the mycelium. When enzymes are produced by submerged fermentation, recovery depends on the enzyme being of extra- or intracellular origin. In the former case, again, solid–liquid separation is required. Cell disruption is required, followed by solid–liquid separation, in the latter case, to remove cell debris. In all cases, the enzyme-rich supernatant proceeds for further steps of purification. A concentration step, usually by ultrafiltration, can be used before further processing or if the enzyme proceeds to the formulation. Besides the target enzyme, the supernatant contains residual soluble and colloidal components from the fermentation medium, including eventually nonproduct enzymes that are produced by the host organism. Depending on the requirements for the application intended, several purification steps may be needed. Yet it should be taken into that these add to production costs and decrease efficiency. Precipitation (viz. with ammonium sulfate) followed by ultrafiltration is common. For a precise application, where a high purity level is needed, chromatographic processes are used (Dodge, 2010; Ramos et al., 2013). Upon recovery and purification, an enzyme concentrate is obtained, which requires a final step of processing and formulation to be delivered in a suitable form for food processing or production and provide convenient shelf life.

18.2.2 Formulation of enzymes for use in food processing

Details on the formulation of enzyme preparations are relatively scarce, yet this is a critical step in producing industrial enzymes, as it often confers the producer a competitive edge. In the case of food enzymes, they can be delivered in either liquid or solid form. The former allows for a simpler dosage, while the latter usually extends the shelf life (Illanes, 2008; Dodge, 2010). During formulation, several ancillary substances are added to the enzyme concentrate as stabilizers and preservatives (Table 18.5). Stabilizers added to

TABLE 18.5 Excipients used in enzyme formulations.

Excipient	Action	Examples and comments
Stabilizer	Maintain protein structure and prevent denaturation	Carbohydrates: dextrose, sucrose, trehalose, xylan Polyols and sugar alcohols: glycerol, mannitol, polyethylene glycols, sorbitol, xylitol
Preservative	Control microbial contamination	Potassium sorbate, sodium benzoate; alternatively, natural inhibitors of microbial growth, namely, plant extracts or peptides are looked after. When at high levels, stabilizers help to control activity of water and hence microbial contamination.
Diluents, carriers, and fillers	Make the enzyme available at proper rate	Starch, anhydrous and spray-dried lactose, gum arabic, maltodextrins, microcrystalline cellulose, diatomaceous earth, gum arabic, alginates, carrageenans, dairy and soy proteins, emulsifiers, and waxes
Binders	Create liquid bridges which form agglomerates from the powder	Gelatin, starch, polyvinylpyrrolidone, high concentrations of sugar

Source: Based on information from Segura et al. (2007); Rajakari et al. (2013); Lohscheidt et al. (2009).

a liquid formulation extend support to maintain a soluble product by preventing aggregation. Once stabilizers and preservatives are added, the liquid formulation is filtered to remove undissolved solids (Iyer and Ananthanarayan, 2008; Dodge, 2010; El-Sherbiny and El-Chaghaby, 2011). Until the 1960s solid formulation consisted almost exclusively of powdered enzyme particles obtained by spray-drying. The enzyme was mixed with stabilizers and diluents, and the solution or suspension was atomized into small droplets and exposed to hot air. This resulted in small particles, often with sizes under 10 μm. Complexity in handling and respiratory allergies led to the quest for dust-free alternatives. Still, spray-dried particles can be included in more structured particles. Currently, a solid formulation can involve granulates, tablets, or immobilization (Aunstrup et al., 1979; Tripathi et al., 2020).

18.2.3 Granulation of enzymes

The granulation of enzymes results from enzyme particles ranging from 425 to 850 μm. These granules can be engineered to encapsulate the enzyme in a uniform matrix or embedded in a core and shell matrix where multiple layers of stabilizers and protective agents are deposited. Several methods can be used to produce enzyme granulates. Still, only a few are suitable for large-scale processes, namely, spray-chilling (prilling), marumerization/spheronization, high shear granulation, and fluid bed.

Prilling involves the incorporation of the enzyme powder in a wax. The mixture is atomized through either nozzles or a rotating disk placed on the top of a tower. The falling droplets cool, solidify, and harden to yield round particles. Although simple and straightforward, the method is limited by the wax's melting temperature and by the risk of agglomeration of granules when exposed to moderate to high temperatures.

In marumerization/spheronization, either enzyme concentrate or powdered enzyme is mixed with suitable excipients to create a dough-like mass extruded through a perforated plate. The resulting cylindrical particles can either be dried in a fluidized bed or converted into small spherical particles using an apparatus known as marumerizer, consisting of a spheronization plate that spins at 500 to 2000 rpm. Finally, the spherical particles are dried in a fluidized bed. The enzyme powder is mixed with a stabilizer and diluent with a plough shear mixer and a high-speed mixer in high shear granulation. Simultaneously, a binder solution is added. As a result of shear forces, small particles are generated that can be coated with wax and dried in a fluidized bed (Chotani et al., 2014). The fluidized bed approach is considered the most adequate to produce low dust granules with high shear.

Moreover, a fluidized bed is the most flexible technique and yields more uniform granules and smoother coatings. In this method, enzyme powder (or concentrate that undergoes spray-drying) is mixed with suitable excipients in a fluidized bed apparatus. If required, spray-coating can also be carried out, allowing sequential layers of the required composition deposition. Examples of food and enzymes formulated as granulates are α-amylases (Duramyl, Termamyl), amyloglucosidase (Amigase), glucose isomerase (Sweetzyme T, Gensweet IGI.), all involved in the production of high-fructose syrups from starch; lipases (Lipozyme TL IM), within the scope of the manufacture of healthy oils; and phytases (Lohscheidt et al., 2009; DiCosimo et al., 2013; Basso and Serban, 2019).

18.2.4 Tablets

Tablets can enable the incorporation of enzymes formulated as dry powders or granulates. These are compressed into the intended format and coated with a suitable component. Tablets can include multiple enzyme activities (Damgaard, 2016).

18.2.5 Immobilization

Immobilization of enzymes consists of the containment of a biomolecule in a given region of space. This containment can be achieved either through physical or chemical methods and can be carrier-less in the latter case (Fig. 18.2). Carriers may be available in particles, fibers, films, and monoliths. Entrapment involves containment within a polymeric network, namely, polyvinyl alcohol, calcium alginate, chitosan, gelatin, and polyacrylamide. Simple, cheap, and often using mild, biocompatible conditions are typically

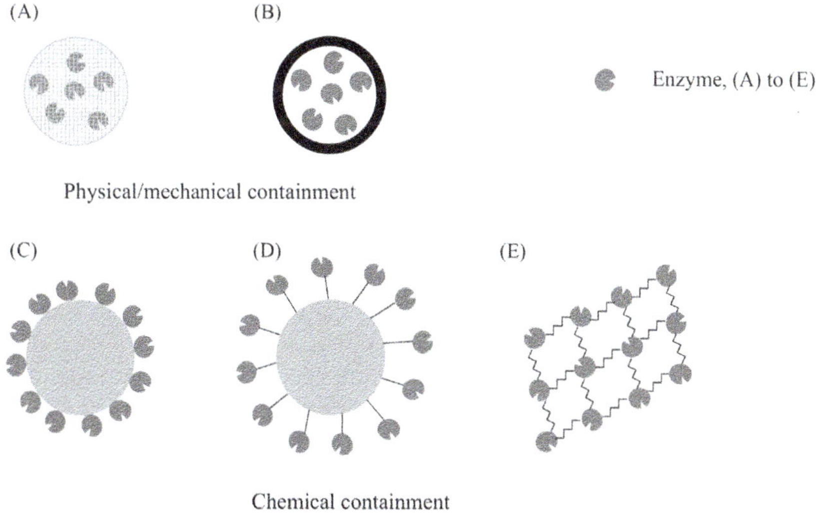

FIGURE 18.2 A pictorial representation of enzyme (E) immobilization methods: (A) entrapment; (B) microencapsulation; (C) adsorption; (D) covalent binding; (E) cross-linking (carrier-less).

associated with diffusion limitations. In microencapsulation, the enzyme is stored within a semipermeable membrane. Mass transfer limitations are mitigated compared to entrapment, but the method is more complex. Adsorption involves interactions between the enzyme and the carrier, namely, ion-exchange resins, activated charcoal, alumina, or celite, through weak forces, namely, hydrogen/hydrophobic/ionic interactions and van der Waals forces. While simple and non-aggressive, adsorption is prone to massive enzyme leakage upon shifts of operational conditions. Covalent binding involves the formation of stable, covalent bonds between the carrier (viz. controlled pore glass, silica, and wood chips) and enzyme residues, but for those essential for catalytic activity, a requisite that may prove not easy to comply with. Currently, preactivated carriers are available that ease the immobilization process. Recently, materials such as graphene/graphene oxide or metal–organic frameworks (MOF) have emerged as promising carriers for enzyme immobilization by both adsorption and covalent binding. The former group features high surface area, high mechanical and thermal endurance, and plenty of surface functional groups that ease dispersion in water. Moreover, it exhibits excellent optical and electrical transmittance, making it suitable for nanoelectronics, for example, biosensors, to assess food quality. The latter consists of microporous crystalline hybrid materials assembled from metal ions and organic linkers. They feature high specific surface areas and porosity, multifunctionality, and stability and have been shown to impart enzymes with a uniform microenvironment. Hence, MOF-immobilized enzymes have often exhibited the most appealing catalytic features as compared to their free counterparts. However, the dissemination of such carriers is currently hampered by cost and availability (Wahab et al., 2020; Sheldon et al., 2021). Cross-linking involves the reaction of enzyme molecules with a bifunctional reagent to yield an insoluble enzyme network, cross-linked enzyme aggregates (CLEAs) requiring no solid carrier, thus potentially triggering biocatalyst productivity ($kg_{product}/kg_{enzyme}$). Typically, glutaraldehyde is used to react with NH_2 groups. Incorporation of magnetic (nano)particles in CLEAs eases the recovery of the biocatalysts, hence enhancing its relevance for both large-scale application and food analysis (Torres-Salas et al., 2011) DiCosimo et al., 2013; Sheldon, 2019; Zhang et al., 2021b). Immobilization allows extended shelf life for the continuous/repeated use of the enzyme and also eases the separation of the biocatalyst from the reaction media, thereby simplifying downstream processing. These features contribute to making processes more cost-effective. However, there is a downside, as immobilization itself has an added cost. Some activities may be lost during the process, namely, when chemical methods are involved, mass transfer limitations may occur, and enzyme leakage may occur. Moreover, the carrier must be discarded once the catalytic activity is lost. It is somehow noteworthy that the first report of immobilized enzymes at a laboratory scale was on a food-related process, as it involved the adsorption of invertase onto charcoal to hydrolyze sucrose to invert sugar syrup. The same reaction is possibly the first commercial application of immobilized enzymes, using bone char as a carrier. Eventually, further developments led to the commercial processes based on invertase adsorbed onto ion-exchange resins or covalently bound to macroporous methacrylate beads, the later allowing for a productivity of 6000 tons (dry weight) of invert sugar syrup per kg of biocatalyst from sugar beet or cane sugar, with a conversion of 90% (Uhlig and Linsmaier-Bednar, 1998; Swaisgood, 2003). Macroporous methacrylate beads were also used to immobilize amyloglucosidase to hydrolyze dextrins to glucose in starch processing. Conversion (94%) was slightly under that achieved with free enzyme (95%–96%) due to diffusion-related hindrances. This minute difference, together with the cost of the enzyme, prevented the immobilized approach from going beyond the pilot-scale (Uhlig and Linsmaier-Bednar, 1998). Although commercial production of invert sugar syrups relies on chemocatalysis, a wide array of immobilization methods has been assayed using this model system, some up to plant scale with invertase adsorbed onto chitosan beads (Serna-Saldivar and Rito-Palomares, 2008; Kotwal and Shankar, 2009). Currently, and following a trend started in the late 1960s/early 1970s, the most significant commercial-scale application of immobilized enzymes within the scope of food industry is by far the isomerization of glucose to fructose for the production of HFS (DiCosimo et al., 2013; Basso and Serban, 2019; Sneha et al., 2019). Some other relevant examples of the use of immobilized enzymes in commercial-scale processes and prospective developments toward such scale in the short term, including one-pot cascade reactions, are given in Table 18.6.

18.2.6 Applications in food industries

18.2.6.1 Starch processing and sweetener production

18.2.6.1.1 Starch industry

Starch is the major form of energy storage in plants. It is essentially a mixture of two polysaccharides, amylose and amylopectin, composed of glucose units. The latter polysaccharide is predominant, accounting for around 65%–75% of the total, depending on the source. The most common is corn, but barley, cassava, rice, sorghum, tapioca, and wheat are also used. Several sugar syrups are produced from starch, namely, dextrins, dextrose, maltose, high fructose and fructose, and hydrogenated derivatives, namely, hydrogenated starch hydrolysates and sorbitol. The production of these goods involves enzymatic hydrolysis, promoted by amylases, glucoamylases and pullulanases. When fructose syrups are aimed at, glucose isomerase is also required. Currently, a starch slurry (around 35% dry matter) is prepared, and a thermostable α-amylase is added (around 0.5 kg/ton) as the pH is adjusted to 5.0–6.5, in the presence of Ca^{2+}. The slurry is pumped through a jet

TABLE 18.6 **Some representative examples on the use of immobilized enzymes in the food industry: current and emerging large-scale processes.**

Application	Comments
Glucose isomerization	Granules of glutaraldehyde cross-linked microbial cells homogenates with glucose isomerase activity (current). Adsorption, entrapment, cross-linking, and covalent binding of the enzymes have been tested, some went into production scale but are currently unavailable (DiCosimo et al., 2013; Basso and Serban, 2019).
Lactose hydrolysis/production of whey hydrolysates/GOS (galacto oligosaccharides) synthesis/lactose-fructose syrup synthesis	β-Galactosidase entrapped within the microcavities of fibers made from cellulose triacetate or adsorbed and cross-linked to food grade resins for lactose hydrolysis/whey hydrolysates (current); covalent binding to porous silica (discontinued). Several other methods involving adsorption, encapsulation, entrapment, cross-linking and covalent binding of the enzymes have been tested for both lactose hydrolysis and GOS synthesis and suggest potential development to plant scale (Swaisgood, 2003; Panesar et al., 2010b; Alnadari et al., 2021; Schulz and Rizvi, 2021; Pottratz et al., 2022). Production of lactose-fructose syrup, a mild sweetener, in a one-pot process using combi-CLEAs comprising β-galactosidase and glucose isomerase (Araya et al., 2019).
Esterifications within the scope of processing fats and oils	Lipase adsorption or covalent binding to Eupergit like resins. Processes for the interesterification of fats and oils are currently at production scale, namely, Betapol, a vegetable fat blend, or Crokvitolt, a set of fats for margarine and baked goods (current). Developments involve the production of human milk fat analogues and substitutes, processing of anhydrous milk fat and of caprine milk, processing of canola oil; or production of semisolid fats (Swaisgood, 2003; Forde and OFagain, 2008; DiCosimo et al., 2013; Basso and Serban, 2019).

Artificial sweeteners (Aspartame) and low-calorie sweeteners	Thermolysin adsorbed onto polymeric resins flowed by cross-linking to produce Aspartame (DiCosimo et al., 2013). D-Allose synthesis: from D-allulose using commercially available immobilized glucose isomerase, Sweetzyme IT (Choi et al., 2021); from D-fructose in one-pot process combining L-rhamnose isomerase and D-psicose 3-epimerase immobilized in anion-exchange resin (ionic binding) and amino resin (covalent binding), respectively (Li et al., 2020). D-Allulose (D-psicose) synthesis with: D-tagatose 3-epimerase immobilized by ionic binding to Chitopearl beads (Takeshita et al., 2000); D-psicose 3-epimerase: adsorbed in graphene dioxide (Dedania et al., 2017); entrapped in nanoflowers composed of the enzyme as organic component and cobalt phosphate as the inorganic component (Zheng et al., 2018); covalently bound to titanium dioxide nanoparticles (Dedania et al., 2020); immobilized in Amino-Epoxide resins by combined ionic/covalent binding and glutaraldehyde cross-linking (Bu et al., 2021); two-step process from glucose with sequential application of glucose isomerase and D-psicose 3-epimerase, each immobilized on *Saccharomyces cerevisiae* spores (Li et al., 2015). Tagatose: synthesis with L-arabinose isomerase immobilized: in calcium alginate beads packed in a fixed-bed reactor (Rai et al., 2021); in cupper-chelate epoxy supports trough covalent binding (Bortone and Fidaleo, 2020); in CLEAs and magnetic CLEAs (de Sousa et al., 2020); in hybrid nanoflowers composed of the enzyme as organic component and manganese phosphate as inorganic component (Rai et al., 2018).
Amino acids (L-aspartate)	Immobilization of *Escherichia coli* aspartase expressed in psychrophilic *Shewanella livingstonensis* Ac10 immobilized in alginate beads (Tajima et al., 2015); microbial (resting) whole cells with aspartase activity either entrapped in polyacrylamide or k-carrageenan cross-linked with glutaraldehyde and hexamethylenediamine or adsorbed to phenolformaldehyde resin; aspartase adsorbed onto weakly basic anion-exchange resin (current) (Wu et al., 2012).

(*Continued*)

TABLE 18.6 (Continued)

Application	Comments
Debittering of fruit juices as alternative to nonspecific, neutral or ion-exchange resins; enhancing aroma in wine	Immobilization of naringinase in a wide array of supports, among them alginate, k-carrageenan, celite, cellulose acetate nanofibers, chitosan, CLEA, magnetic nanoparticles, mesoporous materials, polyvinyl alcohol, zeolites perspective implementation to commercial application (DiCosimo et al., 2013; Nunes et al., 2014; Huang et al., 2017; Carceller et al., 2020; Zheng et al., 2021).
Processing of fruit and vegetable juices and wine	Immobilization of pectinases and glycosidases as CLEAs and in a wide array of supports, among them alginate, celite, chitosan, Eudragit, ion-exchange resins, gelatin, magnetic particles, nylon, silica, ultrafiltration membranes. In particular, the advantages of enzyme immobilization during ultrafiltration of food juices have been highlighted at pilot scale. Perspective implementation to commercial application (DiCosimo et al., 2013; Pagán, 2014; Sojitra et al., 2016; Martín et al., 2019; Ben-Othman and Rinken, 2021; Tavernini et al., 2021).
Synthesis of lactulose	β-galactosidase immobilized in membrane reactors, activated carbon, activated silica gels, CLEA; β-glycosidase immobilized in Amberlite and Eupergit (bench scale) (Wang et al., 2013; Guerrero et al., 2015); coimmobilization of β-glucosidase and glucose isomerase in CLEAs for one-pot cascade reaction (Wilson et al., 2022).

cooker, where steam is injected to increase the temperature to 105oC. Just 5 min of holding time allow for starch's gelatinization (swelling of the granules and breaking up hydrogen bonds), and the slurry is then cooled to 95°C. The slurry is maintained at this temperature for about 2 h, the time required for liquefaction to proceed until the required DE value, usually within 8 to 12 (DE, dextrose equivalent, a measure of the total reducing power of the sugars present, as related to dextrose standard, on a dry mass basis). Further hydrolysis of the liquefied starch, rich in oligosaccharides, proceeds through the action of either glucoamylase and pullulanase or fungal α-amylase/β-amylase. In the former case, pH is adjusted to 4.0–5.0 and incubation proceeds at about 60°C for 72 h, until a DE of 97 or higher is achieved to produce a dextrose syrup. In the latter case, maltose/high maltose syrups are aimed, requiring pH adjustment to 5.5 and incubation periods of 48 h. Dextrose can then be isomerized to fructose. Dextrose syrup is filtered, processed through activated carbon and ion-exchange resins, and concentrated to about 50% dry solids. Magnesium is then added, pH adjusted to about pH 7.7 and the liquor is fed to packed-bed reactors filled with immobilized glucose isomerase and maintained at 60°C. At this temperature, the isomerization of glucose to fructose is governed by the thermodynamic equilibrium between both sugars. Under the conditions used, a mixture of 42% (w/w) fructose, 50% (w/w) glucose, and 8% (w/w) of other saccharides is obtained, which can be referred to as HFS42. Further enrichment in fructose to 55% is required as a sweetener in soft drinks. Thus HFS42 syrup undergoes column chromatography to obtain syrup with 55% fructose, HFS55. This syrup results from the blend of the HFS42 syrup with a fructose-rich syrup (90%) resulting from the chromatographic step.

Some improvements in this process are expected to be implemented shortly due to dedicated efforts within the scope of protein engineering. These led to amylases made available commercially with tailor-improved features to fit in advantageously in the production processes. One such enzyme, LpHera is an α-amylase with intermediate acid stability, allowing for starch liquefaction to be carried out within pH 4.5 to 4.8, unlike the usual pH of 5.0 to 6.5 required due to the range of operation of the current commercial α-amylases. Typically, pH has to be lowered downstream to 4 to 5, for the saccharification step to comply with the range of activity of glucoamylases (Eshra et al., 2014). The range of operation of this α-amylase results in lower demand for chemicals for pH adjustment and ion-exchange resins.

Moreover, the final yield in dextrose from starch processing is increased by 0.2%, on the whole, leading to a more cost-effective process. Within the scope of rendering well-established industrial processes more cost-effective, a microbial β-amylase, Secura, with enhanced thermal stability over plant-derived counterparts, has also been commercially available. A genetically engineered α-amylase from *Geobacillus* sp. displayed high acid resistance and high thermal stability, which was noticeably enhanced upon immobilization in epoxy resins. Thus the half-life at 90°C roughly doubled, as compared with the free form, reaching 205 min, and 90% residual activity remained after six runs of starch saccharification (Zhang et al., 2021a). Also, within the scope of making starch processing more cost-effective, glucoamylase variants were obtained by oligo-directed mutagenesis that displays high catalytic activity at pH 5.0 and about 65°C, moreover displaying high retention of activity within 60°C–75°C. Such features enable the combined use of the variant glucoamylases with α-amylases, allowing the liquefaction and saccharification of starch in a single step (Jing et al., 2014). To minimize unwanted by-product formation and browning reaction while still favoring proper equilibrium shift, isomerization of glucose to fructose should be performed in a slightly acidic environment and under 85°C. Thermostable native/engineered glucose isomerases have accordingly been looked after (Jin et al., 2021a). To enable continuous operation, enzymes that fit in this profile have been recently immobilized on Sepabeads EC-HA (Neifar et al., 2020) and diatomite in the form of whole cells (Jin et al., 2021b). Both formulations were thermostable and under continuous operation space-time yields of 1.08 and 3.84 kg/L.day, respectively, were reported. Still, the former was operated at pH 7.0 and operation was monitored only for 11 days, throughout which fructose yield remained fairly constant, whereas, in the latter, fructose yield decreased noticeably after 11 days. This may have been caused by enzyme leakage, as cracks were observed in the immobilized cells.

18.2.6.1.2 Other sweeteners and prebiotics from fructose and glucose

Invertase is used for the hydrolysis of sucrose into glucose and fructose. The resulting inverted sugar syrup is sweeter than sucrose. It is widely used in confectionery, bakery, and pastry, as it features enhanced moisture-preserving properties and is less prone to crystallization (Manoochehri et al., 2020).

HFSs can also be obtained in a single-step process from inulin, a polyfructan with a terminal glucose residue, used for energy storage by several plants, namely, agave, chicory, and Jerusalem artichoke. Complete inulin hydrolysis can be obtained through the action of *exo*-inulinase, or in combination with *endo*-inulinase that hydrolyze the inner linkages, randomly releasing fructooligosaccharides (FOS) with 1 to 9 units of fructose (Singh et al., 2020). These long-chain FOS are known to play a prebiotic role, as they stimulate the growth of intestinal bifidobacteria, with a concomitant positive, healthy effect in the intestines. Short-chain fructooligosaccharides (scFOS), with up to 4 fructose units and a single glucose unit, are obtained from sucrose by transfructosylation through the action of fructosyltransferase. Again, these scFOS have also a well-identified prebiotic role and as their longer chain counterparts are used as ingredients in functional food such as baking and dairy products, breakfast cereals, frozen desserts; infant formulae, fruit preparations; dietetic products; and sweeteners (Singh and Singh, 2010; Chi et al., 2011; Nobre et al., 2015; Davani-Davari et al., 2019). Oligosaccharides composed mostly of glucose units, glucooligosaccharides (GluOS), have also gained the interest of the consumers, again as an outcome of the prebiotic role. Hence, their production for incorporation in functional foods has also been focused. GluOS synthesis typically relies on sucrose as glycosyl donor and a suitable acceptor, such as lactose, maltose, or even sucrose. The relatively broad range of acceptors allows for a wide diversity of products. GluOS synthesis is catalyzed by glucansucrases (alternansucrases, dextransucrases, mutansucrases, reuternasucrases) that promote successive transfers of glycosyl units onto the oligosaccharide (Daudé et al., 2014; Bivolarski et al., 2018).

Also, alternatives to traditional sweeteners have been looked after within the scope of the growing public interest in functional foods and low caloric sweeteners. Stevioside is a promising candidate, yet this diterpene glycoside features an after-taste bitterness that hampers its application. Such drawback can be eliminated through transglycosylation with glucanotransferase, such as β-cyclodextrin glucanotransferase, pullulanase or β-galactosidase, and a suitable donor, namely, starch, pullulan, or lactose, so that carbohydrates are attached into proper positions of the stevioside molecule (Magomet et al., 2010; Czinkóczky and Németh, 2022).

Growing concerns about the impact of caloric sweeteners on public health have raised interest in low-calorie sweeteners (Shintani, 2019). Examples of these are rare sugars, such as D-psicose (D-allulose), which can be obtained through the isomerization of D-fructose using a suitable 3-epimerase, for example, allulose-3-epimerase (Yoshihara et al., 2017), psicose-3-epimerase (Bu et al., 2021), or tagatose-3-epimerase (Parildi et al., 2021). A different approach was presented by Li and coworkers, who synthesized D-psicose from inulin in a one-pot two-step cascade combining an exoinulinase and a D-allulose-3-epimerase (Li et al., 2020); D-mannose, which can be obtained from D-fructose using D-lyxose isomerase (Zhang et al., 2019). Another example of noncaloric sweetener is isomaltulose, an isomer of sucrose, obtained through sucrose isomerase and patented as Palatinose (Pilak et al., 2020; Liu et al., 2021c).

18.2.6.2 Dairy

Enzymatic processing of milk has a long tradition (Law, 2010). One of the most significant applications involves using rennet to prepare proteases with milk clotting activity, a specific form of coagulation essential in cheese production. Specifically, rennet hydrolyzes k-casein, releasing its terminal hydrophilic region, caseinomacropeptide (CMP). Hydrolysis proceeds, concomitantly with a decrease in milk viscosity, until the micelles aggregate and eventually network into a gel, at which point almost all CMPs have been released into solution (Dalgleish and Corredig, 2012). The gel thus formed undergoes a series of operations where whey is released, ultimately enhancing by 10-fold the concentration of casein, fat, and calcium phosphate, leading to a curd with the high dry matter. The enzymatic-driven coagulation process typically occurs at pH within 6.4–6.6 and temperatures within 30°C–35°C to have adequate control over curd firmness, a feature of interest considering downstream operations in cheese making (Harboe et al., 2010). Depending on the source, the composition of rennet varies. Traditionally, rennet was obtained from the stomach of young calves, where chymosin is vastly dominant over pepsin, in a ratio of 9:1. This was, until quite recently, the standard against which all formulations were matched with. Extracts from older animals have increased pepsin, so their ratio may be up to 1:1. Both enzymes are aspartic proteases, but chymosin is much more specific in action; hence formulations rich in pepsin have an overall higher proteolytic activity. Moreover, pepsins are more sensitive to pH than chymosins. Extracts from sheep, goats, and pigs also provide rennet, but these are far from ideal for clotting milk from cows. Extracts from the macerated and dried stomachs of suckling calves, lambs, or kid-caprine may be used as a source of rennet. This particular rennet is furthermore enriched with lipase, therefore adding a piquancy to the flavor of the cheese and is thus suited for the production of some particular cheeses. Throughout the years, the demand for rennet far exceeded the offer provided by young calves; hence, alternative sources were looked for. These include plants and microbial sources, which were shown to produce coagulants with a set of features that allow their use as rennet alternatives, namely, the ability to clot milk without excessive proteolysis, in a manner akin to chymosin; low proteolytic specificity toward β-casein to prevent bitter taste development in cheese; and a cost comparable or lower than that of traditional rennet (Mistry, 2012). Extracts from *Cynara cardunculus* (L.) cardoon plant have been traditionally used in artisan cheese making in Portugal. This pattern currently endures and allows the production of cheeses like Serra and Serpa. Suitable clotting activity is due to cyprosin and cardosin, two aspartic proteases. The former has been awarded GRAS status (Mercer, 2014, Murlidhar et al., 2017). Additionally, extracts from *Withania coagulans* have also been used in India to produce soft cheeses. Among the wide variety of proteolytic enzymes from microbial sources, only a very few aspartic proteases, particularly those from *Rhizomucor miehei* and *Cryphonectria parasitica* have proved suitable for cheese making, particularly, the former, with a broader range of applications as it presents relatively high milk clotting to proteolytic activity and is heat-labile. The latter features, high overall proteolytic activity, low pH dependency, and heat thermostability make it suitable for the manufacture of cheeses cooked at high temperatures. Otherwise, the high proteolytic activity may result in off-flavors and bitter taste, resulting in high levels of hydrophobic peptides from the C-terminal region of β-casein, if long maturation periods are considered. Still, these fungal proteases are relatively cheap and can comply with the requirements of some relevant market niches, namely, Kosher, Halal, GMO-free products, and vegetarian. Therefore chemical and genetic engineering methods have been implemented to decrease thermal stability and increase the clotting to proteolysis ratio to capitalize on some of these advantages. Some of these successfully engineered proteases are currently commercially available, with particular highlight to variants termed XL, which are more heat sensitive and with lower proteolytic activity than the native enzyme, and to purified forms of variant XL, obtained by chromatography, free of lipase and amylase activity (Harboe et al., 2010; Feijoo-Siota et al., 2014). Still, expression of chymosin in genetically modified microorganisms, GMM, namely, *K. lactis* and *A. niger* and concomitant production in large-scale fermentation, so-called fermentation-produced chymosin, is currently the most widely disseminated alternative for calf rennet. These heterologous chymosine formulations commercially available are from bovine and camelus sources, the latter being considered to display the highest clotting to proteolytic ratio (Harboe et al., 2010; Feijoo-Siota et al., 2014). Details on recent developments involving application research efforts and foreseen developments involving rennet can be found elsewhere (Liu et al., 2021b). Other enzymes are used in cheese making. Thus lipases are used to accelerate cheese ripening and modify flavor due to their lipolytic action that releases fatty acids. Excessive lipolytic action may, however, lead to unwanted odors. Lipases also affect the structure of cheese, eventually allowing for a product with a softer texture (Law, 2010). During the ripening of some cheeses, the gas formation can occur (late-blowing) because of contamination with spores of the Gram-positive bacteria *Clostridium tyrobutyricum*. These can grow within 1°C–45°C and colonize pasteurized goods, producing butyric acid, hence the gas effect. Antimicrobial action is implemented through egg white lysozyme, an enzyme relatively stable in cheese and whey, which hydrolyzes the carbohydrate polymers of alternating *n*-acetyl-glucosamine and N-acetyl-muramic acid. Such polymer, combined with peptide chains, constitutes Gram-positive bacteria's mucopolysaccharide walls. Hence, lysozyme promotes the lysis of the cell wall (Law, 2010).

Lactase or β-galactosidase is another relevant enzyme within the dairy industry. Typical sources of this enzyme are *A. oryzae*, *A. niger*, and *K. lactis*. Lactase promotes the hydrolysis of lactose in milk to its monosaccharides, glucose, and galactose. Thus it allows spanning the market of milk-based products to those suffering from lactose intolerance or lactose maldigestion. On an industrial scale, the process is performed at 35°C and processes typically 8.0 m^3 milk per day with a conversion of 70%–81% (Liese et al., 2006). Aiming to minimize the impact of the enzymatic hydrolysis process in organoleptic properties of milk, the use of cold-active β-galactosidases has been evaluated, featuring its application during transportation and storage (Yao et al., 2019), including in immobilized form (Czyzewska and Trusek, 2021). Moreover, because of lactose hydrolysis, the sweetness of milk is increased as both monosaccharides are sweeter than lactose. The need for adding sugars in the manufacture of flavored milk drinks is minimized, if not avoided (Luzzi et al., 2020). Additionally, lactase is used to process whey, the watery part of the milk resulting from curd formation. A severe environmental hazard is the disposal of whey formed in large amounts. It was traditionally carried out by using it as feed for animals or fertilizers or simply dumping it in sewers or watercourses. Besides current environmental constraints and to make production processes more profitable, there is a growing interest in turning this by-product into a commercially interesting good. The hydrolysis of lactose present in whey results in the formation of sweet syrup, which provides a source of sugar and is used as such in confectionery, dairy desserts, ice creams, soft, as well as in feedstuff (Panesar et al., 2010b; Shukla et al., 2013). The sweetness of sweet syrup can be further enhanced by processing with immobilized glucose isomerase (Weetall and Yaverbaum, 1974). Besides lactose hydrolysis, lactases can also promote the synthesis of galactooligosaccharides (GOS), nondigestible oligosaccharides typically composed of 2–10 galactose units and a terminal glucose unit. GOS have a prebiotic role. Recently, Alnadari and coworkers reported the use of β-glucosidase immobilized in chitosan-functionalized magnetic nanoparticles, to ease recovery and enable recycling as an alternative for the large-scale production of GOS. GOS are also added to infant food formula to mimic human milk oligosaccharides (HMOs). Several studies suggest that blends of GOS with FOS stimulate formula-fed infants' intestinal flora, leading to looser fecal stools. Moreover, Food and Agriculture Organization supports using GOS/FOS blends in infant formula for infants aged 5 months and older (Ackerberg et al., 2012; Faijes et al., 2019). The synthesis of GOS is carried out through a transgalactosylation reaction, where lactose (or other carbohydrates in the mixture) serves as galactosyl acceptor. The reaction is influenced by several environmental parameters, namely, lactose concentration, pH, and temperature, but under optimal conditions, GOS yields of about 40% are achievable. For initial lactose concentrations above 30% (w/v), the influence of this parameter on GOS yield clearly decreases. To overcome the relatively low solubility of lactose, reactions are carried out at temperatures of 40°C and above. GOS production largely uses whey permeate as raw material, where lactose concentration can be adequately adjusted, rather than milk, due to the relatively low concentration of lactose in the latter, 5% (w/v) for cow milk (Torres et al., 2010; Díez-Municio et al., 2014; Xavier et al., 2018; Damin et al., 2021). Still, the production of GOS-enriched milk in a concentration close to that of HMO in human milk and with a low titer in lactose has been successfully performed (Rodriguez-Colinas et al., 2014).

Lactose is also the substrate for the production of lactulose (4-*O*-β-D-galactopyranosyl-D-fructose). The traditional role of lactulose had been as a laxative; however, its prebiotic action has been identified, and its production is thus receiving increased attention. Enzymatic production of this disaccharide has been implemented as an alternative to the chemical alkaline isomerization of lactose to avoid costly and cumbersome downstream processing. Again the process involves transglycosylation from lactose and fructose, using either β-galactosidases or β-glycosidases. The continuous production of lactulose from fructose and lactose in a packed-bed reactor operation with β-galactosidase immobilized in glyoxyl-agarose allowed a maximum product yield of 60% (Guerrero et al., 2019), however still lower than those obtained through chemical synthesis, of about 87% (Wang et al., 2013, Guerrero et al., 2015). An alternative synthetic method, where yields are similar to those obtained chemically, involves using cellobiose-2-epimerase from *Caldicellulosiruptor saccharolyticus* (Park et al., 2017). Given the isomerization ability for the glucose moiety of cellobiose, the enzyme is considered to display functionality to convert lactose (4-*O*-β-D-galactopyranosyl-D-glucose) directly into lactulose (4-*O*-β-D-galactopyranosyl-D-fructose), thus doing away with the need of fructose addition (Kim et al., 2013; Wang et al., 2013). In the quest for novel prebiotics with improved properties, the synthesis of lactulose-based GOS has been implemented, involving either enzymatic route, based on transglycosylation activity of β-galactosidases, or chemoenzymatic methods, where said enzymatic activity is combined with a chemical catalyst that promotes the isomerization of either lactose to lactulose in whey permeate, or of the transgalactosylated cheese whey into the lactulose oligosaccharides (Wang et al., 2013; Padilla et al., 2015; Karim and Aider, 2022).

The role played by lipases in cheese making is extended to milk fats, as these enzymes promote the partial hydrolysis (lipolysis) of triacylglycerols, about 96% of the milk fat, to release fatty acids among those that are located at given positions on the triacylglycerol backbone and produce lipolyzed milk fat, LMF. Typically, most saturated long-chain fatty acids are located in the sn-2 position, whereas short-chain fatty acids are located in the sn-3 position contains. Hence, the region-selective nature of lipases has to be considered for the intended goal. Moreover, the extent of reaction conveys significant different flavors to the resulting product. Therefore lipolysis is carefully controlled, and the reaction is allowed to proceed until either the intended flavor/fragrance or acid degree value is achieved. LMF is incorporated in a wide variety of goods, namely, butter, bakery, and snacks (Fraga et al., 2018). On the other hand, lipases can also perform esterification reactions (*inter*- and *trans*-), acidolysis and alcoholysis, which improve milk's physical and chemical properties. In recent years these efforts have focused on producing triacylglycerols with structures similar to those in human milk fat (Ferreira-Dias and Tecelão, 2014; Ferreira et al., 2021).

Proteases are also used to produce bioactive peptides from milk and whey proteins. The use of digestive enzymes such as chymotrypsin, elastase, pepsin, and trypsin has been reported, yet microbial proteases such as alcalase, subtilisin, and thermolysin are also effective. Moreover, blends from commercial enzyme preparation are also of interest as they allow the production of a larger array of peptides compared with pure preparations (Guha et al., 2021).

Transaminases promote cross-linking both within a protein molecule and between molecules of different proteins. Thus these enzymes are used for conditioning the strength and texture of several products, namely, cheese, particularly by increasing the yield of curd, and yoghurt (Kieliszek and Misiewicz, 2014; Li et al., 2019).

18.2.6.3 Bakery

The bread making process typically involves the synergistic use of α- and β-amylases. As α-amylases hydrolyze starch to dextrins, β-amylases promote further hydrolysis to maltose. Thus, while the former releases low-molecular-chain dextrins from starch, the latter hydrolyzes them to maltose, which can be used as fermentable sugar by yeast. Several advantages result from amylase action: enhanced bread volume and crumb texture and lower dough viscosity. Moreover, the reducing sugars formed to allow for enhanced Maillard reactions, accountable for the crust's browning and intensifying pleasant flavor (Miguel et al., 2013). The combined action of amylases, namely, maltogenic amylases and glucoamylase, can be advantageously used to minimize staling of bread (Amigo et al., 2021). Staling baked goods is a chemical and physical process that reduces their palatability and is often mistaken for a simple dry-out process due to water evaporation. Staling is noticeable through the increase of the firmness of the crumb, decreased elasticity of the crumb, and challenging and leathery appearance of the crust. Staling partly results from the retrogradation of starch. The outcome is the realignment of amylose and amylopectin molecules upon migration of moisture from the starch granules into the interstitial spaces. Retrogradation starts immediately after baking, during which gelatinization of starch occurs, and large amounts of water are absorbed. Amylose retrogrades faster than amylopectin. The partial hydrolysis resulting from enzyme action, preferably performed after gelatinization, significantly alters starch's structure, as the fragments resulting from hydrolysis are too small to retrograde (Else et al., 2013).

Lipases are used to enhance the handling, machinability, and strength of dough; improve bread oven spring; and enhance the volume and crumb structure of white bread through their action on flour lipids or added fat. Thus lipases hydrolyze triacylglycerols into mono- and diacylglycerols, glycerol, and free fatty acids. Besides chemical modifications, the surface-active nature of the reaction products is accountable for the positive action on bread observed. Different generations of lipases have been presented, allowing for an increasingly more comprehensive range of substrates, namely, diacylgalactolipids and phospholipids, and concomitantly, a more extensive array of products (van Oort, 2010; Miguel et al., 2013; Gerits et al., 2014; Huang et al., 2020).

Lipoxigenase catalyzes the oxidation of polyunsaturated fatty acids (PUFA) containing a *cis*, *cis*-1,4-pentadiene moiety to form fatty acid hydroperoxides. Lipoxygenases can be isolated from animal and plant sources. The selectivity depends on the origin of the enzyme, as wheat lipoxygenases act on PUFA in free or monoacyl glycerol, while enzyme from soybean also acts on PUFA in triacylglycerol form. As oxidation occurs, hydroxyl radicals are formed, which react with the yellow carotenoid present in wheat flour and with peptides/proteins present in the dough, with concomitant formation of hydroxyacids, resulting in reducing the yellow color and thus in a whiter crumb. Besides this bleaching effect, the oxidation of thiol of gluten proteins results in rearranging disulfide bonds and cross-linking of tyrosine residues, ultimately leading to enhanced loaf volume. Several oxidases have been used in bread making as an alternative to chemical oxidants, such as potassium bromate or potassium iodate, to enhance dough strength and handling properties and improve the texture and appearance of the baked product. Often referred to within this type of enzyme are glucose oxidase and hexose oxidase. The use of the former is somehow conditioned given the low amount of glucose in dough from cereal flours, the latter with a broader range of substrates being more appealing. Yet the mechanism of action of oxidases is not yet fully established (van Oort, 2010; Miguel et al., 2013; Hayward et al., 2016). Overdosage of oxidase can result in excessive cross-linking, tampering with gas retention, and thus handling of the dough, leading to poor quality (Bonet et al., 2006).

Proteases have been traditionally used in the production of bread and baked goods. The proteolytic activity of these enzymes is used advantageously on both gluten and dough. As an outcome, mixing times are reduced, dough consistency is decreased, and it becomes more uniform. Also, controlled hydrolysis helps regulate gluten strength and makes pulling and kneading easier. Most of these effects are promoted by *endo*-peptidases, since their action has a more noticeable impact on the gluten network and in dough rheology. The action of *exo*-peptidases is more pronounced in flavor and color, as a result of Maillard reactions involving amino acids released and sugars present: Given their effective action and environment-friendly nature, proteases have gradually replaced sodium metabisulfite in dough conditioning (Miguel et al., 2013; Hassan et al., 2014; Heredia-Sandoval et al., 2016).

Transglutaminases are also used in baking, as their cross-linking action over gluten proteins improves the stability and volume of dough and its elasticity and resilience. Transglutaminases promote the formation of an isopeptide bond between the group of γ-carboxamides of glutamine residues (donor) and the primary ε-amine groups of proteins/peptides, acceptors of an acyl residue (Miguel et al., 2013; Kieliszek and Misiewicz, 2014; Ogilvie et al., 2021).

Xylanases are used to break down hemicelluloses, namely arabinoxylans, as the insoluble nature of the latter hampers the formation of the gluten network. As a result of enzyme action, the handling of dough is improved. In addition, the concentration of arabinoxylo-oligosaccharides in bread increases, with a positive impact on human health, given their prebiotic nature (Broekaert et al., 2011; Both et al., 2021).

18.2.6.4 Beer making

The efficiency of the malting process, where a fermentable extract for later yeast action has to be obtained, depends on the addition of exogenous amylolytic, (hemi)-cellulolytic and proteolytic enzymes in a controlled and quantifiable manner. Poor malting ultimately leads to defective fermentation, low alcohol titer, hampered filtration, and low quality and stability of the final product. Thus glucanases are required to break down the cell walls of the grains of cereal (viz. barley, cell wall of which is composed of about 70% of glucans). Xylanases are also included in the process to contribute to the degradation of nonstarch polysaccharides, namely, arabinoxylans, also significantly present in the cell; proteases (*endo-* and *exo-*peptidases) to hydrolyze the large-chain protein molecules of the cereal, ease the assess of amylolytic enzymes to starch, and provide amino acids and small peptides for fermentation, ultimately having also influence in the flavors produced during fermentation. Excess proteolysis tampers with foam stability of the final beer by reducing the level of foam-positive proteins, while deficient proteolysis will also tamper the colloidal stability of beer; α- and β-amylases, amyloglucosidases, pullulanases, and α-glucosidases are required for the process of starch hydrolysis to glucose units (Souppe and Beudeker, 2002; Lalor and Goode, 2010; Blanco et al., 2014; Gomaa, 2018).

Another key enzyme in the process of beer making is α-acetolactate decarboxylase, as it allows the decarboxylation of acetolactate to acetoin. Acetolactate is one of the many flavor compounds produced by yeast during fermentation but, in excess, gives the beer a butterscotch taste. Acetoin is rather tasteless, and enzyme addition speeds up the maturation process (Dulieu et al., 2008).

18.2.6.5 Juices

Enzymes used in the juice industry help separate juice from the fruit/vegetable cells and clarify the juice by degrading pectin and naturally occurring starches that contribute to undesired viscosity, hamper filtration and give the final product a cloudy appearance. Pectin is a generic name for complex structural polysaccharides in fruits and plants, with a backbone of galacturonic acid residues linked by α-1,4 bonds. The side chains of the pectin molecule are composed of sugar residues, namely, arabinose, galactose, and xylose. On the other hand, the carboxyl groups of galacturonic acid are partially esterified by methyl groups. The degradation involves the use of pectinases, a broad designation that encompasses several enzymes: *endo*- and *exo*-polygalacturonases, that promote the hydrolysis of galacturonans in a random or terminal action pattern, respectively, the former decreasing viscosity and the latter releasing galacturonic acid. Pectin methyl esterase hydrolyzes the carboxyl ester bond; endoarabinases promote the endohydrolysis of α-1,5-arabinofuranosidic linkages in 1,5-arabinans, preventing haze formation; *endo*-pectinlyase results in the eliminative cleavage of α-1,4-D-galacturonan methyl ester, decreasing viscosity. Polygalacturonases and pectin methyl esterases are also included in the peeling of citrus fruits. Amylases can also be used to promote the hydrolysis of starch. Cellulases and hemicellulases (xylanases) are used to disaggregate the cell wall (Cautela et al., 2010; Grassin and Coutel, 2010; Tapre and Jain, 2014; García, 2018).

To avoid the bitter taste of citrus juices, these can be processed with naringinase, an enzyme complex composed of α-rhamnosidase and β-glycosidase. Naringinase hydrolyzes naringin, the molecule that conveys the bitter taste: first to prunin (α-L-rhamnosidase) which is then hydrolyzed to naringenin, (β-D-glycosidase) (Puri, 2012; Purewal and Sandhu, 2021).

18.2.6.6 Processing of meat, fish, and seafood

Proteases are used to obtain uniform tender meat by decreasing the amount of connective tissue while retaining myofibrillar proteins. Bromelain, ficin and papain, proteases from plants are typically used for such goals; however, they present some drawbacks: poor selectivity and relatively high thermal stability, preventing their full denaturation during cooking. Microbial proteases, namely, subtilisin and the neutral proteases are more selective and tend to display higher activity at relatively low temperatures, particularly cold-active proteases, and denaturation at cooking temperatures (Bekhit et al., 2014, Mageswari et al., 2017).

The cross-linking action of transglutaminases is widely used in meat, fish, and seafood areas, particularly in the manufacture of restructured meat, as it allows to improve the texture, cohesiveness, and shelf life of goods, namely, sausages, fish protein paste, and other fish raw materials, moreover without the need of phosphate addition, with positive impact in health. Application of *trans*-glutaminases in meat processing enables the use of lower quality materials, namely, collagen, in the production of highly nutritive goods, provided suitable amino acids are added (Ashie and Lanier, 2000; Kieliszek and Misiewicz, 2014; Li et al., 2018b; Erdem et al., 2020).

Proteases, alongside carbohydrases and lipases, are used in the preparation of fish protein hydrolysates (Martosuyono et al., 2019). Given the limited availability and cost of animal and plant proteases, bacterial proteases are mostly used, such as neutral protease and Alcalase, a commercial preparation developed for detergent industry. Enzymatic extraction of protein hydrolysates from cod, capelin, salmon, and tuna, among others, has been reported (Ghaly et al., 2013). Lipases have been used in the hydrolysis of fish oils to promote the enrichment of PUFA (Yan et al., 2012; Aguilera-Oviedo et al., 2021).

18.2.6.7 Wine making

Wine results from the maceration of grapes and processing of the resulting juice; hence, enzymes used for juice processing, namely, pectinases and (hemi)cellulases, are also used, aiming at the same goals. The concerted action of exoglycosidases and β-glycosidases enables the release of molecules accountable for aroma compounds by cleaving the intersugar linkage of glycosides, thus liberating sugars and β-glucosides, thereby allowing for β-glycosidase to release glucose and aromatic aglycone. Care has to be taken with this exogenous activity in the case of red wines, as it may hamper color stability. Urease may also be required to avoid the formation of ethyl carbamate, a potential mild carcinogen formed by the spontaneous reaction of urea and ethanol, as the enzyme hydrolyzes urea to ammonia and carbon dioxide. Glucose oxidase can be added to the must to adjust glucose concentration and thus control ethanol content in wine. Lysozyme can also be used to counter unwanted malolactic fermentation (Gómez-Plaza et al., 2010). Laccases provide an effective alternative to traditional approaches for removing polyphenols in white wines, such as polyvinylpolypyrrolidone and high doses of sulfur dioxide. The enzymatic approach is more selective and prevents, thus the loss of given organoleptic characteristics due to indiscriminate removal of polyphenols (Kunamneni et al., 2008; Ottone et al., 2020).

18.3 Concluding remarks

The use of enzymes has a long tradition within the scope of food industry. Nevertheless, dedicated efforts are continuously improving existing production processes, implementing new ones, and introducing new products. Such measures are necessary to tackle the need for further cost-effective production processes in an increasingly competitive market, the public demand for new and healthier goods, and stringent safety regulations. The use of enzymes in traditional areas has been significantly improved, but novel applications are to be innovated to improve key properties or create novel functionalities. Increased understanding of the mechanisms of enzyme action, development of suitable enzyme formulations, and design of operational conditions can be beneficial for using biocatalysts most advantageously. With the growing public demand for safer and high-quality foods, alongside sustainable and environment-friendly production processes, a set of conditions is gathered for exciting developments related to the use of enzymes in the food sector.

Abbreviations

CAD	computer-assisted design
CAGR	compound annual growth rate
CLEAs	cross-linked enzyme aggregates
CMP	caseinomacropeptide
DE	dextrose equivalent
DNA	deoxyribonucleic acid
FDA	Food and Drug Administration
FOS	fructooligosaccharides
HMOs	human milk oligosaccharides
GluOS	glucooligosaccharides
GMM	genetically modified microorganism
GMO	genetically modified organism
GOS	galactooligosaccharides
GRAS	Generally Regarded as Safe
HFS	high-fructose syrup, high-fructose corn syrup
LMF	lipolyzed milk fat
MOF	metal–organic frameworks
ONPG	*o*-Nitrophenyl-β-D-galactopyranoside
PUFA	polyunsaturated fatty acids
scFOS	short-chain fructooligosaccharides
SSF	solid-state fermentation

References

Ackerberg, T.S., Labuschagne, I.L., Lombard, M.J., 2012. The use of prebiotics and probiotics in infant formula. S. Afr. Fam. Pract. 54, 321–323.

Adrio, J.-L., Demain, A.L., 2010. Recombinant organisms for production of industrial products. Bioengineered Bugs 1, 116–131.

Adrio, J.-L., Demain, A.L., 2014. Microbial enzymes: tools for biotechnological processes. Biomolecules 4, 117–139.

Aguilera-Oviedo, J., Yara-Varón, E., Torres, M., Canela-Garayoa, R., Balcells, M., 2021. Sustainable synthesis of omega-3 fatty acid ethyl esters from monkfish liver oil. Catalysts 11, 100. Available from: https://doi.org/10.3390/catal11010100.

Ahmed, T., Rana, M.R., Zzaman, W., Ara, R., Aziz, M.G., 2021. Optimization of substrate composition for pectinase production from Satkara (*Citrus macroptera*) peel using *Aspergillus niger*-ATCC 1640 in solid-state fermentation. Heliyon 7, e08133. Available from: https://doi.org/10.1016/j.heliyon.2021.e08133.

Akanbi, T.O., Ji, D., Agyei, D., 2020. Revisiting the scope and applications of food enzymes from extremophiles. J. Food Biochem. 44, e13475. Available from: https://doi.org/10.1111/jfbc.13475.

Al-Hagar, O.E.A., Ahmed, A.S., Hassan, I.A., 2015. Invertase production by irradiated *Aspergillus niger* OSH5 using agricultural wastes as carbon source. Microbiol. Res. J. Int. 6, 135–146. Available from: https://doi.org/10.9734/BMRJ/2015/11539.

AL-Sa'ady, A.J.R., 2014. Optimization of invertase production from *Saccharomyces cerevisiae* by solid state fermentation. Curr. Res. Microbiol. Biotechnol. 2, 373–377. Available from: http://crmb.aizeonpublishers.net/content/2014/3/crmb373-377.pdf.

Alcântara, S.R., da Silva, F.L.H., 2014. Solid state fermentation process for polygalacturonase production using cashew apple. Am. J. Chem. Eng. 2, 28–34.

Almanaa, T.N., Vijayaraghavan, P., Alharbi, N.S., Kadaikunnan, S., Khaled, J.M., Alyahya, S. A., 2020. Solid state fermentation of amylase production from *Bacillus subtilis* D19 using agro-residues. J. King Saud. Univ. - Sci. 32 (2), 1555–1561. Available from: https://doi.org/10.1016/j.jksus.2019.12.011.

Alnadari, F., Xue, Y., Almakas, A., Mohedein, A., Samie, A., Abdel-Shafi, M., et al., 2021. Large batch production of galactooligosaccharides using β-glucosidase immobilized on chitosan-functionalized magnetic nanoparticle. J. Food Biochem. 45, e13589. Available from: https://doi.org/10.1111/jfbc.13589.

Amigo, J.M., Olmo, A.D., Engelsen, M.M., Lundkvist, H., Engelsen, S.B., 2021. Staling of white wheat bread crumb and effect of maltogenic α-amylases. Part 3: spatial evolution of bread staling with time by near infrared hyperspectral imaging. Food Chem. 353, 129478. Available from: https://doi.org/10.1016/j.foodchem.2021.129478.

Andersen, C., Borchert, T.V., Nielsen, B.R., 2013. Amylase variants. Patent US 8609811 B2.

Araya, E., Urrutia, P., Romero, O., Illanes, A., Wilson, L., 2019. Design of combined cross-linked enzyme aggregates (combi-CLEAs) of β-galactosidase and glucose isomerase for the one-pot production of fructose syrup from lactose. Food Chem. 288, 102–107. Available from: https://doi.org/10.1016/j.foodchem.2019.02.024.

Ashie, I.N.A., Lanier, T.C., 2000. Transglutaminases in seafood process, Bing. In: Haard, N.F., Simpson, B.K. (Eds.), Seafood Enzymes. Marcel Dekker, NY, USA, pp. 121–145.

Aunstrup, K., Andresen, O., Falch, E.A., Nielsen, T.K., 1979. Production of microbial enzymes. In: 2nd (Ed.) Perlman, D. (Ed.), Microbial Technology, Vol. 1. Microbial Processes, Academic Press, NY, USA, pp. 281–309.

Balakrishnan, M., Jeevarathinam, G., Kumar, S.K.S., Muniraj, I., Uthandi, S., 2021. Optimization and scale-up of α-amylase production by *Aspergillus oryzae* using solid-state fermentation of edible oil cakes. BMC Biotechnol. 21, 33. Available from: https://doi.org/10.1186/s12896-021-00686-7.

Barrios-González, J., 2012. Solid-state fermentation: physiology of solid medium, its molecular basis and applications. Process. Biochem. 47, 175–185. Available from: https://doi.org/10.1016/j.procbio.2011.11.016.

Basso, A., Serban, S., 2019. Industrial applications of immobilized enzymes—a review. Mol. Catal. 479, 110607. Available from: https://doi.org/10.1016/j.mcat.2019.110607.

Bekhit, A.A., Hopkins, D.L., Geesink, G., Bekhit, A.A., Franks, P., 2014. Exogenous proteases for meat tenderization. Crit. Rev. Food Sci. Nutr. 54, 1012–1031. Available from: https://doi.org/10.1080/10408398.2011.623247.

Ben-Othman, S., Rinken, T., 2021. Immobilization of pectinolytic enzymes on nylon 6/6 carriers. Appl. Sci. 11, 4591. Available from: https://doi.org/10.3390/app11104591.

Berikten, D., Kivanc, M., 2014. Optimization of solid-state fermentation for phytase production by Thermomyces lanuginosus using response surface methodology. Prep. Biochem. Biotechnol. 44, 834–848. Available from: https://doi.org/10.1080/10826068.2013.868357.

Bernal, C., Rodríguez, K., Martínez, R., 2018. Integrating enzyme immobilization and protein engineering: an alternative path for the development of novel and improved industrial biocatalysts. Biotechnol. Adv. 36, 1470–1480. Available from: https://doi.org/10.1016/j.biotechadv.2018.06.002.

Bilal, M., Iqbal, H., 2020. State-of-the-art strategies and applied perspectives of enzyme biocatalysis in food sector - current status and future trends. Crit. Rev. Food Sci. Nutr. 60, 2052–2066. Available from: https://doi.org/10.1080/10408398.2019.1627284.

Bivolarski, V., Vasileva, T., Gabriel, V., Iliev, I., 2018. Synthesis of glucooligosaccharides with prebiotic potential by glucansucrase URE 13–300 acceptor reactions with maltose, raffinose and lactose. Eng. Life Sci. 18, 904–913. Available from: https://doi.org/10.1002/elsc.201800047.

Blanco, C.A., Caballero, I., Barrios, R., Rojas, A., 2014. Innovations in the brewing industry: light beer. Int. J. Food Sci. Nutr. 65, 655–660. Available from: https://doi.org/10.3109/09637486.2014.893285.

Bonet, A., Rosell, C.M., Caballero, P.A., Gómez, M., Perez-Munuera, I., Lluch, M.A., 2006. Glucose oxidase effect on dough rheology and bread quality: A study from macroscopic to molecular level. Food Chem. 99, 408–415. Available from: https://doi.org/10.1016/j.foodchem.2005.07.043.

Bornscheuer, U.T., 2013. Protein engineering as a tool for the development of novel bioproduction systems. Adv. Biochem. Eng. Biotechnol. 137, 25–40.

Bortone, N., Fidaleo, M., 2020. Stabilization of immobilized l-arabinose isomerase for the production of d-tagatose from d-galactose. Biotechnol. Prog. 36, e3033. Available from: https://doi.org/10.1002/btpr.3033.

Both, J., Biduski, B., Gómez, M., Bertolin, T.E., Friedrich, M.T., Gutkoski, L.C., 2021. Micronized whole wheat flour and xylanase application: dough properties and bread quality. J. Food Sci. Technol. 58, 3902–3912. Available from: https://doi.org/10.1007/s13197-020-04851-2.

Brandl, J., Andersen, M.R., 2015. Current state of genome-scale modeling in filamentous fungi. Biotechnol. Lett. 37, 1131–1139. Available from: https://doi.org/10.1007/s10529-015-1782-8.

Broekaert, W.F., Courtin, C.M., Verbeke, K., Van de Wiele, T., Verstraete, W., Delcour, J.A., 2011. Prebiotic and other health-related effects of cereal-derived arabinoxylans, arabinoxylan-oligosaccharides, and xylooligosaccharides. Crit. Rev. Food Sci. Nutr. 51, 178–194. Available from: https://doi.org/10.1080/10408390903044768.

Bu, Y., Zhang, T., Jiang, B., Chen, J., 2021. Improved performance of D-psicose 3-epimerase by immobilisation on amino-epoxide support with intense multipoint attachment. Foods 10, 831. Available from: https://doi.org/10.3390/foods10040831.

Carceller, J.M., Galan, J.P.M., Monti, R., Bassan, J.C., Filice, M., Yu, J.H., et al., 2020. Covalent immobilization of naringinase over two-dimensional 2D zeolites and its applications in a continuous process to produce citrus flavonoids and for debittering of juices. ChemCatChem 12, 4502–4511. Available from: https://doi.org/10.1002/cctc.202000320.

Cautela, D., Castaldo, D., Servillo, L., Giovane, A., 2010. Enzymes in citrus juice processing. In: Bayındırlı, A. (Ed.), Enzymes in Fruit and Vegetable Processing. CRC Press, Boca Raton, USA, pp. 197–213.

Chen, X., Li, W., Ji, P., Zhao, Y., Hua, C., Han, C., 2018. Engineering the conserved and noncatalytic residues of a thermostable β – 1,4-endoglucanase to improve specific activity and thermostability. Sci. Rep. 8, 2954. Available from: https://doi.org/10.1038/s41598-018-21246-8.

Chen, Q., Xiao, Y., Shakhnovich, E.I., Zhang, W., Mu, W., 2020. Semi-rational design and molecular dynamics simulations study of the thermostability enhancement of cellobiose 2-epimerases. Int. J. Biol. Macromol. 154, 1356–1365. Available from: https://doi.org/10.1016/j.ijbiomac.2019.11.015.

Chisti, Y., 2010. Fermentation technology. In: Soetaert, W., Vandamme, E.J. (Eds.), Industrial Biotechnology: Sustainable Growth and Economic Success. Wiley-VCH Verlag GmbH & Co. KGaA, Weinheim, Germany, pp. 149–171.

Chi, Z.M., Zhang, T., Cao, T.S., Liu, X.Y., Cui, W., Zhao, C.H., 2011. Biotechnological potential of inulin for bioprocesses. Bioresour. Technol. 102, 4295–4303. Available from: https://doi.org/10.1016/j.biortech.2010.12.086.

Choi, M.N., Shin, K.C., Kim, D.W., Kim, B.J., Park, C.S., Yeom, S.J., et al., 2021. Production of D-allose from D-allulose using commercial immobilized glucose isomerase. Front. Bioeng. Biotechnol. 9, 681253. Available from: https://doi.org/10.3389/fbioe.2021.681253.

Chotani, G., Peres, C., Schuler, A., Moslemy, P., 2014. Bioprocessing technologies. In: Bisaria, V.S., Kondo, A. (Eds.), Bioprocessing of Renewable Resources to Commodity Bioproducts. Wiley, NY, USA, pp. 133–167.

Chowdhury, R., Maranas, C.D., 2020. From directed evolution to computational enzyme engineering — a review. AIChE J. 121, e16847. Available from: https://doi.org/10.1002/aic.16847.

Czinkóczky, R., Németh, Á., 2022. Enrichment of the rebaudioside A concentration in *Stevia rebaudiana* extract with cyclodextrin glycosyltransferase from *Bacillus licheniformis* DSM 13. Eng. Life Sci. 22, 30–39. Available from: https://doi.org/10.1002/elsc.202100111.

Czyzewska, K., Trusek, A., 2021. Encapsulated NOLA™ Fit 5500 lactase—an economically beneficial way to obtain lactose-free milk at low temperature. Catalysts 11, 527. Available from: https://doi.org/10.3390/catal11050527.

Dalgleish, D.G., Corredig, M., 2012. The structure of the casein micelle of milk and its changes during processing. Annu. Rev. Food Sci. Technol. 3, 449–467. Available from: https://doi.org/10.1146/annurev-food-022811-101214.

Dalmaso, G.Z.L., Ferreira, D., Vermelho, A.B., 2015. Marine extremophiles: a source of hydrolases for biotechnological applications. Mar. Drugs 13, 1925–1965. Available from: https://doi.org/10.3390/md13041925.

Damborsky, J., Brezovsky, J., 2014. Computational tools for designing and engineering enzymes. Curr. Op. Chem. Biol. 19, 8–16. Available from: https://doi.org/10.1016/j.cbpa.2013.12.003.

Damgaard, J., 2016. Method for the preparation of an enzyme tablet. Patent Application US 2016/0081356 A1.

Damin, B., Kovalski, F.C., Fischer, J., Piccin, J.S., Dettmer, A., 2021. Challenges and perspectives of the β-galactosidase enzyme. Appl. Microbiol. Biotechnol. 105, 5281–5298. Available from: https://doi.org/10.1007/s00253-021-11423-7.

Das, D., Bhat, R., Selvaraj, R., 2020. Optimization of inulinase production by a newly isolated *Penicillium amphipolaria* strain using solid-state fermentation of hardy sugarcane stems. Biocatal. Agric. Biotechnol. 30, 101875. Available from: https://doi.org/10.1016/j.bcab.2020.101875.

Das, D., Selvaraj, R., Ramananda Bhat, M., 2019. Optimization of inulinase production by a newly isolated strain *Aspergillus flavus* var. flavus by solid state fermentation of *Saccharum arundinaceum*. Biocatal. Agric. Biotechnol. 22, 101363. Available from: https://doi.org/10.1016/j.bcab.2019.101363.

Daudé, D., André, I., Monsan, P., Remaud-Siméon, M., 2014. Successes in engineering glucansucrases to enhance glycodiversification. Carbohydr. Chem. 40, 624–645.

Davani-Davari, D., Negahdaripour, M., Karimzadeh, I., Seifan, M., Mohkam, M., Masoumi, S.J., et al., 2019. Prebiotics: definition, types, sources, mechanisms, and clinical applications. Foods 8, 92. Available from: https://doi.org/10.3390/foods8030092 (Basel, Switzerland).

Deckers, M., Deforce, D., Fraiture, M.A., Roosens, N., 2020. Genetically modified microorganisms for industrial food enzyme production: an overview. Foods 9, 326. Available from: https://doi.org/10.3390/foods9030326. Basel, Switzerland.

Dedania, S.R., Patel, M.J., Patel, D.M., Akhani, R.C., Patel, D.H., 2017. Immobilization on graphene oxide improves the thermal stability and bioconversion efficiency of D-psicose 3-epimerase for rare sugar production. Enzyme: Microb. Technol. 107, 49–56. Available from: https://doi.org/10.1016/j.enzmictec.2017.08.003.

Dedania, S.R., Patel, V.K., Soni, S.S., Patel, D.H., 2020. Immobilization of *Agrobacterium tumefaciens* d-psicose 3-epimerase onto titanium dioxide for bioconversion of rare sugar. Enzyme Microb. Technol. 140, 109605. Available from: https://doi.org/10.1016/j.enzmictec.2020.109605.

Derakhti, S., Shojaosadati, S.A., Hashemi, M., Khajeh, K., 2012. Process parameters study of α-amylase production in a packed-bed bioreactor under solid-state fermentation with possibility of temperature monitoring. Prep. Biochem. Biotechnol. 42, 203–216. Available from: https://doi.org/10.1080/10826068.2011.599466.

Dey, T.B., Banerjee, R., 2012. Hyperactive α-amylase production by Aspergillus oryzae IFO 30103 in a new bioreactor. Lett. Appl. Microbiol. 54, 102–107. Available from: https://doi.org/10.1111/j.1472-765X.2011.03177.x.

de Oliveira, F., Castellane, T., de Melo, M.R., Buzato, J.B., 2022. Preparation of *Aspergillus niger* 426 naringinases for debittering citrus juice utilization of agro-industrial residues. Int. Microbiol. 25, 123–131. Available from: https://doi.org/10.1007/s10123-021-00199-5.

de Oliveira, R.L., da Silva, M.F., Converti, A., Porto, T.S., 2020. Production of β-fructofuranosidase with transfructosylating activity by *Aspergillus tamarii* URM4634 solid-state fermentation on agroindustrial by-products. Int. J. Biol. Macromol. 144, 343–350. Available from: https://doi.org/10.1016/j.ijbiomac.2019.12.084.

de Sousa, M., Gurgel, B.S., Pessela, B.C.;, Gonçalves, L.R., 2020. Preparation of CLEAs and magnetic CLEAs of a recombinant l-arabinose isomerase for d-tagatose synthesis. Enzym. Microb. Technol. 138, 109566. Available from: https://doi.org/10.1016/j.enzmictec.2020.109566.

de Souza, C., Ribeiro, B.D., Coelho, M., 2019. Characterization and application of *Yarrowia lipolytica* lipase obtained by solid-state fermentation in the synthesis of different esters used in the food industry. Appl. Biochem. Biotechnol. 189, 933–959. Available from: https://doi.org/10.1007/s12010-019-03047-5.

DiCosimo, R., McAuliffe, J., Poulose, A.J., Bohlmann, G., 2013. Industrial use of immobilized enzymes. Chem. Soc. Rev. 42, 6437–6474. Available from: https://doi.org/10.1039/C3CS35506C.

Díez-Municio, M., Herrero, M., Olano, A., Moreno, F.J., 2014. Synthesis of novel bioactive lactose-derived oligosaccharides by microbial glycoside hydrolases. Microb. Biotechnol. 7, 315–331. Available from: https://doi.org/10.1111/1751-7915.12124.

Dilipkumar, M., Rajasimman, M., Rajamohan, N., 2013. Enhanced inulinase production by Streptomyces sp. in solid state fermentation through statistical designs. 3 Biotech. 3, 509–515. Available from: https://doi.org/10.1007/s13205-012-0112-2.

Dodge, T., 2010. Production of industrial enzymes. In: Whitehurst, R.J., Van Oort, M. (Eds.), Enzymes in Food Technology, 2nd (Ed.) Wiley-Blackwell, Chichester, UK, pp. 44–58.

Dulieu, C., Moll, M., Boudrant, J., Poncelet, D., 2008. Improved performances and control of beer fermentation using encapsulated alpha-acetolactate decarboxylase and modeling. Biotechnol. Prog. 16, 958–965. Available from: https://doi.org/10.1021/bp000128k.

El-Gendy, M.M.A., 2012. Production of glucoamylase by marine endophytic *Aspergillus* sp. JAN-25 under optimized solid-state fermentation conditions on agro residues. Australian J. Basic. Appl. Sci. 6, 41–54.

El-Gendy, M.M.A.A., Alzahrani, N.H., 2020. solid state fermentation of agro-industrial residues for glucoamylase production from endophytic fungi *Penicillium javanicum* of *Solanum tuberosum* L. J. Microb. Biochem. Technol. 12, 426. Available from: https://doi.org/10.35248/1948-5948.20.12.426.

El-Sherbiny, M.A., El-Chaghaby, G.A., 2011. Storage temperature and stabilizers in relation to the activity of commercial liquid feed enzymes: a case study from Egypt. J. Agrobiol. 28, 129–137.

Else, A.J., Tronsmo, K.M., Niemann, L.-A., Moonen, J.H.E., 2013. Use of an anti-staling enzyme mixture in the preparation of baked bread. Patent Application US 20130059031 A1.

Erdem, N., Babaoğlu, A.S., Poçan, H.B., Karakaya, M., 2020. The effect of transglutaminase on some quality properties of beef, chicken, and turkey meatballs. J. Food Process. Preserv. 44, e14815. Available from: https://doi.org/10.1111/jfpp.14815.

Eshra, D.H., El-Iraki, S.M., Abo Bakr, T.M., 2014. Performance of starch hydrolysis and production of corn syrup using some commercial enzymes. Int. Food Res. J. 21, 815–821.

Faijes, M., Castejón-Vilatersana, M., Val-Cid, C., Planas, A., 2019. Enzymatic and cell factory approaches to the production of human milk oligosaccharides. Biotechnol. Adv. 37, 667–697. Available from: https://doi.org/10.1016/j.biotechadv.2019.03.014.

Farias, M.A., Valoni, E.A., Castro, A.M., Coelho, M.A.Z., 2014. Lipase production by *Yarrowia lipolytica* in solid state fermentation using different agro industrial residues. Chem. Eng. Trans. 38, 301–306. Available from: https://doi.org/10.3303/CET1438051.

Fasim, A., More, V.S., More, S.S., 2021. Large-scale production of enzymes for biotechnology uses. Curr. Op. Biotechnol. 69, 68–76. Available from: https://doi.org/10.1016/j.copbio.2020.12.002.

Feijoo-Siota, L., Blasco, L., Rodríguez-Rama, J.L., Barros-Velázquez, J., Miguel, T., Sánchez-Pérez, A., et al., 2014. Recent patents on microbial proteases for the dairy industry. Recent. Adv. DNA Gene Seq. 8, 44–55. Available from: https://doi.org/10.2174/2352092208666141013231720.

Ferreira-Dias, S., Tecelão, C., 2014. Human milk fat substitutes: advances and constraints of enzyme-catalyzed production. Lipid Technol. 26, 183–185. Available from: https://doi.org/10.1002/lite.201400043.

Ferreira, G.C.A., Silva, J.M., Silva, G.A.R., Ponhozi, I.B., Castro, M.C., Souza, P.M., et al., 2021. Production of human milk fat substitute by enzyme interesterification: a review. Res. Soc. Dev. 10. Available from: https://doi.org/10.33448/rsd-v10i3.13469e36210313469.

Forde, J., OFagain, C., 2008. Immobilized enzymes as industrial biocatalysts. In: Flynne, W.G. (Ed.), Biotechnology and Bioengineering, Nova Science Publishing, NY, USA. pp. 9–36.

Fraga, J.L., Penha, A., da S Pereira, A., Silva, K.A., Akil, E., Torres, A.G., et al., 2018. Use of *Yarrowia lipolytica* lipase immobilized in cell debris for the production of lipolyzed milk fat (LMF. Int. J. Mol. Sci. 19, 3413. Available from: https://doi.org/10.3390/ijms19113413.

García, C.Á., 2018. Application of enzymes for fruit juice processing. In: Rajauria, G., Tiwari, B.K. (Eds.), Fruit Juices. Elsevier, NY, USA, pp. 201–216.

Garuba, E.O., Onilude, A.A., 2020. *Exo*-inulinase production by a catabolite repression-resistant mutant thermophilic *Aspergillus tamarii*-U4 in solid state fermentation. Biotechnol. J. Int. 24, 21–31. Available from: https://doi.org/10.9734/bji/2020/v24i430110.

Gerits, L.R., Pareyt, B., Decamps, K., Delcour, J.A., 2014. Lipases and their functionality in the production of wheat-based food systems. Compr. Rev. Food Sci. Food Saf. 13, 978–989. Available from: https://doi.org/10.1111/1541-4337.12085.

Ghaly, A.E., Ramakrishnan, V.V., Brooks, M.S., Budge, S.M., Dave, D., 2013. Fish processing wastes as a potential source of proteins, amino acids and oils: a critical review. J. Microb. Biochem. Technol. 5, 107–129. Available from: https://doi.org/10.4172/1948-5948.1000110.

Gomaa, A., 2018. Application of enzymes in brewing. J Nutr. Food Sci. Forecast 1, 5. Available from: https://doi.org/10.5281/zenodo.3336203.

Gómez-Plaza, E., Romero-Cascales, I., Bautista-Ortín, A.B., 2010. Use of enzymes for wine production. In: Bayındırlı, A. (Ed.), Enzymes in Fruit and Vegetable Processing. CRC Press, Boca Raton, USA, pp. 215–243.

Grassin, C., Coutel, Y., 2010. Enzymes in fruit and vegetable processing and juice extraction. In: Whitehurst, R.J., van Oort, M. (Eds.), Enzymes in Food Technology, 2nd (Ed.) Wiley-Blackwell Chichester, UK, pp. 236–263.

Guérard, F., Sellos, D., Le Gal, Y., 2005. Fish and shellfish upgrading. Traceability. Adv. Biochem. Engin./Biotechnol. 96, 127–163. Available from: https://doi.org/10.1007/b135783.

Guerrero-Urrutia, C., Volke-Sepulveda, T., Figueroa-Martinez, F., Favela-Torres, E., 2021. Solid-state fermentation enhances inulinase and invertase production by *Aspergillus brasiliensis*. Process. Biochem. 108, 169–175. Available from: https://doi.org/10.1016/j.procbio.2021.06.014.

Guerrero, C., Valdivia, F., Ubilla, C., Ramírez, N., Gómez, M., Aburto, C., et al., 2019. Continuous enzymatic synthesis of lactulose in packed-bed reactor with immobilized *Aspergillus oryzae* β-galactosidase. Bioresour. Technol. 278, 296–302. Available from: https://doi.org/10.1016/j.biortech.2018.12.018.

Guerrero, C., Vera, C., Araya, E., Conejeros, R., Illanes, A., 2015. Repeated-batch operation for the synthesis of lactulose with β-galactosidase immobilized by aggregation and crosslinking. Bioresour. Technol. 190, 122–131. Available from: https://doi.org/10.1016/j.biortech.2015.04.039.

Guha, S., Sharma, H., Deshwal, G.K., Rao, P.S., 2021. A comprehensive review on bioactive peptides derived from milk and milk products of minor dairy species. Food Prod. Process. Nutr. 3, 2. Available from: https://doi.org/10.1186/s43014-020-00045-7.

Harboe, M., Broe, M.L., Qvist, K.B., 2010. The production, action and application of rennet and coagulants. In: Law, B.A., Tamime, A.Y. (Eds.), Technology of Cheesemaking, 2nd (Ed.) Blackwell Publishing Ltd, Oxford, UK, pp. 98–128.

Hassan, A.A., Mansour, E.H., El Bedawey, A.E.-F.A., Zaki, M.S., 2014. Improving dough rheology and cookie quality by protease enzyme. Am. J. Food Sci. Nutr. Res. 1, 1–7.

Hayward, S., Cilliers, T., Swart, P., 2016. Lipoxygenases: from isolation to application. Compr. Rev. Food Sci. Food Saf. 16 (1), 199–211. Available from: https://doi.org/10.1111/1541-4337.12239.

Heerd, D., Yegina, S., Tari, C., Fernandez-Lahore, M., 2012. Pectinase enzyme-complex production by *Aspergillus* spp. in solid-state fermentation: a comparative study. Food Bioprod. Process. 90, 102–110. Available from: https://doi.org/10.1016/j.fbp.2011.08.003.

Heredia-Sandoval, N.G., Valencia-Tapia, M.Y., Calderón de la Barca, A.M., Islas-Rubio, A.R., 2016. Microbial proteases in baked goods: modification of gluten and effects on immunogenicity and product quality. Foods 5, 59. Available from: https://doi.org/10.3390/foods5030059.

Hlima, H.B., Aghajari, N., Ali, M.B., Haser, R., Bejar, S., 2012. Engineered glucose isomerase from *Streptomyces* sp. SK is resistant to Ca2 + inhibition and Co2 + independent. J. Ind. Microbiol. Biotechnol. 39, 537–546. Available from: https://doi.org/10.1007/s10295-011-1061-1.

Huang, Z., Brennan, C.S., Zheng, H., Mohan, M.S., Stipkovits, L., Liu, W., et al., 2020. The effects of fungal lipase-treated milk lipids on bread making. LWT 128, 109455. Available from: https://doi.org/10.1016/j.lwt.2020.109455.

Huang, W., Zhan, Y., Shi, X., Chen, J., Deng, H., Du, Y., 2017. Controllable immobilization of naringinase on electrospun cellulose acetate nanofibers and their application to juice debittering. Int. J. Biol. Macromol. 98, 630–636. Available from: https://doi.org/10.1016/j.ijbiomac.2017.02.018.

Illanes, A., 2008. Enzyme production. In: Illanes, A. (Ed.), Enzyme Biocatalysis: Principles and Applications. Springer Science, Valparaíso, Chile, pp. 57–106.

Iyer, P.V., Ananthanarayan, L., 2008. Enzyme stability and stabilization—aqueous and non-aqueous environment. Process Biochem 43, 1019–1032. Available from: https://doi.org/10.1016/j.procbio.2008.06.004.

Jayaraman, A.B., Kandasamy, T., Venkataraman, D., Meenakshisundaram, S., 2021. Rational design of *Shewanella* sp. L-arabinose isomerase for D-galactose isomerase activity under mesophilic conditions. Enzyme Microb. Technol. 147, 109796. Available from: https://doi.org/10.1016/j.enzmictec.2021.109796.

Jing, G.E., Hua, L., Lee, S.H., Shetty, J.K., Strohm, B.A., Tang, Z., et al., 2014. A process for producing high glucose compositions by simultaneous liquefaction and saccharification of starch substrates. Patent Application WO 2014092961 A1.

Jin, L.Q., Chen, X.X., Jin, Y.T., Shentu, J.K., Liu, Z.Q., Zheng, Y.G., 2021b. Immobilization of recombinant Escherichia coli cells expressing glucose isomerase using modified diatomite as a carrier for effective production of high fructose corn syrup in packed bed reactor. Bioproc. Biosyst. Eng. 44, 1781–1792. Available from: https://doi.org/10.1007/s00449-021-02560-4.

Jin, L.-Q., Jin, Y.-T., Zhang, J.-W., Liu, Z.-Q., Zheng, Y.-G., 2021a. Enhanced catalytic efficiency and thermostability of glucose isomerase from *Thermoanaerobacter ethanolicus* via site-directed mutagenesis. Enzyme Microb. Technol. 152, 109931. Available from: https://doi.org/10.1016/j.enzmictec.2021.109931.

Karim, A., Aider, M., 2022. Production of prebiotic lactulose through isomerisation of lactose as a part of integrated approach through whey and whey permeate complete valorisation: a review. Int. Dairy. J. 126. Available from: https://doi.org/10.1016/j.idairyj.2021.105249Article 105249.

Kaur, S.J., Gupta, V.K., 2017. Production of pectinolytic enzymes pectinase and pectin lyase by *Bacillus subtilis* SAV-21 in solid state fermentation. Ann. Microbiol. 67, 333–342. Available from: https://doi.org/10.1007/s13213-017-1264-4.

Kieliszek, M., Misiewicz, A., 2014. Microbial transglutaminase and its application in the food industry. A review. Folia Microbiol. 59, 241–250. Available from: https://doi.org/10.1007/s12223-013-0287-x.

Kim, Y.S., Kim, J.E., Oh, D.K., 2013. Borate enhances the production of lactulose from lactose by cellobiose 2-epimerase from *Caldicellulosiruptor saccharolyticus*. Bioresour. Technol. 128, 809–812. Available from: https://doi.org/10.1016/j.biortech.2012.10.060.

Kiran, E.U., Trzcinski, A.P., Liu, Y., 2014. Glucoamylase production from food waste by solid state fermentation and its evaluation in the hydrolysis of domestic food waste. Biofuel Res. J. 1, 98–105. Available from: https://doi.org/10.18331/BRJ2015.1.3.7.

Kotwal, S.M., Shankar, V., 2009. Immobilized invertase. Biotechnol. Adv. 27, 311–322. Available from: https://doi.org/10.1016/j.biotechadv.2009.01.009.

Kranthi, V.S., Rao, D.M., Jaganmohan, P., 2012. Protease production by rhizopus stolonifer through solid state fermentation. Cent. Eur. J. Exp. Biol. 1, 113–117.

Kumar, R., Kesavapillai, B., 2012. Stimulation of extracellular invertase production from spent yeast when sugarcane pressmud used as substrate through solid state fermentation. SpringerPlus 1, 81. Available from: https://doi.org/10.1186/2193-1801-1-81.

Kunamneni, A., Plou, F.J., Ballesteros, A., Alcalde, M., 2008. Laccases and their applications: a patent review. Recent. Pat. Biotechnol. 2, 10–24. Available from: https://doi.org/10.2174/187220808783330965.

Lalor, E., Goode, D., 2010. Brewing with enzymes. In: Whitehurst, R.J., van Oort, M. (Eds.), Enzymes in Food Technology, 2nd (Ed.) Wiley-Blackwell Chichester, UK, pp. 163–194.

Law, B.A., 2010. Enzymes in dairy product manufacture. In: Whitehurst, R.J., Van Oort, M. (Eds.), Enzymes in Food Technology, 2nd (Ed.) Wiley-Blackwell, Chichester, UK, pp. 88–101.

Liese, A., Seelbach, K., Buchholz, A., Haberland, J., 2006. Processes. In: Liese, A., Seelbach, K., Wandrey, C. (Eds.), Industrial Biotransformations. Wiley-VCH, Weinheim, Germany, pp. 147–513.

Lincoln, L., More, S.S., 2018. Comparative evaluation of extracellular β-d-fructofuranosidase in submerged and solid-state fermentation produced by newly identified *Bacillus subtilis* strain. J. Appl. Microbiol. 125, 441–456. Available from: https://doi.org/10.1111/jam.13881.

Liu, P., Wu, J., Liu, J., Ouyang, J., 2021a. Engineering of a β-galactosidase from Bacillus coagulans to relieve product inhibition and improve hydrolysis performance. J. Dairy. Sci. 104, 10566–10575. Available from: https://doi.org/10.3168/jds.2021-20388.

Liu, X., Wu, Y., Guan, R., Jia, G., Ma, Y., Zhang, Y., 2021b. Advances in research on calf rennet substitutes and their effects on cheese quality. Food Res. Int. (Ottawa, Ont.) 149, 110704. Available from: https://doi.org/10.1016/j.foodres.2021.110704.

Liu, L., Bilal, M., Luo, H., Zhao, Y., Duan, X., 2021c. Studies on biological production of isomaltulose using sucrose isomerase: current status and future perspectives. Catal. Lett. 151, 1868–1881. Available from: https://doi.org/10.1007/s10562-020-03439-x.

Liu, L., Yang, H., Shin, H.-d, Chen, R.R., Li, J., Du, G., et al., 2013. How to achieve high-level expression of microbial enzymes: strategies and perspectives. Bioengineered 4, 212–223. Available from: https://doi.org/10.4161/bioe.24761.

Li, Q., Gui, P., Huang, Z., Feng, L., Luo, Y., 2018b. Effect of transglutaminase on quality and gel properties of pork and fish mince mixtures. J. Texture Stud. 49, 56–64. Available from: https://doi.org/10.1111/jtxs.12281.

Li, H., Liu, Y., Sun, Y., Li, H., Yu, J., 2019. Properties of polysaccharides and glutamine transaminase used in mozzarella cheese as texturizer and crosslinking agents. LWT 99 (2019), 411–416. Available from: https://doi.org/10.1016/j.lwt.2018.10.011.

Li, Z., Li, Y., Duan, S., Liu, J., Yuan, P., Nakanishi, H., et al., 2015. Bioconversion of d-glucose to d-psicose with immobilized d-xylose isomerase and d-psicose 3-epimerase on Saccharomyces cerevisiae spores. J. Ind. Microbiol. Biotechnol. 42, 1117–1128. Available from: https://doi.org/10.1007/s10295-015-1631-8.

Li, L., Li, W., Gong, J., Xu, Y., Wu, Z., Jiang, Z., et al., 2021. An effective computational-screening strategy for simultaneously improving both catalytic activity and thermostability of α-l-rhamnosidase. Biotechnol. Bioeng. 118, 3409–3419. Available from: https://doi.org/10.1002/bit.27758.

Li, Q., Zhang, G., Du, G., 2022. Production of food enzymes. In: Rai, A.K., Singh, S.P., Pandey, A., Larroche, C., Soccol, C.R. (Eds.), Current Developments in Biotechnology and

Bioengineering. Elsevier, Amsterdam, Netherlands, pp. 139–155. Available from: https://doi.org/10.1016/B978-0-12-823506-5.00015-1.

Li, M., Zhang, H., Li, Y., Hu, X., Yang, J., 2018a. The thermoduric effects of site-directed mutagenesis of proline and lysine on dextransucrase from *Leuconostoc mesenteroides* 0326. Int. J. Biol. Macromol. 107, 1641–1649.

Li, W., Zhu, Y., Jiang, X., Zhang, W., Guang, C., Mu, W., 2020. One-pot production of d-allulose from inulin by a novel identified thermostable exoinulinase from *Aspergillus piperis* and *Dorea* sp. d-allulose 3-epimerase. Process. Biochem. 99, 87–95. Available from: https://doi.org/10.1016/j.procbio.2020.08.021.

Lohscheidt, M., Betz, R., Braun, J., Pelletier, W., Ader, P., 2009. Enzyme Granulate 1 Containing Phytase. Patent Application US 20090274795 A1.

Lutz, S., 2010. Beyond directed evolution - semi-rational protein engineering and design. Curr. Op. Biotechnol. 21, 734–743. Available from: https://doi.org/10.1016/j.copbio.2010.08.011.

Luzzi, G., Steffens, M., Clawin-Rädecker, I., Hoffmann, W., Franz, C.M.A.P., Fritsche, J., et al., 2020. Enhancing the sweetening power of lactose by enzymatic modification in the reformulation of dairy products. Int. J. Dairy. Technol. 73, 502–512. Available from: https://doi.org/10.1111/1471-0307.12681.

Madhavan, A., Arun, K.B., Binod, P., Sirohi, R., Tarafdar, A., Reshmy, R., et al., 2021. Design of novel enzyme biocatalysts for industrial bioprocess: harnessing the power of protein engineering, high throughput screening and synthetic biology. Bioresour. Technol. 325, 124617. Available from: https://doi.org/10.1016/j.biortech.2020.124617.

Madhavan, A., Sindhu, R., Binod, P., Sukumaran, R.K., Pandey, A., 2017. Strategies for design of improved biocatalysts for industrial applications. Bioresour. Technol. 245 (Pt B), 1304–1313. Available from: https://doi.org/10.1016/j.biortech.2017.05.031.

Mageswari, A., Subramanian, P., Chandrasekaran, S., Karthikeyan, S., Gothandam, K.M., 2017. Systematic functional analysis and application of a cold-active serine protease from a novel *Chryseobacterium* sp. Food Chem. 217, 18–27. Available from: https://doi.org/10.1016/j.foodchem.2016.08.064.

Magnuson, B., Munro, I., Abbot, P., Baldwin, N., Lopez-Garcia, R., Ly, K., et al., 2013. Review of the regulation and safety assessment of food substances in various countries and jurisdictions. Food Addit. Contam.: Part. A 30, 1147–1220. Available from: https://doi.org/10.1080/19440049.2013.795293.

Magomet, M., Tomov, T., Somann, T., Abelyan, V.H., 2010. Sweetener and use. Patent US 7807206 B2.

Manning, M.C., Liu, J., Li, T., Holcomb, R.E., 2018. Rational design of liquid formulations of proteins. Adv. Protein Chem. Struct. Biol. 112, 1–59. Available from: https://doi.org/10.1016/bs.apcsb.2018.01.005.

Manoochehri, H., Fayazi, N., Saidijam, M., Taheri, M., 2020. A review on invertase: its potentials and applications. Biocatal. Agric. Biotechnol. 25, 101599. Available from: https://doi.org/10.1016/j.bcab.2020.101599.

Martín, M.C., López, O.V., Ciolino, A.E., Morata, V.I., Villar, M.A., Ninago, M.D., 2019. Immobilization of enological pectinase in calcium alginate hydrogels: a potential biocatalyst for winemaking. Biocatal. Agric. Biotechnol. 2019 (18), 101091. Available from: https://doi.org/10.1016/j.bcab.2019.101091.

Martosuyono, P., Fawzya, Y., Patantis, G., Sugiyono, S., 2019. Enzymatic production of fish protein hydrolysates in a pilot plant scale. Squalen Bull. Mar. Fish. Postharvest Biotechnol. 14, 85–92. Available from: https://doi.org/10.15578/squalen.v14i2.398.

Mazutti, M.A., Zabot, G., Boni, G., Skovronski, A., de Oliveira, D., Di Luccio, M., et al., 2010. Optimization of inulinase production by solid-state fermentation in a packed-bed bioreactor. J. Chem. Technol. Biotechnol. 85, 109–114. Available from: https://doi.org/10.1002/jctb.2273.

Meng, S., Yin, Y., Yu, L., 2019. Exploration of a high-efficiency and low-cost technique for maximizing the glucoamylase production from food waste. RSC Adv. 9, 22980–22986. Available from: https://doi.org/10.1039/C9RA04530A.

Mercer, G.D., 2014. GRAS notification for Cynzime® milk clotting enzyme preparation derived from *Cynara cardunculus* flowers; Fytozimus Biotech, Inc.: Canada.

Miguel, A.S.M., Martins-Meyer, T.S., Figueiredo, E.V.C., Lobo, B.W.P., 2013. Dellamora-Ortiz, 2013. Enzymes in bakery: current and future trends. In: Muzzalupo, I. (Ed.), Food Industry. [Internet]. IntechOpen, London [cited 2022 Mar 09]. Available from: https://www.intechopen.com/chapters/41661 doi:10.5772/53168.

Mikl, M., Dennig, A., Nidetzky, B., 2020. Efficient enzyme formulation promotes *Leloir glycosyltransferases* for glycoside synthesis. J. Biotechnol. 322, 74–78. Available from: https://doi.org/10.1016/j.jbiotec.2020.06.023.

Mistry, V.V., 2012. Chymosin in Cheese Making. In: Simpson, B.K. (Ed.), Food Biochemistry and Food Processing, 2nd. Edition Wiley-Blackwell, Oxford, UK, pp. 223–231.

Molina, M., Cioci, G., Moulis, C., Séverac, E., Remaud-Siméon, M., 2021. Bacterial α-glucan and branching sucrases from GH70 family: discovery, structure–function relationship studies and engineering. Microorganisms 9, 1607. Available from: https://doi.org/10.3390/microorganisms9081607.

Murlidhar, M., Anusha, R., Bindhu, O.S., 2017. Plant-based coagulants in cheese making: review. In: Murlidhar, M., Goyal, M.R., Chavan, R.S. (Eds.), Dairy Engineering: Advanced Technologies and Their Applications, Apple. Academic Press Inc, Oakville, Canada, pp. 4–35.

Neifar, S., Cervantes, F.V., Darenfed, A.B., BenHlima, H., Ballesteros, A.O., Plou, F.J., et al., 2020. Immobilization of the glucose isomerase from *Caldicoprobacter algeriensis* on sepabeads EC-HA and its efficient application in continuous high fructose syrup production using packed bed reactor. Food Chem. 309, 125710. Available from: https://doi.org/10.1016/j.foodchem.2019.125710.

Neifar, M., Maktouf, S., Ghorbel, R.E., Jaouani, A., Cherif, A., 2015. Extremophiles as source of novel bioactive compounds with industrial potential. In: Gupta, V.K., Tuohy, M.G. (Eds.), Biotechnology of Bioactive Compounds: Sources and Applications. John Wiley & Sons, Ltd, Chichester, UK, pp. 245–266.

Nema, A., Patnala, S.H., Mandari, V., Kota, S., Devarai, S.K., 2019. Production and optimization of lipase using *Aspergillus niger* MTCC 872 by solid-state fermentation. Bull. Natl. Res. Cent. 43, 82. Available from: https://doi.org/10.1186/s42269-019-0125-7.

Niyonzima, F.N., 2019. Production of microbial industrial enzymes. Acta Sci. Micobiol 2, 75–89. Available from: https://doi.org/10.31080/ASMI.2019.02.0434.

Nobre, C., Teixeira, J.A., Rodrigues, L.R., 2015. New trends and technological challenges in the industrial production and purification of fructo-oligosaccharides. Crit. Rev. Food Sci. Nutr. 55, 1444–1455. Available from: https://doi.org/10.1080/10408398.2012.697082.

Nunes, M.A.P., Rosa, M.E., Fernandes, P.C.B., Ribeiro, M.H.L., 2014. Operational stability of naringinase PVA lens-shaped microparticles in batch stirred reactors and mini packed bed reactors - one step closer to industry. Bioresour. Technol. 164, 362–370. Available from: https://doi.org/10.1016/j.biortech.2014.04.108.

Ogilvie, O., Roberts, S., Sutton, K., Larsen, N., Gerrard, J., Domigan, L., 2021. The use of microbial transglutaminase in a bread system: a study of gluten protein structure, deamidation state and protein digestion. Food Chem. 340. Available from: https://doi.org/10.1016/j.foodchem.2020.127903Article 127903.

Oh, D.K., 2007. Tagatose: properties, applications, and biotechnological processes. Appl. Microbiol. Biotechnol. 76, 1–8. Available from: https://doi.org/10.1007/s00253-007-0981-1.

Olempska-Beer, Z.S., Merker, R.I., Ditto, M.D., DiNovi, M.J., 2006. Food-processing enzymes from recombinant microorganisms—a review. Regul. Toxicol. Pharmacol. 45, 144–158. Available from: https://doi.org/10.1016/j.yrtph.2006.05.001.

Oliveira, F., Souza, C.E., Peclat, V.R., Salgado, J.M., Ribeiro, B.D., Coelho, M.A.Z., et al., 2017. Optimization of lipase production by *Aspergillus ibericus* from oil cakes and its application in esterification reactions. Food Bioprod. Proc. 102, 268–277. Available from: https://doi.org/10.1016/j.fbp.2017.01.007.

Onilude, A.A., Fadaunsi, I.F., Garuba, E.O., 2012. Inulinase production by *Saccharomyces* sp. in solid state fermentation using wheat bran as substrate. Ann. Microbiol. 62, 843–848. Available from: https://doi.org/10.1007/s13213-011-0325-3.

Ottone, C., Romero, O., Aburto, C., Illanes, A., Wilson, L., 2020. Biocatalysis in the winemaking industry: challenges and opportunities for immobilized enzymes. Compr. Rev. Food Sci. Food Saf. 19, 595–621. Available from: https://doi.org/10.1111/1541-4337.12538.

Padilla, B., Frau, F., Ruiz-Matute, A.I., Montilla, A., Belloch, C., Manzanares, P., et al., 2015. Production of lactulose oligosaccharides by isomerisation of transgalactosylated cheese whey permeate obtained by β-galactosidases from dairy Kluyveromyces. J. Dairy. Res. 82, 356–364. Available from: https://doi.org/10.1017/S0022029915000217.

Pagán, J., 2014. Utilization of enzymes in fruit juice production. In: Falguera, V., Ibarz, A. (Eds.), Juice Processing. Quality, Safety and Value-Added opportunities. CRC Press, Boca Raton, Florida, USA, pp. 151–170.

Pandey, A., Soccol, C.R., Larroche, C., 2008. Current Developments in Solid-state Fermentation. Asiatech Publishers, New Delhi, India.

Panesar, P.S., Chopra, H.K., Marwaha, S.S., 2010a. Fundamentals of enzymes. In: Panesar, P.S., Marwaha, S.S., Chopra, H.K. (Eds.), Enzymes in Food Processing: Fundamentals and Potential Applications. IK International Pvt. Ltd, New Delhi, India, pp. 1–50.

Panesar, P.S., Kumari, S., Panesar, R., 2010b. Potential applications of immobilized β-galactosidase in food processing industries. Enzyme Res. 2010, 473137. Available from: https://doi.org/10.4061/2010/473137.

Parbat, R., Singhal, B., 2011. Production of glucoamylase by *Aspergillus oryzae* under solid state fermentation using agro industrial products. Int. J. Microbiol. Res. 2, 204–207.

Parıldı, E., Kola, O., Özcan, B.D., Akkaya, M.R., Dikkaya, E., 2021. Recombinant D-tagatose 3-epimerase production and converting fructose into allulose. J. Food Process. Preserv. 00, e15508. Available from: https://doi.org/10.1111/jfpp.15508.

Park, A.R., Kim, J.S., Jang, S.W., Park, Y.-G., Koo, B.-S., Lee, H.-C., 2017. Rational modification of substrate binding site by structure-based engineering of a cellobiose 2-epimerase in *Caldicellulosiruptor saccharolyticus*. Microb. Cell Fact. 16, 224. Available from: https://doi.org/10.1186/s12934-017-0841-3.

Pilak, P., Schiefner, A., Seiboth, J., Oehrlein, J., Skerra, A., 2020. Engineering a highly active sucrose isomerase for enhanced product specificity by using a "Battleship" strategy. Chembiochem: Europ. J. Chem. Biol. 21, 2161–2169. Available from: https://doi.org/10.1002/cbic.202000007.

Pires, E.B.E., de Freitas, A.J., Souza, F.Fe, et al., 2019. Production of fungal phytases from agro-industrial byproducts for pig diets. Sci. Rep. 9, 9256. Available from: https://doi.org/10.1038/s41598-019-45720-z.

Pottratz, I., Müller, I., Hamel, C., 2022. Potential and scale-up of pore-through-flow membrane reactors for the production of prebiotic galacto-oligosaccharides with immobilized β-galactosidase. Catalysts 12, 7. Available from: https://doi.org/10.3390/catal12010007.

Purewal, S.S., Sandhu, K.S., 2021. Debittering of citrus juice by different processing methods: a novel approach for food industry and agro-industrial sector. Sci. Hortic. 276, 109750. Available from: https://doi.org/10.1016/j.scienta.2020.109750.

Puri, M., 2012. Updates on naringinase: structural and biotechnological aspects. Appl. Microbiol. and Biotechnol 93, 49–60. Available from: https://doi.org/10.1007/s00253-011-3679-3.

Rai, S.K., Kaur, H., Singh, A., Kamboj, M., Jain, G., Yadav, S.K., 2021. Production of d-tagatose in packed bed reactor containing an immobilized l-arabinose isomerase on alginate support. Biocatal. Agric. Biotechnol. 38, 102227. Available from: https://doi.org/10.1016/j.bcab.2021.102227.

Rai, S.K., Narnoliya, L.K., Sangwan, R.S., Yadav, S.K., 2018. Self-assembled hybrid nanoflowers of manganese phosphate and l-arabinose isomerase: a stable and recyclable nanobiocatalyst for equilibrium level conversion of d-galactose to d-tagatose. ACS Sustain. Chem. Eng. 6, 6296–6304. Available from: https://doi.org/10.1021/acssuschemeng.8b00091.

Rajagopalan, G., Krishnan, C.J., 2010. Hyper-production of alpha-amylase from agro-residual medium with high-glucose in SSF using catabolite derepressed *Bacillus subtilis* KCC103. J. Basic. Microbiol. 50, 336–343. Available from: https://doi.org/10.1002/jobm.200900199.

Rajakari, K., Hotakainen; K., Myllärinen, P., 2013. A liquid enzyme formulation and a process for its preparation. Patent WO2013064736A1.

Ramos, P., Vicente, A.A., Teixeira, J.A., 2013. Biotechnology-derived enzymes for food applications. In: Teixeira, J.A., Vicente, A.A. (Eds.), Engineering Aspects of Food Biotechnology. CRC Press, Boca Raton, USA, pp. 3–20.

Reetz, M.T., Carballeira, J.D., Vogel, A., 2006. Iterative saturation mutagenesis on the basis of B factors as a strategy for increasing protein thermostability. Angew. Chem. Int. (Ed.) 45, 7745–7751. Available from: https://doi.org/10.1002/anie.200602795.

Rodriguez-Colinas, B., Fernandez-Arrojo, L., Ballesteros, A.O., Plou, F.J., 2014. Galactooligosaccharides formation during enzymatic hydrolysis of lactose: towards a prebiotic-enriched milk. Food Chem. 145, 388–394. Available from: https://doi.org/10.1016/j.foodchem.2013.08.060.

Rodríguez-Fernández, D.A., Parada, J.L., Medeiros, A.B.P., de Carvalho, J.C., Lacerda, L.G., Rodríguez-León, J.A., et al., 2013. Concentration by ultrafiltration and stabilization of phytase produced by solid-state fermentation. Process. Biochem. 48, 374–379. Available from: https://doi.org/10.1016/j.procbio.2012.12.021.

Rodríguez-Núñez, K., Bernal, C., Martínez, R., 2021. Immobilized biocatalyst engineering: high throughput enzyme immobilization for the integration of biocatalyst improvement strategies. Int. J. Biol. Macromol. 170, 61–70. Available from: https://doi.org/10.1016/j.ijbiomac.2020.12.097.

Rodríguez, A.P., Leiro, R.F., Trillo, M.C., Cerdán, M.E., Siso, M.I., Becerra, M., 2006. Secretion and properties of a hybrid *Kluyveromyces lactis-Aspergillus niger* β-galactosidase. Microb. Cell Fact. 5, 41. Available from: https://doi.org/10.1186/1475-2859-5-41.

Ruiz, H.A., Rodríguez-Jasso, R.M., Rodríguez, R., Contreras-Esquivel, J.C., Aguilar, C.N., 2012. Pectinase production from lemon peel pomace as support and carbon source in solid-state fermentation column-tray bioreactor. Biochem. Eng. J. 65, 90–95. Available from: https://doi.org/10.1016/j.bej.2012.03.007.

Saito, Y., Oikawa, M., Sato, T., Nakazawa, H., Ito, T., Kameda, T., et al., 2021. Machine-learning-guided library design cycle for directed evolution of enzymes: the effects of training data composition on sequence space exploration. ACS Catal. 11, 14615–14624. Available from: https://doi.org/10.1021/acscatal.1c03753.

Salehmin, M.N.I, Annuar, M.S., Chisti, Y., 2013. High cell density fed-batch fermentations for lipase production: feeding strategies and oxygen transfer. Bioprocess. Biosyst. Eng. 36, 1527–1543. Available from: https://doi.org/10.1007/s00449-013-0943-1.

Sarrouh, B., Santos, T.M., Miyoshi, A., Dias, R., Azevedo, V., 2012. Up-to-date insight on industrial enzymes applications and global market. J. Bioproces. Biotechniq. S 4, 1–10.

Saxena, R., Singh, R., 2011. Amylase production by solid-state fermentation of agro-industrial wastes using Bacillus sp. Braz. J. Microbiol. 42, 1334–1342. Available from: https://doi.org/10.1590/S1517-83822011000400014.

Schulz, P., Rizvi, S.S.H., 2021. Hydrolysis of lactose in milk: current status and future products. Food Rev. Int. Available from: https://doi.org/10.1080/87559129.2021.1983590.

Segura, D., Mygind. P., Tossi, A., Hogenhaug, H.-H., 2007. Antimicrobial peptides. Patent Application US20070259087A1.

Serna-Saldivar, S.R., Rito-Palomares, M.A., 2008. Production of invert syrup from sugarcane juice using immobilized invertase. Patent US7435564B2.

Shanmugaprakash, M., Kumar, V.V., Hemalatha, M., Melbia, V., Karthik, P., 2011. Solid-state fermentation for the production of debittering enzyme naringinase using *Aspergillus niger* MTCC 1344. Eng. Life Sci. 11, 322–325. Available from: https://doi.org/10.1002/elsc.201000128.

Shehata, A.N., El Aty, A.A.A., 2014. Optimization of process parameters by statistical experimental designs for the production of naringinase enzyme by marine fungi. Int. J. Chem. Eng. 2014, 273523. Available from: https://doi.org/10.1155/2014/273523.

Sheldon, R.A., 2019. CLEAs, Combi-CLEAs and 'Smart' Magnetic CLEAs: biocatalysis in a bio-based economy. Catalysts 9, 261. Available from: https://doi.org/10.3390/catal9030261.

Sheldon, R.A., Basso, A., Brady, D., 2021. New frontiers in enzyme immobilisation: robust biocatalysts for a circular bio-based economy. Chem. Soc. Rev. 50, 5850–5862. Available from: https://doi.org/10.1039/d1cs00015b.

Shintani, T., 2019. Food industrial production of monosaccharides using microbial, enzymatic, and chemical methods. Fermentation 5, 47. Available from: https://doi.org/10.3390/fermentation5020047.

Shukla, A., Mishra, K.P., Tripathi, P., 2013. β-galactosidase a novel enzyme for treating dairy waste "an environment friendly approach": a review. Indian. Res. J. Genet. & Biotech. 5, 151–159.

Sicard, P.J., Leleu, J.B., Duflot, P., Drocourt, D., Martin, F., Tiraby, G., et al., 1990. Site-directed mutagenesis applied to glucose isomerase from *Streptomyces violaceoniger* and *Streptomyces olivochromogenes*. Ann. N. Y. Acad. Sci. 613, 371–375. Available from: https://doi.org/10.1111/j.1749-6632.1990.tb18181.x.

Singhania, R.R., Patel, A.K., Pandey, A., 2010. The industrial production of enzymes. In: Soetaert, W., Vandamme, E.J. (Eds.), Industrial Biotechnology: Sustainable Growth and Economic Success. Wiley-VCH Verlag GmbH & Co. KGaA, Weinheim, Germany, pp. 207–225.

Singh, R.S., Chauhan, K., Singh, J., Pandey, A., Larroche, C., 2018. Solid-state fermentation of carrot pomace for the production of inulinase by *Penicillium oxalicum* BGPUP-4. Food Technol. Biotechnol. 56, 31–39. Available from: https://doi.org/10.1713/ftb.56.01.18.5411.

Singh, P., Kumar, S., 2019. Microbial enzyme in food biotechnology. In: Kuddus, M. (Ed.), Enzymes in Food Biotechnology. Academic Press, Cambridge, USA, pp. 19–28. Available from: https://doi.org/10.1016/B978-0-12-813280-7.00002-5.

Singh, B., Kumar, G., Kumar, V., Singh, D., 2021. Enhanced phytase production by *Bacillus subtilis* subsp. subtilis in solid state fermentation and its utility in improving food nutrition. Protein Pept. Lett. 28, 1083–1089. Available from: https://doi.org/10.2174/0929866528666210720142359.

Singh, R.S., Singh, R.P., 2010. Production of fructooligosaccharides from inulin by endoinulinases and their prebiotic potential. Food Technol. Biotechnol. 48, 435–449. Available from: https://hrcak.srce.hr/61714.

Singh, R.S., Singh, T., Hassan, M., Kennedy, J.F., 2020. Updates on inulinases: structural aspects and biotechnological applications. Int. J. Biol. Macromol. 164, 193–210. Available from: https://doi.org/10.1016/j.ijbiomac.2020.07.078.

Singh, R., Singh, A., Sachan, S., 2019. Enzymes used in the food industry: friends or foes? In: Kuddus, M. (Ed.), Enzymes in Food Biotechnology. Academic Press, Cambridge, USA, pp. 827–843. Available from: https://doi.org/10.1016/B978-0-12-813280-7.00048-7.

Singh, R.K., Tiwari, M.K., Singh, R., Lee, J.-K., 2013. From protein engineering to immobilization: promising strategies for the upgrade of industrial enzymes. Int. J. Mol. Sci. 14, 1232–1277. Available from: https://doi.org/10.3390/ijms14011232.

Slivinski, C.T., Machado, A.V.L., Iulek, J., Ayub, R.A., Almeida, M.M., 2011. Biochemical characterisation of a glucoamylase from *Aspergillus niger* produced by solid-state fermentation. Braz. Arch. Biol. Technol. 54, 559–568. Available from: https://doi.org/10.1590/S1516-89132011000300018.

Sneha, H.P.P., Beulah, K.C.C., Murthy, P.S., 2019. Enzyme immobilization methods and applications in the food industry. In: Kuddus, M. (Ed.), Enzymes in Food Biotechnology. Academic Press, Cambridge, USA, pp. 645–658. , 10.1016/b978-0-12-813280-7.00037-2.

Sojitra, U.V., Nadar, S.S., Rathod, V.K., 2016. A magnetic tri-enzyme nanobiocatalyst for fruit juice clarification. Food Chem. 213, 296–305. Available from: https://doi.org/10.1016/j.foodchem.2016.06.074.

Soman, S., Kumarasamy, S., Narayanan, M., Ranganathan, M., 2020. Biocatalyst: phytase production in solid state fermentation by OVAT strategy. Biointerface Res. Appl. Chem. 10, 6119–6127. Available from: https://doi.org/10.33263/briac105.61196127.

Souppe, J., Beudeker, R.F., 2002. Process for the production of alcoholic beverages using maltseed. Patent US 6361808 B1.

Spohner, S.C., Müller, H., Quitmann, H., Czermak, P., 2015. Expression of enzymes for the usage in food and feed industry with *Pichia pastoris*. J. Biotechnol. 202, 118–134. Available from: https://doi.org/10.1016/j.jbiotec.2015.01.027.

Sriprapundh, D., Vieille, C., Zeikus, J.G., 2003. Directed evolution of *Thermotoga neapolitana* xylose isomerase: high activity on glucose at low temperature and low pH. Protein Eng. Des. Sel. 16, 683–690. Available from: https://doi.org/10.1093/protein/gzg082.

Srivastava, N., 2019. Production of food-processing enzymes from recombinant microorganisms. In: Kuddus, M. (Ed.), Enzymes in Food Biotechnology. Academic Press, Cambridge, USA, pp. 739–767. Available from: https://doi.org/10.1016/B978-0-12-813280-7.00043-8.

Steiner, K., Schwab, H., 2012. Recent advances in rational approaches for enzyme engineering. Comput. Struct. Biotechnol. J. 2012 (2), 1–12. Available from: https://doi.org/10.5936/csbj.201209010.

Swaisgood, H.E., 2003. Use of immobilized enzymes in the food industry. In: Whitaker, J.R., Vorgen, A.G.I., Wong, D.W.S. (Eds.), Handbook of Food Enzymology. Marcel Dekker, NY, USA, pp. 185–236.

Tajima, T., Hamada, M., Nakashimada, Y., Kato, J., 2015. Efficient aspartic acid production by a psychrophile-based simple biocatalyst. J. Ind. Microbiol. Biotechnol. 42, 1319–1324. Available from: https://doi.org/10.1007/s10295-015-1669-7.

Takeshita, K., Suga, A., Takada, G., Izumori, K., 2000. Mass production of D-psicose from d-fructose by a continuous bioreactor system using immobilized D-tagatose 3-epimerase. J. Biosci. Bioeng. 90, 453–455. Available from: https://doi.org/10.1016/s1389-1723(01)80018-9.

Tang, X., Luo, T., Li, X., Yang, H., Yang, Y., Li, J., et al., 2018. Application and analysis of *Rhizopus oryzae* mycelia extending characteristic in solid-state fermentation for producing glucoamylase. J. Microbiol. Biotechnol. 28, 1865–1875. Available from: https://doi.org/10.4014/jmb.1805.05023.

Tan, Y., Zhang, Y., Han, Y., Liu, H., Chen, H., Ma, F., et al., 2019. Directed evolution of an α1,3-fucosyltransferase using a single-cell ultrahigh-throughput screening method. Sci. Adv. 5, eaaw8451. Available from: https://doi.org/10.1126/sciadv.aaw8451.

Tapre, A.R., Jain, R.K., 2014. Pectinases: enzymes for fruit processing industry. Int. Food Res. J. 21, 447–453.

Tarafdar, A., Sirohi, R., Gaur, V.K., Kumar, S., Sharma, P., Varjani, S., et al., 2021. Engineering interventions in enzyme production: lab to industrial scale. Bioresour. Technol. 326, 124771. Available from: https://doi.org/10.1016/j.biortech.2021.124771.

Tavernini, L., Aburto, C., Romero, O., Illanes, A., Wilson, L., 2021. Encapsulation of Combi-CLEAs of glycosidases in alginate beads and polyvinyl alcohol for wine aroma enhancement. Catalysts 11, 866. Available from: https://doi.org/10.3390/catal11070866.

Thomas, L., Larroche, C., Pandey, A., 2013. Current developments in solid-state fermentation. Biochem. Eng. J. 81, 146–161. Available from: https://doi.org/10.1016/j.bej.2013.10.013.

Torres, D.P., Goncalves, M., Teixeira, J.A., Rodrigues, L.R., 2010. Galacto-oligosaccharides: production, properties, applications, and significance as prebiotics. Compr. Rev. Food Sci. Food Saf. 9, 438–454. Available from: https://doi.org/10.1111/j.1541-4337.2010.00119.x.

Torres-Salas, P., del Monte-Martinez, A., Cutiño-Avila, B., Rodriguez-Colinas, B, Alcalde, M, Ballesteros, A.O., Plou, F.J., 2011. Immobilized biocatalysts: novel approaches and tools for binding enzymes to supports. Adv. Mater. 23, 5275–5282. Available from: https://doi.org/10.1002/adma.201101821.

Tripathi, B.C., Yadav, P., Sharma, R., 2020. Microbial enzymes in food industry: applications. J. Crit. Rev. 7, 1418–1422.

Uhlig, H., Linsmaier-Bednar, E.M., 1998. Industrial Enzymes and Their Applications. John Wiley & Sons, NY, USA.

Usman, A., Mohammed, S., Mamo, J., 2021. Production, optimization, and characterization of an acid protease from a filamentous fungus by solid-state fermentation. Int. J. Microbiol. 2021. Available from: https://doi.org/10.1155/2021/66859636685963–6685963.

van den Berg, M.A., Roubos, J.A., Parnicová, L., 2010. Enzymes in fruit and vegetable processing: future trends in enzyme discovery, design, production, and application. In: Bayındırlı, A. (Ed.), Enzymes in Fruit and Vegetable Processing. CRC Press, Boca Raton, USA, pp. 341–358.

van Oort, M., 2010. Enzymes in bread making. In: Whitehurst, R.J., van Oort, M. (Eds.), Enzymes in Food Technology, 2nd (Ed.) Wiley-Blackwell Chichester, UK, pp. 103–143.

Velasco-Lozano, S., Volke-Sepulveda, T., Favela-Torres, E., 2012. Lipases production by solid-state fermentation: the case of *Rhizopus homothallicus* in perlite. Methods Mol. Biol. 861, 227–237. Available from: https://doi.org/10.1007/978-1-61779-600-5_14.

Vongsangnak, W., Nielsen, J., 2013. Systems biology methods and developments of filamentous fungi in relation to the production of food ingredients. In: McNeil, B., Archer, D., Giavasis, I., Harvey, L.M. (Eds.), Microbial production of food ingredients, enzymes and nutraceuticals. Woodhead Publishing Ltd, Cambridge, UK, pp. 19–41.

Wahab, R.A., Elias, N., Abdullah, F., Ghoshal, S.K., 2020. On the taught new tricks of enzymes immobilization: an all-inclusive overview. React. Funct. Polym. 152, 104613. Available from: https://doi.org/10.1016/j.reactfunctpolym.2020.104613.

Wang, C., Huang, R., He, B., Du, Q., 2012. Improving the thermostability of alpha-amylase by combinatorial coevolving-site saturation mutagenesis. BMC Bioinforma. 13, 263. Available from: https://doi.org/10.1186/1471-2105-13-263.

Wang, H., Yang, R., Hua, X., Zhao, W., Zhang, W., 2013. Enzymatic production of lactulose and 1-lactulose: current state and perspectives. Appl. Microbiol. Biotechnol. 97, 6167–6180. Available from: https://doi.org/10.1007/s00253-013-4998-3.

Weetall, H., Yaverbaum, S., 1974. Treatment of whey with immobilized lactase and glucose isomerase. Patent US 3852496 A.

Wilson, L., Illanes, A., Ottone, C., Romero, O., 2022. Co-immobilized carrier-free enzymes for lactose upgrading. Curr. Opin. Green. Sustain. Chem. 33, 100553. Available from: https://doi.org/10.1016/j.cogsc.2021.100553.

Wiltschi, B., Cernava, T., Dennig, A., Galindo Casas, M., Geier, M., Gruber, S., et al., 2020. Enzymes revolutionize the bioproduction of value-added compounds: from enzyme discovery to special applications. Biotechnol. Adv. 40, 107520. Available from: https://doi.org/10.1016/j.biotechadv.2020.107520.

Wu, L., Qin, L., Nie, Y., Xu, Y., Zhao, Y.L., 2022. Computer-aided understanding and engineering of enzymatic selectivity. Biotechnol. Adv. 54, 107793. Available from: https://doi.org/10.1016/j.biotechadv.2021.107793.

Wu, B., Szymanski, W., Crismaru, C.G., Ferringa, B.L., Janssen, D.B., 2012. C-N lyases catalyzing addition of ammonia, amines and amides to C = C and C = O bonds. In: Drauz, K., Gröger, H., May, O. (Eds.), Enzyme Catalysis in Organic Synthesis, Third Edition Wiley-VCH, Weinheim, Germany, pp. 749–778. , Enzyme Catalysis in Organic Synthesis.

Xavier, J.R., Ramana, K.V., Sharma, R.K., 2018. β-galactosidase: biotechnological applications in food processing. J. Food Biochem. 42, e12564. Available from: https://doi.org/10.1111/jfbc.12564.

Xiao, Y.Z., Wu, D.K., Zhao, S.Y., Lin, W.M., Gao, X.Y., 2015. Statistical optimization of alkaline protease production from *Penicillium citrinum* YL-1 under solid-state fermentation. Prep. Biochem. Biotechnol. 45, 447–462. Available from: https://doi.org/10.1080/10826068.2014.923450.

Xiong, W., Liu, B., Shen, Y., Jing, K., Savage, T.R., 2021. Protein engineering design from directed evolution to de novo synthesis. Biochem. Eng. J. 174, 108096. Available from: https://doi.org/10.1016/j.bej.2021.108096.

Xu, H., Shen, D., Wu, X.-Q., Liu, Z.-W., Yang, Q.-H., 2014. Characterization of a mutant glucose isomerase from *Thermoanaerobacterium saccharolyticum*. J. Ind. Microbiol. Biotechnol. 41, 1581–1589. Available from: https://doi.org/10.1007/s10295-014-1478-4.

Xu, W., Zhang, W.L., Zhang, T., Jiang, B., Mu, W.M., 2018. L-arabinose isomerases: characteristics, modification, and application. Trends Food Sci. Tech. 78, 25–33. Available from: https://doi.org/10.1016/j.tifs.2018.05.016.

Yan, J., Gui, X., Wang, G., Yan, Y., 2012. Improving. stability and activity of cross-linked enzyme aggregates based on polyethylenimine in hydrolysis of fish oil for enrichment of polyunsaturated fatty acids. Appl. Biochem. Biotechnol. 166, 925–932. Available from: https://doi.org/10.1007/s12010-011-9480-z.

Yao, C., Sun, J., Wang, W., Zhuang, Z., Liu, J., Hao, J., 2019. A novel cold-adapted β-galactosidase from *Alteromonas* sp. ML117 cleaves milk lactose effectively at low temperature. Process. Biochem. 82, 94–101. Available from: https://doi.org/10.1016/j.procbio.2019.04.016.

Yoshihara, A., Kozakai, T., Shintani, T., Matsutani, R., Ohtani, K., Iida, T., et al., 2017. Purification and characterization of d-allulose 3-epimerase derived from *Arthrobacter globiformis* M30, a GRAS microorganism. J. Biosci. Bioeng. 123, 170–176. Available from: https://doi.org/10.1016/j.jbiosc.2016.09.004.

Zhang, Y., He, S., Simpson, B.K., 2018. Enzymes in food bioprocessing – novel food enzymes, applications, and related techniques. Curr. Op. Food Sci. 19, 30–35. Available from: https://doi.org/10.1016/j.cofs.2017.12.007.

Zhang, W.L., Huang, J.W., Jia, M., Guang, C., Zhang, T., Um, W.,M., 2019. Characterization of a novel d-lyxose isomerase from *Thermoflavimicrobium dichotomicum* and its application for d-mannose production. Process. Biochem. 83, 131–136. Available from: https://doi.org/10.1016/j.procbio.2019.05.007.

Zhang, H., Zhai, W., Lin, L., Wang, P., Xu, X., Wei, W., et al., 2021a. In Silico Rational design and protein engineering of disulfide bridges of an α-amylase from *Geobacillus* sp. to improve thermostability. Starch - Stärke 73, 2000274. Available from: https://doi.org/10.1002/star.202000274.

Zhang, Y., Rui, X., Simpson, B.K., 2021b. Trends in nanozymes development vs traditional enzymes in food science. Curr. Op. Food Sci. 37, 10–16. Available from: https://doi.org/10.1016/j.cofs.2020.08.001.

Zhao, G.-Y., Zhou, M.-Y., Zhao, H.-L., Chen, X.-L., Xie, B.-B., Zhang, X.-Y., et al., 2012. Tenderization effect of cold-adapted collagenolytic protease MCP-01 on beef meat at low temperature and its mechanism. Food Chem. 134, 1738–1744. Available from: https://doi.org/10.1016/j.foodchem.2012.03.118.

Zheng, X., Li, Q., Tian, J., Zhan, H., Yu, C., Wang, S., et al., 2021. novel strategy of mussel-inspired immobilization of naringinase with high activity using a polyethylenimine/dopamine co-deposition method. ACS Omega 6, 3267–3277. Available from: https://doi.org/10.1021/acsomega.0c05756.

Zheng, L., Sun, Y., Wang, J., Huang, H., Geng, X., Tong, Y., et al., 2018. Preparation of a flower-like immobilized D-psicose 3-epimerase with enhanced catalytic performance. Catalysts 8, 468. Available from: https://doi.org/10.3390/catal8100468.

Zhu, L., Zhu, Y., Zhang, Y., Li, Y., 2012. Engineering the robustness of industrial microbes through synthetic biology. Trends Microbiol. 20, 94–101. Available from: https://doi.org/10.1016/j.tim.2011.12.003.

Chapter 19

Carbohydrases: a class of all-pervasive industrial biocatalysts

Archana S. Rao[1], Ajay Nair[1], Hima A. Salu[1], K.R. Pooja[1], Nandini Amrutha Nandyal[1], Venkatesh S. Joshi[1], Veena S. More[2], Niyonzima Francois[3], K.S. Anantharaju[4] and Sunil S. More[1]

[1]School of Basic and Applied Sciences, Dayananda Sagar University, Bangalore, Karnataka, India, [2]Department of Biotechnology, Sapthagiri College of Engineering, Bangalore, Karnataka, India, [3]Department of Biotechnologies, Faculty of Applied Fundamental Sciences, INEs Ruhengeri, Rwanda, [4]Department of Chemistry, Dayananda Sagar College of Engineering, Bangalore, Karnataka, India

19.1 Introduction

Carbohydrates are one of the most diverse and abundantly available biomacromolecules—be it vital structural components of the cell, storage molecules, or key players in various biological processes. Carbohydrates are present in different structural forms, from simple sugars such as monosaccharides to oligosaccharides, to complex molecules such as polysaccharides, glycoproteins, glycolipids, etc. Monosaccharides are individual sugar units which form glycosidic linkages to produce disaccharides. Many oligosaccharides join to form a polysaccharide. The majority of carbohydrates are present in polysaccharides such as starch, glycogen, cellulose, hemicelluloses, and chitins in the terrestrial biosphere and as agars, alginates, and carrageenan in the aquatic biosphere (Falkowski et al., 2000).

A diverse set of enzyme regimens involved to biologically accommodate the vast array of carbohydrates in their hydrolysis and synthesis are broadly classified as *carbohydrases*. They catalyze almost all catabolic processes of carbohydrates. The enzymes cleave glycosidic bonds at specific locations and convert the carbohydrate into numerous saccharides, causing polymer fragmentation into smaller units. The glycosidic linkage is broken by a hydrolytic reaction that involves transferring the substrates' components to water.

Some hydrolytic reactions are followed by structural rearrangements of sugar units, resulting in the formation of new saccharides. This is known as *transglycosylation*, where water acts as an acceptor. If aliphatic alcohols are

Biotechnology of Microbial Enzymes. DOI: https://doi.org/10.1016/B978-0-443-19059-9.00018-9
© 2023 Elsevier Inc. All rights reserved.

used instead of water, they may serve as acceptors and lead to new saccharides in suitable conditions.

Several factors affect carbohydrase activity. Some of them are as follows:

- effect of D- and L-configuration,
- configuration of glycosidic linkages,
- size of the molecule,
- conformation of the sugar,
- nature of the linking atom in the glycosidic bond, and
- ring size of sugar (Kulp, 1975).

Carbohydrases are seminal to many metabolic processes, such as digestion (Simpson et al., 2012). Some of the major carbohydrase enzymes include cellulases, amylases, dextranase, glucoamylase, β-galactosidase, pectinases, maltase, xylanase, and many more (Coutinho et al., 2009). Their ability to hydrolyze carbohydrates can be used for various industrial applications, particularly in food industries, starch processing, animal feed, textile, paper industries, biofuel production, agriculture, pharmaceuticals, etc. (Polaina and MacCabe 2007; Olsen, 2000; Himmel et al., 2007; Schafer et al., 2007; Ragauskas et al., 2006). The chapter discusses the classification, sources, production, and assorted application of carbohydrases in numerous fields.

19.2 Classification of carbohydrases

Carbohydrates have one of the highest molecular complexities among polymeric biomolecules. With variations in stereochemistry, structural branching, and monosaccharide diversity, the potential diversity in carbohydrate structures reaches monumental proportions (Laine, 1994). Furthermore, secondary modifications by noncarbohydrate groups such as phosphates, sulfates, acyl esters, and many more groups further increase the structural complexity of the biomolecule group (Laine, 1994; Taylor and Drickamer, 2006). Hence, a vast class of carbohydrases participates in carbohydrate metabolism (Stals et al., 2004).

Based on their function, carbohydrases are classified into glycosyltransferases, transglycolases, glycoside hydrolases (glycosidases), glycoside phosphorylases, polysaccharide lyases, phosphatases, sulfatases, and carbohydrate esterases (Table 19.1) (Brumer, 2010).

19.2.1 Glycosidases

Glycosidases are a class of carbohydrases that catalyze hydrolysis of glycosidic bonds in complex polysaccharide molecules. The reaction leads to low-molecular-weight monosaccharides and oligosaccharides (Divakar, 2013). They are of α- and β-types based on the configuration of the glycosidic

TABLE 19.1 **Classification of carbohydrases and applications.**

Enzyme class	Function	Examples	Industrial applications	References
Glycosidases	Hydrolysis of glycosidic bonds ($\alpha \rightarrow \beta$) ($\alpha \rightarrow \alpha$)	β-Glucosidase, β-galactosidase, amylase, lactase, maltase	Production of maltodextrins, laundry detergents, hydrolysis of lactose in dairy	Brumer (2010)
Glycosyltransferases	Transfer of sugar from donor to acceptor ($\alpha \rightarrow \beta$) ($\alpha \rightarrow \alpha$)	β-1,4-Galactosyltransferase 1, *N*-acetylglucosaminyltransferase	Antistaling agents in baked goods, as biocatalysts, production of cyclodextrins	Ünligil and Rini (2000); Christelle et al. (2005); Han et al. (2014)
Glycosyl phosphorylases	Cleave glycosidic bond and transfer nonreducing end to an inorganic phosphate	Starch phosphorylase, sucrose phosphorylase, maltosyl transferase, glycogen phosphorylase	Development of maltodextrin powered enzymatic fuel cell, production of cellobiose, cellulose, amylose	Puchart (2015)
Polysaccharide lyases	Cleave glycosidic bond in acidic polysaccharides	Pectin lyase, pectate lyase, alginate lyase	Retting of natural fibers, clarification of fruit juices, oil extraction, coffee and tea fermentation	Linhardt et al. (1987); Zheng et al. (2021)
Carbohydrate esterases	Catalyze de-*O* or de-*N*-acetylation by cleaving ester bond	Acetyl xylan esterase, carboxylesterase, chitin deacetylase, cutinase, cinnamoyl esterase	Enhancing animal feedstock, production of food additives, paper and pulp industries	Nakamura et al. (2017); Armendáriz-Ruiz et al. (2018); Kameshwar and Qin (2018)

carbon (Pigman, 1943). Glucosidases catalyze three main reactions: hydrolysis, reverse hydrolysis, and transglycosylation.

Hydrolysis occurs in an aqueous solution, where there is a sufficient amount or an excess of water. It leads to mono- and oligosaccharides (Bojarová and Křen, 2009). Reverse hydrolysis is an equilibrium-controlled reaction. From a polysaccharide molecule, the equilibrium is shifted toward the formation of an alcohol and a glycoside molecule. Transglycosylation is a reaction where the transfer of a glycosyl molecule is facilitated from a donor to an acceptor (Bojarová and Křen, 2009).

19.2.2 Glycosyltransferase

This class of carbohydrases catalyzes the transfer of sugar from an activated donor sugar to saccharide and monosaccharide acceptors (Christelle et al., 2005). The acceptor of a glycosyltransferase can range from a simple monosaccharide homologous to the donor to a component heteropolysaccharide (containing hundreds of glycosidic bonds), like an oligosaccharide, protein, nucleic acid, or a lipid (Weijers et al., 2008). In other words, these enzymes work by forming a glycosidic bond. The mechanism of inverting glycosyltransferases underlies the mode of enzymatic activity (Ünligil and Rini, 2000; Weijers et al., 2008).

19.2.3 Glycosyl phosphorylases

Glycosyl phosphorylases are enzymes that reversibly cleave the glycosidic bonds and transfer the nonreducing end terminal glycoside residue to an inorganic phosphate molecule. These are structurally and mechanically similar to glycosidases (Puchart, 2015).

19.2.4 Polysaccharide lyases

Polysaccharide lyases are a class of carbohydrases which cleave glycosidic bonds in acidic polysaccharides. These polysaccharides are usually found in most plants, animals, and microorganisms. Some common examples are pectins, pectates, components of the extracellular matrix of animals, alginic acid, etc. The complexity of these enzymes can range from forming simple linear homopolymers to producing complex homo-copolymers having branched polymers. Examples of these enzymes are pectin lyase, alginate lyase, pectate lyase, etc. (Linhardt et al., 1987).

19.2.5 Carbohydrate esterases

Carbohydrate esterases catalyze the de-*O* or de-*N*-acylation by removing the esters from carbohydrates, acting on an ester bond that links an acid and an alcohol (Nakamura et al., 2017). The enzyme acts on two classes of substrates:

- Sugar acts as an acid, for example, pectin methyl esters.
- Sugar acts as alcohol, for example, acetylated xylan.

These enzymes catalyze reactions by acetylation–deacetylation reactions (Armendáriz-Ruiz et al., 2018)

19.3 Sources

Microorganisms such as bacteria, yeasts, and molds are the most sought-after source for any enzyme, including carbohydrases. With benefits ranging from ease of isolation, fast growth rate, higher enzyme production, product stability, etc., microorganisms consequentially have become a superior choice for industrial production of enzymes (Simpson et al., 2012; Tufvesson et al., 2010; Dumorne and Severe, 2018; Zhang and Kim, 2010; Kim et al., 2011). Among many microbial sources, those producing extracellular enzymes are preferred more as this simplifies downstream processing and cuts the cost (Simpson et al., 2012; Tufvesson et al., 2010). *Bacillus* sp., *Aeromonas* sp., *Pseudomonas* sp., *Bacteroidetes bacterium*, *Oxalobacteraceae bacterium*, *Paenibacillus* sp, *Staphylococcus* sp., *Aspergillus* spp., *Penicillium notatum*, and *Saccharomyces* spp. are some of the best-known carbohydrates degrading microorganisms (Anzai et al., 1997; Ash et al., 1993; Mendoza et al., 1998).

19.3.1 Marine microorganisms

Marine environments offer great reservoirs to explore microbial sources of industrially important enzymes. Additionally, microbes from the marine ecosystem are adapted to multiple extreme physical factors such as pH, temperature, pressure, salinity, etc. As a result, the enzymes produced by them are extremophilic in nature. Extremozymes isolated from marine environments, such as cellulase, alginate lyase, carrageenase, agarase, etc., have wide industrial applications (Zhang and Kim, 2010; Dumorne and Severe, 2018; Kim et al., 2011; Zilda et al., 2019; Huang et al., 2019).

Marine microorganisms have been often known to survive as a symbiont with marine animals or plants (Minic et al., 2001). These symbionts show high adaptability to ambient factors based on their host and also the niche. One such example is a *Bacillus* sp., isolated from a seaweed which was reported to produce alginate lyase with high temperature (50°C) and pH (up to 9) tolerance. Many other carbohydrases such as amylase, agarase, K-carrageenase etc., were also reported to have been obtained from marine ecosystems (Zilda et al., 2019). *Thermococcus litoralis* and *Pyrococcus furiosus* are good sources for thermostable extracellular pullulanases (Brown and

Kelly, 1993). Enzymes such as thermostable amylases, pullulanase, and glucosidase were reported to be produced from *Thermococcus* (Legin et al., 1997). *Pseudoalteromonas* sp., a symbiont of a krill, is able to produce psychrophilic β-galactosidase (Turkiewicz et al., 2003), *Bacillus* sp., a symbiont with the marine sponge, was able to produce carboxymethyl cellulase, and so on. A certain *Alteromonas* sp., associated with *Laminaria* plant producing intracellular alginate lyases, utilizes alginate as its sole carbon source (Sawabe et al., 1997). Other than the abovementioned marine microbes, *Vibrio* sp., *Pseudoalteromonas* sp., *Pseudoalteromonas* sp., *Aeromonas* sp., and *Halomonas* sp., are other marine microbes known to produce extremophilic carbohydrases (Tirado et al., 2005).

19.3.2 Rumen bacteria

Oligotrichs are a group of rumen microorganisms which are adapted for the plant carbohydrates rich in cellulose and hemicellulose. Hence, the gut microbiota of animals such as cattle, sheep, cows etc. are apt reserves for isolating microbes, which can hydrolyze such polysaccharides. Due to the high consumption of cellulose-rich food by the cattle, many cellulose-digesting rumen bacteria can be isolated from the gut of cattle. Rumen bacteria digest the cellulose fibers of plant cell walls by secreting hydrolytic enzymes. These enzymes are either held close to the bacteria or are firmly adsorbed onto the surface of fibers. This is why cell-free rumen fluid shows significantly less carbohydrase activity. Sheep rumen has been shown to harbor microbes producing cellulases, xylanase, carboxymethyl cellulase, cellobiase, etc. (Francis et al., 1978). *Epidinium ecaudatum*, a ciliate isolated from cows feeding on red clover, exhibited amylase, maltase, hemicellulose, alpha-galactosidase, and pectinase activity (Balley et al., 1962; Wright, 1961).

19.3.3 Genetically modified organisms

When it comes to industrial application, one of the top priorities in enzyme production is to make it cost-effective and keep an above-average production rate. Recombinant DNA technology has thus been instrumental in achieving those above. Studies have shown the use of genetically modified organisms in enhancing the degradation of complex carbohydrates such as ligninocellulose (Schubot et al., 2004; Payne et al., 2013).

A cellulose structure has both crystalline and amorphous regions. Enzymatic hydrolysis is accessible in the amorphous regions over the crystalline regions. The conversion of crystalline to the amorphous region thus enhances the reaction efficiency. Fungi such as *Phanerochaete chrysosporium* and *Neurospora crassa* produce a class of copper-dependent lytic polysaccharide monooxygenases. The enzyme is able to oxidize the C1, C4, and

C6 carbon. Incorporating the gene coding for such enzymes by rDNA technology will be of great use to industries (Vaaje-Kolstad et al., 2010; Horn et al., 2012; Westereng et al., 2011; Phillips et al., 2011; Levasseur et al., 2013; Quinlan et al., 2011).

In some cases, understanding the synergistic action of enzymes is essential. The increased enzyme activity can be accomplished by using complete enzyme systems. For example, an engineered cellulase system comprising exo-cellulase, cbhA + endo-cellulase, and cenA + (*Cellulomonas fimi*) showed eightfold higher glucose production due to synergistic activation (Liu and Yu, 2012).

19.3.4 Fungi and yeasts

Eukaryotic microorganisms such as fungi and yeasts are one of the better choices for the industrial production of enzymes. The posttranslational modification of enzymes benefits the enzyme activity in many ways. Enzyme production using fungi normally requires a solid-state fermentation (SSF) system (Purkarthofer et al., 1993; Pandey et al., 1999). A range of solid substrates such as agricultural wastes and industrial wastes can be efficiently utilized, which helps simplify the process while requiring the minimal cost of production (Raimbault, 1998). Fungi such as *Aspergillus*, *Fusarium*, *Penicillium*, and *Rhizopus*, etc., are identified as potent carbohydrase producers. In a study, *Rhizopus delmar* showed multiple carbohydrase activity ranging from cellulases, xylanases, pectinases, amylase, etc., under SSF (Shruti et al., 2018).

Yeast is one of the leading producers of products of economic importance via fermentation. Ease of large-scale cultivation, cost-effectiveness, and their wide application in industries make them a good source of carbohydrases. Studies found that yeast isolates could degrade two or more polysaccharides by producing amylase, xylanase, and cellulase (Nasr et al., 2014).

19.4 Industrial production of carbohydrase

Industrial techniques for carbohydrase production aim at increasing the yield while reducing the production costs. The industries have developed highly efficient methods for enzyme production. Any enzyme production typically involves pretreatment of substrates, fermentation, and downstreaming processes (Fig. 19.1) (Rodney et al., 1997).

Depending on the substrate being hydrolyzed, different pretreatment methodologies are followed. For example, during the hydrolysis of cellulose, a dilute acid such as sulfuric acid is used. The treatment of biomass with such dilute acids has advantages such as producing a soluble pentose stream that can be separated from particulate residue and increased activity rate due to acid-induced high fiber porosity (Grethlein, 1985). In xylose hydrolysis,

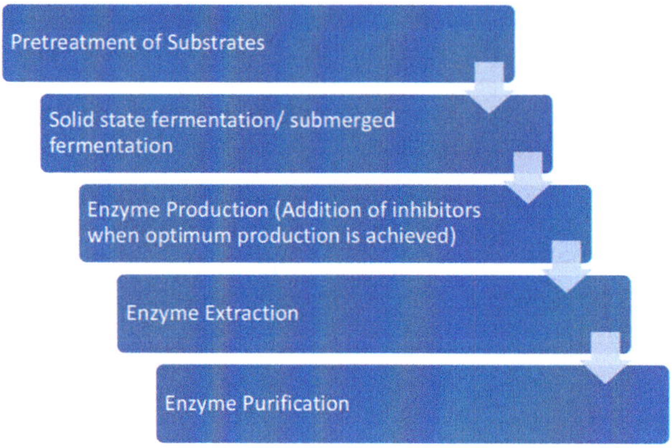

FIGURE 19.1 General process of carbohydrase production.

furfurals are used in pretreating substrates, which kill microorganisms. However, the addition of furfurals increases the acidity, which must be neutralized. This can be achieved by vacuum distillation which involves acid removal in the form of filterable calcium sulfate by the addition of lime (Dunning and Lathrop, 1945).

Fermentation methods such as SSF or submerged fermentation (SmF) are mainly employed in production. SSF is a process used to hydrolyze appropriate solid substrates by selected microorganisms (Leite et al., 2021), whereas SmF is a liquid-state fermentation which involves bacterial action, and the enzyme produced can be easily recovered by filtration or centrifugation (Mussatto and Teixeira, 2010). SSF differs from SmF as it is performed with little or no water, making SSF highly suitable for fungal growth (Leite et al., 2021). Industries prefer SSF over SmF due to many advantages such as lower susceptibility to microbial contamination, high cost-effectiveness, and high enzyme activity rates (Soccol et al., 2017). Various carbohydrases such as cellulases and xylanases can be produced by SSF (Lizardi-Jiménez and Hernández-Martínez, 2017).

Inhibitors can be used to halt the enzyme production once optimal product concentration is obtained. The quantity and nature of these inhibitors depend on the substrate used and the type of prehydrolytic or hydrolytic reaction. For example, acetic acid and its extracts could be used as an inhibitor released when hemicellulose structure is degraded. 5-Hydroxymethyl furfural, levulinic acid, and formic acid formed due to the degradation of sugars act as inhibitors (Ohlson et al., 1984).

Downstreaming process involves extraction of the carbohydrase enzymes. It usually consists of a combination of ion exchange and adsorption chromatography. Industrially, this is performed by simulated moving bed chromatography (Luz et al., 2008). Besides chromatography, cooled crystallization and membrane techniques are also employed (Wang et al., 2016).

19.4.1 Enzyme immobilization

Today's biotechnological industries demand increased enzyme productivity and high shelf life. Enzyme immobilization is a technique in which enzymes are fixed to a support matrix-like inert organic or inorganic materials via physical or chemical means. This results in prolonged availability of the enzyme, thus increasing productivity and reducing the cost. Several physical factors such as the nature of career material, pore size, etc., affect the performance of the immobilized enzymes (Cao, 2006). Different techniques such as adsorption, covalent binding, and entrapment are used for immobilization of carbohydrases such as invertases, galactosidases, glucosidases, invertases, etc. (Fabiano et al., 2013).

19.5 Industrial applications of carbohydrases

Carbohydrases as enzymes have shown potential across various industries such as food, leather, paper, textiles, pharmaceuticals, agriculture, biofuels, etc. Some of the most important include glucosidases and galactosidases, invertases, pectinases, cellulases, glucoamylases, etc. (Table 19.2) (Underkofler et al., 1958).

What advantages do enzymes give in industrial processes?

- *Active under mild conditions*: Enzymes are active under room temperature and neutral pH, which are ideal for industrial production (Underkofler et al., 1958).
- *High specificity*: Enzymes usually catalyze either a single specific reaction or a group of closely related reactions. This makes it easy to find or synthesize enzymes for a particular process (Olsen, 2000).
- *Very high reaction rates*: Enzymes catalyze reactions at a high rate of reaction. They accelerate reactions by a factor of more than a million times. This makes production a faster process, which would otherwise take much longer to complete (Olsen, 2000).
- *High diversity*: There are numerous enzymes available for different tasks in nature, due to their high specificity. Hence, a very wide spectrum of reactions and processes can be carried out (Olsen, 2000).
- *Work at low concentrations*: Enzymes can do all their work at low concentrations. Also, their reaction rates can be easily by adjusting their pH, temperature, and concentrations (Underkofler et al., 1958).
- *Inactivate when a reaction is complete*: When the desired amount of product is reached, these enzymes can be inactivated easily (Underkofler et al., 1958).

TABLE 19.2 Carbohydrases used in industries.

Industry	Enzymes
Baking	α- and β-Amylase Xylanases Hemicellulase
Beverage	Celluloses Hemicelluloses Pectinases Xylanases Amylases Galactosidases
Sweeteners	Amylases Invertases Inulinases Glucosyltransferase
Prebiotics	β-D-Fructosyltransferase β-Fructofuranosidase β-Galactosidase
Biofuels	Amylase Cellulase β-Glucosidase Endoglucanase Cellobiohydrolase Xylulose
Agriculture	Cellulases Amylases Invertases
Dairy	Lactase
Animal feed	Cellulase Hemicellulase Glycosyl hydrolase Xylanase Mannanase
Pharmaceuticals	β-Glucocerebrosidase Sacrosidase Lactase α-Galactosidase
Detergents	Amylases Cellulases
Wastewater treatment	Cellulase Hemicellulase Other polysaccharidases

(*Continued*)

TABLE 19.2 (Continued)

Industry	Enzymes
Paper	Cellulase Hemicellulase Xylanase
Textile	Amylase Cellulase Pectinase Laccase

- *Nontoxic*: This trait is very important when using enzymes in food and beverage industry.

The following table shows the list of enzymes used in various industries:

19.5.1 Enzymes involved in the production of beverages

The beverage industry is one of the major industries producing various beverages such as wine, beer, fruit and vegetable juices, tea, coffee, etc.

19.5.1.1 Cellulases and hemicellulases

Cellulase is made up of three enzymes which are β-1,4-endoglucanase, cellobiohydrolase, and β-glucosidase. Cellulase helps in pressing raw materials to enhance yield in beverage industries (Uzuner and Cekmecelioglu, 2019). Clarification of fruit and vegetable juices is an important step in wine production. In beverages like tea, bringing out more aroma is important. Enzyme cellulase is also useful for such processes (de Souza and Kawaguti, 2021).

19.5.1.2 Amylases

These enzymes are used for breaking down starch molecules which are responsible for cloudiness in the beverage. Starch also decreases the filtration rate and increases membrane fouling, haze, and gelling. Amylase and glucoamylase enzymes are used in beer production to break down starch from cereal grains (Uzuner and Cekmecelioglu, 2019).

19.5.1.3 Xylanases

Xylanases are glycosidases which catalyze *endo*-hydrolysis of 1,4-d-xylosidic linkages in xylan and are involved in xylose production. Along with cellulases, xylanases increase the clarification of fruit juices (Uzuner and Cekmecelioglu, 2019).

19.5.1.4 Pectinases

Pectinases are crucial in the production of beverages. They help in a process called maceration, along with cellulases and hemicellulases. This is the production process of pulps, nectars, and cloudy or transparent juices or their concentrates from fruits. Pectinases also have many roles such as clarification of juices and wines, removal of bitterness from citrus juices, extraction of juices, liquefaction of fruit pulp, and even fermentation of coffee. The main mode of action for pectinase is to break the cell wall of the fruits, which are held together by pectins (de Souza and Kawaguti, 2021; Kantharaj et al., 2017; Uzuner and Cekmecelioglu, 2019).

19.5.2 Enzymes involved in the production of prebiotics

19.5.2.1 Galactooligosaccharides

β-Galactosidase enzymes help produce galactooligosaccharides (GOS) by acting on lactose through transgalactosylation. GOS is the main component of prebiotic food. These enzymes can be extracted from many microbes, such as bifidobacteria and lactic acid bacteria (Contesini et al., 2013; Chourasia et al., 2020; van den Broek et al., 2008).

19.5.2.2 Fructooligosaccharides

These are the prebiotics that are used against colon cancer, reduce cholesterol, and regulate phospholipid and triglyceride levels in serum. These molecules can be synthesized by enzymes such as β-D-fructosyltransferase or β-fructofuranosidase, which hydrolyze sucrose molecules to produce glucose and fructose, and transfer the fructosyl moiety to another FOS or sucrose molecule (Contesini et al., 2013; Chourasia et al., 2020; van den Broek et al., 2008). An added advantage of FOS is that they can be used as calorie-free sweeteners.

19.5.3 Enzymes involved in syrup and isomaltulose production

Demand for sugar and syrup has gone dramatically increased in recent times. Syrups are widely used in food and beverage industries for preparations such as fruit juices, wine, dairy products, etc. They are used as sweeteners or as a source of fermentable sugar (Johnson et al., 2009).

19.5.3.1 Amylolytic enzymes

Good-quality syrups are typically produced using starch as a substrate. Hydrolysis of starch is catalyzed by amylases and amyloglucosidase. 1,4-

α-Glycosidic linkages of amylose and amylopectic chain is hydrolyzed by α-amylase, producing soluble dextrins and oligosaccharides, whereas α-1,4 and α-1,6 linkages of starch are hydrolyzed by amyloglucosidase (Tawil et al., 2011).

The glucose syrups production begins with the liquefaction of starch, followed by saccharification. The first step is catalyzed by α-amylase, which hydrolyzes starch to maltodextrins, whereas latter is catalyzed by glucoamylase where the low dextrose equivalent syrup is completely converted to glucose by glucoamylase (Hobb, 2009). A starch debranching enzyme, pullulanase can also be used in final processing (Roy and Munishwar, 2004).

Industrial production of syrup often uses immobilized enzymes in packed bed bioreactors. Other techniques, such as stirred tanks, bubble columns, airlift bioreactors, and fluidized beds, can be employed. Starch is a large molecule to be immobilized on small pore-sized substrates. A matrix with a large pore size is often a good choice to enhance reaction rate (Cao, 2006). In a bench-scale study, glucose syrup was produced from cassava starch hydrolysis by using stirred tank bioreactor. Immobilization of glucoamylase resulted in a high yield of the syrup.

19.5.3.2 Invertases

Inverted sugar syrup is of great economic value as it is used widely in the food industry. Advantages are its lower freezing point, high hydrophilicity, and it is sweeter than sucrose (Emregul et al., 2006). This syrup is produced by hydrolysis action of invertase on sucrose syrup. Industries prefer immobilized enzyme action due to several advantages such as reusability and increased productivity (Kotwal and Shankar, 2009). Most of the recent studies showed the use of the covalent binding technique to immobilize invertase.

In a particular study, invertase was covalently immobilized using a polyurethane rigid adhesive foam. It was observed that immobilization increased the affinity of the substrate to the enzyme and decreased in km value. A 10-fold decrease in turnover rate, which may be due to restricted diffusion of the substrate was observed (Cadena et al., 2011). In contrast, enhanced conversion of sucrose to high fructose syrup was achieved in a packed bed reactor using thermostable invertase from *Aspergillus awamori*. The enzyme was immobilized using glutaraldehyde by covalent binding on acetic acid solubilized chitosan.

19.5.3.3 Inulinases

Inulin is a plant-based polyfructan. Hydrolysis of inulin is catalyzed by *exo-* and *endo*-inulinase, liberating ultrahigh fructose syrup and inulooligosaccharides (Dodge, 2009; Jiang et al., 2019).

An immobilized endoinulinase isolated from *Aspergillus niger* showed higher production of oligofructose syrup from Jerusalem artichoke juice, a source of inulin (Nguyen et al., 2011).

19.5.3.4 Isomaltulose

Isomaltulose is a monosaccharide having wide applications in food industry. Having a low glycemic index and being nonpathogenic, it is a substitute for sucrose. Isomalt can be prepared from isomaltulose, a sugar alcohol used in many food production industries. Isomaltulose can be industrially produced by the chemical method, transglycosylation, or by biological sources by intracellular glucosyltransferase produced by some bacteria strains.

19.5.4 Enzymes in dairy industry

Lactase or galactosidase catalyzes the hydrolysis of lactose to galactose and glucose. It is used in dairy products to increase taste and solubility and acts as a digestive factor (Emad, 2019). Lactobacilli-produced enzymes could generate glucose from the lactose substrate in lactose-modified medium (Mojumder et al., 2011). β-Galactosidase enhances the overall quality of dairy products by enhancing sweetness, digestibility, etc.

19.5.5 Carbohydrases in animal feed production

Many ingredients present in animal feed are not fully digested by animals. This makes it essential to add enzymes to the feed. To make easy digestion possible, the enzymes and a mix of other necessary vitamins and minerals can be added to the feed (Olsen, 2000).

19.5.5.1 Cell wall degrading enzymes

The main ingredients of animal feed are monocot and dicot grains, the cell walls of which are made of complex polysaccharides and need to be broken down (Fan and Pederson, 2021). The use of cell degrading enzymes reduces the nutrient encapsulating properties of the cell walls. This increases the nutritional value of the feed by significant proportions. The enzymes used are *cellulases*, *hemicellulases*, and *glycosyl hydrolases*.

19.5.5.2 Non-starch polysaccharide degrading enzymes

Most cereals used in animal feed are predominantly wheat, barley, and rye which are not easily degradable due to the presence of non-starch polysaccharides (NSP), which ultimately causes a reduction in nutrient uptake (Olsen, 2000). A mixture of two enzymes—*xylanases and mannases*—is added for this purpose (Zeng et al., 2018). Breakage of NSPs using these enzymes increases the nutrition uptake by animals (Narasimha et al., 2013).

19.5.6 Carbohydrase application in pharmaceutical industries

Carbohydrases, along with proteases and lipases, are widely used in pharmaceutical industries. These enzymes are mainly isolated from bacteria and fungi and are generally recognized to belong to a safe category (Yang et al., 2017). Carbohydrases are mainly used in *enzyme therapy*, where medical conditions such as enzyme deficiencies can be treated. Enzyme therapy uses artificially or naturally synthesized enzymes to treat patients suffering from cystic fibrosis, lactose intolerance, etc. (Cormode et al., 2018). Another enzyme, β-glucocerebrosidase is used to treat patients with Parkinson's disease. Ceredase injection is the commercially available form of the enzyme (Erdem et al., 2018). People who cannot digest sucrose can be treated with enzyme sacrosidase-based drugs, which help them hydrolyze sucrose. Hence, it is useful for patients suffering from congenital sucrase isomaltase deficiency (Lwin et al., 2004). *Saccharomyces cerevisiae* is a good source of enzyme production (Matta et al., 2018). Similarly, lactase and α-galactosidase can be used to treat lactose intolerance (Treem et al., 1999). For such people, lactase-fortified milk is produced, and lactase powder is also made available to reduce bloating and diarrhea caused by lactose intolerance (Kumar et al., 2019; Hertzler et al., 2017). α-Galactosidase is an enzyme that is used to treat indigestion. It hydrolyzes the α-galactosidic residues of sugar substrate. Thereby reducing the bloating and gas caused due to indigestion (Shang et al., 2018).

19.5.7 Carbohydrases involved in detergent

Detergent is a cleansing agent which consists of surfactants and chelating agents. The surfactant is responsible for removing dirt from the soil surface, and the chelating agent binds to the unwanted metal ions of the cleansing solutions. Proteases, amylase, lipases, and cellulases are the enzymes used in the detergent formulation, among which amylase and cellulase are the carbohydrases. Carbohydrases are enzymes which help in the degradation of carbohydrates (Niyonzima and More, 2014).

In medical, the equipment and utensils are cleaned using enzymatic cleaners, including amylases. The benefits are as follows:

- These assist in the removal and breakdown of organic soils at neutral pH.
- Bio-burden gets reduced.
- Limited use of mechanical action in places hard to reach.
- A broader range of material compatibility for delicate instruments.
- Effective at lower concentrations and lower temperatures (Steris healthcare).

When plant amylase was incorporated with detergent, it enhanced the cleansing property of the detergent. 1,4-α-D-glucan glucanohydrolase, catalyzes starch hydrolysis and related polysaccharides. The amylases act on stain containing starch and degrade it into short-chain sugars (Imen et al., 2017). On inhaling the enzyme, there is the possibility of asthma in individuals who are in contact with the detergent (Hole et al., 2000).

α-Amylases are now used in chemical, analytical, pharmaceutical and clinical processes. Amylase also helps in the drainage system of hospitals (Imen et al., 2017). Cellulases present in the detergent act on dust and mud. They are used for cleaning cellulose fiber clothes.

19.5.8 Carbohydrases in wastewater treatment

The huge amounts of excess activated sludge produced from wastewater treatment plants are a massive disposal concern. Although the main part of this sludge is biodegradable, the anaerobic digestion of activated sludge is a rate-limiting process, thereby taking a lot of time. Most of the components of activated sludge can be hydrolyzed using enzymes, effectively saving time and resources (Yin et al., 2016). Sludge dewatering is an important component of this process. It reduces the sludge volume and makes it easier to transport to the disposal site (Houghton et al., 2001). Adding enzymes can degrade extracellular proteins and carbohydrates present in the sludge, improving the dewaterability of the sludge. Enzymes used for this process are *cellulases, hemicellulases and other polysaccharidases*. These enzymes degrade the extracellular polymer aiding in the flocculation process (Houghton et al., 2001).

19.5.9 Agriculture

Carbohydrates are a significant component of many foods and raw materials. Food carbohydrate analytical methods are useful for food quality assurance and product uniformity (Moreno et al., 2014). The applications of carbohydrases in the field of agriculture have been described in the following sections.

19.5.9.1 Soyabean hulls

Soybean hulls are a waste product produced during the processing of soybeans into oil and meal. The main components of soybean hull are cellulose, pectin, and hemicellulose, which are three major plant polysaccharides. It is a low-cost prospective substrate for carbohydrase production since it can induce a wide range of activities that can hydrolyze complicated biomass. Although *Aspergillus* is known for producing carbohydrases, no research has examined and compared the soybean hull induced production of various carbohydrases among *Aspergillus* species and strains (Li et al., 2017).

19.5.9.2 Flaxseed mucilage

Flaxseed mucilage is extracted into aqueous solutions. To lower the concentration of seed coat polysaccharides, whole flax seeds are soaked, followed by a specific end treatment. Soaking seeds in sodium bicarbonate or water solutions or treating them with commercially available carbohydrases (Celluclast K, Viscozyme and Pectinex) reduces the amount of mucilage (Wanasundara and Shahidi, 1997).

19.5.10 Enzymes in textile industry

There is a huge demand for enzymes in textile industry to optimize many processes such as dye production, processing of raw materials, etc. Amylases, cellulases, catalases, pectinases, and laccase are some commonly used enzymes in the textile industry. They help in processes such as removal of starch, degradation of additional hydrogen peroxide, lignin degradation, bleach fabrics, and so on (Kiro, 2012).

19.5.10.1 Amylase

Warping starch paste in textile weaving improves the strength of the fabric in the textile industry. It also reduces string breakage caused by friction, ripping, and static power generation on the thread by softening the area of the thread due to the established regulatory warping. The starch is eliminated from the cloth after weaving, and it is then scrubbed and dyed. α-Amylase is frequently used to remove the starch from the textile (Feitkenhauer,2003).

19.5.10.2 Cellulase

These enzymes are used in the initial processing of raw material and to make a novel variety of fabric. The enzyme was first used in textile industries during the 1980s to give denim a stylish stonewashed look through a process called biostoning. The use of cellulases in the textile industry has several benefits, including ease of handling, especially in wet processing with mild treatment conditions, and also minimization of waste generation (Arja, 2007).

19.5.10.3 Pectinase

Pectinases are indeed a unique enzyme in textile industries (Arja, 2007). Pectins are polysaccharides with branched neutral sugar side chains and a partly methyl esterified (1,4)-linked homogalacturonic acid. They are found in fruits and vegetables and are major elements of the cell wall and middle lamella. Pectinolytic enzymes or pectinases are enzymes that break down pectic compounds into simpler ones (Apoorvi and Vuppu, 2012).

Caustic soda was generally used for removing sizing compounds from cotton. Pectinases, along with cellulases, lipases, amylases, and hemicellulases, can replace caustic soda and are environmentally safe to remove sizing compounds from cotton. Bio-scouring is a revolutionary enzyme-based method for removing noncellulosic contaminants from fiber. Pectinases have been utilized for this purpose without causing any cellulose degradation problems (Mehraj et al., 2013).

19.5.10.4 Laccase

Laccase enzymes have been employed in textile industries for wash-off treatment, dyeing, rove scouring, dye synthesis, finishing, neps removal, printing, bio-bleaching, and effluent treatment, among other things (Rodriguez-Couto et al., 2006).

Laccase has been shown to protect colored or printed textiles from back stains. Laccase, as part of the washing solution, could swiftly bleach discharged dyestuff, reducing the amount of time, energy, and water required to attain satisfactory textile quality. Finishing colored cotton cloth with laccase catalyzed dye bleaching could be beneficial (Vernekar and Lele, 2009).

19.5.11 Carbohydrases involved in biofuel production

Biofuel production and its usage have seen huge demand in recent times. It is considered a sustainable innovation and a low carbon alternative to the existing fossil fuels as it helps in the reduction of the emission of greenhouse gases, one of the major causes of climate change. Biofuel can be produced from biomass such as corn, vegetable oil, liquid animal fats, algae, and other plant sources (Fig. 19.2) (Harish et al., 2020).

Many enzymes, such as glucosidase, lipase, phospholipase, etc., are major players in breaking down complex molecules present in the biomass. Carbohydrases are a set of enzymes responsible for the catabolism of carbohydrates. Lignocellulose is a better choice over starch or other sugars for ethanol production. The advantages of using lignocellulose are effective hydrolysis of cellulose and hemicellulose to simple soluble sugars, effective fermentation, less energy consumption, and cost-effectiveness (Hahn-Hägerdal et al., 2006).

19.5.11.1 Amylase

Amylase, an enzyme, is a major contributor to the breakdown of starch content in the biomass for bio-ethanol production. Microbial enzymes are used to produce amylase for their feasibility and enzyme production (Viktor et al., 2013). Starch is used in biofuel cells, where it is hydrolyzed to oligosaccharide or dextrin, which is then hydrolyzed into glucose by glucoamylase (Yamamoto et al., 2013). Starch-based ethanol production is an emerging solution in the field of biofuel against the existing fossil fuels.

19.5.11.2 Cellulase

Digestion of cellulose, a major constituent of the plant cell wall, is a challenge. Cellulose is a homopolymer linked by β-1,4-glycosidic bond.

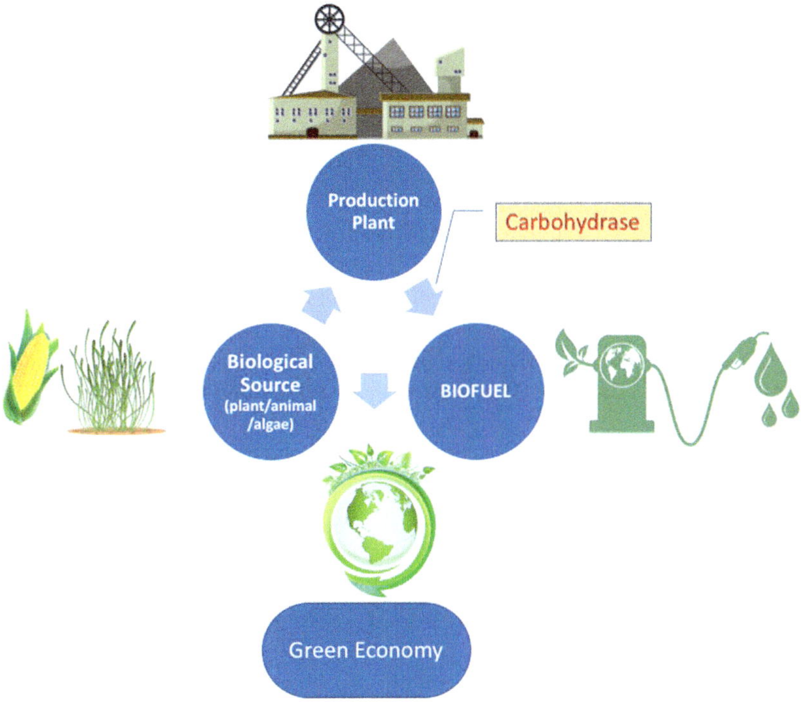

FIGURE 19.2 Production of biofuel from biomass.

The units are bonded by inter and intrapolymer hydrogen bonds into planes. These planes interact by hydrogen bond, van der Waal's interactions, and hydrophobic interactions. Breakdown of the cellulose is by β-glucosidases, cellulase, endoglucanases, and cellobiohydrolases. Due to its inert structure, it is difficult to be accessed organic solvents and water, hence making it challenging to convert cellulose to glucose for liquid fuel production and other value chemicals (Maxim et al., 2011).

Nowadays, C-6 carbohydrates are converted into 5-hydroxymethylfufural (HMF). A derivative of *hexo*se is recognized as an important compound for the production of new products and a replacement for fossil fuel derivatives. HMF is an aromatic alcohol, aromatic aldehyde and a furan ring system (Xiao et al., 2014). Corn residues and sugarcane bagasse are the potential sources of cellulose biofuel.

19.5.11.3 Xylulase

Xylose, is abundantly present in hemicellulose sources. It is the second most abundant sugar in nature. It can be a good source of generating food and fuel. It undergoes hydrolysis to release its respective sugars, xylose and arabinose, for the ethanol fermentation by the microorganism. Naturally occurring xylose-fermenting yeasts can catalyze acid or enzyme hydrolysis of xylan liberating xylulose. Enzymes such as xylose reductase and xylitol dehydrogenase are involved in xylose metabolism (Dodd and Cann, 2009).

19.5.12 Carbohydrases involved in paper industry

Paper is majorly constituted of cellulose fibers (90%–99%). The chemical bonding in cellulose affects paper's physical and chemical characteristics. Cellulose is a linear homopolymer made up of β-anhydroglucose units with the dormant hydroxyl group. β-1,4-glycosidic bonds link anhydroglucose units. Cellulose can form extensive intra and intermolecular hydrogen bonds. Cellulose cannot be hydrolyzed easily as it is insoluble in water and organic solvents for its structure and bonds (Harish et al., 2020; František et al., 2009). Hemicellulose affects the intrinsic fiber properties such as fibrillation, bonding ability, swelling, etc. (Pere et al., 2019).

Cellulases and hemicellulases are the enzymes which help in the degradation of paper. Xylanase degrades xylan and has an impact on the shape of fiber and other sheet properties such as density, tensile strength, two-dimensional formability, etc. (Pere et al., 2019). The degradation initiates when the cellulose degradation takes place homogeneously by hydrolysis, followed by oxidation and crosslinking. Eventually, the saccharides in the paper decrease, leading to the degradation of hemicellulose (František et al., 2009).

19.6 Concluding remarks

The modern industry is heavily reliant on enzymes nowadays, considering their massive inherent potential to speed up the chemical reaction and improve product yield. Among an array of industrially enzymes employed, carbohydrases have been shown to be some of the most sought-after enzymes. The market for carbohydrases ranges from food, beverage, animal feed to pharmaceuticals, textile, etc. All of these applications rely on the fundamental property of carbohydrases to catalyze the breaking down of carbohydrates into simple sugars. The enzyme's ability to hydrolyze diverse carbohydrate molecules stems from the innate potential of being structurally complex. Differences such as structural branching and monosaccharide specific secondary modifications (modifications by chemical moieties such as phosphates, sulfates, acyl esters, etc.) in the structure greatly improve the catalytic potential of carbohydrases. As a direct consequence of its molecular complexity, the industrial techniques devised for the production of carbohydrases have been methodically optimized for increasing product yield. Depending on the carbohydrate molecule, the pretreatment strategies change. To improve the overall shelf life of the enzyme, immobilization techniques affix enzymes onto a matrix that ensures prolonged availability. The only impediment to expanding the potential of carbohydrases is the inability to develop customized carrier systems that will enable the enzyme to be used in contrasting reactor configurations. However, the accomplishment of such an endeavor would ensure the economical use and reuse of the biocatalyst.

Abbreviations

DNA deoxyribo nucleic acid
rDNA recombinant DNA
SSF solid-state fermentation
SmF submerged fermentation
GOS galactooligosaccharide
FOS fructooligosaccharide
NSP non-starch polysaccharides
HMF 5-Hydroxymethylfurfural

References

Anzai, Y., Kudo, Y., Oyaizu, H., 1997. The phylogeny of the genera Chryseomonas, Flavimonas, and Pseudomonas supports synonymy of these three genera. Int. J. Syst. Bacteriol. 47, 249–251.

Apoorvi, C., Vuppu, S., 2012. Microbially derived pectinases: a review. J. Pharm. Biol. Sci. 2 (2), 1–5.

Arja, M.-O., 2007. Chapter 4-Cellulases in the textile industry. In: Polaina, J., MacCabe, A.P. (Eds.), Industrial Enzymes. Springer, pp. 51–63.

Armendáriz-Ruiz, M., Rodríguez-González, A.J., Camacho-Ruiz, R.M., Mateos-Diaz, J.C., 2018. Carbohydrate esterases: an overview. Methods mol. biol. 1835, 39–68.

Ash, C., Priest, F.G., Collins, M.D., 1993. Molecular identi¢cation of rRNA group 3 bacilli (Ash, Farrow, Wallbanks and Collins) using a PCR probe test. Proposal for the creation of a new genus Paenibacillus. Antonie van. Leeuwenhoek 64, 253–260.

Balley, R.W., Clarke, R.T.J., Wright, D.E., 1962. Biochem. J. 83, 517.

Bojarová, P., Křen, V., 2009. Glycosidases: a key to tailored carbohydrates. Trends Biotechnol. 27 (4), 199–209.

Brown, S.H., Kelly, R.M., 1993. Characterization of amylolytic enzymes, having both α-1,4 and α-1,6 hydrolytic activity, from the thermophilic archaea *Pyrococcus furiosus* and *Thermococcus litoralis*. Appl. Environ. Microbiol. 59, 2614.

Brumer, H., 2010. Carbohydrases, second ed John Wiley and Sons.

Cadena, P.G., Vigors, F.N., Silva, R.A., LimaFilho, J.L., Pimental, M.C.B., 2011. Kinetics and bioreactor studies of immobilized invertase on polyurethane rigid adhesive form. Biores. Technol. 102 (2), 513–518.

Cao, L., 2006. Immobilized enzymes: past, present and prospects. Carrier-bound Immobilized Enzymes: Principles, Application and Design. Wiley-VCH Verlag GmbH & Co, KGaA, Weinheim.

Chourasia, R., Phukon, L.C., Singh, S.P., Rai, A.K., 2020. Role of enzymatic bioprocesses for the production of functional food and nutraceuticals. Biomass, Biofuels, Biochemicals 309–334.

Christelle, B., Lenka, S., Charlotte, J., Jaroslav, K., Anne, I., 2005. Structures and mechanisms of glycosyl transferases. Glycobiol 16 (2), 29–37.

Contesini, F.J., de Alencar, F.J., Kawaguti, H.Y., de Barros Fernandes, P.C., de Oliveira Carvalho, P., et al., 2013. Potential applications of carbohydrases immobilization in the food industry. Int. J. Mol. Sci. 14 (1), 1335–1369.

Cormode, D.P., Gao, L., Koo, H., 2018. Emerging biomedical applications of enzyme-like catalytic nanomaterials. Trends Biotechnol. 36, 15–29.

Coutinho, P.M., Rancurel, C., Stam, M., Bernard, T., Couto, F.M., Danchin, E.J.D., et al., 2009. Carbohydrate-active enzymes database: principles and classification of glycosyltransferases. Bioinforma. Glycobiolgy Glycomics: An Introduction. Available from: https://doi.org/10.1002/9780470029619.ch5.

de Souza, T.S.P., Kawaguti, H.Y., 2021. Cellulases, hemicellulases, and pectinases: applications in the food and beverage industry. Food Bioprocess. Technol. 14, 1446–1477.

Divakar, S., 2013. Glycosidases. Enzymatic Transform. 5–21.

Dodd, D., Cann, I.K., 2009. Enzymatic deconstruction of xylan for biofuel production. Glob. Change Biol. Bioenergy 1 (1), 2–17.

Dodge, T., 2009. Production of industrial enzymes. In: Amauri, A.B. (Ed.), Enzymes in Food Technology, 2nd (ed.) Wiley-Blackwell, Hoboken, NJ, USA, pp. 44–58.

Dumorne, K., Severe, R., 2018. Marine enzymes and their industrial and biotechnological applications. Minerva Biotecnologica 30 (4), 113–119.

Dunning, J.W., Lathrop, E.C., 1945. The saccharification of agricultural residues. Indust. Eng. Chern. 37, 24–29.

Emad, A.A. 2019. Application of microbial enzymes in the dairy industry. In: Enzymes in food biotechnology, production, applications and future prospects, 6(1), 17–30.

Emregul, E., Sungur, S., Akbulut, U., 2006. Polyacrylamide gelatine career system used for invertase immobilisation. Food Che 97 (4), 591–597.

Erdem, N., Buran, T., Berber, I., Aydogdu, I., 2018. Enzyme replacement therapy in a Gaucher family. J. Natl. Med. Assoc. 110, 330–333.

Fabiano, J.C., Joelise, de, A.F., Haroldo, Y.K., Pedro, Carlos, de, B.F., et al., 2013. Potential applications of carbohydrases immobilization in the food industry. Int. J. Mol. Sci. 14, 1335–1369.

Falkowski, P., Scholes, R.J., Boyle, E., Canadell, J., Canfield, D., Elser, J., et al., 2000. The Global carbon cycle: a test of our knowledge of earth as a system. Science 290, 291–296.

Fan, Y., Pederson, O., 2021. Gut microbiota in human metabolic health and disease. Nat. Rev. Microbiol. 19, 55–71.

Feitkenhauer, H., 2003. Anaerobic digestion of desizing wastewater: influence of pretreatment and anionic surfactant on degradation and intermediate accumulation. Enzyme Microb. Technol. 33 (2–3), 250–258.

Francis, G.L., Gawthorne, J.M., Storer, G.B., 1978. Factors affecting the activity of cellulases isolated from the rumen digesta of sheep. Appl. Env. Microbiol. 36 (5), 643–649.

František, K., Danica, K., Michal, J., Svetozár, K., 2009. Cellulose degradation in newsprint paper ageing. Polym. Degrad. Stab. 94 (9), 1509–1514.

Grethlein, H.E., 1985. The effect of pore size distribution on the rate of enzymatic hydrolysis of cellulosic substrates. Biotechnol. 3, 155–160.

Hahn-Hägerdal, B., Galbe, M., Gorwa-Grauslund, M.F., Lidén, G., Zacchi, G., 2006. Bio-ethanol – the fuel of tomorrow from the residues of today. Trends Biotechnol. 24 (12), 549–556.

Han, R., Li, J., Shin, H.D., Chen, R.R., Du, G., Liu, L., et al., 2014. Recent advances in discovery, heterologous expression, and molecular engineering of cyclodextrin glycosyltransferase for versatile applications. Biotechnol. Adv. 32 (2), 415–428.

Harish, K.J., Andrew, C., Adisa, A., 2020. Environmental sustainability of biofuels. R. Soc. Publ. 476, 2243.

Hertzler, S., Savaiano, D.A., Dilk, A., Jackson, K.A., Fabrizis, S.N.B., Suarez, L., 2017. Nutrition in the Prevention and Treatment of Disease, 4th Edition Academic Press Books, Elsevier, pp. 875–892.

Himmel, M.E., Ding, S.Y., Johnson, D.K., Adney, W.S., Nimlos, M.R., Brady, J.W., et al., 2007. Biomass recalcitrance: engineering plants and enzymes for biofuels production. Science 315, 804–807.

Hobb, L., 2009. Sweetner from starch. Starch. (Eds.) 3, 797–832.

Hole, A.M., Draper, A., Jolliffe, G., Cullinan, P., Jones, M., Taylor, A.N., 2000. Occupational asthma caused by bacillary amylase used in the detergent industry. Occup. Environ. Med. 57 (12), 840–842.

Horn, S.J., Vaaje-Kolstad, G., Westereng, B., Eijsink, V.G., 2012. Novel enzymes for the degradation of cellulose. Biotechnol. Biofuels. 5, 45.

Houghton, J.I., Quarmby, J., Stephenson, T., 2001. Municipal wastewater sludge dewaterability and the presence of microbial extracellular polymer. Water Sci. Technol. 44 (2–3), 373–379.

Huang, G., Wen, S., Liao, S., Wang, Q., Pan, S., Zhang, R., et al., 2019. Characterization of a bifunctional alginate lyase as a new member of the polysaccharide lyase family 17 from a marine strain BP-2. Biotechnol. Lett. 41 (10), 1187–1200.

Imen, L., Hanen, El, Abed, B., Khemakhem, H., Belghith, F., Ben, A., et al., 2017. Optimization, purification, and starch stain wash application of two new alpha amylases extracted from leaves and stems of *Pergularia tomentosa*. Biomed. Res. Int. 9.

Jiang, R., Qiu, Y., Huang, W., Zhang, Li, Xue, F., Hao, N., et al., 2019. One step bioprocess inulin to product inulo-oligosaccharide using *Bacillus subtilis* secreting an extracellular *endo*-inulinase. Appl. Biochem. Biotechnol. 187, 116–128.

Johnson, R., Padmaga, G., Murthy, S.N., 2009. Comparative production of glucose and high fructose syrup from Cassava and sweet potato roots by direct conversion techniques. Innpv. Food Sci. Emerge. Technol. 10 (4), 616–620.

Kameshwar, A.K.S., Qin, W., 2018. Understanding the structural and functional properties of carbohydrate esterases with a special focus on hemicellulose deacetylating acetyl xylan esterases. Mycology 9 (4), 273–295.

Kantharaj, P., Boobalan, B., Sooriamuthu, S., Mani, R., 2017. Lignocellulose degrading enzymes from fungi and their industrial applications. Int. J. Curr. Res. 9 (21), 1–12.

Kim, J.H., Kim, Y.H., Kim, S.K., Kim, B.W., Nam, S.W., 2011. Properties and industrial applications of seaweed polysaccharides-degrading enzymes from the marine microorganisms. Microbiol. Biotechnol. Lett. 39 (3), 189–199.

Kiro, M., 2012. Application of enzymes in textile industry: a review. 230–239.

Kotwal, S.M., Shankar, V., 2009. Immobilised invertase. Biotechnol. Adv. 27 (4), 311–322.

Kulp, K., 1975. Carbohydrases: enzymes food process. 2, 62–87.

Kumar, R., Henrissat, B., Coutinho, P.M., 2019. Intrinsic dynamic behavior of enzyme:substrate complexes govern the catalytic action of β-galactosidases across clan GH-A. Sci. Rep. 9, 10346.

Laine, R.A., 1994. Glycobiology. A calculation of all possible oligosaccharide isomers both branched and linear yields 1.05 x 10(12) structures for a reducing hexasaccharide: the isomer barrier to development of single-method saccharide sequencing or synthesis systems. Glycobiology. 4, 759–767.

Legin, E., Ladrat, C., Godfroy, A., Barbier, G., Duchiron, F., 1997. Comptes Rendus de. l'Académie des. Sci. - Ser. III - Sci. de la. Vie, 320. Elsevier, p. 893.

Leite, P., et al., 2021. Recent advances in production of lignocellulolytic enzymes by solid-state fermentation of agro-industrial wastes. Curr. Opin. Green Sustain. Chem 27, 1–7.

Levasseur, A., Drula, E., Lombard, V., Coutinho, P.M., et al., 2013. Expansion of the enzymatic repertoire of the CAZy database to integrate auxiliary redox enzymes. Biotechnol. Biofuels. 6, 41.

Linhardt, R.J., Galliher, P.M., Cooney, C.L., 1987. Polysaccharide lyases. Appl. Biochem. Biotechnol. 12, 135–176.

Liu, M., Yu, H., 2012. Co-production of a whole cellulose system in *Escherichia coli*. Biochem. Eng. J. 69, 204–210.

Lizardi-Jiménez, M.A., Hernández-Martínez, R., 2017. Solid state fermentation (SSF): diversity of applications to valorize waste and biomass. Biotechnol. 7, 44.

Li, Q., Loman, A.A., Coffman, A.M., Ju, L.K., 2017. Soybean hull induced production of carbohydrases and protease among *Aspergillus* and their effectiveness in soy flour carbohydrate and protein separation. J. Biotechnol. 248, 35–42.

Luz, D.A., Rodrigues, A.K.O., Silva, F.R.C., Torres, A.E.B., Cavalcante, C.L., Brito, E.S., et al., 2008. Adsorptive separation of fructose and glucose from an agroindustrial waste of cashew industry. Bioresour. Technol. 99, 2455. –246.

Lwin, A., Orvisky, E., Goker-Alpan, O., LaMarca, M.E., Sidransky, E., 2004. Multi-center analysis of glucocerebrosidase mutations in Parkinson disease. Mol. Genet. Metab. 81, 70–73.

Matta, M.C., Vairo, F., Torres, L.C., Schwartz, I., 2018. Could enzyme replacement therapy promote immune tolerance in Gaucher disease type 1? Blood Cell Mol. Dis. 68, 200–202.

Maxim, S., Rajiv, K., Haitao, Z., Steven, H., 2011. Novelties of the cellulolytic system of a marine bacterium applicable to cellulosic sugar production. Biofuels 2 (1), 59–70.

Mehraj, P., Anuradha, K.P., Subbarao, D., 2013. Applications of pectinases in industrial sector K. Int. j. pure appl. sci. 16 (1), 89–95.

Mendoza, M., Meugnier, H., Bes, M., Etienne, J., Freney, J., 1998. Identi¢cation of *Staphylococcus* species by 16S-23S rDNA intergenic spacer PCR analysis. Int. J. Syst. Bacteriol. 48, 1049–1055.

Minic, Z., Simon, V., Penverne, B., Gaill, F., Hervé, G., 2001. Contribution of the Bacterial endosymbiont to the biosynthesis of pyrimidine nucleotides in the deep-sea tube worm *Riftia pachyptila*. J. Biol. Chem. 276, 23777.

Mojumder, N.H.M.R., Akhtaruzzaman, M., Bakr, M.A., Fatema-Tuj-Zohra, 2011. Study on isolation and partial purification of lactase (β-galactosidase) enzyme from *Lactobacillus* bacteria isolated from yogurt. J. Sci. Res. 4 (1), 239–249.

Moreno, F., Javier, M., Luz, S., 2014. Food Oligosaccharides (Production, Analysis and Bioactivity). Wiley. Available from: https://doi.org/10.1002/9781118817360.

Mussatto, S., Teixeira, J., 2010. Lignocellulose as raw material in fermentation processes. Appl. Microbiol. Microb. Biotechnol. 2, 897–907.

Nakamura, A.M., Alessandri, S.N., Igor, P., 2017. Structural diversity of carbohydrate esterases. Biotechnol. Res. Innov. 1 (1), 35–51.

Narasimha, J., Nagalakshmi, D., Viroji, Rao, S.T., 2013. Effect of NSP degrading enzyme supplement on the nutrient digestibility of young chickens fed wheat with different viscosities and triticale. Indian. J. Anim. Nutr. 5 (3), 105–111.

Nasr, S., Soudi, M.R., Salmanian, A.H., 2014. Diversity of secretory carbohydrases among yeasts isolated from environmental samples. volume 1.

Nguyen, Q.D., Judit, M.R., Czukor, B., Hoschke, A., 2011. Continuous production of oligofructose syrup from Jerusalem artichoke juice by immobilised endo-inulinase. Process. Biochem. 46 (1), 298–303.

Niyonzima, F.N., More, S.S., 2014. Detergent-compatible bacterial amylases. Appl. Biochem. Biotechnol. 174 (4), 1215–1232.

Ohlson, I., Trayardh, G., Hahn-Hagerdal, B., 1984. Enzymatic hydrolysis of sodium hydroxide pretreated sallow in an ultra-filtration membrane reactor. Biotechnol. Bioeng. 26, 647–653.

Olsen, H.S., 2000. Enzymes at work: a concise guide to industrial enzymes and their uses, Novo Nordisk A/S, Bagsvaerd, Denmark.

Pandey, A., Seelvakumar, P., Soccol, C.R., Nigam, P., 1999. Solid state fermentation for the production in industrial important enzyme. Enzyme Microb. Technol. 15, 677–682.

Payne, C.M., Resch, M.G., Chen, L., Crowley, M.F., et al., 2013. Glycosylated linkers in multimodular lignocellulose-degrading enzymes dynamically bind to cellulose. Proc. Natl. Acad. Sci. U.S.A. 110, 14646–14651.

Pere, J., Pääkkönen, E., Ji, Y., Retulainen, E., 2019. Influence of the hemicellulose content on the fiber properties, strength, and formability of handsheets. BioRes 14 (1), 251–263.

Phillips, C.M., Beeson, W.T., Cate, J.H., Marletta, M.A., 2011. Cellobiose dehydrogenase and a copper-dependent polysaccharide monooxygenase potentiate cellulose degradation by *Neurospora crassa*. ACS Chem. Biol. 6, 1399–1406.

Pigman, W.W., 1943. Classification of carbohydrases. J. Res. Nat. Bur. 30, 1–9.

Polaina, J., MacCabe, A.P., 2007. Industrial Enzymes: Structure, Function and Applications. Springer, Berlin, Germany, p. 641.

Puchart, V., 2015. Glycoside phosphorylases: structure, catalytic properties and biotechnological potential. Biotechnol. Adv. 33 (2), 261–276.

Purkarthofer, H., Sinner, M., Steiner, W., 1993. Cellulase-free xylanase from Thermomyces langinosus optimization of production in submerged and solid state culture. Enzyme Microb. Technol. 15, 677–682.

Quinlan, R.J., Sweeney, M.D., Leggio, L.L., Otten, H., et al., 2011. Insights into the oxidative degradation of cellulose by a copper metalloenzyme that exploits biomass components. Proc. Natl. Acad. Sci. U.S.A. 108, 15079–15084.

Ragauskas, A.J., Williams, C.K., Davison, B.H., Britovsek, G., Cairney, J., Eckert, C.A., et al., 2006. The path forward for biofuels and biomaterials. Science3 11, 484–489.

Raimbault, M., 1998. General and microbiological aspects of solid substrate fermentation. Electron. J. Biotechnol. 1, 1–15.

Rodney, J., Bothast Badal, C., Saha, 1997. Ethanol production from agricultural biomass substrates. Adv. Appl. Microbiol. 44.

Rodriguez-Couto, S., Toca-HerreraJose, L., 2006. Lacasses in the textile industry. Biotechnol. Mol. Biol. 1 (4), 115–120.

Roy, I., Munishwar, N.G., 2004. Hydrolysis of starch by a mixture of glucoamylase and pullulanase entrapped individually in calcium alginate beads. Enz. Microb. Technol. 34 (1), 26–32.

Sawabe, T., Ohtsuka, M., Ezura, Y., 1997. Isolation and characterisation of new limonoid glycosides from *Citrus unshiu* peels. Carbohydr. Res. 304, 69.

Schafer, T., Borchert, T.W., Nielsen, V.S., Skagerlind, P., Gibson, K., Wenger, K., et al., 2007. Industrial enzymes. White Biotechnol. 105, 59–131.

Schubot, F.D., Kataeva, I.A., Chang, J., Shah, A.K., 2004. Structural basis for the exocellulase activity of the cellobiohydrolase CbhA from *Clostridium thermocellum*. Biochemistry. 43, 1163–1170.

Shang, Q.H., Ma, X.K., Li, M., Zhang, L.H., Piao, X.S., 2018. Effects of alpha galactosidase supplementation on nutrient digestibility, growth performance, intestinal morphology and digestive enzyme activities in weaned piglets. Feed. Sci. Technol. 236, 48–56.

Shruti, P., Nivedita, S., Shweta, H., 2018. Utilization of horticultural waste (Apple Pomace) for multiple carbohydrase production from *Rhizopus delemar* F2 under solid state fermentation. J. Genet. Eng. Biotechnol. 16, 181–189.

Simpson, B.K., Rui, X., Klomklao, S., 2012. Enzymes in food processing. In: Simpson, B.K. (Ed.), Food Biochem and Food Processing, 2nd (ed.) Wiley-Blackwell, Oxford, UK, pp. 181–206.

Soccol, C.R., et al., 2017. Recent developments and innovations in solid state fermentation. Biotechnol. Res. Innov. 1, 52–71.

Stals, I., Sandra, K., Geysens, S., Contreras, R., Van Beeumen, J., Claeyssens, M., 2004. Factors influencing glycosylation of *Trichoderma reesei* cellulases. I: postsecretorial changes of the O- and N-glycosylation pattern of Cel7A. Glycobiology 14 (8), 713–724.

Tawil, G.S., Anders, V.N., Agnes, R.S., Paul, C., Alain, B., 2011. In-depth study of a new highly efficient raw starch hydrolysing alpha amylase from *Rhizomucor* sp. Biomicromol 12, 34–42.

Taylor, M.E., Drickamer, K., 2006. Introduction to Glycobiology, 2nd (ed.) Oxford University Press, New YorK.

Tirado, O., Rosado, W., Govind, N.S., 2005. Characterization of bacteria with carbohydrase activities from tropical ecosystems. J. Mar. Biolog. Assoc. U.K.

Treem, W.R., McAdams, L., Stanford, L., Kastoff, G., Justinich, C., Hyams, J., 1999. Sacrosidase therapy for congenital sucrase-isomalatase deficiency. J. Pediatr. Gastroenterol. Nutr. 28, 137–142.

Tufvesson, P., Lima-Ramos, J., Nordblad, M., Woodley, J.M., 2010. Guidelines and cost analysis for catalyst production in biocatalytic processes. Org. Process. Res. Dev. 15, 266–274. Int. J. Mol. Sci. 2013, 14 1360.

Turkiewicz, M., Kur, J., Bialkowska, A., Cieslinski, H., Kalinowska, H., Bielecki, S., 2003. Antartic Marine bacterium *Pseudoalteromonas* sp. 22b as a source of cold adapted beta galactosidase. Biomol. Eng. 20, 317–335.

Underkofler, L.A., Barton, R.R., Rennert, S.S., 1958. Production of microbial enzymes and their applications. Appl. Environ. Microbiol. 6 (3), 212–221.

Ünligil, U.M., Rini, J.M., 2000. Glycosyltransferase structure and mechanism. Curr. Opin. Struct. Biol. 10 (5), 510–517.

Uzuner, S., Cekmecelioglu, D., 2019. Enzymes in the beverage industry. Enzymes Food Biotechnol. 29–43.

Vaaje-Kolstad, G., Westereng, B., Horn, S.J., Liu, Z., et al., 2010. An oxidative enzyme boosting the enzymatic conversion of recalcitrant polysaccharides. Science. 330, 219–222.

van den Broek, L.A., Hinz, S.W., Beldman, G., Vincken, J.P., Voragen, A.G., 2008. Bifidobacterium carbohydrases-their role in breakdown and synthesis of (potential) prebiotics. Mol. Nutr. Food Res. 52 (1), 146–163.

Vernekar, M., Lele, S., 2009. Laccase: properties and applications. J. Biosci. 4, 1694–1717.

Viktor, M.J., Rose, S.H., van Zyl, W.H., et al., 2013. Raw starch conversion by *Saccharomyces cerevisiae* expressing *Aspergillus tubingensis* amylases. Biotechnol. Biofuels 6, 167.

Wanasundara, P.K.J.P.D., Shahidi, F., 1997. Removal of flaxseed mucilage by chemical and enzymatic treatments. Food Chem. 59 (1), 47–55.

Wang, Y.J., Jiang, X.W., Liu, Z.Q., Jin, L.Q., Liao, C.J., Cheng, X.P., et al., 2016. Isolation of fructose from high-fructose corn syrup with calcium immobilized strong acid cation exchanger: isotherms, kinetics, and fixed-bed chromatography study. Can. J. Chem. Eng. 94, 537–546.

Weijers, C.A.G.M., Franssen, M.C.R., Visser, G.M., 2008. Glycosyltransferase-catalyzed synthesis of bioactive oligosaccharides-. Biotechnol. Adv. 26 (5), 436–456.

Westereng, B., Ishida, T., Vaaje-Kolstad, G., Wu, M., et al., 2011. The putative endoglucanase PcGH61D from *Phanerochaete chrysosporium* is a metal-dependent oxidative enzyme that cleaves cellulose. PLoS One 6.

Wright, D.E., 1961. Bloat in cattle. New Zealand J. Agric. Res. 4, 203.

Xiao, S., Liu, B., Wang, Y., Fang, Z., Zhang, Z., 2014. Efficient conversion of cellulose into biofuel precursor 5-hydroxymethylfurfural in dimethyl sulfoxide–ionic liquid mixtures. Bioresour. Technol. 15 (1), 361–366.

Yamamoto, K., Matsumoto, T., Shimada, S., Tanaka, T., Kondo, A., 2013. Starchy biomass-powered enzymatic biofuel cell based on amylases and glucose oxidase multi-immobilized bioanode. N. Biotechnol. 30 (5), 531–535.

Yang, H., Li, J., Du, G., Liu, L. (Eds.), 2017. Biotechnology of Microbial Enzymes: Production. Biocatalysis and Industrial Applications. Academic Press Books, Elsevier, pp. 151–165.

Yin, Y., Liu, Y.J., Meng, S.J., Kiran, E.U., Liu, Y., 2016. Enzymatic pretreatment of activated sludge, food waste and their mixture for enhanced bioenergy recovery and waste volume reduction via anaerobic digestion. Appl. Energy. 179, 1131–1137.

Zeng, Z.K., Li, Q.Y., Tian, Q.Y., Xu, Y.T., Piao, X.S., 2018. The combination of carbohydrases and phytase to improve nutritional value and non-starch polysaccharides degradation for growing pigs fed diets with or without wheat bran. Anim. Feed. Sci. Technol. 235, 138–145.

Zhang, C., Kim, S.K., 2010. Research and application of marine microbial enzymes: status and prospects. Mar. drugs 8 (6), 1920–1934.

Zheng, L., Xu, Y., Li, Q., et al., 2021. Pectinolytic lyases: a comprehensive review of sources, category, property, structure, and catalytic mechanism of pectate lyases and pectin lyases. Bioresour. Bioprocess. 8, 79.

Zilda, D.S., Yulianti, Y., Sholihah, R.F., Subaryono, S., Fawzya, Y.N., Irianto, H.E., 2019. A novel *Bacillus* sp. isolated from rotten seaweed: identification and characterization alginate lyase its produced. Biodiversitas. J. Biodivers. 20 (4), 1166–1172.

Chapter 20

Role of microbial enzymes in agricultural industry

Prashant S. Arya[1], Shivani M. Yagnik[2] and Vikram H. Raval[1]

[1]*Department of Microbiology and Biotechnology, School of Sciences, Gujarat University, Ahmedabad, Gujarat, India,* [2]*Department of Microbiology, Christ College, Rajkot, Gujarat, India*

20.1 Introduction

Enzymes are highly specific, and they implement biochemical reactions in living organisms, speeding up metabolic reactions without being consumed. They may also be taken from cells and used outside living organisms to carry out processes. Extracellular enzymes play an important role in many aspects of life and can be used to accelerate a variety of economically viable biochemical reactions (Raval et al., 2013; Singh et al., 2016). The utilization of enzymes is not a new concept. They have been finding applications for centuries. The name "enzyme" was coined in 1878 by Wilhelm Kuhne, who tested the capacity of beer yeast to ferment various carbohydrates. It is derived from the Greek terms *en-*, which means inside, and *-zyme*, which means yeast, some things present within the yeast cells.

Enzymes will have a wide range of uses in a variety of industries for a multitude of reasons, including food manufacturing and processing, leather, fabric and textile processing, cosmetics and detergent production, beer brewing, dairy production, meat tenderizing, and paper-making (Raval et al., 2013). As a final point, enzymes are widely used in the environment and agriculture sector, and their use has been steadily increasing for many years. A broad spectrum of enzymes from plants, animals, and microbes plays a vital role in waste management, is involved in an element cycle, and synthesizes precursor metabolites for conversion, transformation, and degradation of such compounds. Some enzymes can function to remove contaminants by triggering them to precipitate and change into other substances. It could also alter the characteristics of particular pollutants to make them more treatment-friendly or assist in converting waste chemicals into value-added goods (Younus, 2019).

Biotechnology of Microbial Enzymes. DOI: https://doi.org/10.1016/B978-0-443-19059-9.00017-7
© 2023 Elsevier Inc. All rights reserved.

Agro-enzymes used as feed additives in animal farms are the most essential from an agricultural standpoint. A wide range of enzymes in agricultural soils ensures an appropriate course of activities at soil–plant–environment interrelations critical to crop development and human and animal feed production. They are frequently employed as indicators of soil health, fertility, and agronomy. Scientists and other stakeholders working in this field need a better knowledge of the main features and activities of soil enzymes (Piotrowska-Długosz, 2019). The present chapter aims to offer an overview of the applications of these microbial enzymes in the agricultural industry.

20.2 Soil and soil bacteria for agriculture

Enzymes are essential for the survival of organisms because they play a critical role in metabolic processes. They are also important in various industries, including agriculture, because they may be used to make products for different purposes. Industries utilize enzyme-producing microorganisms to attain their objectives (Fig. 20.1). Bacteria, fungi, and actinomycetes can produce these enzymes, and industries have genetically altered them to do so (Raval et al., 2014, 2015b). The bacterial genus *Rhizobium*, *Azotobacter*, *Bacillus*, *Clostridium*, and *Pseudomonas*; the fungus genus *Aspergillus*, *Trichoderma*, and *Penicillium*; and the actinomycetes *Streptomyces* and *Cellulomonas* are

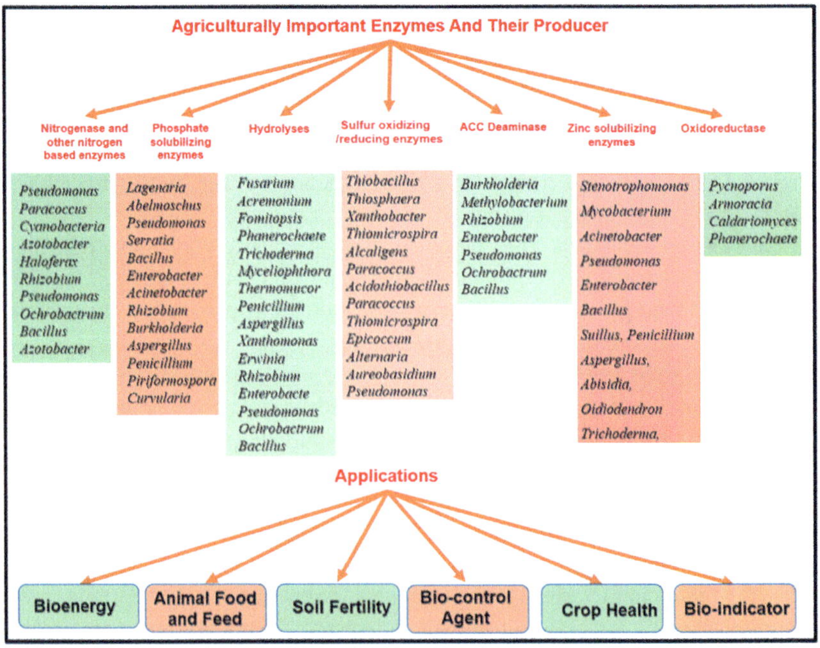

FIGURE 20.1 Agriculturally important enzymes, their producers, and applications.

utilized for this purpose. Many of them are known for their biotechnological potential and are used in agriculture (Piotrowska-Długosz, 2019).

Bacteria are extensively distributed and well recognized for synthesizing agriculturally significant enzymes among all types of microorganisms. *Bacillus cereus*, *Bacillus stearothermophilus*, *Bacillus megaterium*, *Bacillus licheniformis, Bacillus subtilis*, *Bacillus polymyxa*, *Clostridium thermosulfurogenes*, *Hordeum vulgare*, *Paenibacillus chitinolyticus* CKS1, and *Thermoactinomyces* sp. are prominent in producing a range of carbohydratase (amylase, pullulanase, cellulose, pectinase, and other enzymes), proteolytic enzymes, and many more (Raval et al., 2015a). Likewise, several extremophile microbes are involved in agriculture by releasing hydrolytic enzymes that work in various hostile environments. *Thermoplasma*, *Ferroplasma*, *Sulfolobus*, *Leptospirillum*, *Bacillus*, *Halobacillus*, *Humicola*, *Halobacter*, and *Methanocaldococcus* were among the genera intricate in this field (Arya et al., 2022). However, plant-associated bacteria such as *Rhizobium* sp., *Pseudomonas putida, Azospirillum fluorescens, Azospirillum lipoferum, Allorhizobium* sp., *Azorhizobium* sp., *Bradyrhizobium* sp., *Thermomonosporaceae* sp., and *Micromonosporaceae* sp., are not able to release essential enzymes required for plant growth (Fasusi et al., 2021).

Actinobacteria are also recognized to explore a range of plant growth-promoting elements and reduce plant diseases by secreting various compounds such as secondary metabolites. They hence are the key prospects for increasing agricultural production. Researchers who have mentioned the above abilities have identified numerous actinomycetes species, including *Frankia* sp., *Rhodococcus fascians*, *Streptomyces filipinensis*, *Streptomyces olivaceoviridis*, *Streptomyces rimosus*, *Streptomyces rochei*, *Streptomyces albidoflavus*, *Streptomyces thermoautotrophicus*, *Thermoactinomyces thalophilus*, and *Arthrobacter maltophilia* (Husain and Ullah, 2019).

Moreover, the ability of fungi to produce a wide range of extracellular enzymes allows them to break down all types of organic matter, decompose soil components, and manage the carbon and nutrient balance to preserve soil health (Arya et al., 2021). Aside from this feature, the genera *Alternaria*, *Aspergillus*, *Cladosporium*, *Dematium*, *Gliocladium*, *Humicola*, and *Metarhizium* produced hydrolytic enzymes that aid in the synthesis of organic compounds in soil and may thus be required for the preservation of soil organic matter. Many researchers have studied *Aspergillus oryzae*, *Aspergillus awamori*, *Trichoderma harzianum*, *Bispora* sp., *Steccherinum ochraceum*, *Polyporus versicolor*, *Trichoderma reesei*, *Trichoderma longibrachiatum*, *Fusarium venenatum*, *Kluyveromyces marxianus*, and *Penicillium notatum*, which are agriculturally remarkable enzyme-producing fungi, investigated by many researchers (Thapa et al., 2019).

20.3 Microbial enzymes

The agricultural sector is inextricably tied to the soil, an important component of terrestrial ecosystems and the basic matrix for farming. The proper metabolism and biogeochemical reactions are required to support the biochemical cycles of essential nutrients (Bunemann et al., 2018). However, these interactions include microbial populations and their metabolites, such as organic acids and enzymes. Enzymes play a vital role in agriculture because they perform biochemical tasks such as organic matter digestion and regeneration, nutrient turnover, soil structure stabilization, and pollutant degradation (Piotrowska-Długosz, 2019). Agriculture-relevant enzymes are produced by soil microorganisms, plants, and soil animals. The three main types of enzymes are intracellular (found in leaving and proliferating organisms, such as soil dehydrogenases), cell-associated (found in cell and tissue fragments), and free enzymes (Bakshi and Varma, 2010). Extracellularly stabilized enzymatic proteins work independently of normal cell growth and immobilization, protecting enzymes against degradation and denaturation caused by unfavorable environmental conditions and protease. Although stabilized enzymes are less active than free enzymes, their activity is crucial for entire biochemical activities in soil (Arya et al., 2021). To yet, no clear distinction between soil extracellular and intracellular activity has been accomplished due to limited agricultural enzymology technologies. Oxidoreductases, hydrolases, lyases, and transferases are the four kinds of enzymes found in soil, with the first two being the most prevalent (Piotrowska-Długosz, 2019). Table 20.1 describes the most studied agricultural enzymes and their mode of action.

Different soil types include a wide range of enzymes that vary in quantity and quality due to changes in physical, chemical, microbiological, and biochemical properties. Enzymatic activity is generally higher in soils with higher amounts of organic matter, nutrients, and clay and a significant microbial activity than in soils with lower levels of these attributes. As is widely known, clay-organic matter complexes play an important role in maintaining extracellular enzymatic activity in the soil. As enzymes associated with these complexes are more resistant to proteolysis and microbial attack and unfavorable circumstances such as temperature and moisture variations, they maintain their activity (Zimmerman and Ahn, 2010).

20.3.1 Nitro-reductase

Nitrogen is a vital element for life; it is found in proteins, nucleic acids, and many biomolecules and accounts for around 6% of the dry weight of organisms on average. As a result, nitrogen availability restricts microbial and plant development. The distribution of nitrogen is divided into three primary pools: the atmosphere, soils/groundwater, and biomass. The nitrogen cycle refers to the complicated nitrogen exchange between these three pools. Nitrogen may be found in ecosystems as both organic and inorganic molecules.

On the other hand, nitrogen is frequently present in the completely reduced-state inorganic compounds, such as amino, amido, or imino groups

TABLE 20.1 Microbial enzymes, mechanism of action, and their role in the agriculture industry.

Class of enzyme	Enzyme	Mechanism of action	Microorganisms	Role in the agriculture industry	References
Oxidoreductase	Nitrate reductase	Nitrate + NADH + H^+ = nitrite + NAD^+ + H_2O	*Pseudomonas* *Paracoccus* *Serratia* *Bacillus*	The first enzyme of the denitrification process, nitrate reduction 2 (assimilatory).	Younus (2019); Verma et al. (2019); Singh and Gupta (2020)
	Peroxidase	2 phenolic donor + H_2O_2 = 2 phenoxyl radical of the donor + 2 H_2O	*Enterobacter* *Pycnoporus* *Pycnoporus*	Takes part in lignin decomposition Protects cells from oxidative damage by releasing O_2 from H_2O_2.	
	Nitrogenase	8 reduced ferredoxin + 8 H^+ + N_2 + 16 ATP + 16 H_2O = 8 oxidized ferredoxin + H_2 + 2 NH_3 + 16 ADP + 16 phosphate	Chloroperoxidase *Caldariomyces* *Phanerochaete*	Converts the atmospheric, gaseous dinitrogen (N_2) into ammonia (NH_3) by taking part in nitrogen fixation.	
	Catalase	2 H_2O_2 = $2H_2O$ + O_2		Protects cells from oxidative damage by releasing O_2 from H_2O_2.	
	Laccase	Phenolic compounds + O_2 = oxidized phenolic compounds		Waste management, oxidized soil component.	

(*Continued*)

TABLE 20.1 (Continued)

Class of enzyme	Enzyme	Mechanism of action	Microorganisms	Role in the agriculture industry	References
Hydrolases	Acid/alkaline phosphatase	A phosphate monoester + H_2O = phosphate + alcohol	*Burkholderia* *Methylobacterium* *Rhizobium* *Enterobacter* *Pseudomonas* *Ochrobactrum* *Bacillus* *Stenotrophomonas* *Burkholderia* *Mycobacterium* *Acinetobacter* *Pseudomonas* *Enterobacter* *Aureobasidium* *Abisidia* *Suillus* *Aspergillus* *Barnettozyma* *Abisidia* *Torulaspora* *Oidiodendron* *Trichoderma* *Hymenoscyphus*	Organic P compounds are converted into inorganic forms (HPO_4^{-2}, $H_2PO_4^{-}$), which are easily consumed by microorganisms and plants.	Huang et al. (2017); Saikia et al. (2018); Usharani et al. (2019); Mehta et al. (2019); Thapa et al. (2020); Ben Zineb et al. (2020); Fasusi et al. (2021); Arya et al. (2021)
	Arylsulfatase	A phenol sulfate + H_2O = a phenol + sulfate		Catalyze the hydrolysis of aromatic sulfate esters (C–O-SO3) to phenols (R–OH) and sulfate ($SO4^{-2}$), this is important for the mobilization of inorganic S for plant nutrition.	
	Phosphodiesterase I	$R_2NaPO_4 + H_2O = ROH + RNaHPO_4$		Indicator for P cycle, hydrolysis of phosphoric esters, revealed to be a good index of the soil P availability to the plant.	
	Aryl acyl-amidase	Anilide + H_2O = carboxylate + aniline		Hydrolyzes propanol, which is used as a component of herbicide.	
	Arylamidase [-α-aminoacyl-peptide hydrolase (microsomal)]	Valaciclovir-hydrolyzed Acyclovir		Hydrolysis of an N-terminal amino acid from peptides, amides, and arylamines is important in the beginning stages of the soil amino acids mineralization Indicator of soil N mineralization.	

L-Asparaginase and L-glutaminase	Aspargine = aspartic acid + NH_3 Glutamine = glutamic acid + NH_3	Acts on C—N bonds (other than peptide bonds) on respective amino acids release; important in N mineralization to NH_3 to provide plant-available N form.
Amidase	Monocarboxylacid amide + H_2O = monocarboxyl acid + NH_3	Hydrolysis of C—N bonds other than peptide bond in linear amides releasing NH_3, important for N mineralization to provide plant-available N form.
Inorganic diphosphatase (pyrophosphatase)	Diphosphate + H_2O = 2 phosphate (PO_4^{3-})	Indicator of phosphorus transformation, ammonium polyphosphate, an inorganic salt of poly-phosphoric acid and ammonia, is one of the frequently used phosphoric fertilizers.
Cellulases (*endo*-1,4-β-D-glucanase and *exo*-cellobiohydrolase)	Endohydrolysis of (1−4)-β-D-glucosidic linkages in cellulose. D-glucosidic Hydrolysis of (1−4)-β-D-linkages in cellulose releasing cellobiose from the nonreducing ends of the chains	The enzymatic complex involved in the degradation of cellulose, the most abundant polysaccharide found in the biosphere, provides readily available C for soil microorganisms, thus increasing soil microbiological activity, and directly soil fertility.
β-Glucosidase	Hydrolysis of terminal, nonreducing β-D-glucosyl residues with release of β-D-glucose	It is a part of the enzymatic complex involved in the degradation of cellulose.

(*Continued*)

TABLE 20.1 (Continued)

Class of enzyme	Enzyme	Mechanism of action	Microorganisms	Role in the agriculture industry	References
	α-Amylase and β-amylase	α-Amylase: *endo*-hydrolysis of (1-D-4)-α-glucosidic linkages in polysaccharides containing three or more (1–4)-α-linked D-glucose units; β-amylase: hydrolysis of (1-D-Glu-4)-α-glycosidic linkages in polysaccharides to remove successive maltose units from the nonreducing ends of the chains		The amylase system synergistically hydrolyzes starch (glycogen and other poly- and oligosaccharides). The products of this reaction are dextrins, oligosaccharides, maltose, and finally, monosaccharides, like glucose; important in the transformation of plant residues entering the soil.	
	Proteolytic enzymes	Protein + H_2O = amino acid + short peptide		They hydrolyze proteins and peptides and liberate amino acids. The degradation of proteins (proteolysis) is believed to be a limiting step of N-mineralization in soil.	
	Xylanase	Hydrolysis of β-1,4-xylan bonds		Responsible for decomposition of xylan, a polysaccharide found with cellulose in soil.	
	Invertase	$C_{12}H_{22}O_{11} + H_2O = C_6H_{12s}O_6 + C_6H_{12}O_6$		Indicator of carbon transformation, responsible for the breakdown of plant litter in soil; catalyzes the hydrolysis of sucrose to glucose and fructose.	
	Urease	Urea + $H_2O = CO_2 + 2NH_3$		The urease activity is crucial in regulating the N-supply to plants after urea fertilization. Important in a more effective way of managing N fertilizers. Used as an index of N-transformation in soil.	

Transferases	Thiosulfate sulfur transferase (rhodanese)	Thiosulfate ($S_2O_3^{2-}$) + cyanide (CN) = sulfite (SCN^-) + thiocyanide (SO_3^{2-})	*Thiobacillus* *Thiosphaera* *Xanthobacter* *Thiomicrospira* *Alcaligens* *Paracoccus* *Pseudomonas* *Aureobasidium* *Epicoccum* *Penicillium* *Acidithiobacillus*	Indicator for S cycle. Cleaves the S–S bond of thiosulfate, forming S and sulfite, and then S is subsequently oxidized into sulfite by sulfur oxygenase. In contrast, sulfite is further oxidized into sulfate by sulfite oxidase.	Goyal (2019); Zenda et al. (2021)

(Pajares and Ramos, 2019). Microbial enzymatic nitrogen fixation is categorized into abiotic mechanisms (lightning) and biotic (nitrogen fixers). In the abiotic fixation, lightning would have oxidized N_2 with CO_2, and subsequently, NO would have been transformed into soluble nitrosyl hydride (HNO). Some bacteria and archaea carry out this type of biological nitrogen fixing (BNF). Despite a vast atmospheric reservoir, nitrogen bioavailability is mostly dependent on BNF. Ammonia-oxidizing bacteria and archaea are the two major drivers of the global nitrogen cycle (Sun et al., 2021).

Nitrogenase catalyzes the reduction of N_2 to NH_3 and is composed of Fe and Mo-Fe component proteins. It is important in nitrogen fixation because it makes atmospheric nitrogen readily available to plants and other crops for growth. Using a sophisticated nitrogenase enzyme system, BNF converts atmospheric nitrogen to ammonia, a plant-utilizable form. Nitrogenase systems are made up of many components, including (1) nitrogenase reductase, which is an iron protein, and (2) nitrogenase with a metal cofactor, which can be Mo-nitrogenase (the most common), V-nitrogenase, or Fe-nitrogenase. All three nitrogenases are likely linked and are made up of two different metalloproteins known as component 1 or dinitrogenase (Mo-Fe, V-Fe, or Fe-Fe protein) and component 2 or dinitrogenase reductase (Fe protein). Nitrogenase reductase offers strong reducing power, electrons that nitrogenase uses with cofactor to reduce N_2 to NH_3 (Verma et al., 2019). It is estimated that BNF accounts for roughly two-thirds of all nitrogen fixed globally, with the remaining one-third being manufactured industrially. BNF may be classified into several types, including symbiotic BNF that fixes atmospheric nitrogen through mutualistic associations with leguminous plants, nonleguminous plants, and free-living non-symbiotic BNF that may also be endophytes (Bhattacharyya and Jha, 2012).

Another key phase in nitrate assimilation needs nitrate taken up by a transport system and then reduced to ammonium via nitrite by assimilatory nitrate and nitrite reductases (Nas). A molybdo-enzyme catalyzes the two-electron reduction of nitrate to nitrite rather than ammonia. Bacterial Nas are cytoplasmic proteins that vary physically and functionally from the dissimilatory periplasmic nitrate reductases (Nap) and respiratory membrane-bound nitrate reductases (Nar) involved in denitrification (Fariduddin et al., 2018). The ferredoxin or flavodoxin-dependent enzyme found in cyanobacteria, *Azotobacter*, and the archaeon (*Haloferax mediterranei*) and the NADH-dependent enzyme found in heterotrophic bacteria such as *Rhodobacter capsulatus* are the two forms of bacterial Nas. Likewise, nitrate reductase (NR) catalyzes the first process in nitrate assimilation, the reduction of nitrate to nitrite, whereas ferredoxin-nitrite reductase (Fd-NiR) catalyzes the six-electron reduction of nitrite to ammonia, utilizing reduced ferredoxin as the electron donor (Martínez-Espinosa, 2020). Denitrification is required for the full nitrogen cycle, in which nitrate, nitrite, and the gaseous nitric and nitrous oxides serve as alternate terminal acceptors for electron transport phosphorylation, resulting in N_2 as the ultimate product. This reaction is performed progressively by nitrate reductase (Nar and Nap), nitrite reductase (Nir), nitric oxide reductase (NOr), and nitrous oxide reductase (NOr and NOs). Denitrifying species include proteobacteria, halophilic and hyperthermophilic archaea, and even certain fungi, although several bacterial strains such as *Pseudomonas* and *Paracoccus* strains have received the most attention. Ammonia produced from direct absorption or nitrate reduction must be integrated into organic molecules for further metabolism. The GS-GOGAT cycle, which functions in leaves, is the major mechanism of ammonia absorption in higher plants. Even trace amounts of ammonia are absorbed by glutamate dehydrogenase (GDH) (Rajta et al., 2020).

20.3.2 Hydrolases

Hydrolytic enzymes are now gaining popularity due to their uses in various sectors, including agriculture, detergent, textiles, leather, pulp and paper, and food. Hydrolytic enzymes, such as carbohydratase, proteases, lipases, laccase, pectinases, and cellulases, are synthesized by many types of microorganisms, including bacteria, fungi, archaea, and actinomycetes. Carbohydrases are frequent feed enzymes because they improve food absorption by breaking down carbohydrates, whereas proteases hydrolyze peptide bonds in proteins and convert them to peptides and amino acids. The pretreated biomass was exposed to enzymatic hydrolysis using cellulases that generate glucose (Thapa et al., 2020). Similarly, Xylanase, also known as *endo*-β-1,4-xylanase, is a hydrolytic enzyme with necrotizing and enzymatic activity that can activate plant immunity. *Aspergillus niger*, *Aspergillus foetidus*, *Penicillium oxalicum*, *Aspergillus tubingensis*, *Aspergillus terreus*, *Aspergillus fumigatus*, *Trichoderma viride*, and *Trichoderma citrinoviride* may all produce xylanase (Chukwuma et al., 2020).

Additionally, pectinase refers to a class of enzymes that can catalyze the de-esterification, hydrolysis, and trans-elimination of pectin and pectic substances. This enzyme is widely employed in agro-processing, animal feed manufacturing, and agro-waste management (Nayak and Bhushan, 2019). This enzyme has also been shown to have antimicrobial action against infections. Pectinase is produced by fungi such as *Thermomucor indicae seudaticae*, *Penicillium chrysogenum*, *Penicillium glandicola*, *Aspergillus japonicas*, *A. niger*, and *Aspergillus tamari*, and bacteria, for example, *Xanthomonas*, *Erwinia*, and *Pseudomonas*, are utilized for decomposing agriculture wastes (Mohanram et al., 2013).

As we know, cellulose is one of the most prevalent carbohydrate forms in agricultural waste. To handle them, cellulase is necessary to break 1,4-glycosidic bonds. It is the third biggest enzyme consumed globally for animal feed additives, cotton processing, detergent manufacture, juice extraction, and paper recycling (Yarullina et al., 2016). Cellulase enzyme now has found agricultural applications to promote plant development by destroying infections. Cellulase kills pathogens by cleaving the internal link of the glycan chain and providing reducing or nonreducing ends of cello-oligosaccharides for cello-bio-hydrolases, also known as *exo*-glucanases or 1,4-D-glucan-cello-bio-hydrolase (CBH), to attack. Following that, CBH hydrolyzes chain ends to produce cellobiose as the main product (Thapa et al., 2019). *Fusarium chlamydosporum, Acremonium cellulolyticus, Fomitopsis* sp., *Phanerochaete chrysosporium, T. reesei, Myceliophthora thermophila*, and *Penicillium* sp. have been shown to produce this enzyme (Husain, and Ullah, 2019).

20.3.3 1-Aminocyclopropane-1-carboxylic acid deaminase

Plants face biotic and abiotic stressors because of heat, cold, drought, floods, nutrient shortage, heavy metal exposure, phytopathogens, and insect infestations. When stresses surpass a particular level, they have a major impact on agricultural productivity. According to Murali and his coworkers' report, most stressed plants enhanced ethylene production from the precursor, 1-aminocyclopropane-1-carboxylic acid (ACC) (Murali et al., 2021). Ethylene is a plant hormone that regulates several physiological activities, including respiration, nitrogen fixation, and photosynthesis. The rise in the plant hormone ethylene would restrict plant growth and development, and if the ethylene level exceeded the limit, the plant would die (Raghuwanshi and Prasad, 2018). Through hydrolyzing 1-aminocyclopropane-1-carboxylic acid, plant growth-promoting bacteria with ACC deaminase (ACCD) activity play an essential role in managing biotic and abiotic stresses. ACCD is made up of multiple polypeptide chains with monomers with molecular masses ranging from 35 to 42 kDa. ACCD was discovered in soil microorganisms, where it transforms ACC into alpha-ketobutyrate and ammonia (deamination), which is then metabolized by another set of microbes. The cofactor for this enzyme is pyridoxal-5-phosphate. One molecule of ACCD is strongly linked to the pyridoxal phosphate cofactor per subunit. Among the D-amino acids, D-cysteine and D-serine are ACCD substrates, whereas L-serine and L-alanine are ACCD competitive inhibitors (Chandwani and Amaresan, 2022).

ACCD-producing rhizobacteria have been identified in *Burkholderia, Methylobacterium, Rhizobium, Enterobacter, Pseudomonas, Ochrobactrum*, and *Bacillus*, which increased root length, seed germination rate, chlorophyll content, production of antioxidant enzymes, and cellular osmolytes during drought stress. Individual ACCD-producing bacteria, in addition to consortia, eased water stress in the examined plants by solubilizing phosphate, supplying iron and nitrogen, and suppressing infections (Saikia et al., 2018; Singh et al., 2019).

20.3.4 Phosphate-solubilizing enzymes

After nitrogen, phosphorus is the second most important macronutrient for plant metabolism, growth, and development. Phosphorus (P) is one of the most critical elements needed for plant development, and it ranks high among soil macronutrients. Organic phosphate compounds in the agriculture sector are nucleic acids, phospholipids, phosphonates, phytic acid, polyphosphonates, and sugar phosphates (Ben Zineb et al., 2020). Despite the availability of phosphorus in the soil in both organic and inorganic forms, it is generally inaccessible for plant absorption owing to the complexation with metal ions. The use of agrochemicals to meet the need for phosphorus to increase crop output has resulted in a decline in ecosystem and soil health and an imbalance in soil microbiota. Phosphate fertilizers or microbial activity are frequently used to compensate for soil phosphorus deficiency. Phosphorus-deficit in soils is caused by lower overall phosphorus concentrations in the soil as well as fixation of supplied P from chemical fertilizers and other organic sources such as manures. The primary restriction to P availability is its solubilization, as it is fixed in both acidic and alkaline soil (Mehta et al., 2019).

Only phosphate-solubilizing microorganisms (PSMs) can solubilize soil-fixed phosphorus. These bacteria released various enzymes and organic acids into the soil, making phosphorus soluble and accessible to plants. The capacity of these PSMs to solubilize phosphorus varies and is mostly determined by the method used for solubilization, their molecular genetics, and their ability to release phosphorus in soil (Prabhu et al., 2019). Microbial enzymes, such as phosphatase or phosphohydrolase, phytase, phosphonatase, mono- and di-esterase, pyrophosphatases, cellulolytic and ligninolytic enzyme, and CeP lyase, are responsible for organic phosphate solubilization in soil. Phosphatase is one of the most widely secreted enzymes that hydrolyze phosphoric acid into phosphorus (P) ion and a molecule with a free OH group, later removing phosphorus from its substrate (Li et al., 2021). The hydrolysis of ester phosphate bonds results in the release of phosphate ions, converting high-molecular-weight organic phosphate into low-molecular-weight molecules. These enzymes are classified as acidic, neutral, or alkaline phosphatase based on their optimal pH (Zhou et al., 2012). The release of enzymes such as phytase is another way of phosphate solubilization by bacteria mentioned in the literature. This enzyme may liberate P associated with organic molecules in the soil in the form of phytate. The process includes the decomposition of phytate, and the phosphorus released as a result is in a form that plants may use. The plant cannot obtain phosphorus directly from phytate; instead a phosphate-solubilizing bacterium dissolves the phytate and makes phosphorus accessible to the plant for absorption (Ben Zineb et al., 2020). Phosphonatase and CeP lyase hydrolyze phosphonate ester linkages (e.g., phosphoenolpyruvate, phosphonoacetate) and convert them to hydrocarbons and phosphate ions. Phosphatase and cellulolytic enzymes are sometimes required to hydrolyze organic phosphorus or mineralize organic residues and organic materials, respectively. However, understanding enzymes, the soil microbial population, plant phosphorus absorption, exudation by roots, and other rhizospheric activities might be useful in investigating soil P transformations (Huang et al., 2017).

Mainly in the mineralization of phosphorus, phosphodiesterase and phosphor-mono-esterase may function consecutively. Phosphor-mono-esterase is capable of dissociating the phosphate group from phosphate monoester compounds, whereas phosphodiesterase is capable of hydrolyzing the phosphate di-ester link in nucleic acids (Kalsi et al., 2016). D-α-Glycerophosphatase is an uncommon phosphatase isolated and described from *B. licheniformis*. The products of the D-α-glycerophosphatase-catalyzed process were recognized as glycerol and inorganic phosphate. Furthermore, an inorganic phosphatase (pyrophosphate phosphohydrolase) that can hydrolyze pyrophosphate (a fertilizer) to Pi has been found. Phosphates catalyze the dephosphorylation of organic chemical phosphor-esters or phosphor-anhydride linkages (Zhou et al., 2012).

Compared to uninoculated plants in field trials, the fungal strain boosted plant height, leaf length, and fruit number per plant in *Lagenaria siceraria* and *Abelmoschus esculentus*. The phytase production of elite strains *Pseudomonas corrugata* SP77 and *Serratia liquefaciens* LR88 was investigated (Ben Zineb et al., 2020). Many phosphate-solubilizing bacteria, including *Bacillus*, *Pseudomonas*, *Enterobacter*, *Acinetobacter*, *Rhizobium*, *Burkholderia*, and endophytic fungus, including *Aspergillus*, *Penicillium*, *Piriformospora*, and *Curvularia*, have Carbon–phosphorus (C–P) lyase activity (Mehta et al., 2019).

20.3.5 Sulfur-oxidizing and reducing enzymes

Sulfur is a vital plant nutrient that adds to crop output and quality. Sulfur may be found in a wide range of organic and inorganic compounds. The activity of the soil biota, particularly the soil microbial biomass, which has the highest capacity for both mineralization and subsequent alteration of the oxidation state of sulfur, is completely responsible for the movement of sulfur between the inorganic and organic pools (Usharani et al., 2019). These microorganisms produce sulfate, which plants may utilize. At the same time, the acidity caused by oxidation helps to solubilize plant nutrients. It improves alkali soils by producing different enzymes such as sulfatase, sulfur dehydrogenase, sulfite reductase, thiol lyase, desulfurylases, sulfur oxidases, arylsulfatase, sulfohydrolase, rhodanese (thiosulfate reductase), myrosinase, alliinase, dimanganese, sulfate thiohydrolase, ATP sulfurylase, APS reductase sulfate thiol esterase oxidases, reductases, oxidoreductases, transferases, and laccase (Zenda et al., 2021).

The sulfur-oxidizing microorganisms are primarily involved in genera such as *Thiobacillus*, *Thiosphaera*, *Xanthobacter*, *Thiomicrospira*, *Alcaligens*, *Paracoccus*, and *Pseudomonas* (Rai et al., 2020). Certain obligate chemolithotrophs microbes such as *Thiobacillus thioparus*, *T. neapolitanus*, *T. denitrificans*, *T. thiooxidans*, *T. ferrooxidans*, *T. halophilus*, *T. novellus*, *T. acidophilus* (acidophile), *T. aquaesulis*, *T. intermedius*, *Paracoccus denitrificans*, *P. versutus*, *Xanthobacter tagetidis*, *Thiosphaera pantotroph*, and *Thiomicrospira thyasirae*, *Alternaria tenius*, *Aureobasidium pullulans*, *Epicoccum nigrum*, and a variety of *Penicillium* species are also capable of oxidizing elemental sulfur and thiosulfate (Lucheta. and Lambais; 2012, Goyal, 2019). Several obligate chemolithotrophic *beta*- and *gamma*-proteobacteria, such as *Acidithiobacillus*, produce tetrathionate as an intermediate (S4 I), whereas photo and chemolithotrophic Alphaproteobacteria, such as *Paracoccus*, use the "PSO pathway," also known as the Kelly–Friedrich pathway, which is controlled by the sox operon. The sox operon is extensively distributed across the domain bacteria; the appearance and development of monomeric dimanganese-containing protein as well as a sulfur dehydrogenase enzyme complex in the environment that aids in sulfur oxidation and reduction (Lucheta. and Lambais, 2012).

20.3.6 Oxidoreductases

Oxidoreductases can treat chemical wastes comprising phenols, pharmaceuticals, and hormones. This section discusses the properties of three oxidoreductase enzymes: laccases, tyrosinase, and peroxidases, which are the most explored enzymes for treating organic micropollutants (Naghdi et al., 2018). Laccases are the most stable and robust biocatalysts with many uses. It is a copper-containing, organic solvent-resistant oxidoreductase enzyme with the capability to oxidize phenolic and nonphenolic compounds to form dimers, oligomers, and polymers (Singh and Gupta, 2020). Laccase is discovered in a wide variety of organisms, including prokaryotes, fungi, lower eukaryotes, and plants. Laccases are responsible for the biological breakdown of lignin in their natural environment. They have gained a lot of interest from researchers in the last decade due to their widespread application and stability (Varga et al., 2019). Laccase's high versatility, owing to its broad range of substrate specificity, makes it ideal for a wide range of industrial applications, including the paper and pulp industry, bioremediation, food processing, and biodegradation of xenobiotic compounds (Catherine et al., 2016). Since the effectiveness of oxidation is determined by the redox potential difference between the enzyme and the substrate, some organic compounds with lower ionization potential cannot be oxidized directly by laccase. Laccases are classified into several categories based on the source of the enzyme. Laccases from *Pycnoporus sanguineus* CS43 were successful in lab-scale studies against the endocrine-disrupting chemicals, 4-NP and triclosan (TCS) (Varga et al., 2019).

Tyrosinase is another copper-containing enzyme that may be utilized to treat agricultural and medicinal products. Tyrosinases can also react with various phenolic chemicals due to their broad specificity. Polymers are formed throughout the process and can be removed from the effluent via precipitation (Fairhead and Thony-Meyer, 2012).

Peroxidases are oxidoreductases that catalyze oxidative processes using hydrogen peroxide as an electron acceptor. Different peroxidases were isolated from a variety of organisms such as horseradish peroxidase (HRP) from the plant horseradish (*Armoracia rusticana*), chloroperoxidase from *Caldariomyces fumago*, lignin peroxidases, manganese peroxidase (MnP) from *P. chrysosporium*, soybean peroxidase, black radish peroxidase, or turnip peroxide (Reina et al., 2018). HRP is a peroxidase enzyme that is frequently studied enzyme. Horseradish contains a high concentration of heme-containing oxidoreductase enzymes, which may catalyze the oxidation of a wide variety of organic and inorganic compounds utilizing H_2O_2 or other peroxides. The native enzyme (Fe^{3+}) is first oxidized by peroxide in the catalyzed process, producing water as a by-product (Varga et al., 2019).

Apart from that, enzymes such as lignin peroxidase, manganese peroxidase, laccase, versatile peroxidase, glyoxal oxidase, aryl-alcohol oxidase, vanillyl-alcohol oxidase, succinate dehydrogenase, vanillyl-alcohol oxidase, vanillyl-alcohol oxidase, vanillyl-alcohol oxidase, pyranose oxidase, and *p*-benzoquinone reductase are synthesized by a wide range of bacteria and fungi that play an important role in agro-waste management, bioenergy, and in the agro-processing industry (Cadoux and Milton, 2020).

20.3.7 Zinc-solubilizing enzymes

Zinc is a plant micronutrient involved in numerous physiological activities; its deficiency reduces agricultural production. Zinc insufficiency is the most common micronutrient deficiency concern, with virtually all crops and calcareous, sandy, peat, and soils heavy in phosphorus and silicon likely to be deficient. Zinc deficiency can harm plants by limiting development, reducing the number of tillers, causing chlorosis and smaller leaves, extending crop maturity, causing spikelet sterility, and lowering the quality of produced crops. Zn is engaged in the cellular activities of living beings and their involvement in crop production (Fasusi et al., 2021).

Many zinc-solubilizing microorganisms exist, including *Pseudomonas* sp., *Stenotrophomonas maltophilia*, *B. subtilis*, *Burkholderia* sp., *Mycobacterium brisbanense*, *Acinetobacter* sp., *Pseudomonas aeruginosa*, and *Enterobacter aerogen*. In place of more expensive zinc sulfate, such microbial inoculants might be utilized as biofertilizers to solubilize soil zinc accessible in different insoluble forms such as zinc oxide (ZnO), zinc carbonate ($ZnCO_3$), and zinc sulfide (ZnS) (Rehman et al., 2018).

Furthermore, fungi produce enzymes and organic acids that aid in the mobilization of zinc from its insoluble form to a readily available form in soil solution. The zinc-solubilizing ability has been observed in fungi, including *A. pullulans*, *Abisidia cylindrospora*, *Suillus bovines*, *Penicillium simplicissimum*, *Suillus luteus*, *A. niger*, *Barnettozyma californica*, *Penicillium* sp., *Abisidia glauca*, *Paxillus involutus*, *Torulaspora* sp., *Dothideomycetes* sp.,

Oidiodendron maius, *Trichoderma* sp., *Hymenoscyphus ericae*, *Abisidia spinosa*, and *Beauveria caledonica* (Rana et al., 2020). Inoculation with Zn-solubilizing fungus is a promising and environmentally acceptable approach for increasing the bioavailability of native and applied zinc to plants, as well as a viable alternative to chemical fertilizers (Nosheen et al., 2021).

20.4 Microbial enzymes for crop health, soil fertility, and allied agro-industries

Microorganisms play a crucial role in nitrogen, sulfur, phosphorus cycling, and the degradation of organic wastes. Moreover, their enzymes are vital promoters of biological activities and significantly impact soil fertility and crop health. The majority of enzyme activity in the soil comes from microbial sources, including intracellular, cell-associated, and free enzymes. Soil health is maintained by a unique balance of chemical, physical, and biological (including microbial, notably enzyme activity) components. As a result, indices of all of these components are needed to assess soil health. Drought, climate change, insect infestation, pollution, and human exploitation, including agriculture, all require healthy soil for terrestrial ecosystems to stay intact or recover from perturbations (Lee et al., 2020).

The ability of root-colonizing microbes, plant growth promotion, and management of crop health by these microbes is well documented. Even the use as biofertilizers, plant growth regulators, and biotic elicitors are widely known (Mukherjee et al., 2021). They promote plant growth through a variety of mechanisms, including phosphorus solubilization, volatile organic compound production, induction of systemic disease resistance, nitrogen fixation, soil fertility and nutrient uptake, and water stress resistance. On the other hand, recent technological developments have revealed fungal species capable of stimulating subsequent growth and improving soil fertility (Muller and Behrendt, 2021). Soil microorganisms and their enzymes are predominantly used in the agriculture industry to increase crop health, soil fertility, and many more, as described in Fig. 20.1. They act as a bioindicator and biocontrol agent, and their metabolic activity also increases soil fertility. However, their efficacy, substrate specificity, and other features have continuously evolved the applications and their utility.

20.4.1 Crop health (assessment via biocontrol agents)

Harmful bacteria are a significant threat to plant health because they negatively influence crop health, reduce agricultural output, and lower food quality. PSMs with biocontrol activity have emerged as a promising method for reducing pathogen infestation in crops environmentally benign. They are a viable alternative to synthetic phytopathogen control agents such as microbes themselves and their enzymes (Mukherjee et al., 2021). Ganeshan and Manoj

Kumar (2005) discovered that the phosphate-solubilizing *Pseudomonas fluorescens* acts as a biocontrol agent against *Ralstonia solanacearum*, which causes tomato bacterial wilt. These biocontrol agents release hydrolytic enzymes such as lipases, proteases, and α-amylase enzymes that attack pathogen biochemical machinery. Plants treated with these bacteria showed less wilt and enhanced plant growth and development (Ganeshan and Manoj Kumar, 2005).

Phosphorus-solubilizing strains, for example, *Pseudomonas mallei* and *Pseudomonas cepaceae*, improved *Phaseolus vulgaris* growth, yield, photosynthetic efficiency, chlorophyll content, and antioxidant enzyme activity due to enzymes such as catalase, glutathione reductase, proline dehydrogenase, glutathione-S-transferase, and superoxide dismutase. However, genetically engineered *Pseudomonas* biocontrol strains have been developed to promote plant growth and disease resistance in crops (Muller and Behrendt, 2021). Moreover, Rhizobacteria are also capable of controlling plant diseases caused by bacteria and fungus. Induced systemic resistance and the generation of antifungal metabolites help keep the condition at bay. In agriculture, inoculant bacteria are frequently treated to seed coats before they are sowed. Inoculated seeds are more likely to generate big enough rhizobacterial populations in the rhizosphere to benefit the crop significantly (Jiao et al., 2021).

20.4.2 Soil fertility (indicator enzymes)

The importance of agro-enzymes in agriculture may be seen on two levels. In the beginning, they are required for organic matter transformation (both production and decomposition) and nutrient cycling, which significantly impacts soil fertility, productivity, and crop health. Secondly, even though they are highly susceptible to specific agricultural practices and respond quickly to various environmental conditions, their activities are appropriate indicators of soil status. They are eagerly applied to determine the influence of management practices on general soil status, with particular attention to environmental and biological functioning. Enzymatic proteins respond to environmental management techniques by changing soil's physical and chemical attributes, determining microbial biomass, monitoring soil fertility, and determining the impact of numerous elements associated with diversified land use and agricultural management (Lee et al., 2020). Agricultural practices such as the application of inorganic fertilizers, organic amendments, and biofertilizers, tillage, cropping systems, and vegetation cover, irrigation, mulching, pesticide use, urease inhibitors, and environmental pollution caused by the aforementioned human impact (e.g., heavy metals, PAHs) influence soil fertility, soil structure and chemical composition, crop health, crop productivity, crop growth, soil microbial diversity, species structure, and metabolic activation (Cui et al., 2021). Similarly, microbial organic and mineral fertilizer may have a significant influence on soil fertility, microbial diversity, crop health, and their metabolic activity, which was indicated by altering microbial diversity and enzyme activity, such as urease and acid phosphatase, as a result of microbial proliferation or/and enzyme induction (Du et al., 2018).

In investigations devoted to the influence of inorganic nitrogen fertilizer on soil enzymatic activity, contradictory results have been observed. According to certain authors, increasing the quantity of NPK fertilizer can enhance the activity of enzymes such as cellulases, urease, and phosphatases or decrease the activity of catalase and invertase that indirectly increase soil nutritional quality, fertility, and plant growth (Anas et al., 2020). However, enzymatic activity is more commonly stimulated following the combined application of organic and inorganic N fertilization. Similarly, Kumar et al. (2019) discovered that soil enzymes such as dehydrogenase, xylanase, cellulase, phenoloxidase, β-glucosidase, and peroxidase were considerably greater in zero-tilled soils than in tilled soils. Meanwhile, dehydrogenase, alkaline phosphatase, and protease activities were more active in zero-till systems than in conventional tillage systems (Kumar et al., 2019).

20.4.3 Allied agro-industrial applications

Another application of enzymes in agriculture is in-farm animal nutrition, particularly for pigs and poultry. To digest food, all organisms require enzymes. Pigs and poultry, while producing certain enzymes, are unable to digest even 25% of the grain they get as fodder. The lack of certain enzymes, either because the feed constituents contain indigestible antinutritional compounds that inhibit the digestive process or because they lack some unusual enzymes that digest a few compounds present in farm animals, has resulted in the loss of just a few essential nutrients; the attempt to compensate for the appropriate nutritional elements has increased feed costs. Feed enzymes, on the other hand, are an efficient replacement for achieving maximum feed efficiency (Litonina, et al., 2021). In today's animal feed, enzymes that break down fiber, starch, proteins, and phytate are most commonly utilized. Carbohydrases are the most extensively utilized enzymes, accounting for approximately 41.67% of the worldwide feed enzyme market in terms of volume (Mekuriaw and Asmare, 2018). Furthermore, the two essential fiber-decomposing enzymes supplied to animal feed are β-glucanase and xylanase phytase. Xylanases degrade arabinoxylans, protein-degrading enzymes abundant in cereals and their derivatives, whereas β-glucanases deconstruct β-glucans, which are also plentiful in cereals and their products (Vigors et al., 2016).

Feed enzyme technology has been a rapidly expanding field of study. Exogenous enzymes are increasingly added to animal feed to aid digestive enzymes and degrade antinutritive fractions. Exogenous enzyme supplementation of feed provides for better feed utilization and cost savings. Future research should concentrate on developing and testing enzymes that are more suited to the environment found in an animal's digestive system. Following the administration of feed enzymes, these methods will boost the efficiency of animal production even more. Furthermore, the use of feed enzymes can help to prevent pollution in the environment, to improve feed digestion and absorption, as well as to reduce dung production and N and P secretion (Bedford, 2018).

Furthermore, rampant population growth has put enormous strain on energy supplies, and the globe is on the verge of an energy catastrophe. Biomass, particularly agricultural biomass, is a viable solution to the issue and is considered solid waste. Microbial bioconversion of agricultural waste materials is an effective apparatus for utilizing and valorizing agro-industrial wastes. It is not only abundant around the planet, but it also has the potential to become the future generation of fuel. Alternatives to petroleum, diesel, and natural gas include bioethanol, biodiesel, and biogas (Anwar et al., 2014). Agriculture is the primary source of income in most countries throughout the world; in other words, agriculture plays a unique role in the global economy. Crop residues, which is high in cellulose fibers, must be processed by microbial activities (enzymatic action), and each type will have its own set of advantages and disadvantages. Still, in the end, the procedure is not only environmentally benign but also cost-effective. Biomass for energy generation may also be utilized to generate electricity from trash dumps worldwide. Not only would the utilization of agricultural waste address energy shortages, but it will also use up garbage that has been discarded up until now, adding to the annoyance. It will also help manage pollution and provide a cleaner environment because it is environmentally benign (Kaur and Sarao, 2021).

Lignocellulosic agricultural by-products are also a plentiful and inexpensive source of cellulose fibers. Composite, textile, and pulp and paper manufacturers can employ agro-based fibers because of their composition, characteristics, and structure. Furthermore, biofibers may be utilized to manufacture fuel, chemicals, enzymes, and food. Corn, wheat, rice, sorghum, barley, sugarcane, pineapple, bananas, and coconut by-products are the most common agro-based biofibers (Reddy and Yang, 2005). Innovations in biotechnological processes tend to contribute agro-industrial leftovers more economically viable. Due to its widespread availability, it can serve as a good substrate for microbial processes for generating value-added products. Protein-enriched animal feed, enzymes, amino acids, organic acids, and pharmaceutically important compounds have all been attempted to be produced from agricultural residues. A pretreatment method has frequently resulted in better microbial substrate usage. For such bioconversions, solid-state fermentation technology might be a viable option (Pandey et al., 2000).

20.5 Agricultural enzyme market

Agricultural enzymes work as a catalyst, speeding up the chemical reaction that releases nutrients from the soil and makes them available to plant roots. If an agricultural enzyme had not been present, these nutrients would have stayed attached to the soil, inaccessible to the plants. Increased plant output and quality are made possible by adding agricultural enzymes to the feed. Enzymes are commonly employed in agriculture because their use increases crop yields. According to Woese's research, utilizing agricultural enzymes improves photosynthesis by around 20%. Despite rising demand for agricultural enzymes, government regulations may operate as a major stumbling block to industry expansion (marketsandmarkets.com/Market-Reports/agricultural-enzymes-market-180483493.html).

This study segmented the agricultural enzymes market by enzyme type, crop type, application, and geography. Lipases, proteases, carbohydrases, polymerases, and nucleases are the different types of enzymes on the market. The market is divided into three categories based on the application: growth-promoting products, control products, and fertility products. For North America, Europe, Asia-Pacific, and LAMEA, there is a geographic breakdown and in-depth examination of each preceding region. Substantial R&D expenditure by major market participants will stimulate market innovation, resulting in a fierce rivalry. Several new product releases have emerged in the industry due to major R&D spending. Novozymes, for example, released Avantec Amp, an enzyme that boosts agricultural productivity. All of the leading competitors in the agricultural enzymes market are concentrating on increasing crop output, assisting the market's growth. Novozymes, Deepak Fertilizers, Petrochemicals Corporation Ltd., Greenmax Agro Tech, and Agri Life are among the prominent industry competitors profiled in this report (marketsandmarkets.com/Market-Reports/agricultural-enzymes-market-180483493.html).

20.6 Concluding remarks

According to the cited literature, it can be stated that there is still a lot of interest in utilizing enzymes in agriculture. Agro essential enzymes have received particular attention because they are thought to play a key role in soil organic matter transformation and nitrogen element cycling. Understanding the existence and activity of enzymes in the agriculture sector can assist in understanding the transformation of organic matter and nutrients in sustainable soil and agro product management and sustaining agricultural output. Such microbes and their enzymes are commonly utilized in agronomy as accurate soil health, fertility, and production markers and as a biocontrol agent, bioindicator, and agro-processing. Future studies should look at the global, widespread usage of enzymes in the agro-based business and how they might be employed at different stages. This strategy will result in a more accurate portrayal of the microbial enzyme in the agricultural industry, not just for the environment in which it was synthesized. Nowadays, one of the most significant study topics in agro-enzymology is valorization and long-term preservation of agriculture zone enzyme activity.

Abbreviations

ACCD	1-Aminocyclopropane-1-carboxylic acid deaminase
APS reductase	adenylyl-sulfate reductase
BNF	biological nitrogen fixing
CBH	1,4-D-Glucan-cello-bio-hydrolase
Fd-Nir	ferredoxin-nitrite reductase
GDH	glutamate dehydrogenase
GS-GOGAT	glutamate synthase glutamine oxoglutarate aminotransferase
HNO	nitrosyl hydride
HRP	horseradish peroxidase
MnP	manganese peroxidase
NADH	nicotinamide adenine dinucleotide (NAD) + hydrogen (H)
Nar or Nap	nitrate reductases
Nir or Nas	nitrite reductases
NO	nitric oxide
Nor or Nos	nitric oxide reductase
NP	4-Nonylphenol
NPK	nitrogen, phosphorus, potassium
PSMS	phosphate-solubilizing microorganisms
PSO	*Paracoccus* sulfur oxidation
sp	species
TCS	triclosan
$ZnCO_3$	zinc carbonate
ZnO	zinc oxide
ZnS	zinc sulfide

References

Anas, M., Liao, F., Verma, K.K., Sarwar, M.A., Mahmood, A., Chen, Z.L., et al., 2020. The fate of nitrogen in agriculture and environment: agronomic, eco-physiological and molecular approaches to improve nitrogen use efficiency. Biol. Res. 53 (1), 1–20.

Anwar, Z., Gulfraz, M., Irshad, M., 2014. Agro-industrial lignocellulosic biomass a key to unlock the future bio-energy: a brief review. J. Radiat. Res. Appl. Sci. 7 (2), 163–173.

Arya, P.S., Yagnik, S.M., Panchal, R.R., Rajput, K.N., Raval, V.H., 2022. Industrial applications of enzymes from extremophiles. Physiology, Genomics, and Biotechnological Applications of Extremophiles. IGI Global, pp. 207–232.

Arya, P.S., Yagnik, S.M., Rajput, K.N., Panchal, R.R., Raval, V.H., 2021. Understanding the basis of occurrence, biosynthesis, and implications of thermostable alkaline proteases. Appl. Biochem. Biotechnol. 193 (12), 4113–4150.

Bakshi, M., Varma, A., 2010. Soil enzyme: the state-of-art. Soil Enzymology. Springer, Berlin, Heidelberg, pp. 1–23.

Bedford, M.R., 2018. The evolution and application of enzymes in the animal feed industry: the role of data interpretation. Br. Poult. Sci. 59 (5), 486–493.

Ben Zineb, A., Trabelsi, D., Ayachi, I., Barhoumi, F., Aroca, R., Mhamdi, R., 2020. Inoculation with elite strains of phosphate-solubilizing bacteria enhances the effectiveness of fertilization with rock phosphates. Geomicrobiology J. 37 (1), 22–30.

Bhattacharyya, P.N., Jha, D.K., 2012. Plant growth-promoting rhizobacteria (PGPR): emergence in agriculture. World J. Microbiology Biotechnol. 28 (4), 1327–1350.

Bunemann, E.K., Bongiorno, G., Bai, Z., Creamer, R.E., De Deyn, G., de Goede, R., et al., 2018. Soil quality–a critical review. Soil. Biol. Biochem. 120, 105–125.

Cadoux, C., Milton, R.D., 2020. Recent enzymatic electrochemistry for reductive reactions. Chem. Electro. Chem 7 (9), 1974–1986.

Catherine, H., Penninckx, M., Frederic, D., 2016. Product formation from phenolic compounds removal by laccases: a review. Environ. Technol. & Innov. 5, 250–266.

Chandwani, S., Amaresan, N., 2022. Role of ACC deaminase-producing bacteria for abiotic stress management and sustainable agriculture production. Environ. Sci. Pollut. Res. 1–17.

Chukwuma, O.B., Rafatullah, M., Tajarudin, H.A., Ismail, N., 2020. Lignocellulolytic enzymes in biotechnological and industrial processes: a review. Sustainability 12 (18), 7282.

Cui, Y., Wang, X., Wang, X., Zhang, X., Fang, L., 2021. Evaluation methods of heavy metal pollution in soils based on enzyme activities: a review. Soil. Ecol. Lett. 3 (3), 169–177.

Du, C., Abdullah, J.J., Greetham, D., Fu, D., Yu, M., Ren, L., et al., 2018. Valorization of food waste into fertilizer and its field application. J. Clean. Prod. 187, 273–284.

Fairhead, M., Thony-Meyer, L., 2012. Bacterial tyrosinases: old enzymes with new relevance to biotechnology. N. Biotechnol. 29 (2), 183–191.

Fariduddin, Q., Varshney, P., Ali, A., 2018. The perspective of nitrate assimilation and bioremediation in (a non-nitrogen fixing cyanobacterium): an overview. J. Environ. Biol. 39 (5), 547–557.

Fasusi, O.A., Cruz, C., Babalola, O.O., 2021. Agricultural sustainability: microbial biofertilizers in rhizosphere management. Agriculture 11 (2), 163.

Ganeshan, G., Manoj Kumar, A., 2005. *Pseudomonas fluorescens* is a potential bacterial antagonist to control plant diseases. J. Plant. Interact. 1 (3), 123–134.

Goyal, S.C.S., 2019. Sulfur oxidizing fungus: a review. J. Pharmacognosy Phytochemistry 8 (6), 40–43.

Huang, L.M., Jia, X.X., Zhang, G.L., Shao, M.A., 2017. Soil organic phosphorus transformation during ecosystem development: a review. Plant. Soil. 417 (1), 17–42.

Husain, Q., Ullah, M.F., 2019. Biocatalysis. Springer, Cham, https://doi.org/10.1007, pp. 978-3.

Jiao, X., Takishita, Y., Zhou, G., Smith, D.L., 2021. Plant-associated rhizobacteria for biocontrol and plant growth enhancement. Front. Plant. Sci. 12, 420.

Kalsi, H.K., Singh, R., Dhaliwal, H.S., Kumar, V., 2016. Phytases from *Enterobacter* and *Serratia* species with desirable characteristics for food and feed applications. 3 Biotech. 6 (1), 1–13.

Kaur, S., Sarao, L., 2021. Bioenergy from agricultural wastes. Bioenergy Research: Biomass Waste to Energy. Springer, Singapore, pp. 127–147.

Kumar, A., Naresh, R.K., Singh, S., Mahajan, N.C., Singh, O., 2019. Soil aggregation and organic carbon fractions and indices in conventional and conservation agriculture under vertisol soils of sub-tropical ecosystems: a review. Int. J. Curr. Microbiol. App. Sci. 8 (10), 2236–2253.

Lee, S.H., Kim, M.S., Kim, J.G., Kim, S.O., 2020. Use of soil enzymes as indicators for contaminated soil monitoring and sustainable management. Sustainability 12 (19), 8209.

Li, J.T., Lu, J.L., Wang, H.Y., Fang, Z., Wang, X.J., Feng, S.W., et al., 2021. A comprehensive synthesis unveils the mysteries of phosphate-solubilizing microbes. Biol. Rev. 96 (6), 2771–2793.

Litonina, AS., Smirnova, Y.M., Platonov, A.V., Laptev, G.Y., Dunyashev, T.P., Butakova, M.V., 2021. Application of enzyme probiotic drug developed based on microorganisms of the rumen of reindeer (*Rangifer tarandus*) in feeding cows. Regulatory Mechanisms Biosyst. 12 (1), 109–115.

Lucheta, A.R., Lambais, M.R., 2012. Sulfur in agriculture. Rev. Brasileira de. Cienc. do Solo 36 (5), 1369–1379.

Martínez-Espinosa, R.M., 2020. Microorganisms and their metabolic capabilities in the context of the biogeochemical nitrogen cycle in extreme environments. Int. J. Mol. Sci. 21 (12), 4228.

Mehta, P., Sharma, R., Putatunda, C., Walia, A., 2019. Endophytic fungi: role in phosphate solubilization. Advances in Endophytic Fungal Research. Springer, Cham, pp. 183–209.

Mekuriaw, Y., Asmare, B., 2018. Nutrient intake, digestibility and growth performance of Washera lambs fed natural pasture hay supplemented with graded levels of *Ficus thonningii* (Chibha) leaves as a replacement for concentrate mixture. Agriculture & Food Security 7 (1), 1–8.

Mohanram, S., Amat, D., Choudhary, J., Arora, A., Nain, L., 2013. Novel perspectives for evolving enzyme cocktails for lignocellulose hydrolysis in biorefineries. Sustain. Chem. Process. 1 (1), 1–12.

Mukherjee, A., Bhowmick, S., Yadav, S., Rashid, M.M., Chouhan, G.K., Vaishya, J.K., et al., 2021. Re-vitalizing of endophytic microbes for soil health management and plant protection. 3 Biotech. 11 (9), 1–17.

Muller, T., Behrendt, U., 2021. Exploiting the biocontrol potential of plant-associated *Pseudomonads*–a step towards pesticide-free agriculture. Biol. Control. 155, 104538.

Murali, M., Gowtham, H.G., Singh, S.B., Shilpa, N., Aiyaz, M., Niranjana, S.R., et al., 2021. Bio-prospecting of ACC deaminase producing Rhizobacteria towards sustainable agriculture: a special emphasis on abiotic stress in plants. Appl. Soil. Ecol. 168, 104142.

Naghdi, M., Teheran, M., Brar, S.K., Kermanshahi-Pour, A., Verma, M., Surampalli, R.Y., 2018. Removal of pharmaceutical compounds in water and wastewater using fungal oxidoreductase enzymes. Environ. Pollut. 234, 190–213.

Nayak, A., Bhushan, B., 2019. An overview of the recent trends on waste valorization techniques for food wastes. J. Environ. Manag. 233, 352–370.

Nosheen, S., Iqra, A., Yuanda, S., 2021. Microbes as biofertilizers, a potential approach for sustainable crop production. Sustainability 13 (4), 1868.

Pajares, S., Ramos, R., 2019. Processes and microorganisms involved in the marine nitrogen cycle: knowledge and gaps. Front. Mar. Sci. 6, 739.

Pandey, A., Soccol, C.R., Nigam, P., Soccol, V.T., 2000. Biotechnological potential of agro-industrial residues. I: sugarcane bagasse. Bioresour. Technol. 74 (1), 69–80.

Piotrowska-Długosz, A., 2019. Significance of enzymes and their application in agriculture. Biocatalysis. Springer, Cham, pp. 277–308.

Prabhu, N., Borkar, S., Garg, S., 2019. Phosphate solubilization by microorganisms: overview, mechanisms, applications, and advances. Adv. Biol. Sci. Res. 161–176.

Raghuwanshi, R., Prasad, J.K., 2018. Perspectives of rhizobacteria with ACC deaminase activity in plant growth under abiotic stress. Root Biol. 303–321.

Rai, A., Singh, A.K., Mishra, R., Shahi, B., Rai, V.K., Kumari, N., et al., 2020. Sulphur in soils and plants: an overview. Int. Res. J. Pure Appl. Chem. 66–70.

Rajta, A., Bhatia, R., Setia, H., Pathania, P., 2020. Role of heterotrophic aerobic denitrifying bacteria in nitrate removal from wastewater. J. Appl. microbiology 128 (5), 1261–1278.

Rana, K.L., Kour, D., Kaur, T., Devi, R., Yadav, A.N., Yadav, N., et al., 2020. Endophytic microbes: biodiversity, plant growth-promoting mechanisms and potential applications for agricultural sustainability. Antonie Van. Leeuwenhoek 113 (8), 1075–1107.

Raval, V.H., Pillai, S., Rawal, C.M., Singh, S.P., 2014. Biochemical and structural characterization of a detergent-stable serine alkaline protease from seawater haloalkaliphilic bacteria. Process. Biochem. 49 (6), 955–962.

Raval, V.H., Purohit, M.K., Singh, S.P., 2013. Diversity, population dynamics and biocatalytic potential of cultivable and non-cultivable bacterial communities of the saline ecosystems. Marine Enzymes for Biocatalysis. Woodhead Publishing, pp. 165–189.

Raval, V.H., Purohit, M.K., Singh, S.P., 2015a. Extracellular proteases from halophilic and haloalkaliphilic bacteria: occurrence and biochemical properties. Halophiles. Springer, Cham, pp. 421–449.

Raval, V.H., Rawal, C.M., Pandey, S., Bhatt, H.B., Dahima, B.R., Singh, S.P., 2015b. Cloning, heterologous expression and structural characterization of an alkaline serine protease from seawater halo-alkaliphilic bacterium. Ann. Microbiology 65 (1), 371–381.

Reddy, N., Yang, Y., 2005. Biofibers from agricultural byproducts for industrial applications. Trends Biotechnol. 23 (1), 22–27.

Rehman, A., Farooq, M., Naveed, M., Nawaz, A., Shahzad, B., 2018. Seed priming of Zn with endophytic bacteria improves the productivity and grain biofortification of bread wheat. Eur. J. Agron. 94, 98–107.

Reina, R., Garcia-Sanchez, M., Liers, C., García-Romera, I., Aranda, E., 2018. An overview of fungal applications in the valorization of lignocellulosic agricultural by-products: the case of two-phase olive mill wastes. Mycoremediation Environ. Sustainability 213–238.

Saikia, J., Sarma, R.K., Dhandia, R., Yadav, A., Bharali, R., Gupta, V.K., et al., 2018. Alleviation of drought stress in pulse crops with ACC deaminase-producing rhizobacteria isolated from acidic soil of Northeast India. Sci. Rep. 8 (1), 1–16.

Singh, D., Gupta, N., 2020. Microbial Laccase: a robust enzyme and its industrial applications. Biologia 75 (8), 1183–1193.

Singh, S.B., Gowtham, H.G., Murali, M., Hariprasad, P., Lakshmeesha, T.R., Murthy, K.N., et al., 2019. Plant growth-promoting ability of ACC deaminase producing rhizobacteria native to sunflower (*Helianthus annuus* L.). Biocatalysis Agric. Biotechnol. 18, 101089.

Singh, R., Kumar, M., Mittal, A., Mehta, P.K., 2016. Microbial enzymes: industrial progress in 21st century. 3 Biotech. 6 (2), 1–15.

Sun, W., Shahrajabian, M.H., Cheng, Q., 2021. Nitrogen fixation and diazotrophs–a review. Rom. Biotechnol. Lett. 26, 2834–2845.

Thapa, S., Li, H., OHair, J., Bhatti, S., Chen, F.C., Nasr, K.A., et al., 2019. Biochemical characteristics of microbial enzymes and their significance from industrial perspectives. Mol. Biotechnol. 61 (8), 579–601.

Thapa, S., Mishra, J., Arora, N., Mishra, P., Li, H., Bhatti, S., et al., 2020. Microbial cellulolytic enzymes: diversity and biotechnology with reference to lignocellulosic biomass degradation. Rev. Environ. Sci. Bio/Techno. 19 (3), 621–648.

Usharani, K., Naik, D., Manjunatha, R., 2019. Sulphur oxidizing bacteria: oxidation, mechanism and uses-a review.

Varga, B., Somogyi, V., Meiczinger, M., Kovats, N., Domokos, E., 2019. Enzymatic treatment and subsequent toxicity of organic micropollutants using oxidoreductases-a review. J. Clean. Prod. 221, 306–322.

Verma, D.K., Kaur, B., Pandey, A.K., Asthir, B., 2019. Nitrogenase: a key enzyme in microbial nitrogen fixation for soil health. Microbiology for Sustainable Agriculture, Soil Health, and Environmental Protection. CRC Press, pp. 261–294.

Vigors, S., Sweeney, T., O'Shea, C.J., Kelly, A.K., O'Doherty, J.V., 2016. Pigs that are divergent in feed efficiency, differ in intestinal enzyme and nutrient transporter gene expression, nutrient digestibility and microbial activity. Animal 10 (11), 1848–1855.

Yarullina, L.G., Akhatova, A.R., Kasimova, R.I., 2016. Hydrolytic enzymes and their proteinaceous inhibitors in regulation of plant-pathogen interactions. Russian J. Plant. Physiol. 63 (2), 193–203.

Younus, H., 2019. Oxidoreductases: overview and practical applications. Biocatalysis 39–55.

Zenda, T., Liu, S., Dong, A., Duan, H., 2021. Revisiting sulfur—the once-neglected nutrient: its roles in plant growth, metabolism, stress tolerance, and crop production. Agriculture 11 (7), 626.

Zhou, X., Cui, Y., Zhou, X., Han, J., 2012. Phosphate/pyrophosphate and MV-related proteins in mineralization: discoveries from mouse models. Int. J. Biol. Sci. 8 (6), 778.

Zimmerman, A.R., Ahn, M.Y., 2010. Organo-mineral–enzyme interaction and soil enzyme activity. Soil enzymology. Springer, Berlin, Heidelberg, pp. 271–292.

Chapter 21

Opportunities and challenges for the production of fuels and chemicals: materials and processes for biorefineries

Carolina Reis Guimarães[1], Ayla Sant'Ana da Silva[1,2], Daniel Oluwagbotemi Fasheun[1,2], Denise M.G. Freire[2], Elba P.S. Bon[2], Erika Cristina G. Aguieiras[2,3], Jaqueline Greco Duarte[2,4], Marcella Fernandes de Souza[5], Mariana de Oliveira Faber[1,2], Marina Cristina Tomasini[1,2], Roberta Pereira Espinheira[1,2], Ronaldo Rodrigues de Sousa[1,2], Ricardo Sposina Sobral Teixeira[2] and Viridiana S. Ferreira-Leitão[1,2]

[1]*National Institute of Technology, Ministry of Science, Technology and Innovation, Avenida Venezuela, Rio de Janeiro, Brazil,* [2]*Federal University of Rio de Janeiro, Department of Biochemistry, Rio de Janeiro, Brazil,* [3]*Federal University of Rio de Janeiro, Campus UFRJ - Duque de Caxias Prof. Geraldo Cidade, Duque de Caxias, Rio de Janeiro, Brazil,* [4]*SENAI Innovation Institute for Biosynthetics and Fibers, SENAI CETIQT, Rio de Janeiro, Brazil,* [5]*Ghent University, Faculty of Bioscience Engineering, Gent, Belgium*

21.1 Introduction

Global primary energy consumption that increases yearly has been predominantly sourced from fossil fuels. According to the Statistical Review of World Energy (2021), 84% of total consumed energy in 2019 was of fossil origin. Consequently, so far continuous use of fossil resources and carbon dioxide emissions keep building up in the atmosphere, increasing global climate changes. The severity of this scenario has been pushing nations to seek solutions through environmental treaties. One of them is the Paris Agreement, a treaty signed by the United Nations in 2015 aiming to reduce greenhouse gas (GHG) emissions and, by extension, control the increase in global temperature from 2020 onward (Christensen and Olhoff, 2019; Höhne et al., 2019). However, to achieve this goal, it is necessary to change the energy source from fossil to renewables, whereby fuels and chemicals would be produced from renewable materials. This approach has been considered a

Biotechnology of Microbial Enzymes. DOI: https://doi.org/10.1016/B978-0-443-19059-9.00004-9
© 2023 Elsevier Inc. All rights reserved.

viable way to achieve the decarbonization goals established in this agreement and has become a priority in the energy policy strategies of several countries (Straathof, 2014; Straathof and Bampouli, 2017; Saravanan et al., 2020; Lin and Lu, 2021).

Within this scenario, Brazil stands out for its biofuel production and replacement of fossil fuels. Biofuel production in Brazil was started over 45 years ago, in the 1970s, with the National Fuel Alcohol Program (*Pro-álcool*). The program was focused on ethanol production from the sucrose-rich sugarcane juice, known as first-generation (1G) ethanol (Saravanan et al., 2020). Initially, the program predicted the mandatory blend of 20% ethanol (E20) to gasoline or the use of 100% ethanol (E100) in cars with customized engines. Currently, the Brazilian legislation establishes that the government can fix the blend of ethanol to gasoline between 18% and 27%. The percentage is determined by the legislation of each state and the fuel market.

The increased demand for renewable liquid fuels and the production of cars with flex-fuel engines compatible with any ethanol/gasoline blend furthered the ethanol production technology from both the sugarcane juice and the sugarcane biomass, the second-generation (2G) ethanol. In 2019 over 94% of new cars licensed in Brazil were flex-fuel (ANFAVEA – Associação Nacional dos Fabricantes de Veículos Automotoree, 2020), and 69.5% of the entire light vehicle fleet was made up of vehicles with this type of engine (SINDPEÇAS, 2020).

Likewise, despite the well-established role of 1G ethanol in replacing fossil fuels, the 2G ethanol also has a great potential to replace fossil fuels besides aggregating value to an agricultural waste produced in enormous amounts. The 2G technology, however, is rather complex in comparison to the 1G technology as the material to be processed, the lignocellulosic biomass, is a complex, heterogeneous and recalcitrant material composed of cellulose, hemicellulose, and lignin. This complexity imposes a challenging preliminary step for biomass deconstruction and a subsequent costly measure for the enzymatic hydrolysis of the cellulose part of the pretreated biomass. Regardless of its cost, enzymatic hydrolysis is presently the predominant technological choice due to its high specificity, low energy consumption and chemical products, and the generation of less toxic waste (Binod et al., 2011).

Besides bioethanol, biodiesel, another renewable fuel, was implemented in the Brazilian energy matrix by creating the National Program of Production and Use of Biodiesel (PNPB). The program initially provided a mandatory blend of 2% (B2) of biodiesel to diesel until 2008 (Ribeiro and Silva, 2020). Currently, the compulsory combination is 13% (B13), with an expectation of an increase to 15% in the coming years (MME, 2020).

Biodiesel, composed of monoalkyl esters produced from vegetable oils or animal fats, has similar properties to petro-diesel. Its burning lowers emissions of particulates, CO, SOx, and aromatic hydrocarbons (Robles-Medina et al., 2009). Currently, the predominant technology for biodiesel production is alkaline transesterification with methanol. However, this technological option presents several drawbacks, such as generating large amounts of alkaline effluents and their consequent environmental impact, glycerol formation contaminated with the alkaline catalyst, and the challenges regarding catalyst recovery. In addition, this route requires high-quality raw materials, being responsible for some 80% of the final costs of biodiesel production (Demirbas, 2007; Pourzolfaghar et al., 2016; Robles-Medina et al., 2009; Sharma et al., 2008).

Therefore innovative technology for biodiesel production has been focusing on the transition toward biocatalysis using lipases and, more recently, microbial processes. The use of lipases or whole cells in biodiesel production has the advantage of relatively simple downstream processing steps to purify biodiesel and glycerol. In addition, enzymatic processes allow the use of cheaper raw materials, which contain high levels of FFA (free fatty acids) and water, as lipases can catalyze esterification and transesterification reactions (Aguieiras et al., 2013, 2017; Souza et al., 2009; Zeng et al., 2017; Zheng et al., 2020).

Ethanol and biodiesel can also be produced using microalgae, a viable biomass source studied for the production of the third-generation (3G) fuels (Lin and Lu, 2021). The production of ethanol or biodiesel depends on the type of microalgae. Some species of microalgae, when grown under stress, accumulate lipid. These can later be used for the production of biodiesel. Other microalgae species are rich in carbohydrates, which can be used to produce ethanol and other chemical compounds (Souza et al., 2020).

Beyond the liquid fuels from biomass, studies of biogas production, such as hydrogen and methane, should not be neglected. The production of these gases occurs in anaerobic digestion (AD). This technology is considered a viable technology for treating organic waste through the action of microorganisms with subsequent generations of bioenergy and fertilizers (Holm-Nielsen et al., 2009; Mao et al., 2015). Several materials rich in organic matter can be used as raw materials for AD, such as wastes and effluents generated in ethanol and biodiesel production chains (Dawood et al., 2020). In addition, AD reduces the emission of GHGs and pathogens through sanitization, and the final residue could be used as fertilizer (Holm-Nielsen et al., 2009; Mao et al., 2015).

Last but not least, the search for more sustainable and ecological processes based on the circular economy is needed for the most different production chains, not only in the biofuel sector. In the biomass-based chemical industry, biomass can be exploited directly or as a source of platform molecules, such as those obtained by the petrochemical industry (Soccol et al., 2011; Straathof, 2014; Straathof and Bampouli, 2017). According to Straathof (2014), in 2014, there were already at least 22 types of industrial or pilot-scale chemical production processes from biomass, including the synthesis of different classes of compounds: hydrocarbons, alcohols, carbohydrates, carboxylic acids, esters, amines, and amino acids.

In summary, the production of biofuels, bioenergy, and biochemicals from biomass is necessary to achieve the goals of environmental treaties and further the bioeconomy. Considering the Brazilian scenario, the annual generation of agricultural residues, such as lignocellulosic and oleaginous residues, is strategically important to develop sustainable technologies aiming for their processing. Fig. 21.1 summarizes some opportunities for integrating the sugar and oil biorefinery, mainly fostered by biofuel production chains. Nevertheless, with a massive amount of fuels required currently and the need to switch from fossil to renewables, the full implementation and exploitation of biorefineries is still a potential. This chapter will explore the production of bioethanol, biodiesel, biogas, and chemicals, via microbial and enzymatic approaches, from plant and microalgae biomass. Biorefineries in Brazil have been focused on the demand for fuel. This scenario is expected to last until an effective transition toward electric and hydrogen cars is envisaged, which is still being defined in different world markets and Brazil (Noussan et al., 2021). In any scenario, however, Brazil has a favorable and unique condition to give a sizable contribution to the decarbonization of the local and world economy and an enormous potential for establishing a bioeconomy. As an example, but without exhausting the theme, Fig. 21.1 contemplates an integrated sugar and oil biorefinery, envisioning two major processes that use

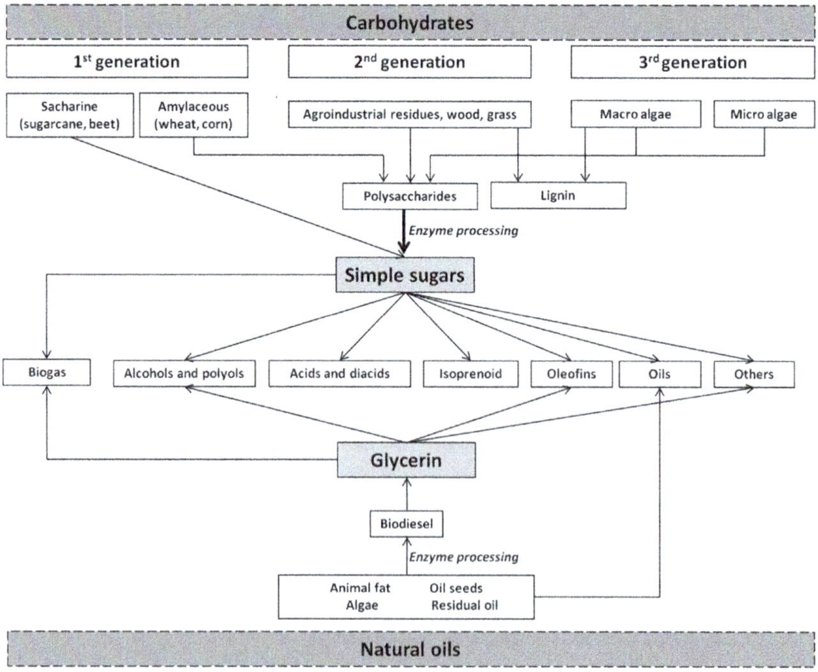

FIGURE 21.1 Sugar and oil biorefinery integration.

renewable raw materials in Brazil, the production of ethanol and biodiesel, generating large volumes of products. It is expected that a natural transition toward the diversification of products in these two main production chains will take place as, while the fuel sector is evolving, the chemical industry has been slower to move. It is still in deficit (ABIQUIM – Química, 2021).

21.2 Brazilian current production and processing of lignocellulosic sugarcane biomass

Around 589 million tons of agricultural residues are produced annually in Brazil, for example, sugarcane straw and bagasse, soybean straw, maize stover and cob, rice straw, and several residues from fruit production (Araújo et al., 2019). Among those, the residues from sugarcane cultivation are still the most represented, resulting from 1G ethanol and sugar production. Indeed, Brazil is the world's largest producer of sugarcane, reaching a harvest of 654.8 million tons in 2020/21, which was used to produce approximately 32.8 billion liters of ethanol and 41.25 million tons of sugar (CONAB, 2020). Most commercial sugarcane varieties have an average bagasse and straw yield of 140 kg each per ton of milled sugarcane (dry basis) (Pippo et al., 2011). Thus, considering these numbers, it is estimated that approximately 183 million tons of bagasse and straw (dry weight) were produced in Brazil's 2020/21 harvest.

A priori, the bagasse was used for energy self-sufficiency of the plants, producing steam in boilers to feed the industry equipment. However, with the development of high-pressure boilers (Dias et al., 2012), electricity cogeneration from sugarcane biomass became a profitable strategic option for bioenergy sources in Brazil (Chandel et al., 2019; Watanabe et al., 2020). The bioelectricity from sugarcane biomass increased from 4% to 7% of Brazil's total electricity between 2010 and 2020 (Bujan, 2020; Watanabe et al., 2020; Brasil, 2020a,b). Over the recent years, the sugarcane straw has become available in large quantities in the field due to the transition from manual to mechanical harvesting, pushed by the elimination of straw burning before sugarcane harvesting (Franco et al., 2018; Gonzaga et al., 2019; Watanabe et al., 2020). Such transition resulted in increased sugarcane straw availability, which represents an excellent opportunity for 2G ethanol production and bioelectricity cogeneration and a new challenge to map the best destination and processing. The agronomic and environmental implications of straw sugarcane removal from the field, the appropriate amount of removal, and their potential for electricity production have recently been studied by the Sugarcane Renewable Electricity Project (SUCRE) (Carvalho et al., 2017; Gonzaga et al., 2019; Sampaio et al., 2019; and Watanabe et al., 2020). Interestingly, considering the use of 50% of the straw, an additional 35 Terawatt-hour (TWh) could be exported to the Brazilian grid yearly (SUCRE, 2017). Besides, as the bioelectricity from sugarcane is generated during sugarcane harvesting in the dry season—when the hydroelectric reservoirs are at low levels– it complements the hydroelectric generation (Carpio and de Souza, 2017).

Although sugarcane biomass bioelectricity generation is a profitable residue management process, it does not hamper the cellulosic ethanol technology implementation. The coproduction (bioelectricity and cellulosic ethanol) following the level of aversion to risk of the sugarcane industry would represent, according to Carpio and de Souza (2017), the best scenario of biomass use. The following topics will focus on cellulosic ethanol production, presenting the main process steps and the perspectives on biomass processing for the production of composites and chemicals.

21.2.1 Cellulosic ethanol: worldwide production and feedstock description

Cellulosic ethanol is an alternative for the diversification of energy sources and decarbonization of the transport sector due to reducing GHG emissions compared to fossil-based fuels and conventional (starch and sugar) ethanol. One of the major challenges is to degrade the highly recalcitrant lignocellulosic biomass into simple carbohydrates at the industrial scale. Among the options available for biomass conversion into simple sugars, enzymatic hydrolysis has been the technology chosen by the commercial plants that produce cellulosic ethanol (Padella et al., 2019).

Cellulosic ethanol technology has been under development for many years, and its commercial production in Brazil only became a reality in 2014. However, since 2015, the production capacity has remained unchanged (127 million liters). The production follows the same pattern, as in 2020/21 the estimated production is 32 million liters, similar to 2019/20 (30 million liters) (Barros, 2020). The production capacity in the European Union (EU) is currently 60 million liters. Although the capacity could increase to 240 million liters in 2020, the actual production is expected to be no more than 50 million liters (Flach et al., 2020). A similar production number was observed in the United States (US). In 2019 15.5 million gallons (60 million liters) of liquid cellulosic biofuel were produced, and according to the projection of the Environmental Protection Agency, the production will continue to be around 15 million gallons (Environmental Protection Agency, 2020).

21.2.2 Lignocellulosic biomass components and biomass-degrading enzymes

21.2.2.1 Composition of lignocellulosic materials

Lignocellulosic biomass is mainly composed of cellulose, hemicellulose, and lignin but can also contain small amounts of pectin, proteins, extractives, and ash. Typically, the distribution of biomass components is 30%–50% cellulose, 20%–35% hemicellulose, and 15%–35% lignin, depending on their origin.

Lignin is an aromatic macromolecule present in all vascular plants and is known to bind physically/chemically to cellulose and hemicelluloses by covalent bonds. Lignin is not evenly distributed in the cell wall; it is absent in the primary cell wall, found in high concentration in the cell corner middle lamella followed by compound middle lamella and in low amount in secondary wall regions (Gellerstedt and Henriksson, 2008; Wang et al., 2017). The presence of lignin in the plant cell wall provides the plant tissue with stiffness, impermeability, and resistance to microbial and mechanical attacks (Grabber, 2005; Mishra et al., 2007). Consequently, lignin imposes one of the main limiting factors on the enzymatic attack of biomass carbohydrates (cellulose and hemicellulose) by hindering the access of enzymes to the substrates and/or promoting unproductive adsorption of cellulases through charged and uncharged interactions (Zhu et al., 2008; Kumar et al., 2012; Wang et al., 2017; Yoo et al., 2020). The influence of lignin on cellulose conversion has recently been discussed in lignocellulosic biomass conversion reviews (Zoghlami and Paës, 2019; da Silva et al., 2020). Lignin is formed by the polymerization of phenyl-propane alcohols, namely, *p*-coumarilic, coniferilic, and synapilic alcohols, which differ in structure, depending on the type of plant (Laurichesse and Avérous, 2014). In coniferous (softwood) trees, lignin consists almost exclusively of coniferilic alcohol, with small amounts of *p*-coumarilic alcohol. In hardwoods, coniferilic and synapilic alcohols are present, while in monocots, such as sugarcane, all three alcohols are lignin precursors (Fengel and Wegener, 1984; Sjöström, 1993; Shindo et al., 2001; Yu et al., 2014).

The hemicelluloses are branched/unbranched heteropolymers with low molecular weight, with a polymerization degree of 80–200 (Peng et al., 2012; Mota et al., 2018a,b). Hemicelluloses can be divided into four major classes: xylans, mannans, xyloglucans, and mixed-linkage β-glucans (Qaseem et al., 2021). Hemicellulose may include pentoses (xylose and arabinose); hexoses (mannose, glucose, galactose); and/or uronic acids (glucuronic and galacturonic acids), which are connected mainly through β-(1→4) glycosidic linkages but also by β-(1→3), β -(1→6), α-(1→2), α-(1→3), and α-(1→6) glycosidic bonds in branches. Other sugars, such as rhamnose and fucose, may also be present in small amounts, and the sugars' hydroxyl groups may be partially replaced by acetyl groups (Gírio et al., 2010). The variety of linkages and branching and the presence of different monomer units contribute to the structural complexity of hemicellulose, which thus requires a complex enzyme system for its degradation (Mota et al., 2018a,b). However, as the hemicelluloses are amorphous and have little physical strength, their hydrolysis is more accessible than the hydrolysis of cellulose. Many studies have reported hemicellulose removal for improving cellulose enzyme accessibility and hydrolysis rate (Leu and Zhu, 2013; Lv et al., 2013; Kruyeniski et al., 2019). Due to the variety of components in its structure, hemicellulose is considered a promising alternative for replacing fossil resources with many important fuels and biopolymers (Isikgor and Becer, 2015; Zhou et al., 2017; Dulie et al., 2021; Qaseem et al., 2021).

Cellulose is the most abundant organic polymer in nature and the major constituent of plant cell walls, with an annual production estimated at 1.5×10^{12} tons (Klemm et al., 2005). Cellulose is defined as a linear polymer consisting of glucose residues connected by type β-(1,4) glycosidic linkages. The formation of β-(1,4) glycosidic bonds requires the adjacent residues to be positioned 180 degrees relative to another, forming cellobiose units (O'Sullivan, 1997). The degree of polymerization (DP) of plant cellulose ranges from 500 to 15,000 D-glucose residues, depending on its location in the primary or secondary cell wall (Albersheim et al., 2011). The linear character observed in cellulose chains allows adjacent chains to be positioned close to each other (Dufrense, 2012). Thus cellulose chains are aligned in strands forming organized fibrils.

Cellulose microfibrils contain crystalline and amorphous regions. The crystalline regions consist of highly ordered cellulose molecules derived from the organization of cellulose chains linked by hydroxyl groups to form intra- and intermolecular hydrogen bonds in different arrangements (Djahedi et al., 2015), while the molecules are less ordered in the amorphous regions (Park et al., 2010). The crystalline regions are more recalcitrant to enzymatic attack, while the amorphous regions are more readily hydrolyzed (Cao and Tan, 2005; Van Dyk and Pletschke, 2012). The cellulose crystallinity index and the DP affect cellulose recalcitrance, as celluloses with low DP are usually more easily hydrolyzed (Meng et al., 2017a,b; Mattonai et al., 2018).

21.2.2.2 Cellulose-degrading enzymes

The typically reported enzyme for complete cellulose degradation into glucose comprises three main activities: cellobiohydrolases (CBHs—EC 3.2.1.91 and EC 3.2.1.176), endoglucanases (EGs—EC 3.2.1.4), and β-glucosidases (BGLs—EC 3.2.1.21) (Chandel et al., 2012; Cao and Tan, 2005; Singh et al., 2021). This cellulase enzyme consortium, classified under the glycoside hydrolase (GH) family, mainly hydrolyzes the glycosidic bond between two or more carbohydrates (cazy.org/Glycoside-Hydrolases.html). According to the currently accepted model of *Trichoderma reesei* cellulase action, EGs randomly hydrolyze internal glycosidic bonds in amorphous regions of cellulose, creating new chain ends and releasing oligosaccharides. CBHs act mainly in the crystalline part of cellulose, removing cellobiose units from the reducing and nonreducing free chain ends. EGs and CBHs perform synergistically, as EGs create new chain ends for CBH action, and CBHs create more substrate for EGs by disrupting the crystalline substrate and/or exposing previously inaccessible less ordered substrates (Al-Zuhair, 2008; Sukumaran et al., 2021). EGs and CBHs are described to have a carbohydrate-binding module, which helps in substrate binding and keeps the catalytic domain (responsible for hydrolysis reaction) closer to the substrate

(Nakamura et al., 2016; Singh et al., 2021; Sukumaram et al., 2021). Interestingly, CBHs from GH family 7 show a processive cellulose degradation or, in other words, work through successive hydrolytic catalytic reactions without dissociation from the cellulose chain (Nakamura et al., 2016; Uchiyama et al., 2020). Ultimately, the cellobiose released by CBHs is hydrolyzed by BGLs to glucose. BGLs can also, to a lesser extent, hydrolyze other small cello-oligosaccharides to glucose (Kostylev and Wilson, 2012). As BGLs are inhibited by their end product (glucose), new, improved commercial cellulase preparations, such as Novozymes' Cellic series and Dupont's Accelerase, present engineered BGLs for increased glucan conversion and reduced product inhibition (Cannella and Jorgensen, 2014). For more information, the effect of sugars and degradation products derived from biomass on enzyme efficiency (including commercial enzymes) and cellulose conversion has recently been reviewed on lignocellulosic biomass conversion under low and high solid conditions (da Silva et al., 2020).

Bacteria and fungi can produce cellulases, but aerobic fungi have been preferred due to their versatile substrate utilization and high production level (Srivastava et al., 2018). Cellulases are in high demand as they are used in numerous industries and occupy the third position in the global enzyme market, after amylases and proteases (Sajith et al., 2016; Singh et al., 2021). Due to their importance, the improvement of cellulolytic enzymatic cocktails has been continuously studied (Karp et al., 2021a,b). In addition to these hydrolytic enzymes, many cellulolytic microorganisms produce enzymes that can degrade crystalline cellulose through an oxidative mechanism of action (Zifcakova and Baldrian, 2012; Dixit et al., 2019). Although studies on cellulose degradation dated back to the early 1950s, it was only in 2010 that studies demonstrated the ability of metalloenzymes, now known as lytic polysaccharide monooxygenases (LPMOs), to disrupt crystalline cellulose and thus complement the GHs by functioning as endoenzymes that cleave crystalline cellulose surfaces, creating new chain ends for CBH action (Laurent et al., 2019; Sukumaram et al., 2021). It has been shown that LPMOs are copper-dependent monooxygenases that oxidize polysaccharides at C1 and/or C4, starting a chain breakage (Harris et al., 2010; Vaaje-Kolstad et al., 2010; Hemsworth et al., 2013). LMPOs were incorporated into commercial cellulases, such as Cellic CTec 2 and CTec3 (Novozymes), within a few years after their discovery, as they were proven to boost cellulose degradation (Horn et al., 2012). The addition of LMPOs to commercial enzyme cocktails adds a new variable to cellulose hydrolysis, as those enzymes require oxygen. To have benefits from LMPOs' boosting activities, processes must be designed to avoid competition with dissolved oxygen, preferably by conducting the hydrolysis step separately from fermentation (Cannella and Jorgensen, 2014). The improvement of enzyme blends containing LPMOs on lignocellulose conversion at high solid conditions has been recently reviewed (da Silva et al., 2020).

21.2.2.3 Hemicellulose-degrading enzymes

Enzymatic depolymerization of hemicellulose is commercially attractive due to the requirement of mild conditions, low formation of toxic degradation products, and its diverse sugar composition. Different plants have different branched/unbranched hemicelluloses, composed of two to six monosaccharides, acetylated or methylated. Due to its heterogeneity, multiple enzymes with distinct specificities acting synergistically and/or sequentially are needed to convert each type of biomass (Juturu and Wu, 2013; Juturu and Wu, 2014; Qaseem et al., 2021). The prerequisite for the conversion of hemicellulose to value-added chemicals is its depolymerization, which may require endoxylanase (EC 3.2.1.8), β-xylosidase (EC 3.2.1.37), α-arabinofuranosidase (EC 3.2.1.55), α-glucuronidase (EC 3.2.1.139), endo-1,4-mannanase (EC 3.2.1.78), β-mannosidases (EC 3.2.1.25), acetyl xylan esterase (EC 3.1.1.72), and feruloyl xylan esterase (EC 3.1.1.73) (Juturu and Wu, 2013; Sukumaram et al., 2021; Qaseem et al., 2021). These enzymes can be classified based on their substrate specificities as GHs or carbohydrate esterases. For example, grasses, such as sugarcane, have arabinoxylan as a major hemicellulose component. Their complete degradation requires an enzyme mixture containing xylanases, which catalyzes the hydrolysis of 1,4-β-d-xylosidic linkages in xylan, β-xylosidases, which catalyzes the hydrolysis of xylo-oligosaccharides and xylobiose into xylose, and α-arabinofuranosidase hydrolyze arabinan side chains attached to the xylan backbone to arabinose (Sorensen et al., 2007; Sweeney and Xu, 2012).

The requirement of hemicellulases for lignocellulosic biomass hydrolysis is strongly dependent on the type of pretreatment used to reduce the biomass recalcitrance. As some pretreatments, as such hydrothermal, remove the hemicellulose content almost completely, the need for hemicellulases is reduced in those cases. However, the trend in new commercial plants for cellulosic ethanol production is to use low-severity pretreatments to reduce capital cost and toxic waste generation (Harris et al., 2014). Pretreatments conducted in lower severity conditions tend to leave more hemicellulose in the biomass material, thus requiring the incorporation of hemicellulases in commercial enzymatic pools, as hemicellulose remains linked to cellulose and act as a physical barrier for cellulose-degrading enzymes. Several studies have shown a significant increase in cellulose hydrolysis or reduced requirement of cellulases by adding hemicellulases in enzyme mixtures for the hydrolysis of pretreated biomass (Tabka et al., 2006; García-Aparicio et al., 2007; Xu et al., 2019)

Thus commercial enzymes for the hydrolysis of lignocellulosic materials, such as the Cellic enzymes series, offer the option of mixing cellulases and hemicellulases for a better hydrolysis response. Although the application sheet of Cellic CTec2 describes the product as a mixture of aggressive cellulases, high levels of BGLs, and hemicellulases, the manufacturer also offers the option of mixing Cellic CTec2 with HTec2, which is described as a product rich in endoxylanases with high specificity to soluble hemicellulose, with a cellulase background. A dose-response test through the addition of Cellic HTec2 to CTec2 in comparison to CTec2-only dose is recommended when pretreatment results in a feedstock that would benefit from additional hemicellulose degradation. Cellic HTec3, which is the new-generation enzyme, is described as a hemicellulase complex that contains endoxylanases and β-xylosidases activities for effective hemicellulose conversion. The manufacturer describes this product as highly effective for the hydrolysis of liquid process streams rich in xylan-oligomers to xylose and the transformation of soluble and insoluble hemicellulose in pretreated biomass slurries.

21.2.3 Perspectives and difficulties of cellulosic ethanol production

The commercial production of cellulosic ethanol has increased in the last few years; nonetheless, it is far behind expected. The processes required to produce cellulosic ethanol are inherently more complex than sugar/starch ethanol production. First of all, before enzymatic hydrolysis can be conducted, the key obstacle of biomass recalcitrance needs to be addressed by performing a pretreatment step to improve the accessibility of cellulose to hydrolytic enzymes. Major efforts have been made toward improving the pretreatment step, and several methods have been developed (e.g., steam explosion, diluted-acid, and hydrothermal pretreatments). The choice of pretreatment technology depends mainly on the biomass type. However, studies and experts agree that the steam explosion is one of the most cost-effective. It removes the lignin and hemicellulose, provides high glucose yields, and results in low environmental impact (Alberts et al., 2016; Chandel et al., 2019). Thus the pretreatment step that was previously considered the main limitation in cellulosic ethanol production is nowadays of lower impact when compared to other factors. A similar pattern is observed in the efforts made to improve the fermentation process, as the development of new yeast strains has been copiously studied during the last few years (Chandel et al., 2019).

Divergently, enzyme production continues to be a critical technology step that still impairs the economic feasibility of cellulosic ethanol production. Although the enzyme cost reduced from \$3.00 per gallon in 2001 to about \$0.40 in 2020, according to the US Department of Commerce, for cellulosic ethanol to reach competitiveness, the enzyme cost should be at least \$0.05 per gallon (Osborne, 2020). Reducing the enzyme cost is incredibly challenging, as in the current market, only one company, Novozymes, produces industrially effective cellulases.

Besides the enzyme costs, the enzymatic hydrolysis step on an industrial scale poses technical issues, as the solids loading can directly affect the ethanol yield. To reach distillation feasibility—ethanol concentrations above 4% (w/w)—(Zacchi and Axelsson, 1989; Varga et al., 2004), the concentration of the glucose syrup must be around 80–100 g/L, and, by extension, the amount of solids required during enzymatic hydrolysis has to be increased. To reach cellulosic ethanol feasibility, high solid enzymatic hydrolysis (at least 15% of solids) is mandatory. However, technical issues are faced when working in this condition. The high viscosity of the media hampers the enzymatic catalysis due to many factors, for example, mass transfer limitation, product inhibition, and water constraint (da Silva et al., 2020).

The beginning of cellulosic ethanol developments occurred when macro- and microenvironmental factors triggered the technology establishment, where oil prices were rising, and the world was facing a financial crisis. The climate change consensus was pressuring governments to seek sustainable energy options. However, the cellulosic ethanol industry faced several challenges due to the dramatic drop in oil prices in the second half of 2014 and uncertainty about the political support for biofuels. Global new investment in renewable power and fuels increased 5% from 2018 to 2019 (Renewables, 2020). However, those investments have been increasing due to the wind and solar power sector, and indeed, the biofuels sector lost 10% of investments from 2018 to 2019. This is a reflex to the lack of effective policies in the vast majority of the countries. Although the global transport sector has the second highest share of total final energy consumption, it remains the sector with the lowest penetration of renewables. It continues to rely heavily on fossil fuels (IEA, 2018). Another important factor that has hindered most investments in this sector is the global COVID-19 pandemic. In Brazil, "the expected negative impact in the Brazilian Otto-cycle fuel consumption (gasoline and ethanol) in the next couple of years forced the Ministry of Mines and Energy to propose the review of compulsory targets."

The price of cellulosic ethanol in the US decreased from $ 5 per gallon in 2001 to about $ 2.65 in 2020 (Osborne, 2020). And the effective communication of the benefits and lessons learned to the public and policy makers will be essential for the full development of this growing industry, as the example of the state of California, which has harsh policies, such as the Low Carbon Fuel Standard over the use of biofuels (The California Air Resources Board, 2021). Another example is the instruments based on RenovaBio Program in Brazil, for example, the Decarbonization Credits (CBio) (Barros, 2020). However, to reach a competitive status with starch ethanol in the US, it is estimated that cellulosic ethanol should cost $1.07 per gallon, which means that more efforts are demanded toward the investments in this sector (Osborne, 2020). Thus renewable-based policies are important to boost the investments in R&D for the development of processes that are still not cost-effective. For this purpose, the focus on the on-site production of enzymes to diminish the enzyme costs and the coproduction of different chemical feedstocks alongside the cellulosic ethanol to increase the economic feasibility of the industrial production of cellulosic ethanol is necessary (Zhao and Liu, 2019; Rosales-Calderon and Arantes, 2019; Chandel et al., 2019).

21.2.4 Enzyme-based initiatives for ethanol production at commercial scale

Although Brazilian initiatives to stimulate cellulosic ethanol production have been made in the last few years (e.g., investments from the Brazilian Development Bank for the construction of cellulosic ethanol facilities), recently, those investments have been drastically reduced mainly by the technical issues in the plants that hindered the full implementation of this technology in Brazil. Indeed, other ethanol production feedstocks have been emerging, such as corn, which is abundant and generally cheap. Since 2014 corn starch ethanol production has been increasing in Brazil to attend to the demand for biofuels required by Brazilian policies during the summer rainy season when sugarcane is unavailable. Since then, its production has been higher than cellulosic ethanol production. According to the Brazilian Corn Ethanol Union (*União Brasileira de Etanol do Milho* - UNEM), the total corn ethanol production in 2020/21 is estimated as 2.5 billion liters, representing an increase of 1.17 billion liters when compared to 2019/20 (Bini, 2020).

Brazil has two commercial-scale facilities for ethanol production from sugarcane biomass: Bioflex, in the state of Alagoas, from Granbio (82 million liters production capacity) and Raizen-Costa Pinto Unit in the State of Sao Paulo (42.2-million liter production capacity). GranBio built its first unit in Alagoas with the technology based on the pretreatment developed by PROESA of the Beta Renewables (company Group M&G). The enzymes are from Novozymes, and the yeasts were developed by DSM (GranBio, 2020). GranBio started operation in September 2014, and due to technical difficulties in the pretreatment stage, they resumed the cellulosic ethanol production operations in 2016 (Susanne, 2016; Kennedy, 2019). Granbio is currently diverting the biomass to the thermoelectric plant, and in 2020 it announced a strategic alliance with NextChem to license its patented technology to produce cellulosic ethanol (GranBio, 2020). Raizen-Costa Pinto Unit is the only one producing cellulosic ethanol at a relatively large scale. Raizen, a company based on the merger of Shell and Cosan, in partnership with Iogen Corporation, in November 2014, began the operation of its biomass-to-ethanol facility in Piracicaba, São Paulo, using Iogen Energy's technology (Green Car Congress, 2014). The company also signed an agreement with Novozymes regarding developing enzyme technology for 2G ethanol production (Novozymes, 2014). In 2016, the company announced reductions in its cellulosic ethanol investment due to low gasoline prices (Jim, 2016). Since 2019 it has been supplying cellulosic ethanol to O Boticário, one of Brazil's biggest beauty and cosmetics companies (Vital, 2021).

The cellulosic ethanol production downturn is worldwide. The plants, Beta Renewables, DowDuPont, Abengoa, and POET-DSM have announced bankruptcy, a break, and switched the energy matrix, respectively (Scott, 2018; Lane, 2017; Voegele, 2016; Bomgardner, 2019). Even though the industrial production of cellulosic ethanol became a reality, it is clear that the cost-effectiveness of the production is still impracticable. Incentives and investments were reduced due to fear of the persistently high cost linked to its production. Thus, solutions for lowering the costs must be developed and tested once the investments return to arise. Indeed, the projections indicate a worldwide demand for ethanol. In this sense, Raízen announced the construction of a second cellulosic ethanol facility by 2023. According to Marcelo Eduardo Martins, the investor relationship vice-president of the group, the new facility will meet the growing international demand for the product (União Nacional da Bioenergia, 2021; Reuters, 2021).

21.2.5 Perspectives on the use of microalgae as sources of fermentable sugars

Algae are simple photosynthetic organisms with no roots, stems, or leaves (Andersen, 2013). They are responsible for half of the world's photosynthetic activity, with a large part occurring in the oceans (Falkowski and Raven, 2007). Due to the great diversity of species, varying between microscopic (microalgae) and macroscopic (macroalgae), there is no consensus on the definition of this group of organisms.

When comparing the use of algae to agricultural crops for obtaining bioproducts, they have several advantages over the latter: (1) high photosynthetic efficiency, (2) requirement of no arable land, (3) no requirement of freshwater, and (4) they do not present seasonality (Sayre, 2010). Moreover, both macro- and microalgae have been shown to grow on wastewater (Ge and Champagne, 2017; Nagarajan et al., 2020), and the CO_2 required for their growth can be obtained from industrial plants, with the flue gas line being diverted to the aqueous culture medium (Cheah et al., 2014). These alternatives would minimize the production costs and offer additional advantages for algae production; however, they are still in the research phase and have not been applied on the industrial scale for biomass production for further recovery of bioproducts.

Regarding the Brazilian prospects for microalgae technology, Brazil has a high global solar irradiation in any region of its territory, 4200–6700 kWh/m^2, which is higher than most countries in the EU at 900–1850 kWh/m^2. Moreover, as a significant part of its territory is located near the equator, Brazil has an even distribution of sunlight throughout the year (Pereira et al., 2017). Recent studies have identified more than 3000 algae species in Brazil, with two endemic genera and 52 endemic species (Bicudo and Menezes, 2010). Therefore Brazil has enormous potential for growing microalgae, a potential thus far hardly explored.

Studies on microalgae as sources of biomolecules for fuel and chemical production have been increasing in recent decades. Historically, using algae as an energetic feedstock dated back to the end of the 1950s and was furthered by the 1970s oil crisis (Chen et al., 2009). From 1978 to 1996, the US Department of Energy invested US$ 25 million in a program on the study of algal fuels, achieving significant advances in this field (Waltz, 2009). As some species of microalgae, when cultivated under metabolic stress, can accumulate higher amounts of lipids as an energy reserve (Brindhadevi et al., 2021; Mata et al., 2010), they have been mainly studied for biodiesel production. However, carbohydrate-rich algae have also been gaining attention in the last decade as a potential source of fermentable sugars to produce ethanol and other biochemicals. Indeed as observed for lipid accumulation, certain microalgae species, under metabolic stress, increase their intracellular carbohydrate content, making them more attractive as a feedstock for obtaining high-concentration sugar syrups (Souza et al., 2020). Common algae, such as species belonging to the *Chlorella* genus, can accumulate between 10% and 80% of carbohydrates depending on the cultivation conditions (Souza et al., 2020, 2017). Table 21.1 presents the biochemical composition of different starch-accumulating microalgae.

TABLE 21.1 Content of carbohydrates (starch content shown in parentheses), lipids, and proteins of several microalgae species.

Species	Carbohydrates (starch) (%)	Lipids (%)	Proteins (%)	References
Arthrospira platensis	~60	~5	~25	Markou et al. (2013)
Chlamydomonas fasciata	n.d. (43.5)	n.d.	30.4	Asada et al. (2012)
Chlamydomonas reinhardtii	59.7 (43.6)	n.d.	9.2	Choi et al. (2010)
Chlorella variabilis	53.9 (37.8)	24.7	19.8	Cheng et al. (2013)
Chlorella vulgaris	50.4 (31.3)	11.6	23.3	Ho et al. (2013)
Dunaliella salina	32	6	57	Becker (2007)
Mychonastes homosphaera	76 (55)	14	6	Mota et al. (2018a,b); Souza et al. (2017)
Chlorella sorokiniana	35 (28)	23	30	Mota et al. (2018a,b); Souza et al. (2017)

n.d., Not determined.

Carbohydrates in microalgae can be found in the cell wall (Scholz et al., 2014; Takeda, 1988) and as intracellular starch (Brányiková et al., 2011; Souza et al., 2017). Since the microalgae cell wall only accounts for a small part of their dry weight, that is between 3%–6% (Scholz et al., 2014; Takeda, 1991), the intracellular starch can be considered the main potential source of fermentable sugars in these microorganisms. Nevertheless, cell walls play an essential part in this technology. They act as a physical barrier that must be ruptured before accessing the intracellular starch to further process the biomass into fermentable sugars (Günerken et al., 2015).

Starch is an insoluble, semicrystalline glucose polymer, the fundamental unit of which is maltose, a glucose dimer with an α-1,4 bond. This polymer consists of two distinct fractions, amylopectin and amylose. Amylose consists of a linear maltose chain without branches, while amylopectin also has α-1,6 bonds, branching its structure (Takeda et al., 1987). Starch molecules stored intracellularly are always found in the form of granules (Buléon et al., 1998), as shown in Fig. 21.2. These granules are formed mainly due to the presence of amylopectin. The chains originating from the branches align and form double helices, giving rise to the crystalline sections of the starch chains, which are insoluble and responsible for the collapse of the structure in granular format. The chain fractions close to the branching points are part of the amorphous sections of the polymer (Imberty et al., 1991).

Due to this crystalline nature, starch in its native form is not easily degraded. However, when heated, the starch undergoes a gelatinization

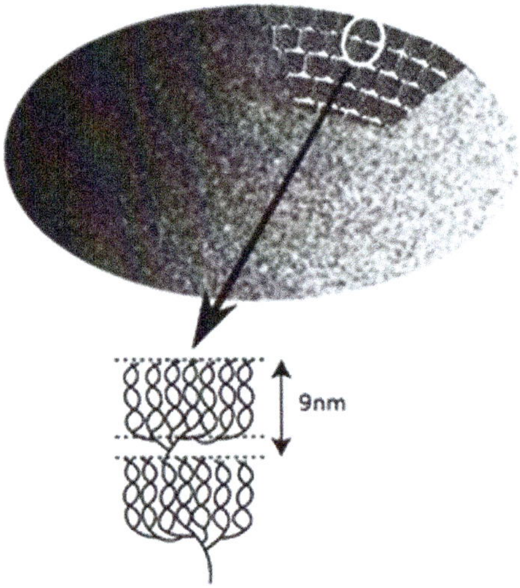

FIGURE 21.2 Starch granules with emphasis on the formation of double helices in the amylopectin chains (Ball et al., 2011).

process (Ratnayake and Jackson, 2008), resulting in the crystalline fractions' disorganization. It becomes more susceptible to amylase hydrolysis (Slaughter et al., 2001). Several factors determine the extent and rate of hydrolysis of starch granules, including granule size and morphology and the ratio of amylose and amylopectin present (Svihus et al., 2005). In general, hydrolysis of amylopectin-rich starches is faster than that of amylose-rich starches (Tester et al., 2006).

For the complete hydrolysis of starch, several enzymes are required. The main ones are α-amylases (EC 3.2.1.1), β-amylases (EC 3.2.1.2), and amyloglucosidases (EC 3.2.1.3), which act on α-1,4 bonds; the first acts internally in the depolymerization of the chains, releasing dextrins, while the others act from the nonreducing terminals, releasing maltose and glucose, respectively. Other enzymes, called isoamylase (EC 3.2.1.68) and pullulanase (EC 3.2.1.41), hydrolyze α-1,6 bonds, acting on the branches and helping the action of the former.

In a well-established corn ethanol industry, corn starch undergoes two subsequent stages of enzymatic hydrolysis to generate glucose syrup. The first stage consists of liquefaction, in which commercial α-amylase from thermoresistant bacteria of the genus Bacillus is used at temperatures of 90°C–110°C; at this temperature, the starch is also gelatinized. Then, there is the saccharification stage, with the action of fungal amyloglucosidases at 60°C–70°C (Sánchez and Cardona, 2008). Although this process is well established, it generates high energy costs. An alternative is the hydrolysis of starch at low temperatures, without the prior need for a gelatinization step. Several amylases have already been described as being able to digest starch *in nature*, without the need for this step; however, in general, more significant amounts of enzymes are required than for the hydrolysis of gelatinized starch (Robertson et al., 2006).

As the microalgal starch processing industry still does not exist, there is no established process for its hydrolysis. The existing research studies vary on using or not a gelatinization step (Souza et al., 2020). Moreover, as mentioned before, there is usually the need for a cell wall disruption step either simultaneously or before the enzymatic hydrolysis of the intracellular starch, which also varies between studies. Table 21.2 shows the different treatments and yields obtained in the main studies published in enzymatic hydrolysis of microalgae for the production of glucose syrup.

As can be seen from Table 21.2, only milling has been used consistently in several studies for cell wall rupture, probably because this mechanical process is independent of cell wall composition and is guaranteed to work as long as enough energy is given during treatment. Even though enzymatic hydrolysis of the cell wall of microalgae has been explored as a method for cell wall disruption, this is highly dependent on the cell wall composition, being genera- or even species-specific (Gerken et al., 2013). For instance, Choi et al. (2010) reported the rupture of the cell wall of *Chlamydomonas*

TABLE 21.2 Reported glucose yields for the enzymatic hydrolysis of starch-rich microalgae [as presented in Souza et al. (2020)].

Microalgae	Treatment	Solids loading	Glucose yield (reaction time)	References
Chlamydomonas reinhardtii	Proteases/ liquefaction	50 g/L	94% (30 min)	Choi et al. (2010)
Chlorella sp.	Milling/ gelatinization	22 g/L	97% (24 h)	Maršálková et al. (2010)
Chlorella vulgaris	Sonication/ autoclave	10 g/L	79% (24 h)	Ho et al. (2013)
Chlorella variabilis	Virus infection	4.25 g/L	43% (120 h)	Cheng et al. (2013)
Chlorella vulgaris	Milling	100 g/L	79% (72 h)	Kim et al. (2014)
Chlorella sp.	Lipid extraction	50 g/L	93% (3 h)	Lee et al. (2015)
Scenedesmus dimorphus	Milling/ gelatinization	20 g/L	96% (1 h)	Chng et al. (2017)
Chlorella vulgaris	Milling	100 g/L	91% (4 h)	de Farias et al. (2018)
Chlorella sorokiniana	Milling	111 g/L	94% (4 h)	Souza et al. (2020)

reinhardtii by treatment with commercial α-amylase rich in proteases since the cell wall of this microalgae is composed of glycoproteins. Another study conducted with several enzymes concluded that the cell wall of *Chlorella vulgaris* was susceptible to the action of chitinase, lysozyme, pectinase, and pectiolase (Gerken et al., 2013).

In addition to the use of commercial enzymes, another approach that has been used is the coculture of microalgae with bacteria or viruses that produce enzymes capable of degrading their cell wall. For instance, Cheng et al. (2013) used virus infection as a strategy to increase the hydrolysis yield of *Chlorella variabilis* starch by 80%. An innovative alternative that is still in the preliminary stage of studies is using enzymes from the algae in a process called autolysis, which occurs naturally during the cell division process. Some enzymes involved in this process have already been identified; however, the only work to induce autolysis in algae was carried out to mitigate the phenomena of red tide and overproliferation (Demuez et al., 2015).

21.3 Technical and economic prospects of using lipases in biodiesel production

21.3.1 Current biodiesel production and perspectives

Biodiesel is composed of monoalkyl esters, commonly methyl and ethyl esters, similar to petro-diesel (Robles-Medina et al., 2009). Biodiesel is produced from vegetable oils or animal fats; its application has gained attention over the years as a potential substitute for crude oil-based diesel due to its implications for global warming by reducing carbon monoxide emissions particulates, hydrocarbons, and sulfates. When blended with petro-diesel, the emission reduction of the resultant fuel depends on the blend's composition. Besides, biodiesel can degrade at a much faster rate than petro-diesel. Biodiesel use has already been implemented in several countries worldwide (Almeida et al., 2021; Robles-Medina et al., 2009). Indonesia, Brazil, the US, and Germany are among the largest biodiesel producers in the world, totaling 24.1 billion liters produced in 2019. In the same year, Indonesia rose to be the world's largest biodiesel producer reaching nearly 8 billion liters. The implementation of the Energy Policy Act of 2005 in the US, which provided tax incentives for certain types of energy, led to an increase in biodiesel production. Future projections estimate production levels of over 1 billion gallons of biodiesel by 2025 (https://www.statista.com/statistics/271472/biodiesel-production-in-selected-countries/#statisticContainer).

In Brazil, biodiesel production and use have been mandatory since the National Program for Use and Production of Biodiesel (PNPB) in 2004, implemented by the law no. 11.097 of January 2005. The PNPB established a minimum percentage of biodiesel blended with crude oil diesel of 2%, named B2, in January 2008, which has increased over time (Ribeiro and Silva, 2020). B11 was adopted in September 2019; although the production capacity was 9.2 million m^3 with forty-two authorized plants in operation, the national production achieved 5.9 million m^3 (64% of the total capacity). However, the production was 10.3% superior to 2018 (ANP, 2020a,b). Currently, Brazil uses B13 with perspectives to reach B14 on March 1, 2022, and B15 on March 1, 2023 (MME, 2020). Considering these perspectives, the Mines and Energy Ministry coordinated a group aiming to attend the Report of Consolidation and Validation Tests and Trials for the Utilization of B15 Biodiesel in Engines Vehicles by the definition of oxidation stability reference established by ANP Resolution no. 798, from August 1, 2019. Moreover, the literature has reported that blends up to B20 are adequate and compatible with most petro-diesel equipment, including storing and distribution (Balat, 2011; MME, 2020).

Nowadays, alkaline transesterification technology broadly uses methanol as an acyl acceptor for biodiesel production. This route presents some drawbacks—generating large amounts of highly alkaline wastewater (with significant environmental impacts that impose additional handling costs), producing impure glycerol by-product (contaminated with alkaline catalyst), and laborious catalyst recovery. Furthermore, the alkaline route requires high-quality and expensive raw materials that represent around 80% of the total biodiesel production costs, hindering the economic competitiveness of this biofuel (Demirbas, 2007; Pourzolfaghar et al., 2016; Robles-Medina et al., 2009; Sharma et al., 2008). Therefore using low-cost feedstocks could make biodiesel more suitable for commercialization.

Lipases have been highlighted as powerful tools for synthesizing esters by biocatalysis to produce several products, including biodiesel. Biocatalysis can circumvent the drawbacks of chemical catalysis, and, thus, several studies focus on developing new technologies in biodiesel production to increase its competitiveness in the international fuel market. Particular attention has been given to biocatalysis using lipases, and, more recently, bioprocesses using whole cells as biocatalysts also emerged as a potential alternative. The use of lipases or whole cells for biodiesel production requires relatively simple downstream processing steps to purify biodiesel. Besides, the by-product glycerol (without catalyst contamination) has better quality and a higher value. Furthermore, enzymatic processes allow the use of cheaper raw materials that contain high FFA and water content, as lipases can catalyze both esterification and transesterification reactions (Aguieiras et al., 2013, 2017; Souza et al., 2009; Zeng et al., 2017; Zheng et al., 2020).

Lipases are versatile biocatalysts and can act in different functional groups and substrates, under several reaction conditions, in various chemical transformations, and with varying transition states. Moreover, lipases show regio-specificity (the distinction between positions 1 and 3 of the triacylglycerols), specificity in terms of fatty acids and alcohol present in the synthesis/hydrolysis, and stereo-specificity. These enzymes also show stability in organic solvents (Kapoor and Gupta, 2012). Lipases produced by multicellular eukaryotes (animals and plants) and microorganisms are typically present in the three domains (Eukarya, Archaea, and Bacteria). Lipases of animal origin are responsible for the metabolism of lipids; in plants, they are present in the energy reserve tissues of several species (Sharma et al., 2001) and can also be found in the seeds of several angiosperms such as castor—*Ricinus communis* (Greco-Duarte et al., 2017), physic nut—*Jatropha curcas* (De Sousa et al., 2010), and others such as sunflower (*Heliantus annuus L.*) and wheat (*Triticum aestivum L.*) (Barros et al., 2010).

Due to their advantages and versatility, lipases are among the most frequently used groups of enzymes in the industrial sector worldwide.

21.3.2 Biocatalytic production of biodiesel

The concepts of lipases and esterases are often mixed up since both enzymes are versatile and catalyze the same diversity of reactions. The definition of esterases compared to lipases is still widely discussed, and over the years, several studies have been carried out to distinguish them. Despite all efforts and criteria of distinction that have been proposed, no consensus has been reached in the scientific community. However, a reorganization was suggested only in a pragmatic way which divides the large group of carboxyl ester hydrolases into lipolytic esterases (LEst or lipases: E.C.: L3.1.1.1) and nonlipolytic esterases (NLEst: E.C.: NL3.1.1.1) (Romano et al., 2015).

A possible difference is in the concept that, by definition, the natural function of lipases is to catalyze the hydrolysis of ester bonds present in long-chain triacylglycerols (TAGs), releasing diacylglycerols (DAGs), monoacylglycerols (MAGs), glycerol, and fatty acids. There is no explicit definition for the term "long-chain" triacylglycerol; however, the "standard substrates" for lipases are believed to be those, the chains of which contain more than eight carbon atoms. It is important to note that lipases can hydrolyze esters of smaller chains, unlike esterases that only hydrolyze esters with chains, the size of which is less than eight carbon atoms (Romano et al., 2015). Thus it can be inferred that lipases can catalyze all reactions catalyzed by esterases, but not all reactions catalyzed by lipases can be catalyzed by esterases. It is a common consensus that lipases are esterases that act on long-chain acylglycerols (Jaeger et al., 1999; Mancheño et al., 2003). Triacylglycerol is just one example of all other compounds that can be substrates for these enzymes, whether of low or high molar mass. The possible lipase substrates vary from amides, thioesters, fatty acids, hydroxy fatty acids, and others.

Despite efforts to separate lipases and esterases into different groups, they are classified by the International Union of Biochemistry and Molecular Biology within the same group, EC 3.1.1. Esterases or, carboxylesterases (EC 3.1.1.1), are systematically classified as carboxylic ester hydrolases and the group name, and lipases or TAGs lipases (EC 3.1.1.3) are systematically classified as triacylglycerol acyl hydrolases. The natural function of these enzymes is, as stated above, the hydrolysis in an aqueous medium of ester bonds releasing fatty acids and alcohol, as shown in Fig. 21.3.

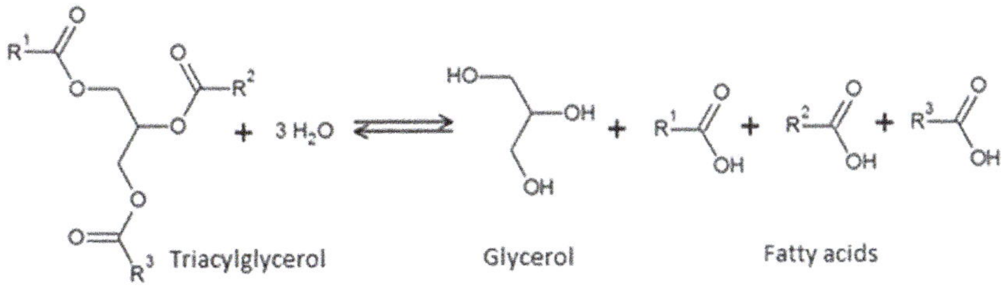

FIGURE 21.3 Hypothetical example of the natural reaction catalyzed by lipases and esterases. Hydrolysis of triacylglycerol is in the direct sense of the reaction producing glycerol and fatty acids as products ($R^1 = \text{or} \neq R^2 = \text{or} \neq R^3 \mid R^1 < 8$ carbon atoms).

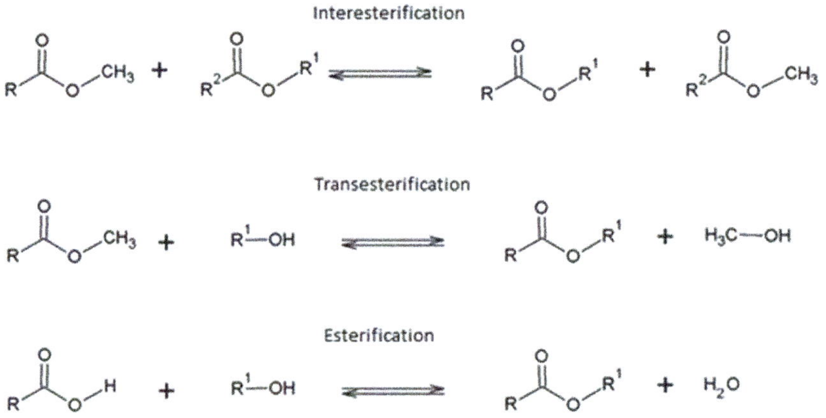

FIGURE 21.4 Example of synthesis reactions performed by lipases and esterases.

However, in water-restricted environments, these enzymes can catalyze reverse synthetic reactions such as interesterification, transesterification, and esterification (Fig. 21.4) (Jaeger and Reetz, 1998; Paques and Macedo, 2006; Sharma et al., 2001).

Most reports on enzymatic biodiesel production in this chapter treated the enzymes as lipases. In industrial production, greater emphasis is given to those enzymes produced by microorganisms since they have a shorter generation time and incredible versatility of operating conditions. Besides, they are easy to manipulate genetically, aiming to improve their production capacity and cultivation conditions, which increases their biotechnological interest (Pereira et al., 2008; Sharma et al., 2001). These enzymes are active in transesterifying vegetable oils and other oils into fatty acid methyl esters (FAME) or fatty acid ethyl esters (FAEE). Moreover, literature also shows the enzymatic transesterification of oils with other acyl acceptors such as 1-propanol or 1-butanol (Ma et al., 2019), 2-propanol (Kumar et al., 2018), and enzymatic interesterification of oils with ethyl acetate (Kovalenko et al., 2015) all known as biodiesel in the literature.

21.3.3 Feedstocks used for biodiesel production

The evaluation of the feedstock to be used has utmost importance since the feedstocks significantly impact the production costs of biodiesel and bring some specificities to the process. Each country has specific quality requirements to produce and commercialize biofuels (Budžaki et al., 2017), and the feedstock needs to be adequate to match these requisites. The use of refined oils can facilitate the achievement of these requirements. However, soybean oil, corn oil, or any oil, the grains of which are included in human feeding implies a discussion about competition between biofuel and food and problems caused by monoculture in the ecosystem. In addition, the use of refined oils reduces the competitiveness of biodiesel compared to crude oil diesel. Therefore the utilization of alternative feedstocks can assure the future of this biofuel. Alternative feedstocks include nonedible and crude vegetable oils, grease and animal fats, microbial oils (e.g., from microorganisms and microalgae), and residual oils (e.g., deodorizer distillate from the process of refinement) (Table 21.3). The availability of these oils in each country will be the main choice point to consider (Aarthy et al., 2014).

Palm and soybean oils are the main feedstocks used for biodiesel production worldwide. In the Brazilian context, through 2019, soybean oil remained the primary raw material for the B100 production, reaching 68.3% of the total, which means an increase of 9% compared to 2018. Other feedstocks, which include canola oil, castor oil, corn oil, palm oil, peanut oil, sesame

TABLE 21.3 Examples of alternative feedstocks (waste oils and nonconventional oils) adopted for biodiesel production.

Nonconventional oils/waste oils	References	Nonconventional oils/waste oils	References
Karanja oil (*Millettia pinnata*)	Kumar et al. (2018)	Rapeseed oil deodorized distillate	Zeng et al. (2017)
Crambe oil (*Crambe abyssinica* Hochst)	Tavares et al. (2018)	Soybean oil deodorized distillate	Zheng et al. (2020); Souza et al. (2009)
Babassu oil (*Orbignya* sp.)	da Rós et al. (2014)	Waste cooking oil	Tacias-Pascacio et al. (2017); Chesterfield et al. (2012); Waghmare and Rathod (2016)
Macauba oil (*Acrocomia aculeata*)	Aguieiras et al. (2014)	Palm fatty acid distillate	Kapor et al. (2017); Aguieiras et al. (2013); Chongkhong et al. (2009)
Garcinia Gummi-gutta oil	Subramani et al. (2018)	Algal oil	Prabakaran et al. (2019)
Neem oil (*Azadirachta indica*)	Aransiola et al. (2012)	Yeast oil	Duarte et al. (2015)
Physic nut oil (*Jathropha curcas* L.)	de Sousa et al. (2010)	Animal fat	Toldra-Reig et al. (2020)

oil, sunflower oil, turnip oil, used frying oil, and other fatty materials, accounted for the second-largest amount used (16.5% of the total), followed by animal fat (14.1% of the total) and cotton oil (1.1%) (ANP, 2020a,b).

The increasing use of waste oils is due to the raw material's cost in biodiesel's total production costs. However, since waste oils contain >1% of FFA, their use in alkaline transesterification requires a pretreatment step to avoid the formation of soaps as by-products, leading to an increase in the total cost. Besides, waste oils also contain high water levels, reducing the yields of the reaction (Budžaki et al., 2017; Sharma et al., 2008). Considering the ability of lipases to convert both FFA and acylglycerols into esters (biodiesel), these biocatalysts become promising alternatives for obtaining biodiesel from these sources, with fewer pretreatment steps.

Feedstocks with high FFA content and lower cost have also been used for enzymatic biodiesel production, with conversions above 90%—crude vegetable oils (Aransiola et al., 2012; Kumar et al., 2018; Prabakaran et al., 2019), waste cooking oils (WCOs) and wastes from vegetable oil refining (Chesterfield et al., 2012; Collaço et al., 2020; Waghmare and Rathod, 2016; Zeng et al., 2017; Zheng et al., 2020), and nonedible oils obtained from many plant species potentially available in a local area (Aguieiras et al., 2014; Arumugam et al., 2018; Subramani et al., 2018; Chen et al., 2020).

Microalgae are another alternative lipid source potentially applicable for biodiesel production. The microalgae lipid content varies from 1% to 70% but can be as high as 90%, depending on species and cultivation conditions. Common algae, including the genus *Chlorella*, present lipid contents between 20% and 50% (Chisti, 2007; Mata et al., 2010). One factor that drove attention to microalgae as a feedstock for biodiesel production is their high oil yield per cultivated area. Palm, the most productive oleaginous land plant, yields 5950 L of oil per hectare, while a microalga with 30% lipid content yields 58,700 L per hectare (Chisti, 2007). Oleaginous organisms as a platform for efficient oleochemical production have been recently highlighted (Khoo et al., 2020; Marcon et al., 2019; Prabakaran et al., 2019).

Some issues have emerged regarding the oil extraction that requires cell disruption and treatment of the proteins, carbohydrates, and other by-products generated during the biosynthesis (Meng et al., 2009). Nevertheless, microbial oils also differ from most vegetable oils in the ratio of polyunsaturated fatty acids, resulting in biodiesel that could be more susceptible to oxidation (Meng et al., 2009; Salimon et al., 2012). However, promising results were obtained using immobilized lipases in enzymatic transesterification of oils extracted from *Chlorella*, with yields of 98% (Li et al., 2007; Tran et al., 2012). Also, crude oil and microbial oil contain other minor components as phospholipid that has shown an inhibitory effect on immobilized lipases (Li et al., 2013; Meng et al., 2009; Nordblad et al., 2016). Li et al. (2013) studied the effect of phospholipids in free lipases from *Aspergillus niger* during biodiesel production. The authors found out that free lipase exhibited better reuse stability, showing that free lipases could be promising for biodiesel production from crude oils.

Besides the challenges, alternative feedstocks for enzymatic biodiesel production open several benefit pathways. The diversification of the oil crops used for biodiesel production contributes to the best use of regional resources and the maintenance of biodiversity. Furthermore, after pressing the seeds to extract the oil, the by-product (cake) can be used as fertilizer for soil enrichment or as a substrate to produce enzymes and other valuable products by solid-state fermentation (SSF). The use of waste oils can diminish the discharge of oil-rich contaminated effluents, reducing the environmental impact caused by improper disposal of such oils. In both cases, the diversification of the sources enables the development of high-value products from low-value raw materials.

21.3.4 Enzymatic routes for biodiesel production

The biocatalytic route (using lipases or whole cells as catalysts) can overcome some alkaline or acid catalysis difficulties in biodiesel synthesis. Lipases can catalyze biodiesel production by different routes—esterification (from FFA), interesterification/transesterification (from tri-, di-, and monoacylglycerols), or hydroesterification (hydrolysis followed by esterification using complex oily matrixes). These enzymatic routes have specific features, as shown in Table 21.4. Besides, biocatalytic processes present gains related to the use of heterogeneous catalysis (enzyme immobilized on solid support acting in liquid reaction media) and milder conditions, which reduce costs and downstream steps.

Esterification is a single-step process in which fatty acids and alcohols react, producing FAME or FAEE, and water. Alkaline catalysis is not adequate for esterification due to the reaction of alkalis with fatty acids (formation of soaps). Acid catalysis makes the process feasible but at rates lower than those observed with alkaline catalysis. Esterification reactions mediated by biocatalysts can be successfully carried out in solvent-free systems, attaining high conversions at intervals lower than 12 h (Talukder et al., 2010; Trentin et al., 2014; Mulalee et al., 2015; Aguieiras et al., 2017; Collaço et al., 2020).

Lipases are susceptible to enzymatic inhibition by short-chain alcohols, the effect of which may become quite significant for the process depending on the enzyme loading adopted. Using a hydrophobic solvent in the reaction media can overcome this limitation. However, reactions with solvents have less volumetric productivity and are less advantageous, considering the complexity of the downstream operations required. In this sense, esterification in solvent-free systems can be performed through controlled feeding of alcohols in the reaction medium. This strategy has been successfully adopted on a lab scale. Another possible strategy is interesterification, in which fatty acids

TABLE 21.4 General considerations about the different routes for biodiesel production.

	Type of catalyst	
	Chemical homogeneous catalysis	**Biocatalysis (immobilized lipases)**
Routes	Mild (alkaline) or high (acid) temperatures. High conversions (>90%). Low-cost catalysts' acid/alkaline wastewater. Complex downstream and purification processes. Catalysts usually cannot be reused.	Mild/low temperatures No-formation of hazardous wastes High conversions (>90%) High-cost biocatalysts. Simple downstream. Biocatalysts can be reused. Possible inhibition of lipases by short-chain alcohols.
Esterification	Applied only for acid catalysts. FFAs of raw material are esterified to biodiesel requirement of raw material with low water content. Corrosion of the equipment by the acid catalyst.	Raw material with high FFA content. Enzymes have less sensibility to the water in the feedstock.
Transesterification	Applied for alkaline catalysts' raw material with low FFA content (<0.5%) and water content. (<0.1%–0.3%). Short reaction time (90 min). The high-quality raw material can represent 70%–95% of the final cost of biodiesel. Crude glycerin contaminated with salts of catalyst neutralization postreaction.	Long reaction time (8–72 h) and accumulation of glycerol on immobilization support. Food-grade glycerol.
Hydroesterification	Use of two different catalysts. Standardization of the raw material—converting all acylglycerols into FFA. Corrosion of the equipment by the acid catalyst.	No limitations related to raw material quality. Lipase used in hydrolysis, esterification (hybrid processes), or both steps (enzyme/enzyme hydroesterification). Food-grade glycerol.

react with an alternative acyl donor (such as methyl and ethyl formate or acetate, among others) rather than methanol or ethanol. Interesterification can produce a coproduct with a higher added value than glycerin (Kashyap et al., 2019; Subhedar and Gogate, 2016). However, the costs related to these acyl donors should be considered.

Transesterification by alkaline catalysis is the most widely adopted process for biodiesel production. This process requires raw materials (triacylglycerols obtained from oilseed plants) with low FFA levels to avoid saponification, a limitation not observed with biocatalysis. In this case, biocatalysis allows greater versatility of raw materials, including residual sources with high content of FFA. Transesterification generates glycerin as a by-product. The increase of the biodiesel volumes obtained via transesterification has caused an economic devaluation of glycerin due to its overproduction. Thus the use of dimethyl carbonate instead of methanol or ethanol has been investigated by some groups generating glyceryl carbonate as a by-product (Su et al., 2007; Zhang et al., 2010; Gharat and Rathod, 2013; Gu et al., 2015). Considering the specificity and selectivity of lipases and their easy separation from the medium (when used in the immobilized form), the glycerin produced has a higher degree of purity (food-grade glycerol) than that obtained by conventional catalysis, with a high sale value.

Hydroesterification has been highlighted in recent years, considering the use of residual raw materials or an increase in the value of the formed glycerin. Hydroesterification is a two-step process involving hydrolysis of acylglycerols (converting them to FFAs) and subsequent esterification. Glycerin is released during hydrolysis by lipases with a low degree of contamination (Meher et al., 2006; Pourzolfaghar et al., 2016), and the resulting fatty acids are esterified with the assistance of lipases to FAME/FAEE. Hydroesterification gains importance to valorize oily residues, such as WCO, which contains a nonstandardized mixture of tri-, di-, and monoacylglycerols and FFAs. The hydrolysis process standardizes the raw material to fatty acids that can be further reacted with methanol or ethanol. Hydroesterification processes using residual or crude oils with high yields have been described in the literature (Watanabe et al., 2007; Talukder et al., 2010; Aguieiras et al., 2014; Bressani et al., 2015; Tacias-Pascacio et al., 2017; Zheng et al., 2020). The use of biocatalysis can also complement the chemical catalysts in a hybrid process—enzymatic hydrolysis of the raw material and posterior chemical esterification (de Sousa et al., 2010) or nonenzymatic hydrolysis followed by lipase esterification (Soares et al., 2013). In addition, another possible strategy consists of applying the enzymatic esterification as a pretreatment to reduce the acidity of the feedstock with posterior alkaline transesterification (Nordblad et al., 2016).

The use of biocatalysts in biodiesel production processes also presents some constraints, mainly related to the enzymatic activity (which impacts reaction rates), operational stability, and reusability (crucial parameters to the economic feasibility of the process). The literature describes great efforts in the last decades to address these issues, including the use of ionic liquids (Sunitha et al., 2007; Arai et al., 2010; de Diego et al., 2011; Liu et al., 2011; Abrahamsson et al., 2015; Merza et al., 2018), deep eutectic solvents (Abbott et al., 2007; Huang et al., 2014; Zhao et al., 2013; Hayyan et al., 2014; Kleiner et al., 2016), supercritical solvents (Madras et al., 2004; Rathore and Madras, 2007; Varma and Madras, 2007; Lee et al., 2011; Taher et al., 2020), reactions assisted by ultrasound or microwaves (Ji et al., 2006; Santos et al., 2009; Kumar et al., 2011; Veljkovic et al., 2012; da Rós et al., 2014; Feiten et al., 2014; Gupta et al., 2020; Sáez-Bastante et al., 2015; Souza et al., 2016), and alternative reactor configuration, mainly fixed-bed reactors (Andrade et al., 2019; Hama et al., 2007; Hama et al., 2011; Nie et al., 2006) with different degrees of feasibility in a large-scale context.

Lipases can catalyze different chemical reactions, such as hydrolysis, esterification/transesterification, amidation, and epoxidation, so it is theoretically possible to integrate biodiesel production in a biorefinery context. Refining processes of vegetable oils (e.g., soybean or palm) generate FFA-rich residual streams. These fatty acids can be converted into esters using lipases, producing biodiesel (in reactions with short-chain alcohols) and/or biolubricants (in reactions with medium/long-chain alcohols or polyalcohols). Residual biomasses, such as straws and fibers, can have their polymeric sugars deconstructed by specific enzymatic processes; the released sugars can also be reacted with fatty acids in reactions catalyzed by lipases to produce surfactants such as SFAE (sugar-fatty acid ester), widely adopted in the food industry. These residual biomasses can also be subjected to anaerobic fermentation processes, producing biohydrogen and biogas (H_2 and CH_4). The variety of compounds present in biomass can also lead to the exploration of compounds potentially applied as antioxidants and additives, enhancing biodiesel (or biolubricants) properties. These theoretical possibilities can be explored, considering all the advantages of enzymatic processes compared to conventional catalysts, and adding value to agro-industrial residues.

21.3.5 Enzymatic biodiesel: state of the art

Although biodiesel has been produced on a large scale in Brazil for at least two decades, no industrial plants adopt enzymatic technologies or produce biodiesel from residual raw materials. Even in global terms, few companies are currently using the biocatalytic route to obtain biodiesel, despite the notable scientific progress in this field and the abundant related literature available. In 2014 the American Blue Sun Energy Company inaugurated a commercial-scale plant with a production capacity of 135,500 m^3 per year of biodiesel using enzymatic biodiesel technology developed by Novozymes (Novozymes, 2014). Moreover, Novozymes developed specific enzymes for biodiesel production—Eversa Transform—which has been tested and applied by different companies worldwide, such as Spain-based Oleofat Trader S.L. and Aemetis Inc. (US), which is currently producing enzymatic biodiesel in an industrial plant in India, using palm oil derivatives as raw material. SRS Biodiesel (US) describes in their portfolio of technologies enzymatic transesterification using solvents to produce biodiesel in different industrial plants in the US. Enzymocore, an Israel-based company, has been developing enzymes for biodiesel production and producing enzymatic biodiesel on a commercial or pilot-scale since 2013 in South Korea, China, Peru, Germany, and Israel (https://enzymocore.com/our-operations).

Despite some potential technical challenges to be overcome—inhibition by methanol/ethanol, accumulation of glycerol on the surface of the immobilization support, and lack of high operational stability and reuse capacity—the main obstacle to the widespread use of the biocatalysis for biodiesel production (as well as for other chemical commodities) is the cost of lipases. The cost of lipases exceeds the expense of other raw materials (including acid or alkaline catalysts) by several magnitude orders. In an economic assessment of enzymatic biodiesel production carried out by Sotoft et al. (2010), refined rapeseed oil cost was estimated at 0.61 euro/kg, while the enzyme cost was estimated at 762.7 euro/kg. Serrano-Arnaldos et al. (2019) studied the economic aspects of long-chain fatty acids' enzymatic esterification; commercial lipase costs were estimated at 1300 euros/kg for Novozym 435 and 70 euros/kg for Lipozyme TL IM, while industrial-grade fatty acids costs were estimated below 10 euros/kg.

Fjerbaek et al. (2009) calculated the productivities from different literature studies. They compared the results with the productivities attained using an alkaline catalyst (NaOH concentration of 1 wt.% based on the weight of oil), which presents a yield of 100 kg biodiesel per kg of catalyst. A 74% higher productivity can be achieved using Novozym 435, considering lipase reuse (100 times) and using a by-product acid from refining vegetable oils as raw material. The study obtained an estimated cost of 0.14 dollars/kg ester in the enzymatic process against 0.006 dollars/kg ester for NaOH. If the cost of lipase was reduced to 44 dollars/kg, or if the enzyme could be reused for at least 6 years, the use of the biocatalyst could become economically viable. According to Nielsen et al. (2008), the maximum cost of the biocatalyst should be the same as a chemical catalyst (25 dollars/ton biodiesel). Thus an enzyme cost of 12–185 dollars/kg could be feasible, depending on the productivity. Sotoft et al. (2010) found that enzymatic biodiesel plants using *tert*-butanol as solvent was economically unfeasible due to the costs of separation and solvent recovery units. Al-Zuhair et al. (2011) simulated an enzymatic biodiesel plant with a production capacity of 1 ton/h, using waste oil (0.02 dollars/kg) as raw material, Novozym 435 as the catalyst, and *tert*-butanol as solvent. The economic study indicated a payback period of 4 years, with a selling price of biodiesel of 0.86 dollars/kg. The economic feasibility of enzymatic biodiesel produced from cooking oil (0.25 euro/kg), using supercritical carbon dioxide as solvent and lipase Lipozyme TL I.M. (estimated cost of 800 euro/kg in this case), resulted in a biodiesel cost of 1.64 euro/L, which is not economically viable (Lisboa et al., 2014). Estimating enzyme reuse for 100 days, the biodiesel cost could be reduced to 0.75 euro/L.

Among the most common optimization studies, approaches in enzymatic biodiesel syntheses determine the lowest enzyme loading. However, such reduced enzyme loadings increase the reaction time, causing potential negative impacts on lipase operational stability in stirred tank reactors. Another approach deals with the immobilization processes and the materials used, commonly polymers, to obtain more stable and active biocatalysts and, at the same time, reduce the costs of the materials used for immobilization. In this case, promising technologies currently study the use of residual materials of plant origin as immobilization support (Girelli et al., 2020). Solid enzymatic preparations, obtained from SSF, in which lipases are spontaneously adsorbed onto the matrix used for SSF without needing extraction and immobilization, are also studied (Aguieiras et al., 2017; Collaço et al., 2020). Another possibility is the study of highly active and stable lipases using engineered microorganisms. These developments deal almost exclusively with the technical aspects involved. Despite notable technological advances, economic considerations are frequently overlooked or underestimated in this field.

21.3.6 Perspectives for enzymatic biodiesel production

In summary, biodiesel has gradually become more relevant in the Brazilian energy matrix, following the same trend observed worldwide. Although biodiesel is already obtained from renewable sources, it is possible to bring it more sustainably than the current methods using biocatalysis. The advantages of biocatalysis include milder reaction conditions (energy savings in the process and lower emissions of GHGs) and minimal production of harmful by-products (savings in wastewater treatment and reduced environmental risks). From a technological point of view, it is possible to produce biodiesel with high yields in different ways. However, the high biocatalysts' costs still represent the most significant obstacle for biocatalysis to be considered in this context.

The versatility of lipases allows agro-industrial waste for biodiesel production, turning waste into valuable products. In association with other enzymatic technologies, lipases can be inserted in a biorefinery concept from oilseed crops, with biodiesel as one of their main targets, adding value to different residual biomasses generated in their processing. These products may be attractive for different productive chains, such as automotive, food, cosmetics, and cleaning products. Strategies for obtaining more active and stable biocatalysts at lower costs have been studied. It is of utmost importance that the efforts of the research groups around the world be evaluated from an economic perspective, aiming to implement such technologies in a large-scale context.

21.4 Perspectives on biomass processing for composites and chemicals production

As mentioned before, Brazil has favorable conditions to contribute to the local and world decarbonization of the economy, besides its enormous potential to establish a bioeconomy. In addition, the fuel sector is changing, and the chemical industry is still in deficit (ABIQUIM – Química, 2021). The use of biomass for chemicals and composites could be a great opportunity.

Studies on biomass to produce commodity chemicals, predominantly derived from fossil feedstocks such as crude oil and natural gas, received a lot of attention due to the global policy to reduce the dependence on fossil carbon sources and GHGs. Several chemicals can be produced via biochemical routes, which comprise enzymatic or microbial conversions. Interestingly, biochemical routes can be explored for both bio-based fossil-based products. A significant example of enzyme-catalyzed commodity production is the hydration of acrylonitrile into acrylamide (Ashina and Suto, 1993; Straathof, 2014). In this case, the substrate is not a renewable material, which demonstrates the versatility of enzymes. However, the focus here is not only on biochemical routes but also on bio-based products, in other words, the use of renewable substrates, specifically lignocellulosic biomass or oleaginous materials, for biorefinery integration considering the biofuels chain in Brazil. Therefore the main available substrates include the carbohydrates: sucrose, cellulose, hemicellulose, and their C6 and C5 monosaccharides; the lipid fractions: triglyceride and derived glycerol and all by-products and residues generated during the biofuel product processes.

In the biomass-based chemical industry, each available biomass substrate can be explored directly or as a source of platform molecules. Many derived products are possible, simultaneously offering a great opportunity and an enormous challenge. Thomas et al. (2010) summarized the direct applications of nanocellulose (nanofibrillated and nanocrystals), which gained considerable attention as a nanoreinforcement for polymer matrices in various industries (medical and health care, packaging, paper and board, composites, printed and flexible electronics, textiles, filtration, rheology modifiers, 3D printing, aerogels and coating films) (Thomas et al., 2010). Few market players were positioned to produce cellulose nanocellulose at commercial or precommercial scale: CelluForce (Canada), Kruger Inc. (Canada), Fiberlean technologies (UK), American Process (USA), Forest Products Laboratory (FPL) (USA), Paper logic, Borregaard (Norway), Innventia (Sweden), Nippon Paper Industries (Japan), Oji Paper Industries (both Japan), Holmen Paper (Sweden), CTP/FCBA (France), and others (Blanco et al., 2018; Ho and Leo, 2021). As an example, CelluForce has a commercial plant capable of producing 300 tons annually of CelluForce NCC, a multifunctional material described to improve the performance of several materials, such as oil and gas, adhesives, paper, cement, plastics, composites, paints, coatings, personal care, health care, food and beverages, and electronics (NanoCrystalline Cellulose, 2021).

Cellulose and hemicelluloses can also be modified with different functional groups to enhance their reactivity and functionality and be used for medicine, food, packing, and many other industries. Recently, Qaseem et al. (2021) reviewed the products manufactured in the industry by direct modification of hemicellulose by copolymerization, amidation, esterification, oxidation, among other reactions, and their application in the food, packing, health care, pharmaceutical, and textile industries. As a common example, modified xylans mixed with other compounds can be used as emulsifiers, wet-end additives, hydrogels, and dispersants in packaging films (Alekina et al., 2014; Xu et al., 2011; Qaseem et al., 2021).

Cellulose and hemicelluloses can initially be hydrolyzed to C6 and C5-sugars, which are versatile building blocks in the carbohydrate-based chemical industry. These sugars can be converted by chemical or enzymatic reactions. However, the most direct and economical way of converting carbohydrates into commodity chemicals is the fermentation process (di Donato et al., 2019; Straathof and Bampouli, 2017). Recent publications reviewed the pathways of producing several chemicals from cellulose (Artz and Palkovits, 2018; Shedon, 2018) and hemicellulose (Qaseem et al., 2021) derived sugars. For example, hydrolysis of hemicellulose could produce many intermediate products, such as C6 and C5-sugars, which can be modified into many high-value polymers such as 5-hydroxymethylfurfural (HMF), xylitol, furfuryl, succinic acid, ethanol, butanediol, butanol, polyhydroxyalkanoates (PHA), and polylactates (PLA), among others (Qaseem et al., 2021).

Interestingly, Straathof and Bampouli (2017) have ranked 58 commodity chemicals based on the economic potential of bio-based products compared to petrochemical production. These authors used a simple model with ethanol as a base case and considered only a few variables: feedstock prices, number of conversion steps, maximum yields per conversion step, and typical feedstock contribution to the product price. The petrochemical industry is based on C2–C4 chemicals (ethylene, propylene, butenes, syngas/methanol) and C6–C8 aromatic compounds (benzene, toluene, xylene) as building blocks to a variety of products. These authors concluded that the production of carbohydrates is not competitive with platform chemicals such as ethylene, propylene, and benzene–toluene–xylene (BTX), as these chemicals can be produced more cheaply from petrochemical resources. Carbohydrates contain a lot of oxygen which is useless for making these base chemicals.

On the other hand, the fine chemical production from carbohydrates can be competitive when production from petrochemicals requires more steps and more oxidation. Among the best-ranked candidates for carbohydrate-based production are adipic acid, acrylic acid, 1,4-butanediol, and methyl methacrylate. These chemicals are relatively oxidized and require several petrochemical conversion steps starting from the base chemicals, thus they can be produced competitively from carbohydrates if theoretical yields are approached, and the processing is efficient.

In a recent review, Rosales-Calderon and Arantes (2019) emphasized the importance of developing a biorefinery, which produces, for example, ethanol and high-value chemicals from lignocellulose as a promising strategy to promote both carbohydrate-based biofuels and carbohydrate-based chemicals industries. These authors reviewed chemicals and materials with a technology readiness level of at least 8, which have reached a commercial scale and could be shortly or immediately integrated into a cellulosic ethanol process. According to Straathof (2014), there are at least 22 types of industrial or pilot-scale processes of bio-based chemicals production by enzyme or cell catalysis, including the synthesis of different classes of compounds: hydrocarbons, alcohols, carbohydrates, carboxylic acids, esters, amines, and amino acids. Tables 21.5 and 21.6 show some examples of platform molecules, products, applications, and stages of development.

New concepts around bio-based chemicals are emerging. Shanks and Keeling (2017) bring out a new concept of the evolution of platform chemicals to bioprivileged molecules, which are bio-based chemical intermediates that can be efficiently converted to various chemicals, including drop-in replacements for petrochemical, as stated above, as well as novel molecules. Novel chemical entities from petrochemical building blocks have received little attention due to a constrained set of alkene and aromatic molecules, leading to a limited number of possible transformations. So, it would be strategic to study novel chemical compounds derived from biological molecules due to the diversity of biomolecule structures and expand the universe of possible molecules beyond the scope of petrochemicals. The authors mention muconic acid, 5-hydroxymethylfurfural, and triacetic acid lactone as examples of bioprivileged molecules.

Differently from the biomass' carbohydrate fraction, lipid fractions can be converted to various relevant compounds with reduced processing steps, mainly if the biocatalytic route is adopted. The most common uses of lipid fraction are food and biofuels (biodiesel). Besides the direct application of triglycerides in edible oils in the food industry, triglycerides' properties can be enhanced, via interesterification or partial hydrolysis, and designed for high-value applications, such as human milk fat and cocoa butter equivalents (Guerrand, 2017; Coelho and Orlandelli, 2020). Nonedible oils, on the other hand, can be extensively explored for chemical production. The hydrolysis of triglycerides generates FFAs that can be esterified with different alcohols, sugars, amines, amides, and amino acids, to obtain an assort of important surfactants and emollients for cosmetics, pharmaceutics, cleaning products,

TABLE 21.5 Examples of products from C6-sugars, their main applications and stage of development (Ansorge-Schumacher and Thum, 2013; Bozell and Peterson, 2010; Straathof, 2014; Straathof and Bampouli, 2017; Taylor et al., 2015; Rosales-Calderon and Arantes, 2019; Markets, 2020).

Product	Application	Development stage
1,2-Propanediol	Heat-transfer fluid, cosmetics	Industrial
1,3-Propanediol	Polymers and cosmetic industries	Industrial
1,4-Butanediol	Synthetic rubber, polymers, solvents, and chemicals	Pilot
1,2-Butanediol	Precursor for polyester polyol or plasticizer; used to produce adhesive resins or as solvent, coolant, refrigerant, hydraulic fluid, or fine chemical raw material	Industrial
1,4-Diaminobutane	Production of Nylon 4,6	Research
1,5-Diaminopentane	Production of polyamides	Pilot
1-Butanol	Drop-in fuel	Industrial
1-Butyrolactone	Solvent	Research
1-Hexanol	Fragrances, plasticizers	Research
1-propanol	Synthesis of n-propyl acetate	Research
2-Aminoethanol	Anticorrosive, detergent, gas sweetening	Research
2-Butanol	Precursor of amines and esters	Research
2,3-Butanediol	Used to manufacture printing inks, perfumes, fumigants, moistening and softening agents, explosives, plasticizers, foods, and pharmaceuticals	Industrial
3-Hydroxybutyric Acid	Monomer of polyhydroxybutyrate	Research
3-Hydroxypropionic acid	Precursor of acrylic acid	Pilot
6-Aminohexanoic acid	Silicones	Research
Acetaldehyde	Chemical intermediate for various compounds	Industrial

(Continued)

TABLE 21.5 (Continued)

Product	Application	Development stage
Acetic acid	Precursor for vinyl acetate, cellulose-based polymers, acetic anhydride, Acetate salts. Used for foam rubber, cable insulation, wood gluing, emulsifiers, cement coatings, and desalination membranes	Industrial
Acetic anhydride	Esterification agent used in the preparation of modified food starch and acetylation of monoglycerides	Industrial
Acetone	Production of acrylic plastics, signs, lighting fixtures and displays, and bisphenol A (BPA), and as a solvent in multiple products, such as paints, cleaning fluids, and adhesives	Industrial
Adipic acid	Production of nylon 6,6	Research
Ascorbic acid	Production of vitamin C	Industrial
Butanol	Used in the manufacture of adhesives, sealant chemicals, paint additives, coating additives, plasticizer, and cleaning products	Industrial
Butanone	Used in paint and glues	Research
Butyric acid	Cellulose acetate butyrate plastics	Pilot
Citric acid	Food/beverage industry	Research
D-Gluconic acid	Solvent for multivalent cations	Industrial
D-Mannitol	Sweetener	Research
Epoxyethane	Production of ethylene glycol	Research
Erythritol	Sweetener	Industrial
Ethanol	Fuel, solvent, beverages	Industrial
Ethyl acetate	Used in the production of inks, adhesives, car care chemicals, plastics, and as synthetic fruit essence, flavor, and perfume in the food industry	Industrial
Ethylene glycol	Manufacture of antifreeze, hydraulic brake fluids, industrial humectants, printer's inks, and in the synthesis of safety explosives, plasticizers, synthetic fibers	Industrial

(Continued)

TABLE 21.5 (Continued)

Product	Application	Development stage
Farnesene	Solvents, surfactants, resins, adhesives	Industrial
Formaldehyde	Resins	Research
Formic acid	Silaging	Research
Glycolic acid	Cosmetics	Research
Glutamic acid	Use as thickener, humectant, cryoprotectant, drug carrier, biodegradable fibers, highly water absorbable hydrogels, biopolymer flocculants, and animal feed additives	Industrial
Hexanoic acid	Fine chemical	Research
Isobutanol	Mobile phase in thin-layer chromatography	Industrial
Isobutyraldehyde	Chemical intermediate for various compounds	Research
Isoprene	Synthetic rubber and thermoplastic elastomer	Pilot
Isopropanol	Solvent and cleaning fluid	Research
Itaconic acid	Synthesis of resins and chemicals	Industrial
Lactic acid	Cosmetics, leather industry	Industrial
L-Aspartic acid/L-arginine	Chemical commodity	Research
L-Glutamic acid	Nylon 6; use as thickener, humectant, cryoprotectant, drug carrier, biodegradable fibers, highly water absorbable hydrogels, biopolymer flocculants, and animal feed additives	Industrial
L-Lysine	Food industry	Industrial
L-Threonine	Chemical commodity	Industrial
L-Valine	Chemical commodity	Research
Malic acid	Acidulant	Research
Methanol	Antifreeze agent, solvent	Pilot
Methyl chloride	Synthesis of polymers	Research

(*Continued*)

TABLE 21.5 (Continued)

Product	Application	Development stage
Phenol	Polycarbonates and resins	Industrial
Propanal	Chemical intermediate for various compounds	Research
Propionic acid	Food preservation	Pilot
Pyruvic acid	Fine chemical	Research
Sorbitol	Used as sweetener, thickener, humectant, excipient, dispersant in food, cosmetic, and toothpaste, and vitamin C synthesis	Industrial
Styrene	Synthesis of polymers	Research
Succinic acid	Building block for polymers; personal care products and food additives to large-volume applications such as biopolymers, plasticizers, polyurethanes, resins, and coatings	Industrial

lubricants, plasticizers, and food additives (Basri et al., 2013; Abdelmoez and Mustafa, 2014; Sarmah et al., 2017). Moreover, to avoid competition between the destination of edible oils for food and chemicals, residual streams of edible oil refining may be used as a source of FFAs that can be converted directly to biofuels, biolubricants, surfactants, and emollients. These approaches have been successfully explored by many different researchers in recent years, including the use of biocatalytic processes (Chaiyaso et al., 2006; Chong et al., 2007; Collaço et al., 2021; Fernandes et al., 2018; Kapor et al., 2017; Marín-Suarez et al., 2019; Musa et al., 2019; Top, 2010; Zhang et al., 2017).

Biocatalyzed processes for oleochemical production have received increasing attention as cleaner alternatives to produce esters for different applications (Ansorge-Schumacher and Thum, 2013; Bozell and Peterson, 2010; Straathof, 2014; Khan and Rathod, 2015; Sarmah et al., 2017). Due to lipase's specificity, a particular class of hydrolytic enzymes, reactions between triglycerides and fatty acids with alcohols (or other acyl acceptors) can be carried out without forming undesirable by-products. This feature improves the quality of products avoiding laborious purification processes,

TABLE 21.6 Examples of products from C5-sugars, fatty acids, and fatty alcohols and glycerol, their main applications and stage of development (Ansorge-Schumacher and Thum, 2013; Bozell and Peterson, 2010; Straathof, 2014; IEA, 2020a,b; METEX, 2020).

Platform molecule	Products	Application	Development stage
C5	Ethene	Polyethylene, ethylene oxide, vinyl chloride	Research
	Isobutene	Synthetic resins, adhesive resins, vitamins	Research
	Ethylene glycol	Antifreeze agent, production of polyester	Research
	Xylitol	Sweetener	Industrial
	Acetic Acid	Vinyl acetate, acetic anhydride, acetate salts	Research
	Furfural	Used in the recovery of lubricants from cracked crude, in the production of specialist adhesives, and as a flavor compound	Industrial
	Glycolic acid	Cosmetics	Research
	2-aminoethanol	Anticorrosive, detergent, gas sweetening	Research
	1,2-Propanediol	Heat-transfer fluid, cosmetics	Research
Fatty acids	Terminal alkenes	Surface-active agents	Research
	FAME	Biodiesel	Industrial
	1-Hexanol	Fragrances	Research
	Glycerol	Biodiesel	Industrial
Fatty alcohols	Butadiene	Synthetic rubbers and plastics	Research
Glycerol	1-Propanol	Synthesis of n-propyl acetate	Research
	1,3-Propanediol	Precursor for polymers	Industrial
	3-Hydroxypropionic acid	Precursor of acrylic acid	Pilot
	Butyric acid	Animal feed	Industrial
	Epichlorohydrin	Intermediate of epoxy resins and other derivatives	Industrial
	Propylene glycol	Heat-transfer fluid, cosmetics	Industrial

besides mild reaction conditions compared to conventional catalysis (Ansorge-Schumacher and Thum, 2013; Khan and Rathod, 2015; Sarmah et al., 2017). Important advances have been observed using lipases for obtaining flavors and aromas esters, emollients, biosurfactants, biofuels, and biolubricants derived from different biomass' lipid fractions. However, the high cost of enzymes is still a drawback for the broad adoption of biocatalysis in this field.

Biomass lipid fraction is extensively adopted for biodiesel production. The most common biodiesel production process is transesterification—a reaction between triglycerides and methanol/ethanol—the by-product of which is glycerol. Thus glycerol is a raw material available in large quantities. Glycerol can be considered as a mini-sugar and consequently used in different fermentative processes. Currently, research is concentrated on 1,3-propanediol (1,3-PDO) production (Bozell and Peterson, 2010; Straathof, 2014; Chen et al., 2018). However, other several products can be produced from glycerol processing, as by catalysis. Bio-based epichlorohydrin (EPI) is a competitive drop-in for oil-based EPI, obtained from glycerol in a process called epicerol (IEA, 2020a,b); propylene glycol can be produced from glycerol hydrogenolysis, offering a reduction in GHG emissions compared to the oil-based process (Kaur et al., 2020).

Concerning the 1,3-PDO industry, there is a trend to replace the chemical processes traditionally performed by Shell and DuPont, with biotechnological ones. Initially, the bio-based 1,3-PDO producers used to apply genetically modified organism (GMO) capable of converting glucose into glycerol and then metabolize glycerol to 1,3-PDO through the fermentation process (Silva et al., 2014). Recently, METEX-NOOVISTA announced a new plant in France to produce 1,3-PDO and butyric acid using glycerin from rapeseed oil as raw material in a GMO-free process (Metex, 2020).

21.5 Biogas/biomethane production

AD has been considered viable for treating organic waste materials and methane production. This process generates agricultural and environmental benefits, such as renewable energy vector production, organic waste treatment, GHG emission reduction, pathogen reduction through sanitation, and improved fertilization efficiency (Kougias and Angelidaki, 2018; Mucha et al., 2019; Atelge et al., 2020). For these reasons, the AD of organic waste has received significant attention worldwide in recent years. The AD process promotes the degradation of organic material into biogas by microorganisms in the absence of oxygen, implying significant advantages, such as low power demand, inexpensive nutrient requirement, moderate and stable sludge production, and high efficiency of both organic matter removal and biogas generation (Rajagopal et al., 2013; Sawatdeenarunat et al., 2019; Zamri et al., 2021). Several organic materials have been used as feedstock for AD,

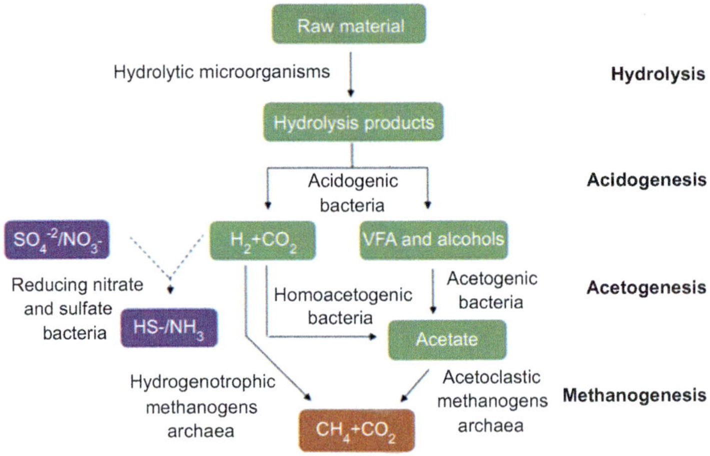

FIGURE 21.5 Anaerobic digestion steps: hydrolysis, acidogenesis, acetogenesis, and methanogenesis. *VFA*, Volatile fatty acid (de Sá et al., 2014).

for instance, lignocellulosic biomass (Ferraro et al., 2020; Weide et al., 2020; Ghimire et al., 2021), municipal solid waste (Chynoweth and Pullammanappallil, 2020; Basinas et al., 2021), animal manure (Weide et al., 2020; Khan and Ahring, 2021), and food processing waste (Andriamanohiarisoamanana et al., 2020; Alrefai et al., 2020), among others. These feedstocks are usually available at small-scale biogas plants, avoiding additional transportation costs and rendering biogas production economically feasible (Naik et al., 2014; Yang et al., 2014).

A consortium of microorganisms generally converts organic materials into biogas through a series of metabolic phases, namely hydrolysis, acidogenesis, acetogenesis, and methanogenesis, in that order (Fig. 21.5). The hydrolysis phase is usually the rate-determining step of the whole AD process. During this phase, undissolved complex organic materials, for example, polysaccharides, proteins, and fats are broken down into simpler organic materials such as sugars, amino acids, and fatty acids. The hydrolytic bacteria involved are usually of the genera *Bacteroides*, *Clostridium*, *Lactobacillus*, *Propionibacterium*, etc. (Deublein and Steinhauser, 2008; Goswami et al., 2016), which secret exoenzymes.

Depending on the composition and structure of the organic matter, the hydrolysis stage may last from a few hours to several days. Hence, many studies have been developed to perform the hydrolysis separately from the other phases of the AD, which allows independent control of the hydrolysis, reduces the time of the process, and increases the availability of the substrates, thereby improving the overall AD process (Menzel et al., 2020). This separation is essential when lignocellulosic biomasses are the primary carbon source. Different biomass pretreatment methods, including physical (ball milling, wet disk, extrusion, microwave irradiation, steam explosion, and liquid hot water); chemical (alkali and acidic); and biological (fungi, bacteria, and enzyme) pretreatments have been reported (Amin et al., 2017; Rajin, 2018; Yu et al., 2019; Ferdeş et al., 2020; Atelge et al., 2020), which may be applied to enhance the first phase of the AD, prior to the subsequent phases.

In the second phase of AD (acidogenesis), acidogenic bacteria (e.g., *Clostridium*, *Paenibacillus*, and *Ruminococcus*) use the simple compounds (monomers of sugars, amino acids, fatty acids) generated from the hydrolysis phase as substrates to produce volatile fatty acids (VFA) (e.g., acetate, propionate, and butyrate), alcohols, H_2, and CO_2. In the third phase (acetogenesis), homoacetogenic bacteria (*Clostridium*) convert H_2 and CO_2 into acetate, while acetogenic bacteria (*Moorella*, *Clostridium*, *Alkaliphilus*, *Caldanaerobacter*, *Thermoanaerobacter*) oxidize VFA and alcohols into acetate. Finally, in the methanogenic phase, methanogenic archaea (acetoclastic and hydrogenotrophic methanogens) convert H_2, CO_2, and acetate into CH_4 and CO_2. The acetoclastic methanogens (Methanosaeta, Methanosarcina) produce methane through acetate decarboxylation, while the hydrogenotrophic methanogens (Methanothermobacter, Methanoculleus, Methanosarcina, Methanoccocus, Methanocaldococcus) produce methane through H_2/CO_2 reduction. If nitrate and sulfate are present in the medium, some species of nitrate- and sulfate-reducing bacteria use H_2 as the electron donor for ammonia and sulfide production, respectively (Chernicharo, 2007; Schnürer and Jarvis, 2009; Soares et al., 2019).

Besides biogas production, the digestate resulting from AD can be applied as biofertilizer in agriculture. However, the quality of the digestate is essential for its acceptance as a replacement for mineral fertilizers in crop production. On a general note, the digestate is rich in several nutrients, principally nitrogen, phosphorous, and potassium, but low in carbon (Mucha et al., 2019; Logan and Visvanathan, 2019; Guilayn et al., 2020). The digestate quality varies with different factors, such as the characteristics of the feedstock material, microbial community, AD operational conditions, and the digestate processing techniques (Mucha et al., 2019; Logan and Visvanathan, 2019; Guilayn et al., 2020). Parameters that allow high-quality digestate are appropriate pH, nutrient and chemical content, and the absence of inorganic impurities and pathological contamination (Mucha et al., 2019).

Different methods such as predigestion, in-vessel cleaning, and postdigestion can be applied to improve the digestate quality without any negative effect on methane yield (Mucha et al., 2019; Logan and Visvanathan, 2019). The predigestion involves various chemical, mechanical, thermal, and enzymatic feedstock pretreatment techniques, while in-vessel cleaning systems are used to remove contaminants from the digester. The postdigestion involves a partial solid–liquid separation of digestate with volume reduction or complete separation of digestate into solid fibers, fertilizer concentrate, and pure water (Mucha et al., 2019; Logan and Visvanathan, 2019). AD digestate could also find further applications in algae cultivation, biopesticides, biosurfactants, biofuels, and biochar production (Mucha et al., 2019; Logan and Visvanathan, 2019; Baştabak and Koçar, 2020; Guilayn et al., 2020).

Several factors can affect the productivity and stability of the anaerobic fermentative system for biogas production, such as temperature, pH, the carbon-to-nitrogen mass ratio (C:N ratio), redox potential, and organic loading rate (OLR), and retention time. Temperature is one of the main factors affecting AD, as it directly influences the CH_4 yield. The growth rate of microorganisms is significantly affected by pH. For example, the growth rate of methanogenic archaea is greatly reduced at pH lower than 6.0 and higher than 8.0 (Mao et al., 2015). The C:N ratio affects the performance of the AD, as the anaerobic bacteria require a balanced nutritional medium for their growth and maintenance of a stable environment. According to the literature, a C:N range of 20–35 was considered to be the optimum condition for AD (Zahan et al., 2018; Kainthola et al., 2020; Ma et al., 2020; Nugraha et al., 2020; Dima et al., 2020).

The redox potential can be used as an indicator of the AD, as the growth of methanogenic archaea requires a low redox potential. This redox potential has been reported to range from −200 to −400 mV (Naik et al., 2014). The stability of the AD is dependent on the OLR and hydraulic retention time (HRT). When the OLR is high, the fermentative system may become unbalanced due to excessive production of volatile acids, leading to inhibition of the process. The same behavior is observed at short HRT. Thus a low OLR and a long HRT provide the best strategy for achieving constant and maximal methane yields (Naik et al., 2014; Mao et al., 2015).

21.5.1 Enzymes applied to improve anaerobic digestion

As discussed earlier, hydrolysis, the first phase of AD, limits the AD process's overall rate. Enzymatic pretreatment is a biological pretreatment method used to improve methane production by speeding up the hydrolysis phase of AD by hydrolytic enzymes such as protease, lipase, α-glucosidase, α-amylase, cellulase, hemicellulases, xylanases, dextranases, and others (Sethupathy et al., 2020).

Hydrolytic enzymes are used to improve the reduction of the particle size of substrates and/or to reduce the crystallinity degree of the biomass. They can be added simultaneously to the substrate and inoculum in the biodigester (one-stage) or initially used to hydrolyze the substrate and then fed into the reactor (two-stage) (Navarro et al., 2020; Xu et al., 2021).

Thus the choice of enzymes and the AD operating conditions depend on the feedstock material's type and composition. For lignocellulosic feedstocks, hydrolysis has been performed using fungal hydrolytic enzyme mixtures (containing cellulase, hemicellulase, xylanase, pectinase, and so on) for a more efficient one-stage or two-stage AD system. For example, Weide et al. (2020) studied the effects of different enzyme mixtures (Cellulase, xylanase, beta-glucanase, endo-pectinase) from *Trichoderma citrinoviride* Bisset on the biomethane yields in single-stage AD of agricultural wastes (silage, straw, and animal manure). They observed an accelerated biomass degradation leading to an increase in methane yield. Specifically, methane yield increased between 8.1% and 21.2% (Weide et al., 2020). However, for the other feedstocks tested, such as grass and maize silage, maize straw, horse and cattle manure, and a mixture of cattle manure and maize silage, the methane yield was only between 0.3% and 6.4% (Weide et al., 2020).

Garcia et al. (2019) evaluated the enzymatic pretreatment of lignocellulosic biomass 24 h before reactor feeding (two-stage condition). They observed an improvement in the degradation performances of the feedstocks, such as sorghum straw and corn cob flour, and an enhancement of over 30% of biogas production (Garcia et al., 2019). Generally, feedstocks with greater degradability can be carried out in one-stage AD without impairing the efficiency of the process, while for recalcitrant feedstocks, the most efficient condition is two-staged (Garcia et al., 2019; Garritano et al., 2017).

Enzymes such as lipases have also been applied in the AD of food waste, sewage sludge, and different animal wastewaters (from dairy and meat processing), which contain large amounts of fats (Meng et al., 2017a,b; Pascale et al., 2019; dos Santos Ferreira et al., 2020; Cheng et al., 2020). These enzymes break down fat to FFAs for better AD performance.

A hydrolytic pool of enzymes can improve the overall AD process, but commercial enzymes represent a high cost to industrial processes. Enzyme extraction from natural sources such as activated sludge serves as an alternative and is cost-effective (Liu and Smith, 2020). Furthermore, purification, concentration, and stabilization/immobilization techniques can also greatly expand the industrial application and increase the economic value of enzymes, developing a potential commercial-scale recovery of hydrolytic enzyme products from waste biomass sources.

21.5.2 Generation and use of biogas/biomethane in Brazil

Biogas produced from the AD process has been presented as an efficient alternative in bioenergy production. Biogas production in the IEA Bioenergy Task 37 member countries is clearly dominated by Germany, with more than 10,000 biogas plants (IEA, 2020a,b). None of the other member countries (Austria, Brazil, Denmark, Finland, France, Norway, Ireland, Korea, Sweden, Switzerland, Netherlands, and the United Kingdom) have more than 1000 biogas plants apiece. The annual biogas production is approximately 120 TWh in Germany, 25 TWh in the United Kingdom, 5 TWh in Brazil, 4 TWh in the Netherlands and France, and less than 3 TWh in the remaining countries (IEA, 2020a,b). In countries such as the United Kingdom, Brazil, and South Korea, biogas produced in landfills is the largest source.

In contrast, landfill gas is only a minor contributor in countries such as Germany, Switzerland, and Denmark. The biogas produced is mainly used for the generation of heat and electricity in most countries, except for Sweden, where approximately half of the produced biogas is used as vehicle fuel (IEA, 2020a,b). Many countries, such as Denmark, Germany, and South Korea showed initiatives and interest in increasing the share of biogas to be used as vehicle fuel in the near future (IEA, 2020a,b).

In Brazil, the potential for biogas production from sugar-energy, animal protein, agricultural production, and sanitation is 57.6, 35.3, 18.1, and 6.1 million Nm^3/day, respectively (ABiogás—Associação Brasileira do Biogás, 2021). The current biogas production is about 3500 Nm^3/day from about 550 biogas plants distributed across the country (ABiogás—Associação Brasileira do Biogás, 2021).

In addition, the energy use of biogas for electrical systems in Brazil remains insufficient. In the first quarter of 2019, the total number of biomass-fueled thermoelectric plants was 559. Only 37 were driven by biogas (approximately 145.6 MW of installed capacity), representing about 0.8% of the total electricity production by biomass (ANEEL—Agência Nacional de Energia Elétrica, 2019). Interestingly, only in 2020, 58 new thermoelectric plants driven by biogas (approximately 49 MW of installed capacity) started operation (ABiogás—Associação Brasileira do Biogás, 2020), making a total of 95 thermoelectric plants caused by biogas, with about 195 MW of installed capacity.

The purification of biogas, through the removal of CO_2, H_2O, H_2S, NH_3, and other impurities, makes it possible to obtain biomethane, which can be used as a substitute for natural gas and as a transportation fuel (Awe et al., 2017; Iglesias et al., 2021). This approach allows the efficient integration of biogas into the energy sector. It is also observed that the industries are intensely interested in this product in Brazil and Africa, Europe, and throughout the Americas (Bley, 2015). Applications of gaseous fuels developed from shale gas in the US have been tendered competitively worldwide.

In Brazil, impacts are already observed on the use of engines relying on 100% natural gas (perfectly replaceable by biomethane), including heavy loads, trucks, and buses. Shale gas has accelerated the arrival of the "Age of Gas" in the world energy matrix and Brazil (Bley, 2015). Currently, the use of biogas as a vehicle fuel is rare. However, the viability of using biomethane as a vehicle fuel has been demonstrated in a project developed by ITAIPU Binacional, the Itaipu Technology Park Foundation, Scania, Haacke Farm, and the International Center on Renewable EnergyBiogas/CIBiogás-ER (IEA, 2015). Currently, there are three compressed biomethane filling stations in Brazil servicing about 110 biomethane-utilizing vehicles (IEA, 2020a,b).

The expansion of the biogas sector in Brazil over the years has been supported by various initiatives, policies, legislations, and research. For example, the 2015 legislation (Resolution No. 8, Jan 30, 2015) developed by the government's National Agency of Petroleum, Natural Gas and Biofuels (ANP) paved the way for the development of biomethane market in Brazil. The legislation was applied to biomethane produced from biodegradable materials originating from agroforestry and organic waste, intended nationwide as a fuel for vehicles, commercial shipping, and residential use. The standard includes obligations regarding quality control to be met by the various economic agents who trade biomethane throughout Brazil (IEA, 2015, 2020a,b). Furthermore, through another legislation (Resolution No. 685, June 29, 2017) by ANP, the rules for approving the quality and specifications of biomethane from landfills, sewage treatment plants for mobility, residential, industrial, and commercial uses were established. In addition, National Biofuel Policy (RenovaBio), recently established by Law No. 13,576/2017 seeks to expand biofuels' production, use, and commercialization, including biomethane, ensuring their competitive participation in the fuel market (IEA, 2020a,b; ANP, 2020a,b).

21.5.3 Hydrogen production

The role of hydrogen (H_2) will be essential to the global decarbonization of energy matrixes worldwide. It is a low-carbon energy vector, and its combustion only generates water as a by-product. The remarkable challenge is the transition from fossil to renewable sources in the hydrogen production chain. According to the Sixth IPCC - Intergovernmental Panel on Climate Change (2021), by 2050, renewable energy must account for 70%–85% of electricity to limit global warming to 1.5°C. Therefore investments in low-carbon technologies need to be a priority. Otherwise, Earth will experience a critical scenario of extreme heat, rising sea levels, and species extinction. Currently, H_2 is produced mostly from fossil sources through thermochemical processes. Therefore sustainable processes for H_2 production, such as biological processes, are of great relevance.

Fermentative microorganisms have been widely explored to produce H_2 along with the feasibility of treating organic waste. Hydrogen is one of the most efficient and cleanest fuels to be inserted in the energy matrix due to its high calorific value of 141.9 MJ/kg (higher than all the biofuels), and the generation of water as the only by-product of its combustion (Reaction 21.1) (Hans and Kumar, 2018; Yang and Wang, 2018).

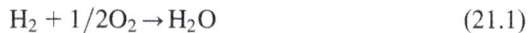

$$H_2 + 1/2O_2 \rightarrow H_2O \tag{21.1}$$

Photo-fermentation and dark fermentation (DF) are the main processes to obtain hydrogen through the biological route. Purple nonsulfur (PNS) bacteria conduct photo-fermentation through organic matter degradation (simple sugars, short organic acids, aromatic compounds, and alcohols) in the presence of light energy (Akhlaghi and Najafpour-Darzi, 2020). PNS bacteria such as *Rhodobacter sphaeroides, Rhodobacter capsulatus, Rhodovulum sulfidophilum, Rhodopseudomonas palustris*, and *Rhodospirillum rubrum* have versatile metabolism dependence on the light intensity, carbon source, and degree of anaerobiosis (Argun and Kargi, 2011). It is important to highlight that simple sugars (hexoses and pentoses) (Reactions 21.2 and 21.3) are less efficient substrates than organic acids such as acetate, lactate, and butyrate (Reactions 21.4–21.6) for H_2 production by PNS bacteria (Akhlaghi and Najafpour-Darzi, 2020).

$$\text{Hexoses:} \quad C_6H_{12}O_6 + 6H_2O \rightarrow 12H_2 + 6CO_2 \tag{21.2}$$

$$\text{Pentoses:} \quad C_5H_{10}O_5 + 5H_2O \rightarrow 10H_2 + 5CO_2 \tag{21.3}$$

$$\text{Acetate:} \quad C_2H_4O_2 + 2H_2O \rightarrow 4H_2 + 2CO_2 \tag{21.4}$$

$$\text{Lactate:} \quad C_3H_6O_3 + 3H_2O \rightarrow 6H_2 + 3CO_2 \tag{21.5}$$

$$\text{Butyrate:} \quad C_4H_8O_2 + 6H_2O \rightarrow 10H_2 + 4CO_2 \tag{21.6}$$

Biological H_2 production through DF allows the use of a variety of organic materials such as lignocellulosic wastes (de Sá et al., 2020; Rena et al., 2020; Wang and Yin, 2018), agro-industrial effluents (Faber and Ferreira-Leitão, 2016; Garritano et al., 2017; Rego et al., 2020), food wastes (Hassan et al., 2020), among others as feedstocks. The use of fermentative microorganisms in converting carbohydrates, lipids, and proteins from biomass into H_2 provides the return of these carbon sources to the productive cycle of biofuel production. In addition, biohydrogen can be produced under room temperature and pressure, resulting in a low energy demand process (Niz et al., 2019; Yang and Wang, 2018).

Dark fermentative microorganisms, such as *Enterobacter* sp., *Bacillus* sp., and *Clostridium* sp., allow a continuous production, fermenting a wide range and more complex substrates without inhibition in the absence of light (Ding; Yang; He, 2016; Elbeshbishy et al., 2017; Yang and Wang, 2018). *Clostridium* is a Gram-positive bacteria, and this genus is the most commonly used microorganism for H_2 production using pure culture. It can metabolize both hexoses (Reactions 21.7 and 21.8) and pentoses (Reactions 21.9 and 21.10), generating acetic and butyric acids and H_2 as products.

Hexoses:

$$C_6H_{12}O_6 + H_2O \rightarrow 2C_2H_4O_2 + 2CO_2 + 4H_2 \tag{21.7}$$

$$C_6H_{12}O_6 + H_2O \rightarrow C_4H_8O_2 + 2CO_2 + 2H_2 \tag{21.8}$$

Pentoses:

$$C_5H_{10}O_5 + 1.67\ H_2O \rightarrow 1.67\ C_2H_4O_2 + 3.33\ H_2 + 1.67\ CO_2 \tag{21.9}$$

$$C_5H_{10}O_5 \rightarrow 0.83\ C_4H_8O_2 + 1.67\ H_2 + 1.67\ CO_2 \tag{21.10}$$

However, a microbial consortium is most often used as inoculum for DF than a pure culture of H_2 producer (Moraes et al., 2019). Mixed culture can stem from a natural source, such as sewage sludge and animal manure, which may not require substrate or equipment sterilization, resulting in an effortless, simple, and cheaper process. Additionally, their microbial diversity can make the system resistant to operational changes (Hiligsmann et al., 2011; Wang et al., 2018).

21.5.4 Sequential production of hydrogen and methane

As a result of DF of organic feedstock applying the mixed culture of microorganisms as an inoculum, H_2 and CO_2 are the major compounds in the gas phase. In contrast, the liquid phase contains acids (mainly acetic and butyric acids), alcohols, and other molecules derived from the metabolism of the raw material. These organic compounds generated in DF can be used as a substrate for CH_4 production by methanogenic archaea. The overall process for obtaining H_2 and CH_4 consists of separate AD into two sequential steps: the acidogenic step leads to H_2 production using a pretreated inoculum, and the methanogenic step culminates in CH_4 production applying the raw mixed culture as an inoculum (Rajendran et al., 2020).

This two-step production is required to avoid using H_2 as a substrate by the hydrogenotrophic methanogenic archaea in the CH_4 production step (Fig. 21.5). Besides, sequential production has the advantage of promoting a stable AD process due to the use of two different optimal pH values, allowing the regulation of organic overloads in the methanogenic step and the increase of overall yield. Two-stage AD has a faster degradation of substrates since it arises in acetogenic steps, resulting in higher productivity than one-stage AD. Another advantage is related to energy recovery. Given the fact that only 10%–20% of the total COD (chemical oxygen demand) in the medium is used for H_2 production, the sequential step promotes the digestion of the most complex compounds for methane production, representing 80%–90% of the COD (Fu et al., 2017; Parthiba et al., 2018; Rajendran et al., 2020).

The setup of the two-stage reactor was initially designed for the high-solid substrates such as agro-industrial residues. Still, it has been used in wastewater treatment systems showing good performance. There is great potential in the treatment of residual biomass by two-stage AD. A crucial factor for the success of this process is influent feed solid concentrations and the evaluation of the characteristics of the substrate. As mentioned, the degradation of the more complex compounds occurs in the fermentative stage (hydrolysis-acidogenesis). Hence, AD in combined reactors provides a retention time of 10–18 days (one-stage, AD is approximately 30 days), and it brings the advantage of reducing the reactor volume by 25%–45%. In addition to less digester volume, the process enables odor control, more significant degradation of VFAs, and provides buffering capacity for the system, which may increase the availability of nutrients for microorganisms (Rajendran et al., 2020). The techno-economic aspects of sequential H_2 and CH_4 production still need to be analyzed to understand market accessibility, success, and profitability.

21.6 Concluding remarks

Sustainability that is linked to the use of renewable materials for industrial production processes is already considered an unavoidable path. Nevertheless, biorefineries, presently, mostly integrate, in the same industrial unit, processes for the production of biofuels, electricity, and heat. It is expected that the experience that has been so far gathered will pave the way for the integrated production of materials and chemicals from renewable raw materials—a new and promising approach to the creation of new and oil-free industries. Although in the current biorefinery context, industrial investments are largely directed toward the production of low-added-value biofuels, such as biodiesel and ethanol, the production of high-added-value products, such as green chemicals and polymeric resins, from residues and by-products derived from biomass processing may improve the material value two- to fourfold. It is also expected that, by extension, six to eight times more jobs could be created as well. The lessening of the environmental impact and the decrease of emissions of GHG by biorefineries will add value to the bio-based industries. It is expected that economic, cultural, and regional dissimilarities will influence the establishment of the renewables-based industry and control its implementation speed. However, it is important to maintain investments in this strategic field to take advantage of the opportunities, which include, with a high degree of protagonism, the area of biocatalysis, both enzymatic and microbial processes, to support a sustainable industry. The latest IPCC report that emphasizes the importance of economic recovery in the post-Covid era also reiterates the importance of local development and the utilization of regional resources, which suits the development of new processes under the biorefinery concept.

Abbreviations

1G	first-generation ethanol
2G	second-generation ethanol
3G	third-generation fuels
AD	anaerobic digestion
ANEEL	National Electric Energy Agency (*Agência Nacional de Energia Elétrica*)
ANP	National Agency of Petroleum, Natural Gas and Biofuels (*Agência Nacional do Petróleo, Gás Natural e Biocombustíveis*)
BGLs	β-Glucosidases
BTX	benzene–toluene–xylene
CBHs	cellobiohydrolases
CBio	Decarbonization Credits
COD	oxygen demand
DAGs	diacylglycerols

DF	dark fermentation
DP	degree of polymerization
EGs	endoglucanases
EPI	epichlorohydrin
EU	European Union
FAEE	fatty acid ethyl esters
FAME	fatty acid methyl esters
FFA	free fatty acids
FPL	Forest Products Laboratory
GH	glycoside hydrolase
GHG	greenhouse gas
GMO	genetically modified organism
HMF	5-Hydroxymethylfurfural
HRT	hydraulic retention time
IPCC	Intergovernmental Panel on Climate Change
LEst	lipolytic esterases or lipases
LPMOs	lytic polysaccharide monooxygenases
MAGs	monoacylglycerols
NLEst	nonlipolytic esterases
OLR	organic loading rate
PHA	polyhydroxyalkanoates
PLA	polylactates
PNPB	National Program of Production and Use of Biodiesel (*Programa Nacional de Produção e Uso de Biodiesel*)
PNS	purple nonsulfur
Pro-álcool	National Fuel Alcohol Program (*Programa Nacional de Álcool Combustível*)
R&D	Research & Development
SSF	solid-state fermentation
SUCRE	Sugarcane Renewable Electricity Project
TAGs	long-chain triacylglycerols
UNEM	Brazilian Corn Ethanol Union (*União Brasileira de Etanol do Milho*)
US	United States
VFA	volatile fatty acid
WCO	waste cooking oil

References

Aarthy, M., Saravanan, P., Gowthaman, M.K., Rose, C., Kamini, N.R., 2014. Enzymatic transesterification for production of biodiesel using yeast lipases: an overview. Chem. Eng. Res. Des. 92, 1591–1601.

Abbott, A.P., Cullis, P.M., Gilbson, M.J., Harris, R.C., Rave, E., 2007. Extraction of glycerol from biodiesel into a eutectic based ionic liquid. Green Chem. 9, 868–872.

Abdelmoez, W., Mustafa, A., 2014. Olechemical industry future through biotechnology. J. Oleo Sci. 63, 545–554.

ABiogás—Associação Brasileira do Biogás, 2020. Novo projetos de 2020 confirmam expansão do setor. Retrospectiva 2020, edicão especial. Available from: <https://abiogas.org.br/abiogasnews-especial-dezembro2020-janeiro2021>.

ABiogás—Associação Brasileira do Biogás, 2021. Biogás no Brasil: conhecendo o mercado no país. Available from: <https://abiogas.org.br/wp-content/uploads/2021/01/Infograficos-Abiogas_D_2021-1.pdf>.

ABIQUIM – Química, 2021. Promovendo Avanços e Protegendo Vidas. Avaiable from: <https://abiquim.org.br/industriaQuimica>.

Abrahamsson, J., Andreasson, E., Hansson, N., Sandström, D., Wennberg, E., Maréchal, M., 2015. A Raman spectroscopic approach to investigate the production of biodiesel from soybean oil using 1-alkyl-3-methylimidazolium ionic liquids with intermediate chain length. Appl. Energy. 154, 763–770.

Aguieiras, E.C.G., Souza, S.L., Langone, M.A.P., 2013. Study of immobilized lipase lipozyme RM in esterification reactions for biodiesel synthesis. Quim. Nova 36, 646–650.

Aguieiras, E.C.G., Cavalcanti-Oliveira, E.D., de Castro, A.M., Langone, M.A.P., Freire, D.M.G., 2014. Biodiesel production from *Acrocomia aculeata* acid oil by (enzyme/enzyme) hydroesterification process: use of vegetable lipase and fermented solid as low-cost biocatalysts. Fuel 135, 315–321.

Aguieiras, E.C.G., de Barros, D.S.N., Sousa, H., Fernandez-Lafuente, R., Freire, D.M.G., 2017. Influence of the raw material on the final properties of biodiesel produced using lipase from *Rhizomucor miehei* grown on babassu cake as biocatalyst of esterification reactions. Renew. Energy. 113, 112–118.

Akhlaghi, N., Najafpour-Darzi, G., 2020. A comprehensive review on biological hydrogen production. Int. J. Hydrog. Energy 45, 22492–22512.

Albersheim, P., Darvill, A., Roberts, K., Sederroff, R., Staehelin, A., 2011. The structural polysaccharides of the cell wall and how they are studied. In: Albersheim, P., Darvill, A., Roberts, K., Sederroff, R., Staehelin, A. (Eds.), Plant Cell Walls. Garland Science, New York, NY, pp. 43–66.

Alberts, G., Ayuso, M., Bauen, A., Boshell, F., Chudziak, C., Gebauer, J.P., et al., 2016. Innovation outlook. Advanced Liquid Biofuels. International Renewable Energy Agency (IRENA), Abu Dhabi, UAE, p. 2016. Available from. Available from: https://www.irena.org//media/Files/IRENA/Agency/Publication/2016/IRENA_Innovation_Outlook_Advanced_Liquid_Biofuels_2016.

Alekina, M., Mikkonen, K.S., Alén, R., Tenkanen, M., Sixta, H., 2014. Carboxymethylation of alkali extracted xylan for preparation of bio-based packaging films. Carbohydr. Polym. 100, 89–96.

Almeida, F.L.C., Travália, B.M., Gonçalves, I.S., Forte, M.B.S., 2021. Biodiesel production by lipase-catalyzed reactions: bibliometric analysis and study of trends. Biofuels, Bioprod. Biorefining .

Alrefai, R., Alrefai, A.M., Benyounis, K.Y., Stokes, J., 2020. An evaluation of the effects of the potato starch on the biogas produced from the anaerobic digestion of potato wastes. Energies 13, 2399.

Al-Zuhair, S., 2008. The effect of crystallinity of cellulose on the rate of reducing sugars production by heterogeneous enzymatic hydrolysis. Bioresour. Technol. 99, 4078–4085.

Al-Zuhair, S., Almenhali, A., Hamad, I., Alshehhi, M., Alsuwaidi, N., Mohamed, S., 2011. Enzymatic production of biodiesel from used/waste vegetable oils: design of a pilot plant. Renew. Energy 36, 2605–2614.

Amin, F.R., Khalid, H., Zhang, H., u Rahman, S., Zhang, R., Liu, G., et al., 2017. Pretreatment methods of lignocellulosic biomass for anaerobic digestion. Amb. Express. 7, 1–12.

Andersen, R.A., 2013. The microalgal cell. Handbook of Microalgal Culture. John Wiley & Sons, Oxford, UK, pp. 1–20.

Andrade, T.A., Martin, M., Errico, M., Christensen, K.V., 2019. Biodiesel production catalyzed by liquid and immobilized enzymes: optimization and economic analysis. Chem. Eng. Res. Des. 141, 1–14.

Andriamanohiarisoamanana, F.J., Yasui, S., Yamashiro, T., Ramanoelina, V., Ihara, I., Umetsu, K., 2020. Anaerobic co-digestion: a sustainable approach to food processing organic waste management. J. Mater. Cycles Waste Manag 22, 1501–1508.

ANEEL—Agência Nacional de Energia Elétrica, 2019. Boletim de Informações Gerenciais. Available from: <https://www.aneel.gov.br/informacoes-gerenciais>.

ANFAVEA – Associação Nacional dos Fabricantes de Veículos Automotoree, 2020. Estatísticas. Available from: <https://www.anfavea.com.br/estatisticas-copiar-3>.

ANP - Agência Nacional do Petróleo, Gás Natural e Biocombustíveis, 2020a. RenovaBio. Available from: <https://www.gov.br/anp/pt-br/assuntos/producao-e-fornecimento-biocombustiveis/renovabio> (accessed 03.14.21).

ANP – Agência Nacional do Petróleo, Gás Natural e Biocombustíveis, 2020b. Anuário estatístico de petróleo, gás natural e biocombustíveis. Available from <http://www.anp.gov.br/publicacoes/anuario-estatistico/5809-anuario-estatistico-2020>. (accessed 03.18.21).

Ansorge-Schumacher, M.B., Thum, O., 2013. Immobilised lipases in the cosmetics industry. Chem. Soc. Ver. 42, 6475–6490.

Arai, S., Nakashima, K., Tanino, T., Ogino, C., Kondo, A., Fukuda, H., 2010. Production of biodiesel fuel from soybean oil catalyzed by fungus whole-cell biocatalysts in ionic liquids. Enzyme Microb. Technol. 46, 51–55.

Aransiola, E.F., Betiku, E., Ikhuomoregbe, D., Ojumu, T.V., 2012. Production of biodiesel from crude neem oil feedstock and its emissions from internal combustion engines. African J. Biotechnol. 11–6178–6186.

Araújo, D.J.C., Machado, A.V., Vilarinho, M.C.L.G., 2019. Availability and suitability of agroindustrial residues as feedstock for cellulose-based materials: Brazil Case Study. Waste Biomass Valorization 10, 2863–2878.

Argun, H., Kargi, F., 2011. Bio-hydrogen production by different operational modes of dark and photo-fermentation: an overview. Int. J. Hydrogen Energy 36, 7443–7459.

Artz, J., Palkovits, 2018. Chemical building blocks from carbohydrates. Curr. Opin. Green Sustain. Chem. 14, 14–18.

Arumugam, A., Thulasidharan, D., Jegadeesan, G.B., 2018. Process optimization of biodiesel production from *Hevea brasiliensis* oil using lipase immobilized on spherical silica aerogel. Renew. Energy. 116, 755–761.

Asada, C., Doi, K., Sasaki, C., Nakamura, Y., 2012. Efficient extraction of starch from microalgae using ultrasonic homogenizer and its conversion into ethanol by simultaneous saccharification and fermentation. Nat. Resour. 03, 175–179.

Ashina, Y., Suto, M., 1993. Development of an enzymatic process for manufacturing acrylamide and recent progress. In: Tanaka, A., Tosa, T., Kobayashi, T. (Eds.), Industrial Application of Immobilized Biocatalysts. Marcel Dekker, New York, NY, pp. 91–107.

Atelge, M.R., Atabani, A.E., Banu, J.R., Krisa, D., Kaya, M., Eskicioglu, C., et al., 2020. A critical review of pretreatment technologies to enhance anaerobic digestion and energy recovery. Fuel 270, 117494.

Awe, O.W., Zhao, Y., Nzihou, A., Minh, D.P., Lyczko, N., 2017. A review of biogas utilization, purification and upgrading technologies. Waste Biomass Valorization 8–267–283.

Balat, M., 2011. Potential alternatives to edible oils for biodiesel production - a review of current work. Energ. Convers Manag 52, 1479–1492.

Ball, S., Colleoni, C., Cenci, U., Raj, J.N., Tirtiaux, C., 2011. The evolution of glycogen and starch metabolism in eukaryotes gives molecular clues to understand the establishment of plastid endosymbiosis. J. Exp. Bot. 62, 1775–1801.

Barros. S., 2020. Biofuels Annual. USDA – United States Department of Agriculture – Foreign Agricultural Service. Available from: <https://apps.fas.usda.gov/newgainapi/api/Report/DownloadReportByFileName?fileName = Biofuels%20Annual_Sao%20Paulo%20ATO_Brazil> (accessed 03.08.20).

Barros, M., Fleuri, L.F., MacEdo, G.A., 2010. Seed lipases: sources, applications and properties - a review. Brazilian J. Chem. Eng 27, 15–29.

Basinas, P., Rusín, J., Chamrádová, K., 2021. Assessment of high-solid mesophilic and thermophilic anaerobic digestion of mechanically-separated municipal solid waste. Environ. Res. 192, 110202.

Basri, M., Rahman, R.N.Z., Salleh, A.B., 2013. Specialty oleochemicals from palm oil via enzymatic syntheses. J. Oil Palm Res 25, 22–35.

Baştabak, B., Koçar, G., 2020. A review of the biogas digestate in agricultural framework. J. Mater. Cycles Waste Manag 22, 1318–1327.

Becker, E.W., 2007. Micro algae as a source of protein. Biotechnol. Adv. 25, 207–210.

Bicudo, C.E.D.E.M., Menezes, M., 2010. Introdução: as algas do Brasil. In: Forzza, R.C. (Ed.), Catálogo de Plantas e Fungos Do Brasil - Vol. 1. I Andrea Jakobsson Estúdio Editorial, Instituto de Pesquisa Jardim Botânico, Rio de Janeiro, pp. 49–60.

Bini, A., 2020. Produção de etanol de milho se firma no país e deve dobrar este ano. Available from.

Binod, P., Janu, K.U., Sindu, R., Pandey, A., 2011. Hydrolysis of lignocellulosic biomass for bioethanol production. Biofuels 229–250.

Blanco, A., Monte, M.C., Campano, C., Balea, A., Merayo, N., Negro, C., 2018. Nanocellulose for industrial use: cellulose nanofibers (CNF), cellulose nanocrystals (CNC), and bacterial cellulose (BC). Nanocellulose for Industrial Use. Handbook of Nanomaterials for Industrial Applications 74–126.

Bley Jr., C., 2015. Biogás: A Energia Invisível, second (ed.) University of São Paulo, São Paulo.

Bomgardner, 2019. POET-DSM to pause cellulosic ethanol production. <https://cen.acs.org/business/biobased-chemicals/POET-DSM-pause-cellulosic-ethanol/97/i46>.

Bozell, J.J., Peterson, G.R., 2010. Technology development for the production of biobased products from biorefinery carbohydrates – the US Department of Energy's "Top 10" revisited. Green Chem. 12, 539–554.

Brányiková, I., Maršálková, B., Doucha, J., Brányik, T., Bišová, K., Zachleder, V., et al., 2011. Microalgae-novel highly efficient starch producers. Biotechnol. Bioeng. 108, 766–776.

Brasil, 2020a 9 (a). Ministério de Minas e Energia. Balanço Energético Nacional. Available from: <https://www.epe.gov.br/sites-pt/publicacoes-dados-abertos/publicacoes/PublicacoesArquivos/publicacao-479/topico-528/BEN2020_sp.pdf>.

Brasil, 2020b (b). Ministério de Minas e Energia. Resenha Energética Brasileira. Oferta e Demanda de Energia. Instalações Energéticas. Energia no Mundo. Available from: <http://antigo.mme.gov.br/documents/36208/948169/Resenha + Energ%C3%A9tica + Brasileira + - + edi%C3%A7%C3%A3o + 2020/ab9143cc-b702-3700-d83a-65e76dc87a9e>.

Bressani, A.P.P., Garcia, K.C.A., Hirata, D.B., Mendes, A.A., 2015. Production of alkyl esters from macaw palm oil by a sequential hydrolysis/esterification process using heterogeneous biocatalysts: optimization by response surface methodology. Bioprocess Biosyst. Eng. 38, 287–297.

Brindhadevi, K., Mathimani, T., Rene, E.R., Shanmugam, S., Chi, N.T.L., Pugazhendhi, A., 2021. Impact of cultivation conditions on the biomass and lipid in microalgae with an emphasis on biodiesel. Fuel 284, 119058.

Budžaki, S., Miljić, G., Tišma, M., Sundaram, S., Hessel, V., 2017. Is there a future for enzymatic biodiesel industrial production in microreactors? Appl. Energy 201, 124–134.

Bujan, J.A.C., 2020. Sugarcane Renewable Energy (SUCRE) Terminal Evaluation Report. Available from: <https://erc.undp.org/evaluation/documents/download/18506>.

Buléon, A., Colonna, P., Planchot, V., Ball, S., 1998. Starch granules: structure and biosynthesis. Biol. Macromol. 23, 85–112.

Cannella, D., Jorgensen, H., 2014. Do new cellulolytic enzyme preparation affect the industrial strategies for high solids lignocellulosic ethanol production. Biotechnol. Bioeng. 111, 59–68.

Cao, Y., Tan, H., 2005. Study on crystal structures of enzyme-hydrolyzed cellulosic materials by X-ray diffraction. Enzyme Microb. Technol. 36, 314–317.

Carpio, L.G.T., de Souza, F.S., 2017. Optimal allocation of sugarcane bagasse for producing bioelectricity and second generation ethanol in Brazil: scenarios of cost reductions. Renew. Energy 111, 771–780.

Carvalho, J.L.N., Nogueirol, R.C., Menandro, L.M.S., Bordonal, R.O., Borges, C.D., Canterella, H., et al., 2017. Agronomic and environmental implications of sugarcane straw removal: a major review. GCB Bioenergy 9, 1181–1195.

Chandel, A.K., Chandrasekhar, G., Silva, M.B., Silva, S.S., 2012. The realm of cellulases in biorefinery development. Crit. Rev. Biotechnol. 32, 187–202.

Chaiyaso, T., H-Kittikun, A., Zimmermann, 2006. Biocatalytic acylation of carbohydrates with fatty acids from palm fatty acid distillates. J. Ind. Microbio. Biotechnol 33, 338–342.

Chandel, A.K., Albarelli, J.Q., Santos, D.T., Chundawat, S.P., Puri, M., Meireles, M.A.A., 2019. Comparative analysis of key technologies for cellulosic ethanol production from Brazilian sugarcane bagasse at a commercial scale. Biofuels, Bioprod. Bioref. 13, 994–1014.

Cheah, W.Y., Show, P.L., Chang, J.-S., Ling, T.C., Juan, J.C., 2014. Biosequestration of atmospheric CO_2 and flue gas-containing CO_2 by microalgae. Bioresour. Technol. 184, 190–201.

Chen, P., Min, M., Chen, Y., Wang, L., Li, Y., Chen, Q., et al., 2009. Review of the biological and engineering aspects of algae to fuels approach. Int. J. Agric. Biol. Eng. 2, 1–30.

Chen, J., Tyagi, R.D., Li, J., Zhang, X., Drogui, P., Sun, F., 2018. Economic assessment of biodiesel production from wastewater sludge. Bioresour. Technol. 253, 41–48.

Chen, H., Ding, M., Li, Y., Xu, H., Li, Y., Wei, Z., 2020. Feedstocks, environmental effects and developmentsuggestions for biodiesel in China. J. Traffic Transp. Eng. 7, 791–807 (Engl. Ed.).

Cheng, Y.S., Zheng, Y., Labavitch, J.M., VanderGheynst, J.S., 2013. Virus infection of *Chlorella variabilis* and enzymatic saccharification of algal biomass for bioethanol production. Bioresour. Technol. 137, 326–331.

Cheng, D., Liu, Y., Ngo, H.H., Guo, W., Chang, S.W., Nguyen, D.D., et al., 2020. A review on application of enzymatic bioprocesses in animal wastewater and manure treatment. Bioresour. Technol. 313, 123683.

Chernicharo, C.A.D.L., 2007. Reatores Anaeróbios, Princípios do tratamento biológico de águas residuárias, vol 5. Universidade Federal do Minas Gerais – UFMG, Belo Horizonte, 239–239.

Chesterfield, D.M., Rogers, P.L., Al-Zaini, E.O., Adesina, A.A., 2012. Production of biodiesel via ethanolysis of waste cooking oil using immobilised lipase. Chem. Eng. J. 207, 701–710.

Chisti, Y., 2007. Biodiesel from microalgae. Biotechnol. Adv. 25, 294–306.

Chng, L.M., Lee, K.T., Chan, D.J.C., 2017. Synergistic effect of pretreatment and fermentation process on carbohydrate-rich *Scenedesmus dimorphus* for bioethanol production. Energy Convers. Manag. 141, 410–419.

Choi, S.P., Nguyen, M.T., Sim, S.J., 2010. Enzymatic pretreatment of *Chlamydomonas reinhardtii* biomass for ethanol production. Bioresour. Technol. 101, 5330–5336.

Chong, F.C., Tey, B.T., Dom, Z.M., Cheong, K.H., Satiawihardja, B., Ibrahim, M.N., et al., 2007. Rice bran lipase catalyzed esterification of palm oil fatty acid disllate and glycerol in organic solvent. Biotechnol. Bioprecess Eng. 12, 250–256.

Chongkhong, S., Tongurai, C., Chetpattananondh, P., 2009. Continuous esterification for biodiesel production from palm fatty acid distillate using economical process. Renew. Energy 34, 1059–1063.

Christensen, J., Olhoff, A., 2019. Lessons from a decade of emissions gap assessments. United Nations Environment Programme, Nairobi. Available from: <https://www.unenvironment.orgesources/emissions-gap-report-10-year-summary, 2019>.

Chynoweth, D.P., Pullammanappallil, P., 2020. Anaerobic digestion of municipal solid wastes. Microbiology of solid waste 71–113.

Coelho, A.L.S., Orlandelli, R.C., 2020. Immobilized microbial lipases in the food industry: a systematic literature review. Crit. Rev. Food Sci. Nutr. 61, 1689–1703.

Collaço, A.C.A., Aguieiras, E.C.G., Santos, J.G., de Oliveira, R.A., de Castro, R.P.V., Freire, D. M.G., 2020. Experimental study and preliminary economic evaluation of enzymatic biodiesel production by an integrated process using co-products from palm (*Elaeis guineensis* Jaquim) industry. Ind. Crops Prod. 157, 112904.

Collaço, A.C.C., Aguieiras, E.C.G., Cavalcanti, E.L.D.C., Freire, D.M.G., 2021. Development of an integrated process involving palm industry co-products for monoglyceride/diglyceride emulsifier synthesis: use of palm cake and for lipase production an palm fatty-acid distillate as raw material. LWT 135, 110039.

CONAB – Campanha Nacional de Abastecimento, 2020, Acompanhamento da Safra Brasileira Cana-de-açúcar. V.7 - SAFRA 2020/21 N.3 - Terceiro levantamento. Available from: <https://www.conab.gov.br/component/k2/item/download/34870_e1c52a336b53ca05c29824831da3c9e9>.

da Rós, P.C.M., Silva, W.C., Grabauskas, D., Perez, V.H., de Castro, H.F., 2014. Biodiesel from babassu oil: characterization of the product obtained by enzymatic route accelerated by microwave irradiation. Ind. Crops Prod. 52, 313–320.

da Silva, A.S., Espinheira, R.P., Teixeira, R.S.S., de Souza, M.F., Ferreira-Leitão, V., Bon, E.P. S., 2020. Constraints and advances in high-solids enzymatic hydrolysis of lignocellulosic biomass: a critical review. Biotechnol. Biofuels 58.

Dawood, F., Anda, M., Shafiullah, G.M., 2020. Hydrogen production for energy: an overview. Int. J. Hydrog. Energy 45, 3847–3869.

de Diego, T., Manjon, A., Lozano, P., Vaultier, M., Iborra, J.L., 2011. An efficient activity ionic liquid-enzyme system for biodiesel production. Green Chem. 13, 444e451.

de Farias, S.C.E., Meneghello, D., Bertucco, A., 2018. A systematic study regarding hydrolysis and ethanol fermentation from microalgal biomass. Biocatal. Agric. Biotechnol. 14, 172–182.

de Sá, L.R.V., Cammarota, M.C., Ferreira-Leitão, V.S., 2014. Produção de hidrogênio via fermentação anaeróbia – aspectos gerais e possibilidade de utilização de resíduos agroindustriais brasileiros. Quim. Nova 37, 857–867.

de Sá, L.R.V., Faber, M.O., da Silva, A.S., Cammarota, M.C., Ferreira-Leitão, V.S., 2020. Biohydrogen production using xylose or xylooligosaccharides derived from sugarcane bagasse obtained by hydrothermal and acid pretreatments. Renew. Energy 146, 2408–2415.

de Sousa, J.S., Cavalcanti-Oliveira, Ed'A., Aranda, D.A.G., Freire, D.M.G., 2010. Application of lipase from the physic nut (*Jatropha curcas* L.) to a new hybrid (enzyme/chemical) hydroesterification process for biodiesel production. J. Mol. Catal. B Enzym 65, 133–137.

Demirbas, A., 2007. Importance of biodiesel as transportation fuel. Energy Policy 35, 4661–4670.

Demuez, M., Mahdy, A., Tomás-Pejó, E., González-Fernández, C., Ballesteros, M., 2015. Enzymatic cell disruption of microalgae biomass in biorefinery processes. Biotechnol. Bioeng. 112, 1955–1966.

Deublein, D., Steinhauser, A., 2008. Biogas from waste and renewable resources. WILEY-VCH Verlag GmbH & Co. KGaA, Weinheim.

di Donato, P., Finori, I., Poli, A., Nicolaus, B., Lama, L., 2019. The production of second generation bioethanol: the biotechnology potential of thermophilic bacteria. J. Clean. Prod 233, 1410–1417.

Dias, M.O., Junqueira, T.L., Cavalett, O., Cunha, M.P., Jesus, C.D., Rossell, C.E., et al., 2012. Integrated vs stand-alone second generation ethanol production from sugarcane bagasse and trash. Bioresour. Technol. 103, 152–161.

Dima, A.D., Pârvulescu, O.C., Mateescu, C., Dobre, T., 2020. Optimization of substrate composition in anaerobic co-digestion of agricultural waste using central composite design. Biomass Bioenergy 138, 105602.

Ding, C., Yang, K.L., He, J., 2016. Biological and fermentative production of hydrogen. Handbook of Biofuels Production: Processes and Technologies: Second Edition Chapter 11, 303–333.

Dixit, P., Basu, B., Puri, M., Tuli, D.K., Mathur, A.S., Barrow, C.J., 2019. A screening approach for assessing lytic polysaccharide monooxygenase activity in fungal strains. Biotechnol. Biofuels 12, 1–16.

Djahedi, C., Berglund, L.A., Wohlert, J., 2015. Molecular deformation mechanisms in cellulose allomorphs and the role of hydrogen bonds. Carbohydr. Polym. 130, 175–182.

dos Santos Ferreira, J., de Oliveira, D., Maldonado, R.R., Kamimura, E.S., Furigo, A., 2020. Enzymatic pretreatment and anaerobic co-digestion as a new technology to high-methane production. Appl. Microbiol. Biotechnol. 104, 4235–4246.

Duarte, S.H., del Peso Hernandez, G.L., Canet, A., Benaiges, M.D., Maugeri, F., Valero, F., 2015. Enzymatic biodiesel synthesis from yeast oil using immobilized recombinant *Rhizopus oryzae* lipase. Bioresour. Technol. 183, 175–180.

Dufrense, A., 2012. Nanocellulose - potential reinforcement in composites. In: John, M.J., Sabu, T. (Eds.), Natural Polymers, Vol. 2: Natural Polymer Nanocomposites. RSC Publishing, London, pp. 1–33.

Dulie, N.W., Woldeyes, B., Demsash, H.D., Jabasingh, A.S., 2021. An insight into the valorization of hemicellulose fraction of biomass into furfural: catalytic conversion and product separation. Waste Biomass Valorization 12, 531–552.

Elbeshbishy, E., Dhar, B.R., Nakhla, G., Hyung-Sool, L., 2017. A critical review on inhibition of dark biohydrogen fermentation. Renew. Sustain. Energy Rev. 79, 656–668.

Environmental Protection Agency, 2020. Renewable Fuel Standard Program: Standards for 2020 and Biomass Based Diesel Volume for 2021 and Other Changes. Available from: <https://www.govinfo.gov/content/pkg/FR-2020-02-06/pdf/2020-00431.pdf>.

Faber, M.O., Ferreira-Leitão, V.S., 2016. Optimization of biohydrogen yield produced by bacterial consortia using residual glycerin from biodiesel production. Bioresour. Technol. 2019, 365–370.

Falkowski, P.G., Raven, J.A., 2007. Aquatic photosynthesis, 2nd (ed.) Princeton University Press, New Jearsey.

Feiten, M.C., Dalla Rosa, C., Treichel, H., Furigo, A., Zenevicz, M.C., de Oliveira, D., et al., 2014. Batch and fed-batch enzymatic hydrolysis of soybean oil under ultrasound irradiation. Biocatal. Agric. Biotechnol. 3, 83–85.

Fengel, D., Wegener, G., 1984. Wood. Chemistry, Ultrastructure, Reactions. Walter de Gruyter, Berlin.

Ferdeş, M., Dincă, M.N., Moiceanu, G., Zăbavă, B.Ş., Paraschiv, G., 2020. Microorganisms and enzymes used in the biological pretreatment of the substrate to enhance biogas production: a review. Sustainability 12, 7205.

Fernandes, K.V., Papadaki, A., da Silva, J.A.C., Fernandes-Lafuente, R., Koutinas, A.A., Freire, D.M.G., 2018. Enzymatic esterification of palm fatty-acid distillate for the production of polyol esters with biolubricant properties. Ind. Crops. Prod. 116, 90–96.

Ferraro, A., Massini, G., Miritana, V.M., Rosa, S., Signorini, A., Fabbricino, M., 2020. A novel enrichment approach for anaerobic digestion of lignocellulosic biomass: Process performance enhancement through an inoculum habitat selection. Bioresour. Technol. 313, 123703.

Fjerbaek, L., Christensen, K.V., Norddahl, B., 2009. A review of the current state of biodiesel production using enzymatic transesterification. Biotechnol. Bioeng. 102, 1298–1315.

Flach, B., Lieberz, S., Bolla, S., 2020. Biofuels Annual. United States Department of Agriculture: Foreign Agriculture Service. European Union.

Franco, H.C.J., Castro, S.G.Q., Sanches, G.M., Kolln, O.T., Bordonal, R.O., Borges, B.M.M.N., et al., 2018. Alternatives to increase the sustainability of sugarcane production in Brazil under high intensive mechanization. In: Singh, P., Tiwari, A.K. (Eds.), Sustainable Sugarcane Production. Waretown. Apple Academic Press, NJ – USA, pp. 350–383.

Fu, S.F., Xu, X.H., Dai, M., Yuan, X.Z., Guo, R.B., 2017. Hydrogen and methane production from vinasse using two-stage anaerobic digestion. Process. Saf. Environ. 107, 81–86.

Garcia, N.H., Benedetti, M., Bolzonella, D., 2019. Effects of enzymes addition on biogas production from anaerobic digestion of agricultural biomasses. Waste Biomass Valorization 10, 3711–3722.

García-Aparicio, M.P., Ballesteros, M., Manzanares, P., Ballesteros, I., González, A., José, N. M., 2007. Xylanase contribution to the efficiency of cellulose enzymatic hydrolysis of barley straw. Appl. Biochem. Biotechnol. 137, 353–365.

Garritano, A.N., de Sá, L.R.V., Aguieiras, E.C.G., Freire, D.M.G., Ferreira-Leitão, V.S., 2017. Efficient biohydrogen production via dark fermentation from hydrolized palm oil mill effluent by non-commercial enzyme preparation. Int. J. Hydrog. Energy. 42, 29166–29174.

Ge, S., Champagne, P., 2017. Cultivation of the marine macroalgae *Chaetomorpha linum* in municipal wastewater for nutrient recovery and biomass production. Environ. Sci. Technol. 51, 3558–3566.

Gellerstedt, G., Henriksson, G., 2008. Lignins: major sources, structure and properties. In: Belgacem, M.N., Gandini, A. (Eds.), Monomers, Polymers and Composites from Renewable Resources. Elsevier, Amsterdam, pp. 201–224.

Gerken, H.G., Donohoe, B., Knoshaug, E.P., 2013. Enzymatic cell wall degradation of *Chlorella vulgaris* and other microalgae for biofuels production. Planta 237, 239–253.

Gharat, N., Rathod, V.K., 2013. Ultrasound assisted enzyme catalyzed transesterification of waste cooking oil with dimethyl carbonate. Ultrason. Sonochem. 20, 900–905.

Ghimire, N., Bakke, R., Bergland, W.H., 2021. Mesophilic anaerobic digestion of hydrothermally pretreated lignocellulosic biomass (Norway Spruce (*Picea abies*)). Processes 9, 190.

Girelli, A.M., Astolfi, M.L., Scuto, F.R., 2020. Agro-industrial wastes as potential carriers for enzyme immobilization: a review. Chemosphere 244, 125368.

Gírio, F.M., Fonseca, C., Carvalheiro, F., Duarte, L.C., Marques, S., Bogel-Lukasik, R., 2010. Hemicelluloses for fuel ethanol: a review. Bioresour. Technol. 101, 4775–4800.

Gonzaga, L.C., Zotelli, L.C., de Castro, S.G.Q., de Oliveira, B.G., Bordonal, R.O., Cantarella, H., et al., 2019. Implications of sugarcane straw removal for soil greenhouse gas emissions in São Paulo State, Brazil. Bioenerg. Res. 12, 843–857.

Goswami, R., Chattopadhyay, P., Shome, A., Banerjee, S.N., Chakraborty, A.K., Mathew, A.K., et al., 2016. An overview of physico-chemical mechanisms of biogas production by microbial communities: a step towards sustainable waste management. 3 Biotech 6, 72.

Grabber, J.H., 2005. How do lignin composition, structure, and cross-linking affect degradability? A review of cell wall model studies. Crop Sci. 45, 820–831.

GranBio, 2020. GranBio e NextChem assinam parceria para desenvolver mercado de etanol celulósico. Available from: <http://www.granbio.com.br/press-releases/granbio-e-nextchem-assinam-parceria-para-desenvolver-mercado-de-etanol-celulosico/>.

Greco-Duarte, J., Cavalcanti-Oliveira, E.D., da Silva, J.A.C., Fernandez-Lafuente, R., Freire, D. M.G., 2017. Two-step enzymatic production of environmentally friendly biolubricants using castor oil: enzyme selection and product characterization. Fuel 202, 196–205.

Green Car Congress, 2014. Iogen and Raízen begin production of cellulosic ethanol in Brazil. Available from: <https://www.greencarcongress.com/2014/12/20141218-iogen.html>. (accessed 01.04.2022).

Gu, J., Xin, Z., Meng, X., Sun, S., Qiao, Q., Deng, H., 2015. Studies on biodiesel production from DDGS-extracted corn oil at the catalysis of Novozym 435/superabsorbent polymer. Fuel 146, 33e40.

Guerrand, D., 2017. Lipases industrial applications: focus on food and agroindustriesLipids of the Future OCL 24, D403.

Guilayn, F., Rouez, M., Crest, M., Patureau, D., Jimenez, J., 2020. Valorization of digestates from urban or centralized biogas plants: a critical review. Rev. Environ. Sci. Biotechnol. 19, 419–462.

Günerken, E., D'Hondt, E., Eppink, M.H.M., Garcia-Gonzalez, L., Elst, K., Wijffels, R.H., 2015. Cell disruption for microalgae biorefineries. Biotechnol. Adv. 33, 243–260.

Gupta, S., Mazumder, P.B., Scott, D., Ashokkumar, M., 2020. Ultrasound-assisted production of biodiesel using engineered methanol tolerant *Proteus vulgaris* lipase immobilized on functionalized polysulfone beads. Ultrason. Sonochem. 68, 105211.

Hama, S., Yamaji, H., Fukumizo, T., Numata, T., Tamalampudi, S., Kondo, A., et al., 2007. Biodiesel fuel production in a packed-bed reactor using lipase-producing *Rhizopus oryzae* cells immobilized within biomass support particles. Biochem. Eng. J. 34, 273–278.

Hama, S., Tamalampudi, S., Yoshida, A., Tamadani, N., Kuratani, N., Noda, H., et al., 2011. Enzymatic packed-bed reactor integrated with glycerol-separating system for solvent-free production of biodiesel fuel. Biochem. Eng. J. 55, 66–71.

Hans, M., Kumar, S., 2018. Biohythane production in two-stage anaerobic digestion system. Int. J. Hydrog. Energy. 44, 17363–17380.

Harris, P.V., Welner, D., McFarland, K.C., Re, E., Poulsen, J.N., Brown, K., et al., 2010. Stimulation of lignocellulosic biomass hydrolysis by proteins of glycoside hydrolase family 61: structure and function of a large, enigmatic family. Biochem 49, 3305–3316.

Harris, P.V., Xu, F., Kreel, N.E., Kang, C., Fukuyama, S., 2014. New enzyme insights drive advances in commercial ethanol production. Curr. Opin. Chem. Biol. 19, 162–170.

Hassan, G.K., Hemdan, B.A., El-Gohary, F.A., 2020. Utilization of food waste for bio-hydrogen and bio-methane production: influences of temperature, OLR, and in situ aeration. J. Mater. Cycles. Waste. 22, 1218–1226.

Hayyan, A., Hashim, M.A., Hayyan, M., Mjalli, F.S., AlNashef, I.M., 2014. A new processing route for cleaner production of biodiesel fuel using a choline chloride based deep eutectic solvent. J. Clean. Prod 65, 246–251.

Hemsworth, G.R., Davies, G.J., Walton, P.H., 2013. Recent insights into copper-containing lytic polysaccharide mono-oxygenases. Curr. Opin. Struct. Biol. 23, 660–668.

Hiligsmann, S., Masset, J., Hamilton, C., Beckers, L., Thonart, P., 2011. Comparative study of biological hydrogen production by pure strains and consortia of facultative and strict anaerobic bacteria. Bioresour. Technol. 102, 3810–3818.

Ho, N.A.D., Leo, C.P., 2021. A review on the emerging applications of cellulose, cellulose derivatives and nanocellulose in carbon capture. Environ. Res. 197, 111100.

Ho, S.H., Li, P.J., Liu, C.C., Chang, J.S., 2013. Bioprocess development on microalgae-based CO2 fixation and bioethanol production using *Scenedesmus obliquus* CNW-N. Bioresour. Technol. 145, 142–149.

Höhne, N., Fransen, T., Hans, F., Bhardwaj, A., Blanco, G., Elzen, M., et al., 2019. Bridging the gap: enhancing mitigation ambition and action at g20 level and globally. An advance chapter of The Emissions Gap Report 2019. United Nations Environment Programme. Nairobi.

Holm-Nielsen, J.B., Seadi, T.A., Oleskowicz-Popielc, P., 2009. The future of anaerobic digestion and biogas utilization. Bioresour. Technol. 100, 5478–5484.

Horn, S.J., Vaaje-Kolstad, G., Westereng, B., Eijsink, V.G.H., 2012. Novel enzymes for the degradation of cellulose. Biotechnol. Biofuels 5, 45.

Huang, Z.L., Yang, T.X., Huang, J.Z., Yang, Z., 2014. Enzymatic production of biodiesel from *Millettia pinnata* seed oil in ionic liquids. Bioenerg. Res. 7, 1519–1528.

IEA—International Energy Agency, 2015. IEA Bioenergy Task 37—Country Reports Summary 2014, first electronic edition produced in 2015. Available from: <https://www.ieabioenergy.com/>.

IEA—International Energy Agency, 2018. World Energy Balances and Statistics, op. cit. note 31; IRENA, IEA and REN21, op. cit. note 1, p. 39; IRENA, "Running on renewables: transforming transportation through renewable technologies," 14 January 2018, <https://irena.org/newsroom/articles/2018/Jan/Running-on-renewables-transforming>.

IEA, 2020a. Bio-based chemicals: a 2020 update. [s.l.] IEA Bioenergy.

IEA—International Energy Agency, 2020b. IEA Bioenergy Task 37—Country Reports Summary 2019, IEA Bioenergy. Available from <http://task37.ieabioenergy.com/>.

Iglesias, R., Muñoz, R., Polanco, M., Díaz, I., Susmozas, A., Moreno, A.D., et al., 2021. Biogas from anaerobic digestion as an energy vector: current upgrading development. Energies 14, 2742.

Imberty, A., Buléon, A., Tran, V., Péerez, S., 1991. Recent advances in knowledge of starch structure. Starch/Stärke 43, 375–384.

IPCC - Intergovernmental Panel on Climate Change, 2021. Report on climate change, desertification, land degradation, sustainable land management, food security, and greenhouse gas fluxes in terrestrial ecosystems.

Isikgor, F.H., Becer, C.R., 2015. Lignocellulosic biomass: a sustainable platform for the production of bio-based chemicals and polymers. Polym. Chem. 6, 4497–4559.

Jaeger, K., Reetz, M.T., 1998. Microbial lipases form versatile tools for biotechnology. TIBTECH 16, 396–403.

Jaeger, K., Dijkstra, B.W., Reetz, M.T., 1999. Molecular biology, three-dimensional structures, and biotechnological applications of lipases. Annu. Rev. Microbiol. 53, 315–351.

Ji, J., Wang, J., Li, Y., Yu, Y., Xu, Z., 2006. Preparation of biodiesel with the help of ultrasonic and hydrodynamic cavitation. Ultrasonics 44, e411–e414.

Jim, L., 2016. Strategic intent: the digest's 2016 multi-slide guide to Raizen, reliance industries. Available from: <http://www.biofuelsdigest.com/bdigest/2016/10/10/strategic-intent-the-digests-2016-multi-slide-guide-to-raizen-reliance-industries/9/>.

Juturu, V., Wu, J.C., 2013. Insight into microbial hemicellulases other than xylanases: a review: microbial hemicellulases other than xylanases. J. Chem. Technol. Biotechnol. 88, 353–363.

Juturu, V., Wu, J.C., 2014. Microbial *Exo*-xylanases: a mini review. Appl. Biochem. Biotechnol. 174, 81–92.

Kainthola, J., Kalamdhad, A.S., Goud, V.V., 2020. Optimization of process parameters for accelerated methane yield from anaerobic co-digestion of rice straw and food waste. Renew. Energy 149, 1352–1359.

Kapoor, M., Gupta, M.N., 2012. Lipase promiscuity and its biochemical applications. Process Biochem 47, 555–569.

Kapor, N.Z.A., Maniam, G.P., Rahim, M.H.A., Yusoff, M.M., 2017. Palm fatty acid distillate as a potential source for biodiesel production-a review. J. Clean. Prod 143, 1–9.

Karp, S.G., Rozhkova, A.M., Semenova, M.V., Osipov, D.O., de Pauli, S.T.Z., Sinitsyna, O.A., et al., 2021a. Designing enzyme cocktails from *Penicillium* and *Aspergillus* species for the enhanced saccharification of agro-industrial wastes. Bioresour. Technol. 330, 124888.

Karp, S.G., Medina, J.D.C., Letti, L.A.J., Woiciechowski, A.L., de Carvalho, J., Schimitt, C.C., et al., 2021b. Bioeconomy and biofuels: the case of sugarcane ethanol in Brazil. Biofuels, Bioprod. Bioref. 15, 889–912.

Kashyap, S.S., Gogate, P.R., Joshi, S.M., 2019. Ultrasound assisted synthesis of biodiesel from karanja oil by interesterification: Intensification studies and optimization using RSM. Ultrason. Sonochem. 50, 36–45.

Kaur, J., Sarma, A.K., Jha, M.K., Gera, P., 2020. Valorisation of crude glycerol to value-added products: perspectives of process technology, economics and environmental issues. Biotechnol. Repo 27, e00487.

Kennedy, H.T., 2019. GranBio to resume ethanol plant commercial operations. Available from: <https://www.biofuelsdigest.com/bdigest/2019/01/25/granbio-to-resume-ethanol-plant-commercial-operations/>.

Khan, N.R., Rathod, V.K., 2015. Enzyme catalyzed synthesis of cosmetic esters and its intensification: a review. Process Biochem 50, 1793–1806.

Khan, M.U., Ahring, B.K., 2021. Improving the biogas yield of manure: effect of pretreatment on anaerobic digestion of the recalcitrant fraction of manure. Bioresour. Technol. 321, 124427.

Khoo, K.S., Chia, W.Y., Tang, D.Y.Y., Show, P.L., Chew, K.W., Chen, W.-H., 2020. Nanomaterials utilization in biomass for biofuel and bioenergy production. Energies 13.

Kim, K.H., Choi, I.S., Kim, H.M., Wi, S.G., Bae, H.J., 2014. Bioethanol production from the nutrient stress-induced microalga *Chlorella vulgaris* by enzymatic hydrolysis and immobilized yeast fermentation. Bioresour. Technol. 153, 47–54.

Kleiner, B., Fleischer, P., Schörken, U., 2016. Biocatalytic synthesis of biodiesel utilizing deep eutectic solvents: a two-step-one-pot approach with free lipases suitable for acidic and used oil processing. Process Biochem 51, 1808–1816.

Klemm, D., Heublein, B., Fink, H.P., Bohn, A., 2005. Cellulose: fascinating biopolymer and sustainable raw material. Angew. Chem. Int. (Ed.) 44, 3358–3393.

Kostylev, M., Wilson, D., 2012. Synergistic interactions in cellulose hydrolysis. Biofuels 3, 61–70.

Kougias, P.G., Angelidaki, I., 2018. Biogas and its opportunities—a review. Front. Environ. Sci. Eng. 12, 1–12.

Kovalenko, G.A., Perminova, L.V., Beklemishev, A.B., Yakovleva, E.Y., Pykhtina, M.B., 2015. Heterogeneous biocatalytic processes of vegetable oil interesterification to biodiesel. Catal. Ind. 7, 73–81.

Kruyeniski, J., Ferreira, P.J., Carvalho, M.G.V.S., Vallejos, M.E., Felissia, F.E., Area, M.C., 2019. Physical and chemical characteristics of pretreated slash pine sawdust influence its enzymatic hydrolysis. Ind. Crops Prod. 130, 528–536.

Kumar, G., Kumar, D., Poonam, Johari, R., Singh, C.P., 2011. Enzymatic transesterification of *Jatropha curcas* oil assisted by ultrasonication. Ultrason. Sonochem. 18, 923–927.

Kumar, L., Arantes, V., Chandra, R., Saddler, J., 2012. The lignin present in steam pretreated softwood binds enzymes and limits cellulose accessibility. Bioresour. Technol. 8, 103–201.

Kumar, D., Das, T., Giri, B.S., Verma, B., 2018. Characterization and compositional analysis of highly acidic karanja oil and its potential feedstock for enzymatic synthesis of biodiesel. New J. Chem. 42, 15593–15602.

Lane, J., 2017. DowDuPont to exit cellulosic biofuels business. Available from <http://www.biofuelsdigest.com/bdigest/2017/11/02/breaking-news-dowdupont-to-exit-cellulosic-ethanol-business>.

Laurent, C.V.F.P., Breslmayr, E., Tunega, D., Ludwig, R., Oostenbrink, C., 2019. Interaction between cellobiose dehydrogenase and lytic polysaccharide monooxygenase. Biochemistry 58, 1226–1235.

Laurichesse, S., Avérous, L., 2014. Chemical modification of lignins: towards biobased polymers. Prog. Polym. Sci. 39, 1266–1290.

Lee, J.H., Kim, S.B., Kang, S.W., Song, Y.S., Park, C., Han, S.O., 2011. Biodiesel production by a mixture of *Candida rugosa* and *Rhizopus oryzae* lipases using a supercritical carbon dioxide process. 2011. Bioresour. Technol. 102, 2105–2108.

Lee, O.K., Oh, Y.K., Lee, E.Y., 2015. Bioethanol production from carbohydrate-enriched residual biomass obtained after lipid extraction of *Chlorella* sp. KR-1. Bioresour. Technol. 196, 22–27.

Leu, S.Y., Zhu, J., 2013. Substrate-related factors affecting enzymatic saccharification of lignocelluloses: our recent understanding. Bioenerg. Res. 6, 405–415.

Li, X., Xu, H., Wu, Q., 2007. Large-scale biodiesel production from microalga Chlorella protothecoides through heterotrophic cultivation in bioreactors. Biotechnol. Bioeng. 98, 764–771.

Li, Y., Du, W., Liu, D., 2013. Effect of phospholipids on free lipase-mediated methanolysis for biodiesel production. J. Mol. Catal. B Enzym 91, 67–71.

Lin, C.Y., Lu, C., 2021. Development perspectives of promising lignocellulose feedstocks for production of advanced generation biofuels: a review. Renew. Sustain. Energy Rev. 136, 110445.

Lisboa, P., Rodrigues, A.R., Martín, J.L., Simões, P., Barreiros, S., Paiva, A., 2014. Economic analysis of a plant for biodiesel production from waste cooking oil via enzymatic transesterification using supercritical carbon dioxide. J. Supercrit. Fluid 85, 31–40.

Liu, Z., Smith, S.R., 2020. Enzyme recovery from biological wastewater treatment. Waste Biomass Valorization.

Liu, Y., Chen, D., Yan, Y., y, C., Xu, L., 2011. Biodiesel synthesis and conformation of lipase from *Burkholderia cepacia* in room temperature ionic liquids and organic solvents. Bioresour. Technol. 102, 10414–10418.

Logan, M., Visvanathan, C., 2019. Management strategies for anaerobic digestate of organic fraction of municipal solid waste: current status and future prospects. Waste Manag. Res. 37, 27–39.

The California Air Resources Board, 2021. Low Carbon Fuel Standard. Available from: <https://ww2.arb.ca.gov/our-work/programs/low-carbon-fuel-standard/about>.

Lv, S., Yu, Q., Zhuang, X., Yuan, Z., Wang, W., Wang, Q., et al., 2013. The influence of hemicellulose and lignin removal on the enzymatic digestibility from sugarcane bagasse. Bioenerg. Res. 6, 1128–1134.

Ma, G., Dai, L., Liu, D., Du, W., 2019. Integrated production of biodiesel and concentration of polyunsaturated fatty acid in glycerides through effective enzymatic catalysis. Front. Bioeng. Biotechnol. 7, 393.

Ma, G., Ndegwa, P., Harrison, J.H., Chen, Y., 2020. Methane yields during anaerobic co-digestion of animal manure with other feedstocks: a meta-analysis. Sci. Total Environ. 728, 138224.

Madras, G., Kumar, R., Modak, J., 2004. Synthesis of octyl palmitate in various supercritical fluids. Ind. Eng. Chem. Res. 43, 7697–7701.

Mancheño, J.M., Pernas, M.A., Martínez, M.J., Ochoa, B., Rúa, M.L., Hermoso, J.A., 2003. Structural insights into the lipase/esterase behavior in the *Candida rugosa* lipases family: crystal structure of the lipase 2 isoenzyme at 1.97A resolution. J. Mol. Biol. 332, 1059–1069.

Mao, C., Feng, Y., Wang, X., Ren, G., 2015. Review on research achievements of biogas from anaerobic digestion. Renew. Sustain. Energy Rev. 45, 540–555.

Marcon, N.S., Colet, R., Bibilio, D., Graboski, A.M., Steffens, C., Rosa, C.D., 2019. Production of ethyl esters by direct transesterification of microalga biomass using propane as pressurized fluid. Appl. Biochem. Biotechnol. 187, 1285–1299.

Markets and MarketsTM, 2020. Available from: https://www.marketsandmarkets.com/.

Marín-Suarez, M., Méndez-Mateos, D., Guadix, A., Guadix, E.M., 2019. Reuse of immobilized lipases in the transesterification of waste fish oil for the production of biodiesel. Renew. Energy 140, 1–8.

Markou, G., Angelidaki, I., Nerantzis, E., Georgakakis, D., 2013. Bioethanol production by carbohydrate-enriched biomass of *Arthrospira* (*Spirulina*) *platensis*. Energies 6, 3937–3950.

Maršálková, B., Širmerová, M., Kuřec, M., Brányik, T., Brányiková, I., Melzoch, K., et al., 2010. Microalgae *Chlorella* sp. as an alternative source of fermentable sugars. Chem. Eng. Trans 21, 1279–1284.

Mata, T.M., Martins, A.A., Caetano, N.S., 2010. Microalgae for biodiesel production and other applications: a review. Renew. Sustain. Energy Rev. 14, 217–232.

Mattonai, M., Pawcenis, D., Seppia, S.D., Łojewska, J., Ribechini, E., 2018. Effect of ball-milling on crystallinity index, degree of polymerization and thermal stability of cellulose. Bioresour. Technol. 270, 270–277.

Meher, L.C., Vidya Sagar, D., Naik, S.N., 2006. Technical aspects of biodiesel production by transesterification - a review. Renew. Sustain. Energy Rev. 10, 248–268.

Meng, X., Yang, J., Xu, X., Zhang, L., Nie, Q., Xian, M., 2009. Biodiesel production from oleaginous microorganisms. Renew. Energy 34, 1–5.

Meng, Y., Luan, F., Yuan, H., Chen, X., Li, X., 2017a. Enhancing anaerobic digestion performance of crude lipid in food waste by enzymatic pretreatment. Bioresour. Technol. 224, 48–55.

Meng, X., Pu, Y., Yoo, C.G., Li, M., Bali, G., Park, D.Y., et al., 2017b. An in-depth understanding of biomass recalcitrance using natural poplar variants as the feedstock. Chem Sus Chem. 10, 139–150.

Menzel, T., Neubauer, P., Junne, S., 2020. Role of microbial hydrolysis in anaerobic digestion. Energies 13, 5555.

Merza, F., Fawzy, A., AlNashef, I., Al-Zuhair, S., Taher, H., 2018. Effectiveness of using deep eutectic solvents as an alternative to conventional solvents in enzymatic biodiesel production from waste oils. Energy Rep. 4, 77–83.

METEX., 2020. METabolic EXplorer ensures future sales of its subsidiary METEX NØØVISTA through a strategic partnership with DSM for the marketing of 1,3 propanediol (PDO) to the cosmetic ingredients market. Available from: <https://www.metabolic-explorer.com/history/ ;https://www.metabolic-explorer.com/2019/12/03/metex-subsidiary-partnership-dsm/>.

Mishra, S.B., Mishra, A., Kaushik, N., Khan, M.A., 2007. Study of performance properties of lignin-based polyblends with polyvinyl chloride. J. Mater. Process. Tech 183, 273–276.

MME, 2020. Ministério de Minas e Energia ratifica apoio ao Programa Nacional de Produção e Uso do Biodiesel e à Política Nacional de Biocombustíveis.

Moraes, B.S., dos Santos, G.M., Delforno, T.P., Fuess, L.T., da Silva, A.J., 2019. Enriched microbial consortia for dark fermentation of sugarcane vinasse towards value-added short-chain organic acids and alcohol production. J. Biosci. Bioeng. 127, 594–601.

Mota, M.F.S., Souza, M.F., Bon, E.P.S., Rodrigues, M.A., Freitas, S.P., 2018a. Colorimetric protein determination in microalgae (Chlorophyta): association of milling and SDS treatment for total protein extraction. J. Phycol. 54, 577–580.

Mota, T.R., Oliveira, D.M., Rogério Marchiosi, O., Ferrarese-Filho Santos, W.D., 2018b. Plant cell wall composition and enzymatic deconstruction. Bioengineering 5, 63–77.

Mucha, A.P., Dragisa, S., Dror, I., Garuti, M., van Hullebusch, E.D., Repinc, S.K., et al., 2019. Re-use of digestate and recovery techniques. Trace Elements in Anaerobic Biotechnologies 181.

Mulalee, S., Srisuwan, P., Phisalaphong, M., 2015. Influences of operating conditions on biocatalytic activity and reusability of Novozym 435 for esterification of free fatty acids with short-chain alcohols: a case study of palm fatty acid distillate. Chinese J. Chem. Eng. 23, 1851–1856.

Musa, H., Kasim, F.H., Gunny, A.A.N., Gopinath, S.C.B., Ahmad, M.A., 2019. Biosynthesis of butyl esters from crude oil palm fruit and kernel using halophilic lipase secretion by *Marinobacter litoralis* SW-45. 3 Biotech 9, 314.

Nagarajan, D., Lee, D.J., Chen, C.Y., Chang, J.S., 2020. Resource recovery from wastewaters using microalgae-based approaches: a circular bioeconomy perspective. Bioresour. Technol. 302, 122817.

Naik, L., Gebreegziabher, Z., Tumwesige, V., Balana, B.B., Mwirigi, J., Austin, G., 2014. Factors determining the stability and productivity of small scale anaerobic digesters. Biomass bioenergy 70, 51–57.

NanoCrystalline Cellulose, 2021. Available from https://www.celluforce.com/en/products/cellulose-nanocrystals/.

Nakamura, A., Tasaki, T., Ishiwata, D., Yamamoto, M., Okuni, Y., Maximilien, A.V.M., et al., 2016. Single-molecule imaging analysis of binding, processive movement, and dissociation of cellobiohydrolase *Trichoderma reesei* Cel6A and its domains on crystalline cellulose. J. Biol. Chem. 291, 22404–22413.

Navarro, R.R., Otsuka, Y., Matsu, K., Sasaki, K., Hori, T., Habe, H., et al., 2020. Combined simultaneous enzymatic saccharification and comminution (SESC) and anaerobic digestion for sustainable biomethane generation from wood lignocellulose and the biochemical characterization of residual sludge solid. Bioresour. Technol. 300, 122622.

Nie, K., Xie, F., Wang, F., Tan, T., 2006. Lipase catalyzed methanolysis to produce biodiesel: optimization of the biodiesel production. J. Mol. Catal. B: Enzym 43, 142–147.

Nielsen, P.M., Brask, J., Fjerbaek, L., 2008. Enzymatic biodiesel production: Technical and economical considerations. Eur. J. Lipid. Sci. Technol. 110, 692–700.

Niz, M., Etchelet, I., Fuentes, L., Etchebehere, C., Zaiat, M., 2019. Extreme thermophilic condition: an alternative for long-term biohydrogen production from sugarcane vinasse. Int. J. Hydrog. Energy 44, 22876–22887.

Nordblad, M., Pedersen, A.K., Rancke-Madsen, A., Woodley, J.M., 2016. Enzymatic pretreatment of low-grade oils for biodiesel production. Biotechnol. Bioeng. 113, 754–760.

Noussan, M., Raimondi, P.P., Scita, R., Hafner, M., 2021. The role of green and blue hydrogen in the energy transition – a technological and geopolitical perspective. Sustainability 13, 298.

Novozymes, 2014. Available from: <http://energy.agwired.com/2014/01/20/blue-sun-opens-most-advanced-biodiesel-plant/>.

Nugraha, W.D., Kusumastuti, V.N., Matin, H.H.A., 2020. Optimization of biogas production in Indonesian region by liquid anaerobic digestion (L-AD) method from rice husk using response surface methodology (RSM). In: IOP Conference Series: Materials Science and Engineering, 845, 012042.

O'Sullivan, C.A., 1997. Cellulose: the structure slowly unravels. Cellulose 4, 173–207.

Osborne, S., 2020. Energy in 2020: assessing the economic efects of commercialization of cellulosic ethanol. Available from <https://www.trade.gov/sites/default/files/2020-12/Energy%20in%202020_Assessing%20the%20Economic%20Effects%20of%20Commercialization%20of%20Cellulosic%20Ethanol.pdf>.

Padella, M., O'Connell, A., Prussi, M., 2019. What is still limiting the deployment of cellulosic ethanol? Analysis of the current status of the sector. Appl. Sci 9, 4523.

Paques, F.W., Macedo, G.A., 2006. Lipases de látex vegetais: propriedades e aplicações industriais. Quim. Nova 29, 93–99.

Park, S., Baker, J.O., Himmel, M.E., Parilla, P.A., Johnson, D.K., 2010. Cellulose crystallinity index: measurement techniques and their impact on interpreting cellulase performance. Biotechnol. Biofuels 3, 1–10.

Parthiba, K.O., Trably, E., Meheriya, S., Bernet, N., Wong, J.W.C., Carrere, H., 2018. Pretreatment of food waste for methane and hydrogen recovery: a review. Bioresour. Technol. 249, 1025–1039.

Pascale, N.C., Chastinet, J.J., Bila, D.M., SantAnna, G.L., Quitério, S.L., Vendramel, S.M.R., 2019. Enzymatic hydrolysis of floatable fatty wastes from dairy and meat food-processing industries and further anaerobic digestion. Water Sci. Technol. 79, 985–992.

Peng, F., Peng, P., Xu, F., Sun, R.-C., 2012. Fractional purification and bioconversion of hemicelluloses. Biotechnol. Adv. 30, 879–903.

Pereira Jr., N., Bon, E.P. da S., Ferrara, M.A., 2008. Séries em Biotecnologia: Tecnologia de Bioprocessos.

Pereira, E.B., Martins, F.R., Gonçalves, A.R., Costa, R.S., Lima, F.J.L., Rüther, R., et al., 2017. Atlas Brasileiro de Energia Solar, 2ed (ed.) INPE, Sao Jose dos Campos.

Pippo, W.A., Luengo, C.A., Alberteris, L.A.M., Garzone, P., Cornacchia, G., 2011. Energy recovery from sugarcane-trash in the light of 2nd generation biofuels. Part 1: current situation and environmental aspects. Waste Biomass Valor 2, 1–16.

Pourzolfaghar, H., Abnisa, F., Daud, W.M.A.W., Aroua, M.K., 2016. A review of the enzymatic hydroesterification process for biodiesel production. Renew. Sustain. Energy Rev. 61, 245–257.

Prabakaran, P., Pradeepa, V., Selvakumar, G., Ravindran, A.D., 2019. Efficacy of Enzymatic transesterification of *Chlorococcum* sp. algal oils for biodiesel production. Waste Biomass Valorization 10, 1873–1881.

Qaseem, M.F., Shaheen, H., Wu, A., 2021. Cell wall hemicellulose for sustainable industrial utilization. Renew. Sustain. Energy Rev. 144, 110996.

Rajagopal, R., Massé, D.I., Singh, G., 2013. A critical review on inhibition of anaerobic digestion process by excess ammonia. Bioresour. Technol. 143, 632–641.

Rajendran, K., Mahapatra, D., Venkatesh, A., 2020. Advancing anaerobic digestion through two-stage processes: current developments and future trends. Renew. Sustain. Energy Rev. 123, 109746.

Rajin, M., 2018. A current review on the application of enzymes in anaerobic digestion. Anaerobic Digestion Processes. Springer, pp. 55–70.

Rathore, V., Madras, G., 2007. Synthesis of biodiesel from edible and non-edible oils in supercritical alcohols and enzymatic synthesis in supercritical carbon dioxide. Fuel 86, 2650–2659.

Ratnayake, W.S., Jackson, D.S., 2008. Starch gelatinization. Advances in Food and Nutrition Research 221–268.

Rego, G.C., Ferreira, T.B., Ramos, L.R., de Menezes, C.A., Soares, L.A., Sakamoto, I.K., et al., 2020. Bioconversion of pretreated sugarcane vinasse into hydrogen: new perspectives to solve one of the greatest issues of the sugarcane biorefinery. Biomass Convers. Bior .

Rena, Zacharia, K.M.B., Yadav, S., Machhirake, N.P., Kim., S.H., Lee, B.D., et al., 2020. Bio-hydrogen and bio-methane potential analysis for production of bio- hythane using various agricultural residues. Bioresour. Technol. 309, 123297.

Renewables, 2020 Global Status Report., 2020. Available from: <https://www.ren21.net/wp-content/uploads/2019/05/gsr_2020_full_report_en.pdf>.

Reuters, 2021. Brazil's Raízen to build second cellulosic ethanol plant – filing. Available from: <https://www.reuters.com/business/energy/brazils-razen-build-second-cellulosic-ethanol-plant-filing-2021-06-25/>. (accessed 01.04.22).

Ribeiro, V.S., Silva, M.A.R., 2020. Biodiesel public policy in Brazil: an analysis in the overview of the public policy cycle. Desenvolvimento Regional em debate 10, 833–861.

Robertson, G.H., Wong, D.W.S., Lee, C.C., Wagschal, K., Smith, M.R., Orts, W.J., 2006. Native or raw starch digestion: a key step in energy efficient biorefining of grain. J. Agric. Food Chem. 54, 353–365.

Robles-Medina, A., González-Moreno, P.A., Esteban-Cerdán, L., Molina-Grima, E., 2009. Biocatalysis: towards ever greener biodiesel production. Biotechnol. Adv. 27, 398–408.

Romano, D., Bonomi, F., de Mattos, M.C., de Sousa Fonseca, T., de Oliveira, M., da, C.F., et al., 2015. Esterases as stereoselective biocatalysts. Biotechnol. Adv. 33, 547–565. Available from: https://doi.org/10.1016/j.biotechadv.2015.01.006.

Rosales-Calderon, O., Arantes, V., 2019. A review on commercial-scale high-value products that can be produced alongside cellulosic ethanol. Biotechnol. Biofuels 12, 240.

Sáez-Bastante, J., Ortega-Román, C., Pinzi, S., Lara-Raya, F.R., Leiva-Candia, D.E., Dorado, M. P., 2015. Ultrasound-assisted biodiesel production from *Camelina sativa* oil. Bioresour. Technol. 185, 116–124.

Sajith, S., Priji, P., Sreedevi, S., Benjamin, S., 2016. An overview on fungal cellulases with an industrial perspective. J. Nutr. Food Sci. 6.

Salimon, J., Salih, N., Yousif, E., 2012. Improvement of pour point and oxidative stability of synthetic ester base stocks for biolubricant applications. Arab. J. Chem. 5, 193–200.

Sampaio, I.L.M., Cardoso, T.F., Souza, N.R., Watanabe, M.D.B., Carvalho, D.J., Bonomi, A., et al., 2019. Electricity production from sugarcane straw recovered through bale system: assessment of retrofit projects. BioEnergy Res 12, 865–877.

Sánchez, Ó.J., Cardona, C.A., 2008. Trends in biotechnological production of fuel ethanol from different feedstocks. Bioresour. Technol. 99, 5270–5295.

Santos, F.F.P., Rodrigues, S., Fernandes, F.A.N., 2009. Optimization of the production of biodiesel from soybean oil by ultrasound assisted methanolysis. Fuel Process. Technol. 90, 312–316.

Saravanan, A.P., Pugazhendhi, A., Mathimani, T., 2020. A comprehensive assessment of biofuel policies in the BRICS nations: implementation, blending target and gaps. Fuel 272, 117635.

Sarmah, N., Revathi, D., Sheelu, G., Rani, K.Y., Sridhar, S., Mehtab, V., et al., 2017. Recent advances on sources and industrial applications of lipases. Biotechnol. Prog. 34.

Sawatdeenarunat, C., Wangnai, C., Songkasiri, W., Panichnumsin, P., Saritpongteeraka, K., Boonsawang, P., et al., 2019. Biogas production from industrial effluents. Biofuels: Alternative Feedstocks and Conversion Processes for the Production of Liquid and Gaseous Biofuels. Academic Press, pp. 779–816.

Sayre, R., 2010. Microalgae: the potential for carbon capture. Bioscience 60, 722–727.

Schnürer, A., Jarvis, Å., 2009. Microbiological Handbook for Biogas Plants Swedish. Swedish Waste Management U2009:03. Swedish Ga, p. 138.

Scholz, M.J., Weiss, T.L., Jinkerson, R.E., Jing, J., Roth, R., Goodenough, U., et al., 2014. Ultrastructure and composition of the *Nannochloropsis gaditana* cell wall. Eukaryot. Cell 13, 1450–1464.

Scott, A., 2018. Chemical & engineering news. Versalis buys Mossi Ghisolfi's biobased businesses. Available from: <https://cen.acs.org/business/biobased-chemicals/Versalis-buys-Mossi-Ghisolfis-biobased/96/i39> (accessed 01.04.22).

Serrano-Arnaldos, M., Montiel, M.C., Ortega-Requena, S., Máximo, F., Bastida, J., 2019. Development and economic evaluation of an eco-friendly biocatalytic synthesis of emollient esters. Bioprocess Biosyst. Eng. 43, 495–505.

Sethupathy, A., Arun, C., Sivashanmugam, Kumar, R.R., 2020. Enrichment of biomethane production from paper industry biosolid using ozonation combined with hydrolytic enzymes. Fuel 279, 118522.

Shanks, B.H., Keeling, P.L., 2017. Bioprivileged molecules: creating value from biomass. Green Chem. 19, 3177.

Sharma, R., Chisti, Y., Banerjee, U.C., 2001. Production purification characterization and applications of lipases. Biotechnol. Adv. 19, 627–662.

Sharma, Y.C., Singh, B., Upadhyay, S.N., 2008. Advancements in development and characterization of biodiesel: a review. Fuel 87, 2355–2373.

Shedon, A.R., 2018. The Road to Biorenewables: Carbohydrates to Commodity Chemicals. ACS Sustainable Chem. Eng 6, 4464–4480.

Shindo, S., Sakakibara, K., Sano, R., Ueda, K., Hases, M., 2001. Characterization of a floricaula/leafy homologue of *Gnetum parvifolium* and its implications for the evolution of reproductive organs in seed plants. Int. J. Plant Sci. 162, 1199–1209.

Silva, G.P.S., Contiero, J., Neto, P.M.A., de Lima, C.J.B., 2014. 1,3-Propanediol: production, applications and biotechnological potential. Quim. Nova 37, 527–534.

SINDPEÇAS – Sindicato Nacional da Insdústria de Componentes para Veículos Automotores, 2020. Avaiable from: <https://www.sindipecas.org.br/sindinews/Economia/2020/RelatorioFrotaCirculante_Abril_2020.pdf>.

Singh, A., Bajar, S., Devi, A., Pant, D., 2021. An overview on the recent developments in fungal cellulase production and their industrial applications. Bioresour. Technol. Rep. 14, 100652.

Sjöström, E., 1993. Wood chemistry, Fundamentals and Applications, second (ed.) Academic Press, San Diego, CA.

Slaughter, S.L., Butterworth, P.J., Ellis, P.R., 2001. Mechanisms of the action of porcine pancreatic α-amylase on native and heat treated starches from various botanical sources. Starch - Adv. Struct. Funct 1525, 110–115.

Soares, D., Pinto, A.F., Gonçalves, A.G., Mitchell, D.A., Krieger, N., 2013. Biodiesel production from soybean soapstock acid oil by hydrolysis in subcritical water followed by lipase-catalyzed esterification using a fermented solid in a packed-bed reactor. Biochem. Eng. J. 81, 15–23.

Soares, L.A., Rabelo, C.A.B.S., Delforno, T.P., Silva, E.L., Varesche, M.B.A., 2019. Experimental design and syntrophic microbial pathways for biofuel production from sugarcane bagasse under thermophilic condition. Renew. Energy 140, 852–861.

Soccol, C.R., Faraco, V., Karp, S., Vandenberghe, L.P.S., Thomaz-Soccol, V., Woiciechowski, A., et al., 2011. Lignocellulosic bioethanol: current status and future perspectives. Biofuels: Alternative Feedstocks and Conversion Processes, Chapter 5. Academic Press, pp. 101–122.

Sorensen, H.R., Pedersen, S., Jorgensen, C.T., Meyer, A.S., 2007. Enzymatic hydrolysis of wheat Arabinoxylan by a recombinant "minimal" enzyme cocktail containing β-xylosidase and novel endo-1,4-β-xylanase and α-L-arabinofuranosidase activities, Biotechnol. Prog., 23. pp. 100–107.

Sotoft, L.F., Rong, B., Christensen, K.V., Norddahl, B., 2010. Process simulation and economical evaluation of enzymatic biodiesel production plant. Bioresour. Technol. 101, 5266–5274.

Souza, M.S., Aguieiras, E.C.G., Da Silva, M.A.P., Langone, M.A.P., 2009. Biodiesel synthesis via esterification of feedstock with high content of free fatty acids. Appl. Biochem. Biotechnol. 154, 253–267.

Souza, L.T.A., Mendes, A.A., Castro, H.F.D., 2016. Selection of Lipases for the synthesis of biodiesel from Jatropha oil and the potential of microwave irradiation to enhance the reaction rate. Biomed Res. Int. 2016.

Souza, M.F., de, Pereira, D.S., Freitas, S.P., Bon, E.P., da, S., Rodrigues, M.A., 2017. Neutral sugars determination in Chlorella: use of a one-step dilute sulfuric acid hydrolysis with reduced sample size followed by HPAEC analysis. Algal Res 24, 130–137.

Souza, M.F., de, Rodrigues, M.A., Freitas, S.P., Bon, E.P., da, S., 2020. Effect of milling and enzymatic hydrolysis in the production of glucose from starch-rich *Chlorella sorokiniana* biomass. Algal Res 50, 101961.

Srivastava, N., Srivastava, M., Mishra, P., Gupta, V.K., Molina, G., Rodriguez-Couto, S., et al., 2018. Applications of fungal cellulases in biofuel production: advances and limitations. Renew. Sust. Energ. Rev. 82, 2379–2386.

Statistical Review of World Energy, 2021. Available from <https://www.bp.com/content/dam/bp/business-sites/en/global/corporate/pdfs/energy-economics/statistical-review/bp-stats-review-2021-full-report.pdf>.

Straathof, A.J.J., 2014. Transformation of biomass into commodity chemicals using en-zymes or cells. Chem. Rev. 114, 1871–1908.

Straathof, A.J.J., Bampouli, A., 2017. Potential of commodity chemicals to become bio-based according to maximum yields and petrochemical prices. Modeling and Analysis: Biobased Commodity Chemicals. Wiley.

Su, E.Z., Zhang, M.J., Zhang, J.G., Gao, J.F., Wei, D.Z., 2007. Lipase-catalyzed irreversible transesterification of vegetable oils for fatty acid methyl esters production with dimethyl carbonate as the acyl acceptor. Biochem. Eng. J. 36, 167–173.

Subhedar, P.B., Gogate, P.R., 2016. Ultrasound assisted intensification of biodiesel production using enzymatic interesterification. Ultrason. Sonochem. 29, 67–75.

Subramani, L., Parthasarathy, M., Balasubramanian, D., Ramalingam, K., 2018. Novel Garcinia gummi-gutta methyl ester (GGME) as a potential alternative feedstock for existing unmodified DI diesel engine. Renew. Energy 125, 568–577.

SUCRE, 2017. Electricity from sugarcane straw can supply 27% of Brazilian household demand. [Eletricidade gerada a partir da palha de cana-de-açúcar pode suprir 27% do consumo residencial no Brasil] <https://pages.cnpem.br/sucre/2017/09/21/eletricidade-geradapartir-da-palha-de-cana-de-acucar-pode-suprir-27-do-consumo-residencialno-brasil/>.

Sukumaram, R.K., Christopher, M., Kooloth-Valappil, P., Sreeja-Raju, A., Mathew, R.M., Sankar, M., et al., 2021. Addressing challenges in production of cellulases for biomass hydrolysis: targeted interventions into the genetics of cellulase producing fungi. Bioresour. Technol. 329, 124746.

Sukumaran, R.K., Christopher, M., Kooloth-Valappil, P., Sreeja-Raju, A., Mathew, R.M., Sankar, M., et al., 2021. Addressing challenges in production of cellulases for biomass hydrolysis: Targeted interventions into the genetics of cellulase producing fungi. Bioresour. Technol. 124746.

Sunitha, S., Kanjilal, S., Reddy, P.S., Prasad, R.B.N., 2007. Ionic liquids as a reaction medium for lipase-catalyzed methanolysis of sunflower oil. Biotechnol. Lett. 29, 1881–1885.

Susanne, R.S., 2016. The latest news and data about ethanol production. Available from: <http://www.ethanolproducer.com/articles/13135/global-cellulosic-ethanol-developments>.

Svihus, B., Uhlen, A.K., Harstad, O.M., 2005. Effect of starch granule structure associated components and processing on nutritive value of cereal starch: a review. Anim. Feed Sci. Technol. 122, 303–320.

Sweeney, A.D., Xu, F., 2012. Biomass converting enzymes and industrial biocatalysts for fuels and chemicals: recent developments. Catalysts 2, 244–263.

Tabka, M.G., Herpoël-Gimbert, I., Monod, F., Asther, M., Sigoillot, J.C., 2006. Enzymatic saccharification of wheat straw for bioethanol production by a combined cellulase xylanase and feruloyl esterase treatment. Enzym. Microb. Technol 39, 897–902.

Tacias-Pascacio, V.G., Virgen-Ortíz, J.J., Jiménez-Pérez, M., Yates, M., Torrestiana-Sanchez, B., Rosales-Quintero, A., et al., 2017. Evaluation of different lipase biocatalysts in the production of biodiesel from used cooking oil: critical role of the immobilization support. Fuel 200, 1–10.

Taher, H., Giwa, A., Abusabiekeh, H., Al-Zuhair, S., 2020. Biodiesel production from *Nannochloropsis gaditana* using supercritical CO_2 for lipid extraction and immobilized lipase transesterification: economic and environmental impact assessments. Fuel Process. Technol. 198, 106249.

Takeda, H., 1988. Classification of *Chlorella* strains by cell wall sugar composition. Phytochemistry 27, 3823–3826.

Takeda, H., 1991. Sugar composition of the cell wall and the taxonomy of *Chlorella* (Chlorophyceae). J. Phycol. 27, 224–232.

Takeda, Y., Hizukuri, S., Takeda, C., Suzuki, A., 1987. Structures of branched molecules of amyloses of various origins, and molar fractions of branched and unbranched molecules. Carbohydr. Res. 165, 139–145.

Talukder, M.M.R., Wu, J.C., Fen, N.M., Melissa, Y.L.S., 2010. Two-step lipase catalysis for production of biodiesel. Biochem. Eng. J. 49, 207–212.

Tavares, F., Da Silva, E.A., Pinzan, F., Canevesi, R.S., Milinsk, M.C., Scheufele, F.B., et al., 2018. Hydrolysis of crambe oil by enzymatic catalysis: an evaluation of the operational conditions. Biocatal. Biotransformation 36, 422–435.

Taylor, R., Alberts, G., Robson, P., Chudziak, C., Bauen, A., 2015. From the Sugar Platform to biofuels and biochemicals. Final Report for the European Commission.

Tester, R.F., Qi, X., Karkalas, J., 2006. Hydrolysis of native starches with amylases. Anim. Feed Sci. Technol. 130, 39–54.

Thomas, P., Duolikun, T., Rumjit, N.P., Moosavi, S., Lai, C.W., Johan, M.R.B., et al., 2010. Comprehensive review on nanocellulose: recent developments, challenges and future prospects. J. Mech. Behav. Biomed. Mater. 110, 103884.

Toldra-Reig, F., Mora, L., Toldra, F., 2020. Trends in biodiesel production from animal fat waste. Appl. Sci 10, 3644.

Top, A.G.M., 2010. Production and utilization of palm fatty acid distillate (PFAD). Lipid Technol 22, 11–13.

Tran, D.T., Chen, C.L., Chang, J.S., 2012. Immobilization of *Burkholderia* sp. lipase on a ferric silica nanocomposite for biodiesel production. J. Biotechnol. 158, 112–119.

Trentin, C.M., Scherer, R.P., Rosa, C.D., Treichel, H., Oliveira, D., Oliveira, J.V., 2014. Continuous lipase-catalyzed esterification of soybean fatty acids under ultrasound irradiation. Bioproc Biosyst Eng 37, 841–847.

Uchiyama, T., Uchihashi, T., Nakamura, A., Watanabe, H., Kaneko, S., Samejima, M., et al., 2020. Convergent evolution of processivity in bacterial and fungal cellulases. Proc. Natl. Acad. Sci. USA 117, 19896–19903.

União Nacional da Bioenergia, 2021. Raízen vai construir sua segunda planta de etanol celulósico. Available from: <https://www.udop.com.br/noticia/2021/06/25/raizen-vai-construir-sua-segunda-planta-de-etanol-celulosico.html> (accessed 01.04. 22).

Vaaje-Kolstad, G., Westereng, B., Horn, S.J., Zhanliang, L., Zhai, H., Sorlie, M., et al., 2010. An oxidative enzyme boosting the enzymatic conversion of recalcitrant polysaccharides. Science 330, 219–222.

Van Dyk, J.S., Pletschke, B.I., 2012. A review of lignocellulose bioconversion using enzymatic hydrolysis and synergistic cooperation between enzymes – factors affecting enzymes, conversion and synergy. Biotechnol. Advan. 30, 1458–1480.

Varga, E., Klinke, H.B., Réczey, K., Thomsen, A.B., 2004. High solid simultaneous saccharification and fermentation of wet oxidized corn stover to ethanol. Biotechnol. Bioeng. 88, 567–574.

Varma, M.N., Madras, G., 2007. Synthesis of biodiesel from castor oil and linseed oil in supercritical fluids. Ind. Eng. Chem. Res. 46, 1–6.

Veljkovic, V.B., Avramovic, C.J.M., Stamenkovic, O.S., 2012. Biodiesel production by ultrasound-assisted transesterification: state of the art and the perspectives. Renew. Sust. Energ. Rev. 16, 1193–1209.

Vital, A., 2021. Raízen passa a fornecer EcoÁlcool à toda linha de perfumaria do Grupo Boticário. Available from: <https://jornalcana.com.br/raizen-passa-a-fornecer-ecoalcool-a-toda-linha-de-perfumaria-do-grupo-boticario/>.

Voegele, E., 2016. Ethanol Producer Magazine. Ocean Park Advisors to market Abengoa's cellulosic ethanol plant. Available from: <http://www.ethanolproducer.com/articles/13537/ocean-park-advisors-to-market-abengoaundefineds-cellulosic-ethanol-plant> (accessed 01.04.22).

Waghmare, G.V., Rathod, V.K., 2016. Ultrasound assisted enzyme catalyzed hydrolysis of waste cooking oil under solvent free condition. Ultrason. Sonochem. 32, 60–67.

Waltz, E., 2009. Biotech's green gold? Nat. Biotechnol. 27, 15–18.

Wang, J., Yin, Y., 2018. Fermentative hydrogen production using various biomass-based materials as feedstock. Renew. Sustain. Energy Rev. 92, 284–306.

Wang, C., Li, H., Li, M., Bian, J., Sun, R., 2017. Revealing the structure and distribution changes of Eucalyptus lignin during the hydrothermal and alkaline pretreatments. Sci. Rep. 7, 593.

Wang, H., Xu, J., Sheng, L., Liu, X., Lu, Y., Li, W., 2018. A review on bio-hydrogen production technology. Int. J. Energy. Res 42, 3442–3453.

Watanabe, Y., Nagao, T., Nishida, Y., Takagi, Y., Shimada, Y., 2007. Enzymatic production of fatty acid methyl esters by hydrolysis of acid oil followed by esterification. JAOCS, J. Am. Oil Chem. Soc. 84, 1015–1021.

Watanabe, M.D.B., Morais, E.R., Cardoso, T.F., Chagas, M.F., Junqueira, T.L., Carvalho, D.J., et al., 2020. A process simulation of renewable electricity from sugarcane straw: techno-economic assessment of retrofit scenarios in Brazil. J. Clean. Prod 254, 120081.

Weide, T., Baquero, C.D., Schomaker, M., Brügging, E., Wetter, C., 2020. Effects of enzyme addition on biogas and methane yields in the batch anaerobic digestion of agricultural waste (silage, straw, and animal manure). Biomass Bioenerg 132, 105442.

Xu, C., Eckerman, C., Smeds, A., Reunanen, M., Eklund, P.C., Sjoholm, R., et al., 2011. Carboxymethylated spruce galactoglucomannans: preparation, characterisation, dispersion stability, water-in-oil emulsion stability, and sorption on cellulose surface. Nordic Pulp & Paper Research Journal .

Xu, C., Zhang, J., Zhang, Y., Guo, Y., Xu, H., Xu, J., et al., 2019. Enhancement of high solids enzymatic hydrolysis efficiency of alkali pretreated sugarcane bagasse at low cellulase dosage by fed-batch strategy based on optimized accessory enzymes and additives. Bioresour. Technol. 292, 121993.

Xu, Q., Luo, T.-Y., Wu, R.-L., Wei, W., Sun, J., Dai, X., et al., 2021. Rhamnolipid pretreatment enhances methane production from two-phase anaerobic digestion of waste activated sludge. Water Res. 194, 116909.

Yang, G., Wang, J., 2018. Various additives for improving dark fermentative hydrogen production: a review. Renewable Sustainable Energy Rev 95, 130–146.

Yang, L., Ge, X., Wan, C., Yu, F., Li, Y., 2014. Progress and perspectives in converting biogas to transportation fuels. Renewable Sustainable Energy Rev 40, 1133–1152.

Yoo, C.G., Meng, X., Pu, Y., Ragauskas, A.J., 2020. The critical role of lignin in lignocellulosic biomass conversion and recent pretreatment strategies: a comprehensive review. Bioresour. Technol. 301, 122784.

Yu, Z., Gwak, K.S., Treasure, T., Jameel, H., Chang, H.M., Park, S., 2014. Effect of lignin chemistry on the enzymatic hydrolysis of woody biomass. Chem Sus Chem 7, 1942–1950.

Yu, Q., Liu, R., Li, K., Ma, R., 2019. A review of crop straw pretreatment methods for biogas production by anaerobic digestion in China. Renewable Sustainable Energy Rev 107, 51–58.

Zacchi, G., Axelsson, A., 1989. Economic evaluation of preconcentration in production of ethanol from dilute sugar solutions. Biotechnol. Bioeng. 34, 223–233.

Zahan, Z., Georgiou, S., Muster, T.H., Othman, M.Z., 2018. Semi-continuous anaerobic co-digestion of chicken litter with agricultural and food wastes: a case study on the effect of carbon/nitrogen ratio, substrates mixing ratio and organic loading. Bioresour. Technol. 270, 245–254.

Zamri, M.F.M.A., Hasmady, S., Akhiar, A., Ideris, F., Shamsuddin, A.H., Mofijur, M., et al., 2021. A comprehensive review on anaerobic digestion of organic fraction of municipal solid waste. Renewable Sustainable Energy Rev 137, 110637.

Zeng, L., He, Y., Jiao, L., Li, K., Yan, Y., 2017. Preparation of biodiesel with liquid synergetic lipases from rapeseed oil deodorizer distillate. Appl. Biochem. Biotechnol. 183, 778–791.

Zhang, L., Sun, S., Xin, Z., Sheng, B., Liu, Q., 2010. Synthesis and component confirmation of biodiesel from palm oil and dimethyl carbonate catalyzed by immobilized-lipase in solvent-free system. Fuel 89, 3960–3965.

Zhang, X., Yu, J., Zeng, A., 2017. Optimization and modeling for the synthesis of sterol esters from deodorizer distillate by lipase-catalyzed esterification. Biotechnol. Appl. Biochem. 64 (2), 270–278.

Zhao, X., Liu, D., 2019. Multi-products co-production improves the economic feasibility of cellulosic ethanol: a case of Formiline pretreatment-based biorefining. Applied Energy 250, 229–244.

Zhao, H., Zhang, C., Crittle, T.D., 2013. Choline-based deep eutectic solvents for enzymatic preparation of biodiesel from soybean oil. J. Mol. Catal. B Enzym 85, 243–247.

Zheng, J., Wei, W., Wang, S., Li, X., Zhang, Y., Wang, Z., 2020. Immobilization of Lipozyme TL 100L for methyl esterification of soybean oil deodorizer distillate. 3 Biotech 10.

Zhou, X., Li, W., Mabon, R., Broadbelt, L.J., 2017. A critical review on hemicellulose pyrolysis. Energy Technol 5, 52–79.

Zhu, L., O'dwyer, J.P., Chang, V.S., Granda, C.B., Holtzapple, M.T., 2008. Structural features affecting biomass enzymatic digestibility. Bioresour. Technol. 99, 3817–3828.

Zifcakova, L., Baldrian, P., 2012. Fungal polysaccharide monooxygenases: new players in the decomposition of cellulose. Fungal Ecol 5, 481–489.

Zoghlami, A., Paës, G., 2019. Lignocellulosic biomass: understanding recalcitrance and predicting hydrolysis. Front. Chem. 7, 874.

Chapter 22

Use of lipases for the production of biofuels

Thais de Andrade Silva[1], Julio Pansiere Zavarise[2], Igor Carvalho Fontes Sampaio[3], Laura Marina Pinotti[2], Servio Tulio Alves Cassini[1,3] and Jairo Pinto de Oliveira[1,3]

[1]*Federal University of Espírito Santo, Vitoria, Espírito Santo, Brazil,* [2]*Federal University of Espírito Santo, São Mateus, Espírito Santo, Brazil,* [3]*Center for Research, Innovation and Development of Espírito Santo, CPID, Cariacica, Espírito Santo, Brazil*

22.1 Introduction

Alternative fuel sources developed worldwide include mainly biodiesel, alcohol, biomass, biogas, and synthetic fuels. Biodiesel can be used directly, with no need for new refueling stations and no modifications to the engines (Srivastava and Prasad, 2000). Biodiesel is renewable, biodegradable, less toxic, and safer for storage and handling because it has a higher flash point, excellent lubricity, and calorific value similar to diesel (Knothe and Van Gerpen, 2010). In addition, its combustion is cleaner than that of mineral diesel, as it contains oxygen and reduced amounts of most emissions (CO_2, CO, SO_x, and particulates) (United States Environmental Protection Agency, 2010).

Chemically, biodiesel is a mono-alkyl fatty acid ester produced by esterification and transesterification of various lipid sources by an acidic, basic, or enzymatic catalyst (Vyas et al., 2010). Among the biodiesel production processes, the most commercially used is alkaline alcoholysis. However, despite the low cost of homogeneous chemical catalysts, this production route has some disadvantages, such as the impossibility of recovering and reusing the catalyst, difficult glycerol recovery, and high energy expenditure (Al-Zuhair et al., 2007). In addition, alkaline transesterification requires that the vegetable oil to be free from moisture and fatty acids (not exceeding 1% FFA), since the base can react with the fatty acids forming soap and water, hampering the separation and purification of biodiesel.

Despite a slower reaction rate, acid-catalyzed esterification is an alternative since acids can use both triglycerides and free fatty acids as substrates,

Biotechnology of Microbial Enzymes. DOI: https://doi.org/10.1016/B978-0-443-19059-9.00016-5
© 2023 Elsevier Inc. All rights reserved.

achieving good conversion levels, according to the literature (Ribeiro et al., 2011). However, acidic catalysis requires high temperatures and the absence of moisture and can compromise the quality of the product because of corrosive action.

In this context, enzymatic transesterification is one of the most promising fields among new technologies for synthesizing high value-added compounds. The chiral nature of enzymes results in the formation of products in a highly stereo- and regioselective manner in neutral and aqueous conditions, with the possibility of developing a high number of catalytic cycles. Furthermore, biocatalysts transform polyfunctionalized and sensitive compounds under mild conditions, unlike the corresponding chemical variants that require severe reaction conditions (Goldbeck, 2008).

Lipases have been widely reported in the literature as effective catalysts for the transesterification of oils. This route has attracted much attention because of the production of high-purity biodiesel: it makes the separation of glycerin simpler than conventional methods do, as there are no saponified by-products and aqueous residues containing glycerin and alcohol from washing to remove soaps. In addition, biocatalytic routes allow the transesterification of a wide variety of oils to be carried out in acidic impurities, such as raw materials with low added value (high content of free fatty acids).

This chapter will discuss the main advances in lipases for the catalysis of biodiesel, the production methods, immobilization strategies, and raw materials used. Moreover, we will give details on the main kinetic parameters as examples of industrial processes in bioreactors. Finally, we will present the perspectives of using lipases for the production of biofuels, taking into account the current limitations and the main challenges to be overcome.

22.2 Lipases

Lipolytic enzymes or lipases (E.C.3.1.1.3) are glycerol ester hydrolases that hydrolyze triacylglycerols (the main components of oils and fats), releasing free fatty acids, glycerol, and mono- and diacylglycerols. These enzymes act at the oil–water interface, increasing enzyme activity, although they also catalyze esterification and transesterification reactions in aqua-restricted media (Shimada et al., 2002).

From an economic and industrial perspective, lipases obtained from microorganisms by fermentation are preferable to those from animal and plant sources because they yield a large amount of product relatively quickly, and raw materials cost less (Zimmer et al., 2009). In addition, microorganisms produce lipases that hydrolyze triglycerides in the extracellular media, facilitating the ingestion of lipids. In these conditions, lipase expression is regulated mainly by environmental factors, for example, as an extracellular response to an environment deprived of nutrients. The presence of fatty acids and other lipids as carbon sources induces the production of these extracellular enzymes. Microorganisms are the most exciting hosts for protein production, and both regulatory and constitutive promoters can be used in the fermentation process (Nielsen, 2013).

Fungi (filamentous and yeast-like) are the preferred sources of lipases for commercial use. These enzymes are usually part of the extracellular metabolism in these organisms, which facilitates their extraction from the fermented medium. Another advantage arises because lipolytic fungi are considered safe microorganisms for manipulation (with exceptions). Their use in the form of immobilized integral lipolytic cells in reaction processes is expected to grow (Colen, 2006). The species with potential for lipase production which have been better described belong to the genera *Rhizopus* sp., *Mucor* sp., *Geotrichum* sp., *Penicillium* sp., and *Aspergillum* sp.

Because of the low yield of the fermentation process, microbial lipases have a high cost of production and purification compared with other hydrolases, such as proteases and carboxylases. However, recent advances in molecular biology have allowed enzyme manufacturers to place highly active microbial lipases on the market at a much more affordable cost. Currently, microbial lipases are produced by several companies, such as Novozymes, Amano, and Gist Brocades, among others. de Castro et al. (2004) published a study on the commercial availability of lipases and listed enzymes from 34 different sources, including 18 from fungi and 7 from bacteria.

Lipases are versatile and robust enzymes: the advantages of using these hydrolyses to produce biodiesel include working in different media in both hydrophilic and hydrophobic solvents. Many lipases show considerable activity in the catalysis of transesterification reactions with long-chain or branched alcohols—a challenging feat when alkaline catalysts are employed; moreover, if the enzyme is immobilized, it can be reused (Ghaly, 2010).

The use of lipases in environmental biocatalysis is in line with the strong tendency of governments to intensify restrictions on environmental pollution. Enzyme treatment techniques have attracted more attention because of stricter environmental regulations and are considered a clean and friendly technology (Gandhi, 1997). Several authors have studied biodiesel production using enzymatic catalysis employing free lipases (Table 22.1).

Although the catalysis using free enzymes has yielded promising results, difficulties in recovering, reusing, and maintaining the activity in different reaction media have hampered the scaling of this application. Therefore different supports have been explored for lipase immobilization processes.

22.2.1 Immobilization of lipases

Immobilizing an enzyme consists of confining it to support that allows contact with the substrate present in the reaction medium but makes it insoluble

TABLE 22.1 Examples of application of free lipases for biodiesel production.

Lipase	Substrate	Time (h)	Yield (%)	References
Aspergillus niger	Palm oil and palm kernel oil	48	90	Kareem et al. (2017)
Candida antarctica	Palm oil	22	94.6	Guo et al. (2020)
Rhizomucor miehei	Macauba oil	8	91.0	Aguieiras et al. (2014)
Burkholderia cepacia	Soy oil	31	92	Soares et al. (2013)
Thermomyces lanuginosus	Soy oil	12	94.3	Rosset et al. (2019)
Rhizopus oryzae	Rubber tree seed oil	48	31	Vipin et al. (2016)
T. lanuginosus	Rapeseed oil	24	97	Firdaus et al. (2016)

or poorly soluble in any medium. This process, carried out to maintain enzymatic activity, can reduce the cost of using enzymes industrially because it ensures that the enzyme is separated from the reaction medium, reused later, or kept in a continuous flow in a reactor (Dalla-Vecchia et al., 2004). Furthermore, many studies indicate that enzymes are more active immobilized than free in the reaction. This occurs because, in a heterogeneous system, the active sites of the enzymes are more readily available when conjugated to a solid support. Thus the use of supports that retain the enzyme, maintaining its catalytic characteristics, can increase the efficiency of the reactions and the possibility of recovery and reuse.

In addition, immobilization can provide other advantages for the enzyme, such as making it more stable to resist significant variations in pH and temperature and conserving its catalytic activity for several reuse cycles. Good support for enzymatic immobilization must have an excellent affinity for proteins, availability of reactive groups, stability, and excellent loading capacity (Pessela et al., 2007).

One lipase most used in studies to produce biodiesel is one obtained from *Candida antarctica*. This lipase of microbial origin is widely used in reactions to produce enantiomerically pure secondary alcohols and carboxylic acid transformations (Kirk and Christensen, 2002). *C. antarctica* lipase is sold immobilized in acrylic resin under the trade name Novozym 435.

Novozym 435 was first used by Nelson et al. (1996) to transesterify fat with high content of free fatty acids to produce biodiesel. Novozym 435 had increased enzyme activity when the secondary alcohol (2-butanol) was used in a solvent-free system. In this experiment, a 96.4% yield was obtained under the following reaction conditions: 0.34 molar of fat, the temperature of 45ºC, alcohol/tallow molar ratio of 3:1, stirring speed of 200 rpm, and reaction time of 16 h.

C. antarctica lipase showed promising activity in the transesterification of soybean oil with methanol (97% yield) (Mittelbach, 1990). However, Rodrigues et al. (2008) showed that the yield decreases proportionally as the alcohol chain increases. Watanabe et al. (2001) used residual oil and lipase from *C. antarctica* immobilized in a column along varying proportions of methanol, observing that the enzyme activity remained unchanged for 100 days.

22.2.2 Immobilization methods and supports

There are several ways to bind enzymes to supports: adsorption, ionic or covalent bonding, and encapsulation, not to mention other more sophisticated immobilizations (Barbosa et al., 2011). Immobilization by physical adsorption is the most straightforward technique in which enzymes adhere to the solid support through low-energy bonds, such as van der Waals, ionic bonds, and hydrogen bonds. Adsorption and covalent bonding methods are considered promising for applications in organic media because they often increase the catalytic activity and stability of the biocatalyst (Dalla-Vecchia et al., 2004).

As for the types of support, polymers are materials that make good enzymatic supports. Synthetic polymers such as acrylic resins show various physical forms and chemical structures. On the other hand, natural polymers such as agarose and chitosan hydrogel are cheaper and can be degraded easily (Mendes et al., 2011). Table 22.2 presents examples of methods to immobilize lipases on different supports.

Nanostructures are an excellent alternative as enzymatic supports (Lei et al., 2009). In particular, iron oxide (Fe_3O_4) magnetic nanoparticles (MNPs) can be used for the heterogeneous support of enzymes. Due to their properties, such as magnetism, large surface area, and excellent resistance to temperature variation, these nanoparticles have been gaining significant attention (Costa et al., 2016).

Magnetic nanomaterial supports are a potential substitute for conventional supports; they bring forth new properties, such as more significant surface area, tolerance to variations of temperature, and pH in different experimental conditions, good chemical reactivity, and strong interactions with enzymes (Lei et al., 2009). The high surface area-to-volume ratio of iron MNPs increases the binding capacity and specificity of enzymes. In addition, MNPs respond to magnetic field, which allows an efficient recovery and reuse of

TABLE 22.2 Methods for immobilizing microbial lipases on different supports.

Lipase	Support	Method	References
Rhizomucor miehei	Acrylic resin	Physical adsorption	De Paola et al. (2009)
Pseudomonas fluorescens	Silica		Salis et al. (2009)
Geotrichum sp.	K_2SO_4 microcrystals		Yan et al. (2011)
Saccharomyces cerevisiae	Mg-Al hydrotalcites		Zeng et al. (2009)
Candida rugosa	Polypropylene		Salis et al. (2008)
Candida antarctica	Activated charcoal		Naranjo et al. (2010)
Thermomyces lanuginosus	Styrene-divinylbenzene copolymer	Covalent bonding	Dizge et al. (2009)
Penicillium camembertii	Epoxy-SiO_2-PVA		Mendes et al. (2011)
T. lanuginosus	Polyurethane foam		Dizge; Keskinler. (2008)
Candida rugosa	Chitosan		Shao et al. (2008)
Pseudomonas fluorescens	Carbon nanotubes		Bartha-Vári et al. (2020)
Candida antarctica	Magnetic nanoparticles		Ashjari et al. (2020), Mehrasbi et al. (2017)

the conjugated enzymes, making the reaction product free of contamination (Yamaura et al., 2003).

However, despite the significant advantages of enzyme support, MNPs present some problems. They tend to form clusters due to anisotropic dipolar attraction, hampering immobilization. In addition, iron oxide undergoes oxidation in biological systems, losing its stability and magnetism (Ma et al., 2006). One way to mitigate these problems is to functionalize nanoparticle surfaces using organic functional groups, such as amine, thiol, aldehyde, carboxyl, and epoxy. Table 22.3 presents a range of studies on the immobilization of lipases, showing the diversity of molecules that can be used to functionalize MNPs. Enzyme immobilization becomes more efficient when MNPs are functionalized as they are more stable and resistant to oxidation (Guo and Sun, 2008).

TABLE 22.3 Methods of immobilizing microbial lipases on paramagnetic supports.

Lipase	Support	Method	References
Bacillus licheniformis; Rhizomucor miehei	Fe_3O_4 and Fe_3O_4-APTS	Physical adsorption and covalent bonding	Badoei-dalfard et al. (2019), Silva et al. (2022)
Candida antarctica			Miao et al. (2018)
Staphylococcus epidermidis	Fe_3O_4 and Fe_3O_4 -citric acid		Patel et al. (2018)
Thermomyces lanuginosus			Sarno and Iuliano (2019)
Burkholderia cepacia	Fe_3O_4 -styrene-divinylbenzene	Physical adsorption	Silva et al. (2018)
T. lanuginosus, Rhizomucor miehei	Fe_3O_4-SiO_2 -GPTMS	Covalent bonding	Ashjari et al. (2020)
Aspergillus niger	Fe_3O_4-SiO_2-APTS/MPTS		Thangaraj et al. (2019)
B. cepacia	Fe_3O_4-SiO_2-CTAB		Karimi (2016)
Candida rugosa	Fe_3O_4-GO-EDC		Xie and Huang (2018)
Pseudomonas cepacia	Fe_3O_4-SiO_2-PEI/PAA		Ahranjani et al. (2020)
Rhizopus oryzae	Fe_3O_4 -Chitosan		Kumar et al. (2013)
C. rugosa	GO-Fe_3O_4, GO-Fe_3O_4 -APTS/MPTS/OTMS	Covalent bonding and electrostatic attraction	Mosayebi et al. (2020)

22.3 Feedstocks

22.3.1 Vegetable oils

Vegetable oils are hydrophobic substances obtained by pressing or extracting oilseeds, consisting mainly of triglycerides with small amounts of mono- and diglycerides and free fatty acids, phospholipids, sterols, water, and other impurities that affect the reaction. When using basic or acid catalysts for transesterification, the separation of the coproduct glycerol becomes difficult. In addition, vegetable oils are liquids at room temperature, highly viscous, with low volatility, polyunsaturated character, and suffer incomplete combustion, making their direct use in engines unfeasible (Felizardo et al., 2006).

Among the various types of vegetable oils used as raw materials for biodiesel production, we mention soybean, peanut, sunflower, and castor bean oil. Among these, soy occupies a prominent position, as most of the oil production in the world comes from this legume. Despite the large production capacity of vegetable oils, Pimentel (1996) reported restrictions on expanding the use of the main cultivated oilseeds as raw material for the production of biofuels, justifying them as follows: competition with food crops and other crops for the use of soil and water; possible environmental impacts resulting from intensive agricultural production of energy crops (erosion, soil, and water contamination with residues of fertilizer, herbicides, and pesticides); and high production costs compared with current fees for the production of fossil fuels (diesel and fuel oil). Suarez et al. (2009) stated that an issue that permeates the use of biomass to produce fuel is the dilemma between food security and energy security.

22.3.2 Animal fats

This raw material is generally cheaper than refined oils because it represents a by-product of the animal agroindustry rather than a primary product. As a result, the demand for this product is lower than for most common vegetable oils. However, in addition to their high content of saturated fatty acids, they have a relatively high melting point, a property that, at low temperatures, can lead to precipitation and poor engine performance. On the positive side, the high content of saturated fatty acid esters ensures that biodiesel derived from animal fats will generally have a higher cetane number than that observed in vegetable oil biodiesel (Knothe et al., 2006).

22.3.3 Oily waste

Residual oils and fats, resulting from domestic, commercial, and industrial processing, are raw materials of great interest because of their high supply potential and low price (Holanda, 2004). The recycling of these types of waste, many of which do not have a commercial purpose, is gaining more and more ground, not simply because they represent low-cost "raw materials" but mainly because such waste has environmental impacts. The organization of efficient collection and purification systems is the most significant limitation to the use of oily waste (La Rovere et al., 2010).

The primary sources for lipid generation are edible oil, ice cream, dairy, tanneries, slaughterhouses, and domestic and restaurant effluents, mainly from fast foods. In the case of effluents originating from industrial activities, the oils and greases present in wastewater are variable, for example, 200–4680 mg/L for the dairy industry and 500–16,000 mg/L for the vegetable oil extraction industries. For the most part, these effluents are not treated or do not receive adequate treatment, which makes them, because of the impact caused, a worrying environmental problem (Mendes et al., 2005).

Regarding domestic sanitary sewage, according to Metcalf (2003), the levels of oils and fats usually observed are in the range of 55–170 mg/L, with an average value of 110 mg/L. These authors also note that the limited amount of oils and greases in wastewater discharged into waterways must represent 15–20 mg/L in the chemical oxygen demand value. In this type of wastewater, the oil and grease contents comprise the sum of the number of oils, greases, waxes, and fatty acids from food residues such as butter, margarine, vegetable oils, and animal fats, in addition to oils derived from lubricants used in industrial establishments, mainly cafeterias and restaurants (Jordão and Pessoa, 2005).

The main components of residual oils and fats are fatty acids that can be free (FFA) or esterified with glycerol in mono-, di-, or triacylglycerides. Triacylglycerols constitute approximately 80% of palmitic, stearic, oleic, and linoleic acid. The most abundant fatty acid is the oleic acid (C18:1) (Oliveira et al., 2014). Furthermore, phosphatides, which are mixed esters of glycerin with fatty acids and phosphoric acid, can also be found. In addition to fatty acid compounds and their derivatives, other lipids, such as sterols, waxes, antioxidants, and vitamins, though present in smaller amounts, make oils and fats a very complex mixture. The physicochemical properties and the reactivity of these mixtures vary enormously depending on their composition, which will define the technical and economic feasibility of their use as raw materials for the production and use of biofuel (Suarez et al., 2009).

It has been recently shown that the lipids in sewage sludge constitute a potential raw material for producing biodiesel. The content of approximately 20% of ether-soluble oils and greases, which can be converted into methyl or ethyl fatty acid esters, has been reported (Dufreche et al., 2007). These authors concluded that the integration of lipid extraction processes in 50% of sewage treatment plants in the United States, and the transesterification of the extracted lipids, could lead to the production of approximately 1.8 billion gallons of biodiesel, which represented, at the time, around 0.5% of the annual demand for petroleum diesel in that country. In addition to reducing the price of biodiesel, as the raw material costs virtually nothing, there is the possibility of helping to solve environmental problems related to the treatment and disposal of sludge.

In Singapore, Li et al. (2012) studied biodiesel production from oily residues of grease traps using intracellular lipase-producing strains of *Serratia marcescens* cloned and expressed in *Escherichia coli* in a solvent-free system with methanol as the esterifying alcohol, obtaining a yield of 97% of biodiesel.

In Canada, Siddiquee and Rohani (2011) analyzed the extraction of lipids and the production of biodiesel from wastewater and obtained yields of over

57% of methyl esters via acid catalysis with oils and greases extracted from primary sludge from an effluent treatment plant. Boocock et al. (1992), while evaluating types of solvents for extracting oils and greases from sewage sludge to produce biofuels, found 18% of lipids per dry weight of sludge, of which 65% were formed by free fatty acids, 7% by glycerides, and 28% by unsaponifiable materials, indicating a potential raw material for the production of biodiesel. Oliveira et al. (2017) evaluated biodiesel production from different sources of oily waste from environmental sanitation (industrial, restaurants, effluent treatment, and septic tanks). The best conversion (96.5%) was obtained with *C. antarctica* lipase (5% w/w oil/24 h) and raw material from the grease trap of restaurants at 50°C and a molar ratio of 1:9 (oil: alcohol).

22.3.4 Microalgae oil and biomass

Microalgae biomass is a versatile substrate for proteins, lipids, carbohydrates, pigments, antioxidants, and other substances with the potential industrial application (de Lima Barizão et al., 2021). In addition, these microorganisms have some advantages over plants, such as cultivation in water, reduced use and pressure on arable land, accelerated growth, and resistance to adverse conditions, not to mention the possibility of using effluents and untreated water as a substrate for its cultivation (Alvarez et al., 2021).

Usually, the energy storage lipids of microalgae are unsaturated and polyunsaturated fatty acids (Lupette and Benning, 2020), which makes them a promising renewable source for obtaining biofuels (biodiesel and bio-oil), thus mitigating CO_2 emissions from fossil fuels (Chhandama et al., 2021; Xue et al., 2021). In addition to their low obtaining cost, biofuels produced from microalgae can totally or partially replace those from plant sources (Chisti, 2008; Correa et al., 2019). In this sense, some progress has been reported in obtaining biodiesel from microalgae lipids/oil subjected to enzymatic catalysis (Table 22.4).

Despite these significant advances, the cost-effective production of biodiesel or bio-oil from microalgae still faces some limitations. First, producing biofuels on a large scale is an expensive process with slow financial return (Hannon et al., 2010), so much so that they cannot compete with fossil fuels yet. Another limitation involves the technologies for lysis and extraction of lipids and other intracellular constituents from microalgae, which often employ toxic compounds and make the process more expensive. Studies on *green* extraction alternatives seek to overcome this limitation (Alam et al., 2021).

On the other hand, it is believed that the association of refinery plants that couple the extraction of lipids for the production of biodiesel and the processing of the microalgae "cake" biomass for other purposes, such as

TABLE 22.4 Microalgae lipids and oil used for biofuel production via enzymatic transesterification.

Lipid/oil source	Lipase source	Yield (%)	References
Isochrysis galbana	Lipase from *Candida antarctica* and *Pseudomonas cepacia* immobilized on SBA-15 mesoporous silica	97.2	Sánchez-Bayo et al. (2019)
Spirulina platensis	Recombinant lipase A from *Pseudomonas aeruginosa* displayed on yeast cell surface	87.6	Raoufi and Mousavi Gargari (2018)
Chlorella vulgaris	*Rhizopus oryzae* lipase in few-layer graphene oxide and Fe3O4	71.2	Nematian et al. (2020)
Nannochloropsis sp., *Nannochloropsis oceanica*	*Hermomyces lanuginosus* lipase	85.1 76.3	He et al. (2020)
C. vulgaris var L3	Lipase B from *C. antarctica* as magnetic cross-linked enzyme aggregates	>90	Picó et al. (2018)
Chlorella sp.	Lipase from *Thermomyces lanuginosus*	81.1	He et al. (2018)
Micractinium sp. IC-76	Cross-linked enzyme aggregates of *Burkholderia cepacia* lipase	92.3	Piligaev et al. (2018)
Nannochloropsis gaditana	Novozym 435 (macroporous acrylic resin)	94.7	Navarro López et al. (2015)
Quad-tailed Scenedesmus	*Candida rugosa* lipase (microporous biosilica polymer)	85.7	Bayramoglu et al. (2015)
Scenedesmus obliquus	*Aspergillus niger* whole cell lipase	90.8	Guldhe et al. (2016)
C. vulgaris	Recombinant *Rhizomucor miehei* lipase	>90	Huang et al. (2015)

animal feed or extraction of bioactive compounds, can improve the economics of the process (Siddiki et al., 2022). Finally, the potential for generating energy (biodiesel) through microalgae biomass (Ravanipour et al., 2021) will continue to be intensively investigated over the next few years to contribute to the worldwide energy matrix in the context of renewable fuel sources.

22.4 Catalytic process

Transesterification parameters are essential to increase the yield of reactions. Below is a review of the main parameters in enzymatic transesterification.

22.4.1 Effect of temperature

Temperature is one of the essential factors in the enzymatic transesterification process. Properly increasing the temperature can raise the reaction rate, whereas too high temperatures can denature and inactivate the enzyme. Furthermore, as the boiling point of methanol is 333.7K (Sharma et al., 2008), higher temperatures will burn the alcohol and reduce yields. At the same time, too low temperatures may reduce the solubility of solutes in the reaction system, which is not conducive to its industrial application (He et al., 2022). Therefore it is essential to investigate the best temperature in the process.

Many studies on enzymatic transesterification use the lipase Novozyme 435 (lipase B from *C. antarctica*). This enzyme has good thermal stability, and it is perhaps the most widely used commercial biocatalyst in both academia and industry (Ortiz et al., 2019). High transesterification yields are achieved at temperatures between 50°C and 60°C (Gharat and Rathod, 2013), although yields of approximately 90% are also obtained at 37°C (Cerveró et al., 2014). Pinotti et al. (2018), when using Novozyme 435, obtained high yields in biodiesel production with the temperature ranging between 30°C and 50°C, with no statistically significant differences. He et al., (2022) also reported positive, constant results between 35°C and 65°C. However, these authors used the Novozym 40086 enzyme.

Studies using enzymes from other microorganisms had better performances at milder temperatures. Pooja et al. (2021) obtained maximum yields (96.4%) in the transesterification of kapok oil at 33°C. They investigated a range between 25°C and 40°C and used the immobilized porcine pancreatic lipase. Pinotti et al. (2021) explored temperatures between 30°C and 50°C with the enzyme of *Mucor miehei* and achieved better yields at temperatures between 30°C and 40°C. Khoobbakht et al. (2020) reported a good performance at 35°C using the *Burkholderia cepacia* enzyme. At 40°C, a good yield was also obtained with the *P. cepacia* enzyme (Salis et al., 2005).

22.4.2 Effect of water content

Water content is one of the critical factors that affect the catalytic activity and stability of lipases. Although a certain amount of water is necessary to preserve the active conformation of the enzyme, excessive water in the reaction medium will cause the hydrolysis of the substrate and decrease the production yield (Babaki et al., 2016). Furthermore, the conformation of the enzyme in reaction media with high water content may be more flexible, which may decrease stability while increasing the lipase's transesterification activity. The amount of water in the reaction can be either detrimental or beneficial for the conversion, depending on the specific enzyme and the conditions applied (Lotti et al., 2015). Water contents ranging between 4% and 30% are recommended when using refined oils (Kaieda et al., 1999).

The literature reports different behaviors regarding the yield of biodiesel production according to the amount of water added. Babaki et al. (2016) tested water contents between 0% and 30% (by oil weight) with immobilized enzymes from *Rhizomucor miehei* and *Thermomyces lanuginosa*. They found that the water content had little effect on biodiesel production when using the *R. miehei* enzyme. However, for the enzyme from *T. lanuginosa*, almost complete conversion (98%) was achieved when testing higher water contents. Lara Pizarro and Park (2003) also obtained better yields with higher water contents with the enzyme from *Rhizopus oryzae*. These authors used vegetable oils in waste-activated bleaching earth (it has approximately 40% of its weight as oil) and tested between 15% and 100% water content (by substrate weight). The best yields were achieved at 75% water content. These authors explained that, as the extracted oil has a high viscosity, a high water content was essential to facilitate the mixing of the substrate and to guarantee a greater oil–water interface area on which *R. oryzae* lipase displays activity. Al-Zuhair et al. (2006) conducted esterification in a two-phase system (*n*-hexane/water) and tested water contents between 0% and 25%; the investigators found that the enzyme is more efficient—from a kinetic point of view—at higher water contents. However, the authors found that the conversion after 1 h was higher at low initial water contents. Lv et al. (2017) used the lipase NS81006 to produce biodiesel and found no apparent difference in the final methyl ester yield in various systems with water content ranging from 3% to 10%. However, there were differences in biodiesel quality: the acid value and the contents of monoglyceride and diglyceride were much lower in the system with lower water content.

Some authors have obtained better yields with lower water contents, such as Zheng et al. (2009), who investigated the transesterification of soybean oil for biodiesel production with water contents between 1% and 6%. The highest biodiesel conversion yield was obtained at 2% water content (based on soybean oil weight). Samukawa et al. (2000) performed methanolysis using the enzyme Novovzym 435 and found that the reaction rate decreased with increasing water contents, having tested contents between 0% and 0.5%. Pooja et al. (2021) studied water contents between 2% and 24%, obtaining better biodiesel production yields at 14.5%.

22.4.3 Effect of acyl acceptor

Short-chain alcohols are employed as acyl acceptors in the transesterification of triacylglycerols for enzymatic biodiesel production. The alcohols most frequently used are methanol, ethanol, propanol, isopropanol, pentanol, and butanol, and it has been suggested that an increase in alcohol concentration drives the formation of biodiesel (Lotti et al., 2015).

Transesterifying renewable oils commonly conduct the industrial-scale synthesis of biodiesel with methanol as one of the substrates in the reaction mixture. This reaction, known as methanolysis, results in the production of fatty acid methyl esters (FAMEs). A combination of methyl/ethyl esters can be obtained when both methanol and ethanol are used as substrates. Still, the formation of methyl esters occurs at a higher velocity when compared with that of ethyl esters (Joshi et al., 2010).

Unlike methanol, ethanol is mainly derived from biomass and has environmental advantages, although propanol or butanol guarantees better miscibility between the alcohol and oil phases (Rodrigues et al., 2008). The predilection for methanol can be due to cost issues (Lotti et al., 2015). Methanol has been reported to reduce the activity of multiple commercial lipases when used at optimal molar ratios by inhibiting or denaturing these enzymes and causing a subsequent reduction in the velocity of biodiesel synthesis (Lotti et al., 2015).

The low activity of lipases under excess methanol is due to this organic solvent's inability to dissolve glycerol, which is adsorbed by the enzyme, blocking the access of substrates to the active site (Guldhe et al., 2015). To overcome methanol inhibition, various alternatives have been put forward, including the stepwise addition of methanol, use of other acyl acceptors, use of solvents, and use of methanol-tolerant lipases (Guldhe et al., 2015).

Methyl acetate can be used as the acyl acceptor to replace methanol, though it causes difficulties in product purification and is more expensive than methanol (Ruzich; Bassi, 2010). Kim et al. (2007) and Modi et al. (2007) concluded that ethyl acetate could be an appropriate acyl acceptor for preparing biodiesel following an enzymatic approach.

In the stepwise addition of methanol, the addition occurs at different time intervals, with part of the alcohol and catalyst being added at the start of each step and the by-products removed at the end (Van Gerpen, 2005). According to Talukder et al. (2010), maintaining methanol concentration at a deficient level by stepwise addition is inappropriate for contraposing the causes of the reduction in the yield of biodiesel production. Tolerance to methanol is an inherent property of some lipases, and, interestingly, it seems to be independent of robustness to other environmental factors, such as temperature and organic solvents (Lotti et al., 2015).

22.4.4 Effect of solvent

The molar excess of alcohol over oil can increase the transesterification yield. Still, it can also inactivate the enzyme when the alcohol is insoluble in reactants/products of the transesterification reaction. The addition of organic solvent to this mixture increases the solubility between triacylglycerols and short-chain alcohols and can protect enzymes from inactivation (Antczak et al., 2009). On the other hand, the presence of solvent can raise the overall production costs, interfering with the separation of biodiesel from by-products/excess reactants in the downstream process, so minimum amounts of solvent must be employed (Guldhe et al., 2015)

According to Han et al. (2005), the organic solvents used for propane biodiesel synthesis are excellent for vegetable oil supercritical CO_2 is a suitable solvent for feedstocks with high relative concentrations of free fatty acids as waste cooking oils. Methanolysis yields are usually low in the absence of a suitable solvent, though the reaction is relatively slower in organic solvents such as hexane (Talukder et al., 2010).

As an alternative, another solvent can be added to methanol in a cosolvent system (Alhassan et al., 2014). The use of heptane as a cosolvent increases the mutual solubility between methanol and triglycerides. It can react readily with methanol even under conditions milder than those without a cosolvent (Tan et al., 2010). Royon et al. (2007) concluded that using t-butanol as a solvent in the enzymatic process would increase the progress of methanolysis, as this solvent can dissolve both methanol and glycerol, which is not a substrate for lipases.

It is well known that, in general, organic solvents with a log P value below two are considered unsuitable for biocatalysts. They can remove the essential water around the lipase structure, destabilizing its active conformation (Nie et al., 2006). Several studies have reported the use of ionic liquids as solvents in the enzymatic production of biodiesel. To Ruzich and Bassi (2010), many ionic liquids are considered nonvolatile and may enhance the operational efficiency of lipases in the transesterification of mixtures of oils and short-chain alcohols.

It is known that the enzymatic synthesis of biodiesel can be carried out in solvent-free systems, which are only a mixture of substrates. In this configuration, the alcohol must be added gradually to the reaction mixture to maintain its concentration at relatively low levels (Shimada et al., 2002). In addition, a set of optimal operating conditions of solubility of alcohols in oils must be ensured in the initial step of alcoholysis processes (Antczak et al., 2009).

22.4.5 Effect of molar ratio

The most important variable affecting the yield of esters is the molar ratio between alcohol and triglycerides (Ganesan et al., 2009). Although the molar ratio of alcohol to oil required by stoichiometry is only three, excess alcohol is often used in the industrial production of biodiesel (Madras et al., 2004). A higher molar ratio—about 6:1 alcohol to oil—is employed to maximize ester production (Ferella et al., 2010; Ma and Hanna, 1999).

A higher molar ratio of alcohol than that required by stoichiometry is often employed to achieve maximum biodiesel production. As transesterification is an equilibrium reaction, a significant excess of alcohol is required to shift the response toward maximum ester yields (Ganesan et al., 2009). The stoichiometry of methanolysis requires three moles of methanol and one mole of triglyceride to produce three moles of FAMEs and one mole of glycerol. Increasing the molar ratio of methanol to oil beyond 6:1 can generally increase the product's yield. However, that is limited by the set of operational conditions of the reaction mixture, such as the presence of cosolvents or the concentration of biocatalyst (Sharma et al., 2008).

However, it is well known that the excess of methanol in a reaction mixture is prejudicial to the progress of methanolysis, as it negatively affects the activity of the lipase by decreasing its stability and depleting the oil (Maceiras et al., 2009). Consequently, alcoholysis should be conducted by adding methanol stepwise, which can reduce the amount of alcohol used because only a fraction of the total amount of alcohol and the lipase is added at the start of each step, at the end of which glycerol is removed (Van Gerpen, 2005; Dizge and Keskinler, 2008).

The molar ratio of alcohol to vegetable oil also interferes with glycerol separation because of the increase in solubility (Ganesan et al., 2009). Therefore alcohol-to-oil ratio is a determinant factor for the purity of biodiesel (Atadashi et al., 2011). In addition, excellent molar ratios of methanol to oil result in increased reaction pressure, which demands the adaption of industrial equipment (Cao et al., 2005).

The molar ratio of methanol to oil in the reaction mixture is an essential factor that must be taken into account for increasing biodiesel yields and reducing the efforts on purification and separation of biodiesel from by-products and excess reactants. It is believed that a lower ratio of methanol to oil would reduce production costs by substantially affecting the price of equipment and that of the separation process (Ting et al., 2008). However, it is essential to emphasize that the molar ratio will depend on the lipase and mainly on the reagents used, which correspond to many vegetable oils and short-chain alcohols.

22.5 Reactors and industrial processes

The most used reactor types for biodiesel production by enzymatic transesterification are the packed bed reactor and the fluidized bed reactor (Amini et al., 2017).

They are usually operated in continuous mode to keep the optimal catalysis variables constant and maximize the use of immobilized enzymes (Zik et al., 2020).

The contact surface between the substrate and the catalyst is excellent in packed bed reactor systems, which are usually arranged in a column through which the reactant solution is pumped. These aspects confer the advantage of a lower substrate–enzyme ratio and, consequently, better performance (Watanabe et al., 2001). In addition, they grant a reduced shear stress operation, increasing the long-term stability of the enzymes (Hama et al., 2011) and facilitating the removal of the glycerol by-product without the need for complex treatments (Watanabe et al., 2000).

On the other hand, in a fluidized bed reactor, enzymes are kept immobilized by recycling the substrate pumped through the system (Guisan, 2006), ensuring lower pressure drops, more uniform flow, and reduced preferential flow channels (Fidalgo et al., 2016). However, despite these advantages, there are not many industrial applications for biodiesel production by enzymatic transesterification with this type of reactor (Amini et al., 2017).

Furthermore, it is known that glycerol accumulation reduces the rate of transesterification, as it directly impacts mass transfer. In this sense, a few real-time glycerol separation system models have been investigated for application in transesterification for biodiesel production in fluidized bed reactors (Fidalgo et al., 2016) and microreactors (Gojun et al., 2021; Šalić et al., 2018).

In this sense, implementing microreactors and flow chemistry (Rial et al., 2019) can provide advantages over conventional systems, such as better surface-to-volume ratio, improved mass and energy transfer, and reduced time (Gojun et al., 2021). However, their implementation has economic limitations, such as the need for expensive state-of-the-art instrumentation to monitor the production of the catalyst.

Although the use of inexpensive raw materials and enzymatic inputs can contribute to reducing the costs of biodiesel production (Cesário et al., 2021; Loh et al., 2021; Pinotti et al., 2021), there still exists a limitation regarding the development of a greater variety of cost-effective methods of enzymatic immobilization for industrial application (Lv et al., 2021). In this sense, some research groups have explored methods of catalytic improvement through ultrasound (Tan et al., 2019) and microwave (Kamel Ariffin and Idris, 2022; Lin et al., 2021; Souza et al., 2016) to improve the production of biodiesel.

Therefore a considerable effort has been made to develop new bioreactor engineering solutions for biodiesel production by enzymatic pathways, expanding the in-depth knowledge of the hydrodynamics and kinetics of enzymes used for this purpose.

22.6 Concluding remarks

The use of lipases as catalysts in biodiesel production has been extensively explored in recent years. The environmental benefits justify this interest due to the biodegradability of the biocatalyst in relation to conventional chemical catalytic systems. In addition, they have lower energy consumption and greater versatility in raw materials, including residual oils with high acidity.

In recent years, many advances have been made in favor of the production of lipases with high activity and stability in organic systems to produce biofuels. As a highlight, we can mention the following:

1. Genetic engineering allowed the expression of highly active lipases in bacteria.
2. Several immobilization strategies were developed, and the use of magnetic supports allowed the reuse of biocatalysts with lower energy expenditure.
3. The exploitation of cheaper potential raw materials, such as inedible oils and fats, residual oils, microalgae biomass, etc., allows for a considerable reduction in the costs of the process and, in addition, the reduction of environmental impacts in the case of the use of oily waste.
4. Considerable advances have been published involving optimizing essential variables of the catalytic process and exploring new technologies such as microwave-assisted transesterification, esterification in pressurized fluids, and transesterification in supercritical fluids.

Despite significant advances in improving conversion and technical aspects, there are still some challenges to be overcome for lipases to be used as biocatalysts in biofuel production processes. Economic considerations should be investigated in more detail, as the enzymatic process is more expensive than conventional chemical catalysis. For biocatalysis, long reaction times and large enzyme loads are required for acceptable yields. It is clear that to meet this challenge, the upstream and downstream processes need to be optimized.

The combination of enzymatic immobilization on low-cost magnetic supports aiming at energy-free recyclability and industrial oily waste as raw material seems to be currently one of the essential guidelines for a sustainable perspective in the future.

References

Aguieiras, E.C.G., Cavalcanti-Oliveira, E.D., de Castro, A.M., Langone, M.A.P., Freire, D.M.G., 2014. Biodiesel production from *Acrocomia aculeata* acid oil by (enzyme/enzyme) hydroesterification process: use of vegetable lipase and fermented solid as low-cost biocatalysts. Fuel 135, 315–321. Available from: https://doi.org/10.1016/j.fuel.2014.06.069.

Ahranjani, E.P., Kazemeini, M., Arpanaei, A., 2020. Green biodiesel production from various plant oils using nanobiocatalysts under different conditions. BioEnergy Res. 13, 552–562. Available from: https://doi.org/10.1007/s12155-019-10022-9.

Al-Zuhair, S., Jayaraman, K.V., Krishnan, S., Chan, W.-H., 2006. The effect of fatty acid concentration and water content on the production of biodiesel by lipase. Biochem. Eng. J. 30, 212–217. Available from: https://doi.org/10.1016/j.bej.2006.04.007.

Al-Zuhair, S., Ling, F.W., Jun, L.S., 2007. Proposed kinetic mechanism of the production of biodiesel from palm oil using lipase. Process. Biochem. 42, 951–960. Available from: https://doi.org/10.1016/j.procbio.2007.03.002.

Alam, M.A., Muhammad, G., Khan, M.N., Mofijur, M., Lv, Y., Xiong, W., et al., 2021. Choline chloride-based deep eutectic solvents as green extractants for the isolation of phenolic compounds from biomass. J. Clean. Prod. 309, 127445. Available from: https://doi.org/10.1016/j.jclepro.2021.127445.

Alhassan, Y., Kumar, N., Bugaje, I.M., Pali, H.S., Kathkar, P., 2014. Co-solvents transesterification of cotton seed oil into biodiesel: effects of reaction conditions on quality of fatty acids methyl esters. Energy Convers. Manag. 84, 640–648. Available from: https://doi.org/10.1016/j.enconman.2014.04.080.

Alvarez, A.L., Weyers, S.L., Goemann, H.M., Peyton, B.M., Gardner, R.D., 2021. Microalgae, soil and plants: a critical review of microalgae as renewable resources for agriculture. Algal Res. 54, 102200. Available from: https://doi.org/10.1016/j.algal.2021.102200.

Amini, Z., Ilham, Z., Ong, H.C., Mazaheri, H., Chen, W.H., 2017. State of the art and prospective of lipase-catalyzed transesterification reaction for biodiesel production. Energy Convers. Manag. 141, 339–353. Available from: https://doi.org/10.1016/j.enconman.2016.09.049.

Antczak, M.S., Kubiak, A., Antczak, T., Bielecki, S., 2009. Enzymatic biodiesel synthesis – key factors affecting efficiency of the process. Renew. Energy 34, 1185–1194. Available from: https://doi.org/10.1016/j.renene.2008.11.013.

Ashjari, M., Garmroodi, M., Amiri Asl, F., Emampour, M., Yousefi, M., Pourmohammadi Lish, M., et al., 2020. Application of multi-component reaction for covalent immobilization of two lipases on aldehyde-functionalized magnetic nanoparticles; production of biodiesel from waste cooking oil. Process. Biochem. 90, 156–167. Available from: https://doi.org/10.1016/j.procbio.2019.11.002.

Atadashi, I.M., Aroua, M.K., Aziz, A.A., 2011. Biodiesel separation and purification: a review. Renew. Energy 36, 437–443. Available from: https://doi.org/10.1016/j.renene.2010.07.019.

Babaki, M., Yousefi, M., Habibi, Z., Mohammadi, M., Yousefi, P., Mohammadi, J., et al., 2016. Enzymatic production of biodiesel using lipases immobilized on silica nanoparticles as highly reusable biocatalysts: effect of water, t-butanol and blue silica gel contents. Renew. Energy 91, 196–206. Available from: https://doi.org/10.1016/j.renene.2016.01.053.

Badoei-dalfard, A., Malekabadi, S., Karami, Z., Sargazi, G., 2019. Magnetic cross-linked enzyme aggregates of km12 lipase: a stable nanobiocatalyst for biodiesel synthesis from waste cooking oil. Renew. Energy 141, 874–882. Available from: https://doi.org/10.1016/j.renene.2019.04.061.

Barbosa, O., Ortiz, C., Torres, R., Fernandez-Lafuente, R., 2011. Effect of the immobilization protocol on the properties of lipase B from *Candida antarctica* in organic media: enantiospecifc production of atenolol acetate. J. Mol. Catal. B Enzym. 71, 124–132. Available from: https://doi.org/10.1016/j.molcatb.2011.04.008.

Bartha-Vári, J.-H., Moisă, M.E., Bencze, L.C., Irimie, F.-D., Paizs, C., Toşa, M.I., 2020. Efficient biodiesel production catalyzed by nanobioconjugate of lipase from *Pseudomonas fluorescens*. Molecules 25, 651. Available from: https://doi.org/10.3390/molecules25030651.

Bayramoglu, G., Akbulut, A., Ozalp, V.C., Arica, M.Y., 2015. Immobilized lipase on microporous biosilica for enzymatic transesterification of algal oil. Chem. Eng. Res. Des. 95, 12–21. Available from: https://doi.org/10.1016/j.cherd.2014.12.011.

Boocock, D.G.B., Konar, S.K., Leung, A., Ly, L.D., 1992. Fuels and chemicals from sewage sludge. Fuel 71, 1283–1289. Available from: https://doi.org/10.1016/0016-2361(92)90055-S.

Cao, W., Han, H., Zhang, J., 2005. Preparation of biodiesel from soybean oil using supercritical methanol and co-solvent. Fuel 84, 347–351. Available from: https://doi.org/10.1016/j.fuel.2004.10.001.

Cerveró, J.M., Álvarez, J.R., Luque, S., 2014. Novozym 435-catalyzed synthesis of fatty acid ethyl esters from soybean oil for biodiesel production. Biomass Bioenergy 61, 131–137. Available from: https://doi.org/10.1016/j.biombioe.2013.12.005.

Cesário, L.M., Pires, G.P., Pereira, R.F.S., Fantuzzi, E., da Silva Xavier, A., Cassini, S.T.A., et al., 2021. Optimization of lipase production using fungal isolates from oily residues. BMC Biotechnol. 21, 65. Available from: https://doi.org/10.1186/s12896-021-00724-4.

Chhandama, M.V.L., Satyan, K.B., Changmai, B., Vanlalveni, C., Rokhum, S.L., 2021. Microalgae as a feedstock for the production of biodiesel: a review. Bioresour. Technol. Rep. 15, 100771. Available from: https://doi.org/10.1016/j.biteb.2021.100771.

Chisti, Y., 2008. Biodiesel from microalgae beats bioethanol. Trends Biotechnol. 26, 126–131. Available from: https://doi.org/10.1016/j.tibtech.2007.12.002.

Colen, G., 2006. Isolamento e seleção de fungos filamentosos produtores de lipases. Federal University of Minas Gerais, Belo Horizonte.

Correa, D.F., Beyer, H.L., Fargione, J.E., Hill, J.D., Possingham, H.P., Thomas-Hall, S.R., et al., 2019. Towards the implementation of sustainable biofuel production systems. Renew. Sustain. Energy Rev. 107, 250–263. Available from: https://doi.org/10.1016/j.rser.2019.03.005.

Costa, V.M., Souza, M.C.M., de, Fechine, P.B.A., Macedo, A.C., Gonçalves, L.R.B., 2016. Nanobiocatalytic systems based on lipase-fe3o4 and conventional systems for isoniazid synthesis: a comparative study. Braz. J. Chem. Eng. 33, 661–673. Available from: https://doi.org/10.1590/0104-6632.20160333s20150137.

Dalla-Vecchia, R., Nascimento, M., da, G., Soldi, V., 2004. Aplicações sintéticas de lipases imobilizadas em polímeros. Quim. Nova 27, 623–630. Available from: https://doi.org/10.1590/S0100-40422004000400017.

de Castro, H.F., Mendes, A.A., dos Santos, J.C., de Aguiar, C.L., 2004. Modificação de óleos e gorduras por biotransformação. Quim. Nova 27, 146–156. Available from: https://doi.org/10.1590/S0100-40422004000100025.

de Lima Barizão, A.C., de Oliveira, J.P., Gonçalves, R.F., Cassini, S.T., 2021. Nanomagnetic approach applied to microalgae biomass harvesting: advances, gaps, and perspectives. Environ. Sci. Pollut. Res. 28, 44795–44811. Available from: https://doi.org/10.1007/s11356-021-15260-z.

De Paola, M.G., Ricca, E., Calabrò, V., Curcio, S., Iorio, G., 2009. Factor analysis of transesterification reaction of waste oil for biodiesel production. Bioresour. Technol. 100, 5126–5131. Available from: https://doi.org/10.1016/j.biortech.2009.05.027.

Dizge, N., Keskinler, B., 2008. Enzymatic production of biodiesel from canola oil using immobilized lipase. Biomass Bioenergy 32, 1274–1278. Available from: https://doi.org/10.1016/j.biombioe.2008.03.005.

Dizge, N., Keskinler, B., Tanriseven, A., 2009. Biodiesel production from canola oil by using lipase immobilized onto hydrophobic microporous styrene–divinylbenzene copolymer. Biochem. Eng. J. 44, 220–225. Available from: https://doi.org/10.1016/j.bej.2008.12.008.

Dufreche, S., Hernandez, R., French, T., Sparks, D., Zappi, M., Alley, E., 2007. Extraction of lipids from municipal wastewater plant microorganisms for production of biodiesel. J. Am. Oil Chem. Soc. 84, 181–187. Available from: https://doi.org/10.1007/s11746-006-1022-4.

Felizardo, P., Neiva Correia, M.J., Raposo, I., Mendes, J.F., Berkemeier, R., Bordado, J.M., 2006. Production of biodiesel from waste frying oils. Waste Manag. 26, 487–494. Available from: https://doi.org/10.1016/j.wasman.2005.02.025.

Ferella, F., Mazziotti Di Celso, G., De Michelis, I., Stanisci, V., Vegliò, F., 2010. Optimization of the transesterification reaction in biodiesel production. Fuel 89, 36–42. Available from: https://doi.org/10.1016/j.fuel.2009.01.025.

Fidalgo, W.R.R., Ceron, A., Freitas, L., Santos, J.C., de Castro, H.F., 2016. A fluidized bed reactor as an approach to enzymatic biodiesel production in a process with simultaneous glycerol removal. J. Ind. Eng. Chem. 38, 217–223. Available from: https://doi.org/10.1016/j.jiec.2016.05.005.

Firdaus, M.Y., Brask, J., Nielsen, P.M., Guo, Z., Fedosov, S., 2016. Kinetic model of biodiesel production catalyzed by free liquid lipase from *Thermomyces lanuginosus*. J. Mol. Catal. B Enzym. 133, 55–64. Available from: https://doi.org/10.1016/j.molcatb.2016.07.011.

Gandhi, N.N., 1997. Applications of lipase. J. Am. Oil Chem. Soc. 74, 621–634. Available from: https://doi.org/10.1007/s11746-997-0194-x.

Ganesan, D., Rajendran, A., Thangavelu, V., 2009. An overview on the recent advances in the transesterification of vegetable oils for biodiesel production using chemical and biocatalysts. Rev. Environ. Sci. Bio/Technology 8, 367–394. Available from: https://doi.org/10.1007/s11157-009-9176-9.

Ghaly, 2010. Production of biodiesel by enzymatic transesterification: review. Am. J. Biochem. Biotechnol. 6, 54–76. Available from: https://doi.org/10.3844/ajbbsp.2010.54.76.

Gharat, N., Rathod, V.K., 2013. Enzyme catalyzed transesterification of waste cooking oil with dimethyl carbonate. J. Mol. Catal. B Enzym. 88, 36–40. Available from: https://doi.org/10.1016/j.molcatb.2012.11.007.

Gojun, M., Šalić, A., Zelić, B., 2021. Integrated microsystems for lipase-catalyzed biodiesel production and glycerol removal by extraction or ultrafiltration. Renew. Energy 180, 213–221. Available from: https://doi.org/10.1016/j.renene.2021.08.064.

Goldbeck, R., 2008. Triagem, produção e avaliação da atividade da enzima lipase a partir de leveduras silvestres. UNICAMP.

Guisan, J.M., 2006. Immobilization of Enzymes as the 21st Century Begins. Springer., pp. 1–13. Available from: https://doi.org/10.1007/978-1-59745-053-9_1.

Guldhe, A., Singh, P., Kumari, S., Rawat, I., Permaul, K., Bux, F., 2016. Biodiesel synthesis from microalgae using immobilized *Aspergillus niger* whole cell lipase biocatalyst. Renew. Energy 85, 1002–1010. Available from: https://doi.org/10.1016/j.renene.2015.07.059.

Guldhe, A., Singh, B., Mutanda, T., Permaul, K., Bux, F., 2015. Advances in synthesis of biodiesel via enzyme catalysis: novel and sustainable approaches. Renew. Sustain. Energy Rev. 41, 1447–1464. Available from: https://doi.org/10.1016/j.rser.2014.09.035.

Guo, Z., Sun, Y., 2008. Characteristics of immobilized lipase on hydrophobic superparamagnetic microspheres to catalyze esterification. Biotechnol. Prog. 20, 500–506. Available from: https://doi.org/10.1021/bp034272s.

Guo, J., Sun, S., Liu, J., 2020. Conversion of waste frying palm oil into biodiesel using free lipase A from *Candida antarctica* as a novel catalyst. Fuel 267, 117323. Available from: https://doi.org/10.1016/j.fuel.2020.117323.

Hama, S., Tamalampudi, S., Yoshida, A., Tamadani, N., Kuratani, N., Noda, H., et al., 2011. Enzymatic packed-bed reactor integrated with glycerol-separating system for solvent-free production of biodiesel fuel. Biochem. Eng. J. 55, 66–71. Available from: https://doi.org/10.1016/j.bej.2011.03.008.

Han, H., Cao, W., Zhang, J., 2005. Preparation of biodiesel from soybean oil using supercritical methanol and CO_2 as co-solvent. Process. Biochem. 40, 3148–3151. Available from: https://doi.org/10.1016/j.procbio.2005.03.014.

Hannon, M., Gimpel, J., Tran, M., Rasala, B., Mayfield, S., 2010. Biofuels from algae: challenges and potential. Biofuels 1, 763–784. Available from: https://doi.org/10.4155/bfs.10.44.

He, S., Lian, W., Liu, X., Xu, W., Wang, W., Qi, S., 2022. Transesterification synthesis of high-yield biodiesel from black soldier fly larvae by using the combination of Lipase Eversa Transform 2.0 and Lipase SMG1. Food Sci. Technol. 42. Available from: https://doi.org/10.1590/fst.103221.

He, Y., Wu, T., Wang, X., Chen, B., Chen, F., 2018. Cost-effective biodiesel production from wet microalgal biomass by a novel two-step enzymatic process. Bioresour. Technol. 268, 583–591. Available from: https://doi.org/10.1016/j.biortech.2018.08.038.

He, Y., Zhang, B., Guo, S., Guo, Z., Chen, B., Wang, M., 2020. Sustainable biodiesel production from the green microalgae nannochloropsis: novel integrated processes from cultivation to enzyme-assisted extraction and ethanolysis of lipids. Energy Convers. Manag. 209, 112618. Available from: https://doi.org/10.1016/j.enconman.2020.112618.

Holanda, A., 2004. Biodiesel e Inclusão Social. Câmara dos Deputados do Brasil, Coordenação de Publicações, Brasília.

Huang, J., Xia, J., Jiang, W., Li, Y., Li, J., 2015. Biodiesel production from microalgae oil catalyzed by a recombinant lipase. Bioresour. Technol. 180, 47–53. Available from: https://doi.org/10.1016/j.biortech.2014.12.072.

Jordão, E.P., Pessoa, C.A., 2005. Tratamento de esgotos domésticos, 3rd (ed.) ABES, Rio de Janeiro.

Joshi, H., Moser, B.R., Toler, J., Walker, T., 2010. Preparation and fuel properties of mixtures of soybean oil methyl and ethyl esters. Biomass Bioenergy 34, 14–20. Available from: https://doi.org/10.1016/j.biombioe.2009.09.006.

Kaieda, M., Samukawa, T., Matsumoto, T., Ban, K., Kondo, A., Shimada, Y., et al., 1999. Biodiesel fuel production from plant oil catalyzed by *Rhizopus oryzae* lipase in a water-containing system without an organic solvent. J. Biosci. Bioeng. 88, 627–631. Available from: https://doi.org/10.1016/S1389-1723(00)87091-7.

Kamel Ariffin, M.F., Idris, A., 2022. Fe2O3/Chitosan coated superparamagnetic nanoparticles supporting lipase enzyme from *Candida antarctica* for microwave assisted biodiesel production. Renew. Energy 185, 1362–1375. Available from: https://doi.org/10.1016/j.renene.2021.11.077.

Kareem, S.O., Falokun, E.I., Balogun, S.A., Akinloye, O.A., Omeike, S.O., 2017. Enzymatic biodiesel production from palm oil and palm kernel oil using free lipase. Egypt. J. Pet. 26, 635–642. Available from: https://doi.org/10.1016/j.ejpe.2016.09.002.

Karimi, M., 2016. Immobilization of lipase onto mesoporous magnetic nanoparticles for enzymatic synthesis of biodiesel. Biocatal. Agric. Biotechnol. 8, 182–188. Available from: https://doi.org/10.1016/j.bcab.2016.09.009.

Khoobbakht, G., Kheiralipour, K., Yuan, W., Seifi, M.R., Karimi, M., 2020. Desirability function approach for optimization of enzymatic transesterification catalyzed by lipase immobilized on mesoporous magnetic nanoparticles. Renew. Energy 158, 253–262. Available from: https://doi.org/10.1016/j.renene.2020.05.087.

Kim, S.-J., Jung, S.-M., Park, Y.-C., Park, K., 2007. Lipase catalyzed transesterification of soybean oil using ethyl acetate, an alternative acyl acceptor. Biotechnol. Bioprocess. Eng. 12, 441–445. Available from: https://doi.org/10.1007/BF02931068.

Kirk, O., Christensen, M.W., 2002. Lipases from *Candida antarctica* : unique biocatalysts from a unique origin. Org. Process. Res. Dev. 6, 446–451. Available from: https://doi.org/10.1021/op0200165.

Knothe, G., Gerpen, J.V., Krahl, J., Ramos, L.P., 2006. Manual do Biodiesel. Editora Blucher, São Paulo.

Knothe, G., Van Gerpen, J. (Eds.), 2010. The Biodiesel Handbook. AOCS Publishing. Available from: https://doi.org/10.1201/9781003040262.

Kumar, V., Jahan, F., Raghuwanshi, S., Mahajan, R.V., Saxena, R.K., 2013. Immobilization of Rhizopus oryzae lipase on magnetic Fe3O4-chitosan beads and its potential in phenolic acids ester synthesis. Biotechnol. Bioprocess. Eng. 18, 787–795. Available from: https://doi.org/10.1007/s12257-012-0793-8.

La Rovere, E.L., Soares, J.B., Oliveira, L.B., Lauria, T., 2010. Sustainable expansion of electricity sector: Sustainability indicators as an instrument to support decision making. Renew. Sustain. Energy Rev. 14, 422–429. Available from: https://doi.org/10.1016/j.rser.2009.07.033.

Lara Pizarro, A.V., Park, E.Y., 2003. Lipase-catalyzed production of biodiesel fuel from vegetable oils contained in waste activated bleaching earth. Process. Biochem. 38, 1077–1082. Available from: https://doi.org/10.1016/S0032-9592(02)00241-8.

Lei, L., Bai, Y., Li, Y., Yi, L., Yang, Y., Xia, C., 2009. Study on immobilization of lipase onto magnetic microspheres with epoxy groups. J. Magn. Magn. Mater. 321, 252–258. Available from: https://doi.org/10.1016/j.jmmm.2008.08.047.

Li, A., Ngo, T.P.N., Yan, J., Tian, K., Li, Z., 2012. Whole-cell based solvent-free system for one-pot production of biodiesel from waste grease. Bioresour. Technol. 114, 725–729. Available from: https://doi.org/10.1016/j.biortech.2012.03.034.

Lin, C.-H., Chang, Y.-T., Lai, M.-C., Chiou, T.-Y., Liao, C.-S., 2021. Continuous biodiesel production from waste soybean oil using a nano-Fe3O4 microwave catalysis. Processes 9, 756. Available from: https://doi.org/10.3390/pr9050756.

Loh, J.M., Amelia, Gourich, W., Chew, C.L., Song, C.P., Chan, E.-S., 2021. Improved biodiesel production from sludge palm oil catalyzed by a low-cost liquid lipase under low-input process conditions. Renew. Energy 177, 348–358. Available from: https://doi.org/10.1016/j.renene.2021.05.138.

Lotti, M., Pleiss, J., Valero, F., Ferrer, P., 2015. Effects of methanol on lipases: Molecular, kinetic and process issues in the production of biodiesel. Biotechnol. J. 10, 22–30. Available from: https://doi.org/10.1002/biot.201400158.

Lupette, J., Benning, C., 2020. Human health benefits of very-long-chain polyunsaturated fatty acids from microalgae. Biochimie 178, 15–25. Available from: https://doi.org/10.1016/j.biochi.2020.04.022.

Lv, L., Dai, L., Du, W., Liu, D., 2017. Effect of water on lipase NS81006-catalyzed alcoholysis for biodiesel production. Process. Biochem. 58, 239–244. Available from: https://doi.org/10.1016/j.procbio.2017.04.033.

Lv, L., Dai, L., Du, W., Liu, D., 2021. Progress in Enzymatic biodiesel production and commercialization. Processes 9, 355. Available from: https://doi.org/10.3390/pr9020355.

Ma, Z., Guan, Y., Liu, H., 2006. Superparamagnetic silica nanoparticles with immobilized metal affinity ligands for protein adsorption. J. Magn. Magn. Mater. 301, 469–477. Available from: https://doi.org/10.1016/j.jmmm.2005.07.027.

Ma, F., Hanna, M.A., 1999. Biodiesel production: a review. Bioresour. Technol. 70, 1–15. Available from: https://doi.org/10.1016/S0960-8524(99)00025-5.

Maceiras, R., Vega, M., Costa, C., Ramos, P., Márquez, M.C., 2009. Effect of methanol content on enzymatic production of biodiesel from waste frying oil. Fuel 88, 2130–2134. Available from: https://doi.org/10.1016/j.fuel.2009.05.007.

Madras, G., Kolluru, C., Kumar, R., 2004. Synthesis of biodiesel in supercritical fluids. Fuel 83, 2029–2033. Available from: https://doi.org/10.1016/j.fuel.2004.03.014.

Mehrasbi, M.R., Mohammadi, J., Peyda, M., Mohammadi, M., 2017. Covalent immobilization of *Candida antarctica* lipase on core-shell magnetic nanoparticles for production of biodiesel from waste cooking oil. Renew. Energy 101, 593–602. Available from: https://doi.org/10.1016/j.renene.2016.09.022.

Mendes, A.A., Castro, H.F., de, Pereira, E.B., Furigo Júnior, A., 2005. Aplicação de lipases no tratamento de águas residuárias com elevados teores de lipídeos. Quim. Nova 28, 296–305. Available from: https://doi.org/10.1590/S0100-40422005000200022.

Mendes, A.A., De Oliveira, P.C., De Castro, H.F., Giordano, R.D.L.C., 2011. Aplicação de quitosana como suporte para a imobilização de enzimas de interesse industrial. Quim. Nova 34, 831–840.

Metcalf, E., 2003. Wastewater Engineering treatment Disposal Reuse, 4th (ed.) McGraw Hill, New York.

Miao, C., Yang, L., Wang, Z., Luo, W., Li, H., Lv, P., et al., 2018. Lipase immobilization on amino-silane modified superparamagnetic Fe3O4 nanoparticles as biocatalyst for biodiesel production. Fuel 224, 774–782. Available from: https://doi.org/10.1016/j.fuel.2018.02.149.

Mittelbach, M., 1990. Lipase catalyzed alcoholysis of sunflower oil. J. Am. Oil Chem. Soc. 67, 168–170. Available from: https://doi.org/10.1007/BF02539619.

Modi, M.K., Reddy, J.R.C., Rao, B.V.S.K., Prasad, R.B.N., 2007. Lipase-mediated conversion of vegetable oils into biodiesel using ethyl acetate as acyl acceptor. Bioresour. Technol. 98, 1260–1264. Available from: https://doi.org/10.1016/j.biortech.2006.05.006.

Mosayebi, M., Salehi, Z., Doosthosseini, H., Tishbi, P., Kawase, Y., 2020. Amine, thiol, and octyl functionalization of GO-Fe3O4 nanocomposites to enhance immobilization of lipase for transesterification. Renew. Energy 154, 569–580. Available from: https://doi.org/10.1016/j.renene.2020.03.040.

Naranjo, J.C., Córdoba, A., Giraldo, L., García, V.S., Moreno-Piraján, J.C., 2010. Lipase supported on granular activated carbon and activated carbon cloth as a catalyst in the synthesis of biodiesel fuel. J. Mol. Catal. B Enzym. 66, 166–171. Available from: https://doi.org/10.1016/j.molcatb.2010.05.002.

Navarro López, E., Robles Medina, A., González Moreno, P.A., Jiménez Callejón, M.J., Esteban Cerdán, L., Martín Valverde, L., et al., 2015. Enzymatic production of biodiesel from *Nannochloropsis gaditana* lipids: influence of operational variables and polar lipid content. Bioresour. Technol. 187, 346–353. Available from: https://doi.org/10.1016/j.biortech.2015.03.126.

Nelson, L.A., Foglia, T.A., Marmer, W.N., 1996. Lipase-catalyzed production of biodiesel. J. Am. Oil Chem. Soc. 73, 1191–1195. Available from: https://doi.org/10.1007/BF02523383.

Nematian, T., Shakeri, A., Salehi, Z., Saboury, A.A., 2020. Lipase immobilized on functionalized superparamagnetic few-layer graphene oxide as an efficient nanobiocatalyst for biodiesel production from *Chlorella vulgaris* bio-oil. Biotechnol. Biofuels 13, 1–15. Available from: https://doi.org/10.1186/s13068-020-01688-x.

Nie, K., Xie, F., Wang, F., Tan, T., 2006. Lipase catalyzed methanolysis to produce biodiesel: optimization of the biodiesel production. J. Mol. Catal. B Enzym. 43, 142–147. Available from: https://doi.org/10.1016/j.molcatb.2006.07.016.

Nielsen, J., 2013. Production of biopharmaceutical proteins by yeast. Bioengineered 4, 207–211. Available from: https://doi.org/10.4161/bioe.22856.

Oliveira, J.P., Antunes, P.W.P., Pinotti, L.M., Cassini, S.T.A., 2014. Physico-chemical characterization of oily sanitary waste and of oils and greases extracted for conversion into biofuels. Quim. Nova 37 (4), 597–602. Available from: https://doi.org/10.5935/0100-4042.20140094.

Oliveira, J.P., Antunes, P.W.P., Santos, A.R., et al., 2017. Transesterification of sanitation waste for biodiesel production. Waste Biomass Valor 8, 463–471. Available from: https://doi.org/10.1007/s12649-016-9581-6.

Ortiz, C., Ferreira, M.L., Barbosa, O., dos Santos, J.C.S., Rodrigues, R.C., Berenguer-Murcia, Á., et al., 2019. Novozym 435: the "perfect" lipase immobilized biocatalyst? Catal. Sci. Technol. 9, 2380–2420. Available from: https://doi.org/10.1039/C9CY00415G.

Patel, U., Chauhan, K., Gupte, S., 2018. Synthesis, characterization and application of lipase-conjugated citric acid-coated magnetic nanoparticles for ester synthesis using waste frying oil. 3 Biotech. 8, 211. Available from: https://doi.org/10.1007/s13205-018-1228-9.

Pessela, B.C.C., Dellamora-Ortiz, G., Betancor, L., Fuentes, M., Guisán, J.M., Fernandez-Lafuente, R., 2007. Modulation of the catalytic properties of multimeric β-galactosidase from *E. coli* by using different immobilization protocols. Enzyme Microb. Technol. 40, 310–315. Available from: https://doi.org/10.1016/j.enzmictec.2006.04.015.

Picó, E.A., López, C., Cruz-Izquierdo, Á., Munarriz, M., Iruretagoyena, F.J., Serra, J.L., et al., 2018. Easy reuse of magnetic cross-linked enzyme aggregates of lipase B from *Candida antarctica* to obtain biodiesel from *Chlorella vulgaris* lipids. J. Biosci. Bioeng. 126, 451–457. Available from: https://doi.org/10.1016/j.jbiosc.2018.04.009.

Piligaev, A.V., Sorokina, K.N., Samoylova, Y.V., Parmon, V.N., 2018. Lipid production by microalga *Micractinium* sp. IC-76 in a flat panel photobioreactor and its transesterification with cross-linked enzyme aggregates of *Burkholderia cepacia* lipase. Energy Convers. Manag. 156, 1–9. Available from: https://doi.org/10.1016/j.enconman.2017.10.086.

Pimentel, M., 1996. Produção de lipases por fungos filamentosos: estudos cinéticos e síntese de ésteres. Universidade Estadual de Campinas.

Pinotti, L.M., Benevides, L.C., Lira, T.S., de Oliveira, J.P., Cassini, S.T.A., 2018. Biodiesel production from oily residues containing high free fatty acids. Waste Biomass Valoriz. 9, 293–299. Available from: https://doi.org/10.1007/s12649-016-9776-x.

Pinotti, L.M., Salomão, G.S.B., Benevides, L.C., Antunes, P.W.P., Cassini, S.T.A., de Oliveira, J.P., 2021. Lipase-catalyzed biodiesel production from grease trap. Arab. J. Sci. Eng. 21–27. Available from: https://doi.org/10.1007/s13369-021-05965-1.

Pooja, S., Anbarasan, B., Ponnusami, V., Arumugam, A., 2021. Efficient production and optimization of biodiesel from kapok (*Ceiba pentandra*) oil by lipase transesterification process: addressing positive environmental impact. Renew. Energy 165, 619–631. Available from: https://doi.org/10.1016/j.renene.2020.11.053.

Raoufi, Z., Mousavi Gargari, S.L., 2018. Biodiesel production from microalgae oil by lipase from *Pseudomonas aeruginosa* displayed on yeast cell surface. Biochem. Eng. J. 140, 1–8. Available from: https://doi.org/10.1016/j.bej.2018.09.008.

Ravanipour, M., Hamidi, A., Mahvi, A.H., 2021. Microalgae biodiesel: a systematic review in Iran. Renew. Sustain. Energy Rev. 150, 111426. Available from: https://doi.org/10.1016/j.rser.2021.111426.

Rial, R., Tahoces, P.G., Hassan, N., Cordero, M.L., Liu, Z., Ruso, J.M., 2019. Noble microfluidic system for bioceramic nanoparticles engineering. Mater. Sci. Eng. C. 102, 221–227. Available from: https://doi.org/10.1016/j.msec.2019.04.037.

Ribeiro, A., Castro, F., Carvalho, J., 2011. Influence of Free Fatty Acid Content in Biodiesel Production on Non-Edible Oils. Wastes: Solutions, Treatments and Opportunities. Routledge.

Rodrigues, R.C., Volpato, G., Wada, K., Ayub, M.A.Z., 2008. Enzymatic synthesis of biodiesel from transesterification reactions of vegetable oils and short chain alcohols. J. Am. Oil Chem. Soc. 85, 925–930. Available from: https://doi.org/10.1007/s11746-008-1284-0.

Rosset, D.V., Wancura, J.H.C., Ugalde, G.A., Oliveira, J.V., Tres, M.V., Kuhn, R.C., et al., 2019. Enzyme-catalyzed production of FAME by hydroesterification of soybean oil using the novel soluble lipase NS 40116. Appl. Biochem. Biotechnol. 188, 914–926. Available from: https://doi.org/10.1007/s12010-019-02966-7.

Royon, D., Daz, M., Ellenrieder, G., Locatelli, S., 2007. Enzymatic production of biodiesel from cotton seed oil using t-butanol as a solvent. Bioresour. Technol. 98, 648–653. Available from: https://doi.org/10.1016/j.biortech.2006.02.021.

Ruzich, N.I., Bassi, A.S., 2010. Investigation of enzymatic biodiesel production using ionic liquid as a co-solvent. Can. J. Chem. Eng. Available from: https://doi.org/10.1002/cjce.20263n/a-n/a.

Šalić, A., Tušek, A.J., Sander, A., Zelić, B., 2018. Lipase catalysed biodiesel synthesis with integrated glycerol separation in continuously operated microchips connected in series. N. Biotechnol. 47, 80–88. Available from: https://doi.org/10.1016/j.nbt.2018.01.007.

Salis, A., Bhattacharyya, M.S., Monduzzi, M., Solinas, V., 2009. Role of the support surface on the loading and the activity of *Pseudomonas fluorescens* lipase used for biodiesel synthesis. J. Mol. Catal. B Enzym. 57, 262–269. Available from: https://doi.org/10.1016/j.molcatb.2008.09.015.

Salis, A., Pinna, M., Monduzzi, M., Solinas, V., 2005. Biodiesel production from triolein and short chain alcohols through biocatalysis. J. Biotechnol. 119, 291–299. Available from: https://doi.org/10.1016/j.jbiotec.2005.04.009.

Salis, A., Pinna, M., Monduzzi, M., Solinas, V., 2008. Comparison among immobilised lipases on macroporous polypropylene toward biodiesel synthesis. J. Mol. Catal. B Enzym. 54, 19–26. Available from: https://doi.org/10.1016/j.molcatb.2007.12.006.

Samukawa, T., Kaieda, M., Matsumoto, T., Ban, K., Kondo, A., Shimada, Y., et al., 2000. Pretreatment of immobilized *Candida antarctica* lipase for biodiesel fuel production from plant oil. J. Biosci. Bioeng. 90, 180–183. Available from: https://doi.org/10.1016/S1389-1723(00)80107-3.

Sánchez-Bayo, A., Morales, V., Rodríguez, R., Vicente, G., Bautista, L.F., 2019. Biodiesel production (FAEEs) by heterogeneous combi-lipase biocatalysts using wet extracted lipids from microalgae. Catalysts 9, 1–15. Available from: https://doi.org/10.3390/catal9030296.

Sarno, M., Iuliano, M., 2019. Highly active and stable Fe3O4/Au nanoparticles supporting lipase catalyst for biodiesel production from waste tomato. Appl. Surf. Sci. 474, 135–146. Available from: https://doi.org/10.1016/j.apsusc.2018.04.060.

Shao, P., Meng, X., He, J., Sun, P., 2008. Analysis of immobilized Candida rugosa lipase catalyzed preparation of biodiesel from rapeseed soapstock. Food Bioprod. Process. 86, 283–289. Available from: https://doi.org/10.1016/j.fbp.2008.02.004.

Sharma, Y.C., Singh, B., Upadhyay, S.N., 2008. Advancements in development and characterization of biodiesel: a review. Fuel 87, 2355–2373. Available from: https://doi.org/10.1016/j.fuel.2008.01.014.

Shimada, Y., Watanabe, Y., Sugihara, A., Tominaga, Y., 2002. Enzymatic alcoholysis for biodiesel fuel production and application of the reaction to oil processing. J. Mol. Catal. B Enzym. 17, 133–142. Available from: https://doi.org/10.1016/S1381-1177(02)00020-6.

Siddiki, S.Y.A., Mofijur, M., Kumar, P.S., Ahmed, S.F., Inayat, A., Kusumo, F., et al., 2022. Microalgae biomass as a sustainable source for biofuel, biochemical and biobased value-added products: an integrated biorefinery concept. Fuel 307, 121782. Available from: https://doi.org/10.1016/j.fuel.2021.121782.

Siddiquee, M.N., Rohani, S., 2011. Experimental analysis of lipid extraction and biodiesel production from wastewater sludge. Fuel Process. Technol. 92, 2241–2251. Available from: https://doi.org/10.1016/j.fuproc.2011.07.018.

Silva, M.V.C., Aguiar, L.G., de Castro, H.F., Freitas, L., 2018. Optimization of the parameters that affect the synthesis of magnetic copolymer styrene-divinilbezene to be used as efficient matrix for immobilizing lipases. World J. Microbiol. Biotechnol. 34, 169. Available from: https://doi.org/10.1007/s11274-018-2553-1.

Silva, T.A., Keijok, W.J., Guimaraes, M.C.C., Cassini, S.T.A., Oliveira, J.P., 2022. Impact of immobilization strategies on the activity and reciclability of lipases in nanomagnetic supports. Sci. Rep. Available from: https://doi.org/10.1038/s41598-022-10721-y.

Soares, D., Pinto, A.F., Gonçalves, A.G., Mitchell, D.A., Krieger, N., 2013. Biodiesel production from soybean soapstock acid oil by hydrolysis in subcritical water followed by lipase-catalyzed esterification using a fermented solid in a packed-bed reactor. Biochem. Eng. J. 81, 15–23. Available from: https://doi.org/10.1016/j.bej.2013.09.017.

Souza, L.T.A., Mendes, A.A., de Castro, H.F., 2016. Selection of lipases for the synthesis of biodiesel from jatropha oil and the potential of microwave irradiation to enhance the reaction rate. Biomed. Res. Int. 2016, 1–13. Available from: https://doi.org/10.1155/2016/1404567.

Srivastava, A., Prasad, R., 2000. Triglycerides-based diesel fuels. Renew. Sustain. Energy Rev. 4, 111–133. Available from: https://doi.org/10.1016/S1364-0321(99)00013-1.

Suarez, P.A.Z., Santos, A.L.F., Rodrigues, J.P., Alves, M.B., 2009. Biocombustíveis a partir de óleos e gorduras: desafios tecnológicos para viabilizá-los. Quim. Nova 32, 768–775. Available from: https://doi.org/10.1590/s0100-40422009000300020.

Talukder, M.M.R., Wu, J.C., Fen, N.M., Melissa, Y.L.S., 2010. Two-step lipase catalysis for production of biodiesel. Biochem. Eng. J. 49, 207–212. Available from: https://doi.org/10.1016/j.bej.2009.12.015.

Tan, K.T., Lee, K.T., Mohamed, A.R., 2010. Effects of free fatty acids, water content and co-solvent on biodiesel production by supercritical methanol reaction. J. Supercrit. Fluids 53, 88–91. Available from: https://doi.org/10.1016/j.supflu.2010.01.012.

Tan, S.X., Lim, S., Ong, H.C., Pang, Y.L., 2019. State of the art review on development of ultrasound-assisted catalytic transesterification process for biodiesel production. Fuel 235, 886–907. Available from: https://doi.org/10.1016/j.fuel.2018.08.021.

Thangaraj, B., Jia, Z., Dai, L., Liu, D., Du, W., 2019. Effect of silica coating on Fe3O4 magnetic nanoparticles for lipase immobilization and their application for biodiesel production. Arab. J. Chem. 12, 4694–4706. Available from: https://doi.org/10.1016/j.arabjc.2016.09.004.

Ting, W.-J., Huang, C.-M., Giridhar, N., Wu, W.-T., 2008. An enzymatic/acid-catalyzed hybrid process for biodiesel production from soybean oil. J. Chin. Inst. Chem. Eng. 39, 203–210. Available from: https://doi.org/10.1016/j.jcice.2008.01.004.

United States Environmental Protection Agency, 2010. EPA [WWW Document]. <http://www.epa.gov/air/particlepollution>.

Van Gerpen, J., 2005. Biodiesel processing and production. Fuel Process. Technol. 86, 1097–1107. Available from: https://doi.org/10.1016/j.fuproc.2004.11.005.

Vipin, V.C., Sebastian, J., Muraleedharan, C., Santhiagu, A., 2016. Enzymatic transesterification of rubber seed oil using *Rhizopus oryzae* lipase. Procedia Technol. 25, 1014–1021. Available from: https://doi.org/10.1016/j.protcy.2016.08.201.

Vyas, A.P., Verma, J.L., Subrahmanyam, N., 2010. A review on FAME production processes. Fuel 89, 1–9. Available from: https://doi.org/10.1016/j.fuel.2009.08.014.

Watanabe, Y., Shimada, Y., Sugihara, A., Noda, H., Fukuda, H., Tominaga, Y., 2000. Continuous production of biodiesel fuel from vegetable oil using immobilized *Candida antarctica* lipase. J. Am. Oil Chem. Soc. 77, 355–360. Available from: https://doi.org/10.1007/s11746-000-0058-9.

Watanabe, Y., Shimada, Y., Sugihara, A., Tominaga, Y., 2001. Enzymatic conversion of waste edible oil to biodiesel fuel in a fixed-bed bioreactor. J. Am. Oil Chem. Soc. 78, 703–707. Available from: https://doi.org/10.1007/s11746-001-0329-5.

Xie, W., Huang, M., 2018. Immobilization of *Candida rugosa* lipase onto graphene oxide Fe 3 O 4 nanocomposite: Characterization and application for biodiesel production. Energy Convers. Manag. 159, 42–53. Available from: https://doi.org/10.1016/j.enconman.2018.01.021.

Xue, J., Balamurugan, S., Li, T., Cai, J.-X., Chen, T.-T., Wang, X., et al., 2021. Biotechnological approaches to enhance biofuel producing potential of microalgae. Fuel 302, 121169. Available from: https://doi.org/10.1016/j.fuel.2021.121169.

Yamaura, M., Camilo, R.L., Moura, E. de, Santos, B.Z., 2003. Preparação de nanopartículas magnéticas silanizadas para utilização em técnicas de separação magnética. Inst. Pesqui. Energéticas e Nucl. - IPEN-CNEN/SP 4.

Yan, J., Yan, Y., Liu, S., Hu, J., Wang, G., 2011. Preparation of cross-linked lipase-coated micro-crystals for biodiesel production from waste cooking oil. Bioresour. Technol. 102, 4755–4758. Available from: https://doi.org/10.1016/j.biortech.2011.01.006.

Zeng, H., Liao, K., Deng, X., Jiang, H., Zhang, F., 2009. Characterization of the lipase immobilized on Mg–Al hydrotalcite for biodiesel. Process. Biochem. 44, 791–798. Available from: https://doi.org/10.1016/j.procbio.2009.04.005.

Zheng, Y., Quan, J., Ning, X., Zhu, L.-M., Jiang, B., He, Z.-Y., 2009. Lipase-catalyzed transesterification of soybean oil for biodiesel production in tert-amyl alcohol. World J. Microbiol. Biotechnol. 25, 41–46. Available from: https://doi.org/10.1007/s11274-008-9858-4.

Zik, N.A.F.A., Sulaiman, S., Jamal, P., 2020. Biodiesel production from waste cooking oil using calcium oxide/nanocrystal cellulose/polyvinyl alcohol catalyst in a packed bed reactor. Renew. Energy 155, 267–277. Available from: https://doi.org/10.1016/j.renene.2020.03.144.

Zimmer, K., Luís Borré, G., da Silva Trentin, D., Woicickoski Júnior, C., Piccoli Frasson, A., de Arruda Graeff, A., et al., 2009. Enzimas microbianas de uso terapêutico e diagnóstico clínico. Rev. Lib. 10, 123–137. Available from: https://doi.org/10.31514/rliberato.2009v10n14.p123.

Chapter 23

Microbial enzymes used in textile industry

Francois N. Niyonzima[1], Veena S. More[2], Florien Nsanganwimana[1], Archana S. Rao[3], Ajay Nair[3], K.S. Anantharaju[4] and Sunil S. More[3]

[1]Department of Math, Science and PE, CE, University of Rwanda, Rwamagana, Rwanda, [2]Department of Biotechnology, Sapthagiri College of Engineering, Bangalore, Karnataka, India, [3]School of Basic and Applied Sciences, Dayananda Sagar University, Bangalore, Karnataka, India, [4]Department of Chemistry, Dayananda Sagar College of Engineering, Bangalore, Karnataka, India

23.1 Introduction

The textile industry is a growing sector that increases the global economy of developed and underdeveloped countries. Substances that were utilized in textile include sodium hydroxide, chlorine, peroxide, and pumice stones. They were used in all fabrics processing steps, including desizing, scouring, and bleaching for removal of impurities and polishing for good-looking fabrics (Araujo et al., 2008; Mojsov, 2011; Kumar and Gunasundari, 2018). Before textile enzymes' use, it was consuming a lot of energy, water, and chemicals, causing environmental pollution due to the release of toxic compounds (such as toxic dyes) into water bodies. However, the utilization of enzymes in textile and fibers processing resulted in eco-friendly, cost-effective, and nontoxic processes. In this way, a lot of money is saved (Mojsov, 2011; Sarkar et al., 2020).

The textile enzymes can be obtained from plant, microbial, and animal sources. However, the microbial textile enzymes are most of the time preferred owing to their secretion in a significant amount, bacteria/fungi can be genetically modified in order to produce textile enzymes in important quantities, and the optimization conditions are cost-effectively and easily optimized. The chief producers of microbial textile enzymes are bacteria such as *Bacillus* and *Streptomyces* species and fungi such as *Trichoderma* and *Aspergillus* species (Sen et al., 2021). The textile enzymes are divided into two types: oxidoreductases and hydrolases, catalyzing redox and hydrolysis biochemical reactions. The oxidoreductases comprise peroxidases, catalases,

Biotechnology of Microbial Enzymes. DOI: https://doi.org/10.1016/B978-0-443-19059-9.00006-2

© 2023 Elsevier Inc. All rights reserved.

ligninases, and laccases, while hydrolases include pectinases, amylases, lipases/esterases, cutinases and cellulases, and proteases (Shen and Smith, 2015). Even if the textile enzymes are very specific in speeding up the reactions, increasing the fabric quality, and reducing water use, pollution, and energy consumption, their prices remain high (Cavaco-Paulo et al., 1996; Pazarlıoğlu et al., 2005). Although textile enzymes can effectively replace conventional substances utilized in the textile industry, their commercialization remains a significant issue since they are produced in insignificant amounts. Thus the search for the microorganisms that may overproduce textile enzymes cost-effectively without causing environmental pollution is necessary. In addition, the search for various, less polluting and inexpensive commercial textile enzymes is also vital (Couto and Toca-Herrera, 2006). The present chapter discusses the optimization of process parameters for microorganism-producing textile enzymes. The microbial enzymes utilized in the textile industries are also highlighted.

23.2 Isolation and identification of microorganism-producing textile enzymes

Microorganism-producing textile enzymes are mostly isolated from soil (Pereira et al., 2005; Chimata et al., 2011; Rajendran et al., 2011; Battan et al., 2012; Khalid-Bin-Ferdaus et al., 2018; Gururaj et al., 2021). *Coriolopsis byrsina* producing textile laccase and *Lentinus* sp. SXS48 secreting lignin and manganese peroxidases were isolated from rotten wood (Gomes et al., 2009). Compost was the best isolation medium for cellulase production by *Thermomonospora* sp. (T-EG). A watery environment is also used as an isolation medium for textile enzymes. For instance, the seawater was the *Nocardiopsis dassonvillei* NRC2aza producing a textile collagenase (Table 23.1). Soil thus harbors uncountable microorganisms, and some of them produce enzymes for textile industries.

Various culture media containing agar are used to isolate and screen bacteria- and fungi-producing microorganisms. For example, cellulose paper agar was used to screen *Thermomonospora* sp. (T-EG), producing textile cellulase (Anish et al., 2007). Abdel-Fattah (2013) screened *N. dassonvillei* NRC2aza-producing collagenase with the help of chitin waste agar. The cutin agar was the isolation medium of choice for the cutinase secretion by *Acinetobacter baumannii* AU10 (Gururaj et al., 2021). *Bacillus pumilus* ASH produced xylanase when xylan agar was used as a screening medium (Battan et al., 2012). For fungi, PDA is often used. For instance, it was used as a screening medium for the amylase by *Aspergillus niger* (Khalid-Bin-Ferdaus et al., 2018), laccases by *Trametes hirsuta* (Pereira et al., 2005) and *C. byrsina* (Gomes et al., 2009), and by lignin/manganese peroxidases by *Lentinus* sp. SXS48 (Gomes et al., 2009). However, starch agar and pectin agar were

TABLE 23.1 Isolation medium and identification of textile enzyme-producing microorganisms.

Enzyme produced	Bacterial or fungal species	Isolation source	Isolation and screening medium	Microbial identification	References
Cellulase	*Thermomonospora* sp. (T-EG)	Compost	Cellulose paper agar	Morphological, chromosomal DNA and cell wall composition aspects	Anish et al. (2007)
Collagenase	*Nocardiopsis dassonvillei* NRC2aza	Seawater	Chitin waste agar	16S rDNA sequencing	Abdel-Fattah (2013)
Cutinase	*Acinetobacter baumannii* AU10	Soil	Cutin agar	16S rDNA sequencing	Gururaj et al. (2021)
Nitrile hydratase	*Rhodococcus* NCIMB 11216	Lab collection	Propionitrile agar	Nucleic acid sequencing	Tauber et al. (2000)
Xylanase	*Bacillus pumilus* ASH	Soil	Xylan agar	16S rDNA sequencing	Battan et al. (2012)
Amylase	*Aspergillus niger* MK 07	Dump yards soil	Starch agar	Microscopic and morphological aspects	Chimata et al. (2011)
Amylase	*A. niger*	Soil	PDA	Microscopic and morphological aspects	Khalid-Bin-Ferdaus et al. (2018)
Laccase	*Trametes hirsuta*	Forest soil	PDA	Morphological and microscopic features	Pereira et al. (2005)
Laccase	*Coriolopsis byrsina*	Rotten wood	PDA	Microscopic and morphological characteristics	Gomes et al. (2009)

(*Continued*)

TABLE 23.1 (Continued)

Enzyme produced	Bacterial or fungal species	Isolation source	Isolation and screening medium	Microbial identification	References
Lignin peroxidase	*Lentinus* sp. SXS48	Rotten wood	PDA	Microscopic and morphological characteristics	Gomes et al. (2009)
Manganese peroxidase	*Lentinus* sp. SXS48	Rotten wood	PDA	Microscopic and morphological aspects	Gomes et al. (2009)
Pectinase	*Fusarium* sp.	Soil	Pectin agar	Microscopic and morphological features	Rajendran et al. (2011)

used to screen amylase and pectinase by *A. niger* MK 07 and *Fusarium* sp., respectively (Chimata et al., 2011; Rajendran et al., 2011) (Table 23.1).

Sometimes, inducers are necessary to screen the desired enzymes. Propionitrile compound was necessary to induce nitrile hydratase production by *Rhodococcus* NCIMB 11216 (Tauber et al., 2000). The occurrence of inducer hydrolysis by the produced textile enzyme is an indication that the isolated microorganisms are producing the desired textile enzyme. This is shown by a clear zone on the solidified medium with agar. Similarly, the carboxy methyl cellulose induced cellulase secretion by *Trichoderma reesei* MTCC 162 (Saravanan et al., 2013). Tetracycline and nystatin are often added to Petri culture plates to inhibit bacterial and fungal species' growth when screening for fungi- and bacteria-producing textile enzymes, respectively (Niyonzima, 2018).

Microorganisms producing textile industrial enzymes are identified at genus and species levels based on microscopic and morphological aspects as in the Bergey's manual (Logan and De Vos, 2009); for instance, *A. niger* (Khalid-Bin-Ferdaus et al., 2018), *T. hirsuta* (Pereira et al., 2005), *C. byrsina* and *Lentinus* sp. SXS48 (Gomes et al., 2009), and *Fusarium* sp. (Rajendran et al., 2011). However, microorganisms are currently identified by nucleic acid sequencing. For instance, 16S rDNA sequencing was used to identify *N. dassonvillei* NRC2aza (Abdel-Fattah, 2013), *A. baumannii* AU10 (Gururaj et al., 2021), and *B. pumilus* ASH (Battan et al., 2012). Likewise, the *Thermomonospora* sp. (T-EG) that produces textile cellulase was identified based on the chromosomal DNA and cell wall composition aspects (Anish et al., 2007) (Table 23.1). After microbial identification, the microbial species exhibiting a higher level of textile enzyme activity are maintained on nutrient potato dextrose agar slants for further usage.

23.3 Production of textile enzymes by bacteria and fungi

For the commercialization of a textile enzyme, the microorganism-producing textile enzyme has to be isolated from the environment, screened with appropriate culture media, identified based on the cultural and molecular aspects, and stored in good conditions. Fermentation with relevant sterilized media has to be carried out after optimization of process parameters. The crude textile enzyme has to be obtained after cell debris removal by centrifugation. The textile enzyme has to be then partially or totally purified and then characterized. After characterization, it has to be applied in various textile processing steps. The produced textile enzyme has to be certified by regulatory bodies before commercialization. Fig. 23.1 shows the various processes to produce and commercialize a microbial textile enzyme.

The production of industrial enzymes, including textile enzymes, has to be inexpensive. One of the nutritional parameters to be taken into account is a substrate. Cost-effective substrates have thus to be utilized during textile

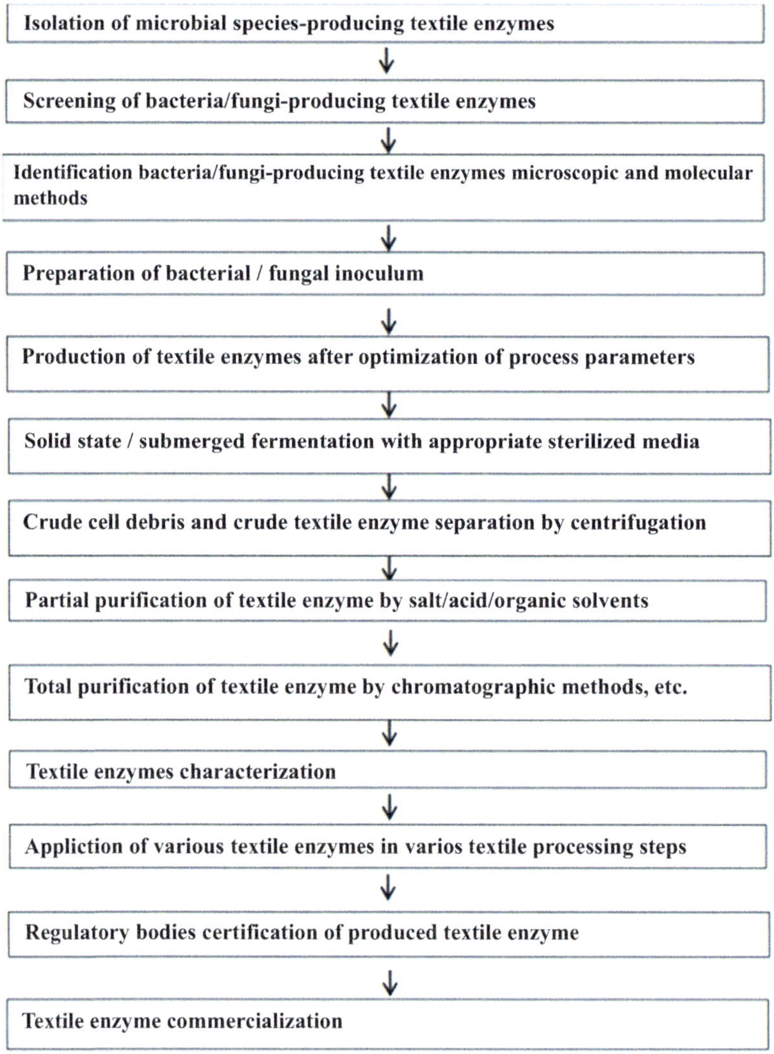
Isolation of microbial species-producing textile enzymes
Screening of bacteria/fungi-producing textile enzymes
Identification bacteria/fungi-producing textile enzymes microscopic and molecular methods
Preparation of bacterial / fungal inoculum
Production of textile enzymes after optimization of process parameters
Solid state / submerged fermentation with appropriate sterilized media
Crude cell debris and crude textile enzyme separation by centrifugation
Partial purification of textile enzyme by salt/acid/organic solvents
Total purification of textile enzyme by chromatographic methods, etc.
Textile enzymes characterization
Appliction of various textile enzymes in varios textile processing steps
Regulatory bodies certification of produced textile enzyme
Textile enzyme commercialization

FIGURE 23.1 Various processes to produce and commercialize a microbial textile enzyme.

enzyme production by bacteria and fungi (Table 23.2). Wheat bran was considered to inexpensively produce textile laccase obtained from *T. hirsuta* (Abadulla et al., 2000; Pereira et al., 2005), cellulase from *Thermomonospora* sp. (T-EG) (Anish et al., 2007), and alpha-amylase from *A. niger* (Khalid-Bin-Ferdaus et al., 2018). Similarly, textile laccase of *C. byrsina* and manganese peroxidase and lignin peroxidase of *Lentinus* sp. SXS48 were cost-effectively produced using rice straw as the best cheap substrate (Gomes et al., 2009). Likewise, chitin wastes and tomato peels were the vital substrates to secrete

TABLE 23.2 Cost-effective substrates utilized during textile enzymes production by bacteria and fungi.

Substrate used	Enzymes produced	Microorganism	References
Chitin wastes	Collagenase	*Nocardiopsis dassonvillei* NRC2aza	Abdel-Fattah (2013)
Rice straw	Laccase	*Coriolopsis byrsina*	Gomes et al. (2009)
Rice straw	Manganese peroxidase	*Lentinus* sp. SXS48	Gomes et al. (2009)
Rice straw	Lignin peroxidase	*Lentinus* sp. SXS48	Gomes et al. (2009)
Tomato peels	Cutinase	*Acinetobacter baumannii* AU10	Gururaj et al. (2021)
Wheat bran	Laccase	*Trametes hirsuta*	Pereira et al. (2005)
Wheat bran	Alpha-amylase	*Aspergillus niger*	Khalid-Bin-Ferdaus et al. (2018)
Wheat bran	Cellulase	*Thermomonospora* sp. (T-EG)	Anish et al. (2007)
Wheat bran flake	Laccase	*T. hirsuta*	Abadulla et al. (2000)

textile collagenase by *N. dassonvillei* NRC2aza (Abdel-Fattah, 2013) and cutinase by *A. baumannii* AU10 (Gururaj et al., 2021), respectively.

Fermentations (solid-state and/or submerged) are utilized to produce textile enzymes in industries. Each procedure has its own advantages and disadvantages. Thus both methods have to be carried out to produce a textile enzyme by a microbial species, and the one with the highest yield has to be considered. For instance, solid-state fermentation was the best method to produce lignin peroxidase and manganese peroxidase using *Lentinus* sp. SXS48 and laccase by *C. byrsina* (Gomes et al., 2009). *N. dassonvillei* NRC2aza and *A. baumannii* AU10 maximally, respectively, produced textile collagenase (Abdel-Fattah, 2013) and cutinase (Gururaj et al., 2021) under solid-state fermentation.

Like solid-state fermentation, submerged fermentation was also used to produce various textile enzymes. For example, the textile laccase was secreted by *T. hirsuta* under submerged conditions (Abadulla et al., 2000). Anish et al. (2007) and Chimata et al. (2011) produced cellulase from *Thermomonospora* sp. (T-EG) and amylase from *A. niger* MK 07, respectively, under submerged fermentation. Likewise, *Fusarium* sp. optimally produced a pectinase under submerged fermentation (Rajendran et al., 2011). Most of the textile enzymes are extracellularly produced by bacteria and fungi. As the production of microbial textile enzymes is expensive, the search for bacteria or fungi able to produce all textile hydrolases and oxidases in a single fermentation medium is necessary. This will allow the various textile steps to be combined. Thus the cost and processing time will be less.

23.4 Process aspect optimization for producing microbial textile enzymes

After identifying microorganism-producing textile enzymes, the physico- and nutritional parameters have to be optimized, one parameter each time, maintaining other constants. Some of these parameters are nitrogen and carbon sources as nutritional factors, as well as incubation time, agitation, pH, inoculum level, temperature, etc., as physicochemical factors. The optimization of parameters for textile enzyme production by bacteria and fungi is prerequisite since the production, in this case, is cost-effective.

23.4.1 Effect of initial pH medium for the secretion of textile enzymes by microorganisms

The culture medium's initial pH for the secretion of textile enzymes is an essential factor in regulating microbial growth (Niyonzima and More, 2013). The optimal pH for the secretion of textile enzymes can be acidic, neutral, or basic (Table 23.3). This will depend on the textile processing step, as some are carried out at basic or acidic pH. For example, the amylases of *Bacillus* species (Haq et al., 2010; Chand et al., 2014) and *A. niger* (Chimata et al., 2011; Sreelaakshmi et al., 2014; Khalid-Bin-Ferdaus et al., 2018) were optimally produced under acidic conditions. A medium of acidic pH was also required to maximally produce laccases by *T. hirsuta* (Abadulla et al., 2000; Pereira et al., 2005) and by *C. byrsina* (Gomes et al., 2009). Optimal pH of 5 and 6 was necessary for textile cellulase and pectinase production by *T. reesei* MTCC 162 (Saravanan et al., 2013) and *Fusarium* sp. (Rajendran et al., 2011), respectively. *Lentinus* sp. SXS48 secreted manganese peroxidase and lignin peroxidase at low acidic pH of 3.5 (Gomes et al., 2009). A neutral pH was needed for the production of textile amylase by *Thermotoga petrophila* gene cloned into *Escherichia coli* (Tauber et al., 2000), cellulase by *Chaetomium globosum* (Chinnamma and Antony, 2015), and nitrile hydratase by *Rhodococcus rhodochrous* NCIMB 11216 (Tauber et al., 2000). An alkaline pH of 8.0 was necessary to optimally produce textile xylanase from *B. pumilus* ASH (Battan et al., 2012), collagenase from *N. dassonvillei* NRC2aza (Battan et al., 2012), and cutinase from *A. baumannii* AU10 (Gururaj et al., 2021). A fungus *Thermomonospora* sp. (T-EG) secreted an alkaline cellulase at an optimum pH of 9.0 (Anish et al., 2007). The variation

TABLE 23.3 Optimum conditions for textile enzymes production by microorganisms.

Bacterial or fungal species	Textile enzyme produced	pH	T (°C)	Agitation (rpm)	Inoculum level (%)	Incubation time	Preferred carbon source	Good nitrogen source	References
Bacterial species									
Acinetobacter baumannii AU10	Cutinase	8	36	150	1	24 h	Sucrose	Gelatin	Gururaj et al. (2021)
Bacillus amyloliquefaciens EMS-6	Amylase	6.5	37	400	8	48 h	Soluble starch	Beef extract, yeast extract, peptone	Haq et al. (2010)
B. amyloliquefaciens UNG-16	Amylase	7	37	200	2	48 h	Glucose	Peptone and $(NH_4)_2SO_4$	Haq et al. (2010)
Bacillus pumilus ASH	Xylanase	8	37	200	2.5	26 h	Wheat bran	Yeast extract, peptone, KNO_3	Battan et al. (2012)
Bacillus sp. KR-8104	Amylase	5.5	45	n s	ns	48 h	Soya powder and chicken excrement	Soya powder and chicken excrement	Chand et al. (2014)
Nocardiopsis dassonvillei NRC2aza	Collagenase	8	37	150	6	6 days	Chitin wastes	Chitin wastes	Abdel-Fattah (2013)
Rhodococcus rhodochrous NCIMB 11216	Nitrile hydratase	7	30	180	2	24 h	Glucose	Propionitrile, yeast extract	Tauber et al. (2000)

(*Continued*)

TABLE 23.3 (Continued)

Bacterial or fungal species	Textile enzyme produced	pH	T (°C)	Agitation (rpm)	Inoculum level (%)	Incubation time	Preferred carbon source	Good nitrogen source	References
Thermotoga petrophila gene cloned into *Escherichia coli*	Amylase	7	22	200	3	24 h	Lactose	Tryptone and yeast extract	Zafar et al. (2019)
Fungal species									
Aspergillus niger	Amylase	5	40	ns	ns	4 days	Wheat starch	Corn	Sreelaakshmi et al. (2014)
A. niger	Amylase	6.2	28	160	ns	4 days	Glucose	NH_4Cl	Khalid-Bin-Ferdaus et al. (2018)
A. niger MK 07	Amylase	5	30	250	10	4 days	Sucrose	Corn steep liquor	Chimata et al. (2011)
Chaetomium globosum	Cellulase	7.2	50	ns	ns	6 days	Wheat bran	Sodium nitrate and Yeast extract	Chinnamma and Antony (2015)
Coriolopsis byrsina	Laccase	3.5	60	ns	3	5 weeks	Rice straw	Rice straw and $(NH_4)_2SO_4$	Gomes et al. (2009)
Fusarium sp.	Pectinase	6	27	ns	ns	48 h	Pectin and sucrose	Tryptone and yeast extract	Rajendran et al. (2011)

Lentinus sp. SXS48	Manganese peroxidase	3.5–4.0	55	ns	3	5 weeks	Rice straw	Rice straw and $(NH_4)_2SO_4$	Gomes et al. (2009)
L. sp. SXS48	Lignin peroxidase	3.5	40	ns	3	5 weeks	Rice straw	Rice straw and $(NH_4)_2SO_4$	Gomes et al. (2009)
Thermomonospora sp. (T-EG)	Cellulase	9	50	200	10	5 days	Cellulose paper powder	Yeast extract, $(NH_4)_2SO_4$, urea	Anish et al. (2007)
Trametes hirsuta	Laccase	5	30	130	ns	5 days	Wheat bran and glucose	Yeast extract and NH_4Cl	Pereira et al. (2005)
T. hirsuta	Laccase	5	30	150	ns	10 days	Wheat bran and glucose	Yeast extract and NH_4Cl	Abadulla et al. (2000)
Trichoderma reesei MTCC 162	Cellulase	5	50	150	10	7 days	Carboxy methyl cellulose	Peptone, urea and yeast extract	Saravanan et al. (2013)

ns: Not specified/not determined.

in textile enzymes at different pH could be attributed to the specificity of fungal/bacterial strains.

23.4.2 Influence of incubation temperature on the production of textile enzymes by microorganisms

The initial optimization has a crucial role in textile enzyme secretion as it regulates the growth of fungi/bacteria (Niyonzima and More, 2013). The optimal incubation temperature observed for bacteria-producing textile enzymes is 30°C–37°C (Table 23.3). For instance, 37°C was optimum for the secretion of xylanase (Battan et al., 2012) and amylase (Haq et al., 2010) by *Bacillus* species, as well as collagenase by *N. dassonvillei* NRC2aza (Abdel-Fattah, 2013). Tauber et al. (2000) reported an optimal temperature of 30°C when producing nitrile hydratase from *R. rhodochrous* NCIMB. A lower and a higher temperature of 22°C and 45°C were needed for the production of bacterial amylases (Chand et al., 2014; Zafar et al., 2019). The fungal textile enzymes are produced in the 30°C–60°C range. For example, the textile cellulase was optimally produced by *Thermomonospora* sp. (T-EG) (Anish et al., 2007), *T. reesei* MTCC 162 (Saravanan et al., 2013), and *C. globosum* (Chinnamma and Antony, 2015) at 50°C. 30°C was the optimum incubation temperature for *T. hirsuta* producing laccase (Abadulla et al., 2000; Pereira et al., 2005) and *A. niger* MK 07 secreting textile amylase (Chimata et al., 2011). Gomes et al. (2009) and Sreelaakshmi et al. (2014) reported 40°C as the optimal incubation temperature for the secretion of lignin peroxidase and amylase by *Lentinus* sp. SXS48 and *A. niger*, respectively. *C. byrsina* optimally secreted a textile laccase at 60°C (Gomes et al., 2009), while *Fusarium* sp. produced a pectinase at 27°C (Rajendran et al., 2011). In general, the yield of textile enzymes at a higher initial incubation temperature is low. This could be attributed to the thermostability of these enzymes used in the textile industry.

23.4.3 Effect of agitation on the secretion of textile enzymes by microorganisms

Some microorganisms producing textile enzymes are shaken in order to produce them in important amounts. Most culture flasks are shaken at 150–200 rpm range (Table 23.3). For instance, 150 rpm was optimum for the secretion of cutinase by *A. baumannii* AU10 (Gururaj et al., 2021), collagenase by *N. dassonvillei* NRC2aza (Abdel-Fattah, 2013), laccase by *T. hirsuta* (Abadulla et al., 2000), and cellulase by *T. reesei* MTCC 162 (Saravanan et al., 2013). Culture flasks of *Bacillus* species were agitated at 200 rpm in order to produce textile amylase and xylanase in an important amount (Haq et al., 2010; Battan et al., 2012). 200 rpm was also necessary to overproduce textile cellulase by *Thermomonospora* sp. (T-EG) (Anish et al.,

2007). A higher and a lower agitation of 400 and 130 rpm were observed for the laccase secretion by *T. hirsuta* (Pereira et al., 2005) and amylase production by *Bacillus amyloliquefaciens* EMS-6 (Haq et al., 2010). Amylase by *A. niger* (Khalid-Bin-Ferdaus et al., 2018) and nitrile hydratase by *R. rhodochrous* NCIMB (Tauber et al., 2000) were produced in a vital amount when the Erlenmeyer flasks were shaken at 160 and 180 rpm, respectively. When a low agitation is considered, there may be less secretion of textile enzymes owing to poor bacterial/fungal growth due to O_2 transfer limitation (Niyonzima et al., 2020).

23.4.4 Influence of inoculum concentration on the production of textile enzymes by microorganisms

The inoculum level utilized for textile enzymes production has to be optimized. A range of 1%–10% was seen for textile enzyme production by bacteria and fungi (Table 23.3). Inoculum size of 1% was used to maximally produce a textile cutinase by *A. baumannii* AU10 (Gururaj et al., 2021). Haq et al. (2010) reported 2% as the optimal inoculum size for amylase production by *B. amyloliquefaciens* UNG-16. The textile amylase from *A. niger* MK 07 (Chimata et al., 2011) and a textile cellulase from *Thermomonospora* sp. (T-EG) (Anish et al., 2007) and *T. reesei* MTCC 162 (Saravanan et al., 2013) were secreted in a significant amount when the inoculum level was 10%. *Lentinus* sp. SXS48 optimally produced lignin peroxidase and lignin peroxidase, and *C. byrsina* produced laccase when the inoculum concentration was 3%. Other optimal inoculum levels noticed were 2.5%, 6%, and 8% for xylanase produced by *B. pumilus* ASH (Battan et al., 2012), collagenase by *N. dassonvillei* NRC2aza (Abdel-Fattah, 2013), and amylase by *Bacillus* sp. (Haq et al., 2010), respectively. The microorganisms producing textile enzymes are usually grown under shaking conditions since conical flask agitation allows proper nutrients available to the organisms.

23.4.5 Effect of initial time on the secretion of textile enzymes by microorganisms

Each microorganism-producing textile enzyme has an optimal incubation time period beyond which its production gets decreased. Most of the bacteria-secreting textile enzymes have 24–48 has optimum incubation time (Table 23.3). For instance, 24 h was the optimal time period for the cutinase production by *A. baumannii* AU10 (Gururaj et al., 2021) and for nitrile hydratase secretion by *R. rhodochrous* NCIMB 11216 (Tauber et al., 2000). *B. pumilus* ASH produced textile xylanase when the incubation period was 26 h (Battan et al., 2012). *Bacillus* species secreted textile enzymes when the incubation time was 48 h (Haq et al., 2010; Chand et al., 2014). Fungal species secrete textile enzymes at higher incubation temperatures compared to bacterial species. Incubation time ranging from 2 to 10 days was seen for most fungi producing fungi. *A. niger* optimally produced textile amylase after 4 days (Chimata et al., 2011; Sreelaakshmi et al., 2014; Khalid-Bin-Ferdaus et al., 2018). Textile cellulases were optimally produced by *Thermomonospora* sp. (T-EG) (Anish et al., 2007), *C. globosum* (Chinnamma and Antony, 2015), and *T. reesei* MTCC 162 (Saravanan et al., 2013) when the incubation time was 5, 6, and 7 days, respectively. A lower and higher incubation time of 2 and 10 days were observed for pectinase production (Rajendran et al., 2011) and laccase secretion by *T. hirsuta* (Abadulla et al., 2000), respectively. At a high incubation time period, less textile enzyme production was observed owing to toxic substances production and/or nutrient exhaustion (Niyonzima and More, 2015).

23.4.6 Influence of carbon sources on the production of textile enzymes by microorganisms

Monosaccharides such as glucose were the best carbon sources for the secretion of textile nitrile hydratase by *R. rhodochrous* NCIMB 11216 (Tauber et al., 2000) and amylases by *B. amyloliquefaciens* (Haq et al., 2010) and *A. niger* (Khalid-Bin-Ferdaus et al., 2018). Similarly, disaccharides such as sucrose were the carbon source of choice for the textile cutinase secretion by *A. baumannii* AU10 (Gururaj et al., 2021) and amylase by *A. niger* MK 07 (Chimata et al., 2011). Lactose was also a carbon source for the production of amylase by *T. petrophila* gene cloned into *E. coli* (Zafar et al., 2019). Likewise, a polysaccharide known as soluble starch produced a textile amylase maximally when *Amyloliquefaciens* EMS was inoculated into the fermentation medium (Haq et al., 2010). Sometimes, inexpensive substrates are used as carbon sources. For instance, wheat bran was considered as a carbon source leading to maximum secretion of xylanase by *B. pumilus* ASH (Battan et al., 2012) and cellulase by *C. globosum* (Chinnamma and Antony, 2015). A similar observation was seen when the rice straw was utilized for the production of laccase by *C. byrsina* and lignin and manganese peroxidases by *Lentinus* sp. SXS48 (Gomes et al., 2009). Saravanan et al. (2013) produced a textile cellulase from *T. reesei* MTCC 162 with carboxy methyl cellulose as the sole carbon source. The cellulose paper powder was also a cheap carbon source for textile cellulase production by *Thermomonospora* sp. (T-EG) (Anish et al., 2007). A combination of carbon sources is sometimes necessary to produce textile enzymes in an important quantity—for example, a mixture of soya powder and chicken excrement for amylase secretion by *Bacillus* sp. KR-8104 (Chand et al., 2014), pectin and sucrose for pectinase secretion by *Fusarium* sp. (Rajendran et al., 2011), and wheat bran and glucose for laccase production by *T. hirsuta* (Abadulla et al., 2000; Pereira et al., 2005) was necessary to overproduce textile enzymes. Carbon sources usually produce textile enzymes when used in minute amounts; however, an enzyme/catabolite repression can be noticed at high concentrations.

23.4.7 Effect of nitrogen sources on the production of textile enzymes by microorganisms

Organic and/or inorganic nitrogen sources are utilized as nutritionals for the microorganisms producing textile enzymes. The best organic nitrogen source for cutinase production was gelatin by *A. baumannii* AU10 (Gururaj et al., 2021). Corn and NH_4Cl are the best organic and inorganic compounds for the secretion of textile amylases by *A. niger* (Sreelaakshmi et al., 2014; Khalid-Bin-Ferdaus et al., 2018). Chimata et al. (2011) reported an important textile amylase production by *A. niger* MK 07 when steep corn liquor was utilized as a nitrogen source. A mixture of nitrogen sources is utilized in most cases. For instance, a combination of peptone and $(NH_4)_2SO_4$ was used by *B. amyloliquefaciens* UNG-16 to produce amylase in a higher amount (Haq et al., 2010). Similarly, a combination of yeast extract and NH4Cl was utilized by *T. hirsuta* for the secretion of textile laccase (Abadulla et al., 2000; Pereira et al., 2005). Gomes et al. (2009) utilized rice straw and $(NH_4)_2SO_4$ to overproduce manganese and lignin peroxidase by *Lentinus* sp. SXS48, and laccase by *C. byrsina*. Cellulase was produced by *T. reesei* MTCC 162 with peptone, urea, and yeast extract (Saravanan et al., 2013) and by *C. globosum* with sodium nitrate and yeast extract (Chinnamma and Antony, 2015). A mixture of yeast extract, peptone, and KNO_3 was appropriate for the production of textile xylanase by *B. pumilus* ASH (Battan et al., 2012), while a mixture of propionitrile and yeast extract was the best for nitrile hydratase secretion by *R. rhodochrous* NCIMB 11216 (Tauber et al., 2000). Textile enzymes are therefore produced significantly when an organic nitrogen source is combined with an inorganic nitrogen source.

23.5 Purification strategies of textile enzymes

The textile enzymes used in the industries need not be always in pure form. This will depend on its use. The crude textile enzyme, partially or totally purified, is used in industries (Table 23.4). However, completely purified textile enzymes have to be used when studying and determining properties of enzymes, such as molecular weight, molecular structure, etc. Laccase from *C. byrsina and enzymes* lignin peroxidase and manganese peroxidase from *Lentinus* sp. SXS48 were used in biobleaching of synthetic dyes as crude textile enzymes (Gomes et al., 2009). Various methods are utilized to purify enzymes. The pectinase from *Fusarium* sp. was only partially purified by acetone precipitation and used as bioscouring agent for cotton fabrics (Rajendran et al., 2011). Similarly, a partial purification by $(NH_4)_2SO_4$ fractionation followed by dialysis was conducted for collagenase of *N.*

TABLE 23.4 Fermentation and purification strategies of textile enzymes.

Microorganism	Textile enzyme produced	Fermentation strategy	Enzyme release	Purification strategies	Textile application	References
Acinetobacter baumannii AU10	Cutinase	Solid-state	Extracellular	$(NH_4)_2SO_4$ precipitation, hydrophobic interaction, and ion-exchange chromatography	Bioscouring of cotton fabric	Gururaj et al. (2021)
A. niger MK 07	Amylase	Submerged	Extracellular	$(NH_4)_2SO_4$ precipitation, size exclusion chromatography	Biodesizing of cotton cloth	Chimata et al. (2011)
Coriolopsis byrsina	Laccase	Solid-state	Extracellular	Filtration under vacuum and centrifugation	Biobleaching of synthetic dyes	Gomes et al. (2009)
Fusarium sp.	Pectinase	Submerged	Extracellular	Acetone precipitation	Bioscouring of cotton fabrics	Rajendran et al. (2011)
Lentinus sp. SXS48	Manganese peroxidase	Solid-state	Extracellular	Filtration under vacuum and centrifugation	Biobleaching of synthetic dyes	Gomes et al. (2009)
L. sp. SXS48	Lignin peroxidase	Solid-state	Extracellular	Filtration under vacuum and centrifugation	Biobleaching of synthetic dyes	Gomes et al. (2009)

Nocardiopsis dassonvillei NRC2aza	Collagenase	Solid-state	Extracellular	$(NH_4)_2SO_4$ fractionation followed by dialysis	Biodyeing of leather	Abdel-Fattah (2013)
Thermomonospora sp. (T-EG)	Cellulase	Submerged	Extracellular	$(NH_4)_2SO_4$ precipitation, cellulose affinity and gel filtration	Biodenim washing/ denim biofinishing	Anish et al. (2007)
Trametes hirsuta	Laccase	Submerged	Extracellular	Acetone precipitation, ultrafiltration, ion-exchange and gel filtration chromatography	Biobleaching of textile dyes	Abadulla et al. (2000)

dassonvillei NRC2aza and was utilized to dyeing a leather (Abdel-Fattah, 2013). However, chromatographic methods are carried out after partial purification. Partial precipitation by acetone and total purification by chromatography (ion-exchange and gel filtration) were conducted to purify the laccase of *T. hirsuta* (Abadulla et al., 2000). Anish et al. (2007) completely purified a textile cellulase from *Thermomonospora* sp. (T-EG) by $(NH_4)_2SO_4$ precipitation, cellulose affinity and gel filtration chromatography. An amylase responsible for a cotton cloth desizing, obtained from *A. baumannii* AU10, was purified by $(NH_4)_2SO_4$ precipitation and size exclusion chromatography (Chimata et al., 2011). $(NH_4)_2SO_4$ precipitation, hydrophobic interaction, and ion-exchange chromatography were combined with purifying a bioscouring cutinase from *A. baumannii* AU10 (Gururaj et al., 2021). Thus a purification strategy will be chosen depending on the application and step to be carried out in textile industry processes.

23.6 Microbial enzymes used in the textile industry

Fabric processing involved various steps catalyzed by textile enzymes. Indeed, after fabric production, the fabric is first biodesized by α-amylases and pectinases. This is followed by the bioscouring where pectinases aided by proteases, xylanases, and lipases are involved. Neutral cellulases then come in for the biostone-washing. The biobleaching is the next step which is carried out with the aid of laccases, catalases, and peroxidases. This step is followed by biodyeing and printing by pectinases and peroxidases. The last step is biopolishing/biofinishing, when acid cellulases are involved. A schematic diagram for treating the raw fabric with various textile enzymes is shown in Fig. 23.2.

23.6.1 Biodesizing by α-amylases

Amylases are the chief hydrolytic textile industrial enzyme, degrading starch into monosugars, maltose and small dextrins, without altering the support fabric material, with an important strength retention of cellulosic fabrics. Biodesizing is the removal of polysaccharide starch from the fabric material by textile amylases. Textile amylases are thus utilized as desizing agents (Araujo et al., 2008; Ahlawat et al., 2009). The biodesizing procedure by textile amylase is carried out in three steps: impregnation, where the enzymic solution gets absorbed by the fabric material at elevated temperature; incubation, where the starch-related size is cleaved by a textile enzyme; and afterwash, where the hydrolytic products get liberated from the fabric support (Kabir and Koh, 2021).

The biodesizing of fabrics by a textile amylase is usually carried at acidic pH of 6.5 and at a temperature varying from 60°C to 80°C (Haq et al., 2010; Chimata et al., 2011; Zafar et al., 2019) (Table 23.5). However, a low temperature for desizing of 30°C–60°C at pH range of 5.5–6.5 was reported by

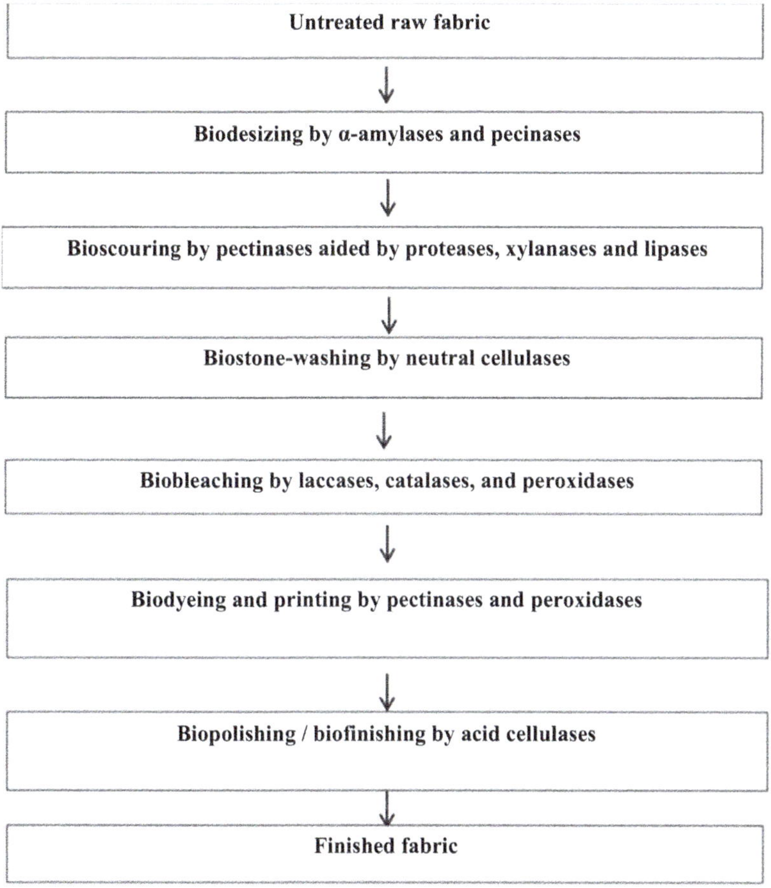

FIGURE 23.2 A schematic representation of treating raw fabric by various textile enzymes.

Gübitz and Cavaco-Paulo(2003). Sometimes, shaking is necessary to favor the process. For instance, Zafar et al. (2019) agitated the containers/flasks at 100 rpm when amylase was desizing a fabric. An optimal desizing by the amylase purified from *B. amyloliquefaciens* was seen at a temperature of 60°C for 60 min when the pH was 6.5 (Haq et al., 2010).The addition of calcium chloride to the *A. niger* MK 07 textile amylase solution was necessary to desize the fabric material at pH 6.5 and 75°C for textile (Chimata et al., 2011). Chand et al. (2014) and Sreelaakshmi et al. (2014) reported an optimal desizing of gray fabric by amylases obtained from *Aspergillus* species when the pH and temperature ranges are 5°C–6°C and 40°C–45°C, respectively. Therefore biodesizing at low temperature by textile amylases is cost-effective as the fabric absorbency is improved, all the impurities are washed away, and less energy is consumed.

TABLE 23.5 Conditions used for various textile processing in industries by enzymes.

Textile process	Textile enzyme	pH	T (°C)	Incubation time	Agitation (rpm)	References
Desizing of cotton cloth	Amylase	6.5	80	1 h	ns	Chimata et al. (2011)
Desizing of gray fabric	Amylase	6.5	60	1 h	ns	Haq et al. (2010)
Textile desizing	Amylase	6.5	85	1 h	100	Zafar et al. (2019)
Textile bleaching process	Catalase	7	50	30 h	ns	Miłek et al. (2014)
Denim washing/biofinishing	Cellulase	8	55	1 h	ns	Anish et al. (2007)
Biopolishing of bleached fabric	Cellulase	4.5–5.5	55	40 min	ns	Uddin (2015)
biopolishing of bleached fabric	Cellulase	7	55	50 min	ns	Uddin (2015)
Denim washing/biofinishing	Humicola cellulase	7	55	1 h	ns	Anish et al. (2007)
Textile dyeing and printing	Laccase	4	55	5 h	20	Yuan et al. (2020)
Bleaching of cotton	laccase	5	50	1 h	ns	Pereira et al. (2005)
Textile dyeing	Laccase	5	30	1 h	40	Abadulla et al. (2000)
Fabric dyeing	Laccase	4.85	30	24 h	ns	Campos et al. (2001)
Acrylic woven fabrics' hydrolysis	Nitrilase	8	40	5 h	ns	Lim et al. (2016)
Cotton scouring and ramie degumming	Pectate lyases	9	55	24 h	ns	Wu et al. (2020)

Cotton/micropoly fabrics' desizing and bioscouring	Pectinase	9.5	65	2 h	50	Ahlawat et al. (2009)
Bioscouring of cotton fabrics	Pectinase	8	40	20 min	ns	Rajendran et al. (2011)
Dyeing of gray wool fabrics	Protease	9	40	45 min	40	Ibrahim and Abd El-Salam (2012)
Denim washing/biofinishing	Trichoderma cellulase	5	55	1 h	ns	Anish et al. (2007)
Desizing of cotton and micropoly fabrics	Xylanase	7	60	90 min	ns	Battan et al. (2012)

23.6.2 Bioscouring by pectinases aided by proteases, cutinases, and lipases

Pectinases are pectic hydrolases, comprising pectin esterases, polygalacturonate lyases, and polygalacturonases (Shahid et al., 2016; Madhu and Chakraborty, 2017). They are important in bioscouring step as they remove pectin, which serves as the cementing and adhesive agent for other impurities. Scouring is a process where noncellulosic materials are eliminated from the cotton/fabric surface. NaOH was utilized in the scouring procedure, but the process is expensive and causes environmental pollution, and industrial workers are exposed to corrosive compounds (Pawar et al., 2002). However, in bioscouring procedure, hydrolytic textile enzymes, namely, pectinases, proteases, cutinases, and cellulases are utilized to selectively digest pectinic, proteinic, cutinic/waxes, and oils/fat materials, respectively, from the cellulosic fibers/textile fabrics (Madhu and Chakraborty, 2017; Rajulapati et al., 2020). The softness, whiteness, higher wettability, undamaged/tearness, weight maintenance, tensile strength, and other properties of the cotton fabric remain intact after bioscouring step. As for desizing step, saving of energy and less pollution (as little biological oxygen demand, total dissolved solids, and chemical oxygen demand are released into the effluent) are also observed. The softness of the fabric material allows for proper dyeing and finishing of textile steps (Madhu and Chakraborty, 2017; Rajulapati et al., 2020).

The bioscouring of cotton fabrics is done with the help of pectinases for 20 min–2 h at 40°C–65°C and at basic pH of 8–9.5 (Table 23.5) (Ahlawat et al., 2009; Rajendran et al., 2011; Wu et al., 2020). A similar range of temperature for bioscouring step was reported by Li et al., 1997. *B. pumilus* BK2 secreted a textile pectate lyase active for cotton fabric bioscouring processes at alkaline pH 8.0 and at 70°C, with a low km of 0.24 g/L (Klug-Santner et al., 2006). The cotton fibers' cuticle and other noncellulosic impurities were degraded by the textile pectinases obtained from *Tetracladium* sp. and *A. niger* (Singh et al., 2020). A textile pectinase which *Bacillus subtilis* SS extracellularly secreted under submerged fermentation was able to bioscoure the fabric at pH 9.5 and 50 rpm for 2 h at 65°C. A maximum bioscouring was achieved when EDTA was added to the enzyme mixture (Ahlawat et al., 2009).

Even if the chief bioscouring enzymes are pectinases, the synergistic effect of pectinases with proteases, cellulases, and cutinases is necessary. For instance, the proteases produced by *Streptomyces* sp. Al-Dhabi-82 were able to decontaminate the cellulosic fibers by digesting the proteinaceous part in the substrate (Al-Dhabi et al., 2020). Cutinases obtained from *A. baumannii* AU10, *Fusarium solani*, *Pseudomonas putida*, and *Streptomyces scabies* were used in the bioscouring processes as they were able to remove wax and cutin impurities at low temperatures (Gururaj et al., 2021; Sen et al., 2021). Similarly, the cutinases purified from *Fusarium oxysporum* increased the absorbency of the bioscoured polyethylene fabric (Kanelli et al., 2015). Lipases purified from fungal species could hydrolyze a polyester during bioscouring studies (Gübitz and Cavaco-Paulo, 2003). Lipases are thus utilized to remove fats/oils/triglycerides from the fabric in scouring-catalyzed step. They may allow these lipidic substances during desizing procedure (Schmid and Verger, 1998; Siddiquee et al., 2014). The supplementation of the nonionic surfactant to the textile lipase improved the scouring process as the lipase penetration is favored, and fibers' surface tension is lowered significantly (Traore and Buschle-Diller, 2000). Kalantzi et al. (2010) reported a bioscouring process improvement (especially absorbency, hydrophilicity, levelness) when a pectinase was coupled with a lipase. *B. pumilus* ASH produced textile xylanase that performed both desizing and scouring the micropoly fabrics and cotton substrates. Thus xylanase has to be exploited for the scouring purpose (Battan et al., 2012). Like lipases, bacterial or fungal esterases are utilized in the textile industry to partially hydrolyze the surfaces of synthetic fiber materials (Araujo et al., 2008).

23.6.3 Biostone-washing by neutral cellulases

Cellulases are hydrolytic enzymes, cleaving the cellulose chains from the interior/middle or from the N or C end (Teeri, 1997). They are used in textile industries to modify cellulosic fibers to improve the quality fabric (Araujo et al., 2008). Indeed, stones (like pumice stones = potassium permanganate + sodium hypochlorite) were used to wash the denim garments, and this resulted in damage to garments/washers/machine, environmental pollution, too much water use, and difficulty in fabrics handling. However, using textile neutral cellulases in biostone-washing/denim washing solved the above-highlighted issues. It leads to the softness, productivity increase, prevent fuzz fibers and fibril formation, back staining, and pilling, as well as good looks of the fabric (Cavaco-Paulo, 1998; Pazarlıoğlu et al., 2005; Pandey et al., 2010; Yu et al., 2013; Kabir and Koh, 2021). Biostone-washing/denim washing is usually carried out with a neutral cellulase at pH 7.0 for 50–60 min at 50°C–55°C (Table 23.5) (Anish et al., 2007; Uddin, 2015). Sarkar et al. (2020) also isolated neutral cellulases active in biostoning processes at pH 6.6–7.0 and at a temperature ranging from 30°C to 60°C. A neutral textile cellulase obtained from *Melanocarpus albomyces* was able to biostone the denim fabric (Das, 2020). A textile cellulase overproduced by *T. reesei* was the best biostoning agent, and no back staining was observed (Cavaco-Paulo, 1998). Therefore the use of neutral cellulase in very small amounts is cost-effective and eco-friendly and replaced many pumice stones in the textile industry.

23.6.4 Biobleaching by laccases, catalases, and peroxidases

Bleaching is an important step in textile industry and is carried out before dyeing and sharp printing as it decolorizes the fabric fibers by removing the natural pigments. Chlorine, hydrogen peroxide, sodium hypochlorite, and oxidizing agents were responsible for bleaching the textile materials, but they cause substantial environmental pollution, fibers' damage, long bleaching time, and a lot of water and energy is utilized (Tavčer et al., 2006; Basto et al., 2007; Shahid et al., 2016; Madhu and Chakraborty, 2017). The oxidative enzymes, namely, catalases, laccases, and peroxidases are currently and directly used to bleach textiles (Araujo et al., 2008).

A textile laccase is utilized for fabric bleaching at pH ranging from 4.0 to 5.0 and temperature varying from 30°C to 55°C under static (Campos et al., 2001; Pereira et al., 2005) or shaking conditions at 20–40 rpm range (Abadulla et al., 2000; Yuan et al., 2020) for 1–5 h (Table 23.5). Many bleaching systems were tested in the textile plants, but the best was laccase/mediator systems (Pereira et al., 2005; Špička and Tavčer, 2013). The laccases decolorize the fabric by oxidizing the phenolic hydroxyl organic groups of flavonoids (Pereira et al., 2005). Basto et al. (2007) reported a bleaching improvement when the laccase was combined with a small amount of hydrogen peroxide in ultrasound energy at 60°C for 30 min at pH 5.0. This excellent bleaching was attributed to the laccase diffusion through the fabric owing to the ultrasound energy (Basto et al., 2007). Some chemical compounds such as tetraacetylethylenediamine (TAED) were reported to be the best activators of bleaching processes in the presence of a textile enzyme (Špička and Tavčer, 2013). The hydrophilicity increase of polyester fabric materials was achieved when the laccase was mediated (Ibrahim and Abd El-Salam, 2012).

Peroxidases or glucose oxidases are also utilized as bleaching agents. With the help of molecular oxygen, they convert monosaccharide glucose to hydrogen peroxide and gluconic acid (Tzanov et al., 2001). Shahid et al. (2016) reported a peroxidase that was able to bleach cotton fabrics, and this glucose oxidase was able to desize and to bioscoure the fabrics. Tear and tensile strengths of linen fabric were improved significantly when a laccase was supplemented with a glucose oxidase, and less water was utilized (Anis et al., 2009). On the other hand, catalases are also used as bleaching agents and act by oxidizing hydrogen peroxide to molecular oxygen and water. They may act at neutral/basic/acidic pH and low temperatures, varying from 20°C to 50°C, and less water and energy are used, and it is an eco-friendly process (Pereira et al., 2005; Basto et al., 2007; Araujo et al., 2008). Catalases produced by *Aspergillus* species were used to bleach fabrics through H_2O_2 excess removal in a cost-effective process (Mojsov, 2011). Miłek et al. (2014) reported a textile bleaching process by catalase at 50°C and pH 7.0 for 30 h. Thus catalases and peroxidases are helping laccases in getting excellent bleaching results.

23.6.5 Biodyeing and printing by pectinases and peroxidases

After biobleaching of fabric materials by textile oxidorectases, biodyeing has to be followed. This step is catalyzed by pectinases which remove dyes and peroxidases which remove excess of the dyes used. This dyeing step is followed by printing the dyed fabric for biopolishing (Kabir and Koh, 2021; Sen et al., 2021). For instance, the dyeing process was optimum after desizing and biobleaching of cotton towels (Ali et al., 2014). Thus if the textile dyeing is adequate, the biofinishing of the fabric is easier.

23.6.6 Biopolishing/biofinishing by acid cellulases

Biopolishing is the procedure where the acidic textile cellulases remove micro-hair and fuzzy fibrils from the fabric material surfaces, after the dyeing step (Araujo et al., 2008). The importance of this process of avoiding fibrillation is that it augments the appearance and good look, the brightness of color, hand touch and feel, fibers' water absorbency, and crystallinity degree (Saravanan et al., 2013; Madhu and Chakraborty, 2017). Chinnamma and Antony (2015) secreted an acidic cellulase from *C. globosum* that act as a biopolishing agent of cotton fabric materials. An acidic textile cellulase biopolished a bleached fabric for 40 min at 55°C in the pH range of 4.5–5.5 (Uddin, 2015). Similarly, *Trichoderma* cellulase biofinished a fabric at pH 5.5 for 1 h at 55°C (Anish et al., 2007). A textile collagenase obtained from *Acinetobacter* sp. also biopolished the wool surface, highlighting its application in the industry of textiles (Abdel-Fattah, 2013). Similar fiber surface modification by textile nitrilases purified from *R. rhodochrous* NCIMB 11216 was observed (Tauber et al., 2000). Ge et al. (2009) reported an eco-friendly textile transglutaminase that repaired the wool damages, favoring dyeing and wettability properties of wool fabric materials.

23.6.7 Use of the mixture of microbial enzymes in textile fabric material processing

A mixture of cellulase and protease was used to bioscoure fiber materials and needed water absorbency, and maximum pectin removal was achieved (Traore and Buschle-Diller, 2000). Similarly, the desired absorbency and wettability during cotton bioscouring were achieved when cellulases, proteases, and pectinases were combined (Karapinar and Sariisik, 2004). Likewise, Agrawal et al. (2008) reported an optimal cotton scouring with pectate lyase and cutinase of *F. solani* at low temperatures. A mixture of textile lipase and protease adequately biohydrolyzed a nylon 66 (Parvinzadeh et al., 2009). With the Taguchi procedure, the mixture of pectinase, protease, and cellulase leads to an optimal bioscouring process (Saravanan et al., 2010). Madhu and Chakraborty (2017) simultaneously bioscoure and bioble-ach a fabric material with a pectinase and a peroxidase under alkaline conditions, and water and energy were enormously saved. Jute fibers were degummed, and cotton fabrics were bioscoured when pectate lyases and pectin methylesterase were screened from *Clostridium thermocellum* when utilized (Rajulapati et al., 2020). A combination of xylanase and pectinase enzymes were used to bioscoure ramie fibers to attain adequate whiteness and brightness cost-effectively (Singh et al., 2020). Although textile enzymes are available, the production of many enzymes by one fungal/bacterial species remains a big challenge as this would make the production cost-effective and the stability between textile enzymes in the presence of protease that may degrade them will be assured.

23.7 Immobilization of textile enzymes

Textile enzymes are sometimes immobilized on adequate support in order to perform efficiently the various fabrics processes. For instance, the pectinase purified from *A. niger* was immobilized on the polyacrylonitrile copolymer membrane, and an effective bioscouring was achieved (Delcheva et al., 2007). Dinçer and Telefoncu (2007) immobilized the textile acid cellulases on the polyvinyl alcohol-coated chitosan beads/functionalized matrices and a maximum thermal stability and action were attained. An improvement in thermal stability was observed when a cellulase was immobilized on matrices, and the enzyme was reused many times (Hirsh et al., 2010). Similar reusability and thermal stability were seen when a cellulase was attached to the modified mesoporous silica (Yin et al., 2013). An optimal bleaching process was achieved after peroxidase immobilization on textile carrier materials (Opwis et al., 2016).

Toxicity alleviation, textile effluents' decolorization, and thermal stability improvements were observed after immobilizing the textile laccase obtained from *T. hirsuta* on the alumina (Abadulla et al., 2000). Likewise, decolorization of textile dyes' effluents was enhanced when a textile laccase was immobilized on strips and greater reusability was noticed (Yuan et al., 2020). Ramie degumming and cotton bioscouring were optimally achieved after immobilizing a textile pectate lyase on the inorganic hybrid nanoflower. This was attributed to the enhancement of the reusability and thermostability of the enzyme (Wu et al., 2020). Although the immobilization of textile enzymes is a cost-effective process and has enormously improved their catalytic stabilities and efficiencies, the search for new immobilization support materials is needed to decrease textile enzyme losses and their durability. The catalytic performance and site have also to be protected by improving the available immobilization methods (Chang et al., 2021).

23.8 Genetic engineering of bacteria- and fungi-producing textile enzymes

Genetic materials of microorganism-producing textile enzymes can be manipulated in order to produce these enzymes in a significant amount. *E. coli*, *S. cerevisiae*, *Bacillus* sp., *Aspergillus* sp., etc., may be utilized as an expression host to produce textile enzymes (Baneyx, 1999; Silbersack et al., 2006; Li et al., 2007). Even if *E. coli* lacks some pathways for the expression of some enzymes, it remains as a microorganism of expression of choice owing to the easy manipulation of its genetic material, cost-effective production of enzyme in huge amounts in a lesser time, etc. (Baneyx, 1999; Li et al., 2007). For the secretion of eukaryotic textile enzymes, *S. cerevisiae* can be utilized in most cases (Araujo et al., 2008). To obtain the microorganisms over producing textile enzymes, molecular or classical procedures can be followed. For instance, using mutagenic agents such as chemicals and UV radiation, desired characteristics of variants were obtained. Through the recombinant DNA method, textile enzymes of interest can be produced after the incorporation of a gene of interest into the best enzyme production host. Textile enzymes can also be engineered (such as site-directed or random mutagenesis) in order to get enzymes with desired aspects, such as thermostability and pH, stability in solvents, surfactants, etc. (Pandey et al., 2010).

Xylanase was genetically engineered. Its stability at high temperatures and at alkaline pH was achieved, and it was used as a bleaching agent (Fenel et al., 2004; Pandey et al., 2010). A textile pectinase that resists at elevated temperature and pH was also engineered and was used as a bioscouring agent at low dose (Solbak et al., 2005). The *B. amyloliquefaciens* UNG-16 was mutated with the help of ethyl methane sulfonate in order to overproduce a textile amylase for gray fabric desizing for 60 min at 60°C and pH 6.5 (Haq et al., 2010). *T. reesei* RUT C-30 is known as a mutant strain to overproduce textile cellulases (Pandey et al., 2010). Zafar et al. (2019) also produced a textile desizing amylase after the expression of the cloned gene into *E. coli*. A textile pectate lyase was overproduced by *Bacillus* sp. RN1 after its expression in *Pichia pastoris*. As it was stable at high temperatures and in acidic and basic conditions, it was utilized in the ramie degumming procedures (Zheng et al., 2020). Even if textile enzymes are somehow produced in a significant with DNA recombinant technology, textile enzymes with desired characteristics that can be commercialized remain a big challenge.

23.9 Manufacturers of some commercial textile enzymes

Big companies worldwide are manufacturing various textile enzymes, even if getting enzymes with desired characteristics remains a challenge. For instance, Novozymes (Denmark) is the chief manufacturer of a textile enzyme, such as Terminox Ultra (commercial name of the catalase), and is used in the bleaching processes (Miłek et al., 2014), *Humicola* cellulase used in the denim washing (Anish et al., 2007), DenLite (a laccase) utilized in denim bleaching (Pereira et al., 2005), Bioprep 3000L (pectate lyase) in bioscouring of gray cotton (Uddin, 2015), and Novoprime Base 268, a laccase used in the denim bleaching (Rodríguez-Couto, 2012). Dehabadi et al. (2011) reported Optisize Next produced by PT. Chemira (Indonesia) for desizing purposes. A textile nitrilase is also commercialized as Cyanovacta Lyase by Novacta Byosystems Ltd. (United Kingdom) and is used in the surface modification of acrylic fibers (Matamá et al., 2007) (Table 23.6).

TABLE 23.6 Manufacturers of some commercial textile enzymes.

Enzyme	Commercial product	Manufacturer	Application/uses	References
Amylase	Optisize Next	PT. Chemira, Indonesia	Desizing	Dehabadi et al. (2011)
Catalase	Terminox Ultra	Novozymes, Denmark	Bleaching processes	Miłek et al. (2014)
Cellulase	Trichoderma cellulase	Novozymes, Denmark	Denim washing/ denim biofinishing	Anish et al. (2007)
Cellulase	Humicola cellulase	Novozymes, Denmark	Denim washing/ denim biofinishing	Anish et al. (2007)
Laccase	DenLite	Novozymes (Novo Nordisk, Denmark)	Denim bleaching/ denim finishing	Pereira et al. (2005)
Laccase	Zytex	Zytex Pvt. Ltd., Mumbai, India	Denim bleaching/ denim finishing	Pereira et al. (2005)
Laccase	Bleach-cut 3S	Chemicals Dyestuffs Ltd., Hong Kong	Bleaching of indigo	Rodríguez-Couto (2012)
Laccase	Ecostonelcc 10	AB Enzymes GmbH, Germany	Bleaching of denim/ indigo dye with a mediator radical	Rodríguez-Couto (2012)
Laccase	Trilite II	Tri-Tex Co. Inc., Canada	Decoloration of indigo dyes in denim wet processing	Rodríguez-Couto (2012)
Laccase	Apcozyme II-S	Apollo Chemical Company, LLC, USA	Bleaching of indigo dye on denim textiles	Rodríguez-Couto (2012)

(*Continued*)

TABLE 23.6 (Continued)

Enzyme	Commercial product	Manufacturer	Application/uses	References
Laccase	Purizyme	Puridet Asia Ltd., Hong Kong	Bleaching of indigo-dyed garments	Rodríguez-Couto (2012)
Laccase	Americos Laccase LTC; Americos Laccase P	Americos Industries Inc., India	Bleaching of denim garments	Rodríguez-Couto (2012)
Laccase	Hypozyme	Condor Speciality Products, USA	Deinking of denim fabrics	Rodríguez-Couto (2012)
Laccase	Lacasa Ultratex	Proenzimas Ltd., Colombia	Decoluration of indigo in denim fabrics	Rodríguez-Couto (2012)
Laccase	Cololacc BB	Colotex Biotechnology Co. Ltd., Hong Kong	Denim finishing applications	Rodríguez-Couto (2012)
Laccase	Novoprime Base 268	Novozymes, Denmark	Denim bleaching	Rodríguez-Couto (2012)
Laccase	Prozyme LAC	Sunson Industry Group Co. Ltd., China	Bleaching of indigo denims with anti-back staining	Rodríguez-Couto (2012)
Laccase	Lava Zyme LITE	DyStar GmbH, Germany	Bleaching of indigo dye	Rodríguez-Couto (2012)
Laccase	IndiStar Color Adjust system	Genencor International Inc., USA	Denim finishing	Rodríguez-Couto (2012)
Nitrilase	Cyanovacta Lyase	Novacta Byosystems Ltd., Hatfield, UK	Surface modification of acrylic fibers	Matamá et al. (2007)
Pectate lyase	Bioprep 3000L	Novozyme, Denmark	Bioscouring of gray cotton	Uddin (2015)

Textile laccase is the most commercialized enzyme compared to others. It was manufactured by many companies. Indeed, various laccases are commercialized as Zytex (Zytex Pvt. Ltd., India) (Pereira et al., 2005), Bleach-cut 3S (Chemicals Dyestuffs Ltd., Hong Kong), Trilite II (Tri-Tex Co. Inc.,

Canada), Apcozyme II-S (Apollo Chemical Company, LLC, United States), Purizyme (Puridet Asia Ltd., Hong Kong), Americos Laccase LTC; Americos Laccase P (Americos Industries Inc., India), Hypozyme (Condor Speciality Products, United States), Lacasa Ultratex (Proenzimas Ltd., Colombia), Cololacc BB (Colotex Biotechnology Co. Ltd., Hong Kong), Prozyme LAC (Sunson Industry Group Co. Ltd., China), Lava Zyme LITE (DyStar GmbH, Germany), and IndiStar Color Adjust system (Genencor International Inc., United States) (Rodríguez-Couto, 2012) (Table 23.6).

23.10 Textile industry effluents' treatment

Textile effluents are in general colored as they harbor various and different chemical compounds, such as surfactants, dyes, phenols, humectants, salts, aromatic amines, dispersants, acids, nitriles, detergents, bases, and oxidants, coming from textile industries. Some of these substances are recalcitrant and nonbiodegradable. It is thus necessary to treat these compounds before reaching the environment like water bodies as they cause mutation, toxicity, and cancer by damaging aquatic organisms or human beings. Chemical, biological, and physical procedures can be used to treat textile effluents; however, the biological method, especially with textile enzymes, was reported to be cost-effective, nontoxic, and environment-friendly. Removal of the textile pollutants/dyes will depend on the pH, chemical concentration, temperature, and reaction time period (Mai et al., 2000; Costa et al., 2002).

Textile effluents were treated with oxidoreductases (such as peroxidases and laccases) to remove chlorinated phenolic substances (Mai et al., 2000). The oxidoreductases, through oxidative coupling, were also utilized to detoxify the organic substances (Karigar and Rao, 2011). Textile effluents resulted from dyeing step were decolorized by a laccase of *T. hirsuta* (Abadulla et al., 2000; Tapia-Tussell et al., 2020). *Trametes trogii* produced a textile laccase that was able to detoxify textile effluent with the help of a mediator (Khlifi et al., 2010). The textile laccases bleached a textile effluent containing aromatic amine and phenolic organic substances with the help of molecular oxygen (Zucca et al., 2016). *Serratia marcescens* and *Phanerochaete chrysosporium* were reported to be the microorganisms of choice in treating effluents containing synthetic and natural dyes. This was attributed to the secretion of laccases and ligninases (such as textile lignin/manganese peroxidases) (Verma and Madamwar, 2003; Asgher et al., 2008). Sarkar et al. (2020) detoxified the effluent with azo dyes rich in nitrogen groups with the help of azoreductase and laccase obtained from bacteria. Even if the decolorization of textile effluent by microorganisms/microbial textile enzymes is cost-effectively advancing, the optimal detoxification mechanisms remain to be elucidated (Chang et al., 2021).

23.11 Concluding remarks

Enzymes are utilized in the textile industries to make the environment safe and the textile manufacturing processes cost-effective. The utilization of textile enzymes in the industry is still at the infant stage as most of the chemicals used are not totally replaced by enzymes, and few textile enzymes are commercialized. To save energy and water, developing an eco-friendly and cost-effective enzymatic process that may combine desizing, scouring, and bleaching at appropriate pH and temperature has to continue. Further studies are needed to get a green economy and meet the sustainable demand for textile enzymes with desirable aspects. The search for commercial textile enzymes that could biomodify natural and synthetic fiber materials is needed. The commercial company managers, genetic materials engineers, and the academicians have to work together to solve the highlighted issues.

References

Abadulla, E., Tzanov, T., Costa, S., Robra, K.H., Cavaco-Paulo, A., Gübitz, G.M., 2000. Decolorization and detoxification of textile dyes with a laccase from *Trametes hirsuta*. Appl. Environ. Microbiol. 66 (8), 3357–336.

Abdel-Fattah, A.M., 2013. Production and partial characterization of collagenase from marine *Nocardiopsis dassonvillei* NRC2aza using chitin wastes. Egyp. Pharm. J. 12 (2), 109–114.

Agrawal, P.B., Nierstrasz, V.A., Warmoeskerken, M.M.C.G., 2008. Role of mechanical action in low-temperature cotton scouring with *F. solani* pisi cutinase and pectate lyase. Enzyme Microb. Technol. 42 (6), 473–482.

Ahlawat, S., Dhiman, S.S., Battan, B., Mandhan, R.P., Sharma, J., 2009. Pectinase production by *Bacillus subtilis* and its potential application in biopreparation of cotton and micropoly fabric. Process. Biochem. 44 (5), 521–526.

Al-Dhabi, N.A., Esmail, G.A., Ghilan, A.K.M., Arasu, M.V., Duraipandiyan, V., Ponmurugan, K., 2020. Characterization and fermentation optimization of novel thermostable alkaline protease from *Streptomyces* sp. Al-Dhabi-82 from the Saudi Arabian environment for eco-friendly and industrial applications. J. King Saud. Univ. Sci. 32 (1), 1258–1264.

Ali, S., Khatri, A., Tanwari, A., 2014. Integrated desizing-bleaching-reactive dyeing process of cotton towel using glucose oxidase enzyme. J. Clean. Prod. 66, 562–567.

Anish, R., Rahman, M.S., Rao, M., 2007. Application of cellulases from an alkalothermophilic *Thermomonospora* sp. in biopolishing of denims. Biotechnol. Bioeng. 96 (1), 48–56.

Anis, P., Davulcu, A., Eren, H.A., 2009. Enzymatic pre-treatment of cotton. Part 2: peroxide generation in desizing liquor and bleaching. Fibres Text. East. Eur. 17 (2), 87–90.

Araujo, R., Casal, M., Cavaco-Paulo, A., 2008. Application of enzymes for textile fibres processing. Biocatal. Biotransfor. 26 (5), 332–349.

Asgher, M., Bhatti, H.N., Ashraf, M., Legge, R.L., 2008. Recent developments in biodegradation of industrial pollutants by white rot fungi and their enzyme system. Biodegradation 19 (6), 771–783.

Baneyx, F., 1999. Recombinant protein expression in *Escherichia coli*. Curr. Opin. Biotechnol. 10, 411–421.

Basto, C., Tzanov, T., Cavaco-Paulo, A., 2007. Combined ultrasound-laccase assisted bleaching of cotton. Ultrason. Sonochem. 14 (3), 350–354.

Battan, B., Dhiman, S.S., Ahlawat, S., Mahajan, R., Sharma, J., 2012. Application of thermostable xylanase of *Bacillus pumilus* in textile processing. Indian. J. Microbiol. 52 (2), 222–229.

Campos, R., Kandelbauer, A., Robra, K.H., Cavaco-Paulo, A., Gübitz, G.M., 2001. Indigo degradation with purified laccases from *Trametes hirsuta* and *Sclerotium rolfsii*. J. Biotechnol. 89 (2–3), 131–139.

Cavaco-Paulo, A., 1998. Mechanism of cellulase action in textile processes. Carbohydr. Polym. 37 (3), 273–277.

Cavaco-Paulo, A., Almeida, L., Bishop, D., 1996. Effects of agitation and endoglucanase pretreatment on the hydrolysis of cotton fabrics by a total cellulase. Text. Res. J. 66 (5), 287–294.

Chand, N., Sajedi, R.H., Nateri, A.S., Khajeh, K., Rassa, M., 2014. Fermentative desizing of cotton fabric using an α-amylase-producing *Bacillus* strain: optimization of simultaneous enzyme production and desizing. Process. Biochem. 49 (11), 1884–1888.

Chang, Y., Yang, D., Li, R., Wang, T., Zhu, Y., 2021. Textile dye biodecolorization by manganese peroxidase: a review. Molecules 26 (15), 4403.

Chimata, M.K., Chetty, C.S., Suresh, C., 2011. Fermentative production and thermostability characterization of α amylase from *Aspergillus* species and its application potential evaluation in desizing of cotton cloth. Biotechnol. Res. Int. 2011, 323–891.

Chinnamma, S.K., Antony, V.A.R., 2015. Production and application of cellulase enzyme for biopolishing of cotton. Int. J. Sci. Technol. Manag. 4 (1), 1606–1612.

Costa, S.A., Tzanov, T., Carneiro, F., Gübitz, G.M., Cavaco-Paulo, A., 2002. Recycling of textile bleaching effluents for dyeing using immobilized catalase. Biotechnol. Lett. 24 (3), 173–176.

Couto, S.R., Toca-Herrera, J.L., 2006. Lacasses in the textile industry. Biotechnol. Mol. Biol. Rev. 1 (4), 115–120.

Das, S., 2020. Bio-polishing of mulberry silk. Indian. J. Fibre Text. Res. 45 (3), 362–365.

Dehabadi, V.A., Opwis, K., Gutmann, J., 2011. Combination of acid-demineralization and enzymatic desizing of cotton fabrics by using industrial acid stable glucoamylases and α-amylases. Starch 63 (12), 760–764.

Delcheva, G., Pishtiyski, I., Dobrev, G., Krusteva, S., 2007. Immobilization of *Aspergillus niger* pectinase on polyacrylonitrile copolymer membrane. Trends Appl. Sci. Res. 2 (5), 419–425.

Dinçer, A., Telefoncu, A., 2007. Improving the stability of cellulase by immobilization on modified polyvinyl alcohol coated chitosan beads. J. Mol. Catal. B Enzym. 45 (1–2), 10–14.

Fenel, F., Leisola, M., Jänis, J., Turunen, O., 2004. A de novo designed N-terminal disulphide bridge stabilizes the *Trichoderma reesei endo*-1, 4-β-xylanase II. J. Biotechnol. 108 (2), 137–143.

Ge, F., Cai, Z., Zhang, H., Zhang, R., 2009. Transglutaminase treatment for improving wool fabric properties. Fibers Polym. 10 (6), 787–790.

Gomes, E., Aguiar, A.P., Carvalho, C.C., Bonfá, M.R.B., Silva, R.D., Boscolo, M., 2009. Ligninases production by *Basidiomycetes* strains on lignocellulosic agricultural residues and their application in the decolorization of synthetic dyes. Braz. J. Microbiol. 40 (1), 31–39.

Gübitz, G.M., Cavaco-Paulo, A., 2003. New substrates for reliable enzymes: enzymatic modification of polymers. Curr. Opin. Biotechnol. 14 (6), 577–582.

Gururaj, P., Khushbu, S., Monisha, B., Selvakumar, N., Chakravarthy, M., Gautam, P., et al., 2021. Production, purification and application of cutinase in enzymatic scouring of cotton fabric isolated from *Acinetobacter baumannii*. Prep. Biochem. Biotechnol. 51 (6), 550–AU561.

Haq, I., Ali, S., Javed, M.M., Hameed, U., Saleem, A., Adnan, F., et al., 2010. Production of alpha amylase from a randomly induced mutant strain of *Bacillus amyloliquefaciens* and its application as a desizer in textile industry. Pak. J. Bot. 42 (1), 473–484.

Hirsh, S.L., Bilek, M.M.M., Nosworthy, N.J., Kondyurin, A., Dos Remedios, C.G., McKenzie, D.R., 2010. A comparison of covalent immobilization and physical adsorption of a cellulase enzyme mixture. Langmuir 26 (17), 14380–14388.

Ibrahim, D., Abd El-Salam, S.H., 2012. Enzymatic treatment of polyester fabrics digitally printed. J. Text. Sci. Eng. 2 (3), 1–4.

Kabir, S.M.M., Koh, J., 2021. Sustainable textile processing by enzyme applications. Biodegradation. Available from: https://doi.org/10.5772/intechopen.97198IntechOpen.

Kalantzi, S., Mamma, D., Kalogeris, E., Kekos, D., 2010. Improved properties of cotton fabrics treated with lipase and its combination with pectinase. Fibres Text. East. Eur. 18 (5), 86–92.

Kanelli, M., Vasilakos, S., Nikolaivits, E., Ladas, S., Christakopoulos, P., Topakas, E., 2015. Surface modification of poly (ethylene terephthalate) (PET) fibers by a cutinase from *Fusarium oxysporum*. Process. Biochem. 50 (11), 1885–1892.

Karapinar, E., Sariisik, M.O., 2004. Scouring of cotton with cellulases, pectinases and proteases. Fibres Text. East. Eur. 12 (3), 79–82.

Karigar, C.S., Rao, S.S., 2011. Role of microbial enzymes in the bioremediation of pollutants: a review. Enzym. Res. Available from: https://doi.org/10.4061/2011/805187.

Khalid-Bin-Ferdaus, K.M., Hossain, M.F., Mansur, S.A., Sajib, S.A., Miah, M.M., Hoque, K.M. F., et al., 2018. Commercial production of alpha amylase enzyme for potential use in the textile industries in Bangladesh. Int. J. Biosci. 13 (4), 149–157.

Khlifi, R., Belbahri, L., Woodward, S., Ellouz, M., Dhouib, A., Sayadi, S., et al., 2010. Decolourization and detoxification of textile industry wastewater by the laccase-mediator system. J. Hazard. Mater. 175 (1–3), 802–808.

Klug-Santner, B.G., Schnitzhofer, W., Vršanská, M., Weber, J., Agrawal, P.B., Nierstrasz, V.A., et al., 2006. Purification and characterization of a new bioscouring pectate lyase from *Bacillus pumilus* BK2. J. Biotechnol. 121 (3), 390–401.

Kumar, P.S., Gunasundari, E., 2018. Sustainable wet processing – an alternative source for detoxifying supply chain in textiles. In: Muthu, S.S. (Ed.), Detox Fashion Sustainable Chemistry and Wet Processing. Springer, Singapore, pp. 37–60.

Li, P., Anumanthan, A., Gao, X.G., Ilangovan, K., Suzara, V.V., Düzgüneş, N., et al., 2007. Expression of recombinant proteins in *Pichia pastoris*. Appl. Biochem. Biotechnol. 142 (2), 105–124.

Li, T., Wang, N., Li, S., Zhao, Q., Guo, M., Zhang, C., et al., 1997. Enzymatic scouring of cotton: effects on structure and properties. Text. Chem. Color. 29 (8), 71–76.

Lim, M., Kim, D., Seo, J., 2016. Enhanced oxygen-barrier and water-resistance properties of poly (vinyl alcohol) blended with poly (acrylic acid) for packaging applications. Polym. Int. 65 (4), 400–406.

Logan, N.A., De Vos, P., 2009. Bacillus. In: 2nd edn De Vos, P., Garrity, G., Jones, D., Krieg, N.R., Ludwig, W., Rainey, F.A., Schleifer, K.-H., Whitman, W.B. (Eds.), Bergey's Manual of Systematic Bacteriology, Vol. 3. Springer, New York, pp. 21–127.

Madhu, A., Chakraborty, J.N., 2017. Developments in application of enzymes for textile processing. J. Clean. Prod. 145, 114–133.

Mai, C., Schormann, W., Milstein, O., Hüttermann, A., 2000. Enhanced stability of laccase in the presence of phenolic compounds. Appl. Microbiol. Biotechnol. 54 (4), 510–514.

Matamá, T., Carneiro, F., Caparrós, C., Gübitz, G.M., Cavaco-Paulo, A., 2007. Using a nitrilase for the surface modification of acrylic fibres. Biotechnol. J. Health Care Nutr. Technol. 2 (3), 353–360.

Miłek, J., Wójcik, M., Verschelde, W., 2014. Thermal stability for the effective use of commercial catalase. Pol. J. Chem. Tech. 16 (4), 75–79.

Mojsov, K., 2011. Application of enzymes in the textile industry; a review', II International Congress, Engineering, Ecology and Materials in the Processing Industry. Jahorina, Bosnia Herzeg. 2, 30–39.

Niyonzima, F.N., 2018. Purification and properties of detergent compatible proteases and lipase from soil microorganisms. Humalaya Publishing Houses, New Delhi.

Niyonzima, F.N., More, S.S., 2013. Optimization of fermentation culture conditions for alkaline protease production by *Scopulariopsis* spp. Appl. Biol. Res. 15 (1), 66–69.

Niyonzima, F.N., More, S.S., 2015. Purification and characterization of detergent compatible alkaline protease from *Aspergillus terreus* gr. 3 Biotech. 5, 61–70.

Niyonzima, F.N., Veena, S.M., More, S.S., 2020. Industrial production and optimization of microbial enzymes. In: Arora, N.A., Mishra, J., Mishra, V. (Eds.), Microbial Enzymes: Roles and Applications in industries. Microorganisms for Sustainability, vol 11. Springer, Singapore, pp. 115–135.

Opwis, K., Kiehl, K., Gutmann, J.S., 2016. Immobilization of peroxidases on textile carrier materials and their use in bleaching processes. Chem. Eng. Trans. 49, 67–72.

Pandey, A., Binod, P., Ushasree, M.V., Vidya, J., 2010. Advanced strategies for improving industrial enzymes. Chem. Ind. Digest. 23, 74–84.

Parvinzadeh, M., Assefipour, R., Kiumarsi, A., 2009. Biohydrolysis of nylon 6, 6 fibers with different proteolytic enzymes. Polym. Degrad. Stab. 94 (8), 1197–1205.

Pawar, S.B., Shah, H.D., Andhorika, G.R., 2002. Man-Made Text. India 45 (4), 133.

Pazarlıoğlu, N.K., Sariişik, M., Telefoncu, A., 2005. Laccase: production by *Trametes versicolor* and application to denim washing. Process. Biochem. 40 (5), 1673–1678.

Pereira, L., Bastos, C., Tzanov, T., Cavaco-Paulo, A., Gübitz, G.M., 2005. Environmentally friendly bleaching of cotton using laccases. Environ. Chem. Lett. 3 (2), 66–69.

Rajendran, R., Sundaram, S.K., Radhai, R., Rajapriya, P., 2011. Bioscouring of cotton fabrics using pectinase enzyme its optimization and comparison with conventional scouring process. Pak. J. Biol. Sci. 14 (9), 519–525.

Rajulapati, V., Dhillon, A., Gali, K., Katiyar, V., Goyal, A., 2020. Green bioprocess of degumming of jute fibers and bioscouring of cotton fabric by recombinant pectin methylesterase and pectate lyases from *Clostridium thermocellum*. Process. Biochem. 92, 93–104.

Rodríguez-Couto, S., 2012. Laccases for denim bleaching: an eco-friendly alternative. Sigma 1, 1–7.

Saravanan, D., Sree Lakshmi, S.N., Raja, K.S., Vasanthi, N.S., 2013. Biopolishing of cotton fabric with fungal cellulase and its effect on the morphology of cotton fibres. Indian. J. Fibre Text. Res. 38, 156–160.

Saravanan, D., Ramanathan, V.A., Karthick, P., Murugan, S.V., Nalankilli, G., Ramachandran, T., 2010. Optimisation of multi-enzyme scouring process using Taguchi methods. Indian. J. Fibre Text. Res. 35, 164–171.

Sarkar, S., Banerjee, A., Chakraborty, N., Soren, K., Chakraborty, P., Bandopadhyay, R., 2020. Structural-functional analyses of textile dye degrading azoreductase, laccase and peroxidase: a comparative in silico study. Electron. J. Biotechnol. 43, 48–54.

Schmid, R.D., Verger, R., 1998. Lipases: interfacial enzymes with attractive applications. Angew. Chem. Int. (Ed.) 37 (12), 1608–1633.

Sen, A., Kapila, R., Chaudhary, S., Nigam, A., 2021. Biotechnological applications of microbial enzymes to replace chemicals in the textile industry-a review. J. Text. Assoc. 82 (2), 68–73.

Shahid, M., Mohammad, F., Chen, G., Tang, R.C., Xing, T., 2016. Enzymatic processing of natural fibres: white biotechnology for sustainable development. Green. Chem. 18 (8), 2256–2281.

Shen, J., Smith, E., 2015. Enzymatic treatments for sustainable textile processing. In: Blackburn, R. (Ed.), Sustainable Apparel: Production, Processing and Recycling. Woodhead Publishing Limited, Cambridge, pp. 119–134.

Siddiquee, A.B., Bashar, M.M., Sarker, P., Tohfa, T.T., Hossan, M.A., Azad, M.I., et al., 2014. Comparative study of conventional and enzymatic pretreatment (scouring & bleaching) of cotton knitted fabric. Int. J. Eng. Technol. 3 (1), 37.

Silbersack, J., Jürgen, B., Hecker, M., Schneidinger, B., Schmuck, R., Schweder, T., 2006. An acetoin-regulated expression system of *Bacillus subtilis*. Appl. Microbiol. Biotechnol. 73 (4), 895–903.

Singh, A., Varghese, L.M., Battan, B., Patra, A.K., Mandhan, R.P., Mahajan, R., 2020. Eco-friendly scouring of ramie fibers using crude xylano-pectinolytic enzymes for textile purpose. Environ. Sci. Pollut. Res. 27 (6), 6701–6710.

Solbak, A.I., Richardson, T.H., McCann, R.T., Kline, K.A., Bartnek, F., Tomlinson, G., et al., 2005. Discovery of pectin-degrading enzymes and directed evolution of a novel pectate lyase for processing cotton fabric. J. Biol. Chem. 280 (10), 9431–9438.

Špička, N., Tavčer, P.F., 2013. Complete enzymatic pre-treatment of cotton fabric with incorporated bleach activator. Text. Res. J. 83 (6), 566–573.

Sreelaakshmi, S.N., Paul, A., Vasanthi, N.S., Sarvanan, D., 2014. Low temperature acidic amylases from *Aspergillus* for desizing of cotton fabrics. J. Text. Inst. 105 (1), 59–66.

Tapia-Tussell, R., Pereira-Patrón, A., Alzate-Gaviria, L., Lizama-Uc, G., Pérez-Brito, D., Solis-Pereira, S., 2020. Decolorization of textile effluent by *Trametes hirsuta* Bm-2 and lac-T as possible main laccase-contributing gene. Curr. Microbiol. 77 (12), 3953–3961.

Tauber, M.M., Cavaco-Paulo, A., Robra, H., Gübitz, G.M., 2000. Nitrile hydratase and amidase from *Rhodococcus rhodochrous* hydrolyze acrylic fibers and granular polyacrylonitriles. Appl. Environ. Microbiol. 66, 1634–1638.

Tavčer, P.F., Križman, P., Preša, P., 2006. Combined bioscouring and bleaching of cotton fibres. J. Nat. Fibers 3 (2–3), 83–97.

Teeri, T.T., 1997. Crystalline cellulose degradation: new insight into the function of cellobiohydrolases. Trends Biotechnol. 15 (5), 160–167.

Traore, M.K., Buschle-Diller, G., 2000. Environmentally friendly scouring processes. Text. Chem. Color. Am. Dyes. Report. 32 (12), 40–43.

Tzanov, T., Calafell, M., Guebitz, G.M., Cavaco-Paulo, A., 2001. Bio-preparation of cotton fabrics. Enzyme Microbiol. Technol. 29, 357–362.

Uddin, M.G., 2015. Effects of biopolishing on the quality of cotton fabrics using acid and neutral cellulases. Text. Cloth. Sustain. 1, 9.

Verma, P., Madamwar, D., 2003. Decolourization of synthetic dyes by a newly isolated strain of *Serratia marcescens*. World J. Microbiol. Biotechnol. 19 (6), 615–618.

Wu, P., Luo, F., Lu, Z., Zhan, Z., Zhang, G., 2020. Improving the catalytic performance of pectate lyase through pectate lyase/$Cu_3(PO_4)_2$ hybrid nanoflowers as an immobilized enzyme. Front. Bioeng. Biotechnol. 8, 280.

Yin, H., Su, Z.L., Shao, H., Cai, J., Wang, X., Yin, H., 2013. Immobilization of cellulase onmodified mesoporous silica shows improved thermal stability and reusability. Afr. J. Microbiol. Res. 7, 3248–3253.

Yuan, H., Chen, L., Cao, Z., Hong, F.F., 2020. Enhanced decolourization efficiency of textile dye Reactive Blue 19 in a horizontal rotating reactor using strips of BNC-immobilized laccase: optimization of conditions and comparison of decolourization efficiency. Biochem. Eng. J. 156, 107501.

Yu, Y., Yuan, J., Wang, Q., Fan, X., Ni, X., Wang, P., et al., 2013. Cellulase immobilization onto the reversibly soluble methacrylate copolymer for denim washing. Carbohydr. Polym. 95 (2), 675–680.

Zafar, A., Aftab, M.N., Iqbal, I., Dind, Z., Saleema, M.A., 2019. Pilot-scale production of a highly thermostable a-amylase enzyme from *Thermotoga petrophila* cloned into *E. coli* and its application as a desizer in textile industry. RSC Adv. 9, 984–992.

Zheng, X., Zhang, Y., Liu, X., Li, C., Lin, Y., Liang, S., 2020. High-level expression and biochemical properties of a thermo-alkaline pectate lyase from *Bacillus* sp. RN1 in *Pichia pastoris* with potential in ramie degumming. Front. Bioeng. Biotechnol. 8, 850.

Zucca, P., Cocco, G., Sollai, F., Sanjust, E., 2016. Fungal laccases as tools for biodegradation of industrial dyes. Biocatalysis 1, 82–108.

Chapter 24

Microbial enzymes in bioremediation

Shivani M. Yagnik[1], Prashant S. Arya[2] and Vikram H. Raval[2]
[1]Department of Microbiology, Christ College, Rajkot, Gujarat, India, [2]Department of Microbiology and Biotechnology, School of Sciences, Gujarat University, Ahmedabad, Gujarat, India

24.1 Introduction

Biogeochemical cycling is one of the most important phenomena on the Earth, which helps in balancing biotic and abiotic energy exchanges directly or indirectly. Although the cycling of compounds in all spheres is a complex and multilevel-dependent continuous cluster of processes, biodegradation is one of the system's most important and unavoidable processes (Pileggi et al., 2020). Biodegradation of complex compounds, whether inorganic or organic, is necessary to replenish the Earth and its resources; it is s a slow, steady, and consistent process of recharging various forms of natural compounds in the smallest possible forms. Such degraded compounds enrich the environment with greater accessibility to life forms, as many organisms can efficiently utilize simpler compounds (Atagana, 2009). A major part of biodegradation is carried out by microbes (Dangi et al., 2019).

Human interventions, industry evolution, and excessive use of novel compounds in almost every sector have drastically increased the load of complex compounds to be degraded by the microbial world. The novelty in the structure and composition of compounds has created a challenge and restricted the degradation patterns through microbes. These tremendous changes created an accumulation of undegraded and nondegradable compounds, which became havoc to all life forms as they are not utilizable or degradable. Furthermore, these compounds prove toxic to almost all life forms, including microbes (Pal et al., 2017). The pollutants are such undesirous compounds that are not degraded at the same pace as their production. Instead, they are getting piled up in aquatic, terrestrial, and air zones and enhance toxicity.

Remediation of air, water, and soil has been a priority task for the world to survive. Various bodies are working hard on the question of restoring

Biotechnology of Microbial Enzymes. DOI: https://doi.org/10.1016/B978-0-443-19059-9.00010-4
© 2023 Elsevier Inc. All rights reserved.

resources with hundreds of remedial methods (Dangi et al., 2019). These are expensive methods, even though countries worldwide spend remarkable budgets from their economies to develop sustainable techniques (Pal et al., 2017). Studies state that bioremediation is an efficient method for removing pollutants. Potential microorganisms are in use to degrade complex molecules to attain more efficacy. Microorganims degrade contaminants with a specific speed by providing optimized physical and chemical conditions, so their pace of degradation is enhanced. The task is challenging and yet to be attained, while the vast volume of research inputs is the ray of hope in achieving the same (Karthigadevi et al., 2021). Solid waste management in major countries is carried out through bioremediation (Sharma and Shukla, 2022). Industrial wastewater treatment gets faster using specific microbial consortia or with the help of specific microbes. Microbial air scrubbers have proven best for restoring air quality (Karthigadevi et al., 2021). Technical expansions are desired in this field with more potent microbes and their enzymes; this will be accountable for better Earth in coming times.

24.2 Robust microbes/superbugs in bioremediation

The process of detoxification through microbial metabolic or enzymatic activity for healthier environment achievement is bioremediation. The process requires a strong army of microbes to sustain and treat toxic and complex compounds (Margesin and Schinner, 2001). Industrialization introduced many novel compounds, known as xenobiotic compounds, to the environment.

24.2.1 Xenobiotic and persistent compounds

The word xenobiotic illustrates "*Xeno*" that means novel and "biotic," that is, organisms. Since microbes are on the Earth, their occurrence depends on various organic and inorganic compounds in different ecosystems with multiple forms of carbon, nitrogen, and hydrogen sources (Dangi et al., 2019). These diversified organisms have vivid metabolic pathways to utilize these naturally occurring compounds. The industrial era has generated various compounds such as persistent organic pollutants, synthetic drugs, dyes, pesticides, and chemical fertilizers (Fig. 24.1), expelled into nature. The synthetic and complex structures of these compounds make them least acceptable for utilization by the living world, as the living forms do not possess a specific metabolic enzymatic pathway for utilization of these compounds. Hence, these are known as xenobiotic compounds. As these are novel to the biotic factor, they are least degradable or nondegradable. These molecules remain in the ecosystem for decades and more; even they are found to be accumulated when entering the tropic levels. Such an accumulated portion retains as it is in the organism throughout its life and passes to the next tropical level upon consumption leading to biomagnification. DDT, lead (Pb), mercury (Hg),

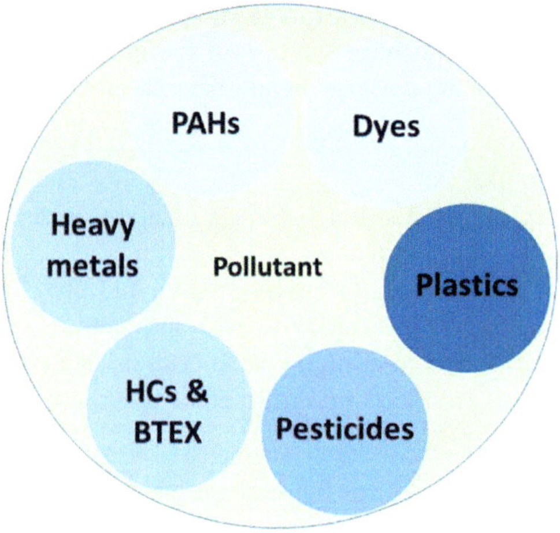

FIGURE 24.1 Example of environmental pollutant.

diclofenac, and many such compounds are found persistent in high amounts of health risk in humans, birds, and animals (Karthigadevi et al., 2021).

The xenobiotic compounds are majorly persistent and toxic to the biotic world. The thinning of an eggshell, decrease in fertility in buffaloes, disturbance in regular reproductive cycles in many birds, kidney failure in eagles, and high carcinogen concentration in humans are a few studied cases of toxicity due to these persistent xenobiotic compounds (Datta et al., 2020). The development of mankind through industrialization cannot be limited, making the constant addition of pollutants to the environment. Detoxification of resources becomes necessary in this situation. There are chemical, physical, photochemical, and biological methods for enhancing the degradation of such xenobiotic pollutants from soil and water. Bioremediation through microorganisms is the most cost-effective and sustainable method (Atagana, 2009). This process requires detailed and practical research outputs. Biotransformed organisms or enzymes of robust organisms can efficiently degrade these pollutants at necessary rates (Raghu et al., 2008).

24.2.2 Robust microbes and their application in bioremediation

Bioremediation can be carried out basically in two ways: in situ and ex situ. Microbes play a vital role in both cases. Industrial wastewater, oil spillages, petroleum polluted soil, chemically polluted agricultural lands, and food and dairy wastewater treatments are carried out by in situ bioremediation techniques. In contrast, domestic solid and liquid waste remediation, nuclear waste treatments, petrochemical industrial waste detoxification, and electronic waste treatments are majorly done by ex situ bioremediation methods worldwide (Dangi et al., 2019).

The pollutants are complex with large molecular weight and are least biodegradable; these properties make the task of bioremediation challenging. The microbes to be used for the purpose should have the following properties:

- The microbe should survive and grow at high levels of pollutants and multiple pollutants.
- The rate of enzyme production for the utilization or degradation of compounds should be compatible.
- That should be active in major environmental and physical changes.
- The organism must work and survive with indigenous organisms too.
- That should be fine active at high concentrations of by-products.
- The by-product(s) produced upon remedial treatment must be least/nontoxic to the environment.

Overall, the microbes used for bioremediation must be robust, known as a superbug. These are used to improve our environment from various perspectives (Margesin, and Schinner, 2001; Datta et al., 2020). Bacteria, fungi, and algae are mainly used as biocleaners. They use various strategies for detoxification according to their metabolic abilities and type of pollutant, such as accumulation, degradation through reduction or oxidation, desorption, and mineralization through aerobic or anaerobic functionalities (Table 24.1).

TABLE 24.1 Microorganisms and their applications in bioremediation.

Name of organism	Algae/ fungi/ bacteria	Application in bioremediation	References
Pseudomonas alcaligenes, Pseudomonas putida	Bacteria	Oil spillage removal, crude oil waste, paints, BTEX compounds biodegradation	Arun et al. (2008)
Geotrichum candidum	Fungi	Phenol and derived compounds	Dragicevic et al. (2010)
Bacillus subtilis	Bacteria	Oil-based products, paints	Kotoky and Pandey (2021)
Myrothecium roridum	Fungi	Industrial dyes' and effluents' treatments	Jasińska et al. (2019)
Phanerochaete chrysosporium	Fungi	Industrial dyes and crude oil degradation	Atagana (2009)
Micrococcus leuteus	Bacteria	Petrochemical and its products degradation	Bodor et al. (2021)

(Continued)

TABLE 24.1 (Continued)

Name of organism	Algae/ fungi/ bacteria	Application in bioremediation	References
Bacillus firmus	Bacteria	Paper and textile industrial waste detoxification	Mandal et al. (2014)
Bacillus macerans	Bacteria	Nitrocompounds, DDT degradation	Pileggi et al. (2020)
Staphylococcus aureus	Bacteria	PAHs compounds' degradation	Perpetuo et al. (2011)
Klebsiella oxytoca	Bacteria	Vinyl compounds	Perpetuo et al. (2011)
Aspergillus niger	Fungi	Organic compounds, crude oil, and heavy metals and degradation	Atagana (2009)
Lipomyces kononenkoae	Fungi (yeast)	BTEX compounds degradation	Sadowsky et al. (2009)
Chlorophyceae	Algae	Insecticides and pesticides	Caceres et al. (2008)
Cyanobacteria	Algae	Naphthalene	Caceres et al. (2008)
Acinetobacter	Bacteria	Petrochemical products	Pal et al. (2017)
Sphingobacter	Bacteria	Herbicides	Pal et al. (2017)

Pseudomonas spp. showed excellent results for crude oil degradation in various studies in situ and ex situ in both conditions (Arun et al., 2008). Solid wastes are detoxified with alternative aerobic and anaerobic degradation with greater efficiencies using defined bacterial consortia (Pileggi et al., 2020). Paints, plastics, and petrochemical compounds are slowly degraded, but the anaerobic biodegradation process obtains efficient results with smaller molecular weight by-products (Datta et al., 2020). Heavy metals (HMs) from soil and wastewater are removed or detoxified efficiently by white-rot fungi, *Acinetobacter* spp., and *Pseudomonas* spp.

24.2.3 Metabolic pathway engineering for high-speed bioremediation

Biodegradation is a prolonged process. Even the known robust organisms take considerably more time, and simultaneously the load of pollutants to be degraded enters the environment comparatively faster. This needs to be balanced with enhancing the rate of degradation. Selecting robust organisms and then modifying their metabolic pathway for enhanced production of a preferred enzyme(s) and effective functioning becomes promising (Dangi et al., 2019).

The degradation of methyl phenols and methyl benzoates is efficiently carried out at a high rate by metabolically engineered *Pseudomonas* spp. B13; various catabolic pathways from diverse organisms were shared and fabricated into *Pseudomonas* spp. B13 (Sharma and Shukla, 2022).

Nickel and cobalt are HMs with higher destructive effects; electronic wastes have increased their amount in water reservoirs. The removal of nickel and cobalt through the increased expression of NiCoT genes of *Escherichia coli* knockout mutant of NiCoT efflux gene (rcnA) gives promising results. The NiCoT genes were conceded from *Rhodopseudomonas palustris* CGA009 (RP) and *Novosphingobium aromaticivorans* F-199 (NA) (Raghu et al., 2008). The degradation pathway of halo-alkane tri-chloro-propane (TCP) to glycerol in *Pseudomonas putida* was reconstructed, and *P. putida* strain KT2440 was designed by inserting genes for the production of epoxide hydrolase, halo-alkane dehalogenase, and halo-alcohol dehalogenase (Gong et al., 2016).

24.3 Role of microbial enzymes

Enzymes are supermolecules that enhance respective reaction even outside the cell. The use of enzymes spurred upon its industrial, agricultural, and medical usage. Emerging science gave the idea of modified microbial enzymatic use for bioremediation improvements as one of the alternative solutions to highly complex compounds (Fig. 24.2). Microbes in consortia and communities show promising results with various metabolic pathways. The metabolism process involves a specific group of enzymes for a specific biochemical reaction. Several enzymes are involved in bioremediation, such as laccases, hydrolases, oxydoreductase, lyase, and hydroxylase. The definite molecule or type of molecules undergoes a suitable metabolic pathway with

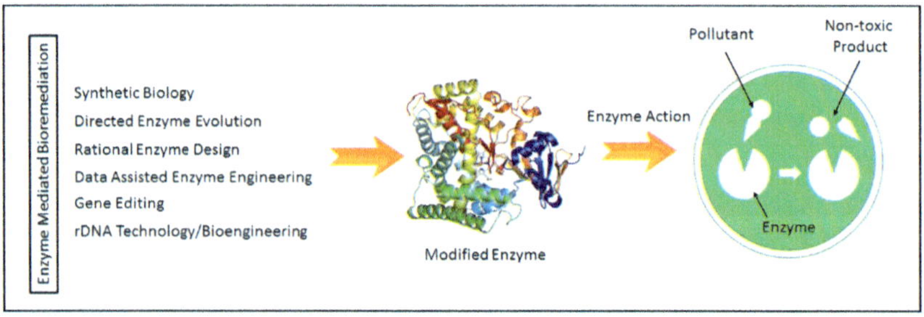

FIGURE 24.2 Enzyme modification for high-speed bioremediation.

a specific set of enzymes for the degradation, and this requires the respective microbe(s) or enzyme(s) (Arya et al., 2022; Sharma and Shukla, 2022).

24.3.1 Dye degradation

Synthetic dyes are used in vast amounts as industrialization has progressed in this era. Cosmetics, paints, leather, plastic stuff, textile (wool, silk, cotton, rayon), food color, and many more such products use synthetic dyes for furnishing. These are the least biodegradable, with larger molecular weights and complex structures. There are various dyes. The chemical structures are classified as acidic dyes, basic dyes, neutral dyes, vat dyes, azo dyes, disperse dyes, sulfur dyes, and reactive dyes (Yagnik et al., 2021a; Saket et al., 2022). Dye producing and dyeing or painting industries of various materials expel dye-polluted water in major water resources creating environmental hazards. The water becomes colored and toxic and even can't be used for agricultural purposes. Synthetic dyes contain benzene and polycyclic groups, majorly with halogen, metal, nonmetal, sulfur, methyl, and/or azo groups. These are recalcitrant compounds, and indigenous microorganisms show their degradation at very slow rates (Saket et al., 2022).

Chemical and physical ways of dye degradation and remediation of dye-polluted water bioremediation with microorganisms have proven cheaper, faster, eco-friendly, and accepted worldwide. Dyes are toxic to microbes also as they consist of complex functional groups, so for bioremediation, potential dye degrading microorganisms are required, which can be seeded in the polluted region to enhance the detoxification through bioremediation (Saravanan et al., 2021). Studies show about 97% of dye degradation at lab scales under optimized conditions by specific microorganisms. These, in actual conditions, give more than 70% removal of dyestuffs. Generally, the concentrations of dye(s) in the effluent or wastewater to be treated are about 300–700 ppm and this can be maximized up to 1100 ppm in some instances (Sarkar et al., 2017). The organism should be robust to withstand these levels of toxicity and can degrade them too. *Bacillus fermis, Virgibacillus marismortui, Proteus mirabilis, Pseudomonas, Alcaligenes faecalis, Aspergillus niger*, and *Arthrobacter soli* are a few examples of potent dye degrading organisms studied, and many are used at industrial levels (Yagnik et al., 2021b).

A community of microorganisms acts on dye pollutant in the environment, and various organisms' consortia studies on dye degradation show promising results. Optimization of environmental and nutritional parameters enhances the bioremediation process of dye-contaminated environments. Oxidase and reductase enzymes are involved in microbial catalysis of dye molecules; the genetically engineered organisms with these enzymes are of interest for the enhanced process. The by-product of bioremediation is of concern in dye degradation with microbes; the by-products may be toxic to plants and animals and get released into the environment. The treated water is generally reused by industry or is majorly used for crops and irrigation. This increases the unavoidable studies on by-product toxicity tests such as protozoal toxicity tests, phytotoxicity tests, and aims tests (Sarkar et al., 2017).

Governing bodies of major countries are dedicating financial and scientific pools in this direction for reuse and restoring usable water supplies to reduce the crisis of water resources. The demand for dyes will remain consistent; the detoxification of such dye-polluted water with sustainable methods and its reuse can fill the gap in water crisis and demands (Saravanan et al., 2021).

24.3.2 Remediation of hydrocarbon and benzene, toluene, ethylbenzene, and xylene compounds

The constantly increasing need for crude oil and its products enhanced soil and water contamination through hydrocarbons. These being large in size and molecular weight are difficult for degradation. The hydrophobic nature of crude oil hinders getting available and adsorbed by any microorganism or plants; they can uptake hydrophilic and water-soluble compounds, while mainly hydrocarbons are the least water-soluble (Margesin and Schinner, 2001). The physical and chemical treatment of such regions is possible with higher costs and partial degradation. These methods release BTEX compounds (benzene, toluene, ethylbenzene, and xylene) into natural resources, which are highly toxic and are a point of environmental and health concerns (Kotoky and Pandey, 2021).

Natural microbial degradation is relatively slow, while NPK sources enhance the degradation rate of indigenous organisms with promising results (Pal et al., 2017). The regions of oil spillage in the marine aquatic environment become hazardous to many aquatic lives, and natural degradation and clearance of sea surface take several months or years, which countably affects marine life. This has been made faster using NPK sources and a few *Pseudomonas* spp.; these bequeathed clear zones on the contaminated area in a few weeks. The nutrients initially enhance the growth of indigenous microbes, which boosts the augmentation rates of hydrocarbons (Margesin and Schinner, 2001).

Extremophiles and their enzymatic actions that are capable of hydrocarbon and petrochemical degradation have been studied (Raval et al., 2018). The hydrocarbon-contaminated regions are mainly in extreme environments such as high or low pH, temperatures, or pressure. Microbes of this environment are already adapted to difficult conditions; studies show their significant role in biodegradation and treatment of hydrocarbon-polluted areas (Raval et al., 2014). Psychrophiles are degraders in seawater regions, especially near oil extraction wells and oil tanker accidents, indigenous halophiles have been studied for oil-polluted desert soil remediation, and thermophiles show enhanced in situ remediation of soil near refineries as they require less water availability which is the point of interest (Margesin and Schinner 2001; Dangi et al., 2019).

The bioavailability is a key step in the bioremediation of hydrocarbons and their compounds. The hydrophobic nature of hydrocarbons makes them nonaccessible for microbes and plants. The cellular membranes can efficiently uptake hydrophilic substances; for this, surfactants are used to attain the process of bioavailability. Surfactants decrease the surface tension of hydrocarbons. With this, they are amphoteric, having both hydrophobic and hydrophilic ends. These properties help surfactants to create micelles (spherical structures with hydrophilic surface and hydrophobic core). The micelles of pollutant molecules can be transported easily through cellular membranes, enhancing bioremediation. Chemical surfactants are toxic to microbes and are harsh to the environment; hence, biosurfactants are used widely for enhanced bioremediation of hydrocarbons and BTEX compounds. Biosurfactants are naturally metabolic products having properties similar to surfactants and are biodegradable. Several biosurfactants produce microbial species which yield molecules such as glycolipids, lipoproteins, lipopolysaccharides, and glycoproteins, which function as biosurfactants (Souza et al., 2014).

24.3.3 Heavy metal remediation

Contamination with HMs is a persistent environmental issue in many nations. Volcanic eruptions, pesticides, ammunition, organic compounds, metal processing, coal ash, paints, industrial waste, domestic and agricultural wastes, and mining operations are all significant causes of HM pollution in soil. Since HMs inflict direct and indirect damage to living creatures and the environment, remediation of polluted soils is critical (Saravanan et al., 2021). For a long time, attempts have been made to address the indefinite persistence of HMs in the soil. Yet, the problem is sustained due to the increased use of HMs and a lack of appropriate methods for their elimination. The toxicity of HMs varies depending on the element. For example, As and Cd are exceedingly hazardous; Hg, Pb, and Ni are somewhat toxic; while Cu, Zn, and Mn are less dangerous in biological systems. HMs, on the other hand, have a variety of toxicological effects on humans, including central nervous system disturbance, renal malfunction, and lung damage (Sharma and Shukla, 2022).

Enzymatic cleanup of HM-polluted locations is known to be accomplished by a variety of bacteria. HM remediation is known for the bacterial and fungal genera *Flavobacterium*, *Pseudomonas*, *Enterobacter*, *Bacillus*, *Micrococcus*, *Arthrobacter*, *Acinetobacter*, *Aspergillus*, *Penicillium*, *Rhizopus*, *Pichia*, *Rhodotorula*, *Saccharomyces*, *Hansenula*, and *Yarrowia* (Verma et al., 2021). Moreover, the capacity of genetically engineered bacteria, such as *Corynebacterium glutamicum*, *Rhodopseudomonas palustris*, *E. coli*, *Mesorhizobium huakuii*, and *Deinococcus radiodurans*, was found to be used in places polluted with HMs (Diep et al., 2018).

One of the promising ways for pollution-free remediation solutions via transformation and demineralization is the bioremediation of HMs by microbial enzymes. Certain microbial enzymes are used to combat the threat of trace metals by reducing them to nontoxic or less damaging components, according to Ojuederie and Babalola (2017). They are master molecules that make it easier to identify contamination before, during, and after cleanup. Mukherjee et al. (2018) isolated *Ralstonia* spp., which can live and withstand a variety of metals, including cadmium, cobalt, zinc, arsenic, nickel, and mercury. Arsenate reductase and mercuric reductase were generated by this species, which reduced As^{+5} to As^{+3} and Hg^{+2} to Hg, respectively.

Furthermore, multiple studies have shown that microbial enzymes can help reduce HM levels in the papermaking business (Nathan et al., 2018). To handle HM-contaminated soils, microbial enzymes such as arylsulfatase, beta-glucosidase, and dehydrogenase have been utilized "in situ," resulting in better soil quality (de Mora et al., 2005). Additionally, oxidative and reductive enzymes employed for metal transformation, such as cupric reductase from *Streptomyces* spp., enhance copper reduction. Likewise, urease from *Sporosarcina pasteurii* is used to biomineralize HMs such as nickel, cobalt, cadmium, copper, and zinc (Saravanan et al., 2021).

24.3.4 Pesticide degradation

Chemical pesticides are widely employed in contemporary agriculture to protect agricultural plants against insects and microbiological pests. Still, the incorrect application of synthetic chemicals poses a risk to nontargeted creatures and the environment surrounding the application locations. In worldwide farmlands, the disposal mechanism of applied pesticides is very diverse; as a result, long-term rehabilitation of pesticide-contaminated landscapes is a major study field for environmentally acceptable management of these pollutants (Sarker et al., 2021). Furthermore, experts advise against physical and chemical pesticide and metabolite cleanup due to time, cost, and sustainability concerns compared to microbial and enzymatic remediation. Microbes from many taxonomic families, such as bacteria, fungus, and algae, have been shown to successfully metabolize pesticides and/or modify their chemical structures, allowing for their breakdown and removing dangerous pesticide residues from polluted environments. Most organic and chemical pesticides are effectively remedied by catalytic enzymes such as oxidases, reductases, and hydrolytic enzymes (Jaiswal et al., 2019).

The breakdown of microbial pesticides is divided into three stages: the initial parent pesticide chemical is changed into nontoxic and soluble compounds in the first phase by reduction, hydrolysis, or oxidation. Several bacterial enzymes have recently been revealed to aid in removing harmful pesticides. According to Sirajuddin et al. (2020), *E. coli* IES-02 carboxylesterase, malathion esterase, and oxidoreductase enzymes may degrade pesticides by 81%. Malathion esterase, for example, converts malathion to malathion monocarboxylic acid during decomposition. Malathion monocarboxy esterase and malathion dicarboxyloxidoreductase eventually convert the chemical to simpler compounds such as succinic acid and ethyl hydrogen formate. Fenamiphos hydrolyzing enzyme, a phosphotriesterase-like enzyme from *Microbacterium esteraromaticum*, has been used to degrade a neurotoxic pesticide called fenamiphos, which is then converted into amino acids by methyltransferases (Logeshwaran et al., 2020). *Bacillus thuringiensis* organophosphate enzymes were shown to be effective at degrading chlorpyrifos, dimethoate, triazophos, and trichloropyridine (TCP) (Ambreen and Yasminm 2020).

Pesticide is broke down by *Trametes versicolor* laccases and other lignin-degrading enzymes—dicofol and chlorpyrifos (Hu et al., 2020). Similarly, enzymes such as cytochrome p450, laccases, and monooxygenases have been widely used to break down xenobiotics such as diuron. The major intermediates generated by the demethylation of diuron are 1-(3, 4-dichlorophenyl)-3-methyl urea (DCPMU) and (3,4-dichlorophenyl) urea. Only DCPMU showed signs of further breakdown by fungal enzymes (Henn et al., 2020).

24.4 Remedial applications for industries

Environmental security is one of the most pressing problems for protecting living species, including humans, from various dangerous chemicals in the environment. To manage and regulate the hazards of environmental pollution, many efforts, legislative activities, as well as scientific and social concerns have been developed and implemented, but it remains a global issue. As a result, accurate, quick, and selective approaches for detecting, screening, and removing contaminants for successful bioremediation procedures are required. Several approaches for bioremediation and their applications are described in Fig. 24.3. In this regard, isolated enzymes or biological systems that produce enzymes, whether as complete cells or immobilized, can be employed as a source for detecting, quantifying, and degrading or transforming pollutants into nonpolluting substances to restore ecological balance (Kumar and Bharadvaja, 2019).

24.4.1 Designing and developing environmental biosensor

Among all methodologies, biosensors are an appropriate and self-contained integrated instrument for the dependable, quantitative, specific, and sensitive detection/measurement of environmental contamination. Biosensors are devices that utilize any biological mechanism in direct contact with a transduction element to detect analytes. The biosensor is not often considered a separate device but rather part of a larger set of instruments. A biosensor's three main components are a biological recognition element, a transducer, and a signal

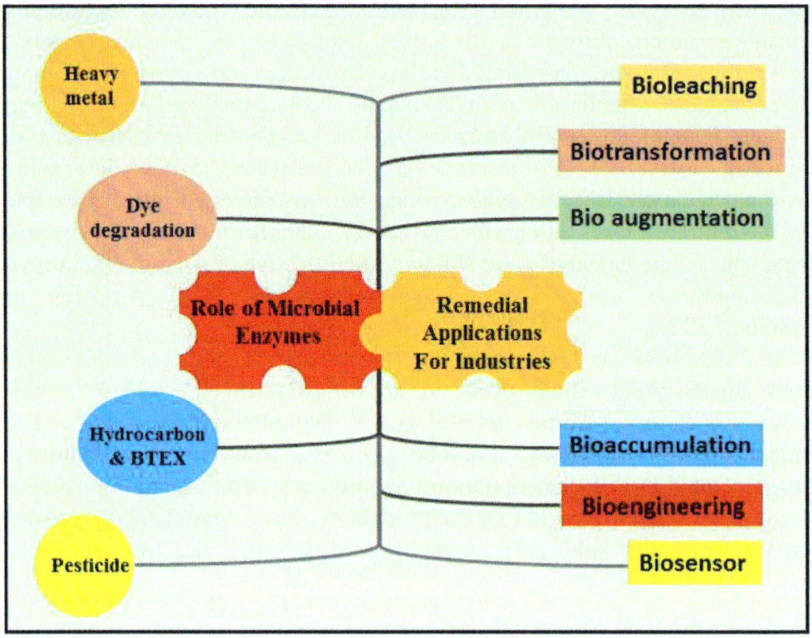

FIGURE 24.3 Applications of enzyme in bioremediation.

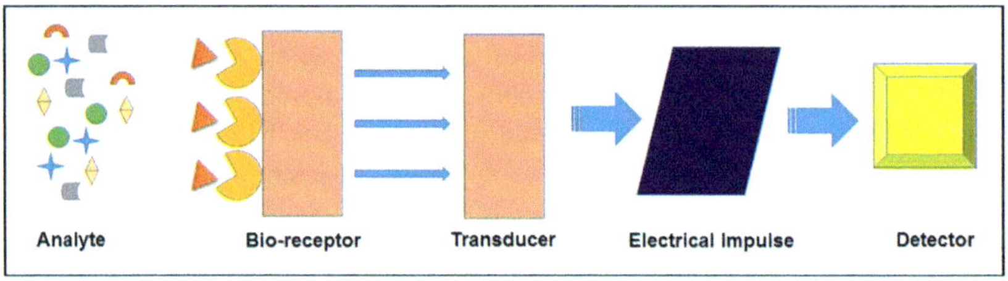

FIGURE 24.4 Basic mechanism of biosensor.

processing system (Gavrilaş et al., 2022). Fig. 24.4 shows a schematic illustration of a biosensor.

The qualities of bioreceptors used in detection techniques (e.g., antibodies, entire microbial cells, enzymes, proteins, or DNA fragments) or the physicochemical nature of transducers employed for toxicant detection is used to further classify biosensors (e.g., optical, thermal, electrochemical, calorimetric, piezoelectrical, etc.). Biological catalyst-based biosensors can detect the presence of particular analytes by monitoring the consumption or synthesis of chemicals such as CO_2, NH_3, H_2O_2, H^+, or O_2, and therefore transducers can identify contaminants and correlate their presence in substrates (Nigam and Shukla, 2015; Reynoso et al., 2022).

The development of biosensors for numerous applications, such as biopharma, food and beverages, biodefense, and environmental analysis, has seen a 10.4% increase in market trends over the last decades. In comparison to biological applications, environmental applications are still in their infancy and face several problems owing to the inherent peculiarities of environmental analysis. Low detection limits, a complex environmental matrix, and analyte specificity are all important factors in developing biosensors for environmental investigation (Cavalcante et al., 2021). Modern biosensors, which were synthesized by various polymers, papers, plastics, and composites to manufacture biosensors, are currently being employed to address this difficulty. These materials are more portable, miniaturized, low cost, user-friendly, environment-friendly, and economically sustainable when they are used (Reynoso et al., 2022).

Many enzymatic biosensors have been created for monitoring water and air contaminants such as Zn, Cu, Cd, Ni, phenol/Cl-phenol, parathion, and diuron. As a necessary by-product of various industries, wastewater contains industrial chemicals such as dyes, aromatic substances, diphenylmethane, chlorine compounds, surfactants, hydrocarbons, metals, and HMs (Cu, Ti, Cr, Cd, As, Zn, Co, Pb, and Hg), as well as pesticides, herbicides, and bactericides (Verma and Rani, 2021). Table 24.2 describes many enzyme-based biosensors.

24.4.1.1 Limitations of biosensors

- The biorecognition elements are hampered by a large load of analytes, resulting in an attenuation response.
- The existence of particles and/or organic elements might cause the sensor signals to be obstructed.
- Reproducibility, repeatability, and the ability to utilize indefinitely.
- Too far, the majority of HM detection biosensors have relied on HM ion detection.
- Environmental variations in pH, mass, temperature, and other factors impact the biosensor's accuracy and dependability.

24.4.2 Immobilization and bioengineering

Bioengineering is the process of transferring coding properties into recombinant proteins to increase the enzymes' catalytic activity, specificity, and stability (Arya et al., 2021). By changing the fundamental structure of enzymes, they boosted their substrate selectivity, pH and temperature stability, stress tolerance, and shelf life. In different phases of the condition, recombinant enzymes have a considerably higher capacity for contaminant breakdown (Akerman-Sanchez and Rojas-Jimenez, 2021). Table 24.3 defines the various modified enzymes via bioengineering that eliminate or reduce the number of

TABLE 24.2 Enzyme in biosensors.

	Analyte	Samples	Transducers	Biosensing elements	References
Heavy metals	Hg, Cd, and As	Industrial effluent	Electrochemical	Urease enzyme	Pal et al. (2009)
	Cd, Cu, and Pb	Synthetic effluents	Electrochemical	Sol–gel-immobilized urease	Ilangovan et al. (2006)
Phenolic compounds	Binary mixtures: phenol/chlorophenol, catechol/phenol, cresol/chloro-cresol and phenol/cresol	Wastewater	Amperometric	Laccase and tyrosinase	Karim and Fakhruddin (2012)
	Phenol, *p*-cresol, *m*-cresol, and catechol	Wastewater	Amperometric	Polyphenol oxidase	
	Phenol	Wastewater	Amperometric	tyrosinase	Silva et al. (2010)
Pesticides	Simazina	Soil and wastewater	Potentiometric	Peroxidase (biocatalytic)	Rodriguez-Mozaz et al. (2006)
	Parathion	Soil	Amperometric	Parathion hydrolase (biocatalytic)	Mostafa (2010)
	Paraoxon	Soil and wastewater	Optical	Alkaline phosphatase	
	Carbaril	Soil and wastewater	Amperometric	Acetylcholinesterase	
Herbicides	2,4-Dichloro-phenoxy acetic acid	Soil	Amperometric	Acetylcholinesterase	Sassolas et al. (2012), Sadowsky et al. (2009)

TABLE 24.3 Bioremediating enzymes modified or produced using bioengineering approach.

Recombinant enzyme	Native microorganism	Engineered microorganism	Expression vector	Pollutants	References
Tetrahydrofuran monooxygenase Polyphenol oxidase	*Pseudonocardia* sp. Strain KT	*Escherichia coli*	PVP55A	Chlorinated solvents Tetrahydrofuran Cyclic ethers 1,4-dioxane	Mihovilovic et al. (2006)
Laccase	*E. coli* K12	*Pichia pastoris* GS5115	pHBM905BD M	Synthetic dye	Li et al. (2007)
Laccase Versatile peroxidase Mn peroxidase Lignin peroxidases	*Saccharomyces cerevisiae* *Trametes versicolor* *Pleurotus. eryngii,* *Phanerochaete chryososporium*	*E. coli* *P. chryososporium*	pETa (þ) Pst2 pPCLACIIIb pPCLIPH8 pPCVPL2 pPCMNP1	Quinine Phenolic component Synthetic dye	Cardona et al. (2011)
Horseradish peroxidase	Horseradish plant	*E. coli* BL21	pET21d pAES30	Endocrine disruptive Chemicals phenolic compounds	Bansal and Kanwar (2013)

contaminants from the environment. Protein engineering has already been identified for high volume and targeted ribonuclide and HM bioremediation. For example, by changing the location of an amino acid in nitrobenzene 1,2-dioxygenase, the oxidation of 2, 6 di-nitro-toluene was boosted, resulting in the release of nitrite and catechol as products (Li et al., 2021b).

Immobilization of proteins can be accomplished via a variety of approaches, such as entrapment, encapsulation, cross-linkage, covalent binding, and adsorption. It preserves enzyme function, increases shelf life, provides a broad surface area for substrate binding, and allows for enzyme recovery and reuse. Because of their regularity and frequent usage in xenobiotic combinations, the immobilized compounds looked financially savvy (Table 24.4) (Shakerian et al., 2020). Immobilized oxidative enzymes are now widely regarded as a green solution to the problem of excessive levels of micropollutants in nature. Laccases and horseradish peroxidase, among other oxidative enzymes, have been utilized often in recent years because they are universal oxidative enzymes that can oxidize a wide range of substances. Compared to free enzymes, immobilized laccase or horseradish peroxidase demonstrated superior stability, reusability, and ease of separation from the reaction mixture, making them more attractive and cost-effective (Yaashikaa et al., 2022).

Kadam and his coworkers (2018) employed a covalent approach to encapsulate Lac on chitosan-functionalized supermagnetic halloysite nanotubes for Dr80 degradation. Lac was covalently immobilized on bacterial cellulose/TiO2-functionalized composite membranes by Li and his team (2017), and it was employed for Reactive Xe3B degradation. To investigate the potential of the entrapped cross-linked enzyme (E-CLE) and E-CLE aggregate (E-CLEA) for Lac immobilization, Fathali et al. (2019) employed carrier-free immobilization techniques of E-CLE and E-CLEA.

Nanozymes or artificial enzymes of the next generation with enzyme properties are enzyme impersonators based on nanoparticles. In physiologic environments, they catalyze substrate adaptation and follow a similar mechanism and kinetics as regular enzymes. Due to remarkable consistency and low cost, nanoenzymes attract much attention from scientists (Kumar and Bharadvaja, 2019). Nanoenzymes can be used to detect a variety of biomolecules and often lack an active site, allowing just a specific substrate to bind for a chemical reaction to occur. Nanoenzymes have a wide range of uses in bioremediation. These enzymes are used to deprive and recognize pollutants such as lignin, organic chemicals, and dyes, among other things (de Souza Vandenberghe et al., 2022).

24.4.3 Biotransformation and bioleaching

Biotransformation, biocomposting, landfarming, biostimulation, bioleaching, bioventing, bioaugmentation, and bioreactor are just a few of the sophisticated

TABLE 24.4 Immobilization of enzymes and their application in bioremediation.

Immobilization method	Compounds	Support materials	Enzymes	References
Adsorption	TC, BPA, Phenol, ARS, HQ, MG, SAs	Modified agarose, chitosan, nanoparticles, magnetic microspheres, metal oxides, graphene aerogel-Zr-MOF, kaolinite	Laccase Peroxidase Lipase Protease Oxidase	Yaashikaa et al. (2022)
Covalent	EDCs, phenol, p-chlorophenol, catechol, synthetic dyes, TCS, DCF, BPA	Polymer containing cyclic carbonate, ceramic membrane, epoxy-functionalized silica, nanofiber, nanotubes, nanoparticle-coated PVDF membrane, micro-biochars	Horseradish peroxidase Laccase Oxidases Hydrolytic enzymes	Shakerian et al. (2020)
Cross-linking	Aromatic compounds, dyes, chemicals	Nanoparticles, nanotubes, ZnO nanowires/ macroporous SiO_2 composite	Lipase Dehydrogenase Laccase Peroxidase	Shakerian et al. (2020); Işık et al. (2021)
Encapsulation	Phenolic compounds, Chemicals, Synthetic dye, BPA, Antibiotics	Phospholipid-templated titania, electrospun fibrous membranes	Laccase Peroxidase Lipase	de Souza Vandenberghe et al. (2022)
Entrapment	BPA, MOs, synthetic dyes Petroleum chemical	Magnetic nanoflowers, Ca-alginate beads, agarose–chitosan hydrogel, chitosan beads	Laccase Oxodase	Imam et al. (2021)

techniques that have been developed for the effective and simple treatment of waste materials. The metabolic conversion of endogenous and xenobiotic substances to more water-soluble molecules is known as biotransformation. A small number of enzymes with broad substrate specificities perform xenobiotic biotransformation, commonly known as drug-metabolizing enzymes (Tak et al., 2022). Absorption, distribution, metabolism, and elimination are phase I processes that include hydrolysis, reduction, and oxidation. Phase 1 and 2 reactions are used to transfer aromatic, aliphatic, heterocyclic, alicyclic, nitrogen- and sulfur-based compounds. *N*-Hydroxylation, dealkylation, reduction, azo, nitro, hydrolysis, ester, amide, hydrazide, and carbamate compounds are just a few of the reactions and pathways involved in biotransformation (Bagiu et al., 2020).

Biotransformation is, therefore, a very crucial step of disposition since it can have a significant impact on the compound's biological action as well as enhance excretion by increasing polarity and hence hydrophilicity. Plant regulators of these items may have been synthesized in goods derived from plants to which agrochemicals, particularly pesticides, have been applied, and they may be detected in food. The same may be said for veterinary drugs and animal products: mammalian metabolism is frequently but not always analogous to human metabolism. Consideration of avian and piscine biotransformation products may be relevant in poultry and fish products. Metabolism in a broad range of species is significant in environmental toxicology because it can lead to human exposure, and bacterial metabolism in bacteria is crucial in the mammalian gut (Li et al., 2021a). This may transfer the acetyl group to another amine molecule or the hydroxy group, resulting in a highly reactive acyloxyarylamine that interacts with proteins and nucleic acids following a rearrangement. High AOX, COD, BOD, suspended particles, toxicity, color, lignin and its derivatives, and chlorinated chemicals that can be biotransformed are all difficulties linked with the pulp and paper industry (Kumar et al., 2021).

Bioleaching, often called microbial leaching or bio-hydrometallurgy, is the method of extracting metals from insoluble ores using biologically existing microbes. In which microorganisms are used to cleanse or detoxify polluted sludge and hazardous materials present in the soil and water to create a safe and habitable environment. One such bioremediation strategy is microbial bioleaching, which uses acidophilic microorganisms (including archaea and bacteria) and fungi to remove hazardous wastes containing heavy or toxic metals, such as electronic wastes, wasted catalysts, steel slag, and sludge (Srichandan et al., 2019). Acidophilic bacteria and fungi acidify sulfuric (H_2SO_4) and organic acids, respectively, and acidophilic microorganisms oxidize ferrous iron (Fe^{2+}) to ferric iron (Fe^{3+}). Such metabolic enzymes react with the hazardous wastes, converting them to nonhazardous forms. Successful investigations on acidophilic bacteria, *Acidithiobacillus ferrooxidans*, *Acidithiobacillus thiooxidans*, and a fungus *Aspergillus niger*, have been published on their function in the bioleaching process. Electronic wastes, spent petroleum catalysts, sludge, and slag are all reduced or decontaminated due to this procedure (Xue and Wang, 2020).

24.5 Concluding remarks

The deposition of environmental pollutants has reached an alarming level in recent years due to urbanization, population growth, and industrial expansion. However, bioremediation is the only environmentally benign answer to this problem. Enzyme-based bioremediation is a viable and cost-effective solution. Researchers looked at a wide spectrum of bacteria from various natural sources to isolate enzymes with biodegradative potential. A broad family of enzymes with bioremediation potential has been identified and described. Enzyme-based bioremediation did not initially appear to be particularly successful due to the small number of enzymes generated by microorganisms in natural circumstances.

Meanwhile, improvements in biosensors, bioengineering, immobilization, and nanomaterial coatings may provide these microbes and their enzymes with more ideal situations for considerable pollution removal. Furthermore, these approaches can significantly improve the catalytic activity, shelf life, and stability of enzymes under stressed conditions. Due to restricted cultivation technologies, the more significant number of microflora in the environment is unknown, potentially having enormous bioremediation potential. As a result, novel enzymes and their roles in bioremediation must be discovered. From either perspective, a systems biology approach including omics methods such as genomics, proteomics, transcriptomics, phenomics, lipidomics, and metabolomics might be useful in studying bacteria and their enzymes for bioremediation. Nonetheless, the methods mentioned above are sufficient for effective bioremediation, and the demand for the most advanced, efficient, and environmentally acceptable solutions necessitates the use of the most advanced technological intervention.

Abbreviations

AOX	adsorbable organically bound halogens
ARS	Alizarin Red S
BOD	biological oxygen demand
BPA	bisphenol A
BTEX	benzene, toluene, ethylbenzene, and xylenes
COD	chemical oxygen demand
DCF	diclofenac
DCPMU	1-(3, 4-Dichlorophenyl)-3-methyl urea
DDT	dichlorodiphenyltrichloroethane
E-CLE	entrapped cross-linked enzyme
E-CLEA	entrapped cross-linked enzyme aggregate

EDCs	endocrine disrupting compounds
HMs	heavy metals
HQ	hydroquinone
MG	malachite green
MOF	metal–organic framework
MOS	microorganisms
NPK	nitrogen (N), phosphorus (P), and potassium (K)
PAHs	polycyclic aromatic hydrocarbons
ppm	parts per million
PVDF	polyvinylidene fluoride
SAs	sulfonamide antibiotics
Spp	species
TC	tetracycline
TCP	trichloro pyridine
TCS	triclosan

References

Akerman-Sanchez, G., Rojas-Jimenez, K., 2021. Fungi for the bioremediation of pharmaceutical-derived pollutants: a bioengineering approach to water treatment. Environ. Adv. 4, 100071.

Ambreen, S., Yasmin, A., 2020. Isolation, characterization and identification of organophosphate pesticide degrading bacterial isolates and optimization of their potential to degrade chlorpyrifos. Int. J. Agriculture Biol. 24 (4), 699–706.

Arun, A., Raja, P.P., Arthi, R., Ananthi, M., Kumar, K.S., Eyini, M., 2008. Polycyclic aromatic hydrocarbons (PAHs) biodegradation by basidiomycetes fungi, *Pseudomonas* isolate, and their cocultures: comparative *in-vivo* and *in-silico* approach. Appl. Biochem. Biotechnol. 151 (2), 132–142.

Arya, P.S., Yagnik, S.M., Rajput, K.N., Panchal, R.R., Raval, V.H., 2021. Understanding the basis of occurrence, biosynthesis, and implications of thermostable alkaline proteases. Appl. Biochem. Biotechnol. 193 (12), 4113–4150.

Arya, P.S., Yagnik, S.M., Panchal, R.R., Rajput, K.N., Raval, V.H., 2022. Industrial applications of enzymes from extremophiles. Physiology, Genomics, and Biotechnological Applications of Extremophiles. IGI Global, pp. 207–232.

Atagana, H.I., 2009. Biodegradation of PAHs by fungi in contaminated soil containing cadmium and nickel ions. Afr. J. Biotechnol. 8 (21).

Bagiu, R.V., Sarac, I., Radu, F., Cristina, R.T., Butnariu, M., Bagiu, I.C., 2020. Chemical transformations of synthetic persistent substances, Bioremediation and Biotechnology, Vol 3. Springer, Cham, pp. 65–103.

Bansal, N., Kanwar, S.S., 2013. Peroxidase (s) in environment protection. Sci. World J. 2013.

Bodor, A., Bounedjoum, N., Feigl, G., Duzs, Á., Laczi, K., Szilágyi, Á., et al., 2021. The exploitation of extracellular organic matter from *Micrococcus luteus* to enhance ex-situ bioremediation of soils polluted with used lubricants. J. Hazard. Mater. 417, 125996.

Caceres, T.P., Megharaj, M., Naidu, R., 2008. Biodegradation of the pesticide fenamiphos by ten different species of green algae and cyanobacteria. Curr. Microbiol. 57 (6), 643–646.

Cardona, F., Orozco, H., Friant, S., Aranda, A., Del Olmo, M., 2011. The *Saccharomyces cerevisiae* flavodoxin-like proteins Ycp4 and Rfs1 play a role in stress response and in the regulation of genes related to metabolism. Arch. Microbiol. 193 (7), 515–525.

Cavalcante, F.T., de A Falcão, I.R., da S Souza, J.E., Rocha, T.G., de Sousa, I.G., Cavalcante, A.L., et al., 2021. Designing of nanomaterials-based enzymatic biosensors: synthesis, properties, and applications. Electrochem 2 (1), 149–184.

Dangi, A.K., Sharma, B., Hill, R.T., Shukla, P., 2019. Bioremediation through microbes: systems biology and metabolic engineering approach. Crit. Rev. Biotechnol. 39 (1), 79–98.

Datta, S., Rajnish, K.N., Samuel, M.S., Pugazlendhi, A., Selvarajan, E., 2020. Metagenomic applications in microbial diversity, bioremediation, pollution monitoring, enzyme and drug discovery. A review. Environ. Chem. Lett. 18 (4), 1229–1241.

de Mora, A.P., Ortega-Calvo, J.J., Cabrera, F., Madejón, E., 2005. Changes in enzyme activities and microbial biomass after "in situ" remediation of heavy metal-contaminated soil. Appl. soil. Ecol. 28 (2), 125–137.

de Souza Vandenberghe, L.P., Junior, N.L., Valladares-Diestra, K.K., Karp, S.G., Siqueira, J.G. W., Rodrigues, C., et al., 2022. Enzymatic bioremediation: current status, challenges, future prospects, and applications. Development in Wastewater Treatment Research and Processes. Elsevier, pp. 355–381.

Diep, P., Mahadevan, R., Yakunin, A.F., 2018. Heavy metal removal by bioaccumulation using genetically engineered microorganisms. Front. Bioeng. Biotechnol. 6, 157.

Dragicevic, T.L., Hren, M.Z., Gmajnić, M., Pelko, S., Kungulovski, D., Kungulovski, I., et al., 2010. Biodegradation of olive mill wastewater by *Trichosporon cutaneum* and *Geotrichum candidum*. Arch. Ind. Hyg. Toxicol. 61 (4), 399–405.

Fathali, Z., Rezaei, S., Faramarzi, M.A., Habibi-Rezaei, M., 2019. Catalytic phenol removal using entrapped cross-linked laccase aggregates. Int. J. Biol. macromolecules 122, 359–366.

Gavrilaş, S., Ursachi, C.Ş., Perţa-Crişan, S., Munteanu, F.D., 2022. Recent trends in biosensors for environmental quality monitoring. Sensors 22 (4), 1513.

Gong, T., Liu, R., Zuo, Z., Che, Y., Yu, H., Song, C., et al., 2016. Metabolic engineering of *Pseudomonas putida* KT2440 for complete mineralization of methyl parathion and γ-hexachlorocyclohexane. ACS Synth. Biol. 5 (5), 434–442.

Henn, C., Arakaki, R.M., Monteiro, D.A., Boscolo, M., da Silva, R., Gomes, E., 2020. Degradation of the organochlorinated herbicide diuron by rainforest basidiomycetes. BioMed. Res. Int. 2020.

Hu, K., Peris, A., Torán, J., Eljarrat, E., Sarrà, M., Blánquez, P., et al., 2020. Exploring the degradation capability of *Trametes versicolor* on selected hydrophobic pesticides through setting sights simultaneously on culture broth and biological matrix. Chemosphere 250, 126293.

Ilangovan, R., Daniel, D., Krastanov, A., Zachariah, C., Elizabeth, R., 2006. Enzyme based biosensor for heavy metal ions determination. Biotechnol. Biotechnol. Equip. 20 (1), 184–189.

Imam, A., Kanaujia, P.K., Ray, A., Suman, S.K., 2021. Removal of petroleum contaminants through bioremediation with integrated concepts of resource recovery: a review. Indian. J. Microbiology 61 (3), 250–261.

Işık, C., Saraç, N., Teke, M., Uğur, A., 2021. A new bioremediation method for removal of wastewater containing oils with high oleic acid composition: *Acinetobacter haemolyticus* lipase immobilized on eggshell membrane with improved stabilities. N. J. Chem. 45 (4), 1984–1992.

Jaiswal, S., Singh, D.K., Shukla, P., 2019. Gene editing and systems biology tools for pesticide bioremediation: a review. Front. Microbiol. 10, 87.

Jasińska, A., Góralczyk-Bińkowska, A., Soboń, A., Długoński, J., 2019. Lignocellulose resources for the *Myrothecium roridum* laccase production and their integrated application for dyes removal. Int. J. Environ. Sci. Technol. 16 (8), 4811–4822.

Kadam, A.A., Jang, J., Jee, S.C., Sung, J.S., Lee, D.S., 2018. Chitosan-functionalized super magnetic halloysite nanotubes for covalent laccase immobilization. Carbohydr. Polym. 194, 208–216.

Karim, F., Fakhruddin, A.N.M., 2012. Recent advances in the development of a biosensor for phenol: a review. Rev. Environ. Sci. Bio/Technology 11 (3), 261–274.

Karthigadevi, G., Manikandan, S., Karmegam, N., Subbaiya, R., Chozhavendhan, S., Ravindran, B., et al., 2021. Chemico-nano treatment methods for the removal of persistent organic pollutants and xenobiotics in water–a review. Bioresour. Technol. 324, 124678.

Kotoky, R., Pandey, P., 2021. The genomic attributes of Cd-resistant, hydrocarbonoclastic *Bacillus subtilis* SR1 for rhizodegradation of benzo (a) pyrene under co-contaminated conditions. Genomics 113 (1), 613–623.

Kumar, L., Bharadvaja, N., 2019. Enzymatic bioremediation: a smart tool to fight environmental pollutants. Smart Bioremediation Technologies. Academic Press, pp. 99–118.

Kumar, A., Saxena, G., Kumar, V., Chandra, R., 2021. Environmental contamination, toxicity profile and bioremediation approach for treatment and detoxification of pulp paper industry effluent. Bioremediation for Environmental Sustainability. Elsevier, pp. 375–402.

Li, Y., Cao, H., Wang, X., Guo, L., Ding, X., Zhao, W., et al., 2021a. Diet-mediated metaorganismal relay biotransformation: health effects and pathways. Crit. Rev. Food Sci. Nutr. 1–19.

Li, T., Gao, Y.Z., Xu, J., Zhang, S.T., Guo, Y., Spain, J.C., et al., 2021b. A recently assembled degradation pathway for 2, 3-dichloronitrobenzene in *Diaphorobacter* spp. strain JS3051. Mbio 12 (4), e02231. -21.

Li, G., Nandgaonkar, A.G., Wang, Q., Zhang, J., Krause, W.E., Wei, Q., et al., 2017. Laccase-immobilized bacterial cellulose/TiO2 functionalized composite membranes: evaluation for photo-and bio-catalytic dye degradation. J. Membr. Sci. 525, 89–98.

Li, X., Wei, Z., Zhang, M., Peng, X., Yu, G., Teng, M., et al., 2007. Crystal structures of *E. coli* laccase CueO at different copper concentrations. Biochemical Biophysical Res. Commun. 354 (1), 21–26.

Logeshwaran, P., Krishnan, K., Naidu, R., Megharaj, M., 2020. Purification and characterization of a novel fenamiphos hydrolyzing enzyme from *Microbacterium esteraromaticum* MM1. Chemosphere 252, 126549.

Mandal, K., Singh, B., Jariyal, M., Gupta, V.K., 2014. Bioremediation of fipronil by a *Bacillus firmus* isolate from soil. Chemosphere 101, 55–60.

Margesin, R., Schinner, F., 2001. Biodegradation and bioremediation of hydrocarbons in extreme environments. Appl. Microbiol. Biotechnol. 56 (5), 650–663.

Mihovilovic, M.D., Bianchi, D.A., Rudroff, F., 2006. Accessing tetrahydrofuran-based natural products by microbial Baeyer–Villiger bio-oxidation. Chem. Commun. 30, 3214–3216.

Mostafa, G.A., 2010. Electrochemical biosensors for the detection of pesticides. Open. Electrochem. J. 2 (1).

Mukherjee, G., Saha, C., Naskar, N., Mukherjee, A., Mukherjee, A., Lahiri, S., et al., 2018. An endophytic bacterial consortium modulates multiple strategies to improve arsenic phytoremediation efficacy in *Solanum nigrum*. Sci. Rep. 8 (1), 1–16.

Nathan, V.K., Rani, M.E., Gunaseeli, R., Kannan, N.D., 2018. Enhanced biobleaching efficacy and heavy metal remediation through enzyme-mediated lab-scale paper pulp deinking process. J. Clean. Prod. 203, 926–932.

Nigam, V.K., Shukla, P., 2015. Enzyme based biosensors for detection of environmental pollutants-a review. J. microbiology Biotechnol. 25 (11), 1773–1781.

Ojuederie, O.B., Babalola, O.O., 2017. Microbial and plant-assisted bioremediation of heavy metal polluted environments: a review. Int. J. Environ. Res. Public. Health 14 (12), 1504.

Pal, P., Bhattacharya, D., Mukhopadhyay, A., Sarkar, P., 2009. The detection of mercury, cadium, and arsenic by the deactivation of urease on rhodinized carbon. Environ. Eng. Sci. 26 (1), 25–32.

Pal, S., Kundu, A., Banerjee, T.D., Mohapatra, B., Roy, A., Manna, R., et al., 2017. Genome analysis of crude oil degrading *Franconibacter pulveris* strain DJ34 revealed its genetic basis for hydrocarbon degradation and survival in an oil-contaminated environment. Genomics 109 (5–6), 374–382.

Perpetuo, E.A., Souza, C.B., Nascimento, C.A.O., 2011. Engineering Bacteria for Bioremediation. IntechOpen.

Pileggi, M., Pileggi, S.A., Sadowsky, M.J., 2020. Herbicide bioremediation: from strains to bacterial communities. Heliyon 6 (12), e05767.

Raghu, G., Balaji, V., Venkateswaran, G., Rodrigue, A., Maruthi Mohan, P., 2008. Bioremediation of trace cobalt from simulated spent decontamination solutions of nuclear power reactors using *E. coli* expressing NiCoT genes. Appl. Microbiol. Biotechnol. 81 (3), 571–578.

Raval, V.H., Bhatt, H.B., Singh, S.P., 2018. Adaptation strategies in halophilic bacteria. Extremophiles. CRC Press, pp. 137–164.

Raval, V.H., Pillai, S., Rawal, C.M., Singh, S.P., 2014. Biochemical and structural characterization of a detergent-stable serine alkaline protease from seawater haloalkaliphilic bacteria. Process. Biochem. 49 (6), 955–962.

Reynoso, E.C., Romero-Guido, C., Rebollar-Pérez, G., Torres, E., 2022. Enzymatic biosensors for the detection of water pollutants. Nanomaterials for Biocatalysis. Elsevier, pp. 463–511.

Rodriguez-Mozaz, S., Lopez de Alda, M.J., Barceló, D., 2006. Biosensors as useful tools for environmental analysis and monitoring. Anal. Bioanal. Chem. 386 (4), 1025–1041.

Sadowsky, M.J., Koskinen, W.C., Bischoff, M., Barber, B.L., Becker, J.M., Turco, R.F., 2009. Rapid and complete degradation of the herbicide picloram by Lipomyces kononenkoae. J Agric Food Chem 1520-5118, 57 (11), 4878–4882. Available from: https://doi.org/10.1021/jf900067f. 19789626.

Saket, P., Mittal, Y., Bala, K., Joshi, A., Yadav, A.K., 2022. Innovative constructed wetland coupled with microbial fuel cell for enhancing diazo dye degradation with simultaneous electricity generation. Bioresour. Technol. 345, 126490.

Saravanan, A., Kumar, P.S., Vo, D.V.N., Jeevanantham, S., Karishma, S., Yaashikaa, P.R., 2021. A review on catalytic-enzyme degradation of toxic environmental pollutants: microbial enzymes. J. Hazard. Mater. 419, 126451.

Sarkar, S., Banerjee, A., Halder, U., Biswas, R., Bandopadhyay, R., 2017. Degradation of synthetic azo dyes of the textile industry: a sustainable approach using microbial enzymes. Water Conserv. Sci. Eng. 2 (4), 121–131.

Sarker, A., Nandi, R., Kim, J.E., Islam, T., 2021. Remediation of chemical pesticides from contaminated sites through potential microorganisms and their functional enzymes: Prospects and challenges. Environ. Technol. Innov. 23, 101777.

Sassolas, A., Blum, L.J., Leca-Bouvier, B.D., 2012. Immobilization strategies to develop enzymatic biosensors. Biotechnol. Adv. 30 (3), 489–511.

Shakerian, F., Zhao, J., Li, S.P., 2020. Recent development in the application of immobilized oxidative enzymes for bioremediation of hazardous micropollutants–a review. Chemosphere 239, 124716.

Sharma, B., Shukla, P., 2022. Futuristic avenues of metabolic engineering techniques in bioremediation. Biotechnol. Appl. Biochem. 69 (1), 51–60.

Silva, L.M.C., Salgado, A.M., Coelho, M.A.Z., 2010. *Agaricus bisporus* as a source of tyrosinase for phenol detection for future biosensor development. Environ. Technol. 31 (6), 611–616.

Sirajuddin, S., Khan, M.A., Qader, S.A.U., Iqbal, S., Sattar, H., Ansari, A., 2020. A comparative study on degradation of complex malathion organophosphate using *Escherichia coli* IES-02 and a novel carboxylesterase. Int. J. Biol. Macromolecules 145, 445–455.

Souza, E.C., Vessoni-Penna, T.C., de Souza Oliveira, R.P., 2014. Biosurfactant-enhanced hydrocarbon bioremediation: an overview. Int. Biodeterior. Biodegrad. 89, 88–94.

Srichandan, H., Mohapatra, R.K., Parhi, P.K., Mishra, S., 2019. Bioleaching: a bioremediation process to treat hazardous wastes. Soil Microenvironment for Bioremediation and Polymer Production. Wiley-Scrivener, pp. 115–129.

Tak, Y., Kaur, M., Tilgam, J., Kaur, H., Kumar, R., Gautam, C., 2022. Microbes assisted bioremediation: a green technology to remediate pollutants. Bioremediation of Environmental Pollutants. Springer, Cham, pp. 25–52.

Verma, S., Bhatt, P., Verma, A., Mudila, H., Prasher, P., Rene, E.R., 2021. Microbial technologies for heavy metal remediation: effect of process conditions and current practices. Clean. Technol. Environ. Policy 1–23.

Verma, M.L., Rani, V., 2021. Biosensors for toxic metals, polychlorinated biphenyls, biological oxygen demand, endocrine disruptors, hormones, dioxin, phenolic and organophosphorus compounds: a review. Environ. Chem. Lett. 19 (2), 1657–1666.

Xue, Y., Wang, Y., 2020. Green electrochemical redox mediation for valuable metal extraction and recycling from industrial waste. Green. Chem. 22 (19), 6288–6309.

Yaashikaa, P.R., Devi, M.K., Kumar, P.S., 2022. Advances in the application of immobilized enzyme for the remediation of hazardous pollutant: A review. Chemosphere 134390.

Yagnik, S., Raval, V.H., Kothari, R.K., Kothari, C.R., 2021a. Efficient dye decolourization by *Virgibacillus marismortui* B6 and its impact on seed germination. GAU Res. J. 46 (3), 135–144.

Yagnik, S., Raval, V.H., Kothari, R.K., Kothari, C.R., 2021b. Recycling and reuse of azo dye polluted water for agriculture through bioremediation-a microbial perspective. A review. GAU Res. J. 46 (3), 117–122.

Chapter 25

The role of microbes and enzymes for bioelectricity generation: a belief toward global sustainability

Lakshana Nair (G)[1,*], Komal Agrawal[1,2,*] and Pradeep Verma[1]
[1]*Bioprocess and Bioenergy Laboratory (BPEL), Department of Microbiology, Central University of Rajasthan, Bandarsindri, Kishangarh, Ajmer, Rajasthan, India,* [2]*Department of Microbiology, School of Bioengineering and Biosciences, Lovely Professional University, Phagwara, Punjab, India*

25.1 Introduction

With the rise of globalization and industrialization, economic stability and the quality of human life have increased. The world has developed into a space where daily life is easier and hassle-free. With the technological advancements and the rise in consumer product demand, the industries and production units have expanded to enhance production rates. This leads to an increase in the economy and job opportunities. Also, the discoveries and production strategies produced high-quality materials at a low production rate.

The side effects of the new production strategies may simultaneously cause many environmental problems. The increasing usage of chemicals and similar additives in many industries leads to pollution in the environment at a hazardous level. The untreated effluents from the industries contain many toxic substances, which, when released into the environment, cause harm to the ecological niches (Shindhal et al., 2021). The increased exposure to such chemicals may also lead to many environmental and health hazards. Conventional techniques such as physical and chemical methods to manage wastes are available. Their high cost and lower efficiency make them economically unviable (Abdel-Shafy and Mansour, 2018) and constitute a significant problem globally. We face another significant problem nowadays: the depletion of sustainable energy sources such as fossil fuels used to produce electricity and other

*Both authors have contributed equally.

Biotechnology of Microbial Enzymes. DOI: https://doi.org/10.1016/B978-0-443-19059-9.00001-3
© 2023 Elsevier Inc. All rights reserved.

energy refineries. Due to the depleted levels of such sources, the future availability of electricity and related energies in coming future has become a global concern (Armaroli and Balzani, 2007).

Thus bio-based methodologies are currently being explored to which microbes can be of outstanding contributions. The action of microbes in bioremediation, biorefinery, production of bioelectricity, etc. has been well exploited in the current scenario (Venkata Mohan et al., 2016). Enzymes such as laccases, xylanases, etc., produced by microbes, especially fungi, can be used to treat effluents or pollutants from the industries. Thus they can also be regarded as green tools or green catalysts. Besides their role in bioremediation, they can also be used in biofuel, biosensor, fiber board synthesis, clinical, textile industry, food, cosmetics, and many more (Agrawal et al., 2018). The biomass substrates such as cellulose, lignin, hemicellulose, etc., can be used to produce various value-added compounds. Using biorefinery methods, biomass can be converted to biofuels (Carvalheiro et al., 2008). Additionally, wastewater from industries containing hexavalent chromium, agro-waste, azo dyes, etc. can be used as a substrate in a microbial fuel cell (MFC) to produce electricity (Chaturvedi and Verma, 2016). These methods utilizing microbes are of low cost and high efficiency.

Hence, the present chapter discusses microorganisms and microbial enzymes used in various industrial sectors. The principle behind the utilization of microbes in MFCs is also discussed. The utilization of wastewater for the working of MFC with its bonus of mitigation is also explained. Lastly, the limitations and prospects are highlighted to understand its potential future role and possible opportunities.

25.2 Bioresources: biorefinery

Global energy is mainly met by the intense search and exploitation of fossil resources. The negative impacts of using fossil fuels on the environment/climate forced scientists to develop greener and cleaner resources (Kumari and Singh, 2018). Among the emerging alternatives was biomass use, which gave rise to a fresh concept called biorefinery or biomass refinery (Rivas et al., 2019). Biorefinery is explained as a sustainable processing of biomass into an array of products which includes food, feed, fuels, chemicals, etc. This shows similarity to a petroleum-based refinery. It relies on the total valorization of biomass to ensure a steady supply of energy and critical value-added chemicals besides creating enough job opportunities (de Jong et al., 2012; Kumar and Verma, 2020). The biomass includes cellulose, hemicellulose, lipids, proteins, lignin, pectin, starch, extractives, ash, etc.

Biomasses, known for their complex structure, require complex hydrolytic enzymes to break them down (Dutta and Wu, 2014; Kumar and Verma, 2020; Sari et al., 2015; Shi et al., 2010; Singh et al., 2019). The hydrolytic enzymes used in such a way include cellulases, hemicellulases, ligninolytic enzymes, lipases, lytic polysaccharide monooxygenases (LPMOs), pectinases (Escamilla-Alvarado et al., 2016), xylanases, etc. The enzymes and the microorganisms from which they are processed are shown in Table 25.1. Many living organisms

TABLE 25.1 Different microorganisms producing hydrolytic enzymes.

Family of hydrolytic enzymes	Major organisms	Microorganisms	References
Ligninolytic enzymes	Bacteria	*Bacillus aryabhattai*	Paz et al. (2020)
		Aeromonas hydrophila	Bharagava et al. (2018)
		Klebsiella pneumoniae	Gaur et al. (2018)
		Raoultella ornithinolytica	Falade et al. (2017)
		Ensifer adhaerens	Falade et al. (2017)
		Bacillus albus MW407057	Kishor et al. (2021)
		Stenotrophomonas sp.	Olajuyigbe et al. (2018)
		Streptomyces sp. S6	Riyadi et al. (2020)
		Thermobifida fusca	Chen et al. (2016)
		Ureibacillus thermosphaericus	Nakamura and Kurosawa (2021)
	Fungus	*Bjerkandera adusta*	Tripathi et al. (2012)
		Lentinus squarrosulus	Tripathi et al. (2012)
		Panus tigrinus	Ruqayyah et al. (2013)
		Pleurotus ostreatus	Ozcirak Ergun and Ozturk Urek (2017)
		Leptosphaerulina sp.	Copete-Pertuz et al. (2019)
		Pleurotus florida	Illuri et al. (2021)
		Trametes polyzona	Lueangjaroenkit et al. (2018)
		Coriolopsis byrsina	Agrawal et al. (2021)
		Lentinula edodes	Cai et al. (2017)
		Phanerochaete chrysosporium	Ansari et al. (2016)

(*Continued*)

TABLE 25.1 (Continued)

Family of hydrolytic enzymes	Major organisms	Microorganisms	References
Laccases	Fungus	*Aspergillus flavus* PUF5	Ghosh and Ghosh (2017)
		Lepista nuda	Zhu et al. (2016)
		Myrothecium verrucaria	Agrawal et al. (2019)
		Panus strigellus	Cardoso et al. (2018)
		Trametes versicolor	Atilano-Camino et al. (2020)
		Lentinus tigrinus	Zavarzina et al. (2018)
		L. squarrosulus Mr13	Mukhopadhyay and Banerjee (2015)
		Phlebia brevispora BAFC 633	Fonseca et al. (2015)
		Stropharia aeruginosa	Daroch et al. (2014)
		Sclerotinia sclerotiorum	Moţ et al. (2012)
Cellulases	Bacteria	*Caldicellulosiruptor changbaiensis* sp. nov.	Bing et al. (2015)
		Anoxybacillus sp. 527	Liang et al. (2010)
		Geobacillus thermodenitrificans	Priya et al. (2016)
		Bacillus anthracis	Duza and Mastan (2015)
		Ochrobactrum anthropi	Duza and Mastan (2015)
		Bacillus cereus	Tabssum et al. (2018)
		Bacillus licheniformis NCIM 5556	Shajahan et al. (2016)
		Cellulosimicrobium cellulans	Song and Wei (2010)
		Micrococcus sp.	Mmango-Kaseke et al. (2016)
		Vibrio xiamenensis	Gao et al. (2012)

	Fungus	*Aspergillus nidulans*	De Assis et al. (2015)
		*Aspergillus oryzae*ITCC-4857.01	Begum et al. (2009)
		Acremonium sp.	De Almeida et al. (2011)
		Aspergillus fumigates	Ang et al. (2013)
		Aspergillus niger	Dos Santos et al. (2016)
		Rhizopus sp.	Dos Santos et al. (2016)
		Chaetomium globosporum	Yadav et al. (2019)
		Neocallimastix patriciarum	Wang et al. (2011)
		Penicillium chrysogenum	Chinedu et al. (2011)
		Trichoderma asperellum	Ezeilo et al. (2019)
		Trichoderma harzianum	Chinedu et al. (2011)
Xylanases	Bacteria	*Arthrobacter* sp.	Khandeparker and Jalal (2015)
		Lactobacillus sp.	Khandeparker and Jalal (2015)
		Bacillus sp.	Irfan et al. (2016)
		Streptomyces sp.	Rosmine et al. (2017)
		Anoxybacillus kamchatkensis NASTPD13	Yadav et al. (2018)
		C. cellulans CKMX1	Walia et al. (2015)
		Geobacillus sp.	Bhalla et al. (2015)
		Micrococcus sp.	Mmango-Kaseke et al. (2016)
		Arthrobacter oxidans KQ11	Ren et al. (2019)
		Bacillus amyloliquefaciens	Kumar et al. (2017)

(*Continued*)

TABLE 25.1 (Continued)

Family of hydrolytic enzymes	Major organisms	Microorganisms	References
	Fungus	*Pichia stipitis*	Ding et al. (2018)
		Aureobasidium pullulans	Yegin (2017)
		Aspergillus foetidus	Cunha et al. (2018)
		Thermomyces lanuginosus VAPS24	Kumar and Shukla (2018)
		Myceliophthora thermophila	Wang et al. (2015)
		Penicillium citrinum	Saha and Ghosh (2014)
		Cladosporium oxysporum	Guan et al. (2016)
		T. asperellum	Sridevi et al. (2017)
		Fusarium sp. BVKT R2	Ramanjaneyulu et al. (2017)
		Thermoascus aurantiacus	Ping et al. (2018)
LPMOs	Bacteria	*T. fusca*	Kruer-Zerhusen et al. (2017)
		B. thuringiensis	Zhang et al. (2015)
		Streptomyces ambofaciens	Valenzuela et al. (2017)
		B. licheniformis	Courtade et al. (2015)
		Teredinibacter turnerae	Fowler et al. (2019)
		Serratia marcescens	Yang et al. (2017)
		B. amyloliquefaciens	Gregory et al. (2016)
		Listeria monocytogenes	Paspaliari et al. (2015)
		Streptomyces coelicolor	Tanghe et al. (2017)
		Photorhabdus luminescens	Munzone et al. (2020)

	Fungus	*M. thermophila* C1	Frommhagen et al. (2016)
		Gloeophyllum trabeum	Kojima et al. (2016)
		Neurospora crassa	Petrović et al. (2019)
		P. chrysosporium	Wu et al. (2013)
		Podospora anserina	Bennati-Granier et al. (2015)
		Botrytis cinerea	Zarattini et al. (2021)
		Thielavia australiensis	Calderaro et al. (2020)
		Heterobasidion irregulare	Liu et al. (2017)
		Fusarium graminearum	Nekiunaite et al. (2016)
		Malbranchea cinnamomea	Hüttner et al. (2019)
Amylases	Bacteria	*Bacillus species*	Ajayi and Fagade (2006)
		Chromohalobacter sp. TVSP 101	Prakash et al. (2009)
		Bacillus caldolyticus	Schwab et al. (2009)
		B. licheniformis	Vaseekaran et al. (2010)
		Caldimonas taiwanensis	Chen et al. (2005)
		Anoxybacillus ayderensis FMB1	Matpan Bekler et al. (2021)
		Anoxybacillus amylolyticus	Paola Di Donato (2014)
		Geobacillus thermoleovorans	Mehta and Satyanarayana (2013)
		Rhodothermus marinus	Hamed et al. (2017)
		Thermomyces dupontii	Wang et al. (2019a)

(*Continued*)

TABLE 25.1 (Continued)

Family of hydrolytic enzymes	Major organisms	Microorganisms	References
	Fungus	*Aspergillus awamori*	Prakasham et al. (2007)
		Aspergillus fumigatus	Ratnasri et al. (2014)
		T. harzianum	Mohamed et al. (2011)
		Penicillium expansum MT-1	Erdal and Taskin (2010)
		Streptomyces cheonanensis	Naragani et al. (2015)
		Streptomyces clavifer	Megead Yassien (2012)
		Streptomyces griseus	Lakshmi et al. (2020)
		Cryptococcus flavus	Galdino et al. (2011)
		A. niger (BTM-26)	Abdullah et al. (2014)
		Laceyella sacchari TSI-2	Shukla and Singh (2015)
Pectinases	Bacteria	*Bacillus subtilis*	Ahlawat et al. (2009)
		Bacillus mojavensis I4	Ghazala et al. (2015)
		Chryseobacterium indologenes Strain SD	Roy et al. (2018)
		B. licheniformis	Rehman et al. (2015)
		Bacillus tequilensis	Chiliveri et al. (2016)
		Penicillium oxalicum	Neagu et al. (2012)

	Fungus	*Thermomucor indicae-seudaticae* N31	Martin et al. (2010)
		A. niger	Akhter et al. (2011)
		Filobasidium capsuligenum	Merín et al. (2014)
		Aspergillus parvisclerotigenus KX928754	Satapathy et al. (2021)
		P. chrysogenum	Banu et al. (2010)
		Penicillium griseoroseum	Teixeira et al. (2011)
Lipases	Bacteria	*Galactomyces geotrichum* mafic-0601	Wang et al. (2019a)
		Pseudomonas fluorescens KE38	Adan Gökbulut and Arslanoğlu (2013)
		Pseudomonas putida	Fatima and Khan (2015)
		B. subtilis PCSIRNL-39	Mazhar et al. (2017)
		Pseudomonas aeruginosa JCM5962(T)	Sachan et al. (2018)
		Chromobacterium viscosum	Taipa et al. (1995)
		Pseudomonas glumae	Taipa et al. (1995)
		Acinetobacter sp. strain SU15	Ugras and Uzmez (2016)
		B. tequilensis (F7)	Verma et al. (2020)
		S. griseus	Vishnupriya et al. (2010)

(*Continued*)

TABLE 25.1 (Continued)

Family of hydrolytic enzymes	Major organisms	Microorganisms	References
	Fungus	*A. awamori* BTMFW032	Basheer et al. (2011)
		Penicillium restrictum	De Azeredo et al. (2007)
		Candida rugosa	Domínguez De María et al. (2006)
		Fusarium solani	Kanmani et al. (2013)
		Penicillium verrucosum	Kempka et al. (2008)
		Rhizopus homothallicus	Mateos Diaz et al. (2006)

ranging from plants, animals, and microorganisms generally produce such enzymes for their nutrition and day-to-day metabolic activities, making them ubiquitous (Lynd et al., 2002; Robinson, 2015). The microorganisms' fast growth, easy regulation, and certain other properties make them a favorable option for hydrolytic enzyme production (Gurung et al., 2013). Such an alternative is not only greener and pollution-free but also is considered to be an efficient method in industries. There are also studies regarding laccases, a multicopper oxidase from microorganisms, which is also known as the green tool due to the various applications (Xenakis et al., 2016). The different enzymes used and their merits and demerits are discussed in the chapter.

25.3 Hydrolytic enzymes and their applications in various sectors

Hydrolytic enzymes used in various sectors are presented in the following sections.

25.3.1 Ligninolytic enzymes

The first step before producing biofuel from biomass requires removing lignin, the process known as biological delignification or pretreatment (Moreno et al., 2012). Selective degradation of lignin is performed using ligninolytic enzymes of microbial origin, and it does not affect hemicellulose or cellulose (Moreno et al., 2015). It is well known that microbial delignification is a time-consuming process and is thus considered one of the major concerns of its usage in the biofuel industries. It usually takes about 13 days to complete but may extend up to 40–50 days, depending on the microorganisms employed (Lu et al., 2010; Niladevi, 2009; Wan and Li, 2011). Establishing the correlation among enzyme production, lignin degradation, glucose yield, and other parameters was difficult in cases where microorganisms were applied directly (Salvachúa et al., 2011; Yamagishi et al., 2011). Studies show that chemical-assisted pretreatment before the delignification process improves the overall duration and efficiency of the process. The use of acid- and alkali-assisted treatment enhances the delignification rates and improves the glucose yield and the subsequent outcome of ethanol production (Khuong et al., 2014; López-Abelairas et al., 2013; Martín-Sampedro et al., 2017).

25.3.2 Laccases

The ability of laccases to oxidize an extended range of aromatic and nonaromatic compounds makes it a preferable option for the pretreatment of lignocellulosic biomass (LCB) for bioethanol or bioenergy production (Agrawal et al., 2018; Tabka et al., 2006). Since this approach replaces the existing and conventional method of chemical treatments, it is considered to be a greener or environmentally friendly approach (Agrawal et al., 2018). Studies on laccase put forward its profound applications in biofuel cells due to its ability to oxidize phenols and reduce oxygen to water. During this reduction, a four-electron transfer occurs at high redox potential, making it a useful application in fuel cells' cathode compartment (Zheng et al., 2008). The laccases are currently applied in the bioremediation of chlorophenols, polycyclic aromatic hydrocarbons, lignin-related structures, organophosphorus compounds, phenols, azo dyes, etc. (Saratale et al., 2011; Viswanath et al., 2014). Other than these functions, they are known to metabolize dichlorodiphenyltrichloroethane content (present in the soil) and also biodegrade 2,4-dichlorophenol (Bhattacharya et al., 2009; Zhao et al., 2010). The other applications of laccases include the decolorization, and detoxification of effluent discharge from various industries like food, textile, pulp/paper, plastic, etc. (Chandra and Chowdhary, 2015; Viswanath et al., 2014). The clinical approach regarding laccases involves the development of biosensors and the removal of clinically generated pollutants from the environment. The clinically generated pollutants affecting the nonsteroidal antiinflammatory drugs usually found in the aquatic habitats are known to damage the liver and kidneys of animals consuming water contaminated with drugs (Lloret et al., 2013). Studies report the anti-proliferative activity of laccase strains from *Agrocybe cylindracea* and *Inonotus baumii* against cancer cells, inhibitory activity against HIV-1 reverse transcriptase by laccase strain from *A. cylindracea* (Hu et al., 2011a; Sun et al., 2014).

25.3.3 Cellulases

The applications of cellulase enzymes are mainly recorded in fields including the textile, food, beverages, animal feed, biofuels, etc. (Drahansky et al., 2016). Cellulose, one of the world's most abundant polymers, forms monosaccharides on enzymatic hydrolysis, a process or technology which is recognized as of high importance to fulfilling visions like "circular bioeconomy" or "lignocellulosic feedstock based biorefinery" (Kumar and Verma, 2020). The monosaccharides produced as a result of the process can be used either for the synthesis of SCP (single-cell protein) or can be made to undergo fermentation to produce ethanol, which is also a representation of its role as an alternate and sustainable source of energy (Kuhad et al., 2010; Sukumaran et al., 2005). During the process of biorefining, the enzymatic hydrolysis using cellulases is done right next to the delignification step, collaborating with other hydrolytic proteins (Rajak and Banerjee, 2016).

Cellulases are widely used in food, animal feed, textile, pulp and paper, research and development, beer and wine, agricultural, biofuel, waste management, pharmaceutical industries, etc. In food industries, cellulases are generally applied for the hydrolysis of the cell wall components to reduce the viscosity and maintain the texture of fruit juice extracts. It also has applications in the food coloring industry as a food coloring agent, alters the sensory properties of fruits, vegetables, etc., oils from olives, soups, etc., and reduces food spoilage. Cellulases in the food industry have significant contributions to improving extraction, clarification, stabilization, etc., of fruit, vegetable juices, beverage processing, etc. (Basak et al., 2021; Bhat, 2000). Cellulases are used to pre-treat agricultural silage and grain feed for partial hydrolysis of the lignocellulosic materials in the animal feed industry. This helps in improving the nutritional quality of the feed, increases the weight gain of poultry, decreases pathogenic bacterial colonization in the large intestine of the animal-consuming feed (prepared by cellulose treatment), etc. In paper and pulp industries, it plays a major role in producing good quality paper wherein it helps to debark the biomass and remove the lignocellulosic part of the matter. It also helps in biobleaching and deinking, thereby helping to recycle paper. Cellulases are employed to produce better-quality alcohol in the brewery industries (Basak et al., 2021). In textile industries, it has various uses, including biopolishing and improved absorbance properties of textile fibers, biostoning of jeans, softening garments, removing excess dye from fabrics, etc. It also restores the color brightness and improves fabric quality. In agricultural industries, cellulase has a major role in the plant pathogen and disease control and improves soil quality. Also, it improves root system and seed germination rates and reduces the dependence of plants on mineral fertilizers (Kuhad et al., 2011). In pharmaceutical industries, a digestive enzyme product containing cellulose, digestin, can be used by humans to increase the digestibility of cellulose-containing food. Celluloses, hemicelluloses, etc., from fungal systems, can also be used for the increased hydrolysis of cellulose, hemicellulose, beta-glucan polymers, etc., present in the food (Gupta et al., 2013).

25.3.4 Xylanases

Xylan is known as the second most abundant polymer globally. The hydrolysis of cellulases generates pentoses plentily, which can be converted into bioethanol or other valuable compounds like xylooligosaccharides (XOS) (Gírio et al., 2010). An important example includes xylanases. Xylanases aid in selective xylan removal from biomass, leading to biomass fiber swelling. This improves the porosity of hemicellulases present in biomass for the better functioning of other hydrolytic enzymes, including enzymes like cellulases (Hu et al., 2011b).

Xylanases are mainly employed to convert LCB into fuel-grade ethanol. It has already become a world priority to produce environment-friendly renewable energy. Its availability at a reasonable price for the transportation sector is also a bonus. Microbial xylanases are known to play important roles in the saccharification of LCB (Basit et al., 2018; Choudhary, 2014; Hu et al., 2011b; Ramanjaneyulu et al., 2017). Xylanases are also used for biopulping and biobleaching in pulp and paper industries (Walia et al., 2017). The application of xylanases in textile industries includes desizing, scouring, bleaching, etc. Its other applications include animal feed industries, bread making, food industries, solid waste management, biofuel industries, etc. (Dhiman et al., 2008; Mandal, 2015).

The XOS can be applied in various sectors such as biotechnology, pharmaceutical, food, feed industries, etc. (Chang et al., 2017). They are produced by either the enzymatic or chemical processing of xylan. Since these are neither hydrolyzed nor absorbed in the gastrointestinal tract, it plays a vital role as prebiotics. They can selectively stimulate the growth of important gastrointestinal microorganisms that regulate the human digestive health and livestock. (Collins and Gibson, 1999; Roberfroid, 1997; Samanta et al., 2015; Vázquez et al., 2000). XOS, when used as a feed alternative, helps in cholesterol reduction, inhibits starch retro-gradation, improves the bioavailability of calcium, etc. This improves the nutritional and sensory properties of the feed. It also guards the intestine of livestock against the onslaught of pathogenic microflora and hence is also considered a preventive medicine (Motta et al., 2013; Samanta et al., 2015; Voragen, 1998). Reports also show the role of XOS in phytopharmaceutical and feed applications due to its growth regulatory activity in aquaculture and poultry (Bhardwaj et al., 2019). Owing to its immunomodulatory, anticancerous, antimicrobial, antioxidant, antiallergic, antiinflammatory, and antihyperlipidemic functions, XOS has wide applications in the pharmaceutical sectors (Aachary and Prapulla, 2011; Chen et al., 2012;Gupta et al., 2018; Kallel et al., 2015; Li et al., 2010).

25.3.5 Amylases

Amylases find useful applications in various industries, such as food, pharmaceutical, detergent, paper, and textile (Kathiresan and Manivannan, 2006). They make up 25% of the world's enzymes (Mojsov, 2012). They are the second type of enzymes most utilized in detergent formulations, and almost 90% of all liquid detergents constitute amylases (Gurung et al., 2013). They also play a major role in biofuel production from starchy biomass. Here, the biomass is first grounded to form a pulp, further treated with amylases for liquefaction. It also generates a soluble form of the maltodextrin oligosaccharides (Lee et al., 2013).

It is applied to produce corn, maltose, and glucose syrups in the food industry and juices and alcohol fermentation (Gopinath et al., 2017). Another major application of amylases in the food industries is to increase the shelf life due to their antistaling effect (Gupta et al., 2003; Van Der Maarel et al., 2002). In the textile industry, amylases are employed for the desizing process. In contrast, in the paper industry, it is used for the modification of starch of coated paper, which is known to make the paper smooth and strong and also enhance its writing quality (de Souza and Magalhães, 2010; Gupta et al., 2003; Van Der Maarel et al., 2002).

25.3.6 Pectinases

Pectinases used to break down protein-rich biowastes such as wastes from the vegetable and fruit industry are also known for extracting vegetable oils which could be transesterified to form biofuels later on (Biz et al., 2014; Iconomou et al., 2010). The commercial extraction of vegetable oils utilizes hexane-like synthetic carcinogenic chemicals, which are very dangerous. So as a safe and greener alternative, a combination of pectinolytic enzymes with cellulases and hemicellulases can be employed to extract oils from the parts of plants (Kashyap et al., 2001). Pectinases also have applications in the fruit and textile industries. Acidic pectinases are known to minimize the cloudiness and bitterness of fruit juice, and alkaline pectinases are known for the retting and degumming of fiber crops, good quality paper production, coffee and tea fermentation, oil extractions, and treatment of pectic wastewater, etc. in the textile industry (Kashyap et al., 2001).

25.3.7 Lytic polysaccharide monooxygenases

The LPMOs are copper enzymes and can catalyze the glycosidic bond cleavage and degrade polysaccharides (Johansen, 2016; Sato et al., 2020). Belonging to auxiliary families, they assist other hydrolytic enzymes such as cellulase, hemicellulose, chitinase, amylases, etc., for their smooth performance in their fields of application and mediate the oxidative cleavage of polysaccharides. Blending LPMOs with several commercially available hydrolytic cocktails improves their hydrolytic potential (Hu et al., 2015; Müller et al., 2015, 2018; Salwan and Sharma, 2020). The action of LPMOs is on the crystalline regions, at the sites where the endoglucanases cannot bind. It eventually leads to the cleavage of the internal inaccessible bonds (Harris et al., 2014; Johansen, 2016). They are also known to improve the cellulase, which otherwise is a recalcitrant and crystalline cellulosic substrate (Eibinger et al., 2014, 2017). LPMO-containing cellulosic cocktails are known to increase the rate of saccharification. The LPMOs are also considered to have an immense potential for their use in the complete or absolute valorization of a broader biomass range (Müller et al., 2018).

25.3.8 Lipases

Lipases constitute the third largest fraction of the group of enzymes consumed in the market and are of great importance in the oleochemical refineries (Escamilla-Alvarado et al., 2016; Uçkun Kiran et al., 2014). Conversion of glycerol (a by-product from both biodiesel industry and saponification in soap industries) to high-value phenolic structured compounds leads to broad applications in pharmaceutical, cosmetic, food industries, etc. The properties of lipases, including requirements for milder reaction, high selectivity, lower energy consumption, regiospecificity, etc., make the enzymatic conversion mediated by lipase more preferable over the chemical mediated catalysis of glycerol into phenolic antioxidants (Wang et al., 2019b). Studies conducted by Karmee et al. (2018) put forward the use of lipases from the different microbial origins to synthesize biodiesel from spent coffee oil.

25.4 Bioelectricity and microbial electrochemical system

Toxic waste disposal and generation of electrical power are two significant problems faced in today's world. These toxic wastes cause many negative environmental impacts (Balat, 2008; Mmereki et al., 2014). Many strategical approaches were developed to treat these toxic wastes, but they, in turn, also used chemicals for the treatment of waste (El-Bestawy et al., 2005). Major pollutants are toxic dyes, heavy metals, etc. The dye wastes are generated from industries, including textile, paper, plastics, leather tanning, food, agricultural research, hair coloring, light-harvesting arrays, photoelectrochemical cells, etc. (Forgacs et al., 2004; Ji et al., 2013; Rangabhashiyam et al., 2013; Ziessel et al., 2013). Some of the known synthetic dyes used in the industries are azo dyes, anthraquinone dyes, sulfur dyes, indigoid dyes, triphenylmethyl (trityl) dyes, and phthalocyanine derivatives (Ertugay and Acar, 2017; Kahloul et al., 2019; Khataee et al., 2017; Mateos Diaz et al., 2006; Nguyen et al., 2016).

It is common knowledge that the industrial revolution resulted from man's discovery of the extraction and usage of metals from the earth's crust. These metals have a high impact on the industrial sector. Some of the industries which involve the utilization or generation of heavy metals include metal processing (smelters), mining, burning of fossil fuel (coal, diesel, petrol, etc.), manufacturing of high-tension wires, nuclear power plants, the processing of crude petroleum (in refineries), and in industrial units for plastics, textiles, paper, microelectronics, and wood preservative production (Arruti et al., 2010; Sträter et al., 2010; Tchounwou et al., 2012). Similarly, the heavy metal contamination is not the result of anthropogenic activities alone. Many natural phenomena, including corrosion of the Earth's crust, weathering, volcanic eruptions, etc., are known to release heavy metals into the environment (Tchounwou et al., 2012). The heavy metals, whichever source they come from when exposed to living organisms, are found to be affected with an instant (irritation, itching, nausea, vomiting, dysfunction of exposed organs) to long-term impact (carcinogenic effect and even genetic manipulations) (Jaishankar et al., 2014). Heavy metals and dye-based pollution are both relevant environmental issues. The recalcitrant nature of both leads them to have a long half-life, and resistance to several treatment processes, such as physical and chemical processes, also raises concerns (Ali et al., 2019; Bi et al., 2006)

The major conventional methods include physical methods such as adsorption, filtration, and chemical methods such as ion exchange, oxidation, ozonation, chemical precipitation, coagulation or flocculation, photochemical treatment, and electrochemical treatment (Kumar et al., 2021). The conventional method of treatment of waste is costly and causes harm to the environment (Campos et al., 2016). Bioremediation approaches are evolving as one of the most sustainable approaches for treating waste. A microbial electrochemical system (MES) is one such approach. MES is considered a promising technology wherein the wastewater polluted with heavy dye metals is treated along with the simultaneous generation of electricity (Agrawal et al., 2019; Bagchi and Behera, 2020; Chaturvedi and Verma, 2016) (Table 25.2). Since energy requirements are ever-increasing worldwide and the depletion of fossil resources, the MES systems are a great hope for today's world. It reduces the pollution rates caused due to the combustion of fossil fuel and proves to be a reliable, clean, and efficient process that does not produce any toxic by-products (Logan, 2004). The working of such cells and mitigation of wastes are discussed in the following section.

25.4.1 Working of the microbial fuel cell

The MFC/MES system works by the principle of conversion of chemical energy to electrical energy. Here, the chemical energy is produced by converting organic or inorganic waste materials into ATP (adenosine triphosphate) via sequential reactions. During these sequential reactions, the generation of an electric current occurs due to the transfer of electrons to terminal electron acceptors (Chaturvedi and Verma, 2016).

An MFC generally constitutes two main compartments, that is, the anode and the cathode compartments, both of which are separated by a cationic membrane. It is in the anode compartment where the microbes reside and the site where they metabolize the organic compounds such as glucose, which also functions as the electron donor. Both electrons and protons generated during the metabolism of organic compounds aid in generating electricity. Electrons generated in the anode compartment are transferred to the anode surface, from where it moves to the cathode through the electrical circuit. The proton generated migrates via the electrolyte and then through the cationic membrane to the cathode chamber (Chaturvedi and Verma, 2016). The electron and proton consumption are both done in the cathode by reducing a soluble electron acceptor such as oxygen or hexacyanoferrate, acidic permanganate, etc. (Rabaey et al., 2004; You et al., 2006). A load is placed in between the anode and cathode compartments to harness electrical power (Allen and Bennetto, 1993). A demonstration of the working of an MFC is given in Fig. 25.1.

TABLE 25.2 The various microorganisms employed in microbial fuel cell

Microorganisms	References
Proteus vulgaris	Bennetto et al. (1985)
Shewanella putrefaciens	Kim et al. (1999)
Geobacter metallireducens	Bond et al. (2002)
Geothrix fermentans	Nevin and Lovley (2002)
Rhodoferax ferrireducens	Chaudhuri and Lovley (2003)
Desulfobulbus propionicus	Holmes et al. (2004)
Geothrix species	Lovley (2006)
Geobacter psychrophilus	
Desulfuromonas acetoxidans	
Geopsychrobacter electrodiphilus	
Klebsiella pneumoniae	Menicucci et al. (2006)
Escherichia coli	Zhang et al. (2006)
Pichia anomala	Prasad et al. (2007)
Brevibacillus sp. PTH1	Pham et al. (2008)
Pseudomonas sp.	
Synechocystis sp. PCC 6803	Logan (2009)
Pseudomonas aeruginosa	Zuo et al. (2008)
Geobacter sulfurreducens	Ishii et al. (2008)
Shewanella oneidensis	Watson (2009)
Rhodopseudomonas palustris	Xing et al. (2008)
Ochrobactrum anthropi	Zuo et al. (2008)
K. pneumoniae strain L17	Liu et al. (2009a); Liu et al. (2009b)
Clostridium acetobutylicum	Logan (2009)
Desulfovibrio desulfuricans	
E. coli strain K-12	Zheng and Nirmalakhandan (2010)
S. oneidensis strain 14063	Fernando et al. (2012)
Streptomyces enissocaesilis KNU (K strains)	Hassan et al. (2012)
Nocardiopsis sp. KNU (S Strain)	Hassan et al. (2012); Walter et al. (2015)
Synechococcus leopoliensis	
Thiobacillus ferrooxidans	Ulusoy and Dimoglo (2018)

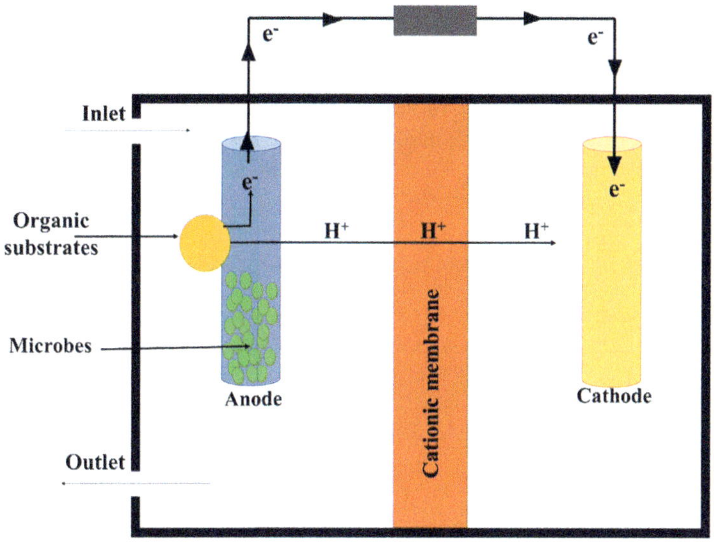

FIGURE 25.1 Demonstration of the working of an MFC. *MFC*, Microbial fuel cell.

The MFC is of two types based on the number of chambers. They are single-chambered and dual-chambered MFCs. As the name suggests, the dual-chambered MFC contains two chambers, that is, anionic and cationic chambers, separated by a cationic membrane. The single-chambered MFC has only a single chamber (Chaturvedi and Verma, 2016). Instead of using platinum, a high-cost metal in the cathode, a cheaper alternative by the use of microorganisms as catalysts to assist electron transfer in the cathode was put forward by studies. This alternative is also known to increase the cathode performance. There are also other options as alternatives to terminal electron acceptors in biocathode. They include compounds such as hexavalent chromium, nitrate, sulfate, iron, selenate, manganese, arsenate, fumarate, urinate, carbon dioxide, etc. (Stams et al., 2006; Wang et al., 2008). This draws light on a potential approach for wastewater treatment using biocathode since the presence of various terminal electron acceptors is found to be present in them like recalcitrant wastes such as azo dyes (Sun et al., 2011a; Sun et al., 2011b).

MFCs can utilize wastewaters from different industrial sources such as potato-producing units, starch processing, swine farms, food processing, meat-packing industry, etc. (Gil et al., 2003; Heilmann and Logan, 2006; Min et al., 2005; Oh and Logan, 2005; Pham et al., 2008; Rabaey et al., 2005). Solid agricultural wastes, including corn stover, carbohydrates, etc. (Scott and Murano, 2007; Zuo et al., 2006), have also been tested as fuel after their pretreatment. Studies show that these organic sources affect the power output of MFCs, and as a consequence, power density usually varies from 1 to 3600 mW/m^2, with most values lying between 10 and 1000 mW/m^2.

It was Potter in 1911, who, for the first time, employed fungi *Saccharomyces cerevisiae* and bacteria such as *Escherichia coli* for power generation in MFC. Studies demonstrate the efficiency of mixed cultures or microbial consortia to be robust and more productive than the pure strains used in the technology. Additionally, it also makes the extraction process of microbes from their natural resources easier (Ha et al., 2008).

25.4.2 Use of wastes for electricity generation

The proper functioning or working of an MFC significantly depends on the type of raw materials employed for the metabolism process by microbes (Logan and Regan, 2006a, 2006b). The use of waste materials in place of the raw materials or substrates used is deemed an eco-friendly method of mitigation of waste. Examples of the uncommon wastes and pollutants utilized in the cells include hexavalent chromium, selenite, agro-wastes, azo dyes, anthraquinone dyes, cadmium, copper, mercury, thallium, arsenic, nitrate, etc. (Kumar et al., 2021). The sources of such pollutants are undoubtedly from various industries. Hexavalent chromium is a major waste product released from the industries involving electroplating, metallurgy, leather tanning, wood preservatives, etc. (Humphries et al., 2004). The selenite used is a waste product from glass manufacturing and electronic industries (Chaturvedi and Verma, 2016). The agro-wastes used in the MFCs are generally from various agricultural operations such as farming, poultry processing industries, slaughterhouses, etc. (Kaewkannetra et al., 2009; Peters et al., 2001), as well as the nitrate from the excessive use of nitrogen-based fertilizers and animal wastes (Chebotareva and Nyokong, 1997). Many reports suggest that the retrieval of costly metals from wastewater or industrial effluents could be done using this technology. Studies showed that the results for the retrieval of gold and silver involving this method showed better results than the conventional ones in terms of both yields and cost efficiency (Kalathil et al., 2013; Tao et al., 2012; Wang and Ren, 2013). The working of MFCs utilizing wastewater is demonstrated in Fig. 25.2.

25.4.3 Hydrolytic enzymes in microbial fuel cell

Laccases with their high redox potential can oxidize phenols through electron transfer, and hence in the MFCs, these can be harnessed in the cathode compartment (Zebda et al., 2012). The fuel cell coated with laccases can degrade a range of recalcitrant molecules and enables an application for the treatment of a wide range of industrial effluents (Mani et al., 2018). The advantages of using enzyme-coated MFCs include (1) the economic feasibility of the enzyme coat, (2) minimal regulation of temperature, pH, etc. (since it is known to run even at room temperature and natural pH), (3) null

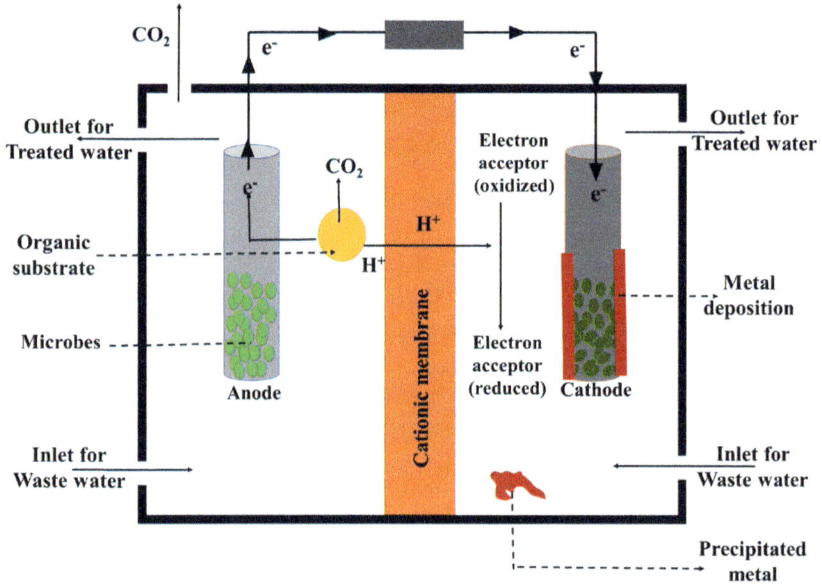

FIGURE 25.2 Schematic representation of the treatment of wastewater using MFC. *MFC*, Microbial fuel cell.

requirement of electro-catalysts, and (4) zero greenhouse gas emission (Agrawal et al., 2018; Nazaruk et al., 2008).

25.5 Limitations and their possible solutions in biorefinery and bioelectricity generation

In association with the production, application, and recovery of hydrolytic enzymes, there are some limitations; including production cost, enzyme stability corresponding to their environment, and recovery of the enzyme or its reuse (Chapman et al., 2018; Singh et al., 2016). The choice of cheap substrates can overcome such limitations, optimization of the production protocols, enzyme recovery by using a suitable support system, improvement of the production potential of the hydrolytic enzymes in plants, genetic engineering, metagenomics approach, etc. The new advances in biorefining based on hydrolytic enzymes include the cell surfacing engineering of yeast, the development of integrated biorefineries, the development of consolidated processing, etc. (Kumar and Verma, 2020).

Some major bottlenecks in the MES system include the low power density obtained from wastewater utilization compared to pure carbon sources, high cost of pure carbon sources, scaling up of MFC, etc. The low power density is also a roadblock to the application of MFC in wastewater treatment. The high cost of carbon sources also makes it economically nonfeasible. Similar is the case of the cost of membranes used in MFCs. The scaling-up process makes the MFCs bulkier and reduces the power density (Chaturvedi and Verma, 2016).

25.6 Prospects

A bio-based refinery is an ideal and eco-friendly approach to industries, where enzymes and related products from living organisms are utilized. The use of microorganisms for this approach is considered feasible and cost-effective in addition to its being environment-friendly. Industrialization using these hydrolytic enzyme usages at a large scale can result in the efficient bioremediation of the environment in a greener and cleaner way. To overcome the present limitations faced in the industries regarding the usage of enzymes, further studies should be done to discover and design better strains and efficient enzymes. Advanced biotechnological approaches, such as metagenomics, genetic cell surface, metabolic engineering, next-generation sequencing, etc., are some of the methods that can be employed.

The chapter suggests that the MES application on a large scale can itself be a self-sustainable wastewater treatment strategy. Its simultaneous removal of toxic wastes along with the production of electricity makes it a good option for bioremediation. This method can be used to generate electricity in homes and help people of developing countries produce electricity, as it doesn't require an enormous infrastructure power plant. The limitations faced by the current technologies can be overcome by research and the discovery and design of better and more efficient MFCs or assisted technologies.

25.7 Concluding remarks

Hydrolytic enzymes play an important role in bio-based refineries. Factors such as the enhanced rate of enzyme production, low production cost, high specificity and activity, and the ability to withstand extreme industrial conditions or environments are usually preferred for large-scale applications. For this, the properties of enzymes should be improved, coupled with the reduced number of steps to reduce the cost of biorefineries. The MES is an eco-friendly approach to the mitigation of toxic wastewater using microbes with the simultaneous production of electricity. This method also faces limitations, and studies are ongoing to overcome these and design better systems and discover efficient microorganisms. Various biotechnological methods can be employed to produce and discover novel strains of efficient microorganisms to be applied in both biorefinery and MES processes. When applied on a large scale, both these methods described in the chapter are believed to reduce pollution and synthesize sustainable energy with reduced cost. Further development of these systems will aid in a safer and cleaner approach to industrial applications as well as energy production.

Abbreviations

ATP	adenosine triphosphate
LCB	lignocellulosic biomass
LPMOs	lytic polysaccharide monooxygenases
MES	microbial electrochemical system
MFCs	microbial fuel cells
SCP	single-cell protein
XOS	xylooligosaccharides

References

Aachary, A.A., Prapulla, S.G., 2011. Xylooligosaccharides (XOS) as an emerging prebiotic: microbial synthesis, utilization, structural characterization, bioactive properties, and applications. Compr. Rev. Food Sci. Food Saf. 10 (1), 2–16. Available from: https://doi.org/10.1111/j.1541-4337.2010.00135.x.

Abdel-Shafy, H.I., Mansour, M.S.M., 2018. Solid waste issue: sources, composition, disposal, recycling, and valorization. Egypt. J. Pet. 27 (4), 1275–1290. Available from: https://doi.org/10.1016/j.ejpe.2018.07.003.

Abdullah, R., Shaheen, N., Iqtedar, M., Naz, S., Iftikhar, T., 2014. Optimization of cultural conditions for the production of alpha amylase by *Aspergillus niger* (BTM-26) in solid state fermentation. Pak. J. Botany 46 (3), 1071–1078.

Adan Gökbulut, A., Arslanoğlu, A., 2013. Purification and biochemical characterization of an extracellular lipase from psychrotolerant *Pseudomonas fluorescens* KE38. Turkish J. Biol. 37 (5), 538–546. Available from: https://doi.org/10.3906/biy-1211-10.

Agrawal, K., Bhardwaj, N., Kumar, B., Chaturvedi, V., Verma, P., 2019. Process optimization, purification and characterization of alkaline stable white laccase from *Myrothecium verrucaria* ITCC-8447 and its application in delignification of agroresidues. Int. J. Biol. Macromolecules 125, 1042–1055. Available from: https://doi.org/10.1016/j.ijbiomac.2018.12.108.

Agrawal, K., Chaturvedi, V., Verma, P., 2018. Fungal laccase discovered but yet undiscovered. Bioresour. Bioprocess. 5 (1). Available from: https://doi.org/10.1186/s40643-018-0190-z.

Agrawal, N., Kumar, V., Shahi, S.K., 2021. Biodegradation and detoxification of phenanthrene in in vitro and in vivo conditions by a newly isolated ligninolytic fungus *Coriolopsis byrsina* strain APC5 and characterization of their metabolites for environmental safety. Environ. Sci. Pollut. Res. 1–16. Available from: https://doi.org/10.1007/s11356-021-15271-w.

Ahlawat, S., Dhiman, S.S., Battan, B., Mandhan, R.P., Sharma, J., 2009. Pectinase production by *Bacillus subtilis* and its potential application in biopreparation of cotton and micropoly fabric. Process. Biochem. 44 (5), 521–526. Available from: https://doi.org/10.1016/j.procbio.2009.01.003.

Ajayi, O.A., Fagade, O.E., 2006. Growth pattern and structural nature of amylases produced by some *Bacillus* species in starchy substrates. Afr. J. Biotechnol. 5 (5), 440–444. Available from: http://www.academicjournals.org/AJB.

Akhter, N., Morshed, M.A., Uddin, A., Begum, F., Sultan, T., Azad, A.K., 2011. Production of pectinase by *Aspergillus niger* cultured in solid state media. Int. J. Biosci. 1 (1), 33–42. Available from: http://www.innspub.net.

Ali, H., Khan, E., Ilahi, I., 2019. Environmental chemistry and ecotoxicology of hazardous heavy metals: environmental persistence, toxicity, and bioaccumulation. J. Chemistry, Cd 1–14. Available from: https://doi.org/10.1155/2019/6730305.

Allen, R.M., Bennetto, H.P., 1993. Microbial fuel-cells - electricity production from carbohydrates. Appl. Biochem. Biotechnol. 39–40 (1), 27–40. Available from: https://doi.org/10.1007/BF02918975.

Ang, S.K., Shaza, E.M., Adibah, Y.A., Suraini, A.A., Madihah, M.S., 2013. Production of cellulases and xylanase by *Aspergillus fumigatus* SK1 using untreated oil palm trunk through solid state fermentation. Process. Biochem. 48 (9), 1293–1302. Available from: https://doi.org/10.1016/j.procbio.2013.06.019.

Ansari, Z., Karimi, A., Ebrahimi, S., Emami, E., 2016. Improvement in ligninolytic activity of *Phanerochaete chrysosporium* cultures by glucose oxidase. Biochemical Eng. J. 105, 332–338. Available from: https://doi.org/10.1016/j.bej.2015.10.007.

Armaroli, N., Balzani, V., 2007. The future of energy supply: challenges and opportunities. Angew. Chem. - Int. (Ed.) 46 (1–2), 52–66. Available from: https://doi.org/10.1002/anie.200602373.

Arruti, A., Fernández-Olmo, I., Irabien, Á., 2010. Evaluation of the contribution of local sources to trace metals levels in urban PM2.5 and PM10 in the Cantabria region (Northern Spain). J. Environ. Monit. 12 (7), 1451–1458. Available from: https://doi.org/10.1039/b926740a.

Atilano-Camino, M.M., Álvarez-Valencia, L.H., García-González, A., García-Reyes, R.B., 2020. Improving laccase production from *Trametes versicolor* using lignocellulosic residues as cosubstrates and evaluation of enzymes for blue wastewater biodegradation. J. Environ. Manag. 275 (July). Available from: https://doi.org/10.1016/j.jenvman.2020.111231.

Bagchi, S., Behera, M., 2020. Assessment of heavy metal removal in different bioelectrochemical systems: a review. J. Hazard. Toxic Radioact. Waste 24 (3), 04020010. Available from: https://doi.org/10.1061/(asce)hz.2153-5515.0000500.

Balat, H., 2008. Contribution of green energy sources to electrical power production of Turkey: a review. Renew. Sustain. Energy Rev. 12 (6), 1652–1666. Available from: https://doi.org/10.1016/j.rser.2007.03.001.

Banu, A.R., Devi, M.K., Gnanaprabhal, G.R., Pradeep, B.V., Palaniswamy, M., 2010. Production and characterization of pectinase enzyme from *Penicillium chrysogenum*. Indian. J. Sci. Technol. 3 (4), 377–381.

Basak, P., Adhikary, T., Das, P., Shee, M., Dutta, T., Biswas, S., et al., 2021. Cellulases in paper and pulp, brewing and food industries: principles associated with its diverse applications. Current Status and Future Scope of Microbial Cellulases. Elsevier, pp. 275–293.

Basheer, S.M., Chellappan, S., Beena, P.S., Sukumaran, R.K., Elyas, K.K., Chandrasekaran, M., 2011. Lipase from marine *Aspergillus awamori* BTMFW032: Production, partial purification and application in oil effluent treatment. N. Biotechnol. 28 (6), 627–638. Available from: https://doi.org/10.1016/j.nbt.2011.04.007.

Basit, A., Liu, J., Miao, T., Zheng, F., Rahim, K., Lou, H., et al., 2018. Characterization of two *endo*-β-1, 4-xylanases from *Myceliophthora thermophila* and their saccharification efficiencies, synergistic with commercial cellulase. Front. Microbiol. 9 (FEB), 1–11. Available from: https://doi.org/10.3389/fmicb.2018.00233.

Begum, F., Absar, N., Alam, M.S., 2009. Purification and characterization of extracellular cellulase from *A. oryzae* ITCC-4857.01. J. Appl. Sci. Res. 5 (10), 1646–1651.

Bennati-Granier, C., Garajova, S., Champion, C., Grisel, S., Haon, M., Zhou, S., et al., 2015. Substrate specificity and regioselectivity of fungal AA9 lytic polysaccharide monooxygenases secreted by *Podospora anserina*. Biotechnol. Biofuels 8 (1), 1–14. Available from: https://doi.org/10.1186/s13068-015-0274-3.

Bennetto, H.P., Delaney, G.M., Mason, J.R., Roller, S.D., Stirling, J.L., Thurston, C.F., 1985. The sucrose fuel cell: efficient biomass conversion using a microbial catalyst. Biotechnol. Lett. 7 (10), 699–704.

Bhalla, A., Bischoff, K.M., Sani, R.K., 2015. Highly thermostable xylanase production from a thermophilic *Geobacillus* sp. strain WSUCF1 utilizing lignocellulosic biomass. Front. Bioeng. Biotechnol. 3 (JUN), 16. Available from: https://doi.org/10.3389/fbioe.2015.00084.

Bharagava, R.N., Mani, S., Mulla, S.I., Saratale, G.D., 2018. Degradation and decolourization potential of a ligninolytic enzyme producing *Aeromonas hydrophila* for crystal violet dye and its phytotoxicity evaluation. Ecotoxicol. Environ. Saf. 156, 166–175. Available from: https://doi.org/10.1016/J.ECOENV.2018.03.012.

Bhardwaj, N., Kumar, B., Verma, P., 2019. A detailed overview of xylanases: an emerging biomolecule for current and future prospective. Bioresour. Bioprocess. 6 (1). Available from: https://doi.org/10.1186/s40643-019-0276-2.

Bhat, M.K., 2000. Cellulases and related enzymes in biotechnology. Biotechnol. Adv. 18 (5), 355–383. Available from: https://doi.org/10.1016/S0734-9750(00)00041-0.

Bhattacharya, S.S., Karmakar, S., Banerjee, R., 2009. Optimization of laccase mediated biodegradation of 2,4-dichlorophenol using genetic algorithm. Water Res. 43 (14), 3503–3510. Available from: https://doi.org/10.1016/j.watres.2009.05.012.

Bi, X., Feng, X., Yang, Y., Qiu, G., Li, G., Li, F., et al., 2006. Environmental contamination of heavy metals from zinc smelting areas in Hezhang County, western Guizhou, China. Environ. Int. 32 (7), 883–890. Available from: https://doi.org/10.1016/j.envint.2006.05.010.

Bing, W., Wang, H., Zheng, B., Zhang, F., Zhu, G., Feng, Y., et al., 2015. *Caldicellulosiruptor changbaiensis* sp. nov., a cellulolytic and hydrogen-producing bacterium from a hot spring. Int. J. Syst. Evolut. Microbiol. 65 (1), 293–297. Available from: https://doi.org/10.1099/ijs.0.065441-0.

Biz, A., Farias, F.C., Motter, F.A., De Paula, D.H., Richard, P., Krieger, N., et al., 2014. Pectinase activity determination: an early deceleration in the release of reducing sugars throws a spanner in the works!. PLoS ONE 9 (10). Available from: https://doi.org/10.1371/journal.pone.0109529.

Bond, D.R., Holmes, D.E., Tender, L.M., Lovley, D.R., 2002. Electrode-reducing microorganisms that harvest energy from marine sediments. Science 295 (5554), 483–485. Available from: https://doi.org/10.1126/science.1066771.

Cai, Y., Gong, Y., Liu, W., Hu, Y., Chen, L., Yan, L., et al., 2017. Comparative secretomic analysis of lignocellulose degradation by *Lentinula edodes* grown on microcrystalline cellulose, lignosulfonate and glucose. J. Proteom. 163, 92–101. Available from: https://doi.org/10.1016/j.jprot.2017.04.023.

Calderaro, F., Keser, M., Akeroyd, M., Bevers, L.E., Eijsink, V.G.H., Várnai, A., et al., 2020. Characterization of an AA9 LPMO from *Thielavia australiensis*, TausLPMO9B, under industrially relevant lignocellulose saccharification conditions. Biotechnol. Biofuels 13 (1), 1–17. Available from: https://doi.org/10.1186/s13068-020-01836-3.

Campos, J.L., Valenzuela-Heredia, D., Pedrouso, A., Val Del Río, A., Belmonte, M., Mosquera-Corral, A., 2016. Greenhouse gases emissions from wastewater treatment plants: minimization, treatment, and prevention. J. Chem. 2016. Available from: https://doi.org/10.1155/2016/3796352.

Cardoso, B.K., Linde, G.A., Colauto, N.B., do Valle, J.S., 2018. *Panus strigellus* laccase decolorizes anthraquinone, azo, and triphenylmethane dyes. Biocatalysis Agric. Biotechnol. 16, 558–563. Available from: https://doi.org/10.1016/j.bcab.2018.09.026.

Carvalheiro, F., Duarte, L.C., Gírio, F.M., 2008. Hemicellulose biorefineries: a review on biomass pretreatments. J. Sci. & Ind. Res. 67, 849–864.

Chandra, R., Chowdhary, P., 2015. Properties of bacterial laccases and their application in bioremediation of industrial wastes. Environ. Sciences: Process. Impacts 17 (2), 326–342. Available from: https://doi.org/10.1039/c4em00627e.

Chang, S., Chu, J., Guo, Y., Li, H., Wu, B., He, B., 2017. An efficient production of high-pure xylooligosaccharides from corncob with affinity adsorption-enzymatic reaction integrated approach. Bioresour. Technol. 241, 1043–1049. Available from: https://doi.org/10.1016/j.biortech.2017.06.002.

Chapman, J., Ismail, A.E., Dinu, C.Z., 2018. Industrial applications of enzymes: recent advances, techniques, and outlooks. Catalysts 8 (6), 20–29. Available from: https://doi.org/10.3390/catal8060238.

Chaturvedi, V., Verma, P., 2016. Microbial fuel cell: a green approach for the utilization of waste for the generation of bioelectricity. Bioresour. Bioprocess. 3 (1). Available from: https://doi.org/10.1186/s40643-016-0116-6.

Chaudhuri, S.K., Lovley, D.R., 2003. Electricity generation by direct oxidation of glucose in mediatorless microbial fuel cells. Nat. Biotechnol. 21 (10), 1229–1232. Available from: https://doi.org/10.1038/nbt867.

Chebotareva, N., Nyokong, T., 1997. Metallophthalocyanine catalysed electroreduction of nitrate and nitrite ions in alkaline medialnJ. Appl. Electrochem. 27 (8), 975–981. Available from: https://doi.org/10.1023/A:1018466021838.

Chen, C.Y., Lee, C.C., Chen, H.S., Yang, C.H., Wang, S.P., Wu, J.H., et al., 2016. Modification of lignin in sugarcane bagasse by a monocopper hydrogen peroxide-generating oxidase from *Thermobifida fusca*. Process. Biochem. 51 (10), 1486–1495. Available from: https://doi.org/10.1016/j.procbio.2016.07.009.

Chen, H.H., Chen, Y.K., Chang, H.C., Lin, S.Y., 2012. Immunomodulatory effects of xylooligosaccharides. Food Sci. Technol. Res. 18 (2), 195–199. Available from: https://doi.org/10.3136/fstr.18.195.

Chen, W.M., Chang, J.S., Chiu, C.H., Chang, S.C., Chen, W.C., Jiang, C.M., 2005. *Caldimonas taiwanensis* sp. nov., a amylase producing bacterium isolated from a hot spring. Syst. Appl. Microbiol. 28 (5), 415–420. Available from: https://doi.org/10.1016/j.syapm.2005.02.008.

Chiliveri, S.R., Koti, S., Linga, V.R., 2016. Retting and degumming of natural fibers by pectinolytic enzymes produced from *Bacillus tequilensis* SV11-UV37 using solid state fermentation. SpringerPlus 5 (1). Available from: https://doi.org/10.1186/s40064-016-2173-x.

Chinedu, S.N., Okochi, V.I., Omidiji, O., 2011. Cellulase production by wild strains of *Aspergillus niger, Penicillium chrysogenum and Trichoderma harzianum* grown on waste cellulosic materials. Ife J. Sci. 13 (1), 57–62.

Choudhary, J., 2014. Enhanced saccharification of steam-pretreated rice straw by commercial cellulases supplemented with xylanase. J. Bioprocess. Biotech. 04 (07). Available from: https://doi.org/10.4172/2155-9821.1000188.

Collins, M.D., Gibson, G.R., 1999. Probiotics, prebiotics, and synbiotics: approaches for modulating the microbial ecology of the gut. Am. J. Clin. Nutr. 69 (5), 1052–1057. Available from: https://doi.org/10.1093/ajcn/69.5.1052s.

Copete-Pertuz, L.S., Alandete-Novoa, F., Plácido, J., Correa-Londoño, G.A., Mora-Martínez, A.L., 2019. Enhancement of ligninolytic enzymes production and decolourising activity in *Leptosphaerulina* sp. by co-cultivation with *Trichoderma viride* and *Aspergillus terreus*. Sci. Total. Environ. 646, 1536–1545. Available from: https://doi.org/10.1016/j.scitotenv.2018.07.387.

Courtade, G., Balzer, S., Forsberg, Z., Vaaje-Kolstad, G., Eijsink, V.G.H., Aachmann, F.L., 2015. 1H, 13C, 15N resonance assignment of the chitin-active lytic polysaccharide monooxygenase BlLPMO10A from *Bacillus licheniformis*. Biomolecular NMR Assign. 9 (1), 207–210. Available from: https://doi.org/10.1007/s12104-014-9575-x.

Cunha, L., Martarello, R., De Souza, P.M., De Freitas, M.M., Barros, K.V.G., Filho, E.X.F., et al., 2018. Optimization of xylanase production from *Aspergillus foetidus* in soybean residue. Enzyme Res. 2018. Available from: https://doi.org/10.1155/2018/6597017.

Daroch, M., Houghton, C.A., Moore, J.K., Wilkinson, M.C., Carnell, A.J., Bates, A.D., et al., 2014. Glycosylated yellow laccases of the basidiomycete *Stropharia aeruginosa*. Enzyme Microb. Technol. 58–59, 1–7. Available from: https://doi.org/10.1016/j.enzmictec.2014.02.003.

De Almeida, M.N., Guimarães, V.M., Bischoff, K.M., Falkoski, D.L., Pereira, O.L., Gonçalves, D.S.P.O., et al., 2011. Cellulases and hemicellulases from endophytic *Acremonium* species and its application on sugarcane bagasse hydrolysis. Appl. Biochem. Biotechnol. 165 (2), 594–610. Available from: https://doi.org/10.1007/s12010-011-9278-z.

De Assis, L.J., Ries, L.N.A., Savoldi, M., Dos Reis, T.F., Brown, N.A., Goldman, G.H., 2015. *Aspergillus nidulans* protein kinase A plays an important role in cellulase production. Biotechnol. Biofuels 8 (1), 1–20. Available from: https://doi.org/10.1186/s13068-015-0401-1.

De Azeredo, L.A.I., Gomes, P.M., Sant'Anna, G.L., Castilho, L.R., Freire, D.M.G., 2007. Production and regulation of lipase activity from *Penicillium restrictum* in submerged and solid-state fermentations. Curr. Microbiol. 54 (5), 361–365. Available from: https://doi.org/10.1007/s00284-006-0425-7.

de Jong, E., Higson, A., Walsh, P., Wellisch, M., 2012. Bio-based chemicals value added products from biorefinery. IEA Bioenergy I Task. 42 Biorefinery, 1–33.

de Souza, P.M., de Oliveira Magalhães, P., 2010. Application of microbial α-amylase in industry - a review. Braz. J. Microbiol. 41 (4), 850–861. Available from: https://doi.org/10.1590/s1517-83822010000400004.

Dhiman, S.S., Sharma, J., Battan, B., 2008. Industrial applications and future prospects of microbial xylanases: a review. BioResources 3 (4), 1377–1402. Available from: https://doi.org/10.15376/biores.3.4.1377-1402.

Ding, C., Li, M., Hu, Y., 2018. High-activity production of xylanase by *Pichia stipitis*: purification, characterization, kinetic evaluation and xylooligosaccharides production, Int. J. Biol. Macromol., 117. Elsevier B.V. Available from: https://doi.org/10.1016/j.ijbiomac.2018.05.128.

Domínguez De María, P., Sánchez-Montero, J.M., Sinisterra, J.V., Alcántara, A.R., 2006. Understanding *Candida rugosa* lipases: an overview. Biotechnol. Adv. 24 (2), 180–196. Available from: https://doi.org/10.1016/j.biotechadv.2005.09.003.

Dos Santos, T.C., Filho, G.A., De Brito, A.R., Pires, A.J.V., Bonomo, R.C.F., Franco, M., 2016. Production and characterization of cellulolytic enzymes by *Aspergillus niger* and *Rhizopus* sp. By solid state fermentation of prickly pear. Rev. Caatinga 29 (1), 222–233. Available from: https://doi.org/10.1590/1983-21252016v29n126rc.

Drahansky, M., Paridah, M., Moradbak, A., Mohamed, A., Owolabi, F., Asniza, M., et al., 2016. Microbial cellulases: an overview and applications. Intech i 13. Available from: https://doi.org/10.5772/57353.

Dutta, S., Wu, K.C.W., 2014. Enzymatic breakdown of biomass: Enzyme active sites, immobilization, and biofuel production. Green. Chem. 16 (11), 4615–4626. Available from: https://doi.org/10.1039/c4gc01405g.

Duza, M.B., Mastan, S.A., 2015. Optimization studies on cellulase production from *Bacillus anthracis* and *Ochrobactrum anthropic* (YZ1) isolated from soil. Int. J. Appl. Sci. Biotechnol. 3 (2), 272–284. Available from: https://doi.org/10.3126/ijasbt.v3i2.12616.

Eibinger, M., Ganner, T., Bubner, P., Rošker, S., Kracher, D., Haltrich, D., et al., 2014. Cellulose surface degradation by a lytic polysaccharide monooxygenase and its effect on cellulase hydrolytic efficiency. J. Biol. Chem. 289 (52), 35929–35938. Available from: https://doi.org/10.1074/jbc.M114.602227.

Eibinger, M., Sattelkow, J., Ganner, T., Plank, H., Nidetzky, B., 2017. Single-molecule study of oxidative enzymatic deconstruction of cellulose. Nat. Commun. 8 (1), 4–10. Available from: https://doi.org/10.1038/s41467-017-01028-y.

El-Bestawy, E., Hussein, H., Baghdadi, H.H., El-Saka, M.F., 2005. Comparison between biological and chemical treatment of wastewater containing nitrogen and phosphorus. J. Ind. Microbiol. Biotechnol. 32 (5), 195–203. Available from: https://doi.org/10.1007/s10295-005-0229-y.

Erdal, S., Taskin, M., 2010. Production of α-amylase by *Penicillium expansum* MT-1 in solid-state fermentation using waste loquat (Eriobotrya japonica Lindley) kernels as substrate. Rom. Biotechnol. Lett. 15 (3), 5342–5350.

Ertugay, N., Acar, F.N., 2017. Removal of COD and color from Direct Blue 71 azo dye wastewater by Fenton's oxidation: kinetic study. Arab. J. Chem. 10, S1158–S1163. Available from: https://doi.org/10.1016/j.arabjc.2013.02.009.

Escamilla-Alvarado, C., Pérez-Pimienta, J.A., Ponce-Noyola, T., Poggi-Varaldo, H.M., 2016. An overview of the enzyme potential in bioenergy-producing biorefineries. J. Chem. Technol. Biotechnol. 92 (5), 906–924.

Ezeilo, U.R., Lee, C.T., Huyop, F., Zakaria, I.I., Wahab, R.A., 2019. Raw oil palm frond leaves as cost-effective substrate for cellulase and xylanase productions by *Trichoderma asperellum* UC1 under solid-state fermentation. J. Environ. Manag. 243 (April), 206–217. Available from: https://doi.org/10.1016/j.jenvman.2019.04.113.

Falade, A.O., Eyisi, O.A.L., Mabinya, L.V., Nwodo, U.U., Okoh, A.I., 2017. Peroxidase production and ligninolytic potentials of fresh water bacteria *Raoultella ornithinolytica* and *Ensifer adhaerens*. Biotechnol. Rep. 16, 12–17. Available from: https://doi.org/10.1016/J.BTRE.2017.10.001.

Fatima, H., Khan, N., 2015. Production and partial characterization of lipase from *Pseudomonas putida*. Fermentation Technol. 4 (1). Available from: https://doi.org/10.4172/2167-7972.1000112.

Fernando, E., Keshavarz, T., Kyazze, G., 2012. Enhanced bio-decolourisation of acid orange 7 by *Shewanella oneidensis* through co-metabolism in a microbial fuel cell. Int. Biodeterior. Biodegrad. 72, 1–9. Available from: https://doi.org/10.1016/j.ibiod.2012.04.010.

Fonseca, M.I., Fariña, J.I., Sadañoski, M.A., D'Errico, R., Villalba, L.L., Zapata, P.D., 2015. Decolorization of Kraft liquor effluents and biochemical characterization of laccases from *Phlebia brevispora* BAFC 633. Int. Biodeterior. Biodegrad. 104, 443–451. Available from: https://doi.org/10.1016/j.ibiod.2015.07.014.

Forgacs, E., Cserháti, T., Oros, G., 2004. Removal of synthetic dyes from wastewaters: a review. Environ. Int. 30 (7), 953–971. Available from: https://doi.org/10.1016/j.envint.2004.02.001.

Fowler, C.A., Sabbadin, F., Ciano, L., Hemsworth, G.R., Elias, L., Bruce, N., et al., 2019. Discovery, activity and characterisation of an AA10 lytic polysaccharide oxygenase from the shipworm symbiont *Teredinibacter turnerae*. Biotechnol. Biofuels 12 (1), 1–11. Available from: https://doi.org/10.1186/s13068-019-1573-x.

Frommhagen, M., Koetsier, M.J., Westphal, A.H., Visser, J., Hinz, S.W.A., Vincken, J.P., et al., 2016. Lytic polysaccharide monooxygenases from *Myceliophthora thermophila* C1 differ in substrate preference and reducing agent specificity. Biotechnol. Biofuels 9 (1), 1–17. Available from: https://doi.org/10.1186/s13068-016-0594-y.

Galdino, A.S., Silva, R.N., Lottermann, M.T., Álvares, A.C.M., Moraes, L.M.P., De, et al., 2011. Biochemical and structural characterization of amy1: An alpha-amylase from *Cryptococcus flavus* expressed in *Saccharomyces cerevisiae*. Enzyme Res. 2011 (1). Available from: https://doi.org/10.4061/2011/157294.

Gao, Z.M., Xiao, J., Wang, X.N., Ruan, L.W., Chen, X.L., Zhang, Y.Z., 2012. *Vibrio xiamenensis* sp. nov., a cellulase-producing bacterium isolated from mangrove soil. Int. J. Syst. Evolut. Microbiol. 62 (8), 1958–1962. Available from: https://doi.org/10.1099/ijs.0.033597-0.

Gaur, N., Narasimhulu, K., Pydi Setty, Y., 2018. Extraction of ligninolytic enzymes from novel *Klebsiella pneumoniae* strains and its application in wastewater treatment. Appl. Water Sci. 8 (4), 1–17. Available from: https://doi.org/10.1007/s13201-018-0758-y.

Ghazala, I., Sayari, N., Romdhane, M., Ben, Ellouz-Chaabouni, S., Haddar, A., 2015. Assessment of pectinase production by *Bacillus mojavensis* I4 using an economical substrate and its potential application in oil sesame extraction. J. Food Sci. Technol. 52 (12), 7710–7722. Available from: https://doi.org/10.1007/s13197-015-1964-3.

Ghosh, P., Ghosh, U., 2017. Statistical optimization of laccase production by *Aspergillus flavus* PUF5 through submerged fermentation using agro-waste as cheap substrate. Acta Biologica Szegediensis 61 (1), 25–33.

Gil, G.C., Chang, I.S., Kim, B.H., Kim, M., Jang, J.K., Park, H.S., et al., 2003. Operational parameters affecting the performance of a mediator-less microbial fuel cell. Biosens. Bioelectron. 18 (4), 327–334. Available from: https://doi.org/10.1016/S0956-5663(02)00110-0.

Gírio, F.M., Fonseca, C., Carvalheiro, F., Duarte, L.C., Marques, S., Bogel-Łukasik, R., 2010. Hemicelluloses for fuel ethanol: a review. Bioresour. Technol. 101 (13), 4775–4800. Available from: https://doi.org/10.1016/j.biortech.2010.01.088.

Gopinath, S.C.B., Anbu, P., Arshad, M.K.M., Lakshmipriya, T., Voon, C.H., Hashim, U., et al., 2017. Biotechnological processes in microbial amylase production. BioMed. Res. Int. 2017. Available from: https://doi.org/10.1155/2017/1272193.

Gregory, R.C., Hemsworth, G.R., Turkenburg, J.P., Hart, S.J., Walton, P.H., Davies, G.J., 2016. Activity, stability and 3-D structure of the Cu(II) form of a chitin-active lytic polysaccharide monooxygenase from: *Bacillus amyloliquefaciens*. Dalton Trans. 45 (42), 16904–16912. Available from: https://doi.org/10.1039/c6dt02793h.

Guan, G.Q., Zhao, P.X., Zhao, J., Wang, M.J., Huo, S.H., Cui, F.J., et al., 2016. Production and Partial Characterization of an Alkaline Xylanase from a Novel Fungus *Cladosporium oxysporum*. BioMed. Res. Int. 2016. Available from: https://doi.org/10.1155/2016/4575024.

Gupta, P.K., Agrawal, P., Hedge, P., Akhtar, M.S., 2018. Xylooligosaccharides and their anticancer potential: an update, Anticancer Plants: Natural Products and Biotechnological Implements, 2. Springer, pp. 255–271. Available from: https://doi.org/10.1007/978-981-10-8064-7_11.

Gupta, R., Gigras, P., Mohapatra, H., Goswami, V.K., Chauhan, B., 2003. Microbial α-amylases: a biotechnological perspective. Process. Biochem. 38 (11), 1599–1616. Available from: https://doi.org/10.1016/S0032-9592(03)00053-0.

Gupta, R., Mehta, G., Deswal, D., Sharma, S., Jain, K.K., Singh, A., et al., 2013. Cellulases and their biotechnological applications. Biotechnol. Environ. Manag. Resour. Recovery 1–315. Available from: https://doi.org/10.1007/978-81-322-0876-1.

Gurung, N., Ray, S., Bose, S., Rai, V., 2013. A broader view: microbial enzymes and their relevance in industries, medicine, and beyond. BioMed. Res. Int. 2013. Available from: https://doi.org/10.1155/2013/329121.

Ha, P.T., Tae, B., Chang, I.S., 2008. Performance and bacterial consortium of microbial fuel cell fed with formate. Energy Fuels 22 (1), 164–168. Available from: https://doi.org/10.1021/ef700294x.

Hamed, M.B., Karamanou, S., Ólafsdottir, S., Sofia, J., Basílio, M., Simoens, K., et al., 2017. Large-scale production of a thermostable Rhodothermus marinus cellulase by heterologous secretion from *Streptomyces lividans*. Microb. Cell Factories 1–12. Available from: https://doi.org/10.1186/s12934-017-0847-x.

Harris, P.V., Xu, F., Kreel, N.E., Kang, C., Fukuyama, S., 2014. New enzyme insights drive advances in commercial ethanol production. Curr. Opin. Chem. Biol. 19 (1), 162–170. Available from: https://doi.org/10.1016/j.cbpa.2014.02.015.

Hassan, S.H.A., Kim, Y.S., Oh, S.E., 2012. Power generation from cellulose using mixed and pure cultures of cellulose-degrading bacteria in a microbial fuel cell. Enzyme Microb. Technol. 51 (5), 269–273. Available from: https://doi.org/10.1016/j.enzmictec.2012.07.008.

Heilmann, J., Logan, B.E., 2006. Production of electricity from proteins using a microbial fuel cell. Water Environ. Res. 78 (5), 531–537. Available from: https://doi.org/10.2175/106143005x73046.

Holmes, D.E., Bond, D.R., Lovley, D.R., 2004. Electron transfer by desulfobulbus propionicus to Fe(III) and graphite electrodes. Appl. Environ. Microbiol. 70 (2), 1234–1237. Available from: https://doi.org/10.1128/AEM.70.2.1234-1237.2004.

Hu, D.D., Zhang, R.Y., Zhang, G.Q., Wang, H.X., Ng, T.B., 2011a. A laccase with antiproliferative activity against tumor cells from an edible mushroom, white common *Agrocybe cylindracea*. Phytomedicine 18 (5), 374–379. Available from: https://doi.org/10.1016/j.phymed.2010.07.004.

Hu, J., Arantes, V., Saddler, J.N., 2011b. The enhancement of enzymatic hydrolysis of lignocellulosic substrates by the addition of accessory enzymes such as xylanase: Is it an additive or synergistic effect? Biotechnol. Biofuels 4 (October). Available from: https://doi.org/10.1186/1754-6834-4-36.

Hu, J., Chandra, R., Arantes, V., Gourlay, K., Susan van Dyk, J., Saddler, J.N., 2015. The addition of accessory enzymes enhances the hydrolytic performance of cellulase enzymes at high solid loadings. Bioresour. Technol. 186, 149–153. Available from: https://doi.org/10.1016/j.biortech.2015.03.055.

Humphries, A.C., Nott, K.P., Hall, L.D., Macaskie, L.E., 2004. Continuous removal of Cr(VI) from aqueous solution catalysed by palladised biomass of *Desulfovibrio vulgaris*. Biotechnol. Lett. 26 (19), 1529–1532. Available from: https://doi.org/10.1023/B:BILE.0000044457.80314.4d.

Hüttner, S., Várnai, A., Petrović, D.M., Bach, C.X., Kim Anh, D.T., Thanh, V.N., et al., 2019. Specific xylan activity revealed for AA9 lytic polysaccharide monooxygenases of the thermophilic fungus *Malbranchea cinnamomea* by functional characterization. Appl. Environ. Microbiol. 85 (23). Available from: https://doi.org/10.1128/AEM.01408-19.

Iconomou, D., Arapoglou, D., Israilides, C., 2010. Improvement of phenolic antioxidants and quality characteristics of virgin olive oil with the addition of enzymes and nitrogen during olive paste processing. Grasas y. Aceites 61 (3), 303–311. Available from: https://doi.org/10.3989/gya.064809.

Illuri, R., Kumar, M., Eyini, M., Veeramanikandan, V., Almaary, K.S., Elbadawi, Y.B., et al., 2021. Production, partial purification and characterization of ligninolytic enzymes from selected basidiomycetes mushroom fungi. Saudi J. Biol. Sci. 28 (12), 7207–7218. Available from: https://doi.org/10.1016/J.SJBS.2021.08.026.

Irfan, M., Asghar, U., Nadeem, M., Nelofer, R., Syed, Q., 2016. Optimization of process parameters for xylanase production by *Bacillus* sp. in submerged fermentation. J. Radiat. Res. Appl. Sci. 9 (2), 139–147. Available from: https://doi.org/10.1016/j.jrras.2015.10.008.

Ishii, S., Watanabe, K., Yabuki, S., Logan, B.E., Sekiguchi, Y., 2008. Comparison of electrode reduction activities of *Geobacter sulfurreducens* and an enriched consortium in an air-cathode microbial fuel cell. Appl. Environ. Microbiol. 74 (23), 7348–7355. Available from: https://doi.org/10.1128/AEM.01639-08.

Jaishankar, M., Tseten, T., Anbalagan, N., Mathew, B.B., Beeregowda, K.N., 2014. Toxicity, mechanism and health effects of some heavy metals. Interdiscip. Toxicol. 7 (2), 60–72. Available from: https://doi.org/10.2478/intox-2014-0009.

Ji, Z., He, M., Huang, Z., Ozkan, U., Wu, Y., 2013. Photostable p-type dye-sensitized photoelectrochemical cells for water reduction. J. Am. Chem. Soc. 135 (32), 11696–11699. Available from: https://doi.org/10.1021/ja404525e.

Johansen, K.S., 2016. Lytic polysaccharide monooxygenases: the microbial power tool for lignocellulose degradation. Trends Plant. Sci. 21 (11), 926–936. Available from: https://doi.org/10.1016/j.tplants.2016.07.012.

Kaewkannetra, P., Imai, T., Garcia-Garcia, F.J., Chiu, T.Y., 2009. Cyanide removal from cassava mill wastewater using *Azotobactor vinelandii* TISTR 1094 with mixed microorganisms in activated sludge treatment system. J. Hazard. Mater. 172 (1), 224–228. Available from: https://doi.org/10.1016/j.jhazmat.2009.06.162.

Kahloul, M., Chekir, J., Hafiane, A., 2019. Dye removal using keggin polyoxometalates assisted ultrafiltration: characterization and UV visible study. Arch. Environ. Prot. 45 (4), 30–39. Available from: https://doi.org/10.24425/aep.2019.130239.

Kalathil, S., Lee, J., Cho, M.H., 2013. Gold nanoparticles produced in situ mediate bioelectricity and hydrogen production in a microbial fuel cell by quantized capacitance charging. ChemSusChem 6 (2), 246–250. Available from: https://doi.org/10.1002/cssc.201200747.

Kallel, F., Driss, D., Chaabouni, S.E., Ghorbel, R., 2015. Biological activities of xylooligosaccharides generated from garlic straw xylan by purified xylanase from *Bacillus mojavensis* UEB-FK. Appl. Biochem. Biotechnol. 175 (2), 950–964. Available from: https://doi.org/10.1007/s12010-014-1308-1.

Kanmani, P., Karthik, S., Aravind, J., Kumaresan, K., 2013. The Use of Response Surface Methodology as a Statistical Tool for Media Optimization in Lipase Production from the Dairy Effluent Isolate *Fusarium solani*. ISRN Biotechnol. 2013, 1–8. Available from: https://doi.org/10.5402/2013/528708.

Karmee, S.K., Swanepoel, W., Marx, S., 2018. Biofuel production from spent coffee grounds via lipase catalysis. Energy Sources, Part. A: Recovery, Util. Environ. Eff. 40 (3), 294–300. Available from: https://doi.org/10.1080/15567036.2017.1415394.

Kashyap, D.R., Vohra, P.K., Chopra, S., Tewari, R., 2001. Applications of pectinases in the commercial sector: a review. Bioresour. Technol. 77 (3), 215–227. Available from: https://doi.org/10.1016/S0960-8524(00)00118-8.

Kathiresan, K., Manivannan, S., 2006. α-Amylase production by *Penicillium fellutanum* isolated from mangrove rhizosphere soil. Afr. J. Biotechnol. 5 (10), 829–832. Available from: https://doi.org/10.5897/AJB.

Kempka, A.P., Lipke, N.L., Da Luz Fontoura Pinheiro, T., Menoncin, S., Treichel, H., Freire, D.M.G., et al., 2008. Response surface method to optimize the production and characterization of lipase from *Penicillium verrucosum* in solid-state fermentation. Bioprocess. Biosyst. Eng. 31 (2), 119–125. Available from: https://doi.org/10.1007/s00449-007-0154-8.

Khandeparker, R., Jalal, T., 2015. Xylanolytic enzyme systems in *Arthrobacter* sp. MTCC 5214 and *Lactobacillus* sp. Biotechnol. Appl. Biochem. 62 (2), 245–254. Available from: https://doi.org/10.1002/bab.1253.

Khataee, A., Kayan, B., Gholami, P., Kalderis, D., Akay, S., 2017. Sonocatalytic degradation of an anthraquinone dye using TiO2-biochar nanocomposite. Ultrason. Sonochem. 39, 120–128. Available from: https://doi.org/10.1016/j.ultsonch.2017.04.018.

Khuong, L.D., Kondo, R., De Leon, R., Kim Anh, T., Shimizu, K., Kamei, I., 2014. Bioethanol production from alkaline-pretreated sugarcane bagasse by consolidated bioprocessing using *Phlebia* sp. MG-60. Int. Biodeterior. Biodegrad. 88, 62–68. Available from: https://doi.org/10.1016/j.ibiod.2013.12.008.

Kim, B.H., Kim, H.J., Hyun, M.S., Park, D.H., 1999. Direct electrode reaction of Fe(III)-reducing bacterium, *Shewanella putrefaciens*In J. Microbiol. Biotechnol. 9 (2), 127–131.

Kishor, R., Saratale, G.D., Saratale, R.G., Romanholo Ferreira, L.F., Bilal, M., Iqbal, H.M.N., et al., 2021. Efficient degradation and detoxification of methylene blue dye by a newly isolated ligninolytic enzyme producing bacterium *Bacillus albus* MW407057. Colloids Surf. B: Biointerfaces 206 (May), 111947. Available from: https://doi.org/10.1016/j.colsurfb.2021.111947.

Kojima, Y., Várnai, A., Ishida, T., Sunagawa, N., Petrovic, D.M., Igarashi, K., et al., 2016. A lytic polysaccharide monooxygenase with broad xyloglucan specificity from the brown-rot fungus *Gloeophyllum trabeum* and its action on cellulose-xyloglucan complexes. Appl. Environ. Microbiol. 82 (22), 6557–6572. Available from: https://doi.org/10.1128/AEM.01768-16.

Kruer-Zerhusen, N., Alahuhta, M., Lunin, V.V., Himmel, M.E., Bomble, Y.J., Wilson, D.B., 2017. Structure of a *Thermobifida fusca* lytic polysaccharide monooxygenase and mutagenesis of key residues. Biotechnol. Biofuels 10 (1), 1–12. Available from: https://doi.org/10.1186/s13068-017-0925-7.

Kuhad, R.C., Gupta, R., Singh, A., 2011. Microbial cellulases and their industrial applications. Enzyme Res. 2011 (1). Available from: https://doi.org/10.4061/2011/280696.

Kuhad, R.C., Mehta, G., Gupta, R., Sharma, K.K., 2010. Fed batch enzymatic saccharification of newspaper cellulosics improves the sugar content in the hydrolysates and eventually the ethanol fermentation by Saccharomyces cerevisiae. Biomass Bioenergy 34 (8), 1189–1194. Available from: https://doi.org/10.1016/j.biombioe.2010.03.009.

Kumar, B., Verma, P., 2020. Enzyme mediated multi-product process: A concept of bio-based refinery. Ind. Crop. Products 154 (May), 112607. Available from: https://doi.org/10.1016/j.indcrop.2020.112607.

Kumar, B., Agrawal, K., Verma, P., 2021. Microbial electrochemical system: a sustainable approach for mitigation of toxic dyes and heavy metals from wastewater. J. Hazard. Toxic Radioact. Waste 25 (2), 04020082. Available from: https://doi.org/10.1061/(asce)hz.2153-5515.0000590.

Kumar, S., Haq, I., Prakash, J., Singh, S.K., Mishra, S., Raj, A., 2017. Purification, characterization and thermostability improvement of xylanase from *Bacillus amyloliquefaciens* and its application in pre-bleaching of kraft pulp. 3 Biotech. 7 (1), 1–12. Available from: https://doi.org/10.1007/s13205-017-0615-y.

Kumar, V., Shukla, P., 2018. Extracellular xylanase production from *T. lanuginosus* VAPS24 at pilot scale and thermostability enhancement by immobilization. Process. Biochem. 71, 53–60. Available from: https://doi.org/10.1016/j.procbio.2018.05.019.

Kumari, D., Singh, R., 2018. Pretreatment of lignocellulosic wastes for biofuel production: a critical review. Renew. Sustain. Energy Rev. 90 (March), 877–891. Available from: https://doi.org/10.1016/j.rser.2018.03.111.

Lakshmi, S.A., Shafreen, R.M.B., Balaji, K., Ibrahim, K.S., Shiburaj, S., Gayathri, V., et al., 2020. Cloning, expression, homology modelling and molecular dynamics simulation of four domain-containing α-amylase from *Streptomyces griseus*. J. Biomol. Struct. Dyn. 39 (6), 2152–2163. Available from: https://doi.org/10.1080/07391102.2020.1745282.

Lee, B.H., Yan, L., Phillips, R.J., Reuhs, B.L., Jones, K., Rose, D.R., et al., 2013. Enzyme-synthesized highly branched maltodextrins have slow glucose generation at the mucosal α-glucosidase level and are slowly digestible in vivo. PLoS ONE 8 (4). Available from: https://doi.org/10.1371/journal.pone.0059745.

Li, T., Li, S., Du, L., Wang, N., Guo, M., Zhang, J., et al., 2010. Effects of haw pectic oligosaccharide on lipid metabolism and oxidative stress in experimental hyperlipidemia mice induced by high-fat diet. Food Chem. 121 (4), 1010–1013. Available from: https://doi.org/10.1016/j.foodchem.2010.01.039.

Liang, Y., Feng, Z., Yesuf, J., Blackburn, J.W., 2010. Optimization of growth medium and enzyme assay conditions for crude cellulases produced by a novel thermophilic and cellulolytic bacterium, *Anoxybacillus* sp. 527. Appl. Biochem. Biotechnol. 160 (6), 1841–1852. Available from: https://doi.org/10.1007/s12010-009-8677-x.

Liu, B., Olson, Å., Wu, M., Broberg, A., Sandgren, M., 2017. Biochemical studies of two lytic polysaccharide monooxygenases from the white-rot fungus *Heterobasidion irregulare* and their roles in lignocellulose degradation. PLOS ONE . Available from: https://doi.org/10.1371/journal.pone.0189479.

Liu, L., Li, F.B., Feng, C.H., Li, X.Z., 2009a. Microbial fuel cell with an azo-dye-feeding cathode. Appl. Microbiol. Biotechnol. 85 (1), 175–183. Available from: https://doi.org/10.1007/s00253-009-2147-9.

Liu, Z., Liu, J., Zhang, S., Su, Z., 2009b. Study of operational performance and electrical response on mediator-less microbial fuel cells fed with carbon- and protein-rich substrates. Biochem. Eng. J. 45 (3), 185–191. Available from: https://doi.org/10.1016/j.bej.2009.03.011.

Lloret, L., Eibes, G., Moreira, M.T., Feijoo, G., Lema, J.M., 2013. On the use of a high-redox potential laccase as an alternative for the transformation of non-steroidal anti-inflammatory drugs (NSAIDs). J. Mol. Catal. B: Enzym. 97, 233–242. Available from: https://doi.org/10.1016/j.molcatb.2013.08.021.

Logan, B.E., 2004. Extracting hydrogen and electricity from renewable resources. Environ. Sci. Technol. 38 (9).

Logan, B.E., 2009. Exoelectrogenic bacteria that power microbial fuel cells. Nat. Rev. Microbiol. 7 (5), 375–381. Available from: https://doi.org/10.1038/nrmicro2113.

Logan, B.E., Regan, J.M., 2006a. Electricity-producing bacterial communities in microbial fuel cells. Trends Microbiol. 14 (12), 512–518. Available from: https://doi.org/10.1016/j.tim.2006.10.003.

Logan, B.E., Regan, J.M., 2006b. Microbial fuel cells—challenges and applications. Environ. Sci. Technol. 5172–5180. Available from: https://doi.org/10.1016/B978-0-444-53563-4.10002-1.

López-Abelairas, M., Álvarez Pallín, M., Salvachúa, D., Lú-Chau, T., Martínez, M.J., Lema, J. M., 2013. Optimisation of the biological pretreatment of wheat straw with white-rot fungi for ethanol production. Bioprocess. Biosyst. Eng. 36 (9), 1251–1260. Available from: https://doi.org/10.1007/s00449-012-0869-z.

Lovley, D.R., 2006. Bug juice: Harvesting electricity with microorganisms. Nat. Rev. Microbiol. 4 (7), 497–508. Available from: https://doi.org/10.1038/nrmicro1442.

Lu, C., Wang, H., Luo, Y., Guo, L., 2010. An efficient system for pre-delignification of gramineous biofuel feedstock in vitro: application of a laccase from *Pycnoporus sanguineus* H275. Process. Biochem. 45 (7), 1141–1147. Available from: https://doi.org/10.1016/j.procbio.2010.04.010.

Lueangjaroenkit, P., Teerapatsakul, C., Chitradon, L., 2018. Morphological characteristic regulation of ligninolytic enzyme produced by *Trametes polyzona*. Mycobiology 46 (4), 396–406. Available from: https://doi.org/10.1080/12298093.2018.1537586.

Lynd, L.R., Weimer, P.J., Zyl, W.H., Van, Isak, S., 2002. Microbial cellulose utilization : fundamentals and biotechnology. Microbiol. Mol. Biol. Rev. 66 (3), 506–577. Available from: https://doi.org/10.1128/MMBR.66.3.506.

Mandal, A., 2015. Review on microbial xylanases and their applications. Int. J. Life Sci. 4 (3), 178–187.

Mani, P., Kumar, V.T.F., Keshavarz, T., Sainathan Chandra, T., Kyazze, G., 2018. The role of natural laccase redox mediators in simultaneous dye decolorization and power production in microbial fuel cells. Energies 11 (12), 1–12. Available from: https://doi.org/10.3390/en11123455.

Martin, N., Guez, M.A.U., Sette, L.D., Da Silva, R., Gomes, E., 2010. Pectinase production by a Brazilian thermophilic fungus *Thermomucor indicae-seudaticae* N31 in solid-state and submerged fermentation. Microbiology 79 (3), 306–313. Available from: https://doi.org/10.1134/S0026261710030057.

Martín-Sampedro, R., López-Linares, J.C., Fillat, Ú., Gea-Izquierdo, G., Ibarra, D., Castro, E., et al., 2017. Endophytic fungi as pretreatment to enhance enzymatic hydrolysis of olive tree pruning. BioMed. Res. Int. 2017. Available from: https://doi.org/10.1155/2017/9727581.

Mateos Diaz, J.C., Rodríguez, J.A., Roussos, S., Cordova, J., Abousalham, A., Carriere, F., et al., 2006. Lipase from the thermotolerant fungus *Rhizopus homothallicus* is more thermostable when produced using solid state fermentation than liquid fermentation procedures. Enzyme Microb. Technol. 39 (5), 1042–1050. Available from: https://doi.org/10.1016/j.enzmictec.2006.02.005.

Matpan Bekler, F., Güven, K., Gül Güven, R., 2021. Purification and characterization of novel α-amylase from *Anoxybacillus ayderensis* FMB1. Biocatal. Biotransformation 39 (4), 322–332. Available from: https://doi.org/10.1080/10242422.2020.1856097.

Mazhar, H., Abbas, N., Ali, S., Sohail, A., Hussain, Z., Ali, S.S., 2017. Optimized production of lipase from *Bacillus subtilis* PCSIRNL-39. Afr. J. Biotechnol. 16 (19), 1106–1115. Available from: https://doi.org/10.5897/ajb2017.15924.

Megead Yassien, M.A., 2012. Improved production, purification and some properties of α-amylase from *Streptomyces clavifer*. Afr. J. Biotechnol. 11 (80), 14603–14611. Available from: https://doi.org/10.5897/ajb12.1790.

Mehta, D., Satyanarayana, T., 2013. Biochemical and molecular characterization of recombinant acidic and thermostable raw-starch hydrolysing α-amylase from an extreme thermophile *Geobacillus thermoleovorans*. J. Mol. Catal. B: Enzym. 85–86, 229–238. Available from: https://doi.org/10.1016/j.molcatb.2012.08.017.

Menicucci, J., Beyenal, H., Marsili, E., Veluchamy, R.A., Demir, G., Lewandowski, Z., 2006. Procedure for determining maximum sustainable power generated by microbial fuel cells. Environ. Sci. Technol. 40 (3), 1062–1068. Available from: https://doi.org/10.1021/es051180l.

Merín, M.G., Mendoza, L.M., Morata de Ambrosini, V.I., 2014. Pectinolytic yeasts from viticultural and enological environments: novel finding of *Filobasidium capsuligenum* producing pectinases. J. Basic. Microbiol. 54 (8), 835–842. Available from: https://doi.org/10.1002/jobm.201200534.

Min, B., Kim, J.R., Oh, S.E., Regan, J.M., Logan, B.E., 2005. Electricity generation from swine wastewater using microbial fuel cells. Water Res. 39 (20), 4961–4968. Available from: https://doi.org/10.1016/j.watres.2005.09.039.

Mmango-Kaseke, Z., Okaiyeto, K., Nwodo, U.U., Mabinya, L.V., Okoh, A.I., 2016. Optimization of cellulase and xylanase production by *Micrococcus* species under submerged fermentation. Sustainability (Switz.) 8 (11), 1–15. Available from: https://doi.org/10.3390/su8111168.

Mmereki, D., Li, B., Meng, L., 2014. Hazardous and toxic waste management in Botswana: practices and challenges. Waste Manag. Res. 32 (12), 1158–1168. Available from: https://doi.org/10.1177/0734242X14556527.

Mohamed, S.A., Azhar, E.I., Ba-Akdah1, M.M., Tashkandy, N.R., Kumosani, T.A., 2011. Production, purification and characterization of α-amylase from *Trichoderma harzianum* grown on mandarin peel. Afr. J. Microbiol. Res. 5 (9). Available from: https://doi.org/10.5897/ajmr10.890.

Mojsov, K., 2012. Microbial α-amylase and their industrial applications: a review. Int. J. Manag. IT Eng. 2 (10), 583–609.

Moreno, A.D., Ibarra, D., Alvira, P., Tomás-Pejó, E., Ballesteros, M., 2015. A review of biological delignification and detoxification methods for lignocellulosic bioethanol production. Crit. Rev. Biotechnol. 35 (3), 342–354. Available from: https://doi.org/10.3109/07388551.2013.878896.

Moreno, M.L., Piubeli, F., Bonfá, M.R.L., García, M.T., Durrant, L.R., Mellado, E., 2012. Analysis and characterization of cultivable extremophilic hydrolytic bacterial community in heavy-metal-contaminated soils from the Atacama Desert and their biotechnological potentials. J. Appl. Microbiol. 113 (3), 550–559. Available from: https://doi.org/10.1111/j.1365-2672.2012.05366.x.

Moţ, A.C., Pârvu, M., Damian, G., Irimie, F.D., Darula, Z., Medzihradszky, K.F., et al., 2012. A "yellow" laccase with "blue" spectroscopic features, from *Sclerotinia sclerotiorum*. Process. Biochem. 47 (6), 968–975. Available from: https://doi.org/10.1016/j.procbio.2012.03.006.

Motta, F.L., Andrade, C.C.P., Santana, M.H.A., 2013. A review of xylanase production by the fermentation of xylan: classification, characterization and applications. Sustainable Degradation of Lignocellulosic Biomass: Techniques, Applications and Commercialization. Sciencedirect, pp. 251–275. Available from: https://www.sciencedirect.com/science/article/pii/B9780444635037000061.

Mukhopadhyay, M., Banerjee, R., 2015. Purification and biochemical characterization of a newly produced yellow laccase from *Lentinus squarrosulus* MR13. 3 Biotech. 5 (3), 227–236. Available from: https://doi.org/10.1007/s13205-014-0219-8.

Müller, G., Chylenski, P., Bissaro, B., Eijsink, V.G.H., Horn, S.J., 2018. The impact of hydrogen peroxide supply on LPMO activity and overall saccharification efficiency of a commercial cellulase cocktail. Biotechnol. Biofuels 11 (1), 1–17. Available from: https://doi.org/10.1186/s13068-018-1199-4.

Müller, G., Várnai, A., Johansen, K.S., Eijsink, V.G.H., Horn, S.J., 2015. Harnessing the potential of LPMO-containing cellulase cocktails poses new demands on processing conditions. Biotechnol. Biofuels 8 (1), 1–9. Available from: https://doi.org/10.1186/s13068-015-0376-y.

Munzone, A., El Kerdi, B., Fanuel, M., Rogniaux, H., Ropartz, D., Réglier, M., et al., 2020. Characterization of a bacterial copper-dependent lytic polysaccharide monooxygenase with an unusual second coordination sphere. FEBS J. 287 (15), 3298–3314. Available from: https://doi.org/10.1111/febs.15203.

Nakamura, S., Kurosawa, N., 2021. Decomposition of Rice chaff using a cocultivation system of *Thermobifida fusca* and *Ureibacillus thermosphaericus*. Proceedings 66 (1), 31. Available from: https://doi.org/10.3390/proceedings2020066031.

Naragani, K., Muvva, V., Munaganti, R., Bindu, B., 2015. Studies on optimization of amylase production by *Streptomyces cheonanensis* VUK-A isolated from mangrove habitats. J. Adv. Biol. & Biotechnol. 3 (4), 165–172. Available from: https://doi.org/10.9734/jabb/2015/18025.

Nazaruk, E., Smoliński, S., Swatko-Ossor, M., Ginalska, G., Fiedurek, J., Rogalski, J., et al., 2008. Enzymatic biofuel cell based on electrodes modified with lipid liquid-crystalline cubic phases. J. Power Sources 183 (2), 533–538. Available from: https://doi.org/10.1016/j.jpowsour.2008.05.061.

Neagu, D.A., Destain, J., Thonart, P., Socaciu, C., 2012. Effects of different carbon sources on pectinase production by *Penicillium oxalicum*. Bull. Univ. Agric. Sci. Veterinary Med. Cluj-Napoca - Agriculture 69 (2), 327–333. Available from: https://doi.org/10.15835/buasvmcn-agr:8781.

Nekiunaite, L., Petrović, D.M., Westereng, B., Vaaje-Kolstad, G., Hachem, M.A., Várnai, A., et al., 2016. FgLPMO9A from *Fusarium graminearum* cleaves xyloglucan independently of the backbone substitution pattern. FEBS Lett. 3346–3356. Available from: https://doi.org/10.1002/1873-3468.12385.

Nevin, K.P., Lovley, D.R., 2002. Mechanisms for accessing insoluble Fe(III) oxide during dissimilatory Fe(III) reduction by *Geothrix fermentans*. Appl. Environ. Microbiol. 68 (5), 2294–2299. Available from: https://doi.org/10.1128/AEM.68.5.2294-2299.2002.

Nguyen, T.A., Fu, C.C., Juang, R.S., 2016. Biosorption and biodegradation of a sulfur dye in high-strength dyeing wastewater by *Acidithiobacillus thiooxidans*. J. Environ. Manag. 182, 265–271. Available from: https://doi.org/10.1016/j.jenvman.2016.07.083.

Niladevi, K.N., 2009. Ligninolytic enzymes. Biotechnology for Agro-Industrial Residues Utilisation: Utilisation of Agro-Residues. Springer, Dordrecht, pp. 397–414. Available from: https://doi.org/10.1007/978-1-4020-9942-7.

Oh, S.E., Logan, B.E., 2005. Hydrogen and electricity production from a food processing wastewater using fermentation and microbial fuel cell technologies. Water Res. 39 (19), 4673–4682. Available from: https://doi.org/10.1016/j.watres.2005.09.019.

Olajuyigbe, F.M., Fatokun, C.O., Oyelere, O.M., 2018. Biodelignification of some agro-residues by *Stenotrophomonas* sp. CFB-09 and enhanced production of ligninolytic enzymes. Biocatal. Agric. Biotechnol. 15, 120–130. Available from: https://doi.org/10.1016/j.bcab.2018.05.016.

Ozcirak Ergun, S., Ozturk Urek, R., 2017. Production of ligninolytic enzymes by solid state fermentation using *Pleurotus ostreatus*. Ann. Agrarian Sci. 15 (2), 273–277. Available from: https://doi.org/10.1016/j.aasci.2017.04.003.

Paola Di Donato, I.F., 2014. Use of agro waste biomass for α-amylase production by *Anoxybacillus amylolyticus*: purification and properties. J. Microb. Biochem. Technol. 06 (06). Available from: https://doi.org/10.4172/1948-5948.1000162.

Paspaliari, D.K., Loose, J.S.M., Larsen, M.H., Vaaje-Kolstad, G., 2015. *Listeria monocytogenes* has a functional chitinolytic system and an active lytic polysaccharide monooxygenase. FEBS J. 282 (5), 921–936. Available from: https://doi.org/10.1111/febs.13191.

Paz, A., Costa-Trigo, I., Oliveira, R.P., de, S., Domínguez, J.M., 2020. Ligninolytic enzymes of endospore-forming *Bacillus aryabhattai* BA03. Curr. Microbiol. 77 (5), 702–709. Available from: https://doi.org/10.1007/s00284-019-01856-9.

Peters, D., Ngai, D.D., An, D.T., 2001. Agro-processing waste assessment in Peri-urban Hanoi. In Scientist and Farmer (pp. 451–457).

Petrović, D.M., Várnai, A., Dimarogona, M., Mathiesen, G., Sandgren, M., Westereng, B., et al., 2019. Comparison of three seemingly similar lytic polysaccharide monooxygenases from *Neurospora crassa* suggests different roles in plant biomass degradation. J. Biol. Chem. 294 (41), 15068–15081. Available from: https://doi.org/10.1074/jbc.RA119.008196.

Pham, T.H., Boon, N., Aelterman, P., Clauwaert, P., De Schamphelaire, L., Vanhaecke, L., et al., 2008. Metabolites produced by *Pseudomonas* sp. enable a Gram-positive bacterium to achieve extracellular electron transfer. Appl. Microbiol. Biotechnol. 77 (5), 1119–1129. Available from: https://doi.org/10.1007/s00253-007-1248-6.

Ping, L., Wang, M., Yuan, X., Cui, F., Huang, D., Sun, W., et al., 2018. Production and characterization of a novel acidophilic and thermostable xylanase from *Thermoascus aurantiacus*.

Int. J. Biol. Macromolecules 109, 1270–1279. Available from: https://doi.org/10.1016/j.ijbiomac.2017.11.130.

Potter, M.C., 1911. Electrical effects accompanying the decomposition of organic compounds. Proceedings of the Royal Society of London. Series B, Containing Papers of a Biological Character 571 (84), 260–276. Available from: https://doi.org/10.1098/rspb.1911.0073.

Prakash, B., Vidyasagar, M., Madhukumar, M.S., Muralikrishna, G., Sreeramulu, K., 2009. Production, purification, and characterization of two extremely halotolerant, thermostable, and alkali-stable α-amylases from *Chromohalobacter* sp. TVSP 101. Process. Biochem. 44 (2), 210–215. Available from: https://doi.org/10.1016/j.procbio.2008.10.013.

Prakasham, R.S., Subba Rao, C., Sreenivas Rao, R., Sarma, P.N., 2007. Enhancement of acid amylase production by an isolated *Aspergillus awamori*. J. Appl. Microbiol. 102 (1), 204–211. Available from: https://doi.org/10.1111/j.1365-2672.2006.03058.x.

Prasad, D., Arun, S., Murugesan, M., Padmanaban, S., Satyanarayanan, R.S., Berchmans, S., et al., 2007. Direct electron transfer with yeast cells and construction of a mediatorless microbial fuel cell. Biosens. Bioelectron. 22 (11), 2604–2610. Available from: https://doi.org/10.1016/j.bios.2006.10.028.

Priya, I., Dhar, M.K., Bajaj, B.K., Koul, S., Vakhlu, J., 2016. Cellulolytic activity of *Thermophilic bacilli* isolated from Tattapani Hot Spring Sediment in North West Himalayas. Indian J. Microbiol. 56 (2), 228–231. Available from: https://doi.org/10.1007/s12088-016-0578-4.

Rabaey, K., Boon, N., Siciliano, S.D., Verhaege, M., Verstraete, W., 2004. Biofuel cells select for microbial consortia that self-mediate electron transfer. EurekaMag 70 (9), 5373–5382. Available from: https://doi.org/10.1128/AEM.70.9.5373.

Rabaey, K., Clauwaert, P., Aelterman, P., Verstraete, W., 2005. Tubular microbial fuel cells for efficient electricity generation. Environ. Sci. Technol. 39 (20), 8077–8082. Available from: https://doi.org/10.1021/es050986i.

Rajak, R.C., Banerjee, R., 2016. Enzyme mediated biomass pretreatment and hydrolysis: a biotechnological venture towards bioethanol production. RSC Adv. 6 (66), 61301–61311. Available from: https://doi.org/10.1039/c6ra09541k.

Ramanjaneyulu, G., Sridevi, A., Seshapani, P., Ramya, A., Dileep Kumar, K., Praveen Kumar Reddy, G., et al., 2017. Enhanced production of xylanase by *Fusarium* sp. BVKT R2 and evaluation of its biomass saccharification efficiency. 3 Biotech. 7 (5), 1–17. Available from: https://doi.org/10.1007/s13205-017-0977-1.

Rangabhashiyam, S., Anu, N., Selvaraju, N., 2013. Sequestration of dye from textile industry wastewater using agricultural waste products as adsorbents. J. Environ. Chem. Eng. 1 (4), 629–641. Available from: https://doi.org/10.1016/j.jece.2013.07.014.

Ratnasri, P.V., Lakshmi, B.K.M., Ambika Devi, K., Hemalatha, K.P.J., 2014. Isolation, characterization of *Aspergillus fumigatus* and optimization of cultural conditions for amylase production. Int. J. Res. Eng. Technol. 03 (02), 457–463. Available from: https://doi.org/10.15623/ijret.2014.0302080.

Rehman, H.U., Siddique, N.N., Aman, A., Nawaz, M.A., Baloch, A.H., Qader, S.A.U., 2015. Morphological and molecular based identification of pectinase producing *Bacillus licheniformis* from rotten vegetable. J. Genet. Eng. Biotechnol. 13 (2), 139–144. Available from: https://doi.org/10.1016/j.jgeb.2015.07.004.

Ren, W., Liu, L., Gu, L., Yan, W., Feng, Y.L., Dong, D., et al., 2019. Crystal structure of GH49 dextranase from *Arthrobacter oxidans* KQ11: identification of catalytic base and improvement of thermostability using semirational design based on B-factors. J. Agric. Food Chem. 67 (15), 4355–4366. Available from: https://doi.org/10.1021/acs.jafc.9b01290.

Rivas, S., Vila, C., Alonso, J.L., Santos, V., Parajó, J.C., Leahy, J.J., 2019. Biorefinery processes for the valorization of *Miscanthus* polysaccharides: from constituent sugars to platform chemicals. Ind. Crop. Products 134 (March), 309–317. Available from: https://doi.org/10.1016/j.indcrop.2019.04.005.

Riyadi, F.A., Tahir, A.A., Yusof, N., Sabri, N.S.A., Noor, M.J.M.M., Akhir, F.N.M.D., et al., 2020. Enzymatic and genetic characterization of lignin depolymerization by *Streptomyces* sp. S6 isolated from a tropical environment. Sci. Rep. 10 (1), 1–9. Available from: https://doi.org/10.1038/s41598-020-64817-4.

Roberfroid, M.B., 1997. Health benefits of non-digestible oligosaccharides. In: Kritchevsky, Bonfield (Eds.), Dietary Fiber in Health and Disease. Plenum Press, pp. 211–219.

Robinson, P.K., 2015. Enzymes: principles and biotechnological applications. Essays Biochem. 59, 1–41. Available from: https://doi.org/10.1042/BSE0590001.

Rosmine, E., Sainjan, N.C., Silvester, R., Alikkunju, A., Varghese, S.A., 2017. Statistical optimisation of xylanase production by estuarine *Streptomyces* sp. and its application in clarification of fruit juice. J. Genet. Eng. Biotechnol. 15 (2), 393–401. Available from: https://doi.org/10.1016/j.jgeb.2017.06.001.

Roy, K., Dey, S., Uddin, M.K., Barua, R., Hossain, M.T., 2018. Extracellular pectinase from a novel bacterium *Chryseobacterium indologenes* strain SD and its application in fruit juice clarification. Enzyme Res. 2018, 1–7. Available from: https://doi.org/10.1155/2018/3859752.

Ruqayyah, T.I.D., Jamal, P., Alam, M.Z., Mirghani, M.E.S., 2013. Biodegradation potential and ligninolytic enzyme activity of two locally isolated *Panus tigrinus* strains on selected agro-industrial wastes. J. Environ. Manag. 118, 115–121. Available from: https://doi.org/10.1016/j.jenvman.2013.01.003.

Sachan, S., Iqbal, M.S., Singh, A., 2018. Extracellular lipase from *Pseudomonas aeruginosa* JCM5962(T): isolation, identification, and characterization. Int. Microbiol. 21 (4), 197–205. Available from: https://doi.org/10.1007/s10123-018-0016-z.

Saha, S.P., Ghosh, S., 2014. Optimization of xylanase production by *Penicillium citrinum* xym2 and application in saccharification of agro-residues. Biocatal. Agric. Biotechnol. 3 (4), 188–196. Available from: https://doi.org/10.1016/j.bcab.2014.03.003.

Salvachúa, D., Prieto, A., López-Abelairas, M., Lu-Chau, T., Martínez, Á.T., Martínez, M.J., 2011. Fungal pretreatment: an alternative in second-generation ethanol from wheat straw. Bioresour. Technol. 102 (16), 7500–7506. Available from: https://doi.org/10.1016/j.biortech.2011.05.027.

Salwan, R., Sharma, V., 2020. Fungal lytic polysaccharide monooxygenases in biofuel production from agricultural waste. Recent Developments in Bioenergy Research. Elsevier BV. Available from: https://doi.org/10.1016/b978-0-12-819597-0.00008-8.

Samanta, A.K., Jayapal, N., Jayaram, C., Roy, S., Kolte, A.P., Senani, S., et al., 2015. Xylooligosaccharides as prebiotics from agricultural by-products: production and applications. Bioact. Carbohydr. Diet. Fibre 5 (1), 62–71. Available from: https://doi.org/10.1016/j.bcdf.2014.12.003.

Saratale, R.G., Saratale, G.D., Chang, J.S., Govindwar, S.P., 2011. Bacterial decolorization and degradation of azo dyes: a review. J. Taiwan. Inst. Chem. Eng. 42 (1), 138–157. Available from: https://doi.org/10.1016/j.jtice.2010.06.006.

Sari, Y.W., Syafitri, U., Sanders, J.P.M., Bruins, M.E., 2015. How biomass composition determines protein extractability. Ind. Crop. Products 70, 125–133. Available from: https://doi.org/10.1016/j.indcrop.2015.03.020.

Satapathy, S., Soren, J.P., Mondal, K.C., Srivastava, S., Pradhan, C., Sahoo, S.L., et al., 2021. Industrially relevant pectinase production from *Aspergillus parvisclerotigenus* KX928754 using apple pomace as the promising substrate. J. Taibah Univ. Sci. 15 (1), 347–356. Available from: https://doi.org/10.1080/16583655.2021.1978833.

Sato, K., Chiba, D., Yoshida, S., Takahashi, M., Totani, K., Shida, Y., et al., 2020. Functional analysis of a novel lytic polysaccharide monooxygenase from *Streptomyces griseus* on cellulose and chitin. Int. J. Biol. Macromolecules 164, 2085–2091. Available from: https://doi.org/10.1016/j.ijbiomac.2020.08.015.

Schwab, K., Bader, J., Brokamp, C., Popović, M.K., Bajpai, R., Berovič, M., 2009. Dual feeding strategy for the production of α-amylase by *Bacillus caldolyticus* using complex media. N. Biotechnol. 26 (1–2), 68–74. Available from: https://doi.org/10.1016/j.nbt.2009.04.005.

Scott, K., Murano, C., 2007. Microbial fuel cells utilising carbohydrates. J. Chem. Technol. & Biotechnol. 82 (1), 92–100. Available from: https://doi.org/10.1002/jctb.

Shajahan, S., Moorthy, I.G., Sivakumar, N., Selvakumar, G., 2016. Statistical modeling and optimization of cellulase production by *Bacillus licheniformis* NCIM 5556 isolated from the hot spring, Maharashtra, India. J. King Saud. Univ. - Sci. 29 (3), 302–310. Available from: https://doi.org/10.1016/j.jksus.2016.08.001.

Shi, P., Tian, J., Yuan, T., Liu, X., Huang, H., Bai, Y., et al., 2010. *Paenibacillus* sp. Strain E18 bifunctional xylanase-glucanase with a single catalytic domain. Appl. Environ. Microbiol. 76 (11), 3620–3624. Available from: https://doi.org/10.1128/AEM.00345-10.

Shindhal, T., Rakholiya, P., Varjani, S., Pandey, A., Ngo, H.H., Guo, W., et al., 2021. A critical review on advances in the practices and perspectives for the treatment of dye industry wastewater. Bioengineered 12 (1), 70–87. Available from: https://doi.org/10.1080/21655979.2020.1863034.

Shukla, R.J., Singh, S.P., 2015. Characteristics and thermodynamics of α-amylase from thermophilic actinobacterium, *Laceyella sacchari* TSI-2. Process. Biochem. 50 (12), 2128–2136. Available from: https://doi.org/10.1016/j.procbio.2015.10.013.

Singh, R.S., Singh, T., Pandey, A., 2019. Microbial enzymes-an overview. Biomass, Biofuels, Biochemicals: Advances in Enzyme Technology. Elsevier B.V. Available from: https://doi.org/10.1016/B978-0-444-64114-4.00001-7.

Singh, R., Kumar, M., Mittal, A., Mehta, P.K., 2016. Microbial enzymes: industrial progress in 21st century. 3 Biotech. 6 (2). Available from: https://doi.org/10.1007/s13205-016-0485-8.

Song, J.M., Wei, D.Z., 2010. Production and characterization of cellulases and xylanases of *Cellulosimicrobium cellulans* grown in pretreated and extracted bagasse and minimal nutrient medium M9. Biomass Bioenerg. 34 (12), 1930–1934. Available from: https://doi.org/10.1016/j.biombioe.2010.08.010.

Sridevi, A., Ramanjaneyulu, G., Suvarnalatha Devi, P., 2017. Biobleaching of paper pulp with xylanase produced by *Trichoderma asperellum*. 3 Biotech. 7 (4), 1–9. Available from: https://doi.org/10.1007/s13205-017-0898-z.

Stams, A.J.M., De Bok, F.A.M., Plugge, C.M., Van Eekert, M.H.A., Dolfing, J., Schraa, G., 2006. Exocellular electron transfer in anaerobic microbial communities. Environ. Microbiol. 8 (3), 371–382. Available from: https://doi.org/10.1111/j.1462-2920.2006.00989.x.

Sträter, E., Westbeld, A., Klemm, O., 2010. Pollution in coastal fog at Alto Patache, Northern Chile. Environ. Sci. Pollut. Res. 17 (9), 1563–1573. Available from: https://doi.org/10.1007/s11356-010-0343-x.

Sukumaran, R.K., Singhania, R.R., Pandey, A., 2005. Microbial cellulases—production, applications and challenges. J. Sci. Ind. Res. 64 (11), 832–844.

Sun, J., Bi, Z., Hou, B., Cao, Y.Q., Hu, Y.Y., 2011a. Further treatment of decolorization liquid of azo dye coupled with increased power production using microbial fuel cell equipped with an aerobic biocathode. Water Res. 45 (1), 283–291. Available from: https://doi.org/10.1016/j.watres.2010.07.059.

Sun, J., Chen, Q.J., Zhu, M.J., Wang, H.X., Zhang, G.Q., 2014. An extracellular laccase with antiproliferative activity from the Sanghuang mushroom *Inonotus baumii*. J. Mol. Catal. B: Enzym. 99, 20–25. Available from: https://doi.org/10.1016/j.molcatb.2013.10.004.

Sun, J., Hu, Y.Y., Hou, B., 2011b. Electrochemical characteriztion of the bioanode during simultaneous azo dye decolorization and bioelectricity generation in an air-cathode single chambered microbial fuel cell. Electrochim. Acta 56 (19), 6874–6879. Available from: https://doi.org/10.1016/j.electacta.2011.05.111.

Tabka, M.G., Herpoël-Gimbert, I., Monod, F., Asther, M., Sigoillot, J.C., 2006. Enzymatic saccharification of wheat straw for bioethanol production by a combined cellulase xylanase and feruloyl esterase treatment. Enzyme Microb. Technol. 39 (4), 897–902. Available from: https://doi.org/10.1016/j.enzmictec.2006.01.021.

Tabssum, F., Irfan, M., Shakir, H.A., Qazi, J.I., 2018. RSM based optimization of nutritional conditions for cellulase mediated saccharification by *Bacillus cereus*. J. Biol. Eng. 12 (1), 1–10. Available from: https://doi.org/10.1186/s13036-018-0097-4.

Taipa, M.A., Liebeton, K., Costa, J.V., Cabral, J.M.S., Jaeger, K.E., 1995. Lipase from *Chromobacterium viscosum*: biochemical characterization indicating homology to the lipase from *Pseudomonas glumae*. Biochimica et. Biophysica Acta (BBA)/Lipids Lipid Metab. 1256 (3), 396–402. Available from: https://doi.org/10.1016/0005-2760(95)00052-E.

Tanghe, M., Danneels, B., Last, M., Beerens, K., Stals, I., Desmet, T., 2017. Disulfide bridges as essential elements for the thermostability of lytic polysaccharide monooxygenase LPMO10C from *Streptomyces coelicolor*. Protein Eng. Des. Sel. 30 (5), 401–408. Available from: https://doi.org/10.1093/protein/gzx014.

Tao, H.C., Gao, Z.Y., Ding, H., Xu, N., Wu, W.M., 2012. Recovery of silver from silver(I)-containing solutions in bioelectrochemical reactors. Bioresour. Technol. 111, 92–97. Available from: https://doi.org/10.1016/j.biortech.2012.02.029.

Tchounwou, P.B., Yedjou, C.G., Patlolla, A.K., Sutton, D.J., 2012. Molecular, clinical and environmental toxicicology. Volume 3: environmental toxicology, Molecular, Clinical and Environmental Toxicology, 101. Springer. Available from: https://doi.org/10.1007/978-3-7643-8340-4.

Teixeira, J.A., Gonçalves, D.B., de Queiroz, M.V., De Araújo, E.F., 2011. Improved pectinase production in *Penicillium griseoroseum* recombinant strains. J. Appl. Microbiol. 111 (4), 818–825. Available from: https://doi.org/10.1111/j.1365-2672.2011.05099.x.

Tripathi, A., Upadhyay, R.C., Singh, S., 2012. Extracellular ligninolytic enzymes in *Bjerkandera adusta* and *Lentinus squarrosulus*. Indian. J. Microbiol. 52 (3), 381–387. Available from: https://doi.org/10.1007/s12088-011-0232-0.

Uçkun Kiran, E., Trzcinski, A.P., Ng, W.J., Liu, Y., 2014. Enzyme production from food wastes using a biorefinery concept. Waste Biomass Valoriz. 5 (6), 903–917. Available from: https://doi.org/10.1007/s12649-014-9311-x.

Ugras, S., Uzmez, S., 2016. Characterization of a newly identified lipase from a lipase-producing bacterium. Front. Biol. 11 (4), 323–330. Available from: https://doi.org/10.1007/s11515-016-1409-z.

Ulusoy, I., Dimoglo, A., 2018. Electricity generation in microbial fuel cell systems with *Thiobacillus ferrooxidans* as the cathode microorganism. Int. J. Hydrog. Energy 43 (2), 1171–1178. Available from: https://doi.org/10.1016/j.ijhydene.2017.10.155.

Va'zquez, M.J., Alonso, J.L., Domínguez, H., Parajo, J.C., 2000. Xylooligosaccharides : manufacture and xylooligosaccharides : manufacture and applications. Trends Food Sci. & Technol. 11 (11), 387–393.

Valenzuela, S.V., Ferreres, G., Margalef, G., Pastor, F.I.J., 2017. Fast purification method of functional LPMOs from *Streptomyces ambofaciens* by affinity adsorption. Carbohydr. Res. 448, 205–211. Available from: https://doi.org/10.1016/j.carres.2017.02.004.

Van Der Maarel, M.J.E.C., Van Der Veen, B., Uitdehaag, J.C.M., Leemhuis, H., Dijkhuizen, L., 2002. Properties and applications of starch-converting enzymes of the α-amylase family. J. Biotechnol. 94 (2), 137–155. Available from: https://doi.org/10.1016/S0168-1656(01)00407-2.

Vaseekaran, S., Balakumar, S., Arasaratnam, V., 2010. Isolation and identification of a bacterial strain producing thermostable α-amylase. Tropical Agric. Res. 22 (1), 1–11.

Venkata Mohan, S., Nikhil, G.N., Chiranjeevi, P., Nagendranatha Reddy, C., Rohit, M.V., Kumar, A.N., et al., 2016. Waste biorefinery models towards sustainable circular bioeconomy: critical review and future perspectives. Bioresour. Technol. 215, 2–12. Available from: https://doi.org/10.1016/j.biortech.2016.03.130.

Verma, S., Kumar, R., Kumar, P., Sharma, D., Gahlot, H., Sharma, P.K., et al., 2020. Cloning, characterization, and structural modeling of an extremophilic bacterial lipase isolated from saline habitats of the Thar Desert. Appl. Biochem. Biotechnol. 192 (2), 557–572. Available from: https://doi.org/10.1007/s12010-020-03329-3.

Vishnupriya, B., Sundaramoorthi, C., Kalaivani, M., Selvam, K., 2010. Production of lipase from *Streptomyces griseus* and evaluation of bioparameters. Int. J. ChemTech Res. 2 (3), 1380–1383.

Viswanath, B., Rajesh, B., Janardhan, A., Kumar, A.P., Narasimha, G., 2014. Fungal laccases and their applications in bioremediation. Enzyme Res. 2014, 1–21. Available from: https://doi.org/10.1155/2014/163242.

Voragen, A.G.J., 1998. Technological aspects of functional food-related carbohydrates. Trends Food Sci. Technol. 9 (8–9), 328–335. Available from: https://doi.org/10.1016/S0924-2244(98)00059-4.

Walia, A., Guleria, S., Mehta, P., Chauhan, A., Parkash, J., 2017. Microbial xylanases and their industrial application in pulp and paper biobleaching: a review. 3 Biotech. 7 (1), 1–12. Available from: https://doi.org/10.1007/s13205-016-0584-6.

Walia, A., Mehta, P., Guleria, S., Shirkot, C.K., 2015. Improvement for enhanced xylanase production by *Cellulosimicrobium cellulans* CKMX1 using central composite design of response surface methodology. 3 Biotech. 5 (6), 1053–1066. Available from: https://doi.org/10.1007/s13205-015-0309-2.

Walter, X.A., Greenman, J., Taylor, B., Ieropoulos, I.A., 2015. Microbial fuel cells continuously fuelled by untreated fresh algal biomass. Algal Res. 11, 103–107. Available from: https://doi.org/10.1016/j.algal.2015.06.003.

Wan, C., Li, Y., 2011. Effectiveness of microbial pretreatment by *Ceriporiopsis subvermispora* on different biomass feedstocks. Bioresour. Technol. 102 (16), 7507–7512. Available from: https://doi.org/10.1016/j.biortech.2011.05.026.

Wang, G., Huang, L., Zhang, Y., 2008. Cathodic reduction of hexavalent chromium [Cr(VI)] coupled with electricity generation in microbial fuel cells. Biotechnol. Lett. 30 (11), 1959–1966. Available from: https://doi.org/10.1007/s10529-008-9792-4.

Wang, H., Ren, Z.J., 2013. A comprehensive review of microbial electrochemical systems as a platform technology. Biotechnol. Adv. 31 (8), 1796–1807. Available from: https://doi.org/10.1016/j.biotechadv.2013.10.001.

Wang, J., Liu, Y., Guo, X., Dong, B., Cao, Y., 2019a. High-level expression of lipase from *Galactomyces geotrichum* mafic-0601 by codon optimization in *Pichia pastoris* and its application in hydrolysis of various oils. 3 Biotech. 9 (10), 1–10. Available from: https://doi.org/10.1007/s13205-019-1891-5.

Wang, J., Wu, Y., Gong, Y., Yu, S., Liu, G., 2015. Enhancing xylanase production in the thermophilic fungus *Myceliophthora thermophila* by homologous overexpression of Mtxyr1. J. Ind. Microbiol. Biotechnol. 42 (9), 1233–1241. Available from: https://doi.org/10.1007/s10295-015-1628-3.

Wang, T.Y., Chen, H.L., Lu, M.Y.J., Chen, Y.C., Sung, H.M., Mao, C.T., et al., 2011. Functional characterization of cellulases identified from the cow rumen fungus *Neocallimastix patriciarum* W5 by transcriptomic and secretomic analyses. Biotechnol. Biofuels 4, 1–16. Available from: https://doi.org/10.1186/1754-6834-4-24.

Wang, Y.chuan, Zhao, N., Ma, J.wen, Liu, J., Yan, Q.juan, Jiang, Z.qiang, 2019b. High-level expression of a novel α-amylase from *Thermomyces dupontii* in *Pichia pastoris* and its application in maltose syrup production. Int. J. Biol. Macromol. 127, 683–692. Available from: https://doi.org/10.1016/j.ijbiomac.2019.01.162.

Watson, V., 2009. *Shewanella oneidensis* MR-1 compared to mixed cultures for electricity production in four different microbial fuel cell configurations. May, 53. https://etda.libraries.psu.edu/paper/9056/5117

Wu, M., Beckham, G.T., Larsson, A.M., Ishida, T., Kim, S., Payne, C.M., et al., 2013. Crystal structure and computational characterization of the lytic polysaccharide monooxygenase GH61D from the Basidiomycota fungus *Phanerochaete chrysosporium*. J. Biol. Chem. 288 (18), 12828–12839. Available from: https://doi.org/10.1074/jbc.M113.459396.

Xenakis, A., Zoumpanioti, M., Stamatis, H., 2016. Enzymatic reactions in structured surfactant-free microemulsions. Curr. Opinion Colloid. Interface Sci. 22, 41–45. Available from: https://doi.org/10.1016/j.cocis.2016.02.009.

Xing, D., Zuo, Y., Cheng, S., Regan, J.M., Logan, B.E., 2008. Electricity generation by *Rhodopseudomonas palustris* DX-1. Environ. Sci. Technol. 42 (11), 4146–4151. Available from: https://doi.org/10.1021/es800312v.

Yadav, M., Singh, A., Balan, V., Pareek, N., Vivekanand, V., 2019. Biological treatment of lignocellulosic biomass by *Chaetomium globosporum*: process derivation and improved biogas production. Int. J. Biol. Macromol. 128, 176–183. Available from: https://doi.org/10.1016/j.ijbiomac.2019.01.118.

Yadav, P., Maharjan, J., Korpole, S., Prasad, G.S., Sahni, G., Bhattarai, T., et al., 2018. Production, purification, and characterization of thermostable alkaline xylanase from *Anoxybacillus kamchatkensis* NASTPD13. Front. Bioeng. Biotechnol. 6 (MAY). Available from: https://doi.org/10.3389/fbioe.2018.00065.

Yamagishi, K., Kimura, T., Watanabe, T., 2011. Treatment of rice straw with selected *Cyathus stercoreus* strains to improve enzymatic saccharification. Bioresour. Technol. 102 (13), 6937–6943. Available from: https://doi.org/10.1016/j.biortech.2011.04.021.

Yang, Y., Li, J., Liu, X., Pan, X., Hou, J., Ran, C., et al., 2017. Improving extracellular production of *Serratia marcescens* lytic polysaccharide monooxygenase CBP21 and *Aeromonas veronii* B565 chitinase Chi92 in *Escherichia coli* and their synergism. AMB. Express 7 (1). Available from: https://doi.org/10.1186/s13568-017-0470-6.

Yegin, S., 2017. Xylanase production by *Aureobasidium pullulans* on globe artichoke stem: Bioprocess optimization, enzyme characterization and application in saccharification of lignocellulosic biomass. Prep. Biochem. Biotechnol. 47 (5), 441–449. Available from: https://doi.org/10.1080/10826068.2016.1224245.

You, S., Zhao, Q., Zhang, J., Jiang, J., Zhao, S., 2006. A microbial fuel cell using permanganate as the cathodic electron acceptor. J. Power Sources 162 (2 SPEC. ISS.), 1409–1415. Available from: https://doi.org/10.1016/j.jpowsour.2006.07.063.

Zarattini, M., Corso, M., Kadowaki, M.A., Monclaro, A., Magri, S., Milanese, I., et al., 2021. LPMO-oxidized cellulose oligosaccharides evoke immunity in *Arabidopsis* conferring resistance towards necrotrophic fungus *B. cinerea*. Commun. Biol. 4 (1), 1–13. Available from: https://doi.org/10.1038/s42003-021-02226-7.

Zavarzina, A.G., Lisov, A.V., Leontievsky, A.A., 2018. The role of ligninolytic enzymes laccase and a versatile peroxidase of the white-rot fungus *Lentinus tigrinus* in biotransformation of soil humic matter: comparative in vivo study. J. Geophys. Res. Biogeosci. 123 (9), 2727–2742. Available from: https://doi.org/10.1029/2017JG004309.

Zebda, A., Gondran, C., Cinquin, P., Cosnier, S., 2012. Glucose biofuel cell construction based on enzyme, graphite particle and redox mediator compression. Sens. Actuators, B: Chem. 173, 760–764. Available from: https://doi.org/10.1016/j.snb.2012.07.089.

Zhang, H., Zhao, Y., Cao, H., Mou, G., Yin, H., 2015. Expression and characterization of a lytic polysaccharide monooxygenase from *Bacillus thuringiensis*. Int. J. Biol. Macromol. 79, 72–75. Available from: https://doi.org/10.1016/j.ijbiomac.2015.04.054.

Zhang, T., Cui, C., Chen, S., Ai, X., Yang, H., Shen, P., et al., 2006. A novel mediatorless microbial fuel cell based on direct biocatalysis of *Escherichia coli*. Chem. Commun. 21, 2257–2259. Available from: https://doi.org/10.1039/b600876c.

Zhao, J., Mou, Y., Shan, T., Li, Y., Zhou, L., Wang, M., et al., 2010. Antimicrobial metabolites from the endophytic fungus *Pichia guilliermondii* Isolated from *Paris polyphylla var. yunnanensis*. Molecules 15 (11), 7961–7970. Available from: https://doi.org/10.3390/molecules15117961.

Zheng, W., Zhou, H.M., Zheng, Y.F., Wang, N., 2008. A comparative study on electrochemistry of laccase at two kinds of carbon nanotubes and its application for biofuel cell. Chem. Phys. Lett. 457 (4–6), 381–385. Available from: https://doi.org/10.1016/j.cplett.2008.04.047.

Zheng, X., Nirmalakhandan, N., 2010. Cattle wastes as substrates for bioelectricity production via microbial fuel cells. Biotechnol. Lett. 32 (12), 1809–1814. Available from: https://doi.org/10.1007/s10529-010-0360-3.

Zhu, M., Zhang, G., Meng, L., Wang, H., Gao, K., Ng, T., 2016. Purification and characterization of a white laccase with pronounced dye decolorizing ability and HIV-1 reverse transcriptase inhibitory activity from *Lepista nuda*. Molecules 21 (4), 1–16. Available from: https://doi.org/10.3390/molecules21040415.

Ziessel, R., Ulrich, G., Haefele, A., Harriman, A., 2013. An artificial light-harvesting array constructed from multiple Bodipy dyes. J. Am. Chem. Soc. 135 (30), 11330–11344. Available from: https://doi.org/10.1021/ja4049306.

Zuo, Y., Maness, P.C., Logan, B.E., 2006. Electricity production from steam-exploded corn stover biomass. Energy Fuels 20 (4), 1716–1721. Available from: https://doi.org/10.1021/ef060033l.

Zuo, Y., Xing, D., Regan, J.M., Logan, B.E., 2008. Isolation of the exoelectrogenic bacterium *Ochrobactrum anthropi* YZ-1 by using a U-tube microbial fuel cell. Appl. Environ. Microbiol. 74 (10), 3130–3137. Available from: https://doi.org/10.1128/AEM.02732-07.

Chapter 26

Discovery of untapped nonculturable microbes for exploring novel industrial enzymes based on advanced next-generation metagenomic approach

Shivangi Mudaliar[1,*], Bikash Kumar[1,2,*], Komal Agrawal[1,3] and Pradeep Verma[1]

[1]*Bioprocess and Bioenergy Laboratory (BPEL), Department of Microbiology, Central University of Rajasthan, Bandarsindri, Kishangarh, Ajmer, Rajasthan, India,* [2]*Department of Biosciences and Bioengineering, Indian Institute of Technology Guwahati, Guwahati, Assam, India,* [3]*Department of Microbiology, School of Bioengineering and Biosciences, Lovely Professional University, Phagwara, Punjab, India*

26.1 Introduction

In 1998 the concept of metagenomics was introduced to have a better idea of the microbial diversities that cannot be cultured. It is very well known that the microbial diversities we are studying at present are just 0.01% of what we can have. If we could identify those untapped nonculturable microbes, many folds of information related to microbial diversity could be revealed. These unculturable microbes represent an unlimited resource for developing novel products (Schmeisser et al., 2007). Also, microbial diversity analysis could provide us with all necessary information about their population and functions during changing environments. There are two fundamental approaches to assessing the functional diversity of microbes—culture-dependent and culture-independent methods. In culture-dependent methods, the characterization of microorganisms for their biochemical, morphological,

* Both authors have contributed equally.

Biotechnology of Microbial Enzymes. DOI: https://doi.org/10.1016/B978-0-443-19059-9.00020-7
© 2023 Elsevier Inc. All rights reserved.

and physiological properties is done by cultivating isolates in the laboratory, while in the culture-independent technique, with the help of sequencing, the study is done where no isolation of microbes and culturing are needed (Srivastava et al., 2018). They are being studied under the heading of metagenomics. Metagenomics enables us to explore these uncultured microbes with a better understanding. It is widely used to exploit these untapped microbes to reveal their hidden potential (Bilal et al., 2018).

Commercial sectors, pharmaceuticals, and food sectors have an essential role in microbial enzymes in their production. This is time-consuming and laborious and requires large-scale production, formulations, and purification with minimum effect on the environment. The activity and stability of microbial enzymes are more than that of plants and animal sources, making them suitable for various purposes (Tatta et al., 2022). The only possible reason that made us opt for this metagenomic approach is the difficulty of culturing multiple microorganisms in the laboratory (Kamble and Vavilala, 2018). This exploitation of microbes will help explore some novel enzymes and compounds that can be used to develop medicine and diagnostic tools. With these discoveries, this approach could impact the economic success of industries as well. The genetic material for screening can be extracted from various habitats such as soil, seawater, wastewater, and the human gut. Then they could be analyzed to know the relationship with the community, how they function together, and their survival. This helps us to study the whole community in a single run. Unlike the conventional approach, this method has three basic steps: first is the extraction of samples from the environment. With the help of transformation metagenomic library is created, and finally, the screening of desired clones is done to identify the desired product (Ahmad et al., 2019). In the current chapter, we will discuss the need, problems associated with nonculturable microbes, and how this could be overcome with the help of a metagenomic approach.

26.2 Need for nonculturable microbe study

Earth is a massive pool of biodiversity, but less than 1% can be cultured from this existing pool, and the rest are unculturable. The 99% of unculturable microbes can be an excellent source of novel enzymes and can be used in various industrial applications, especially in enzymatic sectors. In this developing world, the need and demand for resources are increasing with the increasing population (Bilal et al., 2018). To meet the demands, the culturable microbes have been exploited a lot with the help of biotechnological tools (Kirubakaran et al., 2020). But these activities lead to the deterioration of natural resources and the environment. Industries, pharmaceutical, food, agriculture, and energy sectors depend on enzymes for their production. The microbial enzyme system compensates this growing demand for enzymes. But this system still faces problems such as the unavailability of enzymes that could work well in extreme conditions and the absence of potential enzyme sources. These problems and increasing enzyme demand can be solved by culturing these nonculturable microbes as they could be a great source of potential compounds. These are found in almost every part of the Earth, from extreme to moderate environments, thus offering great potential in novel enzyme production. The enzymes extracted from extremophilic microbes could be thermostable with broad application. Also, studies on these microbes could better understand the origin of earliest life because many preserved fossils were found from these extremophiles (Debnath et al., 2019).

The unculturable microbes and their study will help in revealing the relationship between the individual and their environment. These microbes are not unculturable, but they just need some special conditions for their growth and development. Cultivating these unculturable microbes would be of great importance to the whole community and the entire ecosystem (Montgomery, 2020). Unculturable microbes could be a great source of some unknown and novel compounds that will remain as such if not cultured. They offer an excellent source for some new unidentified enzymes of industrial application. The need to endorse white biotechnology that requires the production of the novel enzyme has drawn attention to these nonculturable microbes and their potential. In this, metagenomics has appeared as an alternative to conventional screening methods and helped reveal the genetic structure of the whole microbial community and the potential enzymes isolated from them. Enzymes have become an environment-friendly and sustainable approach in industrial sectors. The present need is to dig more and more microbial enzymes to develop enzymatic industries. This combination of metagenomics and nonculturable microorganism can elevate the finding of superior novel microbial enzymes that can tolerate harsh conditions with several industrial applications (Ahmad et al., 2019).

26.3 Problems associated with nonculturable microbial studies

It is still impossible to identify the diversified species of bacteria and archaea using conventional cultivation methods, thus remaining unculturable. Many new dimensions and novel compounds could be produced from these unculturable microbes. Several culture-independent techniques could be used to characterize polymerase chain reaction (PCR), cloning, and sequencing of genes. With the help of these techniques, many novel products have been identified.

But still, the problem remains there, that is, the uncultivability of microbes. Many bacterial species have not been identified or overlooked in the cultural analyses because of their slow growth. Also, some bacteria have resilient growth on conventional media and are very specific in their requirements such as pH conditions, temperatures, or salt content. In mixed cultures, competition for resources could be a possible reason for resistance to growth.

Other than this, some bacterium produces bacteriocin that inhibits the growth of other bacteria in the culture. The struggle with these unculturable microbes could be because of the lack of metabolic signals and interaction with an unrelated environment. Many efforts are being made to overcome all these problems by devising culturing methods for these unculturable microbes (Vartoukian et al., 2010). Many microorganisms cannot grow on regularly used laboratory media, and these microorganisms are viable. Still, they are not culturable, and they are commonly known as VBNC (viable but nonculturable cells). VBNC is considered an adaptive state that is acquired by the organism to survive under extreme conditions such as high temperature, high salinity, nutrient starvation, pH change, osmotic stress, heavy metals, UV radiations, or chlorination. VBNC cannot be cultured by conventional methods, they have metabolic activity, but it is reduced. Many of the microorganisms are in the VBNC state because of the dominancy of the surrounding environment, and hence, they remain unculturable.

26.3.1 Relationship with coexisting microbes

In the environment, every single organism interacts with its surroundings. Interaction is very important for their survival. Not a single organism can exist alone, and so do these unculturable microbes also keeps interacting with their neighboring species. The interaction could be intra- or interspecific. All unculturable microbes in aquatic, terrestrial, and extremophiles show interactions with the nearby like *Planctomycetes* DNA were found associated with some other bacterial species suggesting a symbiotic (both the bacteria are getting benefits from each other) cooperation between them. Also, bacterial groups, for example, *Proteobacteria* and *Bacteroidetes* are found in association and are dependent on each other.

This also shows that the growth of these microbes is dependent on each other, and in the absence of one, the other growth could be affected. And thus their culture in pure culture would be difficult, and this could be a possible reason for their uncultivability (Kaboré et al., 2020).

26.4 Culture-independent molecular-based methods

Many attempts have been made to culture these unculturable microorganisms. Many methods have recovered unculturable microbes from populated habitats such as soil or aquatic systems (Vartoukian et al., 2010). Most bacteria are unculturable and remain in the VBNC stage. They urgently need to study their molecular and functional diversity as they are unculturable. This strongly suggests having a culture-independent technique for the characterization of these untapped microorganisms to investigate microbial communities. To obtain information, many molecular approaches are being implemented along with the omics strategies such as metagenomic, metaproteomic, and metabolomics (Gnaneswar Gude, 2015). The steps for the culture-independent methodology are discussed in the following sections.

26.4.1 Isolation of sample DNA

The isolation can be done by two methods: direct and indirect extraction. In the direct extraction method, detergents and enzymes are used to lyse the cell of the samples directly. The destructive nature of these solvents results in the inappropriate size of DNA and poor yield, and thus they are not a good fit for the construction of the metagenomic library. Unlike the direct method, physical separation of cells is done in the indirect extraction method, followed by lysis and subsequent steps. Despite drawbacks, mostly direct extraction methods are used because they have a good recovery rate, 10–100 times better than the indirect extraction method (Ahmad et al., 2019).

26.4.2 Metagenomic library construction

For creating a metagenomic library, an appropriate host-vector system is needed. The vector is designed according to the need of the research. For efficient cloning and screening of genes, selecting a proper host strain is very important. The most common and preferred hosts for library construction in *Escherichia coli*, followed by *Pseudomonas* and *Streptomyces* sp. (Ahmad et al., 2019). *E. coli* is a great host strain for cloning purposes for the enzymes extracted from Gram-negative bacteria but is not suitable for Gram-positive GC-rich bacteria (Kimura, 2018). The screening of a metagenomic library is technically a difficult task. A few commonly used screening methods are summarized in the following sections.

26.4.2.1 Sequence- and function-based screening

Sequence-based screening is simply based on the PCR approach, where the desired gene is detected using a DNA probe (Kimura, 2018). At the same time, the function-based screening is based on differentiating the biochemical activities of biocatalyzed reactions. Based on different functions, various new genes are identified (Kimura, 2018).

26.4.2.2 Substrate-induced gene-expression screening

Both the abovementioned techniques are time-consuming and laborious; hence, the technique SIGEX was developed to overcome these limitations. In this, the expression of green fluorescent protein (GFP) is induced when clones produce specific types of secondary metabolites, and with the help of fluorescent-activated cell sorting (FACS), the positive clones are separated. The synergy of the culture-independent approach and the omics provides an opportunity to develop novel products. The discovery of the novel enzymes is the latest example of this synergy and the successful implementation of omics. Not only based on DNA, but at every step of metabolism, one can identify the microbe. The different omics tools are metagenomics (study based on DNA), metatranscriptomics (based on RNA), metaproteomics (based on protein), and metabolomics (based on metabolites).

26.4.3 Metagenomics

Also termed "environmental genomics" and "community genomics," metagenomics helps in elucidating the genomic materials of microbes. This method needs live microbes only, and no culture is needed. The concept of metagenomics goes a step further, focusing on taxonomic diversity by creating a metagenomic library that links phylogenetic profiles and emphasizes the functional profile of the uncultivable microorganisms. Microbial samples from a particular habitat are taken directly and are cloned for genomic analysis by metagenomics. A library is created to provide all the vital information about the microbes and their capabilities. All this information could be helpful in the further broad application of those particular microbes. These organisms are a rich source of novel compounds that could be of industrial importance, and their production could be enhanced by using this approach (Table 26.1).

26.4.4 Metatranscriptomics

A new subdiscipline within the metagenomic thoroughly analyzes the transcriptomes. This method identifies the transcriptionally active microbial population and transcribed genes (Zarraonaindia et al., 2013). The active metabolic pathways are identified with the help of the gene expression, and the abundance of functional genesis analyzed (Srivastava et al., 2018) (Table 26.1).

26.4.5 Metaproteomic

It is also called community proteogenomic. This technique determines the post-transcriptionally regulated and translated proteins. At a specific point in time, the characterization of protein takes place in the metaproteomic. Tracking the functional genes as well as those proteins which are formed under stress conditions could help in identifying the microbial ecology. In combination with mass spectrometry (Ms), one-dimensional and two-dimensional gel electrophoresis can be used for the separation of protein. This approach could be used with soil or groundwater to extract protein and can be analyzed (Srivastava et al., 2018). All these techniques enable us to extract information at different molecular stages, and the synergistic approach can help us better understand the microbial environment. Overall, these techniques are proving to be helpful in the discovery of novel compounds for industrial application and could be sustainable (Srivastava et al., 2018).

TABLE 26.1 List of culture-independent genomic analysis tools and enzymes discovered.

Methods	Approaches	Enzymes identified/ discovered	Microorganism	Substrates	Temperature	References
Metagenomics	Next-generation sequencing	Cellulase (GH9)	*Thermobifida fusca*	Phosphoric acid swollen cellulose	100°C	Stepnov et al. (2019)
		Pectinase	–	Polygalacturonic acid	60°C	Singh et al. (2012)
		Tannase	*Clostridium butyricum*	Methyl gallate, propyl gallate, hexyl gallate	37°C	Ristinmaa et al. (2022)
		Protease	*Psychrobacter* sp.	Tributyrin	4°C–18°C	Perfumo et al. (2020)
Metatranscriptomics	RNA-sequencing	Cellulase	–	Carboxymethyl cellulose	40°C–50°C	He et al. (2019)
		Xylanase (GH11)	*Trichoderma reesei*	–	50°C	Yi et al. (2021)
Metaproteomic	LC-Ms, MALDI	Lipase	–	*p*-Nitrophenyl butyrate	55°C–60°C	Sander et al. (2021)
		Chitosanase	*Bacillus cereus*	Chitin, chitosan	25°C–60°C	Liang et al. (2014)

26.5 Different approaches for metagenomic analysis of unculturable microbes

The conventional needs for isolation and cultivation are being bypassed with the help of the metagenomic approach. The samples are directly taken for genomic DNA isolation and can be used for studying the hidden truths. There are two types of approaches for the analysis, which are discussed in the following sections.

26.5.1 Sequence-based screening

In sequence-based screening, primers and probes are used. The important bioactive compounds or some known genes coding for enzymes are used for designing specific primers and probes. PCR amplification followed by sequencing is done for these genes and subsequently detected. For cross-verification, these genes can be cloned into various expression systems (Datta et al., 2020). This strategy got the attention when an uncultivated gamma-proteobacterium, a new variant of a light-driven proton pump, was isolated by Kimura (2018). Some computational approaches are also required when the bulky metagenomic sequence is taken with these laboratory methods. With the highly complex microbial community, a new sequencing approach (454-pyrosequencing) should also be used (Madhavan and Sindhu, 2017). This method uses two strategies to sequence the target genes: PCR-based sequencing helps in reconstructing the route of the evolution of the desired compound with the change in the ecosystem, while the hybridization-based sequencing reads the intensity of hybridization signals and positive expression levels of the desired transcript can be acquired (Mahapatra et al., 2020) (Fig. 26.1).

26.5.2 Function-based screening

The sequencing and biochemical analysis characterize the isolated clones with the desired trait. It can identify both novel and previously identified enzymes by discriminating their biochemical activities. The gene clusters essential for the trait are secured. The clones with the desired trait are identified in this analysis, followed by sequencing and biochemical analysis. And with the help of fluorescence microscopy, one can identify the metagenomic clones as they will produce quorum sensing inducers, and cells will produce a GFP (Kimura, 2006). There are different function-based screening strategies as follows: (1) direct detection of phenotypes, (2) heterologous compatibility, and (3) substrate-induced gene expression.

26.5.2.1 Direct detection of phenotypes

In this method, the detection of biocatalyst can be done with the use of dyes and the substrate of targeted enzymes. These enzymes are attached to chromophores which help in visual detection. This method is best for the

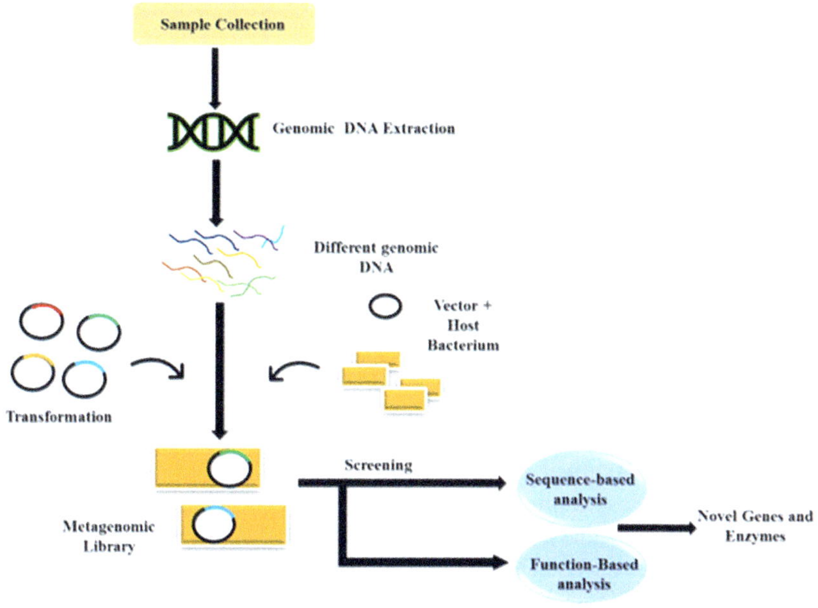

FIGURE 26.1 A diagrammatical representation of metagenomic analysis of DNA samples.

identification of novel genes that encode novel enzymes for industrial applications (Madhavan and Sindhu, 2017).

26.5.2.2 Heterologous compatibility

This method uses a complemented host and requires the presence of target genes for selective growth (Madhavan and Sindhu, 2017). The compatibility between cloned sequence and the host cells is checked to determine the success rate of heterologous expression. The major determining factors in heterologous compatibility are phylogenetic origin, promoters in gene library, sequence composition, resistance mechanism, and toxicity of products formed (Mahapatra et al., 2020).

26.5.2.3 Substrate-induced gene expression

It is a new approach to novel gene cloning and expression. A library is created with a cloning host, and restriction digested metagenomic clones, and an expression vector in a culture medium (Madhavan and Sindhu, 2017). In this method, a reporter gene is found, that is, GFP. When the clones produce some specific types of metabolites, the expression of GFP is induced. The positive clones can be separated from the others with the help of FACS, which helps in providing rapid screening of the library (Mahapatra et al., 2020). A modified version of SIGEX is PIGEX, that is, product-induced gene expression developed by Uchiyama and Miyazaki. A transcriptional activator was introduced that is highly sensitive to product detection. A high-throughput screening was developed by Williamson named METREX to screen the small active molecules produced by metagenomic clones. The advantage of this approach is that there is no need for sequence data to identify the biological activity of microbes (Madhavan and Sindhu, 2017).

26.6 Next-generation sequencing and metagenomics

Next-generation sequencing (NGS) has come up with considerable abilities in the last few years (Verma and Gazara, 2021). NGS is a high-throughput sequencing method where billions of nucleic acid fragments can be independently and simultaneously sequenced (Slatko et al., 2018). NGS involves splitting the genome of organisms into several small fragments that generate small reads/sequences ranging from a hundred to thousands of bases in length. The computational approaches are used to assemble these short fragments into a single genome by overlapping sequence reads stitched together to form a long sequence called a contig. These contigs often consist of gaps aligned to a reference database resulting in the identification of the organism (Henson et al., 2012, Lee, 2019) (Fig. 26.2). Although the generation of

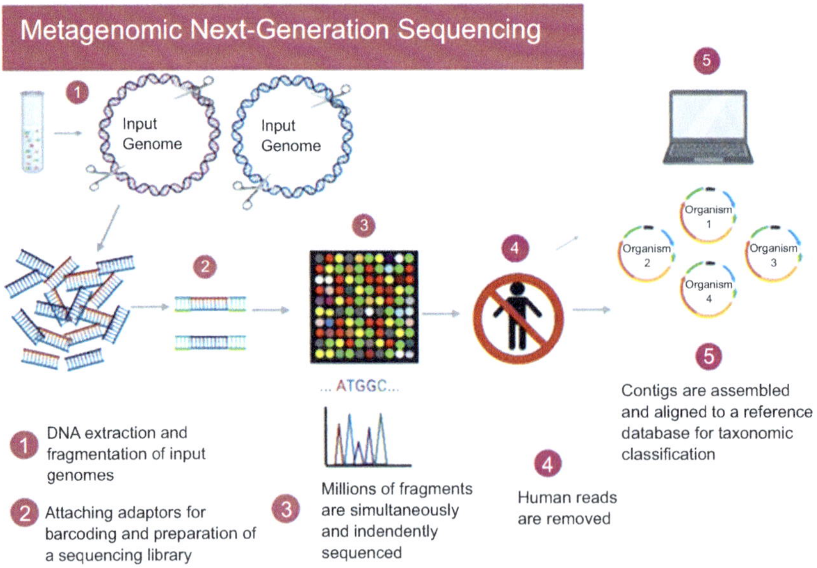

FIGURE 26.2 Workflow for metagenomic next-generation sequencing. (1) Extraction and fragmentation of genomic DNA. (2) Attaching adapters for barcoding and preparation of library sequences. (3) Sequencing of these short fragments of DNA independently and simultaneously. (4) Removal of the human-related DNA sequence read. (5) These assemblages of contigs of long DNA stretches from shorter, overlapping sequences. Alignment of these contigs to a reference database for taxonomic classification (Lee, 2019, https://asm.org/Articles/2019/November/Metagenomic-Next-Generation-Sequencing-How-Does-It.)

longer high-fidelity sequences of reads/contigs can be considered an ideal method for sequencing, platforms that result in shorter reads are usually less costly and can overlap smaller reads, making them more accurate (Lee, 2019). The NGS is in massive contrast to the conventional sequencing method, such as Sanger sequencing (dideoxynucleotide chain termination sequencing), which processes one nucleotide sequence per reaction. For example, NGS helps in sequencing a bacterial genome in a very short time compared to several years by the conventional approach (Gupta and Verma, 2019; Kulski, 2016; Lee, 2019; Liu et al., 2012; Pareek et al., 2011).

Metagenomic NGS can help identify the mixed population of microbes and in what proportions they exist. The ability of NGS to simultaneously identify nucleic acids from multiple taxa makes it a powerful platform for studying unculturable microbes in an easy and fast manner (Forbes et al., 2017). Different NGS platforms are developed by leading companies such as Illumina, Oxford Nanopore, Ion torrent, Pacific Biosciences, Roche 454, Beijing Genomics Institute/BGI, etc. Some of the platforms and their efficiency are tabulated in Table 26.2.

NGS, together with metagenomics, can explore all those uncultivable novel compounds that could be exploited for enhancing production in the industries and other enzymatic-based systems (Wilson and Piel, 2013). The development of NGS has revolutionized the exploration of unculturable microorganisms from different habitats (Dubey et al., 2022). Many potential novel enzymes of industrial importance have been discovered from various environmental samples such as soil, air, insects, extreme environment, and animal gut with the help of metagenomics (Datta et al., 2020). With the advancements in sequencing techniques, the metagenomic industry has also flourished. The replacement of single genome assembly models with new genomic pipelines has helped in extracting a lot of information from those untapped microorganisms (Lapidus and Korobeynikov, 2021). The conventional techniques were bulky, time-consuming, and laborious, and high-end portable computers were required for monitoring.

With the advancements in this technique, the cost has been brought down significantly. In environmental and pathogen diagnosis, this metagenomic NGS (mNGS) approach is feasible. Multiple databases such as NCycDB, GOLD, HVPC, and Meta Biome have been introduced (Datta et al., 2020). That has helped in fast and high-throughput screening of unculturable microbes.

26.6.1 Benefits of metagenomic next-generation sequencing

The unbiased hypothesis-free diagnostic method is the biggest strength of the mNGS, unlike its conventional counterpart where it depends on primers for targeted polymerase chain amplification and their identification (Lee, 2019). The universal primers or broad range primers are considered in many cases for conventional sequencing. Still, these primers could not cover a broad range of metagenomes as they are dependent on the specific primers of

TABLE 26.2 Comparison of some of the next-generation sequencing platforms (https://genohub.com/ngs-instrument-guide/).

Platform	Instrument	Unit	Reads/unit[a]	Max read length	Read type	Error type	Highlights
Illumina	HiSeq X	Lane	375,000,000	2×150 bp	PE	Substitution	Greatest throughput and number of reads.
Illumina	NovaSeq-S4	Lane	2,500,000,000	1×300 bp, 2×150 bp	SR & PE	Substitution	NovaSeq is Illumina's latest high-output instrument designed for research labs that can't afford the capital costs of the HiSeq X without having any application restrictions.
PacBio	PacBio Sequel	SMRT Cell	187,500	20,000 bp	SR	Indel	First desktop instrument that delivers $\sim 7 \times$ more reads as compared to its predecessor.
Oxford Nanopore	PromethION with kit 12 chemistry	Flow cell	None	Typically 6–20 Kbp	SR	Indel & substitution	Largest device that can run and analyze data in real time for 1–48 independent flow cells.Best suited for high-throughput sequencing for large whole genomes and transcriptomes or population-scale sequencing.
Ion Torrent	Proton I Chip	Chip	60,000,000	200 bp	SR	Indel	Ion instrument with highest throughput. Compared to MiSeq, it has greater number of reads but shorter read lengths.
Roche 454	GS FLX 1 PTP	1 PTP	700,000	450 bp	SR	Indel	Long read lengths make it ideal for sequencing of small genomes.
BGI	DNBSEQ-T7	Lane	5,000,000,000	2×150 bp	PE	–	–

[a]*PE, Pair end; SR, single-read sequencing technology.*

conserved 16s ribosomal RNA and internal transcribed spacers for amplification of nucleic acids that can be later classified into bacteria, archaea, or fungi using bioinformatics tools (Peng et al., 2021; Clarridge 2004).

While diagnosing a microbial consortium and polymicrobial infections via molecular identification, universal primers often pose problems (Clarridge, 2004; Maher-Sturgess et al., 2008; Raja et al., 2017). For example, using 16s sequencing for polymicrobial infections, multiple base calls can be observed per nucleotide resulting in a mixed nucleotide chromatogram that is very difficult to interpret (Kommedal et al., 2008). Several computational methods are available for the prediction of the identified organism, but these are not in standard use throughout the laboratories, thus supporting NGS of 16s polymicrobial sample (Lee, 2019; Weinstock et al., 2016).

26.7 Application of unculturable microbes and significance of next-generation metagenomic approaches

Sustainable resources are depleting and increasing the demand for basic resources, and an attempt to solve this problem is being made with the help of a metagenomic approach. There are various environmental problems for which we need sustainable solutions that have increased the demand for enzymes in industries. A diversified application of enzymes discovered by metagenomic analysis in the food and pharmaceutical industry is found. There are numerous applications in which metagenomics is used, such as phylogenetic assortment, development of an effective bioremediation system, identifying the role of the microbial community in agriculture, and analysis of the human microbiome (Kirubakaran et al., 2020). Some of the metagenomic applications are discussed in the following sections and are represented diagrammatically in Fig. 26.3.

FIGURE 26.3 A diagrammatical representation of applications of metagenomics in various sectors.

26.7.1 Agricultural applications

With the growing population, it is difficult to meet the demand for food day by day. Another problem that makes it difficult is the pathogen attack, especially in Asia and Africa regions. The attack causes a huge loss of crops, and the pathogen is very well adapting themselves, so they develop resistance against the insecticides quickly. In this, the metagenomic approach helped in monitoring these plant pathogens. It will also help in identifying the diversity of the viruses, thus their early detection is possible. The metagenomic method has helped in identifying unculturable viruses and also revealed data on how they affect plants' pathogen routes of transmission and their habitat. Also, enzymes such as cellulase play an important role in agriculture by increasing crop yield and soil fertility and providing defense against plant pathogens. With the help of microbes such as *Trichoderma* and *Aspergillus*, the soil fertility is enhanced (Tatta et al., 2022).

26.7.2 Clinical diagnosis

All the diagnostic methods pose some limitations that hinder the diagnosis of pathogens. Many of the pathogens are unculturable in lab conditions. Several techniques for diagnosis are time-consuming and have many demerits. Here metagenomic approach could be employed as it offers great potential. With the advent of NGS, the metagenomic approach has helped in detecting almost all types of pathogens (Datta et al., 2020). Duan et al. (2021) demonstrated mNGS analysis and culture-based assay of blood, bone marrow, bronchoalveolar lavage fluid, cerebrospinal fluid, nasal swab, pleural effusion, pus, sputum, and tissue collected from 109 adult patient's samples for diagnosis of infectious disease. The study suggested that the mNGS was significantly more sensitive (67%) than that of the culture-based method (23.6%, $P < .001$). Similarly Chen et al. (2021) demonstrated the mNGS of bronchoalveolar lavage samples to diagnose pulmonary infectious pathogens. Qian et al. (2020) performed a diagnosis of ventriculitis and meningitis using mNGS of cerebrospinal fluids. The most interesting fact is that mNGS confirmed the presence of ventriculitis and meningitis-associated pathogens in 22 cases earlier confirmed as negative using conventional methods. Thus mNGS can help in the early detection of diseases that can help in providing timely treatment and also plays a vital role in preventing disease outbreaks.

26.7.3 Xenobiotic degradation

Compounds that are released into the environment due to human activity and are persistent for a very long time are known as xenobiotics. These xenobiotics may be present anywhere; the longer the persistence higher the chances of microorganisms to adapt accordingly. The exposure of xenobiotics to microbes may cause mutations that may be positive or negative; the positive mutations may help in the degradation of xenobiotics. They may acquire novel genes that encode for the degradation of these compounds and could be of great industrial importance. Many have reported that few microbial species are efficient in degrading some of these compounds such as polyaromatic, polychlorinated, and polyester molecules. The better the understanding of the xenobiotic activity, the more benefit the clinical and industrial sectors will get.

26.7.4 Industrial applications

Microbial metagenomics has helped in identifying several different enzymes of industrial use. The most commercial type of enzyme is the amylase, which used to break starch into its monomeric form. Various fungi, bacteria, and archaea produce this enzyme and catalyze the starch breakdown. The metagenomic approach has helped in finding an amylase enzyme that can tolerate high temperatures and has broad applications. Similarly, another group of enzymes, lipases used for the hydrolysis of lipids, were obtained from different microbial sources (bacteria, yeast) and was used for different industrial purposes such as the detergent industry and food industry, for aroma compound production in the cosmetic industry (Sarmah et al., 2018). Many more enzymes have been metagenomically identified and are being used in industries. With the help of metagenomics, new variants of these enzymes with better efficiency were discovered and are being significantly used in industries (Datta et al., 2020). Some of the commonly used enzymes in industries discovered by metagenomics are as follows.

26.7.4.1 Lipases

To date, a total of 80 positive clones for lipase have been reported through metagenomics from various sources. One isolated from a deep-sea hypersaline environment. They were found to be active at alkaline pH and showed higher activities under high pressure. From the soil sample of the vegetable garden, an enzyme named pyrethroid-hydrolyzing esterase was discovered that has potential application in insecticide production. Many thermostable and organic solvent esterases were discovered with the help of this technique (Ngara and Zhang, 2018).

26.7.4.2 Tannase

The main application of this enzyme is as a clarifying agent in wine, beer, and fruit juices. Tannase is also used in the manufacturing of feed, food, and tea. The large-scale production of it is still restricted because of less knowledge of the enzyme, and here the metagenomic method has helped a lot in gaining the information. With the help of metagenomics and immobilization, a novel tannase (Tan410) was discovered from a soil sample. The enzyme had a higher optimum temperature and more alkaline pH optimum, and high storage capabilities, which made it a good option for industrial application (Chen et al., 2014).

26.7.4.3 *Proteases*

It is used for the catalysis of proteins and protein compounds. With the help of metagenomics, many proteases have been discovered. A sample from goatskin was taken and screened for protease activity, and a protease with high alkalinity tolerance was identified. Similarly, soil samples from Death Valley and Gobi Desert with the metagenomic technique helped in screening two serine proteases with different thermophilic profiles (Ngara and Zhang, 2018) (Table 26.3).

26.7.5 Bioeconomy

The recent trend following the principle of "waste to wealth and value addition" puts forward the idea of the development of bioenergy to achieve the bioeconomy. Bioeconomy can be defined as the usage of waste materials such as food, bioenergy, and other bioproducts to produce renewable materials of biological origin, such as land and water. It is a tool for investment as well as for innovation and for having sustainability. But still, there is no proper definition for bioeconomy. The main goal of the bioeconomy includes three main visions: (1) to highlight the importance of research in biotechnology and commercial application; (2) to signify the conservation of ecological processes such as soil, water, and biodiversity; and (3) to emphasize on the vision in the research and development of various bio-based raw materials from different sectors.

In this era of increasing demands for energy where all the conventional sources are depleting, this intensifies the need for research on looking for sustainable means for energy generation. Nowadays, the concept of the unculturable majority that is impacting the culturable microbes is gaining pace. Tapping unculturable microbes can open wide aspects of opportunity for enhancing the bioenergy sector, which is the major player in bioeconomy. Bioeconomy is the new term that is dependent on bio-based materials for growth. The bioeconomy is a large area covered by every sector of society, thus the development of each of these sectors directly or indirectly benefits the growth of the bioeconomy. The metagenomic is the driving path for sustainable development and bioeconomy (Pandey and Singhal, 2021).

The metagenomic tool help in identifying the complete information from various sources through gene sequencing. The development and advancements in NGS comprise a metagenomic approach that helps in-depth insight into the diversity of microbes. Metagenome has enriched our knowledge of microbiome and has opened several paths for development. By serving different sectors, metagenomics directly or indirectly has its hands on different

TABLE 26.3 Enzymes discovered through a metagenomic approach and their applications.

Enzymes	Source	Applications	Microorganisms	References
Lipases	Intertidal flat of the yellow sea in Korea	Fatty acid production, pharmaceutical	*Photobacterium lipolyticum*	Ryu et al. (2006)
Glycosyl hydrolase	Cow rumen	Biomass saccharification, textile, and pulp and paper processing	*Prevotella bryantii*	Palackal et al. (2007)
Protease	Wastewater	Industrial	*Bacillus licheniformis*	Hmidet et al. (2009)
Lactase	Geothermal spring in the northern Himalayas	Dairy industry, also in the pharmaceuticals and food industry	*Meiothermus ruber*	Gupta et al. (2012)
Tannase	Soil	Beverage industry, tea industry	–	Yao et al. (2014)
DNA polymerase	Hot spring	PCR, sequencing	–	Moser et al. (2012)
Amylase	Wastewater	Detergent processing	*B. licheniformis*	Hmidet et al. (2009)
Cellulase	Anaerobic beer	Food processing, chemical, textile, biofuel	–	Yang et al. (2016)
Lignocellulose degrading enzyme	Porcupine microbiome	Wood and biofuel	*Bacteroides* sp.	Thornbury et al. (2019)
Chitinase	Soil	Agriculture purpose	*Streptomyces* sp.	Hjort et al. (2014)
Hemicellulase	Degrading wheat straw	Wood and biofuel	*Klebsiella* sp.	Maruthamuthu et al. (2016)
Exosialidase	Hot spring	Biofuel and dairy	*Caldilinea aerophile*, *Thermomicrobium roseum*	Chuzel et al. (2018)
Asparaginase	Soil	Pharmaceutical and food processing	–	Kumar et al. (2018)

sectors of the economy, such as biorefinery which has nearly more than 50% of its share in the global bioeconomy, marine bioeconomy, and agricultural bioeconomy, and health. The need to fulfill the growing energy demand and the urge to attain sustainability have turned us into a bio-based economy. The whole soul of a bio-based economy is bioprocess which is in a cyclic manner, connected. The problems with this remain there, which is the main barrier to this development. Computational analysis with the help of omics is showing potential for identifying novel enzymes and products sustainably. Also, it is well applicable for the production of value-added products. The bioeconomy is a great tool for addressing today's problems and meeting the demands of society by giving novel and innovative products. This opens an opportunistic path to attain a circular bioeconomy for the coming generations (Pandey and Singhal, 2021).

26.8 Concluding remarks

The industrialization has made humans harness enzymes from microbes, and the yield from conventional methods could be consumable as microbes keep modifying their characteristics day by day. This has created a need to opt for the new advanced techniques for isolating the metagenomic approach that could provide us with an unlimited benefit from these untapped microbes in different areas. Several novel and valuable products can be produced from these unculturable microbes. The advancements in the metagenomic approach could help us create a more sustainable environment, improved agricultural production, and a sustainable bioeconomy. Many enzymes remain unexpressed in laboratory conditions and need some advanced techniques to be expressed. Synthetic biology shows a promising role in enhancing these effects for overcoming these. Like in the biofuel generation from lignocellulosic biomass, NGS and metagenomic have identified and exploited several novel enzymes from unculturable microbes. Many untouched aspects are to be dealt with in the coming future to explore more about unculturable microbes with advancements in metagenomics. This approach could better understand the uncultured microbes in the environment and their possible application in the coming future.

Conflict of interest

All the authors approve for the submission and do not have any conflict of interest to declare.

Abbreviations

FACS fluorescent-activated cell sorting
GFP green fluorescent protein
Ms mass Spectrometry

NGS next-generation sequencing
SIGEX substrate-induced gene-expression screening
VBNC viable but nonculturable

References

Ahmad, T., Singh, R.S., Gupta, G., Sharma, A., Kaur, B., 2019. Metagenomics in the search for industrial enzymes. Biomass, Biofuels, Biochemicals: Adv. Enzyme Technol. Available from: https://doi.org/10.1016/B978-0-444-64114-4.00015-7.

Bilal, T., Malik, B., Hakeem, K.R., 2018. Metagenomic analysis of uncultured microorganisms and their enzymatic attributes. J. Microbiol. Methods 155, 65–69. Available from: https://doi.org/10.1016/j.mimet.2018.11.014.

Chen, Y., Feng, W., Ye, K., Guo, L., Xia, H., Guan, Y., et al., 2021. Application of metagenomic next-generation sequencing in the diagnosis of pulmonary infectious pathogens from bronchoalveolar lavage samples. Front. Cell. Infect. Microbiol. 11, 168. Available from: https://doi.org/10.3389/fcimb.2021.541092.

Chen, R., Li, C., Pei, X., Wang, Q., Yin, X., Xie, T., 2014. Isolation an Aldehyde Dehydrogenase Gene from Metagenomics Based on Semi-nest Touch-Down PCR. Indian J. Microbiol. 54, 74–79. Available from: https://doi.org/10.1007/s12088-013-0405-0.

Chuzel, L.Ã., Ganatra, M.B., Rapp, E., Henrissat, B., Taron, C.H., 2018. Functional metagenomics identifies an exosialidase with an inverting catalytic mechanism that defines a new glycoside hydrolase family (GH156). J. Biol. Chem. 293, 18138–18150. Available from: https://doi.org/10.1074/jbc.RA118.003302.

Clarridge III, J.E., 2004. Impact of 16S rRNA gene sequence analysis for identification of bacteria on clinical microbiology and infectious diseases. Clin. Microbiol. Rev. 17 (4), 840–862. Available from: https://doi.org/10.3389/fmicb.2021.613791.

Datta, S., Rajnish, K.N., Samuel, M.S., Pugazlendhi, A., Selvarajan, E., 2020. Metagenomic applications in microbial diversity, bioremediation, pollution monitoring, enzyme and drug discovery. A review. Environ. Chem. Lett. 18, 1229–1241. Available from: https://doi.org/10.1007/s10311-020-01010-z.

Debnath, T., Kujur, R.R.A., Mitra, R., Das, S.K., 2019. Diversity of microbes in hot springs and their sustainable use. Microbial Diversity in Ecosystem Sustainability and Biotechnological Applications. Springer, pp. 159–186. Available from: https://doi.org/10.1007/978-981-13-8315-1_6.

Duan, H., Li, X., Mei, A., Li, P., Liu, Y., Li, X., et al., 2021. The diagnostic value of metagenomic next- generation sequencing in infectious diseases. BMC Infect. Dis. 21 (1), 1–13. Available from: https://doi.org/10.1186/s12879-020-05746-5.

Dubey, A., Malla, M.A., Kumar, A., 2022. Role of next-generation sequencing (NGS) in understanding the microbial diversity. Molecular Genetics and Genomics Tools in Biodiversity Conservation. Springer, Singapore, pp. 307–328. Available from: https://doi.org/10.1007/978-981-16-6005-4_16.

Forbes, J.D., Knox, N.C., Ronholm, J., Pagotto, F., Reimer, A., 2017. Metagenomics: the next culture-independent game changer. Front. Microbiol. 8, 1069. Available from: https://doi.org/10.3389/fmicb.2017.01069.

Gnaneswar Gude, V., 2015. A new perspective on microbiome and resource management in wastewater systems. J. Biotechnol. Biomater. 05. Available from: https://doi.org/10.4172/2155-952x.1000184.

Gupta, N., Verma, V.K., 2019. Next-generation sequencing and its application: empowering in public health beyond reality. Microbial Technology for the Welfare of Society. Springer, Singapore, pp. 313–341. Available from: https://doi.org/10.1007/978-981-13-8844-6_15.

Gupta, R., Govil, T., Capalash, N., Sharma, P., 2012. Characterization of a glycoside hydrolase family 1 β-galactosidase from hot spring metagenome with transglycosylation activity. Appl. Biochem. Biotechnol. 168, 1681–1693. Available from: https://doi.org/10.1007/s12010-012-9889-z.

He, B., Jin, S., Cao, J., Mi, L., Wang, J., 2019. Metatranscriptomics of the Hu sheep rumen microbiome reveals novel cellulases. Biotechnol. Biofuels 12, 1–15. Available from: https://doi.org/10.1186/s13068-019-1498-4.

Henson, J., Tischler, G., Ning, Z., 2012. Next-generation sequencing and large genome assemblies. Pharmacogenomics 13 (8), 901–915. Available from: https://doi.org/10.2217/pgs.12.72.

Hjort, K., Presti, I., Elväng, A., 2014. Bacterial chitinase with phytopathogen control capacity from suppressive soil revealed by functional metagenomics. Appl. Microbiol. Biotechnol. 2819–2828. Available from: https://doi.org/10.1007/s00253-013-5287-x.

Hmidet, N., El-Hadj Ali, N., Haddar, A., Kanoun, S., Alya, S.K., Nasri, M., 2009. Alkaline proteases and thermostable α-amylase co-produced by *Bacillus licheniformis* NH1: characterization and potential application as detergent additive. Biochem. Eng. J. 47, 71–79. Available from: https://doi.org/10.1016/j.bej.2009.07.005.

Kaboré, O.D., Godreuil, S., Drancourt, M., 2020. Planctomycetes as host-associated bacteria: a perspective that holds promise for their future isolations, by mimicking their native environmental niches in clinical microbiology laboratories. Front. Cell. Infect. Microbiol. 10, 1–19. Available from: https://doi.org/10.3389/fcimb.2020.519301.

Kamble, P., Vavilala, S.L., 2018. Discovering novel enzymes from marine ecosystems: A metagenomic approach. Bot. Mar. 61, 161–175. Available from: https://doi.org/10.1515/bot-2017-0075.

Kimura, N., 2006. Metagenomics: access to unculturable microbes in the environment. Microbes Env. 21, 201–215. Available from: https://doi.org/10.1264/jsme2.21.201.

Kimura, N., 2018. Novel biological resources screened from uncultured bacteria by a metagenomic method. Metagenomics: Perspectives, Methods, and Applications. Elsevier Inc. Available from: https://doi.org/10.1016/B978-0-08-102268-9.00014-8.

Kirubakaran, R., AruIJothi, K.N., Revathi, S., Shameem, N., Parray, J.A., 2020. Emerging priorities for microbial metagenome research. Bioresour. Technol. Rep. 11, 100485. Available from: https://doi.org/10.1016/j.biteb.2020.100485.

Kommedal, Ø., Karlsen, B., Sæbø, Ø., 2008. Analysis of mixed sequencing chromatograms and its application in direct 16S rRNA gene sequencing of polymicrobial samples. J. Clin. Microbiol. 46 (11), 3766–3771. Available from: https://doi.org/10.1128/JCM.00213-08.

Kulski, J.K., 2016. Next-generation sequencing—an overview of the history, tools, and "Omic" applications, Next Generation Sequencing - Advances, Applications and Challenges, 10. IntechOpen, p. 61964. Available from: https://doi.org/10.5772/61964.

Kumar, J., Balakrishna, A., Aneesh, P., Kavitha, T., 2018. Characterization of a novel asparaginase from soil metagenomic libraries generated from forest soil. Biotechnol. Lett. 40, 303–308. Available from: https://doi.org/10.1007/s10529-017-2470-7.

Lapidus, A.L., Korobeynikov, A.I., 2021. Metagenomic data assembly–the way of decoding unknown microorganisms. Front. Microbiol. 12, 653. Available from: https://doi.org/10.3389/fmicb.2021.613791.

Lee, R., 2019. Metagenomic Next Generation Sequencing: How Does It Work and Is It Coming to Your Clinical Microbiology Lab? American Society of Microbiology. Available from: https://asm.org/Articles/2019/November/Metagenomic-Next-Generation-Sequencing-How-Does-It.

Liang, T.W., Chen, Y.Y., Pan, P.S., Wang, S.L., 2014. Purification of chitinase/chitosanase from *Bacillus cereus* and discovery of an enzyme inhibitor. Int. J. Biol. Macromol. 63, 8–14. Available from: https://doi.org/10.1016/j.ijbiomac.2013.10.027.

Liu, L., Li, Y., Li, S., Hu, N., He, Y., Pong, R., et al., 2012. Comparison of next-generation sequencing systems. J. biotechnol. biomed. 2012, Volume 2012 |Article ID 251364 |. Available from: https://doi.org/10.1155/2012/251364.

Madhavan, A., Sindhu, R., 2017. Metagenome analysis : a powerful tool for enzyme bioprospecting. Appl. Biochem. Biotechnol. 636–651. Available from: https://doi.org/10.1007/s12010-017-2568-3.

Mahapatra, G.P., Raman, S., Nayak, S., Gouda, S., Das, G., Patra, J.K., 2020. Metagenomics approaches in discovery and development of new bioactive compounds from marine actinomycetes. Curr. Microbiol. 77, 645–656. Available from: https://doi.org/10.1007/s00284-019-01698-5.

Maher-Sturgess, S.L., Forrester, N.L., Wayper, P.J., Gould, E.A., Hall, R.A., Barnard, R.T., et al., 2008. Universal primers that amplify RNA from all three flavivirus subgroups. Vir. J. 5 (1), 1–10. Available from: https://doi.org/10.1186/1743-422X-5-16.

Maruthamuthu, M., Jiménez, D.J., Stevens, P., Elsas, J.D.Van, 2016. A multi-substrate approach for functional metagenomics-based screening for (hemi) cellulases in two wheat straw-degrading microbial consortia unveils novel thermoalkaliphilic enzymes. BMC Genomics 1–16. Available from: https://doi.org/10.1186/s12864-016-2404-0.

Montgomery, B.L., 2020. Lessons from microbes: what can we learn about equity from unculturable bacteria. mSphere 5. Available from: https://doi.org/10.1128/msphere.01046-20.

Moser, M.J., DiFrancesco, R.A., Gowda, K., Klingele, A.J., Sugar, D.R., Stocki, S., et al., 2012. Thermostable DNA polymerase from a viral metagenome is a potent RT-PCR enzyme. PLoS One 7. Available from: https://doi.org/10.1371/journal.pone.0038371.

Ngara, T.R., Zhang, H., 2018. Recent Advances in Function-based Metagenomic Screening. Genomics, Proteom. Bioinforma. 16, 405–415. Available from: https://doi.org/10.1016/j.gpb.2018.01.002.

Palackal, N., Lyon, C.S., Zaidi, S., Luginbühl, P., Dupree, P., Goubet, F., et al., 2007. A multifunctional hybrid glycosyl hydrolase discovered in an uncultured microbial consortium from ruminant gut. Appl. Microbiol. Biotechnol. 74, 113–124. Available from: https://doi.org/10.1007/s00253-006-0645-6.

Pandey, M., Singhal, B., 2021. Metagenomics: adding new dimensions in bioeconomy. Biomass Convers. Biorefinery . Available from: https://doi.org/10.1007/s13399-021-01585-9.

Pareek, C.S., Smoczynski, R., Tretyn, A., 2011. Sequencing technologies and genome sequencing. J. Appl. Genet. 52 (4), 413–435. Available from: https://doi.org/10.1007/s13353-011-0057-x.

Peng, X., Wilken, S.E., Lankiewicz, T.S., Gilmore, S.P., Brown, J.L., Henske, J.K., et al., 2021. Genomic and functional analyses of fungal and bacterial consortia that enable lignocellulose breakdown in goat gut microbiomes. Nat. Microbiol. 6 (4), 499–511. Available from: https://doi.org/10.1038/s41564-020-00861-0.

Perfumo, A., Freiherr von Sass, G.J., Nordmann, E.L., Budisa, N., Wagner, D., 2020. Discovery and characterization of a new cold-active protease from an extremophilic bacterium via comparative genome analysis and in vitro expression. Front. Microbiol. 11, 1–12. Available from: https://doi.org/10.3389/fmicb.2020.00881.

Qian, L., Shi, Y., Li, F., Wang, Y., Ma, M., Zhang, Y., et al., 2020. Metagenomic next-generation sequencing of cerebrospinal fluid for the diagnosis of external ventricular and lumbar drainage-associated ventriculitis and meningitis. Front. Microbiol. 3096. Available from: https://doi.org/10.3389/fmicb.2020.596175.

Raja, H.A., Miller, A.N., Pearce, C.J., Oberlies, N.H., 2017. Fungal identification using molecular tools: a primer for the natural products research community. J. Nat. Prod. 80 (3), 756–770. Available from: https://doi.org/10.1021/acs.jnatprod.6b01085.

Ristinmaa, A.S., Coleman, T., Cesar, L., Langborg Weinmann, A., Mazurkewich, S., Brändén, G., et al., 2022. Structural diversity and substrate preferences of three tannase enzymes encoded by the anaerobic bacterium *Clostridium butyricum*. J. Biol. Chem. 298, 101758. Available from: https://doi.org/10.1016/j.jbc.2022.101758.

Ryu, H.S., Kim, H.K., Choi, W.C., Kim, M.H., Park, S.Y., Han, N.S., et al., 2006. New cold-adapted lipase from *Photobacterium lipolyticum* sp. nov. that is closely related to filamentous fungal lipases. Appl. Microbiol. Biotechnol. 70, 321–326. Available from: https://doi.org/10.1007/s00253-005-0058-y.

Sander, D., Yu, Y., Sukul, P., Schäkermann, S., Bandow, J.E., Mukherjee, T., et al., 2021. Metaproteomic discovery and characterization of a novel lipolytic enzyme from an Indian hot spring. Front. Microbiol. 12, 1–15. Available from: https://doi.org/10.3389/fmicb.2021.672727.

Sarmah, N., Revathi, D., Sheelu, G., Yamuna Rani, K., Sridhar, S., Mehtab, V., et al., 2018. Recent advances on sources and industrial applications of lipases. Biotechnol. Prog. 34, 5–28. Available from: https://doi.org/10.1002/btpr.2581.

Schmeisser, C., Steele, H., Streit, W.R., 2007. Metagenomics, biotechnology with non-culturable microbes. Appl. Microbiol. Biotechnol. 75, 955–962. Available from: https://doi.org/10.1007/s00253-007-0945-5.

Singh, R., Dhawan, S., Singh, K., Kaur, J., 2012. Cloning, expression and characterization of a metagenome derived thermoactive/thermostable pectinase. Mol. Biol. Rep. 39, 8353–8361. Available from: https://doi.org/10.1007/s11033-012-1685-x.

Slatko, B.E., Gardner, A.F., Ausubel, F.M., 2018. Overview of next-generation sequencing technologies. Curr. Protoc. Mol. Biol, 122(1) e59. Available from: https://doi.org/10.1002/cpmb.59.

Srivastava, N., Gupta, B., Gupta, S., Danquah, M.K., Sarethy, I.P., 2018. Analyzing functional microbial diversity: an overview of techniques. an overview of techniques. Microbial Diversity in the Genomic Era. Elsevier Inc. Available from: https://doi.org/10.1016/B978-0-12-814849-5.00006-X.

Stepnov, A.A., Fredriksen, L., Steen, I.H., Stokke, R., Eijsink, V.G.H., 2019. Identification and characterization of a hyperthermophilic GH9 cellulase from the Arctic Mid-Ocean Ridge vent field. PLoS One 14, 1–19. Available from: https://doi.org/10.1371/journal.pone.0222216.

Tatta, E.R., Imchen, M., Moopantakath, J., Kumavath, R., 2022. Bioprospecting of microbial enzymes: current trends in industry and healthcare. Appl. Microbiol. Biotechnol. 1813–1835. Available from: https://doi.org/10.1007/s00253-022-11859-5.

Thornbury, M., Sicheri, J., Slaine, P., Getz, L.J., Finlayson-Trick, E., Cook, J., et al., 2019. Characterization of novel lignocellulose-degrading enzymes from the porcupine microbiome using synthetic metagenomics. PLoS One . Available from: https://doi.org/10.1371/journal.pone.0209221.

Vartoukian, S.R., Palmer, R.M., Wade, W.G., 2010. Strategies for culture of “unculturable” bacteria. FEMS Microbiol. Lett. 309, 1–7. Available from: https://doi.org/10.1111/j.1574-6968.2010.02000.x.

Verma, S., Gazara, R.K., 2021. Next-generation sequencing: an expedition from workstation to clinical applications. Translational Bioinformatics in Healthcare and Medicine. Academic Press, pp. 29–47. Available from: https://doi.org/10.1016/B978-0-323-89824-9.00003-3.

Weinstock G., Goldberg B., Ledeboer N., Rubin E., Sichtig H., Geyer C., 2016. Applications of clinical microbial next-generation sequencing. Report on an American Academy of Microbiology Colloquium held in Washington, DC, in April 2015. <https://www.ncbi.nlm.nih.gov/books/NBK513764/?report = reader>

Wilson, M.C., Piel, J., 2013. Metagenomic approaches for exploiting uncultivated bacteria as a resource for novel biosynthetic enzymology. Chem. & Biol. 20 (5), 636–647. Available from: https://doi.org/10.1016/j.chembiol.2013.04.011.

Yang, C., Xia, Y., Qu, H., Li, A.D., Liu, R., Wang, Y., et al., 2016. Biotechnology for Biofuels Discovery of new cellulases from the metagenome by a metagenomics - guided strategy. Biotechnol. Biofuels 1–12. Available from: https://doi.org/10.1186/s13068-016-0557-3.

Yao, J., Chen, Q., Zhong, G., Cao, W., Yu, A., Liu, Y., 2014. Immobilization and characterization of tannase from a metagenomic library and its use for removal of tannins from green tea infusion. J. Microbiol. Biotechnol. 24, 80–86. Available from: https://doi.org/10.4014/jmb.1308.08047.

Yi, Y., Xu, S., Kovalevsky, A., Zhang, X., Liu, D., Wan, Q., 2021. Characterization and structural analysis of a thermophilic GH11 xylanase from compost metatranscriptome. Appl. Microbiol. Biotechnol. 105, 7757–7767. Available from: https://doi.org/10.1007/s00253-021-11587-2.

Zarraonaindia, I., Smith, D.P., Gilbert, J.A., 2013. Beyond the genome: community-level analysis of the microbial world. Biol. Philos. 28, 261–282. Available from: https://doi.org/10.1007/s10539-012-9357-8.

Index

Note: Page numbers followed by "*f*" and "*t*" refer to figures and tables, respectively.

A

B

D

E

F

H

Q

R

S

T

U

V

W

X

Y

Z

CPI Antony Rowe
Eastbourne, UK
January 24, 2023